Desk Encyclopedia of
PLANT AND
FUNGAL VIROLOGY

DESK ENCYCLOPEDIA OF PLANT AND FUNGAL VIROLOGY

EDITOR-IN-CHIEF

Dr MARC H V VAN REGENMORTEL

ELSEVIER

AMSTERDAM • BOSTON • HEIDELBERG • LONDON • NEW YORK • OXFORD
PARIS • SAN DIEGO • SAN FRANCISCO • SINGAPORE • SYDNEY • TOKYO
Academic Press is an imprint of Elsevier

ACADEMIC PRESS

Academic Press is an imprint of Elsevier
Linacre House, Jordan Hill, Oxford, OX2 8DP, UK
525 B Street, Suite 1900, San Diego, CA 92101-4495, USA

British Library Cataloguing in Publication Data
A catalogue record for this book is available from the British Library

Library of Congress Cataloguing in Publication Data
A catalogue record for this book is available from the Library of Congress

ISBN: 978-0-12-375148-5

For information on all Elsevier publications
visit our website at books.elsevier.com

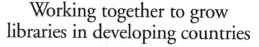

EDITOR-IN-CHIEF

Marc H V van Regenmortel PhD
Emeritus Director at the CNRS,
French National Center for Scientific Research,
Biotechnology School of the University of Strasbourg,
Illkirch, France

ASSOCIATE EDITORS

Dennis H Bamford, Ph.D.
Department of Biological and Environmental Sciences
and Institute of Biotechnology, Biocenter 2,
P.O. Box 56 (Viikinkaari 5),
00014 University of Helsinki,
Finland

Charles Calisher, B.S., M.S., Ph.D.
Arthropod-borne and Infectious Diseases Laboratory
Department of Microbiology, Immunology and Pathology
College of Veterinary Medicine and Biomedical Sciences
Colorado State University
Fort Collins
CO 80523
USA

Andrew J Davison, M.A., Ph.D.
MRC Virology Unit
Institute of Virology
University of Glasgow
Church Street
Glasgow G11 5JR
UK

Claude Fauquet
ILTAB/Donald Danforth Plant Science Center
975 North Warson Road
St. Louis, MO 63132

Said Ghabrial, B.S., M.S., Ph.D.
Plant Pathology Department
University of Kentucky
201F Plant Science Building
1405 Veterans Drive
Lexington
KY 4050546-0312
USA

Eric Hunter, B.Sc., Ph.D.
Department of Pathology and Laboratory Medicine, and
Emory Vaccine Center
Emory University
954 Gatewood Road NE
Atlanta Georgia 30329
USA

Robert A Lamb, Ph.D., Sc.D.
Department of Biochemistry,
Molecular Biology and Cell Biology
Howard Hughes Medical Institute
Northwestern University
2205 Tech Dr.
Evanston
IL 60208-3500
USA

Olivier Le Gall
IPV, UMR GDPP, IBVM,
INRA Bordeaux-Aquitaine, BP 81,
F-33883 Villenave d'Ornon Cedex
FRANCE

Vincent Racaniello, Ph.D.
Department of Microbiology
Columbia University
New York, NY 10032
USA

David A Theilmann, Ph.D., B.Sc., M.Sc
Pacific Agri-Food Research Centre
Agriculture and Agri-Food Canada
Box 5000, 4200 Highway 97
Summerland
BC V0H 1Z0
Canada

H Josef Vetten, Ph.D.
Julius Kuehn Institute, Federal Research Centre for
Cultivated Plants (JKI)
Messeweg 11-12
38104 Braunschweig
Germany

Peter J Walker, B.Sc., Ph.D.
CSIRO Livestock Industries
Australian Animal Health Laboratory (AAHL)
Private Bag 24
Geelong
VIC 3220
Australia

PREFACE

The *Desk Encyclopedia of Plant and Fungal Virology* is the fourth in a series of four volumes that reproduces many entries that appeared in the third edition of the *Encyclopedia of Virology*, edited by Brian W J Mahy and Marc H V van Regenmortel, published by Academic Press/Elsevier in 2008.

It consists of 85 chapters that highlight recent advances in our knowledge of the viruses that infect plants and fungi. The first section of the book, comprising 10 chapters, discusses general topics in plant virology such as the movement of viruses in plants, the transmission of plant viruses by vectors, antiviral defense mechanisms in plants, and the development of virus-resistant transgenic plants. A chapter is devoted to viroids.

The second section of 48 chapters presents an overview of the properties of a selection of 20 well-studied plant viruses, 23 plant virus genera and a few larger groups of plant viruses.

The third section of 12 chapters describes the most economically important virus diseases of cereals, legumes, vegetable crops, fruit trees, and ornamentals. This section is abundantly illustrated and should be very useful to plant pathologists and all those interested in viral infections in plants. The last section of 15 chapters describes the major groups of viruses that infect fungi.

As all the chapters initially appeared in an encyclopedia, little prior specialized knowledge is required to follow the material that is presented. When used in conjunction with the first volume of the series, which is devoted to *General Virology* and describes the structure, replication, molecular biology, and general properties of viruses, this volume could form the basis of an introductory course on virology, suitable for students of plant sciences.

Marc H V van Regenmortel

CONTRIBUTORS

G Adam
Universität Hamburg, Hamburg, Germany

M J Adams
Rothamsted Research, Harpenden, UK

N K van Alfen
University of California, Davis, CA, USA

M Bar-Joseph
The Volcani Center, Bet Dagan, Israel

Y Bigot
University of Tours, Tours, France

S Blanc
INRA–CIRAD–AgroM, Montpellier, France

J F Bol
Leiden University, Leiden, The Netherlands

L Bos
Wageningen University and Research Centre (WUR),
Wageningen, The Netherlands

C Bragard
Université Catholique de Louvain, Leuven, Belgium

J N Bragg
University of California, Berkeley, CA, USA

J K Brown
The University of Arizona, Tucson, AZ, USA

J Bruenn
State University of New York, Buffalo, NY, USA

J J Bujarski
Northern Illinois University, DeKalb, IL, USA and Polish
Academy of Sciences, Poznan, Poland

J Burgyan
Agricultural Biotechnology Center, Godollo, Hungary

M I Butler
University of Otago, Dunedin, New Zealand

P Caciagli
Istituto di Virologia Vegetale – CNR, Turin, Italy

T Candresse
UMR GDPP, Centre INRA de Bordeaux, Villenave
d'Ornon, France

S Chakraborty
Jawaharlal Nehru University, New Delhi, India

I-R Choi
International Rice Research Institute, Los Baños,
The Philippines

P D Christian
National Institute of Biological Standards and Control,
South Mimms, UK

T Dalmay
University of East Anglia, Norwich, UK

C J D'Arcy
University of Illinois at Urbana-Champaign, Urbana,
IL, USA

W O Dawson
University of Florida, Lake Alfred, FL, USA

P Delfosse
Centre de Recherche Public-Gabriel Lippmann,
Belvaux, Luxembourg

M Deng
University of California, Berkeley, CA, USA

C Desbiez
Institut National de la Recherche Agronomique (INRA),
Station de Pathologie Végétale, Montfavet, France

R G Dietzgen
The University of Queensland, St. Lucia, QLD, Australia

S P Dinesh-Kumar
Yale University, New Haven, CT, USA

L L Domier
USDA–ARS, Urbana, IL, USA

L L Domier
USDA-ARS, Urbana-Champaign, IL, USA

T W Dreher
Oregon State University, Corvallis, OR, USA

K C Eastwell
Washington State University – IAREC, Prosser, WA, USA

H Edskes
National Institutes of Health, Bethesda, MD, USA

A Engel
Ntioal Institutes of Health, Bethesa, MD, USA

D Kryndushkin
National Institutes of Health, Bethesda, MD, USA

J Engelmann
INRES, University of Bonn, Bonn, Germany

R Esteban
Instituto de Microbiología Bioquímica CSIC/University de Salamanca, Salamanca, Spain

R Esteban
Instituto de Microbiología Bioquímica CSIC/University of Salamanca, Salamanca, Spain

R-X. Fang
Chinese Academy of Sciences, Beijing, People's Republic of China

D Fargette
IRD, Montpellier, France

C M Fauquet
Danforth Plant Science Center, St. Louis, MO, USA

B A Federici
University of California, Riverside, CA, USA

S A Ferreira
University of Hawaii at Manoa, Honolulu, HI, USA

R Flores
Instituto de Biología Molecular y Celular de Plantas (UPV-CSIC), Valencia, Spain

M Fuchs
Cornell University, Geneva, NY, USA

T Fujimura
Instituto de Microbiología Bioquímica CSIC/University de Salamanca, Salamanca, Spain

T Fujimura
Instituto de Microbiología Bioquímica CSIC/University of Salamanca, Salamanca, Spain

D Gallitelli
Università degli Studi and Istituto di Virologia Vegetale del CNR, Bari, Italy

F García-Arenal
Universidad Politécnica de Madrid, Madrid, Spain

R J Geijskes
Queensland University of Technology, Brisbane, QLD, Australia

S A Ghabrial
University of Kentucky, Lexington, KY, USA

M Glasa
Slovak Academy of Sciences, Bratislava, Slovakia

Y Gleba
Icon Genetics GmbH, Weinbergweg, Germany

D Gonsalves
USDA, Pacific Basin Agricultural Research Center, Hilo, HI, USA

M M Goodin
University of Kentucky, Lexington, KY, USA

T J D Goodwin
University of Otago, Dunedin, New Zealand

A-L Haenni
Institut Jacques Monod, Paris, France

J Hamacher
INRES, University of Bonn, Bonn, Germany

R M Harding
Queensland University of Technology, Brisbane, QLD, Australia

D Hariri
INRA – Département Santé des Plantes et Environnement, Versailles, France

P A Harries
Samuel Roberts Noble Foundation, Inc., Ardmore, OK, USA

E Hébrard
IRD, Montpellier, France

B I Hillman
Rutgers University, New Brunswick, NJ, USA

S A Hogenhout
The John Innes Centre, Norwich, UK

T Hohn
Institute of Botany, Basel university, Basel, Switzerland

J S Hong
Seoul Women's University, Seoul, South Korea

A O Jackson
University of California, Berkeley, CA, USA

P John
Indian Agricultural Research Institute, New Delhi, India

P Kazmierczak
University of California, Davis, CA, USA

C Kerlan
INRA, UMR1099 BiO3P, Le Rheu, France

C Kerlan
Institut National de la Recherche Agronomique (INRA), Le Rheu, France

V Klimyuk
Icon Genetics GmbH, Weinbergweg, Germany

R Koenig
Biologische Bundesanstalt für Land- und Forstwirtschaft, Brunswick, Germany

R Koenig
Institut für Pflanzenvirologie, Mikrobiologie und biologische Sicherheit, Brunswick, Germany

G Konaté
INERA, Ouagadougou, Burkina Faso

H D Lapierre
INRA – Département Santé des Plantes et
Environnement, Versailles, France

H Lecoq
Institut National de la Recherche Agronomique (INRA),
Station de Pathologie Végétale, Montfavet, France

B Y Lee
Seoul Women's University, Seoul, South Korea

D-E Lesemann
Biologische Bundesanstalt für Land- und Forstwirtschaft,
Brunswick, Germany

D J Lewandowski
The Ohio State University, Columbus, OH, USA

G Loebenstein
Agricultural Research Organization, Bet Dagan, Israel

S A Lommel
North Carolina State University, Raleigh, NC, USA

G P Lomonossoff
John Innes Centre, Norwich, UK

S A MacFarlane
Scottish Crop Research Institute, Dundee, UK

V G Malathi
Indian Agricultural Research Institute, New Delhi, India

S Marillonnet
Icon Genetics GmbH, Weinbergweg, Germany

G P Martelli
Università degli Studi and Istituto di Virologia vegetale
CNR, Bari, Italy

G P Martelli
Università degli Studi and Istituto di Virologia Vegetale del
CNR, Bari, Italy

G P Martelli
Università degli Studi, Bari, Italy

D P Martin
University of Cape Town, Cape Town, South Africa

L McCann
National Institutes of Health, Bethesda, MD, USA

M Meier
Tallinn University of Technology, Tallinn, Estonia

R G Milne
Istituto di Virologia Vegetale, CNR, Turin, Italy

F J Morales
International Center for Tropical Agriculture,
Cali, Colombia

T J Morris
University of Nebraska, Lincoln, NE, USA

B Moury
INRA – Station de Pathologie Végétale, Montfavet, France

J W Moyer
North Carolina State University, Raleigh, NC, USA

E Muller
CIRAD/UMR BGPI, Montpellier, France

T Nakayashiki
National Institutes of Health, Bethesda, MD, USA

M S Nawaz-ul-Rehman
Danforth Plant Science Center, St. Louis, MO, USA

R S Nelson
Samuel Roberts Noble Foundation, Inc., Ardmore,
OK, USA

D L Nuss
University of Maryland Biotechnology Institute, Rockville,
MD, USA

A Olspert
Tallinn University of Technology, Tallinn, Estonia

R A Owens
Beltsville Agricultural Research Center, Beltsville,
MD, USA

M S Padmanabhan
Yale University, New Haven, CT, USA

P Palukaitis
Scottish Crop Research Institute, Dundee, UK

R T M Poulter
University of Otago, Dunedin, New Zealand

F Qu
University of Nebraska, Lincoln, NE, USA

B C Ramirez
CNRS, Paris, France

D V R Reddy
Hyderabad, India

M H V van Regenmortel
CNRS, Illkirch, France

P A Revill
Victorian Infectious Diseases Reference Laboratory,
Melbourne, VIC, Australia

L Rubino
Istituto di Virologia Vegetale del CNR, Bari, Italy

E Ryabov
University of Warwick, Warwick, UK

E P Rybicki
University of Cape Town, Cape Town, South Africa

K H Ryu
Seoul Women's University, Seoul, South Korea

H Sanfaçon
Pacific Agri-Food Research Centre, Summerland, BC, Canada

C Sarmiento
Tallinn University of Technology, Tallinn, Estonia

K Scheets
Oklahoma State University, Stillwater, OK, USA

J E Schoelz
University of Missouri, Columbia, MO, USA

P D Scotti
Waiatarua, New Zealand

D N Shepherd
University of Cape Town, Cape Town, South Africa

F Shewmaker
National Institutes of Health, Bethesda, MD, USA

P A Signoret
Montpellier SupAgro, Montpellier, France

T L Sit
North Carolina State University, Raleigh, NC, USA

P Sreenivasulu
Sri Venkateswara University, Tirupati, India

J Stanley
John Innes Centre, Colney, UK

J Y Suzuki
USDA, Pacific Basin Agricultural Research Center, Hilo, HI, USA

N Suzuki
Okayama University, Okayama, Japan

G Szittya
Agricultural Biotechnology Center, Godollo, Hungary

M Taliansky
Scottish Crop Research Institute, Dundee, UK

S Tavantzis
University of Maine, Orono, ME, USA

J E Thomas
Department of Primary Industries and Fisheries, Indooroopilly, QLD, Australia

S A Tolin
Virginia Polytechnic Institute and State University, Blacksburg, VA, USA

L Torrance
Scottish Crop Research Institute, Invergowrie, UK

S Tripathi
USDA, Pacific Basin Agricultural Research Center, Hilo, HI, USA

E Truve
Tallinn University of Technology, Tallinn, Estonia

M Tsompana
North Carolina State University, Raleigh, NC, USA

J K Uyemoto
University of California, Davis, CA, USA

A M Vaira
Istituto di Virologia Vegetale, CNR, Turin, Italy

H J Vetten
Federal Research Centre for Agriculture and Forestry (BBA), Brunswick, Germany

R B Wickner
National Institutes of Health, Bethesda, MD, USA

S Winter
Deutsche Sammlung für Mikroorganismen und Zellkulturen, Brunswick, Germany

N Yoshikawa
Iwate University, Ueda, Japan

S K Zavriev
Shemyakin and Ovchinnikov Institute of Bioorganic Chemistry, Russian Academy of Sciences, Moscow, Russia

CONTENTS

SECTION III: PLANT VIRUS DISEASES

SECTION IV: FUNGAL VIRUSES

GENERAL TOPICS

Movement of Viruses in Plants

P A Harries and R S Nelson, Samuel Roberts Noble Foundation, Inc., Ardmore, OK, USA

Glossary

Ancillary viral proteins Virus-encoded proteins that do not meet the definition of a movement protein, but are required for virus movement.

Intercellular movement Movement between two cells.

Intracellular movement Movement within a single cell.

Microfilaments A component of the cytoskeleton formed from polymerized actin monomers.

Microtubules A component of the cytoskeleton composed of hollow tubes formed from α–β tubulin dimers.

Molecular chaperones A family of cellular proteins that mediate the correct assembly or disassembly of other polypeptides.

Movement protein Virus-encoded proteins that can transport themselves cell to cell, bind RNA, and increase the size exclusion limits of plasmodesmata.

Phloem Vascular tissue that transports dissolved nutrients (e.g., sugars) from the photosynthetically active leaves to the other parts of the plant. In most plants there is only one phloem class, but for some plant families this tissue is divided into two classes: (1) internal phloem (internal or adaxial to xylem) and (2) external phloem (external or abaxial to xylem).

Systemic movement Movement through vascular tissue to all parts of the plant.

Viroid A plant pathogen containing nucleic acid that encodes no proteins.

Xylem Vascular tissue that transports water and minerals through the plant.

Introduction

In order for a plant virus to infect its host systemically, it must be capable of hijacking the host's cellular machinery to replicate and move from the initially infected cell. Plant viruses require wounding, usually by insect or fungal vectors or mechanical abrasion, for an infection to begin. Once inside a cell, the virus initiates transcription (DNA viruses) and translation and replication (DNA and RNA viruses) activities. Some of these viral products are required for virus movement and often interact with host factors (proteins or membranes) to carry out this function.

Virus movement in plants can be broken down into three distinct steps: (1) intracellular movement, (2) intercellular movement, and (3) systemic movement. Intracellular movement refers to virus movement to the periphery of a cell and includes all metabolic activities necessary to recycle the host and viral constituents required for the continued transport of the intracellular complex. Intercellular movement refers to virus movement between cells. In order for a plant virus infection to spread between cells, viruses must move through specific channels in the cell wall, called plasmodesmata (PD), that connect neighboring cells. Once intracellular and intercellular movement is established, the virus can invade the vascular cells of the plant and then spread systemically through the open pores of modified PD within the sugar-transporting phloem sieve elements. Upon delivery by the phloem to a tissue distant from the original infection site, virus exits the vasculature and resumes cell-to-cell movement via PD in the new tissue. Although it will not be discussed further in this article, it is important to know that a few viruses utilize the water-transporting xylem vessels for systemic transport.

When contemplating plant virus movement it is critical to understand that each virus movement complex varies in viral and host factor composition over time as it travels within and between cell types. In addition, individual viruses often utilize unique host factors to support their movement. The diverse and dynamic nature of virus movement complexes makes it difficult to summarize plant virus movement in a simple unified model. However, there is evidence that some stages of virus movement, although carried out by apparently unrelated host or virus proteins, do have functional convergence.

Virus movement in plants has been studied with a wide range of virus genera, including, but not limited to, tobamoviruses, potexviruses, hordeiviruses, comoviruses, nepoviruses, potyviruses, tombusviruses, tospoviruses, and geminiviruses. In this article we do not review virus movement by all plant viruses, but rather focus on model viruses within genera that provide the most information on the subject. We review what is currently known about the three steps of virus movement in plants and attempt to convey the complexity of movement mechanisms utilized by members of different virus genera. However, we also highlight recent findings indicating that irrespective of the presence of seemingly unrelated host or viral factors, functional similarities exist for some aspects of movement displayed by viruses from different genera.

Intracellular Movement

Intracellular movement is necessary to deliver the virus genome to PD for cell-to-cell transmission. This has been an understudied area, as researchers have only recently had the ability to label and observe the movement of viral proteins and RNA in near-real-time conditions. Early studies relied on static images of immunolabeled viral proteins from light and transmission electron microscopes to determine their intracellular location. While a few of these studies related the intracellular location of the viral protein to the stage of infection, most did not and thus the importance of the intracellular location for virus movement was not understood. Other early studies of virus movement relied on the mutation of specific viral genes in virus genomic clones and the assessment of the intercellular movement of the resulting mutant virus, through the presence of local (representing intracellular and intercellular movement) or systemic (representing intracellular, intercellular and vascular movement) disease. Although these genetic experiments often determined which viral proteins were important for virus intercellular or systemic movement, they could not determine whether the mutation prevented intracellular or intercellular movement, both outcomes being visually identical. In more recent studies, fusion of viral proteins with fluorescent reporter genes such as the green fluorescent protein (GFP) have given researchers a powerful method to observe both the intracellular movement and final subcellular destination of many viral proteins in near-real-time conditions. However, it is important not to over-interpret movement studies using GFP since GFP maturation for fluorescence emission takes hours and thus the visible movement and position of the GFP or GFP:viral protein fusion may not reflect early movement activity. Additionally, the level of GFP within the movement form of the virus may be too low to detect during critical phases of movement.

Although intracellular virus movement in plants is just beginning to be elucidated, it is clear that specific viral proteins regulate this activity. Chief among these are the virally encoded movement proteins (MPs), named to indicate their genetically determined requirement for intercellular virus movement. MPs are defined based upon three functional characteristics: their (1) association with, or ability to increase, the size exclusion limit (SEL) of PD; (2) ability to bind to single-stranded RNA (ssRNA); and (3) ability to transport themselves or viral RNA cell to cell. Based upon these defining characteristics, a number of proteins have been classified as MPs (**Table 1**). Many viral MPs have similar sequences indicating a shared evolutionary history. However, a considerable number have no obvious sequence similarity between them. The absence of a shared sequence for these MPs suggests convergent evolution for movement function by unrelated predecessor proteins. MPs often interact with host proteins that modify their amino acid backbone (e.g., through phosphorylation) or host proteins associated with intracellular trafficking (e.g., cytoskeletal or vesicle-associated proteins) (**Table 1**). However, the role of MPs in intracellular movement remains largely unknown because technical limitations have prevented visualizing movement of individual viral RNA or DNA associated with MPs in real time. In addition, it is becoming clear that ancillary viral proteins (**Table 1**), which do not fulfill the classical definition of an MP, are essential for virus movement. These proteins are often associated with membranes or cytoskeletal elements and thus likely function primarily for intracellular virus movement. The interaction of MPs with host factors and the impact of the ancillary viral proteins on intracellular virus movement are discussed in detail in the following section. Models for intracellular virus movement of particular genera of viruses are presented based on some of this information (**Figure 1**).

Host factors and intracellular virus movement

Host proteins shown to interact with viral MPs include kinases, chaperones, nuclear-localized proteins (often transcription co-activators), and proteins that are associated with or are core components of the cytoskeletal or vesicle trafficking systems (**Table 1**). In addition, some MPs have been shown to associate with host membranes.

For geminiviruses, whose DNA genomes replicate in the nucleus, it is not surprising that nuclear factors may be necessary to transport viral genetic components required for virus replication into or out of the nucleus. For RNA viruses, however, there must be other reasons for an interaction between a nuclear protein and viral MP since these viruses are replicated in the cytoplasm. Some of the nuclear host proteins are non-cell-autonomous factors (e.g., HiF22) and thus it has been suggested that their interaction with MPs may inadvertently aid in virus intracellular and intercellular movement. It is also possible that MP and nuclear protein interactions occur to prevent transcription of host defense proteins or enhance transcription of host proteins necessary for virus movement, either within the infected cell or after transport to uninfected cells at the infection front.

The discovery over 10 years ago that tobacco mosaic virus (TMV) MP associates with microtubules (MTs) and microfilaments (MFs) was the first evidence that the host cytoskeleton might be involved in virus movement in plants. Although results from early studies indicated that disruption of MT arrays or their association with TMV MP could inhibit TMV movement, later studies suggested this was not so. Disruption of MT arrays with pharmacological agents or by tubulin transcript knock-down using virus-induced gene silencing had no effect on TMV movement or MP localization. Other work showed that the association of the MP with MTs happened late in

Table 1 Proteins necessary for the cell-to-cell movement of plant viruses

Virus	MP[a]	Ancillary viral proteins[b]	Host protein interactors with MP
Tobacco mosaic virus, Tomato mosaic virus	**30 kDa**	126 kDa	Actin, tubulin, MPB2C, PME, KELP, MBF1, calreticulin
Red clover necrotic mosaic virus	**35 kDa**		
Groundnut rosette virus	ORF4		
Cowpea chlorotic mottle virus	3a		
Brome mosaic virus	**3a**	CP	
Cucumber mosaic virus	**3a**	CP	NtTLP1
Bean dwarf mosaic virus	BC1	BV1	
Tobacco etch virus	CP	CI, HC-Pro, VPg	
Barley stripe mosaic virus	TGBp1	TGBp2, TGBp3	
Potato virus X	**TGBp1**	TGBp2, TGBp3 + CP	TIPs
Cowpea mosaic virus	**48 kDa**	CP	
Cauliflower mosaic virus	**38 kDa**	CP	MPI7, PME
Turnip crinkle virus	**p8 + p9**		Atp8
Tomato bushy stunt virus	**P22**		HFi22, REF
Potato leaf roll virus	17 kDa		
Tomato spotted wilt virus	**NS$_m$**		DnaJ-like, At4/1
Beet necrotic yellow vein virus	TGBp1	TGBp2, TGBp3, p14	
Grapevine fanleaf virus	2B	CP	Knolle, actin, tubulin
Rice yellow stunt virus	P3		
Rice dwarf virus	Pns6		
Southern bean mosaic virus	ORF1		
Turnip yellow mosaic virus	69 kDa		
Alfalfa mosaic virus	**P3**	CP	
Prunus necrotic ringspot virus	3a		
Tobacco rattle virus	**29 kDa**		
Soil-borne wheat mosaic virus	37 kDa		
Peanut clump virus	P51	P14, P17	
Potato mop top virus	**TGBp1**	TGBp2, TGBp3	
Commelina yellow mottle virus	N-term 216 kDa		
Beet yellows virus	p6, Hsp70h, p64	CPm, CP	
Rice stripe virus	Pc4		
Apple stem grooving virus	36 kDa		
Raspberry bushy dwarf virus	39 kDa		

[a]Regular (i.e., no bold) font indicates marginal classification as MP because the protein either has not been fully tested or has some but not all of the functions classically associated with MP (see text for definition).
[b]Necessary for viral cell-to-cell movement.

infection, probably after virus movement had occurred. Also, during time-course studies it was determined that the MP disappeared during late stages of infection. This finding, in combination with the discovery that a mutant virus expressing a functionally enhanced MP with limited affinity for MTs moved cell to cell better than the parental virus, led to the idea that the association of MP with MTs is critical for MP degradation rather than to aid virus cell-to-cell movement. Further support for this idea came from the finding that the *Nicotiana tabacum* host protein, MPB2C, binds to MP and promotes its accumulation at MTs, yet acts as a negative effector of MP cell-to-cell transport. The role of MTs during TMV movement remains to be fully understood, but at this time it appears that they are more involved with MP degradation or compartmentalization than with virus movement (**Figure 1(a)**).

In contrast to the large body of work focusing on the role of the MT–MP interaction in TMV movement, studies on the role of the MFs in the movement of TMV and other viruses have only recently been published. It was demonstrated that intracellular movement of TMV viral replication complexes (VRCs; large multiprotein complexes comprised of host and viral factors) and cell-to-cell spread of the virus were blocked by MF inhibitors (pharmacological and transcript silencing agents). VRCs were later determined to physically traffic along MFs (**Figure 1(a)**). The interaction of TMV VRCs with MFs may be mediated by the TMV 126 kDa protein (a protein containing helicase, methyltransferase, and RNA silencing suppressor domains), since expression of a 126 kDa protein:GFP fusion in the absence of the virus results in fluorescent protein bodies that, like VRCs, traffic along MFs. VRC association with MFs may be mediated through a direct interaction of the 126 kDa protein with MFs or through an intermediary cell membrane. MFs are known to associate with membranes in plant cells and

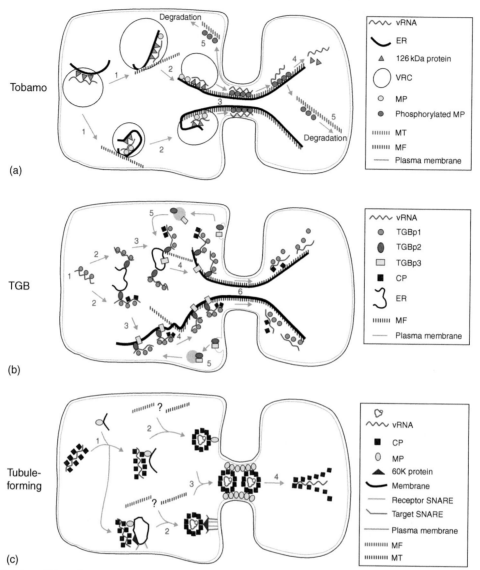

Figure 1 Models for cell-to-cell movement of plant viruses using tobamovirus triple gene blocks (TGBs), or tubule-forming strategies. (a) Viral 126 kDa protein binds both viral RNA (vRNA) and endoplasmic reticulum (ER) forming a cytoplasmic body in the cell termed a VRC. MP associates with the ER and possibly the vRNA within the VRC (step 1). VRCs associated with microfilaments (MFs) traffic toward plasmodesmata (PD; step 2). Here we show an indirect association of the VRC with actin mediated by the ER, but it is also possible that this interaction is mediated directly by the viral 126 kDa protein or MP. At the PD, vRNA is released from its association with the 126 kDa protein and is transported through the PD in association with MP (step 3). Phosphorylation of the MP occurs either within the cytoplasm, the cell wall, or both, and likely regulates transport to and through PD and subsequent translation of the vRNA in the new cell (steps 3 and 4). MP is degraded in the later stages of infection, likely via association with MTs and delivery to specific cellular sites of degradation (step 5). (b) Progeny vRNA binds to TGB protein 1 (TGBp1; step 1). The TGBp1/vRNA complex, either in the presence or absence of coat protein (CP, depending on virus genus), then binds TGBp2 attached to the ER to form a movement-competent ribonucleoprotein complex (RNP, step 2). The RNP then interacts with TGBp3, either directly or indirectly, to be positioned near the PD (steps 3 and 4). RNPs associate with actin MFs likely through an interaction with TGBp2, which may be responsible for transport to the PD. Following delivery of vRNA to the PD, TGBp2 and TGBp3 are likely recycled via an endocytic pathway (step 5). vRNA is actively transported through the PD via an unknown mechanism (step 6), although TGBp1 or CP may be involved, and is released from associated proteins in the next cell to allow replication to initiate. (c) CP-bound vRNA associates with the MP (itself associated with a membrane of unknown origin, step 1). The complex then moves, either as a vesicle directed to the PD through targeting proteins such as those from the SNARE family, or through other unknown targeting signals to the cell periphery (step 2). Interaction between SNAREs and virus may be mediated by a viral 60K protein for cowpea mosaic virus. The requirement for the cytoskeleton in transport of MP–vRNA complex is unclear since the nepovirus, grapevine fanleaf virus, requires cytoskeletal elements for proper delivery of its MP to the cell wall while cowpea mosaic virus does not. At or near the PD, the vesicular or nonvesicular membranes fuse with the plasma membrane and the attached MP directs the CP-associated vRNA through the PD (step 3). The vRNA is then released into the next cell to initiate virus replication and movement (step 4). Reproduced from Nelson RS (2005) Movement of viruses to and through plasmodesmata. In: Oparka KJ (eds.) *Plasmodesmata*, 1st edn., pp. 188–211. Oxford: Blackwell, with permission from Blackwell.

the 126 kDa protein binds to an integral membrane host protein, TOM1. However, the MP of TMV is long known to bind actin and associate with membranes, so the relative importance of the TMV 126 kDa protein or MP for directing intracellular VRC movement is unclear (**Figure 1(a)**).

Recently, MFs were demonstrated to co-localize with ancillary proteins required for movement of the hordeivirus, potato mop-top virus (PMTV). PMTV encodes a conserved group of proteins termed the triple gene block (TGB) that are required for cell-to-cell virus movement (see **Table 1**). Two of these TGB proteins (TGBp2 and TGBp3) co-localize with motile granules that are dependent upon the endoplasmic reticulum (ER)–actin network for intracellular movement. In addition, the TGBp2 protein from the potexvirus, potato virus X (PVX), localizes to MFs in what are likely ER-derived vesicles (**Figure 1(b)**). The association of TGBp2 from potexviruses and the 126 kDa protein from TMV with MFs and their requirement for successful virus movement provide an elegant example of convergent evolution since TGBp2 and 126 kDa protein have no sequence identity.

The role of the cytoskeleton in the intracellular transport of some plant viruses is unclear. Cowpea mosaic virus (CPMV), for example, does not require the host cytoskeleton for the formation of tubular structures containing MP on the surface of protoplasts. These tubular structures are similar to the tubules formed in modified PD (that likely do not contain cytoskeleton) which are necessary for intercellular movement of this virus (**Figure 1(c)**). The role of tubules in intercellular transport of CPMV is discussed in next section. Work with grapevine fanleaf virus (GFLV), another tubule-forming virus, has revealed the possibility that this virus may be targeted to the PD by membrane vesicle SNARE (v-SNARE)-mediated trafficking. The MP of GFLV co-immunoprecipitates with KNOLLE, a target SNARE (t-SNARE). The 60 kDa protein of CPMV has been shown to bind a SNARE-like protein. t-SNAREs such as KNOLLE act as specific receptors for targeted delivery of Golgi-derived vesicles to sites where fusion with the plasma membrane will occur. Thus, it is possible that the SNARE trafficking machinery delivers viral proteins (and possibly associated viral RNA) to the plasma membrane near PD (**Figure 1(c)**).

There is evidence that, following movement of the viral RNA to the PD, some viral factors involved in this movement may be recycled for further use. The TGBp2 and TGBp3 proteins from PMTV localize to endocytic vesicles as evidenced by labeling with FM4-64 dye, a marker for internalized plasma membrane (**Figure 1(b)**). Additionally, TGBp2 co-localizes with Ara7, a marker for early endosomes. The functional significance of this endocytic association of viral proteins remains to be determined and it is not known whether proteins from other viruses may also traffic in the host cell's endocytic pathway.

Intercellular Movement

Following intracellular movement to the cell periphery, the virus must then move through PD in order to spread into neighboring cells. PDs are plasma membrane-lined aqueous tunnels connecting the cytoplasm of adjacent cells. An inner membrane, termed the desmotubule, is a tubular form of the ER and is an extension of the cortical ER. PDs can be subdivided structurally into simple (containing a single channel) or branched (containing multiple channels) forms. The SELs of PD are increased by the disruption of MFs indicating a role for actin in PD gating and indeed both actin and myosin have been observed in PD. Thus, it is possible that cytoskeletal-mediated transport of viral components results in direct delivery of virus to and passage through PD.

Protein movement through PD is dependent on the developmental stage of the PD. For example, free GFP (27 kDa) moves through simple but not branched PDs. Branched PDs generally have smaller SELs than simple PDs and the presence of more branched PDs in mature photosynthate-exporting (source) versus immature photosynthate-importing (sink) leaves represents a developmental change that limits transport of macromolecules through PD. This developmental change also affects the localization pattern for some viral MPs. For example, both cucumber mosaic virus and TMV MPs are observed predominantly or solely within branched PD in source leaves and not simple PD in sink leaves. The TMV MP expressed in transgenic plants, however, is sufficient to complement the movement of an MP-deficient TMV mutant in both source and sink leaves. Thus, the presence of MP in the central cavity of branched PD in source leaves may not represent a site of function for the MP, but rather the final deposition of inactive MP. Although it is possible that the level of MP binding in simple PD is below the detection limits of the current technology, questions remain about where and how the MP functions in virus movement.

A clue to TMV MP function during virus movement comes from findings showing that a TMV MP–viral RNA complex could not establish an infection in protoplasts, but could do so when introduced into plants. It was suggested that a change in the phosphorylation state of the MP at the cell wall was necessary to weaken the binding between the MP and viral RNA, thereby allowing translation of the viral genome and initiation of infection in the next cell. Indeed, a PD-associated protein kinase has been identified that phosphorylates TMV MP. Thus, the protein kinase in the cell wall may be necessary to end the involvement of MPs in virus movement and release the viral RNA for translation in the new cell (**Figure 1(a)**). Also, considering that there are additional phosphorylation sites on the TMV MP besides those targeted by the PD-associated protein kinase, it is likely that proper sequential phosphorylation of this protein is necessary to allow it to function in both intracellular and intercellular virus movement. For potyviruses, the eukaryotic elongation factor, eIF4E, appears to modulate both virus accumulation, likely by affecting translation, and cell-to-cell movement. Thus, as for TMV, virus accumulation and movement may be linked activities.

Chaperones of host or viral origin may be required for PD translocation of some MPs. Host-encoded calreticulin modulates TMV intercellular movement and co-localizes with TMV MP in PD. The virus-encoded virion-associated protein (VAP) of cauliflower mosaic virus (CaMV) binds MP through coiled-coil domains and co-localizes with MP on CaMV particles within PD. The mechanism by which a molecular chaperone can support intercellular virus movement is illustrated by the virally encoded Hsp70 chaperone homolog (Hsp70h) of beet yellows virus. Hsp70h requires MFs to target it to the PD. The Hsp70h is a component of the filamentous capsid and its ATPase activity is required for virus cell-to-cell movement. These findings led to a model where Hsp70h mediates virion assembly and, once localized to the PD, actively translocates the virion from cell to cell via an ATP-dependent process. The idea that viral proteins may actively participate in plasmodesmal translocation of virus is further supported by the finding that the NTPase activity of the TGBp1 helicase from PMTV is necessary for its translocation to neighboring cells and that the coat protein (CP) of PVX, necessary for virus cell-to-cell movement, has ATPase activity (**Figure 1(b)**). It has also been found that the helicase domain of the TMV 126 kDa protein is required for cell-to-cell movement. In these cases it seems likely that the helicase activity is necessary to remodel viral RNA, thereby easing passage of the virus through PD.

Tubule-forming viruses have adopted another strategy for intercellular movement whereby virus-induced tubules span modified PD that lack a desmotubule in order to transmit capsids from cell to cell (**Figure 1(c)**). Such capsid-containing tubules are known to be composed, at least in part, of MP and have been identified for a number of viruses, including commelina yellow mottle virus, CaMV, CPMV, and tomato ringspot virus.

Although the tubule-forming viruses modify PD differently than those utilizing classical MPs, it was recently determined that the tubule-forming MP from tomato spotted wilt virus can functionally substitute for the non-tubule-forming TMV MP to support TMV movement. This is likely another example where two proteins with no sequence identity and therefore no apparent evolutionary relationship have independently evolved to functionally support movement of viruses.

Systemic Movement

Some viruses are limited to the phloem of plants (i.e., phloem-limited viruses) and require inoculation, often by aphids, directly to vascular cells for infection. Systemic movement of a non-phloem-limited virus through vascular tissue, however, requires that the virus moves from nonvascular cells into veins. Veins are defined as major or minor based on their structure, location, branching pattern, and function (**Figure 2**). Whether major or minor, each vein contains many different cell types with greatly differing structures. Within *N. tabacum*, minor veins include phloem parenchyma, xylem parenchyma, and companion cells, along with sieve elements and xylem vessels (**Figure 2**). All of these cells have distinct structures and locations within the vein which present unique regulatory sites for virus entry. Between plant species, companion cell morphologies vary greatly with an obvious difference being the number of PDs between these cells and other vascular cells. This difference is functionally related to the type of photosynthate transport system exhibited by the plant (i.e., apoplastic versus symplastic). In addition, bundle sheath cells, which have their own unique position and structure surrounding the minor veins, must be considered as potential regulators of virus movement. These complex cell types are difficult to study because it is problematic to directly access or isolate them.

Recently, studies have been conducted that conclusively indicate which veins, minor or major, can serve as entry sites for rapid systemic infection. Using surgical procedures to isolate specific veins and TMV or CPMV modified to express GFP as a reporter, it was determined that either major or minor veins in leaves of *Nicotiana benthamiana* and *Vigna unguiculata* can be invaded directly and serve as inoculum sources for systemic infection. In addition, for TMV, direct infection of cells in transport veins in stems yielded a systemic infection. Considering that major and transport veins do not have terminal endings bounded by nonvascular tissue, it is likely that virus entered these veins by passing through bundle sheath cells and interior vein cells.

Virus transport and accumulation are regulated within vascular tissue. In plants that have internal and external phloem, potyviruses and tobamoviruses accumulate in specific tissue depending on the tissue's position relative to the inoculated leaf. In the inoculated leaf and the stem below, virus accumulates in the external phloem, whereas in the stem and leaf veins above the inoculated leaves, virus accumulates in the internal phloem.

Exit of PVX, TMV, and CPMV from vascular cells in sink tissues only occurs from major veins. For a growing number of viruses, however, exit occurs from both major and minor veins indicating that there is not a uniform exit strategy for all viruses.

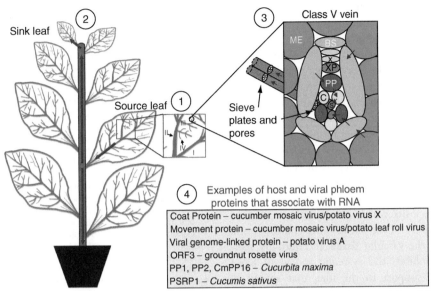

Figure 2 (1) Viral infection of a source leaf occurs by intercellular movement of the virus into the vasculature (class I–V veins indicated). (2) Virus travels through the phloem from the source leaf throughout the plant (red arrows) and exits vascular tissue to resume cell-to-cell movement in sink tissue. (3) In order to enter the phloem of a class V vein, a non-phloem-limited virus must travel through mesophyll cells (ME), bundle sheath cells (BS), and possibly phloem parenchyma cells (PP) before entering the companion cells (C) and finally the sieve elements (S). Movement through SEs requires passage of virus through pores within the sieve plates. A minority of viruses move through the xylem (X). (4) Examples of host and viral proteins that have been identified in phloem and that associate with RNA are indicated.

The virus and host factors that control systemic virus accumulation are becoming better understood, mostly through genetic studies. Virus factors include CP, some MPs, and some nonstructural proteins such as the 126 kDa protein of TMV. Although CPs are often necessary for systemic infection, it is clear that for some viruses, such as groundnut rosette virus, a CP is not present and the virus still produces a systemic infection in the host. Also, for viruses that normally require the CP for systemic infection, the loss of the CP through mutation or deletion may still allow systemic movement of the virus in specific hosts. Lastly, viroids, which do not encode any proteins, can systemically infect plants. These results indicate that although a capsid may be required to protect viral RNA for systemic transport in some hosts, other viral or host proteins can functionally mimic the CP and allow systemic infection.

MP function during systemic infection has, in one case, been uncoupled from its role during intra- and intercellular transport. Some point mutations in the red clover necrotic mosaic virus MP still allow intercellular movement, but prevent systemic movement. Additional support that MPs function to allow systemic movement comes from studies with the 17 kDa MP of potato leafroll virus, a phloem-limited virus. This MP, when expressed from within an infectious virus sequence in transgenic plants, is uniquely localized to PD connecting the companion cells with sieve elements, even though virus accumulated in both vascular and nonvascular cells. Thus, the PD between companion cells and sieve elements may be

uniquely recognized by this MP to allow the virus to only invade vascular tissue. More recently, it has been shown that a host factor, CmPP16, that is thought to function by forming ribonucleoprotein complexes with phloem transcripts has sequence similarity with viral MPs. Thus, some MPs may function to protect RNA while in transit through the phloem.

Other viral proteins such as the 2b protein of CMV, p19 of TBSV, and the 126 kDa protein of TMV have been linked to supporting systemic movement of their respective viruses. Considering that all of these proteins are suppressors of gene silencing, it is possible that this activity is related to their function in supporting systemic movement. It is known that a member of the plant silencing pathway, specifically, the RNA-dependent RNA polymerase, RDR6, functions in sink tissue (e.g., the shoot apex) by responding to incoming signals for RNA silencing. RDR6 has also been shown to control virus accumulation in systemic, but not inoculated, leaves. Thus, it is possible that viral suppressor activity could function to specifically allow systemic accumulation of viruses.

Host factors that modulate virus systemic spread either support or restrict this activity. A protein methylesterase (PME) is involved in both intercellular and systemic movement of TMV. For systemic movement, PME is essential for virus to exit into nonvascular tissue of the uninoculated leaves. A phloem protein from cucumber, p48, was found to interact with CMV capsids and may function to protect the capsid during transport.

Host factors that restrict virus systemic movement include the restricted TEV movement (RTM) proteins, which are expressed only in phloem-associated cells and accumulate in sieve elements. RTM1 is related to the lectin, jacalin, while RTM2 has a heat shock protein motif. RTM1 may function in a plant defense pathway within the veins, although the jacalin-like proteins have not been previously linked to virus defense. RTM2 may function as a chaperone to prevent unfolding of a transport form of the virus within the sieve elements. A third protein that serves as a negative regulator is a cadmium-induced glycine-rich protein, cdiGRP. This protein does not act directly to restrict systemic movement. Instead, it induces callose deposits which are thought to restrict intercellular transport of the virus. This could prevent exit of virus from the vascular tissue. Interestingly, cadmium treatment inhibits the systemic spread of RNA silencing, lending support to the idea that spread of specific viruses affected by cadmium treatment (i.e., TMV and turnip vein clearing virus) is functionally similar to that of a host silencing signal.

See also: Bromoviruses; Carmovirus; Citrus Tristeza Virus; Cucumber Mosaic Virus; Furovirus; Luteoviruses; Nepovirus; Plant Resistance to Viruses: Engineered Resistance; Plant Resistance to Viruses: Geminiviruses; Potexvirus; Tobacco Mosaic Virus; Tobamovirus; Tobravirus; Tombusviruses; Tospovirus; Umbravirus; Viroids; Virus Induced Gene Silencing (VIGS).

Further Reading

Boevink P and Oparka KJ (2005) Virus–host interactions during movement processes. *Plant Physiology* 138: 1815–1821.

Derrick PM and Nelson RS (1999) Plasmodesmata and long-distance virus movement. In: van Bel AJE and van Kesteren WJP (eds.) *Plasmodesmata: Structure, Function, Role in Cell Communication,* 1st edn., pp. 315–339. Berlin: Springer.

Gilbertson RL, Rojas MR, and Lucas WJ (2005) Plasmodesmata and the phloem: Conduits for local and long-distance signaling. In: Oparka KJ (ed.) *Plasmodesmata,* 1st edn., pp. 162–187. Oxford: Blackwell.

Heinlein M and Epel BL (2004) Macromolecular transport and signaling through plasmodesmata. *International Review of Cytology* 235: 93–164.

Lewandowski DJ and Adkins S (2005) The tubule-forming NSm protein from *Tomato spotted wilt virus* complements cell-to-cell and long-distance movement of *Tobacco mosaic virus* hybrids. *Virology* 342: 26–37.

Lucas WJ (2006) Plant viral movement proteins: Agents for cell-to-cell trafficking of viral genomes. *Virology* 344: 169–184.

Morozov SY and Solovyev AG (2003) Triple gene block: Modular design of a multifunctional machine for plant virus movement. *Journal of General Virology* 84: 1351–1366.

Nelson RS (2005) Movement of viruses to and through plasmodesmata. In: Oparka KJ (ed.) *Plasmodesmata,* 1st edn., pp. 188–211. Oxford: Blackwell.

Nelson RS and Citovsky V (2005) Plant viruses: Invaders of cells and pirates of cellular pathways. *Plant Physiology* 138: 1809–1814.

Oparka KJ (2004) Getting the message across: How do plant cells exchange macromolecular complexes? *Trends in Plant Science* 9: 33–41.

Rakitina DV, Kantidze OL, Leshchiner AD, *et al.* (2005) Coat proteins of two filamentous plant viruses display NTPase activity *in vitro. FEBS Letters* 579: 4955–4960.

Requena A, Simón-Buela L, Salcedo G, and García-Arenal F (2006) Potential involvement of a cucumber homolog of phloem protein 1 in the long-distance movement of cucumber mosaic virus particles. *Molecular Plant Microbe Interactions* 19: 734–746.

Roberts AG (2005) Plasmodesmal structure and development. In: Oparka KJ (ed.) *Plasmodesmata,* 1st edn., pp. 1–32. Oxford: Blackwell.

Scholthof HB (2005) Plant virus transport: Motions of functional equivalence. *Trends in Plant Science* 10: 376–382.

Silva MS, Wellink J, Goldbach RW, and van Lent JWM (2002) Phloem loading and unloading of cowpea mosaic virus in *Vigna unguiculata. Journal of General Virology* 83: 1493–1504.

Verchot-Lubicz J (2005) A new cell-to-cell transport model for potexviruses. *Molecular Plant Microbe Interactions* 18: 283–290.

Waigman E, Ueki S, Trutnyeva K, and Citovsky V (2004) The ins and outs of nondestructive cell-to-cell and systemic movement of plant viruses. *Critical Reviews in Plant Sciences* 23: 195–250.

Vector Transmission of Plant Viruses

S Blanc, INRA–CIRAD–AgroM, Montpellier, France

Glossary

Fitness The relative ability of an individual (or population) to survive and reproduce in a given environment.

Helper component (HC)-transcomplementation An HC encoded by a viral genome X mediates the vector transmission of a virus particle containing a viral genome Y.

Horizontal transmission The transmission of a virus, parasite, or other pathogen from one individual to another within the same generation, as opposed to vertical transmission.

Pierce-sucking insects Insects adapted to sap or blood feeding, with the mouthparts transformed into long chitin needles that can pierce and penetrate tissues and allow pumping up their content.

Quasispecies Ensemble of mutant viral genomes constituting a viral population.
Vector Organism acquiring a pathogen on an infected host and inoculating it in a new healthy one.
Vertical transmission The transmission of a pathogen from the parent(s) to the offspring, usually through the germline.

Introduction

Viruses are intracellular parasites diverting the host cellular machinery for their own replication and offspring particles production. As such, they most often negatively affect the hosting cells, sometimes even killing them, and hence repeatedly and unavoidably face the problem of moving on and colonizing new healthy and potent 'territories'. Within a single host, viruses can both diffuse from cell to cell and be transported on longer distances by the vascular system. While animal viruses use membrane fusions (if enveloped) or membrane receptors to penetrate healthy cells, plant virus entry during the host invasion is always resulting from a passage through 'tunnels' traversing the cell wall, called plasmodesmata, and ensuring a cytoplasm continuity between adjacent cells. Any viral population can grow this way only until the physical limits of the host are reached. Then, a critical passage in the 'outside world' separating two compatible hosts has to be successfully achieved. Because animals are motile and often come in contact, some associated viruses can directly access either blood or permissive tissues of a healthy host and operate a cell entry resembling that involved during invasion of single hosts. However, a most frequently adopted strategy relies on additional organisms, capable of sampling the virus population within an infected host, transporting, disseminating, and efficiently inoculating infectious forms of this virus within host population. Such organisms are designated vectors, giving rise to the term vector transmission. Vector transmission is found frequently in animal viruses and, presumably due to stable hosts and to the need of covering considerable distances between them, has been adopted by the vast majority of plant viruses. Each virus species is submitted to different ecological conditions; hence, an impressive complexity of host–virus–vector interactions has been unraveled over a century of research efforts. The object of this chapter is to synthesize the knowledge available at present in the field of vector transmission of viruses, with a special emphasis on plant viruses, where a great diversity of strategies have been discovered and documented. Indeed, the numerous patterns of vector transmission described for plant viruses include all those reported in animal viruses and many more.

Plant Virus Vectors

Any organism that is creating a break into the cell wall, either for penetrating a plant or simply for feeding on it, and that is capable of covering the distance between two separated plants, can possibly be used as a vector by viruses, for traveling through space and time. Vectors have been described in groups of organisms as diverse as parasitic fungi, nematodes, mites, and most importantly insects (**Table 1**). The pattern of virus uptake, preservation, transport, dissemination, and inoculation can be very different, due to the specific biology of all three (plant, virus, and vector) partners. However, viruses transmitted by 'pierce-sucking' insects are quantitatively predominant, and the classification established for their various modes of transmission is widely used as a reference for comparison with others. For this reason, hemipteran insect transmission will be described first in details and succinctly compared later on with that by other types of vectors.

Transmission of Plant Viruses by Insects

History of the Classification of the Different Modalities of Transmission

The transmission of plant viruses has been investigated for over a century with the most common vectors being sap-feeding insects with pierce-sucking mouth parts, particularly aphids, and also whiteflies, leafhoppers, planthoppers, and mealybugs. Pioneer studies have demonstrated the complexity and diversity of the interactions between plant viruses and their insect vectors. Even as late as the 1950s, scientists, using the tools at hand, were merely measuring quantitative traits such as the time required for virus acquisition on infected plants and the time during which the virus remained infectious within the vector. Three categories were then defined: (1) the nonpersistent viruses, acquired within seconds and retained only a few minutes by their vectors; (2) the semipersistent viruses, acquired within minutes to hours and retained during several hours; and (3) the persistent viruses that require minutes to hours for acquisition, and can be retained for very long periods, often until the death of the vector. It is important to note that, though the classification and terminology have changed in the last decades, these categories are still used by a number of authors, and thus often encountered in the literature.

In an early study on 'nonpersistent' viruses, the transmission of potato virus Y was abolished by chemical (formaldehyde) or ultraviolet (UV) treatments of the extremity of the stylets of live viruliferous aphids, demonstrating that infectious virus particles were retained there. It was first believed that the transmission of 'nonpersistent' viruses could be assimilated to mechanical transmission, stemming from nonspecific contamination of the stylets, the

Table 1   Vectors and mode of transmission in families of plant viruses

Family[a]	Vector	Mode of vector-transmission[b]
Bromoviridae genus *Alfamovirus*	Aphids	Noncirculative capsid strategy
Bromoviridae genus *Cucumovirus*	Aphids	Noncirculative capsid strategy
Bromoviridae genus *Ilarvirus*	Thrips	?
Bromoviridae genus *Oléavirus*	?	?
Bromoviridae genus *Bromovirus*	Beetle	?
Bunyaviridae	Thrips, planthopper	Circulative propagative
Caulimoviridae	Aphid, mealybug, leafhopper	Noncirculative helper strategy
Circoviridae	Aphid	Circulative nonpropagative
Closteroviridae	Aphid, whitefly, mealybug	Noncirculative
Comoviridae genus *Comovirus*	Beetle	?
Comoviridae genus *Fabavirus*	Aphid	Noncirculative
Comoviridae genus *Nepovirus*	Nematode	Noncirculative capsid strategy
Geminiviridae	Leafhopper, whitefly	Circulative nonpropagative[c]
Luteoviridae	Aphid	Circulative nonpropagative
Partiviridae	?	?
Potyviridae genus *Potyvirus*	Aphid	Noncirculative helper strategy
Potyviridae genus *Ipomovirus*	Whitefly	Noncirculative
Potyviridae genus *Macluravirus*	Aphid	Noncirculative
Potyviridae genus *Rymovirus*	Mite	Noncirculative
Potyviridae genus *Tritimovirus*	Mite	Noncirculative
Potyviridae genus *Bymovirus*	Fungus	Circulative
Reoviridae	Planthopper, leafhopper	Circulative propagative
Rhabdoviridae	Leafhopper, aphid	Circulative propagative
Sequiviridae	Aphid, leafhopper	Noncirculative helper strategy
Tombusviridae	Fungus	Noncirculative

[a]The families are broken down to the genus level when they contain genera with totally different vectors and mode of transmission.
[b]The helper or capsid strategies (see **Table 2**) are mentioned when experimentally demonstrated for at least one of the member species. When no complement is added to either circulative or noncirculative, it reflects the lack of further information.
[c]For at least one member species (*Tomato yellow leaf curl virus*, TYLCV), replication within the vector is still being debated.
The noncirculative viruses, or assimilated as discussed in the text, are in blue. The circulative viruses, or assimilated as described in the text, are in green.

vector acting simply as a 'flying needle'. Consistent with this was the repeated demonstration that 'nonpersistent' viruses are lost upon moulting of the viruliferous vectors. Later on, the hypothesis of virus uptake during sap ingestion and inoculation during putative regurgitation led to a change from vectors as flying needles to vectors as 'flying syringes', the virus–vector relationship still being considered as nonspecific (**Figure 1**). It is interesting that, while in plant viruses recent data unequivocally convinced the scientific community that the situation is much more complex, likely involving specific receptors in vectors for specific virus species (see the next section), in animal viruses very few studies are available at present and this mode of transmission is still referred to as 'mechanical vector transmission'. The prime conclusion from these experiments is that the virus–vector association occurs externally, on the cuticle lining the food or salivary canal in the insect stylets. Because semipersistent viruses are also lost upon vector moulting, their association with the vector was also proposed to be external, likely in the stylets, though a possible location 'upstream', on the cuticle lining the anterior gut of the insect, was also proposed in some cases.

In sharp contrast, many persistent viruses were observed within the vector body by electron microscopy, in various organs and tissues, indicating an internal association with the vectors. Such viruses were shown to pass through the gut epithelium into the hemolymph and join the salivary glands to be ejected together with saliva (**Figure 1**). A latent period of hours to days after acquisition, during which the virus cannot be efficiently inoculated, is consistent with the time needed for completing this cycle within the vector body. Moreover, microinjection of purified persistent viruses within the insect hemolymph subsequently resulted in efficient transmission to new healthy plants, proving that virus within the vector body can get out and be inoculated to host plants.

Altogether, these results prompted a revision of the classification of the modes of transmission in the late 1970s, based on qualitative criteria and still valid today (**Table 2**). The non- and semipersistent viruses were grouped in a new category designated 'noncirculative viruses', and the persistent viruses were named 'circulative viruses'. While circulative animal viruses (arboviruses) in fact infect their vectors where they efficiently replicate, some circulative plant viruses can seemingly operate their cycle in the vector body without any cell infection and replication. Hence, the category 'circulative' has been broken down into the two subcategories: 'propagative'

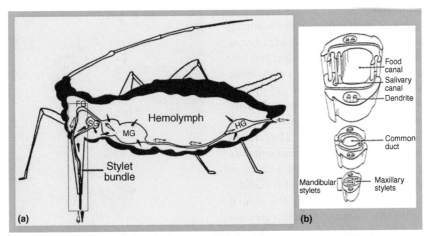

Figure 1 Different routes of plant viruses in their aphid vectors. (a) The white arrows represent the ingestion of circulative viruses, whereas the black arrows materialize their cycle within the aphid body, and inoculation in a new host plant. The red square area indicates the region of the anterior feeding system, where noncirculative viruses are retained in their vectors. (b) Cross sections of the stylet bundle illustrating the inner architecture of maxillary stylets which defines interlocking structures, food canal and salivary canal, fused at the distal extremity into a single common duct, where most noncirculative viruses are thought to be retained (see text). Adapted from Taylor CE and Robertson WM (1974) Electron microscopy evidence for the association of tobacco severe etch virus with the Maxillae in Myzuspersical (Sulz.). *Journal of Phytopathology* 80: 257–266.

Table 2   Different modes of plant virus transmission by insects with pierce-sucking mouth parts

	Circulative		Noncirculative	
Transmission modes[a]	Propagative	Nonpropagative	Capsid strategy	Helper strategy
Acquisition time[b]	Minutes to hours	Minutes to hours	Seconds to hours	Seconds to hours
Retention time[c]	Days to months	Days to months	Minutes to hours	Minutes to hours
Inoculation time[d]	Minutes to hours	Minutes to hours	Seconds to minutes	Seconds to minutes
Association with vectors[e]	Internal	Internal	External	External
Replication in vectors	Yes	No	No	No
Requirement of an HC[f]	No	No	No	Yes

[a]These modes of transmission were established and are widely accepted for virus transmission by pierce-sucking insects. As discussed in the text, they sometimes also apply to other types of vector.
[b]The length of time required for a vector to efficiently acquire virus particles upon feeding on an infected plant.
[c]The length of time during which the virus remains infectious within its vector, after acquisition.
[d]The length of time required for a vector to efficiently inoculate infectious virus particles to a new healthy plant.
[e]Internal means that the virus enters the inner body of its vector, passing through cellular barriers. External means that the virus binds the cuticle of the vector and never passes through cellular barriers.
[f]A helper component (HC) is involved in cases where the virus particles do not directly recognize vectors, acting as a molecular bridge between the two.

and 'nonpropagative'. The various families and/or genera of plant viruses and their associated vectors and modes of transmission are listed in **Table 1**.

During the last decades, the implementation of molecular and cellular biology has provided invaluable tools for studying the molecular mechanisms of virus–vector interaction. The data currently available for each category are summarized in the following subsections.

Circulative Transmission

Logically, circulative viruses are ingested by vectors, while feeding on infected plants. Some viruses are limited to phloem tissues, which the insect vector can reach within minutes to hours depending on the species, which explains the long feeding period required for their acquisition. As schematized in **Figure 1(a)**, the viruses cross the mid- or hindgut epithelium, are released into the hemolymph, and can then adopt various pathways to traverse the salivary glands, and be released in their lumen, wherefrom they will be inoculated upon salivation into healthy hosts. During this basic cycle , the virus encounters and must overcome diverse cellular barriers, where the existence of specific virus–vector interaction has long been established experimentally, though specific receptors have not been identified so far.

Propagative transmission

Propagative transmission of plant viruses is the homolog of that of arboviruses in vertebrates. Members of the virus families *Rhabdoviridae*, *Reoviridae*, *Bunyaviridae*, and the genus *Marafivirus* are transmitted this way. In compatible virus–vector associations, after infecting the gut epithelium, virus particles are released in the hemocoel cavity and colonize various organs and tissues of the vector, including, ultimately, the salivary glands. The viruses can either diffuse in the hemolymph and concomitantly infect different organs, or follow a constant pattern of spread from organ to organ, as demonstrated for rhabdoviruses which move in (and spread from) the central nervous system. Within the vector, all these cases are very similar to genuine infection of an insect host; hence, it is difficult to decide whether insects are proper vectors of propagative viruses, or should rather be considered as alternative hosts. Apart from the genus *Marafivirus*, all propagative viruses are in families that comprise viruses that infect animals, suggesting that they might have evolved from insect viruses by secondarily acquiring the capacity to replicate in plants.

Nonpropagative transmission

This particular association between insect vectors and plant viruses is reasonably well understood only for members of some plant virus species in the family *Luteoviridae*. Such viruses have developed original mechanisms of viral transport, both when passing through gut and salivary gland barriers, and when traveling into the hemocoel cavity.

The cycle of luteoviruses within their vector body involves specific ligand-receptor-like recognition at the cell entry of both the gut epithelium and the salivary glands. While viral ligands are known to be structural proteins of the coat and extension thereof, very little is known about the putative counterpart receptors on the cell membranes of the vectors. Despite this lack of full understanding of the molecular process, many electron microscopy and molecular studies have determined in detail the route of luteovirus particles within the vector, across cellular layers (**Figure 2**). Once the virus reaches either the apical membrane of the gut epithelium, or the basal membrane of the accessory salivary gland cells, and attaches to the specific receptors, it provokes invagination of the plasmalemma, forming small coated virus-containing vesicles. Soon after budding, the coated vesicles deliver the virus particles to a larger uncoated membrane endosomal compartment. Interestingly, luteoviruses mostly escape the route of degradation of internalized material ending in lysosomes. Instead, the virus particles become concentrated in the endosomes, and *de novo* elongated uncoated vesicles are repacked, transporting the viruses to the basal or apical membrane,

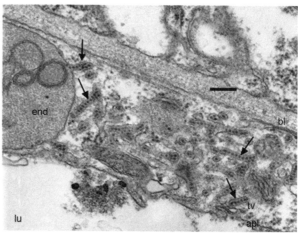

Figure 2 Transcytosis of cucurbit aphid-born yellows virus (family *Luteoviridae*) in hindgut cell of the aphid vector *Myzus persicae*. Luteovirus particles present in the gut lumen (lu) are internalized from the apical plasmalemma (apl) and transported to the basal lamina (bl) in a complex pattern involving different vesicular structures, described in the text. A network of uncoated tubular vesicles is visible (tv indicated by arrows), sometimes connected to the endosome (end). The bar represents 100 nm. The photograph is provided by Catherine Reinbold and Véronique Brault (INRA, Colmar, France). Reproduced from Blanc S (2007) Virus transmission – getting out and in. In: Waigmann E and Heinlein M (eds.) *Viral Transport in Plants*, pp. 1–28. Berlin: Springer, with kind permission of Springer Science and Business Media.

in gut and accessory salivary gland cells, respectively. The elongated vesicles, which contain rows of virions (**Figure 2**), finally fuse with plasma membranes and release the virus either into the hemocoel cavity or into the lumen of the salivary ducts. Despite these extensive searches of luteovirus particles in their insect vectors, by several independent groups of scientists, no particles have been observed in any organ other than the gut or the accessory salivary glands. Furthermore, classical monitoring of the viral titers within the vector, for assessing viral replication, failed to demonstrate an increase over time. Consequently, it is generally accepted that virus particles, either included in membrane vesicles or suspended in the hemolymph, never come in contact with the cell cytoplasm, thus precluding any possibility of viral replication.

Virus transfer into the hemolymph is believed to occur by passive diffusion. However, the possible impact of the insect immune system at this step of the virus life cycle is often discussed. A study demonstrated that a major protein of the aphid hemolymph, the symbionin, was mandatory for efficient luteovirus transmission. The symbionin is a homolog of the *Escherichia coli* chaperone GroEL, secreted in aphids by endosymbiotic bacteria of the genus *Buchnera*. Eliminating symbiotic bacteria, and thus symbionin, by antibiotic treatments significantly reduced the aphid efficiency as a vector. Consistently, direct evidence of a physical interaction between symbionin and luteovirus particles has been detected in several viral species, and virus mutants deficient in symbionin binding are poorly transmitted. Two hypotheses can explain the positive action of symbionin and are still debated, since no direct proof could be experimentally obtained: (1) it exhibits protective properties, masking the virus to the immune system and maintaining its integrity during transfer through the hostile hemolymph environment, or (2) its putative chaperon activity ensures correct folding facilitating transfer into the salivary glands.

A similar phenomenon was later demonstrated for other circulative nonpropagative viruses (in the family *Geminiviridae*) by other vectors (whiteflies), suggesting that symbionin participation may be a general phenomenon in this mode of transmission.

Noncirculative Transmission of Plant Viruses

This mode of transmission is by far the most frequently encountered in plant viruses and concerns over 50% of the viral species described to date. Noncirculative viruses do not enter the body of their vectors. They simply attach to receptor sites located externally on the cuticle lining the anterior part of the digestive tract, most often the alimentary/salivary canals within the stylets or the foregut region, and wait until the vector has moved to another plant, where they contrive to be released to initiate a new infection. It has often been proposed that difference between non- and semipersistent viruses, in acquisition and retention time within the vector, was due to a differential location of the binding sites, the former being retained in the stylets and the latter higher up in the foregut. This distinction, however, is not experimentally supported at present, and these two categories should be considered with caution (they are no longer included in **Table 2**). Because noncirculative viruses are adsorbed on the vector cuticle during sap ingestion, and released in many cases during salivation, the favored but still hypothetical location for the retention sites is at the distal extremity of the maxillary stylets, where a single common salivary/food canal is serving for both sap uptake and saliva ejection (**Figure 1(b)**). The best-studied cases of noncirculative transmission are those of viruses in the genera *Cucumovirus*, *Potyvirus*, and *Caulimovirus*, transmitted by various aphid species. Mutagenesis studies of the viral proteins interacting with the vector's mouthparts have clearly indicated that some single-amino-acid substitutions can specifically abolish the virus transmissibility by certain vector species but not by others. Such results, obtained both with cucumber mosaic virus and cauliflower mosaic virus, demonstrate a very specific virus–vector recognition, and strongly suggest the existence of a receptor in the insect mouthparts. Unfortunately, here again and despite the quantitative importance of noncirculative transmission in plant viruses, the viral ligands are well characterized (see below), but the receptors on the insect cuticle stand as the major black box in this field of research, their chemical nature and even their precise location remaining largely hypothetical.

The viral protein motifs directly involved in the attachment to vector putative receptors have been characterized in a number of cases. The frequent occurrence of both transmissible and nontransmissible isolates in the same virus species has greatly facilitated the identification of viral gene regions involved in vector transmission, by simple sequencing of viral genomes, and reverse genetic approaches have been successfully used for definitive confirmation. These investigations clearly revealed two distinct viral strategies for controlling the molecular mechanisms of virus–vector association (**Figure 3**), which represent the currently preferred subcategories in noncirculative transmission (**Table 2**).

The helper strategy

The development of artificial feeding of insect vectors, through stretched parafilm membranes, made it possible to assess the transmission of purified virus particles. The primary striking result was that, in most cases, purified virions are not transmissible, suggesting the requirement of an additional component that is probably eliminated during purification. Sequential feeding on plants infected with transmissible and nontransmissible isolates of such virus species showed that the former produced a compound that could be acquired by the vector, and subsequently mediated the transmission of the nontransmissible isolate. This compound, first described in potyviruses and caulimoviruses, was designated helper component (HC). In both cases, the HC was later purified and demonstrated to be a viral nonstructural protein produced upon plant cell infection, named P2 in caulimoviruses and HC-Pro in potyviruses. P2 and HC-Pro are responsible for specific recognition of the vector. They can be acquired on their own by the aphid and bind to putative receptors within its mouthparts. They exhibit a separated domain, which specifically binds homologous virus particles, thus creating a molecular bridge between virus and vector (**Figure 3**). It is important to realize that HCs can be acquired on their own by the vector, subsequently scavenging virions from homologous species acquired in other locations of the same plant, or even in other plants. This phenomenon extending the assistance between related viral genomes during vector transmission has been termed HC-transcomplementation (**Figure 3**), and may have some important implications in the population genetics and evolution of a virus species (discussed below).

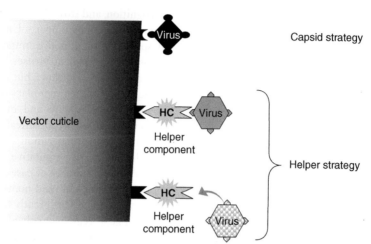

Figure 3 Two molecular strategies for virus–vector interaction in noncirculative transmission of plant viruses. Both strategies allow the retention of virus particles in the vector mouth parts or foregut on putative receptors located at the surface of the cuticular lining. In the capsid strategy, a motif of the coat protein is able to directly bind to the vector's receptor. In the helper strategy, virus–vector binding is mediated by a virus-encoded nonstructural protein (HC), which creates a reversible 'molecular bridge' between the two. HC can be acquired alone, prior to the virion, and thereby allows HC-transcomplementation. In this case, an HC encoded by a genome X (for instance that encapsidated in the gray virion) can subsequently assist in the transmission of a genome Y of the same population, encapsidated in the dotted virion. This possible sequential acquisition of HC and virion is symbolized by the arrow. Reproduced from Froissart R, Michalakis Y, and Blanc S (2002) Helper component-transcomplementation in the vector transmission of plant viruses. *Phytopathology* 92: 576–579, with permission from The American Phytopathological Society.

Capsid Strategy

However, for some viruses transmitted by aphids, particularly in the genera *Cucumovirus, Alfamovirus,* and *Carlavirus,* purified viral particles are readily acquired and transmitted by the vector. This indicated unequivocally that the coat protein of members of the species must be capable of direct attachment to vector receptors (**Figure 3**). An experiment confirming this conclusion was the demonstration that tobacco mosaic virus, which is not transmitted by any vector, can be transmitted by aphids when its RNA genome is encapsidated *in vitro* into the coat protein of an aphid-transmissible cucumovirus. Recently, the amino acid positions in the coat protein of cucumber mosaic virus, involved in binding to the putative receptors of aphids, have been identified. Their substitution differentially affects the efficiency of transmission by distinct aphid species, definitely confirming that no HC is required for virus/vector interaction.

Transmission by Beetles

The only cases of insect transmission diverting clearly from one of the patterns detailed above is that described in insect with 'biting-chewing' mouthparts, the best documented examples being beetles, though other insects have a similar feeding behavior.

In contrast to insect with 'pierce-sucking' mouthparts, the beetle feeding is damaging tissues and killing cells, implying an obligate translocation of deposited viruses toward adjacent live cells where they can initiate infection.

Beetles have long been reported to acquire and retain viruses from numerous virus species, usually those with highly stable virus particles, and to release them together with the regurgitation of disrupted plant material, lubricating the mouthparts during feeding on new host plants. Surprisingly, however, only some rare species are actually successfully transmitted. A high amount of RNase activity has been found in the beetle regurgitation liquid, and shown to block infection by non-beetle-transmissible viruses. Hence, those viral species that are efficiently transmitted are likely capable of translocation in the vascular system, and/or transfer to unwounded cells, away from the RNase activity. Some virus particles have been detected within the hemolymph of their beetle vector, only seconds after acquisition on infected plants. No correlation between the presence of virions in the hemolymph and success of transmission has yet been established, precluding a possible comparison with the circulative or noncirculative transmission described above.

Transmission by Noninsect Vectors

Although quantitatively less important, noninsect vectors have been identified in several plant virus species. Some are arthropod mites with a morphology distantly related to that of insects, others are totally unrelated organisms such as nematodes or even fungi. However, in many instances, a comparison with the mode of transmission defined in **Table 2** remains possible, demonstrating that this classification can be used as a general basis applicable to all cases of vector transmission.

Transmission by Mites

The transmission of plant viruses by mites has been far less studied, and the molecular and cellular mechanisms of virus–vector relationships remain poorly understood. Several different viral species have been studied and the results suggest the existence of different types of interactions with their respective vectors. This variety is illustrated by two examples: (1) the presence of massive amounts of bromegrass mosaic virus (BMV, family *Bromoviridae*) particles within cells of the gut epithelium of the insect *Eriophyes tulipae* seems to indicate that the virus replicates within the vector, resembling circulative-propagative transmission; (2) wheat streak mosaic virus (WSMV, family *Potyviridae*) appears to be noncirculative. WSMV was recently demonstrated to interact with its mite vectors through an HC-Pro homolog, thus likely suggesting a helper strategy. Further investigations on additional viral species are required to evaluate the diversity of the modes of transmission found with mite vectors, but it seems likely that they will be closely related to those described for insect vectors in **Table 2**.

Transmission by Nematodes

The transmission by nematode vectors is particularly interesting. Such vectors are moving very slowly in the soil (one or a few meters per year), and cannot disseminate the transmitted virus over long distances. Instead, they retain virus particles, for several months or even years, and usually transmit them in the same location but from plants of a growing season *n* to plants of the growing season *n* + 1. Consequently, nematode transmission is resulting in a time travel for the virus, rather than a space travel.

Nematodes feed by piercing root cells with protractile stylets, constituting the anterior part of their feeding apparatus, and by ingesting the cell content together with viruses when host plants are infected. In all cases described in the literature, the virus particles are adsorbed on the cuticle lining the foregut and are presumably released from there when inoculated in a new host plant. Since viruses are lost upon moulting of the nematode, the virus–vector association is external and this mode of transmission can be assimilated to the noncirculative transmission. Interestingly, the two subcategories of noncirculative transmission also appear to be represented in nematode transmission. Indeed, among the viruses belonging to two viral genera using nematodes as vectors, *Nepovirus* and *Tobravirus*, viruses in the former genus have been shown to directly attach to the retention sites through their coat protein (capsid strategy), whereas in the latter genus, viruses produce a nonstructural protein which allows transcomplementation (see the helper

strategy in **Figure 3**). As in other virus–vector associations, the viral proteins involved in vector recognition were identified, but the putative receptor in the nematodes remains unknown.

Transmission by Fungi

Parasitic fungi are coming in contact with their host plants under the form of motile zoospores, which can digest the root cell wall and penetrate into the cytoplasm, from where they will colonize the whole plant. Two different patterns of virus transmission exist. Some viruses, for instance in the genus *Bymovirus*, are present within the fungus cytoplasm early during formation of the future zoospores in infected plant cells. They will remain inside the zoospore until its cytoplasm is injected in the next host cell. In other cases, for instance in cucumber necrotic virus (genus *Tombusvirus*), the best-studied example of a fungus-transmitted virus, virions are specifically retained at the surface of the zoospore envelope, and inoculated into the plant upon cell wall digestion and fungal penetration. In this case, the receptors of the vector were partially characterized. They have been shown to be distributed at the surface of the zoospore of *Olpidium bornovanus*, and their chemical nature was identified as a glycoprotein.

Altogether, the transmission of plant viruses by fungi can tentatively be regarded as homologous to the categories noncirculative, for those viruses associated to the external coat of the zoospore, and circulative for those internalized in its cytoplasm. Unfortunately, not enough information is available to definitely decide whether viral replication is effective within the fungus, and hence whether both propagative and nonpropagative subcategories are present.

Transmission and Evolution

One aspect of vector transmission that is rarely considered in the literature is its evident impact on the population genetics and evolution of viruses. Indeed, while vector transmission is always regarded as an excellent means of maintenance and spread within host populations, it also implies that the virus is able to solve certain problems. First of all, one must bear in mind that a virus population rapidly accumulates very numerous mutations and develops a swarm of genome variants generally designated 'quasispecies'. In such a genetic context, the virus sample collected, transported, and inoculated by the vector will be highly variable. This sample, which gives rise to a new population after each round of vector transmission, will obviously condition the composition of the viral lineage, and thus its evolution. In this regard, two key parameters must be considered:

1. The distribution of viral variants within the host plant determines what is actually accessible for sampling by

the vector. If variants are homogeneously distributed within the whole plant, several can be acquired together by the vector. In contrast, if they are physically isolated in different cells, organs, or tissues, only one or few will be collected.

2. The pattern of the virus acquisition by the vector can dramatically change the composition of the transmitted viral sample. When a given vector collects the virus all at once in a single location, the diversity of genome variants transmitted tends to be restricted. On the other hand, collecting the viral sample bits by bits upon successive probing in different locations of the infected host will increase the genetic diversity of the pool of viral genomes transmitted.

It has been repeatedly demonstrated experimentally that successive genetic bottlenecks imposed in viral lineages are drastically decreasing the mean fitness of RNA virus populations, sometimes even driving them toward extinction. One can then easily conceive that viruses will develop strategies to avoid or compensate for genetic bottlenecks during their life cycle. It is highly probable that different strategies of vector transmission found in plant viruses have different impacts on the fitness and evolution of viruses, through the different pools of genome variants that are collected and inoculated. For instance, in the helper strategy, it has been suggested that HC-transcomplementation (**Figure 3**) allows the vector to sample viruses in several steps and in several locations of the host, thus opening the genetic bottleneck, and perhaps attenuating its detrimental effect.

Future Prospects

From a mechanistic standpoint, future advances will mainly concern the characterization of the vector receptors that are used by viruses to ensure successful transmission. Indeed, none of these molecules have been identified so far, in any vector species and category of transmission. Apart from the example of the fungus-transmitted CNV described above, not even their chemical nature and precise location have been established.

From an evolutionary standpoint, two aspects have been mostly neglected in spite of their undeniable importance: (1) the size of the virus sample transmitted by a given vector remains largely unknown in most cases, (2) the distribution of the genome variants, constituting the swarm of mutants of the viral quasispecies, within different cells, organs, and tissues of the host plant is largely unresolved, precluding any evaluation of what genetic diversity from a given viral population is actually accessible to the vectors.

Further Reading

Blanc S (2007) Virus transmission – getting out and in. In: Waigmann E and Heinlein M (eds.) *Viral Transport in Plants*, pp. 1–28. Berlin: Springer.

Froissart R, Michalakis Y, and Blanc S (2002) Helper component-transcomplementation in the vector transmission of plant viruses. *Phytopathology* 92: 576–579.

Gray S and Gildow FE (2003) Luteovirus–aphid interactions. *Annual Review of Phytopathology* 41: 539–566.

Ng JC and Falk BW (2006) Virus–vector interactions mediating non-persistent and semi-persistent transmission of plant viruses. *Annual Review of Phytopatholy* 44: 183–212.

Power AG (2000) Insect transmission of plant viruses: A constraint on virus variability. *Current Opinion in Plant Biology* 3(4): 336–340.

Rochon D, Kakani K, Robbins M, and Reade R (2004) Molecular aspects of plant virus transmission by olpidium and plasmodiophorid vectors. *Annual Review of Phytopathology* 42: 211–241.

Taylor CE and Robertson WM (1974) Electron microscopy evidence for the association of tobacco severe etch virus with the Maxillae in Myzuspersical (Sulz.). *Journal of Phytopathology* 80: 257–266.

Whitfield AE, Ullman DE, and German TL (2005) Tospovirus–thrips interactions. *Annual Review of Phytopathology* 43: 459–489.

Diagnostic Techniques: Plant Viruses

R Koenig and D-E Lesemann, Biologische Bundesanstalt für Land- und Forstwirtschaft, Brunswick, Germany
G Adam, Universität Hamburg, Hamburg, Germany
S Winter, Deutsche Sammlung für Mikroorganismen und Zellkulturen, Brunswick, Germany

Glossary

Cylindrical inclusions (CI) CI are induced in infected cells by all viruses belonging to the family *Potyviridae*. The term pinwheel inclusion is also widely used, because it describes the typical appearance of the CI in cross sections. CI are composed of one virus-encoded 66–75 kDa nonstructural protein which aggregates to monolayer sheets forming a complicated structure in which several curved sheets are attached to a central tubule. Details of the CI architecture are specific for the virus species and can serve as an additional identification feature of the virus.

Immunosorbent techniques (e.g., enzyme-linked immunosorbent assay, immunosorbent electron microscopy) Techniques in which a viral antigen is trapped on a solid matrix by means of specific antibodies that are bound to the matrix by adsorpiton.

Plant virus vectors Plant viruses can be transmitted specifically by various vector organisms, for example, aphids, mites, nematodes, or plasmodiophorid protozoans (*Olpidium* sp., *Polymyxa* sp.).

Viroplasm Cytoplasmic inclusions induced by members of the *Caulimoviridae*, *Rhabdoviridae*, *Reoviridae*, and *Bunyaviridae* are about the size of nuclei and are not bound by any membrane. The viroplasm consists of amorphous and/or fibrillar material and may or may not enclose immature or mature virus particles. It is generally assumed that viroplasms are the site of virus synthesis.

Introduction

Diagnostic techniques for plant viruses are indispensable for at least three major sets of applications:

1. the identification and classification of viruses associated with 'new' plant diseases (primary diagnosis);
2. the routine detection of known viruses, for example, in
   - indexing programs designed to provide healthy plant propagation material (seeds, tubers, grafting material, etc.);
   - breeding programs designed to select virus-resistant plants; and
   - epidemiological investigations designed to monitor the spread of viruses into new areas and in quarantine tests designed to prevent this spread;
3. investigations on the functional, pathogenic, and structural properties of a virus, for example, on
   - its spread within a plant or a vector;
   - changes produced in the ultrastructure of infected cells;
   - intracellular localization of the virions and of nonstructural viral proteins;
   - the time course of the expression of various parts of the viral genome; and
   - the surface structure of virus particles, for example, the accessibility or nonaccessibility to antibodies of various parts of coat protein in the virions.

The primary diagnosis usually requires the application of a number of different techniques, whereas for routine diagnosis only one well-adapted technique is employed. For studying additional properties of a virus, in order to understand its pathogenic effects, one or several techniques may be necessary. The most important diagnostic techniques for plant viruses are based on their morphological, serological, molecular, and pathogenic properties, quite often on a combination of them.

Techniques Used for the Identification and Classification of Viruses Associated with New Plant Diseases (Primary Diagnosis)

Inspection of Symptoms

The first step in attempts to identify the causal agent of a new plant disease is usually a careful inspection of the type of symptoms present in the natural host. **Figure 1** shows some typical virus symptoms. Mosaics of green and yellow (chlorotic) tissue areas (**Figure 1(a)**) or of green and brownish (necrotic) ones (**Figure 1(c)**) or irregular yellow spots (mottling) (**Figure 1(b)**) may be seen on leaves, flowers, or fruits. Flowers may show irregular discolorations (flower break) (**Figure 1(d)**). More or less severe growth reduction (stunting) and/or malformations of either the whole plants or of individual parts, such as flowers and tubers, are also often observed.

Adsorption Electron Microscopy

A quick means to detect the presence of viruses – preferentially in symptom-showing parts of a plant – is the electron microscopical examination of a crude sap extract. In adsorption electron microscopy (**Figure 2(a)**), the virus particles are adsorbed to support films on electron microscope grids and are negatively stained, for example with uranyl acetate. Rod-shaped, filamentous, or isometric particles are readily distinguished from normal plant constituents. More than 30 different morphological types of viruses (examples in **Figure 3**) have been recognized. Particle morphology may serve as a guide for further identification, for example, by means of serology or molecular analyses (see below). Unfortunately, many viruses with very different pathogenic and other properties may be morphologically indistinguishable. Another severe limitation of this test is its rather low sensitivity. Thus, viruses occurring in low concentrations may not be detected. These drawbacks may be overcome in the immunoelectron microscopical tests described in the following sections.

Immunosorbent Electron Microscopy

In immunosorbent electron microscopy (ISEM) (**Figures 2(b)**, **4(a)**, and **4(b)**), the electron microscope

Figure 1 Typical symptoms induced by plant viruses: (a) mosaic symptoms on abutilon, (b) mottling symptoms on zucchini fruits, (c) chlorotic and necrotic stripe mosaic on barley leaves, (d) flower break on lilies.

grids carrying a support film are coated with antibodies specific for viruses having the same morphology as the ones found in the electron microscopical adsorption test or with antibodies to viruses being suspected to be present in the plant material under investigation. Over a thousand times more virus particles can be detected on such antibody-coated grids than on uncoated grids or on grids coated with antibodies to an unrelated virus. This ISEM test is a very versatile technique. Its sensitivity and the broadness of cross-reactivities detected can be modulated by the length of time during which the virus-containing fluids are in contact with the antibody-coated grids. When the incubation time is short (15 min or less), mainly the homologous and very closely related viruses are trapped. However, after incubation overnight, the efficiency of virus trapping increases and more distantly related viruses are also detected. Very small amounts of antisera are needed (1 μl of crude antiserum can support 200 ISEM tests) and the quality of the antisera need not necessarily be high, since antibodies to normal host constituents do not interfere with virus-specific reactions and low-titered antisera can also be used. Since ISEM tests are labor intensive – they are not suitable for the routine detection of viruses in large numbers of samples.

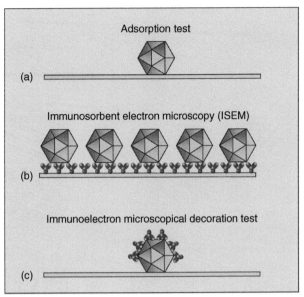

Figure 2 Schematic representation of various electron microscopical tests. For detailed explanations, see text. Adapted with permission from Koenig R and Lesemann D-E (2001) Plant virus identification. In: *Encyclopedia of Life Sciences*. Copyright 2001, © John Wiley & Sons Limited.

Immunoelectron Microscopical Decoration Tests

In the immunoelectron microscopical decoration test (**Figures 2(c)**, **4(c)**, and **4(d)**), the reaction between free antibodies and viruses which have been immobilized on electron microscope support films either by adsorption or serological binding can be visualized as antibody coating of particles ('decoration'). At low antiserum dilutions (e.g., 1:50), homologous antisera produce a dense coating. Reactions with heterologous antisera may be recognized by a much weaker coating efficiency. The closeness of serological relationships may be assessed by comparing the dilution end points (titers) of antisera with homologous and heterologous viruses. The more refined immunogold-labeling technique uses antibodies that are labeled with colloidal gold particles having diameters between 5 and 20 nm. The binding of small amounts of antibodies to individual epitopes, for example, at the extremities of rod-shaped or filamentous virus particles (**Figure 4(d)**), can be visualized by means of this technique.

The immunoelectron microscope decoration tests are the most reliable means for verifying reactions between viruses and antibodies. Mixed infections by morphologically indistinguishable viruses are readily detected (**Figure 4(c)**). The presence of antibodies can be recognized even on a single virus particle. Thus, viruses occurring at low concentrations, as often happens in field samples, can nevertheless be analyzed directly. The detectability of viruses occurring at low concentration may be increased when the decoration test is preceded by an ISEM step. Antisera even with high host plant reactivities can be used successfully, because the specific antibody attachment to virus particles is directly visualized. The technique is not suited for detecting small antigenic differences between viruses and since it is labor intensive and cannot be automated, it is not suitable for routinely testing large numbers of samples.

Observations on Host Plant Reactions

The transmission of new viruses to experimental host plants is another helpful method to identify and characterize suspected 'new' viruses. Commonly used indicator and propagation hosts are *Chenopodium* species, for example, *C. quinoa* (**Figure 5**), and *Nicotiana* species, for example, *Nicotiana benthamiana*. These plant species are susceptible to a great number of plant viruses, but by far not to all of them. Host range studies may be helpful in finding a suitable host for propagating a new virus and for differentiating it from other similar viruses. In experimental hosts, viruses may reach higher concentrations than in their natural hosts and they may often be purified more easily from them, because natural hosts frequently contain tannins, glycosides, slimy substances, and other inhibitory compounds. Virus purification is necessary for the production of specific antisera which are needed for many diagnostic procedures, such as the above-mentioned immunoelectron microscopical techniques and many others, some of which will be described in the following sections.

Nucleotide Sequence Analyses

Nucleotide sequence analyses are the most reliable means to identify a virus either as a 'new' one or as a 'new' strain of

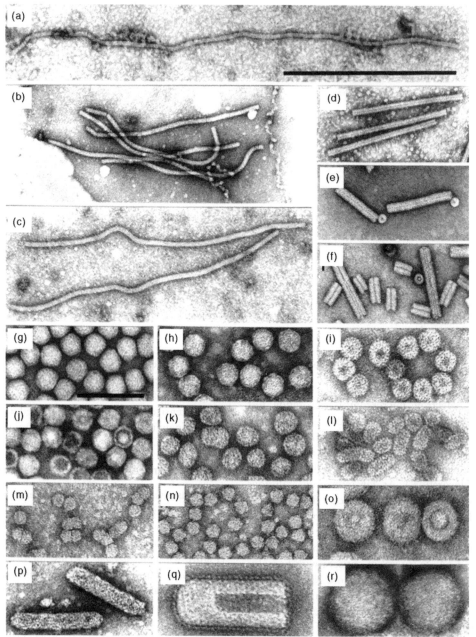

Figure 3 Various morphological types of plant viruses. Filamentous and rod-shaped particles: (a) beet yellows virus (genus *Closterovirus*), (b) potato virus X (genus *Potexvirus*), (c) potato virus Y (genus *Potyvirus*), (d) tobacco mosaic virus (genus *Tobamovirus*), (e): barley stripe mosaic virus (genus *Hordeivirus*), (f) tobacco rattle virus (genus *Tobravirus*). Isometric particles: (g) maize chlorotic mottle virus (genus *Machlomovirus*) , (h) barley yellow dwarf virus (genus *Luteovirus*), (i) cucumber mosaic virus (genus *Cucumovirus*), (j) arabis mosaic virus (genus *Nepovirus*). Irregular isometric to bacilliform particles: (k) apple mosaic virus (genus *Ilarvirus*), (l) alfalfa mosaic virus (genus *Alfamovirus*). Geminate or isometric 18 nm diameter particles: (m) tomato yellow leaf curl virus (genus *Begomovirus*), (n) faba bean necrotic yellows virus (genus *Nanovirus*). Isometric or bacilliform particles: (o) cauliflower mosaic virus (genus *Caulimovirus*), (p) cacao swollen shoot virus (genus *Badnavirus*) . Bullet-shaped or spherical particles: (q) Laelia red leaf spot-type virus (genus *Rhabdovirus*), (r) tomato spotted wilt virus (genus *Tospovirus*). The magnification bar shown in (a) equals 500 nm; the same magnification is also used in (b) to (f). The magnification bar shown in (g) equals 100 nm; the same magnification is also used in (h) to (r). Adapted with permission from Koenig R and Lesemann D-E (2001) Plant virus identification. In: *Encyclopedia of Life Sciences.* Copyright 2001, © John Wiley & Sons Limited.

a known virus. These analyses are more time-consuming than the above-described immunoelectron microscope tests. The latter can provide preliminary information concerning the putative genus or family membership of a new virus within 1–2 days. This preliminary information may serve as a valuable basis for further molecular studies, because it allows the design of broad specificity primers for cDNA synthesis and amplification of genome portions by means of the polymerase chain reaction (PCR). Such primers will be derived from highly conserved genome

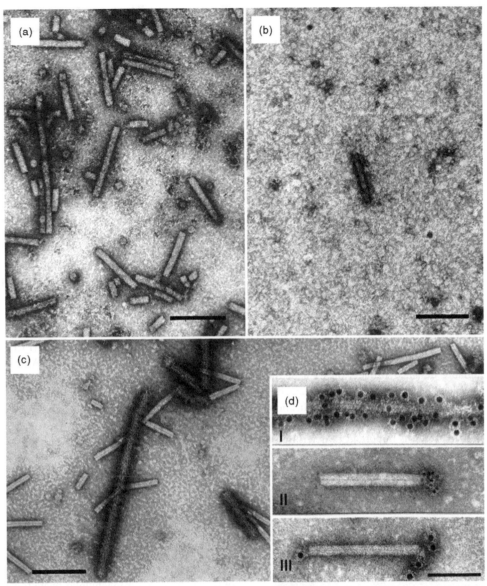

Figure 4 Immunoelectron microscopical analyses of virus-containing fluids: (a) Many virus particles are trapped within 15 min on a grid which had been coated with the homologous antiserum. (b) Only one particle is adsorbed nonspecifically from the same extract as in (a) on a grid coated with serum from a nonimmunized rabbit. (c) Detection of a mixed infection with two morphologically indistinguishable viruses: only those virus particles are decorated for which the antiserum is specific. (d) Binding of monoclonal antibodies to virus particles visualized by immunogold labeling (10 nm gold particles in the upper and lower parts and 5 nm gold particles in the middle part of the figure, respectively). In the upper part of the figure, the antibodies are bound along the entire surface of the particle, and in the middle and lower parts of the figure, on one or both extremities of the particles, respectively. Scale = 200 nm (a–c); 100 nm (d). Adapted with permission from Koenig R and Lesemann D-E (2001) Plant virus identification. In: *Encyclopedia of Life Sciences*. Copyright 2001, © John Wiley & Sons Limited.

areas in the genera or families suggested by electron microscope studies.

Genome sequences of viruses with circular DNAs can be amplified directly by means of isothermic rolling-circle amplification (RCA) techniques using random primers and the DNA polymerase of *Bacillus subtilis* bacteriophage Phi29. This enzyme in addition to polymerase activity possesses strand-displacement activity, thus allowing circular DNA to be replicated to a nearly unlimited extent. Since random primers can be used, no previous sequence information is necessary. Virus variants or viruses in mixed infections can be recognized after cleavage with restriction endonucleases.

For RNA viruses, reverse transcription of the genome into cDNA is necessary. The PCR products obtained with broad specificity primers can often be sequenced directly without the need of time-consuming and costly cloning procedures. Several commercial companies offer this service. The sequences are compared with those of other viruses in the respective taxonomic group that can be obtained from gene banks, for example, 'Pubmed' or the 'Descriptions of Plant Viruses'. The latter as well as the 'ICTVdB Index of Viruses' offer additional information on previously described viruses.

If no information on the possible genus or family membership of a new virus is available from initial electron microscope studies, the double-stranded RNA (dsRNA) found as a replication intermediate in plants infected with RNA viruses may serve as a template for cDNA synthesis. Reverse transcription may be initiated by a 'universal' primer containing a sequence of about 20 defined nucleotides on its 5′ end and a 3′ tail consisting of six variable nucleotides that may bind randomly to various internal sites in the denatured dsRNA. The cDNAs obtained are rendered double-stranded and are amplified by means of PCR using a second primer that contains only the sequence of the 20 defined nucleotides forming the 5′ end of the 'universal' primer. The many different PCR products obtained will form a smear after electrophoresis in an agarose gel. The more slow-moving fraction of this smear is eluted from the gel. It contains the larger-sized PCR products that are cloned into a suitable vector. Nucleotide sequences are determined with clones containing differently sized inserts. Gaps between individual portions of a sequence may be bridged up by means of PCRs using suitably designed primers on each of these sequences. Complete sequences may be obtained by means of various RACE techniques (rapid amplification of cDNA ends). Since genome recombinations are frequently found with plant viruses, it is important to determine the full-length or almost-full-length nucleotide sequences for newly described viruses in order to allow comparisons to be made along the entire length of the genome.

Techniques That are Especially Useful for the Routine Detection of Plant Viruses

Most of the techniques that are presently used for the routine detection of plant viruses are either based on the use of enzyme-labeled antibodies or on the amplification of portions of the viral genome by means of the PCR or other nucleic acid amplification techniques (see below). Tests based on symptom formation on indicator plants to which the viruses are transmitted either mechanically or by grafting have been used extensively as diagnostic tools in earlier times, but they are increasingly replaced by molecular or serological techniques which provide results within 1–2 days rather than after weeks or months. In addition, these latter techniques do not require extensive greenhouse or outdoor space, are more specific, do not require the propagation of unwanted pathogens, and – very importantly – can be readily automated.

Enzyme-Linked Immunosorbent Assay

Enzyme-linked immunosorbent assay (ELISA), especially its double antibody sandwich form (DAS ELISA), is presently probably the most widely used technique for the routine detection of plant viruses. The principle of DAS ELISA is outlined in **Figure 6**. Virus particles present even in low concentrations in crude sap extracts of diseased plants are trapped by specific antibodies that have been immobilized on the surface of multiwell polystyrene plates. The trapped particles are detected by means of enzyme-labeled antibodies. The enzyme attached to these antibodies is able to convert an unstained substrate into a brightly colored reaction product. Alkaline phophatase is used most commonly for antibody labeling in plant virus work. One molecule of alkaline phosphatase attached to an antibody molecule can dephosphorylate large numbers of molecules of the colorless p-nitrophenylphosphate to yield the bright yellow p-nitrophenol. Both the initial trapping step and the dye-production step make the test more than a 1000-fold

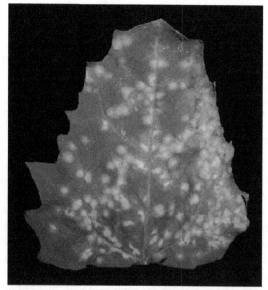

Figure 5 Chlorotic local lesions produced by a plant virus (beet necrotic yellow vein virus) on a leaf of the test plant *Chenopodium quinoa*.

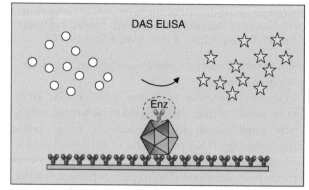

Figure 6 Schematic representation of the principle of DAS ELISA. For a detailed explanation, see text.

more sensitive than the previously used precipitin tests which relied on the detection of a visible precipitate formed as a result of the antigen/antibody reaction. Triple antibody sandwich ELISAs are used especially in work with monoclonal antibodies (MAbs), because the latter often lose their activity when labeled directly with an enzyme. The binding of mouse or rat MAbs is detected by means of enzyme-labeled antibodies from rabbits specific for mouse or rat immunoglobulins, respectively. This requires an additional assay step, but often results in increase in sensitivity of virus detection and a broadening of the range of serologically related viruses which can be detected.

Tissue Print Immunoblotting

In the tissue print immunoblotting assay (**Figure** 7), the surfaces of freshly cut virus-infected plant organs, for example, of leaves or roots, are firmly pressed on a nitrocellulose or nylon membrane. Viruses contained in the sap become adsorbed to this membrane and are detected by means of antibodies labeled with an enzyme that is able to convert a colorless soluble substrate into a colored insoluble product. For alkaline phospatase, 5'-bromo-4-chloro-3-indolylphosphate (BCIP) is commonly used as a substrate. Together with 4-nitrotetrazoliumchloride blue (NTB), its dephosphorylation products form an intense blue deposition of diformazan. The binding of unlabeled antibodies produced in a rabbit can be detected by means of enzyme-labeled antibodies produced against rabbit immunogloblins in a different animal species, such as goats. Small portions of a leaf are sufficient for the routine detection of a virus. The test also allows studies of the distribution of a virus in large organs, such as sugar beet tap roots.

PCR and Other Nucleic Acid Amplification Techniques

PCR techniques using specific primers have become very popular for the routine detection of plant viruses. They are in general more sensitive than serological techniques and hence can be used even for the detection of many viruses in woody plants, for example, in fruit trees, for which ELISA is usually not sensitive enough.

With RNA viruses, transcription of the RNA into cDNA prior to PCR is necessary (reverse transcription PCR, RT-PCR). Viral RNAs can be extracted from infected tissues by means of various commercially available kits which often are based on the selective binding of RNAs to silica gel-containing membranes and microspin technology. Alternatively, virus particles can be trapped from plant sap by means of specific antibodies which have been immobilized, as in ELISA, on a polystyrene surface, for example, in an Eppendorf tube. In this immunocapture RT-PCR (IC-RT-PCR), the nucleic acid is released from the virus by heat treatment and reverse-transcribed (**Figure** 8). Fragments of the generated cDNA are amplified and converted to dsDNA by means of a heat-stable DNA polymerase in the presence of deoxynucleotide triphosphates (dNTPs) and specific upstream and downstream primers flanking the sequence. The formation of correctly sized PCR products is checked by means of electrophoresis in agarose gels which are stained with ethidium bromide.

As mentioned earlier, viruses with circular DNA can readily be detected by means of the isothermic RCA technique using the DNA polymerase of *B. subtilis* bacteriophage Phi29. RT-LAMP (loop-mediated isothermal amplification) might be a promising isothermic alternative to PCR that can be used for RNA viruses.

In multiplex PCRs, several viruses can be detected simultaneously using pairs of primers specific for different viruses. The primers are designed in a way that each virus in a mixture yields a differently sized PCR product. The sensitivity of conventional PCRs can be further increased by means of a second amplification step using nested primers that anneal to internal sequences in the first PCR product which may be present in a concentration too low for direct detection.

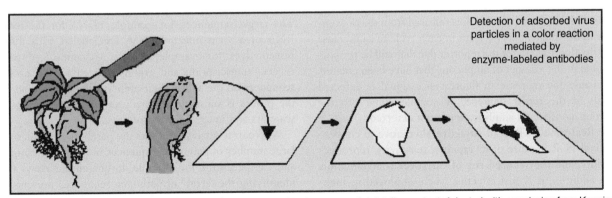

Figure 7   Schematic representation of the tissue print immunoblotting assay – for details, see text. Adapted with permission from Koenig R and Lesemann D-E (2001) Plant virus identification. In: *Encyclopedia of Life Sciences*. Copyright 2001, © John Wiley & Sons Limited.

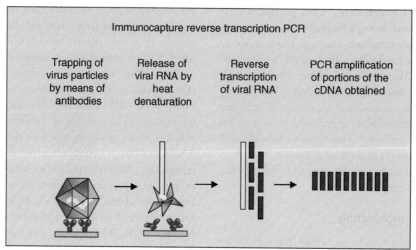

Figure 8 Schematic representation of the principle of IC-RT-PCR – for details, see text. Adapted with permission from Koenig R and Lesemann D-E (2001) Plant virus identification. In: *Encyclopedia of Life Sciences*. Copyright 2001, © John Wiley & Sons Limited.

The technical simplicity of PCR has made it a routine procedure in many laboratories, including virus-diagnostic laboratories. However, despite the ease of the method, careful interpretation of the results and positive, internal, and negative controls are extremely important in PCR analyses. The reason is that samples may easily become contaminated during working procedures by minute amounts of cDNAs or PCR products that may be present, for example, in aerosols in laboratories. UV irradiation is a useful means to decontaminate working areas.

Real-Time PCR

Real-time PCR combines the amplification and identification of target sequences in one complex reaction. Between the upstream and downstream amplification, primers most commonly a TaqMan probe is annealed specifically to the target sequence. This probe carries a fluorophore quencher dye attached to its 3′ end and a fluorpophore reporter dye attached to its 5′ end. During PCR, the polymerase replicates the template to which the probe is bound. Due to the 5′ exonuclease activity of the enzyme, the probe annealed to the target sequence is then cleaved and the reporter is thus released from the proximity of the quencher. This leads in each cycle to an increase in the fluorescence of the reporter dye that will be proportional to the number of amplicons that have been created. Because the increase in fluorescence signal is detected only if the target sequence is complementary to the probe, nonspecific amplification is not detected.

Real-time PCR is an attractive alternative to conventional PCR because of its rapidity, sensitivity, reproducibility, and the reduced risk of carryover contamination. In addition, it is cost effective when employed in high-throughput diagnostic work where it can be conveniently automated. As other PCR-based techniques, real-time PCR permits the differentiation of strains or pathotypes that differ in a few nucleotides only. With the possibility of quantification, real-time PCR is a highly versatile molecular technique for routine virus detection and identification.

Microarrays

The simultaneous identification of a number of viral nucleotide sequences, theoretically comprising all possible virus targets in a given sample, is possible by means of microarray analysis, a technique originally designed for gene expression studies with large numbers of nucleic acid sequences. With the availability of rapidly increasing viral sequence data, sequences conserved in all members of a genus or, alternatively, sequences highly specific for one or a few isolates can be defined and used in such hybridization analysis. PCR products or nowadays rather oligonucleotides less than 50 nt in length complementary to informative conserved or discriminatory sequences in virus genomes are printed on glass slides. Several hundred to a few thousand oligonucleotide spots which act as capture probes are compiled on a so-called DNA chip. The target sequences, for example, cDNA for the total RNA of a virus-infected plant, are labeled with fluorescent dyes, for example, cyanine 3/cyanine 5, during reverse transcription and are hybridized to the probe sequences on the slide. Their ability or failure to bind to the probes is subsequently analyzed using specific laser scanners and image analysis software.

Microarray analysis permits the parallel detection of a large number of pathogen sequences in one plant sample and can be useful, for example, in quarantine assays for identifying the 'array' of pathogen sequences present in one sample. The technique is commonly used for the detection and genotyping of human viruses, for example,

HIV and hepatitis virus C. Its application in plant virology, however, is still at the level of proof of principle and its use for high-throughput testing is still limited by the high cost of chip production and target labeling. Also, preamplification steps may be necessary to enrich target sequences present in low concentrations.

Techniques That are Especially Useful for Studying Functional, Physical, and Structural Properties of Plant Viruses

Western Blotting

The correct sizes and the time course of the expression of structural and nonstructural proteins of a virus can be determined by means of Western blotting. In this technique, the proteins present in a plant extract are first separated by means of sodium dodecyl sulfate (SDS) polyacrylamide electrophoresis and are then transferred electrophoretically to a nitrocellulose or nylon membrane where the viral proteins are detected by means of enzyme-labeled antibodies as described above for tissue print immunoassay.

Epitope Mapping by Investigating the Binding of MAbs to Overlapping Synthetic Oligopeptides and Subsequent Immunoelectron Microscopical Studies

Epitopes, that is, binding sites of viral coat proteins recognized by antibodies, can be identified by studying the ability or inability of MAbs to bind to a series of overlapping synthetic oligopeptides corresponding to the amino acid sequence of the viral protein. The binding sites of those MAbs which have reacted with one or several of the oligopeptides are subsequently checked on the native virus particles by means of immunoelectron microscopical decoration tests using the sensitive immunogold-labeling technique (see above). By this means, information can be obtained on the accessibility or nonaccessibility of antibody-binding sites in the intact virus particles (**Figure 4(d)**).

Methods for Studying the Distribution of a Virus within a Plant or a Vector

The distribution of a virus inside a plant or a vector can be investigated macroscopically or by means of a light microscope using the above-mentioned tissue print immunoassay or by means of antibodies labeled with fluorescent dyes. Alternatively, the coding sequences for the green fluorescent protein (GFP) or other fluorescent tags can be attached to the coding sequences for structural or nonstructural viral proteins, provided that infectious cDNA clones of a virus are available. Investigations on the movement of a tagged virus or its tagged nonstructural proteins, such as of movement proteins, between different cell compartments are best performed by means of confocal laser scan microscopy (CLSM), which produces blur-free high-resolution images of thick specimens at various depths as well as computer-generated three-dimensional (3-D) reconstructions.

Studies on the Ultrastructure of Infected Host Cells by Means of Conventional Electron Microscopy

Alterations of the fine structure in infected host cells are studied by means of conventional electron microscopy in ultrathin sections that are obtained from plant tissues after chemical fixation, dehydration, and embedding in resin-like polymers such as epoxides or methacrylates (Epon, Araldite, Spurr's resin, LR White, or LR Gold). Cryosectioning after rapid freezing procedures is used less frequently. Cytological alterations are easily detected in ultrathin sections from well-selected tissue samples. The preparation of the sections, however, is time consuming and requires special equipment and expertise. Also, not all plant viruses induce characteristic cellular alterations.

Cytological alterations appearing as unusual inclusions may be recognized in infected cells also by means of light microscopy, especially after appropriate staining. However, details of their fine structure and the complete spectrum of cellular alterations can be analyzed only in ultrathin sections. Tissues from plant organs showing macroscopically visible symptoms like chlorosis and necrosis may show pathological alterations of various cell organelles, for example, chloroplasts, mitochondria, peroxisomes, nuclei or membranes of the endoplasmic reticulum. These alterations may be indicative of the damaging physiological effects induced by the infection in incompletely adapted host plants, rather than of virus-specific effects on cytology. However, virus-specific alterations may also be recognized in infected cells. They are most clearly seen in cells of hosts which do not produce pronounced macroscopically visible symptoms (latent infections).

Characteristic inclusions may be formed by accumulated virus particles (**Figure 9(a)**), nonstructural viral proteins (e.g., in the typical cylindrical inclusions induced by potyviruses; **Figure 9(b)**), proliferated host membrane elements, or they may be composed of complex aggregates of one or more of these components with host cell constituents. The particular structure and composition of such inclusions may be characteristic for a virus. Members of the Sindbis-like supergroup of positive-stranded single-strand RNA (ssRNA) viruses may induce virus-specific or genus-specific membrane-associated vesicles (**Figure 9(c)**), whereas members of the picorna virus-like supergroup may induce accumulations of free vesicles. Both types of vesicles contain dsRNA and

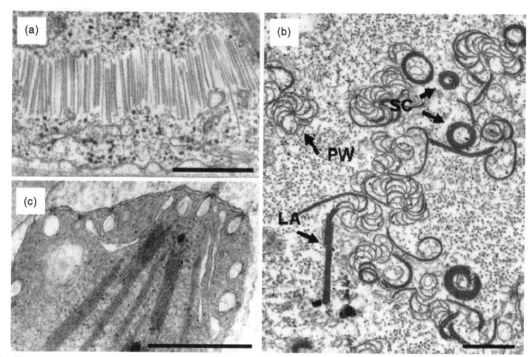

Figure 9 Electron-microscopically visible cytological alterations due to virus infections. (a) Plate-like cytoplasmic aggregate formed by the particles of a rod-shaped virus (cross section) – the highly regular *in vivo* arrangement with particle ends in register has been disturbed during fixation. Scale = 500 nm. (b) Ultrathin cross sections of pinwheels (PWs), scrolls (SCs), and laminated aggregates (LAs) formed by a nonstructural protein of a potyvirus. (c) Flask-shaped vesicles associated with the peripheral membrane of a chloroplast typical for tymovirus infections; the fibrillar content of the vesicles suggests the presence of dsRNA. Scale = 500 nm. Adapted with permission from Koenig R and Lesemann D-E (2001) Plant virus identification. In: *Encyclopedia of Life Sciences*. Copyright 2001, © John Wiley & Sons Limited.

are believed to be associated with viral RNA replication. Viruses with genomic nucleic acids other than positive-stranded ssRNA may induce viroplasm-like inclusions presumably involved in virus replication.

The fine structure of virus particle aggregates is usually determined by the morphology of the viruses (see **Figure 9(a)**) and is, thus, mostly specific for a virus genus. Occasionally, however, the particle arrangement may be specific for a certain virus strain. Inclusions composed of nonstructural viral proteins may exhibit very diverse and sometimes virus-specific shapes and fine structures (see **Figure 9(b)**). Proliferated membrane inclusions and membrane-associated vesicles are also species or genus specific (see **Figure 9(c)**).

Outlook: New and Neglected Techniques with Promise for Plant Virus Work

New technical developments that are promising for plant virus work include the use of liquid arrays and Padlock probes. A useful technique for RNA quantification that has been widely neglected by plant virologists is the RNase protection assay.

Liquid Arrays

Instead of applying antibodies to different viruses as an array to a planar surface, differently colored plastic beads are coated with different antibodies and serve as a basis for a miniaturized DAS ELISA in which the second antibody is labeled with a green fluorescent dye. In ELISA plates, mixtures of such beads carrying different antibodies allow the simultaneous testing of up to 100 different antigens. The assay is demultiplexed by means of a flow-through fluorescence reader that identifies by means of a laser beam on the basis of the color of a bead the antibody that has been used for coating the bead. Subsequently, a second laser beam shows how much fluorescent antibody has been bound to a specific type of bead because of the presence of the antigen in the testing material.

Padlock Probes

Padlock probes are DNA molecules whose 3′ and 5′ ends are complementary to adjacent parts of the target sequence (**Figure 10**). They become circularized in a ligation reaction when they fit the target sequence completely. In addition, they contain forward and backward primer-binding regions (P1 and P2) which can be

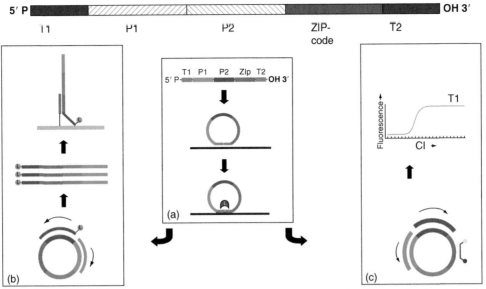

Figure 10 The Padlock probe, structure, and application for nucleic acid-based detection. The top drawing shows the construction of a Padlock probe. It consists of the two target-specific regions T1 and T2, the primer-binding regions P1 and P2, and the ZipCode, which serves as identification sequence. The probe is hybridized to the target and in case of a complete fit it is circularized (a). Subsequently the probe can serve either as a template for PCR amplification using fluorescent-labeled primers (b). In this case, the labeled amplicons are hybridized via the ZipCode to oligonucleotide arrays for differential detection (b). Alternatively, the circularized probe can be used for quantitative PCR with a TaqMan probe directed against the ZipCode (c). Courtesy of C. D. Schoen and P. J. M. Bonants, Plant Research International BV, Wageningen, The Netherlands.

used for amplification using unlabeled (**Figure 10(c)**) or labeled primers (**Figure 10(b)**). The probes also contain a ZipCode region which can be used either for hybridizing the labeled Padlock amplicons to oligonucleotide arrays (**Figure 10(b)**) or as binding region for a TaqMan probe or a molecular beacon in quantitative real-time PCR (**Figure 10(c)**).

RNase Protection Assay

These tests offer a simple and cheap method to quantify RNA without PCR. They are especially useful when several RNAs have to be quantified simultaneously. Discrete-sized complementary RNAs are labeled, for example, by radioactive isotopes, and are hybridized in solution with the target RNAs to yield labeled dsRNAs. After RNase digestion of the unhybridized probe and target RNAs, the labeled dsRNAs are separated electro-phoretically in a denaturing polyacrylamide gel and are visualized and quantified according to the label used.

Further Reading

Abdullahi I, Koerbler M, Stachewicz H, and Winter S (2005) The 18S rDNA sequence of *Synchytrium endobioticum* and its utility in microarrays for the simultaneous detection of fungal and viral pathogens of potato. *Applied Microbiology and Biotechnology* 68: 368–375.

Baulcombe DC, Chapman SN, and Santa Cruz S (1995) Jellyfish green fluorescent protein as a reporter for virus infections. *Plant Journal* 7: 1045–1053.

Boonham N, Fisher T, and Mumford RA (2005) Investigating the specificity of real time PCR assays using synthetic oligonucleotides. *Journal of Virological Methods* 130: 30–35.

Boonham N, Walsh K, Smith P, et al. (2003) Detection of potato viruses using microarray technology: Towards a generic method for plant viral disease diagnosis. *Journal of Virological Methods* 108: 181–187.

Bystrika D, Lenz O, Mraz I, et al. (2005) Oligonucleotide-based microarray: A new improvement in microarray detection of plant viruses. *Journal of Virological Methods* 128: 176–182.

Commandeur U, Koenig R, Manteuffel R, et al. (1993) Location, size and complexity of epitopes on the coat protein of beet necrotic yellow vein virus studied by means of synthetic overlapping peptides. *Virology* 198: 282–287.

Francki RIB, Milne RG, and Hatta T (1985) *Atlas of Plant Viruses* Vols. I and II. Boca Raton, FL: CRC Press.

Froussard P (1992) A random-PCR method (rPCR) to construct whole cDNA library from low amounts of RNA. *Nucleic Acids Research* 20: 2900.

Haible D, Kober S, and Jeske H (2006) Rolling circle amplification revolutionizes diagnosis and genomics of geminiviruses. *Archives of Virology* 135: 9–16.

Koenig R and Lesemann D-E (2001) Plant virus identification. *Encyclopedia of Life Sciences*. London: Nature Publishing Group.

Rott ME and Jelkmann W (2001) Characterization and detection of several filamentous viruses of cherry: Adaptation of an alternative cloning method (DOP-PCR), and modification of an RNA extraction profile. *European Journal of Plant Pathology* 107: 411–420.

Tzanetakis IE, Keller KE, and Martin RR (2005) The use of reverse transcriptase for efficient first- and second-strand cDNA synthesis from single- and double-stranded RNA templates. *Journal of Virological Methods* 124: 73–77.

van Regenmortel MHV, Bishop DHL, Fauquet CM, Mayo MA, Maniloff J, and Calisher CH (1997) Guidelines to the demarcation of virus species. *Archives of Virology* 142: 1505–1518.

Varga A and James D (2006) Use of reverse transcription loop-mediated isothermal amplification for the detection of plum pox virus. *Journal of Virological Methods* 138: 184–190.

Webster CG, Wylie SJ, and Jones MGK (2004) Diagnosis of plant viral pathogens. *Current Science* 86: 1604–1607.

Relevant Websites

http://www.pri.wur.nl – Characterization identification and detection, Plant Research International – Research, Wageningen UR.

http://www.csl.gov.uk – Diagnostic Protocols, Special Interest, Central Science Laboratory, York.

https://www.invitrogen.com – D-LUX Assays, Invitrogen.

http://www.whatman.com – FTA Cards (Product of Whatman).

http://www.ncbi.nlm.nih.gov – ICTVdB Index of Viruses; PubMed, NCBI.

http://www.pcrstation.com – Inverse PCR, PCR Station.

http://www.brandeis.edu – LATE-PCR, Wangh Laboratory of Brandeis University.

http://www.luminexcorp.com/technology – Liquid Arrays Technology, Luminex.

http://www.ambion.com – Nuclease Protection Assays: The Basics, Ambion, Inc.

http://www.dpvweb.net – Plant Virus and Viroid Sequences (compiled 11 Jul. 2007), Descriptions of Plant Viruses.

http://pathmicro.med.sc.edu – Real Time PCR Tutorial (updated 7 Dec. 2006), Microbiology and Immunology On-line, University of South Carolina.

http://aptamer.icmb.utexas.edu – The Ellington Lab Aptamer Database.

Plant Antiviral Defense: Gene Silencing Pathway

G Szittya, Agricultural Biotechnology Center, Godollo, Hungary

T Dalmay, University of East Anglia, Norwich, UK

J Burgyan, Agricultural Biotechnology Center, Godollo, Hungary

Introduction

One of the most striking biological discoveries of the 1990s is RNA silencing, which is an evolutionarily conserved gene-silencing mechanism among eukaryotes. Initially, it was identified in plants as a natural defense system thought to have evolved against molecular parasites such as viruses and transposons. However, it has become clear that RNA silencing also plays a very important role in plant and animal development by controlling gene expression.

RNA silencing operates through diverse pathways; however it uses a set of core reactions, which are triggered by the presence of long, perfect or imperfect double-stranded (ds) RNA molecules that are cleaved by Dicer or Dicer-like (DCL), an RNase III-type enzyme, into short 21–26 nucleotide (nt) long molecules, known as short interfering RNAs (siRNA) and microRNAs (miRNA). These small RNAs are then incorporated into a protein complex called RNA-induced-silencing complex (RISC) containing members of the Argonaute (AGO) family. RISCs are guided by the short RNAs to target RNAs containing complementary sequences to the short RNAs and this interaction results in cleavage or translational arrest of target RNAs. In fungi, nematodes, and plants, the silencing signal is thought to be amplified by RNA-dependent RNA polymerases (RDRs); however, RDRs have not been identified in other systems.

Recent discoveries have shown that the RNA-silencing machinery also affects gene function at the level of genomic DNA. The first observation of RNA-guided epigenetic modification was the RNA-directed DNA methylation (RdDM) of chromosomal DNA in plants resulting in covalent modifications of cytosins. RdDM was shown to require a dsRNA that was processed into short 21–24 nt RNAs and it also required some of the components of the silencing machinery such as RDR, DCL, and AGO proteins. RNA-silencing-mediated heterochromatin formation is also an epigenetic process and results in covalent modifications of histones (usually methylation of k9H3) and has been reported in fission yeast, animals, and plants.

It is now clear that RNA silencing is diverse and is involved in a variety of biological processes that are essential in maintaining genome stability, development, and adaptive responses to biotic and abiotic stresses.

Diverse Gene-Silencing Pathways in Plants

Most of the current knowledge about RNA silencing in plants has arisen from analysis of the model plant *Arabidopsis thaliana*. To date, there are several different, yet partially overlapping gene-silencing pathways known in plants. These pathways operate through the production of small RNAs and these small RNA molecules guide the effector complexes to homologous sequences where they exert their effect. Small RNAs are diverse and are involved in a variety of phenomena. They can be classified on the basis of their biogenesis and function as: microRNAs (miRNA), *trans*-acting siRNAs (ta-siRNA), natural *cis*-antisense transcript-derived siRNAs (nat-siRNA), heterochromatin siRNAs (hc-siRNA), and *cis* acting siRNAs. In *Arabidopsis* the existence of multiple paralogs of RNA-silencing-associated proteins such as ten AGOs, six RDRs and four DCLs may explain the diversification of small RNAs and gene-silencing pathways.

MicroRNA Pathway

miRNAs are mainly 21 nt long and regulate endogenous gene expression during development and environmental adaptation. Important features of miRNAs are that one miRNA duplex is produced from one locus and that they are encoded by miRNA genes that are distinct from the genes that they regulate. Primary miRNAs (pri-miRNAs) are transcribed by RNA polymerase II (Pol II) and possess 5′ caps and poly(A) tails. These long single-stranded primary transcripts with an extensive fold-back structure are cleaved during a stepwise process in the cell nucleus by DCL1. DCL1 interacts with HYL1 (a double-stranded RNA binding protein) to produce the mature miRNA duplex (miRNA:miRNA*). Both strands of the miRNA duplex are methylated (specific to plants) at the 2′ hydroxyl group of the 3′ terminal ribose by HEN1 (an S-adenosyl methionoine (SAM)-binding methyltransferase). Methylation protects miRNAs from polyuridylation and probably from degradation at the 3′ end. The plant exportin-5 homolog HASTY exports the miRNA duplexes to the cytoplasm, where the miRNA strand is incorporated into the RISC complex and the miRNA* strand is degraded.

Plant miRNAs have near-perfect complementarity to their target sites and they bind to these sites to guide the cleavage of target mRNAs. AGO1 is part of the plant RISC, since plant miRNAs are associated with AGO1 and immuno-affinity-purified AGO1 cleaves miRNA targets *in vitro*.

In *Arabidopsis* miRNAs also have a role in RNA-directed DNA methylation, a mechanism generally associated with siRNA-mediated gene silencing. It was shown that PHABULOSA (PHB) and PHAVOLUTA (PHV) transcription factor coding sequences are heavily methylated several kilobases downstream of the miRNA-binding site. It was proposed that DNA methylation of PHB/PHV occurs *in cis* and depends on the ability of the miR165/166 to bind to the transcribed RNAs.

Currently, more than 100 miRNAs have been identified in *Arabidopsis*, but many more are likely to exist. The comparison of miRNAs from different plant species has shown that many of them are conserved; however, there are species-specific miRNAs. These 'young' miRNA genes were proposed to evolve recently and show a high degree of homology to the target genes even beyond the mature miRNA sequence. The list of the currently known plant miRNAs can be found in miRBase.

Trans-Acting siRNA Pathway

ta-siRNAs are 21 nt long endogenous siRNAs and they require components of both the miRNA and *cis*-acting siRNA pathways for their biogenesis. They are produced from noncoding (nc) TAS genes transcribed by Pol II. TAS ncRNAs contain an miRNA-binding site and are cleaved by an miRNA-programmed RISC. The production of cleaved TAS ncRNA therefore requires DCL1, HYL1, HEN1, and AGO1. The cleaved TAS ncRNA is converted into dsRNA by SGS3 (coiled-coil putative Zn²⁺ binding protein) and RDR6. This long dsRNA is then cleaved in a phased dicing reaction by DCL4 to generate 21 nt ta-siRNAs. The phase of the dicing reaction is determined by the initial miRNA cleavage site. Ta-siRNAs are methylated by HEN1 and guide mRNA cleavage through the action of AGO1 or AGO7 RISC. Some of the ta-siRNAs regulate juvenile-to-adult transition but other functions remain to be identified.

Natural *cis*-Antisense Transcript-Derived siRNA Pathway

Ten percent of the *Arabidopsis* genome contains overlapping gene pairs, also known as natural *cis*-antisense gene pairs. Nat-siRNA is a recently discovered class of endogenous siRNAs that are derived from natural *cis*-antisense transcripts regulating salt tolerance in *Arabidopsis*. It was reported that both 24 and 21 nt siRNAs (nat-siRNAs) were generated from a locus consisting of two overlapping antisense genes *P5CDH* and *SRO5*; *P5CDH* mRNA is expressed constitutively, whereas *SRO5* is induced by salt stress. When both transcripts are present, a dsRNA can form by the annealing of *P5CDH* and *SRO5* transcripts, which in turn initiates the biogenesis of a 24 nt nat-siRNA by the action of DCL2, RDR6, SGS3, and NRPD1A (RNA polymerase IVa subunit). This 24 nt nat-siRNA guides the cleavage of the constitutively expressed *P5CDH* transcript and sets the phase for a series of secondary, phased 21 nt nat-siRNAs produced by DCL1. The function of these secondary nat-siRNAs is unclear since the *P5CDH* transcript is also downregulated in their absence (in *dcl1* background) as well. However, in wt plants 21 nt nat-siRNAs mediate the cleavage of *P5CDH* transcript. The presence of many overlapping genes in eukaryotic genomes suggests that other nat-siRNAs may be generated from other natural *cis*-antisense gene transcripts.

Heterochromatin siRNA Pathway

The large majority of endogenous small RNAs is derived from transposons and repeated sequences. To explain their biogenesis, it is proposed that tandem repeats or multiple copies of transposable elements generate dsRNAs. Analysis of mutant plants suggests that these molecules are produced by one of the plant RNA-dependent RNA polymerases (RDR2). The long dsRNAs are cleaved by DCL3 to produce 24 nt long hc-siRNAs that are methylated by HEN1. The accumulation of hc-siRNAs also requires NRPD1a and NRPD2, putative DNA-dependent RNA polymerases collectively called PolIV. The PolIV complex is proposed to be involved in an amplification loop together with the *de novo* methyl-transferases (DRMs), the maintenance methyl-transferase (MET1), and the SWI2/SNF2 chromatin remodeling factor DDM1 that are also required for hc-siRNA production. The resulting hc-siRNAs are incorporated into an AGO4-containing complex which then guides DNA methylation and heterochromatin formation. An overview of these and other proteins with roles in *Arabidopsis* small RNA pathways is shown in **Table 1**.

cis-Acting siRNAs

Exogenous nucleic acid invaders such as viruses and transgenes trigger the production of *cis*-acting siRNAs, which are both derived from and target these molecular invaders. This pathway was the first RNA-silencing pathway identified in plants more than 15 years ago. Since this pathway is activated upon virus infection, it was proposed to be an RNA-based immune system against molecular parasites.

In plants, the *cis*-acting siRNA pathway can be used to induce sequence-specific silencing of endogenous mRNA through post-transcriptional gene silencing (PTGS). For this, sequences homologous to the endogenous gene are expressed ectopically in sense, antisense, or inverted-repeat orientation causing S-PTGS, AS-PTGS, and IR-PTGS, respectively. During IR-PTGS, IR transgenes direct the production of long dsRNAs which are processed to 21 and 24 nt siRNAs and methylated by HEN1. Twenty-one nt siRNAs are produced by DCL4 and guide AGO1 RISC to cleave the homologous endogenous mRNAs. Most likely, DCL3 produces 24 nt siRNAs that mediate DNA or histone modification at homologous loci. In *Arabidopsis* forward-genetic screens have identified many components of the S-PTGS pathway. During S-PTGS single-stranded transgene RNAs can be converted into dsRNAs by the combined action of RDR6, SGS3, SDE3 (putative DEAD box RNA helicase), and possibly WEX (RNase D $3'-5'$ exonuclease). The resulting dsRNA is likely to be processed by DCL4 to 21 nt siRNAs, which are then methylated by HEN1. These siRNAs are incorporated into AGO1 RISC and guide the cleavage of homologous mRNAs. It is likely that this pathway also operates as an RNA-based immune system against viruses, where the dsRNA could be a replication intermediate or an internal secondary structure feature of the viral RNA. However, it is very likely that different viruses activate different siRNA pathways.

Gene Silencing and Antiviral Defense

The phenomenon that plants are able to recover from virus infection was first described almost 80 years ago. During recovery the initially infected leaves and first systemic leaves showed severe viral symptoms. However, the upper leaves were symptom free and became immune to the virus. As a consequence the plant became resistant to secondary infection against the inducing virus or close relatives. The discovery of recovery led to the concept of cross-protection. Cross-protection is a type of induced resistance and its basis is that prior infection with one virus provides protection against closely related viruses. Although recovery and cross-protection were described a very long time ago, it was not clear until recently how they operated. However, cross-protection demonstrated that plants harbor an adaptive antiviral defense system. During the last decade it became clear that this adaptive antiviral defense system was similar to RNA silencing and suggested that RNA silencing served as a natural defense system against viruses.

Successful virus infection in plants requires replication in the infected cells, cell-to-cell movement, and long-distance spread of the virus. However, plants try to protect themselves against virus infection and RNA silencing was evolved to detect and degrade foreign RNA molecules. The vast majority of plant viruses have RNA genomes with dsRNA replication intermediates and also with short imperfect hairpins in their genome that are ideal targets of the RNA-silencing mechanism. Moreover, plant viruses with DNA genome also serve as a target for the antiviral RNA-silencing pathway. In these cases, during replication the DNA is transcribed bi-directionally and overlapping transcripts form dsRNA and induce RNA silencing. It has become evident that plant viruses are potent inducers and targets of RNA silencing. In addition, as a part of co-evolution, plant viruses have developed silencing suppressor proteins to defend themselves against RNA silencing.

Many different types of plant viruses exist. They have different nucleic-acid content, genome organization, replication strategies, and have very different host ranges. However, it seems to be a common feature that the virus infection of plants results in the accumulation of viral siRNAs, yet the size profile of viral siRNAs varies depending on the virus. It has also been demonstrated that plants, mutant in different components of the RNA-silencing pathway, showed different responses to virus infection depending on the virus suggesting that different siRNA-mediated pathways are activated in plants in response to infection by different viruses.

Dicer-Like Proteins in Antiviral Defense

Arabidopsis encodes four DCL enzymes. DCL1 produces both miRNAs and siRNAs, while the three other DCLs

Table 1 Plant proteins involved in small RNA pathways

Gene	Alternative name	Protein name	Gene code	Biochemical activity and function	Pathway
AGO1	DND	ARGONAUTE 1	At1g48410	Ribonuclease activity; RNA slicer	miRNA ta-siRNA cis-acting siRNA
AGO4		ARGONAUTE 4	At2g27040	Unknown; RNA-directed DNA methylation	hc-siRNA
AGO7	ZIPPY	ARGONAUTE 7	At1g69440	Ribonuclease activity; RNA slicer	ta-siRNA
DCL1	CAF/SIN1/SUS1/EMB76	DICER-LIKE 1	At1g01040	RNase III; miRNA synthesis	miRNA nat-siRNA
DCL2		DICER-LIKE 2	At3g03300	RNase III; 22 nt or 24 nt siRNA synthesis	nat-siRNA
DCL3		DICER-LIKE 3	At3g43920	RNase III; 24 nt siRNA	hc-siRNA cis-acting siRNA
DCL4		DICER-LIKE 4	At5g20320	RNase III; 21 nt siRNA	ta-siRNA cis-acting siRNA
DDM1	SOM	DECREASE IN DNA METHYLATION 1	At5g66750	SWI2/SNF2-like chromatin remodeling enzyme; maintenance of CG DNA and histone methylation	hc-siRNA
DRM1		DOMAINS REARRANGED METHYLASE 1	At5g15380	Cytosine DNA methyltransferase; de novo DNA methylation and maintenance of asymmetric methylation of DNA sequences	hc-siRNA
DRM2		DOMAINS REARRANGED METHYLASE 2	At5g14620	Cytosine DNA methyltransferase; de novo DNA methylation and maintenance of asymmetric methylation of DNA sequences	hc-siRNA
HEN1		HUA ENHANCER 1	At4g20910	S-adenosyl methionine(SAM)-binding methyltransferase; miRNA and siRNA methylation	miRNA ta-siRNA nat-siRNA hc-siRNA cis-acting siRNA
HST		HASTY	At3g05040	Exportin; miRNA transport	miRNA

Continued

Table 1   Continued

Gene	Alternative name	Protein name	Gene code	Biochemical activity and function	Pathway
HYL1	DRB1	HYPONASTIC LEAVES 1	At1g09700	dsRNA biding; assisting in efficient and precise cleavage of pri-miRNA through interaction with DCL1	miRNA
MET1	DDM2/RTS2	METHYLTRANSFERASE 1	At5g49160	Cytosine DNA methyltransferase; maintenance of CG DNA methylation	hc-siRNA
NRPD1a	SDE4	NUCLEAR DNA-DEPENDENT RNA POLYMERASE IVa	At1g63020	DNA-dependent RNA polymerase; 24 nt siRNA production and siRNA-directed DNA methylation	nat-siRNA hc-siRNA
NRPD2a	DRD2	NUCLEAR DNA-DEPENDENT RNA POLYMERASE IVb	At3g23780	DNA-dependent RNA polymerase; 24 nt siRNA production and siRNA-directed DNA methylation	hc-siRNA
RDR2		RNA-DEPENDENT RNA POLYMERASE 2	At4g11130	RNA-dependent RNA polymerase; endogeneous 24 nt siRNA production	hc-siRNA
RDR6	SDE1/SGS2	RNA-DEPENDENT RNA POLYMERASE 6	At3g49500	RNA-dependent RNA polymerase; 21 nt and 24 nt siRNA biogenesis, natural virus resistance	ta-siRNA nat-siRNA cis-acting siRNA
SDE3		SILENCING DEFECTIVE 3	At1g05460	RNA helicase; Unclear	nat-siRNA cis-acting siRNA
SGS3	SDE2	SUPPRESSOR OF GENE SILENCING 3	At5g23570	Unknown; RNA stabilizer	ta-siRNA nat-siRNA cis-acting siRNA
WEX		WERNER SYNDROME-LIKE EXONUCLEASE	At4g13870	RNaseD 3′-5′ exonuclease; Unclear	cis-acting siRNA

produce only siRNAs. During the biogenesis of endogenous small RNAs DCL1 produces predominantly 21 nt miRNAs and 21 nt nat-siRNAs. DCL2 synthesizes a 24 nt nat-siRNA, DCL3 generates 24 nt hc-siRNA, and DCL4 produces 21 nt ta-siRNA and *cis*-acting siRNA. Analysis of Arabidopsis *dcl* mutants has revealed that there are some redundancies among the four DCL enzymes.

When a virus infects a plant, the virus-derived dsRNA or, as was shown with tombusvirus infection, the partially self-complementary structures of its genomic RNA, become substrates of DCLs and viral infection results in the accumulation of viral siRNAs. However, it has been shown that the size profile of viral siRNAs varies depending on the virus. Many viruses trigger the generation of 21 nt long viral siRNAs while others induce the production of both 21 and 24 nt siRNAs, with some viruses generating mostly 24 or 22 nt long viral siRNAs.

In *Arabidopsis* cucumber mosaic virus (CMV) siRNAs accumulate as 21 nt long species, tobacco rattle virus (TRV) produces virus-specific 21 and 24 nt siRNAs and turnip crinkle virus (TCV) produces only 22 nt long viral siRNAs. However, these viral siRNA size profiles were altered in different *Arabidopsis dcl* mutants. For example, in *dcl4* the accumulation of 21 nt CMV siRNAs was undetectable, but 22–24 nt viral siRNAs were present. Further analysis of *dcl* mutants has revealed that CMV 21 nt siRNAs are produced by DCL4, whereas the TRV 21 and 24 nt long siRNAs are generated by DCL4 and DCL3, respectively. However, the DCL3-dependent 24 nt long siRNAs are neither necessary nor sufficient to mediate defense against TRV. Thus, the DCL3-dependent RNA-silencing pathway alone cannot limit TRV infection. It was also shown that TCV 22 nt siRNAs are produced by DCL2 and they are sufficient to direct destruction of the virus. Further analysis has revealed that both DCL4 and DCL2 mediate TCV RNA silencing, with DCL2 providing redundant viral siRNA processing function, likely because TCV suppresses the activity of DCL4. It was concluded that viral siRNAs are produced by DCL4; however, DCL2 can substitute for DCL4 when the activity of the latter is reduced or inhibited by viruses. Furthermore, it was proposed that DCL4 might be the first line of antiviral defense, with DCL2-mediated activity coming to the fore when DCL4 is deactivated by a viral silencing suppressor protein. Whether all viral siRNAs are produced primarily by DCL4 from viruses with RNA genomes remains to be determined.

DNA viruses are also targets of RNA silencing, and virus-derived siRNAs have been detected in plants infected with DNA geminiviruses. *Arabidopsis* infected with Cabbage leaf curl begomovirus (CaLCuV), a geminivirus, generates 21, 22, and 24 nt viral siRNA of both polarities, with the 24 nt species being a predominant size-class. The accumulation of three different size-classes of viral siRNAs suggests that more than one DCL is involved in the biogenesis of these RNAs. Genetic evidence has shown that DCL3 is required to generate the 24 nt viral siRNAs and DCL2 is necessary for the production of a substantial fraction of the 22 nt viral siRNAs. Other yet unidentified DCLs or combination of DCLs generate the 21 nt and the remaining of 22 nt viral siRNAs. It has been shown that at least two RNA-silencing pathways are involved in DNA virus–plant interactions.

HEN1

HEN1 encodes an S-adenosyl methionine(SAM)-binding methyl-transferase that methylates the $2'$-OH of the $3'$-terminal nucleotide of miRNAs and siRNAs. It is suggested that all known endogenous small RNAs in *Arabidopsis* are methylated by HEN1. Infection of *Arabidopsis* with CaLCuV shows that the $3'$-terminal nucleotide of viral siRNA is methylated by HEN1. However, mutations in HEN1 gene do not increase plant susceptibility to DNA viruses. Conversely, during the infection of a cytoplasmic RNA virus (oilseed rape mosaic tobamovirus), a major fraction of virus-derived siRNAs are not methylated in *Arabidopsis*. Moreover, it was also demonstrated that viral RNA-silencing suppressors interfere with miRNA methylation in *Arabidopsis*. However, the understanding of the exact biochemical process of viral siRNA methylation and its biological relevance together with the effect of viral silencing suppressors remains a challenge in plant virus interaction.

RISC

The existence of viral siRNA-programmed antiviral RISC that cleaves viral target RNA is widely accepted. However, in theory, DCL-mediated dicing of the viral genome would be sufficient to repress virus infection. Recently, it was shown that efficient accumulation of DCL-3-dependent (in *dcl2–dcl4* double mutant) TRV-derived 24 nt siRNAs was not sufficient to mediate viral defense against TRV. This may argue for an antiviral RISC to promote defense against TRV. In plants, AGO1 was shown to have RISC activity and it was physically associated with miRNAs, ta-siRNAs, and *cis*-acting siRNAs. However, the AGO1 slicer did not contain virus-derived siRNAs when *Arabidopsis* plants were infected with three different viruses (CMV, TCV, tobacco mosaic virus (TMV)). Unfortunately, the experiment did not examine more viruses and had not tested the effect of the viral suppressors on RISC assembly. Furthermore, no other viral siRNA-containing RISC complex has been identified, yet. However, the Arabidopsis genome encodes 10 AGOs and it is possible that some of them may have diversified to become the antiviral slicer of RISC.

The Role of RNA-Dependent RNA Polymerase 6 in Virus Infection

Plants contain many *RDR* genes, for example, the model plant *Arabidopsis* has six RDR paralogs. *Arabidopsis* RDR6 is necessary for ta-siRNA, nat-siRNA, and S-PTGS pathways and in some cases also in antiviral defense. A likely biochemical role of RDR6 in RNA silencing is to produce dsRNA that is cleaved by DCLs. *Arabidobsis rdr6* mutants showed hypersusceptibility to CMV, a cucumovirus, but not to TRV, a tobravirus, TMV, a tobamovirus, turnip vein clearing virus (TVCV), also a tobamovirus, TCV, a carmovirus, and turnip mosaic virus (TuMV), a potyvirus. In *Nicotiana benthamiana*, a species often used in plant virus research, RDR6 plays a role in defense against potato virus X (PVX), a potexvirus, potato virus Y (PVY), a potyvirus but is not involved in resistance against TMV, TRV, TCV, and CMV. The role of NbRDR6 in viral defense against PVX has been studied in detail. During PVX infection, NbRDR6 has no effect on replication, cell-to-cell movement, and virus-derived siRNA accumulation. However, NbRDR6 is implicated in systemic RNA silencing. In plants, there is a mobile silencing signal that spreads through the plasmodesmata and phloem. The exact nature of the silencing signal is unknown, but it is likely to be a 24 nt siRNA species. Both RDR6 and SDE3 are involved in the long-distance spread of the silencing signal generated during PVX infection. During infection NbRDR6 prevents meristem invasion by PVX and is required for the activity but not for the production of a systemic silencing signal. A model for NbRDR6 function in antiviral defense against PVX is that RDR6 uses the incoming silencing signal to produce dsRNA precursors of secondary siRNAs which mediate RNA silencing as an immediate response to slow down the systemic spread of the virus. However, it remains obscure why *rdr6* mutant plants are not hypersusceptible to all tested viruses.

SDE3

Arabidopsis SDE3 encodes a DEAD box putative RNA helicase. SDE3, in addition to RDR6, is involved in long-range but not cell-to-cell signaling of RNA silencing. *Arabidopsis* SDE3, like RDR6, is implicated in defense against CMV but has no effect on TMV, TCV, and TRV infection. It is very likely that they act together on the same RNA-silencing pathway during viral defense.

Many proteins have been identified with a role in RNA silencing. However, only a few have known function in viral defense. A simplified model summarizes the antiviral role of RNA silencing during virus infection (**Figure 1**). In the inoculated cell the entering viruses start to replicate using the virus-encoded replicase protein (Viral RdRP). During the replication cycle the virus forms dsRNA replication intermediates. The plant antiviral defense system recognizes the dsRNA and DCL4 uses it as a substrate to produce virus-derived 21 nt siRNAs. The viral siRNAs may be methylated at the 3′-nucleotide by HEN1; however, the extent and general appearance of methylation on viral siRNAs remains undetermined. Virus-derived dsRNAs are diced mainly through DCL4. However, in some cases the invading virus-encoded silencing suppressor protein can repress DCL4 activity. In these cases other DCLs with redundant function become the major viral siRNA producer. Another source of virus-derived siRNAs is the partially self-complementary structures of the viral genomic RNA. These secondary structures can also be the substrate of a DCL. According to the size of these viral siRNAs, they may be DCL4 products. However, further investigation is needed to determine their exact origin. The viral siRNAs from both origins are unwound and one strand of the siRNA duplex is incorporated into RISC. The siRNA-loaded RISC recognizes and destroys complementary target viral RNAs. In some cases RDR6 and SDE3 are implicated in antiviral defense by slowing down the accumulation of certain viruses in systemically infected leaves.

Many viral genomes encode silencing suppressors that can inhibit different points of the plant defense system. They can repress HEN1 function or bind to the siRNA duplexes and sequester them from RISC assembly.

Environmental Influence on Gene-Silencing Pathways

It has long been known that plant–virus interactions are strongly modified by environmental factors, especially by temperature. Higher temperature is frequently associated with milder symptoms, low virus content of the infected plants, and rapid recovery from virus disease. In contrast, at cold air temperatures plants become more susceptible to virus infection, develop severe symptoms, and have higher virus content. The effect of temperature on virus-induced RNA silencing has been tested by infecting *N. benthamiana* plants with cymbidium ringspot virus (CymRSV) and Cym19stop virus (an RNA-silencing suppressor mutant of the same virus) and growing the infected plants at different temperatures. It has been found that CymRSV symptoms were attenuated at 27 °C and that the attenuated symptoms were associated with reduced virus level and that the amount of virus-derived siRNAs gradually increased with rising temperature. The CymRSV-encoded p19 protein acts as a viral suppressor of RNA silencing; thus, Cym19stop-infected plants recover from viral infection at a standard temperature (21 °C). However, at low temperature (15 °C) Cym19stop-infected plants display strong viral symptoms, and the level of virus-derived siRNAs is dramatically reduced. At low temperature, RNA silencing fails to protect the plants, even when the virus lacks a silencing

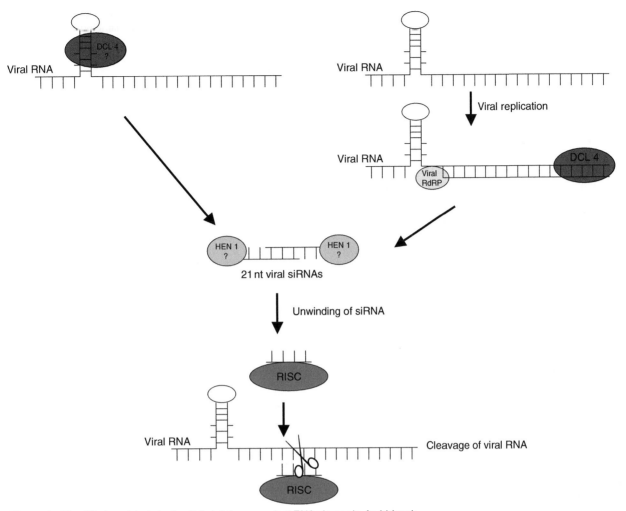

Figure 1 Simplified model of plant antiviral defence against RNA viruses in *Arabidopsis*.

suppressor. Therefore, RNA-silencing-mediated plant defense is temperature dependent and temperature regulates this defense pathway through the control of siRNA generation.

In *N. benthamiana*, NbRDR6 participates in the antiviral RNA silencing pathway that is stimulated by rising temperatures, suggesting that the function of NbRDR6 may be closely related to the temperature sensitivity of the RNA-silencing pathway. Importantly, it has also been demonstrated that temperature does not influence the accumulation of miRNAs, thus ensuring normal development at lower temperatures.

Viral Counterdefense: Silencing Suppressors

RNA silencing in plants prevents virus accumulation. To counteract this defense mechanism, most plant viruses express silencing suppressor proteins. More than a dozen silencing suppressors have been identified from different types of RNA and DNA viruses. These proteins probably evolved independently in different virus groups, because they are structurally diverse and no obvious sequence homology has been detected between distinct silencing suppressors. An interesting feature common to many viral silencing suppressors is that they were initially identified as pathogenicity or host range determinants. Suppressor activity has been identified in structural as well as nonstructural proteins involved in almost every viral function.

Viral silencing suppressor proteins operate through a variety of mechanisms. The molecular basis for suppressor activity has been proposed for several viruses, including p19 of tombusviruses, HC-Pro of potyviruses, p21 of closteroviruses, and B2 protein of flock house virus, a nodavirus infecting both plants and animals. Currently, the suppressor activity of p19 has been studied in most detail. The high-resolution crystal structures of two different tombusvirus p19 proteins, combined with molecular and biochemical data, have indicated precisely how silencing is blocked. It has been found that a tail-to-tail p19 homodimer specifically binds to siRNAs and recognizes their characteristic size, thus sequestering the products of Dicer and suppressing the silencing effect. Other studies have demonstrated that the molecular basis of silencing suppression of p21 and HC-Pro is also siRNA sequestration. B2 has been shown to bind to dsRNAs and inhibit siRNA formation. Analysis of many silencing suppressor proteins that bind dsRNA, either size selectively or size independently, has suggested that dsRNA binding is a general plant viral silencing suppression strategy which has evolved independently many times.

Viral infection leads to various symptoms, and many viral silencing suppressors have been previously described as pathogenicity determinants. Overexpression of viral silencing suppressors can result in developmental abnormalities in plants because they affect miRNA accumulation and function. Whether viral silencing suppressors inhibit the miRNA pathway because the miRNA and siRNA pathways share common components or because the miRNA pathway directly or indirectly influence virus infection is unknown.

Application of Virus-Induced Gene Silencing

Replicating plant viruses are both strong inducers and targets of the plant RNA-silencing mechanism. This strong RNA-silencing-inducing ability of plant viruses is used in virus-induced gene silencing (VIGS). VIGS is an RNA-silencing-based technique used to reduce the level of expression of a gene of interest and study the function of the knocked-down gene. Full-length viral clones can be modified to carry a fragment of an endogenous gene of interest and these are known as VIGS vectors. DsRNA of the inserted fragment is generated during viral replication and mediates the silencing of the target gene. Both RNA and DNA viral genomes have been successfully developed into VIGS vectors and have been used in reverse genetics studies in many different plants. The most widely used VIGS vectors are based on TRV, TMV, and cabbage leaf curl geminivirus (CbLCV).

See also: Plant Resistance to Viruses: Engineered Resistance; Virus Induced Gene Silencing (VIGS).

Further Reading

Akbergenov R, Si-Ammour A, Blevins T, *et al.* (2006) Molecular characterization of geminivirus-derived small RNAs in different plant species. *Nucleic Acids Research* 34: 462–471.

Baulcombe D (2004) RNA silencing in plants. *Nature* 431: 356–363.

Bouche N, Lauressergues D, Gasciolli V, and Vaucheret H (2006) An antagonistic function for Arabidopsis DCL2 in development and a new function for DCL4 in generating viral siRNAs. *EMBO Journal* 25: 3347–3356.

Brodersen P and Voinnet O (2006) The diversity of RNA silencing pathways in plants. *Trends in Genetics* 22: 268–280.

Chan SW, Henderson IR, and Jacobsen SE (2005) Gardening the genome: DNA methylation in *Arabidopsis thaliana*. *Nature Reviews Genetics* 6: 351–360.

Deleris A, Gallego-Bartolome J, Bao J, Kasschau KD, Carrington JC, and Voinnet O (2006) Hierarchical action and inhibition of plant Dicer-like proteins in antiviral defense. *Science* 313: 68–71.

Herr AJ (2005) Pathways through the small RNA world of plants. *FEBS Letters* 579: 5879–5888.

Qu F and Morris TJ (2005) Suppressors of RNA silencing encoded by plant viruses and their role in viral infections. *FEBS Letters* 579: 5958–5964.

Silhavy D and Burgyan J (2004) Effects and side-effects of viral RNA silencing suppressors on short RNAs. *Trends in Plant Science* 9: 76–83.

Szittya G, Silhavy D, Molnar A, *et al.* (2003) Low temperature inhibits RNA silencing-mediated defence by the control of siRNA generation. *EMBO Journal* 22: 633–640.

Vaucheret H (2006) Post-transcriptional small RNA pathways in plants: Mechanisms and regulations. *Genes and Development* 20: 759–771.

Vazquez F (2006) Arabidopsis endogenous small RNAs: Highways and byways. *Trends in Plant Science* 11: 460–468.

Voinnet O (2005) Induction and suppression of RNA silencing: Insights from viral infections. *Nature Reviews Genetics* 6: 206–220.

Voinnet O (2005) Non-cell autonomous RNA silencing. *FEBS Letters* 579: 5858–5871.

Wang MB and Metzlaff M (2005) RNA silencing and antiviral defense in plants. *Current Opinion in Plant Biology* 8: 216–222.

Relevant Website

http://microrna.sanger.ac.uk – miRBase::Sequences.

Virus-Induced Gene Silencing (VIGS)

M S Padmanabhan and S P Dinesh-Kumar, Yale University, New Haven, CT, USA

Glossary

Functional genomics The study of genes with respect to the role they play within biological processes.

High-throughput screening A method to efficiently test a large number of putative genes to identify candidates that may regulate a specific biological process.

Hypersensitive response A defense response in plants that initiates cell death to restrict the growth of pathogen.
Knockdown Reduction in the expression of a gene.
Knockout Complete inhibition of gene expression.
Reverse genetics Approaches used to define the function of a gene or sequence of DNA within the context of the organism.
T-DNA Transfer DNA from *Agrobacterium tumefaciens*, a vector routinely used for transforming plants.

Introduction

Virus-induced gene silencing (VIGS) is an RNA silencing-based technique used for the targeted downregulation of a host gene through the use of a recombinant virus. The term VIGS was originally coined to describe the phenomenon of recovery of a host plant from viral infection. Today, its usage is predominantly in reference to a tool for turning down host gene expression, especially in plants. In principle, a plant gene of interest can be silenced by infecting the plant with a viral vector that has been modified to express a nucleic acid sequence homologous to the host gene. Viral infection and synthesis of double-stranded RNA (dsRNA) viral replication intermediates initiate the innate RNA silencing or post-transcriptional gene silencing (PTGS) pathway, ultimately leading to the degradation of host transcript. The potency, specificity, and speed of PTGS have thus been harnessed to create an efficient gene knockout system. The ability of VIGS to rapidly initiate silencing and generate mutant phenotypes without the use of laborious transgenic approaches has made it the tool of choice for characterizing genes, especially in plant species where conventional analytical techniques have had limited success.

PTGS and the Principle behind VIGS

The term virus-induced gene silencing was first used by A. van Kammen to describe the process by which some plants were able to recover from viral infection, coincident with the loss of viral RNA from the infected tissue. The recovery process was soon understood to be associated with the plant inherent RNA-silencing mechanism which is an evolutionarily conserved antiviral system. Most plant viruses are RNA viruses and produce dsRNA intermediates during replication. In contrast to host messenger RNA, the viral dsRNA is made up of relatively long stretches of complementary RNA strands. The host identifies these as foreign and triggers the RNA-silencing pathway. The plant recovery is thus the successful outcome of the actions of this surveillance system.

The phenomenon of RNA silencing, initially observed in plants where it was called PTGS, has since been known to operate in almost all eukaryotic species including fungi, worms, flies, and mammals. In fungi, the process is called quelling while in animal systems it is known as RNAi (RNA interference). In plants and other eukaryotes, the mechanism of RNA silencing of viruses is highly conserved. The dsRNA which is generally synthesized by a viral polymerase or in some cases a host-encoded RNA-dependent RNA polymerase is recognized and cleaved by Dicer, a ribonuclease III (RNase III)-like enzyme. This results in the production of 21–24 nt duplex molecules called short-interfering RNAs (siRNA's). In an ATP-dependent manner, the siRNA's are denatured and one strand is exclusively incorporated into a multi-subunit nuclease complex called the RNA-induced silencing complex (RISC). Within the RISC, siRNA serves as a guide to recognize and base pair with the homologous RNA (in this case the viral RNA), eventually leading to degradation of the target RNA. This occurs via an RNase H-like cleavage mechanism directed by the RISC. One of the key factors that make PTGS such a potent system of host defense is that once initiated, it can maintain the silencing effect by relaying a diffusible silencing signal throughout the plant. As the virus moves within the host tissue, those cells that have received the silencing signal are primed to recognize the viral RNA and initiate its degradation.

PTGS and its efficiency against viral genomes can be exploited to create recombinant viral vectors that serve as tools for gene silencing. Viruses carrying segments of host genes, when used to infect plants, would produce siRNA's specific to the host mRNA. RISC-mediated degradation of target host mRNAs would ultimately lead to downregulation of gene expression. The infected plant would thus have a phenotype similar to a loss-of-function mutant of the gene of interest. The efficiency of viruses in initiating silencing was first shown in plants infected with a recombinant tobacco mosaic virus (TMV) carrying a fragment of the plant gene *phytoene desaturase* (*PDS*), a regulator of carotenoid biosynthesis. Infected *Nicotiana benthamiana* plants showed degradation of the host PDS mRNA and resultant alterations in the pigment synthesis pathway led to significant photobleaching in the leaves. The potential of this gene knockout system was immediately recognized and has since been applied to more than ten plant species. The *PDS* gene now serves as the conventional gene used for testing the efficacy of a virus in inducing silencing (**Figure 1**).

The Development of VIGS as a Tool for Functional Genomics

Several VIGS vectors have been developed over the past 8 years (**Table 1**). The choice of VIGS vector plays a key

Figure 1   Silencing of the *PDS* gene (phytoene desaturase) in *N. benthamiana* using TRV-based VIGS system. *Nicotiana benthamiana* plants infected with the empty vector TRV-VIGS do not have a noticeable phenotype while TRV-*NbPDS*-infected plants show photobleaching, a hallmark of PDS silencing.

Table 1   Selected list of plant viruses used as VIGS vectors for gene silencing

Virus	Genus	Host	Method of infection
Tobacco mosaic virus	*Tobamovirus*	*N. benthamiana*	Inoculation of infectious viral RNA
Potato virus X	*Potexvirus*	*N. benthamiana*, potato	Inoculation of infectious viral RNA; agro-inoculation
Tobacco rattle virus	*Tobravirus*	*N. benthamiana*, *Arabidopsis thaliana*, tomato, pepper, petunia, potato, poppy	Agro-inoculation
Pea early browning virus	*Tobravirus*	Pea	Agro-inoculation
Satellite Tobacco mosaic virus (with helper TMV)	*Tobamovirus*	Tobacco	Inoculation of infectious viral RNA
Bean pod mottle virus	*Comovirus*	Soybean	Inoculation of infectious viral RNA
Tomato golden mosaic virus (TGMV)	*Geminivirus*	*N. benthamiana*	Microprojectile bombardment
Cabbage leaf curl virus (CaLCuV)	*Geminivirus*	*A. thaliana*	Microprojectile bombardment
African cassava mosaic virus (ACMV)	*Geminivirus*	Cassava	Microprojectile bombardment
Barley stripe mosaic virus	Hordeivirus	Barley, wheat	Inoculation of infectious viral RNA
Fescue strain of Brome mosaic virus (F-BMV)	Bromovirus	Rice, maize, barley	Inoculation of infectious viral RNA

role in efficient silencing and there are many factors to be considered when choosing the virus to be used for VIGS. Ideally, the virus must produce little or no symptoms during infection, thereby facilitating easy visualization and interpretation of the mutant phenotype. It must induce persistent silencing. Viruses with strong silencing suppressors are avoided since they can interfere with the establishment of silencing. It is advantageous to have infectious cDNA clones of the virus for cloning purposes and the virus must retain infectivity after insertion of foreign DNA. The virus should also show uniform spread, infect most cell types including the meristem, and preferably show a broad host range.

RNA Virus-Based Vectors

TMV, an RNA virus belonging to the genus *Tobamovirus*, was the earliest viral vector to be used for VIGS but the severity of symptoms in susceptible tissue and inability to invade the meristem limited its application. A PVX (potato virus X)-based vector produced milder infection symptoms but displayed a narrower host range and did not infect the growing points of infected plants. Tobacco rattle virus (TRV) was shown to overcome all these disadvantages and currently is the most widely used VIGS vector, especially in solanaceous hosts. It can infect all cell types including the meristem and so can be used to investigate the role of genes in early developmental processes. The silencing is more persistent when compared to other viral vectors and it is documented to infect a wide range of hosts.

TRV is a plus-sense RNA virus with a bipartite genome. RNA-1 encodes the viral replicase and movement proteins, while RNA-2 contains genes for the coat protein and two nonstructural proteins, one of which is needed for vector transmission. Since the latter two genes are dispensable for viral replication and spread in plants, RNA-2 could be modified to replace the nonstructural genes with a multiple cloning site for insertion of host-derived gene sequences. The host sequence to be inserted into the viral vector could theoretically be as small as 23 nt (the size of an siRNA) but in practice the ideal size range is between 300 and 1500 bp.

There are many ways of introducing the virus into the plant system and these include rubbing the plants with infectious viral transcripts or biolistic delivery. However, the most effective means appear to be incorporation of the viral cDNA into *Agrobacterium tumefaciens*-based plant transformation vectors (T-DNA vectors) under the control of constitutive promoters followed by infiltration of the *Agrobacterium* cultures into the host. This ensures not just ease in the inoculation procedure but also efficient transformation and high levels of viral RNA production within plant cells. For the TRV system, VIGS can be induced by simply mixing individual *Agrobacterium* cultures containing RNA-1 and modified RNA-2 and then co-infiltrating into young plants using a needle-less syringe. Alternatively, the cultures can also be vacuum-infiltrated or sprayed onto the plant.

The appearance of VIGS-related phenotype depends on a number of factors including the plant species used, the age of the plant, and environmental factors like temperature and humidity. Under optimal conditions in *N. benthamiana* plants, TRV-induced silencing phenotypes can be visualized as early as 1 week. TRV is also one of the few viruses that have been modified into a highly efficient cloning and expression system for use in large-scale functional genomics screens. One of the most useful features of TRV vectors is their ability to induce VIGS in a number of solanaceous hosts like *N. benthamiana*, tomato, potato, pepper, petunia, poppy (Eudicot species), and the model system *Arabidopsis thaliana* (family Brassicaceae).

A number of other viruses have been adopted for VIGS in species where TRV has not been efficacious (**Table 1**). For tobacco plants, researchers have developed the satellite virus-induced silencing system (SVISS), which is a two-component system that uses a modified satellite TMV (that acts as the inducer of silencing) and a helper TMV-U2 virus (that promotes viral replication). Most of the VIGS vectors thus far described have been designed for dicotyledonous species though a majority of the economically important food crops such as rice, maize, wheat, and barley are monocots. The RNAi machinery has been shown

to function in monocots and hence research in recent years has focused on finding ideal monocot VIGS vectors. The hordeivirus barley stripe mosaic virus (BSMV) was the first virus adapted for VIGS in barley and subsequently applied to wheat. A specific strain of brome mosaic virus isolated from tall fescue (F-BMV) was cloned, characterized, and modified for use as a VIGS vector. F-BMV-based vectors were shown to effectively silence genes from rice, barley, and some cultivars of maize. The successful adaptation of VIGS technology in monocots will, in the future, aid in the functional characterization of genes in these otherwise recalcitrant systems.

DNA Virus-Based Vectors

DNA viruses belonging to the family *Geminiviridae* like tobacco golden mosaic virus (TGMV) and cabbage leaf curl virus (CaLCuV) have been successfully employed for silencing and serve as good alternatives to RNA viruses. CaLCuV was also the first VIGS vector to be used to knockdown gene expression in *Arabidopsis*. Geminiviruses are highly conserved in their genetic sequence and so, in theory, these viruses can be quickly adapted for VIGS based on information from one virus. Other beneficial properties of geminiviruses are that they are genetically more stable than RNA-based vectors, can invade meristematic tissue, and infect a wide range of plants including economically important crops. In fact, begomoviruses like African cassava mosaic virus (ACMV) and pepper huasteco yellow vein virus (PHYVV) have been successfully employed as VIGS vectors in cassava and pepper. As a caveat, it must be noted that most of the geminiviruses used so far appear to induce some symptoms of infection and thus care must be taken when interpreting VIGS phenotypes.

Applications of VIGS in Aiding Plant Gene Characterization

There have been numerous examples validating VIGS as an effective method for loss-of-function studies in plant systems. The preeminent use of this technique has been in studying plant–pathogen interactions, especially in deciphering plant defense pathways. Resistance against pathogens in many plants is dictated by the presence of specific *R* genes and each gene exclusively initiates defense against one race of a given pathogen. Numerous studies have used VIGS to study genes that are activated or regulated in *R* gene pathways. These include *N* gene-mediated resistance to TMV, *Rx1*-mediated resistance to PVX, *Mla13*-mediated resistance to powdery mildew, *Cf-4* regulated resistance to *Cladosporium fulvum*, and *Pto*-mediated resistance to *Pseudomonas syringae* in tomato. In some of these studies, cDNA libraries inserted into VIGS vectors were screened to identify genes that were essential for the *R* gene-activated hypersensitive response (HR) which is a hallmark of plant defense. These studies in parallel identified common components in disease resistance like the chaperone protein HSP90, mitogen-activated protein (MAP) kinase pathway proteins, regulators of the protein degradation pathway, and TGA and WRKY transcription factors. The use of VIGS has also aided in the identification of novel pathways that play a role in defense or pathogenesis. For instance, autophagy, an evolutionarily conserved programmed cell death response, was found to be important for limiting cell death induced during viral infection. Similarly, P58[IPK], a plant ortholog of the inhibitor of mammalian dsRNA-activated protein kinase (PKR), was shown to be recruited by TMV to maintain infectivity. While the function of PKR in defense has been well characterized in animal systems, this was the first study showing a role for this protein in plant viral pathogenesis.

Meristem invasion by viruses like TRV make them ideal for use in silencing genes that may be important for cell proliferation, tissue differentiation, and flower development. Studies on floral genes have mainly been limited to model systems like *Arabidopsis* and petunia, but with the advent of VIGS it is now possible to test floral homeotic genes in a wide range of hosts and analyze conservation or divergence of gene functions between species. As a case study, in solanaceous plants, the ortholog of *Arabidopsis* and *Antirrhinum* floral homeotic gene *AP3/DEFICIENS* was silenced and found to produce a phenotype similar to that seen in *Arabidopsis*. The TRV system was also used to initiate VIGS in tomato fruits showing its potential in studying fruit development. An *N. benthamiana* cystein protease, calpain, was shown to play a key role in cell differentiation and organogenesis using the TRV-VIGS system. Similarly, the role of proliferating cell nuclear antigen (PCNA) was investigated with gemini-virus-derived vectors. The primary advantage of using VIGS for such studies is not just that it expands the study of plant growth and differentiation into nonhost systems, but also that it overcomes problems of embryo lethality or sterility which would be unavoidable if true mutants of these genes were being studied.

Among its other applications, VIGS has been adopted for analyzing biosynthetic pathways (sterol synthesis and pigment synthesis), characterizing genes involved in stress response or hormone response, organelle biogenesis, and to look at transport of proteins within the cell (especially nuclear transport mediated by different importins).

Advantages of Using VIGS as a Tool for Functional Genomics

In plants, the conventional methodology for gene function analysis has been to use insertional mutagenesis to shut down expression of the gene of interest and study the consequence of its loss to the plant. The T-DNA plasmid of *Agrobacterium*, which is capable of inserting into the plant genome, has for a long time been one of the tools of choice for disrupting plant genes. An alternative approach has been to use transposons. While being powerful tools for reverse genetics, they came with a number of limitations. Laborious and time-consuming screens had to be carried out to search for T-DNA insertions or transposon tagged lines. Loss of genes essential in the early developmental stages led to embryo lethality while functionally redundant genes rarely showed a discernible phenotype when knocked out and so these genes could not be identified in screens. These techniques were also not amenable for genomic-scale studies because of the difficulty of achieving genome saturation and problems with multiple insertions. The advent of RNA-silencing technology meant that it was possible to disrupt the function of a specific gene by the introduction of gene-specific self-complementary RNA, also known as hairpin RNA (hpRNA). This process still required plant transformation and screening to identify silenced plants. Many of the constraints imposed by these methodologies have been overcome with the discovery of VIGS and its subsequent use a tool.

The biggest advantage with this technology is the short time needed to go from gene sequence to functional characterization of its knockdown. While screening for a potential T-DNA knockout would take at least a few months, with VIGS it can be scaled down to 2 weeks. Thus relationships between genes and phenotypes/functions can be established quickly. This factor alone has catapulted VIGS into one of the most widely accepted and popular tools for gene analysis. VIGS is not as labor intensive as the other techniques since it does not involve transformation of plant tissue. Application of the virus and induction of VIGS are easy, especially after incorporation of viral sequences into intermediate plant transformation vectors. VIGS can, in general, be induced at any stage during a plant's life cycle – thus it is still possible to characterize genes that may otherwise be lethal when knocked out during early stages of development. One can overcome redundancy by carefully choosing the gene sequence to be inserted into the viral vector so as to maximize (or minimize) the silencing of related genes. Many VIGS vectors can be used on multiple host species and this can prove useful for testing gene functions against different genetic backgrounds. These vectors have accelerated the characterization of genes in many nonmodel hosts like *N. benthamiana*, petunias, potato, soybean, and even poplars where conventional techniques achieved very limited success. Finally, as has been shown with TRV and PVX, VIGS vectors can be adapted for use in high-throughput functional genomic screens.

Limitations of VIGS

It must be noted that, like all other techniques, VIGS also has its limitations. Above all, it can only induce a transient silencing response. In almost all cases, we observe an eventual recovery of the host from the viral infection and subsequent replenishment of gene transcript levels. Due to the nature of the silencing phenomenon, even during the peak silencing stages, one cannot be assured of complete knockout of transcript levels. In spite of using weak or attenuated viruses, it is not possible to avoid some of the host gene expression changes that are associated with the viral infection and these factors must be taken into consideration while interpreting the results. Ideally, therefore, VIGS should be used in conjunction with or validated by other techniques of gene function analysis. The use of recombinant viruses also requires greater precautionary measures and care must be taken to avoid accidental transmission of the modified or infectious virus.

Conclusions

With the sequencing of numerous genomes completed or near completion, we have in our hands a vast amount of genetic data that is to be deciphered, genes that need to be characterized, and pathways that are to be elucidated. The need of the hour is for powerful functional analysis tools and in the last few years VIGS has proven to be the breakthrough technology that aids in rapid and robust gene characterization. Plant gene analysis has traditionally been carried out in few model systems whose genomes have been sequenced and are also easily amenable to transformation techniques. VIGS is one of the few tools that can be applied on a broad spectrum of plants, since most plants are susceptible to viral infections. VIGS will therefore play a crucial role in promoting gene analysis in many nonmodel systems including those that have duplicated genomes or do not have a sequenced genome. A plethora of new viruses are being added to the list of VIGS vectors and it is exciting to note the addition of vectors for monocots like rice, wheat, and maize. Small grain cereals are among the world's most important food crops but gene discovery and annotation in these systems have been carried out on only a small scale mainly due to the lack of tools for reverse genetics. The adaptation of VIGS in monocots will therefore play an important role

in gene analysis and annotation in these economically important crops.

Within the animal kingdom, RNA-silencing technology has been harnessed to carry out genomewide silencing screens. In human cells, *Drosophila*, and *Caenorhabditis elegans*, dsRNA or hpRNA's designed to systematically target a majority of predicted genes have been screened to identify novel players in cell growth, proliferation, and cell death. We now have all the tools necessary to carry out similar whole genome functional discovery studies in plants. The successful application of VIGS in two completed plant genome sequences, *Arabidopsis* and rice, opens up the possibility of carrying out systematic loss-of-function studies in these systems. Many plant genomes have been partially annotated in the form of expressed sequenced tags (EST) libraries, which serve as inventories of expressed genes. These EST libraries can also serve as ideal resources for use in VIGS-mediated high-throughput screening. In the postgenomics era, VIGS will prove to be a technique that will help accelerate the conversion of genomic data to functionally relevant information and contribute to our understanding of the molecular processes occurring within plants.

See also: Plant Antiviral Defense: Gene Silencing Pathway.

Further Reading

Baulcombe DC (1999) Fast forward genetics based on virus-induced gene silencing. *Current Opinion in Plant Biology* 2: 109–113.

Burch-Smith T, Miller J, and Dinesh-Kumar SP (2003) PTGS Approaches to large-scale functional genomics in plants. In: Hannon G (ed.) *RNAi: A Guide to Gene Silencing*, p. 243. New York: Cold Spring Harbor Laboratory Press.

Burch-Smith TM, Anderson JC, Martin GB, and Dinesh-Kumar SP (2004) Applications and advantages of virus-induced gene silencing for gene function studies in plants. *Plant Journal* 39: 734–746.

Carrillo-Tripp J, Shimada-Beltran H, and Rivera-Bustamante R (2006) Use of geminiviral vectors for functional genomics. *Current Opinion in Plant Biology* 9: 209–215.

Dinesh-Kumar SP, Anandalakshmi R, Marathe R, Schiff M, and Liu Y (2003) Virus-induced gene silencing. *Methods in Molecular Biology* 236: 287–294.

Ding XS, Rao CS, and Nelson RS (2007) Analysis of gene function in rice through virus-induced gene silencing. *Methods in Molecular Biology* 354: 145–160.

Lu R, Martin-Hernandez AM, Peart JR, Malcuit I, and Baulcombe DC (2003) Virus-induced gene silencing in plants. *Methods* 30: '296–303.

Robertson D (2004) VIGS vectors for gene silencing: Many targets, many tools. *Annual Review of Plant Biology* 55: 495–519.

Plant Resistance to Viruses: Engineered Resistance

M Fuchs, Cornell University, Geneva, NY, USA

Introduction

Plant viruses can cause severe damage to crops by substantially reducing vigor, yield, and product quality. They can also increase susceptibility to other pathogens and pests, and increase sensitivity to abiotic factors. Detrimental effects of viruses can be very costly to agriculture. Losses of over $1.5 billion are reported in rice in Southeast Asia, $5.5 million in potato in the United Kingdom, $63 million in apple in the USA, and $2 billion in cassava worldwide. Also, grapevine fanleaf virus (GFLV) alone is responsible for losses of over $1.0 billion in grapevines in France.

A number of strategies are implemented to mitigate the impact of plant viruses. Quarantine measures and certification of seeds and propagation stocks limit the introduction of virus diseases in virus-free fields and areas. Cultural practices, rouging, control of vectors, and cross-protection based on mild virus strains or benign satellite RNA reduce the transmission rate of viruses in areas where epidemics can be problematic. However, the ideal and most effective approach to control viruses relies on the use of resistant crop cultivars. Host resistance genes have been extensively exploited by traditional breeding techniques. In recent years, cloning, sequencing, and functional characterization of some plant resistance genes have provided new insights into their structure and effect on virus multiplication at the cell and plant level. Notwithstanding, host resistance genes have been identified for a few viruses and only a limited number of commercial elite crop cultivars exhibit host resistance to viruses.

The advent of biotechnology through plant transformation and the application of the concept of pathogen-derived resistance (PDR) has opened new avenues for the development of resistant crop cultivars, in particular when resistant material with desired horticultural characteristics has not been developed by conventional breeding or when no host resistance sources are known. The past two decades have witnessed an explosion in the development of virus-resistant transgenic plants. Some crop plants expressing viral genes or other antiviral factors have been tested successfully in the field and a few have been commercialized. Extensive field evaluation experiments and commercial releases have demonstrated the benefits offered by virus-resistant transgenic plants to agriculture. Risk assessment studies have been conducted to address environmental safety issues associated with the release of transgenic plants expressing viral genes. Field experiments have provided a level of certainty of extraordinarily limited, if any, hazard to the environment. Also, studies on RNA silencing have provided new insights into the molecular and cellular mechanisms underlying engineered resistance to viruses in plants. Overall, the safe deployment of virus-resistant transgenic plants has become an important strategy to implement effective and sustainable control measures against major virus diseases.

Development of Virus-Resistant Transgenic Plants

Historical Perspectives

Over the past two decades, biotechnology expanded the scope of innovative approaches for virus control in plants by providing new tools to engineer resistance. The first approach to confer resistance to viruses in plants resulted from the development of efficient protocols for plant transformation and the application of the concept of PDR. This concept was initially conceived to engineer resistance by transferring and expressing a dysfunctional pathogen-specific molecule into the host genome in order to inhibit the pathogen. Therefore, a segment of the virus' own genetic material was hypothesized to somehow protect a plant against virus infection by acting against the virus itself.

The first demonstration of PDR in the case of plant viruses was with tobacco mosaic virus (TMV) in 1986. Transgenic tobacco plants expressing the TMV coat protein (CP) gene either did not display symptoms or showed a substantial delay in symptom development upon inoculation with TMV particles. Early experiments suggested a direct correlation between expression level of TMV CP and the degree of resistance. This type of PDR was known as CP-mediated resistance. It was successfully applied against numerous plant viruses. In the early 1990s, transgenic expression of the viral replicase domain was also shown to induce resistance. This approach was referred to as replicase-mediated resistance. Meanwhile, other virus-derived gene constructs, including sequences encoding movement protein, protease, satellite RNA, defective interfering RNAs, or noncoding regions, were also described to confer virus resistance in plants. It soon became evident that more or less any viral sequence could provide some level of resistance to virus infection when expressed in transgenic plants. Interestingly, in many cases, resistance was shown to require transcription of the virus-derived transgene rather than expression of a protein product. Subsequently, at the turn of the twenty-first century, tremendous progress has been made at unraveling the molecular and cellular mechanisms underlying virus resistance.

The discovery of RNA silencing as a key mechanism in antiviral defense in plants and regulation of gene expression in eukaryotes has been a major scientific breakthrough.

Resistance to viruses has also been achieved by transforming susceptible plants with antiviral factors other than sequences derived from viral genomes. Production of antibodies or antibody fragments, $2'-5'$ oligoadenylate synthase, ribosome inactivating proteins, double-stranded RNA (dsRNA)-specific RNases, dsRNA-dependent protein kinases, cystein protease inhibitors, nucleoprotein-binding interfering aptamer peptides, and pathogenesis-related proteins has been reported to protect plants against virus infection.

Over the past two decades, many plant species and crop plants have been successfully engineered for resistance against numerous viruses with diverse taxonomic affiliation and various mode of transmission. The majority of virus-resistant transgenic plants result from the application of PDR. Virus-derived transgene constructs include full-length, untranslated, and truncated coding and noncoding fragments, in sense or antisense orientation. The CP gene is the most commonly used viral sequence segment to engineer resistance.

Control of Virus Diseases

Virus resistance is evaluated in transgenic plants upon mechanical inoculation, agro-infiltration, grafting, or vector-mediated infection and expressed as immunity, restricted infection, delay in the onset of disease symptoms, or recovery. As an example, squash expressing the CP genes of zucchini yellow mosaic virus (ZYMV), watermelon mosaic virus (WMV), and cucumber mosaic virus (CMV) are highly resistant to mechanical inoculation with a mixture of these three viruses while control plants inoculated with only one of these three viruses are readily infected and exhibit severe symptoms (**Figure 1(a)**).

Some transgenic crop plants, such as cereal, vegetable, legume, flower, forage, and fruit crops, expressing virus-derived gene constructs or other antiviral factors have been tested under field conditions (**Table 1**). So far, a total of 58 virus species that belong to 26 distinct genera have been the target of field resistance evaluation in 37 different plant species.

In the USA, virus resistance represented 10% (1242 of 11 974) of the total applications approved for field releases between 1987 and early August 2006. Other target traits were herbicide tolerance (30%), insect resistance (26%), improved product quality (19%), agronomic properties (9%), and fungal resistance (6%). In the USA, field tests with virus-resistant transgenic crops started in 1988 and accounted for 13–37% of the approved releases from 1988 to 1999.

A high level of resistance to virus infection has been well documented for numerous transgenic crops

expressing viral genes even under conditions of very stringent disease pressure and vector-mediated virus infection. For example, in the case of virus-resistant squash, transgenic cultivars expressing the CP gene of ZYMV and WMV are highly resistant to mixed infection by these two aphid-transmitted potyviruses (**Figure 1(b)**). Transgenic plants have a lush canopy and vigorous growth, whereas control plants show a chlorotic and distorted canopy, and stunted growth (**Figure 1(f)**). Transgenic plants produce fruits of marketable quality throughout the growing season (**Figure 1(h)**), whereas most fruits of control plants are malformed and discolored (**Figure 1(g)**), hence not marketable (**Figure 1(e)**). Remarkably, transgenic squash plants expressing the CP gene of ZYMV, WMV, and CMV exhibit identical levels of resistance to mechanical inoculation (**Figure 1(c)**) and aphid-mediated infection (**Figure 1(d)**) by these three viruses under field conditions in which no insecticides were sprayed to control aphid vector populations. Some of the 37 transgenic crops that have been field-tested or commercialized so far are also engineered for resistance to viruses that are not mentioned in **Table 1**. Similarly, in addition to the crops listed in **Table 1**, other virus-resistant transgenic plant species, including cabbage, cassava, grapevine, ryegrass, and petunia, among others, will likely be extensively field-tested in the future based on their promising performance under greenhouse conditions.

A few transgenic crops expressing viral CP genes have been commercially released in the USA and the People's Republic of China (**Table 1**). In the USA, virus resistance represented 10% (9 of 87) of the petitions granted for deregulation by the regulatory authorities until August 2006. Squash expressing the CP gene of ZYMV and WMV received exemption status in 1994 and was released thereafter. This was the first disease-resistant transgenic crop to be commercialized in the USA. Squash expressing the CP gene of ZYMV, WMV, and CMV has been deregulated and commercialized in 1995. Subsequently, numerous squash types and cultivars have been developed by crosses and back crosses with the two initially deregulated lines. The adoption of virus-resistant squash cultivars is steadily increasing since 1995. In 2004, the adoption rate was estimated to be 10% (2500 ha) across the country with an average rate of 20% in the states of Florida, Georgia, and New Jersey. Papaya expressing the CP gene of papaya ringspot virus (PRSV) was deregulated in 1998 and commercialized in Hawaii. PRSV-resistant papaya was the first transgenic fruit crop to be commercially released in the USA. Its adoption rate has been very high from the start of its release in 1998 because PRSV had devastated the papaya industry in Hawaii. Transgenic papaya cultivars were planted on more than half of the total acreage (480 ha) in 2004. Several potato cultivars expressing the replicase gene of potato leafroll virus (PLRV) or the CP gene of potato virus Y (PVY) were deregulated in 1998 and 2000.

Figure 1   Reaction of transgenic squash ZW-20 expressing the coat protein (CP) gene of zucchini yellow mosaic virus (ZYMV) and watermelon mosaic virus (WMV), and transgenic squash CZW-3 expressing the CP gene of ZYMV, WMV, and cucumber mosaic virus (CMV) to virus infection. (a) Resistance of transgenic CWZ-3 to mechanical inoculation with ZYMV, WMV, and CMV (upper right), and susceptibility of nontransgenic plants to single infection by mechanical inoculation of CMV (lower right), WMV (upper left), and ZYMV (lower left); (b) fields of transgenic ZW-20 (foreground) and nontransgenic plants (background) surrounded by a border row of mechanically inoculated nontransgenic plants; (c) transgenic ZW-20 (right) and nontransgenic (left) plants mechanically inoculated with ZYMV and WMV; (d) transgenic CZW-3 (right) and nontransgenic (left) plants subjected to aphid-mediated inoculation of ZYMV and WMV; (e) close-up of a fruit from transgenic ZW-20 (left) and fruits from nontransgenic plants (five fruits on the right), all subjected to aphid-mediated inoculation of ZYMV and WMV; (f) nontransgenic squash mechanically inoculated with ZYMV and WMV (left) and transgenic ZW-20 (right) exposed to aphid inoculation of ZYMV and WMV; and fruit production of (g) nontransgenic and (h) transgenic squash following aphid-mediated inoculation of ZYMV and WMV.

Table 1 Examples of virus-resistant transgenic crops that have been tested in the field or commercially released

Crop		Resistance to	
Category/common name	Scientific name	Virus	Genus
Cereals			
Barley	*Hordeum vulgare*	Barley yellow dwarf virus	*Luteovirus*
Canola	*Brassica napus*	Turnip mosaic virus	*Potyvirus*
Corn	*Zea mays*	Maize dwarf mosaic virus	*Potyvirus*
		Maize chlorotic dwarf virus	*Waikavirus*
		Maize chlorotic mottle virus	*Machlomovirus*
		Sugarcane mosaic virus	*Potyvirus*
Oat	*Avena sativa*	Barley yellow dwarf virus	*Luteovirus*
Rice	*Oryza sativa*	Rice stripe virus	*Tenuivirus*
		Rice hoja blanca virus	*Tenuivirus*
Wheat	*Triticum aestivum*	Barley yellow dwarf virus	*Luteovirus*
		Wheat streak mosaic virus	*Tritimovirus*
Ornamentals			
Chrysanthemum	*Chrysanthemum indicum*	Tomato spotted wilt virus	*Tospovirus*
Dendrobium	*Encyclia cochleata*	Cymbidium mosaic virus	*Potexvirus*
Gladiolus	*Gladiolus* sp.	Bean yellow mosaic virus	*Potyvirus*
Fruits			
Grapefruit	*Citrus paradisi*	Citrus tristeza virus	*Closterovirus*
Grapevine	*Vitis* sp.	Grapevine fanleaf virus	*Nepovirus*
Lime	*Citrus aurantifolia*	Citrus tristeza virus	*Closterovirus*
Melon	*Cucumis melo*	Cucumber mosaic virus	*Cucumovirus*
		Papaya ringspot virus	*Potyvirus*
		Squash mosaic virus	*Comovirus*
		Watermelon mosaic virus	*Potyvirus*
		Zucchini yellow mosaic virus	*Potyvirus*
Papaya[a]	*Carica papaya*	Papaya ringspot virus[a]	*Potyvirus*
Pineapple	*Ananas comosus*	Pineapple wilt-associated virus	*Ampelovirus*
Plum	*Prunus domestica*	Plum pox virus	*Potyvirus*
Raspberry	*Rubus idaeus*	Raspberry bushy dwarf virus	*Idaeovirus*
		Tomato ringspot virus	*Nepovirus*
Strawberry	*Fragaria* sp.	Strawberry mild yellow edge virus	*Potexvirus*
Tamarillo	*Cyphomandra betacea*	Tamarillo mosaic virus	*Potyvirus*
Walnut	*Juglans regia*	Cherry leafroll virus	*Nepovirus*
Watermelon	*Citrullus lanatus*	Cucumber mosaic virus	*Cucumovirus*
		Watermelon mosaic virus	*Potyvirus*
		Zucchini yellow mosaic virus	*Potyvirus*
		Papaya ringspot virus	*Potyvirus*
Forage			
Alfalfa	*Medicago sativa*	Alfalfa mosaic virus	*Alfamovirus*
Grass			
Sugarcane	*Saccharum* sp.	Sugarcane mosaic virus	*Potyvirus*
		Sugarcane yellow leaf virus	*Polerovirus*
		Sorghum mosaic virus	*Potyvirus*
Legumes			
Bean	*Phaseolus vulgaris*	Bean golden mosaic virus	*Begomovirus*
Clover	*Trifolium repens*	Alfalfa mosaic virus	*Alfamovirus*
Groundnut	*Arachis hypogaea*	Peanut clump virus	*Pecluvirus*
		Groundnut rosette virus	*Umbravirus*
Pea	*Pisum sativum*	Alfalfa mosaic virus	*Alfamovirus*
		Bean leafroll virus	*Luteovirus*
		Bean yellow mosaic virus	*Potyvirus*
		Pea enation mosaic virus	*Umbravirus*
		Pea seed-borne mosaic virus	*Potyvirus*
		Pea streak virus	*Carlavirus*
Peanut	*Arachis hypogaea*	Tomato spotted wilt virus	*Tospovirus*
		Groundnut rosette assistor virus	Unassigned
		Peanut stripe virus	*Potyvirus*
Soybean	*Glycine max*	Soybean mosaic virus	*Potyvius*

Continued

Table 1   Continued

Crop		Resistance to	
Category/common name	Scientific name	Virus	Genus
		Bean pod mottle virus	Comovirus
		Southern bean mosaic virus	Sobemovirus
Vegetables			
Cucumber	*Cucumis sativus*	Cucumber mosaic virus	Cucumovirus
		Papaya ringspot virus	Potyvirus
		Squash mosaic virus	Comovirus
		Watermelon mosaic virus	Potyvirus
		Zucchini yellow mosaic virus	Potyvirus
Lettuce	*Lactuca sativa*	Lettuce mosaic virus	Potyvirus
		Lettuce necrotic yellows virus	Tenuivirus
Pepper[b]	*Capsicum*	Cucumber mosaic virus[b]	Cucumovirus
		Tobacco etch virus	Potyvirus
		Potato virus Y	Potyvirus
Potato	*Solanum tuberosum*	Potato virus A	Potyvirus
		Potato virus X	Potexvirus
		Potato virus Y	Potyvirus
		Potato leafroll virus	Polerovirus
		Tobacco rattle virus	Tobravirus
		Tobacco vein mottling virus	Potyvirus
Squash[a]	*Cucurbita pepo*	Cucumber mosaic virus[a]	Cucumovirus
		Papaya ringspot virus	Potyvirus
		Squash mosaic virus	Comovirus
		Watermelon mosaic virus[a]	Potyvirus
		Zucchini yellow mosaic virus[a]	Potyvirus
Sugar beet	*Beta vulgaris*	Beet necrotic yellow vein virus	Benyvirus
		Beet western yellows virus	Polerovirus
Sweet potato	*Ipomea batatas*	Sweet potato feathery mottle virus	Potyvirus
Tomato[b]	*Solanum lycopersicum*	Beet curly top virus	Curtovirus
		Cucumber mosaic virus[b]	Cucumovirus
		Tobacco mosaic virus	Tobamovirus
		Tomato mosaic virus	Tobamovirus
		Tomato spotted wilt virus	Tospovirus
		Tomato yellow leaf curl virus	Begomovirus

[a]Commercially released in the USA.
[b]Commercially released in the People's Republic of China.

However, soon after their release, these potato lines were withdrawn from the market due to food processor rejection. In the People's Republic of China, tomato and pepper resistant to CMV through expression of the virus CP gene have been released.

Benefits of Virus-Resistant Transgenic Plants

Virus-resistant transgenic plants offer numerous benefits to agriculture, particularly in cases where genetic sources of resistance have not been identified or are not easy to transfer into elite cultivars by traditional breeding approaches due to genetic incompatibility or links to undesired traits. This is well illustrated for PRSV-resistant papaya and GFLV-resistant grapevine. In such cases, engineered resistance may be the only viable option to develop virus-resistant cultivars. Engineered resistance may also be the only approach to develop cultivars with multiple sources of resistance by pyramiding several virus-derived gene constructs. This is the case for squash resistant to CMV, ZYMV, and WMV. Commercial releases have demonstrated the stability and durability of the engineered resistance over more than a decade in the case of squash and papaya. Benefits of virus-resistant transgenic plants are also of economic importance because yields are increased and quality of crop products improved. For example, virus-resistant transgenic squash allowed growers to restore their initial yields in the absence of viruses with a net benefit of $19 million in 2004 in the USA. Also, after the release of PRSV-resistant papaya cultivars, papaya production has reached similar level than before PRSV became epidemic in Hawaii, with a $4.3 million net benefit over a 6-year period in Hawaii. Further, benefits are of epidemiological importance with virus-resistant transgenic plants limiting virus infection rates by restricting challenge viruses, reducing their titers, or inhibiting their replication and/or cell-to-cell or systemic movement. Therefore, lower virus levels reduce the frequency of acquisition by vectors and subsequent transmission within and between fields. Consequently, virus epidemics are substantially limited. Recently, it has been shown that transgenic squash resistant to ZYMV and WMV do not serve as virus source for secondary spread. In addition, benefits of virus-resistant transgenic plants are of environmental importance because chemicals directed to control virus vector populations are reduced or not necessary. Restricting the reliance on chemicals directed to arthropod, fungal, plasmodiophorid, and nematode vectors of plant viruses is important for sustainable agriculture. Finally, benefits of virus-resistant transgenic plants are of social importance. In Hawaii, growing papaya was not viable anymore prior to the release of PRSV-resistant transgenic papaya despite huge efforts to eradicate infected trees in order to limit the propagation of the virus. The impact of PRSV was so severe that some growers abandoned their farms and had to find new jobs. PRSV-resistant transgenic papaya saved the papaya industry and strengthened the social welfare of local communities in Hawaii. Altogether, benefits of virus-resistant transgenic crops are important not only to facilitate their adoption by growers but also to counterbalance real and potential environmental and health risks.

Mechanisms Underlying Resistance

Two types of mechanisms underlying engineered virus resistance have been described, one requiring expression of a transgene protein product (protein-mediated resistance) and the other depending on the expression of transgene transcripts (RNA silencing).

Protein-Mediated Resistance

Resistance to TMV in tobacco plants was initially related to the expression level of the viral CP. Resistance could be overcome by high doses of inoculum and was not very effective against virus RNA inoculation. Interference with an early step in the virus infection cycle, that is, disassembly of TMV particles, has been hypothesized to explain CP-mediated resistance, maybe by inhibition of viral uncoating or reduction of protein–protein interactions between transgene-expressed CP and challenge virus. It would be interesting to analyze whether RNA-mediated resistance is also active in transgenic plants for which the CP has been suspected as an elicitor of the engineered resistance.

RNA Silencing and Post-Transcriptional Gene Silencing

In the early 1990s, the use of untranslatable virus CP transgenes was shown to confer resistance to virus infection, indicating that resistance was RNA rather than protein-mediated. In addition, resistance was correlated to actively transcribed transgenes but low steady-state levels of transgene RNA, indicating a sequence-specific post-transcriptional RNA-degradation system. Remarkably, RNA degradation was shown to target transgene transcripts and viral RNA. This was the first indication of the occurrence of post-transcriptional gene silencing (PTGS) as a manifestation of RNA silencing, a mechanism that inhibits gene expression in a sequence specific manner.

RNA silencing operates through diverse pathways but relies on a set of core reactions that are triggered by dsRNAs, which are processed into RNA duplexes of 21–24 bp in length, called small interfering RNAs (siRNAs), by the enzyme Dicer and its homologs, which have RNAseIII, helicase, dsRNA-binding, Duf283, and PIWI-Argonaute-Zwille (PAZ) domains. One siRNA strand is incorporated into a large multi-subunit ribonucleoprotein complex called the RNA-induced silencing complex (RISC) upon ATP-dependent unwinding and guides the complex to degrade cellular single-stranded RNA molecules that are identical in nucleotide sequence to the siRNA such as challenge viral RNAs.

RNA silencing is a host defense mechanism against viruses that is triggered by dsRNA and targets the destruction of related RNAs. During virus infection, dsRNA intermediates of viral RNA replication, or dsRNA produced from viral RNAs by host polymerases, or partially self-complementary structures of viral genomic RNAs become substrates for the production of siRNAs by Dicer-like (DCL) enzymes. Plants produce at least four different variants of DCLs of which DCL2 and DCL4 provide a hierarchical antiviral defense system.

Viruses have the ability to not only trigger but also suppress RNA silencing as counterdefense mechanism. Indeed, viruses have evolved silencing-suppressor proteins that minimize PTGS. These proteins act in different ways and on different steps of the silencing pathways. Some viral suppressors bind to siRNA duplexes and prevent their incorporation into RISC. Others act directly on the enzymes or cofactors of this defense pathway. It seems that plants have duplicated DCLs as a response to the functional complexity of viral suppressors.

It is anticipated that RNA silencing mechanisms will be devised extensively in the future to create virus-resistant transgenic crop plants. Indeed, resistance can be very strong if transgene transcripts form a hairpin structure containing a substantial base-paired stem, if challenge viruses have extended nucleotide sequence identity with the transgene, and if infection with unrelated viruses carrying silencing-suppressor genes does not interfere with RNA silencing.

Environmental Safety Issues

The insertion and expression of virus-derived genes in plants has raised concerns on their environmental safety.

Since the early 1990s, considerable attention has been paid to potential environmental risks associated with the release of virus-resistant transgenic crops. Potential risks relate to the occurrence and outcomes of heterologous encapsidation, recombination, and gene flow. However, it is important to keep in perspective that these phenomena are known to occur in conventional plants, in particular for heterologous encapsidation and recombination, in the case of co-infection. Therefore, it is critical to determine if they occur in transgenic plants beyond base line events in conventional plants.

Heterologous Encapsidation

Heterologous encapsidation refers to the encapsidation of the genome of a challenge virus by the CP subunits from another virus, upon co-infection or expressed in a transgenic plant containing a viral CP gene. Heterologous encapsidation is a particular case of functional complementation, that is, a transgene might help a challenge virus in another function. Since the CP carries determinants for pathogenicity and vector specificity, among other features, the properties of field viruses may change. For example, it is conceivable that an otherwise vector-nontransmissible virus could become transmissible through heterologous encapsidation in a transgenic plant and infect otherwise nonhost plants. Such changes in vector specificity will be a single generation, not a permanent, event because they will not be perpetuated in the virus genome progeny. Heterologous encapsidation is known to occur in conventional plants subjected to mixed virus infection. Therefore, it is not too surprising that it was documented with herbaceous transgenic plants in the laboratory. In contrast, heterologous encapsidation has not been found to detectable level in transgenic vegetable plants – expressing virus CP gene that were tested extensively in the field over several years at different locations. The only exception was with transgenic squash expressing the CP gene of WMV for which a low rate of transmission of an aphid non-transmissible strain of ZYMV was documented. However, transmission of this ZYMV strain did not reach epidemic proportion. Also, heterologous encapsidation is unlikely to occur in RNA-silenced plants since expression of the transgene does not result in the accumulation of a detectable protein product. Altogether, compelling evidence has shown that heterologous encapsidation in transgenic plants expressing virus CP genes is of limited significance and should be considered negligible in regard to adverse environmental effects.

Recombination

Recombination refers to template switching between viral transgene transcripts and the genome of a challenge virus. Resulting recombinant viruses may have chimerical genomic molecules consisting of a segment from the challenge viral genome and another segment from viral transgene transcripts. It is argued that recombinant viruses may have identical biological properties as their parental lineages or new biological properties such as changes in vector specificity, expanded host range, and increased pathogenicity. Since recombination alters the genome of challenge viruses, new properties of chimera viruses will be stably transmitted to and perpetuated within the virus progeny. Comprehensive studies have documented the occurrence of recombination in transgenic plants expressing viral genes. The stringency of selective pressure applied to the challenge virus has been shown to be a critical factor in the recovery of recombinant viruses. Conditions of high selective pressure enhance the creation of recombinant viruses. In contrast, limited, if any, recombinant viruses are found to detectable level under conditions of low or no selective pressure. It is important to keep in perspective that the latter conditions prevail under field conditions in which plants are infected by functional, not defective, viruses. So far, no recombination event has been found to detectable level in CP gene-expressing transgenic perennial plants that were tested in experimental fields over a decade. Therefore, the significance of recombination in transgenic plants expressing viral genes is very limited in regard to adverse environmental effects. Furthermore, recombination is unlikely to occur in RNA-silenced plants because transgene expression often results in no detectable RNA, especially so when noncoding sequences are used as antiviral constructs.

Gene Flow

Gene flow refers to the pollen-driven movement of *trans-genes* from a virus-resistant transgenic plant into a non-transgenic compatible recipient plant, for example, a wild relative. Hybrids resulting from gene flow can acquire and express virus-derived transgenes, and become resistant to the corresponding viruses. Subsequently, plants acquiring viral resistance traits can have a competitive advantage, exhibit increased fitness, and eventually become more invasive, maybe as noxious weeds. Movement of viral transgene constructs through pollen flow has been documented from virus-resistant transgenic squash into a wild squash relative under experimental field conditions. Hybrids between transgenic and wild squash exhibited increased fitness under conditions of intense disease pressure. In contrast, under conditions of low disease pressure, no difference was observed between hybrids and wild squash in terms of growth and reproductive potential. Since viruses do not limit the size and dynamics of wild squash populations in natural habitats, it is anticipated that gene flow with virus-resistant transgenic squash will be of limited significance. It remains to be seen if increased fitness will provide hybrids with a competitive edge that could eventually lead to enhanced weediness. Altogether, compelling evidence suggest that gene flow with virus-resistant transgenic squash should not be perceived more risky than the equivalent situation with virus-resistant conventional squash.

Gene flow can also occur from a virus-resistant transgenic plant into a compatible conventional plant. Although of negligible biological impact, this phenomenon can be essential for economical reasons such as organic production and export to countries that have not deregulated transgenic crops. Worth noting is the fact that the coexistence of transgenic and conventional papaya in spatiotemporal proximity is a reality in Hawaii.

Real versus Perceived Risks

It is important to discriminate perceived and real risks associated with virus-resistant transgenic plants. Field environmental safety assessment studies have provided strong evidence of limited or no environmental risks, indicating that issues associated with virus-resistant transgenic plants are substantially less significant than initially predicted. To fully grasp the significance of environmental risks, the situation in the absence of transgenic plants needs to be taken into account and considered as base line information. So far, there is no compelling evidence to indicate that transgenic plants expressing viral genes increase the frequency of heterologous encapsidation or recombination beyond background rates. Similarly, there is little evidence, if any, to infer that transgenic plants expressing viral genes alter the properties of existing virus populations or create new viruses that could not arise naturally in conventional plants subjected to multiple virus infection. Also, the consequence of gene flow is perceived identical with transgenic and conventional virus-resistant plants. Therefore, there seems to be increasing scientific evidence that, while initially perceived as major concern, heterologous encapsidation, recombination, and gene flow with transgenic plants expressing viral genes are not deemed hazardous to the environment. Nevertheless, a case-by-case approach is recommended to make sound decisions on the safe release of virus-resistant transgenic plants. Communicating scientific facts on the safety of transgenic plants expressing viral genes is critical to distinguish real and perceived risks, help counterbalance risks and benefits, and assist regulatory authorities in the decision-making process for the release of virus-resistant transgenic crops.

Conclusions

Engineered resistance has expanded the scope of innovative approaches for virus control by providing new tools to develop resistant crop cultivars and increasing

opportunities to implement effective and sustainable management strategies. Many advances have been made on the development of virus-resistant transgenic plants over the past two decades. Since its validation with TMV in tobacco plants in 1986, the concept of PDR has been applied successfully against a wide range of viruses in many plant species. Studies on RNA silencing have provided new insights into virus–host interactions. They also shed light on the molecular and cellular mechanisms underlying engineered resistance in plants expressing virus-derived gene constructs. The exploitation of the sequence-specific antiviral pathways of RNA silencing has facilitated the design of transgene constructs for more predictable resistance. Broader spectrum resistance against multiple virus strains should also be achievable more easily by designing virus transgenes in highly conserved sequence regions and applying RNA silencing. It is anticipated that RNA silencing mechanisms will be devised more extensively in the future to create virus-resistant crop plants.

The past two decades have witnessed an explosion in the development of virus-resistant transgenic plants. Further, virus resistance has been extensively evaluated under field conditions. So far, 37 different plant species have been tested successfully in the field for resistance to 58 viruses. Despite remarkable progress, only a limited number of virus-resistant transgenic crops have been released so far for commercial use. Based on their efficacy at controlling virus diseases and high adoption rate by growers, more virus-resistant transgenic crops are likely to reach the market in the future.

There is no doubt that innovative and sustainable control strategies are needed for virus diseases to mitigate their impact on agriculture. Lessons from field experiments with various transgenic crops engineered for virus resistance and the commercial release of virus-resistant squash, papaya, tomato, and pepper have conclusively demonstrated that benefits outweigh by far risks to the environment and human health. A timely release and adoption of new virus-resistant transgenic crops is desirable, in particular in regions where viruses are devastating.

See also: Papaya Ringspot Virus; Plant Antiviral Defense: Gene Silencing Pathway; Plant Virus Diseases: Economic Aspects; Tobamovirus; Vector Transmission of Plant Viruses.

Further Reading

Baulcombe D (2004) RNA silencing in plants. *Nature* 431: 356–363.
Bendahmane M and Beachy RN (1999) Control of tobamovirus infections via pathogen-derived resistance. *Advances in Virus Research* 53: 369–386.
Fuchs M, Chirco EM, and Gonsalves D (2004) Movement of coat protein genes from a virus-resistant transgenic squash into a free-living relative. *Environmental Biosafety Research* 3: 5–16.

Fuchs M, Chirco EM, McFerson J, and Gonsalves D (2004) Comparative fitness of a free-living squash species and free-living x virus-resistant transgenic squash hybrids. *Environmental Biosafety Research* 3: 17–28.

Fuchs M and Gonsalves D (2007) Safety of virus-resistant transgenic plants two decades after their introduction: Lessons from realistic field risk assessment studies. *Annual Review of Phytopathology* 45: 173–220.

Fuchs M, Klas FE, McFerson JR, and Gonsalves D (1998) Transgenic melon and squash expressing coat protein genes of aphid-borne viruses do not assist the spread of an aphid nontransmissible strain of cucumber mosaic virus in the field. *Transgenic Research* 7: 449–462.

Gonsalves C, Lee DR, and Gonsalves D (2004) *Transgenic Virus-Resistant Papaya: The Hawaiian 'Rainbow' was Rapidly Adopted by Farmers and Is of Major Importance in Hawaii Today.*

Gonsalves D (1998) Control of papaya ringspot virus in papaya: A case study. *Annual Review of Phytopathology* 36: 415–437.

Gonsalves D, Gonsalves C, Ferreira S, *et al.* (2004) *Transgenic Virus-Resistant Papaya: From Hope to Reality for Controlling Papaya ringspot virus in Hawaii.* http://www.apsnet.org/online/feature/ringspot.

Lindbo JA and Dougherty WG (2005) Plant pathology and RNAi: A brief history. *Annual Review of Phytopathology* 43: 191–204.

MacDiarmid R (2005) RNA silencing in productive virus infections. *Annual Review of Phytopathology* 43: 523–544.

Powell AP, *et al.* (1986) Delay of disease development in transgenic plants that express the tobacco mosaic virus coat protein gene. *Science* 232: 738–743.

Qu F and Morris TJ (2005) Suppressors of RNA silencing encoded by plant viruses and their role in viral infections. *FEBS Letters* 579: 5958–5964.

Sanford JC and Johnston SA (1985) The concept of parasite-derived resistance-deriving resistance genes from the parasite's own genome. *Journal of Theoretical Biology* 113: 395–405.

Tepfer M (2002) Risk assessment of virus-resistant transgenic plants. *Annual Review of Phytopathology* 40: 467–491.

Vigne E, Bergdoll M, Guyader S, and Fuchs M (2004) Population structure and genetic diversity within *Grapevine fanleaf virus* isolates from a naturally infected vineyard: Evidence for mixed infection and recombination. *Journal of General Virology* 85: 2435–2445.

Voinnet O (2005) Induction and suppression of RNA silencing: Insights from viral infections. *Nature Reviews Genetics* 6: 206–221.

Waterhouse PM, Wang MB, and Lough T (2001) Gene silencing as an adaptive defense against viruses. *Nature* 411: 834–842.

Plant Resistance to Viruses: Geminiviruses

J K Brown, The University of Arizona, Tucson, AZ, USA

Glossary

Bipartite virus A virus having a genome comprising two segments of nucleic acid.

Consensus sequence An identical sequence, usually shared by a group of close virus relatives.

Hemiptera The insect order that contains plant feeding insects such as aphids, leafhoppers, planthoppers, and whiteflies.

Monopartite virus A virus having a genome comprising one segment of nucleic acid.

Reassortment The process by which nonhomologous viral components interact during an infection cycle. Naturally reassorted viruses are thought to be more fit than their homologous counterparts and so arise by natural selection.

Recombination The process by which nucleic acid sequences are exchanged between two or more nucleic acid molecules during replication.

Satellite A segment of nucleic acid that associates with a 'helper' virus and utilizes it for replication and transmission. Satellites typically share little to no sequence homology with the helper virus.

Vector A biological organism that transmits plant viruses from one host to another, usually an arthropod or nematode.

Introduction

Geminiviruses

For perspective, it is important to recognize that the family *Geminiviridae* was established only in 1978. This is an unusual family among plant viruses in that their genomes are circular, single-stranded DNA. When the group was originally established, fewer than 20 possible members were recognized. Even so, among the whitefly-transmitted group, now recognized as the genus *Begomovirus* (originally subgroup III), additional associated diseases had been described, but definitive characterization for most of them was delayed until tools for detecting and identifying them were developed.

Geminiviruses are composed of circular, single-stranded DNA genome that is packaged within a pair of quasi-isometric particles, hence, the name 'geminate', or twinned particles (**Figure 1**). Geminiviruses are divided into four genera based on the genome organization and biological properties. One of the most striking differences between the four genera is their restriction to either monocots or dicots (eudicots), and a highly specific relationship with the respective arthropod vector, treehoppers, leafhoppers, or a whitefly, all of which are classified in the order Hemiptera, suborder Homoptera, and which feed on phloem sap using a slender stylet.

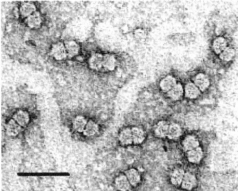

Figure 1 Typical geminate particles (~20 × 32 nm²) by transmission electron microscopy. Courtesy of C. Fauquet.

The four genera

Geminiviruses with monopartite genomes are transmitted by leafhopper vectors to monocotyledonous plants and are placed in the genus *Mastrevirus*. *Maize streak virus* is the type species. Geminiviruses with monopartite genomes that are transmitted to dicotyledonous plants by leafhopper vectors are classified in the genus *Curtovirus* with *Beet curly top virus* as the type species. The third genus, *Topocuvirus*, was recently established and has a single member, *Tomato pseudo-curly top virus*. Tomato pseudo-curly top virus (TPCTV) is monopartite, infects dicotyledonous plants, and has a treehopper vector. The fourth and largest genus, *Begomovirus*, houses the majority of the species in the family. Begomoviruses are transmitted exclusively by the whitefly *Bemisia tabaci* (Gennadius) complex to dicotyledonous plants, and *Bean golden yellow mosaic virus* is recognized as the type species. They can have either a monopartite or bipartite genome. For bipartite viruses, the two components are referred to as DNA-A and DNA-B. The B component is thought to have arisen from a duplication of DNA-A followed by the loss of all but the two genes involved in cell-to-cell movement and nuclear localization of the virus during replication.

Viral Genome and Protein Functions

Begomoviruses have a circular, single-stranded DNA (ssDNA) genome that replicates through double-stranded DNA (dsDNA) intermediates by a rolling-circle mechanism. The genomes of begomoviruses are approximately 2.6–2.8 kbp in size. Bipartite genomes share a common region (CR) of approximately 200 nt that is highly conserved among cognate components of a viral species, while the analogous region in monopartite viruses is referred to as the large intergenic region (LIR). The CR/LIR contain modular *cis*-acting elements of the origin of replication (*ori*) and promoter elements.

Four to six open reading frames (ORFs) capable of encoding proteins >10 kDa in size are present on the A component or monopartite genome. The viral capsid protein (CP) is encoded by the ORF (AV1), which is the most highly conserved begomovirus gene. The CP of begomoviruses is required for encapsidation of ssDNA and for whitefly-mediated transmission. The CP is also necessary for systemic spread of all monopartite and for some bipartite viruses, depending upon the degree of virus–host adaptation. The AC1 ORF encodes REP, a replication initiation protein, and specificity is mediated through sequence-specific interactions with *cis*-acting elements of the *ori*. The AC2 ORF encodes a transcription factor protein required for rightward gene expression and this protein also functions to suppress gene silencing imposed by the plant host. The AC3 ORF encodes a replication enhancer protein. Movement and systemic infection in bipartite viruses is accomplished by B component ORFs BV1 and BC1, which are responsible for nuclear transport of ssDNA and cell-to-cell movement functions, respectively. For monopartite viruses, several ORFs are involved in movement functions, including the analogous V1 (CP), V2, and the C4 ORF, which exert host-specific effects on symptom severity, virus accumulation, and systemic movement.

Satellite DNAs of Begomoviruses

Certain monopartite begomoviruses are associated with a nonviral circular, ssDNA referred to as a DNA-β type satellite. SatDNAs associate with a 'helper' begomovirus on which they depend for replication, encapsidation, and whitefly-mediated transmission. This type of satDNA comprises all nonviral sequences except for a 9 nt stretch that mimics the sequence of the viral origin of replication; hence, when replication initiates, the DNA-β is recognized as template and is replicated together with the begomoviral genome. In turn, the DNA-β satellite encodes a protein that has experimentally been shown to localize to the nucleus of the cell where it is thought to suppress host-induced gene silencing, an innate defense response that is intended to protect the plant from virus infection.

In the absence of the satellite, the 'helper' virus cannot sustain infection of the host and so if the helper virus becomes separated from its satellite, the virus would quickly become extinct. This relationship seems to imply that the virus itself has lost the capacity to be sufficiently pathogenic and so is dependent upon the satellite for its survival. It is interesting to speculate about the evolutionary significance of this type of interaction, in light of the hypothesis that after a time most pathogens and their hosts 'adjust' to one another, reaching a sort of evolutionary equilibrium. How these 'helper' begomoviruses became disabled and subsequently rescued by such a nonviral sequence is not known. And why they associate only with monopartite genomes, even though they have had ample opportunity for contact with bipartite begomoviruses, which also occur in the Old World, is not understood. It is thought by some that a monopartite, compared to bipartite, genome organization is the more primitive condition. However, it is not clear why satDNAs did not also co-evolve with the most ancient bipartite lineages.

Diversification and Host Shifts

There are over 389 complete geminiviral genome sequences, and at least 200 full-length DNA-β satellite sequences available in public databases, indicating the widespread nature of these viruses, discovered only recently. Geminiviruses appear readily adaptable to new hosts, and have proven remarkably capable of 'host-shifting'. Although the specific coding or noncoding regions of the genome that undergo change during host adaptation have not been identified, geminivirus genomes seem capable of undergoing quite rapid evolutionary change, for example, sometimes in a single growing season. The evidence that such rapid change occurs is taken from the observation that very few virus sequences in endemic species match identically with their closest relatives in a crop species, and many diverge by 10% or more. In only a very few instances have geminiviruses in an endemic species been found in a counterpart cultivated host, and then the virus is known to have emerged only recently as a pathogen of the crop plant. These observations suggest that the development of sustainable, resistant varieties to abate geminivirus diseases may prove difficult if these viruses are likewise capable of diversifying rapidly and overcome resistance genes. Moreover, the members of this virus family are well known for their ability to undergo recombination allowing them to diversify and evolve into more fit species. This phenomenon is best exemplified by the begomoviruses, some of which have been shown to contain sequence fragments that closely match sequences in several extant begomovirus genomes.

Diversification in bipartite viruses also appears to occur through reassortment of the DNA components based on sequence analysis. Under laboratory conditions, mixtures of certain heterologous DNA-A and DNA-B components have been shown to work cooperatively to infect a plant, and some of these 'pseudo-recombinants or reassortants' can infect host species that at least one of the parent viruses cannot. Because geminiviruses apparently do not 'cross-protect' against one another or undermine one another in the event of a mixed infection, it is clear that more than a species can infect the same plant. Experimental evidence has demonstrated that two begomoviruses can occupy the same nucleus in the plant host, a circumstance that would facilitate diversification processes and readily yield new and emerging viral species.

Symptomatology on Respective Major Hosts

Symptoms of infection caused by geminiviruses vary depending on whether the host is a monocot or a dicot, characterized as having distinct arrangements of vascular tissues and parallel or branching leaf vein patterns, respectively (**Figures 2(a)–2(d)**). Even so, geminivirus-infected plants exhibit unique symptom phenotypes that are often readily recognizable to the trained eye.

Monocots infected with mastreviruses often exhibit interveinal chlorosis that takes on the appearance of foliar streaking. Leaves may develop splotchy yellow or white patterns between the veins, and the veins may be distorted and cleared and plant vigor is reduced leading to stunted growth. Commonly, the symptoms develop in leaves closest to the growing tip when infection occurred.

A pattern of symptom development in the newest growth also is typical of begomoviruses, curtoviruses, TPCTV, and several unclassified, monopartite Old World geminiviruses. Symptoms in the leaves of dicots range from mosaic to yellow and bright yellow mosaics or sectored patterns, and leaf curling. Infected plants often develop shortened internodes and become dramatically stunted. Certain viruses have the effect of causing profuse flowering, while at the same time reducing fruit set and fruit size of those that are set. Certain geminiviruses also cause outgrowths on the top or underside of leaves, referred to as enations.

Economic Importance

Viruses of the family *Geminiviridae* are considered emergent pathogens because they have become recently increasingly important in a large number of agriculturally important crops ranging from cereals such as maize and wheat, to species grown for fiber, like cotton and kenaf, and to a number of annual and perennial ornamental plants.

Figure 2 (a) Bean plants infected with the New World, bipartite bean calico mosaic virus showing typical mosaic symptoms; (b) tomato plant infected with the Old World, monopartite tomato yellow leaf curl virus; (c) symptoms of the leafhopper-transmitted beet curly top virus infecting a pumpkin plant; (d) maize leaves showing characteristic symptoms of the Old World monopartite, monocot infected with maize streak virus. (c) Courtesy of R. Larsen. (d) Reproduced by permission of E. Rybicki.

Other viruses in this family are primarily yield-limiting pathogens of vegetable and root crop plants that are important as staples and for vitamins and nutritional variety. Yield losses in infected crops range from 20% to 100%, depending on the virus–host combination and the growth stage of the plant at time of infection. Such extensive reductions in crop productivity often result in conditions of malnutrition and in the loss of precious incomes, particularly in developing countries that have become reliant on exporting their crops.

Geminiviruses are distributed throughout the Tropics, subtropics, and mild climate regions of the temperate zones, occurring on all continents where food, fiber, and ornamental crops are cultivated. Geminiviruses also infect a wide range of endemic plant species, with which they have likely co-evolved, based on the evidence that they appear to cause little damage. This would not be surprising given their predicted longstanding association with uncultivated eudicot species. Thus the endemic species serve as a melting pot for native viruses that find their way into cultivated plants, particularly recently, as practices in many locations have shifted to monoculture production. In the subtropics, it is not uncommon to grow crops year-round. These practices enabled by extensive irrigation projects worldwide have resulted in sustained, high levels of virus inoculum and vector populations in the environment, and more frequent outbreaks of disease, particularly since the 1970s.

Molecular Diagnosis of the Virus and Its Homology with Other Viruses

For most plant viruses, serology has traditionally been the method of choice for identification. However, virus-specific antisera for geminiviruses are not generally available. This is in part because geminiviruses can be difficult to purify. Further, begomoviruses in particular are not good antigens, and a small amount of contaminating plant protein in the purified preparation outcompetes the viral epitopes during immunization. Even when antibodies are available, they are useful primarily for detection, not identification, owing to the high genus-wide conservation of the viral coat proteins. However, the greatest factor underlying the popular use of molecular diagnostics for begomoviruses is that they contain a circular, ssDNA genome that yields a dsDNA intermediate from replication, which is highly amenable to linearization with an endonuclease to yield a dsDNA fragment that can be cloned. Second, shortly after the geminiviruses were discovered, polymerase chain reaction (PCR) was invented and the two clearly went hand in hand.

Commonly, detection is accomplished using PCR primers designed to be broad spectrum (usually degenerate) or virus specific. Because substantial information on begomovirus genome organization and gene function is now available, as are numerous genomic and CP gene sequences, it is possible to scan partial or entire viral sequences and develop hypotheses regarding the prospective utility of a particular ORF or nontranslated region for inferring phylogenetic relationships. Sequences associated with functional domains of viral polypeptides, nucleotide sequence motifs or regulatory elements that are conserved within the genus or groups of species or strains, or sequences that are highly variable have been explored for this purpose.

Identification can be accomplished by determining the DNA sequence of the viral coat protein (tentative identification) and ultimately the complete DNA-A or monopartite DNA component. More recently the method, rolling-circle amplification, has been introduced as a non-sequence-specific approach for amplifying multimers of a viral genome, which can subsequently be cleaved to a full-length monomer using restriction endonucleases, allowing the sequence to be determined.

Sequence alignments are conveniently presented as cladograms to illustrate predicted relationships and distances can be calculated to estimate divergence. Phylogenetic analysis of the complete genome (monopartite) and A component (bipartite) viral sequences or the CP gene nucleotide sequences indicate strict clustering of begomoviruses by extant geographic origin, and not by monopartite or bipartite genome organization, crop of origin, or host range. Certain Old World viruses may be somewhat more divergent than the majority of New World viruses, perhaps because they have been more isolated by geographic barriers and/or patterns of human travel, suggesting fewer opportunities for interactions between viruses in different regions, while at the same time, great potential for localized diversity. All begomoviruses in the New World have bipartite genomes and many are distributed as overlapping geographic clusters, suggesting a lesser role of geographic barriers, fewer genotypes that have diverged in isolation, and, consequently, a more confounded evolution (**Figure 3**).

The genera in this family are discriminated based on numerical taxonomy in which a percentage nucleotide identity is used to classify them based on relatedness. Within each genus the working cutoff for species versus strains have been determined empirically by comparing the frequencies of percentage divergence across many sequences. For example, for the begomoviruses, a species is presently recognized when it is less than 89% identical to another, a value that has been shown to correlate nicely with biological differences. Species in the genera *Mastrevirus* and *Curtovirus* are classified as separate species when the genome sequence is found to share less than 75% and 89% nt identity, respectively.

All geminiviruses contain a conserved consensus sequence of *c.* 30 nt in length in this region. This sequence contains the T/AC cleavage site at which rolling-circle replication is initiated, and this conserved feature led to early speculation that recombination might occur

frequently at this site. In addition, virus-specific, directly repeated sequences that bind the viral replicase-associated protein (REP) during replication initiation are also present in this region and are considered to provide useful clues with respect to the likelihood that the DNA-A and DNA-B components are cognate, or of the same virus. The directly repeated element in bipartite and most monopartite genomes contains two 4–5 nt directly repeated sequences separated by two to four 'spacer' nucleotides which act in a virus-specific manner to bind REP. As a result, these elements constitute informative indicators of viral genealogy useful in defining species and strains when traced through viral lineages. It is not entirely clear how much plasticity in sequence or spacing can be tolerated at this site without interfering with replication. Finally, these repeated sequences in the CR have been explored to make predictions about which bipartite genomes might feasibly interact *in trans* and reassort either with a close relative, or a more distantly related virus that harbors the same or similar iteron sequences and REP binding ability.

Recently Adopted Management Approaches

Management of diseases caused by geminiviruses requires a multifaceted strategy. This involves reduction of vector population levels in the field, typically by insecticide applications and/or encouragement of natural enemies in refuge areas such as predators or parasites of the vector. Sanitation in and around the field in which virus-infected plants, including endemic weeds, are removed after the crop is harvested and before it is planted are common practices. Implementation of a host-free period has proven relatively effective in reducing inoculum levels to a minimum so that when a crop is planted, it achieves near-mature growth before becoming infected with virus. Plants that become infected when they are partially mature usually experience less damage compared to virus infection early after planting. Implementation of a host-free period is accomplished by having a voluntary or an enforced period of time when no virus or vector hosts may be grown for a designated period of time usually ranging from 4 to 8 weeks, across an entire region. This is often an intangible goal given the requirement for complete cooperation. In all of the above examples, a single viruliferous vector can feasibly transmit virus to susceptible young plants, and so the most desirable control strategy is the use of resistant or tolerant crop varieties. For the most part, such varieties are not available. Recently, several resistant cassava, maize, and tomato varieties have been developed as the result of widespread efforts to provide genetically tolerant germplasm. There also is an interest in developing and employing genetically engineered resistance using virus-derived resistance

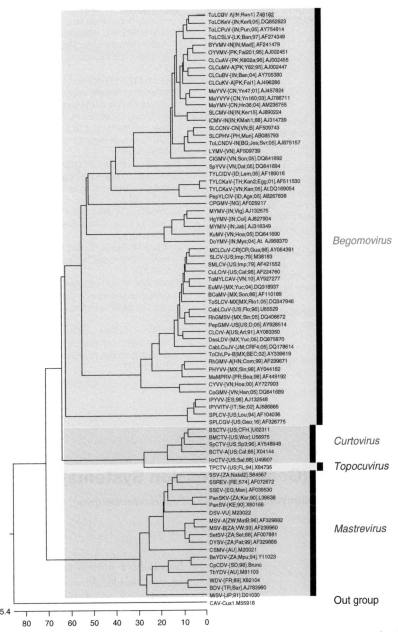

Figure 3 Phylogenetic tree of the family *Geminiviridae* showing the four major virus clades that serve to classify geminiviruses in one of four genera. The viruses are listed in the 8th ICTV Report. Courtesy of C. Fauquet.

approaches such as RNA interference; however, the reluctance on the parts of many to accept this strategy has set back progress, which likely would have been brought to fruition for the control of at least certain geminivirus-incited crop diseases.

The recognition that begomoviruses are capable of rapid divergence employing multiple mechanisms underscores the need for accurate molecularly based methods that permit detection and tracking of biologically significant variants. Molecular approaches will continue to be combined to achieve knowledge about viral biology, ecol-

ogy, and molecular and cellular aspects of the infection cycle. Databases containing genome sequences for geminiviruses now permit comparisons for establishing identity, taxonomic relationships, and interpreting disease patterns. Accurate molecular epidemiological information will assist plant breeding and genetic engineering efforts in developing disease-resistant varieties by enabling the selection of timely and relevant viral species (and variants) for germplasm screening and as sources of the most conserved viral transgene sequences, respectively, for sustainable disease control.

See also: Beet Curly Top Virus; Emerging Geminiviruses; Maize Streak Virus; Mungbean Yellow Mosaic Viruses; Tomato Leaf Curl Viruses from India.

Further Reading

Arguello-Astorga GR, Guevara-Gonzalez RG, Herrera-Estrella LR, and Rivera-Bustamante RF (1994) Geminivirus replication origins have a group-specific organization of iterative elements: A model for replication. *Virology* 203: 90–100.

Bisaro DM (1994) Recombination in geminiviruses: Mechanisms for maintaining genome size and generating genomic diversity. In: Paszkowski J (ed.) *Homologous Recombination and Gene Silencing in Plants*, pp. 39–60. Dordrecht, The Netherlands: Kluwer.

Brown JK (2001) The molecular epidemiology of begomoviruses. In: Khan JA and Dykstra J (eds.) *Trends in Plant Virology*, pp. 279–316. New York: Haworth Press.

Brown JK (2007) *The Bemisia tabaci* complex: Genetic and phenotypic variability drives begomovirus spread and virus diversification. Plant Disease APSNet Feature Article December–January 2006–07. http://www.apsnet.org/online/feature/btabaci/ (accessed February 2008).

Brown JK and Czosnek H (2002) Whitefly transmitted viruses. In: Plumb RT (ed.) *Advances in Botanical Research*, pp. 65–100. New York: Academic Press.

Dry IB, Krake LR, Rigden JE, and Reizan AM (1997) A novel subviral agent associated with a geminivirus: The first report of a DNA satellite. *Proceedings of the National Academy of Sciences, USA* 94: 7088–7093.

Goodman RM (1977) Single-stranded DNA genome in a whitefly-transmitted virus. *Virology* 83: 171–179.

Hanley-Bowdoin L, Settlage SB, Orozco BM, Nagar S, and Robertson D (1999) Geminiviruses: Models for plant DNA replication, transcription, and cell cycle regulation. *Critical Reviews in Plant Sciences* 18: 71–106.

Harrison BD and Robinson DJ (1999) Natural genomic and antigenic variation in whitefly-transmitted geminiviruses (begomoviruses). *Annual Review of Phytopathology* 37: 369–398.

Hou YM and Gilbertson RL (1996) Increased pathogenicity in a pseudorecombinant bipartite geminivirus correlates with intermolecular recombination. *Journal of Virology* 70: 5430–5436.

Idris AM and Brown JK (2005) Evidence for interspecific recombination for three monopartite begomoviral genomes associated with tomato leaf curl disease from central sudan. *Archives of Virology* 150: 1003–1012.

Lazarowitz SG (1992) Geminiviruses: Genome structure and gene function. *Critical Reviews in Plant Sciences* 11: 327–349.

Padidam M, Sawyer S, and Fauquet CM (1999) Possible emergence of new geminiviruses by frequent recombination. *Virology* 265: 218–225.

Sanderfoot AA and Lazarowitz SG (1996) Getting it together in plant virus movement: Cooperative interactions between bipartite geminivirus movement proteins. *Trends in Cell Biology* 6: 353–358.

Saunders K, Norman A, Gucciardo S, and Stanley J (2004) The DNA beta satellite component associated with *Ageratum* yellow vein disease encodes an essential pathogenicity protein (BC1). *Virology* 324: 37–47.

Plant Virus Vectors (Gene Expression Systems)

Y Gleba, S Marillonnet, and V Klimyuk, Icon Genetics GmbH, Weinbergweg, Germany

Glossary

Magnifection Process for heterologous protein expression in plants that relies on transient amplification of viral vectors delivered to multiple areas of plant body (systemic delivery) by *Agrobacterium*.

Noncompeting vectors Viral vectors capable of co-expressing recombinant proteins in the same cell throughout the plant body.

Introduction

Plant viral vectors have been and are still being developed to take advantage of the unique capability of viruses to extremely rapidly redirect most of the biosynthetic resources of a cell for the expression of a (usually) single nonhost protein – the viral coat protein (CP) (and hopefully, a protein of interest as well). Efficient viral vectors should allow to bypass the limitations of other expression methods, such as the low yield obtained with stable transgenic plants or standard transient expression methods, and the long time necessary for development of stable transgenic lines. Most of the progress achieved so far has been with RNA viruses, and the most advanced vectors have been built using just a handful of plant viruses such as tobacco mosaic virus (TMV), potato virus X (PVX), alfalfa mosaic virus (AMV), and cowpea mosaic virus (CPMV). Tremendous technical progress has been made over the past few years in the development of production processes and industrial plant hosts. In particular, novel developments include the design of vectors that are not just simple carbon copies of wild-type viruses carrying a heterologous coding sequence, but that, instead, have been 'deconstructed' by delegating some rate-limiting functions to agrobacteria or the plant host, thus allowing for a more efficient, versatile, controlled, and safe process.

Viruses have to carry out a number of processes to complete the viral replication cycle. These include host infection, nucleic acid amplification/replication, protein translation, assembly of mature virions, cell-to-cell spread, long-distance spread, reprogramming of the host biosynthetic processes, suppression of host-mediated gene silencing, etc. Viral vectors, in contrast, do not necessarily have to be able to perform all of these functions, but they have to perform a new function – high-level expression of a heterologous sequence.

Two different strategies can be used to develop a viral vector/host process.

Historically, the first approach that was developed, starting in 1990, consisted of designing vectors that were, in essence, wild-type viruses engineered to express an additional sequence – the gene of interest (approach also known as the 'full-virus' vector strategy). These vectors were expected to perform all of the functions that a virus normally undertakes, from the infection step to the formation of complete infectious viral particles. However, addition of a heterologous sequence reduced the efficiency of many of the normal steps of the replication cycle, such as high-level amplification and systemic movement. Moreover, these vectors did not necessarily provide the maximal level of expression for the protein of interest that one might have expected. Nevertheless, most of the practical results reported in the literature have been obtained with these 'first-generation' viral vectors.

The design of the next-generation viral vectors reflects an approach that admits inherent limitations of the viral replication process applied to heterologous gene expression. Rather than mimic the design of the wild-type virus, the goal is to 'deconstruct' the virus, and eliminate the functions that are either limiting (e.g., too species-specific, or rate-limiting) or undesired (such as the ability to create functional infectious viral particles), and rebuild the process, by either delegating the missing functions to the host (genetically modified to provide those functions in *trans*) or replacing them with analogous functions that are not derived from a virus (the 'deconstructed virus' vector strategy). In these 'second-generation' expression systems, elements of the viral machinery such as RNA/DNA amplification and cell-to-cell movement are integrated along with nonviral processes such as replicon formation via *Agrobacterium*-mediated delivery of T-DNAs encoding the viral vector or via activation from a plant chromosomal DNA encoding a pro-replicon or pro-virus.

First-Generation Virus Vectors

Under the 'full-virus' scenario, delivery of the replicons is provided by infecting the host with a mature viral particle or with a full copy of the viral DNA/RNA. The vector is essentially a functional virus that, in addition to the viral sequence, contains a coding sequence of interest under control of a strong viral promoter such as the CP subgenomic promoter (**Figure 1(a)**). Several efficient vectors have been made, primarily based on RNA viruses such as TMV and PVX. Vectors based on TMV were developed in several steps. Earlier versions simply used a duplicated CP subgenomic promoter for expression of the gene of interest, but this led to instability of the vector. Later versions replaced the duplicated subgenomic promoter by sequences derived from a different but phylogenetically related virus. Finally, different 3′ untranslated sequences from a range of related viruses were screened to obtain a vector expressing the gene of interest efficiently.

An alternative expression strategy consists of expressing the protein of interest as a fusion to the CP (**Figure 1(a)**). In this case, only an essential part of the protein of interest, such as an immunogenic epitope, is fused to the CP since larger fusions usually eliminate viral particle formation and systemic movement ability.

Large-scale production with either type of vector can be achieved by spraying plants in the field or a greenhouse with a mixture of viral particles and carborundum. Depending on the efficiency of the vector and its ability to move systemically, 2–3 weeks are required for plants to achieve maximal transfection/expression.

During the last few years, improvements have been made to this first-generation vectors. Among the most interesting experiments coming from recent work is the improvement of viral vectors by directed evolution. It was found that cell-to-cell movement of the vectors described above was not as high as with wild-type viruses due to lower expression of the movement protein. The strategy employed was to use DNA shuffling to improve the movement protein gene. This work was very successful and resulted in mutant movement protein genes working better than the native version in the viral vector context, essentially in tobacco, where engineered vectors normally do not work very well. This work suggests that other parts of the viral vectors might be improved by DNA shuffling as well.

In general, first-generation vectors were very successful and were able to provide expression of heterologous protein at the levels close to 10% of total soluble protein or over 1 g recombinant protein kg^{-1} of fresh leaf biomass. Several pharmaceutical proteins made using such vectors have actually reached the stage of clinical trial.

However, despite these successes, first-generation viral vectors have several limitations. The yield that can be obtained with these vectors varies greatly depending on the protein to be expressed. In most cases, only relatively small proteins are expressed efficiently, any proteins larger than 30 kDa are usually poorly expressed. Moreover, yield of heterooligomeric proteins requiring two or more subunits expressed from the same vector are extremely low. With CP fusions, the yield can be as high

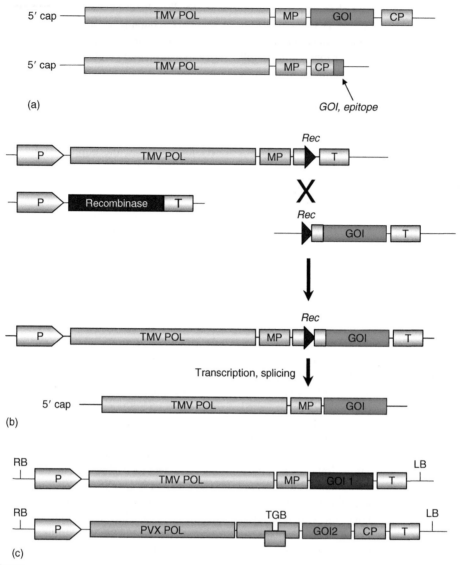

Figure 1 Continued

as $2-4\,\mathrm{g\,kg^{-1}}$ of plant biomass; however, usually only short (up to 25 amino acids) epitopes are tolerated.

Host Improvement

To survive in nature, plants have naturally evolved resistance mechanisms against pathogen attacks. It is therefore expected that plants could be engineered to be more susceptible to viruses and therefore better hosts for expression of heterologous sequences using viral vectors. A dilemma is that plants susceptible to viral vectors and therefore good hosts for protein expression will also have poor agronomic characteristics.

One important element of plant defense mechanisms against viruses is the ability of plants to induce post-transcriptional silencing. A large amount of work in

this area has shown that not only plants are able to defend themselves against viruses, but that viruses have counterattacked by evolving suppressors of silencing. All these elements uncovered by basic research in virology offer elements that can be incorporated in expression strategies (with or without viral replication) to boost the level of protein expression. Another important field of research consists of approaches that aim at silencing/superactivating some plants genes that are, directly or indirectly, involved in viral replication or spread.

A typical host for viral research is *Nicotiana tabacum*, the host plant of the most-studied plant virus – tobacco mosaic virus. The most commonly used plant in virology is however the wild Australian species, *N. benthamiana*, which is unusually sensitive to a very wide array of viruses. Despite its poor agronomic characteristics, *N. benthamiana*

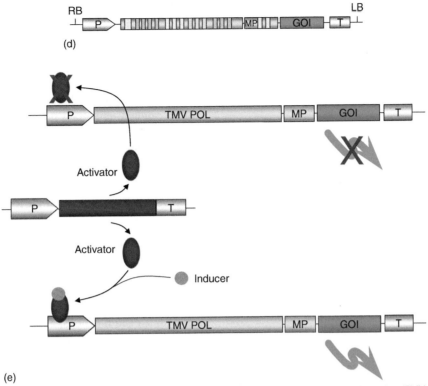

Figure 1 Plant viral vectors currently in use. (a) Typical first-generation RNA expression vectors based on TMV virus. (b) Second-generation vectors. Pro-vector system for rapid *in planta* assembly of viral replicons from T-DNAs delivered by *Agrobacteria*. (c) DNA vectors optimized for *Agrobacterium*-mediated delivery and expression of heterooligomers (IgG antibodies). (d) DNA vectors optimized for *Agrobacterium*-mediated delivery. (e) Systems relying on the regulated release of a replicon from a plant chromosome upon chemical induction. The viral components of the vector backbones are in yellow or in light green and include RNA-dependent RNA polymerase (TMV POL or PVX POL), movement protein (MP), coat protein (CP), triple gene block (TGB); the genes of interest encoding recombinant protein are in green or pink (GOI); the site-specific recombinase (REC) and its recognition sites (*Rec*) are in blue; multiple colored segments on the vector optimized for *Agrobacterium* delivery are plant introns; activator gene and its product are in red; promoter and terminator sequences are in gray.

can be grown very well in the greenhouse under controlled conditions and could therefore even be considered for industrial production.

One drawback of using plants as hosts for manufacturing therapeutic proteins is that plant post-translational modifications are not exactly identical to those made by animal/human cells (although, fortunately, not extremely different either). In particular, the enzymatic machinery responsible for N-glycosylation differs in plants and animals. More specifically, plant cells add some specific sugars (core-bound xylose and α-1,3-fucose), which might be immunogenic if protein containing these were administered to humans. In contrast, plant cells do not add other sugars such as terminal β-1,4-galactose residues or sialic acid, which are added to proteins made in animal cells. Fortunately, significant work has been made to engineer plant hosts, including tobacco and *N. benthamiana*, to provide more 'human-like' or even 'designer' post-translational modifications – in particular, low amounts of xylose and α-1,3-fucose as well as galactosyl terminal residues added to the core sugars. Such host plants will be useful for all

vector types, whether of the first generation or vectors developed later.

Second-Generation Vectors

This strategy reflects attempts by engineers to part with the inherent limitations of the viral machinery, while keeping the useful viral elements. Some of the elements that can be used outside of the integrated virus system are the 'molecular machines' that provide for infectivity, amplification/replication, cell-to-cell movement, assembly of viral particles, suppression (shutoff) of the synthesis of plant proteins, silencing suppression, systemic spread, etc. Some of these elements are less efficient than others. For example, the ability of the virus to infect the host is low and requires some type of mechanical injury to the plant, or an insect as a vector. Systemic spread is a process that is usually very species-specific, easily impaired as a result of genetic manipulation. Replication/amplification ability, on the other hand, is a central element of viral vectors that is valuable as an 'amplifier'. Fortunately, replication is a relatively robust and species-independent mechanism.

Two basic applications, namely rapid expression in small or large quantities and large-scale industrial production of proteins in plants, each require different design of the expression strategies.

Vectors for Specific Application Areas

Plant Virus Vectors as Research Tools

Many different expression systems have been developed based on the backbones of entirely different viruses, and relying on different modifications of the core viral design. However, since the purpose of using such vectors is often rapid and high-throughput expression, and since only small (usually milligram) amounts of protein are usually required, many existing vectors do not contain the viral component(s) providing systemic movement.

An important and limiting step for the use of viral vectors is infection of the plants with the viral replicons. Use of *Agrobacterium* to deliver a copy of the viral vector encoded on the T-DNA ('agrodelivery') provides an excellent solution since *Agrobacterium* is an extremely efficient vector. All the steps that are necessary for the conversion of the 'agrodelivered' T-DNA into a functional DNA or RNA replicon have been shown to occur in plants. 'Agroinfection' has been used for many years. It is also often much more efficient than mechanical inoculation using viral particles, and is definitely more efficient than using DNA or RNA as infectious molecules. In case of RNA viruses, agroinoculation also represents a very inexpensive alternative to *in vitro* transcription to convert the DNA vector into an infectious RNA.

The use of agrodelivery for inoculation of RNA-viruses vectors also provides entirely new opportunities for vector engineering, since the vector is delivered to the plant cell as a DNA molecule, which can be manipulated prior to conversion into an RNA replicon. In particular, it has been shown that several T-DNAs transiently co-delivered from one or more agrobacteria can be efficiently recombined *in planta* by site-specific recombinases such as Cre or the φc31 integrase. The RNA obtained after transcription of the recombined DNA molecule can then be cleaned of the recombination sites by the nuclear RNA splicing machinery, provided that the recombination sites are engineered by flanking by intron sequences (**Figure 1(b)**).

Using such elements, a process has been developed that, in essence, allows simple and inexpensive '*in vivo*' engineering of RNA vectors by simply mixing various combinations of agrobacteria containing different components of a viral vector and co-infiltrating these mixtures into plants. Such an approach allows, for example, to rapidly assemble and express a variety of different protein fusions such as fusions of the protein of interest to targeting or signal peptides, binding domains, various coding sequences, purification domains, or cleavage sites. All of these expression experiments are done by simple mixing of pre-fabricated bacteria strains harboring the desired vector modules and a strain containing the gene-specific module. Depending on the viral vectors and the specific protein studied, milligram quantities of protein can be expressed in various plant compartments or as different fusions, within just a few days.

Vectors for Expression of Heterooligomeric Proteins

The first transgenic plants expressing full-size human antibodies were made around 1990. Despite these early positive results, low yields have prevented the widespread use of this technology for commercial applications. For example, although stably transformed plants express correctly folded and functional antibodies of immunoglobulin of the IgG and IgA classes, yields are generally very low (1–25 mg kg^{-1} of plant biomass). Moreover, the time necessary to generate the first gram(s) of such antibody material is longer than 2 years.

Transient systems, on the other hand, allow production of research quantities of material much faster. However, as for transgenic plants, transfection systems relying on *Agrobacterium*-mediated delivery of standard nonreplicating expresssion cassettes also provide only low yield of expressed proteins. The level of expression of such systems can be increased by using suppressors of silencing, but yields are still inferior to expression using viral vectors and are not readily scalable. Initial attempts to combine the speed of expression of transient systems with the high level of expression of viral vectors failed due to the inability of viral vectors to co-express two genes within one cell at high level; engineering of vectors for expression of two genes from one viral vector led to drastically reduced expression level of both genes; co-infiltration of two vectors expressing a separate gene also failed because competition between vectors resulted in either one or the other protein expressed in individual cells, but not of the two proteins within the same cell (which would be required for expression of a functional IgG antibody).

Recently a solution to this problem was found by cloning the two genes to be expressed in two separate viral vectors that were constructed on the backbone of two different noncompeting viruses such as, for example, TMV and PVX. Unlike vectors that are built on the same backbone, 'noncompeting viral vectors' were shown to be able to efficiently co-infect and replicate within the same cell. Therefore, high level of functional full-size monoclonal antibodies (Mabs) of the IgG class can now be obtained in plants by co-infiltrating two vectors containing the heavy- and the light-chain separately in each vector. This strategy was shown to work with different Mabs of the IgG1 and IgG2 classes; the molecules were found to be fully functional, and the first gram of material could be produced in less than 2 weeks after cloning in the vector (**Figure 1(c)**).

Expression Vectors for Industrial-Scale Protein Production: Transient Systems (Magnifection)

Industrial production requires viral amplification and expression in large numbers of plants (in a greenhouse or field) and, to obtain maximal yield, in as many tissues and leaves as possible within each plant. For achieving this goal, a first solution was provided by using vectors capable of systemic spread (the 'full-virus' strategy). Although successful, this strategy had many limitations: large inserts (larger than 1 kbp) could not be efficiently expressed, not only because of less-efficient amplification within each cell, but also because viral vectors containing large inserts are less able to move in systemic tissue without recombining; systemic vectors also never infect all harvestable parts of the plant (e.g., the lower leaves are usually not infected); the process is asynchronous as it invades different leaves in a sequential order. Moreover, for viral vectors designed to express protein fragments as CP fusions, only short epitopes (20 amino acids or less) could be successfully expressed until recently.

A simple technical solution was found to bypass these limitations; a viral vector lacking the ability to move systemically (with a simpler design, and more stable) is delivered to the entire plant using *Agrobacterium*-mediated delivery by vacuum-infiltration of the entire plant. This eclectic technology called 'magnifection' combines advantages of three biological systems: the speed and expression level/yield of a virus, the transfection efficiency and systemic delivery of an *Agrobacterium*, and the posttranslational capabilities and low production cost of a plant. Such a process can be inexpensively performed on an industrial scale and does not require genetic modification of plants, and is therefore potentially safer (it also does not generate infectious virions as the full-vector strategy does) and more compatible with the current industrial infrastructure.

For magnifection to become an efficient tool, technical difficulties had however to be addressed. Indeed, initial vectors based on the backbones of plus-sense RNA viruses such as TMV could not be delivered to plant cells efficiently by *Agrobacterium* delivery. The first attempts were made with vectors built on the backbone of TMV and led to a rate of infectivity estimated at one successful transfection event per 10^8 agrobacteria. Later experiments using viral vectors based on different tobamoviral strains (infecting both crucifers and *Nicotiana* plants) were more success-

ful, but clearly not optimal. Analysis of the early infection steps initiated by *Agrobacterium* delivery of T-DNAs encoding RNA viral vectors suggested that a bottleneck for the formation of active replicons may be the low ability of the primary transcript to leave the nucleus. The RNAs of cytoplasmic RNA viruses such as TMV normally never enter the plant cell nucleus at any point during the entire replication cycle and have therefore not evolved to be in contact with the cell RNA-processing machinery and to be exported to the cytosol. *In silico* analysis of the viral genome structure with programs designed to assess whether an unknown sequence is likely to be coding or noncoding (by comparison to expressed nuclear plant genes, using neural networks), showed that the viral RNA is unlikely to be recognized as a correct coding sequence and suggests that it will probably be processed abnormally or even degraded before it is able to reach the cytosol. By changing the codon usage of at least portions of the viral RNA, the viral genome can be modified to look more 'exon-like' (conversion of T-rich sequences to more GC-rich sequences, and removal of putative cryptic splice sites). Such changes have been useful to improve export of functional RNAs to the cytosol, but addition of plant introns improved the process even more significantly. The resultant synthetic T-DNA templates, when delivered as DNA precursors using *Agrobacterium*, provided efficient processing of the DNA information into active amplicons in almost all (>93%) cells of infiltrated tobacco (*Agrobacterium* infiltrated at optical density (OD) of 0.7), a 10^3-fold improvement compared with nonmodified vectors, and an up to 10^7-fold improvement over nonoptimized DNA templates reported in the first publications. Improved vectors lead to one successful infection event per 10–20 infiltrated agrobacteria (**Figure 1(d)**).

Using these vectors, a simple fully scalable protocol for heterologous protein expression in plants has been designed that does not require transgenic plants, but instead relies on transient amplification of viral vectors delivered to the entire plant using *Agrobacterium*. Entire plants are infiltrated (although the same procedure can be performed on detached mature leaves) with a highly diluted (up to 10^{-4} dilution) suspension of bacteria carrying a proviral amplicon on the T-DNA. The combination of vacuum infiltration/agroinfection can therefore be considered as replacing the conventional viral functions of primary infection and systemic movement. Viral vector-controlled amplification and cell-to-cell (short distance) spread is performed by the replicon, as is the case with standard vectors. Depending on the vector used, the host organism, and the initial density of bacteria, the magnifection process takes from 4 to 10 days and, for well-expressed proteins, results in level of up to 5 g recombinant protein kg^{-1} of fresh leaf biomass or over 50% of total soluble protein. Furthermore, since the viral vector lacks a CP gene (for tobamoviruses, the CP gene is needed only for plant infection or systemic movement, and is therefore not needed for magnification), it can express longer genes (up to 2.3 kbp inserts or up to 80 kDa proteins). Infiltration of plants/detached leaves with bacteria can be performed in several ways, one simple process being vacuum infiltration by immersing whole aerial parts of plants in a bacterial suspension and applying a weak vacuum (approximately minus 0.8–1.0 bar) for 10–30 s. This process can be performed on a large scale and is expected to be commercially viable for production of up to 1 ton of recombinant protein per year. Industrial-scale production will require containment of infiltrated plants which contain agrobacteria (**Figures 2** and **3**).

Expression Vectors for Industrial-Scale Protein Production: Transgenic Systems

A radically different approach to establish an industrial expression system would be to produce a stably transformed plant that would contain an inactive or repressed copy of the viral replicon inserted on one of the host's chromosomes. Induction of viral replication could be activated at the time of the operator's choosing by a specific treatment. Obviously, establishing such a system would slow down the development phase by several months/years, in order to obtain stable transformants and select well-performing lines. However, this delay would be largely compensated by the ease of production that the system would provide and by the resulting low production costs.

Several strategies are possible for the design of an inactive replicon, and specific strategies will of course depend on the type of viral vector used. A viral vector might be encrypted by insertion of additional sequences within the vector, or inversion of part of the vector, and activation could be provided by regulated expression of a recombinase. Alternatively, a replicon could be kept inactive by silencing by the host plant, and activated by regulated expression of a silencing suppressor. Finally, a replicon could be kept in an inactive form by fusion to a promoter that is inactive in the absence of a specific activator or that is repressed in the presence or absence of a particular inducer.

The inducing treatment could be provided in many different ways. A simple strategy would be to induce release of the active replicon by genetic hybridization, which would combine in the same plant both the transgene locus encoding the replicon as well as the transgene(s) that control unencryption (e.g., a silencing suppressor or a recombinase that is developmentally regulated). Alternatively, hybridization could also be used to bring together genomes of multipartite viruses, thus providing for replication of RNAs that carry the transgene of interest but cannot replicate in the absence of the master RNA expressing the necessary RNA-dependent RNA polymerase.

A more versatile solution would be provided by spraying a small chemical inducer on an engineered plant. One technical challenge for establishing such a system is the design of an effective chemical switch that provides a tight control and can be effectively used for an industrial process.

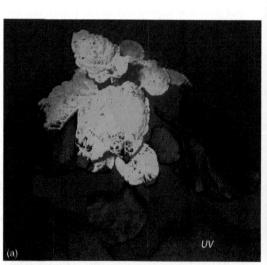

Figure 2   Plant transfection with (a) first- and (b) second- generation vectors; GFP expression in *Nicotiana benthamiana*, photographed under UV light.

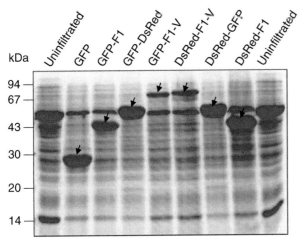

Figure 3 Expression of proteins and protein fusions using second-generation vectors; Coomassie-stained polyacrylamide protein gel showing protein profiles in crude extract from uninfected leaf tissue as well as tissues infected with genes encoding fluorescent protein GFP or protein fusions involving GFP or DsRed as fusion components. The arrow indicates the band with heterologous protein. Molecular weight ladder in kDa is shown on the left side.

One specific issue in dealing with chemically inducible promoters for industrial purpose is the availability of small chemical molecules that are commercially available and safe. Among those available today, one should mention ethanol (used as a part of ethanol-inducible system), the commercial insecticide methoxyfenozide, which acts as an agonist for the ecdysone receptor, and tetracycline antibiotics.

Several groups are now working on the development of protocols relying on chemically inducible promoters for induction of both DNA and RNA viral vectors. As one example, an inducible geminivirus viral vector system based on bean yellow dwarf virus has been described. In this system, treatment with ethanol results in expression of the replicase, which induces release and replication of the gene-specific replicon and leads to expression of the gene of interest. Ethanol induction was reported to result in up to 80-fold increase of the mRNA levels and up to 10-fold increase of the translation product. In another example, an inducible system that relies on the estrogen-inducible promoter was shown to work efficiently in tobacco cell suspensions. Upon induction, a modified tomato mosaic virus expressing green fluorescent protein (GFP) instead of its CP amplified and expressed the gene of interest at high levels (10% of total soluble protein). In contrast, neither viral RNA nor GFP were detectable in uninduced cells (**Figure 1(e)**).

Despite these positive results, it is clear that the full potential of inducible viral vectors systems has still not been completely reached. Given the progress in this field, it is however expected that fully effective technology processes will be available within the next few years.

Vectors for Manufacturing 'Nanoscale' Materials

Due to their relatively simple macromolecular organization and very high accumulation titers, tobamoviruses provide an extremely cheap source of biopolymers than can be manufactured rapidly and under very simple conditions. TMV, the best-known tobamovirus, is also one of the most extensively studied viruses. It has a positive-sense RNA genome encoded in a single 6.4 kbp RNA molecule. The genome encodes four proteins, including the 17.5 kDa CP, the most abundant viral product and the only component of the TMV capsid. Over 2100 copies of CP fully protect the single-stranded viral RNA, resulting in rigid rod-shaped viral particles with a length of 300 nm and a diameter of 18 nm and a molecular mass of 40 000 kDa. The genomic RNA is packaged inside of a 2 nm-wide canal formed by the assembled CP capsid. Ninety-five percent of the mass of TMV particles consists of the CP. TMV accumulates to levels of up to $10 \, \text{g kg}^{-1}$ of leaf biomass, and, therefore, the CP represents the most abundant individual protein that can be harvested from plants. The virions can be purified industrially using simple 'low-tech' protocols.

Due to the ability of the CP to polymerize *in vivo* and *in vitro* and the high stability and defined size of the assembled virions, the CP represents a potentially promising biopolymer feedstock for a number of applications in nanobiotechnology. The CP, the basic element of the 'viral polymer' (the viral particle) can be modified in two different ways.

One strategy consists of attaching novel chemical moieties to the viral particles *in vitro* by chemical modification of reactogenic groups exposed on the surface of the virus or in the inner cavity. One such reactogenic group, lysine, allows biotinylation of the capsid. In the case of the TMV CP, which does not contain a lysine, a randomized library was made and screened to introduce this desired amino acid at an externally located position, and this without affecting the formation of viral particles. The protein/peptide fusion component was then expressed independently as a streptavidin fusion, and used to decorate the TMV particles.

Another strategy for modification of the viral core monomer, the CP, consists of engineering the viral vector to express a protein fusion. Since the structure of many viral particles has been determined at atomic resolution, fusion proteins can be designed in such a way that the added sequences are predicted to lie at the surface of the assembled viral particle. For TMV, fusions can theoretically be made either at the N- or the C-terminal ends of the CP since both ends are exposed on the surface of the viral particle. For other viruses, short peptide epitopes can be inserted in surface-exposed loops. However, constraints such as the size or the pI of the inserted sequence are limiting the type of epitopes that can be successfully expressed using this strategy.

So far, extensive work by numerous groups to create new products based on protein fusions has met with limited success. Since the main goal of these studies was the design of new vaccines by surface display of immunotopes, only short inserts were extensively analyzed. For example, the limited size of peptides that could be fused to the CP of TMV without preventing virion assembly has restricted these systems to expression of 20-amino-acid (aa) or shorter peptide immunogens and of one peptidehormone only. However, it has recently been found that much longer polypeptides of up to 133 aa (for a fully functional fragment of protein A) can be displayed on the surface of TMV as C-terminal fusions to the CP, provided a flexible linker is being used. These nanoparticles coated with protein A allow purification of Mabs with a recovery yield of 50% and higher than 90% purity. The extremely dense packing of protein A on the nanoparticles confers an immunoadsorbent feature with a binding capacity of 2 g Mab per g. This characteristic, combined with the high level of expression of the nanoparticles (more than $3 \, \mathrm{g \, kg}^{-1}$ of leaf biomass), provides a very inexpensive self-assembling matrix that could meet industrial criteria for a single-use immunoadsorbent for antibody purification.

The field of nanomaterials built around plant viruses as scaffolds is still *in statu nascendi*, and we believe that it will offer multiple practical solutions not only to specific problems such as mentioned above for downstream processing of antibodies and other proteins, but also to entirely new technology processes that are difficult to imagine today.

Conclusions

Although there are currently no products or medicines in the market produced using viral vectors, compared to other expression platforms utilizing plants, this technology has made tremendous progress since its inception approximately 15 years ago (**Figure 4**), and it is simply a matter of time before products made with viral vectors become available. This progress has come as a result of the extensive studies that revealed the structure, functions, and molecular organization of plant viruses. In 1955, fascinated by the elegant simplicity of TMV's morphology, F. Krick supposedly exclaimed: 'Even a child could make a virus'. Today, some 50 years later, we tend to agree with him.

Acknowledgments

The authors wish to thank Dr. Anatoliy Girich, Icon Genetics, Halle, Germany, for the materials used in **Figure 3**.

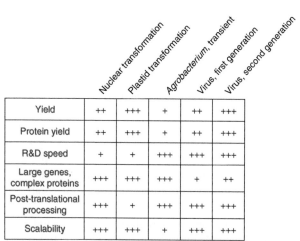

	Nuclear transformation	Plastid transformation	Agrobacterium, transient	Virus, first generation	Virus, second generation
Yield	++	+++	+	++	+++
Protein yield	++	+++	+	++	+++
R&D speed	+	+	+++	+++	+++
Large genes, complex proteins	+++	+++	+++	+	++
Post-translational processing	+++	+	+++	+++	+++
Scalability	+++	+++	+	+++	+++

Figure 4 Strengths and weaknesses of different plant expression platforms. The platforms include nuclear transformation, plastid transformation, *Agrobacterium*-mediated transient expression, and the two viral expression platforms. The parameters that are most essential for expression of recombinant proteins, in particular biopharmaceuticals, include yield of recombinant protein, relative yield as a percentage of total soluble protein, speed of research and development supported by the platform, ability to express large genes and manufacture complex proteins, post-translational processing capabilities and scalability. The best performance is assigned three crosses, the worst is assigned one cross.

Further Reading

Canizares MC, Nicholson L, and Lomonossoff GP (2005) Use of viral vectors for vaccine production in plants. *Immunology and Cell Biology* 83: 263–270.

Donson J, Kearney CM, Hilf ME, and Dawson WO (1991) Systemic expression of a bacterial gene by a tobacco mosaic virus-based vector. *Proceedings of the National Academy of Sciences, USA* 88: 7204–7208.

Gleba Y, Marillonnet S, and Klimyuk V (2004) Engineering viral expression vectors for plants: The 'full virus' and the 'deconstructed virus' strategies. *Current Opinion in Plant Biology* 7: 182–188.

Ma JKC, Drake PMW, and Christou P (2003) The production of recombinant pharmaceutical proteins in plants. *Nature Genetics* 4: 794–805.

Mallory AC, Parks G, Endres VB, *et al.* (2002) The amplicon-plus system for high-level expression of transgenes in plants. *Nature Biotechnology* 20: 622–625.

Marillonnet S, Thoeringer C, Kandzia R, Klimyuk V, and Gleba Y (2005) Systemic *Agrobacterium tumefaciens*-mediated transfection of viral replicons for efficient transient expression in plants. *Nature Biotechnology* 23: 718–723.

Pogue GP, Lindbo JA, Garger SJ, and Fitzmaurice WP (2002) Making an ally from an enemy: Plant virology and the new agriculture. *Annual Review of Phytopathology* 40: 45–74.

Porta C and Lomonosoff GP (2002) Use of viral replicons for the expression of genes in plants. *Biotechnology & Genetic Engineering Reviews* 19: 245–291.

Smith M, Lindbo JA, Dillard-Telm S, *et al.* (2006) Modified tobacco mosaic virus particles as scaffolds for display of protein antigens for vaccine applications. *Virology* 348: 475–488.

Yusibov V, Shivprasad S, Turpen TH, Dawson W, and Koprowski H (1999) Plant viral vectors based on tobamoviruses. *Current Topics in Microbiology and Immunology* 240: 81–94.

Vaccine Production in Plants

E P Rybicki, University of Cape Town, Cape Town, South Africa

Glossary

Immunogenicity Ability of a substance to stimulate the acquired immune response in a recipient.
Pharming The production of pharmaceutically relevant substances in plants.
Vaccine Substance which elicits an immune reaction which protects against infection by a natural pathogen.

Introduction

The concept of plant-produced vaccines, and in particular viral vaccines for animals and humans, is both recent – the first vaccine-relevant protein was made in plants only in 1989 – and controversial, with dire misgivings still being expressed concerning a number of issues. The central planks of the justification for plant-produced pharmaceuticals in general, and for the production of vaccines in particular, have been that their production in plants is both safe and potentially very cheap. A strong additional argument has been that the use of food plants will allow edible vaccines to be locally produced where they are needed most – in the developing world. Thus, the kind of argument that has developed is that plant production of especially viral vaccines allows the possibility of producing large amounts of high-grade vaccines for very low cost (1) for diseases where there is a high burden of preventable disease, (2) where existing vaccines are expensive, (3) where there are no vaccines for 'orphan diseases', and (4) where therapeutic vaccination could be very effective in treating disease.

The need for low-cost vaccines is borne out by the fact that at least 3 million people die every year of vaccine-preventable diseases, yet it takes many years on average for any new vaccine's price to come down to levels (<US$ 1 per dose) where it can be incorporated into the extended programme of immunization (EPI) bundle.

As with any maturing technology, however, problems have come to light along with proofs of concept, and the current view of the field is very different to the early one. Thus, concerns over genetic 'contamination' of food crops have largely blunted the prospect of edible viral vaccines delivered via unprocessed food plants; regulatory concerns – both for production and registration – have tempered the optimism of early years concerning widespread and rapid adoption of the technology; production via transgenic plants has, on the whole, proved to be less reliable and lower-yielding than was hoped; immunogenicity of model antigens delivered orally has often proved to be disappointingly low; and, most importantly, it has been realised that costs of production are far less important in determining the final price of a vaccine than was generally assumed.

However, there have been notable successes as well: a number of animal model systems have been thoroughly tested, and protection against disease has been obtained; production systems have to some extent been worked out, and the first products have been licenced; transient expression systems are proving to have great potential for even large-scale production; and, most significantly, large pharmaceutical companies are increasingly interested in the prospects of making high-value products in plants.

The increasing exposure of the concept to public view has also brought its share of problems: it was after all the public outcry over 'contamination' of food plants with vaccine protein genes that led to the involvement of regulatory agencies in the issue, and a moratorium on the use of food crops for vaccine production. Moreover, public acceptance of human vaccines made in plants is by no means assured, and the anti-genetically modified (anti-GM) lobby has already made an issue of the means of production. Together with these sobering developments has come the realization that high-value products not meant for use in humans are probably the first best use of the advantages of plant production of biopharmaceuticals. For example, it is probably a far better idea to use plants to produce proteins and other macromolecules as reagents or components of diagnostic kits than it is to produce far more regulated vaccine components – as well as being a much quicker route to financial gain.

While the initially determinedly rosy picture of very cheap, edible vaccines against orphan diseases and developing country scourges has not been realized, there is now an altogether more realistic consensus view on how to proceed with the application of this still very promising technology to human and animal health. This article covers the historical development of the concept of plant-produced viral vaccines and relevant expression systems, as well as central issues affecting production, purification, and application of such vaccines.

History

The first publication on the production of vaccine-relevant proteins in plants was in 1989, when A Hiatt and colleagues at the Scripps Institute in La Jolla produced mouse-derived monoclonal antibodies (MAbs) in transgenic tobacco (see **Table 1**). This was soon followed in 1992 by the first of a series of papers – in this case involving hepatitis B virus (HBV) surface antigen (HBsAg) – by the group of Charles J. Arntzen and Hugh Mason on the production of human viral vaccine candidate antigens in transgenic plants. While they did not at the time hold out the promise of edible vaccines, they did conclude that plants did produce antigens that were equivalent to

Table 1   Landmarks in plant-produced vaccine research

Year	Landmark
1989	Expression of immunoglobulin genes in plants
1992	Expression of hepatitis B surface antigen in transgenic plants
1995	Oral immunization with a plant-produced recombinant bacterial antigen
1996	Plant expression of Norwalk virus capsid protein and its oral immunogenicity
1998	Use of a recombinant plant monoclonal secretory antibody in humans
1998	LT-B gene in potatoes protects mice against LT
1999	Oral immunization with potato-produced VP60 protein protects against hemorrhagic disease virus
1999	Protection against FMDV by vaccination with foliar extracts of rTMV-infected plants expressing FMDV VP1
2000	Human responses to Norwalk virus vaccine delivered in transgenic potatoes
2001	Oral immunogenicity of oral HPV L1 co-administered with *E. coli* enterotoxin mutant R192G or CpG DNA
2002	Plant-derived HPV 16 E7 oncoprotein induces immune response and specific tumor protection in mice
2002	Boosting of a DNA measles immunization with an oral plant-derived measles virus vaccine
2003	Production of HPV-16 VLPs in transgenic potato and tobacco and oral immunogenicity
2003	Expression of HPV-16 VLPs in transgenic tobacco and parenteral immunogenicity
2003	Oral immunogenicity of HPV-11 VLPs expressed in potato
2003	Oral immunization with rotavirus VP7 in transgenic potatoes induces high titers of mucosal neutralizing IgA
2004	Conformational analysis of hepatitis B surface antigen fusions in an *Agrobacterium*-mediated transient expression system
2004	Oral TGEV vaccine in maize seed boosts lactogenic immunity in swine
2004	Mucosally targeted HIV gp41-derived fusion protein with CTB elicits transcytosis-blocking Abs
2005	Magnifection – TMV-based transient agroinfection platform
2005	Immunogenicity as a boost in humans of potatoes containing HBsAg
2005	Development of a plant-based vaccine for measles
2006	Proof of concept of a plant-produced papillomavirus vaccine in a rabbit model
2006	A plant-derived edible rotavirus subunit vaccine is stable for over 50 generations in transgenic plants

those produced in more conventional systems, and they held promise as low-cost vaccine production systems. Concurrent with these advances was a rapid and parallel improvement in the understanding of mucosal immunity and the fact that the mucosa-associated lymphoid tissue (MALT) – and, in particular, the gut-associated lymphoid tissue (GALT) – was probably the premier 'organ' of the entire adaptive immune response. It was perhaps inevitable that these two should come together in the concept of 'edible vaccines', where vaccine antigens would be introduced into people and animals in food, given that most human and animal viruses gain entry to their hosts via mucosal surfaces.

The first proof of concept for edible plant-produced vaccines was performed in the mid-1990s using a bacterial protein – *Escherichia coli* heat-labile enterotoxin (LT-B) – but this was soon followed by work on HBsAg, secretory antibodies, Norwalk virus capsid protein, and a *Vibrio cholerae* enterotoxin subunit vaccine (CTB). These studies laid important ground in terms of demonstrating conclusively that plant-produced vaccine antigens were effectively homologous to those produced by more conventional routes, and that oral dosing or gavage of experimental animals elicited significant immune responses, in the absence of replication of the agent. However, the first demonstration that a plant-produced viral antigen was protective against disease came from an animal model system and the production of the foot-and-mouth disease virus (FMDV) structural protein VP1: mice that were parenterally immunized with a crude extract of transgenic *Arabidopsis thaliana* plants were protected against viral challenge. Another development along these lines in 1999 was the proof that injection of rabbits with rabbit hemorrhagic disease virus (RHDV) VP60, produced in transgenic plants, could protect them against virus challenge.

Another important development in the mid-1990s was the application of transient plant virus-based expression systems to the expression of vaccines. The first experimental vaccines produced were chimeras of tobacco mosaic virus (TMV) capsid protein fused to malarial peptides; by the late 1990s, a HIV-1 gp41 peptide fused to cowpea mosaic virus capsids had been tested in mice both parenterally and orally. However, it was again an animal model which provided the best proof of concept, and again with FMDV VP1: production of whole VP1 in plants infected with recombinant TMV and parenteral vaccination of mice with leaf extracts allowed successful protection against live challenge with FMDV. This work provided some of the first evidence that full-length foreign proteins could successfully be produced using a plant virus, in amounts that were sufficient to allow immunization using only crude extracts. However, the technology was limited in that the choice of appropriate vectors was limited, as was the choice of host plants, and vectors were often unstable and did not express large proteins very well.

By the late 1990s, it was apparent that many of the early fears voiced by vaccinologists and immunologists about oral delivery of vaccines 'tolerizing' the recipients to the vaccine – and thereby abolishing an immune response to it – were essentially groundless. The basis for a good response to an orally delivered vaccine was that it should be particulate, and that some form of adjuvant may be required. This was borne out by work with human papillomavirus (HPV) virus-like particles produced in a baculovirus expression system, where the number of particles necessary for a strong systemic humoral and mucosal response to orally introduced VLPs in mice could be reduced by an order of magnitude by co-administration with CpG DNA or mutant *Escherichia coli* LT-B (LT R192G). This work, together with reports on oral immunization with respiratory syncytial virus (RSV) F protein and HBsAg, showed that strong Th1-type responses could be elicited by such treatment. The state of the art in mucosal immunity at the time can be summarized as follows: (1) the mucosal immune system consisted of different compartments (such as gut, respiratory, genital mucosa) which intercommunicated, so that, for example, intranasal immunization gave significant immunity at the vaginal mucosa; (2) oral vaccination elicited antigen-specific mucosal secretory IgA as well as systemic immunity, and CD8+ T-cell responses could be elicited. While co-administration of adjuvants such as LT-B, CTB, CpG DNA, and saponins considerably improved mucosal responses and allowed lower dosages, fusion of nonpolymerizing polypeptides to self-assembling adjuvant entities such as LT-B and CTB also allowed significant improvements in immunogenicity.

By 2000, the first human trial results of Norwalk virus capsid protein made in potato had shown that the vaccine was both tolerated and immunogenic, with almost all volunteers developing a significant immune response. Shortly afterward, the Arntzen group showed that a plant material containing HBsAg given orally was an effective prime for a parenteral boost with HBsAg. In 2002, the first successful combination of DNA prime–plant-produced protein boost vaccination in mice with a measles envelope glycoprotein vaccine showed that plant-produced vaccines could be used in very effective combination with DNA vaccines. However, yields of virus vaccine proteins in transgenic plants are low, and oral immunization even with concentrated extracts is not very successful. In contrast, production of antibodies in plants has been more successful and many different kinds of immunoglobulins with various specificities have been produced.

Plant tissues used for production of viral vaccines include leaf and stem tissues of tobaccos of various varieties, *Arabidopsis*, alfalfa, spinach, and potatoes; of aquatic weeds such as *Lemna* spp. (duckweed); seeds of tobaccos, beans, and maize; fruits like tomatoes and strawberries; carrots; single-cell cultures of *Chlorella* and *Chlamydomonas*; suspension cell cultures of tobacco and other plants; hairy root cultures derived from various plants via *Agrobacterium rhizogenes*; and transformed chloroplasts of a variety of plant species. Experience gained from these various systems showed that there was no way of predicting whether or not a given DNA sequence would express protein at a reasonable level, or that the protein would be stable. Although chloroplast expression sometimes gave high yields, this system was not suitable for glycosylated proteins. Single-cell cultures offered few advantages over conventional fermentation/or cell culture techniques, and it seemed that accumulation of proteins in seeds was preferable to expression in green tissue, because of easier purification and higher accumulation levels.

A major setback to the development of plant-produced vaccines occurred in 2002 when soybean and maize harvests in two states in the USA were found to contain maize seeds from 'volunteer' plants engineered to express transmissible gastroenteritis virus (TGEV) capsid protein. The company involved – ProdiGene – had proprietary technology allowing high levels of accumulation of foreign proteins in especially maize seeds, and had potential oral vaccine products including a hepatitis B vaccine, an LT-B vaccine to treat *E. coli* infections in humans, as well as other products under development with large commercial partners. ProdiGene was fined and forced to clean up the seed by the US Department of Agriculture, which has since issued guidelines to prevent a recurrence. The fallout from these incidents has led to an effective moratorium on the development of 'pharmed' products in food crops including vaccine pharming. Another limitation became obvious in 2005 when the Mason–Arntzen group at Arizona State University released results of a small human trial of a potato-based HBsAg vaccine. In this trial, volunteers previously immunized parenterally with a conventional vaccine ate doses of potato containing ~850 µg of antigen and 63% of them developed increased serum anti-HBsAg titers. However, the immunogenicity was still disappointingly low, despite the ingestion of relatively large doses of antigen.

Growing public unease over the use of transgenic plants for biopharmaceutical production, with the attendant risks of 'contamination' of crop plant gene pools, has prompted the development of other technologies for producing such materials in plants. One such technology is the *Agrobacterium tumefaciens*-based transient expression system: this allows very high levels of expression without the uncertainties inherent in the regeneration and propagation of transgenic plants, and was being used very successfully to produce antibody molecules by 2004. Its main advantage is that the simultaneous expression of a large number of constructs can very rapidly be investigated. Indeed, the Mason group used transient 'agroexpression' very productively in 2004 to explore the conformation and chimeric fusion properties of HBsAg.

By 2005, another major advance in expression technology was announced, which combined the advantages of agroinfection and of viral vectors: this was the 'magnifection' technology of ICON Genetics, which was based on the introduction of recombinant DNA into plants by agroinfection and the subsequent amplification of gene expression by the recombinational reconstitution of a self-replicating TMV-based RNA genome.

Recently, serious reservations have been expressed by researchers and companies about the regulatory acceptance of plant-produced vaccines and other therapeutics, for human and even for animal use. The observed reluctance of large companies to buy into the technology, or to fund it, is probably due to the lack of appropriate regulatory structures and some conservatism in regulatory agencies concerning plant products. However, there is no reason to believe that existing frameworks and regulations could not be adapted to handle plant-produced vaccines: this was demonstrated in 2006 when Dow AgroSciences registered a plant-produced vaccine against Newcastle disease virus (NDV), which affects poultry. This was the first plant-based vaccine registered in the US, and the first to be licenced by the USDA's Animal and Plant Health Inspection Service. The HN envelope glycoprotein of NDV was produced in suspension-cultured tobacco cells, thus avoiding all the production and regulatory problems associated with whole plant production. In the same year, the Centre for Genetic Engineering and Biotechnology (CIGB) in Cuba registered a tobacco-produced MAb for use in purification of their commercially produced HBV vaccine: this development obviates the need for MAb production in ascitic fluid from mice, and considerably reduces the cost of manufacture of a vaccine which is distributed worldwide.

The introduction in mid-2006 of Merck's yeast-produced Gardasil anti-human papillomavirus (HPV) L1 major capsid protein VLP vaccine was hailed as a major development in cancer prevention, and the vaccine is predicted to be a 'blockbuster'. However, at a cost of US$ 360 per course of three doses, it will not make inroads into limiting cervical cancer incidence in developing countries. In 2003, three groups published simultaneous accounts of plant production of HPV L1 proteins. While the antigens were correctly folded and assembled, the yields were low and oral and parenteral immunogenicities were also low: however, these problems were circumvented for HPV-16 L1 in early 2007 by a combination of codon optimization and intracellular targeting, and excellent parenteral immunogenicity was achieved.

Obstacles to Vaccine Pharming

The following obstacles impede the development and potential application of plant-based vaccines, especially in developing countries:

1. increasingly stringent regulatory barriers in developed countries which deter commercial development of new vaccines generally, and vaccines depending on new technology in particular;
2. declining commercial interest in producing vaccines due to negligible return on vaccines needed in developing countries (people who do need vaccines cannot afford them and have to rely on governments and/or aid agencies for making them available);
3. earlier expectations from the technology have not been fulfilled;
4. many developing countries have little or no infrastructure for the approval of experimental drugs and vaccines, or for GM crops;
5. there is widespread and profound suspicion in developing countries of 'GM products', usually fueled by developed country activists;
6. the 'guinea pig' argument: why should developing country populations be used to test products not used in developed countries, where existing but expensive 'best practice' products are already available? and
7. production of vaccine antigens in plants may not significantly decrease the cost of the final vaccines, given that downstream processing and other costs remain the same.

Most of these obstacles could possibly be overcome: countries could pass relevant legislation; regulatory bodies could adopt suitable guidelines; publicity/and information campaigns could change public perceptions, both in the developed as well as in the developing world. The 'guinea pig' argument could be countered by testing products in developed countries at the same time as in developing countries.

However, a more serious obstacle is that conventional pharma industry is not engaging with plant production technology primarily because it is seen as technology in search of a product, when the technology already available is adequate to the task. Moreover, it is a serious misconception that using plants to produce vaccines would lower prices significantly, as cost of materials and processing is only a minor component of total retail vaccine cost, that is, about 24%. Thus, even a significantly lower cost of vaccine material would not change the cost of the product much, especially as processing costs would remain the same however cheaply one could produce raw vaccine ingredients.

The Way Forward

Vaccine pharming has moved on toward maturity in an encouraging way in recent years, as the dogma of 'cheap oral vaccines' has to some extent shifted to an acceptance that processed and possibly injectable products may be preferred to raw or crudely processed oral vaccines. Indeed, as new needle-free developments in vaccine delivery such as oral and nasal and transdermal delivery are becoming increasingly popular, plant-made vaccines may find a niche, especially in products that need less processing than those intended for parenteral delivery.

It is possible that practitioners will come to accept that success will not be achieved by trying to bring cheap vaccines and therapies for poor people, but by producing generics. For example, it may be a better idea to concentrate on HPV vaccines, which currently are very expensive and have only a few remaining years of patent protection, rather than on HBV vaccines, which are out of patent, are produced conventionally by several high-volume producers, and currently sell for low enough prices to be included in EPI packages.

Another important area to focus on is animal vaccines: these are necessarily very low cost for meat animals such as poultry, sheep, and cattle, and animal model systems have shown very satisfactorily that viral and other plant-produced vaccines could be very useful. The development times are shorter and regulatory hurdles are lower than for similar human vaccines, and edible or minimally processed oral vaccines are probably far more feasible for animals than humans. Major targets could be the already-tested FMDV vaccine for cattle, sheep, pigs, and even wild animals; Newcastle disease and other chicken and poultry viruses; H5N1 and other influenza viruses in domestic and wild fowl; and African horsesickness virus which is currently threatening to emerge into Europe.

It is possible that expression of viral antigens in, for example, maize seed could allow very cheap production of orally dosable vaccines which could be stored for years in a dried form for very little expense, to be pulled out when and if the need for emergency barrier immunization occurred.

It is important for the development of plant-made vaccines that several large companies are interested in the technology, and have done a lot of thorough investigation of its potential. It is especially significant that the first product – Dow's Newcastle disease vaccine – has run the regulatory course and was accepted in 2006.

What is probably needed for the successful future application of pharmed products, especially in developing countries, is a highly effective set of plant-derived products – vaccines and therapeutics – which have been approved by regulatory bodies such as the US Food and Drug Administration (FDA) and European Medicines Agency (EMEA). This will lead to public acceptance and a more welcoming stance by industry, both of which are essential if the technology is to survive.

See also: Plant Virus Vectors (Gene Expression Systems); Vector Transmission of Plant Viruses.

Further Reading

Daniell H (2006) Production of biopharmaceuticals and vaccines in plants via the chloroplast genome. *Biotechnology Journal* 1: 1071–1079.

Gerber S, Lane C, Brown DM, *et al.* (2001) Human papillomavirus virus-like particles are efficient oral immunogens when coadministered with *Escherichia coli* heat-labile enterotoxin mutant R192G or CpG DNA. *Journal of Virology* 75: 4752–4760.

Gleba Y, Klimyuk V, and Marillonnet S (2005) Magnifection – a new platform for expressing recombinant vaccines in plants. *Vaccine* 23: 2042–2048.

Kirk DD and Webb SR (2005) The next 15 years: Taking plant-made vaccines beyond proof of concept. *Immunology and Cell Biology* 83: 248–256.

Mason HS, Lam DM, and Arntzen CJ (1992) Expression of hepatitis B surface antigen in transgenic plants. *Proceedings of the National Academy of Sciences, USA* 89: 11745–11749.

Thanavala Y, Huang Z, and Mason HS (2006) Plant-derived vaccines: A look back at the highlights and a view to the challenges on the road ahead. *Expert Review of Vaccines* 5: 249–260.

Voinnet O, Rivas S, Mestre P, and Baulcombe D (2003) An enhanced transient expression system in plants based on suppression of gene silencing by the p19 protein of tomato bushy stunt virus. *Plant Journal* 33: 949–956.

Viroids

R Flores, Instituto de Biología Molecular y Celular de Plantas (UPV-CSIC), Valencia, Spain
R A Owens, Beltsville Agricultural Research Center, Beltsville, MD, USA

Glossary

Catalytic RNA RNA molecules that are able to catalyze, in a protein-free medium, specific reactions involving the formation or breakage of covalent bonds. In nature, these reactions are usually transesterifications (self-cleavage and ligation) affecting the catalytic RNA itself.

Hammerhead structure The conserved secondary/tertiary structure shared by the smallest class of natural ribozymes. Most have been found in one or both strands of certain viroid and viroid-like satellite RNAs where they mediate self-cleavage of multimeric intermediates arising from replication through a rolling-circle mechanism.

Ribozyme RNA motif responsible for the catalytic activity of certain RNA molecules. In nature, they are found embedded within catalytic RNAs.

Introduction

Viroids are the smallest known agents of infectious disease – small (246–401 nt), highly structured, circular, single-stranded RNAs that lack detectable messenger RNA activity. While viruses have been described as 'obligate parasites of the cell's translational system' and supply some or most of the genetic information required for their replication, viroids can be regarded as 'obligate parasites of the cell's transcriptional machinery'. Thus far, viroids are known to infect only plants.

The first viroid disease to be studied by plant pathologists was potato spindle tuber. In 1923, its infectious nature and ability to spread in the field led Schultz and Folsom to group potato spindle tuber disease with several other 'degeneration diseases' of potatoes. Nearly 50 years were to elapse before Diener's demonstration in 1971 that the molecular properties of its causal agent, potato spindle tuber viroid (PSTVd), were fundamentally different than those of conventional plant viruses.

Genome Structure

Efforts to understand how viroids replicate and cause disease without the assistance of any viroid-encoded polypeptides have prompted detailed analysis of their structure. Viroids possess rather unusual properties for single-stranded RNAs (e.g., a pronounced resistance to digestion by ribonuclease and a highly cooperative thermal denaturation profile), leading to an early realization that they might have an unusual higher-order structure.

To date, the complete sequences of 29 distinct viroid species plus a large number of sequence variants have been determined (**Table 1**). All are single-stranded circular RNAs containing 246–401 unmodified nucleotides. Theoretical calculations and physicochemical studies indicate that PSTVd and related viroids assume a highly base-paired, rod-like conformation *in vitro* (**Figure 1**). Pairwise sequence comparisons suggest that the series of short double helices and small internal loops that comprise this so-called 'native' structure are organized into five domains whose boundaries are defined by sharp differences in sequence similarity.

The 'central domain' is the most highly conserved viroid domain and contains the site where multimeric PSTVd RNAs are cleaved and ligated to form circular progeny. The 'pathogenicity domain' contains one or more structural elements which modulate symptom expression, and the relatively small 'variable domain' exhibits the greatest sequence variability between otherwise closely related viroids. The two 'terminal domains' appear to play an important role in viroid replication and evolution. Although these five domains were first identified in PSTVd, apple scar skin viroid (ASSVd) and related viroids also contain a similar domain arrangement. Certain viroids such as *Columnea* latent viroid (CLVd), Australian grapevine viroid (AGVd), and *Coleus blumei* viroid 2 (CbVd 2) appear to be 'mosaic molecules' formed by exchange of domains between two or more viroids infecting the same cell. RNA rearrangement/recombination can also occur within individual domains, leading, in coconut cadang-cadang (CCCVd) and citrus exocortis (CEVd) viroids, to duplications of the right terminal domain plus part of the variable domain. This domain model is not shared by avocado sunblotch (ASBVd) and related viroids.

Much less is known about viroid tertiary structure, especially *in vivo* where these molecules almost certainly accumulate as ribonucleoprotein particles. UV-induced cross-linking of two nucleotides within a loop E motif in the central domain of PSTVd provided the first definitive evidence for such tertiary interactions. Similar UV-sensitive structural elements have also been discovered in a number of other RNAs including 5S eukaryotic rRNA, adenovirus VAI RNA, and the viroid-like domain of the hepatitis delta virus genome. Loop E forms during the conversion of multimeric PSTVd RNAs into monomers. The ability of ASBVd-related RNAs to undergo spontaneous self-cleavage mediated by hammerhead ribozymes as well as the presence of pseudoknots critical for infectivity in some other members of the *Avsunviroidae* (**Figure 1**) provide additional evidence for the functional importance of viroid tertiary structure.

Classification

Based upon differences in the structural and functional properties of their genomes, viroids species are assigned to one of two taxonomic families (see **Table 1**). Members of the family *Pospiviroidae* (type member PSTVd) have a rod-like secondary structure that contains five structural–functional domains and several conserved motifs. Most members of the family *Avsunviroidae* (type member ASBVd), in contrast, appear to adopt a branched conformation, and multimeric RNAs of all family members behave as catalytic RNAs and undergo spontaneous self-cleavage (**Figure 1**). Differences in their sites of replication also support this classification scheme; that is, PSTVd and ASBVd replicate in the nucleus and the

Table 1 Classification of viroids of known nucleotide sequence

Family[a]	Genus[a]	Name	Abbreviation	Nucleotides[b]
Pospiviroidae	Pospiviroid	Chrysanthemum stunt	CSVd	354–356
		Citrus exocortis	CEVd	368–375 (463–467)
		Columnea latent	CLVd	370–373
		Iresine	IrVd	370
		Mexican papita	MPVd	359–360
		Potato spindle tuber	PSTVd	356–361 (341)
		Tomato apical stunt	TASVd	360–363
		Tomato chlorotic dwarf	TCDVd	360
		Tomato planta macho	TPMVd	359–360
	Cocadviroid	Citrus viroid IV	CVd-IV	284
		Coconut cadang-cadang	CCCVd	246–247 (287–301)
		Coconut tinangaja	CTiVd	254
		Hop latent	HLVd	256
	Hostuviroid	Hop stunt[c]	HSVd	294–303
	Apscaviroid	Apple dimple fruit	ADFVd	306,307
		Apple scar skin[d]	ASSVd	329–334
		Australian grapevine	AGVd	369
		Citrus bent leaf	CBLVd	315,318
		Citrus dwarfing	CDVd	294,297
		Grapevine yellow speckle 1	GVYSVd 1	366–368
		Grapevine yellow speckle 2[e]	GYSVd 2	363
		Pear blister canker	PBCVd	315,316
	Coleviroid	Coleus blumei 1	CbVd 1	248–251
		Coleus blumei 2	CbVd 2	301,302
		Coleus blumei 3	CbVd 3	361–364
Avsunviroidae	Avsunviroid	Avocado sunblotch	ASBVd	246–251
	Pelamoviroid	Chrysanthemum chlorotic mottle	CChMVd	398–401
		Peach latent mosaic	PLMVd	335–351
	Elaviroid	Eggplant latent	ELVd	332–335

[a]Classification follows scheme proposed by Flores *et al.* (see VIII Report of the International Committee on Taxonomy of Viruses) with minor modifications. The nucleotide sequences of blueberry mosaic, burdock stunt, *Nicotiana glutinosa* stunt, pigeon pea mosaic mottle, and tomato bunchy top viroids are currently unknown; consequently, these viroids have not been assigned to specific genera. Whether apple fruit crinkle and citrus viroid original source should be considered variants or new viroid species of genus *Apscaviroid* is pending.
[b]Sizes of variants containing insertions or deletions arising *in vivo* are shown in parentheses.
[c]Includes cucumber pale fruit, citrus cachexia, peach dapple, and plum dapple viroids.
[d]Includes pear rusty skin and dapple apple viroids.
[e]Formerly termed grapevine viroid 1B.

chloroplast, respectively, and the same appears to occur for other members of each family. Each family is subdivided into genera according to certain demarcating criteria. Groups of sequence variants that show >90% sequence identity in pairwise comparisons and share some common biological property are arbitrarily defined as viroid species. *In vivo*, each viroid species is actually a 'quasispecies', that is, a collection of closely related sequences subject to a continuous process of variation, competition, and selection. There is phylogenetic evidence for an evolutionary link between viroids and other viroid-like subviral RNAs (**Figure 2**).

Host Range and Transmission

All viroids are mechanically transmissible, and most are naturally transmitted from plant to plant by man and his tools. Individual viroids vary greatly in their ability to infect different plant species. PSTVd can replicate in about 160 primarily solanaceous hosts, while only two members of the family Lauraceae are known to support ASBVd replication. HSVd has a particularly wide host range that includes herbaceous species as well as woody perennials. Many natural hosts are either vegetatively propagated or crops that are subjected to repeated grafting or pruning operations. PSTVd, ASBVd, and CbVd1 are vertically transmitted through pollen and/or true seed, but the significance of this mode of transmission in the natural spread of disease is unclear. PSTVd can be encapsidated by the coat protein of potato leafroll virus (PLRV, a polerovirus) as well as velvet tobacco mottle virus (VTMoV, a sobemovirus), and epidemiological surveys suggest that PLRV facilitates viroid spread under field conditions.

Commonly used techniques for the experimental transmission of viroids include the standard leaf abrasion

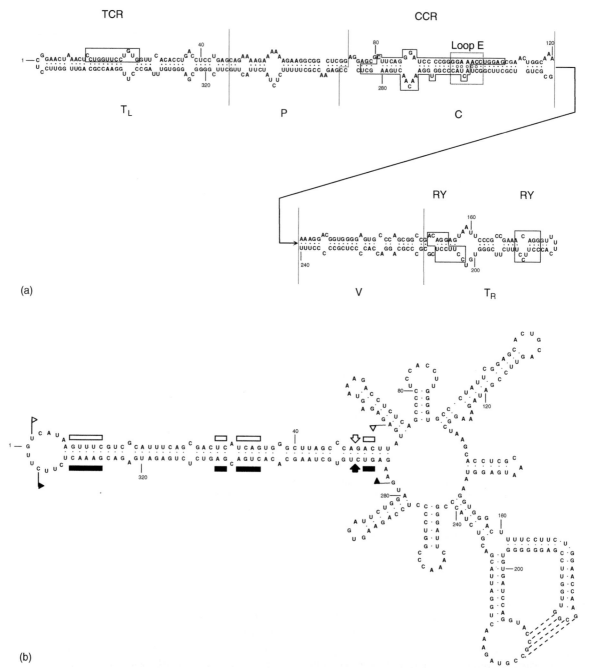

Figure 1 (a) The rod-like secondary structure of PSTVd (intermediate strain) showing the five domains characteristic of members of the family *Pospiviroidae:* the terminal left (T$_L$), pathogenicity (P), central (C), variable (V), and terminal right (T$_R$). The central conserved region (CCR) is located within the C domain and contains a UV-sensitive loop E motif with noncanonical base pairs (denoted by circles). The T$_L$ domains of genera *Pospiviroid* and *Apscaviroid* contain a terminal conserved region (TCR), while those of genera *Hostuviroid* and *Cocadviroid* contain a terminal conserved hairpin (not shown). The T$_R$ may also contain 1–2 copies of a protein-binding RY motif. (b) The branched secondary structure of PLMVd (reference variant). Plus and minus self-cleavage domains are indicated by flags, nucleotides conserved in most natural hammerhead structures by bars, and the self-cleavage sites by arrows. Black and white symbols refer to plus and minus polarities, respectively. Nucleotides involved in a pseudoknot are indicated by broken lines. Redrawn with modifications from Gross HJ, Domdey H, Lossow C, *et al.* (1978) Nucleotide sequence and secondary structure of potato spindle tuber viroid. *Nature* 273: 203–208; Hernández C and Flores R (1992) Plus and minus RNAs of peach latent mosaic viroid self-cleave *in vitro* through hammerhead structures. *Proceedings of the National Academy of Sciences, USA* 89: 3711–3715; Bussière F, Ouellet J, Côté F, Lévesque D, and Perreault JP (2000) Mapping in solution shows the peach latent mosaic viroid to possess a new pseudoknot in a complex, branched secondary structure. *Journal of Virology* 74: 2647–2654.

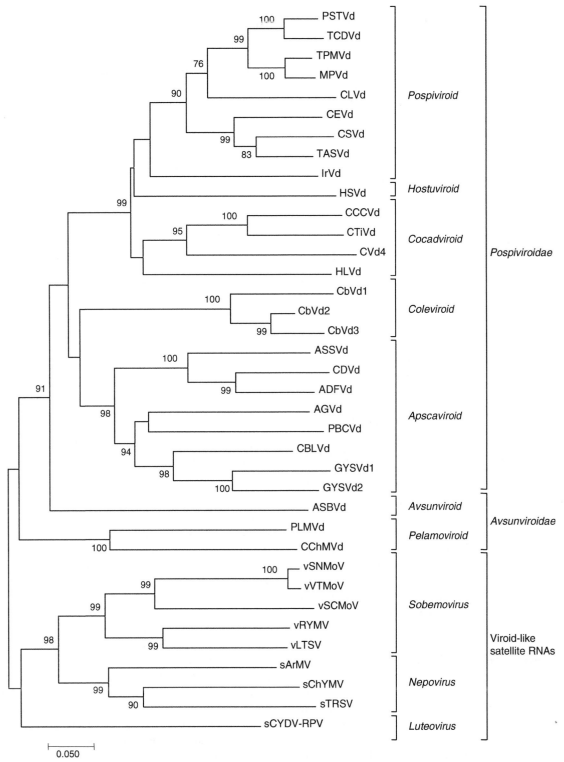

Figure 2 Neighbor-joining phylogenetic tree obtained from an alignment manually adjusted to take into account local similarities, insertions/deletions, and duplications/rearrangements described in the literature for viroid and viroid-like satellite RNAs. Bootstrap values were based on 1000 random replicates (only values >70% are shown). Viroid abbreviations are those used in **Table 1**. Viroid-like satellite RNAs: lucerne transient streak virus (sLTSV); rice yellow mottle virus (sRYMV); subterranean clover mottle virus (sSCMoV); *Solanum nodiflorum* mottle virus (sSNMoV)); velvet tobacco mottle virus (sVTMoV); tobacco ringspot virus (sTRSV); *Arabis* mosaic virus (sArMV); chicory yellow mottle virus (sChYMV); cereal yellow dwarf virus-RPV (sCYDV-RPV). Adapted from Elena SF, Dopazo J, de la Peña M, Flores R, Diener TO, and Moya A (2001) Phylogenetic analysis of viroid and viroid-like satellite RNAs from plants: A reassessment. *Journal of Molecular Evolution* 53: 155–159.

methods developed for conventional viruses, 'razor slashing' methods in which phloem tissue in the stem or petiole is inoculated via cuts made with a razor blade previously dipped into the inoculum, and, in the case of CCCVd, high-pressure injection into folded apical leaves. Viroids can also be transmitted by either plant transformation or 'agroinoculation' during which a modified *Agrobacterium tumefaciens* Ti plasmid is used to introduce full-length viroid-complementary DNA into the potential host cell. Either technique can overcome the marked resistance of some hosts to mechanical inoculation. Identification of the molecular mechanism(s) that determine viroid host range remains an important research goal.

Symptomatology

Viroids and conventional plant viruses induce a very similar range of macroscopic symptoms. Symptom expression is usually optimal at the same relatively high temperatures (30–33 °C) that promote viroid replication. Stunting and leaf epinasty (a downward curling of the leaf lamina resulting from unbalanced growth within the various cell layers) are considered the classic symptoms of viroid infection. Other commonly observed symptoms include vein clearing, veinal discoloration or necrosis, and the appearance of localized chlorotic/necrotic spots or mottling in the foliage. Symptoms may also be expressed in flowers and bark, and fruits or tubers from viroid-infected plants may be abnormally shaped or discolored. Viroid infection of certain citrus rootstock/scion combinations may result in tree dwarfing (**Figure 3**). Viroid infections are often latent and rarely kill the host.

Viroid infections are also accompanied by a number of cytopathic effects – chloroplast and cell wall abnormalities, the formation of membranous structures in the cytoplasm, and the accumulation of electron-dense deposits in both chloroplasts and cytoplasm. Metabolic changes include dramatic alterations in growth regulator levels.

Geographic Distribution

Although PSTVd, HSVd, CEVd, and ASBVd are widely distributed throughout the world, other viroids have never been detected outside the areas where they were first reported. Several factors may contribute to this variation in distribution pattern. Among the crops most affected by viroid diseases are a number of valuable woody perennials such as grapes, citrus, various pome and stone fruits, and hops. Propagation and distribution of improved cultivars is highly commercialized, with the result that many cultivars are now grown worldwide. The international exchange of plant germplasm also continues to increase at a rapid rate. In both instances, the

large number of latent (asymptomatic) hosts facilitates viroid spread.

Epidemiology and Control

Viroid diseases pose a potential threat to agriculture, and several are of considerable economic importance. Ready transmission of PSTVd by vegetative propagation, foliar contact, and true seed or pollen continues to pose a serious threat to potato germplasm collections and breeding programs. Coconut cadang-cadang has killed over 30 million palms in the Philippines since it was first recognized in the early 1930s. While many viroids were first detected in ornamental or crop plants, most viroid diseases are thought to result from chance transfer from endemically infected wild species to susceptible cultivars. Several lines of circumstantial evidence are consistent with this hypothesis:

1. The experimental host ranges of several viroids include many wild species, and these wild species often tolerate viroid replication without the appearance of recognizable disease symptoms.
2. Although co-evolution of host and pathogen is often accompanied by appearance of gene-for-gene vertical resistance, no useful sources of resistance to PSTVd infection have been identified in the cultivated potato.
3. Viroids and/or viroid-related RNAs closely related to TPMVd and CCCVd have been detected in weeds and other wild vegetation growing near fields containing viroid-infected plants.

Growers and plant pathologists are unlikely to have simply overlooked diseases with symptoms as severe as those of chrysanthemum stunt or cucumber pale fruit, two diseases first reported after World War II. Large-scale monoculture of genetically identical crops and the commercial propagation/distribution of many cultivars are two comparatively modern developments which would facilitate the development of serious disease problems following the chance transfer of viroids from wild hosts to cultivated plants. Viroid diseases may also arise by transfer between cultivated crop species. For example, pears provide a latent reservoir for ASSVd; likewise, while HSVd infections of grapes are often symptomless, this viroid causes severe disease in hops. In both instances, the two crops are often grown in close proximity.

Because no useful sources of natural resistance to viroid disease are known, diagnostic tests continue to play a key role in efforts to control viroid diseases. Since viroids lack a protein capsid, the antibody-based techniques used to detect many plant viruses are not applicable. Tests based upon their unique molecular properties have largely supplanted biological assays for

Figure 3 (a) Symptoms of PSTVd infection in Rutgers tomato approximately 4 weeks after inoculation of cotyledons with PSTVd strains causing mild, intermediate, and severe symptoms. (b) Symptoms of ASBVd infection in avocado fruits and leaves. (c) Viroid-induced dwarfing of citrus growing on susceptible rootstocks: All trees in the block were graft-inoculated with CDVd shortly after transfer to the field; only one tree (right foreground) escaped infection. Note the difference in height.

viroid detection. Problems with viroid bioassays include the length of time required for completion (weeks to years) and difficulties in detecting mild or latent strains. Several rapid (1–2 day) protocols involving polyacryl-amide gel electrophoresis (PAGE) under denaturing conditions take advantage of the circular nature of viroids. Using these protocols, nanogram amounts of viroid can be unambiguously detected without the use of radioactive isotopes. In recent years, diagnostic procedures based upon nucleic acid hybridization or the polymerase chain reaction (PCR) are being widely used. The simplest methods involve the hybridization of a nonradioactively labeled viroid-complementary DNA or RNA probe to viroid samples that have been bound to a solid support followed by colorimetric or chemiluminescent detection of the resulting DNA–RNA or RNA–RNA hybrids. Such conventional 'dot blot' assays can detect picogram amounts of viroids using clarified plant sap or tissue prints rather than purified nucleic acid as the viroid source. PCR-based protocols are finding increasing acceptance in those cases where either this level of sensitivity is inadequate or a number of closely related viroids are present in the same sample.

Molecular Biology

Although devoid of messenger RNA activity, viroids replicate autonomously and cause disease in a wide variety of plants. Much has been learned about the molecular biology of viroids and viroid–host interaction over the past 25 years, but the precise nature of the molecular signals involved remains elusive. A series of questions first posed by Diener summarizes the many gaps in our current understanding of the biological properties of these unusual molecules:

1. What molecular signals do viroids possess (and cellular RNAs evidently lack) that induce certain DNA-dependent RNA polymerases to accept them as templates for the synthesis of complementary RNA molecules?
2. What are the molecular mechanisms responsible for viroid replication? Are these mechanisms operative in uninfected cells? If so, what are their functions?
3. How do viroids induce disease? In the absence of viroid-specified proteins, disease must arise from direct interaction(s) of viroids (or viroid-derived RNA molecules) with host-cell constituents. Infections by PSTVd and ASBVd induce RNA silencing (see below).
4. What determines viroid host range? Are viroids restricted to higher plants, or do they have counterparts in animals?
5. How did viroids originate?

Replication

A variety of multimeric plus- and minus-strand RNAs have been detected by nucleic acid hybridization in viroid-infected tissues. Based on their analysis, viroid replication has been proposed to proceed via a 'rolling circle' mechanism that involves reiterative transcription of the incoming plus circular RNA to produce a minus-strand RNA template. ASBVd and related viroids utilize a symmetric replication cycle in which the multi-meric minus strand is cleaved to unit-length molecules and circularized before serving as template for the synthesis of multimeric plus strands. PSTVd and related viroids utilize an asymmetric cycle in which the multimeric minus strand is directly transcribed into multimeric plus strands. In both cases, the multimeric plus strands are cleaved to unit-length molecules and circularized.

A diversity of host-encoded enzymes have been implicated in viroid replication. Low concentrations of α-amanitin specifically inhibit the synthesis of both PSTVd plus and minus strands in nuclei isolated from infected tomato, strongly suggesting the involvement of DNA-dependent RNA polymerase II, transcribing an RNA template, in the replication of PSTVd and related viroids. In nuclear extracts, transcription of the PSTV plus strand by RNA polymerase II starts in the left terminal loop; furthermore, incubation of active replication complexes containing CEVd with a monoclonal antibody directed against the carboxy-terminal domain of RNA polymerase II results in the immunoprecipitation of both CEVd plus- and minus-strand RNAs. Mature PSTVd plus strands accumulate in the nucleolus and the nucleoplasm, while *in situ* hybridization indicates that minus-strand RNAs are confined to the nucleoplasm. The identity of the polymerase(s) responsible for replication of members of the family *Avsunviroidae* in the chloroplast is less certain. ASBVd synthesis is resistant to tagetitoxin, strongly indicating the involvement of a nuclear-encoded chloroplastic RNA polymerase. Initiation sites for both ASBVd plus- and minus-strand synthesis have been mapped to the AU-rich terminal loops of their respective native structures.

In vitro evidence indicates that specific cleavage of multimeric PSTVd plus-strand RNAs requires (1) rearrangement of the conserved central region to form a branched structure containing a GNRA tetraloop and (2) the action of one or more host-encoded nucleases. Other less-efficient processing sites can also be used *in vivo*. Plus- and minus-strand RNAs of ASBVd and related viroids, in contrast, undergo spontaneous self-cleavage through hammerhead ribozymes to form linear monomers (**Figure 4**). Addition of certain chloroplast proteins acting as RNA chaperones facilitates this hammerhead ribozyme-mediated self-cleavage reaction. The final step in viroid replication is the ligation of linear monomers to form mature circular progeny. Plant cells are known to contain RNA ligase activities which can act upon the 5′ hydroxyl and 2′,3′ cyclic phosphate termini formed by either cleavage pathway.

Movement

Upon entering a potential host cell, viroids must move to either the nucleus (*Pospiviroidae*) or chloroplast (*Avsunviroidae*) before beginning replication. Available data

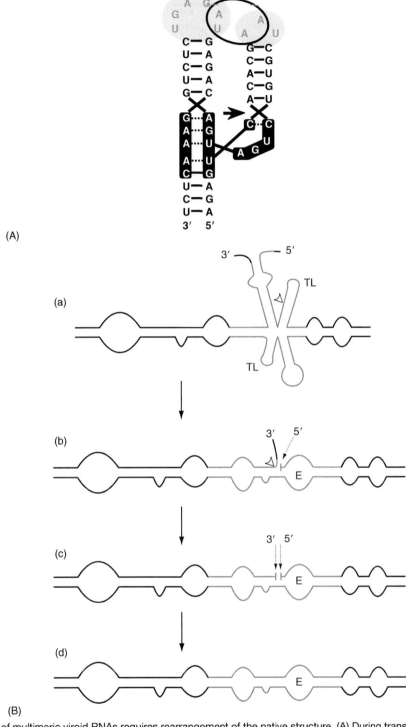

(A)

(B)

Figure 4 Cleavage of multimeric viroid RNAs requires rearrangement of the native structure. (A) During transcription, the strands of both polarities of members of the family *Avsunviroidae* can fold into hammerhead structures (here illustrated for the hammerhead of the PLMVd plus RNA) and self-cleave accordingly. Nucleotides conserved in most natural hammerhead structures are on a black background, and the self-cleavage site is denoted by an arrow. A circle delimits the presumed tertiary interaction between terminal loops enhancing the catalytic activity. Watson-Crick and noncanonical base pairs are represented by continuous and discontinuous lines, respectively. After self-cleavage, the RNA adopts a new conformation favoring ligation (or self-ligation). (B) Processing of a longer-than-unit-length plus PSTVd RNA transcript in a potato nuclear extract. The central conserved region of the substrate for the first cleavage reaction (a) contains a tetraloop (denoted TL). After dissociation of the 5′ segment from the cleavage site, the new 5′ end refolds and is stabilized by formation of a UV-sensitive loop E (b), while the 3′ end partially base-pairs with the lower strand. Single-stranded nucleotides at the 3′ end are then cleaved between positions 95 and 96 (c), and ligation of the 5′ and 3′ termini (d) results in formation of mature circular progeny. From Baumstark T, Schröder ARW, and Riesner D (1997) Viroid processing: Switch from cleavage to ligation is driven by a change from a tetraloop to a loop E conformation. *EMBO Journal* 16: 599–610.

suggest that PSTVd enters the nucleus as a ribonucleo-protein complex formed by the interaction of cellular proteins with specific viroid sequence or structural motifs. VirP1, a bromodomain-containing protein isolated from tomato, has a nuclear localization signal and binds to the terminal right domain of PSTVd. Proteins such as TFIIIA and ribosomal protein L5 that bind to the loop E motif may also be involved in viroid transport into the nucleus. How ASBVd or other members of the family *Avsunviroidae* enter and exit the chloroplast is currently unknown.

To establish a systemic infection, viroids leave the initially infected cell – moving first from cell to cell and then long distances through the host vasculature. Upon injection into symplasmically isolated guard cell in a mature tomato leaf, fluorescently labeled PSTVd RNA does not move. Injection into interconnected mesophyll cells, in contrast, is followed by rapid cell-to-cell movement through the plasmodesmata. Long-distance movement of viroids, like that of nearly all plant viruses, occurs in the phloem where it follows the typical source-to-sink pattern of photoassimilate transport. Viroid movement in the phloem almost certainly requires formation of a ribonu-cleoprotein complex, possibly involving a dimeric lectin known as phloem protein 2 (PP2), the most abundant protein in phloem exudate. Movement of PSTVd in the phloem appears to be sustained by replication in supporting cells and is tightly regulated by developmental and cellular factors. For example, *in situ* hybridization reveals the presence of PSTVd in vascular tissues underlying the shoot apical meristem of infected tomato, but entry into the shoot apical meristem itself appears to be blocked. Another important control point for PSTVd trafficking is the bundle sheath–mesophyll boundary in the leaf. By disrupting normal pattern of viroid movement, it may be possible to create a plant that is resistant/immune to viroid infection.

Pathogenicity

Sequence comparisons of naturally occurring PSTVd and CEVd variants as well as infectivity studies with chimeric viroids, constructed by exchanging the pathogenicity domains of mild and severe strains of CEVd, have clearly shown that the pathogenicity domain in the family *Pospiviroidae* contains important determinants of symptom expression. Symptom expression is also affected by the rate of viroid replication, and sequence changes in the variable domain have been shown to regulate progeny titers in infected plants. Studies with TASVd revealed the presence of a third pathogenicity determinant in the left terminal loop. Also, a single U/A change position 257 in the central domain of PSTVd results in the appearance of severe stunting and a 'flat top' phenotype. In the family *Avsunviroidae*, determinants of pathogenicity have been mapped to either a tetraloop capping a hairpin stem in

chrysanthemum chlorotic mottle viroid (CChMVd) or an insertion that folds into a hairpin also capped by a tetra-loop in peach latent mosaic viroid (PLMVd).

The ability of novel viroid chimeras to replicate and move normally from cell to cell implies certain basic similarities between their structures *in vitro* and *in vivo* but provides no information about the nature of the molecular interactions responsible for symptom development. Until recently, it was widely assumed that the mature viroid RNA was the direct pathogenic effector. Just like viruses, however, viroid replication is also accompanied by the production of a variety of small (21–26 nt) RNA molecules. The role of these small interfering RNAs (siRNAs) in viroid pathogenicity is not yet clear, but the inverse relationship between accumulation levels of the mature viroid RNAs and the corresponding siRNAs for members of the family *Avsunviroidae* suggest that the latter may regulate the titer of the former. Also, recovery of tomato plants from the symptoms of severe PSTVd infections is preceded by the accumulation of PSTVd-specific siRNA.

Viroid infections are accompanied by quantitative changes in a variety of host-encoded proteins. Certain of these are 'pathogenesis-related' proteins whose synthesis or activation is part of a general host reaction to biotic or abiotic stress, but others appear to be more specific. In tobacco, PSTVd infection results in the preferential phosphorylation of a host-encoded 68 kDa protein that is immunologically related to an interferon-inducible, dsRNA-dependent mammalian protein kinase of similar size. The human kinase is differentially activated by PSTVd strains of varying pathogenicity *in vitro*, while infection of tomato by intermediate or severe strains of PSTVd induces the synthesis of PKV, a dual-specificity, serine/threonine protein kinase. Broad changes in host gene expression following PSTVd infection have been detected by complementary DNA macroarray analysis.

Host Range

Possibly as a result of its involvement in the cleavage/ligation of progeny RNA, nucleotides in the central domain of PSTVd and related viroids appear to play an important role in determining host range. For example, a single nucleotide substitution in the loop E motif results in a dramatic increase in the rate of PSTVd replication in tobacco. The biological properties of CLVd also suggest that this domain contains one or more host-range determinants. CLVd appears to be a natural mosaic of sequences present in other viroids; phylogenetic analysis (see **Figure 2**) suggests that it can be considered to be a PSTVd-related viroid whose conserved central domain has been replaced by that of HSVd. Like HSVd (but not PSTVd or related viroids), CLVd can replicate and cause disease in cucumber.

Origin and Evolution

Much of the early speculation about viroid origin involved their possible origin as 'escaped introns' (i.e., descent from normal host RNAs). More recently, however, viroids have been proposed to represent 'living fossils' of a pre-cellular RNA world that assumed an intracellular mode of existence sometime after the evolution of cellular organisms. The presence of ribozymes in members of the *Avsunviroidae* strongly supports this view.

The inherent stability of viroids and viroid-like satellite RNAs (structurally similar to viroids but functionally dependent on helper viruses) which arises from their small size and circularity would have enhanced the probability of their survival in primitive, error-prone RNA self-replicating systems and assured their complete replication without the need for initiation or termination signals. Most viroids (but not satellite RNAs or random sequences of the same base composition) also display structural periodicities with repeat units of 12, 60, or 80 nt. The high error rate of prebiotic replication systems may have favored the evolution of polyploid genomes, and the mechanism of viroid replication (i.e., rolling-circle transcription of a circular template) provides an effective means of genome duplication.

Viroids and viroid-like satellite RNAs all possess efficient mechanisms for the precise cleavage of their oligomeric replication intermediates to form monomeric progeny. PSTVd and related viroids appear to require proteinaceous host factor(s) for cleavage, but others (members of the family *Avsunviroidae* and viroid-like satellite RNAs) contain ribozymes far smaller and simpler than those derived from introns. Thus, ASBVd and the other self-cleaving viroids may represent an evolutionary link between viroids and viroid-like satellite RNAs. No viroid is known to code for protein, a fact that is consistent with the possibility that viroids are phylogenetically older than introns.

Phylogenetic evidence for an evolutionary link between viroids and other viroid-like subviral RNAs has been presented by Elena *et al.* (see **Figure 2**). Among several subviral RNAs possibly related to viroids is carnation small viroid-like RNA, a 275 nt circular molecule with self-cleaving hammerhead structures in both its plus and minus strands that has a DNA counterpart. This novel retroviroid-like element shares certain features with both viroids and a small RNA transcript from newt.

Further Reading

Baumstark T, Schröder ARW, and Riesner D (1997) Viroid processing: Switch from cleavage to ligation is driven by a change from a tetraloop to a loop E conformation. *EMBO Journal* 16: 599–610.

Bussière F, Ouellet J, Côté F, Lévesque D, and Perreault JP (2000) Mapping in solution shows the peach latent mosaic viroid to possess a new pseudoknot in a complex, branched secondary structure. *Journal of Virology* 74: 2647–2654.

Diener TO (1979) *Viroids and Viroid Diseases.* New York: Wiley-Interscience.

Ding B, Itaya A, and Zhong X (2005) Viroid trafficking: A small RNA makes a big move. *Current Opinion in Plant Biology* 8: 606–612.

Elena SF, Dopazo J, de la Peña M, Flores R, Diener TO, and Moya A (2001) Phylogenetic analysis of viroid and viroid-like satellite RNAs from plants: A reassessment. *Journal of Molecular Evolution* 53: 155–159.

Flores R, Hernandez C, Martinez de Alba AE, Daros JA, and DiSerio F (2005) Viroids and viroid–host interactions. *Annual Review of Phytopathology* 43: 117–139.

Flores R, Randles JW, Owens RA, Bar-Joseph M, and Diener TO (2005) Viroids. In: Fauquet CM, Mayo MA, Maniloff J, Desselberger U, and Ball AL (eds.) *Virus Taxonomy: Eighth Report of the International Committee on Taxonomy of Viruses*, pp. 1145–1159. San Diego, CA: Elsevier Academic Press.

Gross HJ, Domdey H, Lossow C, *et al.* (1978) Nucleotide sequence and secondary structure of potato spindle tuber viroid. *Nature* 273: 203–208.

Hadidi A, Flores R, Randles JW, and Semancik JS (eds.) (2003) *Viroids.* Collingwood: CSIRO Publishing.

Hernández C and Flores R (1992) Plus and minus RNAs of peach latent mosaic viroid self-cleave *in vitro* through hammerhead structures. *Proceedings of the National Academy of Sciences, USA* 89: 3711–3715.

Hull R (2002) *Matthews' Plant Virology,* 4th edn. New York: Academic Press.

VIRUSES AND VIRUS GENERA

Alfalfa Mosaic Virus

J F Bol, Leiden University, Leiden, The Netherlands

Glossary

Agroinfiltration Infiltration of plant leaves with a suspension of Agrobacterium tumefaciens for transient expression of genes from a T-DNA vector.

T-DNA vector Plasmid with the T-DNA (transferred DNA) sequence of the Ti plasmid of Agrobacterium tumefaciens, which is transferred to plant cells where it can be transiently expressed or become integrated in the plant genome.

History

Alfalfa mosaic virus (AMV) was identified in 1931 by Weimer as the causal agent of an economically important disease in alfalfa. Purification of AMV around 1960 showed that virus preparations contained bacilliform particles of different length. Fractionation of these particles by sucrose-gradient centrifugation revealed that the four major components each contained a specific type of RNA, termed RNAs 1, 2, 3, and 4. Initially, an analysis of the biological activity of the viral nucleoproteins and RNAs resulted in a puzzle. A mixture of the three largest viral particles was fully infectious, but a mixture of RNAs 1, 2, and 3, purified from these particles, was not. At the RNA level, RNA 4 was required in the inoculum to initiate infection. In 1971 it became clear that the AMV genome consisted of RNAs 1, 2, and 3, and that a mixture of these genomic RNAs became infectious only after addition of coat protein (CP) or its subgenomic messenger, RNA 4. High-affinity binding sites for CP were identified in the AMV RNAs in 1972 and could be localized to the 3' termini of the RNAs in 1978. From 1975 onwards it became clear that, similar to AMV, ilarviruses required CP to initiate infection. Moreover, CPs of AMV and ilarviruses could bind to the 3' termini of each other's RNAs, and could be freely exchanged in the initiation of infection. Currently, AMV is the type species of the genus *Alfamovirus* and is classified together with the genus *Ilarvirus* in the family *Bromoviridae*.

Taxonomy and Classification

In addition to the genera *Alfamovirus* and *Ilarvirus*, the family *Bromoviridae* contains the genera *Bromovirus, Cucumovirus,* and *Oleavirus*. The tripartite RNA genomes of bromo- and cucumoviruses are infectious as such and do not require CP to initiate infection. Although a mixture of the genomic RNAs of AMV and ilarviruses has a low intrinsic infectivity, initiation of infection is stimulated approximately 1000-fold by addition of CP or RNA 4. This phenomenon has been termed 'genome activation'. CP of bromo- and cucumoviruses is unable to initiate infection by AMV or ilarvirus genomic RNAs. Oleaviruses have been studied in less detail but are believed to resemble bromo- and cucumoviruses in their replication strategy. The 3' termini of the RNAs of bromo- and cucumoviruses contain a tRNA-like structure (TLS) that can be charged with tyrosine, whereas the RNAs of AMV and ilarviruses cannot be charged with an amino acid. Many isolates of AMV are known which are closely related by serology and nucleotide sequence similarity. The complete nucleotide sequence of AMV isolate 425 (Leiden [L] and Madison [M] isolates) has been determined, and partial sequences of the Strasbourg (S) and yellow spot mosaic virus (Y) isolates have been published. The available data indicate that sequence similarity between isolates is over 95%.

Host Range and Economic Significance

AMV occurs worldwide. Strains of this virus have been found in natural infections of about 150 plant species representing 22 families. The experimental and natural host ranges include over 600 species in 70 families. Recently, 68 *Arabidopsis thaliana* ecotypes were analyzed for their susceptibility to AMV infection. Thirty-nine ecotypes supported both local and systemic infection, 26 ecotypes supported only local infection, and three ecotypes could not be infected. Although AMV infects mostly herbaceous plants, several woody hosts are included in its natural host range. AMV is an economically important pathogen in alfalfa and sweet clover and may affect pepper, pea, tobacco, tomato, soybean, and celery. Worldwide, AMV causes calico mosaic in potato but its economic importance in this crop is limited.

Particle Structure and Composition

Figure 1 shows an electron micrograph of AMV. The four major classes of particles in AMV preparations are called bottom component (B), middle component (M), top component *b* (Tb), and top component *a* (Ta). B, M, and Tb are bacilliform and contain the genomic RNAs 1, 2, and 3, respectively. Ta contains two molecules of the subgenomic

RNA 4 and can be subdivided into bacilliform Ta-b and spheroidal Ta-t particles. The bacilliform particles are all 19 nm wide and have lengths of 56 nm (B), 43 nm (M), 35 nm (Tb), and 30 nm (Ta-b). The RNAs are encapsidated by a single type of CP which in the case of strain AMV-L has a length of 220 amino acids (mol. wt 24 280). In solution, AMV CP occurs as dimers which under appropriate conditions of pH and ionic strength form a $T = 1$ icosahedral structure built from 30 dimers. This structure can be crystallized and has been studied by X-ray diffraction, cryoelectron microscopy, and image reconstruction methods. The CP was found to have the canonical eight-stranded β-barrel fold with the N- and C-terminal arms as extended chains. Dimer formation in the $T = 1$ particle is based on the clamping of the C-terminal arms of the subunits.

From particle weight measurements and analysis of electron microscopic images, it has been concluded that the number of CP monomers in the major viral components is $60 + (n \times 18)$, n being 10 (B, 240 subunits), 7 (M, 186 subunits), 5 (Tb, 150 subunits), or 4 (Ta, 132 subunits). By gel electrophoresis at least 13 minor components have been resolved which probably represent other n values and contain monomers of genomic RNAs, multimers of genomic RNAs or RNA 4, or specific degradation products of RNA 3. Although details of the arrangement of the protein monomers have not been established, electron microscopical studies indicate that the cylindrical parts of the bacilliform particles have a hexagonal surface lattice with dimers of CP associated with the twofold symmetry axes. The cylindrical part is believed to be capped by two halves of an icosahedron by changing the axes from sixfold

symmetry in the cylinder into axes of fivefold symmetry. Neutron scattering data suggested that the capsid structure of spheroidal Ta-t particles is represented by a deltahedron with 52-point group symmetry built from 120 subunits. The percentage of RNA in the virions decreases from 16.3 in B to 15.2 in Ta-b. The buoyant density in CsCl of the major components fixed with formaldehyde varies from 1.366 (Ta) to 1.372 (B) g cm^{-3}. The protein shell has an inner radius of 6.5 nm and an outside radius of 9.4 nm. The RNA is uniformly packed within the 6.5 nm radial limit, occupies about 20% of the interior volume available, and slightly penetrates the protein shell. The particles are mainly stabilized by protein–RNA interactions. The RNA is easily accessible to ribonucleases A and T1 through holes in the protein shell. At slightly alkaline pH the particles unfold reversibly.

Genome Structure

The genome structure of the L isolate of AMV is shown in **Figure 2**. RNAs 1 and 2 encode the replicase proteins P1 and P2, respectively. P1 contains an N-terminal methyltransferase-like domain and C-terminal helicase-like domain, whereas P2 contains a polymerase-like domain. RNA 3 is dicistronic and encodes the 5′-proximal movement protein (MP) gene and the 3′-proximal CP gene. MP and CP are both required for cell-to-cell

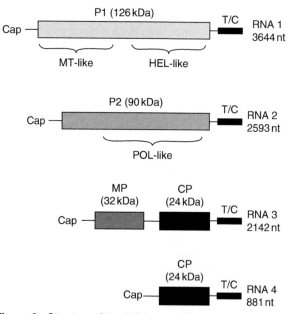

Figure 2 Structure of the AMV genome (isolate 425). RNAs 1 and 2 encode the replicase proteins P1 and P2; RNA 3 encodes the movement protein (MP) and coat protein (CP). CP is translated from subgenomic RNA 4. The bar labeled T/C represents the 3′-terminal 112 nt of the RNAs, which can adopt either a tRNA-like structure (TLS) or a structure with a high affinity for CP (CPB).

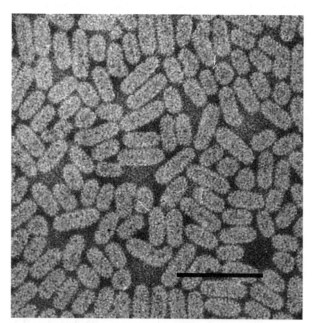

Figure 1 Electron micrograph of alfalfa mosaic virus. Scale = 100 nm.

transport of the virus. CP is translated from RNA 4, which is identical in sequence to the 3'-terminal 881 nucleotides (nt) of RNA 3. The length of the intercistronic region in RNA 3 is 52 nt including the leader sequence of RNA 4 of 36 nt. At the 5' end, all four AMV RNAs are capped. The organization of the leader sequence of RNA 3 varies between strains. For strains M, S, L, and Y the length of this leader sequence is 240, 313, 345, and 391 nt, respectively. The increased length of the last three strains is due to the presence of direct repeats of 56 (S), 75 (L), or 149 (Y) nt. At their 3' termini, the viral RNAs contain a homologous sequence of 145 nt. The 3' 112 nt of this sequence can adopt two mutually exclusive conformations, one representing a strong CP binding site (CPB) and the other representing a TLS resembling the TLS of bromo- and cucumoviruses. In the CPB structure, the 112 nt sequence consists of four hairpins (designated A, B, C, and D from 3' to 5' end) flanked by tetranucleotide sequences AUGC (or a UUGC derivative). Base-pairing between a four nucleotide sequence in the loop of the 5'-proximal hairpin D and a four nucleotide sequence in the stem of the 3'-proximal hairpin A results in a pseudoknot interaction that generates the TLS structure. Structures similar to the AMV CPB and TLS conformations have been identified at the 3' termini of the RNAs of prunus necrotic ring-spot ilarvirus. Although the TLS of AMV and ilarviruses cannot be charged with an amino acid, the 3' end of AMV RNAs is specifically recognized by the host enzyme that adds CCA to the 3' termini of cellular tRNAs (CTP/ATP:tRNA nucleotidyl transferase). The CCA-adding enzyme was able to adenylate the AMV TLS structure *in vitro*, but not the CPB structure. The AMV CPB and TLS structures have been shown to be required for translation of viral RNA and minus-strand promoter activity, respectively.

Interaction between Viral RNA and Coat Protein

In vitro, the 3'-untranslated region (UTR) of AMV RNAs can bind several dimers of CP. A minimal CP binding site consists of the 3'-terminal 39 nt of the RNAs with the structure 5'-AUGC-[hairpin B]-AUGC-[hairpin A]-AUGC-3'. This structure contains two overlapping binding sites for the N-terminal peptides of the two subunits of a CP dimer. The consensus binding site in the RNA is UGC-[hairpin]-RAUGC (in which R is a purine). In addition to dimers of native CP, N-terminal peptides of CP can bind the 39 nt RNA fragment in a 2:1 ratio. The N terminus of CP contains basic residues at positions 5, 6, 10, 13, 16, 17, 25, and 26, but only arginine-17 appeared to be critical for binding of CP to RNA. This Arg residue is part of a Pro-Thr-x-Arg-Ser-x-x-Tyr (PTxRSxxY) RNA binding domain conserved among AMV and ilarvirus CPs.

The complex of the 3'-terminal 39 nt RNA fragment and a peptide corresponding to the N-terminal 26 amino acids of CP has been crystallized and its structure was solved to 3 Å resolution. Co-folding altered the structure of both the peptide and the RNA. In the co-crystal, hairpins A and B are oriented at approximately right angles and each hairpin is extended by 2 bp formed between nucleotides from adjacent AUGC sequences. If the AUGC motifs in the 39 nt fragment are numbered 1–3, starting from the 3' end of the RNA, hairpin B is extended by a duplex formed by base-pairing between the U and C residue of motif 3 and the A and G residue of motif 2. Similarly, hairpin A is extended by a duplex formed by base-pairing between the U and C residue of motif 2 and the A and G residue of motif 1. The two peptides in the co-crystal each form an α helix with residues 12–26 ordered in peptide 1 and residues 9 to 26 ordered in peptide 2. The data provided insight into the role of the PTxRSxxY-motif in RNA binding. *In vitro* selection of RNA fragments with a high affinity for full-length AMV CP from a pool of randomized RNAs yielded fragments that maintained the unusual inter-AUGC base pairs observed in the crystal structure, although the primary sequences diverged from the wild-type RNA.

Translation of Viral RNA

Extension of the 3' end of AMV genomic RNAs with an artificial poly(A)-tail increased the basal level of infectivity of these RNAs 50-fold, compared to a 1000-fold increase caused by binding of CP to the RNAs. A role of CP in translation of viral RNAs was investigated by extension of the 3' end of a luciferase reporter RNA with the AMV 3'-UTR [Luc-AMV], a poly(A)-tail [Luc-poly(A)], or a plasmid-derived sequence [Luc-control]. Transfection of plant protoplasts with these transcripts in the absence or presence of AMV CP showed that CP did not affect translation of Luc-control or Luc-poly(A), but stimulated translational efficiency of Luc-AMV 40-fold. Moreover, CP had only a minor effect on the half-life of Luc-AMV. GST-pull-down assays and Far Western blotting revealed that AMV CP specifically interacted with the eIF4G-subunit from the eIF4F initiation factor complex from wheat germ (and with the eIFiso4G-subunit from the eIFiso4F complex that is present in plants). eIF4F consists of the helicase eIF4A, the cap-binding protein eIF4E, and the multifunctional scaffold protein eIF4G. **Figure 3** illustrates that translational efficiency of cellular messengers is strongly enhanced by the formation of a closed-loop structure, due to interactions of the poly(A) binding protein (PABP) with the poly(A)-tail and with the eIF4G subunit of the eIF4F complex. Due to its binding to the CPB structure at the 3' end of viral

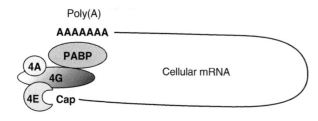

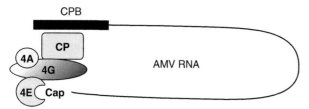

Figure 3 Model for the role of coat protein in translation of AMV RNAs. Translation of cellular mRNAs is strongly enhanced by the formation of a closed-loop structure by interactions of the poly(A) binding protein (PABP) with the 3′ poly(A) tail and with the eIF4G subunit of the eIF4F complex of initiation factors bound to the 5′ cap structure (upper panel). AMV coat protein (CP) enhances translational efficiency of viral RNAs 40-fold by binding to the CP binding site (CPB) at the 3′ end of the viral RNAs. The finding that CP also interacts with eIF4G indicates that CP mimics the function of PABP in formation of the closed-loop structure (lower panel).

RNAs and its interaction with eIF4G, AMV CP could stimulate translation of viral RNAs by a similar mechanism (**Figure 3**).

Protoplasts transfected with wild-type AMV RNA 4 accumulated CP at a detectable level, although accumulation was 100-fold lower than in productively infected protoplasts. However, translation of the RNA 4 mutant R17A was below the detection level. This mutant encodes CP with arginine-17 replaced by alanine, and this CP is unable to bind to the 3′ end of AMV RNAs. Translation of mutant R17A could be rescued to wild-type levels by expression in *trans* of CP that was functional in RNA binding. Also, translation of mutant RNA 4 was rescued to wild-type levels by replacing the 3′-UTR of this mutant by the 3′-UTR of brome mosaic virus (BMV) RNA 4. BMV is the type species of the genus *Bromovirus* in the family *Bromoviridae*. Apparently, the 3′-UTR of AMV requires binding of CP to stimulate translation, whereas the BMV 3′-UTR stimulates translation independently of CP, possibly by the binding of host factors.

Replication of Viral RNA

In tobacco, AMV replication complexes are associated with vacuolar membranes. Template-dependent replicase preparations have been purified from infected plants, which specifically accept exogenous AMV RNA as template. These preparations have been used to study viral minus-strand and plus-strand RNA synthesis *in vitro*.

The identification in 1978 of high-affinity binding sites for CP at the 3′-termini of AMV RNAs led to the hypothesis that binding of CP was required to permit initiation of minus-strand RNA synthesis. However, all experiments done to test this hypothesis showed that CP is not required for the synthesis of AMV minus-strand RNA *in vivo* or *in vitro*: (1) expression of AMV RNAs 1 and 2 from a T-DNA vector in agroinfiltrated leaves results in wild-type levels of minus-strand RNA synthesis in the complete absence of CP; (2) a CP-free replicase, purified from these agroinfiltrated leaves, transcribes AMV RNAs *in vitro* into minus-strand RNAs as efficiently as does the replicase purified from infected leaves; (3) in an *in vitro* replicase assay, addition of CP inhibits minus-strand RNA synthesis; (4) mutation of the 3′-terminal AUGC-motifs in AMV RNAs interferes with binding of CP *in vitro* but does not affect minus-strand RNA synthesis by the purified viral replicase. However, transcription of such mutant RNA by purified AMV replicase is no longer inhibited by CP.

An analysis of sequences in AMV RNAs which direct minus-strand RNA synthesis by the purified viral replicase *in vitro* showed that the entire 3′-terminal homologous sequence of 145 nt is required for minus-strand promoter activity. This promoter consists of the TLS structure formed by the 3′-terminal 112 nt and a hairpin structure, termed hairpin E, between nt 112 and 145 from the 3′ end of the RNAs. Hairpin E is probably the primary element recognized by the viral replicase, and the TLS serves to direct the bound replicase to the very 3′ end of the template. If the TLS structure is disrupted by mutations affecting the pseudoknot interaction or if the TLS is completely deleted, hairpin E directs initiation of minus-strand synthesis to a position located 5′ from the hairpin. In the absence of the TLS, the mechanism of RNA synthesis directed by hairpin E is very similar to that directed by the subgenomic promoter hairpin. This subgenomic promoter is located in minus-strand RNA 3 and directs plus-strand RNA 4 synthesis. The finding that the subgenomic promoter hairpin could be replaced by hairpin E without loss of infectivity illustrates the functional equivalence between the two hairpins.

Although a knockout mutation of the CP gene does not affect AMV minus-strand RNA synthesis in infected protoplasts or in agroinfiltrated leaves, such a mutation results in a 100-fold drop in the accumulation of viral plus-strand RNAs. Initially, it was proposed that this reflected a role of CP in *de novo* synthesis of plus-strand RNA. The observation that CP stimulated RNA 4 synthesis on a minus-strand RNA 3 template by the purified replicase *in vitro* supported this hypothesis. However, this observation could not be reproduced in later experiments. Recently, it was shown that expression of CP in agroinfiltrated leaves not only enhances the accumulation of replication-competent RNAs but also the accumulation of replication-defective viral RNAs. Thus, it is possible that the stimulation of the accumulation of plus-strand RNAs by CP in infected cells reflects a role of CP in protection of the RNAs from degradation.

RNA 3 can be replicated in *trans* by P1 and P2 proteins expressed from replication-defective RNAs 1 and 2 in agroinfiltrated tobacco leaves or by P1 and P2 expressed from nuclear transgenes in transgenic P12 tobacco. However, RNAs 1 and 2 are unable to make use of these transiently or transgenically expressed replicase proteins and are dependent for replication on their encoded proteins in *cis*. This requirement in *cis* explains why RNA 3 can initiate replication in protoplasts of transgenic P12 plants without a requirement for CP in the inoculum, whereas initiation of replication of RNA 1 or RNA 2 in this system requires CP to permit efficient translation of these RNAs into their encoded proteins. Moreover, replication of RNAs 1 and 2 is strictly coordinated. If RNAs 1, 2, and 3 replicate in plant cells in the presence of a replicase transiently expressed from a T-DNA vector, a mutation in RNA 1 that is lethal to the function of the encoded P1 protein blocks the replication of both mutant RNA 1 and wild-type RNA 2, but does not affect the replication of RNA 3 by the transiently expressed replicase. Similarly, a mutation in RNA 2 that affects the conserved GDD polymerase sequence in the P2 protein blocks replication of RNAs 1 and 2 in this system, but not replication of RNA 3. Apparently, replication of RNA 1 controls replication of RNA 2 and vice versa.

Virus Encapsidation and Movement

Expression of mutant RNAs 1 and 2 with their 3'-UTRs deleted and wild-type RNA 3 in agroinfiltrated leaves results in replication of RNA 3 by the transiently expressed replicase proteins and encapsidation of all three viral RNAs into virions. This indicates that the high-affinity binding sites for CP in the 3'-UTRs of RNAs 1 and 2 are dispensable for encapsidation of these RNAs. Possibly, assembly of virions initiates on internal CP binding sites that have been identified in AMV RNAs.

The RNA 3 encoded MP and CP are both required for cell-to-cell movement of AMV in infected plants. The MP is able to form tubular structures on the surface of infected protoplasts, suggesting that virus movement in plants involves transport of virus particles through tubules, which traverse the cell wall through modified plasmodesmata. Such a mechanism has been found for viruses of the genera *Comovirus*, *Caulimovirus*, and *Badnavirus*. However, tubular structures filled with virus particles have not been observed in AMV infected leaf tissue. Moreover, a CP mutant has been reported that is defective in the formation of virions but is able to move cell-to-cell

at a reduced level. Possibly, AMV moves cell-to-cell as viral ribonucleoprotein complexes that structurally differ from virions.

Role of Coat Protein in the AMV Replication Cycle

A natural infection with AMV will be initiated by viral particles B, M, and Tb, containing the genomic RNAs 1, 2, and 3, respectively. Inoculation of plants in the laboratory with a mixture of these three genomic RNAs results in very low, barely detectable levels of infection and infectivity of this mixture is increased 1000-fold by addition of four to ten molecules of CP per RNA molecule. Alternatively, infectivity can be increased by addition of RNA 4, the subgenomic messenger for CP. It is assumed that AMV infection starts with the three genomic RNAs, each associated with one or more dimers of the viral CP. These RNA/CP complexes can be generated by partial disassembly of particles B, M, and Tb in inoculated cells, by mixing purified viral RNAs and CP *in vitro*, or by translation of RNA 4 in the inoculated cells and subsequent binding of *de novo* synthesized CP to the viral RNAs. Infection can be initiated independently of CP when the three genomic RNAs are extended with an artificial 3'-terminal poly(A) tail or are transcribed in the plant nucleus from viral cDNA by polymerase II. It has been shown that binding of CP to the 3' end of AMV RNAs strongly stimulates translation of the RNAs in plant cells, and that CP specifically binds to the eIF4G subunit of the eIF4F complex of plant initiation factors. Based on these observations, it has been proposed that the role of CP in the inoculum is to stimulate translation of AMV (and ilarvirus) RNAs by mimicking the function of the poly(A) binding protein in translation of cellular mRNAs (**Figure 3**). A similar function in translation of nonpolyadenylated viral RNA has been reported for the NSP3 protein of rotaviruses (family *Reoviridae*). Efficient translation of AMV RNA 4 in protoplasts requires the ability of its encoded CP to bind to the 3' end of its own messenger. Available data support the notion that after inoculation of plants with AMV RNAs 1, 2, 3, and 4, RNA 4 is initially translated with low efficiency until translation of this messenger is stimulated by its own translation product.

In vitro, binding of CP to the 3' end of AMV RNAs blocks minus-strand RNA synthesis by the purified viral replicase. It is possible that early in infection the binding of CP to the 3' end of the RNAs not only promotes translation but also blocks premature initiation of replication to prevent a collision between ribosomes and replicase, traveling along the RNA in opposite directions. However, this possibility is not supported by experimental evidence. The switch from translation to replication requires dissociation of parental CP from the 3' end of the RNAs to allow the formation of the TLS structure, as the TLS conformation has been shown to be required for minus-strand promoter activity. Replicase proteins translated from inoculum RNAs 1 and 2 may target the viral RNAs to vacuolar membranes where replication complexes are formed and may bind to minus-strand promoter hairpin E (between nt 112 and 145 from the 3′ end of these RNAs) to promote dissociation of CP. Another possibility is that embedding of viral ribonucleoprotein complexes in the vacuolar membrane promotes dissociation of CP.

The mechanism of the switch from minus-strand to plus-strand RNA synthesis is not yet clear. Although the promoter for plus-strand subgenomic RNA 4 synthesis is structurally similar to hairpin E in the minus-strand promoter, the promoters for plus-strand genomic RNA synthesis have not yet been characterized. The requirement of CP for efficient accumulation of plus-strand RNA in infected cells may reflect a role of CP in protection of the RNAs from degradation, but a role in plus-strand RNA synthesis has not yet been ruled out. There is growing evidence for AMV and other viruses in the family *Bromoviridae* that many steps in the viral replication cycle are tightly linked.

Virus–Host Relationships

The AMV group is a large conglomerate of strains infecting a high number of susceptible hosts. This accounts for the tremendous range of symptoms displayed by AMV-infected plants. Mutations in the coat protein gene and 5′-UTR of RNA 3 have been shown to affect symptom formation in tobacco. Cytological modifications in AMV-infected plants occur only in cells of organs showing symptoms. In these cells fragmentation of the ground cytoplasm and an increased accumulation of membrane-bound vesicles has been observed. Sometimes the lamellar system of chloroplasts is affected and invaginations of the nuclear membrane have been reported. The MP protein has been localized in the middle lamella of walls of those cells that had just been reached by the infection front and in which viral multiplication had just begun. The P1 protein is exclusively localized at the tonoplast, whereas P2 was found both at the tonoplast and other locations in the infected cell. Virus particles are mainly found in the cytoplasm with a few records of particles in the nucleus. Depending on the strain, different types of intracellular aggregates of virus particles may occur.

Transmission

AMV is easily transmissible manually. Field spread occurs predominantly by aphid transmission. At least 15 aphid species are known to transmit the virus in the stylet-borne or nonpersistent manner. Acquisition of the virus occurs within 10–30 s and is followed by immediate transmission without a latent period. The ability to continue transmission is lost by the aphid within 1 h. The variability of individual aphid species in their capacity to transmit different AMV strains suggests a specific virus–vector relationship which is probably governed by the structural properties of the CP. Seed transmission of AMV has been reported for alfalfa and seven other plant species with rates of transmission varying from 0.1% to 50%. Transmission of the virus between plants by parasitic dodder has been observed with five *Cuscuta* species.

Epidemiology and Control

Although there have been reports on resistance and tolerance of alfalfa to AMV, control of the virus in this crop can be done mainly by using virus-free seed and avoiding reservoir hosts of the virus. Because the virus occurs naturally in many different plant species, this is practically impossible.

Tobacco plants transformed with the CP gene of AMV were found to be highly resistant to the virus when infection was done by mechanical inoculation. Resistance to transmission of virus by aphids has not yet been tested. The resistance was clearly protein mediated as plants with the highest level of CP accumulation were the most resistant. Plants with the highest level of CP were resistant to infection with inocula consisting of either viral particles or RNAs, whereas plants with lower levels of CP were resistant to infection with particles only. A mutation in the transgene that affected the N-terminal sequence of the encoded CP destroyed resistance to the wild-type virus but the mutant transgene conferred resistance to virus expressing the mutant CP.

See also: Ilarvirus.

Further Reading

Balasubramaniam M, Ibrahim A, Kim B-S, and Loesch-Fries S (2006) *Arabidopsis thaliana* is an asymptomatic host of alfalfa mosaic virus. *Virus Research* 121: 215–219.

Bol JF (2005) Replication of alfamo- and ilarviruses: Role of the coat protein. *Annual Review of Phytopathology* 43: 39–62.

Boyce M, Scott F, Guogas LM, and Gehrke L (2006) Base-pairing potential identified by *in vitro* selection predicts the kinked RNA backbone observed in the crystal structure of the alfalfa mosaic virus RNA–coat protein complex. *Journal of Molecular Recognition* 19: 68–78.

Guogas LM, Filman DJ, Hogle JM, and Gehrke L (2004) Cofolding organizes alfalfa mosaic virus RNA and coat protein for replication. *Science* 306: 2108–2111.

Hull R (1969) Alfalfa mosaic virus. *Advances in Virus Research* 15: 365–433.

Jaspars EMJ (1985) Interaction of alfalfa mosaic virus nucleic acid and protein. In: Davies JW (ed.) *Molecular Plant Virology*, vol.1, pp. 155–230. Boca Raton, FL: CRC Press.

Krab IM, Caldwell C, Gallie DR, and Bol JF (2005) Coat protein enhances translational efficiency of alfalfa mosaic virus RNAs and interacts with the eIF4G component of initiation factor eIF4F. *Journal of General Virology* 86: 1841–1849.

Kumar A, Reddy VS, Yusibov V, *et al.* (1997) The structure of alfalfa mosaic virus capsid protein assembled as a $T = 1$ icosahedral particle at 4.0-Å resolution. *Journal of Virology* 71: 7911–7916.

Olsthoorn RCL, Haasnoot PC, and Bol JF (2004) Similarities and differences between the subgenomic and minus-strand promoters of an RNA plant virus. *Journal of Virology* 78: 4048–4053.

Olsthoorn RCL, Mertens S, Brederode FT, and Bol JF (1999) A conformational switch at the 3′ end of a plant virus RNA regulates viral replication. *EMBO Journal* 18: 4856–4864.

Allexivirus

S K Zavriev, Shemyakin and Ovchinnikov Institute of Bioorganic Chemistry, Russian Academy of Sciences, Moscow, Russia

History

The genus *Allexivirus* of plant viruses belongs to a new family of plant viruses, *Flexiviridae*, which also includes the genera *Capillovirus, Carlavirus, Citrivirus, Foveavirus, Potexvirus, Trichovirus, Vitivirus,* and *Mandarivirus.* The first publication on an allexivirus as a member of a new plant virus group appeared in 1992 after analysis of the complete genome structure of a filamentous virus isolated from shallot plants in the Institute of Agricultural Biotechnology (Moscow, Russia). The virus thus discovered and described was named shallot virus X (ShVX). Allexiviruses acquired their name by fusion of the host (family Alliaceae) and the type member (ShVX) names. The allexivirus-associated diseases are usually mild and in many cases symptomless. Very often, allexiviruses persist in the infected plants as a part of multiple infection induced by viruses with similarly restricted host ranges. Since allexiviruses do not cause serious plant diseases (as well as the majority of related carlaviruses), they do not attract special attention of phytopathologists, but they are quite interesting for taxonomy studies, genome structure, and relationships between genera, species, and subspecies of distinct viruses.

Taxonomy and Classification

Table 1 lists definite and possible or insufficiently characterized members of the genus *Allexivirus.* Practically, all viruses have been included in the group according to the characteristics of their genome structure (primarily, a unique virus-specific protein encoded by ORF4) as well as the serological relatedness to ShVX and the common host range.

Table 1   Definite and tentative species of the genus *Allexivirus*

Species in the genus	GenBank accession no	
Garlic mite-borne filamentous virus	[X98991, AY390254]	GarMbFV
Garlic virus A	[AB010300, F478197]	GarVA
Garlic virus B	[AB010301, F543829]	GarVB
Garlic virus C	[AB010302, D49443]	GarVC
Garlic virus D	[AB010303, AF519572, L38892]	GarVD
Garlic virus E	[AJ292230]	GarVE
Garlic virus X	[AJ292229, U89243]	GarVX
Shallot virus X (397[a])	[M97264, L76292]	ShVX
Tentative species in the genus		
Garlic mite-borne latent virus		GarMbLV
Onion mite-borne latent virus		OMbLV
Shallot mite-borne latent virus		ShMbLV

[a]Number of the CMI/AAB Plant Virus Description (see http://www.dpvweb.net).

Poorly characterized viruses serologically related to ShVX have been found in several *Allium* species, as well as in tulip and narcissus plants, but it is unclear whether these should be regarded as strains of ShVX or as distinct viruses. The serological relationships of ShVX to other well-characterized allexiviruses or to the viruses in other genera have not yet been studied.

Viral Structure and Composition

Virions

Virions of allexiviruses are highly flexible filamentous particles, about 800 nm in length and 12 nm in diameter. They resemble potyviruses in size and closteroviruses in flexibility and cross-banded substructure (**Figure 1**).

Virus preparations obtained from individual *Allium ascalonicum* plants infected only with ShVX were found to contain virions of two morphological types: (1) typical allexivirus particles with characteristic cross-striation; and (2) thinner (6 nm), more flexible, and aggregation-prone particles. The latter were minor in most preparations, but constituted the bulk of the viral population in about 5% of plants. Particles of both types were formed of homologous genomic RNAs and serologically close coat proteins (CPs). Sequence analysis of the 2500 3'-terminal nucleotides of the genomic RNA of type 1 and 2 particles showed 87% homology; the amino acid sequences of the coat and the 15 kDa (OPF6) proteins had a single difference each, whereas the 42 kDa (ORF4) proteins had about 15% changes.

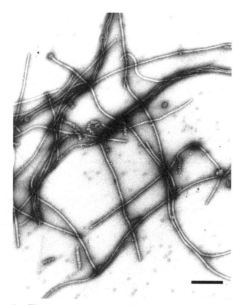

Figure 1   Electron micrograph of ShVX virions. Scale = 200 nm. Courtesy of V. Vishnichenko.

Nucleic Acid

Virions contain a single molecule of linear single-stranded RNA (ssRNA), about 9.0 kb in length, with a 3' poly(A) tract. ShVX RNA preparations, besides genomic ssRNA, contain molecules of 1.5 kb double-stranded RNA (dsRNA), whose genesis and function(s) are unknown. The complete nucleotide sequences of the genomic RNA of ShVX, garlic viruses A (GarVA), C (GarVC), E (GarVE), and X (GarVX), and the partial sequences of the RNA of garlic miteborne filamentous virus (GarMBFV) and garlic virus B (GarVB) and D (GarVD) have been determined.

Proteins

Virions are composed of a single polypeptide of 28–37 kDa. It was reported that the ORF6 protein is a minor component in ShVX virions.

Physicochemical and Physical Properties

The sedimentation constant of the ShVX virion is about 170S (0.1 M Tris-HC1, pH 7.5 at 20 °C). The buoyant density in CsCl is 1.33 g cm^{-3}.

Transmission

Allexiviruses are supposed to be mite-borne; the only vector known is the eriophyd mite *Aceria tulipae*, which was proved to transmit GarVC and GarVD. All allexiviruses are manually transmissible by sap inoculation of healthy host plants. None could be transmitted by aphids or any other insects.

Genome Organization and Replication

The genome organization of allexiviruses resembles that of carlaviruses, with the major exception of an 'additional' ORF between the triple gene block (TGB) and the CP ORF. The genomic RNA of allexiviruses contains six large ORFs and noncoding sequences of about 100 nucleotides at the 5' terminus and about 100 nucleotides followed by a poly(A) tail at the 3' terminus (**Figure 2**). Determination of the complete genomic RNA sequence of several allexiviruses proves their genome organization to be almost identical. The type member, ShVX, codes for polypeptides of 195, 26, 11, 42, 28, and 15 kDa, respectively, from the 5' end to the 3' end. Gene arrangement of other incompletely sequenced allexiviruses is similar.

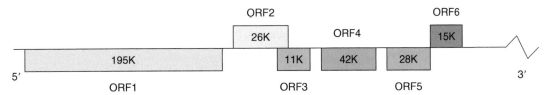

Figure 2 Genome organization of shallot virus X (genus *Allexivirus*).

The 195 kDa polypeptide is a viral RNA polymerase, and most probably it is the only virus-encoded protein required for replication. In comparison of the amino acid sequences of methyltransferase, helicase, or RNA-dependent RNA polymerase (RdRp) motifs, these conserved domains of allexiviruses are most similar to those of potexviruses. The 26 and 11 kDa proteins are similar to the first two proteins encoded by the TGB of potexviruses and carlaviruses and are probably involved in cell-to-cell movement of the virus. There is a coding sequence for a small (7–8 kDa) TGB protein but it lacks the initiation AUG codon. The 42 kDa polypeptide has no significant homology with any proteins known, though it is expressed in plants infected with ShVX in relatively large amounts. Immunoelectron microscopy using polyclonal antisera against the recombinant ShVX 42 kDa protein showed reaction with certain regions of the virions, with the immune complexes nonuniformly distributed along the particle. The 42 kDa protein was supposed to act as a cofactor to provide proper interaction of the CP with the genomic RNA in the virion assembly process. The 28 kDa polypeptide is the CP. In polyacrylamide gel electrophoresis (PAGE) it migrates as a 32–36 kDa protein, which can be due to its high hydrophilicity evident from amino acid sequence. The 15 kDa protein is similar to the 11–14 kDa proteins encoded by the 3′-proximal ORFs of carlaviruses, and has a zinc-binding-finger motif with affinity for nucleic acids. The exact function of this polypeptide is unknown, but it may be involved in virion assembly.

At least two subgenomic mRNAs (sgRNA) are used for translating the 5′-distal ORFs: one for the movement proteins (ORF2 and ORF3) and the other for the CP (ORF5) and the 3′-distal nucleic acid-binding protein (ORF6). Although the ORF4 protein may be expressed in the infected plants in relatively large amounts, no evidence is available for an ORF4-specific sgRNA. It is suggested that the ORF4-encoded protein is translated from the sgRNA for the ORF2 and ORF3 proteins.

Geographical Distribution

Allexiviruses have been identified in Russia, Japan, France, Germany, the UK, The Netherlands, Korea, China, Taiwan, Thailand, and Argentina. Most probably, they are all distributed across the world, especially in the regions where bulbous plants are widespread.

Phylogenetic Information

Phylogenetic trees based on the nucleotide sequence depend on the genome region used for analysis. In the family *Flexiviridae*, the allexiviruses in general occupy an intermediate position in putative phylogeny between carlaviruses, potexviruses, and mandariviruses.

See also: Capillovirus, Foveavirus, Trichovirus, Vitivirus; Carlavirus; Flexiviruses; Vegetable Viruses.

Further Reading

Adams MJ, Antoniw JF, Bar-Joseph M, *et al.* (2004) The new plant virus family Flexiviridae and assessment of molecular criteria for species demarcation. *Archives of Virology* 149: 1045–1060.

Arshava NV, Konareva TN, Ryabov EV, and Zavriev SK (1995) The 42K protein of shallot virus X is expressed in the infected *Allium* plants. *Molecular Biology* (Russia) 29: 192–198.

Barg E, Lesemann DE, Vetten HJ, and Green SK (1994) Identification, partial characterization and distribution of viruses infecting crops in south and south-east Asian countries. *Acta Horticulturae* 358: 251–258.

Chen J, Chen J, and Adams MJ (2001) Molecular characterization of a complex mixture of viruses in garlic with mosaic symptoms in China. *Archives of Virology* 146: 1841–1853.

Kanyuka KV, Vishnichenko VK, Levay KE, Kondrikov DYu, Ryabov EV, and Zavriev SK (1992) Nucleotide sequence of shallot virus X RNA reveals a 5′-proximal cistron closely related to those of potexviruses and a unique arrangement of the 3′-proximal cistrons. *Journal of General Virology* 73: 2553–2560.

Song SI, Song JT, Kim CH, Lee JS, and Choi YD (1998) Molecular characterization of the garlic virus X genome. *Journal of General Virology* 79: 155–159.

Sumi S, Tsuneyoshi T, and Furutani H (1993) Novel rod-shaped viruses isolated from garlic, possessing a unique genome organization. *Journal of General Virology* 74: 1879–1885.

Van Dijk P and van der Vlugt RA (1994) New mite-borne virus isolates from Rakkyo, shallot and wild leek species. *European Journal of Plant Pathology* 100: 269–277.

Vishnichenko VK, Konareva TN, and Zavriev SK (1993) A new filamentous virus in shallot. *Plant Pathology* 42: 121–126.

Vishnichenko VK, Stelmashchuk VY, and Zavriev SK (2002) The 42K protein of the Shallot virus X participates in formation of virus particles. *Molecular Biology* (Russia) 36: 1080–1084.

Yamashita K, Sakai J, and Hanada K (1996) Characterization of a new virus from garlic (*Allium sativum* L.), garlic mite-borne virus. *Annals of Phytopathology Society of Japan* 62: 483–489.

Banana Bunchy Top Virus

J E Thomas, Department of Primary Industries and Fisheries, Indooroopilly, QLD, Australia

Glossary

Cell-cycle link protein A plant virus protein which most probably subverts the cell-cycle control of the host, forcing cells into DNA synthesis or S phase favorable to viral replication.

Circulative transmission Mode of transmission whereby the virus is acquired from plant sieve tubes, via the insect's stylet, and traverses a number of specific barriers as it passes from the hindgut, into the hemocoel and then to the salivary glands, from where it can be reinjected into a plant during feeding. It is characterized by a latent period before which re-transmission cannot occur and by retention of infectivity by the vector for periods ranging from several days to the entire life span of the insect. The virus does not replicate in the insect.

Nuclear shuttle protein A virus-encoded protein that transports the viral single-stranded DNA, as a complex with the viral movement protein, to and from its site of replication in the nucleus and into adjacent cells.

Introduction

Banana bunchy top disease is the most economically important virus disease of banana and plantain (*Musa* spp.) worldwide, due to its devastating effect on crop yield, and the importance of banana and plantain as both a staple food and a major export commodity in much of the developing world. The causal agent is banana bunchy top virus (BBTV). Edible bananas are derived from wild progenitors including *M. acuminata*, *M. balbisiana*, and *M. schizocarpa*, which have a center of origin in the South and Southeast Asian–Australasian region, and it is likely that BBTV also originated within this area. BBTV has a multipartite circular single-stranded (ssDNA) genome, encapsidated in small isometric virions, and is transmitted by the banana aphid, *Pentalonia nigronervosa*.

Disease Symptoms

The symptoms of bunchy top disease in banana are characteristic, especially in the Cavendish subgroup of cultivars, and easily distinguished from all other virus diseases of banana. Plants can become infected at any stage of growth, and the initial appearance of symptoms can depend on the manner of infection. BBTV is systemic within the banana plant, and following aphid inoculation, symptoms do not appear until at least two more new leaves have been produced (bananas produce single new leaves sequentially from a basal meristem). The first symptoms comprise a few dark green streaks and dots on the lower part of the lamina and on the petiole, becoming more general on subsequent leaves. These streaks form hooks as they enter the midrib, and are best viewed from the underside of the leaf, with transmitted light (**Figure 1**). However, these dark streaks can be rare or absent in some cultivars. Successive leaves become shorter and narrower, and have a brittle lamina with upturned, chlorotic, ragged margins. Leaves fail to emerge fully, giving the plant a bunched appearance (**Figure 2**). Plants derived from infected planting material

Figure 1   Dark green dot-dash, hooking and vein clearing fleck symptoms on a Cavendish banana leaf

Figure 2   Young BBTV-infected banana plants, showing stunting, and successively shorter narrow leaves with upturned, chlorotic margins. Leaves have failed to emerge fully, giving the plant a bunched appearance.

(suckers, bits) develop severe symptoms from the first leaf to emerge.

Infected plants seldom produce a bunch, though if infected late in the current cropping cycle, a small, distorted bunch may result. With very late infections, the only symptoms to appear in the current season may be a few dark streaks on the tips of the flower bracts. No fruit is produced in subsequent years, and plants generally die within a couple of years.

Histological examination suggests that BBTV is restricted to the phloem tissue, which shows hypertrophy and hyperplasia and a reduction in the development of the fibrous sclerenchyma sheaths surrounding the vascular bundles. The cells surrounding the phloem contain abnormally large numbers of chloroplasts, resulting in the macroscopic dark green streak symptom.

Using RNA probes and polymerase chain reaction (PCR), it has been demonstrated that BBTV replicates briefly at the site of aphid inoculation, then moves down the pseudostem to the basal meristem, subsequently infecting the newly formed leaves, the corm, and the roots. The virus apparently does not replicate in leaves formed prior to infection, consistent with the lack of symptoms on these leaves and an inability to recover the virus from them via the aphid vector.

From Taiwan, symptomless strains of BBTV, and mild strains that produce only limited vein clearing and dark green streaks have been reported. Also, some plants of the Cavendish subgroup cultivar Veimama from Fiji have been observed to initially show severe symptoms, then to recover and display few if any symptoms.

Host Range

Confirmed hosts of BBTV are confined to the family Musaceae. Known susceptible hosts include *Musa* species, cultivars in the Eumusa and Australimusa series of edible banana and *Ensete ventricosum*. Susceptible *Musa* species include *M. balbisiana*, *M. acuminata* ssp. *banksii*, *M. textilis*, *M. velutina*, *M. coccinea*, *M. jackeyi*, *M. ornata*, and *M. acuminata* ssp. *zebrina*. There are some reports of hosts outside the Musaceae, though in independent tests none has been confirmed.

Causal Agent

The viral nature of bunchy top disease was established by C. J. P. Magee, in Australia, in the 1920s. However, it was not until 1990 that the virus particles were first isolated, in part due to lack of a suitable herbaceous experimental host, its low titer in infected plants, and its restriction to the phloem in the fibrous vascular tissue of *Musa*. BBTV has icosahedral particles, 18–20 nm in diameter (**Figure 3**), a

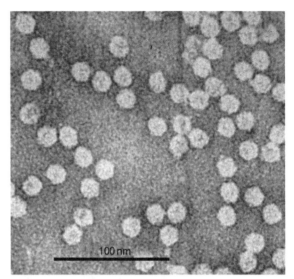

Figure 3 Electron micrograph of BBTV virions, negatively contrasted with 1% ammonium molybdate.

buoyant density of 1.29–1.30 in cesium sulfate, and a sedimentation coefficient of 46S.

BBTV has a multicomponent genome comprising six transcriptionally active components (**Figure 4**; **Table 1**), each *c.* 1 kbp in size. DNA-R encodes the master replication initiation protein (Rep), which can initiate and terminate replication, and a potential second internal ORF of unknown function, for which a transcript only has been identified. All other components are monocistronic; DNA-S encodes the coat protein, DNA-C the cell-cycle link protein, DNA-M the cell-to-cell movement protein, DNA-N the nuclear shuttle protein, and the function of the protein potentially encoded by DNA-U3 is unknown. The untranslated regions of all six components share two areas of high sequence identity, both concerned with the rolling-circle mode by which BBTV replicates. A stem–loop common region (CR-SL) of 69 nt includes a nonanucleotide sequence (TAT/GTATTAC) which is shared between plant-infecting circular ssDNA nanovirids and geminivirids, and which is the site of the origin of viral replication. It also contains iterated sequences, thought to act as recognition sites for the replication initiation protein. The major common region (CR-M) varies between 66 and 92 nt in length among the various components. Virions contain a heterogeneous population of DNA primers, *c.* 80 nt in length, which bind to the CR-M and prime complementary strand synthesis.

The coat protein gene of BBTV is highly conserved, with a maximum difference of 3% at the amino acid level between isolates. No serological differences have been detected between any isolates using polyclonal or monoclonal antibodies.

The intergenic regions of all six integral components of BBTV have been shown to have promoter activity. The highest activity in banana embryonic cells was shown

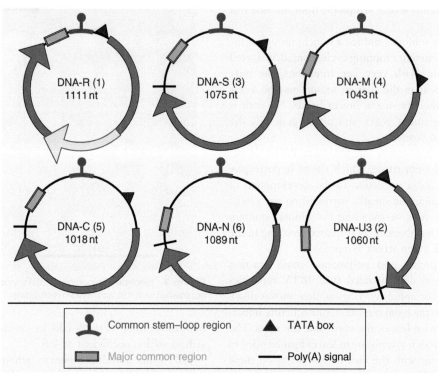

Figure 4 Diagram of the integral genome components of BBTV, showing transcribed ORFs (black arrows) and other main genome features. Reproduced from Vetten HJ, Chu PWG, Dale JL, *et al*. (2005) *Nanoviridae*. In: Fauquet CM, Mayo MA, Maniloff J, Desselberger U, and Ball LA (eds.) *Virus Taxonomy: Eighth Report of the International Committee on Taxonomy of Viruses*, p. 349, figure 3. San Diego, CA: Elsevier Academic Press, with permission from Elsevier.

Table 1   Size and function of the six integral components of BBTV

Genome component	Size (nt)	Size of encoded protein(s) (kDa)	Function
DNA-R	1111	(1) 33.6	(1) Replication initiation protein
		(2) 5.2[a]	(2) Unknown
DNA-S	1075	20.1	Capsid protein
DNA-M	1043	13.7	Movement protein
DNA-C	1018	19.0	Cell-cycle link protein
DNA-N	1089	17.4	Nuclear shuttle protein
DNA-U3	1060	10.4[b]	Unknown

[a]Expression has not been demonstrated for the transcript of this internal ORF.
[b]Most isolates of BBTV do not encode a functional ORF on DNA-U3, and where present there is no evidence for its expression.

by promoters from DNA-C and DNA-M, components thought to be intrinsic to the infection process. Studies on the DNA-N promoter demonstrated that expression in banana embryonic cells is limited to phloem tissue, consistent with the circulative mode of transmission by the aphid vector and observations by Magee in the 1920s that histological effects were confined to the phloem and phloem parenchyma cells. Circulative transmission by aphid vectors usually involves specific feeding on the phloem tissue for virus acquisition. The CR-M and CR-SL are not essential for promoter activity. All essential elements of the promoter are located 3′ of the stem–loop and within 239 bp of the translation start codon and within this region are an ASF-1

motif (TGACG), a hexamer motif (ACGTCA), *rbc*S-I-box (GATAAG), G-boxcore, and the TATA box, all associated with promoter activity in other genomes.

Some isolates of BBTV from Taiwan and Vietnam contain additional Rep-encoding DNAs that are thought to be capable only of self-replication and to behave like satellite molecules. These molecules have a CR-SL, though the stem sequence is not conserved with the six integral DNA components. Unlike DNA-R, the putative satellites lack the internal ORF, their TATA boxes are 5′ of the stem loop, and they generally lack the CR-M. Interestingly, the amino acid sequences of satellite Reps BBTV-S1, BBTV-S2, and BBTV-Y1 are actually more closely related phylogenetically to the Reps encoded by nanovirids outside the genus *Babuvirus* than they are to DNA-R of BBTV. Similar molecules have been detected by Southern hybridization in isolates from the Philippines, Tonga, and Western Samoa, but not from Australia, Egypt, Fiji, and India.

Taxonomy and Phylogenetic Relationships

BBTV is a member of the genus *Babuvirus*, in the family *Nanoviridae*. Other members of this family are classified in the genus *Nanovirus*, and include faba bean necrotic yellows virus, milk vetch dwarf virus, and subterranean clover stunt virus.

BBTV isolates worldwide fall into two broad phylogenetic groups (**Figure 5**), called the South Pacific group (isolates from Australia, Fiji, Hawaii, Western Samoa, Tonga, India, Pakistan, Burundi, Egypt, Malawi) and the Asian group (China, Philippines, Japan, Vietnam, Taiwan). Sequence differences between the two groups have been demonstrated across all genome components. Most striking is the variation in the CR-M of DNA-R, where the mean sequence difference between the groups was 30%, and up to a maximum of 55% for individual

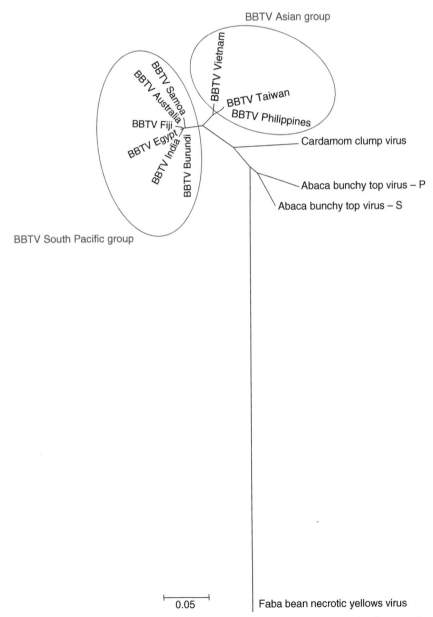

Figure 5 Phylogenetic tree based on the amino acid sequences of babuvirus Rep proteins, using the nanovirus faba bean necrotic yellows virus as an outgroup. The Asian and South Pacific groups of BBTV are circled in blue and red, respectively.

pairs of isolates. Within group sequence diversity is much greater for the Asian group, suggesting a longer evolutionary period and possible evolutionary origin for the virus. The presence of the virus in many countries from the South Pacific group can be traced to introductions within the last century. A Taiwanese isolate (TW4) causing only mild symptoms was recently identified. Interestingly, it apparently lacks DNA-N and is clearly a chimeric isolate, having DNA-M of the Asian group and DNA-S, -C, and -U3 of the South Pacific group. It also contains two DNA-R molecules, one from each group.

Recently, two new viruses have been identified which are clearly distinct members of the genus *Babuvirus* (**Figure 5**). Abaca bunchy top virus has homologous components for all six genome components of BBTV and also infects *Musa* in Southeast Asia. The sequence and organization of DNA-R of cardamom clump virus (syn. cardamom bushy dwarf virus), associated with Foorkey disease of cardamom in India, also suggests that it is also a genome component of a novel babuvirus.

Geographical Distribution

Historical Records and Possible Origins

The origin of BBTV is unclear. The first records were from Fiji in 1889, and though reports and photographs clearly indicate that it was present at least 10 years prior, soon after the establishment of an export industry based on Cavendish cultivars, evidence suggests that the disease did not originate there. Other early records are from Egypt (1901 – source unknown) and Australia and Sri Lanka (both in 1913, and probably from planting material imported from Fiji).

The wild progenitors of modern edible bananas originated in the South and Southeast Asian–Australasian region, and the Cavendish cultivars of international trade, associated with early outbreaks of bunchy top disease, are thought to have originated in Vietnam. These factors lead to speculation that BBTV also originated and evolved in this region. The recent discovery of a chimeric BBTV isolate from Taiwan containing genome components of both geographic groups could support this hypothesis.

Current Distribution

BBTV has a widespread, but scattered distribution in many of the banana-growing countries of the Asia-Pacific regions and Africa, but at present is not found in the Americas. In some countries, the occurrence is localized, probably due to geographic isolation. For example, in Australia, it is present in southern Queensland and northern New South Wales, but not in the major production area of north Queensland. Banana bunchy top disease has been recorded in the following countries, and for those

marked with asterisk (*), the presence of the virus has been confirmed by serological or molecular assays:

Asia. China*, Japan (Okinawa*, Bonin Is.), India*, Indonesia*, Iran*, Myanmar, Malaysia (Sarawak*), Pakistan*, Philippines*, Sri Lanka*, Taiwan*, Vietnam*.

Pacific. Australia*, Fiji*, Guam, Kiribati (formerly Gilbert Is.), New Caledonia*, Samoa (American, Western*), Tonga*, Tuvalu (formerly Ellice Is.), USA (Hawaii*), Wallis Is.

Africa. Angola, Burundi*, Central African Republic, Congo* Democratic Republic of the Congo (formerly Zaire), Egypt*, Gabon*, Malawi*, Rwanda, Zambia*.

Virus Transmission

The banana aphid (*Pentalonia nigronervosa*) has a worldwide distribution, and in 1925 was shown to be a vector of BBTV. It remains the only known insect vector of this virus. Hosts of the aphid include species in the family Musaceae and several related families, including Araceae, Heliconiaceae, Strelitzeaceae, and Zingiberaceae, though the aphid shows a degree of host preference and can be difficult to transfer between host species. On banana, they commonly colonize the base of the pseudostem at soil level and for several centimeters below the soil surface, beneath the outer leaf sheaths and newly emerging suckers.

Transmission by *P. nigronervosa* is of the persistent, circulative, nonpropagative type, and individuals from areas where the virus is endemic and from where it is absent, both transmit the virus with equal efficiency. There is a minimum acquisition access period of 4 h, a minimum latent period of a few hours, and a minimum inoculation access period of 15 min. Aphids retain infectivity, after removal from a virus source, for at least 20 days and probably for life. Both nymphs and adults can acquire the virus, though more efficiently by the former, and reported transmission rates for individual aphids are in the range 46–67%. There is no evidence for transmission of BBTV to parthenogenetic offspring or for replication of the virus in the aphid.

BBTV is also efficiently transmitted in vegetative planting material, both conventional corms, corm pieces (bits) and suckers, and through micropropagation. All meristems from an infected corm will eventually become infected.

Epidemiology and Control of BBTV

Epidemiology

Outbreaks of bunchy top disease can have a devastating effect on banana production, especially industries based on Cavendish cultivars. Production in Fiji fell by more than 80% from 1892 to 1895, primarily due to bunchy top disease. By 1925, nearly 10 years since the introduction of

BBTV to Australia, the banana industry in northern New South Wales and southern Queensland had collapsed, with most plantations affected and production decreased by 90–95%. Magee noted at the time "It would be difficult for anyone who has not visited these devastated areas to visualize the completeness of the destruction wrought in such a short time by a plant disease."

More recently, a severe outbreak of banana bunchy top disease occurred in Pakistan. From 1991 to 1992, production area fell by 55% and total production by 90%, as a direct result of the disease. The disease has also recently appeared in Hawaii, New Caledonia, Angola, Zambia, and Malawi.

The epidemiology of banana bunchy top disease is simplified by the occurrence of a single insect vector species and a limited host range for the virus, usually cultivated or feral edible bananas. Long-distance spread is usually via infected planting material, and local spread via aphids and planting material.

Analysis of actual outbreaks of bunchy top disease in commercial banana plantations in Australia showed that the average distance of secondary spread by aphids was only 15.5–17.2 m, with nearly two-thirds of new infections less than 20 m from the nearest source of infection and 99% less than 86 m. Isolation of new plantations has a marked effect on reducing the risk of infection. New plantations situated adjacent to affected plantations had an 88% chance of recording infections in the first year. This was reduced to 27% if the plantations were separated by 50–1000 m, and to 5% if separated by more than 1000 m. The disease latent period (i.e., period from inoculation of a plant until an aphid can transmit the virus from this plant to another) is equivalent to the time taken for 3.7 new leaves to emerge from the plant. The actual time varies seasonally.

Control

Control strategies were devised by C. J. P. Magee, in the 1920s, and these measures still form the basis of the very successful control program in Australia today. The two major elements of the strategy are (1) exclusion of the disease from unaffected and lightly affected areas and (2) eradication of infected plants from both lightly and heavily affected areas.

These measures require the participation of all growers and are unlikely to succeed if left to the goodwill of growers alone and are thus enforced by legislation in Australia. The measures include:

- registration of all banana plantations;
- establishment of quarantine zones;
- restrictions on the movement and use of planting material;
- regular inspections of all banana plantations for bunchy top;

- prompt destruction of all infected plants; and
- ongoing education and extension programs for growers.

When adopted, these measures allowed the complete rehabilitation of the Australian banana industry. Occasionally, total eradication of BBTV from a district has been achieved, but in most cases, incidence has been reduced to very low, manageable levels. Such successful control of bunchy top is rarely achieved in other countries, in most cases due to an inability to enforce an organized control program across whole districts.

Detection Assays

Polyclonal and monoclonal antibodies to BBTV are used in enzyme-linked immunosorbent assay (ELISA) to detect the virus in field and tissue culture plants and can detect the virus in single viruliferous aphids. PCR was shown to be about a thousand times more sensitive than ELISA or dot blots with DNA probes. Substances in banana sap inhibitory to PCR can be circumvented by simple extraction procedures or by immunocapture PCR.

BBTV has been detected in most parts of the banana plant, including the leaf lamina and midrib, pseudostem, corm, meristems, roots, fruit stalk, and fruit rind.

Resistance

There are no confirmed reports of immunity to BBTV in *Musa*. However, it has frequently been observed that there are differences in susceptibility between cultivars to both field and experimental infection. Edible bananas have diploid, triploid, or tetraploid genomes containing, predominantly, elements of the *M. acuminata* (A) or *M. balbisiana* (B) genomes. Cultivars in the Cavendish subgroup (AAA genome), which dominates the international export trade, and many other A genome cultivars are highly susceptible and show severe disease symptoms. By contrast, Gros Michel (AAA) displays resistance to the disease under both experimental inoculation and field conditions. Compared with highly sensitive cultivars such as Cavendish, the cultivar is less susceptible to aphid inoculation, contains a lower level of virions in infected plants, and symptoms are less severe and develop more slowly. These factors may contribute to a reduced rate of aphid transmission and field spread in plantations of Gros Michel, and introduction of this cultivar may explain the partial recovery of the Fijian banana industry after devastation of the Cavendish-based industry in the early 1900s. Field observations and glasshouse inoculations suggest that some B genome-containing cultivars are less susceptible to infection and/or display more limited symptoms, but this needs to be further investigated.

Despite concerted efforts to generate transgenic resistance to BBTV, no successful glasshouse or field results have yet been reported. However, some promising strategies are being developed at Queensland University of Technology involving the following steps.

1. Transdominant negative strategies to interfere with replication, by constitutive overexpression of mutated Rep proteins, are employed. Single mutations in either of two motifs involved with rolling-circle replication render the Rep inactive, and in transient assays with constitutive overexpression, virus replication is significantly reduced, but not abolished.

2. Rep-activated cell death is carried out using a so-called suicide gene. DNA-R intergenic sequence is cloned within an intron and flanked by a split barnase gene construct. The suicide gene is only activated in the presence of the Rep protein, resulting in cell death and containment of the virus.

See also: Nanoviruses.

Further Reading

Allen RN (1987) Further studies on epidemiological factors influencing control of banana bunchy top disease, and evaluation of control measures by computer simulation. *Australian Journal of Agricultural Research* 38: 373–382.

Bell KE, Dale JL, Ha CV, Vu MT, and Revill PA (2002) Characterisation of Rep-encoding components associated with banana bunchy top nanovirus in Vietnam. *Archives of Virology* 147: 695–707.

Burns TM, Harding RM, and Dale JL (1995) The genome organization of banana bunchy top virus: Analysis of six ssDNA components. *Journal of General Virology* 76: 1471–1482.

Dugdale B, Becker DK, Beetham PR, Harding RM, and Dale JL (2000) Promoters derived from banana bunchy top virus DNA-1 to -5 direct vascular-associated expression in transgenic banana (*Musa* spp.). *Plant Cell Reports* 19: 810–814.

Geering ADW and Thomas JE (1997) Search for alternative hosts of banana bunchy top virus in Australia. *Australasian Plant Pathology* 26: 250–254.

Hafner GJ, Harding RM, and Dale JL (1995) Movement and transmission of banana bunchy top virus DNA component one in bananas. *Journal of General Virology* 76: 2279–2285.

Hu J-M, Fu H-C, Lin C-H, Su H-J, and Yeh H-H (2007) Reassortment and concerted evolution in banana bunchy top virus genomes. *Journal of Virology* 81: 1746–1761.

Hu JS, Wang M, Sether D, Xie W, and Leonhardt KW (1996) Use of polymerase chain reaction (PCR) to study transmission of banana bunchy top virus by the banana aphid (*Pentalonia nigronervosa*). *Annals of Applied Biology* 128: 55–64.

Karan M, Harding RM, and Dale JL (1994) Evidence for two groups of banana bunchy top virus isolates. *Journal of General Virology* 75: 3541–3546.

Magee CJ (1953) Some aspects of the bunchy top disease of banana and other *Musa* spp. *Journal and Proceedings of the Royal Society of New South Wales* 87: 3–18.

Magee CJP (1927) *Investigation on the Bunchy Top Disease of the Banana. Bulletin No. 30.* Melbourne: Council for Scientific and Industrial Research.

Thomas JE and Dietzgen RG (1991) Purification, characterization and serological detection of virus-like particles associated with banana bunchy top disease in Australia. *Journal of General Virology* 72: 217–224.

Thomas JE and Iskra-Caruana ML (2000) Bunchy top. In: Jones DR (ed.) *Diseases of Banana, Abaca and Enset*, pp. 241–253. Wallingford: CABI Publishing.

Thomas JE, Smith MK, Kessling AF, and Hamill SD (1995) Inconsistent transmission of banana bunchy top virus in micropropagated bananas and its implication for germplasm screening. *Australian Journal of Agricultural Research* 46: 663–671.

Vetten HJ, Chu PWG, Dale JL, *et al.* (2005) *Nanoviridae.* In: Fauquet CM, Mayo MA, Maniloff J, Desselberger U, and Ball LA (eds.) *Virus Taxonomy, Eighth Report of the International Committee on Taxonomy of Viruses*, pp. 343–352. San Diego, CA: Elsevier Academic Press.

Wu R-Y and Su H-J (1990) Purification and characterization of banana bunchy top virus. *Journal of Phytopathology* 128: 153–160.

Barley Yellow Dwarf Viruses

L L Domier, USDA-ARS, Urbana-Champaign, IL, USA

Glossary

Hemocoel The primary body cavity of most arthropods that contains most of the major organs and through which the hemolymph circulates.
Hemolymph A circulatory fluid in the body cavities (hemocoels) and tissues of arthropods that is analogous to blood and/or lymph of vertebrates.

Introduction

Barley yellow dwarf (BYD) is the most economically important virus disease of cereals, and is found in almost every grain growing region in the world. Widespread BYD outbreaks in cereals were noted in the United States in 1907 and 1949. However, it was not until 1951 that a virus was proposed as the cause of the disease. The causal agents of BYD are obligately transmitted by aphids, which probably delayed the initial classification of BYD as a virus disease. Subsequently, BYD was shown to be caused by multiple viruses belonging to the species barley yellow dwarf virus (BYDV) and cereal yellow dwarf virus (CYDV). Depending on the virulence of the virus strain, infection may contribute to winter kill in regions with harsh winters, induce plant stunting, inhibit root growth, reduce or prevent heading, or increase plant susceptibility to opportunistic pathogens and other stresses. Yield losses to wheat in the United States alone are estimated at 1–3% annually, exceeding 30% in certain regions in epidemic years. The effects of BYD in barley and oats typically are more severe than in wheat; sometimes resulting in complete crop losses. The existence of multiple strains of viruses that are transmitted in strain-specific manner has made BYDV and CYDV model systems to study interactions between viruses and aphid vectors in the circulative transmission of plant viruses. In addition, the compact genomes of the viruses have provided useful insights into the manipulation of host translation machinery by RNA viruses.

Taxonomy and Classification

The viruses that cause BYD are members of the family *Luteoviridae*, and were first grouped because of their common biological properties. These properties included persistent transmission by aphid vectors, the induction of yellowing symptoms in grasses, and serological relatedness. Different viruses are transmitted more efficiently by different species of aphids, a fact that was originally used to distinguish the viruses. Around 1960, the viruses were separated into five 'strains' (now recognized as distinct species) based on their primary aphid vector(s). BYDVs transmitted most efficiently by *Sitobion* (formerly *Macrosiphum*) *avenae* were assigned the acronym MAV, for *Macrosiphum avenae* virus. Similarly, viruses transmitted most efficiently by *Rhopalosiphum maidis* and *Rhopalosiphum padi* were assigned the acronyms RMV and RPV, respectively. Viruses transmitted most efficiently by *Schizaphis graminum* were assigned the acronym SGV. Finally, vector-nonspecific viruses, that is, viruses transmitted efficiently by both *R. padi* and *S. avenae* were assigned the acronym PAV.

Based on genome organization and predicted amino acid sequence similarities, BYDV-MAV, -PAS, and -PAV have been assigned to the genus *Luteovirus*, and CYDV-RPS and -RPV to the genus *Polerovirus*. The RNA-dependent RNA polymerases (RdRps) encoded by open reading frames (ORFs) 1 and 2 of BYDVs resemble those of members of the family *Tombusviridae* (**Figure 1**). In contrast, the predicted amino acid sequence of the RdRps encoded by ORFs 1 and 2 of CYDVs resemble those of viruses in the genus *Sobemovirus*. The two polymerase types are distantly related in evolutionary terms.

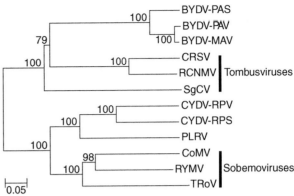

Figure 1   Phylogenetic relationships of the predicted amino acid sequences of RNA-dependent RNA polymerases (RdRps; ORFs 1 and 2) of barley yellow dwarf viruses (BYDVs) and cereal yellow dwarf viruses (CYDVs) and members of the genus *Sobemovirus* and family *Tombusviridae*. The RdRps of BYDV-MAV, -PAS, and -PAV are more similar to those of members of the family *Tombusviridae* (carnation ringspot virus (CRSV), red clover necrotic mosaic virus (RCNMV), and saguaro cactus virus (SgCV)) than to those of CYDVs. The RdRps of CYDV-RPS and -RPV and potato leaf roll virus (PLRV, type member of the genus *Polerovirus*) are more similar to those of members of the genus *Sobemovirus* (cocksfoot mottle virus (CoMV), rice yellow mottle virus (RYMV), and turnip rosette virus (TRoV)) than to those of the BYDVs. The resulting consensus tree from 1000 bootstrap replications is shown. The numbers above each node indicate the percentage of bootstrap replicates in which that node was recovered.

For this reason, viruses for which RdRp sequences have not been determined (BYDVs GPV, RMV, and SGV) have not been assigned to a genus. These observations suggest that the genomic RNAs of BYDVs and CYDVs resulted from recombination between RNAs expressing a common set of structural and movement proteins and RNAs expressing two different sets of replication proteins. Because of these differences, it has been suggested that BYDVs should be placed in the family *Tombusviridae* and CYDVs in the genus *Sobemovirus*.

Virion Properties and Composition

All BYDVs and CYDVs have nonenveloped icosahedral particles with diameters of 25–28 nm (**Figure 2**). Capsids are composed of major (22 kDa) and minor (65–72 kDa) coat proteins (CPs), which is formed by a carboxy-terminal extension to the major CP called the readthrough domain (RTD). According to X-ray diffraction and molecular mass analysis, virions consist of 180 protein subunits, arranged in $T = 3$ icosahedra. Virus particles do not contain lipids or carbohydrates, and have sedimentation coefficients $s_{20,w}$ (in Svedberg units) that range from 115–118S. Buoyant densities in CsCl are approximately $1.4\,\mathrm{g\,cm^{-3}}$. Virions are moderately stable, insensitive to freezing, and are insensitive to treatment with chloroform or nonionic detergents, but are disrupted by prolonged treatment with high concentrations of salts.

The single encapsidated genomic RNA molecule is single-stranded, positive-sense, and lacks a 3'-terminal poly(A) tract. A small protein (VPg) is covalently linked to the 5'-terminus of CYDV RNAs. CYDV-RPV also encapsidates a 322-nucleotide satellite RNA that accumulates to high levels in the presence of the helper virus. Complete genome sequences have been determined for BYDV-MAV, -PAS, and -PAV and CYDV-RPS and -RPV (**Table 1**). For several viruses, notably BYDV-PAV, genome sequences have been determined from multiple isolates.

Genome Organization and Expression

Genomic RNAs of BYDVs and CYDVs for which complete nucleotide sequences are available contain five to six ORFs (**Figure 3**). ORFs 1, 2, 3, and 5 are shared among all BYDVs and CYDVs. BYDVs lack ORF0. Genomic sequences of

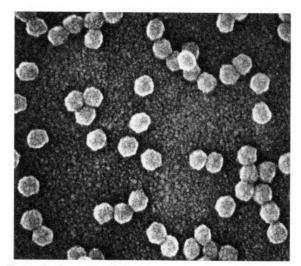

Figure 2 Scanning electron micrograph of barley yellow dwarf virus-PAV particles, magnified 200 000×. Virions are *c*. 25 nm in diameter, hexagonal in appearance, and have no envelope.

some BYDVs contain one or two small ORFs, ORFs 6 and 7, downstream of ORF5. In CYDVs, ORFs 0 and 1 and ORFs 1 and 2 overlap by more than 600 nucleotides. In BYDVs, ORF1 overlaps ORF2 by less than 50 nucleotides. In BYDV and CYDV genome sequences, ORF4 is contained within ORF3. An amber (UAG) termination codon separates ORFs 3 and 5.

BYDVs and CYDVs have relatively short 5' and intergenic noncoding regions. ORFs 2 and 3 are separated by about 200 nucleotides in BYDVs and CYDVs. The lengths of noncoding sequences downstream of ORF5 are very different between BYDVs and CYDVs. BYDV-PAV contains over 860 nucleotides downstream of ORF5 compared to just 170 nucleotides for CYDV-RPV.

The expression of BYDV-PAV RNA has been studied in detail and has revealed a complex set of RNA–RNA and RNA–protein interactions that are employed to express and replicate the virus genome. Less experimental data are available for CYDVs. However, expression and replication strategies and gene functions can be inferred from those of closely related poleroviruses, particularly beet western yellows virus (BWYV) and potato leaf roll virus (PLRV). ORFs 0, 1, and 2 are expressed directly from genomic RNAs. Downstream ORFs are expressed from subgenomic RNAs (sgRNAs) that are transcribed from internal initiation sites by virus-encoded RdRps from negative strand RNAs and are 3'-coterminal with the genomic RNA. Since the initiation codon for ORF0 of CYDVs is upstream of that of ORF1, translation of ORF1 is initiated by 'leaky scanning' in which ribosomes bypass the AUG initiation codon of ORF0 and continue to scan the genomic RNA until they reach the initiation codon of ORF1. The protein products of ORF2 are expressed only as a translational fusion with the product of ORF1. At a low frequency during the expression of ORF1, translation continues into ORF2 through a −1 frameshift that produces a large protein containing sequences encoded by both ORFs 1 and 2 in a single polypeptide. In BYDV-PAV, frameshifting between ORFs 1 and 2 is dependent upon the interaction of RNA sequences close to the site of frameshifting and a

Table 1   Viruses causing barley yellow dwarf in cereals

Genus	Virus (alternative name)	Abbreviation
Luteovirus	Barley yellow dwarf virus-MAV	BYDV-MAV
	Barley yellow dwarf virus-PAS	BYDV-PAS
	Barley yellow dwarf virus-PAV	BYDV-PAV
	(Barley yellow dwarf virus-RGV) (rice giallume)	
Polerovirus	Cereal yellow dwarf virus-RPS	CYDV-RPS
	Cereal yellow dwarf virus-RPV	CYDV-RPV
Unassigned	Barley yellow dwarf virus-GAV	BYDV-GAV
	Barley yellow dwarf virus-GPV	BYDV-GPV
	Barley yellow dwarf virus-RMV	BYDV-RMV
	Barley yellow dwarf virus-SGV	BYDV-SGV

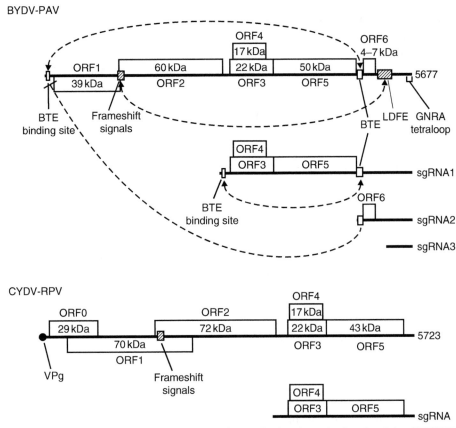

Figure 3 Genome organizations of barley yellow dwarf virus-PAV (BYDV-PAV) and cereal yellow dwarf virus-RPV (CYDV-RPV). Individual open reading frames (ORFs) are shown as staggered open boxes. The predicted sizes of the protein products are indicated. The genome-linked protein (VPg) attached to the 5′ terminus of CYDV RNA is indicated by a solid circle. Based on homology to other viruses ORF0 encodes a silencing suppressor and ORFs 1 and 2 encode replication-related proteins. ORFs 3 and 5 encode the major coat protein and readthrough domain, respectively. ORF4 encodes a protein required for virus cell-to-cell movement. The BYDV translation enhancer (BTE) facilitates translation initiation of BYDV-PAV genomic RNA and subgenomic RNA1 (sgRNA1). In both BYDV-PAV and CYDV-RPV, ORF2 is expressed as a translational fusion with the product of ORF1 via a −1 frameshift. In BYDV-PAV, frameshifting requires interaction between the 5′ frameshift signals and the long-distance frameshift element (LDFE). Dashed lines indicate long-distance RNA–RNA interactions.

long-distance frameshift element (LDFE) located 4000 nucleotides downstream in the 3′ noncoding region of genomic RNAs. Mutations that disrupt the interactions between these two distal regions suppress frameshifting and abolish RNA replication.

ORFs 3, 4, and 5 are expressed from sgRNA1, the 5′ terminus of which is located about 200 nucleotides upstream of ORF3, and extends to the 3′ terminus of the genome. BYDVs produce a second sgRNA that contains ORF6. BYDV-PAV also produces a third sgRNA, which does not appear to encode a protein. ORF3 is translated from the 5′ terminus of sgRNA1. ORF4 of BYDVs and CYDVs, which encodes a 17 kDa protein, is contained within ORF3 and is expressed from the same sgRNA as ORF3 through a leaky scanning mechanism much like that used to express ORF1 of CYDVs. In BYDVs and CYDVs, ORF5 is expressed only as a translational fusion with the products of ORF3 by readthrough of the UAG termination codon at the end of ORF3. This produces a protein with the

product of ORF3 at its amino terminus and the product of ORF5 at its carboxyl terminus.

While genomic RNAs of CYDVs contain 5′ VPgs that interact with translation initiation factors, BYDV-PAV RNA contains only a 5′ phosphate. Unmodified 5′ termini usually are recognized poorly for translation initiation. To circumvent this problem, a short sequence located in the noncoding region just downstream of ORF5 in the BYDV-PAV genome, called the BYDV translation enhancer (BTE), interacts with sequences near the 5′ termini of the genomic and subgenomic RNAs to promote efficient cap-independent translation initiation.

Functions BYDV and CYDV proteins have been ascribed based on homology to virus proteins with known functions and mutational characterization of protein coding regions. Similarity to proteins encoded by BWYV and PLRV suggests that the 28–29 kDa proteins encoded by ORF0 of CYDVs are inhibitors of post-transcriptional gene silencing (PTGS). PTGS is an innate and highly adaptive antiviral defense found in all eukaryotes that is activated by double-stranded RNAs (dsRNAs), which are produced during virus replication. The ORF1-encoded proteins of CYDVs contain the VPg and a chymotrypsin-like serine protease that is responsible for the proteolytic processing of ORF1-encoded polyproteins. The protease cleaves the ORF1 protein in *trans* to liberate the VPg, which is covalently attached to genomic RNAs. ORF2s of BYDVs and CYDVs, which are expressed as translational fusions with the product of ORF1, have coding capacities of 59–72 kDa and predicted amino acid sequences that are very similar to known RdRps and hence likely represent the catalytic portion of the viral replicase.

ORF3 encodes the major 22 kDa CP for both BYDVs and CYDVs. ORF5 has a coding capacity of 43–50 kDa, which is expressed only as a translational fusion with the product of ORF3 when translation reads through the termination codon at the end of ORF3 and continues through to the end of ORF5. The ORF5 portion of this readthrough protein has been implicated in aphid transmission and virus stability. Recombinant viruses that do not express ORF5 produce virions assembled from the major CP alone, which are not transmitted by aphid vectors and are less efficient in systemic infection of host plants than wild-type viruses. The amino-terminal portions of ORF5 proteins are highly conserved among BYDVs and CYDVs while the carboxyl termini are much more variable.

ORF4 of both BYDVs and CYDVs is contained within ORF3 and encodes a 17 kDa protein. Viruses that contain mutations in ORF4 are able to replicate in isolated plant protoplasts, but are deficient or delayed in systemic movement in whole plants. Hence, proteins encoded by ORF4 are thought to facilitate intra- and intercellular virus movement.

Some BYDV genomic sequences contain small ORFs (ORF6) downstream of ORF5. The predicted sizes of the proteins expressed by ORF6 range from 4 to 7 kDa. The predicted amino acid sequences of the proteins encoded by ORF6 are poorly conserved among BYDV-PAV isolates. Repeated attempts to detect protein products of ORF6 have been unsuccessful. In addition, BYDV-PAV genomes into which mutations have been introduced that disrupt ORF6 translation are still able to replicate in protoplasts. Based on these observations, it has been concluded that ORF6 is not translated *in vivo*.

Host Range and Transmission

BYD-causing viruses infect over 150 species of annual and perennial grasses in five of the six subfamilies of the Poaceae. The feeding habits of vector aphids have a major impact on the host ranges of virus species. Hence the number of species naturally infected by the viruses is much lower than the experimental host range.

As techniques for infecting plants with recombinant viruses have improved, the experimental host ranges of BYDVs and CYDVs have been expanded to include plants on which aphid vectors would not normally feed. For example, BYDV-PAV and CYDV-RPV have been shown to infect *Nicotiana* species when inoculated using *Agrobacterium tumefaciens* harboring binary plasmids containing infectious copies of the viruses, which had not been described previously as experimental hosts for the viruses.

Viruses that cause BYD are transmitted in a circulative strain-specific manner by at least 25 aphid species. Circulative transmission of the viruses is initiated when the piercing–sucking mouthparts of aphids acquire viruses from sieve tubes of infected plants during feeding. Aphids that do not probe into and feed from the vascular tissues of infected plants do not transmit the viruses. The virions of BYDVs and CYDVs travel up the stylet, through the food canal, and into the foregut (**Figure 4**). After 12–16 h, virions then are actively transported across the cells of the hindgut into the hemocoel in a process that involves receptor-mediated endocytosis of the viruses and the formation of tubular vesicles that transport viruses

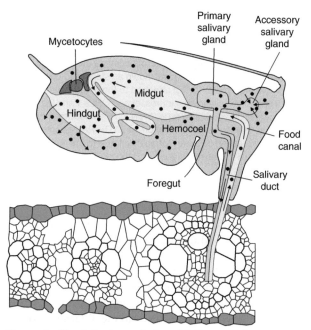

Figure 4 Circulative transmission of BYDVs and CYDVs by vector aphids. While feeding from sieve tubes of an infected plant, an aphid (shown in cross section) acquires virions, which travel up the stylet, through the food canal, and into the foregut. Virions are actively transported across cells of the hindgut into the hemocoel. Virions then passively diffuse through the hemolymph to the accessory salivary gland where they are again actively transported into the lumen of the gland. Once in the salivary gland lumen, the virions are expelled with the saliva into the vascular tissue of host plants. Viruses that are not transmitted by a particular species of aphid often accumulate in the hemocoel, but do not traverse the membranes of the accessory salivary gland.

through epithelial cells and into the hemocoel. Virions then passively diffuse through the hemolymph to the accessory salivary gland where virions must pass through the membranes of accessory salivary gland cells in a similar type of receptor-mediated transport process to reach the lumen of the gland. The accessory salivary gland produces a watery saliva, containing few or no enzymes, that is thought to prevent phloem proteins from clogging the food canal. Once in the salivary gland lumen, virions are expelled with the watery saliva into vascular tissues of host plants. Typically hindgut membranes are much less selective than those of the accessory salivary glands. Consequently, viruses that are not transmitted by a particular species of aphid often are transported across gut membranes and accumulate in the hemocoel, but do not traverse the membranes of the accessory salivary gland. The specificity of aphid transmission and gut tropism has been linked to the RTD of the minor capsid protein. Even though large amounts of virions can accumulate in the hemocoel, there is no evidence for virus replication in their aphid vectors. Aphids may retain the ability to transmit virus for several weeks.

Genetic and biochemical studies have been conducted to identify aphid determinants of strain-specific transmission of BYDV-MAV and BYDV-PAV. Protein–protein and protein–virus interaction experiments were used to isolate two proteins from heads of vector aphids that bind BYDV-MAV that were not detected in nonvector aphids. These two proteins are good candidates for the cell-surface receptors that are thought to be involved in strain-specific transport of viruses into accessory salivary gland lumens. In addition, endosymbiotic bacteria that reproduce in specialized cells called mycetocytes in abdomens of aphids express chaperonin-like proteins that bind BYDV particles and the amino-terminal region of recombinant BYDV-PAV RTD proteins. However, the role of these proteins in aphid transmission is unclear since they are found in both vector and nonvector aphid species. Interactions of virus particles with these proteins seem to be essential for persistence of the viruses in aphids. The proteins may protect virus particles from degradation by aphid immune systems.

Replication

Like other viruses of the family *Luteoviridae*, BYDVs and CYDVs infect and replicate in sieve elements and companion cells of the phloem and occasionally are found in phloem parenchyma cells (**Figure 5**). BYDVs and CYDVs induce characteristic ultrastructural changes in infected cells. BYDV-MAV, -PAV, and -SGV induce single-membrane-bound vesicles in the cytoplasm near plasmodesmata early in infection. Subsequently, filaments are observed in nuclei, and virus particles are first observed in the cytoplasm. In contrast, BYDV-RMV and CYDV-RPV induce double-membrane-bound vesicles in the cytoplasm that are continuous with the endoplasmic reticulum. Later, filaments and tubules form in the cytoplasm, and BYDV-RMV and CYDV-RPV particles are first observed in nuclei.

The subcellular location of viral RNA replication has not been determined unequivocally. However, early in infection, negative-strand RNAs of BYDV-PAV are first detected in nuclei and later in the cytoplasm, which suggests that at least a portion of the BYDV-PAV replication occurs in the nucleus. A nuclear location for replication is supported by the observation that the movement protein encoded by ORF4, which also binds single-stranded RNA, localizes to the nuclear envelope and is associated with virus RNA in nuclei of infected cells. Synthesis of negative-strand RNA, which requires tetra-loop structures at the $3'$ end of BYDV-PAV genomic RNAs, is detected in infected cells before the formation of virus particles. Because tetraloops have been implicated in RNA–protein interactions, these structures could be binding and/or recognition sites for BYDV replication proteins. BYDV-PAV sgRNAs are synthesized by internal initiation of RNA synthesis on negative-strand RNAs from three dissimilar subgenomic promoters. Late in infection, the BTE near the $5'$ terminus of BYDV-PAV sgRNA2 inhibits translation from genomic RNA, which may promote a switch from translation to replication and packaging of genomic RNAs. In addition to genomic RNAs, CYDV-RPV replicates a satellite RNA by a rolling-circle mechanism that generates multimeric satellite RNAs that self-cleave to unit length.

Virus–Host Relationships

Visible symptoms induced by BYDVs and CYDVs vary greatly depending on the host and strain of the virus. The most common symptoms are stunting and chlorosis. While some infected plants display no obvious symptoms, most BYDVs and CYDVs induce characteristic symptoms that include stunting, leaves that become thickened, curled or serrated, and yellow, orange or red leaf discoloration, particularly of older leaves of infected plants. These symptoms result from phloem necrosis that spreads from inoculated sieve elements and causes symptoms by inhibiting translocation, slowing plant growth, and inducing the loss of chlorophyll. Symptoms may persist, may vary seasonally, or may disappear soon after infection. Temperature and light intensity often affect symptom severity and development. In addition, symptoms can vary greatly with different virus isolates or strains and with different host cultivars. Yield losses caused by BYD are difficult to estimate because the viruses are so pervasive and symptoms often are overlooked or attributed to

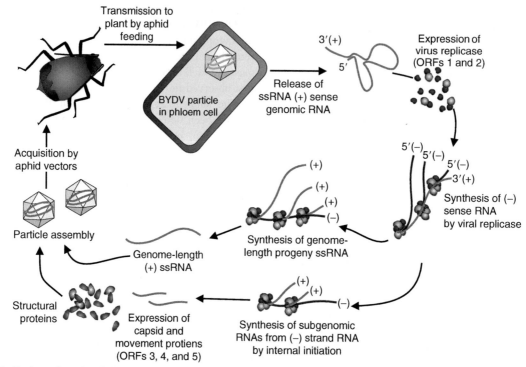

Figure 5 Barley yellow dwarf virus-PAV life cycle. Virus particles are deposited in sieve elements by aphid vectors. By a yet unknown process, single-stranded messenger-sense genomic RNA is released from virus particles and translated by host translation machinery, which is facilitated by long-distance RNA–RNA interactions. Open reading frames (ORFs) 1 and 2, which encode the viral replicase, are expressed first because of their proximity to the 5′ termini of genomic RNAs. Virus encoded replicase then synthesizes negative-sense RNAs that are used as templates for the production of new full-length positive-sense genomic RNAs and subgenomic RNAs. Production of subgenomic RNAs results in synthesis of structural and cell-to-cell movement proteins. Subgenomic RNA2 suppresses translation from genomic RNAs, furthering the switch from early to late gene expression. Full-length positive-sense genomic RNAs and structural proteins then assemble into virions in cells of phloem tissues where they can be ingested by aphid vectors to start the process again.

other agents. In Australia alone, losses in barley production have been valued at over 100 million US dollars annually. Plants infected with BYD at early developmental stages suffer the most significant yield losses, which often are linearly correlated with the incidence of virus infection.

Epidemiology

BYD infections have been reported from temperate, subtropical, and tropical regions of the world. Even though the incidence of infections of individual viruses varies from year to year and can differ among annual and perennial hosts, BYDV-PAV usually is the most prevalent of the viruses causing BYD in small grains worldwide followed by CYDV-RPV or BYDV-MAV. The remaining BYD-causing viruses are typically much less prevalent. BYDVs and CYDVs must be reintroduced into annual crops each year by their aphid vectors. Alate, that is, winged aphids may transmit viruses from local cultivated, volunteer, or weed hosts. Alternatively, alate aphids may

be transported into crops from distant locations by wind currents. These vectors may bring the virus with them, or they may first have to acquire virus from locally infected hosts. In temperate regions of Europe and North America moderate and long-distance migration of viruliferous aphids is important to development of BYD epidemics. In Australasia, and other regions with Mediterranean climates, alate aphids usually transmit viruses from relatively close infected plants. Secondary spread of the viruses is often primarily by apterous, that is, wingless aphids. The relative importance of primary introduction of viruses by alate aphids and of secondary spread of viruses by apterous aphids in disease severity varies with the virus, aphid species, crop, and environmental conditions.

Diagnosis

Accurate diagnosis of infections has been important in understanding the transmission and epidemiology of BYDVs and CYDVs and developing control strategies for BYD. Because BYD symptoms resemble those caused by other biotic and abiotic factors, visual diagnosis is unreliable and other methods have been developed. Initially, infectivity, or biological, assays were used to diagnose infections. In bioassays, aphids are allowed to feed on infected plants and then are transferred to indicator plants. These techniques have also been used to determine vector specificities of viruses causing BYD and to identify viruliferous vector aphids in epidemiological studies. These techniques are very sensitive, but they can require several weeks for symptoms to develop on indicator plants. The viruses causing BYD are strongly immunogenic, which has facilitated development of genus- and even strain-specific antibodies that have been used extensively in BYD diagnosis. Because the viruses causing BYD are present in infected tissues at very low levels, mice have been used to produce monoclonal antibodies against the viruses. Mice typically require much less viral antigen per immunization than rabbits, and hybridoma cell lines that produce monoclonal antibodies can be stored for extended periods and used for many years, which further reduces the amount of antigen needed to produce diagnostic antibodies. Techniques have also been developed to detect viral RNAs from infected plant tissues by reverse transcription polymerase chain reaction, which can be more sensitive and discriminatory than serological diagnostic techniques. Even so, serological tests are the most commonly used techniques for the detection of infections because of their simplicity, speed, and relatively low cost.

Control

Planting of insecticide-treated seeds that protect emerging seedlings from aphid infestation has been shown to reduce losses caused by BYD in North America, Australasia and Africa. Foliar applications of insecticides on older plants typically have been less effective. Alternatively, planting of tolerant or resistant cereals has proved to be a much more cost-effective and sustainable management strategy for BYD. Breeding programs have successfully integrated genes conferring high levels of tolerance into barley and oat and to a lesser extent in wheat. Even though a limited number of single genes for BYD resistance/tolerance have been identified in cultivated barley and rice, in most instances, tolerance to BYD is conditioned by multiple genes in a quantitative fashion, which has made moving BYD tolerance into new plant lines challenging. Particularly in barley, molecular markers have begun to facilitate the process of breeding for BYD tolerance. Because of a lack of effective single-gene resistance in cultivated wheat, some researchers have moved BYD resistance genes from wheat grasses (*Thinopyrum intermedium* and *Thinopyrum ponticum*) into wheat, which have provided high levels of resistance. The lack of naturally occurring resistance in cereals to BYD has made transgene-mediated resistance

very attractive. Even though the expression of CP sequences in transgenic plants has conferred resistance in several other plant-virus systems, it has not provided significant resistance to BYD in barley, oat, or wheat. In contrast, transgenic barley and oat plants have been produced that express either intact or inverted copies of BYDV-PAV replicase genes, which conferred high levels of resistance to BYDV-PAV and closely related viruses.

In many small grain growing regions, viruliferous aphids arrive at similar times each spring and fall even though sizes of the aphid populations can vary significantly from year to year. In these areas, it is sometimes possible to plant crops so that young, highly susceptible plants are not in the field when the seasonal aphid migrations occur. However, crops planted later typically do not yield as well as those planted early in the growing season. Consequently, growers must weigh the probability of obtaining higher yields against possible yield losses caused by BYD. In some instances, biological control agents such as predatory insects and parasites have reduced aphid populations significantly.

See also: Cereal Viruses: Wheat and Barley; Luteoviruses.

Further Reading

Crasta OR, Francki MG, Bucholtz DB, *et al.* (2000) Identification and characterization of wheat-wheatgrass translocation lines and localization of barley yellow dwarf virus resistance. *Genome* 43: 698–706.

D'Arcy CJ and Burnett PA (1994) *Barley Yellow Dwarf Virus, Forty Years of Progress.* St. Paul, MN: American Phytopathological Society Press.

Falk BW, Tian T, and Yeh HH (1999) Luteovirus-associated viruses and subviral RNAs. *Current Topics in Microbiology and Immunology* 239: 159–175.

Figueira AR, Domier LL, and Darcy CJ (1997) Comparison of techniques for detection of barley yellow dwarf virus PAV-IL. *Plant Disease* 81: 1236–1240.

Gray S and Gildow FE (2003) Luteovirus–aphid interactions. *Annual Review of Phytopathoogy* 41: 539–566.

Koev G, Mohan BR, Dinesh-Kumar SP, *et al.* (1998) Extreme reduction of disease in oats transformed with the 5' half of the barley yellow dwarf virus PAV genome. *Phytopathology* 88: 1013–1019.

Miller WA, Liu SJ, and Beckett R (2002) Barley yellow dwarf virus: *Luteoviridae* or *Tombusviridae*? *Molecular Plant Pathology* 3: 177–183.

Miller WA and Rasochova L (1997) Barley yellow dwarf viruses. *Annual Review of Phytopathology* 35: 167–190.

Miller WA and White KA (2006) Long-distance RNA–RNA interactions in plant virus gene expression and replication. *Annual Review of Phytopathology* 44: 447–467.

Nass PH, Domier LL, Jakstys BP, and D'Arcy CJ (1998) *In situ* localization of barley yellow dwarf virus PAV 17-kDa protein and nucleic acids in oats. *Phytopathology* 88: 1031–1039.

Ordon F, Friedt W, Scheurer K, *et al.* (2004) Molecular markers in breeding for virus resistance in barley. *Journal of Applied Genetics* 45: 145–159.

Zhu S, Kolb FL, and Kaeppler HF (2003) Molecular mapping of genomic regions underlying barley yellow dwarf tolerance in cultivated oat (*Avena sativa* L.). *Theoretical and Applied Genetics* 106: 1300–1306.

Beet Curly Top Virus

J Stanley, John Innes Centre, Colney, UK

Glossary

Enation Virus-induced swelling of the plant tissue.

Endoreduplication DNA replication in the absence of cell division.

Etiology The cause of a disease.

Hemolymph The insect circulatory system.

Hyperplasia Unregulated cell division.

Hypertrophy Cell enlargement.

Nonpropagative transmission The virus does not replicate in the insect.

RFLP Restriction fragment length polymorphism, an analytical tool to distinguish DNA viruses.

Transovarial transmission Transmission of virus through the insect egg.

History

Curly top disease of sugar beet was first reported in the USA in 1888. With the establishment and growth of the sugar beet industry, it was soon realized that the disease was widespread throughout the western states where it also affected a variety of crops including tomato, bean, and potato. The disease spread eastward during the middle of the twentieth century and similar diseases were reported to occur in western Canada, northern Mexico, Brazil, Argentina, the Caribbean Basin, Turkey, and Iran, although in most cases the etiology still remains to be established. A relationship between the disease and the beet leafhopper *Circulifer tenellus* (Baker) (Homoptera: Cicadellidae) was established in 1909. Although the leafhopper transmission characteristics suggested viral etiology, the low level of accumulation of the pathogen in plants hampered its isolation, and it was not until 1974 that a virus was eventually isolated from tobacco and shown to adopt a twinned particle morphology characteristic of a geminivirus infection. The viral genomic component was eventually cloned from viral DNA isolated from symptomatic sugar beet and characterized by sequence analysis in 1986. Infectivity studies using the cloned genomic component and transmission of the clone progeny by *C. tenellus* confirmed that the disease was caused by the geminivirus beet curly top virus (BCTV).

Taxonomic Classification and Phylogenetic Relationships

Members of the family *Geminiviridae* are divided into the genera *Begomovirus*, *Curtovirus*, *Topocuvirus*, and *Mastrevirus* on the basis of genome organization, host range, and insect vector characteristics. Originally collectively referred to as strains of *Beet curly top virus*, the genus *Curtovirus* currently includes BCTV as the type species and *Beet mild curly top virus*, *Beet severe curly top virus*, *Horseradish curly top virus*, and *Spinach curly top virus* (**Figure 1**). The names given to beet mild curly top virus (BMCTV) and beet severe curly top virus (BSCTV) reflect the symptom severity of these viruses in sugar beet although their phenotypes vary in other hosts. Species demarcation within the genus is based primarily on an 89% nucleotide sequence identity threshold for the entire genomic DNA while other biological factors such as host range, pathogenicity, and replication compatibility help to distinguish species and variants. The organization of BCTV complementary-sense genes resembles that of whitefly-transmitted begomoviruses while the virion-sense coat protein gene is more closely related to that of leafhopper-transmitted mastreviruses, suggesting that a curtovirus ancestor may have evolved by recombination between members of distinct genera. Phylogenetic analysis of horseradish curly top virus (HrCTV) suggests that a recombination event involving a region within the complementary-sense genes has also occurred between an ancestral curtovirus and begomovirus. Phylogenetic analysis of

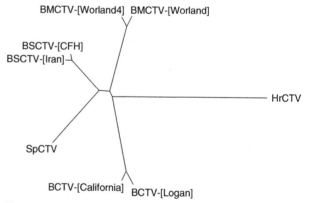

Figure 1 Phylogenetic relationships between members of the genus *Curtovirus*. The unrooted tree is based on the alignment of the complete nucleotide sequences of isolates of beet curly top virus (BCTV), beet mild curly top virus (BMCTV), beet severe curly top virus (BSCTV), horseradish curly top virus (HrCTV), and spinach curly top virus (SpCTV).

spinach curly top virus (SpCTV) indicates that its complementary-sense genes most closely resemble those of BMCTV and BSCTV while its virion-sense genes are more closely related to those of HrCTV, suggesting that it may have originated by recombination between members of the genus. However, because recombination frequently occurs between geminiviruses and undoubtedly plays a major role in their evolution, it is extremely difficult to establish an exact lineage for a particular virus and whether it represents the parent or progeny of a recombination event.

Geographic and Seasonal Distribution

Curly top disease occurs widely throughout arid and semiarid regions of western USA that are favored by the leafhopper vector, and it is here that the epidemiology of the disease has been most closely monitored. Early investigations of host range, symptom induction, and virulence suggested that curly top disease was caused by a complex of viral pathogens, a notion borne out by the identification of distinct species and variants with overlapping geographic distributions. Adult leafhoppers overwinter on perennial weeds from which they acquire the virus. Viruliferous adults migrate to cultivated areas during spring where they undergo several generations while feeding on weeds and crops to which they transmit the disease. In autumn, the adult leafhoppers migrate back to their overwintering grounds. Using restriction fragment length polymorphism (RFLP) analysis, a field survey of sugar beet growing in Texas during 1994 showed BSCTV to be the predominant species. Both BMCTV and BSCTV were detected in a more comprehensive survey of sugar beet from California, Colorado, Idaho, New Mexico, Oregon, Washington, Wyoming, and Texas in 1995. Plants frequently contained genotypic variants of these species and occasionally maintained mixed infections of both species. Although BCTV was detected in samples maintained in laboratories and nurseries for analytical purposes, it was not recovered from field samples. This might reflect limitations of the sampling and screening procedures although a change in the population structure in which BCTV is no longer prevalent in the field seems likely. This is supported by a more recent survey of beet, pepper, and tomato as well as native weed species growing in California between 2002 and 2004, which again indicated the presence of BMCTV and BSCTV and the absence of BCTV. HrCTV was isolated in 1990 from horseradish originating from Illinois that exhibited brittle root disease, and SpCTV was isolated in 1996 from spinach growing in southwest Texas. The lack of RFLP patterns diagnostic of these two species in other surveys suggests they may have rather restricted geographic distributions.

A variant of BSCTV is also associated with curly top disease in the Mediterranean Basin, implying a common origin for the New and Old World viruses. Interbreeding experiments indicated that the leafhopper populations in these two parts of the world are closely related, although leafhopper genetic diversity in the Old World is much greater than in the USA. Coupled with the fact that sugar beet is native to Europe, it has been suggested that disease originated in the Old World and was introduced into the USA from the Mediterranean Basin as a result of movement of infected plants and accompanying viruliferous insects.

Genome Organization and Gene Expression

In common with all geminiviruses, BCTV has a genome of circular single-stranded DNA (ssDNA) that is encapsidated in twinned quasi-isometric particles approximately 20×30 nm in size. The BCTV genome comprises a single component of $c.$ 3000 nt that encodes seven genes distributed between the virion-sense (V) and complementary-sense (C) strands (also referred to as rightward (R) and leftward (L) strands in the literature), separated by an intergenic region (IR) that contains the origin of replication. A nonanucleotide motif (TAATATTAC) that is highly conserved in geminiviruses is located within the IR, between inverted repeat sequences with the potential to form a stem–loop structure. Short repeat sequences, termed iterons, occur upstream of the stem–loop (**Figure 2**).

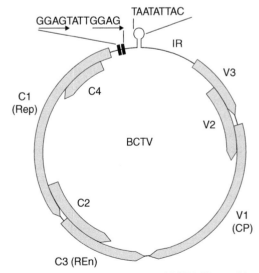

Figure 2 The genome organization of BCTV. The position and orientation of virion-sense (V) and complementary-sense (C) ORFs are shown in relation to the intergenic region (IR) that contains the invariant nonanucleotide motif TAATATTAC and iterons (repeat sequences indicated by arrows) involved in the initiation of viral DNA replication. CP, coat protein; Rep, replication-associated protein; REn, replication-enhancer protein.

The complementary-sense ORFs encode the replication-associated protein (Rep) required for the initiation of viral DNA replication (ORF C1), a replication-enhancer protein (REn) (ORF C3), and two proteins that contribute to viral pathogenicity (ORFs C2 and C4), all of which may be considered to be required during the early stages of infection. The virion-sense ORFs encode proteins that are required later in infection, namely the coat protein (ORF V1) and proteins involved in the regulation of the relative levels of viral ssDNA and double-stranded DNA (dsDNA) (ORF V2) and virus movement (ORF V3).

The identification of BCTV virion- and complementary-sense transcripts is consistent with a bidirectional transcription strategy. The two most abundant complementary-sense transcripts are similar in size to their begomovirus counterparts, the larger transcript mapping across all four ORFs and the smaller transcript across ORFs C2 and C3. However, the most abundant transcripts map across the virion-sense ORFs, downstream of two consensus eukaryotic promoter sequences, and are suitably positioned for the expression of ORFs V2 and V3, while a less abundant transcript maps across ORF V1. Precisely how the viral proteins are expressed from these transcripts is not yet understood although the overlapping nature of both the virion-sense and complementary-sense ORFs may provide a means for their temporal control during the infection cycle.

Replication

Fractionation of viral DNA forms by a combination of chromatography and two-dimensional gel electrophoresis has identified BCTV intermediates consistent with both rolling-circle and recombination-dependent replication strategies. At the onset of infection, the viral ssDNA is uncoated and converted to a circular dsDNA intermediate by host factors. During rolling-circle replication, Rep interacts with iterons located in the IR and introduces a nick into the virion-sense strand of the dsDNA within the nonanucleotide motif (TAATATT$^{\downarrow}$AC). By analogy with the replication strategy of begomoviruses, Rep then forms a covalent bond with the 5' terminus of the nicked virion-sense strand. However, as Rep does not have polymerase activity, the 3' terminus must be extended by a host polymerase. The full-length nascent strand is nicked and relegated by Rep to produce circular ssDNA that either reenters the replication cycle or is encapsidated. BCTV ORF C3 mutants produce severe symptoms and accumulate to wild-type levels in *Nicotiana benthamiana*, although they induce only a mild infection in sugar beet associated with significantly reduced levels of viral DNA accumulation, consistent with the proposed role for REn in viral DNA replication. Interestingly, HrCTV does not contain an intact ORF C3.

The interaction between Rep and the origin of replication, defined by the composition of the N-terminal region of Rep and the iteron sequence, is highly specific and determines whether distinct curtovirus species and variants are compatible for *trans*-replication. For example, BCTV Rep can only functionally interact with its own iteron (GGAGTATTGGAG; **Figure 2**) and not those of BMCTV and BSCTV (GGTGCTATGGGAG and GGTGCTTTGGGTG, respectively). Conversely, BMCTV and BSCTV Reps will not functionally interact with the BCTV iteron. However, the BMCTV and BSCTV iterons are sufficiently similar to allow mutual Rep recognition and, hence, *trans*-replication compatibility. Comparison of the Rep N-terminal sequences has identified amino acid residues conserved between BMCTV and BSCTV but differing in BCTV that may participate in iteron recognition, but this awaits experimental confirmation. While replication incompatibility between *cis*- and *trans*-acting elements serves to maintain the integrity of a particular species, this constraint may be overcome by recombination whereby the functional module comprising the 5' terminus of the Rep ORF and the origin of replication is exchanged between incompatible viruses, as has been suggested to have occurred during the evolution of SpCTV.

Although replication intermediates consistent with a rolling-circle strategy for BCTV replication have been observed, the production of other viral DNA forms can be explained readily by a recombination-dependent replication mechanism. Recombination-dependent replication is generally considered to occur later in the infection cycle than rolling-circle replication whereby a ssDNA byproduct recombines with homologous sequences within a circular or multimeric dsDNA template to initiate replication, a process mediated by a recombination protein and often triggered by a double-stranded break. It remains to be established which, if any, viral proteins contribute to this process, although it is possible that Rep helicase activity could participate in both replication strategies. The nick site for rolling-circle replication has been shown to be a recombinational hot spot. However, recombination-dependent replication may explain the propensity of geminiviruses to undergo recombination throughout their genomes.

Subgenomic-sized viral DNAs of diverse size (ranging from 800 to 1800 nt) and complexity are rapidly produced from the cloned BCTV genomic component in sugar beet, tomato, and *N. benthamiana*. Deletions occur within all ORFs and only the IR is retained, consistent with its participation in viral DNA replication. Field-affected sugar beet frequently contains such defective DNAs, indicating that they are also produced under natural conditions. They occur as both single- and double-stranded DNA forms but it is not known if they are encapsidated or have the ability to move systemically in the plant. However, in view of their rapid appearance, it is likely that they are produced *de novo* in every infected cell. Short repeat sequences of 2–6 nt at the deletion boundaries may specify the deletion endpoints and suggest that the defective DNAs are produced by recombination or as a consequence of errors during replication. Many geminiviruses produce subgenomic-sized DNAs, several of which, including those associated with BCTV, have been shown to interfere with infection. Hence, they may have a biological role in modulating pathogenicity of the helper virus, thus conferring a selective advantage by slowing down infection of the plant to encourage leafhopper feeding and virus transmission.

Virus Movement and Insect Transmission

The coat protein is a multifunctional protein that is targeted to the nucleus where virions accumulate, and is essential for systemic infection and leafhopper transmission. BMCTV coat protein mutants that retain the ability to form virions (mainly N-terminal mutations) are generally able to produce a systemic infection while those unable to form virions (mainly C-terminal mutations) cannot, despite

being competent for replication. This serves to reinforce the proposal that long-distance movement of the virus throughout the plant occurs in the form of virions, consistent with the observation that high levels of virus-like particles accumulate in the phloem. It further suggests that C-terminal amino acids play an important role in determining the structure of the coat protein during virion assembly. ORF V2 mutants accumulate high levels of dsDNA and greatly reduced levels of ssDNA compared to the wild-type virus, implicating V2 protein in the regulation of the relative levels of these viral DNA forms. The function of V2 protein may be to ensure that ssDNA is available in sufficient amounts for encapsidation and systemic movement in the latter stages of the infection cycle. ORF V3 mutants remain competent for replication but produce only sporadic and asymptomatic infections in sugar beet and *N. benthamiana* associated with low levels of viral DNA accumulation, suggesting a role for V3 protein in virus movement.

All curtoviruses are transmitted in nature by the leafhopper *C. tenellus* (**Figure 3**) in a persistent circulative manner, and there is no evidence to suggest that there are any significant differences in transmission characteristics between curtovirus species. *Circulifer opacipennis* is

Figure 3   (a) Adult *Circulifer tenellus* (Baker) leafhopper. Symptoms of BCTV infection in (b) sugar beet and (c) *Nicotiana benthamiana*. Plants show upward leaf roll and vein-swelling symptoms typical of curly top disease.

reported to be an additional vector of the disease in the Mediterranean Basin, but this leafhopper does not occur in the USA. The leafhopper feeds primarily in the phloem from where it rapidly acquires the virus. Using a PCR-based approach, BMCTV has been detected in the digestive tract of the insect after an acquisition access period (AAP) of 1 h, in the hemolymph after 3 h, and the salivary glands after 4 h, from where it is reintroduced into plants through saliva during feeding. This is consistent with earlier reports estimating the minimum time between virus acquisition and the leafhopper becoming infective (the latent period) to be 4 h. The minimum feeding time necessary for transmission of the virus can be as short as 1 min. Depending on the length of the AAP, leafhoppers can remain infective for most of their lifetime, although the amount of virus they retain and their ability to transmit the disease decline with time when maintained on plants that are not hosts for the virus, implying that transmission is nonpropagative. There is no evidence for transovarial transmission of the virus.

Several lines of evidence demonstrate that coat protein composition defines leafhopper transmission specificity. First, BMCTV coat protein mutants that are unable to produce virions are not transmitted by *C. tenellus*. Second, recombinant virus in which the coat protein coding sequence of the whitefly-transmitted begomovirus African cassava mosaic virus has been replaced with that of BCTV produces virions that are transmissible by the leafhopper. Finally, the autonomously replicating nanovirus-like DNA-1 component, normally associated with the whitefly-transmitted begomovirus ageratum yellow vein virus, can be maintained in systemically infected sugar beet by BCTV and is encapsidated in BCTV coat protein, conferring on DNA-1 the ability to be transmitted by *C. tenellus*. This implies that surface features of the virion interact in a highly specific manner during circulative transmission, possibly by receptor-mediated endocytosis during virus acquisition across the gut wall and movement into the salivary glands.

BCTV and other members of the genus are generally poorly transmitted by mechanical inoculation, reflecting their phloem limitation. Infection can be achieved using a fine-gauge needle (pin-pricking) to introduce virus or viral DNA inoculum directly into the phloem, although infection rates are dramatically improved using either biolistic delivery under pressure or *Agrobacterium tumefaciens*-mediated inoculation (agroinoculation) of cloned DNA. Agroinoculation exploits the ability of *A. tumefaciens* to introduce binary vector T-DNA, containing partially or tandemly repeated cloned genomic DNA, into the plant cell nucleus. To initiate an infection, circular viral DNA is resolved from the T-DNA, either by recombination between the repeat sequences or by a more efficient replicative mechanism if two copies of the origin of replication are present.

Host Range and Pathogenesis

In contrast to most other geminiviruses which tend to have a limited host range, BCTV is reported to have an extremely wide host range confined to dicotyledonous plants, including over 300 species in 44 plant families, particularly members of the Chenopodiaceae, Compositae, Cruciferae, Leguminosae, and Solanaceae. Although the wide host range is not in doubt, these largely early observations were conducted using diseased plants of undefined etiology and, hence, the exact host range of BCTV remains to be confirmed. BMCTV, BSCTV, and SpCTV also have wide host ranges, although HrCTV has an atypically narrow host range. Curly top disease was originally described in association with agricultural crops showing severe symptoms, and virus tends to accumulate to higher levels in sugar beet and tomato compared to weeds. However, hosts include a wide variety of annual and perennial weed species which act as a reservoir for the disease and in which the virus frequently produces a mild or asymptomatic infection.

The products of BCTV complementary-sense ORFs C2 and C4 have been implicated in pathogenesis. ORF C2 mutants induce severe symptoms and accumulate to wild--type virus levels in sugar beet and *N. benthamiana*. However, plants infected with the mutants have a greater propensity for recovery from severe symptoms than those infected with wild-type virus, suggesting that C2 protein suppresses a plant stress or defense mechanism. Consistent with this, transgenic *N. benthamiana* and *N. tabacum* plants constitutively expressing C2 protein exhibit enhanced susceptibility to BCTV infection, manifested by a decrease in the inoculum concentration required to elicit an infection and a reduction in latent period before the onset of infection, although they do not develop more severe symptoms or accumulate higher levels of viral DNA than nontransformed plants. The transgenic plants are also more susceptible to the begomovirus tomato golden mosaic virus (TGMV) and the unrelated RNA virus tobacco mosaic virus, indicating that C2 protein impacts on a nonspecific host response. C2 protein has been shown to suppress post-transcriptional gene silencing (PTGS) in *N. benthamiana*, and in this respect resembles its positional counterpart (AC2) in the begomovirus TGMV. However, the two proteins share only limited homology and BCTV C2 protein does not appear to have transcriptional activator activity for virion-sense gene expression shown by its TGMV counterpart. Despite this, both BCTV C2 and TGMV AC2 proteins bind to, and inactivate, adenosine kinase (ADK), an enzyme required for $5'$-adenosine monophosphate (AMP) synthesis. ADK activity is also reduced in BCTV-infected plants and transgenic plants expressing C2 protein. Regulation of adenosine levels by ADK plays a key role in the control of intermediates required for methylation, providing a possible link between C2 protein activity and suppression of transcriptional gene silencing (TGS) of the viral genome. Both BCTV C2 and TGMV AC2 proteins also bind to, and inhibit, SNF1-related nucleoside kinase. Reduction of SNF1 expression in *N. benthamiana* results in an enhanced susceptibility phenotype resembling that associated with the expression of C2 and AC2 proteins, while SNF1 overexpression produces plants exhibiting enhanced resistance. This activity is not directly linked to silencing suppression, suggesting that C2 protein inhibition of SNF-1 activity affects a distinct plant defense pathway.

BCTV typically induces upward curling of the leaves associated with vein clearing and the development of enations on the lower surface of veins (**Figure 3**). *In situ* localization studies have shown that BCTV is tightly phloem-limited in sugar beet and *N. benthamiana*. Although the virus cannot access apical meristematic tissues, it is possible that it exploits undifferentiated cambium cells in the vascular bundles. Infected tissues exhibit hyperplasia and hypertrophy within the phloem and adjacent parenchyma, and the dividing cells differentiate into sieve-like elements. In older infected tissues, the affected phloem cells eventually become necrotic and collapse. Pathogenic effects occur in developing tissues only after mature sieve elements have developed, suggesting that the virus moves from source to sink cells with the flow of metabolites. Infected sugar beet leaves accumulate enhanced levels of sucrose, attributed to impaired transport resulting from disruption of the phloem. This is associated with a reduction in chlorophyll content, reduced activity of key photosynthetic enzymes and concomitant reduction in the rate of photosynthesis, and altered turgor pressure that causes an increase in mesophyll cell size.

Geminiviruses control the plant cell cycle to produce an environment suitable for their replication. This is achieved by the interaction of Rep with the plant homolog of retinoblastoma-related tumor-suppressor protein (pRBR), which relieves the constraint imposed by pRBR binding to E2F transcription factor and allows the expression of E2F-responsive genes. In most cases, infected cells enter S phase and may undergo endoreduplication but do not divide. In contrast, BCTV infection induces cell division within vascular tissues, attributed to the action of C4 protein. Sugar beet infected with ORF C4 mutants remain asymptomatic and *N. benthamiana* shows an altered phenotype, and neither host develops hyperplasia of the phloem and enations. Furthermore, ectopic expression of C4 protein in transgenic *N. benthamiana* results in abnormal plant development and tumorigenic growths. Consistent with this phenotype, C4 protein interacts with BIN2, a negative regulator of transcription factors in the brassinosteroid signaling pathway that controls cell division and tissue development. The reason for the induction of cell division remains obscure, particularly as the hyperplastic tissues often do not contain detectable levels of virus.

The host range of BCTV includes *Arabidopsis thaliana* which represents an important resource for the study of virus–host interactions. Most susceptible ecotypes become stunted and develop enations on affected tissues although the Sei-O ecotype is hypersusceptible to BCTSV and develops callus-like structures containing high levels of the virus, suggesting virus-induced hormonal imbalance. Symptom development correlates with enhanced expression of the cell cycle gene *cdc2* and small auxin-upregulated RNA gene (*saur*). Hence, disruption of the phloem in infected tissues may affect auxin transport, causing localized increases that result in cell proliferation.

Disease Control

As young plants are most susceptible to infection, a significant reduction in disease incidence may be achieved by protecting seedlings using either a physical barrier to prevent leafhopper access or by timing the emergence of the crop to avoid predicted leafhopper spring migration from the overwintering grounds. Breeding resistant varieties of sugar beet has been very successful although the basis of resistance is not understood. Plants remain susceptible to infection but are sufficiently productive, particularly when used in combination with an integrated pest-management scheme. Breeding program have provided better uniformity of resistance with improved yield and sugar content, which has reinvigorated the sugar beet industry in the western USA. Resistant bean varieties are also available, although breeding for resistance to curly top disease in tomato has not been so successful. Systemic insecticides, for example, soil treatment with imidacloprid and dimethoate foliar sprays, have been used to control the leafhopper vector on crops and on weeds located in the breeding areas, both with some success, particularly when applied to young susceptible plants, although this approach is both costly and damaging to the environment. Parasites and predators of the leafhopper are numerous, but it has proved difficult to assess their impact on leafhopper populations under natural conditions.

The subgenomic-sized DNAs produced during BCTV infection are known to adversely affect virus proliferation, for which reason they are termed defective interfering (DI) DNAs. Transgenic *N. benthamiana* plants containing an integrated tandem copy of the DI DNA are less susceptible to BCTV infection. The DI DNA is mobilized from the transgene and amplified to high levels during infection. Competition between the helper virus and DI DNA results in a significant reduction in both viral DNA accumulation and symptom severity. DI DNA amplification is dependent on *trans*-replication by the helper virus, governed by the specific interaction between Rep and the origin of replication. However, as BMCTV and BSCTV are the predominant species in the western USA and are mutually competent for *trans*-replication, genetic modification of sugar beet with the appropriate DI DNA transgene may provide a viable alternative for the production of resistant plants.

See also: Plant Resistance to Viruses: Geminiviruses; Plant Virus Diseases: Fruit Trees and Grapevine; Plant Virus Diseases: Ornamental Plants; Vector Transmission of Plant Viruses.

Further Reading

Bennet CW (1971) *American Phytopathology Society Monograph No. 7: The Curly Top Disease of Sugar Beet and Other Plants.* St. Paul, MN: American Phytopathology Society.

Esau K and Hoefert LL (1978) Hyperplastic phloem in sugarbeet leaves infected with the beet curly top virus. *American Journal of Botany* 65: 772–783.

Hanley-Bowdoin L, Settlage SB, Orozco BM, Nagar S, and Robertson D (1999) Geminiviruses: Models for plant DNA replication, transcription, and cell cycle regulation. *Critical Reviews in Plant Sciences* 18: 71–106.

Stanley J, Bisaro DM, Briddon RW, *et al.* (2005) Geminiviridae. In: Fauquet CM, Mayo MA, Maniloff J, Desselberger U,, and Ball LA (eds.) *Virus Taxonomy: Eighth Report of the International Committee on Taxonomy of Viruses,* pp. 301–326. San Diego, CA: Elsevier.

Stanley J, Markham PG, Callis RJ, and Pinner MS (1986) The nucleotide sequence of an infectious clone of the geminivirus beet curly top virus. *EMBO Journal* 5: 1761–1767.

Benyvirus

R Koenig, Institut für Pflanzenvirologie, Mikrobiologie und biologische Sicherheit, Brunswick, Germany

Glossary

Benyvirus Siglum derived from *Beet necrotic yellow vein virus.*
Rhizomania, root beardedness (root madness) Extensive proliferation of often necrotizing secondary rootlets at the expense of the main tap root.

History

Beet necrotic yellow vein virus is the type species of the genus *Benyvirus.* Originally it was classified as a possible member of the tobamovirus group, because its rod-shaped particles resemble those of tobamoviruses. In 1991, the fungus-transmitted rod-shaped viruses, all of which have multi-partite genomes, were separated from the monopartite tobamoviruses to form a new group, named furovirus group. Soil-borne wheat mosaic virus (SBWMV) became the type member of this new group; beet necrotic yellow vein virus (BNYVV) and rice stripe necrosis virus (RSNV) were listed as possible members. Molecular studies performed in the following years revealed that the genome organization of many of these furoviruses greatly differed from that of SBWMV. Eventually, four new genera were created, that is, the genus *Furovirus* with *Soil-borne wheat mosaic virus* as the type species, the genus *Benyvirus* with *Beet necrotic yellow vein virus* as the type species, the genus *Pomovirus,* and the genus *Pecluvirus.* The genus *Benyvirus* presently comprises two definitive species, viz. *Beet necrotic yellow vein virus* and *Beet soil-borne mosaic virus,* and two tentative species, viz. RNSV and burdock mottle virus (BdMV).

Host Ranges and Diseases

BNYVV is the causal agent of rhizomania, one of the most damaging diseases of sugar beet. Fodder beets, Swiss chard (*Beta vulgaris* var. *cicla*), red beets, and spinach may also become infected naturally. Infections of sugar beet are mainly confined to the root system. Susceptible sugar beet varieties show an extensive proliferation of nonfunctional, necrotizing secondary rootlets, a condition described by the names 'root beardedness', 'root madness', or rhizomania. The tap roots are stunted, their shape tends to be constricted, and their sugar content is low. A brownish discoloration of the vascular system is often seen. Leaves may become pale and have an upright position. Under dry conditions, wilting is often observed due to the disturbances in the root system. The upper parts of the plants are invaded only rarely by the virus. In infected leaves, the veins turn yellow, and occasionally become necrotic, a condition after which the virus has been named. In susceptible varieties, rhizomania may cause yield losses of 50% and more. Like other diseases caused by soil-borne organisms, rhizomania often occurs in patches, especially in recently infested fields. BNYVV can be mechanically transmitted to many species in the Chenopodiaceae and to some species in the Aizoacea (e.g., *Tetragonia expansa*), Amaranthaceae, and Caryophyllaceae. *Beta macrocarpa* becomes infected systemically. Recently obtained isolates also infect *Nicotiana benthamiana* systemically, causing slight mottling and growth reduction.

The symptoms caused by beet soil-borne mosaic virus (BSBMV) on its natural host *B. vulgaris* are more variable than those produced by BNYVV. Infected roots may remain symptomless or show rhizomania-like symptoms. Infections of the upper parts of the plants occur somewhat more frequently than with BNYVV. Leaves may develop vein banding or faint mottling. When plants are dually infected with BSBMV and BNYVV, foliar symptoms appear more frequently. In general, BSBMV causes much less damage to sugar beets than BNYVV. BSBMV can be transmitted mechanically to *Chenopodium quinoa*, *Chenopodium album*, and *Tetragonia tetragonioides*, all of which become infected locally, and to *Beta maritima*, which becomes infected systemically.

RSNV was first described in 1983 in West Africa (Ivory Coast) as the causal agent of a long-known disease named 'rice crinkling disease'. The percentage of infection was found to vary greatly according to the year and to the variety. In 1991, the virus was first noticed in Columbia where it causes a severe disease of rice named 'entorchamiento' that is characterized by seedling death, foliar striping, and severe plant malformation. The virus can be transmitted mechanically to *C. quinoa*, *C. album*, and *Chenopodium amaranticolor* where it produces local lesions, but not to rice. *Nicotiana benthamiana* does not become infected.

BdMV has been isolated from the leaves of naturally infected *Arctium lappa* L. Leaves as well as roots of this plant are common vegetables in Japan. The virus often occurs in mixed infections with other viruses and produces only mild symptoms which often become masked in older plants. BdMV can be transmitted mechanically to *C. quinoa*, *Chenopodium murale*, *Nicotiana clevelandii*, and *Nicotiana rustica*, which are infected systemically, and to *B. vulgaris* var. *cicla* and var. *rapa*, *Spinacia oleracea*, *Cucumis sativus*, and *Tetragonia expansa*, which are infected only locally, some of them only with difficulty.

Geographic Distribution and Epidemiology

BNYVV is now found worldwide in sugar beet-growing areas. Molecular analyses have revealed the existence of different genotypes (A type, B type, P type) that cannot be distinguished serologically, but their RNAs differ in *c.* 1–5% of their nucleotides. The A type is most common and occurs in Southern, Eastern, and parts of Northern Europe, in the USA, and – with a number of nucleotide exchanges – also in East Asia. The B type is prevalent in Germany, France, and other central European countries. B-type-like BNYVV is also found in East Asian countries. The P type, which usually has a fifth RNA species and causes especially severe symptoms, has only been found in limited areas of France (Pithiviers), of the UK, and in Kazakhstan. In the UK, a new RNA5-containing virus source (FF) has recently been described. Its RNA5

occupies an intermediate position between East Asian and European forms of RNA5. In the UK, where rhizomania was observed later than in most other European countries, a larger number of different BNYVV types have now been identified than in any other country in the world. Originally it has been assumed that rhizomania, which was first observed in Italy in 1952, has spread from that country to central and later to Northern Europe. It is unlikely, however, that all BNYVV types have originated in Italy where only the A type has so far been observed. It seems more likely that naturally infected local hosts in different geographic regions may harbor various types of BNYVV which may be transmitted to beets only with difficulty, perhaps due to vector/host compatibility problems or to their occurrence in different habitats. BNYVV is not commonly found in weeds in sugar beet fields. Once a transmission from a native host to sugar beet has been successful, further spread in beet-growing areas by machinery, irrigation, or infested soil may be very rapid.

BSBMV is widely distributed in the USA, but has so far not been found in other parts of the world. Single-strand conformation analyses of polymerase chain reaction (PCR) products indicated that this virus is genetically much more variable than BNYVV. This may explain the variability in the symptoms produced by this virus. RSNV occurs throughout West Africa and several countries in South America, that is, Colombia, Ecuador, Brazil, and Panama. It is assumed that it has been introduced to these countries from West Africa with seed contaminated at the surface with RSNV-harboring *Polymyxa graminis* resting spores. BdMV has been reported only from Japan.

Transmission

In nature, the assigned species of the genus *Benyvirus*, viz. *Beet necrotic yellow vein virus* and *Beet soil-borne mosaic virus*, are transmitted by the zoospores of *Polymyxa betae*, a soil-borne ubiquitous plasmodiophorid protozoan. *P. betae* and the related *P. graminis* that has always been found in natural RSNV infections had formerly been considered to belong to the fungi. *Polymyxa*-transmitted viruses (or their RNAs?) are taken up by the plasmodia of the vector in infected root cells, but there is no evidence that they multiply in the vector. The multinucleate plasmodia which are separated from the host cytoplasm by distinct cell walls may either develop into zoosporangia from which secondary viruliferous zoospores are released within a few days. Alternatively, the plasmodia may form cystosori which act as resting spores and may survive in the soil for many years even under extreme conditions. They may be distributed on agricultural equipment, by irrigation, or even by blowing of wind, and upon germination they release primary zoospores transmitting the virus. Zoospores inject their contents into the cytoplasm of root cells where new plasmodia are formed. With the furovirus SBWMV, immunolabeling and *in situ* RNA hybridization tests have recently revealed the presence of viral movement protein and RNA but not of viral coat protein in the resting spores of *P. graminis*. This might suggest that the vector does not transmit virions but rather a ribonucleoprotein complex possibly formed by the viral movement protein and the viral RNAs. It remains to be shown whether benyviruses are transmitted in a similar manner. The RNA-binding ability of P42 movement protein and further properties of the other two benyviral movement proteins as well as the additional involvement in the transmission process of two transmembrane regions found in the coat protein readthrough proteins of benyviruses and other *Polymyxa*-transmitted viruses are described in the section 'Organization of the genome and properties of the encoded proteins'.

RSNV-harboring cystosori may be carried in soil adhering to the surface of rice seeds, but true seed transmission has not been observed. The natural mode of transmission of BdMV is unknown. *P. betae* and the aphid species *Myzus persicae* and *Macrosiphum gobonis* failed to transmit a BdMV isolate obtained from leaves of *Arctium lappa*.

Under experimental conditions, benyviruses are readily transmitted mechanically from test plants to test plants. However, attempts to transmit BNYVV mechanically from sugar beet rootlets to test plants may not always be successful.

Control

Polymyxa-transmitted plant viruses may survive in soil in the long-living resting spores of the vector for many years, probably even decades. The diseases caused by these viruses are, therefore, more difficult to control than those caused by, for example, insect-transmitted viruses. Chemical control of these soil-borne diseases, for example, by soil treatment with methyl bromide, is neither efficient nor acceptable for economic and ecological reasons. Growing resistant or tolerant varieties currently represents the only practical and environmentally friendly means to lower the impact of theses diseases on yield. BNYVV-tolerant or partially resistant sugar beet varieties are now available which enable good yields also on infested fields. Their resistance is based mainly on the *Rz1* gene from a sugar beet line ('Holly' resistance) and the *Rz2* gene from the *Beta maritima* line WB42. In some locations, there are indications of a breakdown of the *Rz1*-mediated resistance. Genes which would confer immunity against BNYVV to sugar beet have not been found so far. Vector resistance has been detected in *Beta patellaris* and *Beta procumbens*, but attempts to develop agronomically acceptable sugar beet cultivars resistant to *Polymyxa* have so far failed. The degree of resistance or tolerance to RSNV differs in different rice cultivars. A high degree of resistance is found in *Oryza*

glaberrima. The genes responsible for this resistance have been transferred to cultivated rice in breeding programs.

Genetically modified sugar beets expressing various portions of the BNYVV genome have been shown to be highly resistant to BNYVV, but have not yet been utilized commercially. Genome recombinations have not been observed when A-type BNYVV coat protein gene-expressing beets were grown in soil infested with B-type BNYVV-carrying *P. betae*.

Particle Properties and Relations of Particles with Cells

Benyvirus virions are nonenveloped rods which have a helical symmetry (**Figure 1**). The diameter of the particles is *c.* 20 nm and they usually show several length maxima ranging from *c.* 80 to 400 nm depending on the RNA species encapsidated. Additional length maxima may be due to end-to-end aggregation or breakage of particles. The right-handed helix of BNYVV particles has a 2.6 nm pitch with an axial repeat of four turns involving 49 subunits of the *c.* 21 kDa major coat protein which consists of 188 amino acids. Each coat protein subunit occupies four nucleotides on the RNAs of BNYVV. The 75 kDa coat protein readthrough protein (**Figure 2**) has been detected by immunogold-labeling on one end of particles in freshly extracted plant sap. It is believed to act as a minor coat protein which initiates the encapsidation process.

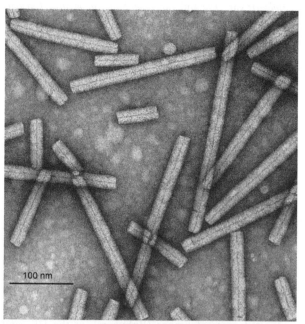

100 nm

Figure 1 Particles of *Beet necrotic yellow vein virus* in a purified preparation negatively stained with uranyl acetate. Courtesy D. E. Lesemann and J. Engelmann, BBA, Braunschweig, Germany.

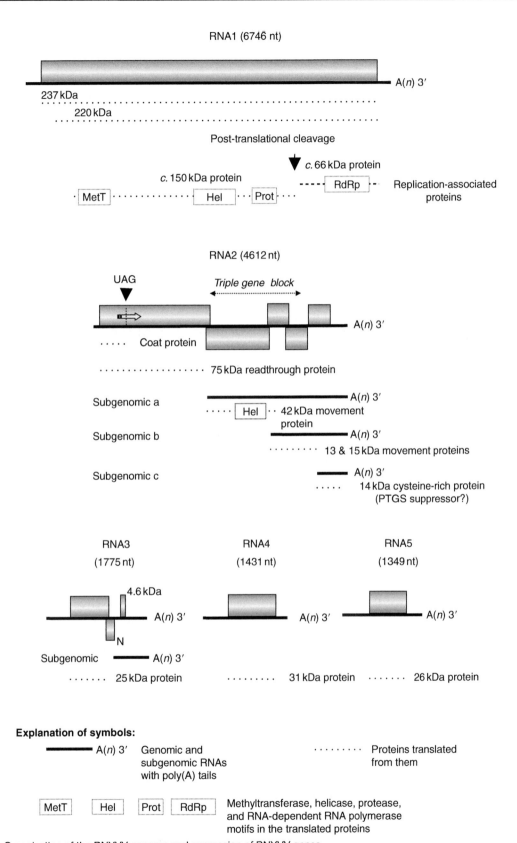

Figure 2 Organization of the BNYVV genome and expression of BNYVV genes.

Green fluorescent protein (GFP)-labeled particles of BNYVV were shown to localize to the cytoplasmic surface of mitochondria early during infection, but later they are relocated to semi-ordered clusters in the cytoplasm. In ultrathin sections, BNYVV particles are found scattered throughout the cytoplasm or occur in aggregates. More or less dense masses of particles arranged in parallel or angle-layer arrays may be formed. Membranous accumulations of endoplasmic reticulum may also be found.

Antigenic Properties

Benyviruses are moderately to strongly immunogenic. Polyclonal antibodies have been obtained for BNYVV, BSBMV, and RSNV. BNYVV and BSBMV are only very distantly related serologically – some antisera may fail to detect this relationship. Particles of RSNV and BdMV do not react with antisera to BNYVV. Monoclonal antibodies to BNYVV have been used for diagnostic purposes and for determining the accessibility of stretches of the coat protein amino acid chain on the virus particles. The C-terminal amino acids 182–188 which are readily cleaved off by treatment with trypsin are exposed along the entire surface of the particles whereas amino acids 42–51 and 156–121 are accessible only on one end of the particles; amino acids 125–140 are located inside and are exposed only after disrupting the particles. Antibody single-chain fragments (scFv's) have been expressed in *Escherichia coli* and *N. benthamiana*.

Nucleic Acid Properties and Interrelationships between Benyviruses

Benyvirus genomes consist of two to five molecules of linear positive-sense single-stranded RNAs (ssRNAs) which terminate in a 3′ poly(A) tail. In naturally infected sugar beets, the BNYVV genome consists of four and in some isolates five RNAs of *c.* 6.7, 4.6, 1.8, 1.4, and 1.3 kb (**Figure 2**). Complete or partial nucleotide sequences have been determined for several isolates. After repeated mechanical transfers to local lesion hosts, BNYVV isolates often contain only partially deleted forms of their RNAs 3, 4, and/or 5, or they lose these small RNAs altogether. RNA1 and RNA2 contain all the genetic information necessary to enable multiplication, encapsidation, and cell-to-cell movement on local lesion hosts (see below); the additional presence of RNA3 and RNA4, however, is essential for vector-mediated transmission and disease development in sugar beet roots. RNA5 which is found only in limited areas increases the virulence of the virus.

Four genomic RNA species have also been identified for BSBMV and RSNV, but only two for BdMV. The BdMV isolate might possibly have lost additional small RNAs after prolonged cultivation on *C. quinoa* as do isolates of BNYVV. The complete sequence has been published for all four RNAs of an isolate of BSBMV. The complete sequences of RNA1 (*c.* 7.0 kb) and RNA2 (*c.* 4.3 kb) of BdMV and a partial sequence of RSNV RNA1 (2239 nt) have also been determined but not yet released.

Available data indicate that the genome organization is similar for all four viruses. Highest nucleotide sequence identities are found in the RNA1-encoded replication-associated proteins (*c.* 84% for the pair BNYVV/BSBMV) and the RNA2-encoded second triple gene block (TGB) proteins (*c.* 82% for the pair BNYVV/BSBMV; *c.* 50% for the pair BNYVV/BdMV), but the lowest ones in the RNA2-encoded cysteine-rich proteins, which are presumably RNA-silencing suppressors (*c.* 38% for the pair BNYVV/BSBMV; *c.* 20% for the pair BNYVV/BdMV).

Organization of the Genome and Properties of the Encoded Proteins

The genome organization of BNYVV is outlined in **Figure 2**. BNYVV RNA1 contains one large open reading frame (ORF) for a replication-associated protein which is cleaved autocatalytically by a papain-like proteinase. In *in vitro* systems, its translation may start either at the first AUG at position 154 or at a downstream AUG at position 496. The resulting proteins of 237 and 220 kDa, respectively, contain in their N-terminal part methyltransferase motifs (MetT), in their central part helicase (Hel) and papain-like protease motifs (Prot), and in their C-terminal part RNA-dependent RNA polymerase (RdRp) motifs. BNYVV RNA2 contains six ORFs, viz. the coat protein gene which is terminated by a suppressible UAG stop codon, the coat protein readthrough protein gene, a TGB coding for proteins of 42, 13, and 15 kDa involved in viral movement, and a gene coding for a 14 kDa cysteine-rich protein, a putative RNA silencing suppressor. The N-terminal part of the 75 kDa coat protein readthrough protein is apparently necessary for initiating encapsidation (see section 'Particle properties and relations of particles with cells'), the C-terminal part for enabling transmission by *P. betae*. A KTER motif in positions 553–556 of the 75 kDa coat protein readthrough protein is essential for efficient transmission of the virus by *P. betae*. After prolonged cultivation on a local lesion host, the area containing this motif may be lost. Computer analyses have revealed the presence of two complementary transmembrane domains (TM1 and TM2) in the coat protein readthrough proteins not only of benyviruses, but also of the likewise *Polymyxa*-transmitted furoviruses and pomoviruses and in the P2 proteins of the *Polymyxa*-transmitted bymoviruses. The second domain is absent or disrupted in deletion mutants which are not vector transmitted. The TM helices are apparently tightly packaged with ridge/groove arrangements between the two helices and strong electrostatic associations. It has been suggested that they facilitate the movement of the virus across the membrane surrounding the plasmodia in plant cells. Benyvirus TM1 and TM2 helices are identical in length, each consisting of 23 amino acids. The E in the KTER motif is highly conserved in position −1 of TM 2.

Three subgenomic RNAs are derived from BNYVV RNA2. The first TGB protein (p42) is translated from subgenomic RNA2a, the second and third TGB proteins (P13 and P15) from the bicistronic subgenomic RNA2b, and the 14 kDa cysteine-rich protein from subgenomic RNA2c (**Figure 2**). The three TGB proteins can be functionally substituted in *trans* by the 30 kDa movement protein of tobacco mosaic virus or by the three TGB proteins of peanut clump virus when these are supplied together but not when they are substituted for their BNYVV counterparts one by one. This suggests that highly specific interactions among cognate TGB proteins are important for their function and/or stability *in planta*. The N-terminal part of the first (42 kDa) TGB protein has nucleic acid-binding activity; its C-terminal part contains helicase motifs. P42 labeled with GFP on its N-terminus is targeted by the second and third TGB proteins (P15 and P13) to punctuate bodies associated with plasmodesmata. It has been speculated that P15 and P13 provide a docking site for a P42–viral RNA complex at the plasmodesmata where P42 alters the plasmodesmatal size exclusion limit and potentiates transit of viral RNA.

The typical rhizomania symptoms in beet are produced only in the presence of BNYVV RNA3. The 25 kDa protein (P25) encoded by this RNA acts as a movement protein in beet roots and *Beta macrocarpa* and is also responsible for the production of bright yellow rather than pale green local lesions on *C. quinoa*. P25 has an amino acid tetrad in positions 67–70 which is highly variable in virus isolates originating from different geographical areas, especially in A-type BNYVV. A connection between the composition of this tetrad and the severity of symptoms in beets has been discussed, but not yet proven. P25 is found in the cytoplasm as well as in the nuclei of infected cells. A nuclear localization signal has been identified on its amino acids 57–62 and a nuclear export sequence within the region of amino acids 169–183. The four valine residues in positions 169, 172, 175, and 178 are all necessary for nucleocytoplasmic shuttling. Mutants that lack this shuttling ability fail to produce yellow lesions on *C. quinoa*. P25 has also a zinc-finger motif in positions 73–90. The small N gene (**Figure 2**) is not detectably expressed from full-length RNA3, but it is translationally activated by the deletion of upstream sequences which positions it closer to the 5′ end of RNA3. Expression of the N gene induces the formation of necrotic lesions in BNYVV infections and also when it is expressed by another plant virus (i.e., cauliflower mosaic virus). The function of the RNA3-encoded 4.6 kbp

ORF, which is presumably expressed from an abundantly present subgenomic RNA, is not known. RNA5 enhances symptom expression in sugar beet. Scab-like symptoms rather than root proliferation have been observed with a recombinant virus containing RNA5, but lacking RNA3. Addition of P-type RNA5 to RNA1 and -2 results in the formation of necrotic local lesions in *C. quinoa*. The RNA5-encoded P26 apparently occurs mainly in the nucleus as suggested by studies with fluorescent fusion proteins. BNYVV RNA4 greatly increases the transmission rate of the virus by *P. betae*. RNA3 and RNA4 or RNA5 and RNA4 may act in a synergistic way. Some authors have classified BNYVV RNA5 as a satellite RNA and RNAs 3 and 4 as satellite-like RNAs.

Diagnosis

Symptoms caused by benyviruses in the field can easily be confused with those produced by other causes, for example, nematodes, soil-borne fungi, or nitrogen deficiency. Enzyme-linked immunosorbent assay, immunoelectron microscopy, and PCR techniques, such as immunocapture reverse transcription-PCR, are useful for diagnosing benyvirus infections in plant parts likely to be infected (e.g., root beards of sugar beets). Infestation of soil may be detected by means of bait plants which are tested for the presence of virus by the above techniques. PCR also allows the detection of genome deletions and specific primers can be designed to differentiate the various BNYVV types (A type, B type, etc.), which are undistinguishable serologically. Even more detailed information is obtained by sequence analyses of PCR products. The sensitivity of BNYVV detection has been increased by using nested primers and real-time PCR.

Similarities and Dissimilarities with Other Taxa

The morphology of benyviruses resembles that of other rod-shaped viruses, that is, of furoviruses, pecluviruses, pomoviruses, hordeiviruses, tobraviruses, and tobamoviruses. The CPs of these viruses have a number of conserved residues, for example, RF and FE in their central and C-terminal parts, respectively, which are presumably involved in the formation of salt bridges. The benyvirus genomes – with the possible exception of the BdMV genome – consist of at least four RNA species, whereas the tobamoviruses have monopartite, the furoviruses, pecluviruses, and tobraviruses bipartite, and the pomoviruses and hordeiviruses tripartite genomes. The fact that the RNAs of benyviruses are polyadenylated differentiates them from the RNAs of the other viruses listed above. Benyviruses have a single large ORF on their RNA1: it codes for a polypetide which is cleaved post-translationally to yield two replication-associated proteins (**Figure 2**). This also differentiates the benyviruses from the other viruses listed above which have their replication-associated proteins encoded by two ORFs located either on two different RNA species (hordeiviruses) or on the same RNA where an ORF-1 is terminated by a leaky stop codon and extends into an ORF-2 (tobamoviruses, tobraviruses, furoviruses, pecluviruses). Benyviruses like pomo-, peclu-, and hordeiviruses, but unlike furo-, tobamo-, and tobraviruses, have their movement function encoded on a TGB. Sequence identities in the first and second TGB-encoded proteins reveal affinities not only to pomo- and hordeiviruses, but also to potex- and carlaviruses. The methyltransferase, helicase, and RdRp motifs in the putative replication-associated proteins of benyviruses show a higher degree of similarity to those of hepatitis virus E (genus *Hepevirus*) and rubella virus (family *Togaviridae*) than to those of other rod-shaped plant viruses.

See also: Cereal Viruses: Rice; Furovirus; Pomovirus; Vector Transmission of Plant Viruses.

Further Reading

Adams MJ, Antoniw JF, and Mullins JG (2001) Plant virus transmission by plasmodiophorid fungi is associated with distinctive transmembrane regions of virus-encoded proteins. *Archives of Virology* 146: 1139–1153.

Erhardt M, Vetter G, Gilmer D, et al. (2005) Subcellular localization of the triple gene block movement proteins of beet necrotic yellow vein virus by electron microscopy. *Virology* 340: 155–166.

Harju VA, Skelton A, Clover GR, et al. (2005) The use of real-time RT-PCR (TaqMan) and post-ELISA virus release for the detection of beet necrotic yellow vein virus types containing RNA 5 and its

comparison with conventional RT-PCR. *Journal of Virological Methods* 123: 73–80.

Koenig R and Lennefors BL (2000) Molecular analyses of European A, B and P type sources of beet necrotic yellow vein virus and detection of the rare P type in Kazakhstan. *Archives of Virology* 145: 1561–1570.

Koenig R and Lesemann D-E (2005) Benyvirus. In: Fauquet CM, Mayo M, Maniloff J, Desselberger U,, and Ball LA (eds.) *Virus Taxonomy: Eighth Report of the International Committee on Taxonomy of Viruses*, pp. 1043–1048. San Diego, CA: Elsevier Academic Press.

Lee L, Telford EB, Batten JS, Scholthof KB, and Rush CM (2001) Complete nucleotide sequence and genome organization of beet soil-borne mosaic virus, a proposed member of the genus *Benyvirus*. *Archives of Virology* 146: 2443–2453.

Link D, Schmidlin L, Schirmer A, et al. (2005) Functional characterization of the beet necrotic yellow vein virus RNA-5-encoded p26 protein: Evidence for structural pathogenicity determinants. *Journal of General Virology* 86: 2115–2125.

Meunier A, Schmit JF, Stas A, Kutluk N, and Bragard C (2003) Multiplex reverse transcription-PCR for simultaneous detection of beet necrotic yellow vein virus, beet soil-borne virus, and beet virus Q and their vector *Polymyxa betae* KESKIN on sugar beet. *Applied and Environmental Microbiology* 69: 2356–2360.

Ratti C, Clover GR, Autonell CR, Harju VA, and Henry CM (2005) A multiplex RT-PCR assay capable of distinguishing beet necrotic yellow vein virus types A and B. *Journal of Virological Methods* 124: 41–47.

Rush CM (2003) Ecology and epidemiology of benyviruses and plasmodiophorid vectors. *Annual Review of Phytopathology* 41: 567–592.

Rush CM, Liu HY, Lewellen RT, and Acosta-Leal R (2006) The continuing saga of rhizomania of sugar beets in the United States. *Plant Disease* 90: 4–15.

Tamada T (2002) Beet necrotic yellow vein virus. *AAB Descriptions of Plant Viruses* 391. http://www.dpvweb.net/dpv/showadpv.php?dpvno=391 (accessed July 2007).

Valentin C, Dunoyer P, Vetter G, et al. (2005) Molecular basis for mitochondrial localization of viral particles during beet necrotic yellow vein virus infection. *Journal of Virology* 79: 9991–10002.

Vetter G, Hily JM, Klein E, et al. (2004) Nucleo-cytoplasmic shuttling of the beet necrotic yellow vein virus RNA-3-encoded p25 protein. *Journal of General Virology* 85: 2459–2469.

Ward L, Koenig R, Budge G, et al. (2007) Occurrence of two different types of RNA 5-containing beet necrotic yellow vein virus in the UK. *Archives of Virology* 152: 59–73.

Bromoviruses

J J Bujarski, Northern Illinois University, DeKalb, IL, USA and Polish Academy of Sciences, Poznan, Poland

Introduction

The family *Bromoviridae* represents one of the most important families of plant viruses, infecting a wide range of herbaceous plants, shrubs, and trees. Several of them are responsible for major epidemics in crop plants. The *Bromoviridae* include the spherical icosahedral viruses with a tripartite positive-sense RNA genome. Since these viruses usually accumulate to a high level in the infected tissue, they have been a convenient subject of molecular

studies. The type members of different genera, such as cucumber mosaic virus (CMV), brome mosaic virus (BMV), and alfalfa mosaic virus (AMV), constitute excellent molecular models for basic research on viral gene expression, RNA replication, virion assembly, and the role of cellular genes in basic virology.

The genus *Bromovirus* (**Table 1**) comprises not only the best-characterized RNA viruses of the family such as brome mosaic virus (BMV), broad bean mottle virus (BBMV), or cowpea chlorotic mottle virus (CCMV), but

Table 1 Main characteristics of the RNA genome in members of the genus *Dromovirus*

Species	Particle size (nm)	RNA1[a]	RNA2	RNA3	3′end[b]	sgRNAs[c]/DI RNAs[d]
BMV	26	3234	2865	2117	tRNA	2/yes
BBMV	26	3158	2799	2293	tRNA	1/yes
CCMV	26	3171	2774	2173	tRNA	1/n.a.
CYBV	26	3178	2720	2091	n.a.	1/n.a.
MYFV	25	n.a.	n.a.	n.a.	n.a.	1/n.a.
SBLV	28	3252	2898	2213	n.a.	1/n.a.

[a]Sequences of RNAs 1 through 3 are available in GenBank.
[b]For these bromoviruses tRNA-like structures were analyzed.
[c]Shows the number of identified sgRNAs.
[d]n.a. – not analyzed.

also melandrium yellow fleck virus (MYFV), spring beauty latent virus (SBLV), and cassia yellow blotch virus (CYBV). BMV infects cereal grains and numerous members of the Graminae, and is commonly distributed throughout the world. BBMV and CCMV infect legume species, while other bromoviruses infect selected members of several plant families. Interestingly, BMV, MYFV, and SBLV were found to infect *Arabidopsis thaliana*, a model plant host.

Bromoviruses are mostly related to cucumoviruses, the agriculturally important genus of *Bromoviridae*. Both bromo- and cucumoviruses share such properties like the molecular and genetic features of their tripartite RNA genome, the number of encoded proteins, and similar virion structure. The computer-assisted comparisons of amino acid sequences revealed similarity among their RNA replication proteins. More broadly, the replication factors share amino acid sequence similarity within the alphavirus-like replicative superfamily of positive-strand RNA viruses, which includes numerous plant and important animal/human viruses.

Phylogeny and Taxonomy of *Bromoviridae*

The current taxonomy divides the family *Bromoviridae* into five genera, named after their best-known members: *Alfamovirus* (one member, type virus: *Alfalfa mosaic virus*, AMV), *Bromovirus* (five members, type virus: *Brome mosaic virus*), *Cucumovirus* (three members, type member: *Cucumber mosaic virus*), *Ilarvirus* (16 viruses, type virus: *Tobacco streak virus*), and *Oleavirus* (one member: *Olive latent virus 2*, OLV-2). However, the evaluation of the phylogenetic relationships based on integrating information on the entire proteome of viruses of the *Bromoviridae* revealed that (1) AMV should be considered a true ilarvirus instead of forming a distinct genus *Alfamovirus*; (2) one virus, pelargonium zonate spot virus (PZSV), should probably constitute a new genus; (3) the genus *Ilarvirus* should be divided into three subgroups; and (4) the exact location of OLV-2 within the *Bromoviridae* remains unresolved.

Virion Properties and Structure

The bromovirus virions are nonenveloped, possess a $T = 3$ icosahedral symmetry, measure about 28 nm in diameter, and are composed of 180 molecules of a 20 kDa coat protein (CP) clustered into 12 pentamers and 20 hexamers. Three types of icosahedral particles of identical diameters (28 nm) and similar sedimentation coefficients (c. 85 S) encapsidate different RNA components: RNA1 (mol. wt. c. 1.1×10^6), RNA2 (mol. wt. c. 1.0×10^6), and RNA3 plus sgRNA4 (mol. wt. c. 0.75×10^6 and 0.3×10^6). The crystal structures of both BMV and CCMV have been resolved showing very similar organization (**Figure 1**). The structure of CCMV serves as a model bromovirus structure. Similar to other icosahedral viruses, its CP subunits, folded into a β-barrel core, are oriented vertically to the capsid surface and organized within both the protruding pentameric and hexameric capsomers. The interactions among hydrophobic amino acid residues stabilize the capsomers, and the hexameric subunits are further stabilized via interactions between N-terminal portions, where six short β-strands form a hexameric tubule called β-hexamer. Mutational analysis demonstrated that this structure was not required for virion formation but modulated virus spread *in planta*. In addition, the capsomers are held together by interactions through C-terminal portions that extend radially from the capsid. The C-termini are anchored between the β-barrel core and the N-proximal loop, and this interaction might be responsible for initiation of assembly of CCMV capsids.

The molecular replacement using the CCMV structure as a model revealed that distinct portions of the BMV capsids can also form the pentameric and the hexameric capsomers. The CP has the canonical β-barrel topology with extended N-terminal polypeptides where a significant fraction of the N-terminal peptides is cleaved. Overall, the virion appears to assemble loosely among the hexameric capsomers. This is likely responsible for virion swelling at neutral pH. The structure also coordinates metal (Mg^{2+}) ions. Interestingly, $T = 1$ imperfect

20 Å

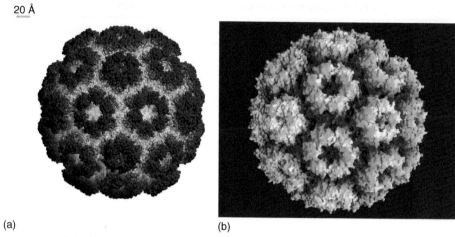

(a)   (b)

Figure 1 Surface structure of the (a) BMV and (b) CCMV capsids. The hexameric and the pentameric structural elements are visible. See also text for further description (a, b). From the virion picture collection at the web-site of the Institute for Molecular Virology at the University of Wisconsin–Madison). Reproduced by permision of Institute for Molecular Virology.

icosahedral particles of BMV virions can be created by treatment of $T = 3$ virus, with particles composed of a loose arrangement of reoriented pentameric capsomers.

The single-stranded viral RNA is located inside the capsid, as a separate torus-shaped subshell, where the basic N-terminal amino acids of the CP interact with the RNA and neutralize the phosphate groups. Other sites of RNA interaction localize to the internally proximal basic amino acids of the CP subunits. CCMV capsid undergoes well-studied reversible structural transitions where shifting pH from 5.0 to 7.0 causes capsid expansion. The swollen forms can be completely disassembled at 1 M salt concentration. Some CP mutations can further stabilize the capsids by a new series of bonds that are resistant to high ionic strength. Other mutations in the CP subunits show that capsid geometry is flexible and may adapt to new requirements as the virus evolves. Capsids of bromoviruses are also stabilized by metals. In CCMV, there are 180 unique metal-binding sites that coordinate five amino acids from two adjacent CP subunits. Some metals have less affinity for the binding sites than the others. The biological significance of capsid swelling is not completely understood.

The detailed knowledge about the structure of bromoviral virions has found applications in nanotechnology. The pH-dependent structural transitions of CCMV capsids provide opportunities for the reversible pH-dependent gating, useful during the regulation of size-constrained biomimetic mineralization. The interior surface of CCMV capsids can be genetically engineered to act as a ferritin surrogate that spatially constrains the formation of iron oxide nanoparticles, whereas the exterior surfaces have been functionalized for ligand presentation. The latter can be tuned by asymmetric reassembly of differentially functionalized CCMV CP subunits. The electrostatically driven adsorption behavior of CCMV on Si and amine-functionalized Si as well as the fabrication of multilayer CCMV films have been reported.

Genome Organization and Expression

High-resolution density gradients as well as gel electrophoresis under denaturing conditions reveal three classes of BMV virion particles, each encapsidating separate plus-sense components of the viral genome. One class encapsidates one copy of RNA1, another class one copy of RNA2, and the third class one copy of RNA3 and one copy of subgenomic RNA4 (**Figure 2**). All three virion classes must infect a cell to initiate viral infection. All BMV RNAs are 5′-capped with RNA1 and RNA2 coding for viral replicase proteins 1a and 2a, respectively, while RNA3 encodes for two proteins, the 5′ movement protein 3a and the 3′ CP. The CP is translated from the 3′ subgenomic RNA4. Recent studies reveal that 3a may be translated from another 5′ subgenomic RNA3a (sgRNA3a). *In vitro* data demonstrate that sgRNA3a is a more efficient RNA template for translation of 3a than full-length RNA3, suggesting a separation of translation (sgRNA3a and RNA4) from replication (whole RNA3) functions.

The purified bromovirus genomic RNAs are directly infectious. By using recombinant DNA technology, the *in vitro* transcripts from complete cDNA clones of RNAs 1, 2, and 3 can be synthesized, and such combined RNAs are also infectious to the bromovirus hosts. In addition, transient expression of DNA constructs carrying bromovirus cDNAs adjacent to the transcription promoter can directly induce bromovirus infection.

RNA Replication

Replication of bromovirus RNAs completely relies on RNA templates without known DNA intermediates. Both positive and negative RNA components accumulate in the infected tissue but the plus strands reach a 100-fold excess

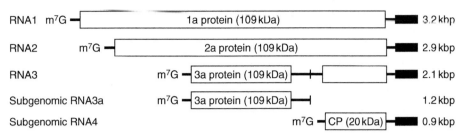

Figure 2 Molecular organization of a typical bromovirus RNA genome. The RNA components 1 and 2 encode two replicase polypeptides (1a and 2a) while RNA3 encodes the movement protein (3a) and the coat protein (CP). The open reading frames are boxed and labeled. The 3′ terminal sequences, which are common among all four RNA components, are marked as black solid boxes on the right. The oligo(A) tract is shown as a small vertical rectangle within the intercistronic region of RNA3 or at the 3′ end of sgRNA3a. The existence of sgRNA3a has been observed, so far, only for BMV.

over the minus strands. In addition to whole plants, bromovirus RNAs can replicate in single cells (protoplasts) isolated from various plants (including nonhosts) or even in yeast cells. The complete BMV RNA replication cycle occurs in yeast, demonstrating the presence of compatible host factors.

Only viral proteins 1a and 2a, but not proteins 3a or CP, are required from bromovirus RNA replication cycle, as demonstrated for BMV. These proteins form the active RNA replicase complex that localize to the endoplasmic reticulum membranes called spherules. Computer-assisted comparisons reveal two large domains in protein 1a and a central domain in protein 2a that are conserved with viral proteins of other members of the alphavirus-like superfamily (**Figure 2**). Protein 2a is actual RdRp enzyme whereas 1a has both helicase and methyltransferase activities, and both function cooperatively inside the spherular structures.

Bromoviral RdRp enzyme preparations can be extracted from virus-infected plants or from yeast cells. These activities are more or less virus RNA-specific and in case of BMV, the enzyme can copy full-length bromoviral RNAs and can generate *in vitro* both sgRNAs (sgRNA4 and sgRNA3a) on (−)-RNA3 templates. The well-studied promoter of (−)-strand synthesis comprises the last 200 nt of the 3′ noncoding region in all genomic BMV RNAs. In addition, it participates in several tRNA-specific activities, including adenylation or aminoacylation of the 3′ CAA terminus. The promoters of (+)-strand synthesis have also been mapped to the 5′ proximal noncoding region in BMV RNAs.

The sgp promoter in (−)-strand of BMV RNA3 of CCMV RNA3 includes a core region that binds the RdRp complex, an upstream polyU tract, and an A/U-rich enhancer, both increasing the level of RNA4 synthesis (transcription). Apparently the sgp is responsible not only for the initiation of sgRNA4 but also for premature termination of sgRNA3a (as demonstrated recently for BMV).

The RNA encapsidation signals have been mapped for BMV RNAs. The 3′ tRNA-like structure has been shown to function as a nucleating element of CP subunits. In addition, a *cis*-acting, position-dependent 187 nt region present in the 3a open reading frame (ORF) is essential for efficient packaging of RNA3. The co-packaging of sgRNA4 is contingent upon both RNA replication and translation of CP.

Homologous and Nonhomologous Recombination of Bromoviral RNAs

Genetic RNA recombination in plant RNA viruses has been first demonstrated by noting the crossovers between BMV RNAs during infection. Both homologous and nonhomologous crossovers have been reported to occur during BMV replication cycle and the restoration of CCMV infectious virus from genomic RNA3 fragments has been reported.

Most importantly, homologous crossing-over has been demonstrated between the same BMV RNA components, and a high-frequency hot spot was observed within the sgp region in BMV RNA3. This type of crossing-over is analogous to the well-studied meiotic DNA crossing-over in DNA-based organisms.

Bromoviruses are capable of generating the defective-interfering (DI) RNAs. In particular, strains of BBMV accumulate RNA2-derived deletion variants that tend to exacerbate the severity of symptoms. The *de novo* generation of DI-like RNAs was demonstrated during serial passages of BBMV in broad bean, and the importance of sequence features for BBMV DI RNA accumulation has been demonstrated. In BMV, both replicating and nonreplicating truncated RNA2-derived artificial DI RNAs have been shown to interfere with accumulation of BMV RNAs.

Host Factors Involved in Replication of Bromoviral RNAs

Numerous results, including the role of tRNA-specific enzymes, the composition of BMV RdRp preparations, or the dependence of bromovirus infectivity on the host plant, all suggest possible functions of cellular host factors during replication of bromovirus RNAs. More recently, the use of powerful yeast genetics has demonstrated a direct involvement of several host genes in BMV replication, including those related to RNA, protein, or membrane modification pathways. Also, some of the bromoviruses, including BMV or SBLV, have been shown to infect *Arabidopsis thaliana*, and preliminary data reveal the role of host genes in BMV replication in *Arabidopsis*.

Transmission and Virus–Host Relationship

Mechanical inoculation has been used to efficiently transmit bromoviruses from plant to plant. Also in the field, mechanical transmission by human activity can spread bromoviruses among plant hosts. In general, seed transmission has not been reported in case of bromoviruses, although there is one report of seed transmission of BBMV. Beetles can reproducibly but inefficiently transmit bromoviruses. Aphids are generally negative in transmission whereas inefficient transmission of BMV by nematodes has been reported.

Bromoviruses have restricted host range for systemic spread, with BMV generally infecting monocotyledonous plants whereas BBMV and CCMV infect dicotyledonous plants. BMV requires the CP for both the cell-to-cell and long-distance movement in barley. The interactions between BMV CP and certain host genes (e.g., an oxidoreductase) have been reported during viral infection in barley. The use of interstrain pseudorecombinants

has demonstrated the involvement of RNA3 genetic information in virus spread and in symptom formation for BMV and CCMV, and during the breakage of CCMV resistance in soybean. Certain changes in RNAs 1 and 2 also confirmed their roles as determinants of systemic spread and symptom formation in BMV and CCMV. Similarly, the genome exchange experiments suggest that the adaptations of bromoviruses to their hosts rely on movement protein and on both the replicase proteins. Also, various strains of BBMV cause different symptom intensity in their host plants. However, there is no direct correlation between bromovirus yield and symptom severity, but rather the symptoms seem to be associated with changes in the CP gene or in the subgenomic promoter.

See also: Alfalfa Mosaic Virus; Cucumber Mosaic Virus; Ilarvirus.

Further Reading

Fauquet CM, Mayo MA, Maniloff J, Desselberger U,, and Ball LA (eds.) (2005) *Virus Taxonomy: Eighth Report of the International Committee on Taxonomy of Viruses,* 2nd edn., 1162pp. San Diego, CA: Elsevier: Academic Press.
Figlerowicz M and Bujarski JJ (1998) RNA recombination in brome mosaic virus, a model plus strand RNA virus. *Acta Biochimica Polonica* 45(4): 847–868.
Johnson JE and Speir JA (1997) Quasi-equivalent viruses: A paradigm for protein assemblies. *Journal of Molecular Biology* 269: 665–675.
Noueiry AO and Ahlquist P (2003) Brome mosaic virus RNA replication: Revealing the role of the host in RNA virus replication. *Annual Review of Phytopathology* 41: 77–98.
Wooley RS and Kao CC (2004) Brome mosaic virus. Descriptions of Plant Viruses, No. 405. http://www.dpvweb.net/dpv/showdpv.php?dpvno=405 (accessed Jul. 2007).

Cacao Swollen Shoot Virus

E Muller, CIRAD/UMR BGPI, Montpellier, France

Introduction

Cacao swollen shoot virus (CSSV) is a member of the genus *Badnavirus*, family *Caulimoviridae*. The importance and diversity of badnaviruses has only been recognized relatively recently and particularly on tropical plants, due to progress made in molecular diagnostic techniques.

In addition to CSS3V, other badnaviruses commonly reported are banana streak virus (BSV) on banana, dioscorea alata bacilliform virus (DaBV) on yam, taro

bacilliform virus (TaBV) on taro, sugarcane bacilliform virus (SCBV) on sugarcane, citrus mosaic bacilliform virus (CMBV) on citrus species, and pineapple bacilliform virus (PBV) on pineapple, but these viruses also infect various ornamental plants such as Aucuba, Commelina, Kalanchoe, Yucca, Mimosa, and Schefflera. Some of the above are only tentative species of the genus *Badnavirus* due to insufficient molecular data. There are regularly new reports of badnaviruses on other plants.

CSSV is naturally transmitted to cacao (*Theobroma cacao*) in a semipersistent manner by several mealybug species, the vector of most badnaviruses. Cacao swollen shoot disease occurs in all the main cacao-growing areas of West Africa, where it has caused enormous damage. It was highlighted for the first time in Ghana in 1922 but was described and named in 1936, then in Nigeria in 1944, in Ivory Coast in 1946, in Togo in 1949, and in Sierra Leone in 1963. The disease was also described in Trinidad and Tobago but currently seems to have disappeared. Cacao swollen shoot disease is present in Sri Lanka and in Indonesia (Java and North Sumatra).

Host Range and Symptomatology

Experimental host range is limited to species of the families of Sterculiaceae, Malvaceae, Tiliaceae, and Bombaceae but the principal host of the virus is *Theobroma cacao*. Symptoms are mostly seen in leaves, but stem and root swellings, as well as pod deformation also occur. In some varieties of cocoa, particularly Amelonado cocoa, reddening of primary veins and veinlets in flush leaves is characteristic (**Figure 1(a)**). This red vein banding later disappears. There can be various symptoms on mature leaves, depending on the cocoa variety and virus strain. These symptoms can include: yellow clearing along main veins; tiny pin-point flecks to larger spots; diffused flecking; blotches or streaks. Chlorotic vein flecking or banding is common and may extend along larger veins to give angular flecks.

Stem swellings may develop at the nodes, internodes, or shoot tips. These may be on the chupons, fans, or branches (**Figure 1(a)**). Many strains of CSSV also induce root swellings. Stem swellings result from the abnormal proliferation of xylem, phloem, and cortical cells.

Infected trees may suffer from partial defoliation initially due to the incompletely systemic nature of the infection. Ultimately, in highly susceptible varieties, severe defoliation and dieback occurs.

Smaller, rounded to almost spherical pods may be found on trees infected with severe strains. Occasional green mottling of these pods is seen and their surface may be smoother than the surface of healthy pods. Various isolates were described in Ghana as in Togo with a variability and a gradation in the type of symptoms observed.

A few avirulent strains occur in limited, widely scattered outbreaks, usually inducing stem swellings only, and having little effect, if any, on growth or yield. Moreover, there are periods of remission during which symptoms are not visible.

The natural host range of the virus includes *Ceiba pentandra*, *Cola chlamydantha*, *Cola gigantea* var. *glabrescens*, *Sterculia tragacantha*, and *Adansonia digitata* with associated symptoms of transient leaf chlorosis or conspicuous leaf chlorosis.

Transmission

Fourteen species of mealybugs (*Pseudococcidae* spp.), including *Planococcoides njalensis*, *Planococcus citri*, *Planococcus kenyae*, *Phenacoccus hargreavesi*, *Planococcus* sp. *Celtis*, *Pseudococcus concavocerrari*, *Ferrisia virgata*, *Pseudococcus*

Figure 1 (a) Symptoms of red-vein banding observed on young flush leaves of cocoa tree in Togo, Kloto area. (b) Symptoms of swellings on chupons of cocoa tree, in Togo, Litimé area. (c) Symptoms of swollen shoot observed on *Theobroma cacao* plantlets four months after agroinoculation with the Togolese Agou 1 isolate. From left to right: A plantlet inoculated with the wild strain *Agrobacterium tumefaciens* LBA4404 and two plantlets inoculated with the recombinant *A. tumefaciens* bacteria LBA4404 (pAL4404, pBCPX2) containing CSSV insert.

longispinus, *Delococcus tafoensis*, and *Paraputo anomalus*, have been reported to transmit CSSV. Only the nymphs of the first, second, and third larval stages and the adult females are able to transmit the virus. The virus does not multiply in the vector and is not transmitted to its progeny. CSSV is transmitted neither by seed nor by pollen. CSSV can infect cocoa at any stage of plant growth. The virus is transmitted experimentally to susceptible species by grafting, particle bombardment, by agro-infection using transformed *Agrobacterium tumefaciens* and with difficulty by mechanical inoculation. Seedlings usually produce acute red vein banding within 20–30 days and, 8–16 weeks later, swellings on shoots and tap roots (**Figure 1(c)**).

As this disease appeared soon after the introduction of cacao to West Africa, it is likely that CSSV came from indigenous hosts. The species *Adansonia digitata*, *Ceiba pentandra*, *Cola chlamydantha*, and *Cola gigantea* were identified as reservoir hosts in Ghana.

Economic Importance

CSSV is a serious constraint to cocoa production in West Africa, particularly in Ghana. Severe strains of this virus can kill susceptible cocoa trees within 2–3 years. They affect Amelonado cocoa, widely considered to give the best-quality cocoa, more seriously than Upper Amazon cocoa and hybrids which have been selected for resistance to the virus.

The disease was first recognized in 1936 but almost certainly occurred in West Africa in the 1920s. Estimates of annual yield losses due to this virus vary from about 20 000 tonnes to approximately 120 000 tonnes of cocoa from the Eastern Region of Ghana alone. The average annual loss between 1946 and 1974 in Ghana was estimated to be worth over £3 650 000.

Attempts at CSSV control in Ghana have required substantial financial and manpower inputs. The 'cutting-out' policies in place in Ghana since the early 1940s which attempted to control cocoa swollen shoot disease resulted in the removal of over 190 million trees up to 1988. Over ten million infected trees which still required 'cutting out' were identified in the field by 1990.

CSSV is currently confined to West Africa, Sri Lanka, and Sumatra. The disease does not pose a real economic problem for the culture of cocoa in Sri Lanka and Indonesia. However, the devastation which has occurred in West Africa has serious implications for germplasm movement. International attempts at crop improvement are hampered by the need to index cocoa germplasm for this virus, particularly if the germplasm is to be moved to where highly susceptible varieties are grown. Although seed transmission of this virus is not known to occur, there is often the need to move germplasm as stem cuttings, which must then pass through intermediate quarantine for indexing.

Molecular Characterization

Characteristics of the Virions

The viral origin was shown in 1939. CSSV possesses small nonenveloped bacilliform particles and a double-stranded DNA genome of 7–7.3 kbp. The bacilliform particles measure 130 nm × 28 nm in size and have been shown by dot–blot hybridization to occur in the cytoplasm of phloem companion and xylem parenchyma cells.

Description of Full-Length Sequence

Molecular characterization of CSSV had a boost in 1990, thanks to the improvement of the techniques of purification.

The first complete sequence of a CSSV isolate (Agou1 from Togo) was determined in 1993. Five putative open reading frames (ORFs) are located on the plus strand of the 7.16 kbp CSSV genome. ORF1 encodes a 16.7 kDa protein whose function is not yet determined. The ORF2 product is a 14.4 kDa nucleic acid-binding protein. ORF3 codes for a polyprotein of 211 kDa which contains, from its amino- to carboxyl-terminus, consensus sequences for a cell-to-cell movement protein, an RNA binding domain of the coat protein, an aspartyl proteinase, a reverse transcriptase (RTase), and an RNase H. The last two ORFs X (13 kDa) and Y (14 kDa) overlap ORF3 and encode proteins of unknown functions. CSSV, CMBV, and TaBV are the only badnaviruses which, to date, are known to code for more than three ORFs, CMBV, and TaBV encode six and four ORFs respectively.

Variability of the CSSV

A more extensive study of the molecular aspects of CSSV variability is relevant for three reasons. First, the knowledge of molecular variability will allow the improvement and the validation of a PCR diagnostic test for better virus-indexing procedures. Second, the variability of the virus must be taken into account for resistance screening of new cocoa varieties to CSSV. Finally, a better understanding of the genetic diversity of CSSV in West Africa and elsewhere will in turn help to provide a better understanding of the development of the epidemics and their eradication, and of the evolution of viral populations.

Badnaviruses are highly variable at both the genomic and serological level, a feature which complicates the development of both molecular and antibody-based diagnostic tests. Moreover, CSSV isolates were for a long time classified according to the variability of the symptoms expressed on *T. cacao*, but it is not known if there is a correspondence between this variability of symptoms and the intrinsic molecular variability of the virus.

A first study analyzed the molecular variability of the area of the ORF3 coding for N-terminal coat protein for

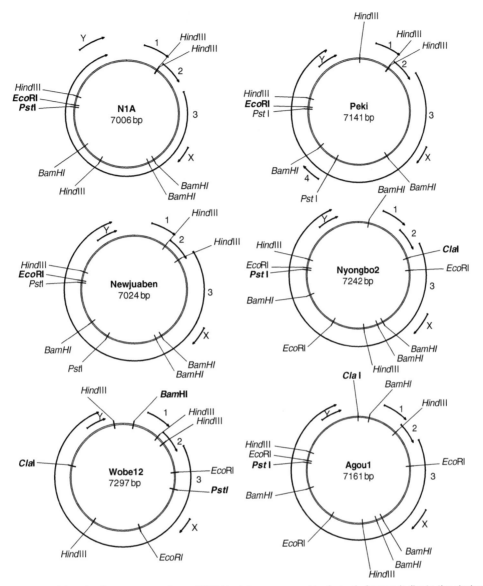

Figure 2 Organization of the circular genomes of new CSSV isolates compared to Agou 1. Arrows indicate the deduced ORFs 1–4, X and Y capable of encoding proteins larger than 9 kDa. *Bam*HI, *Cla*I, *Eco*RI, *Hind*III, and *Pst*I restriction sites are shown. Unique restriction sites are in bold. Reprinted from Muller E and Sackey S (2005) Molecular variability analysis of five new complete cacao swollen shoot virus genomic sequences. *Archives of Virology* 150: 53–66, with permission from Springer Wien.

1A-like isolates from Ghana and recently the analysis of new whole sequences made it possible to have a better idea of the variability of the CSSV.

Figure 2 presents the circular genome maps of the six isolates of CSSV sequenced to date, and these isolates come from Togo (Agou1, Nyongbo2 from Kloto area, and Wobe12 from Litimé area) and Ghana. The size of the total genome varies from 7024 bp for isolate N1A to 7297 bp for the isolate Wobe12. The overall organization of ORFs is the same for all the isolates but differs by the partial disappearance of the ORFX for the three Ghanaian isolates, and by the existence of an additional ORF for the Peki isolate, which codes for a putative protein of 11 kDa. **Table 1** shows the nucleotide and amino acid sequence percentage identity and similarity between the different ORFs of isolates. ORF1 is the most conserved ORF coding region (81 synonymous mutations versus 101 mutations among the six isolates) and the amino acid dissimilarity of pairwise combinations of CSSV isolates ranges from 1.4% to 8.4%. The maximum amino acid dissimilarity between isolates is 25.5% for ORF2, 23.8% for ORF3, and slightly higher for ORFY (33.6%). The value of overall variability calculated for ORF3 corresponds to a succession of more or less variable regions that can be identified on alignment. In agreement with the division of the ORF3 polyprotein in three regions, we observed that the first region (amino acids 1–350), which corresponds to the movement protein region, is highly conserved as

Table 1 Nucleotide sequence identity and amino acid sequence similarity of pairwise combinations of CSSV isolates for the ORFs 1, 2, 3, X, and Y

percentage of nucleotide sequence identity
percentage of amino acid sequence similarity

	ORF1					
	Ag1	*Nb2*	*NJ*	*Peki*	*N1A*	*w12*
Agou1	100	98.8	80	80.9	80.7	95.6
	100	**98.6**	**93.7**	**94.4**	**93**	**97.2**
Nyongbo2		100	80.2	81.1	80.9	94.9
		100	**93.7**	**94.4**	**93**	**97.2**
New Juaben			100	97.2	96.7	80.9
			100	**97.9**	**97.9**	**92.3**
Peki				100	97.7	81.4
				100	**97.2**	**93.7**
N1A					100	80.7
					100	**91.6**
Wobe12						100
						100

	ORF2					
	Ag1	*Nb2*	*NJ*	*Peki*	*N1A*	*w12*
Agou1	100	88.3	73.3	73.6	73.6	70.7
	100	**89**	**82.8**	**82.1**	**82.1**	**74.5**
Nyongbo2		100	81.8	81.8	81.8	77.6
		100	**89.7**	**88.3**	**88.3**	**76.5**
New Juaben			100	96.6	96.3	76.7
			100	**96.6**	**96.6**	**77.9**
Peki				100	96.6	76.5
				100	**95.9**	**77.2**
N1A					100	77
					100	**78.5**
Wobe12						100
						100

	ORF3					
	Ag1	*Nb2*	*NJ*	*Peki*	*N1A*	*w12*
Agou1	100	98.2	85.5	84.7	81.8	74.4
	100	**98.2**	**90.6**	**90.2**	**86.5**	**79.8**
Nyongbo2		100	85.1	84.4	81.5	74.2
		100	**90.3**	**89.8**	**86.3**	**79.7**
New Juaben			100	96.4	93.3	73.3
			100	**96.2**	**93.6**	**79.5**
Peki				100	91.8	73.5
				100	**92**	**80**
N1A					100	70.4
					100	**76.2**
Wobe12						100
						100

	ORFX					
	Ag1	*Nb2*	*NJ*	*Peki*	*N1A*	*w12*
Agou1	100	97.1	58.7	58.4	59.5	40.1
	100	**92.9**	**41.9**	**45.3**	**42.7**	**22.4**
Nyongbo2		100	58.7	58.1	60.1	41.3
		100	**41.9**	**43.6**	**44.4**	**22.4**
New Juaben				96	96.3	47.3
				89	**91.2**	**14.3**
Peki				100	94.1	46.9
				100	**89**	**20.5**
N1A					100	51.3
					100	**20.5**
Wobe12						100
						100

	ORFY					
	Ag1	*Nb2*	*NJ*	*Peki*	*N1A*	*w12*
Agou1	100	98.5	89.3	87.3	89.1	71
	100	**97.7**	**92.4**	**87.8**	**89.3**	**69.5**
Nyongbo2		100	88.3	86.3	88	71
		100	**90.1**	**85.5**	**87**	**71**
New Juaben			100	92.9	94.9	71.8
			100	**90.1**	**92.4**	**69.5**
Peki				100	93.4	69.7
				100	**88.5**	**65.6**
N1A					100	70.7
					100	**66.4**
Wobe12						100
						100

Ag1, Agou1; Nb2, Nyongbo2; NJ, New Juaben; w12, Wobe12.

Reproduced from Muller and a Sockey (2005) Molecular variability analysis of five new complete cacao swollen shoot virus genomic sequences. *Archives of Virology* 150: 53–66, with permission from Springer Wien.

already observed for other pararetroviruses. Region 2 is far less conserved (particularly amino acids 370–500 and 1010–1070). Region 3 has an intermediary level of variability. ORFX is the least conserved ORF.

Maximum nucleotide sequence variability between pairwise combinations of complete genomic sequences of CSSV isolates was 29.4% (between Wobe12-CSSV and Peki-CSSV).

The alignments of the nucleotide and protein sequences of the ORFs (**Table 1**) and the phylogenetic trees built from these sequences make it possible to separate the isolates into three groups according to their geographical origin, rather than their aggressiveness (CSSV-N1A are CSSV-Peki are considered as mild isolates).

A study of sequence variability was made on the level of the first region of the ORF3 on isolates taken in two areas different from Togo having distinct epidemiologic histories. The swollen shot disease was observed for

the first time in 1949 in Togo in the area of Kloto, and then spread only in this area. It is only toward the end of the 1990s that the disease started to be observed in Litimé. The phylogenetic tree built from the sequences of isolates of these two areas distinguishes three groups (A, B, and C), as does the analysis based on the whole sequences (**Figure 3**). One of these groups, A, containing Wobe12 is more distantly related to the others and contains only isolates coming from the area of Litimé, Togo. Moreover, the amplified sequence of isolates from group A (724 bp instead of the 721 bp amplified for the other isolates) code for an additional amino acid. Group A of isolates cannot therefore originate from the area of Kloto, infected before and probably originate from Ghana. In Kloto, only one group of isolates is present (C), whereas in the area of Litimé the three groups of diversity coexist. It is probable that the group B of isolates of Kloto, that are present also in Litimé, come from a contamination of Litimé by isolates of Kloto.

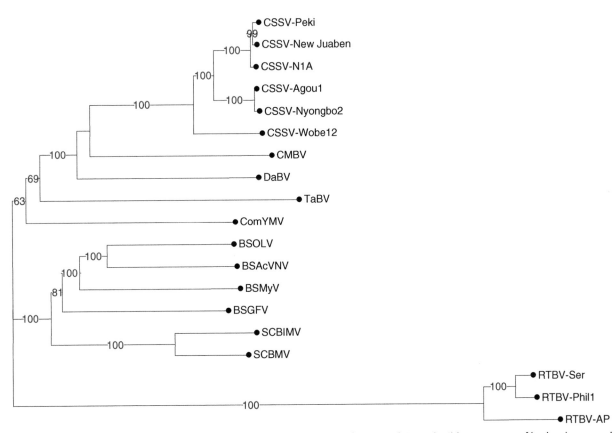

Figure 3 Neighbor-joining tree generated by the Darwin 4 program based on complete nucleotide sequences of badnaviruses and rice tungro bacilliform virus (RTBV). Numbers at the nodes of the branches represent percentage bootstrap values (1000 replicates) when superior to 60. The GenBank accession numbers of sequences are L14546 (CSSV-Agou1), AJ534983 (CSSV-Nyongbo2), AJ608931 (CSSV-New Juaben), AJ609019 (CSSV-Peki), AJ609020 (CSSV-NIA), AJ781003 (CSSV-Wobe12), AJ002234 (banana streak OL virus – BSOLV), AY805074 (banana streak mysore virus – BSMyV) AY750155 (banana streak acuminate vietnam virus – BSAcVNV), AY493509 (banana streak GF virus – BSGFV), X52938 (commelina yellow mottle virus – comYMV), AF347695 (citrus mosaic bacilliform virus – CMBV), X94576 (dioscorea alata bacilliform virus – DaBV), M89923 (sugarcane bacilliform Mor virus – SCBMV), AJ277091 (sugarcane bacilliform IM virus – SCBMIV), AF357836 (taro bacilliform virus – TaBV), AJ292232 (RTBV-AP), O76470 (RTBV-Ser), X57924 (RTBV-Phill).

Homology with Other Viruses

A phylogenetic tree was built from all the available full-length sequences of badnaviruses (**Figure 3**). CSSV isolates are closer to each other than to badnaviruses and to members of the genus *Tungrovirus*. In addition, among the badnaviruses sequenced to date, CSSV has the closest relationship with CMBV whose host is also a tree crop. The majority of the hosts of the badnavirus are indeed plants with vegetative multiplication.

However, one group of BSV-like sequences recently detected in Uganda are closer to CSSV sequences than to other BSV sequences and one sequence found integrated in *Musa acuminata* germplasm has good homology with CSSV.

The maximum variability level observed between the six CSSV isolates is slightly higher than the one observed between the three rice tungro bacilliform virus (RTBV) isolates but much lower than between the various isolates of BSV sequenced until now.

Molecular Epidemiology

A systematic molecular characterization of viral isolates should make it possible to study the molecular epidemiology of CSSV. Moreover, studies on CSSV in adventitious plants or insect vectors would make it possible to better understand the epidemiologic factors that lead to the rapid expansion of the disease in some plots.

Diagnosis

Serological Diagnostics

The virus is not strongly immunogenic; however, several antisera have been raised and shown to react with CSSV. Enzyme-linked immunosorbent assay (ELISA) has been used to detect CSSV but high background values were obtained and difficulties were found in detecting CSSV in plants suspected of having only a low virus titer. Immunosorbent electron microscopy has been used for detection and comparison of some isolates of CSSV in Ghana. Both of these techniques cannot detect latent infection.

The virobacterial agglutination (VBA) test has been found to be a useful test for detecting CSSV in leaf tissue from infected trees. Using this assay, CSSV can be detected in trees showing symptoms as well as in infected, but symptomless, trees. The immunocapture polymerase chain reaction (IC-PCR) technique was adapted to the only detection of CSSV-1A isolates from Ghana.

There is considerable strain variation among the many recognized CSSV isolates, some of which react only weakly with certain antisera. Using monoclonal antibodies, four serotypes of CSSV have been distinguished by ELISA analysis of 31 samples of the virus from different geographical locations in Ghana. The efficiency of serological diagnosis depends on the use of polyvalent antiserum able to detect all serotypes.

PCR Diagnosis

For a reliable PCR diagnosis, it is necessary to design primers from conserved regions of the genome. Until recently, only few badnaviruses were sequenced in full, and only the end of the ORF3 which contains the conserved motifs coding for RTase and RNaseH made it possible to obtain primers useful for diagnosis. PCR-based diagnosis is able to detect CSSV not only in symptomless leaves of symptomatic plants, but also in symptomless plants as early as one week post-inoculation.

The alignment of the six full-length sequences of CSSV allowed conserved regions to be identified and polyvalent primers to be designed for the diagnosis of CSSV. These primers can detect all the isolates tested to date from Togo and Ghana. These primers are located in the first part of the ORF3 and amplify a fragment from 721 to 724 bp, including from isolates having only 70% nucleotide identity (CSSV-Wobe12 and CSSV-Peki). The extraction method is based on a buffer containing MATAB (mixed alkyltrimethylammonium bromide), which reduces the co-extraction of PCR-inhibitory substances, as well as being less costly and generates a higher DNA yield.

These primers for CSSV diagnosis may require modifications as more sequence data from other CSSV isolates are generated. As the sensitivity of diagnosis is better when young symptomatic host plant leaves are tested, it is recommended that validation of the diagnostic test is done using this type of material in the first instance, especially when testing isolates from different areas.

Control of the Disease

Control of the swollen shoot disease based only on 'cutting-out' campaigns has not been successful due to several factors including political and socioeconomic problems. The better strategy for dealing with the disease is to develop a combination of control measures in an integrated approach. Moreover, this approach should implicate cocoa farmers as much as possible.

Intermediate quarantine facilities are at present hampered by the lack of suitable indexing methods for the virus. The early PCR diagnosis could be easily tested in intermediate quarantine facilities and compared with the grafting procedure used for indexation on Amelonado cocoa seedlings.

Mild strain protection is a possibility which is being investigated. Mild strains which appear to confer some protection against the severe strains are available and are being tested in the field in Ghana. However, the degree of protection afforded so far is not sufficient and further research is necessary.

The possibility of isolating new cocoa plantings from infected cocoa by using barriers of CSSV-immune crops could be considered. These crops would form a barrier to the movement of vectors. Examples of possible barrier crops include oil palm, coffee, cola, and citrus. The 'cutting out' of the adventitious plants of the type Commelina and Taro could be a possible strategy if it is confirmed that these plants harbor CSSV.

The use of resistant cocoa is advocated as many of the new hybrids available in West Africa do have some resistance to CSSV and because it seems to be the most sustainable method. Replanting with resistant cacao trees, however, requires the installation of a protocol of effective screening for resistance. Severe isolates representative of the different molecular groups should be used as well as a suitable screening method. However, a standardized inoculation method is not yet available because particle bombardment is difficult to develop on a large scale and agroinoculation needs biosafety confinement. The screening for CSSV resistance for two types of severe isolates has been initiated in 2003 in Togo.

See also: Caulimoviruses: General Features; Caulimoviruses: Molecular Biology.

Further Reading

Brunt AA (1970) *CMI/AAB Descriptions of Plant Viruses No 10. Cacao Swollen Shoot Virus.* Wellesbourne: Association of Applied Biologists.

CABI, Crop Protection Compendium(2002) *Cacao Swollen Shoot Virus.* Wallingford: CABI Publishing.

Castel C, Amefia YK, Djiekpor EK, Partiot M, and Segbor A (1980) Le swollen shoot du cacaoyer au Togo. Les différentes formes de viroses et leurs conséquences économiques. *Café, Cacao, Thé* 24(2): 131–146.

Fauquet C, Mayo M, Maniloff J, Desselberger U,, and Ball L (eds.) (2005) *Virus Taxonomy: Eighth Report of the International Committee on Taxonomy of Viruses.* San Diego, CA: Elsevier Academic Press.

Hagen LS, Jacquemond M, Lepingle A, Lot H, and Tepfer M (1993) Nucleotide sequence and genomic organization of cacao swollen shoot virus. *Virology* 196(2): 619–628.

Hagen LS, Lot H, Godon C, Tepfer M, and Jacquemond M (1994) Infection of *Theobroma cacao* using cloned DNA of cacao swollen shoot virus and particle bombardment. *Molecular Plant Pathology* 84: 1239–1243.

Lot H, Djiekpor E, and Jacquemond M (1991) Characterization of the genome of cacao swollen shoot virus. *Journal of General Virology* 72: 1735–1739.

Muller E and Sackey S (2005) Molecular variability analysis of few new complete cacao swollen shoot virus genomic sequence. *Archives of Virology* 150: 53–66.

Capillovirus, Foveavirus, Trichovirus, Vitivirus

N Yoshikawa, Iwate University, Ueda, Japan

Glossary

Alfavirus A genus in the family *Togaviridae.* The type species is *Sindbis virus.*

Helical symmetry A form of capsid structure found in many RNA viruses in which the protein subunits which interact with the nucleic acid form a helix.

Methyltransferase Enzyme activity involved in capping of viral mRNAs.

Movement protein A virus-encoded protein which is essential for the cell-to-cell movement of the virus in plant tissues.

Semipersistent manner The relationship between a plant virus and its arthropod vector which is intermediate between nonpersistent manner and persistent manner. It has the features of short acquisition feed and no latent period found in nonpersistent manner, but the vector remains able to transmit the virus for periods of hours to days which is longer than the nonpersistent manner.

Stem grooving Deformation of the normally smooth surface of a trunk cased by its furrowing.

Stem pitting A plant disease characterized by the formation of larger or smaller depression in the old wood, between the phloem and the xylem of the tree trunk.

Subgenomic RNA A species of RNA less than genomic length found in infected cells. Viral genomic RNA codes for several proteins but all the 3′ open reading frames will effectively be closed for translation. The formation of subgenomic RNAs overcomes this problem as each species has a different cistron at its 5′ end, thus opening it for translation.

Genus *Capillovirus*

The genus *Capillovirus* contains three species – *Apple stem grooving virus* (ASGV, the type species), *Cherry virus A* (CVA), and *Lilac chlorotic leafspot virus* (LiCLV) – and a tentative species – Nandina stem pitting virus (NSPV) (**Table 1**). Citrus tatter leaf virus (CTLV) from citrus and lily is indistinguishable from ASGV from Rosaceae fruit trees biologically, serologically, in genome organization, and in nucleotide sequence, and these days CTLV is regarded as an isolate of ASGV.

Biological Properties

ASGV occurs worldwide in Rosaceae fruit trees, including apple, European pear, Japanese pear, Japanese apricot and cherry, and it is usually symptomless. However, the virus causes stem grooving, brown line, and graft union abnormalities in Virginia Crab, and it causes topworking disease of apple trees grown on Mitsuba kaido (*Malus sieboldii*) in Japan. ASGV is also widespread in citrus, and it induces bud union abnormalities of citrus trees on trifoliate orange. It also infects lily. CVA occurs in cherry trees in Germany and Japan, but it probably is not associated with any disease. LiClV and NSPV occur in England and the United states, respectively.

No vectors of any of these viruses have been reported. ASGV has been known to be transmitted through seeds to progeny seedlings of lily (1.8%) and *Chenopodium quinoa* (2.5–60%).

Particle Structure

Virions are flexuous filamentous particles 670–700 nm long and 12 nm in diameter, with obvious cross-banding and helical symmetry and a pitch of *c.* 3.8 nm (**Figure 1**). Virus particles are composed of a linear positive-sense ssRNA, 6.5–7.4 kbp, and a single polypeptide species of M_r 24–27 kDa. The 3′ terminus of the RNA has poly A-tail and the 5′ terminus probably has a cap structure.

Genome Organization and Replication

The complete nucleotide sequences (6496 bases) of the single RNA genome of three ASGV isolates have been determined: isolate P-209 from apple, and isolates L and Li-23 from lilies. Identities of the nucleotide sequences are 82.9% (P-209/L), 83.0% (P-209/Li-23), and 98.4% (L/Li-23). The genomic RNA has the same structural organization and two overlapping open reading frames (ORFs) in the positive strand (**Figure 2**). ORF1 (bases 37–6341) encodes a 241–242 kDa polyprotein (2105 aa) containing the consensus motifs of methyltransferase (Met), papain-like protease (P-pro), nucleotide triphosphate-binding helicase (Hel), RNA polymerase (Pol), and coat protein (CP) in the C-terminal region. The protein has homologies with putative polymerase of the 'alphavirus-like' supergroup of RNA viruses. ORF2 (bases 4788–5747) encodes a 36 kDa putative movement protein (320 aa). A region (amino acid (aa) position 1585–1868) of the ORF1-encoded protein between the polymerase and the CP, that encodes ORF2 in another frame has none of the other functional motifs found in other known plant virus genomes. This region (designated the V-region) shows high variability among isolates and sequence variants. The genome organization of CVA is composed of a 266 kDa polyprotein (ORF1) and a 52 kDa protein (ORF2) located within ORF1. ORF1 encodes the CP (24 kDa) in the

Table 1 Virus species in the genus *Capillovirus, Foveavirus, Trichovirus, Vitivirus* in the family *Flexiviridae*

Genus	Species	Sequence accession numbers
Capillovirus	Apple stem grooving virus (ASGV) Cherry virus A (CVA) Lilac chlorotic leafspot virus (LiCLV) Nandina stem pitting virus (NSPV)[a]	D14995, D16681, D16368, D14455, AB004063 X82547
Foveavirus	Apple stem pitting virus (ASGV) Apricot latent virus (ApLV) Rupestris stem pitting-associated virus (RSPaV)	D21829, AB045731, D21828 AF057035 AF026278, AF057136
Trichovirus	Apple chlorotic leaf spot virus (ACLSV) Cherry mottle leaf virus (CMLV) Grapevine berry inner necrosis virus (GINV) Peach mosaic virus (PMV)	M58152, D14996, X99752, AJ243438 AF170028 D88448
Vitivirus	Grapevine virus A (GVA) Grapevine virus B (GVB) Grapevine virus D (GVD) Heracleum latent virus (HLV) Grapevine virus C (GVC)[a]	X75433, AF007415 X75448 Y07764 X79270

[a]Tentative species.

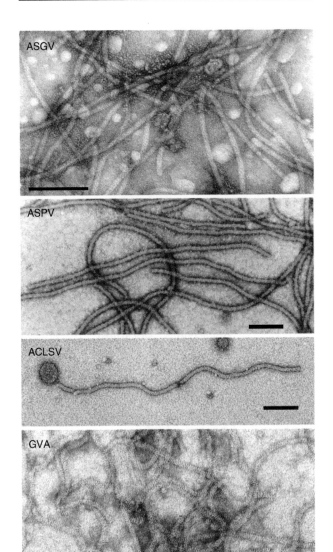

Figure 1 Electron micrographs of particles of *Apple stem grooving virus* (ASGV), the type species of the genus *Capillovirus*; *Apple stem pitting virus* (ASPV), the type species of the genus *Foveavirus* (courtesy of H. Koganezawa); *Apple chlorotic leaf spot virus* (ACLSV), the type species of the genus *Trichovirus*; and *Grapevine virus A* (GVA), the type species of the genus *Vitivirus* (courtesy of J. Imada). Scale = 100 nm.

C-terminal region. The overall nucleotide sequence identity between CVA and ASGV is 57.6%.

Although the ASGV-CP is located in the C-terminal region of the ORF1-encoded polyprotein and genomic RNA directs the synthesis of a polypeptide of *c*. 200 kDa as a major product in *in vitro* translation, and is immuno-precipitated by antiserum to virus particle preparations, the following evidence suggests that the CP is expressed from a subgenomic RNA. Analysis of double-stranded (ds) RNA from infected *C. quinoa* tissues indicates that

all ASGV isolates tested contain five virus-specific dsRNAs (6.5, 5.5, 4.5, 2.0, and 1.0 kbp). The 6.5 kbp species represents the double-stranded form of the full-length genome, whereas the 2.0 and the 1.0 kbp species may be the double-stranded forms of subgenomic RNAs coding for the putative movement protein and the CP, respectively. The 5.5 and 4.5 kbp species are thought to be 5′ co-terminal with the genome. The size of the *Escherichia coli*-expressed protein corresponding to the C-terminal region of the ORF1-encoded protein, which starts with the methionine at aa position 1869, agrees with that of the CP. The single-stranded subgenomic RNAs for movement protein (MP) and CP have also been reported in infected tissues.

In infected *C. quinoa* leaves, the particles occur singly or as aggregates in the cytoplasm of mesophyll and phloem parenchyma cells, suggesting that the replication of the genome and the assembly of the particles may occur in the cytoplasm, although no virus-specific inclusion bodies, such as pinwheels, viroplasmas, or vesicles, have been observed.

Serology

Polyclonal antisera were prepared in rabbits against purified virus or CP expressed in *E. coli*. Enzyme-linked immunosorbent assay was used to detect the virus in fruit trees. Monoclonal antibodies were produced in mice against an isolate from citrus and three selected monoclonal antibodies reacted with all isolates tested, including nine isolates from citrus trees in Japan, four isolates from citrus in the USA, six isolates from Chinese citrus, and isolates from lily, apple, and Japanese apricot. ASGV is serologically unrelated to all known virus species in the genus *Capillovirus*.

Strains and Genome Heterogeneity

Many isolates have been reported from apple, Japanese pear, European pear, Japanese apricot, lily, and citrus plants, but most have not been characterized. Some isolates have been differentiated only on symptomatology. Virus isolates from apple, Japanese pear, and European pear trees comprise at least two to four variants that differ considerably from each other in nucleotide sequence. The composition of sequence variants within a tree differs among leaves from different branches, showing that each sequence variant is distributed unevenly within an individual tree.

Genus *Foveavirus*

The genus *Foveavirus* consists of three species: *Apple stem pitting virus* (ASPV, the type species), *Apricot latent virus* (ApLV), and *Rupestris stem pitting-associated virus* (RSPaV) (**Table 1**).

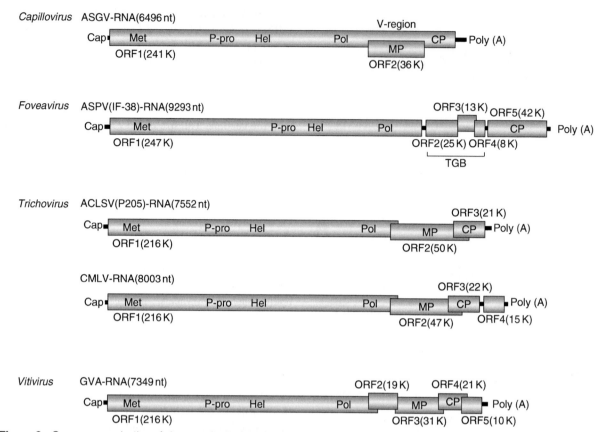

Figure 2 Gemone organization of virus species in the genera _Capillovirus, Foveavirus, Trichovirus,_ and _Vitivirus._ ASGV, apple stem grooving virus; ASPV, apple stem pitting virus; ACLSV, apple chlorotic leaf spot virus; CMLV, cherry mottle leaf virus; GVA, grapevine virus A; Met, methyltransferase; P-pro, papain-like protease; Hel, nucleotide triphosphate-binding helicase; Pol, RNA-dependent RNA polymerase; CP, coat protein; MP, movement protein; and V-region, variable region.

Biological Properties

ASPV is one of the causative agents of apple topworking disease in Japan and induces lethal decline in apple trees grown on _Malus sieboldii_ (Regel) Rehd. rootstock. The virus is usually latent in apple cultivars and is distributed widely in many apple trees. Pear vein yellows is also thought to be caused by this virus. ApLV infects apricot, peach, and sweet cherry, and may be the causal agents of the peach asteroid spot disease and of the peach sooty ringspot diseases. RSPaV is thought to be an agent of rupestris stem pitting, that is probably the most common component of the Rugose wood complex on grapevines. No vectors have been reported for viruses in the genus _Foveavirus,_ and the viruses are probably spread in nature by graft-transmission.

Particle Structure

ASPV (isolate B-39) has flexuous filamentous particles, approximately 800 nm in length and 12–15 nm in width (**Figure 1**). Virus particles readily form end-to-end aggregates with four prominent peaks appearing at 800, 1600, 2400, and 3200 nm in length. ASPV is comprised of a single species of RNA of M_r 3.1×10^6 and a major CP of M_r 48 kDa.

Genome Organization and Replication

The genomes of two ASPV isolates (PA66 and IF 38) have been completely determined and found to consist of 9306 and 9237 nucleotide (nt), respectively, excluding the 3' poly A-tail. The base composition of ASPV genome is 27.6% A, 20.0% C, 23.4–23.8% G, and 28.6–29.0% T. Analysis of the putative ORFs of the nucleotide sequence in both positive and negative strands showed that ASPV genome contains five ORFs in the positive-strand, encoding proteins with M_r's of 247K (ORF1), 25K (ORF2), 13K (ORF3), 7–8K (ORF4), and 42–44K (ORF5) (**Figure 2**). The 5'-noncoding regions of the genome have been reported to be 33 nt (PA66) and 60 nt (IF38). ORF1 encodes a protein (247K) with motifs associated with Met, P-pro, Hel, and Pol (**Figure 2**). ORF2 encodes a protein (25K) containing a helicase motif (GSGKT, aa positions 31–35). ORF3 and ORF4 encode proteins with 13K and 7–8K, respectively. The ORFs 2, 3, and 4 constitute the triple gene block (TGB) found in allexi-, carla-, potex-, and mandariviruses. ORF5 encodes a protein with an M_r of 42–44K, and the sequence of this protein contains the conserved amino acids (R and D) potentially involved in a salt bridge formation that is typical of the CPs of rod-shaped and filamentous plant viruses. The 3'-noncoding region of both genomes of PA66 and IF38 isolates consists of 132 nucleotides, excluding the poly A-tail. The comparison of the nucleotide sequence of the PA66 genome with that of the IF38 genome showed a high level of divergence (76% identity) between the two isolates. Comparisons of aa sequences of five proteins between PA66 and IF38 show the identities of 87% (247K), 94% (25K), 87% (13K), 77% (7–8K), and 81% (42–44K). A hypervariable region was found between the MET and P-Pro domains in the 247K protein. This hypervariable region between MET and P-Pro domains of the 247K protein was also found in the apple chlorotic leaf spot virus (ACLSV) 216K protein, indicating that this region has not undergone severe evolutionary constraint. The 25K protein was mostly conserved between the two isolates. Another striking variability between PA66 and IF38 was found in the N-terminal region of ASPV-CP, in which there were many deletions compared with PA66. This resulted in an aa number of IF38-CP which is 18 aa fewer than that of PA66.

The genome of RSPaV consists of 8726 nt excluding the 3' poly A-tail and has five potential ORFs on its positive strand which have the capacity to code for the replicase (ORF1; 244 K), the TGB (ORF2–4; 24K, 13K, and 8K), and the CP (ORF5; 28K). The identities of aa sequences of five proteins between RSPaV and ASPV (PA66) are 39.6% (ORF1), 38.0% (ORF2), 39.3% (ORF3), 27.1% (ORF4), and 31.3% (ORF5). Partial sequences of the 3'-terminal regions of ApLV have been reported.

In Northern blot hybridization analysis of virus-specific dsRNAs and ssRNAs in ASPV (IF38)-infected tissues by use of three negative-sense RNA probes complementary to nt positions 6207–6683 (ORF1 region), nt positions 6678–7447 (ORF2 region), nt positions 8717–9293 (CP and the 3'-noncoding regions) of the IF38 genome showed the presence of five dsRNAs (9, 7.5, 6.5, 2.6, and 1.6 kbp) and three ssRNAs (9, 2.6, and 1.6 kbp) in infected tissues (**Figure 3**). The slowest migrating RNA (*c.* 9 kbp) was equivalent to that of the ASPV genome, and other two RNAs (2.6 and 1.6 kbp) are thought to be subgenomic RNAs of ORF2–ORF4 proteins and CP (ORF5), respectively.

In electron microscopy of infected leaves, virus particles were found in mesophyll, epidermal, and vascular parenchyma cells. The particles were observed often as large aggregates in the cytoplasm, but not in the vacuole, the nucleus, or in other cellular organelles.

Strains and Genome Heterogeneity

Restriction fragment length polymorphism (RFLP) analysis of the cDNA clones from four ASPV isolates from apple (B39, B12, IG39, and IF38) showed that three different patterns were found in cDNA clones from isolate B39, two patterns from isolates B12 and IG39, and one pattern from isolate IF38. Sequence analysis of the 3'-terminal 600 nt of the genomes indicated that sequence identities among cDNA clones showing the same RFLP pattern were more than 99%, and in contrast, considerable sequence variations were found among the cDNA clones showing a different RFLP pattern. These results indicated that isolate B39 is composed of at least three sequence variants (SVs), and isolates B12 and IG39 at least two SVs. The nucleotide sequence identity among nine isolates and SV including PA66 was 77.9–95.7% in the 3'-terminal 600 nt regions. The aa number of CP varied from 396 to 415 among nine isolates and SV, and their identities ranged from 69.4% to 92.7%.

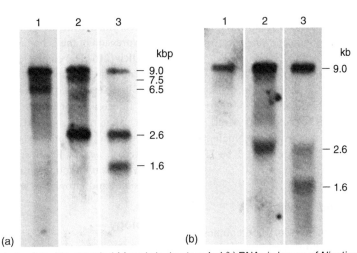

Figure 3 Northern blot analysis of double-stranded (a) and single-stranded (b) RNAs in leaves of *Nicotiana occidentalis* infected with apple stem pitting virus (IF-38) using three different RNA probes specific to different sequences within the genomic RNA. Lane 1, a probe complementary to nt positions 6207–6683 (ORF1 region); lane 2, nt positions 6678–7447 (ORF2 region); and lane 3, nt positions 8717–9293 (CP and the 3'-noncoding regions).

Genus *Trichovirus*

The genus *Trichovirus* consists of four viral species: *Apple chlorotic leaf spot virus* (ACLSV, the type species), *Cherry mottle leaf virus* (CMLV), *Grapevine berry inner necrosis virus* (GINV), and *Peach mosaic virus* (PcMV) (**Table 1**).

Biological Properties

ACLSV occurs in woody plants of the family Rosaceae including apple, pear, peach, plum, cherry, apricot, and prune. Though ACLSV infection is symptomless in most commercial apple varieties, the virus causes topworking disease of apple grown on Maruba kaido (*Malus prunifolia* var. *ringo*) rootstocks in Japan. ACLSV may cause plum bark split, plum pseudopox, pear ring pattern mosaic, and apricot pseudopox disease. CMLV occurs naturally in sweet cherry, peach, and apricot, and causes mottle leaf disease of cherry in some regions of North America. GINV is the causal virus of grapevine berry inner necrosis disease, one of the most important virus diseases of grapevine in Japan.

The viruses can be transmitted by mechanical inoculation, grafting, and through propagating materials. No vectors have been reported for ACLSV. On the other hand, CMLV and GINV are transmitted by the peach bud mite *Eriophyes insidiosus* and the grape erineum mite *Colomerus vitis*, respectively.

Particle Structure

Virus particles are very flexuous filaments 680–780 nm long and 9.25–12 nm in width (**Figure 1**), with obvious cross-banding and helical symmetry; the pitch of the helix is about 3.47–3.8 nm. ACLSV has a buoyant density of 1.27 g ml^{-1} in CsSO$_4$ gradients, and the particles are degraded in CsCl gradients. The A$_{260}$/A$_{280}$ of purified ACLSV preparation is 1.85–1.89. The particles require the presence of divalent cations to maintain the integrity of the quaternary structure.

Virus particles are composed of a linear positive-sense ssRNA, 7.2–8.0 kbp, and a single polypeptide species of M_r 21–22 kDa. The genomic RNA is about 5% of particle weight. The 3′ terminus of the RNA has a poly A-tail, and the 5′ terminus probably has a cap structure. Nucleotide base ratios are 32% A, 18% C, 23% G, and 27% U for ACLSV-RNA.

Genome Organization and Replication

The complete nucleotide sequences of the genomes of ACLSV (four isolates from plum, apple, and cherry), CMLV, and GINV have been determined. The genomes of ACLSV consist of 7549–7555 nt in length excluding the 3′ poly A-tail. The complete genomes of CMLV and GINV are 8003 nt and 7243 nt long excluding the poly A tail, respectively.

The genomic RNAs of ACLSV and GINV have three slightly overlapping ORFs in the positive strand (**Figure 2**). ORF1 encodes a replication-associated protein of M_r 214–216 kDa containing the consensus motifs of Met, P-pro, Hel, and Pol of the alphavirus-like supergroups of ssRNA. ORF2 encodes a putative MP of M_r 39–50 kDa. The 50 kDa protein encoded by ORF2 of ACLSV genome is an MP, which has the following characteristics: (1) it is localized to plasmodesmata of infected and transgenic plant cells; (2) it can spread from the cells that initially produce it into neighboring cells; (3) it enables cell-to-cell trafficking of green fluorescent protein (GFP) when the 50 kDa protein and GFP are co-expressed in leaf epidermis; (4) it induces the production of tubular structures protruding from the surface of protoplast; and (5) it binds to single-stranded nucleic acids. Additionally, transgenic plants expressing the 50 kDa protein can complement the systemic spread of mutants of an infectious cDNA clone that are defective in ORF2. ORF3 encodes a 21–22 kDa CP.

The genomic RNA of CMLV has four putative ORFs encoding 216 kDa (ORF1), 47 kDa (ORF2), 22 kDa (ORF3), and 15 kDa (ORF4) proteins (**Figure 2**). The ORF4 protein that is not present in the genomes of ACLSV and GINV may be a nucleic acid-binding protein because similarities have been found between ORF4 of CLMV and ORF5 of grapevine virus B.

In analysis of the dsRNAs of ACLSV-infected plants, six dsRNA species of *c.* 7.5, 6.4, 5.4, 2.2, 1.1, and 1.0 kbp were found in infected tissues. The 7.5 kbp species corresponding to the full-length genome may be a replicative form of the ACLSV genome, and the 2.2 and 1.1 kbp species are thought to be the double stranded forms of subgenomic RNAs coding for the MP and the CP, respectively. The 6.4 and 5.4 kbp species were found to be 5′ co-terminal with the genomic RNA. In the model for the expression of the genome of ACLSV, only the 5′-proximal ORF (216 kDa) coding for the putative viral replicase is translated directly, and the two other ORFs encoding MP and CP are expressed through two subgenomic RNAs.

In infected plant leaves, ACLSV and GINV particles occur as aggregates in the cytoplasm. ACLSV was also observed in nucleoplasm of a mesophyll cell in the infected *C. quinoa*. Replication is presumed to occur in the cytoplasm although no virus-specific inclusions, such as vesicles and viroplasm, were observed.

Serology

No serological relationships were found among ACLSV, CMLV, and GINV. Polyclonal and monoclonal antibodies (MAbs) of CMLV cross-reacted with all isolates of PcMV tested, indicating that CMLV and PcMV are closely related viruses. MAbs against ACLSV particles were produced and used to investigate the antigenic structure of the virus. Epitope studies using these MAbs identified seven independent antigenic domains in ACLSV particles.

Strains

Many isolates have been reported from apple, cherry, peach, plum, and prune. Several strains could be differentiated serologically as well as by symptoms in indicator plants. Sequence comparisons among isolates indicate large molecular variability, that is, sequence conservation rates vary between 77.4% and 99.4%, with most of the isolates differing by 10–20% from any other given isolates.

Genus *Vitivirus*

The genus *Vitivirus* consists of four viral species – *Grapevine virus A* (GVA, the type species), *Grapevine virus B* (GVB), *Grapevine virus D* (GVD), and *Heracleum latent virus* (HLV) – and a tentative species – Grapevine virus C (**Table 1**).

Biological Properties

GVA, GVB, and GVD naturally infect only grapevines and induce severe diseases, the rugose wood complex of grapevine, characterized by pitting and grooving of the wood. HLV occurs in hogweed (*Heracleum* sp. family Apiaceae) without causing obvious symptoms. Herbaceous plant species, *Nicotiana benthamiana* and *N. occidentalis*, are used for propagation hosts of GVA and GVB, respectively, and *Chenopodium* species are used for diagnostic and propagation species of HLV.

All virus species are transmitted by mechanical inoculation. Transmission by grafting and dispersal through propagating materials is common with infected grapevines. GVA and GVB are transmitted in nature by several species of the pseudococcid mealybug genera *Planococcus* and *Pseudococcus* in a semipersistent manner, whereas HLV is transmitted from naturally infected hogweed plants by the aphids in a semipersistent manner, which depends on a helper virus present in naturally infected plants. No seed transmission of HLV is found in *C. quinoa*, chervil, coriander, or hogweed.

Particle Structure

Virus particles are flexuous filaments 725–825 nm long and 12 nm in width, showing obvious cross-banding and helical symmetry with a pitch of 3.5 nm. Particles sediment as a single peak in sucrose or Cs_2SO_4 density gradients with a sedimentation coefficient (S_{20w}) of about 96S and a buoyant density in Cs_2SO_4 of 1.24 g cm^{-3} for HLV. The A_{260}/A_{280} of purified virus preparation is 1.5–1.52 for GVB and HLV.

Virus particles are composed of a linear positive-sense ssRNA, 7.3–7.6 kbp, and a single polypeptide species of M_r 18–21.5 kDa. The genomic RNA is about 5% of the particle weight. The 3′ terminus of the RNA has a poly A tail, and the 5′ terminus probably has a cap structure. The overall A+U and G+C content of GVB-RNA is 53.5% and 46.5%, respectively.

Genome Organization and Replication

The complete nucleotide sequences of the genomes of GVA and GVB have been determined. The genomes of GVA and GVB consist of 7349 and 7598 nt in length excluding the 3′-poly A tail, respectively.

The genomic RNAs of GVA and GVB have five slightly overlapping ORFs in the positive strand (**Figure 2**). ORF1 of both viruses encodes a replication-associated protein of M_r 194–195 kDa containing the consensus motifs of methyltransferase, papain-like protease, nucleotide triphosphate-binding helicase, and RNA polymerase of the alphavirus-like supergroups of ssRNA. ORF2 encodes a protein of M_r 19 kDa (GVA) or 20 kDa (GVB) that does not show any significant sequence homology with protein sequences from the databases. The biological function of these proteins has not been determined yet. ORF3 encodes an MP with M_r 31 kDa (GVA) and 36.5 kDa (GVB), possessing the G/D motif of the '-30K' superfamily movement protein. ORF4 encodes a CP with M_r 21.5 kDa (GVA) and 21.6 kDa (GVB). ORF5 encodes a protein with M_r 10 kDa (GVA) and 14 kDa (GVB). A 14 kDa protein encoded by GVB-ORF5 shares weak homologies to proteins with nucleic acid-binding properties.

The gene expression strategy may be based on proteolytic processing of ORF1 protein and subgenomic RNA production for translation of ORF2 to ORF5 proteins. In analysis of the dsRNAs of infected plants, four major dsRNA bands with a size of 7.6, 6.48, 5.68, and 5.1 kbp for GVA and GVD, and 7.6, 6.25, 5.03, and 1.97 kbp for GVB were found in infected tissues. In *N. benthamina* plants infected with GVA, three 3′-terminal subgenomic RNAs of 2.2, 1.8, and 1.0 kbp were also detected and thought to serve for the expression of ORF2, ORF3, and ORF4, respectively.

In electron microscopy of leaves infected with GVA and GVB, tonoplast-associated membranous vesicles containing finely fibrillar materials were found in phloem parenchyma cells, and virus particles occurred only in phloem tissues. Thus, replication may occur in the cytoplasm, possibly in association with vesicles protruding from the tonoplast of phloem cells.

Serology

In immunoelectron microscopy, the antiserum against GVB-NY clearly decorated homologous (GVB-NY) and heterologous (GVB-CAN) virus particles, but did not

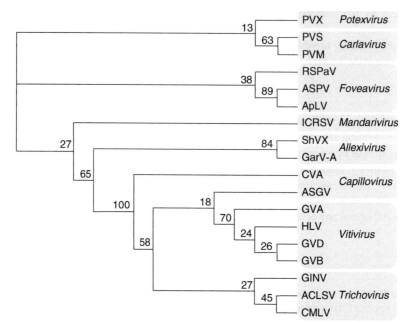

Figure 4 Dendrogram showing the relationships among virus species in the family *Flexiviridae* using the aa sequences of the coat protein. The tree was produced and bootstrapped using CLUSTAL W. PVX, potato virus X; PVS, potato virus S; PVM, potato virus M; RSPaV, ruspestris stem pitting-associated virus; ASPV, apple stem pitting virus; ApLV, apricot latent virus; ICRSV, Indian citrus ringspot virus; ShVX, shallot virus X; GarV-A, garlic virus A; CVA, cherry virus A; ASGV, apple stem grooving virus; GVA, grapevine virus A; HLV, heracleum latent virus; GVD, grapevine virus D; GVB, grapevine virus B; GINV, grapevine berry inner necrosis virus; ACLSV, apple chlorotic leaf spot virus; CMLV, cherry mottle leaf virus.

decorate GVA particles, indicating no serological relationships between GVA and GVB.

Phylogenetic Relationships

Sequences from viruses within the genus *Foveavirus* clustered into a branch different from other TGB-containing viruses (allexi, carla, potex, and mandariviruses) (**Figure 4**). Sequences for viruses within the genera *Capillovirus, Trichovirus,* and *Vitivirus* also clustered into branches different from each other (**Figure 4**).

See also: Allexivirus; Carlavirus; Flexiviruses; Plant Virus Diseases: Fruit Trees and Grapevine; Potexvirus.

Further Reading

Adams MJ, Antoniw JF, Bar-Joseph, *et al.* (2004) The new plant virus family *Flexiviridae* and assessment of molecular criteria for species demarcation. *Archives of Virology* 149: 1045–1066.

Adams MJ, Accotto GP, Agranovsky AA, *et al.* (2005) Family *Flexiviridae.* In: Fauquet CM, Mayo MA, Maniloff J, Desselberger U,, and Ball LA (eds.) *Virus Taxonomy Eighth Report of the International Committee on Taxonomy of Viruses,* pp. 1089–1124. San Diego, CA: Elsevier Academic Press.

Boscia D, Savino V, Minafra A, *et al.* (1993) Properties of a filamentous virus isolated from grapevines affected by corky bark. *Archives of Virology* 130: 109–120.

German S, Candresse T, Lanneau M, *et al.* (1990) Nucleotide sequence and genome organization of apple chlorotic leaf spot closterovirus. *Virology* 179: 1104–1112.

Jelkmann W (1994) Nucleotide sequence of apple stem pitting virus and of the coat protein of a similar virus from pear associated with vein yellows disease and their relationship with potex- and carlaviruses. *Journal of General Virology* 75: 1535–1542.

Koganezawa H and Yanase H (1990) A new type of elongated virus isolated from apple trees containing the stem pitting agent. *Plant Disease* 74: 610–614.

Magome H, Yoshikawa N, Takahashi T, Ito T, and Miyakawa T (1997) Molecular variability of the genomes of capilloviruses from apple, Japanese pear, European pear, and citrus trees. *Phytopathology* 87: 389–396.

Martelli GP and Jelkmann W (1998) *Foveavirus,* a new plant virus genus. *Archives of Virology* 143: 1245–1249.

Martelli GP, Minafra A, and Saldarelli P (1997) *Vitivirus,* a new genus of plant viruses. *Archives of Virology* 142: 1929–1932.

Meng B, Pang S-Z, Forsline PL, McFerson JR, and Gonsalves D (1998) Nucleotide sequence and genome structure of grapevine rupestris stem pitting associated virus-1 reveal similarities to apple stem pitting virus. *Journal of General Virology* 79: 2059–2069.

Minafra A, Saldarelli P, and Martelli GP (1997) Grapevine virus A: Nucleotide sequence, genome organization, and relationship in the *Trichovirus* genus. *Archives of Virology* 142: 417–423.

Yanase H (1974) Studies on apple latent viruses in Japan. *Bulletin of the Fruit Tree Research Station, Japan, Series* C1: 47–109.

Yoshikawa N (2000) Apple stem grooving virus. AAB Descriptions of Plant viruses 376. http://www.dpvweb.net/dpv/showdpv.php?dpvno=376 (accessed January 2008).

Yoshikawa N (2001) Apple chlorotic leaf spot virus. AAB Descriptions of Plant viruses 386. http://www.dpvweb.net/showdpv.php?dpvno=386 (accessed January 2008).

Yoshikawa N, Sasaki E, Kato M, and Takahashi T (1992) The nucleotide sequence of apple stem grooving capillovirus geneme. *Virology* 191: 98–105.

Carlavirus

K H Ryu and B Y Lee, Seoul Women's University, Seoul, South Korea

Glossary

Symptomless Having no apparent symptoms of disease.
Triple gene block (TGB) A specialized evolutionarily conserved gene module of three partially overlapping ORFs involved in the cell-to-cell and long-distance movement of some plant viruses.

Introduction

The genus *Carlavirus* derived from the name of the type species *Carnation latent virus* is one of nine genera in the family *Flexiviridae* and contains a large number of members and tentative members. The carlavirus-infected natural host plants usually have very mild symptoms or remain symptomless, and thus the term 'latent' appears in the names of many species of the genus *Carlavirus*. The natural host range of individual species is restricted to a few plant species. This tendency toward mild or latent (symptomless) infection, a characteristic feature of carlaviruses, has led to carlaviruses and carlavirus-associated diseases failing to attract any special attention from phytopathologists and phytovirologists. However, many members of the genus have been identified in the last few decades and most of them have been associated with more serious diseases when plants are co-infected with other viruses. Most species are transmitted by aphids in a nonpersistent manner but two species (*Coupea mild mottle virus* and *Melon yellowing-associated virus*, CPMMV and MYaV) are transmitted by whiteflies (*Bemisia tabaci*).

Taxonomy and Classification

The genus *Carlavirus* belongs to the family *Flexiviridae*. The type species of the genus is *Carnation latent virus*. The name carlavirus was derived from the type species. The genus contains a large number of members. In total, 68 members of the genus *Carlavirus* (39 definitive species and 29 tentative species) are listed in **Table 1**. Some members listed as species and all members listed as tentative species have no sequence data, and therefore the taxonomic status of these members is yet to be verified.

Carlaviruses are either closely or distantly related to each other and these relationships have been confirmed by phylogenetic analysis of some viral proteins. The phylogenetic tree analysis of sequenced carlaviruses is presented in **Figure 1**.

Species demarcation criteria in this genus include the following:

1. distinct species have less than about 72% nucleotides or 80% amino acids identical between their coat protein (CP) or replicase genes;
2. distinct species are readily differentiated by serology, and strains of individual species are often distinguishable by serology;
3. distinct species do not cross-protect in infected common host plant species, and each distinct species usually has a specific natural host range and distinguishable experimental host ranges.

Virus Structure and Composition

Virions are slightly flexuous filaments that are not enveloped, and measure 610–700 nm in length and 12–15 nm in diameter. Nucleocapsids appear longitudinally striated. Virions have helical symmetry with a pitch of 3.3–3.4 nm. Virion M_r is about 60×10^6 with a nucleic acid content of 5–7%. Virions contain a monopartite, linear, single-stranded, positive-sense RNA that has a size range of 7.4–9.1 kbp in length. The $3'$ terminus has a poly(A) tract and the $5'$ terminus occasionally has a methylated nucleotide cap, or a monophosphate group. CP subunits are of one type, with size 31–36 kDa. The genome is encapsidated in 1600–2000 copies of CP subunit. Purified preparations of carlaviruses such as potato virus S and helinium virus S (PVS, HVS) contain small amounts of encapsidated subgenomic RNAs.

Physicochemical Properties

Virus particles in purified preparations have a sedimentation coefficient of 147–176 S. Isoelectric point of virions is about pH 4.5. The UV absorbance spectra of carlaviruses have maxima at 258–260 nm and minima at 243–248 nm, with A_{max}/A_{min} ratios of 1.1–1.3, and the A_{260}/A_{280} ratio of purified preparation is 1.08–1.40. Physical properties of viruses in this genus are: thermal inactivation point (TIP) 50–85 °C, longevity *in vitro* (LIV) 1–21 days, and dilution endpoint (DEP) 10^{-2}–10^{-7}. Infectivity of sap does not change on treatment with diethyl ether. Infectivity is

Table 1   Virus species in the genus *Carlavirus*

Mode of transmission	Virus species name	Abbreviation	Accession number
Species			
Aphid-transmitted	*American hop latent virus*	AHLV	
	Blueberry scorch virus	BlScV	NC_003499
	Cactus virus 2	CV-2	
	Caper latent virus	CapLV	
	Carnation latent virus	CLV	X52627, AJ010697
	Chrysanthemum virus B	CVB	NC_009087
	Cole latent virus	CoLV	AY340584
	Dandelion latent virus	DaLV	
	Daphne virus S	DVS	AJ620300
	Elderberry symptomless virus	ESLV	
	Garlic common latent virus	GarCLV	AB004805
	Helenium virus S	HVS	D10454
	Honeysuckle latent virus	HnLV	
	Hop latent virus	HpLV	NC_002552
	Hop mosaic virus	HpMV	AB051109
	Hydrangea latent virus	HdLV	
	Kalanchoe latent virus	KLV	AJ293570-1
	Lilac mottle virus	LiMoV	
	Lily symptomless virus	LSV	NC_005138
	Mulberry latent virus	MLV	
	Muskmelon vein necrosis virus	MuVNV	
	Nerine latent virus	NeLV	DQ098905
	Passiflora latent virus	PLV	NC_008292
	Pea streak virus	PeSV	AF354652, AY037925
	Potato latent virus	PotLV	AY007728
	Potato virus M	PVM	NC_001361
	Potato virus S	PVS	NC_007289
	Red clover vein mosaic virus	RCVMV	
	Shallot latent virus	SLV	NC_003557
	Sint-Jan's onion latent virus	SJOLV	
	Strawberry pseudo mild yellow edge virus	SPMYEV	
Whitefly-transmitted	*Cowpea mild mottle virus*	CPMMV	DQ444266
	Melon yellowing-associated virus	MYaV	AY373028
Unknown vector	*Aconitum latent virus*	AcLV	NC_002795
	Narcissus common latent virus	NCLV	NC_008266
	Poplar mosaic virus	PopMV	NC_005343
	Potato rough dwarf virus	PRDV	AJ250314
	Sweetpotato chlorotic fleck virus	SPCFV	NC_006550
	Verbena latent virus	VeLV	AF271218
Tentative species			
Aphid-transmitted	Anthriscus latent virus	AntLV	
	Arracacha latent virus	ALV	
	Artichoke latent virus M	ArLVM	
	Artichoke latent virus S	ArLVS	
	Butterbur mosaic virus	ButMV	
	Caraway latent virus	CawLV	
	Cardamine latent virus	CaLV	
	Cassia mild mosaic virus	CasMMV	
	Chicory yellow blotch virus	ChYBV	
	Coleus vein necrosis virus	CVNV	DQ915963
	Cynodon mosaic virus	CynMV	
	Dulcamara virus A	DuVA	
	Dulcamara virus B	DuVB	
	Eggplant mild mottle virus	EMMV	
	Euonymus mosaic virus	EuoMV	
	Fig virus S	FVS	
	Fuchsia latent virus	FLV	
	Gentiana latent virus	GenLV	
	Gynura latent virus	GyLV	
	Helleborus mosaic virus	HeMV	

Continued

Table 1 Continued

Mode of transmission	Virus species name	Abbreviation	Accession number
	Hydrangea chlorotic mottle virus	HCMV	DQ412999
	Impatiens latent virus	ILV	
	Lilac ringspot virus	LiRSV	
	Narcissus symptomless virus	NSV	NC_008552
	Plantain virus 8	PIV-8	
	Potato virus P	PVP	DQ516055
	Prunus virus S	PruVS	
	Southern potato latent virus	SoPLV	
	White bryony mosaic virus	WBMV	

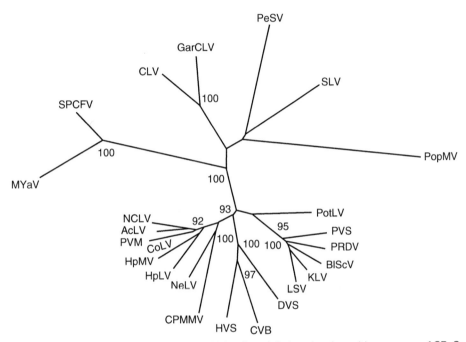

Figure 1 Phylogenetic tree of carlaviruses derived from multiple-aligned deduced amino acid sequences of CP. See **Table 1** for abbreviations of carlavirus names.

retained by the virions despite being deproteinized by proteases and by phenols or detergents.

The CPs of the carlaviruses are partly degraded during purification and storage of virus preparations. The carlavirus virions are susceptible to and can be broken down by denaturing agents such as sodium dodecyl sulfate (SDS), urea, guanidine hydrochloride, acetic acid, and alkali.

Genome Structure and Gene Expression

The genome is a single-stranded RNA (ssRNA), 7.4–9.1 kbp in size, which contains six open reading frames (ORFs), encoding the replicase, three putative protein components of movement proteins called triple gene block (TGB), the CP, and a putative nucleic acid-binding regulatory protein,

from the 5′–3′-end in that order (**Figure 2**). In genome organization, the genus *Carlavirus* particularly resembles the genera *Allexivirus, Foveavirus, Mandarivirus,* and *Potexvirus* but is distinguished from them by its six ORFs and a large replication protein. The genome RNA is capped at the 5′ end and polyadenylated at the 3′ end. The 223 kDa viral replicase ORF is translated from the full-length genomic RNA. For expression of its 3′-proximal viral genes, the virus utilizes at least two subgenomic RNAs. The genome structure of potato virus M (PVM) (8553 nt) is represented in **Figure 2**. PVM RNA has 75 nt 5′ untranslated region (UTR) at the 5′ terminus and 70 nt 3′ UTR followed by a poly(A) tail at the 3′ terminus and intergenic UTRs of 38 and 21 nt between three large blocks of coding sequences. Carlavirus-infected plants contain double-stranded RNA (dsRNA) whose molecular mass corresponds to that of genomic RNA.

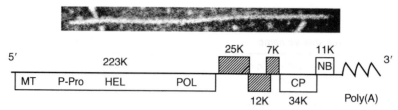

Figure 2 Particle morphology and genome organization of potato virus M (PVM), showing a typical genome structure for the genus *Carlavirus* in the family *Flexiviridae*. The 5′-proximal one large ORF encodes an RNA-dependent RNA polymerase (viral replicase), three overlapping ORFs encode the putative MPs (TGBs), and two ORFs encode CP and nucleic acid binding protein (NB). Motifs in the replicase are methyltransferase (MT), papain-like protease (P-Pro), helicase (HEL), and RNA-dependent RNA polymerase (POL).

ORF1 encodes a 223 kDa polypeptide that is presumed to be the viral replicase. It contains motifs of methyltransferase, a papain-like protease, helicase, and RNA-dependent RNA polymerase. With blueberry scorch virus (BLScV) and probably other carlaviruses, the 223 kDa viral replicase is proteolytically processed by papain-like protease activity, which results in about 30 kDa being removed. ORFs 2–4 form the TGB encoding polypeptides of 25, 12, and 7 kDa, respectively, which facilitate virus cell-to-cell movement. ORF5 encodes the 34 kDa CP, and ORF6 encodes a cysteine-rich 11 kDa protein. The function of the 3′-proximal 11 kDa polypeptide has yet to be elucidated, but its ability to bind nucleic acid and agroin-filtration-mediated transient expression studies suggest that it may facilitate vector transmission, may be involved in viral RNA transcription, or may be a viral pathogenicity determinant for plant defense system.

Putative promoters for the subgenomic RNA synthesis have been identified for several carlaviruses, C/UUUAGGU, 19–43 residues upstream from both putative subgenomic RNA initiation sites.

Complete genomic sequences have recently been obtained for some carlaviruses: BLScV (8512 nt), chrysanthemum virus B (CVB) (8870 nt), daphne virus S (DVS) (8739 nt), hop latent virus (HpLV) (8612 nt), lily symptomless virus (LSV) (8394 nt), PVM (8553 nt), PVS (8478 nt), shallot latent virus (SLV) (8363 nt), aconitum latent virus (AcLV) (8657 nt), poplar mosaic virus (PopMV) (8741 nt), passiflora latent virus (PLV) (8386 nt), narcissus symptomless virus (NSV) (8281 nt), sweetpotato chlorotic fleck virus (SPCFV) (9104 nt), and narcissus common latent virus (NCLV) (8539 nt).

Infectious cDNA clones have been reported for two carlaviruses, BlScV and PopMV.

Viral Transmission

Carlaviruses are spread by mechanical contact, by horticultural and agricultural equipment, and transmitted in a nonpersistent or semipersistent manner by aphids with varying efficiencies, or by whiteflies. Some members are not transmitted by mechanical inoculation. Seed transmission may occur with some legume-infecting carlaviruses, but is not common. Transmission by root grafts may occur in the case of PopMV. Those viruses that infect vegetatively propagated hosts persist in the propagated parts.

Cytopathology

The distribution of carlaviruses in the infected plant is not tissue specific. Virions are usually found in the cytoplasm, or sometimes in chloroplasts or in mitochondria of infected tissues, or may occur in membrane-associated bundle-like or plate-like aggregates. Inclusions are present in infected cells of some members. They occur as crystals in the cytoplasm, as amorphous X-bodies, membranous bodies, and viroplasms and they sometimes contain virions.

Host Range

Infection by most members of this genus is symptomless in the natural host. The individual natural host ranges are, usually narrow, although a few species can infect a wide range of experimental hosts.

Symptomatology

Carlavirus infection often results in no apparent symptoms in natural hosts. Symptoms vary cyclically or seasonally or may disappear soon after infection. If symptoms are evident, more severe symptoms such as mosaics appear in early stages of infection. Some carlaviruses such as PVM, PVS, and BlScV cause diseases that are of economic importance on their own; however, most of them are associated with more serious diseases when the plants are co-infected with other viruses.

Serology

Carlavirus virions are good immunogens. Serological relationships among carlaviruses may be close or distant or sometimes undetectable; some species are serologically interrelated but others apparently distinct. A number of carlaviruses are more or less closely interrelated serologically, with serological differentiation indices (SDIs) ranging from about 3.5 to 6.5.

Geographical Distribution

Some carlaviruses are found wherever their natural hosts are grown, but the geographical distribution of many species is restricted to only certain parts of the world. Those infecting vegetatively propagated crops are usually widely distributed. Most species commonly occur in temperate regions, but whitefly transmitted carlaviruses are restricted to tropical and subtropical regions.

Viral Epidemiology and Control

Most carlavirus-associated diseases are usually very mild or symptomless and few efforts are made to control them. However, crops such as potatoes, certain legumes, and blueberries that may contain more damaging carlaviruses require suitable control measures. Seed potatoes and vegetatively propagated materials must be screened continuously to certify a virus-free status. Rapid removal of infected plants is particularly important for plants associated with aphid- or whitefly-vectored carlaviruses. PVS is the only carlavirus for which transgenic plants (potato and *Nicotiana debney*), which are resistant to virus infection, have been reported so far.

See also: Allexivirus; Capillovirus, Foveavirus, Trichovirus, Vitivirus; Flexiviruses; Plant Virus Vectors (Gene Expression Systems); Potexvirus; Vector Transmission of Plant Viruses.

Further Reading

Adams MJ, Antoniw JF, Bar-Joseph M, *et al.* (2004) The new plant virus family *Flexiviridae* and assessment of molecular criteria for species demarcation. *Archives of Virology* 149: 1045–1060.

Fauquet CM, Mayo MA, Maniloff J, Desselberger U,, and Ball LA (eds.) (2005) *Virus Taxonomy, Classification and Nomenclature of Viruses, Eighth Report of the International Committee on the Taxonomy of Viruses*, p. 1101. San Diego, CA: Elsevier Academic Press.

Hillman BI and Lawrence DM (1994) Carlaviruses. In: Singh RP, Singh US,, and Kohmoto K (eds.) *Pathogenesis and Host Specificity in Plant Diseases*, vol. 3, 35pp. New York: Plenum.

Lee BY, Min BE, Ha JH, Lee MY, Paek KH, and Ryu KH (2006) Genome structure and complete sequence of genomic RNA of Daphne virus S. *Archives of Virology* 151: 193–200.

Martelli GP, Adams MJ, Kreuze JF, and Dolja VV (2007) Family Flexiviridae: A case study in virion and genome plasticity. *Annual Review of Phytopathology* 45. 73–100.

Carmovirus

F Qu and T J Morris, University of Nebraska, Lincoln, NE, USA

Taxonomy, Classification and Evolutionary Relationships

Carmovirus is one of eight genera in the family *Tombusviridae*. Members of this family all have icosahedral virions of about 30 nm in diameter with $T = 3$ symmetry that consists of 180 coat protein (CP) subunits of about 38–43 kDa and a single-stranded (ss) RNA genome ranging in size from 4.0 to 4.8 kbp. Carmoviruses share recognizable yet varied sequence similarity with members of other genera of *Tombusviridae*.

Carmoviruses contain a single-component positive-sense genome of about 4.0 kbp. The genome, as exemplified by turnip crinkle virus (TCV) in **Figure 1**, consists of five definitive open reading frames (ORFs) which encode proteins of about 28, 88, 8, 9, and 38 kDa from the 5′ to the 3′ end, respectively. Some carmoviruses such as hibiscus chlorotic ringspot virus (HCRSV) have additional ORFs of unknown functions. The virions are icosahedral in symmetry and consist of 180 CP subunits of approximately 38 kDa. The genus name is derived from the first member of the genus to be sequenced, carnation mottle virus (CarMV). Much more detailed knowledge about virus structure and genome function is, however, known for TCV because its crystal structure has been determined and it was the first carmovirus for which infectious transcripts were produced from a cDNA clone of the genome.

To date, the nucleotide sequences of 12 definitive carmoviruses have been determined (see **Table 1**). These sequenced members share similar morphological and physicochemical properties with about a dozen other viruses listed in **Table 1** that are recognized as species or tentative species depending on the level of detail of the molecular characterization of the viral genomes. Various carmoviruses are sufficiently distant from each other that

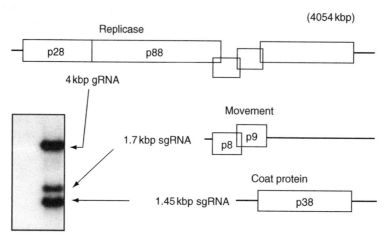

Figure 1   Genome organization of a typical carmovirus as represented by turnip crinkle virus (TCV). The boxes represent open reading frames with the sizes of the encoded proteins indicated within the boxes, in kilodaltons. The two proteins required for replication (p28 and p88) are translated from the 4 kbp genomic RNA with the p88 protein translated by readthrough of an amber codon at the end of the p28 gene. Two small proteins involved in cell-to-cell movement (p8 and p9) are translated from a 1.7 kbp subgenomic RNA. The viral coat protein (p38) is translated from a 1.45 kbp subgenomic RNA. The panel on the left is a Northern blot showing the typical pattern of accumulation of genomic and subgenomic RNAs in carmovirus-infected cells.

Table 1   Virus members in the genus *Carmovirus*

Sequenced viruses	
Angelonia flower break virus	(AnFBV)
Cardamine chlorotic fleck virus	(CCFV)
Carnation mottle virus	(CarMV)
Cowpea mottle virus	(CPMoV)
Galinsoga mosaic virus	(GaMV)
Hibicus chlorotic ringspot virus	(HCRSV)
Japanese iris necrotic ring virus	(JINRV)
Melon necrotic spot virus	(MNSV)
Pea stem necrosis virus	(PSNV)
Pelargonium flower break virus	(PFBV)
Saguaro cactus virus	(SCV)
Turnip crinkle virus	(TCV)
Unsequenced viruses	
Ahlum water-borne virus	(AWBV)
Bean mild mosaic virus	(BMMV)
Cucumber soil-borne virus	(CSBV)
Weddel water-borne virus	(WWBV)
Viruses assigned to tentative species	
Blackgram mottle virus	(BMoV)
Calibrachoa mottle virus	(CbMV)
Elderberry latent virus	(ELV)
Glycine mottle virus	(GMoV)
Narcissus tip necrosis virus	(NTNV)
Plaintain virus 6	(PIV-6)
Squash necrosis virus	(SqNV)
Tephrosia symptomless virus	(TeSV)

they do not cross-react in standard RNA hybridization or serological tests.

Carmoviruses share properties with viruses belonging to other genera of the family *Tombusviridae*. Their particle structure and CP sequences are closely related to tombus-, aureus-, diantho-, and avenaviruses. Their RNA-dependent RNA polymerase (RdRp) genes share significant homology with viruses of the following genera: *Machlomovirus, Panicovirus, Necrovirus, Aureusvirus*, and *Tombusvirus*. Carmovirus RdRp genes also share similarity with more distantly related viruses outside of *Tombusviridae*, such as umbraviruses and luteoviruses. In a broader context, phylogenetic comparisons of viral RNA polymerase genes have identified the *Tombusviridae* as a representative plant virus cluster for one of three RNA virus supergroups with relatedness to animal viruses of *Flaviviridae* and small RNA phage (*Leviviridae*).

Distribution, Host Range, Transmission, and Economic Significance

Carmoviruses occur worldwide and are generally reported to cause mild or asymptomatic infections on relatively restricted natural host ranges. Most accumulate to high concentrations in infected tissues and are mechanically transmitted. Beetle transmission has been reported for some carmoviruses as has transmission in association with soil and/or irrigation water, and in some cases in association with fungal zoospores.

A number of carmoviruses have been identified in association with ornamental hosts and have been widely distributed in such hosts by vegetative propagation. CarMV is the most noteworthy, being widespread in cultivated carnations, and recognized as one of the more important components of viral disease complexes in this crop worldwide. It accumulates to high concentrations without producing severe symptoms and spreads primarily by contact transmission and vegetative propagation. It has a broad experimental host range that includes over 30 species in 15 plant families. Pelargonium flower break virus (PFBV) is widespread in vegetatively propagated *Pelargonium* species causing disease in association with other viruses. The incidence of narcissus tip necrosis virus (NTNV) in narcissus cultivars and HCRSV in hibiscus primarily reflects distribution of infected nursery stock.

Numerous small RNA viruses have been reported to naturally infect cucurbits causing significant disease problems. Several of these viruses are recognized tombusviruses while others such as melon necrotic spot virus (MNSV) and cucumber soil-borne virus (CSBV) have been identified as carmoviruses based on sequence and genome organization properties. MNSV occurs worldwide in greenhouse cucurbits and is both soil and seed transmitted, while CSBV has been primarily restricted to infrequent outbreaks around the Mediterranean. Both have been reported to be transmitted in association with the fungus *Olpidium bornovanus*.

Several carmoviruses have been discovered in natural leguminous hosts, with glycine mottle virus (GMoV) being potentially the most important, causing serious disease losses in legumes in Africa. Bean mild mosaic virus (BMMV) has been reported to be a latent virus widely distributed in bean cultivars in El Salvador, and backgram mottle virus (BMoV) has been found in *Vigna* species in Asia. Beetle vectors have been identified for these viruses, but seed transmission may also be an important factor in their distribution.

TCV is neither common nor widespread in nature in spite of the fact that it is reportedly beetle transmitted. It has a relatively wide experimental host range in some 20 plant families including experimentally useful species such as *Arabidopsis* and *Brassica* in which it accumulates to extremely high concentrations, often approaching a level equivalent to 0.5% of the fresh weight of the plant tissue. Cardamine chlorotic fleck virus (CCFV) was first discovered in the Mount Kosiusko alpine region of Australia in *Cardamine lilacina*, a wild perennial of the family Brassicaceae. It has also been shown to infect *Arabidopsis* and other Brassicaceae species.

Other carmoviruses have been isolated worldwide from natural hosts with little apparent disease and are presumably of little agricultural concern. These include tephrosia symptomless virus (TeSV) from legumes in Kenya, GMoV from glycine in Australia, saguaro cactus virus (SCV) from saguaro cactus in Arizona galinsoga mosaic virus (GaMV) from potato weed in Australia, and plaintain virus 6 (PlV-6) from plantain weed in England. The infrequent isolation of these genetically similar viruses in remote locations around the world has prompted the speculation that ancestor carmoviruses may have been introduced into their natural hosts well before the last Ice Age and have since co-evolved in isolation in their diverse host plants.

Virion Structure and Assembly

Four carmoviruses (TCV, CarMV, cowpea mottle virus (CPMoV), and HCRSV) have now been analyzed in structural detail, with TCV being the first and most thoroughly studied by high-resolution X-ray crystallography. The detailed information about CP structure and intersubunit interactions established that TCV and tomato bushy stunt tombusvirus (TBSV) show marked structural conservation. In this regard, the common structural features shared by other members of the family *Tombusviridae* have been primarily deduced from alignment of the amino acid sequence of the coat proteins of TBSV and TCV. TCV consists of a $T = 3$ icosahedral capsid of 180 subunits of the 38 kDa CP. The individual CP subunit folds into three distinct domains typical of CP subunit of tombusviruses. The relatively basic N-terminal R domain extends into the interior of the virus particle and presumably interacts with viral RNA. The R domain is connected by an arm to the S domain which constitutes the virion shell. The S domain is attached through a hinge to the P domain which projects outward from the virion surface. The protein subunits are believed to form dimers in solution and during assembly.

TCV is also the only carmovirus on which detailed *in vitro* assembly studies have been performed. The virion has been shown to dissociate at elevated pH and ionic strength to produce a stable RNA–CP complex (rp-complex) and free CP subunits. Reassembly under physiological conditions in solution could be demonstrated using the isolated rp-complex and the soluble CP subunits. This rp-complex, consisting of six CP subunits tightly attached to viral RNA, could be generated *in vitro* and was shown to be important in selective assembly of TCV RNA. A model for assembly was proposed in which three sets of dimeric CP interact with a unique site on the viral RNA to form an initiation complex to which additional subunit dimers could rapidly bind. Preliminary characterization of the 'origin of assembly' for this virus identified two possible sites based on the identification of RNA fragments in the rp-complex protected from RNase digestion by CP. Further *in vivo* studies narrowed the assembly origin site to a bulged hairpin-loop of 28 nt within a 180 nt region at the $3'$ end of the CP gene.

Genome Structure

Complete nucleotide sequences have been determined for 12 carmoviruses as listed in **Table 1**. Comparative studies of the deduced ORFs revealed that all of these viruses encode a similar set of genes that are closely related and in the same gene order as illustrated for TCV in **Figure 1**. The genome organization of the carmoviruses is quite compact with most of the identified

ORFs overlapping each other. Both the product of the most 5′ proximal ORF (26–28 kDa) and its readthrough product (86–89 kDa) are essential for replication of the TCV genome. The 3′ proximal gene encodes the viral CP which varies from 37 to 42 kDa for the different viruses. All of the sequenced carmoviruses characteristically encode two small ORFs in the middle of the genome that both have been shown in TCV to be indispensable for cell-to-cell movement (movement proteins or MP). Although the genome organizations of all sequenced carmoviruses are quite similar, there are some unique features evident in the individual carmoviruses. For example, the two small central ORFs in MNSV (p7a and p7b) are connected by an in-frame amber codon that could result in the production of a 14 kDa fusion protein of the two ORFs by a readthrough mechanism. The CPMoV as well as HCRSV are predicted to encode a sixth ORF nested within the 3′ proximal CP gene. HCRSV has also been reported to encode another novel protein that is nested within the RdRp gene.

The 5′ end of the genome is not capped. The 5′ noncoding region varies from 34 nt in CPMoV to 88 nt in MNSV. No extensive sequence homology was observed within this region. The 3′ noncoding region of carmoviruses varies from about 200 to 300 nt in length and possesses neither a poly-A tail nor a tRNA-like structure.

Replication and Gene Expression

Carmoviruses replicate to very high concentrations in protoplasts, with the genomic RNA accumulating to levels approaching that of the ribosomal RNAs. Upon infection of susceptible plants, carmoviruses transcribe two 3′ coterminal subgenomic RNAs (sgRNAs) for expression of the MP and CP genes. The smaller sgRNA (*c.* 1.5 kbp) is the mRNA for CP. The larger sgRNA (*c.* 1.7 kbp) presumably functions as the mRNA for the two MP genes utilizing a leaky scanning mechanism. Results involving transgenic expression of the p8 and p9 gene products of TCV in *Arabidopsis* plants have demonstrated that both of them are essential for viral cell-to-cell movement and that they function by in *trans*-complementation in the same cell.

Carmoviruses are thought to replicate through a (−) strand intermediate because virus-specific double-stranded RNAs (dsRNAs) corresponding in size to the genomic RNA and sgRNAs characteristically accumulate in infected plant tissue. The product of the 5′ proximal ORF (p28 in TCV) and its readthrough product (p88 in TCV) are the only virus-encoded components of the polymerase complex. When expressed from two separate mRNAs, p28 and p88 complemented in *trans* to enable the genome replication. The only host factor found to augment TCV replication so far is the eukaryotic translation initiation factor 4G (eIF4G) of *Arabidopsis*, presumably through more efficient translation of viral genes.

Recently, a membrane-containing extract prepared from evacuolated protoplasts of uninfected *Arabidopsis* plants has been shown to faithfully produce both genomic and subgenomic RNAs from a full-length TCV RNA template. Such an extract should be useful in characterizing viral as well as host elements required for virus replication.

Recent studies have demonstrated that, besides the virus-encoded proteins, a plethora of structural elements in the viral RNA play critical roles in the genome replication of carmoviruses. These elements are located throughout the entire viral genome and occur in both the (+) and (−) strands. Their roles range from promoters, enhancers, or repressors for RNA replication and transcription, enhancers for translation, and specificity determinants for virus assembly. The current knowledge of these RNA structures suggests that different secondary structural motifs, some of them mutually exclusive, are formed at the different stages of virus multiplication, and the highly coordinated nature of their formation ensures optimal utilization of the compact viral genome.

Satellites, Defective-Interfering RNAs

TCV is the only carmovirus in which replication of associated small subviral RNAs in infected plants has been characterized, and the situation for this virus is curiously complex. TCV infections are associated with defective interfering (DI) RNAs derived totally from the parent genome (e.g., RNA G of 342–346 nt), satellite RNAs of nonviral origin (e.g., RNAs D, 194 nt and F, 230 nt), and chimeric RNAs (e.g., RNA C of 356 nt) with a 5′ region derived from sat RNA D and a 3′ region derived from the 3′ end of the TCV genome. All three types of small RNAs depend on the helper virus for their replication and encapsidation within the infected plant. The different satellite and DI RNAs have been shown to affect viral infections in different ways. Both RNA C and G intensify viral symptoms while interfering with the replication of the helper virus, while RNAs D and F seem to produce no detectable effects on either expression of symptoms or helper virus replication.

Virus–Host Interaction

Considerable progress has been made in the last decade in our understanding of the molecular mechanisms of virus–plant interactions with significant contributions coming from studies utilizing TCV and its host plant *Arabidopsis*. Most notably, one ecotype of *Arabidopsis thaliana* has been determined to be resistant to TCV infection and the resistance gene (*HRT*) and its encoded protein have been characterized. Additional host factors that contribute to the resistance response have also been elucidated including the identification of a novel transcription factor (TIP) whose interaction with TCV CP is critical for the initiation of the resistance response. The determination that TCV CP is targeted by HRT resistance protein suggests that TCV CP plays a key role in combating antiviral defense mechanisms of the plant host, in addition to being required for virus assembly. Interestingly, recent studies have also established that TCV CP is a strong suppressor of RNA silencing, a host defense mechanism that targets invading RNA. RNA silencing is a potent defense mechanism that is conserved in nearly all eukaryotic organisms. Accordingly, TCV CP is an effective silencing suppressor in both plant and animal cells. Silencing suppressor activity has also been associated with several other carmovirus CPs, establishing this activity as a conserved feature of the structural subunit.

TCV CP is also important in satellite RNA C interactions in the host plant. Normally, the presence of RNA C results in symptom intensification in TCV infections. However, when the TCV CP ORF is either deleted or replaced by the CCFV CP ORF, RNA C attenuates symptoms caused by the helper virus suggesting that CP either downregulates the replication of RNA C or enhances its own competitiveness.

Finally, the replicase gene has also been implicated in the symptom modification by satellite RNA C by two independent groups. The 3' end of the TCV genome, a sequence common in TCV RNA, RNA C, and DI RNA G, was also suggested to be a symptom determinant. Environmental conditions also affect the extent of resistance of *Arabidopsis* plants to TCV. In conclusion, it is clear that recent intensive studies of TCV have established this small RNA virus as an ideal model for unraveling the complicated processes involved in viral pathogenesis.

See also: Legume Viruses; Luteoviruses; Machlomovirus; Necrovirus; Plant Virus Diseases: Ornamental Plants; Tombusviruses.

Further Reading

Hacker DL, Petty ITD, Wei N, and Morris TJ (1992) Turnip crinkle virus genes required for RNA replication and virus movement. *Virology* 186: 1–8.

Kachroo P, Yoshioka K, Shah J, Dooner HK, and Klessig DF (2000) Resistance to turnip crinkle virus in *Arabidopsis* is regulated by two host genes, is salicylic acid dependent but NPR1, ethylene and jasmonate independent. *Plant Cell* 12: 677–690.

Komoda K, Naito S, and Ishikawa M (2004) Replication of plant RNA virus genomes in a cell-free extract of evacuolated plant protoplasts. *Proceedings of the National Academy of Sciences, USA* 101: 1863–1867.

Lommel SA, Martelli GP, Rubino L, and Russo M (2005) *Tombusviridae*. In: Fauquet CM, Mayo MA, Maniloff J, Desselberger U, and Ball LA (eds.) *Virus Taxonomy: Eighth Report of the International Committee on Taxonomy of Viruses*, pp. 907–936. San Diego, CA: Elsevier Academic Press.

Morris TJ and Carrington JC (1988) Carnation mottle virus and viruses with similar properties. In: Koenig R (ed.) *The Plant Viruses: Polyhedral Virions with Monopartite RNA Genomes*, vol. 3, pp. 73–112. New York: Plenum.

Nagy PD, Pogany J, and Simon AE (1999) RNA elements required for RNA recombination function as replication enhancers *in vitro* and *in vivo* in a plus-strand RNA virus. *EMBO Journal* 18: 5653–5665.

Qu F and Morris TJ (2005) Suppressors of RNA silencing encoded by plant viruses and their role in viral infections. *FEBS Letters* 579: 5958–5964.

Ren T, Qu F, and Morris TJ (2000) HRT gene function requires interaction between a NAC protein and viral capsid protein to confer resistance to turnip crinkle virus. *Plant Cell* 12: 1917–1925.

Russo M, Burgyan J, and Martelli GP (1994) Molecular biology of *Tombusviridae*. *Advances in Virus Research* 44: 381–428.

Yoshii M, Nishikiori M, Tomita K, *et al.* (2004) The *Arabidopsis cucumovirus multiplication 1* and 2 loci encode translation initiation factors 4E and 4G. *Journal of Virology* 78: 6102–6111.

Caulimoviruses: General Features

J E Schoelz, University of Missouri, Columbia, MO, USA

Glossary

Agroinoculation An inoculation technique in which *Agrobacterium tumefaciens* is used to deliver a full-length infectious clone of a virus into a plant cell.

Plasmodesmata A narrow channel of cytoplasm that functions as a bridge between two plant cells to facilitate movement of macromolecules.

Reverse transcriptase An enzyme that utilizes an RNA template for synthesis of DNA.

Ribosome shunt A translational mechanism in which ribosomes enter the RNA at the 5' end and scan for a short distance before being translocated to a downstream point.

Semipersistent transmission Vector acquires the virus in minutes to hours, and can transmit to other plants for hours after the initial feeding.

Transgene A gene introduced into an organism through any one of a number of genetic engineering techniques.

Introduction

The members of the family *Caulimoviridae* are plant viruses that replicate by reverse transcription of an RNA intermediate and whose virions contain circular, double-stranded DNA (dsDNA). They replicate by reverse transcription, but unlike the true retroviruses, integration into the host chromosomes is not required for completion of their replication cycle. The circular DNA encapsidated in virions is not covalently closed, as it contains at least one discontinuity in each DNA strand, and these discontinuities occur as a consequence of the replication strategy of the virus. There are six genera in this family and they can be divided into two groups based on virion morphology; the members of the genera *Caulimovirus*, *Petuvirus*, *Cavemovirus*, and *Soymovirus* contain viruses that form icosahedral particles that are largely found within amorphous inclusion bodies in the cell (**Figures 1(a)** and **1(b)**). In contrast, the members of the genera *Badnavirus* and *Tungrovirus* form bacilliform particles that are not associated with inclusion bodies (**Figure 1(c)**). Their virions are found in the cytoplasm either individually or clustered in palisade-like arrays.

Cauliflower mosaic virus (CaMV), a member of the type species *Cauliflower mosaic virus* of the genus *Caulimovirus*, was the first of the plant viruses to be shown to contain dsDNA in its icosahedral virion by Shepherd in 1968. This discovery led to an extended investigation into its replication strategy throughout the 1970s and early 1980s, culminating in Pfeiffer and Hohn's study in 1983 that showed that CaMV replicates through reverse transcription of an RNA intermediate. Perhaps because the genome of CaMV is composed of dsDNA, it was also the first of the plant viruses to be completely sequenced and cloned into bacterial plasmids in an infectious form.

In the early 1980s, CaMV was thought to have some promise as a vector for foreign genes in plants. However, the effort to convert CaMV into a vector was scaled back when it was shown that the virus genome could tolerate only small insertions of up to a few hundred basepairs of DNA. A few small genes, such as dihydrofolate reductase and interferon, were eventually expressed in plants via a CaMV vector, but other viruses have been shown to be much more versatile as vectors for foreign genes. Although the caulimoviruses have had only limited utility as plant virus vectors, they continue to have a great impact on plant biotechnology. The 35S promoter of CaMV is capable of directing a high level of transcription in most types of plant tissues. This promoter was used to drive expression of one of the first transgenes introduced into transgenic plants, and it is still widely used for expression of transgenes for both research and commercial applications. The promoter regions of several other caulimoviruses have also been evaluated for expression of transgenes in both monocots and dicots, and they can be good alternatives to the CaMV 35S promoter.

The feature that was used initially to characterize the caulimoviruses was the presence of circular dsDNA encapsidated into icosahedral virions of approximately 50 nm diameter. However, as icosahedral viruses continued to be isolated from a variety of hosts and their genomes sequenced, it was discovered that their genome structure diverged from that of CaMV. These differences were significant enough for the International Committee on Taxonomy of Viruses (ICTV) to create three new genera in its VIII ICTV Report. Consequently, *Soybean chlorotic mottle virus*, *Cassava vein mosaic virus*, and *Petunia vein clearing virus* became the type species of the genera *Soymovirus*, *Cavemovirus*, and *Petuvirus*, respectively.

For many years, only plant viruses that had icosahedral virions of 50 nm in diameter were thought to have genomes composed of dsDNA. However, in 1990, Lockhart showed that the bacilliform virions of commelina yellow mottle virus (CoYMV) contained circular, dsDNA, and *Commelina yellow mottle virus* became the type species for the genus *Badnavirus*. Lockhart showed that nucleic acid isolated from the bacilliform virions was resistant to RNase and degraded by DNase. Furthermore, DNA treated with S1 nuclease revealed that the CoYMV genome contained at

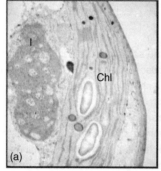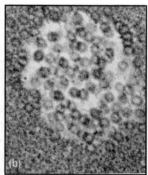

Figure 1 Inclusion bodies and virions of the family *Caulimoviridae*. (a) Amorphous inclusion body of CaMV (I) adjacent to a chloroplast (Chl). Individual virions can be seen in the vacuolated regions of the inclusion body. (b) Icosahedral virions of CaMV visualized within an inclusion body. (c) Purified bacilliform virions of a badnavirus. (a) Reproduced with permission from the *Encyclopedia of Virology*, 2nd edn. (c) Photo courtesy of Ben Lockhart (University of Minnesota).

least two single-stranded discontinuities, a hallmark of the plant viruses that replicate by reverse transcription. In addition to CoYMV, Lockhart also purified DNA from banana streak virus, kalanchoe top-spotting virus (KTSV), and canna yellow mottle virus, providing evidence that they should also be placed in the new genus *Badnavirus*. There are now 18 species in the genus *Badnavirus* (**Table 1**), many of which cause economically important diseases in the tropics. However, new badnaviruses have been characterized since the publication of the VIII Report of the ICTV, so this number will almost certainly be revised upward in the near future.

Soon after the discovery that CoYMV was a DNA virus, the bacilliform component of rice tungro disease was also shown to contain circular, dsDNA. Rice tungro is the most important virus disease of rice in South and Southeast Asia, with annual losses approaching $1.5 billion dollars. The disease is caused by a complex of an RNA virus, called rice tungro spherical virus (RTSV) coupled with the bacilliform dsDNA virus called rice tungro bacilliform virus (RTBV). The genus *Tungrovirus* is distinguished from the genus *Badnavirus* because RTBV has one more open reading frame (ORF) than CoYMV and overall, the RTBV genome has only 20–25% nucleotide sequence identity with members of the genus *Badnavirus*. There is only one species in the genus *Tungrovirus*, although several isolates of the species *Rice tungro bacilliform virus* have been collected and sequenced.

Taxonomy and Classification

The family *Caulimoviridae* consists of six genera (**Table 1**), and they can be conveniently divided into two groups based on virion morphology. Viruses with icosahedral virions (**Figure 1(b)**) include the genera *Caulimovirus*, *Petuvirus*, *Soymovirus*, and *Cavemovirus*. Viruses with bacilliform virions (**Figure 1(c)**) include the genera *Badnavirus* and *Tungrovirus*. Genera can be further distinguished because of differences in genome organization and nucleotide differences in common genes such as the reverse transcriptase. All members of the family *Caulimoviridae* infect only plants. There are no animal or insect viruses in this family.

Virion Structure and Composition

The viruses in the genera *Caulimovirus*, *Cavemovirus*, *Soymovirus*, and *Petuvirus* form nonenveloped, isometric particles that vary in size from 43 to 50 nm. The virion is composed of 420 subunits with a $T = 7$ structure. The viruses in the genera *Badnavirus* and *Tungrovirus* form nonenveloped, bacilliform particles that are 30 nm in width, but can vary in length from 60 to 900 nm. The length most commonly observed is 130 nm. Their

structure is based on an icosahedron, in which the ends are formed from pentamers and the tubular section is made up of hexamers.

The virions of the family *Caulimoviridae* contain a single, circular dsDNA that is 7.2–8.3 kbp in length. Complete and partial nucleotide sequences of members of the family *Caulimoviridae* are listed in **Table 1**. Only a single accession number is given for CaMV and RTBV, although multiple strains of each have been sequenced.

Genome Organization and Expression

Several genome features are common to all members of the family *Caulimoviridae*, in addition to their circular dsDNA genomes. For example, the dsDNA is not covalently closed; it has at least two discontinuities, but may have up to four. Furthermore, all of the ORFs are found on only one of the DNA strands and all of the viruses have at least one strong promoter that drives the expression of a terminally redundant mRNA. In addition, all of the viruses have a reverse transcriptase, and in most cases it is located downstream from the coat protein. Several active sites can be identified within the reverse transcriptase protein, including a protease, the core reverse transcriptase, and RNaseH activity. The reverse transcriptase of PVCV is distinguished from all other members of the family *Caulimoviridae* because it has the core features of an integrase function, in addition to other functions. The most significant difference among the caulimoviruses concerns the arrangement and number of ORFs, and this is illustrated in an examination of the genome organization and expression strategies of CaMV and CoYMV (**Figure 2**).

The genome of CaMV is approximately 8000 bp in size and it contains three single-stranded discontinuities (**Figure 2(a)**). One discontinuity occurs in the negative-sense DNA strand and, by convention, this is the origin of the DNA sequence. Two other discontinuities occur in the positive-sense strand, one at nucleotide position 1600 and a second at approximately nucleotide position 4000. The virus genome consists of seven ORFs. ORF1 encodes a protein necessary for cell-to-cell movement (P1). ORF2 and -3 (proteins P2 and P3) are both required for aphid transmission. P2 is responsible for binding to the aphid stylet, whereas P3 is a virion-associated protein. To complete the bridge for aphid transmission of virions, the C-terminus of P2 physically interacts with the N-terminus of P3. Recent evidence also indicates that P3 may have an additional role in cell-to-cell movement. ORF4 encodes the coat protein, and ORF5 encodes a reverse transcriptase that also has protease and RNaseH domains. With the exception of PVCV, none of the reverse transcriptases of the *Caulimoviridae* have evidence for an integrase function. The ORF6 product (P6) was originally described as the major inclusion body protein, but it also has a function as a

Table 1   Virus members in the family *Caulimoviridae*

Genus	Virus	Abbreviation	Host (family)	Geographic distribution	Representative accession number
Caulimovirus	Carnation etched ring virus	CERV	Carnation (Caryophyllaceae)	Worldwide	X04658
	Cauliflower mosaic virus[a]	**CaMV**	***Brassica* sp. (Crucifereae)**	**Worldwide**	**V00140**
	Dahlia mosaic virus	DMV	Dahlia (Compositae)	Worldwide	
	Figwort mosaic virus	FMV	Figwort (Scrophulariaceae)	USA	X06166
	Horseradish latent virus	HRLV	Horseradish (Crucifereae)	Denmark	
	Mirabilis mosaic virus	MiMV	Mirabilis (Nyctaginaceae)	USA	AF454635
	Strawberry vein banding virus	SVBV	Strawberry (Rosaceae)	Worldwide	X97304
	Thistle mottle virus	ThMoV	Thistle (Compositae)	Europe	
	Aquilegia necrotic mosaic virus[b]	ANMV	Columbine (Ranunculaceae)	East Asia, Japan	
	Plantago virus 4[b]	PlV-4	Plantain (Plantaginaceae)	United Kingdom	
	Sonchus mottle virus[b]	SMoV	*Sonchus* sp. (Asteraceae)		
Petuvirus	**Petunia vein clearing virus**	**PVCV**	**Petunia (Solanaceae)**	**Worldwide**	**U95208**
Soymovirus	Blueberry red ringspot virus	BRRSV	Blueberry (Ericaceae)	USA	AF404509
	Peanut chlorotic streak virus	PCSV	Peanut (Leguminosae)	India	U13988
	Soybean chlorotic mottle virus	**SbCMV**	**Soybean (Leguminosae)**	**Japan**	**X15828**
	Cestrum yellow leaf curling virus[b]	CmYLCV	*Cestrum* sp. (Solanaceae)	Italy	AF364175
Cavemovirus	**Cassava vein mosaic virus**	**CsVMV**	**Cassava (Euphorbiaceae)**	**Brazil**	**U59751**
	Tobacco vein clearing virus	TVCV	*Nicotiana* sp. (Solanaceae)	Worldwide	AF190123
Badnavirus	Aglaonema bacilliform virus	ABV	*Aglaonema* sp. (Araceae)	Southeast Asia	
	Banana streak GF virus	BSGFV	Banana (Musaceae)	Worldwide	AY493509
	Banana streak MYsore virus	BSMyV	Banana (Musaceae)	Worldwide	AY805074
	Banana streak OL virus	BSOLV	Banana (Musaceae)	Worldwide	AJ002234
	Cacao swollen shoot virus	CSSV	*Theobroma* sp. (Sterculiaceae)	Africa	NC001374
	Canna yellow mottle virus	CaYMV	Canna (Cannaceae)	Japan, USA	
	Citrus mosaic virus	CMBV	*Citrus* sp. (Rutaceae)	India	AF347695
	Commelina yellow mottle virus	**ComYMV**	***Commelina* (Commelinaceae)**	**Caribbean**	**X52938**
	Dioscorea bacilliform virus	DBV	Yam (Dioscoreaceae)	Africa	X94576
	Gooseberry vein banding-associated virus	GVBAV	Ribes sp. (Grossulariaceae)	Worldwide	AF298883
	Kalanchoe top-spotting virus	KTSV	*Kalanchoe* (Crassulaceae)	UK, USA	AY180137
	Piper yellow mottle virus	PYMoV	Pepper (Piperaceae)	Brazil, India, Asia	
	Rubus yellow net virus	RYNV	Raspberry (Rosaceae)	Eurasia, USA	AF468454
	Schefflera ringspot virus	SRV	*Schefflera* (Araliaceae)	Worldwide	
	Spiraea yellow leaf spot virus	SYLSV	Spiraea (Rosaceae)		AF299074
	Sugarcane bacilliform IM virus	SCBIMV	Sugarcane (Poaceae)	USA, Cuba, Morocco	AJ277091
	Sugarcane bacilliform Mor virus	SCBMV	Sugarcane (Poaceae)	USA, Cuba, Morocco	M89923
	Taro bacilliform virus	TaBV	Taro (Araceae)	South Pacific	AF357836
	Aucuba bacilliform virus[b]	AuBV	*Aucuba* sp. (Garryaceae)	Japan, UK	

Continued

Table 1 Continued

Genus	Virus	Abbreviation	Host (family)	Geographic distribution	Representative accession number
	Mimosa bacilliform virus[b]	MBV	Mimosa sp. (Fabaceae)	USA	
	Pineapple bacilliform virus[b]	PBV	Pineapple (Bromeliaceae)	Australia	Y12433
	Stilbocarpa mosaic bacilliform virus[b]	SMBV	Stilbocarpa (Araliaceae)	Subantarctic islands	AF478691
	Yucca bacilliform virus[b]	YBV	Yucca (Agavaceae)	South America, Italy	AF468688
Tungrovirus	**Rice tungro bacilliform virus**	**RTBV**	**Rice (Poaceae)**	**East Asia, China**	**X57924**

[a]The entry for type member of each genus is highlighted in bold font.
[b]Tentative member.

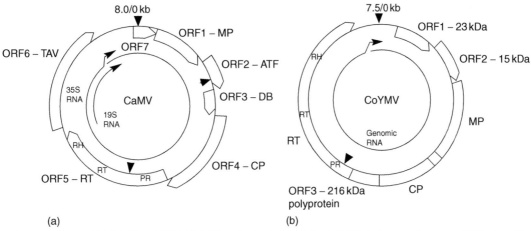

Figure 2 Genomic maps of (a) CaMV and (b) CoYMV. The single-stranded discontinuity in the negative-sense DNA strand is indicated by the triangle outside of the circle, whereas single-stranded discontinuities in the positive-sense DNA strand are indicated by triangles inside the circle. The mRNAs for each virus are represented by the inner circles and the 3' end of the RNA is indicated by the arrowhead. The functions for each of the ORFs are MP, cell-to-cell movement protein; ATF, aphid transmission factor; DB, DNA-binding protein, which also has role in aphid transmission; CP, coat protein; RT, reverse transcriptase; TAV, translational transactivator and major inclusion body protein. The domains within the reverse transcriptase are: PR, protease; RT, reverse transcriptase domain; and RH, ribonucleaseH activity. ORF3 of CoYMV encodes a 216 kDa polyprotein that is cleaved to produce the MP, CP, and RT proteins.

translational transactivator (TAV). It physically interacts with host ribosomes to reprogram them for reinitiation of translation of the polycistronic 35S RNA. Furthermore, P6 is an important symptom and host range determinant. No protein product has been found for ORF7. Its nucleotide sequence appears to play a regulatory role in aligning ribosomes for translation of other CaMV gene products.

Two transcripts are produced from the CaMV genomic DNA. The 19S RNA serves as the mRNA for P6, whereas the terminally redundant 35S RNA serves as a template for reverse transcription and as a polycistronic mRNA for ORFs 1–5. One feature common to many of the caulimoviruses is the ribosomal shunt mechanism of translation. The ribosomal shunt has undoubtedly evolved over time to compensate for the complexity and length of the leader sequence of the genomic RNA. In the case of CaMV, this leader sequence is approximately 600 bp in length and it

contains up to nine short ORFs that vary in size from 9 to 102 nt. The complexity of the CaMV 35S RNA leader sequence is bypassed through the formation of a large stem–loop structure, which allows ribosomes to bypass most of the leader and to initiate translation at ORF7. CaMV utilizes several other strategies for expression of the 35S RNA, including splicing and reprogramming of host ribosomes by the TAV for reinitiation of translation.

The genome of CoYMV is approximately 7500 bp in size, and the circular DNA contains two single-stranded discontinuities, one in each strand (**Figure 2(b)**). As with CaMV, the discontinuity in the negative-sense DNA strand serves as the origin of the DNA sequence, whereas the discontinuity in the positive-sense strand is found at approximately nucleotide position 4800. CoYMV encodes a single transcript that is terminally redundant, with 120 nucleotides reiterated on the 5' and 3' ends. The virus genome consists of three ORFs, encoding proteins of 23, 15, and 216 kDa, respectively. Both the 23 and 15 kDa proteins are associated with the virions. The 23 kDa protein contains a coiled-coil motif and by analogy with the P3 protein of CaMV, may be necessary for cell-to-cell movement. The 216 kDa P3 protein is a polyprotein that contains motifs for a movement protein, coat protein, aspartate protease, reverse transcriptase, and RNase H.

Replication

Caulimoviruses replicate by reverse transcription of an RNA intermediate, but integration of the viral DNA into host chromosomes is not required to complete the replication process. **Figure 3** illustrates the replication of an icosahedral caulimovirus, but the same steps are applicable for the bacilliform viruses. After virions enter a plant cell, the viral DNA becomes unencapsidated and is transported into the nucleus. The viral DNA contains two to four single-stranded discontinuities, which form as a consequence of the reverse transcription process. Once in the nucleus, the single-stranded discontinuities are covalently closed and the DNA is associated with histones to form a minichromosome. The host RNA polymerase II is responsible for synthesizing an RNA that is terminally redundant; the sequence on the 5′ end is reiterated on the 3′ end. In the

case of CaMV, the terminal redundancy is 180 nt in size, whereas the terminal redundancy in the CoYMV RNA is 120 nt. The terminally redundant RNA is transported out of the nucleus into the cytoplasm where it can either serve as a template for translation of viral proteins or as a template for reverse transcription. Reverse transcription is thought to occur in nucleocapsid-like particles.

First strand DNA synthesis is primed by a methionine (Met) tRNA that binds to a complementary sequence near the 5′ end of the terminally redundant RNA. In the case of CaMV, the Met tRNA binds to a sequence approximately 600 nt from the 5′ end of its 35S RNA. The virally encoded reverse transcriptase synthesizes DNA up to the 5′ end of the terminally redundant RNA and the RNase H activity of the reverse transcriptase degrades the 5′ end of the RNA. The terminal redundancies present in the genomic RNA provide a mechanism for a template switch, as the reverse transcriptase is able to switch from the 5′ end of the genomic RNA to the same sequences on the 3′ end of the genomic RNA and continue to synthesize the first strand of DNA.

The RNase H activity of the reverse transcriptase degrades the viral RNA template and small RNA fragments serve as the primers for second strand DNA synthesis. The RNA fragments bind to guanosine-rich tracts present in the first DNA strand, and these priming sites determine the positions of the discontinuities in the

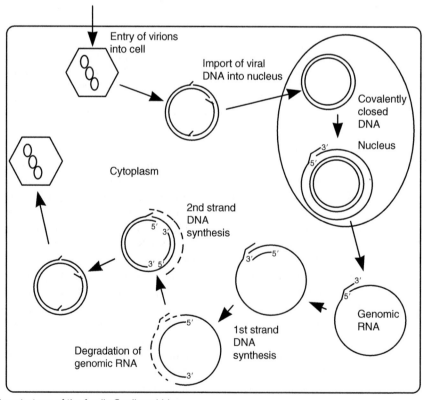

Figure 3 Replication strategy of the family *Caulimoviridae*.

second DNA strand. A second template switch is required to bridge the gap from the 5′ to the 3′ end of the first DNA strand. The completion of the second DNA strand results in the formation of a dsDNA molecule that contains the characteristic single-stranded discontinuities.

Integration of Some Petuviruses, Cavemoviruses, and Badnaviruses into Host Chromosomes

Although caulimoviruses do not integrate into the host as part of their replication strategy, the sequences of several viruses have been detected in the genomes of their hosts. Furthermore, the integrated copies of several of these viruses, the banana streak viruses, TVCV and PVCV, can be activated to yield episomal infections. All three viruses can form virions once episomal infections are initiated, but they differ in their capacity to be transmitted to other plants. Banana streak virus can be transmitted by mealybugs, whereas PVCV is transmitted only by grafting. TVCV and PVCV are transmitted vertically, through seed, but this probably involves only the integrated copies of the virus rather than the episomal forms.

The sites of integration are complex, as the viral sequences have undergone rearrangements, and multiple copies are arranged in tandem. Banana streak virus is integrated at two loci in *Musa* chromosomes, TVCV is integrated at multiple loci in *Nicotiana edwardsonii* chromosomes, and PVCV is integrated into four loci in *Petunia hybrida* chromosomes. The mechanism that results in episomal infections remains to be elucidated, but there are some features common to all three viruses. First, the infections arise in interspecific hybrids. Consequently, each parent plant must contribute some factor to activate the integrated virus. Second, the infections arise after the hybrid has been exposed to some sort of stress. PVCV is activated when plants are exposed to water or nutrient stress, or the plants are wounded. TVCV infections arise in *N. edwardsonii* in the winter months in the greenhouse and are thought to be related to changes in light quality or duration. Integrated banana streak virus sequences are activated in otherwise healthy *Musa* species when they are subjected to tissue culture. In each case, the virus is likely released through a series of recombination events or through reverse transcription of an RNA template.

Portions of other DNA viruses have been detected in plant chromosomes, but they have not been associated with episomal infections. This has led to speculation that plant DNA viruses might be capable of recombination in every infected plant, likely through a mechanism involving nonhomologous recombination. However, since episomal forms of the caulimoviruses and badnaviruses have not been found in tissues fated for seed formation, the integrated forms would not appear in the next generation

of plants. Consequently, the banana streak viruses, TVCV, and PVCV must have gained access to germline cells at some point such that their integrants would be passed through seed. This is likely to have been a relatively rare event.

Transmission and Host Range

Most members of the genus *Caulimovirus* are transmitted by as many as 27 species of aphids in a semipersistent manner. The virions are acquired rapidly by the aphid and can be transmitted immediately upon acquisition. Virions can be maintained by the aphid for as little as 5 h up to 3 days, but are not retained after the aphid molts and are not passed on to aphid progeny. The protein products encoded by ORF2 (P2) and ORF3 (P3) are both required for aphid transmission. P3 forms a tetramer that binds to both the virion and to the C-terminus of P2. P2 is responsible for the binding of this complex to the aphid, as the N-terminus of P2 binds to a site in the aphid foregut. No insect vectors have been identified for species in the genera *Petuvirus*, *Soymovirus*, or *Cavemovirus*.

None of the virions of the icosahedral viruses in the genera *Caulimovirus*, *Petuvirus*, *Soymovirus*, or *Cavemovirus* are transmitted through seed or pollen. Most are transmitted after mechanical inoculation, but there are a few exceptions, as PVCV, TVCV, blueberry red ringspot virus, and strawberry vein banding virus cannot be mechanically inoculated. In addition, most of these viruses are transmitted by grafting. In particular, dahlia mosaic virus and blueberry red ringspot virus infections in the field may be initiated through vegetative propagation or through grafting of infected plant material.

Badnaviruses are transmitted primarily by mealybugs, but a few are transmitted by aphids. Badnaviruses transmitted by mealybugs include the banana streak virus species, the sugarcane bacilliform viruses, cacao swollen shoot virus, dioscorea bacilliform virus, kalanchoe top spotting virus (KTSV), piper yellow mottle virus, taro bacilliform virus, and schefflera ringspot virus. These viruses are transmitted in a semipersistent manner and can be retained after molts, but do not multiply in the mealybug and are not transmitted to progeny. The badnaviruses shown to be transmitted by aphids include gooseberry vein banding-associated virus, rubus yellow net virus, and spiraea yellow leafspot virus.

Other modes of transmission of the badnaviruses vary with the species. For example, CoYMV and KTSV can be mechanically inoculated, cacao swollen shoot virus and piper yellow mottle virus are mechanically inoculated with some difficulty, and rubus yellow net virus and taro bacilliform virus have not been shown to be mechanically inoculated. The preferred method for inoculation of infectious badnavirus clones is agroinoculation. Some of the badnaviruses are transmitted through seed, including the banana streak virus, KTSV, and mimosa bacilliform virus. KTSV is very efficiently transmitted by seed, with transmission rates from 60% to 90%, and is also transmitted in pollen.

RTBV is dependent on the RNA virus, RTSV, for its transmission. RTSV is vectored by a leafhopper, but causes only very mild symptoms. RTBV is only transmitted by the leafhopper in the presence of RTSV, but is responsible for the severe symptoms associated with the rice tungro disease. RTBV is not mechanically transmitted or carried in seed or pollen.

The host range of most caulimoviruses is fairly narrow, as in nature they generally infect plants within a single family (**Table 1**). Their experimental host range may extend to members of one or two other families, but in many instances, they may only infect a single genus of plants. There are a few exceptions. For example, the sugarcane bacilliform viruses have a broader host range than most badnaviruses, as they can infect *Sorghum*, *Rottboellia*, *Panicum*, rice, and banana. A limited host range can also be associated with similar limitations in the geographic distribution of the virus. For example, soybean chlorotic mottle virus has only been recovered from a few samples in Japan. Perhaps the virus with the smallest geographic distribution is stilbocarpa mosaic bacilliform virus, which has only been found on a single, small island in the Subantarctic, midway between Tasmania and Antarctica. Other viruses, such as CaMV, carnation etched ring virus, and dahlia mosaic virus are found worldwide, wherever their hosts are grown. Furthermore, the distribution of viruses that originate from integrated copies in their host's genomes, PVCV, TVCV, and the banana streak virus species, are also closely aligned with the locations of their hosts. Interestingly, the icosahedral viruses of the caulimoviruses tend to infect hosts in temperate climates, whereas the bacilliform viruses of the badnavirus group are more likely to infect hosts in tropical or subtropical climates.

Virus–Host Relationships

Caulimoviruses induce a variety of systemic symptoms in their hosts, from chlorosis, streaking, and mosaics, to necrosis. The best-characterized pathogenicity determinant is the P6 protein of CaMV, as it has been shown to play a key role in the formation of chlorotic symptoms in turnips. This virulence function was first associated with P6 through gene-swapping experiments between CaMV isolates. It was confirmed when P6 was transformed into several species of plants, and in most cases, they exhibited virus-like symptoms. The P6 protein is also responsible for triggering systemic cell death in *Nicotiana clevelandii*, as well as a non-necrotic resistance response in *N. glutinosa*, and a hypersensitive resistance response in *N. edwardsonii*.

One feature that distinguishes the icosahedral viruses from the bacilliform viruses is that the former have the capacity to aggregate into amorphous inclusion bodies (**Figure 1(a)**), whereas the latter do not form inclusions. The inclusion bodies formed by the icosahedral viruses are not bound by a membrane, can range in size from 5 to 20 μm, and occur in virtually all types of plant cells. The inclusions can be visualized with a light microscope in strips of epidermal tissue that has been stained with phloxine B. Close examination of CaMV inclusion bodies by electron microscopy reveals that there are actually two types. One type contains many vacuoles and consists of an electron-dense, granular matrix that is composed primarily of the P6 protein (**Figure 1(a)**). A second, electron translucent type is made up of the P2 protein, a protein required for aphid transmission. Both types of inclusions are thought to have a role in the biology of the virus. The vacuolated inclusion bodies may be considered pathogen organelles, as they are thought to serve as the sites for replication of the viral nucleic acid, as well as translation of the 35S RNA and assembly of the virions. The electron translucent inclusions are considered to have a role in aphid transmission.

A second feature characteristic of the caulimoviruses is that the plasmodesmata of infected cells are enlarged enough to accommodate the 50 nm virions, as electron micrographs have revealed the presence of CaMV and CoYMV virions in the enlarged plasmodesmata. For both CaMV and CoYMV, the alteration in size is mediated by their proteins required for cell-to-cell movement. In infected protoplasts, the CaMV P1 protein has been shown to induce the formation of tubular structures that extend away from the protoplast surface. It is hypothesized that virions are assembled in the cell and then are escorted to the enlarged plasmodesmata by the cell-to-cell movement proteins.

See also: Caulimoviruses: Molecular Biology.

Further Reading

Calvert LA, Ospina MD, and Shepherd RJ (1995) Characterization of cassava vein mosaic virus: A distinct plant pararetrovirus. *Journal of General Virology* 76: 1271–1278.

Geering ADW, Olszewski NE, Harper G, *et al.* (2005) Banana contains a diverse array of endogenous badnaviruses. *Journal of General Virology* 86: 511–520.

Harper G, Hull R, Lockhart B, and Olszewski N (2002) Viral sequences integrated into plant genomes. *Annual Review of Phytopathology* 40: 119–136.

Hasagewa A, Verver J, Shimada A, *et al.* (1989) The complete sequence of soybean chlorotic mottle virus DNA and the identification of a novel promoter. *Nucleic Acids Research* 17: 9993–10013.

Hay JM, Jones MC, Blakebrough ML, *et al.* (1991) An analysis of the sequence of an infectious clone of rice tungro bacilliform virus, a plant pararetrovirus. *Nucleic Acids Research* 19: 2615–2621.

Hohn T and Fütterer J (1997) The proteins and functions of plant
    pararetroviruses: Knowns and unknowns. *Critical Reviews in Plant
    Sciences* 16: 133–161.
Hull R (1996) Molecular biology of rice tungro viruses. *Annual Review of
    Phytopathology* 34: 275–297.
Hull R, Geering A, Harper G, Lockhart BE, and Schoelz JE (2005)
    Caulimoviridae. In: Fauquet CM, Mayo MA, Maniloff J, Desselberger
    U,, and Ball LA (eds.) *Virus Taxonomy: Eighth Report of the
    International Committee on Taxonomy of Viruses*, pp. 385–396. San
    Diego, CA: Elsevier Academic Press.
Lockhart BEL (1990) Evidence for a double-stranded circular DNA
    genome in a second group of plant viruses. *Phytopathology* 80:
    127–131.
Pfeiffer P and Hohn T (1983) Involvement of reverse transcription in the
    replication of cauliflower mosaic virus: A detailed model and test of
    some aspects. *Cell* 33: 781–789.

Qu R, Bhattacharya M, Laco GS, *et al.* (1991) Characterization of the
    genome of rice tungro bacilliform virus: Comparison with Commelina
    yellow mottle virus and caulimoviruses. *Virology* 185: 354–364.
Richert-Pöggeler KR and Shepherd RJ (1997) Petunia vein-clearing
    virus: A plant pararetrovirus with the core sequences for an integrase
    function. *Virology* 236: 137–146.
Schoelz JE, Palanichelvam K, Cole AB, Király L, and Cawly J (2003)
    Dissecting the avirulence and resistance components that comprise
    the hypersensitive response to cauliflower mosaic virus in *Nicotiana*.
    In: Stacey G and Keen N (eds.) *Plant–Microbe Interactions*, vol. 6,
    pp. 259–284. St. Paul, MN: The American Phytopathological Society.
Shepherd RJ, Wakeman RJ, and Romanko RR (1968) DNA in
    cauliflower mosaic virus. *Virology* 36: 150–152.
Skotnicki ML, Selkirk PM, Kitajima, *et al.* (2003) The first subantarctic
    plant virus report: Stilbocarpa mosaic bacilliform badnavirus (SMBV)
    from Macquarie Island. *Polar Biology* 26: 1–7.

Caulimoviruses: Molecular Biology

T Hohn, Institute of Botany, Basel university, Basel, Switzerland

Viruses of the Genus *Caulimovirus*

Several viruses have been assigned to the genus *Caulimo-virus*, characterized by their content of open circular DNA, their icosahedral capsid, and their arrangement of seven open reading frames (ORFs) (**Table 1**). They have a narrow host range, usually restricted to one of the plant families. The genus *Caulimovirus* is one of six genera of the family *Caulimoviridae*, or plant pararetroviruses, which include three more genera of icosahedral viruses, *Soymovirus*, *Cavemovirus*, and *Petuvirus*, and two genera of bacilliform viruses, *Badnavirus* and *Tungrovirus*. Besides the differences in capsid structure of the two categories, the genera are distinct in the details of their genome arrangement and expression strategies (**Figure 1**).

Although members of the family *Caulimoviridae* (plant pararetroviruses), unlike retroviruses, do not integrate obligatorily into the host genome, 'illegitimately' integrated

sequences have been found for several genera to date, that is, caulimo, petu-, cavemo-, badna-, and tungroviruses. These have been named commonly 'endogenous plant pararetroviruses' (EPRVs). EPRVs can be found in the pericentromeric region of chromosomes, are passively replicated together with the host DNA, and are inherited from generation to generation.

Properties of the Virion and Inclusion Bodies

Members of the genus *Caulimovirus* are icosahedral ($T = 7$), 45–50 nm in diameter and sediment between 200 and 250S. The main capsid components of Cauliflower mosaic virus (CaMV) are proteins with mobilities of 37 and 44 kDa on sodium dodecyl sulfate polyacrylamid electrophoresis. They are derived from ORF IV. The capsid is further

Table 1 Caulimoviruses species

Species	Abbr.	Host range	Sequence
Cauliflower mosaic virus	CaMV	Cruciferae	V00141; X02606; J02046
Carnation etched ring virus	CERV	Caryophyllaceae	X04658
Figwort mosaic virus	FMV	Scrophulariaceae	X06166
Strawberry vein banding virus	SVBV	Rosaceae	X97304
Horseradish latent virus	HRLV	Cruciferae	AY534728 - 33
Dahlia mosaic virus	DMV	Compositae	
Mirabilis mosaic virus	MiMV	Nyctaginacea	NC_004036
Thistle mottle virus	ThMoV	*Cirsium arvense*	

Note that *Blueberry ringspot virus* and *Cestrum yellow leaf curl virus*, originally classified as caulimoviruses, are not included because they belong to the genus *Soymovirus*.

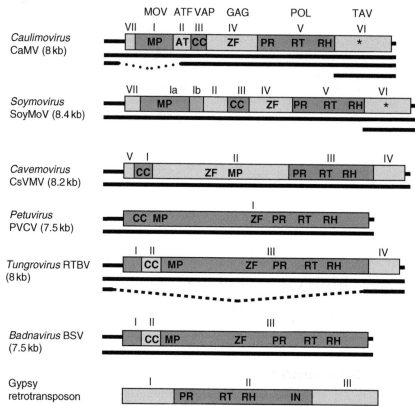

Figure 1 Open reading frame (ORF) arrangement in *Caulimovirus* and the other genera of the family *Caulimoviridae*. Examples of each of the six genera are shown, as well as the retrotransposon 'Gypsy' for comparison. For CaMV, the ORFs code for dispensable unknown function (VII), movement protein (I, MOV), aphid transmission factor (II, ATF), virion-associated protein (III, VAP), capsid protein precursor (IV, GAG), Pol polyprotein (V, POL), and transactivator/viroplasmin (VI, TAV). Although the number of final proteins might be similar for the other genera, the number of original ORFs differs, with soymoviruses having the most ORFs and petuviruses the least, namely only one original ORF, meaning more work for the protease to produce individual proteins. Key motifs are conserved in the family *Caulimoviridae*, movement protein motif (MP), coiled-coil domain (CC), zinc-finger (ZF), protease domain (PR), reverse transcriptase domain (RT), RNase H domain (RH), transactivator domain (*). In contrast to retroviruses and retrotransposons, an integrase domain (IN) is missing in members of the family *Caulimoviridae*.

decorated with 420 subunits of loosely bound 15 kDa virus-associated protein (VAP) derived from ORF III. In addition, CaMV particles contain minor amounts of other polypeptides: protease and reverse transcriptase derived from ORF V, and host casein kinase II. In cells infected with CaMV or a number of other caulimoviruses large numbers of virus particles accumulate in typical stable inclusion bodies, the matrix of which is the virus-encoded transactivator/viroplasmin (TAV) protein of 62 kDa, derived from ORF VI. Inclusion bodies are visible in the light microscope and can be stained with phloxin.

Properties of the Genome

The genome of the members of the family *Caulimoviridae* is built from 7200 to 8200 bp. It exists (1) in infected plant nuclei as supercoiled DNA bound to nucleosomes (minichromosome), (2) in infected cytoplasm as RNA with a ~180 nt redundancy, and (3) in virus particles as open

circular dsDNA. The open form is due to nicks at specific sites in both strands with short 5′ overlaps (**Figure 2**), remnants of the reverse transcription process, which are apparently removed in the infected nucleus. The sequences of several of the caulimoviruses are known (**Table 1**). They contain seven ORFs in the order VII (dispensable unknown function), I (cell-to-cell movement), II (aphid transmission factor), III (VAP), IV (capsid protein precursor), V (protease, reverse transcriptase, RNase H), and VI (multifunctional protein, transactivator of translation/viroplasmin).

Properties of the Proteins

Movement Protein

CaMV and probably all the other members of the family *Caulimoviridae* with icosahedral capsid move from cell to cell as particles through tubular structures spanning the cell walls between adjacent cells. In CaMV the 37 kDa

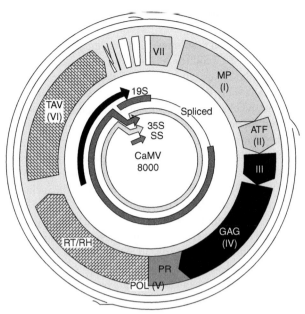

Figure 2 Map of cauliflower mosaic virus. On the outside the open circular double-stranded viral DNA with its overhangs is symbolized. The main circle shows the arrangement of the open reading frames (ORFs). ORF I encodes movement protein (MP or MOV), ORF II aphid transmission factor (ATF), ORF III the VAP, ORF IV the capsid protein precursor (GAG, in analogy to the corresponding protein in retroviruses), ORF V the POL polyprotein consisting of protease (PR), reverse transcriptase (RT), and RNase H (RH), and ORF VI the transactivator/viroplasmin (TAV). On the inside, the primary transcripts and the essential spliced RNAs are shown.

movement protein (MOV), coded for by ORF I, is responsible for the formation of these tubules and also provides their main component.

Aphid Transmission Factor

In general, aphids transmit viruses of the genus *Caulimovirus* and probably other genera of icosahedral (but not bacilliform) Caulimoviruses. In the case of CaMV, ORF II codes for the 18 kDa aphid transmission factor (ATF). Some of the viral inclusion bodies in CaMV-infected cells consist mainly of ATF. ATFs interact with the virus particles via their VAP and with tubulin and also with the cuticulum lining the tip the aphid's stylet. ATF mutants can still be transmitted by mechanical inoculation and by aphids that had previously be in contact with wild-type ATF either by feeding on plants infected with a related virus or on nutrient solutions supplied with artificially produced ATF.

Virion-Associated Protein

The VAP provides for coupling between virions and the movement and insect transmission factors and is therefore required for both insect transmission and cell-to-cell movement.

MOV, ATF, and VAP have no counterpart in retroviruses, reflecting the differences in cell-to-cell movement and infection routes used by animal and plant viruses. However, the movement proteins of plant pararetroviruses are related to a corresponding class of genes in most other plant viruses. In animals, viruses usually move from cell to cell by endocytosis and budding, while in plants, passage through modified plasmodesmata is used.

Capsid Proteins and Their Precursor

The product of CaMV ORF IV has a molecular weight of 56 kDa and an electrophoretic mobility of 80 kDa. It is flanked by very acidic regions, which are removed from the mature capsid proteins, which have a mobility corresponding to proteins of 44 and 37 kDa, although their molecular weight is probably smaller. The sequence of p37 is included within p44 and thus antibodies against p37 react also with p44. p44, but not p37, is phosphorylated. It is not known whether both proteins are required for full infection, or whether only one of them is the true functional capsid protein. It might be that p44 is the mature form and p37 a degradation product. Alternatively, p37 might be the mature form and p44 still a precursor. Both these proteins include a Zn-finger motif, which is implicated in RNA binding. Furthermore, large stretches of basic amino acids at the C-termini of both proteins constitute DNA- and RNA-binding motifs.

The precapsid protein has a nuclear localization signal which is masked by the N-terminal acidic domain. Upon virus assembly and removal of the acidic regions, the NLS becomes exposed and virus particles are transported to the nuclear pore where they release the DNA into the nucleus.

POL Polyprotein, Protease, Reverse Transcriptase/RNase H

CaMV ORF V corresponds to the POL ORF of retroviruses. It produces a polyprotein which is cleaved during virus production into a 15 kDa aspartic proteinase and a 60 kDa reverse transcriptase/RNase H. It is not known whether the RNase H functions as part of the 60 kDa protein or whether it is released by another cleavage reaction. One has to assume that the proteinase also functions as part of the polyprotein, that is, by releasing itself. Members of the family *Caulimoviridae* do not have an integrase and consequently their DNA is not or only accidentally integrated into the host chromatin.

Transactivator of Translation/Viroplasmin

The 62 kDa transactivator of TAV, encoded by ORF VI, has no significant homologies to any other known viral or host genes. It has been implicated in virus assembly, reverse transcription, host range determination, symptom severity, control of polycistronic translation, and seclusion of virus functions and silencing suppression. Perhaps as a consequence of the role of the ORF VI protein in translation, all CaMV proteins including foreign proteins carried by a CaMV vector can be found within the inclusion bodies. A subdomain in its center has been assigned to the translational transactivation and other domains have been found to unspecifically bind single- and double-stranded RNA, respectively. Furthermore, the protein interacts with capsid protein. TAV also has properties of a nuclear shuttle protein and might be involved in viral RNA transport.

Virus Stability

Both, the CaMV inclusion bodies and the virus particles, are very stable. Dissolution of the inclusion bodies to obtain virus particles requires treatment with 1 M urea over a long period. Virus particles aggregate at acidic pH and disintegrate at very high pH (0.1 M NaOH). Virus particles can also be disintegrated by boiling in sodium dodecyl sulfate/dithiothreitol or by treatment with proteinase K and phenol extraction. During these treatments, VAP is released and therefore usually escapes detection as virus component.

Genome Replication

As in retroviruses, replication occurs by production of a terminally redundant RNA by transcription in the nucleus and its reverse transcription in the cytoplasm. Also as in retroviruses, and unlike in hepadnaviruses, a tRNA primer (met-initRNA) is used as the primer for (−)-strand DNA synthesis from the RNA template. The RNA template is digested by RNase H following the reverse transcriptase. However, oligo(G) stretches are resistant to this digestion and remain bound to the (−)-strand DNA where they act as primers for (+)-strand DNA synthesis. The number of (+)-strand synthesis initiation events varies between one and three major events in different members of the family *Caulimoviridae* and even within strains from a particular virus. Minor initiation events also occur. The synthesis of both the strands overshoots, creating short overhangs that can only be repaired by repair exonuclease and ligase after transport into the nucleus. Accordingly, the packaged viral DNA is open circular, while the nuclear one is closed and supercoiled. The latter interacts with histones and forms minichromosomes. Caulimovirus DNA does not integrate into the host chromatin obligatorily. The template for transcription is the supercoiled circular DNA.

Transcription

Transcription in caulimoviruses is unidirectional. All caulimoviruses produce a transcript covering the total genome plus about 180 nt, such that the RNA is terminally redundant. This RNA is called 35S RNA due to its sedimentation behavior. The terminal redundancy is caused by the polyadenylation signal located on the circular DNA 180 bp downstream of the transcription start site being ignored by the polymerase at its first passage (see below). CaMV-like caulimoviruses produce a second transcript, the 19S RNA, covering ORF VI and encoding the inclusion body protein/translation transactivator.

The 35S promoter of CaMV, FMV, and probably other caulimoviruses is very strong and quasi-constitutive, that is, it is expressed in nearly all types of cells and at all developmental stages. This constitutivity is caused by a number of different enhancer elements, each with some specificity for certain cell types. Some of the corresponding transcription factors have been identified, for example, ASF-1, ASF-2, and CAF recognizing a TGACG motif, a GATA motif, and CA-rich region, respectively.

RNA Processing

Full-length RNA made from the circular DNA template includes a polyadenylation signal. If this is used on the first encounter, a short-stop RNA of 180 nt is formed; if used on the second encounter full-length terminally redundant RNA is formed. The role of the short-stop RNA, if any, is not known.

The CaMV polyadenylation signal consists of an AAUAAA sequence, which determines the cleavage site 13 nt downstream of it. Occasionally cryptic signals are used in addition. In contrast to the animal system, single point mutations in the AAUAAA signal are partially tolerated by the plant system. Polyadenylation enhancers are located upstream of the AAUAAA signal and not downstream as in most animal cases. In CaMV and FMV, these elements are tandem repeats of UAUUUGUA.

In addition to the primary CaMV 35S transcript, alternatively spliced versions of it have been detected. By removing an intron extending from close to the end of the leader to a position within ORF II, an mRNA is created in which ORF III is the first ORF and ORF IV the second. Additional splicing events were found using the same splice acceptor but using donors within ORF I. These led to ORF I–ORF II in-frame fusions, the function of which is not known. Whether the ratio of spliced to unspliced RNA is controlled, for example by nuclear export of unspliced RNA as in human immunodeficiency virus, is not known.

Translation Mechanism

The 35S RNA of caulimoviruses and also its spliced derivatives serve as polycistronic mRNAs for the viral proteins. The ORFs of these viruses closely follow each other, are opened by efficiently recognized start codons, and usually also contain internal start codons. In plant protoplasts and in transgenic plants most of the downstream ORFs are poorly expressed in the absence of TAV interestingly the only ORF that is translated from a monocistronic mRNA. Transactivation activity has been demonstrated for CaMV and FMV, and probably resides also in the corresponding ORFs of the other CaMV-like and PCSV-like caulimoviruses. Transactivation is thought to be based on a reinitiation mechanism.

Another special feature of the caulimovirus (and also the badnavirus) 35S RNAs is their 600–700-nt-long leader, which is rather large for eukaryotic RNAs. These leaders contain several small ORFs and include a large hairpin structure. These are features that usually make an RNA a poor messenger, since initiation factors and/or 40S ribosomes initiate scanning at the cap of an RNA moving in the 3' direction until an AUG codon is encountered and translation proper begins, whereas translation from AUGs further downstream is precluded. To overcome this problem caulimoviruses employ a shunting mechanism for translation initiation, whereby scanning is initiated at the cap as usual, but the scanning complex bypasses (shunts) most of the leader to reach the first longer ORF. This mechanism is unlike the internal initiation mechanism used by enteroviruses.

Protein Processing

Many of the caulimovirus proteins are processed, that is, in CaMV-like (and by analogy probably the other) caulimoviruses the precapsid protein is cleaved at both ends removing its very acidic termini. This process might be coupled to virus assembly and maturation. The capsid protein is both methylated and glycosylated. The Pol protein is cleaved to yield the aspartic proteinase and reverse transcriptase/RNase H. At least some of these cleavages occur by the action of the viral proteinase. In contrast, the ORF III product is cleaved by a host cysteine proteinase.

Silencing and Silencing Suppression

Another level of expressional control is provided by transcriptional and post-transcriptional silencing, which lead to strong inhibition of transcription and destabilization of the transcript, respectively. Silencing is viewed as a plant defense mechanism directed against viruses and transposons. On the other hand, viruses have developed silencing suppressors to counteract this plant defence strategy. The response of *Brassica napus* to systemic infection with cauliflower mosaic virus that results first in enhancement followed by subsequent suppression of viral gene expression in parallel with changes in symptom formation can be explained by the battle being waged between silencing suppression and silencing.

Silencing is autocatalytic and systemic, and hence the silencing of CaMV can also lead to the silencing in *trans* of transgenes driven by the CaMV 35S promoter. In a special case, herbicide resistance in oilseed rape conferred by expression of a 35S promoter-driven bialaphos tolerance transgene can be silenced due to the host response to CaMV infection.

See also: Caulimoviruses: General Features; Legume Viruses; Plant Virus Vectors (Gene Expression Systems); Virus Induced Gene Silencing (VIGS).

Further Reading

Blanc S (2002) Caulimoviruses. *Virus–Insect–Plant Interactions.* New York: Academic Press.

Covey SN and Al Kaff NS (2000) Plant DNA viruses and gene silencing. *Plant Molecular Biology* 43(special issue): 307–322.

Goldbach R and Hohn T (1996) Plant viruses as gene vectors. In: Bryant JA (ed.) *Methods in Plant Biochemistry 10b: Molecular Biology,* pp. 103–120. San Diego: Academic Press.

Haas M, Geldreich A, Bureau M, et al. (2005) The open reading frame VI product of cauliflower mosaic virus is a nucleocytoplasmic protein: Its N terminus mediates its nuclear export and formation of electron-dense viroplasms. *Plant Cell* 17: 927–943.

Hohn T (2007) Plant virus transmission from the insect point of view. *Proceedings of the National Academy of Sciences, USA* 104: 17905–17906.

Hohn T, Park HS, Guerra-Peraza O, et al. (2001) Shunting and controlled reinitiation: The encounter of cauliflower mosaic virus with the translational machinery. *Cold Spring Harbor Symposia on Quantitative Biology* 66: 269–276.

Hohn T and Richert-Poeggeler K (2006) Caulimoviruses. In: Heffron KL (ed.) *Recent Advances in DNA Virus Replication,* pp. 289–319. Kerala: Research Signpost Transworld.

Kiss-László Z and Hohn T (1996) Pararetro- and retrovirus RNA: Splicing and the control of nuclear export. *Trends in Microbiology* 4: 480–485.

Lam E (1994) Analysis of tissue specific elements in the CaMV 35S promoter. In: Nover L (ed.) *Results and Problems in Cell Differentiation, #20,* pp. 181–196. Berlin: Springer.

Plisson C, Uzest M, Drucker M, et al. (2005) Structure of the mature P3-virus particle complex of cauliflower mosaic virus revealed by cryo-electron microscopy. *Journal of Molecular Biology* 346: 267–277.

Rothnie HM, Chapdelaine Y, and Hohn T (1994) Pararetroviruses and retroviruses: A comparative review of viral structure and gene expression strategies. *Advances in Virus Research* 14: 1–67.

Ryabova L, Pooggin MM, and Hohn T (2005) *Translation Reinitiation and Leaky Scanning in Plant Viruses. Virus Research (Translational Control during Virus Infection).* Oxford: Elsevier.

Staginnus C and Richert-Poeggeler KR (2006) Endogenous pararetroviruses: Two-faced travelers in the plant genome. *Trends in Plant Science* 11: 485–491.

Citrus Tristeza Virus

M Bar-Joseph, The Volcani Center, Bet Dagan, Israel
W O Dawson, University of Florida, Lake Alfred, FL, USA

Glossary

Cross-protection Prevention of the symptomatic phase of a disease by prior inoculation with a mild or nonsymptomatic isolate of the same virus.
Defective RNA Subviral RNA molecules lacking parts of the genome while maintaining the signals enabling their synthesis by the viral replication system.
Genome The complete genetic information encoded in the RNA of the virus, including both translated and nontranslated sequences.
Subgenomic RNA Shorter than full-length genomic RNA molecules produced during the replication process sometimes to allow translation of open reading frames (ORFs).
Transgenic pathogen-derived resistance Plants genetically transformed to harbor viral or other pathogen-derived sequences expected to confer resistance against related virus isolates.

History

Over the last 70 years citrus tristeza virus (CTV) has killed, or rendered unproductive, millions of trees throughout most of the world's citrus-growing areas and hence it is rightfully considered as the most important virus of citrus, the world's largest fruit crop, hence the name 'tristeza' which means 'sadness' in Spanish and Portuguese. However, as with many other disease agents, the actual damages of CTV infections and their timings varied considerably at different periods and geographical regions. The origins of CTV infections remain unknown; the virus however existed for centuries in Asia as an unidentified disease agent, but growers in these areas adapted citrus varieties and rootstock combinations with resistance or tolerance to CTV infections. Out of Asia, citrus production moved to the Mediterranean region, primarily through the introduction of fruit and propagation of seed which does not transmit CTV to the resulting plantings; hence, the new cultivation areas starting from seed sources remained free of the virus for centuries. However, the improvement in maritime transportation, allowing long-distance transport of rooted citrus plants, led to the outbreak of the deadly citrus root rot disease caused by *Phytophthora* sp., which was managed by adapting the more tolerant sour orange

rootstock. The considerable horticultural advantages offered by this rootstock coincided with the large-scale expansion of plantings throughout the Mediterranean and American countries, and as a result citrus production in many areas was almost entirely based on a single rootstock. Consequently, this decision had grave effects when CTV pandemics swept throughout the world, causing 'quick decline' (death) of trees on this rootstock. Although the CTV problems were first noticed in South Africa and Australia where by the end of the nineteenth century the growers had found that sour orange was an unsuitable rootstock, it was only in the 1930s that the extent of this deadly disease problem manifested itself first in Argentina and shortly later in Brazil and California, with the death of millions of trees.

The virus-like nature of the tristeza-related diseases was demonstrated experimentally by graft and aphid transmission of the disease agent in 1946; however, description of CTV properties began with the seminal findings of Kitajima *et al.* of thread-like particles (TLPs) associated with tristeza-infected trees. These unusually long and thin particles presented a challenging problem for isolation (purification), which was an essential step toward virus characterization. Development of effective purification methods and biophysical characterization of TLP and similar viruses led to their assignment to the closteroviruses, a group of elongated viruses. The association of infectivity with the TLP-enriched preparations was demonstrated first by Garnsey and co-workers; however, the unequivocal completion of Koch's postulates for TLPs was completed only in 2001 with the mechanical infection of citrus plants with TLPs obtained from RNA transcripts of an infectious CTV cDNA clone amplified through serial passaging in *Nicotiana benthamiana* protoplasts.

Developments of TLP purification methods paved the way to antibody preparations and to improved CTV diagnosis by enzyme-linked immunosorbent assay (ELISA), which revealed considerable serological identity among CTV isolates. Later, RNA extracts from CTV-enriched particle preparations were used for molecular cloning of DNA molecules, which when used as probes demonstrated considerable genomic variation among CTV isolates. Nucleic acid probes also demonstrated that plants infected with CTV contained many defective RNAs (dRNAs) in addition to the normal genomic and subgenomic RNAs. Infected plants also contain unusually large amounts of dsRNAs corresponding to the genomic and subgenomic RNAs and the dRNAs. Because of the ease of purifying these abundant dsRNAs, they often have been the template of choice for producing cDNAs. The advent of cDNA cloning of CTV led to the sequencing of its genome and to description of dRNAs.

Taxonomy and Classification

CTV belongs to the genus *Closterovirus*, family *Closteroviridae*. The family *Closteroviridae* contains more than 30 plant viruses with flexuous, filamentous virions, with either mono- or bipartite (one tripartite) single-stranded positive-sense RNA genomes. A recent revision of the taxonomy of the *Closteroviridae* based on vector transmission and phylogenetic relationships using three proteins highly conserved among members of this family (a helicase, an RNA-dependent RNA polymerase, and a homolog of the HSP70 proteins) led to the demarcation of three genera: the genus *Closterovirus* including CTV and other aphid-borne viruses with monopartite genomes; the genus *Ampelovirus* comprising viruses with monopartite genomes transmitted by mealybugs; and the genus *Crinivirus* that includes white-fly-borne viruses with bipartite or tripartite genomes. Viruses of this family are all phloem-limited viruses. Among the hallmarks of the group is the presence of a conserved five-gene module that includes the four proteins involved in assembly of virions, the major (CP) and minor (CPm) coat proteins, p61, and p64, a homolog of the HSP70 chaperon. In addition, infected phloem-associated cells have clusters of vesicles considered specific to the cytopathology of closteroviruses. Although the evolution of the three genera of the family *Closteroviridae* is unknown, its is interesting to note that dRNAs of CTV analogous to each of the two RNA segments of the crinivirus genome were found in infected plants raising the possibility that criniviruses arose from a closterovirus ancestor.

Geographic Distribution

Citrus is a common fruit crop in areas with sufficient rainfall or irrigation from the equator to about 41° of latitude north and south. CTV is now endemic in most of the citrus-growing areas, with only a few places in the Mediterranean basin and Western USA still remaining free of CTV infections. The spread during the last decade of the brown citrus aphid (*Toxoptera citricida*) to Florida and its recent spread to parts of Portugal and Spain are now threatening most of the remaining CTV-free areas of North America and in the Mediterranean basin. One important aspect of CTV's geographic distribution is that some areas that have endemic isolates of CTV still do not have the most damaging stem-pitting isolates. With the spread of the vector comes the threat of extremely severe CTV isolates which so far have not spread to North

America and Mediterranean countries, where only the milder CTV isolates were established in the past.

Host Range and Cytopathology

CTV infects all species, cultivars, and hybrids of *Citrus* sp., some citrus relatives such as *Aeglopsis chevalieri*, *Afraegle paniculata*, *Fortunella* sp., *Pamburus missionis*, and some intergeneric hybrids. Species of *Passiflora* are the only nonrutaceous hosts, infected both naturally and experimentally. The CTV decline strains are associated with the death of the phloem near the bud union, resulting in a girdling effect that may cause the overgrowth of the scion at the bud union, loss of feeder roots, and thus drought sensitivity, stunting, yellowing of leaves, reduced fruit size, poor growth, dieback, wilting, and death. However, other virulent and damaging CTV strains cause stem-pitting (SP), which results in deep pits in the wood under depressed areas of bark and are often associated with severe stunting and considerably reduced fruit production. The seedling-yellow reaction (SY) which includes severe stunting and yellowing on seedlings of sour orange, lemon, and grapefruit is primarily a disease of experimentally inoculated plants but might also be encountered in the field in top-grafted plants.

CTV infection is closely restricted to the phloem tissues which often show strongly stained cells, termed as chromatic. Electron microscopy of infected cells shows that they are mostly filled with fibrous inclusions, consisting of aggregates of virus particles and of membranous vesicles differing in tonicity and containing a fine network of fibrils.

The CTV Virions

The CTV virions are long flexuous particles, 2000 nm long and 10–12 nm in width. The virions have a helical symmetry with a pitch of 3–4 nm, about 8–9 capsids per helix turn, and a central hole of 3–4 nm. Unlike the virions of other elongated plant viruses that possess cylindrical nucleocapsids made of a single coat protein (CP), the CTV virions consist of bipolar helices with a long body and a short tail. Immunoelectron microscopy showed antibodies to CPm attached to only one end of the virions, with the major part (>97%) of the virion encapsidated by CP. Interestingly, the 'tail' corresponds to 5' end region of the viral genome and the particle tails of other closteroviruses have been associated with small amounts of p61 and p65.

The CTV Genome

Figure 1 shows a schematic presentation of the 19.3 kbp RNA of CTV. Generally, the CTV genome is divided almost equally into two parts, the 5' part consisting of

ORF1a and -1b harboring the viral replication machinery and the 3′ half harboring ten ORFs encoding a range of structural proteins and other gene products involved in virion assembly and host and vector interactions. CTV replicons containing only ORF1a and -1b plus the 5′ and 3′ nontranslated ends (**Figure 1**) fulfill all the requirements for efficient replication in protoplasts.

Interestingly, while the sequences of the 3′ half of all CTV isolates that have been sequenced are 97% and 89% identical when comparing the 3′ nontranslated regions (NTRs) and the rest of the 3′ halves, respectively, the 5′ half sequences often differ considerably. For instance, the isolates T36 and VT show only 60% and 60–70% identities for their 5′ NTR and the remaining 5′ halves, respectively. The considerable deviation of strain T36 from that of the VT group suggests that T36 might have resulted from a recent recombination event involving the 3′ part of a VT-like isolate and the 5′ half derived from a different closterovirus. The common finding of numerous recombination events both within and between CTV isolates and multiple dRNAs supports such a possibility.

The Nontranslated Regions

The remarkable feature of CTV isolates are the close identities (97%) of their 3′ NTR primary sequences and the considerable divergence of their 5′ NTR sequences. Computer-assisted calculations, however, suggest that not only the 3′ NTRs of different isolates fold into similar predicted structures, but surprisingly the dissimilar sequences of the 5′ NTRs also are predicted to fold into

similar secondary structures. The 3′ replication signal of several CTV isolates was mapped to 230 nt within the NTR and predicted to fold into a secondary structure composed of ten stem-and-loop (SL) structures. Three of these SL regions and a terminal 3′ triplet, CCA, were essential for replication. Replacement of T36 3′ NTR with 3′ NTRs from other strains allowed replication, albeit with slightly less efficiency, confirming the significance of the primary sequence of this part. The plus-strand sequence of the 5′ NTRs from different strains are predicted to form similar secondary structures consisting of two stem loops (SL1 and SL2) separated by a short spacer region, that were essential for replication. These structures were shown to be *cis*-acting elements involved in both replication and initiation of assembly by CPm.

Genome Organization and Functions

In infected cells, the 12 ORFs of CTV (**Figure 1**) are expressed through a variety of mechanisms, including proteolytic processing of the polyprotein, translational frameshifting, and production of ten 3′-coterminal subgenomic RNAs. The first two mechanisms are used to express proteins encoded by the 5′ half of the genome while the third mechanism is used to express ORFs 2–11. The ORF1a encodes a 349 kDa polyprotein containing two papain-like protease domains plus methyltransferase-like and helicase-like domains. Translation of the polyprotein could also continue through the polymerase-like domain (ORF1b) by a +1 frameshift. These proteins along with the signals at the 5′ and 3′ ends of the genome are the

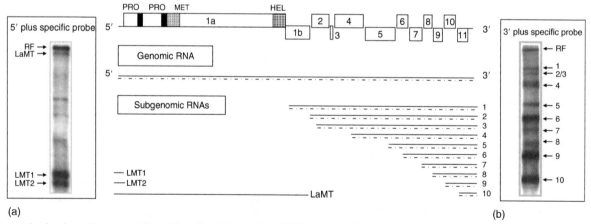

(a)   (b)

Figure 1   A schematic presentation of the citrus tristeza virus (CTV) genomic (g) and subgenomic (sg) RNAs. The putative domains of papain-like proteases (PRO), methyltransferase (MT), helicase (HEL), and the various open reading frames (ORFs) with their respective numbers are indicated. Lines shown on the left and right side below the genomic map indicate 5′ large molecular single stranded (ss) subgenomic (sg) RNA (LaMT) and two low molecular weight ssRNAs (LMT1 and LMT2) and a nested set of 3′ co-terminal sgRNAs, respectively. Inset (a), shows Northern blot hybridization of dsRNA-enriched extracts from a citrus plant infected with the VT strain of CTV using riboprobes specific to the 5′ end of the viral RNA. Note the presence of the large replicative form (RF) molecules (upper band), LaMT, and two abundant LMT molecules. Inset (b), Hybridization of dsRNA with the 3′ probe. Note the hybridization bands of the different 3′ sgRNAs indicated by arrows.

minimal requirements for replication of the RNA. The function of p33 (ORF2) is unknown, but is required for infection of a subset of the viral host range. The next five gene products from the 3' genes include the unique signature block characteristic of closteroviruses, which consist of the small, 6 kDa hydrophobic protein (ORF3), 65 kDa cellular heat-shock protein homolog (HSP70h, ORF4), 61 kDa protein (ORF5), and the tandem pair of p27 (CPm, ORF6) followed by p25 (CP, ORF7). The four latter proteins are required for efficient virion assembly. The small hydrophobic p6 is a single-span transmembrane protein not required for virus replication or assembly but for systemic invasion of host plants. CP is also a suppressor of RNA silencing. The function of p18 (ORF8) and p13 (ORF9) is unknown. Protein p20 (ORF10) accumulates in amorphous inclusion bodies of CTV-infected cells and has been shown to be a suppressor of RNA silencing. p23 (ORF11) has no homolog in other closteroviruses but is a multifunctional protein that (1) binds cooperatively both single-stranded and dsRNA molecules in a non-sequence-specific manner; (2) contains a zinc-finger domain that regulates the synthesis of the plus- and minus-strand molecules and controls the accumulation of plus-strand RNA during replication; (3) is an inducer of CTV-like symptoms in transgenic *C. aurantifolia* plants; (4) is a potent suppressor of intracellular RNA silencing in *Nicotiana tabacum* and *N. benthamiana*; and (5) controls the level of genomic and subgenomic negative-stranded RNAs.

The CTV Subgenomic (sg) RNAs

The replication of CTV involves the production of a large number of less than full-length RNAs. These include ten 3' coterminal sg mRNAs, and ten negative-stranded sgRNAs corresponding to the ten 3' sg mRNAs, plus ten 5'-coterminal sgRNAs that apparently are produced by termination just 5' of each of the ten ORFs (**Figure 1**). The amounts of different sg mRNAs vary with the highest levels for sg mRNAs p23 and p20, located at the distal 3' end. Infected cells also contain abundant amounts of two other small 5' coterminal positive-stranded sgRNAs of ~600 and 800 nt designated as low molecular weight tristeza (LMT).

Defective RNAs

Most CTV contain dRNAs, which consist of the two genomic termini, with extensive internal deletions. CTV dRNAs accumulate abundantly even when their genomes contain less than 10% of the viral genome (**Figure 2**). For convenience, the dRNAs were divided into six groups, class I molecules contained different sizes of 5' and 3' sequences. Some dRNAs of this class contained direct repeats of 4–5 nt flanking or near their junction sites, supporting the possibility that they were generated through a replicase-driven template-switching mechanism.

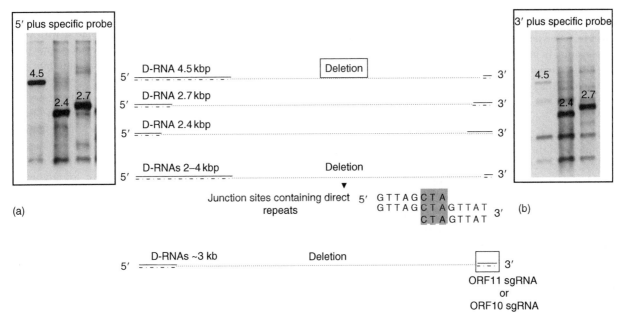

Figure 2 A diagram of class I CTV defective (d) RNAs (=D-RNA) with three different sizes and class II CTV-dRNAs that show 3' moieties of the size and structure of the full-length sgRNA of ORF11. Insets (a) and (b) show Northern hybridization of class I dsRNAs from three different CTV-VT subisolates, hybridized with riboprobes specific to the 5' and 3' ends, respectively, of the genomic RNA of CTV. Note the intense bands resulting from the hybridization of the two probes with the abundantly present dRNAs. Lower part is a diagrammatic presentation of direct repeats at the junction sites of some class I dRNAs. The bottom part shows a schematic presentation of a class II dRNA, with a 3' terminus corresponding to the ORF11 sgRNA of CTV-VT.

Class II molecules showed 3′ moieties of the size and structure of the full-length sgRNA of ORF11 (**Figure 2**, lower part). An extra C at the junction sites of several dRNAs of this class corresponded for the extra G reported for 3′ of minus strands from sgRNAs and RF molecules. Class III molecules are large (*c.* 12 kbp) encapsidated dRNAs which are infectious when used to inoculate *N. benthamiana* protoplasts. Class IV consists of dRNAs that retained all or most of the ten 3′ ORFs, analogous to crinivirus RNA 2. Class V includes double recombinants with identical 5′ regions of 948 nt followed with an internal ORF2 sequence and 3′ parts of different sizes, and class VI molecules are closely similar to class I with variable regions from the 5′ and 3′ with junction sites comprised of non-CTV inserts of 14–17 nt.

The biological roles of CTV dRNAs remain obscure as no specific associations have been established for any of the dRNA. Examination of the dRNAs from Alemow plants infected with SY and non-SY inducing isolates revealed mostly a major single dRNA of 4.5 or 5.1 kbp in non-SY plants and two different dRNAs of 2.4 or 2.7 kbp in SY plants. These results suggested the possibility that large dRNAs might play a role in suppression of SY symptoms.

Transmission

Although the virus is phloem limited, mechanical inoculation can be done experimentally at relatively low efficiency by slashing the trunks of small citrus trees with blades containing sap extracts. In commercial groves, the virus is spread naturally by aphid vectors and by vegetative propagation in infected budwood. Long-distance spread of the virus, particularly from country to country, had been through the long-distance transfer of plant propagation material.

Several aphid species including *Aphis gossypii*, *A. spiraecola*, *A. craccivora*, and *T. citricida* transmit CTV semipersistently. The rate of transmissibility varies considerably between different virus isolates and different aphid species. The brown citrus aphid (*T. citricida*) is the most efficient vector, followed by *A. gossypii*. Despite the less efficient transmission by *A. gossypii* and the absence of brown citrus aphid in the Mediterranean region and California, mild strains are spread in these areas. Recently, the brown citrus aphid invaded Central America and Florida, and was also reported to be in the island of Madeira and Portugal.

CTV Control Measures

Strategies to control CTV have varied at different locations and periods and included (1) quarantine and budwood certification programs to prevent the introduction of CTV; (2) costly and ambitious eradication programs to contain situations of virus spread; (3) the use of CTV-tolerant rootstocks and mild (or protective) strain cross-protection, often also named preimmunization; (4) breeding for resistance; and (5) attempts to obtain resistance by genetic engineering. Mild-strain cross-protection has been widely applied for millions of citrus trees in Australia, Brazil, and South Africa to protect against stem pitting of sweet orange and grapefruit trees. However, mild-strain cross-protection has not provided field protection against CTV isolates causing quick decline of trees on the sensitive sour orange rootstock.

Useful resistance to CTV has been found in a citrus relative, *Poncirus trifoliate*. This resistance has been mapped and shown to be controlled primarily by a single dominant gene, *Ctv.* Although the region containing this gene has been identified and sequenced, a specific gene has not been identified by transformation into susceptible citrus with the resistance phenotype.

CTV Diagnosis

The outcome of CTV infections varies considerably depending on the virulence of the prevailing virus isolates and the sensitivity of infected citrus varieties and root stock combinations. Hence, the need of effective means of CTV diagnosis are of utmost importance. Biological indexing of the disease by grafting sensitive citrus indicators has been the definitive assay, although it is costly and time consuming. It has largely been replaced with more rapid immunoassays. ELISA has been widely practiced for almost 30 years for a variety of CTV sanitation programs. Development of recombinant antigens considerably advanced diagnostic possibilities and production of monoclonal antibodies has allowed for more precise differentiation of isolates. PCR and combinations of immunocapture PCR allow more sensitive CTV diagnosis. However, despite the progress in development of better diagnostic tools, none is effective for predicting the biological properties of new CTV isolates.

Economic Costs of CTV

Control strategies varied at different geographic regions and periods, depending on the extent to which CTV was spreading within the newly infested areas, the sensitivity of the specific local varieties, and the economical availability of alternative crops to replace diseased groves. Most citrus-growing countries that managed to remain free of CTV did so by enforcing sanitation practices to prevent virus spread. These included certification schemes aimed to ensure that citrus nurseries propagate and distribute to growers only CTV-free planting material. Other areas had far more costly and ambitious programs of eradication. These efforts were mostly only temporarily successful, mainly because of lack of long-term grower and governmental commitment to such costly operation. In areas where CTV was endemic, alternative strategies to enable continued commercial citrus production in the presence of CTV were developed. Indeed, long before the viral nature of the tristeza disease was realized, Japanese citrus growers were grafting CTV-tolerant Satsuma mandarins on the cold-tolerant and CTV-resistant trifoliate orange (P. trifoliate) rootstocks, thus enabling them to produce quality fruits despite endemic CTV infections. Similarly, the change of rootstocks from the 'quick declining' sour orange rootstocks to rough lemon rootstocks, and to a less extent to mandarin rootstocks, allowed the South African growers to produce good crops of oranges despite the presence of the most efficient aphid vector and infections with severe CTV strains. Similarly, half a century later the Brazilian citrus industry that was completely decimated during 1940s, when all trees were grafted on sour orange rootstocks was saved by adopting CTV-tolerant rootstocks and mild-strain cross-protection.

Historically, the costs of CTV epidemics were estimated to be of the order of tens and even hundreds of millions of dollars. These estimates, however, varied depending on actual market value of the lost production capacity, alternative uses of the land, and the time needed for the CTV-tolerant replants to enter production. There were also indirect costs resulting in poorer performance and/or sensitivity of some CTV-tolerant rootstocks used to replace the widely adapted sour orange rootstocks to other citrus diseases such as citrus blight and citrus sudden death.

Attempts to Apply Transgenic Resistance to CTV

Conventional breeding for CTV resistance is a long, difficult, and inconvenient process mainly because most citrus varieties are hybrids of unknown parentage. Hence, the considerable interest in applying pathogen-derived resistance (PDR) to render both citrus rootstocks and varieties tolerant to stem-pitting isolates and decline causing CTV isolates. Yet, literally hundreds of independent transformations with a range of different types of configuration of CTV sequences have resulted in failure to obtain durable protection against CTV. These failures are especially frustrating since the control of severe stem- pitting isolates is needed in several major citrus production areas. The lack of RNA silencing against CTV in citrus might be due to more effective suppression by the combination of the three suppressors of this virus.

See also: Bromoviruses; Plant Virus Diseases: Fruit Trees and Grapevine.

Further Reading

Bar-Joseph M, Garnsey SM, and Gonsalves D (1979) The *Closteroviruses*: A distinct group of elongated plant viruses. *Advances in Virus Research* 25: 93–168.

Bar-Joseph M, Marcus R, and Lee RF (1989) The continuous challenge of citrus tristeza virus control. *Annual Review of Phytopathology* 27: 291–316.

Dawson WO (in press) Molecular genetics of *Citrus tristeza virus*. In: Karasev AV and Hilf ME (eds.) *Citrus Tristeza Virus and Tristeza Diseases*. St. Paul, MN: APS Press.

Karasev AV (2000) Genetic diversity and evolution of closteroviruses. *Annual Review of Phytopathology* 38: 293–324.

Martelli GP, Agranovsky AA, Bar-Joseph M, *et al.* (2002) The family *Closteroviridae* revised. *Archives of Virology* 147: 2039–2044.

Cowpea Mosaic Virus

G P Lomonossoff, John Innes Centre, Norwich, UK

Introduction

Cowpea mosaic virus (CPMV) is the type member of the genus *Comovirus* which includes 13 additional members in the family *Comoviridae*. CPMV was first isolated from an infected cowpea (*Vigna unguiculata*) plant in Nigeria in 1959. Subsequently, it has been found to occur in Nigeria, Kenya, Tanzania, Japan, Surinam, and Cuba. While its natural host is cowpea, it can infect other legumes, and *Nicotiana benthamiana* has proven to be an extremely valuable experimental host. In nature, CPMV is usually transmitted by leaf-feeding beetles, especially by members of the Chrysomelidae. CPMV has also been reported to be transmitted by thrips and grasshoppers. The beetle vectors can acquire the virus by feeding for as little as one minute and can retain and transmit the virus for a period of days or weeks. The virus does not, however, multiply in the insect vector. Experimentally, CPMV is mechanically transmissible.

In Nigeria, infection of cowpeas with CPMV causes a considerable reduction in leaf area, flower production, and yield. Infected plant cells show a number of characteristic cytological changes. These include the appearance of viral particles, a proliferation of cell membranes and vesicles in the cytoplasm, and a variety of modifications to plasmodesmata.

Physical Properties of Viral Particles

Viral particles can reach a yield of up to $2\,g\,kg^{-1}$ of fresh cowpea tissue and can be readily purified by polyethylene glycol precipitation and differential centrifugation. The particles are very stable with a thermal inactivation point in plant sap of 65–75 °C and a longevity in sap of 3–5 days at room temperature. Once purified, the particles can be stored for prolonged periods at 4 °C. The ease with which virus particles can be propagated, purified, and stored has undoubtedly contributed to the early popularity of CPMV as an object of study.

CPMV preparations consist of nonenveloped isometric particles, 28 nm in diameter, which can be separated on sucrose density gradients into three components, designated top (T), middle (M), and bottom (B), with sedimentation coefficients of 58S, 95S, and 115S, respectively. The three components have identical protein compositions, containing 60 copies each of a large (L) and small (S) coat protein, with sizes 42 and 24 kDa, respectively, as calculated from the nucleotide sequence. The discovery, in 1971, that CPMV particles contained equimolar amounts of two different polypeptides suggested that the capsids had an architecture more similar to the animal picornaviruses than to other plant viruses of known structure. This provided an early clue as to the common origins of plant and animal viruses.

The difference in the sedimentation behavior of the three centrifugal components of CPMV lies in their RNA contents. Top components are devoid of RNA, while mid-

dle and bottom components each contain single molecules of positive-strand RNA of 3.5 and 6.0 kbp, respectively. The two RNA molecules were originally termed middle (M) and bottom (B) component RNA after the component from which they were isolated. However, more recently they have been referred to as RNA-2 and RNA-1, respectively. The three-component nature of CPMV preparations is summarized in **Figure 1**. The determination of the component structure of the virus, and particularly the relationship between this and infectivity, was important in establishing the principle that plant viruses frequently have divided genomes, the individual components of which are separately encapsidated.

Because of their differing RNA contents, the three components of CPMV also differ in density and can hence be separated by isopycnic centrifugation on cesium chloride gradients. While T and M components give single bands of densities of 1.30 and $1.41\,g\,ml^{-1}$, respectively, B component gives two bands of 1.43 and $1.47\,g\,ml^{-1}$, the proportion of the denser band increasing under alkaline conditions. This increase in density results from an increase in capsid permeability, which allows the exchange of the polyamines present in B components (where they serve to neutralize the excess negative charges from RNA-1) for cesium ions.

CPMV preparations are not only centrifugally heterogeneous but can also be separated in two forms, fast and slow, electrophoretically. Both electrophoretic forms contain all three centrifugal components. The proportion of the two electrophoretic forms in a given virus preparation varies both with the time after infection at which the virus was isolated and the age of the preparation itself. Conversion of one form to the other is caused by loss of 24 amino acids from the C-terminus of the S protein.

Viral Structure

X-ray crystallographic studies on CPMV, as well as the related comoviruses bean pod mottle virus (BPMV) and red clover mottle virus (RCMV), have provided a detailed picture of the arrangement of the two viral coat proteins in the three-dimensional structure of the particle. Overall, the virions are icosahedral, with 12 axes of fivefold and

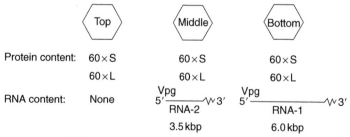

Figure 1   The three-component nature of CPMV indicating the protein and RNA content of each component.

20 axes of threefold symmetry, and resemble a classic $T = 3$ particle. The two coat proteins taken together consist of three distinct β-barrel domains, two being derived from the L and one from the S protein. Thus, in common with the $T = 3$ viruses, each CPMV particle is made up of 180 β-barrel structures. The S protein, with its single domain, is found at the fivefold symmetry axes and therefore occupies a position analogous to that of the A-type subunits in $T = 3$ particles (**Figure 2**). The N- and C-termini domains of the L protein occur at the threefold axes and occupy the positions equivalent to those of the C- and B-type subunits of a $T = 3$ particle, respectively (**Figure 2**). This detailed analysis confirmed the earlier suggestion that CPMV particles are structurally homologous to those of picornaviruses, with the N- and C-terminal domains of the L protein being equivalent to viral protein VP2 and viral protein VP3, respectively, and the S protein being equivalent to viral protein VP1 (**Figure 2**). However, CPMV particles are structurally less complex than those of picornaviruses. The L and S subunits lack the extended N- and C-termini found in VP2, VP3, and VP1 of picornaviruses and there is no equivalent of VP4. Moreover, CPMV subunits lack the relatively large insertions between the strands of β-sheet, sequences that form the major antigenic determinants of picornaviruses. No RNA is visible in either the M or B components of CPMV, in contrast to the situation found with BPMV where segments of ordered RNA could be detected in middle components.

a positive-strand RNA virus, a mixture of the genomic RNAs within the particles can also be used to initiate an infection. However, RNA-1 is capable of independent replication in individual plant cells but this leads to the establishment of gene silencing, rather than a productive infection, in the absence of RNA-2. Both genomic RNAs have a small basic protein (VPg) covalently linked to their 5′ termini and both are polyadenylated at their 3′ ends. The elucidation of the overall structure of the RNA segments once more underscored the similarity between CPMV and picornaviruses. However, unlike picornaviruses, the VPg is linked to the viral RNA via the β-hydroxyl group of its N-terminal serine residue rather than via a tyrosine. The VPg is not required for the viral RNAs to be infectious.

The complete nucleotide sequences of both genomic RNAs were reported in 1983, making CPMV one of the first RNA plant viruses to be completely sequenced. The length of the RNAs are 5889 and 3481 nucleotides, for RNA-1 and RNA-2, respectively, excluding the poly(A) tails and the full sequences appear in GenBank under accession numbers NC_003549 and NC_003550. The two genomic RNAs have no sequence homology apart from that at the 5′ and 3′ termini. Full-length infectious cDNA clones of both RNAs of CPMV have been constructed, allowing the genome to be manipulated. This was an important development since it allowed both reverse genetic experiments to be undertaken and for the virus to be used for biotechnological applications.

Genome Structure

Both M and B (but not T) components of a virus preparation are essential for infection of whole plants. As CPMV is

Expression of the Viral Genome

Both genomic RNAs contain a single long open reading frame, which occupies over 80% of the length of the RNA.

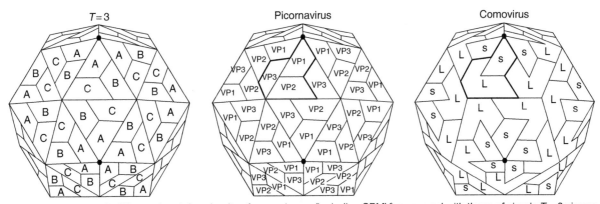

Figure 2 Arrangement of the coat protein subunits of comoviruses (including CPMV) compared with those of simple $T = 3$ viruses and picornaviruses. The asymmetric unit of $T = 3$ viruses contains three β-barrels contributed by three coat protein subunits with identical amino acid sequences (labeled A, B, and C). The asymmetric unit of a picornavirus also contains three β-barrels, but in this case each is contributed by a different coat protein (labeled viral protein VP1, viral protein VP2, and viral protein VP3). The comovirus capsid is similar to that of a picornavirus except that two of the β-barrels (corresponding to viral protein VP2 and viral protein VP3) are fused to give the L protein. Reproduced from Lomonossoff GP, Shanks M, et al. (1991) Comovirus capsid proteins: Synthesis, structure and evolutionary implications. *Proceedings of the Phytochemical Society of Europe* 32: 76.

A combination of *in vitro* translation and protoplast studies has unraveled the basic mechanism of gene expression of the virus. Both RNAs of CPMV are expressed through the synthesis and subsequent cleavage of large precursor polyproteins. This was the first example of a plant virus using this strategy for the expression of its genome, a strategy that was subsequently shown to be used by a number of other plant viruses. On RNA-1, initiation of translation occurs at the first AUG encountered on the sequence (at position 207) and results in the synthesis of a protein of approximately 200 kDa (the 200K protein). This initial product undergoes rapid cotranslational autoproteolysis to give proteins with apparent sizes of 32 and 170 kDa (the 32K and 170K proteins). The 170K protein undergoes further cleavages to give the range of virus-specific proteins shown in **Figure 3**. *In vitro* translation studies using mutant RNA-1 molecules have shown that all the cleavages occur most efficiently in *cis*. The 170K product can initially be cleaved at three different sites to give three different combinations of secondary cleavage products, 58K + 112K, 60K + 110K, and 84K + 87K. *In vitro*, and probably also *in vivo*, the 60K and 110K are stable and do not undergo further cleavage reactions. This is particularly curious in the case of the 110K

protein as it contains both the 24K proteinase domain and a cleavage site. By contrast, the 112K and 84K proteins do undergo further cleavages. The end products of the cleavage pathway of the 170K protein are, from N- to C-terminus, the 58K protein, the VPg, the 24K proteinase, and the 87K protein.

Initiation of translation of RNA-2 occurs at two different positions on the RNA and results in the synthesis of two carboxy coterminal proteins, the 105K and 95K proteins (**Figure 3**). This double initiation phenomenon, which occurs as a result of 'leaky scanning', is found with the RNA-2 molecules of all comoviruses. In the case of CPMV, synthesis of the 105K protein is initiated from an AUG at position 161 while initiation from an AUG at position 512 directs the synthesis of the 95K protein. CPMV RNA-2 has an additional AUG (position 115) upstream of both these initiation sites but this feature is not conserved in the RNA-2 molecules of other comoviruses. Both RNA-2-encoded primary translation products are cleaved by the RNA-1-encoded proteolytic activity to give either the 58K or the 48K protein (depending on whether it is the 105K or 95K protein that is being processed) and the two viral coat proteins. Processing of the RNA-2-encoded polyproteins, at least

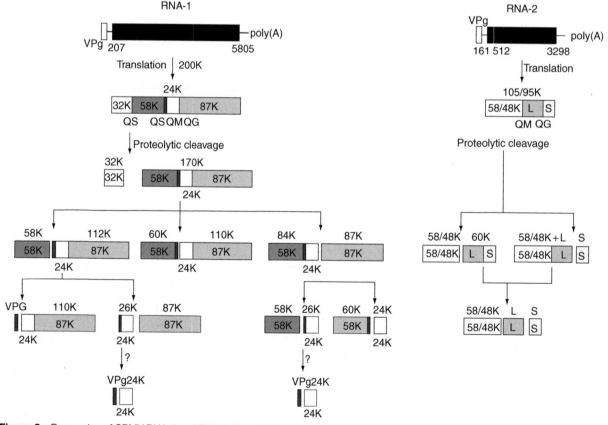

Figure 3 Expression of CPMV RNA-1 and RNA-2. Both RNAs contain a single long open reading frame, which is processed to yield a series of proteins. The positions of the initiation and termination codons and the dipeptides at the proteolytic processing sites are shown.

at the site between the 48K and L coat protein, has been shown to require the presence of the 32K protein as well as the 24K proteinase.

Functions of the Viral Proteins

Functions have been ascribed to most of the regions of the polyproteins encoded by both RNA-1 and RNA-2 of CPMV. In most cases, however, it is not certain at what stage(s) in the cleavage pathway they manifest their activity.

In the case of RNA-1, the 32K protein, which is rapidly cleaved from the N-terminus of the 200K primary translation product, is a cofactor which modulates the activity of the virus-encoded protease. As described earlier, the presence of the 32K protein is required for the cleavage of the RNA-2-encoded 105K and 95K proteins but is not essential for the cleavage of the RNA-1-encoded 170K protein. It does, however, seem to play a role in determining the rate at which cleavage of the 170K protein occurs. When mutant RNA-1 molecules carrying deletions in the region encoding the 32K protein are translated *in vitro*, the rate of processing of the 170K protein is greatly increased, indicating that the 32K protein acts as an inhibitor of processing. This inhibition may be achieved through the interaction of the 32K with the 58K domain of the 170K protein. The mechanism by which the 32K protein enables the 24K proteinase to cleave in *trans* is unclear.

The RNA-1-encoded 58K protein is associated with cell membranes and contains a nucleotide-binding motif. The 60K protein (**Figure 3**), containing the amino acid sequence of the 58K protein linked to VPg, is involved in rearrangements in the endoplasmic reticulum of CPMV-infected cells and acts in concert with the 32K protein. The 24K protein is the virus-encoded proteinase that carries out all the cleavages on both the RNA-1- and RNA-2-encoded polyproteins. Its proteolytic activity has been shown to be expressed in a number of the processing intermediates that contain its sequence. Indeed, it is not known whether the free form of the protein has any biological significance. Although the proteinase contains a cysteine at its active site, it is structurally related to serine proteases, such as trypsin, rather than cellular thiol proteases, such as papain. In this regard, it is similar to the 3C proteinases of picornaviruses. All comoviral cleavage sites identified so far have glutamine (Q) residue at the −1 position. The enzyme encoded by a given comovirus is specific for the polyproteins encoded by that virus and is unable to cleave the polyproteins from other comoviruses either in *cis* or in *trans*.

The 87K protein is believed to contain the virus-encoded RNA-dependent RNA polymerase (RdRp) activity since it contains the G-D-D sequence motif found in all such enzymes. It also has amino acid sequence homology to the $3D^{Pol}$ polymerases encoded by picornaviruses. However, when replication complexes capable of elongating nascent RNA chains were isolated from CPMV-infected cowpea plants, they were found to contain the 110K protein (**Figure 3**), consisting of the sequence of 87K protein linked to the 24K proteinase, rather than the free 87K protein.

In the case of RNA-2, the 48K protein, derived from processing of the 95K protein, is involved in potentiating the spread of the virus from cell to cell. This protein is found in tubular structures that are formed in the plasmodesmata of infected cells. Tubules extending into the culture medium can also be seen in protoplasts either infected with CPMV or transiently expressing the 48K protein. Virus particles can be seen within these tubules when protoplasts are infected with CPMV but not when only 48K protein is expressed. At present, no definite role has been assigned to the 58K protein, which is produced by processing of the 105K protein. Mutants in which translation of the 105K protein is disrupted replicate poorly, if at all. In light of these observations, it has been suggested that the 105K protein may play a role in the replication of RNA-2. Apart from containing many hydrophobic and aromatic amino acids, the approximately 10 kDa of protein present in the 58K but absent from the 48K protein is not conserved between comoviruses. The viral coat proteins are required to enable capsids to be formed. As well as protecting the genomic RNAs, capsid formation is essential for the virus to be able to spread from cell to cell through modified plasmodesmata and long-distance movement also requires capsid formation. An additional function in suppressing gene silencing is also provided by the C-terminal region of the S protein.

Replication

CPMV replicates to high level in infected cells. Replication is believed to involve the initial transcription of the incoming positive-sense RNA into minus-strands followed by initiation and synthesis of new plus-strands from the recently formed minus-strands. It has been shown that the 5′ ends of both the plus- and minus-strands are covalently linked to the VPg, suggesting that this protein has an essential role in the initiation of RNA synthesis. There also appears to be a tight linkage between the translation of the viral RNAs and their replication.

Replication of the viral RNAs has been shown to occur in the membraneous cytopathological structures, which are formed in the cytoplasm of cells during infection through the action of the RNA-1-encoded 32 and 60K proteins. Both CPMV-specific double-stranded replicative form (RF) RNA and an enzyme activity capable of completing nascent RNA strands can be isolated from such structures. Purified preparations of the enzyme

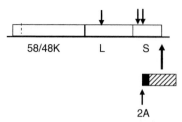

Figure 4 Structure of CPMV RNA-2-based vectors used to express heterologous peptides and proteins in plants. Positions where epitopes have been inserted into the sequence of the coat proteins are shown by black arrows. The position where foreign proteins (shown hatched) have been inserted into RNA-2 is indicated. The FMDV 2A sequence is shown as a black box at the N-terminus of the foreign protein.

activity contain the RNA-1-encoded 110K protein and two host-encoded proteins of 68 and 57 kDa. However, at present, no enzymatic activity capable of initiating RNA synthesis *in vitro* has been described.

Relationships with Other Viruses

Together with the genera *Nepovirus* and *Fabavirus*, the genus *Comovirus* belongs to the family *Comoviridae*. Within the family, the greatest affinity is between the genera *Comovirus* and *Fabavirus*. On a wider scale, consideration of genome structure and organization, translational strategy, and amino acid homologies between the virus-encoded proteins has led to grouping the family *Comoviridae* with the families *Potyviridae* and *Picornaviridae* as members of picorna-like superfamily of viruses. Members of this superfamily are all nonenveloped positive-strand RNA viruses with 3′ polydenylated genomic RNAs, which have a protein (VPg) covalently linked to their 5′ ends. All members of the supergroup have a similar mode of gene expression, which involves the synthesis of large precursor polyproteins and their subsequent cleavage by a virus-encoded proteinase. The members of the superfamily all contain similar gene order, membrane-bound protein-VPg-proteinase-polymerase (see **Figure 3**) and share significant amino acid sequence homology in the membrane-bound proteins, the proteinases, and polymerase coding regions. Comovirus capsids are also clearly structurally related to those of picornaviruses (**Figure 2**).

Use in Biotechnology

CPMV has been extensively used as a vector for the expression of foreign peptides and proteins in plants.

To date, all vectors have involved modifications to RNA-2 (**Figure 4**). In the first instance, antigenic peptides (epitopes) were genetically fused to exposed loops on the surface of the viral capsids. The resulting chimeric virus particles (CVPs) could be propagated in plants and the modified virions purified. When injected into experimental animals, CVPs can elicit the production of antibodies against the inserted epitope and in a number of instances can confer protective immunity against the pathogen from which the epitope was derived. This was a significant breakthrough as it represented the first instance where protection against an animal pathogen was conferred by material produced from a plant virus-based vector.

In an alternative approach, the sequence encoding an entire heterologous polypeptide has been fused to the C-terminus of the RNA-2-encoded polyprotein via a 2A catalytic peptide derived from foot-and-mouth disease virus (FMDV). The inclusion of the 2A sequence promotes efficient release of the foreign polypeptide from the polyprotein (**Figure 4**). This system has been used to express antibody derivatives in cowpea plants and crude plant extracts containing the antibodies have been shown to be capable of passively immunizing newborn pigs against challenge with the porcine coronavirus, transmissible gastroenteritis virus.

Recent developments in the use of comoviruses in biotechnology include the creation of combined transgene/viral vector systems based on CPMV, and the use of CPMV particles in bionanotechnology.

See also: Nepovirus; Plant Virus Diseases: Economic Aspects.

Further Reading

Cañizares MC, Lomonossoff GP, and Nicholson L (2005) Development of cowpea mosaic virus-based vectors for the production of vaccines in plants. *Expert Review of Vaccines* 4: 687–697.

Gergerich RC and Scott HA (1996) Comoviruses: Transmission, epidemiology, and control. In: Harrison BD and Murant AF (eds.) *The Plant Viruses 5: Polyhedral Virions and Bipartite RNA Genomes*, 77p. New York: Plenum.

Goldbach RW and Wellink J (1996) Comoviruses: Molecular biology and replication. In: Harrison BD and Murant AF (eds.) *The Plant Viruses 5: Polyhedral Virions and Bipartite RNA Genomes*, 35p. New York: Plenum.

Lin T and Johnson JE (2003) Structure of picorna-like plant viruses: Implications and applications. *Advances in Virus Research* 62: 167–239.

Lomonossoff GP, Shanks M, *et al.* (1991) Comovirus capsid proteins: Synthesis, structure and evolutionary implications. *Proceedings of the European Phytochemical Society* 32: 76.

Cucumber Mosaic Virus

F García-Arenal, Universidad Politécnica de Madrid, Madrid, Spain
P Palukaitis, Scottish Crop Research Institute, Dundee, UK

Glossary

Chlorotic Yellowing or light green symptoms induced by virus infection affecting chlorophyll accumulation.

Cross-protection Inhibition of systemic virus accumulation and disease by prior inoculation of plants with a mild or symptomless strain of the same virus.

Filiformism Narrowing of the leaf blade, often leading to symptoms referred to as shoestring.

Pseudorecombination Reassortment of the genomic RNAs of two or more strains of a multipartite RNA virus to generate novel combinations of the full complement of genomic RNAs.

Satellite RNA A subviral RNA genome dependent on a helper virus for both replication and encapsidation.

Tonoplast The membrane surrounding the central vacuole of plant cells.

History

Cucumber mosaic virus (CMV) was first described as a disease of cucurbits in 1916 by Doolittle in Michigan and Jagger in New York. The virus can infect a large number of indicator plant species and has been isolated from over 500 naturally infected species. Cross-protection was used in the 1930s to discriminate isolates of CMV with differences in phenotypes or host range (strains). CMV was not purified reliably until the middle 1960s. Later serology and hybridization technology were used to detect and differentiate two major subgroups of CMV. The nucleotide sequence and the genome organization of one strain of each CMV subgroup were determined between 1984 and 1990, while biologically active cDNA clones of several CMV strains were developed in the early 1990s. The major functions of each of the five encoded proteins have been assigned, although each protein is also involved in other host–virus relationships.

Taxonomy and Classification

Isolates of CMV are heterogeneous in symptoms, host range, transmission, serology, physicochemical properties, and nucleotide sequence of the genomic RNAs. On the basis of different criteria (e.g., serological typing, peptide mapping of the coat protein, sequence similarity of their genomic RNA) CMV isolates can be classified into two major subgroups, now named subgroup I and subgroup II. The percentage identity in the nucleotide sequence between pairs of isolates belonging to each of these subgroups ranges from 69% to 77%, depending on the pair of isolates and the RNA segment compared, dissimilarity being highest for RNA2. Nucleotide sequence identity among isolates within a subgroup is above 88% for subgroup I and above 96% for subgroup II, indicating a higher heterogeneity of subgroup I. Analysis of the open reading frames (ORFs) and 5′ noncoding regions of RNA3 of subgroup I isolates shows a group of closely related isolates forming a monophyletic cluster, named subgroup IA; the rest of subgroup I isolates are included in the nonmonophyletic group IB. Analyses of RNA2 show that subgroups IA and IB constitute monophyletic groups, while analyses of RNA1 show no clear division into groups IA and IB. Hence, the different genomic segments have followed different evolutionary histories. Cross-protection occurs between strains from all subgroups. Isolates of subgroup I and II can be distinguished using monoclonal antibodies, and isolates from subgroup IA, IB, and II can be distinguished by reverse transcriptase polymerase chain reaction (RT-PCR). CMV isolates differ from isolates of the other two cucumovirus species, *Tomato aspermy virus* (TAV) and *Peanut stunt virus*, having only 50–67% nucleotide sequence identity, depending on the RNA and isolates being compared.

Geographic Distribution

CMV isolates have a worldwide distribution, having been reported from both temperate and tropical regions. Most reported isolates belong to subgroup I. Subgroup II isolates are found more frequently in cooler areas or seasons of temperate regions. This has been associated with lower temperature optima for *in planta* virus accumulation shown for the few isolates characterized for this property. Most isolates in subgroup IB have been reported from East Asia, which is presumed to be the origin of this subgroup. Subgroup IB isolates also have been reported from other areas, for example, the Mediterranean region, California, Brazil, and Australia. Those in the Mediterranean could have been introduced recently from East Asia.

Host Range and Propagation

The host range of the collective isolates of CMV is over 1300 species in more than 500 genera of over 100 families, with new hosts reported each year. Some recently described strains from new hosts have lost the ability to infect many of the typical hosts of CMV. This may be a general feature for adaptation to unusual hosts. CMV infects most of the major horticultural crops as well as many weed species; the latter act as reservoirs for the virus. Infection of various indicator plants was used to differentiate CMV from other viruses, since unlike most other viruses of cucurbitaceous or solanaceous hosts, CMV could infect representative species of both families. These include cucumber (*Cucumis sativus*), tomato (*Solanum lycopersicum*), and tobacco (*Nicotiana tabacum*), all systemic hosts of CMV, as well as cowpea (*Vigna unguiculata*) and *Chenopodium quinoa*, which limit CMV infection to the inoculated leaves, although there are legume strains that will infect cowpea systemically. Most isolates of CMV are best propagated in squash (marrow) (*Cucurbita pepo*), tobacco, *N. clevelandii*, or *N. glutinosa*.

Virus Structure and Properties

CMV has icosahedral particles 29 nm in diameter, which sediment as a single component with an $S_{20,w}$ of 98.6–104c (c being virus concentration in $mg\,ml^{-1}$). Particles are built of 180 capsid protein subunits arranged with $T = 3$ quasisymmetry, contain about 18% RNA, and have an extinction coefficient at 260 nm ($1\,mg\,ml^{-1}$, 1 cm light path) of 5.0. RNA1 and RNA2 are encapsidated in different particles, whereas RNA3 and RNA4 are probably packaged together in the same particle, some particles may contain three molecules of RNA4. Thus, virus preparations contain at least three different types of particles, but with similar morphology and sedimentation properties. Virus particles also contain low levels of the RNA species designated RNA4A, RNA5, and RNA6. There is a limit to the size of encapsidated RNAs, those larger than RNA1 are not encapsidated *in vivo*. CMV particles are stabilized by RNA–protein interactions, and no empty particles are formed. Particles disrupt at high neutral chloride salt concentrations or at low sodium dodecyl sulfate concentrations; biologically active particles can be reassembled by lowering the salt concentration or removing the sodium dodecyl sulfate. Particles are stable at pH 9.0 and do not swell at pH 7.0, an important difference with bromoviral particles.

The structure of CMV particles has been resolved at 3.2 Å by X-ray crystallography. The $T = 3$ lattice of capsids is composed of 60 copies of three conformationally distinct subunits designated A, B, and C that form trimers with quasi-threefold symmetry. There are

20 hexameric capsomers of B and C subunits with quasi-sixfold symmetry and 12 pentameric capsomers formed by A subunits with fivefold symmetry. The exterior radius along the quasi-sixfold axes is 144 Å, the RNA is tightly packaged against the protein shell and leaves a hollow core of about 110 Å along the threefold axes. The protein subunit has a β-barrel structure, with the long axis of the β-barrel domain oriented roughly in a radial direction. The N-terminal 22 amino acids of the capsid protein are needed for particle assembly. This region is positively charged and, in the B and C subunits, forms amphipatic helices that run parallel to the quasi-sixfold axes. There is an external region of negative electrostatic potential that surrounds the fivefold and quasi-sixfold axes and locates above regions of positive potential which extend to cover nearly homogeneously the inner surface of capsids, where interaction with encapsidated RNA occurs. Electrostatic distributions in CMV particles explain the physicochemical conditions required for particle stability.

Genome Organization

The genome of CMV consists of five genes distributed over three, single-stranded, positive-sense, capped, genomic RNAs: RNA1 (3.3–3.4 kb) encodes the *c.* 111 kDa 1a protein. RNA2 (3.0 kb) encodes the 98 kDa 2a protein, as well as the 13–15 kDa 2b protein, which is translated from a 630–702 nt subgenomic RNA designated RNA 4A that is co-terminal with the 3′ end of RNA2. The ORF expressing the 2b protein overlaps with the ORF encoding the 2a protein, but in a +1 reading frame. RNA3 encodes the 30 kDa 3a protein, as well as the 25 kDa 3b protein, which is expressed from a 1010–1250 nt subgenomic RNA designated RNA4 that is co-terminal with the 3′ end of RNA3. The 224–338 nt 3′ nontranslated regions of all three genomic and both subgenomic RNAs are highly conserved, forming a tRNA-like structure as well as several pseudoknots. The 5′ nontranslated regions of RNA1 (95–98 nt) and RNA2 (78–97 nt) are more conserved in sequence with each other than with those of RNA3 (96–97 nt or 120–123 nt). CMV also produces an RNA5 of unknown function, which is co-terminal with the 3′ nontranslated regions of RNA1 and RNA2. RNA4A and RNA5 are only encapsidated by subgroup II strains. CMV particles also encapsidated a low level of tRNAs, which have been reported in the literature as CMV RNA6. Although rarely reported, some strains of CMV can also encapsidate defective RNAs derived from CMV RNA3.

Satellite RNAs

CMV can also support satellite RNAs varying in size from 333 to 405 nt. These satellite RNAs are dependent upon CMV as the helper virus for both their replication and encapsidation, but have sequence similarity to the CMV RNAs limited to no more than 6–8 contiguous nt. More than 100 satellite variants have been found associated with over 65 isolates of CMV from both of the CMV subgroups. These satellite RNAs usually reduce the accumulation of the helper viruses and on most hosts also reduce the virulence of CMV. However, this attenuation of disease is not due to competition between the helper virus and the satellite RNA for a limited amount of replicase or capsid protein. Some CMV satellite RNAs can also be replicated and packaged by strains of the cucumovirus TAV, although these satellite RNAs do not attenuate the symptoms induced by TAV. Certain satellite RNAs in some selected hosts can enhance the disease induced by CMV. In the case of tomato plants infected by CMV and certain satellites, this has led to systemic necrosis observed in the fields of several Mediterranean countries. This necrosis is actually caused by sequences of the complementary-sense satellite RNA produced in large quantities during satellite RNA replication.

Genetics

The CMV 1a and 2a proteins encode proteins that replicate the virus, but which also function in promoting virus movement in several host species. The 2b protein is an RNA-silencing suppressor protein that antagonizes the salicylic acid defense pathway and also influences virus movement in some hosts. The 3a protein is the major movement protein of the virus and is essential for both cell-to-cell as well as long-distance (systemic) movement. The 3b protein is the sole viral capsid protein and is also required for cell-to-cell and long-distance movement, although the ability to form virions is not a requirement for movement. All of the CMV-encoded proteins are RNA-binding proteins. Viral RNAs from different strains and subgroups can be exchanged to form novel viruses, allowing mapping of some phenotypes to specific RNAs. Higher-resolution mapping requires the use of biologically active cDNA clones for generating chimeras and site-directed mutants. By these approaches, the following functions have been delimited to specific RNAs, with some mapped to specific nucleotide changes: hypersensitive response in tobacco, rapid local and systemic movement in squash, seed transmission in legumes, temperature-sensitive replication in melon; and replication of satellite (all in RNA1); hypersensitive response versus systemic infection in cowpea, suppression of gene silencing, and host range and pathology responses (all in RNA2); limited movement between epidermal cells, systemic movement in cucurbits, virion assembly, hypersensitive response on *Nicotiana* sp., local and systemic infection in cucurbits or in maize, symptom responses, and aphid transmission (all in RNA3).

Replication

CMV replication takes place on the vacuole membrane (the tonoplast). Replication involves the 1a and 2a proteins of the virus and presumably several host proteins. The purified CMV replicase contains the 1a and 2a proteins, as well as a host protein of *c.* 50 kDa of unknown function. The 1a protein of CMV contains a putative N-terminal proximal methyltransferase domain believed to be involved in capping of the RNAs, as well as a putative C-terminal proximal helicase domain, presumed to be required for the unwinding of the viral RNAs during replication. The 1a protein has been found to be able to bind to several tonoplast intrinsic proteins, although what roles these have in virus replication has not been established. The N-terminal region of the 2a protein interacts with the C-terminal region of the 1a protein *in vivo* and *in vitro*. Phosphorylation of the 2a protein prevents interaction with the 1a protein. The C-terminal half of the 2a protein contains conserved domains found in RNA polymerases and therefore together with the 1a protein forms the core of the CMV replicase.

CMV replication is initiated by the binding of the tonoplast-associated replicase to the tRNA-like structure and various pseudoknots present in the $3'$ nontranslated region of the positive-sense CMV RNAs. Minus-sense RNA is then synthesized from each of the genomic RNAs and the synthesized minus-sense RNA acts as a template for synthesis of new plus-stranded genomic RNAs. The minus-sense RNA2 and RNA3 also serve as the templates for the synthesis of the two plus-sense subgenomic RNAs (4 and 4A), through recognition of the subgenomic promoters present on the minus-sense RNAs. The subgenomic RNAs are not themselves replicated, but defective RNAs and satellite RNAs are replicated by the CMV replicase. Differences in the relative levels of accumulation of the various CMV genomic RNAs and satellite RNAs have been observed in different host species, which probably is due to host-specific differences in template copying.

Movement

CMV moves cell to cell via plasmodesmata between cells until it reaches the vasculature, when the virus moves systemically via the phloem. The viral RNAs move as a nucleoprotein complex between cells involving the 3a movement protein and some involvement by the capsid protein. This movement does not involve interactions with microtubules. No specific plant proteins have yet been identified as being involved in cell-to-cell movement. CMV appears to move between epidermal cells as well as from epidermal cells down to mesophyll cells toward vascular cells. The 2b protein also influences the path of virus movement, since without the 2b protein, the virus moves preferentially to and between mesophyll cells. The virus replicates in all of these cell types, but not in the sieve elements of the vasculature. Virion assembly may take place inside sieve elements from RNAs and capsid protein moving from neighboring vascular cells. Virion assembly is necessary in some, but not all, species for systemic infection. A 48 kDa phloem protein (PP1) from cucumber phloem exudates interacts with CMV particles *in vitro* and increases virus particle stability. How the virus moves from the vasculature back to mesophyll and epidermal cells is unknown. The virus moves from plant to plant either by transmission by the aphid vectors of CMV, or in some cases via seed transmission, at a low but variable frequency.

Pathology

The symptoms induced by CMV are not generally specific to CMV, but rather reflect sets of host responses to viral pathogens. Therefore, symptoms such as light green–dark green mosaics, generalized chlorosis, stunting, leaf filiformism, and local chlorotic or necrotic lesions associated with various strains of CMV are not specific to CMV, but can also be elicited by other viruses in the same plant species. Some strains of CMV can induce a bright yellow chlorosis in some *Nicotiana* species. This can be due to either specific amino acid changes in the viral capsid protein, or the presence of a chlorosis-inducing satellite RNA. A white-leaf disease of tomato also was due to the effects of a particular satellite RNA, as well as systemic necrosis and lethal necrosis disorders seen in the field of several Mediterranean countries. Some pathogenic responses, such as local lesions versus systemic infection in cowpea have been mapped to two amino acid changes in RNA 2, while other pathogenic responses have been mapped to sequences present in more than one viral RNA molecule, and to different viral sequences for different strains, indicating that several interactions are involved in the elicitation of some symptom responses.

Cytopathology associated with infection by CMV includes viral inclusions within the cytoplasm and vacuoles, and as membrane-bound clusters in sieve elements. In some cases, angular inclusions corresponding to virus crystals can be seen in vacuoles by staining and light microscopy. CMV infection usually also leads to proliferation of cytoplasmic membranes, which originate from the plasma membrane, the endoplasmic reticulum, or the tonoplast. Effects on the nucleus or nucleolus (e.g., vacuolation), usually due to virion accumulation, have been observed with some strains of CMV. Similarly, some strains have caused effects on mitochondria or chloroplasts. Yellowing strains, in particular, show effects on chloroplasts development leading to smaller and rounded chloroplasts that have fewer grana and starch granules. These various effects appear to be host specific. How virus–plant interactions lead to cytopathic effects remains unknown; however, as the virus expands from the initial site of infection, rings or zones of responses occur, in which the expression patterns of numerous plant genes are altered in a spatial- and temporal-specific manner.

Transmission

Seed transmission of CMV has been reported in many plant species, with efficiencies varying from less than 1% up to 50%. Virus may be present in the embryo, endosperm, and seminal integuments, as well as in pollen. RNA1, and possibly protein 1a, affects the efficiency of seed transmission.

Horizontal transmission of CMV is vectored by aphids in a nonpersistent manner. Over 80 species of aphids have been reported to transmit CMV, *Aphis gossypii* and *Myzus persicae* being two efficient and most studied vectors. Transmission efficiency depends on several factors, particularly the specific combination of virus isolate and aphid species, and the accumulation of particles in the source leaf. Differences in transmissibility of various isolates by different aphid species are determined solely by the virus coat protein and amino acid determinants for transmission have been mapped. The amino acid positions that determine transmission are either exposed on the outer surface of capsids (e.g., amino acid position 129, on the β strand H-I loop) or lay in the inner surface (e.g., position 162). Hence, the effect could be through direct interaction with the aphid mouth parts or by affecting particle stability. Loss of transmissibility upon repeated mechanical passage seems to be rare, perhaps because of tradeoffs with particle stability, but particle stability is not always correlated with transmissibility. Transmission efficiency also depends on the host plant, as shown by the resistance to transmission on melon genotypes containing the *Vat* gene, which does not confer resistance to the aphid or the virus, but impairs aphid transmission.

Ecology and Epidemiology

CMV infects a wide range of food crops, ornamentals, and wild plant species. Economic losses in crops are highest in field-grown vegetables and ornamentals, pasture legumes, and banana. In recent times, CMV has caused severe epidemics in many crops, including necrosis of tomato in Italy, Spain, and Japan; mosaic and heart rot of banana worldwide; mosaic of melons in California and Spain; mosaic of pepper in Australia and California; and mosaic of lupins and other legumes in Australia and the USA. In crops in which seed transmission is effective (e.g., pasture or fodder legumes), the primary inoculum for epidemics may be the seedlings from infected seeds. In most vegetable and ornamental crops, seed transmission does not occur or is negligible, and the primary inoculum must come from outer sources as other crops or weeds, which should be near the crop as aphid transmission is nonpersistent. In the absence of crops during unfavorable seasons, infected perennial weeds or crops, and infected seeds from weeds act as reservoir inoculum. Seed transmission has been shown to be important in several weed species from different regions. In spite of its general broad host range, there is evidence of host adaptation or preference for some CMV strains, which might have important consequences for inoculum flows among host species. Also, the dynamics of virus infection may differ largely in weeds and crops within a region, indicating that the relevant reservoirs and inoculum sources for crops need not be the most frequently infected weeds. In banana, secondary spread of infection within the crop is ineffective for most strains, and alternative hosts are both primary and secondary inoculum sources for epidemics.

Variation and Evolution

In agreement with the high mutation rates of RNA genomes, populations of CMV derived from biologically active cDNA clones were found to be genetically diverse. When the cDNA-derived population was passaged in different hosts, the amount of genetic diversity depended on the host species. Genetic diversity has been shown to be countered by genetic drift associated to population bottlenecks during systemic colonization of the host and, probably also, during host-to-host transmission. Sequence analyses have shown different evolutionary constraints for the different viral proteins, which show different evolutionary dynamics. A second source of genetic variation is the exchange of RNA sequences by recombination or by reassortment of genomic segments. Experimentally, recombination has been shown to occur between the 3' nontranslated region of the genomic RNAs, with recombinants being up to 11% of the population. Recombinants in the 3' nontranslated region may have an increased fitness in some hosts, as shown for isolates infecting alstroemeria. Recombination in the RNA3 also was frequent in mixed infection between CMV and TAV. Recombination between CMV strains or CMV and TAV strains seems to be facilitated by stem–loop structures in the RNA. In spite of abundant evidence for frequent recombination, analyses of the genetic structure of field populations of CMV show that recombinant RNAs are not frequent, and that selection operates against most recombinants. A second mechanism of genetic exchange is reassortment of genomic

segments also called pseudorecombination. Reassortants exchanging any genomic segment have been obtained between different CMV strains, which multiply efficiently under experimental conditions. Natural reassortants also have been described, and reassortment may have played an important role in the evolution of CMV, as suggested by the different phylogenies obtained for each genomic RNA. In field population, reassortant isolates are rare and seem to be selected against, as is the case for recombinants. Evidence for selection against genotypes originating by genetic exchange suggests co-adaptation of the different viral genes that, when disrupted, results in a decreased fitness. Analyses of the population structure of CMV in Spain and California indicate a metapopulation structure, with local extinctions and recolonizations, which suggests that population bottlenecks occur, probably associated with unfavorable seasons for the host plants and/or the aphid vectors. Interestingly, this is not the case for the population structure of the satellite RNA. Analyses in Italy and Spain during epidemics of CMV plus satellite RNAs have shown that the satellite RNAs have an undifferentiated population. The different population structure for CMV and its satellite RNAs indicates that the satellite RNAs expanded as a molecular parasite on the CMV population, rather than satellite expansion was linked to a particular CMV isolate.

Control

Control of CMV can be achieved by planting of resistant crops, but resistance in many crop species is often not available to a broad range of CMV strains. The use of insecticides to control the aphid vectors of the virus has met with only limited success, since the virus is transmitted in a nonpersistant manner and thus before the aphids would have been killed by the insecticide, it would have transmitted the virus. Rather, insecticides are used to reduce aphid numbers and thus reduce the progressive spread of infections. Since many species of weeds act as reservoirs for the virus and many of these are asymptomatic hosts, it is important to remove these from the borders of fields to eliminate the source of infectious material that could then be spread by aphids. This also applies to removal of infected crop plants during the growing season. Others sources of resistance include the use of transgenic plants expressing either protein-mediated or RNA silencing-mediated resistance. Although most of these approaches have led to resistance to only members of one of the two major subgroups, the use of pyramiding of viral segments from different subgroups offers the promise of obtaining a broad spectrum resistance to CMV together with other viruses infecting the same crop species. Transgenic expression of satellite RNAs has also been used to confer resistance to CMV. This has met with success, but has raised concerns about using a virulent pathogen, as did the use of mild strains of CMV for cross-protection against severe strains.

See also: Ilarvirus.

Further Reading

Edwardson JR and Christie RG (eds.) (1991) Cucumoviruses. *CRC Handbook of Viruses Infecting Legumes*, pp. 293–319. Boca Raton, FL: CRC Press.

Kaper JM and Waterworth HE (1981) Cucumoviruses. In: Kurstak E (ed.) *Handbook of Plant Virus Infections and Comparative Diagnosis*, pp. 257–332. New York: Elsevier/North-Holland.

Palukaitis P and Garcia-Arenal F (2003) Cucumoviruses. *Advances in Virus Research* 62: 241–323.

Palukaitis P and García-Arenal F (2003) Cucumber mosaic virus. In: Antoniw J and Adams M (eds.) *Description of Plant Viruses*. DPV400, Rothamstead Research, UK: Association of Applied Biologists. http://www.dpvweb.net/dpv/showdpv.php?dpvno=400 (accessed September 2007).

Palukaitis P, Roossinck MJ, Dietzgen RG, and Francki RIB (1992) Cucumber mosaic virus. *Advances in Virus Research* 41: 281–348.

Flexiviruses

M J Adams, Rothamsted Research, Harpenden, UK

Glossary

Triple gene block (TGB) protein A set of three proteins in overlapping reading frames that function together in facilitating cell-to-cell movement of plant viruses.

Introduction

The creation of the family *Flexiviridae* was approved by the International Committee on the Taxonomy of Viruses (ICTV) in 2004 to include a number of existing genera and some unassigned species of plant viruses. The family is so named because its members have flexuous, filamentous virions, and there are a number of characteristics that distinguish it from the families *Closteroviridae* and *Potyviridae*, the two other families of plant viruses that have virions of similar morphology (**Table 1**).

Main Characteristics

The chief characteristics shared by all members of the family are considered in this section.

Virion Morphology

Virions are flexuous filaments, 12–13 nm in diameter, and usually 500–1000 nm in length, depending on the genus. They are constructed from helically arranged coat protein subunits in a primary helix with a pitch of about 3.4 nm and nine to ten subunits per turn. Each particle contains a single molecule of the full-length genomic RNA. Smaller 3′ terminal subgenomic RNAs may also be encapsidated in some species, sometimes in shorter virions. In negative contrast electron microscopy the virions usually appear to be cross-banded. A typical electron micrograph is shown in **Figure 1** (for red clover vein mosaic virus, genus *Carlavirus*) but in the genera *Trichovirus* and *Vitivirus*, the virions appear to be rather more flexuous and hence resemble more those of the family *Closteroviridae*.

Table 1 Characteristics of the families of filamentous plant viruses[a]

Family	Genome size (kbp)	RNAs[b]	ORFs	Translation strategy[c]	5′ end VPg[d]	3′-polyA
Closteroviridae[e]	15–20	1, 2, or 3	10+	Subgenomic	No	No
Flexiviridae	5.5–9.5	1	2–6	Subgenomic	No	Yes
Potyviridae	9–12	1 or 2	1 or 2	Polyprotein	Yes	Yes

[a]All these have ssRNA genomes that encode in the virion (positive) sense.
[b]Numbers of genomic RNA components.
[c]Proteins are translated either from a set of nested subgenomic mRNAs or are cleaved by virus-encoded proteases from a polyprotein precursor.
[d]Presence or absence of a genome-linked protein covalently bound to the 5′ terminus of the RNA.
[e]Members of the family *Closteroviridae* have distinctive, tailed virions. The tail is constructed from a second, minor, coat protein.

Genome

All members have a monopartite, positive-sense, single-stranded RNA (ssRNA) genome in the size range 5.5–9.5 kbp, depending on the species. This forms about 5–6% of the virion by weight. The RNA is capped at the 5′ terminus and there is a 3′ polyA tail.

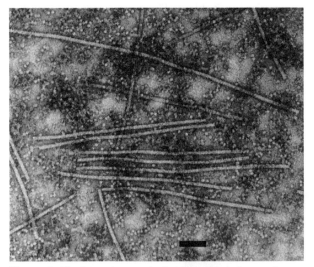

Figure 1 Electron micrograph of virions of red clover vein mosaic virus (genus *Carlavirus*), negatively stained. Scale = 100 nm.

Replication and Translation

In all members, there is a short 5′ nontranslated region followed by a single large alpha-like replication protein (145–250 kDa). This contains (in this order) conserved methyltransferase, helicase, and RNA-dependent RNA polymerase (RdRp) motifs and appears to be the only protein required for replication, which is believed to take place in the cytoplasm. Translation of the smaller open reading frames (ORFs) downstream occurs from one or more 3′ terminal subgenomic mRNAs. In some species, there is a 'leaky scanning' mechanism that ensures that two or three different proteins are translated from the same mRNA.

Genomic Organization

Genomic organization varies between genera but there are up to six ORFs ordered from 5′ to 3′:

1. a large alpha-like replication protein (see above);
2. one or more cell-to-cell movement proteins (MPs): in some genera there is a single MP of the '30K' superfamily (distantly related to the 30 kDa MP of tobacco mosaic virus (genus *Tobamovirus*)) while in the others there is a set of three overlapping proteins known as a 'triple gene block' (TGB);
3. a single capsid (coat) protein (CP) (21–45 kDa);
4. a final ORF (in some viruses only), which is thought to have RNA-binding properties and is probably a

Table 2 Genera and unassigned species included within the family *Flexiviridae*

Genus and unassigned species[a]	Virion length (nm)	ORFs	MP type[b]	Rep (kDa)	CP (kDa)
Potexvirus (28)	470–580	5	TGB	146–191	22–27
Mandarivirus (1)	650	6	TGB	187	36
Allexivirus (8)	~800	6	TGB	175–194	26–28
Carlavirus (35)	610–700	6	TGB	226–238	31–40
Foveavirus (3)	800+	5	TGB	244–247	28–45
Capillovirus (3)	640–700	2 or 3	30K	214–241	25–27
Vitivirus (4)	725–785	5	30K	195–196	18–23
Trichovirus (4)	640–760	3 or 4	30K	216–217	21–22
Citrivirus (1)[c]	960	3	30K	227	41
Banana mild mosaic virus (virus: BanMMV)	580	5	TGB	205	27
Banana virus X (virus: BanVX)	?	5	TGB	nd[d]	24
Cherry green ring mottle virus (virus: CGRMV)	1000+	5	TGB	230	30
Cherry necrotic rusty mottle virus (virus: CNRMV)	1000+	5	TGB	232	30
Potato virus T (virus: PVT)	640	3	30K	nd	24
Sugarcane striate mosaic-associated virus (virus: SCSMaV)	950	5	TGB	222	23
Possible species					
Botrytis virus X (BotV-X)	720	5	na[e]	158	44

[a]Numbers of species recognized within each genus are shown in parentheses (2006 data).
[b]Either a set of three overlapping proteins (TGB) or a single protein of the '30K' superfamily (phylogenetically related to the 30 kDa MP of *Tobacco mosaic virus*, genus *Tobamovirus*).
[c]Genus proposed in 2005 for the species *Citrus leaf blotch virus* (virus: CLBV).
[d]Not determined (only partial sequence data available).
[e]Not applicable (no MP encoded).

suppressor of gene silencing; it partially overlaps the 3′ end of the CP gene in some species.

Biological Properties

Biologically, the viruses are fairly diverse. They have been reported from a wide range of herbaceous and woody mono- and dicotyledonous plant species but the host range of individual members is usually limited. Natural infections by members of the genera *Capillovirus, Citrivirus, Foveavirus, Mandarivirus, Trichovirus,* and *Vitivirus* are mostly or exclusively of woody hosts. Many of the viruses have relatively mild effects upon their host. All species can be transmitted by mechanical inoculation, often readily. Many of the viruses have no known invertebrate or fungal vectors but allexiviruses and some trichoviruses are thought to be mite borne, most carlaviruses are transmitted naturally by aphids in a nonpersistent

manner, and a range of vectors have been reported for different vitiviruses. Aggregates of virus particles accumulate in the cytoplasm but there are generally no distinctive cytopathic structures.

Component Genera and Species

Some properties of the nine genera included in the family and of six species that are not assigned to any genus are summarized in **Table 2**. These are mostly distinguished by the number of ORFs, the type of MP encoded, and the sizes of their virions and replication protein genes. In some genera, there are additional ORFs not present in other members of the family. These distinctions based on genome organization are also supported by phylogenetic analyses (see below). Genome organization is summarized in **Figure 2**.

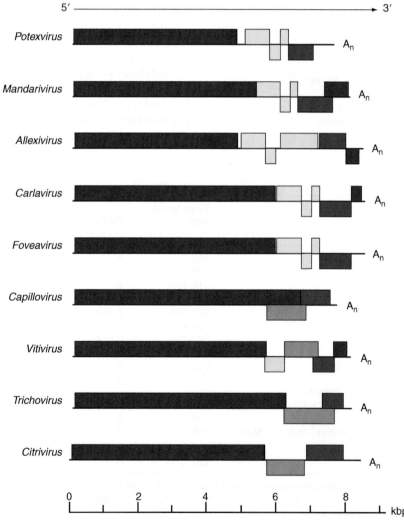

Figure 2 Diagram showing genome organization for the component genera of the family *Flexiviridae*. Blocks represent predicted ORFs. Replication proteins are shown in dark blue, '30K-like' MPs in orange, TGB proteins in yellow, CPs in red, and RNA-binding proteins in purple. ORFs of unknown function are shown in light blue.

A recently described mycovirus, botrytis virus X, shares many of the characteristics of the family and its CP and replication protein are phylogenetically related to members of the family *Flexiviridae*, in particular to members of the genera *Potexvirus* and *Allexivirus*. It is likely that a new genus will be proposed to accommodate this species in due course.

Phylogenetic Relationships

The relationships detected between the genera within the family and those between members of the family *Flexiviridae* and other viruses are dependent on the genome region used for analysis. This probably indicates that the viruses have arisen from reassortment of entire genes (or blocks of genes). These relationships are summarized below.

Replication Protein

A dendrogram based on the nucleotide sequence of the replication proteins (**Figure 3**) supports the taxonomic division of the family *Flexiviridae* into genera and falls into two major parts. One part contains the genera *Allexivirus*, *Mandarivirus*, and *Potexvirus* together with botrytis virus

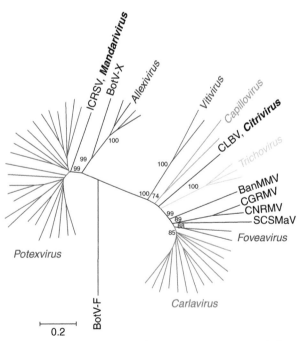

Figure 3 Phylogenetic tree of the replication proteins of all sequenced members of the family *Flexiviridae* with botrytis virus X and botrytis virus F. For abbreviations, see **Table 2**. The neighbor-joining tree is based on the codon-aligned nucleotide sequences. Values on the major branches show the percentage of trees in which this grouping occurred after bootstrapping the data (10 000 replicates). The scale bar shows the number of substitutions per base.

X These proteins are the smaller ones in the family (145–195 kDa) and the plant-infecting members all have a TGB. The second part of the tree contains those viruses with larger proteins (195–250 kDa). It includes all those with a '30K'-like MP (*Capillovirus*, *Citrivirus*, *Trichovirus*, and *Vitivirus*), together with the TGB-containing species that make up the genera *Carlavirus* and *Foveavirus* and some unassigned viruses that have foveavirus-like genome organization.

Most studies of 'longer distance' relationships between viruses have been based upon the motifs within the replication protein as these are relatively conserved within ssRNA viruses. Such studies consistently place the family *Flexiviridae* within a larger group ('alpha-like' or 'Sindbis-like' with 'type III' RdRp) that includes the family *Tymoviridae* (plant-infecting viruses that have isometric particles) and, more distantly, the families *Togaviridae*, *Bromoviridae*, *Closteroviridae,* and the rod-shaped plant viruses including the genus *Tobamovirus*. Phylogenetic analysis of the entire replication protein shows that this protein in members of the *Tymoviridae* is related to that found in the genera *Allexivirus, Mandarivirus*, and *Potexvirus* but distinct from that found in the remaining members of the family *Flexiviridae* (**Figure 4**). This might suggest that the *Tymoviridae* and *Flexiviridae* could be placed within the same family or that the current family *Flexiviridae* could be divided into two separate families. However, the relationships between other genes (see below) and differences in virion morphology and cytopathology suggest that the current classification makes better biological sense. The replication protein of the mycovirus botrytis virus F also appears to belong to the *Flexiviridae–Tymoviridae* group but falls outside any of the major groups and is probably best left unclassified at present.

Movement Protein

'30K'-like MPs occur in the genera *Capillovirus*, *Citrivirus*, *Trichovirus*, and *Vitivirus* and also in the unassigned potato virus T. The '30K' superfamily proteins are largely recognized by structural similarities rather than by primary sequence comparisons. Nonetheless, phylogenetic analysis shows that those from the family *Flexiviridae* are related to one another (although *Vitivirus* is more distant) and that there are some similarities to the corresponding proteins in the rod-shaped plant virus genera *Furovirus*, *Tobravirus,* and *Tobamovirus* (**Figure 5**).

TGB Proteins

Within the family, the TGB proteins (for those genera that have them) are related in much the same way as their CPs. A TGB module is also found in members of the genera *Benyvirus*, *Hordeivirus*, *Pecluvirus*, and *Pomovirus*. However, the first and third proteins of the TGB of the

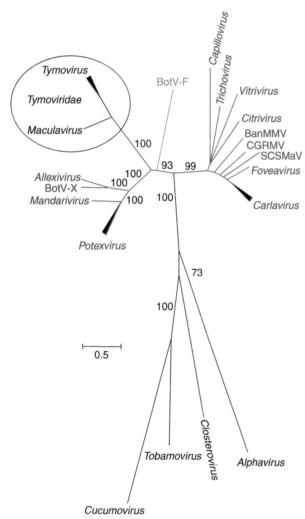

Figure 4 Phylogenetic tree of the replication proteins of selected members of the families *Flexiviridae* (in red) and *Tymoviridae* (in blue) with other related viruses. For abbreviations, see **Table 2**. The neighbor-joining tree is based on the aligned peptide sequences. Values on the major branches show the percentage of trees in which this grouping occurred after bootstrapping the data (500 replicates). The scale bar shows the number of substitutions per base.

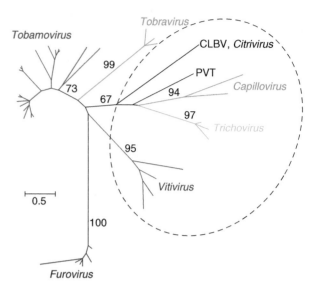

Figure 5 Phylogenetic tree of the '30K-like' MPs of the family *Flexiviridae* (in dotted oval) and the corresponding proteins of rod-shaped plant viruses. For abbreviations, see **Table 2**. The neighbor-joining tree is based on the aligned peptide sequences. Values on the major branches show the percentage of trees in which this grouping occurred after bootstrapping the data (500 replicates). The scale bar shows the number of substitutions per base.

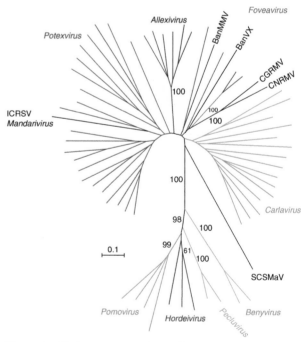

Figure 6 Phylogenetic tree of the TGB proteins of the family *Flexiviridae* and related rod-shaped plant viruses. For abbreviations, see **Table 2**. The neighbor-joining tree is based on the codon-aligned nucleotide sequences of all three proteins. Values on the major branches show the percentage of trees in which this grouping occurred after bootstrapping the data (10 000 replicates). The scale bar shows the number of substitutions per base.

rod-shaped viruses are considerably larger than the corresponding proteins in the family *Flexiviridae* and clearly form a separate grouping in phylogenetic analyses (**Figure 6**). In some species, notably all members of the genus *Allexivirus*, a sequence similar to that coding for the third TGB protein can be recognized but there is no standard AUG start codon and it is not known whether the protein is expressed.

Coat Protein

A phylogenetic tree of the CP sequences shows that most viruses with a '30K'-like MP in the genera *Capillovirus*, *Trichovirus*, and *Vitivirus* are only distantly related to those in the remainder of the family but that citrus leaf blotch virus, the single member of the genus *Citrivirus*, groups with a well-defined group containing all the TGB-containing members (**Figure 7**).

It is difficult to detect any significant relationship between the CPs of members of the *Flexiviridae* and those of other viruses beyond the modest similarity that occurs between the CPs of all viruses with filamentous particles.

RNA-Binding Protein

Putative RNA-binding proteins are found in the genera *Allexivirus, Carlavirus, Mandarivirus,* and *Vitivirus*. The proteins are clearly related and share a strictly conserved core

amino acid sequence $GxSxxAxR/KRRAx_5Cx_{1-2}Cx_{5-13}C$. In phylogenetic analyses, the proteins from each genus cluster separately. The proteins probably act as silencing suppressors and are related to those of the closteroviruses (p23 of citrus tristeza virus and homologs in other species) (**Figure 8**).

General Comments

The limitations of phylogenetic analysis are demonstrated by the complex relationships within and outside the family that are revealed when different genes are used in the analysis (**Table 3**). In particular, there are big differences in the relative positions of the genera *Carlavirus, Citrivirus,* and *Foveavirus* to the rest of the family. The single virus in the genus *Mandarivirus* consistently groups with the genus *Potexvirus*, but it has a larger CP and an RNA-binding protein related to those of the genera *Allexivirus* and *Carlavirus*. Outside the family, the replication proteins are related to those found in the *Tymoviridae* and to botrytis virus F. By contrast, the movement protein(s), whether TGB or '30K'-like, are clearly related to those of the rod-shaped viruses, which are much more distantly related in their replication proteins.

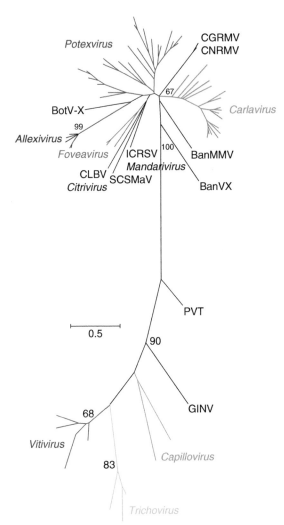

Figure 7 Phylogenetic tree of the CPs of all sequenced members of the family *Flexiviridae*. For abbreviations, see **Table 2**. The neighbor-joining tree is based on the aligned peptide sequences. Values on the major branches show the percentage of trees in which this grouping occurred after bootstrapping the data (500 replicates). The scale bar shows the number of substitutions per base.

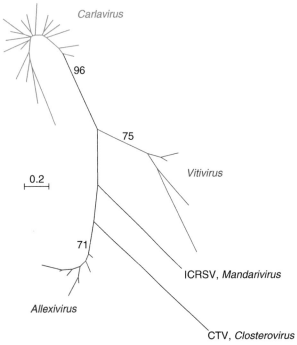

Figure 8 Phylogenetic tree of the nucleic acid-binding proteins of members of the family *Flexiviridae* and citrus tristeza virus (CTV, genus *Closterovirus*). The neighbor-joining tree is based on the aligned peptide sequences. Values on the major branches show the percentage of trees in which this grouping occurred after bootstrapping the data (500 replicates). The scale bar shows the number of substitutions per base.

Table 3 Summary of the main phylogenetic relationships within the family *Flexiviridae*

Genus	Rep[a]	MP[b]	CP[c]	RNABP[d]
Potexvirus	1	TGB	1	No
Mandarivirus	1	TGB	1	Yes
Allexivirus	1	TGB	1	Yes
(Botrytis virus X)	1	na[e]	1	No
Carlavirus	2	TGB	1	Yes
Foveavirus+[f]	2	TGB	1	No
Capillovirus	2	30K	2	No
Vitivirus	2	30K	2	Yes
Trichovirus	2	30K	2	No
Citrivirus	2	30K	1	No

[a]Type of replication protein.
[b]Either a set of three overlapping proteins (TGB) or a single protein of the '30K' superfamily.
[c]Type of coat protein.
[d]Presence of RNA-binding protein.
[e]Not applicable (no protein encoded).
[f]Including the unassigned members of the family: *Banana mild mosaic virus, Banana virus X, Cherry green ring mottle virus, Cherry necrotic rusty mottle virus,* and *Sugarcane striate mosaic-associated virus.*

See also: Allexivirus; Capillovirus, Foveavirus, Trichovirus, Vitivirus; Carlavirus; Carmovirus; Citrus Tristeza Virus; Potexvirus; Tobamovirus; Tymoviruses.

Further Reading

Adams MJ, Antoniw JF, Bar-Joseph M, *et al.* (2004) The new plant virus family *Flexiviridae* and assessment of molecular criteria for species demarcation. *Archives of Virology* 149: 1045–1060.

Howitt RLJ, Beever RE, Pearson MN, and Forster RLS (2001) Genome characterization of Botrytis virus F, a flexuous rod-shaped mycovirus resembling plant 'potex-like' viruses. *Journal of General Virology* 82: 67–78.

Howitt RLJ, Beever RE, Pearson MN, and Forster RLS (2006) Genome characterization of a flexuous rod-shaped mycovirus, Botrytis virus X, reveals high amino acid identity to genes from plant 'potex-like' viruses. *Archives of Virology* 151: 563–579.

Koonin EV and Dolja VV (1993) Evolution and taxonomy of positive-strand RNA viruses: Implication of comparative analysis of amino acid sequences. *Critical Reviews in Biochemistry and Molecular Biology* 28: 375–430.

Martelli G, Adams MJ, Kreuze JF, and Dolja VV (2007) Family *Flexiviridae*: A case study in virion and genome plasticity. *Annual Review of Phytopathology* 45: 73–100.

Melcher U (2000) The '30K' superfamily of viral movement proteins. *Journal of General Virology* 81: 257–266.

Morozov SY and Solovyev AG (2003) Triple gene block: Modular design of a multifunctional machine for plant virus movement. *Journal of General Virology* 84: 1351–1366.

Wong S-M, Lee K-C, Yu H-H, and Leong W-F (1998) Phylogenetic analysis of triple gene block viruses based on the TGB 1 homolog gene indicate a convergent evolution. *Virus Genes* 16: 295–302.

Furovirus

R Koenig, Institut für Pflanzenvirologie, Mikrobiologie und biologische Sicherheit, Brunswick, Germany

Glossary

Furovirus Siglum derived from fungus-transmitted rod-shaped virus.

History

Soil-borne wheat mosaic virus (SBWMV) is the type species of the genus *Furovirus.* Originally it was classified as a possible member of the tobamovirus group, because its rod-shaped particles resemble those of the tobamoviruses. In 1991, the fungus-transmitted rod-shaped viruses which all have several genome segments were separated from the tobamoviruses with a monopartite genome to form a new group, named furovirus group, with SBWMV as the type member. Molecular studies performed in the following years revealed that the genome organization of many of these furoviruses greatly differed from that of SBWMV.

Eventually four new genera were created, that is, the genus *Furovirus* with SBWMV as the type species, the genus *Benyvirus* with *Beet necrotic yellow vein virus,* the genus *Pomovirus* with *Potato mop-top virus,* and the genus *Pecluvirus* with *Peanut clump virus* as type species, respectively. The genus *Furovirus* presently comprises five species, that is, SBWMV, *Soil-borne cereal mosaic virus* (SBCMV), *Chinese wheat mosaic virus* (CWMV), *Oat golden stripe virus* (OGSV), and *Sorghum chlorotic spot virus* (SrCSV). SBCMV was simultaneously and independently described by two different working groups who had suggested the names European wheat mosaic virus (EWMV) and soil-borne rye mosaic virus (SBRMV), respectively. These names are no longer in use.

Host Ranges, Diseases, and Geographic Distribution

SBWMV has been known since the early 1920s. It causes mosaic, stunting, and severe losses of yield in winter wheat in the USA where it is widely distributed in the central parts. As with other soil-borne diseases the symptoms often occur in patches in the fields. SBWMV may also naturally infect barley. A deviating strain of SBWMV which differs from the type strain considerably in its nucleic acid sequences, but not in the proteins translated from them has been observed to be rapidly spreading in upper New York State since 1998. Diseases on wheat with symptoms similar to those described for SBWMV are caused by SBCMV in Europe, CWMV in China and by the distantly related Japanese strain (jap) of SBWMV in Japan. A virus closely related to SBWMV jap has recently been isolated from barley in France. SBWMV jap has also been obtained from barley in Japan. SBCMV is now widely distributed in Europe where it infects mainly wheat in Italy, France, and England, but mainly rye in Germany, Denmark, and Poland. SBWMV has been detected on a single field in Germany for the first time in 2002 with apparently no tendency to spread. OGSV has been detected in oats at various sites in Britain, France, and the USA (North Carolina). In oats it induces conspicuous chlorotic striping of leaves, but it fails to infect wheat. SBWMV, SBCMV, CWMV, and OGSV may be transmitted mechanically (sometimes only with difficulty) to some *Chenopodium* and *Nicotiana* species. SrCSV was isolated in 1986 from a single sorghum line from a breeder's plot in Kansas/USA. It is readily mechanically transmissible to maize where it produces a bright yellow mosaic and elongated ringspot symptoms several weeks after inoculation. Local infections are produced on mechanically inoculated *Chenopodium quinoa*, *C. amaranticolor*, and *N. clevelandii*. Attempts to transmit the virus back to *Sorghum bicolor* or to winter wheat either mechanically or by growing plants in soil in which infected sorghum was growing were unsuccessful.

Plants infected naturally by furoviruses are often also infected by bymoviruses, that is, wheat spindle streak mosaic virus in Europe and North America, wheat yellow mosaic virus in East Asia or oat mosaic virus in Europe and the USA. The symptoms caused by these bymoviruses are very similar to those caused by the furoviruses, and both the furoviruses and the bymoviruses are transmitted by *Polymyxa graminis* (see below).

Transmission in Nature and Long-Distance Movement in Infected Plants

Furoviruses, with the possible exception of SrCSV, are soil-borne and transmitted by the zoospores of *Polymyxa graminis*, a ubiquitous plamodiophorid protozoan formerly considered to belong to the fungi. *Polymyxa*-transmitted viruses (or their RNAs?) are taken up by the plasmodia of the vector in infected root cells, but there is no evidence that they multiply in the vector.

The multinucleate plasmodia which are separated from the host cytoplasm by distinct cell walls may either develop into zoosporangia from which secondary viruliferous zoospores are released within a few days. Alternatively, the plasmodia may form cystosori which act as resting spores and may survive in the soil for many years even under extreme conditions. They may be distributed on agricultural equipment, by irrigation or even by wind blow. Upon germination they release primary zoospores transmitting the virus. Zoospores inject their contents into the cytoplasm of root cells where new plasmodia are formed. Zoospores treated with antisera to SBWMV or resting spores treated with 0.1 N NaOH or HCl retained their ability to transmit virus into plants indicating that the infectious viral material was carried inside. Immunolabeling and *in situ* RNA hybridization studies have revealed the presence of SBWMV movement protein and RNA but not of SBWMV coat protein in the resting spores of *P. graminis*. This might suggest that the vector does not transmit SBWMV in form of its particles but rather as a ribonucleoprotein complex possibly formed by the movement protein and the viral RNAs. The furovirus movement protein belongs to the '30 K' superfamily of movement proteins which are known to mediate cell-to-cell and vascular transport of viruses by binding viral nucleic acids and carrying them through plasmodesmata and through the vasculature. There is also strong evidence that two transmembrane regions found in the coat protein readthrough proteins of furoviruses are involved in the transmission process.

Immunogold-labeling studies have suggested that SBWMV uses the xylem in order to move from infected roots to the leaves. It may enter primary xylem elements before cell death occurs and then move upward in the plant after the xylem has matured into hollow vessels. There is also evidence for lateral movement between adjacent xylem vessels.

Control

Polymyxa-transmitted plant viruses may survive in soil for decades in the long-living resting spores of the vector. The diseases caused by these soil-borne viruses are, therefore, much more difficult to control than those caused, for instance, by insect-transmitted viruses. Chemical control, for example, by soil treatment with methyl bromide, is neither efficient nor acceptable for economic and ecological reasons. Growing resistant or tolerant varieties currently represent the only practical and environmentally friendly means to lower the impact of these diseases on yield. Immunity to the wheat-infecting furoviruses has not been found. The tolerant varieties which have been developed contain high virus levels in the root system. The partial resistance in some varieties may be overcome when the plants are grown at temperatures >23 °C. Cereal genotypes with resistance to *P. graminis* have so far not been identified.

Particle Properties

Furovirus particles are nonenveloped hollow rods which have a helical symmetry (**Figure 1**). The diameter of the particles is *c.* 20 nm and in leaves of freshly infected field grown plants in the spring the predominant lengths are *c.* 280–300 and 140–160 nm. Due to internal deletions in the coat protein readthrough protein genes shorter particles arise later in season in naturally infected plants and in laboratory isolates which may outcompete the 140–160 nm particles. The single coat protein species of furoviruses has a molecular mass of *c.* 20 kDa.

Antigenic Properties

Serological relationships exist between all five furoviruses. Monoclonal antibodies (MAb) prepared in several laboratories were either species or even isolate specific or allowed broad-range detection of several or all furoviruses except for SrCSV that was not included in such studies. MAbs to SBWMV have also been used to determine the accessibility of stretches of the coat protein amino acid chain on the virus particles. The C-terminus of the coat protein is apparently exposed along the length of the particles and is readily removed by treatment with trypsin.

Nucleic Acid Properties and Differentiation of Furoviruses

Furovirus genomes consist of two molecules of linear positive-sense ssRNAs. Their complete or almost complete nucleotide sequences have been determined for all five furoviruses described so far. They are 5′ capped and terminate in a functional tRNA-like structure with an anticodon for valine. This tRNA-like structure is preceded in the 3′ untranslated region by an upstream hairpin and an upstream pseudoknot domain (UPD)

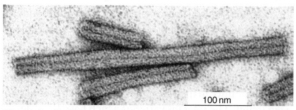

Figure 1 Particles of *Soil borne wheat mosaic virus* in a purified preparation negatively stained with uranyl acetate. Courtesy of Dr. D.-E. Lesemann and Dr. J. Engelmann, BBA Braunschweig.

with two to seven possible pseudoknots. The genome organization is more or less identical for all furoviruses although there are considerable differences in nucleotide sequences (**Figure 2**). The percentages of sequence identities range from *c.* 60% to 75% for RNA1 and from *c.* 50% to 80% for RNA2 of different furovirus species. With strains of CWMV and SBCMV percentages of sequence identities are >94% for RNA1 and >85% for RNA2. Classification on the basis of sequence dissimilarities would suggest that SBWMV jap might be considered to be a separate species rather than a strain of SBWMV. However, the fact that in reassortment experiments a mixture of RNA1 of SBWMV jap and RNA2 of the type strain of SBWMV yielded an infectious progeny and the observation that the biological properties of all wheat-infecting furoviruses are very similar has been taken by some researchers as evidence that the type and the Japanese strains of SBWMV as well as CWMV and SBCMV should all be regarded as distantly related strains of the same species. As with viruses in other genera the decision whether related viruses should be considered as different strains of one virus or as separate species is not always easy, because these categories are men-made and many border-line cases exist in nature.

Organization of the Genome and Properties of the Encoded Proteins

Furoviral RNA1 codes for two N-terminally overlapping, presumably replication-associated proteins and for a *c.* 37 kDa movement protein (**Figure 3**). The shorter of the two replication-associated proteins contains methyltransferase and helicase motifs and the longer one, in addition, the RNA-dependent RNA polymerase motifs (**Figure 3**). The green fluorescent protein (GFP)-labeled 37 kDa protein of SBWMV, as opposed to its GFP-labeled coat protein, was shown to move from cell to cell in leaves of wheat but not in those of tobacco (a nonhost) and, similar to other viral movement proteins, to accumulate in the cell wall of both SBWMV-infected wheat leaves and transgenic wheat plants expressing the 37 kDa protein.

Furoviral RNA2 codes for the *c.* 20 kDa coat protein, a *c.* 84 kDa coat protein readthrough protein and a *c.* 19 kDa cysteine-rich protein. A 24 kDa protein found in *in vitro* translation experiments and also *in planta* has been shown to be coat protein with a 40-amino-acid N-terminal extension initiated at a conserved CUG codon 120 nt upstream of the AUG initiation codon for the coat protein gene. Neither the coat protein with this N-terminal extension nor the coat protein readthrough protein are required for SBWMV virion formation and systemic infection. SBWMV with full-sized RNA2 is found in naturally infected wheat in winter or early spring. Prolonged cultivation of field-infected plants, virus propagation by

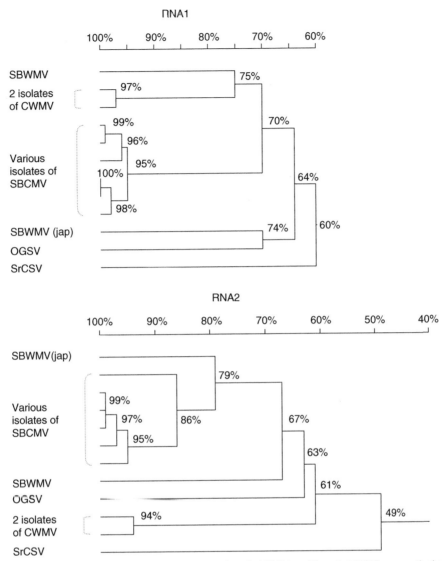

Figure 2   Percentages of nucleotide sequence identities among furoviral RNA1 and furoviral RNA2, respectively.

repeated mechanical inoculations or growth at elevated temperatures result in spontaneous deletions in the coat protein readthrough domain which may vary in size from 519 to 1030 nt. Computer analyses have revealed the presence of two complementary transmembrane domains (TM1 and TM2) in the coat protein readthrough proteins of furoviruses and also in those of the likewise *Polymyxa*-transmitted benyviruses and pomoviruses and in the P2 proteins of the *Polymyxa*-transmitted bymoviruses. The TM helices are apparently tightly packaged with ridge/groove arrangements between the two helices and strong electrostatic associations. It has been suggested that they facilitate the movement of the virus across the membrane surrounding the plasmodia in the plant cells. Nontransmissible deletion mutants lack the second transmembrane region partially or completely. The cysteine-rich 19 kDa

protein of SBWMV is a suppressor of post-transcriptional gene silencing.

The products of the two 3′ proximal ORFs on RNA1 and RNA2, that is, the movement protein and the cysteine-rich protein, have not been found in *in vitro* translation experiments. It is assumed that they are expressed from subgenomic RNAs.

Diagnosis

Symptoms caused by furoviruses may differ depending on environmental conditions, such as moisture and temperature. Nutrient deficiencies, winter injury or other viruses (especially bymoviruses), may produce symptoms that are easily confused with those caused by wheat-infecting

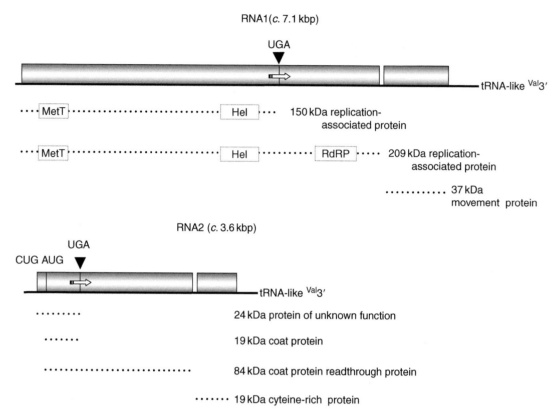

Figure 3 Organization of furoviral genomes (illustrated for SBWMV). The arrows in the open reading frames (boxed) indicate a readthrough translation due to a leaky stop codon.

furoviruses. Serological tests such as the enzyme-linked immunosorbent assay (ELISA), tissue print immunoassay, and immunoelectron microscopy as well as polymerase chain reaction (PCR) techniques allow a reliable detection. Real-time reverse transcriptase-PCR (RT-PCR) and PCR assays based on TaqMan chemistry have been developed for the detection and quantitation not only of SBCMV but also of its vector *P. graminis*. Real-time assays were found to be a 1000 times more sensitive than ELISA for the quantitation of SBCMV, and a 100 times more sensitive than conventional PCR for the quantitation of *P. graminis*.

Similarities and Dissimilarities with Other Taxa

The morphology of furoviruses resembles that of other rod-shaped viruses, that is, of benyviruses, pecluviruses, pomoviruses, hordeiviruses, tobraviruses, and tobamoviruses. The CPs of all these viruses have a number of conserved residues, for example, RF and FE in their central and C-terminal parts, respectively, which are presumably involved in the formation of salt bridges. Furoviruses, like pecluviruses and tobraviruses but unlike the other aforementioned viruses, have bipartite genomes. They differ from pecluviruses by having their movement function encoded on a single ORF rather than a triple

gene block. They also differ from pecluviruses as well as from tobraviruses by having a coat protein readthrough protein gene. The gene for their cysteine-rich protein is located on RNA2, whereas with pecluviruses and tobraviruses it is located on RNA1. The cysteine-rich proteins of furoviruses, pecluviruses, and tobraviruses act as suppressors of post-transcriptional gene silencing and are phylogenetically interrelated, whereas those of beny- and pomoviruses are surprisingly unrelated to them. Furoviruses have two N-terminally overlapping replication-associated proteins (**Figure 2**) which in their amino acid sequences and structure are related to those of pomoviruses, pecluviruses, tobraviruses, and tobamoviruses. The 37 kDa movement protein of the furoviruses which belongs to the '30 K' superfamily of movement proteins relates them to the dianthoviruses and the tobamoviruses, whereas the valine-accepting tRNA-like structures on the 3′ ends of their RNAs relate them to the tymoviruses. In other properties, such as particle morphology and mode of transmission, dianthoviruses and tymoviruses are very different from furoviruses. Furoviruses like tobamoviruses have an upper pseudoknot domain in the 3′ untranslated regions of their genomic RNAs.

See also: Benyvirus; Cereal Viruses: Wheat and Barley; Pecluvirus; Pomovirus; Vector Transmission of Plant Viruses.

Further Reading

Adams MJ, Antoniw JF, and Mullins JG (2001) Plant virus transmission by plasmodiophorid fungi is associated with distinctive transmembrane regions of virus-encoded proteins. *Archives of Virology* 146: 1139–1153.

An H, Melcher U, Doss P, *et al.* (2003) Evidence that the 37 kDa protein of soil-borne wheat mosaic virus is a virus movement protein. *Journal of General Virology* 84: 3153–3163.

Diao A, Chen J, Gitton F, *et al.* (1999) Sequences of European wheat mosaic virus and oat golden stripe virus and genome analysis of the genus furovirus. *Virology* 261: 331–339.

Driskel BA, Doss P, Littlefield LJ, Walker NR, and Verchot-Lubicz J (2004) Soilborne wheat mosaic virus movement protein and RNA and wheat spindle streak mosaic virus coat protein accumulate inside resting spores of their vector, *Polymyxa graminis*. *Molecular Plant–Microbe Interactions* 17: 39–48.

Goodwin JB and Dreher TW (1998) Transfer RNA mimicry in a new group of positive-strand RNA plant viruses, the furoviruses: Differential aminoacylation between the RNA components of one genome. *Virology* 246: 170–178.

Hariri D and Meyer M (2007) A new furovirus infecting barley in France closely related to the Japanese soil-borne wheat mosaic virus. *European Journal of Plant Pathology* 118: 1–10.

Kanyuka K, Ward E, and Adams MJ (2003) *Polymyxa graminis* and the cereal viruses it transmits: A research challenge. *Molecular Plant Pathology* 4: 393–406.

Koenig R, Bergstrom GC, Gray SM, and Loss S (2002) A New York isolate of *Soil-borne wheat mosaic virus* differs considerably from the Nebraska type strain in the nucleotide sequences of various coding regions but not in the deduced amino acid sequences. *Archives of Virology* 147: 617–625.

Koenig R, Pleij CW, and Huth W (1999) Molecular characterization of a new furovirus mainly infecting rye. *Archives of Virology* 144: 2125–2140.

Miyanishi M, Roh SH, Yamamiya A, Ohsato S, and Shirako Y (2002) Reassortment between genetically distinct Japanese and US strains of *Soil-borne wheat mosaic virus*: RNA 1 from a Japanese strain and RNA 2 from a US strain make a pseudorecombinant virus. *Archives of Virology* 147: 1141–1153.

Ratti C, Budge G, Ward L, *et al.* (2004) Detection and relative quantitation of soil-borne cereal mosaic virus (SBCMV) and *Polymyxa graminis* in winter wheat using real-time PCR (TaqMan). *Journal of Virological Methods* 122: 95–103.

Shirako Y, Suzuki N, and French RC (2000) Similarity and divergence among viruses in the genus *Furovirus*. *Virology* 270: 201–207.

Te J, Melcher U, Howard A, and Verchot-Lubicz J (2005) Soil-borne wheat mosaic virus (SBWMV) 19K protein belongs to a class of cysteine rich proteins that suppress RNA silencing. *Virology Journal* 2: 18.

Torrance L and Koenig R (2005) Furovirus. In: Fauquet CM, Mayo M, Maniloff J, Desselberger U, and Ball LA (eds.) *Virus Taxonomy: Eighth Report of the International Committee on Taxonomy of Viruses*, pp 1027–1032. San Diego, CA: Elsevier Academic Press.

Verchot J, Driskel BA, Zhu Y, Hunger RM, and Littlefield LJ (2001) Evidence that soil-borne wheat mosaic virus moves long distance through the xylem in wheat. *Protoplasma* 218: 57–66.

Yamamiya A and Shirako Y (2000) Construction of full-length cDNA clones to Soil-borne wheat mosaic virus RNA 1 and RNA 2, from which infectious RNAs are transcribed *in vitro*: Virion formation and systemic infection without expression of the N-terminal and C-terminal extensions to the capsid protein. *Virology* 277: 66–75.

Ilarvirus

K C Eastwell, Washington State University – IAREC, Prosser, WA, USA

Glossary

Cross-protection A virus-infected plant often exhibits resistance to infection by strains of the same or closely related viruses. The degree of cross-protection has been used as a measure of the relatedness of viruses.

Dicistronic An RNA that encodes two gene products, often with a noncoding region separating them.

Recovery A phenomenon in which a virus-infected plant displays severe symptoms after initial infection, but then symptoms decline; this is often associated with a decline in virus concentration.

Subgenomic RNA A segment of RNA generated from a genomic RNA molecule through an internal promoter sequence. Since eukaryotic systems only efficiently translate the open reading frame located at the 5′ terminus of mRNA, this permits efficient translation of downstream open reading frames.

Suppressor of virus-induced gene silencing A molecule produced through virus infection that mitigates the natural plant defense system.

Virus-induced gene silencing A natural plant defense system that uses small fragments of double-stranded RNA to direct the RNA-degrading or silencing enzymes to a target RNA molecule. Double-stranded RNA results from the synthesis of complementary minus- and plus-sense RNA sequences during virus replication.

Taxonomy

The genus *Ilarvirus* was officially recognized in 1975. It is currently placed within the family *Bromoviridae* that also includes the genera *Alfamovirus*, *Bromovirus*, *Cucumovirus*, and *Oleavirus*. The quasi-isometric nature of ilarvirus particles and their relative instability in the absence of antioxidants are reflected in the sigla which originates

Table 1   Virus members of the genus *Ilarvirus*

Virus (abbreviation) Synonyms	GenBank accession numbers used for comparison		
	RNA1	RNA2	RNA3
Subgroup 1			
Tobacco streak virus (TSV)	NC_003844	NC_003842	NC_003845
Annulus orae virus			
Asparagus stunt virus			
Bean red node virus			
Black raspberry latent virus			
Datura quercina virus			
Nicotiana virus 8			
Nicotiana virus vulnerans			
Sunflower necrosis virus			
Tractus orae virus			
Parietaria mottle virus (PMoV)	NC_005848	NC_005849	NC_005854
Tentative members			
Blackberry chlorotic ringspot virus (BCRSV)	DQ091193	DQ091194	DQ091195
Grapevine angular mosaic virus	—	—	—
Strawberry necrotic shock virus (SNSV)	—	AY743591	AY363228
Subgroup 2			
Asparagus virus 2 (AV-2)	—	—	X86352
Asparagus latent virus			
Asparagus virus C			
Citrus leaf rugose virus (CiLRV)	NC_003548	NC_003547	NC_003546
Citrus crinkly leaf virus			
Citrus variegation virus (CVV)	—	—	U17389
Citrus psorosis virus complex (infectious variegation component)			
Elm mottle virus (EMoV)	NC_003569	NP_619575	NC_003570
Hydrangea mosaic virus (HdMV)			AF172965
Lilac streak mosaic virus			
Lilac white mosaic virus			
Spinach latent virus (SpLV)	NC_003808	NC_003809	NC_003810
GE 36 virus			
Tulare apple mosaic virus (TAMV)	NC_003833	NC_003834	NC_003835
Subgroup 3			
Apple mosaic virus (ApMV)	NC_003464	NC_003465	NC_003480
Birch line pattern virus			S78319[a]
Birch ringspot virus			AAL84586[b]
Dutch plum line pattern virus			AAN01243[c]
Hop virus A			
Horse chestnut yellow mosaic virus			
Mild apple mosaic virus			
Mountain ash variegation virus			
Plum (Dutch) line pattern virus			
Severe apple mosaic virus			
Blueberry shock virus (BlShV)	—	—	—
Humulus japonicus latent virus (HJLV)	NC_006064	NC_006065	NC_006066
Humulus japonicus virus			
Prunus necrotic ringspot virus (PNRSV)	NC_004362	NC_004363	NC_004364
Cherry rugose mosaic virus			CAB37309[d]
Currant (red) necrotic ringspot virus			
Danish plum line pattern virus			
European plum line pattern virus			
Hop virus B			
Hop virus C			
North American plum line pattern virus			
Peach ringspot virus			
Plum (Danish) line pattern virus			
Plum (European) line pattern virus			
Plum (North American) line pattern virus			
Plum line pattern virus			
Prunus ringspot virus			

Continued

Table 1 Continued

Virus (abbreviation) Synonyms	GenBank accession numbers used for comparison		
	RNA1	RNA2	RNA3
Red currant necrotic ringspot virus			
Rose chlorotic mottle virus			
Rose line pattern virus			
Rose mosaic virus			
Rose vein banding virus			
Rose yellow mosaic virus			
Rose yellow vein mosaic virus			
Sour cherry necrotic ringspot virus			
Subgroup 4			
Prune dwarf virus (PDV)	PDU57648	AF277662	L28145
Cherry chlorotic ringspot virus			
Chlorogenus cerasae virus			
Peach stunt virus			
Prunus chlorotic necrotic ringspot virus			
Sour cherry yellows virus			
Subgroup 5			
American plum line pattern virus (APLPV)	NC_003451	NC_003452	NC_003453
Peach line pattern virosis virus			
Plum (American) line pattern virus			
Plum line pattern virus			
Prunus virus 10			
Subgroup 6			
Fragaria chiloensis latent virus (FCiLV)	NC_006566	NC_006567	NC_006568
Lilac ring mottle virus (LiRMoV)	—	—	U17391

[a]ApMV-G strain, CP sequence only.
[b]ApMV-hop strain, CP sequence only.
[c]ApMV-Fuji strain, CP sequence only.
[d]PNRSV-MRY1 isolate, CP sequence only.

from the descriptive phrase: isometric labile ringspot viruses. Other distinguishing features of the genus are the distribution of the genome over three plus-sense RNA segments and the absolute requirement for the coat protein to activate the RNA genome to initiate infection. These traits are shared with the genus *Alfamovirus*.

Experimental host ranges of ilarviruses are quite broad, although many have a relatively narrow range of natural hosts consisting of woody plants. Woody hosts may require several years after initial infection for the virus to uniformly infect the plant. The ringspot symptom for which the genus is named is a common symptom of many virus–host combinations. Infections by ilarviruses frequently result in a shock phase during which severe chlorotic or necrotic rings form, followed by a period of recovery with greatly reduced symptoms. This recovery phenomenon has now been described for many virus groups. No inclusion bodies have been described in tissue infected with members of the genus *Ilarvirus*.

Members of the genus *Ilarvirus* are currently divided into six subgroups (**Table 1**). Historically, this separation was based solely on serological reactivity, but it is generally supported by comparison of the amino acid sequences of structural proteins (**Figure 1**). The exception is the

coat protein sequence of *Humulus japonicus* latent virus (HJLV) of subgroup 3 which is most similar to that of prune dwarf virus (PDV) of subgroup 4 rather than other members of subgroup 3. *Alfalfa mosaic virus* (AMV) is the monotypic member of the *Alfamovirus* genus; the sequence of the coat protein encoded by AMV also segregates with HJLV and PDV in this analysis. Analyses of amino acid sequences of the nonstructural proteins encoded by all three RNA segments are consistent with current subgroup designations.

In addition to the currently recognized members of the genus *Ilarvirus*, many synonyms have appeared in the literature, most of which refer to specific strains of virus species. These are listed in **Table 1**. The taxonomy of ilarvirus isolates that infect hop plants (*Humulus lupulus* L.) is still evolving. Viruses designated as hop virus C, NRSV-intermediate, and NRSV-HP-2 are related serologically to prunus necrotic ringspot virus (PNRSV) and to apple mosaic virus (ApMV) while hop virus A, apple mosaic virus-hop, and prunus necrotic ringspot virus-HP-1 are serologically related to ApMV only. Analyses of coat protein sequences of viruses obtained from hop plants of diverse geographic origins indicate that both serotypes are strains of ApMV.

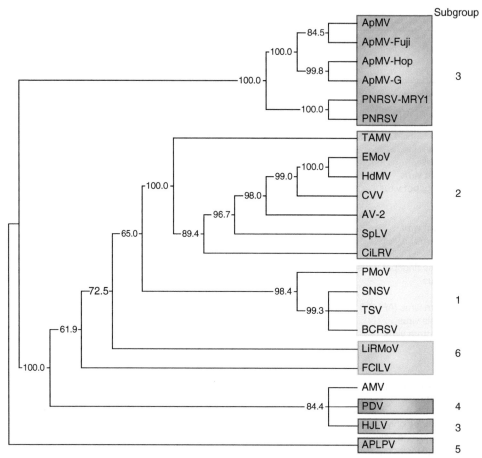

Figure 1 Cladogram illustrating the sequence relationships of ilarviruses based on the deduced amino acid sequence of the coat protein encoded by ORF3b of RNA3. Numbers at branch points indicate the percentage of bootstrap replicates in which that branch occurs. Subgroup designations are derived primarily from serological as well as molecular data and are indicated to the right of the shaded boxes. The sequence of the alfamovirus alfalfa mosaic virus is included in the analysis. See **Table 1** for abbreviations.

Particle Structure and Genome Organization

The virions of ilarviruses are generally quasi-isometric (**Figure 2**), and the size of particle can vary within a species in relation to RNA content; particles range in diameter from 20 to 35 nm. Some ilarviruses also produce a small proportion of bacilliform particles. Virus particles contain subunits of a single structural protein (coat protein). The molecular mass of the coat protein subunits in the range of 19–30 kDa, with the majority of members having coat protein subunits within the rather narrow range of 24–26 kDa. The entire genome of ilarviruses is distributed over three RNA molecules; each is encapsidated separately. However, preparations of virions contain a fourth RNA species that is 3′-co-terminal with RNA3; this subgenomic RNA encodes the coat protein. Recently, a fifth RNA species corresponding to the 3′-untranslated region (UTR) of RNAs 1–3 was found to be encapsidated in virions of PNRSV. This is believed to represent the chance encapsidation of partially degraded RNA

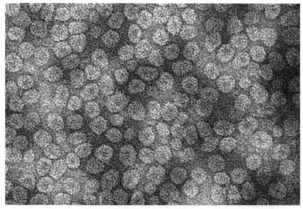

Figure 2 Electron micrograph of negative-stained particles of prune dwarf virus. In addition to quasi-isometric particles 20 and 23 nm in diameter, preparations from some ilarviruses including prune dwarf virus also contain bacilliform particles 19 × 33 nm and 19 × 38 nm.

that is frequently found within purified virus particles. All of the major RNA species bear a 5′-7-methyl-G cap structure.

The relationship between the genera *Alfamovirus* and *Ilarvirus* is particularly close. The unique dependence on coat protein for infectivity of viral RNA is shared by members of both genera. Moreover, the coat proteins of AMV and some ilarviruses can reciprocally activate infectivity of RNA of the other. This phenomenon has led some investigators to suggest that the genera *Ilarvirus* and *Alfamovirus* should be combined. In as much as these viruses also share several features in common with other genera of the family *Bromoviridae*, the two genera remain separate. Studies of AMV have contributed much of the basic information from which the structure and replication of ilarviruses were determined and will be included here in subsequent comparisons of ilarviruses.

The type member of the genus *Ilarvirus* is *Tobacco streak virus* (TSV). The total genome of the WC isolate of TSV consists of 8622 nucleotides distributed over three RNA segments of 3491, 2926, and 2205 nucleotides (**Figure 3**). The largest segment of RNA, RNA1, contains a single open reading frame (ORF1) encoding a protein of 123.4 kDa. This protein is the replicase and contains the methyltranferase and helicase motifs. RNA2 contains ORF2a encoding a protein of 91.5 kDa. This protein is the RNA-dependent RNA-polymerase based on the presence of conserved RNA-binding motifs. RNA2 of TSV and other members of subgroups 1 and 2 contain a second potential overlapping ORF, ORF2b; AMV and ilarviruses of the remaining subgroups apparently lack this second ORF on RNA2. The function of the ORF2b product with a molecular mass of 22.4 kDa has not been determined. The dicistronic organization of RNA2 detected in ilarvirus subgroups 1 and 2 is similar to that of members of the genus *Cucumovirus*; the putative product of ORF2b of cucumoviruses has been proposed to be a key element in suppressing virus-induced gene silencing. The sequence of *Fragaria chiloensis* latent

virus (FCiLV) of ilarvirus subgroup 6 is unique in that its RNA2 has the potential to encode a second small protein located toward the 5′ end of RNA2 (nucleotide positions 350–856 of RNA2). However, the putative product of this ORF bears no significant similarity to protein 2b of the cucumoviruses, nor other proteins thought to be encoded by ilarviruses.

RNA3 of ilarviruses is dicistronic with two potential ORFs. ORF3a encodes a protein of 31.7 kDa with cell-to-cell movement function (MP) and ORF3b encodes the structural coat protein (CP) with a molecular mass of 26.2 kDa. RNA3 of FCiLV contains a third potentially functional ORF, whose putative product does not have any sequence similarity to proteins expressed by other ilarviruses. It is unknown if this ORF is expressed *in vivo*.

Virus Replication

Members of this genus require an 'activated' RNA genome to initiate infection, that is, viral RNA must form a complex with the coat protein to be infectious. This is a distinguishing feature of ilarvirus and alfamovirus replication that differentiates them from other genera in the family *Bromoviridae*. Activation of the genome can be accomplished either through the co-inoculation of viral RNA with coat protein, or by the inclusion of subgenomic RNA4 that is then translated to produce the activating coat protein *in vivo*. This activation can be accomplished through heterologous complex formation, that is, the coat protein from one ilarvirus can bind to and potentiate the replication of RNA from several other but not all members of the genus *Ilarvirus*. Furthermore, the coat protein of AMV can activate the genome of several ilarviruses and vice versa. More recently, it has been demonstrated that RNA3 is able to fulfil the role of either coat protein or RNA4, but with much reduced efficiency. The presence of RNA4 or coat protein increases the efficiency of infection up to 1000-fold relative to the RNA activation achieved by RNA3. This is the result of very inefficient expression of the coat protein gene directly from the dicistronic RNA3.

The mechanism by which genome activation occurs is the result of the specific interaction of the coat protein with sequences located at the 3′ terminus of viral RNA. All ilarviruses for which sequences are known possess sequences at the 3′ terminus of genomic RNAs that are predicted to assume secondary structures that play a role in regulating the relative efficiency of the RNA species for transcription versus translation. Indeed, in the presence of coat protein, transcription of viral RNA is inhibited and translation is favored. The 3′-UTR of each RNA segment contains a series of conserved (A/U) (U/A/G)GC motifs that occur as single-stranded regions at the base of stem-loop structures that are recognized by and bound to coat protein molecules (**Figures 4** and **5**). The sequence and

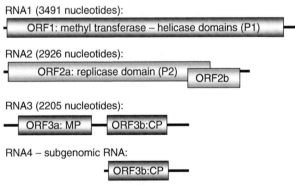

RNA1 (3491 nucleotides):

ORF1: methyl transferase – helicase domains (P1)

RNA2 (2926 nucleotides):

ORF2a: replicase domain (P2)   ORF2b

RNA3 (2205 nucleotides):

ORF3a: MP   ORF3b:CP

RNA4 – subgenomic RNA:

ORF3b:CP

Figure 3 Genome organization of tobacco streak virus (TSV). The positive-sense RNA genome of the WC isolate of TSV is distributed between three segments. ORF2b is present only in genus *Ilarvirus* subgroups 1 and 2. ORF3a encodes a protein (MP) associated with virus movement and ORF3b encodes that coat protein (CP). The subgenomic RNA4 is transcribed from an internal promoter in the intercistronic region of minus-sense RNA3; RNA4 is 3′ co-terminal with RNA3. ORF2b is present only in genus *Ilarvirus* subgroups 1 and 2.

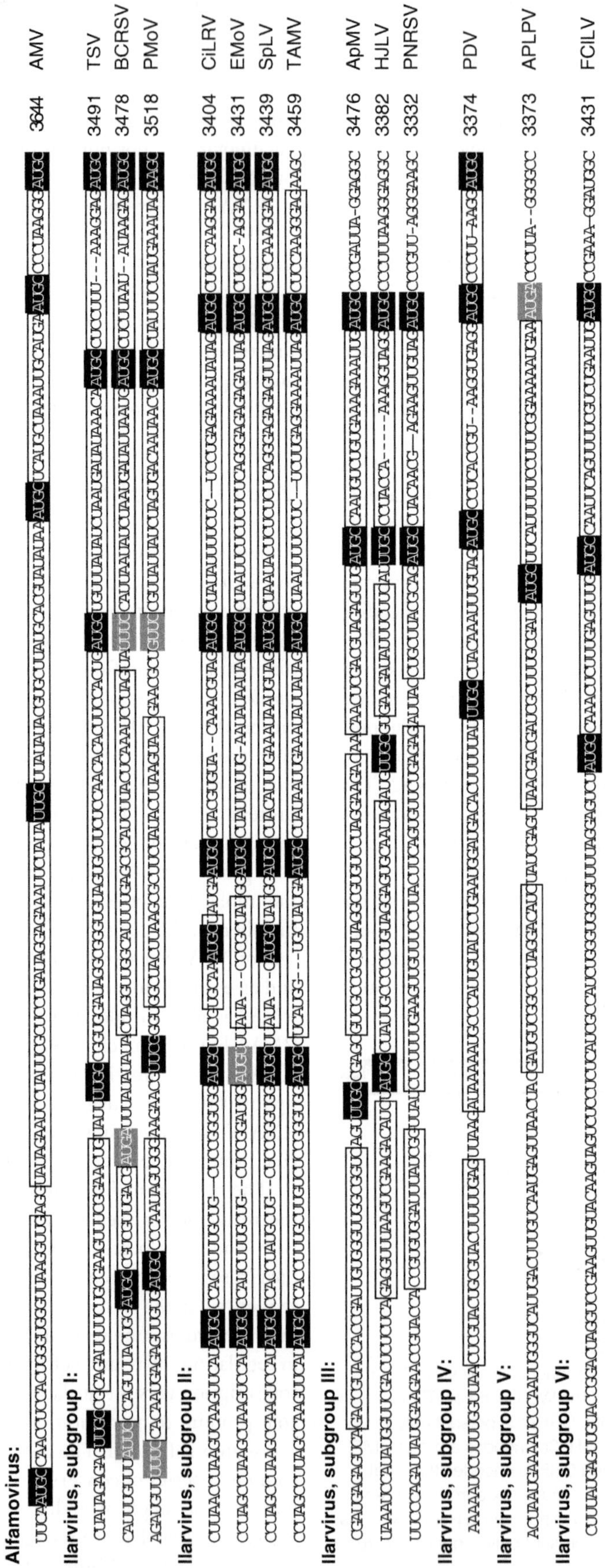

Figure 4 The 3'-untranslated region of RNA1 of ilarviruses contains a repeated (A/U) (U/A/G)GC sequence. The repeated single-stranded regions are indicated by black highlighting. Corrupted copies of (A/U) (U/A/G)GC motifs are highlighted in gray. Potential stem–loop structures are indicated by boxes.

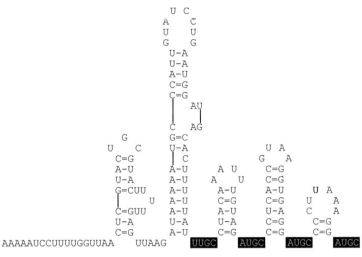

Figure 5 The 3′-untranslated region of RNA1 of ilarviruses contains a repeated (A/U) (U/A/G)GC sequence. These conserved regions are single-stranded motifs (black highlighting) that appear between base-paired stem-loops and are critical for coat protein binding and potentially for replicase binding. The secondary structure of the 3′ terminus of prune dwarf virus RNA1 illustrates the potential stem-loops configuration with intervening conserved single-stranded motifs.

the size of the stem–loop structures vary considerably when all of the ilarviruses are considered. However, within each subgroup of ilarvirus, the overall length and secondary structures of the binding region(s) are well conserved. In the case of TSV, the (A/U) (U/A/G)GC motif is repeated five times flanking four potential stem–loop structures within the 260 nucleotides located at the 3′ terminus of RNA1. This region may represent two coat protein binding sites as has been demonstrated for AMV. Ilarviruses belonging to the subgroups 4, 5, and 6 possess shorter discernable recognition sequences and may contain single coat protein binding sites. Deletion analyses of AMV RNA4 have shown that the first and third 3′-most stem loops are critical for coat protein binding and hence translation, whereas the other stem loops play a role in enhancing efficient transcription but are not strictly required for that function. There is an absolute requirement for the single-stranded (A/U) (U/A/G)GC motifs in virus function.

The role of the RNA–coat protein complex was elucidated through studies with recombinant viruses. The 3′-UTR and its ability to form a complex with coat protein subunits is functionally analogous to a poly-A tail; a poly-A tail of 40–80 residues can experimentally compensate for the absence of activating coat protein molecules. The formation of a RNA–coat protein complex or the addition of a poly-A tail both act to stimulate the translation of viral RNA resulting in the synthesis of proteins needed for virus replication. In eukaryotes, translation efficiency of mRNA is enhanced through the formation of a loop formed by the interaction of a poly-A binding protein (PABP) with poly-A, and then subsequent binding of this complex to the 5′-cap structure of RNA in combination with other cellular initiation factors. It has been suggested that the complex formed between viral RNA and coat protein mimics

the complex formed by the PABP and the poly-A tail. This proposal is further strengthened by the observation that the RNA–coat protein complex binds elongation factors associated with cellular mRNA translation. Based on data from AMV, it appears that the binding of coat protein to 3′-terminal sequences of viral RNA not only facilitates translation, but also effectively blocks transcription of the RNA.

The essential complex formation between the coat protein and RNA has been examined in detail to determine critical features of the coat protein. Limited digests indicate that the N-terminus of the coat protein is primarily responsible for RNA binding. Inspection of the amino acid sequences at the N-terminus of the coat protein reveals a high percentage of basic amino acid residues that are consistent with a nucleic acid binding function. More detailed RNA protection assays and competitive binding studies suggest that a consensus sequence represented by (Q/K/R) (P/N)TXRS(R/Q) (Q/N/S) (W/F/Y)A is involved in coat protein binding to the 3′-UTR of AMV and ilarvirus genomic RNA (**Figure 6**). Although this functional domain is not strictly conserved in all ilarviruses, a core motif centered on the critical arginine residue is generally well preserved in all ilarviruses and AMV, suggesting a key role in virus function. Despite significant variation in primary sequence, the striking similarity in protein characteristics across all ilarviruses and AMV may account for the reciprocity in the ability of coat protein from several ilarviruses and AMV to activate the virus genome in heterologous combinations.

In addition to the functional domain of the coat protein described above, additional features of the coat protein may contribute to RNA binding. The critical basic arginine residue and its flanking residues are

Alfamovirus:

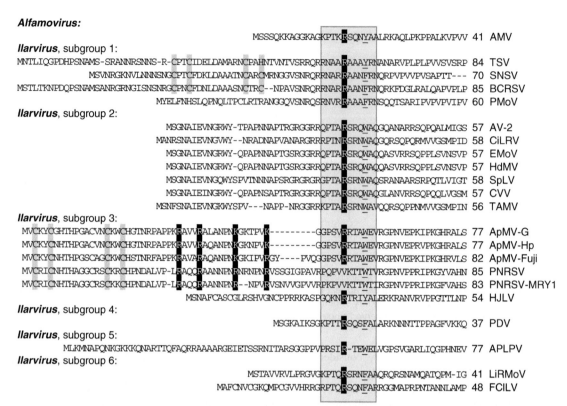

Figure 6 Alignment of the N-terminus of the coat proteins of ilarviruses. The box shaded in red indicates the putative RNA binding region with the consensus sequence (Q/K/R) (P/N)TXRS(R/Q) (Q/N/S) (W/F/Y)A and the central arginine residue highlighted in black. It has been demonstrated that in prunus necrotic ringspot virus, four arginine residues upstream of the conserved binding site are essential for RNA binding; these are highlighted with black, as are comparable residues in the apple mosaic virus coat protein sequence. Members of subgroups 1 and 3 also contain a putative zinc finger that may be involved in RNA binding; the cysteine/histidine residues potentially involved in chelating zinc cations are highlighted in gray. See **Table 1** for abbreviations.

preceded by a putative zinc finger in the coat proteins of members of subgroups 1 and 3, but not in other subgroups of ilarviruses, nor in AMV (**Figure 6**). Zinc fingers are typically involved in nucleic acid binding. Cysteine and/or hisitidine residues are critical elements of the motif represented by $CN_2CN_{10}(C/H)N_2(C/H)$; the cysteine and histidine residues are the functional amino acids capable of chelating zinc cations. The participation or requirement of this putative zinc finger in virus replication is uncertain, particularly since it is not present in all ilarviruses.

The RNA-binding region of the coat protein that is conserved in other ilarviruses appears to be absent or at least highly diverged in PNRSV. In particular, an arginine expected to be critical for RNA binding is absent in the coat protein of PNRSV in this central conserved position, although a basic lysine residue is located nearby. Moreover, electrophoretic shift assay suggests that other basic residues are involved in RNA binding. Four arginine residues of PNRSV coat protein have been experimentally determined to be involved in RNA binding and are located towards the N-terminus of the site of the conserved motif present in other ilarviruses (**Figure 6**). A similar distribution of four basic residues is observed

in ApMV in addition to the conserved (Q/K/R) (P/N) TXRS(R/Q) (Q/N/S) (W/F/Y)A motif.

RNA folding predictions suggest that the 3′-UTR of some ilarviruses can fold into complex tRNA-like structures or pseudoknots that do not accommodate coat protein binding. By analogy to information obtained from AMV, the pseudoknot conformation favors transcription and hence synthesis of viral RNAs. Mutational analysis indicates that stem–loop structures located just upstream of the coat protein binding site as well as the ability to form a tRNA-like pseudoknot are required for transcription and synthesis of minus-sense RNA. Thus, the relative abundance of RNA in the pseudoknot conformation relative to that bound to coat protein provides a feedback mechanism to coordinate phases of virus replication. Translation to form nonstructural proteins is favored in the presence of coat protein, while transcription as part of the replication process is favored in the absence of the coat protein.

In addition to a requirement for regulating the use of positive-sense RNA for translation versus transcription, is the requirement for coordinating the rates of synthesis of positive-sense versus minus-sense RNA molecules. The virus replication cycle requires the switch from minus-sense RNA synthesis to the repetitive transcription into positive-sense progeny viral RNA. The genomic features responsible for this transition are not well understood. However, recent experiments with transformed plants suggest that coat protein is not involved in this process. Secondary structural features of the minus-sense RNAs are believed to regulate the synthesis of plus-sense RNA.

The subgenomic RNA encoding the coat protein, RNA4, is transcribed from a minus-sense copy of RNA3. In brome mosaic virus, the intercistronic region between the ORFs of RNA3 contains a subgenomic promoter region consisting of sequences enriched for A/U flanking the core promoter and a poly-U tract that enhances transcription. However, the intercistronic regions of AMV and the ilarviruses are considerably shorter than that of brome mosaic virus. Moreover, although still slightly enriched in U residues, the poly-U tract and strong bias for U are absent. The transcriptional start site and a strong promoter for the synthesis of positive-sense subgenomic RNA4 occurs immediately 3' of the position corresponding to the termination codon of ORF3a. The subgenomic promoter of AMV has been characterized; a critical component of the promoter region is a three-nucleotide loop on a base-paired stem that bears a 'bulge' at the fifth nucleotide from the loop along the base-paired stem. The stem sequence below the bulge is less critical for promoter activity. Potential stem–loop structures are evident in the intercistronic regions between ORF3a and ORF3b of ilarviruses and these may fulfill the same critical role as those in AMV and brome mosaic virus.

Based on studies of transgenic plants conducted with AMV, replication of RNA1 and RNA2 are coordinated in that a mutation leading to changes in the encoded protein of either RNA product prevents the replication of both. Moreover, this requirement is for a *cis*-encoded function since transgenic expression of the wild-type protein does not rescue the mutated RNA. In contrast, the replication of RNA3 can be supported by RNA1 and RNA2 gene products expressed in *trans*. This is a critical distinction highlighting the independence of RNA3 replication from that of RNA1 and RNA2. During the natural infection cycle, binding of the coat protein in the inoculum to the 3' terminus of RNA1 and RNA2 is sufficient to facilitate their translation into a replication complex in the cytoplasm of the infected cell. This is accomplished by the interaction of the coat protein–RNA complex with host initiation factors. However, replication of viral RNA can only be initiated once the coat protein is dissociated from the 3' of the RNA molecules, allowing the 3' terminus to assume the tRNA-like pseudoknot conformation. Regulation of this dissociation process is unknown. However, it has been speculated that the association of the replication complex with membrane structures could displace the coat protein, or proteolysis of the coat protein could result in its removal from RNA. Alternatively, the coat protein could be displaced by binding of the replicase to stem–loop structures upstream of the coat protein binding sites. Replication then proceeds through synthesis of minus-sense RNA and the replication of progeny plus-sense RNA segments, including the synthesis of the plus-sense subgenomic RNA encoding the coat protein. It has been demonstrated that both the coat protein and the movement protein are required for systemic infection. It appears that the infective agent moves through modified plasmodesmata as a protein–RNA complex rather than as intact virions.

Epidemiology and Control

One or more modes of transmission have been demonstrated for each member of the genus *Ilarvirus*. All ilarviruses can be transmitted by mechanical inoculation to experimental hosts, although some with difficulty. Ilarviruses tend to have a wide experimental host range, but many have woody plants as their primary natural hosts. The role of mechanical transmission between woody plants is likely small. However, succulent hosts such as hop and asparagus plants may be more prone to field spread through mechanical transmission associated with horticultural practices.

The dissemination of viruses through vegetative propagation is one of the most critical factors affecting the distribution of ilarviruses that infect perennial crops. Budding and grafting are important means of transmission of ilarviruses that infect woody plants such as fruit trees and roses. These plants are routinely budded onto rootstock to allow rapid expansion of a particular clone into hundreds or thousands of plants. Other hosts such as hop or asparagus plants can be vegetatively propagated by stem cuttings and/or crown division. In both cases, there is high likelihood that virus will be present in the progeny plants. Chance grafting of root systems is also an important means of transmission of some viruses of perennial crops.

Many ilarviruses are transmitted through seed when the seed-bearing plant is infected. Transmission of pollen-borne virus occurs to seedlings with variable frequency, and transmission of virus from pollen to the pollinated plant has also been demonstrated for PNRSV, PDV, and TSV. Seed and pollen transmission poses a particularly high risk in fruit trees since seedling rootstocks are frequently used in production and virus infection can originate from the rootstock, the scion, or both. The most effective means of control is through the use of virus-tested material during the establishment of plantings. This approach is implemented through programs where plants are propagated from virus-tested plants and grown under conditions that minimize re-infection by viruses, including the pollen-borne ilarviruses. Many industries rely on informal virus control programs whereas others have developed official certification programs. The use of spatial separation of the propagation source material from potentially infected host plants is a basic strategy to minimize the introduction of virus through pollen. The risk associated with seed transmission is further reduced by screening samples of seed lots for the presence of virus. Where seed is produced from perennial plants such as *Prunus* species and asparagus, routine virus testing and rouging of mother plantings is implemented to limit the distribution of virus through seed.

The association of virus transmission with pollen raises concern about the role of bees in virus epidemiology. In the case of PNRSV and PDV, it has been shown that the virus is detectable by serological assays on honeybees emerging from hives weeks after exposure to flowers with virus-laden pollen. Pollen from hives is no longer viable at this stage, but the virus is infectious. The ability of this virus to infect trees then depends on whether the virus is transmitted to the tree by fertilization or by some other physical process. This issue remains largely unresolved. Holding bees in hives for several days before allowing them to enter a new orchard has been suggested as one means of minimizing this risk.

It has been demonstrated that TSV is transmitted to natural hosts in a process known to be facilitated by thrips. The thrips species demonstrated to transmit TSV include *Frankliniella occidentalis* (Pergande), *Frankliniella schultzei* (Trybom), *Microcephalothrips abdominalis* (Crawford), *Thrips tabaci* (Lindeman), and *Thrips parvispinus* (Karny). The association of thrips with pollen transmission of TSV is one of the major factors leading to the increased importance of this virus in vegetable production in many areas of the world. The thrips facilitate transmission by mechanical abrasion of the leaf cells, allowing pollen-associated virus particles to enter cells and initiate infection. Pollen-to-leaf transmission has been demonstrated, but leaf-to-leaf transmission has not. There have been many attempts to confirm the role of thrips in the transmission of ilarviruses other than TSV. Other ilarviruses have been transmitted from pollen to the leaves of experimental hosts in a thrips-mediated manner, but transmission to natural hosts by thrips has not been convincingly demonstrated. Unlike AMV, there is no evidence of aphid transmission of any members of the genus *Ilarvirus*.

Most ilarviruses have a limited natural host range. This feature in combination with a narrow temporal window for transmission imposed by the short flowering period of many hosts means that plant-to-plant spread of ilarviruses is relatively slow. TSV is an exception where a wide natural host range sustains sources of infection throughout the year. This may explain the growing impact of TSV on production of vegetables and legumes such as soybean. Agronomic strategies such as removing perennial weeds that would otherwise provide a green bridge for the thrips

vectors and the virus has helped reduce the carryover of virus inoculum in a limited number of situations. Since rainfall abates thrips populations, in some areas, sowing seed so that germination will occur during periods of increased rain also helps reduce crop losses caused by TSV. As with the ilarviruses that infect fruit trees and asparagus, the use of certified, virus-tested seed is critical for effective virus control.

Rugose mosaic is a serious debilitating disease of sweet cherry (*Prunus avium* L.) that is caused by a severe strain of PNRSV. The long latent period between infection and visible expression of disease symptoms complicates efficient control strategies. In this case, trees can be inoculated prophylactically with less severe strains of PNRSV. The inoculated plant is then resistant to subsequent infection by the severe virus strain. This phenomenon is termed cross-protection. In combination with accurate diagnosis and prompt inoculum removal, cross-protection is an effective tool in controlling rugose mosaic disease.

See also: Alfalfa Mosaic Virus; Bromoviruses; Plant Antiviral Defense: Gene Silencing Pathway; Plant Virus Diseases: Fruit Trees and Grapevine; Vector Transmission of Plant Viruses; Vegetable Viruses.

Further Reading

Ansel-McKinney P and Gehrke L (1998) RNA determinants of a specific RNA-coat protein peptide interaction in alfalfa mosaic virus: Conservation of homologous features in ilarvirus RNAs. *Journal of Molecular Biology* 278: 767–785.

Ansel-McKinney P, Scott SW, Swanson M, Ge X, and Gehrke L (1996) A plant viral coat protein RNA binding consensus sequence contains a crucial arginine. *EMBO Journal* 15: 5077–5084.

Aparicio F, Vilar M, Perez-Payá E, and Pallás V (2003) The coat protein of *Prunus* necrotic ringspot virus specifically binds to and regulates the conformation of its genomic RNA. *Virology* 313: 213–223.

Bol JF (2005) Replication of Alfamo- and Ilarviruses: Role of the coat protein. *Annual Review of Phytopathology* 43: 39–42.

Garrett RG, Cooper JA, and Smith PR (1985) Virus epidemiology and control. In: Francki RIB (ed.) *The Plant Viruses, Vol 1:Polyhedral Virions with Tripartite Genomes*, ch. 9, pp. 269–297. New York: Plenum.

Greber RS, Teakle DS, and Mink GI (1992) Thrips-facilitated transmission of prune dwarf and *Prunus* necrotic ringspot viruses from cherry pollen to cucumber. *Plant Disease* 76: 1039–1041.

Haasnoot PCJ, Brederode FTh, Olsthoorn RCL, and Bol JF (2000) A conserved hairpin structure in *Alfamovirus* and *Bromovirus* subgenomic promoters is required for efficient RNA synthesis *in vitro*. *RNA* 6: 708–716.

Li WX and Ding SW (2001) Viral suppressors of RNA silencing. *Current Opinion in Biotechnology* 12: 150–154.

Mink GI (1983) The possible role of honeybees in long-distance spread of *Prunus* necrotic ringspot virus from California into Washington sweet cherry orchards. In: Plumb RT and Thresh JM (eds.) *Plant Virus Epidemiology: The Spread and Control of Insect-Borne Viruses*, pp. 85–91. Oxford: Blackwell Scientific Publications.

Roossinck MJ, Bujarski MJ, Ding SW, et al. (2005) Bromoviridae. In: Fauquet CM, Mayo MA, Maniloff J, Desselberger U, and Ball LA (eds.) Virus Taxonomy: Eighth Report of the International Committee on Taxonomy of Viruses, pp. 1049–1058. San Diego, CA: Elsevier Academic Press.

Scott SW, Zimmerman MT, and Ge X (2003) Viruses in subgroup 2 of the genus Ilarvirus share both serological relationships and characteristics at the molecular level. Archives of Virology 148: 2063–2075.

Swanson MM, Ansel-McKinney P, Houser-Scott F, Yusibov V, Loesch-Fries LS, and Gehrke L (1998) Viral coat protein peptides with limited sequence homology bind similar domains of alfalfa mosaic virus and tobacco streak virus RNAs. Journal of Virology 72: 3227–3234.

Xin H-W, Ji L-H, Scott SW, Symons RH, and Ding S-W (1998) Ilarviruses encode a cucumovirus-like 2b gene that is absent in other genera within the Bromoviridae. Journal of Virology 72: 6956–6959.

Luteoviruses

L L Domier, USDA–ARS, Urbana, IL, USA
C J D'Arcy, University of Illinois at Urbana-Champaign, Urbana, IL, USA

Glossary

Hemocoel The primary body cavity of most arthropods that contains most of the major organs and through which the hemolymph circulates.
Hemolymph A circulatory fluid in the body cavities (hemocoels) and tissues of arthropods that is analogous to blood and/or lymph of vertebrates.

Introduction

Viruses of the family *Luteoviridae* (luteovirids) cause economically important diseases in many monocotyledonous and dicotyledonous crop plants, including barley, wheat, potatoes, lettuce, legumes, and sugar beets. Yield reductions as high as 30% have been reported in epidemic years, although in some cases crops can be totally destroyed. Diseases caused by the viruses were recorded decades and even centuries before they were associated with the causal viruses. In many cases, the stunted, deformed, and discolored plants that result from luteovirid infection were thought to be the result of abiotic factors, such as mineral imbalances or stressful environmental conditions, or of other biotic agents. This, along with their inabilities to be transmitted mechanically, delayed the initial association of the symptoms with plant viruses. For example, curling of potato leaves was first described in Lancashire, UK, in the 1760s, but was not recognized as a specific disease of potato until 1905 and to be caused by an aphid-transmitted virus until the 1920s. The causal agent, potato leaf roll virus (PLRV), was not purified until the 1960s. Similarly, widespread disease outbreaks in cereals, probably caused by barley yellow dwarf virus (BYDV), were noted in the United States in 1907 and 1949. In 1951, a virus was proposed as the cause.

Other diseases caused by luteovirids, like sugarcane yellow leaf, which is caused by sugarcane yellow leaf virus (ScYLV), were not described until the 1990s.

Taxonomy and Classification

Members of the family *Luteoviridae* were first grouped because of their common biological properties. These properties included persistent transmission by aphid vectors and the induction of yellowing symptoms in many infected host plants. 'Luteo' comes from the Latin *luteus*, which translates as yellowish. All luteovirids have small (*c.* 25 nm diameter) icosahedral particles, composed of one major and one minor protein component and a single molecule of positive-sense single-stranded RNA of approximately 5600 nt in length.

The family *Luteoviridae* is divided into three genera – *Luteovirus*, *Polerovirus* (derived from potato leaf roll), and *Enamovirus* (derived from pea enation mosaic) – based on the arrangements, sizes, and phylogenetic relationships of the predicted amino acid sequences of the open reading frames (ORFs). In some plant virus families, a single gene can be used to infer taxonomic and phylogenetic relationships. Within the family *Luteoviridae*, however, different taxonomic relationships can be predicted depending on whether sequences of the replicase (ORF2) or coat protein (CP; ORF3) genes are analyzed (**Figure 1**). ORFs 1 and 2 of the luteoviruses are most closely related to the polymerase genes of viruses of the family *Tombusviridae*, while ORFs 1 and 2 of the poleroviruses and enamoviruses are related to those of the genus *Sobemovirus*. These polymerase types are distantly related in evolutionary terms. Consequently, it has been suggested that luteovirid genomic RNAs arose by recombination between ancestral genomes containing the CP genes characteristic of the family *Luteoviridae* and genomes containing either of the two

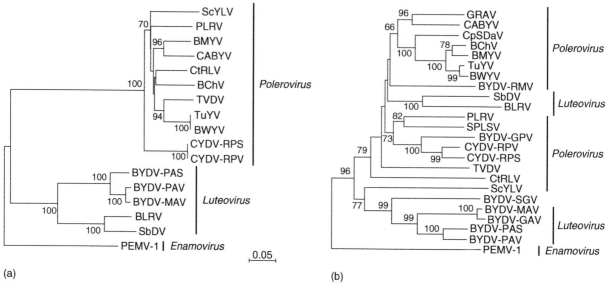

Figure 1   Phylogenetic relationships of the predicted amino acid sequences of the (a) RNA-dependent RNA polymerase (ORF2) and (b) major capsid protein (ORF3). When predicted amino acid sequences from ORF2 are used to group virus species the genera form three distinct groups. Using predicted amino acid sequences from ORF3, species of the genera *Luteovirus* and *Polerovirus* are intermingled in the tree. The resulting consensus trees from 1000 bootstrap replications are shown. The numbers above each node indicate the percentage of bootstrap replicates in which that node was recovered. For virus abbreviations, see **Table 1**.

polymerase types. For taxonomic purposes, the polymerase type has been the primary determinant in assigning a virus to a genus. For this reason, viruses for which only CP sequences have been determined have not been assigned to a genus. The current members of the family are listed in **Table 1**. The genus *Luteovirus* contains five species, and the *Polerovirus* genus has nine species. The genus *Enamovirus* contains a single virus, pea enation mosaic virus 1 (PEMV-1). The family also contains 11 virus species that have not been assigned to a genus. Of these, recently determined sequences of genomic RNAs of BYDV-GAV and carrot red leaf virus (CtRLV) suggest that BYDV-GAV is a strain of BYDV-PAV and that CtRLV is a unique species in the genus *Polerovirus*.

Virion Properties

The sedimentation coefficients $S_{20,w}$ (in Svedberg units) for luteoviruses and poleroviruses range from 106S to 127S. Buoyant densities in CsCl are approximately 1.40 g cm^{-3}. The particles formed as result of the mixed infections by PEMV-1 and PEMV-2 sediment as two components. The $S_{20,w}$ are 107–122S for B components (PEMV-1) and 91–106S for T components (PEMV-2, an umbravirus). Virions are moderately stable and are insensitive to treatment with chloroform or nonionic detergents, but are disrupted by prolonged treatment with high concentrations of salts. Luteovirus and polerovirus particles are insensitive to freezing.

Virion Structure and Composition

All members of the *Luteoviridae* have nonenveloped icosahedral particles with diameters of 25–28 nm (**Figure 2**). Capsids are composed of major (21–23 kDa) and minor (54–76 kDa) CPs, which contain a C-terminal extension to the major CP called the readthrough domain (RTD). According to X-ray diffraction and molecular mass analysis, virions consist of 180 protein subunits, arranged in $T = 3$ icosahedra. Virus particles do not contain lipids or carbohydrates.

Virions contain a single molecule of single-stranded positive-sense RNA of 5300–5900 nt. The RNAs do not have a 3′ terminal poly(A) tract. A small protein (VPg) is covalently linked to the 5′ end of polerovirus and enamovirus genomic RNAs. Cereal yellow dwarf virus RPV (CYDV-RPV) also encapsidates a 322 nt satellite RNA that accumulates to high levels in the presence of the helper virus. Complete genome sequences have been determined for 17 members of the *Luteoviridae* (**Table 1**). For several viruses, genome sequences have been determined from multiple isolates.

Genome Organization and Expression

Genomic RNAs of luteovirids contain five to eight ORFs (**Figure 3**). ORFs 1, 2, 3, and 5 are shared among all members of the *Luteoviridae*. Luteoviruses lack ORF0. Enamoviruses lack ORF4. The luteo- and polerovirus

Table 1   Virus members in the family *Luteoviridae*

Genus	Virus	Abbreviation	Accession number
Luteovirus	Barley yellow dwarf virus – MAV	BYDV-MAV	NC_003680[a]
	Barley yellow dwarf virus – PAS	BYDV-PAS	NC_002160
	Barley yellow dwarf virus – PAV	BYDV-PAV	NC_004750
	Bean leafroll virus	BLRV	NC_003369
	Soybean dwarf virus	SbDV	NC_003056
Polerovirus	Beet chlorosis virus	BChV	NC_002766
	Beet mild yellowing virus	BMYV	NC_003491
	Beet western yellows virus	BWYV	NC_004756
	Cereal yellow dwarf virus – RPS	CYDV-RPS	NC_002198
	Cereal yellow dwarf virus – RPV	CYDV-RPV	NC_004751
	Cucurbit aphid-borne yellows virus	CABYV	NC_003688
	Potato leafroll virus	PLRV	NC_001747
	Turnip yellows virus	TuYV	NC_003743
	Sugarcane yellow leaf virus	ScYLV	NC_000874
Enamovirus	Pea enation mosaic virus 1	PEMV-1	NC_003629
Unassigned	Barley yellow dwarf virus – GAV	BYDV-GAV	NC_004666
	Barley yellow dwarf virus – GPV	BYDV-GPV	L10356
	Barley yellow dwarf virus – RMV	BYDV-RMV	Z14123
	Barley yellow dwarf virus – SGV	BYDV-SGV	U06865
	Carrot red leaf virus	CtRLV	NC_006265
	Chickpea stunt disease associated virus	CpSDaV	Y11530
	Groundnut rosette assistor virus	GRAV	Z68894
	Indonesian soybean dwarf virus	ISDV	
	Sweet potato leaf speckling virus	SPLSV	
	Tobacco necrotic dwarf virus	TNDV	
	Tobacco vein distorting virus	TVDV	AJ575129

[a]Accession numbers beginning with NC_ represent complete genomic sequences.

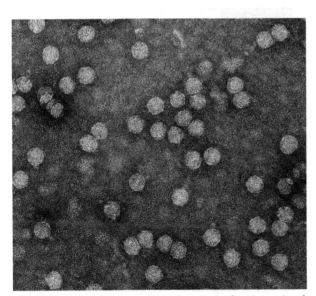

Figure 2   Transmission electron micrograph of soybean dwarf virus particles magnified 240 000×. Virions (stained with uranyl acetate) are *c.* 25 nm in diameter, hexagonal in appearance, and have no envelope.

genomes contain one or two small ORFs, ORFs 6 and 7, within or downstream of ORF5. An additional ORF, ORF8, has been discovered in ORF1 of PLRV. In the enamo- and poleroviruses ORF0 overlaps ORF1 by more than 600 nt, which also overlaps ORF2 by more than 600 nt. In the luteoviruses, ORF1 overlaps ORF2 by less than 50 nt. In all luteo- and polerovirus genome sequences (except for cucurbit aphid-borne yellows virus (CABYV) and GRAV), ORF4 is contained within ORF3. A single, in-frame amber (UAG) termination codon separates ORF5 from ORF3.

Luteovirids have relatively short 5' and intergenic noncoding sequences. The first ORF is preceded by 21 nt in CABYV RNA and 142 nt in soybean dwarf virus (SbDV) RNA. ORFs 2 and 3 are separated by 112–200 nt of noncoding RNA. There is considerable variation in the length of sequence downstream of ORF5, which ranges from 125 nt for CYDV-RPV to 650 nt for SbDV.

Luteovirids employ an almost bewildering array of strategies to express their compact genomes. ORFs 0, 1, 2, and 8 are expressed directly from genomic RNA. Downstream ORFs are expressed from subgenomic RNAs (sgRNAs) that are transcribed from internal initiation sites by virus-encoded RNA-dependent RNA polymerases (RdRps) from negative-strand RNAs and are 3'-co-terminal with the genomic RNA. Since the initiation codon for ORF0 of polero- and enamoviruses is upstream of that of ORF1, translation of ORF1 is initiated by 'leaky scanning' in which ribosomes bypass

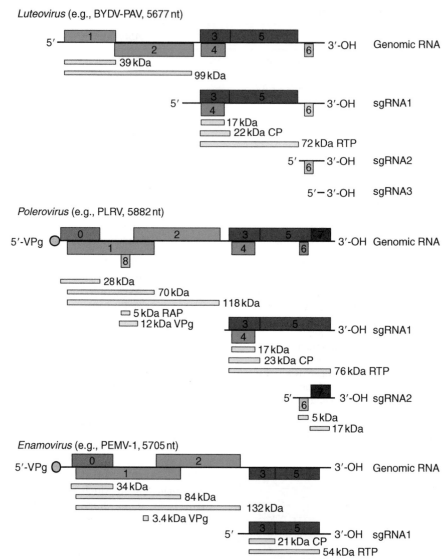

Figure 3 Maps of the virus genomes of genera in the family *Luteoviridae*. Individual ORFs are shown with open boxes. The ORFs are staggered vertically to show the different reading frames occupied by each ORF. The yellow boxes indicate protein products with the predicted sizes listed to the right of each. The polyproteins encoded by ORF1 of enamo- and poleroviruses contain the protease and the genome-linked protein (VPg). The predicted amino acid sequences of proteins encoded by ORF2 are similar to RNA-dependent RNA polymerases. ORF3, which encodes the major coat protein, is separated from ORF5 by an amber termination codon. ORF4, when present, is contained within ORF3 and encodes a protein required for virus cell-to-cell movement.

the AUG of ORF0 and continue to scan the genomic RNA until they reach the ORF1 AUG. The protein products of ORF2 are expressed as a translational fusion with the product of ORF1. At a low but significant frequency during the expression of ORF1, translation continues into ORF2 through a −1 frameshift that produces a large protein containing sequences encoded by both ORFs 1 and 2 in a single polypeptide. ORF8, which has only been identified in PLRV, resides entirely within ORF1 in a different reading frame and encodes a 5 kDa replication-associated protein. To express ORF8, sequences within the ORF fold into a structure called an internal ribosome entry site (IRES), which recruits ribosomes to initiate translation about 1600 nt downstream of the 5′ terminus of PLRV RNA.

ORFs 3, 4, and 5 are expressed from sgRNA1, the 5′ terminus of which is located about 200 nt upstream of ORF3 at the end of ORF2, and extends to the 3′ terminus of the genome. Luteo- and poleroviruses produce a second sgRNA that expresses ORFs 6 and 7. Luteoviruses produce a third sgRNA, which does not appear to encode a protein. ORF3 is translated from the 5′ terminus of sgRNA1. ORF4 of luteo- and poleroviruses, which encodes a 17 kDa protein, is contained within ORF3, and is expressed from the same sgRNA as ORF3 through a leaky scanning mechanism much like that used to express ORF1 in polero- and enamoviruses. In all luteovirids, ORF5 is expressed only as a translational fusion with the products of ORF3 by readthrough of the

UAG stop codon at the end of ORF3. This produces a protein with the product of ORF3 at its N-terminus and the product of ORF5 at its C-terminus.

While enamo- and polerovirus RNAs contain 5′ VPgs that interact with translation initiation factors, luteovirus RNAs contain only a 5′ phosphate. Unmodified 5′ termini are recognized poorly for translation initiation. To circumvent this problem, a short sequence located in the noncoding region just downstream of ORF5 in the BYDV-PAV genome acts as a potent enhancer of cap-independent translation by interacting with sequences near the 5′ termini of the genomic and sgRNAs to promote efficient translation initiation.

Research into the functions of the proteins encoded by luteovirids has shown that the 28–34 kDa proteins encoded by ORF0 are effective inhibitors of post-transcriptional gene silencing (PTGS). PTGS is an innate and highly adaptive antiviral defense found in all eukaryotes that is activated by double-stranded RNAs (dsRNAs), which are produced during virus replication. Consequently, viruses that contain mutations in ORF0 show greatly reduced accumulations in infected plants.

The ORF1-encoded proteins of enamo- and poleroviruses contain the VPg and a chymotrypsin-like serine protease that is responsible for the proteolytic processing of ORF1-encoded polyproteins. The protease cleaves the ORF1 proteins in *trans* to liberate the VPg, which is covalently attached to genomic RNA. The protein expressed by ORF8 of PLRV is required for virus replication. Luteovirid ORF2s have a coding capacity of 59–67 kDa for proteins that are very similar to known RdRps and hence likely represent the catalytic portion of the viral replicase.

ORF3 encodes the major CP of the luteovirids, which ranges in size from 21 to 23 kDa. ORF5 has a coding capacity of 29–56 kDa. However, ORF5 is expressed only as a translational fusion with the product of ORF3 when, about 10% of the time, translation does not stop at the end of ORF3 and continues through to the end of ORF5. The ORF5 portion of this readthrough protein has been implicated in aphid transmission and virus stability. Experiments with PLRV and BYDV-PAV have shown that the N-terminal region of the ORF5 readthrough protein determines the ability of virus particles to bind to proteins produced by endosymbiotic bacteria of aphid vectors. Interactions of virus particles with these proteins seem to be essential for persistence of the viruses in aphids. Nucleotide sequence changes within ORF5 of PEMV-1 abolish aphid transmissibility. The N-terminal portions of ORF5 proteins are highly conserved among luteovirids while the C-termini are much more variable.

The luteo- and polerovirus genomes possess an ORF4 that is contained within ORF3 and encodes proteins of 17–21 kDa. Viruses that contain mutations in ORF4 are able to replicate in isolated plant protoplasts, but are deficient or delayed in systemic movement in whole plants. Hence, the product of ORF4 seems to be required for movement of the virus within infected plants. This hypothesis is supported by the observation that enamoviruses lack ORF4. While luteo- and poleroviruses are limited to phloem and associated tissues, the enamovirus PEMV-1 is able to move systemically through other plant tissues in the presence of PEMV-2, which under natural conditions invariably coexists with PEMV-1.

Some luteo- and polerovirus genomes contain small ORFs within and/or downstream of ORF5. In luteoviruses, no protein products have been detected from these ORFs in infected cells. BYDV-PAV genomes that do not express ORF6 are still able to replicate in protoplasts. The predicted sizes of the proteins expressed by ORFs 6 and 7 of PLRV are 4 and 14 kDa, respectively. Based on mutational studies, it has been proposed that these genome regions may regulate transcription late in infection.

Evolutionary Relationships among Members of the *Luteoviridae*

Viruses in the family *Luteoviridae* have replication-related proteins that are similar to those in other plant virus families and genera. The luteovirus replication proteins encoded by ORFs 1 and 2 resemble those of members of the family *Tombusviridae*. In contrast, polymerases of poleroviruses and enamoviruses resemble those of viruses in the genus *Sobemovirus*. The structural proteins of some sobemoviruses also are similar to the major CP of luteovirids. Using an X-ray crystallography-derived structure of virions of the sobemovirus rice yellow mottle virus, which shares a CP amino acid sequence similarity of 33% with PLRV, it was possible to predict the virion structure of PLRV and other luteovirids.

Host Range and Transmission

Several luteovirids have natural host ranges largely restricted to one plant family. For example, BYDV and CYDV infect many grasses, BLRV infects mainly legumes, and CtRLV infects mainly plants in the family Apiaceae. Other luteovirids infect plants in several or many different families. For example, beet western yellows virus (BWYV) infects more than 150 species of plants in more than 20 families. As techniques for infecting plants with recombinant viruses have improved, the experimental host ranges of viruses have been expanded to include plants on which aphid vectors would not normally feed. For example, BYDV, CYDV, PLRV, and SbDV have been shown to infect *Nicotiana* species that had not been described previously as experimental hosts for the viruses when inoculated biolistically with viral RNA or using

Agrobacterium tumefaciens harboring binary plasmids containing infectious copies of the viruses. These results suggest that feeding preferences of vector aphids play important roles in defining luteovirid host ranges.

Luteovirids are transmitted in a circulative manner with varying efficiencies by at least 25 aphid species. With the exception of the enamovirus PEMV-1, members of the family *Luteoviridae* are transmitted from infected plants to healthy plants in nature only by the feeding activities of specific species of aphids. There is no evidence for replication of the viruses within aphid vectors. *Myzus persicae* is the most common aphid vector of luteovirids that infect dicots. Several different species of aphids transmit luteovirids that infect monocotyledenous plants (BYDV and CYDV) in a species-specific manner.

Circulative transmission of the viruses is initiated when aphids acquire viruses from sieve tubes of infected plants during feeding. The viruses travel up the stylet, through the food canal, and into the foregut (**Figure 4**). The viruses then are actively transported across the cells

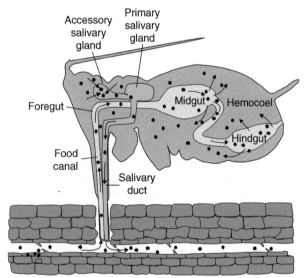

Figure 4   Circulative transmission of viruses of the family *Luteoviridae* by vector aphids. While feeding from sieve tubes of an infected plant, an aphid (shown in cross section) acquires virus particles, which travel up the stylet, through the food canal, and into the foregut. The virions are actively transported across cells of the posterior midgut and/or hindgut into the hemocoel in a process that involves receptor-mediated endocytosis. Virions then passively migrate through the hemolymph to the accessory salivary gland where they are again transported by a receptor-mediated process to reach the lumen of the gland. Once in the salivary gland lumen, the virions are expelled with the saliva into the vascular tissue of host plants. Aphids can retain the ability to transmit virus for several weeks. Hindgut membranes usually are much less selective than those of the accessory salivary glands, which is why viruses that are not transmitted by a particular species of aphid often accumulate in the hemocoel, but do not traverse the membranes of the accessory salivary gland.

of the alimentary tract into the hemocoel in a process that involves receptor-mediated endocytosis of the viruses and the formation of tubular vesicles that transport the viruses through the epithelial cells and into the hemocoel. Luteovirids are acquired at different sites within the gut of vector aphids. PLRV and BWYV are acquired in the posterior midgut. BYDV, CYDV, and SbDV are acquired in the hindgut. CABYV is taken up at both sites. Viruses then passively migrate through the hemolymph to the accessory salivary gland where the viruses must pass through the membranes of the accessory salivary gland cells in a similar type of receptor-mediated transport process to reach the lumen of the gland. Once in the salivary gland lumen, viruses are expelled with saliva into vascular tissues of host plants. Since large amounts of virus can accumulate in the hemocoel of aphids, they may retain the ability to transmit virus for several weeks. Typically hindgut membranes are much less selective than those of the accessory salivary glands. Consequently, viruses that are not transmitted by a particular species of aphid often are transported across gut membranes and accumulate in the hemocoel, but do not traverse the membranes of the accessory salivary gland.

The RTD of the minor capsid protein plays a major role in aphid transmission of luteovirids. The RTD interacts with symbionin produced by endosymbiotic aphid-borne bacteria, which may protect virions from degradation by the aphid immune system. The specificity of aphid transmission and gut tropism has been linked to the RTD in multiple luteovirids.

Unlike other luteovirids, PEMV-1 can be transmitted by rubbing sap taken from an infected plant on a healthy plant, in addition to being transmitted by aphids. This difference in transmissibility is dependent on its multiplication in cells co-infected with PEMV-2, but aphid transmissibility can be lost after several mechanical passages.

Replication

Luteovirids infect and replicate in sieve elements and companion cells of the phloem and occasionally are found in phloem parenchyma cells. PEMV-1 is able to move systemically into other tissues in the presence of PEMV-2. Virus infections commonly result in cytopathological changes in cells that include formation of vesicles containing filaments and inclusions that contain viral RNA and virions. The subcellular location of viral RNA replication has not been determined unequivocally. However, early in infection, negative-strand RNAs of BYDV-PAV are first detected in the nucleus and later in the cytoplasm, which suggests that at least a portion of luteovirus replication occurs in the nucleus. Synthesis of negative-strand RNA, which requires tetraloop structures at the 3′ end of BYDV-PAV genomic RNAs, is detected in infected cells before the formation of virus particles. Late in infection, BYDV-PAV sgRNA2 inhibits translation from genomic RNA, which may promote a switch from translation to replication and packaging of genomic RNAs.

Virus–Host Relationships

While some infected plants display no obvious symptoms, most luteovirids induce characteristic symptoms that include stunting, leaves that become thickened, curled or brittle, and yellow, orange, or red leaf discoloration, particularly of older leaves of infected plants. These symptoms result from phloem necrosis that spreads from inoculated sieve elements and causes symptoms by inhibiting translocation, slowing plant growth, and inducing the loss of chlorophyll. Symptoms may persist, may vary seasonally, or may disappear soon after infection. Temperature and light intensity often affect symptom severity and development. In addition, symptoms can vary greatly with different virus isolates or strains and with different host cultivars.

Yield losses caused by luteovirids are difficult to estimate because the symptoms often are overlooked or attributed to other agents. US Department of Agriculture specialists estimated that yield losses from BWYV, BYDV, and PLRV infections were over $65 million during the period 1951–60. Plants infected at early stages of development by luteovirids suffer the most significant yield losses, which often are linearly correlated with the incidence of virus infection.

Epidemiology

Luteovirid infections have been reported from temperate, subtropical, and tropical regions of the world. Some of the viruses are found worldwide, such as BWYV, BYDV, and PLRV. Others have more restricted distribution, such as tobacco necrotic dwarf virus, which has been reported only from Japan, and groundnut rosette assistor virus, which has been reported only in African countries south of the Sahara.

Most luteovirids infect annual crops and must be reintroduced each year by their aphid vectors. Some viruses are disseminated in infected planting material, like PLRV where infected potato tubers are the principal source of inoculum for new epidemics. Consequently, programs to produce clean stock are operated around the world to control these viruses. Alate, that is, winged, aphid vectors may transmit viruses from local cultivated, volunteer, or weed hosts. Alternatively, alate aphids may be transported into crops from distant locations by wind currents. These vectors may bring the virus with them, or they may first have to acquire virus from locally infected hosts. The agronomic impact of luteovirid diseases depends both on meteorological events that favor movement and reproduction of vector aphids and susceptibility of the crop at the time of aphid arrival. Only aphid species that feed on a particular crop plant can transmit virus. Aphids that merely probe briefly to determine a plant's suitability will not transmit the viruses. Secondary spread of the viruses is often primarily by apterous, that is, wingless, aphids. The relative importance of primary introduction of virus by alate aphids and of secondary spread of virus by apterous aphids in disease severity varies with the virus, aphid species, crop, and environmental conditions.

Some members of the family *Luteoviridae* occur in complexes with other members of the family or with other plant viruses. For example, BYDV and CYDV often are found co-infecting cereals; BWYV and SbDV are often found together in legumes; and PLRV is often found co-infecting potatoes with potato virus Y and/or potato virus X. Some other plant viruses depend on luteovirids for their aphid transmission, such as the groundnut rosette virus, carrot mottle virus, and bean yellow vein banding virus (all umbraviruses), which depend on groundnut rosette assistor virus, carrot red leaf virus, and PEMV-1, respectively.

Diagnosis

An integral part of controlling luteovirid diseases is accurate diagnosis of infection. Because symptoms caused by luteovirids often resemble those caused by other biotic and abiotic factors, visual diagnosis is unreliable and other methods have been developed. Initially, infectivity, or biological, assays were used to diagnose infections. These techniques also have been used to identify species of vector aphids and vector preferences. In bioassays, aphids are allowed to feed on infected plants and then are transferred to indicator plants. These techniques are very sensitive, but can require several weeks for symptoms to develop on indicator plants. The strong immunogenicity of luteovirids has facilitated development of very specific and highly sensitive serological tests that can discriminate different luteovirids and sometimes even strains of a single virus species. Poly- and monoclonal antibodies for virus detection are produced by immunizing rabbits and/or mice with virus particles purified from infected plants. Techniques also have been developed to detect viral RNAs from infected plant tissues by reverse transcription-polymerase chain reaction (RT-PCR), which can be more sensitive and discriminatory than serological diagnostic techniques. Even so, serological tests are the most commonly used techniques for the detection of infections because of their simplicity and speed.

Control

Because methods are not available to cure luteovirid infections after diagnosis, emphasis has been placed on reducing losses through the use of tolerant or resistant plant cultivars and/or on reducing the spread of viruses by controlling aphid populations. Many luteovirids are transmitted by migrating populations of aphids that occur at similar times each year. For those virus–aphid combinations, it is sometimes possible to plant crops so that young, highly susceptible plants are not in the field when the seasonal aphid migrations occur. Insecticides have been used in a prophylactic manner to reduce crop losses. While insecticide treatments do not prevent initial infections, they can greatly limit secondary spread of aphids and therefore of viruses. In some instances biological control agents such as predatory insects and parasites have reduced aphid populations significantly. Genes for resistance or tolerance to infection by luteovirids have been identified in most agronomically important plant species infected by the viruses. For BYDV, PLRV, and SbDV, transgenic plants that express portions of the virus genomes have been produced through DNA-mediated transformation. In some cases, the expression of these virus genes in transgenic plants confers higher levels of virus resistance than resistance genes from plants.

See also: Barley Yellow Dwarf Viruses; Cereal Viruses: Wheat and Barley; Sobemovirus; Tombusviruses.

Further Reading

Brault V, Perigon S, Reinbold C, *et al.* (2005) The polerovirus minor capsid protein determines vector specificity and intestinal tropism in the aphid. *Journal of Virology* 79: 9685–9693.

Falk BW, Tian T, and Yeh HH (1999) Luteovirus-associated viruses and subviral RNAs. *Current Topics in Microbiology and Immunology* 239: 159–175.

Gray S and Gildow FE (2003) Luteovirus–aphid interactions. *Annual Review of Phytopathology* 41: 539–566.

Hogenhout SA, van der Wilk F, Verbeek M, Goldbach RW, and van den Heuvel JF (2000) Identifying the determinants in the equatorial domain of Buchnera GroEL implicated in binding potato leafroll virus. *Journal of Virology* 74: 4541–4548.

Lee L, Palukaitis P, and Gray SM (2002) Host-dependent requirement for the potato leafroll virus 17-kDa protein in virus movement. *Molecular Plant–Microbe Interactions* 10: 1086–1094.

Mayo M, Ryabov E, Fraser G, and Taliansky M (2000) Mechanical transmission of potato leafroll virus. *Journal of General Virology* 81: 2791–2795.

Miller WA and White KA (2006) Long-distance RNA–RNA interactions in plant virus gene expression and replication. *Annual Review of Phytopathology* 44: 447–467.

Moonan F, Molina J, and Mirkov TE (2000) Sugarcane yellow leaf virus: An emerging virus that has evolved by recombination between luteoviral and poleroviral ancestors. *Virology* 269: 156–171.

Nass PH, Domier LL, Jakstys BP, and D'Arcy CJ (1998) In situ localization of barley yellow dwarf virus PAV 17-kDa protein and nucleic acids in oats. *Phytopathology* 88: 1031–1039.

Nixon PL, Cornish PV, Suram SV, and Giedroc DP (2002) Thermodynamic analysis of conserved loopstem interactions in P1–P2 frameshifting RNA pseudoknots from plant *Luteoviridae*. *Biochemistry* 41: 10665–10674.

Pfeffer S, Dunoyer P, Heim F, *et al.* (2002) P0 of beet western yellows virus is a suppressor of posttranscriptional gene silencing. *Journal of Virology* 76: 6815–6824.

Robert Y, Woodford JA, and Ducray-Bourdin DG (2000) Some epidemiological approaches to the control of aphid-borne virus diseases in seed potato crops in Northern Europe. *Virus Research* 71: 33–47.

Taliansky M, Mayo MA, and Barker H (2003) *Potato leafroll virus*: A classic pathogen shows some new tricks. *Molecular Plant Pathology* 4: 81–89.

Terradot L, Souchet M, Tran V, and Giblot Ducray-Bourdin D (2001) Analysis of a three-dimensional structure of potato leafroll virus coat protein obtained by homology modeling. *Virology* 286: 72–82.

Thomas PE, Lawson EC, Zalewski JC, Reed GL, and Kaniewski WK (2002) Extreme resistance to potato leafroll virus in potato cv. Russet Burbank mediated by the viral replicase gene. *Virus Research* 71: 49–62.

Machlomovirus

K Scheets, Oklahoma State University, Stillwater, OK, USA

History

Maize chlorotic mottle virus (MCMV) was initially found in maize (*Zea mays*) and sorghum (*Sorghum bicolor*) fields in Peru in 1973. In 1976, the virus appeared in Kansas (USA) maize fields, alone and as part of a synergistic disease. In 1978, the second component of the synergistic disease, corn lethal necrosis (CLN), was identified as any maize-infecting virus of the family *Potyviridae*. MCMV and CLN spread to Nebraska, and both MCMV and CLN appeared in Mexico in 1982. On the island of Kauai, Hawaii (USA), a severe outbreak of MCMV appeared in the winter seed nurseries in 1989–90. MCMV is endemic in Peru and along the Kansas–Nebraska border. In 2004, MCMV and CLN were first detected in Thailand where MCMV continues to spread.

Taxonomy and Classification

The only known member of the genus *Machlomovirus* is the species *Maize chlorotic mottle virus*. Its inclusion in the family *Tombusviridae* is based on the high degree of homology of the encoded viral replicase. The genome organization of the single viral RNA of MCMV is most similar to panicum mosaic virus (PMV), species *Panicum mosaic virus*, genus *Panicovirus*, family *Tombusviridae*. Key taxonomic features of the genus *Machlomovirus* are a unique open reading frame (ORF) at the 5′ end of the genome that largely overlaps the pre-readthrough portion of the viral replicase gene, and a readthrough ORF preceding and overlapping the capsid protein (CP) gene near the 3′ end of the genome.

Virion Structure and Properties

Transmission electron microscopy of negatively stained MCMV virions shows a smooth sphere or hexagonal structure approximately 28 nm in diameter (**Figure 1**). The virion consists of a 4437 nt single-stranded RNA surrounded by 25.1 kDa CP subunits that lack the protruding domain found on CPs of viruses from many genera in the family *Tombusviridae*. Sequence similarity to the CPs of PMV, tobacco necrosis virus genus *Necrovirus*, family *Tombusviridae*, and southern bean mosaic virus genus *Sobemovirus* suggest that MCMV is a $T = 3$ icosahedral virion with 180 copies of its CP in the viral shell. The estimated weight of the virion is 5.95 mDa. Purified MCMV has a sedimentation coefficient ($s_{20,w}$) of 109S and a buoyant density in CsCl of 1.365 g ml^{-1}. The thermal inactivation point in maize sap is 85–90 °C. The virion is stable at pH 5.

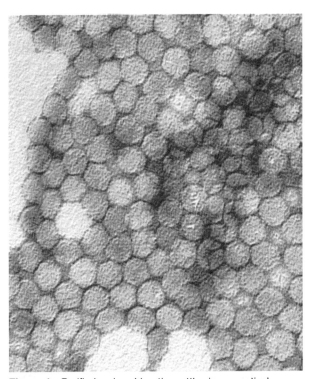

Figure 1 Purified maize chlorotic mottle virus negatively stained with uranyl acetate.

Genome Structure and Organization

The genome organization of the genus *Machlomovirus* is based on the complete sequence of MCMV. The plus-sense RNA is 4437 nt long and contains seven overlapping ORFs that encode proteins of 7 kDa or larger (**Figure 2**). The RNA was reported to have no poly(A) tail and an ^{m7}G cap at the 5′ end. However, in common with other viruses in the family, the encoded MCMV replicase does not have any motifs characteristic of the methyltransferase domain found in viral replicases of capped RNA viruses. None of the other MCMV-encoded proteins contain a methyltransferase domain, so it is likely that the RNA in MCMV, like in other members of the family, is in fact uncapped. The 5′ untranslated region (UTR) is 117 nt long, and ORF1 encodes a 32 kDa highly acidic protein (estimated pI = 3.83). ORF2 begins 19 nt downstream and encodes a 50 kDa highly basic protein (estimated pI = 10.59) so the migration of these two proteins in sodium dodecyl sulfate (SDS) polyacrylamide gels is likely to be anomalous. Suppression of the UAG stop codon of ORF2 would produce a 111 kDa protein. A cluster of four ORFs is encoded in the 3′ third of the viral RNA downstream of the transcription start site for the 1467-nt-long subgenomic RNA1 (sgRNA1). ORF4 encodes a 7.5 kDa protein (p7a), and suppression of its UGA stop codon would produce a 31 kDa protein. The second AUG of sgRNA1 begins ORF7 which encodes the viral CP. ORF6 was identified by similarity of its gene product p7b to small peptides encoded in similar locations on carmoviruses, necroviruses, and PMV, and it begins with a noncanonical start codon. *In vitro* translation of MCMV virion RNA in rabbit reticulocyte lysate produces p32, p50, p111, and p25. The two 7 kDa peptides and p31 were not detected in rabbit reticulocyte lysate translations. The 3′ UTR is 343 nt long and encodes a 337-nt-long sgRNA (sgRNA2).

Replication

The replication strategy of MCMV has not been completely determined, but inoculation of maize protoplasts with transcripts from wild type and mutant versions of an infectious cDNA has provided some information. Transcripts with mutations in the 3′ third of the genome that stop expression of one or more of the proteins encoded on sgRNA1 are capable of replication. Additionally, mutations just upstream of the sgRNA1 transcription start site that stop expression of sgRNA1 but do not alter the sequence of p111 are capable of replication, indicating that none of the proteins encoded on sgRNA1 are necessary for replication. Based on the replication mechanisms of other tombusvirus family members it is likely that after virion disassembly, MCMV viral RNA is translated to produce the viral

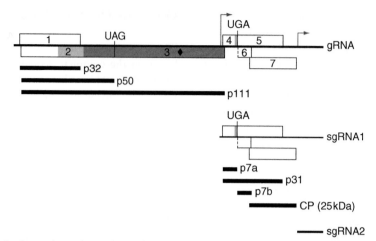

Figure 2 Genome organization and protein products of maize chlorotic mottle virus. The seven ORFs are marked as boxes on the genomic RNA and subgenomic RNA1 (sgRNA1) in each reading frame. The two suppressible stop codons are indicated (UAG and UGA), and the noncanonical start codon for ORF6 is marked with a dashed line. The proteins are indicated as heavy black bars beneath their mRNAs. The dark blue box marks the region encoding high protein sequence similarity to replicases of other members of the family *Tombusviridae* with a monopartite genome, and the light blue box indicates the area encoding protein similarity only to PMV replicase. The diamond marks the location encoding the 'GDD' motif. The green box in ORF4 marks the coding region for a conserved peptide sequence also found in PMV, carmoviruses, and necroviruses. Bent red arrows mark the sgRNA transcription start sites. sgRNA2 does not contain any significant ORFs.

replicase which then synthesizes the negative strand of genomic RNA after recognizing sequences and structures located at the viral 3′ terminus that have sequence and structural similarities to the promoters of carmoviruses. The complementary strand is then used as template for synthesis of progeny viral RNA strands. sgRNA synthesis mechanisms differ between genera in the family *Tombusviridae*, and it is not known which mechanism is used by MCMV. sgRNA1 synthesis may initiate by replicase binding internally to the sgRNA promoter on the genomic complementary strand. Alternatively, occasional premature termination of viral complementary strand synthesis at a specific location may produce separate complementary strand copies of sgRNA1 that are used as templates to synthesize many copies of sgRNA1. Although sgRNA2 accumulates in infected maize plants and inoculated protoplasts, its function and method of transcription are not known.

Geographic and Seasonal Distribution

Originally MCMV was found only in the Western Hemisphere. Peru, Argentina, Mexico, and the USA have reported infections, and it is possible that MCMV is present in other countries in the Western Hemisphere anywhere the beetle vectors are found in maize-growing areas. In Peru, MCMV is found in northern coastal areas where maize is grown as a seasonal crop as well as in the southern coastal areas where it is grown year-round. MCMV is endemic in several valleys in the Department of Lima. In the continental USA, where maize is a

seasonal crop, MCMV was initially found in Kansas near the Nebraska border, then spread along the Republican and Big Blue River Valleys. The virus continued spreading along river valleys and is endemic in Nebraska and Kansas. MCMV initially appeared in the state of Guanajuato, Mexico, and has spread from there to additional nearby states. After its appearance in Kauai, Hawaii, MCMV has not remained as a problem. In Thailand, maize production is year-round, and MCMV was first detected in the province of Saraburi. It is now found in Saraburi, Nakohn Ratchasrima, Tak, and Petchaboon provinces.

Host Range and Virus Propagation

Susceptibility has been tested for 73 plant species (19 dicots) representing 35 genera, and only members of the family Poaceae are hosts. Although some cultivars of hexaploid wheat (*Triticum aestivum*), durum wheat (*Triticum durum*), barley (*Hordeum vulgare*), and sorghum (*S. bicolor*) are susceptible to infection, maize is the only crop in which MCMV infection is of economical importance, with many susceptible cultivars in all maize types (sweet, floury, field, and popcorn). MCMV can be readily propagated by mechanical inoculation in the inbred lines N28Ht, N28, and Oh28, but no reliable local lesion host has been identified.

Genetics

Viral RNA or uncapped transcripts of a full-length cDNA are sufficient for replication in protoplasts and infection of plants. The first ORF encodes a highly acidic 32 kDa protein with no similarity to any other protein in databases. The location of ORF1 suggests that it is a protein needed early in the viral life cycle, but its function is not known. ORF2 and ORF3 comprise a 50 kDa protein and a readthrough protein of 111 kDa containing the 'GDD' motif found in almost all RNA-dependent RNA polymerases (**Figure 2**). Sequence similarities of the carboxyl end of p50 and the p111 readthrough region to viral replicase proteins of other members of the family *Tombusviridae* suggest that these are similarly involved in replication of MCMV. p50 is much larger than the pre-readthrough proteins encoded by most of the monopartite viruses in the same family (27–33 kDa) and the sequence similarity only extends to the carboxyl third of p50. The similarity between p50 and the 48 kDa protein from PMV encompasses about 60% of the carboxyl end. Along with other members of the family, no regions identifiable as a helicase domain or a methyltransferase domain are found in p111. Like PMV and oat chlorotic stunt virus (genus *Avenavirus*), MCMV produces a single sgRNA to express a cluster of ORFs in the 3' third of the genome, and none of the proteins encoded in this region are required for replication. The genes for p7a and p7b are upstream of the CP ORF, similar to the location of small ORFs in PMV, necroviruses, and carmoviruses. p7a and p7b have similar hydrophobic/hydrophilic characteristics and some sequence similarity to the peptides encoded in the corresponding locations in PMV, necroviruses, and carmoviruses, with the greatest sequence similarity occurring in the C-terminal region of p7a. These small peptides are required for cell-to-cell movement for PMV, necroviruses, and carmoviruses, so it is likely that p7a and p7b have similar functions. Database searches with the protein encoded in ORF5 do not identify any proteins with related sequence. The most 3' ORF encodes the viral CP.

Serologic Relationships and Variability

The complete relationship of MCMV isolates from countries where MCMV has been found on a noticeable scale has not been determined. Two serotypes from Kansas (K1 and K2) and one from Peru have been compared, and they can be differentiated by agar double-diffusion assays. Most isolates from Nebraska and Kansas are similar to MCMV-K1, which is the source used for the MCMV sequence in GenBank (X14736) and the infectious transcript cDNA. Sequence comparisons have been done using reverse transcription-polymerase chain reaction (RT-PCR) amplification of the CP shell domain ORF from 47 isolates collected from Nebraska/Kansas, Hawaii, and Peru. Alignment of the 200 bp fragments indicates that each geographic region can be identified by a predominant unique genotype. The Hawaiian isolates show the least intrapopulational divergence, consistent with the recent appearance of the virus in Hawaii before the sample collection period. The Hawaiian isolates are more closely related to Peru isolates than to those from Nebraska/Kansas. The complete sequence of the CP gene of MCMV isolated in Thailand (AY587605) shows 96% identity to MCMV-K1 at the nucleotide level and 97% identity at the protein level. When the corresponding 200 nt region is compared to the data in the larger study, the Thai sequence groups with Nebraska/Kansas sequences. These data suggest that Peru was the source for MCMV in Hawaii while the source of virus in Thailand was Nebraska/Kansas.

Transmission

MCMV is readily transmitted by mechanical inoculation of leaves or roots and by vascular puncture inoculation of seeds. Six species of beetles of the family *Chrysomelidae* found in the continental USA can transmit MCMV: the southern corn rootworm beetle (*Diabrotica undecimpunctata howardi*), the northern corn rootworm beetle (*Diabrotica barberi*), the western corn rootworm beetle (*Diabrotica virgifera virgifera*), the cereal leaf beetle and larvae (*Oulema melanopus*), the corn flea beetle (*Chaetocnema pulicaria*), and the flea beetle (*Systena frontalis*). The southern corn rootworm beetle is the vector in Mexico. In Peru, *Diabrotica viridula* (adult and larvae) and *Diabrotica decempunctata sparsella* transmitted MCMV-P. MCMV was spread by a large infestation of thrips (*Frankliniella williamsi*) in Hawaii. MCMV-K1 epidemiology indicates a soil and water connection, so it is possible that MCMV is soil borne, water borne, or transmitted by a fungus as is seen with various other viruses in the tombusvirus family. It is possible that the different major isolates of MCMV are predominantly transmitted by either a soil route or insects. Seed transmission occurs at a very low rate (0.008–0.04%).

Epidemiology

MCMV reappears in previously infected fields in Nebraska and Kansas, and there is little indication that the virus spreads very rapidly during the maize-growing season, suggesting that beetles are not the major source of viral spread there. MCMV does not overwinter in grasses surrounding the fields, suggesting that soil is the reservoir for future infections. Although it was hypothesized that infected crop residue might be a food source for beetle larvae which would infect new seedlings, it was later shown that MCMV loses its infectivity over the winter, and its loss in infectivity correlates with drying. Additionally, larvae do not feed on dead infected plant material. Infected fields are in river valleys and likely to be irrigated suggesting a water-borne mechanism. In the regions of Peru where maize is planted year-round, infection patterns correlate with emergence and spread of the beetle vectors. The mechanism of spread in Thailand has not yet been determined. The large geographic jumps in MCMV outbreaks suggest that initial infections occur via seed transmission. Although all the known beetle vectors are native to the New World, the western corn rootworm beetle was introduced to Serbia in 1992 and has since spread to additional European countries. This increases the potential for an outbreak of MCMV in Europe if it is introduced there via seeds.

Pathogenicity

MCMV infections can reduce yields of maize in experimental plots by up to 59%, but natural infections are seldom as severe. Average losses of 10–15% of sweet and floury maize crops occur in Peru. The more devastating disease CLN is caused by a synergistic interaction with potyviruses. CLN has been reported in Peru, Kansas, Nebraska, Mexico, and Thailand. In Kansas and Nebraska, maize dwarf mosaic virus-A (MDMV-A; genus *Potyvirus*), sugarcane mosaic virus MD-B (SCMV MD-B) (genus *Potyvirus*), and wheat streak mosaic virus (WSMV) (genus *Tritimovirus*) are found in natural CLN infections. In Mexico and Thailand, the potyvirus is usually SCMV MD-B. CLN has caused crop losses as high as 91%. If both viruses are present when plants are young, they initially develop a chlorotic mottle, then become extremely chlorotic and stunted. Leaves become necrotic starting at the leaf margins, then necrosis spreads inward causing rapid plant death. If the viral combination infects the plants at later stages, plants may appear green and healthy except for early drying of husks, producing ears that are fully developed except for wrinkled and shriveled seeds. As well as the synergistic increase in symptoms in CLN, the concentration of MCMV increases dramatically similarly to most synergisms involving potyviruses. With SCMV MD-B present, MCMV concentrations averaged more than fivefold higher than in singly infected plants, and in WSMV/MCMV synergisms the MCMV levels increased up to 11-fold higher than in singly infected plants. Although most potyviral synergisms do not cause an increase in potyvirus concentrations, the WSMV/MCMV synergism increased the WSMV infection rate and caused a two- to threefold increase in WSMV concentrations.

Pathology and Histopathology

Disease symptoms vary depending on maize genotype and age at infection. The mildest symptoms are a light chlorotic mottle on leaves with little effect on corn yield. In more susceptible varieties, these symptoms progress to chlorotic stripes parallel to the midvein that may coalesce forming elongated chlorotic blotches that may further turn necrotic. Severe symptoms include necrosis,

stunting, decreased male inflorescence with few spikes, decreased number of ears, and malformed and partially filled ears. Plants infected at earlier stages usually have stronger symptoms and lower yields than later infections. MCMV has been found in all parts of the plant including leaves, stem, roots, cob, silk, pollen, ear sheaves, bracts, and in all seed parts (embryo, cotyledon, endosperm, and pericarp). Electron microscopy of cross sections of MCMV-infected maize showed virus-like particles in xylem vessels, often filling the entire vessel lumen. Some xylem tubes were filled with an electron dense matrix that appeared to be virus-like particles embedded in viroplasm. Some parenchyma cells contained vacuolated viroplasms. Chloroplasts were highly disorganized.

Prevention and Control

MCMV-resistant maize lines are being developed, and perennial diploid teosinte (*Zea diploperennis*) is a source for additional resistance genes. If maize lines resistant to either MCMV or the potyviral components of CLN are used, the potential for CLN outbreaks is diminished. Interestingly, a transgenic maize line expressing the CP gene from SCMV MD-B showed fewer disease symptoms when inoculated singly with SCMV MD-B, MDMV-A, and MCMV, as well as showing fewer symptoms when inoculated with MCMV + SCMV-MD-B or MCMV + MDMV-A. In Kansas and Nebraska, crop rotation with soybeans or MCMV-resistant sorghum in fields previously infected with MCMV decreases the incidence of MCMV in the following year. In Kauai, Hawaii, MCMV was essentially eliminated as a detectable disease by completely destroying all maize on the island and not replanting for 6 months. This was followed by including a 90 day fallow period each year for 2 years, and has continued with a 60 day fallow period every year.

Evolution

Based on genome organization and protein sequence similarities, MCMV is most closely related to PMV and vice versa. MCMV CP and the putative viral replicase proteins are markedly most similar to the PMV equivalents, and p7b is most similar to the 6.6 kDa peptide encoded by PMV. Both MCMV and PMV encode four overlapping ORFs on large sgRNAs that have similar start codon characteristics; the second ORFs begin with noncanonical start codons, the CP ORFs begin with the second AUGs, and a fourth ORF overlaps the CP ORFs. Additionally, the host range for both viruses is restricted to monocots. Since there is no similarity in the proteins encoded by the ORFs overlapping the CP ORFs, it suggests that MCMV and PMV evolved from a virus lacking those ORFs and the 5′ ORF encoding p32. The origins of the p32 ORF and p31 readthrough region are unknown. MCMV and the other members of the family *Tombusviridae* belong to virus supergroup II based on their viral replicases along with umbraviruses, a subgroup of luteoviruses, hepatitis C virus, pestiviruses, flaviviruses, and the positive-strand RNA coliphages.

See also: Carmovirus; Cereal Viruses: Maize/Corn; Necrovirus; Tombusviruses.

Further Reading

Batten JS and Scholthof K-BG (2004) Genus *Panicovirus*. In: Lapierre H and Signoret PA (eds.) *Viruses and Virus Diseases of Poaceae (Gramineae)* pp. 411–412. Paris: INRA.

Castillo J and Hebert TT (1974) A new virus disease of maize in Peru. *Fitopatologia* 9: 79–84.

Chiemsombat P, Larprom A, See-Tou W, and Patarapuwadol S (2006) Mixed infection of maize chlorotic mottle virus and sugarcane mosaic virus in sweet corn. In: Pohsoong T (ed.) *Proceedings of the Second Kasetsart University Corn and Sorghum Research Program Workshop*, pp. 214–219. Nakhon Nayok: Thailand Kasetsart University.

Gordon DT, Bradfute OE, Gingery RE, Nault LR, and Uyemoto JK (1984) Maize chlorotic mottle virus. In: *CMI/AAB Description of Plant Viruses*, No. 284. Kew, UK: Commonwealth Mycological Institute.

Lommel SA, Martelli GP, Rubino L, and Russo M (2005) Genus *Machlomovirus*. In: Fauquet CM, Mayo MA, Maniloff J, Desselberger U,, and Ball LA (eds.) *Virus Taxonomy: Eighth Report of the International Committee on Taxonomy of Viruses*, pp. 932–934. San Diego, CA: Academic Press.

Scheets K (2000) Maize chlorotic mottle machlomovirus expresses its coat protein from a 1.47-kb subgenomic RNA and makes a 0.34-kb subgenomic RNA. *Virology* 267: 90–101.

Scheets K (2004) Maize chlorotic mottle. In: Lapierre H and Signoret PA (eds.) *Viruses & Virus Diseases of Poaceae (Gramineae)*, pp. 642–644. Paris: INRA.

Shafer KS (1992) *Molecular Evolution of Maize Chlorotic Mottle Virus: Isolates from Nebraska, Hawaii and Peru*. MD Thesis, University of Nebraska, Lincoln.

Maize Streak Virus

D P Martin, D N Shepherd, and E P Rybicki, University of Cape Town, Cape Town, South Africa

Glossary

Agroinfection Technique used for the infection of host plants with cloned virus genomes involving transfer of virus DNA into the nuclei of host cells by the bacterium *Agrobacterium tumefaciens*.

Agroinoculation See agroinfection.

Bicistronic Contains two protein-coding regions within a single mRNA transcript.

Capsomer A subunit of the mature virus particle containing an ordered series of polymerized coat protein molecules.

Cicadulina spp. A group of leafhopper species involved in transmission of MSV.

C-ori Complementary strand (i.e., the half of the DNA duplex that is not packaged into virus particles) origin of replication.

CP Coat protein. The only protein component of the virus particle also believed to be involved in nuclear trafficking and cell-to-cell movement of viral DNA.

LIR Long or large intergenic region containing the origin of virion strand replication and gene promoters.

MP Movement protein. A small (*c.* 10 kDa) protein believed to be involved in intercellular virus movement via plasmodesmata.

Oviposition Egg-laying. To oviposit means to lay eggs.

Plastochron The time interval between successive leaf primordia, or the attainment of a certain stage of leaf development.

Rep Replication-associated protein involved in the initiation of virion strand replication.

RepA A truncated version of Rep with a unique C-terminal domain believed to be involved in regulation of host and/or virus gene expression.

SIR Short or small intergenic region containing gene polyadenylation signals and the origin of complementary strand replication.

Viruliferous A state in which an MSV vector species is carrying and is capable of transmitting the virus.

V-ori Virion strand (i.e., the half of the DNA duplex that is packaged into virus particles) origin of replication.

Introduction

Maize streak virus (MSV) is the causal agent of maize streak disease (MSD), the most serious viral disease of maize in Africa. It is a major contributor to the continent's food security problems and is endemic throughout Africa south of the Sahara. It is also found on the Indian Ocean islands of Madagascar, Mauritius, and La Réunion. There is no obvious barrier to spread of the virus outside of this region; hence it should be considered as a serious potential problem for other as yet unaffected maize-growing areas.

History and Taxonomy

"The disorder of the mealie plant, locally described as 'Mealie Blight', 'Mealie Yellows', or 'Striped Leaf Disease', belongs to a group of plant troubles arising from obscure causes…" was how MSD was first described by Claude Fuller in 1901 in Natal, South Africa. Fuller mistakenly attributed the disease to a soil disorder, but in retrospect it is quite clear that the 'mealie (a local word for maize or corn) variegation' he described and drew in minute detail can be attributed to MSV.

The first milestone in MSD research was reached in 1924, when H. H. Storey determined that a virus obligately transmitted by leafhopper species of the genus *Cicadulina* (**Figure 1**) was the causal agent of MSD. Storey named the virus 'maize streak virus', and was also the first to determine both the genetic basis of MSV transmission by *Cicadulina mbila*, and that resistance to MSD in maize could be inherited.

When MSV particles were first purified in 1974 they were found to have a novel twinned quasi-icosahedral (geminate) shape (**Figure 2**), from which the name 'geminivirus' was derived. This was followed by the unexpected discovery in 1977 that geminivirus particles contain circular single-stranded DNA (ssDNA), a genome type never before observed in plant viruses. These novel characteristics led to the proposal of a new virus group – the geminiviruses – consisting of MSV and other viruses with geminate particle morphology and ssDNA genomes. *Maize streak virus* is now recognized as the type species of the genus *Mastrevirus*, in the family *Geminiviridae*.

Host Range and Symptoms

While most notorious for the yield losses it causes when infecting maize, MSV also infects over 80 other grass species including the economically important crops wheat, barley, and rye. In susceptible maize and grass genotypes, the virus first causes symptoms between 3 and 7 days after inoculation. These first appear as almost circular pale spots of 0.5–2 mm diameter in the lowest exposed portions of the youngest leaves. Later, fully emerged symptomatic leaves show veinal streaks from a few millimeters long to the entire length of the leaf and between 0.5 and 3 mm wide. These streaks often fuse laterally and symptomatic leaves may become >95% chlorotic (**Figure 2**).

Plants are worst affected when infected within a few days of coleoptile emergence; symptoms only develop above the site of inoculation on newly emerging leaves. Susceptible varieties may display severe stunting as well as very severe streaking, and cob development may be abolished. Yield losses can reach 100% when susceptible varieties are infected early.

Of the nine major MSV strains so far identified (designated MSV-A through MSV-I; (**Figure 3**)), only MSV-A produces economically significant infections in maize. The 'grass-adapted' MSV strains (MSV-B, -C, -D, -E, -F, -G, -H, and -I) differ from MSV-A types by 5–25% in nucleotide sequence, and produce substantially milder symptoms in maize than do MSV-A viruses and are often incapable of producing symptomatic infections in MSV-resistant maize genotypes. They may also have distinct but overlapping host ranges.

Diversity and Evolution

MSV is closely related to the other distinct 'African streak' mastreviruses, panicum streak virus, sugarcane streak virus, sugarcane streak Egypt virus, and sugarcane streak Reunion virus, with which it shares ~65% genome sequence

Figure 1 The leafhopper vector of MSV, *Cicadulina mbila* Naudé. Photograph courtesy of Dr. Benjamin Odhiambo, Kenyan Agricultural Research Institute (KARI).

Figure 2 MSD symptoms on a maize leaf: note characteristic veinal streaks. Photograph courtesy of Dr. Frederik Kloppers, PANNAR (Pty) Ltd., Greytown, KwaZulu-Natal, South Africa.

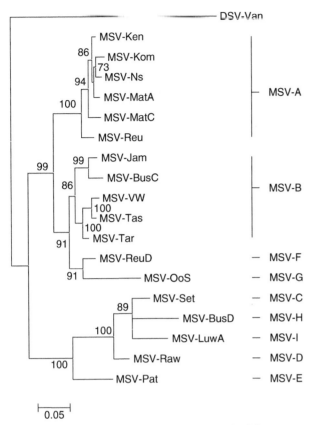

Figure 3 Phylogenetic relationships between the full genome sequences of different MSV strains. The tree is constructed using the maximum-likelihood method (HKY model transition and transversion weight determined from the data and 100 bootstrap replicates) and numbers associated with branches indicate degrees of bootstrap support for those branches. Branches with less than 70% support have been collapsed and the genome sequence of a digitaria streak virus (DSV) from Vanuatu is included as an outgroup. Only viruses in the MSV-A group have been isolated from maize. All viruses in the other groups have been isolated from wheat, barley, or wild grass species.

identity. It is, however, most similar to, although not necessarily more closely related to, an isolate of digitaria streak virus from the Pacific island of Vanuatu, with which it shares ~67% genome sequence identity (**Figure 3**).

The full genome nucleotide sequences of MSV-A isolates display relatively low degrees of diversity, with any two MSV-A isolates obtained from anywhere in Africa invariably having genome sequences that are more than 97% identical. MSV isolates from La Réunion share ~95% identity with mainland isolates. Given that maize is a crop introduced at multiple points into Africa and its neighboring islands less than 400 years ago, and that the virus is not seed-borne, this may indicate either that the rate at which MSV-A is evolving is fairly slow, or that continent-wide spread and dominance of new MSV genotypes with enhanced fitness is very rapid. Experimental assessments of the rates at which the MSV-A-type isolate MSV-Reu evolves when maintained in a susceptible maize

genotype, a resistant maize genotype, and a non-maize host (*Coix lacryma-jobi*) are, respectively, 9.5×10^{-5}, 17.3×10^{-5}, and 26.5×10^{-5} nucleotide substitutions per site per year. These evolutionary rates are relatively low when compared with RNA viruses and imply that genome-wide only one nucleotide becomes fixed every ~1.5–4 years, depending on the selection pressures exerted by the host species. Despite these low nucleotide fixation rates, it has been demonstrated that certain artificially induced mutations will revert to wild-type states at an unexpectedly high rate of approximately 36 substitutions per site per year. This combination of a low evolution rate and high mutation rate implies that MSV is currently occupying a fitness peak in maize. Despite the virus' apparent evolutionary sluggishness it may therefore be able to adapt very quickly and break any inbred or transgenic MSV resistance traits it is exposed to. In any case, the fact that there is only a very narrow range of sequence variants that can cause severe disease in maize over the whole geographical range of the virus, compared to the wide range of sequence variation in viruses adapted to other grasses, indicates that there is a high degree of sequence selection by the host genotype.

Transmission

In nature, MSV and other African streak viruses are neither seed nor contact transmissible and rely instead on transmission by cicadellid leafhoppers in the genus *Cicadulina* (including among others *C. mbila*, *C. storeyi*, *C. bipunctella zeae*, *C. latens*, and *C. parazeae*). Of these *C. mbila* is considered the most important MSV vector as it is the most widely distributed. Also, a greater proportion of *C. mbila* individuals are capable of transmitting the virus than is found with other *Cicadulina* species. The virus may be acquired by leafhoppers at any developmental stage in less than 1 h of feeding with a minimum acquisition time of 15 s. A latent period within the vector during which the virus cannot be transmitted lasts between 12 and 30 h at 30 °C. Once this latent period is over (signaled by the appearance of virus within the leafhopper's body fluids) the virus can be transmitted within 5 min of feeding. The so-called viruliferous leafhoppers can then transmit the virus for the rest of their lives.

Particle Structure

MSV particles (**Figure 4**) consist of two incomplete icosohedra with a $T = 1$ surface lattice, comprising 22 pentameric capsomers each containing five coat protein (CP) molecules. Particle dimensions are 38 nm × 22 nm with the 110 CP molecules in each virion packaging a ~2690 nt covalently closed mostly single-stranded circular DNA genome. The packaged DNA has annealed to it

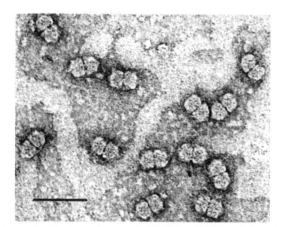

Figure 4 Electron micrograph of MSV purified from infected maize, showing particles of 18 nm × 30 nm stained with uranyl acetate. Scale = 50 nm. Photograph courtesy of Kassie Kasdorf; copyright EP Rybicki.

a complementary ∼80 nt sequence believed to act as a primer for complementary strand synthesis following infection and uncoating.

Genome Organization

As with all other mastreviruses discovered to date, the MSV genome encodes four proteins: a movement protein (MP) and a CP in the virion sense, and a replication-associated protein (Rep) and a regulatory protein (RepA) that is expressed from the same transcript as Rep in the complementary sense. MSV genomes also contain two intergenic regions: these are a short or small one (SIR), and a long or large one (LIR), which are the complementary sense strand and virion sense strand origins of replication, respectively (**Figure 5**).

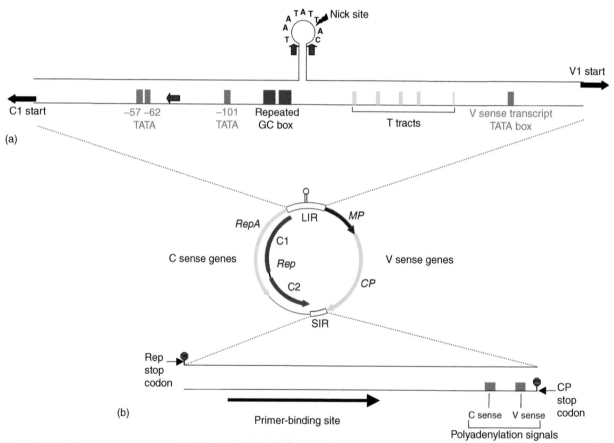

Figure 5 A schematic representation of the MSV LIR (a) and SIR (b), shown in context with the MSV genome. In (a) the main features of the MSV LIR are shown. These include a stem–loop structure with the loop's nonanucleotide sequence conserved among all geminiviruses and other rolling-circle systems. The site at which Rep introduces an endonucleolytic nick to initiate virion strand replication is shown. Iterated sequences (iterons) are shown in the V sense, with blue arrows indicating their location in the LIR. Iterons are potentially specific Rep-recognition sequences via which Rep may bind to the LIR. 5′ of the stem–loop is a repeated GC-box, which binds host transcription factors. A series of T tracts 3′ of the stem–loop may be involved in DNA bending of this region of the LIR. TATA boxes 5′ and 3′ of the stem–loop are potential C sense and V sense transcription initiation sites, respectively. In (b), the main features of the SIR include polyadenylation signals for V and C sense transcripts, and a primer-binding site on the plus strand. An ∼80 bp DNA primer-like molecule, encapsidated with the viral genome and annealed to this site, is thought to be involved in initiating complementary strand replication. Both the MSV LIR and SIR are essential for viral replication.

The Long Intergenic Region

Besides containing divergent RNA polymerase II-type promoters and other transcriptional regulatory features necessary for the expression of the complementary and virion sense genes, the LIR also contains sequence elements that are essential for replication. The most striking of these is an inverted repeat sequence that is capable of forming a stable hairpin loop structure. All geminiviruses sequenced to date have the highly conserved nonanucleotide sequence TAATATT↓AC within the loop sequence of similar hairpin structures: this sequence contains the virion sense strand origin of replication (V-ori;↓).

A sequence 6–12 nt long occurring in all known mastreviruses between the TATA box that directs rep/repA transcription and the repA initiation codon is directly repeated in the stem near the V-ori hairpin, and is probably involved in Rep and/or RepA binding during replication.

The hairpin and two GC boxes on the 5' side of the stem also forms part of an upstream activator sequence (UAS) required for efficient CP expression. The GC boxes bind nuclear factors to the UAS and are known as the rightward promoter element (rpe1).

The Short Intergenic Region

The MSV SIR occurs between the termination codons of the CP and rep genes (**Figure 5**) and contains the polyadenylation and termination signals of the virion and complementary sense transcripts. The SIR also contains the origin of complementary strand synthesis (C-ori). A small 80-nt-long primer-like molecule is bound to the SIR of encapsidated MSV DNA and, at the onset of an infection, probably enables synthesis of double-stranded DNA (dsDNA) replicative forms (RFs) of the genome from newly uncoated virion strand DNA.

The Complementary Sense Genes (rep and repA)

The replication initiator/associated protein (Rep) is the only MSV gene product that is absolutely required for virus replication. In mastreviruses, repA (the C1 open reading frame (ORF)) and the C2 ORF (also called repB), respectively, encode the N- and C-terminal portions of Rep (**Figure 5**). Beginning at the same transcription initiation site, two C sense transcripts (1.5 and 1.2 kbp in size) are produced during MSV infections. Splicing of the larger transcript removes an intron, which permits expression of full-length Rep from the two ORFs.

It is very probable, although as yet unproven in vivo, that RepA is translated from both the unspliced 1.5 and 1.2 kbp C sense transcripts. If expressed in infected cells, MSV RepA would have the same N-terminal 214-amino-acid sequence

as Rep, but would have a different C-terminus. RepA is possibly a multifunctional protein that modifies the nuclear environment to favor viral replication.

The N-terminal portions of Rep and RepA contain three conserved amino acid sequence motifs commonly found in replication-associated proteins of many extremely diverse rolling-circle replicons. Other significant landmarks include a plant retinoblastoma-related protein (pRBR) interaction motif, via which RepA but not Rep binds to the host pRBR which is involved in host cell cycle regulation, and oligomerization domains via which Rep and RepA bind to other Rep/RepA molecules (Rep activities are influenced by the aggregation state of Rep and/or RepA homo- and heterooligomers). It is also likely that the approximately 100 N-terminal amino acids of both Rep and RepA are involved in binding of the proteins to the viral LIR.

The C-terminal portion of Rep contains a dNTP-binding domain with motifs similar to those found in proteins with kinase and helicase activities. The dNTP-binding domain also sits within a region with similarity to the DNA-binding domains of the myb-related class of plant transcription factors: this domain may be functional in the induction of virus and/or host gene transcription.

The C-terminal portion of RepA, which is different from that of Rep, contains another potential transactivation domain also possibly involved in regulation of virus and/or host gene expression. A second domain within the C-terminal portion of RepA possibly interacts with host proteins involved in developmental regulation.

The Virion Sense Genes (MP and CP)

Transcription of the MSV virion (V) sense genes is directed by two TATA boxes within the LIR 26 and 214 nucleotides 5' of the MP start codon. Each TATA box directs the production of different-sized transcripts, both of which terminate at the same place. Splicing of an intron within the MP portion of V sense transcripts appears to be an important determinant of relative MP and CP expression levels. Whereas CP is expressed from both long and short, spliced and unspliced V sense transcripts, MP is most likely only expressed from unspliced long transcripts.

The MP is post-translationally modified and contains a hydrophobic domain that may either facilitate its interaction with host cell membranes or be involved in homo- or heterooligomerization with the CP.

The N-terminal ~100 amino acids of the CP contain both nuclear localization signals and a sequence-nonspecific dsDNA- and ssDNA-binding domain. The CP and MP interact with one another and it is possible that this interaction is involved in trafficking of naked and/or packaged virus DNA from nuclei through nuclear pores, to the cell periphery, through plasmodesmata, and into the nuclei of neighboring cells.

Molecular Biology of MSV

Replication

As with other geminiviruses, MSV replicates by a rolling-circle mechanism (rolling-circle replication, or RCR; (**Figure 6**)) within the nuclei of infected cells. Replication may also occur by recombination-dependent mechanisms but this has not been conclusively demonstrated for MSV. As with other rolling-circle replicons, MSV replication is discontinuous with virion strand replication being initiated from the hairpin structure in the LIR and complementary strand synthesis being initiated from a short 80 nt primer-like molecule synthesized on the SIR of newly replicated virion strands.

Particle Assembly and Movement

Besides being the primary location of replication, the nucleus is also the site of virus particle assembly. CP molecules in the nucleus nonspecifically bind virion strands released during RCR (there is no known encapsidation signal in mastrevirus genomes), arresting the synthesis of new RF DNAs. Viral ssDNA molecules are packaged into particles that aggregate to form large paracrystalline nuclear inclusions. Crystalline arrays of MSV particles have also been detected outside nuclei within physiologically active phloem companion cells, and inside the vacuoles of dead and dying cells within chlorotic lesions. These lesions are caused by an as yet unexplained degeneration of chloroplasts in infected cells.

The mechanistic details of MSV cell-to-cell movement are still obscure, but it seems to involve an interaction between the CP, MP, and viral DNA. Besides requiring the coordinated interactions of viral gene products and DNA, the successful movement of MSV genomes from infected to uninfected cells is strongly dependent on the extent of plasmodesmatal connections between neighboring cells. Also, in maize it appears as though certain cell types are more sensitive to MSV infection than others. For example, in maize leaves the virus infects all photosynthetic cell types (e.g., mesophyll and bundle sheath cells) but despite abundant plasmodesmatal connections between photosynthetic, epidermal, and parenchyma cells, MSV is only rarely detectable in the latter two cell types.

It is unknown whether systemic movement of geminiviruses within plants simply relies on normal cell-to-cell movement to deliver genomic DNA into the phloem, or whether viral DNA is specifically packaged for long-distance transport. It is possible that cell-to-cell movement might involve unencapsidated ss- or dsDNA but that long-distance movement in the phloem might require encapsidation.

Long-distance movement of MSV within infected plants occurs via phloem elements and it is believed that MSV is incapable of invading the root apical, shoot apical, and reproductive meristems due to the absence of developed vasculature in these tissues. Thus, the virus is not found in tissue which develops into gametes and is not seed-borne. It also does not appear to travel down plants from the site of inoculation to older tissue.

Within the shoot apex where most productive MSV replication occurs, MSV first enters developing leaves at approximately plastochron five. While the virus is restricted to the developing leaf vasculature before plastochron 12, it is likely that the development of metaphloem elements at approximately plastochron 12 provides an opportunity for the virus to escape the vasculature into the photosynthetic cells of the leaf. Metaphloem develops with the abundant plasmodesmatal connections required for efficient loading of photoassimilates once the leaf emerges from the whorl. Before emergence, however, the developing photosynthetic tissues are still net importers of photoassimilates and the virus most likely moves into these cells through their plasmodesmatal connections with the metaphloem.

On the leaves, the pattern of chlorotic streak-like lesions that characterizes MSV infections is directly correlated with the pattern of virus accumulation within the leaves and the virus can only be acquired by leafhoppers from these lesions. The degree of chlorosis that occurs within lesions can differ between MSV isolates and is related to the severity of chloroplast malformation that occurs in infected photosynthetic cells.

Control of MSD

Although effective control of MSD in cultivated crops is possible with the use of carbamate insecticides, and it is possible to avoid the worst of leafhopper infestations by varying planting dates, the fact that small farmers cannot generally use these options means that the development and use of MSV-resistant crop genotypes is probably the best way to minimize the impact of MSD on African agriculture. MSV resistance is associated with up to five separate alleles conferring a mixture of both dominant and recessive traits, none of which is sufficient by itself. Despite great successes achieved in the development of MSV-resistant maize genotypes that tolerate infection without significant yield loss, there has been only limited success in the field. For example, severe infections of the so-called MSV-tolerant genotypes can occur when they are grown under environmental conditions different from those in which the plants were selected, meaning each distinct geographical growing area requires specific varieties to be bred for maximum resistance. Another problem facing breeders is that natural genetic resistance is not usually associated with desirable agronomic traits such as good yield and it can therefore be difficult to transfer resistance traits without also transferring undesirable

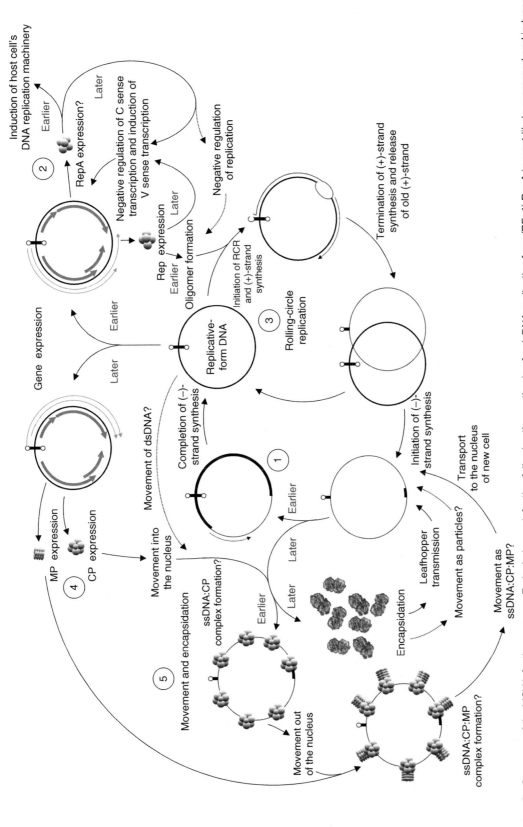

Figure 6 Summary of the MSV infection process. Early during an infection following the synthesis of a dsDNA replicative form (RF; 1) RepA is most likely expressed and induces a cellular state in which viral DNA replication can occur (2). Rep is also expressed early and RCR begins (3). At a later point in the infection process, following genome amplification and possibly Rep and/or RepA induction of the V sense promoter, MP and CP are expressed (4), and movement and encapsidation occur (5). Represented here is movement of unencapsidated ssDNA, but it should be noted that it is possible that dsDNA and/or encapsidated ssDNA may also be moved either cell to cell or systemically within the phloem of plants. Whereas the involvement of MSV CP and MP in movement has been demonstrated, the mechanics of the process are obscure. While the probable timing of events is indicated, it is unlikely, for example, that absolutely no MP and CP expression occurs during the earlier stages of the infection process. ssDNA is represented by blue lines, dsDNA by bold black lines, and RNA by orange lines.

characteristics. Moreover, the number of alleles involved means that successful breeding takes years for each release. Even in the absence of any predictive modeling of sporadic MSD outbreaks, most farmers would still prefer to gamble on the use of higher-yielding MSV-sensitive genotypes. Efforts are currently underway to introduce MSV resistance traits into commercial maize genotypes by genetic engineering. This technique has the advantage of enabling the direct transfer of single-gene resistance, without linkage to undesirable characteristics, to many different breeding lines suited to different environmental conditions. However, up till now this strategy has been limited by negative public perception of genetically modified organisms, and the expensive and time-consuming risk assessment necessary to ensure a safe feed and food product.

MSD Epidemiology

There are loose correlations between MSD incidence and both environmental conditions and agricultural practices. Environmental influences on MSD epidemiology are mostly driven by a strong correlation between rainfall and leafhopper population densities. For example, drought conditions followed by irregular rains at the beginning of growing seasons tend to be associated with severe MSD outbreaks. Also, maize planted later in the growing season tends to get more severely infected than that planted at the beginning of the season, probably due to steady increases in leafhopper numbers and inoculum sources over the course of the season. As is the case with most insect-borne virus diseases, however, the incidence of MSD is erratic. Whereas MSD can devastate maize production in some years, in others it has only a negligible effect. The reason for this is that apart from MSD epidemiology being strongly dependent on environmental variables it is also the product of extremely complex interactions between the various MSV leafhopper vector and host species, and an as yet unknown number of virus strains.

Leafhopper Vectors

Serious MSD outbreaks are absolutely governed by leafhopper acquisition and movement of severe MSV isolates from infected plants (wild grasses or crop plants) to sensitive, uninfected crop plants. The distance that MSV spreads from a source of inoculum is determined by the movement behavior of leafhoppers. Distinct long- and short-distance flight morphs have been detected among certain *Cicadulina* populations. It is believed that the long-flight morphs are a migratory form and as such they may play an important part in the rapid long-distance spread of virulent MSV variants. Migratory movement is more common in certain *Cicadulina* species than in others and it is probably influenced by environmental conditions.

The dynamics of primary infection following leafhopper invasion of a susceptible maize crop are influenced by leafhopper population densities, the proportions of viruliferous individuals in populations, and the virus titer within these individuals. Disease spread within individual maize fields is apparently linear when only a few viruliferous leafhoppers are involved in transmission, but becomes exponential once the number of insects exceeds one individual per three plants.

Plant Hosts

Although attempts to understand the dynamics of MSD epidemics have focused primarily on vector population dynamics and behavior, an important component of MSD epidemiology is the population density, turnover, and demographics of the over 80 grass species that are both MSV and vector hosts. Because *Cicadulina* species favor certain annual grass hosts for mating and oviposition, the species composition of grass populations that vary seasonally in any particular area will directly influence leafhopper population densities and feeding behaviors in that area.

The species composition and age distribution of grasses (including cultivated crops) in an area may also affect the amount of MSV inoculum available for transmission in that area. While MSV infects at least 80 of the 138 grass species that leafhoppers feed on, both the susceptibility of these grasses to MSV infection and the severity of symptoms that occur following their infection may be strongly influenced by a number of factors. While sensitivity to infection can vary substantially from species to species, it can also vary within a species with genotype and plant age at the time of inoculation: for example, plants from many species, including maize, generally become more resistant to MSV infection with age, thereby reducing the inoculum available for transmission to other plants.

The Virus

While efforts are underway to promote the widespread cultivation of MSV-resistant maize in Africa, surprisingly little is known about the MSV populations that will confront these new genotypes. Although to date nine major MSV strain groupings have been discovered, it is unknown whether any other than the maize-adapted MSV-A strain play an important role in the epidemiology of MSD. MSV-B, -C, -D, and -E isolates only produce very mild symptoms in MSV-sensitive maize genotypes and are therefore unlikely to pose any significant direct threat to maize production. Mixed MSV-A and -B infections have, however, been detected in nature and there is also strong evidence of recombination occurring between these strains. It is therefore possible that MSV-B, and possibly other MSV strains, may indirectly influence

MSD epidemiology through recombination with MSV-A-type viruses. Recombination has been linked with the emergence of a number of geminivirus diseases and it is quite conceivable that it may have already contributed to the emergence of MSV and may also eventually contribute to the evolution of MSV genotypes with elevated virulence in resistant maize varieties.

It seems highly probable that MSV isolates, strains, and even distinct MSV-related mastreviruses travel with the leafhoppers as a 'swarm' of virus types through a variety of grasses, both perennial and annual, each of which has a virus genotype or group of genotypes most suited to replication in it. Thus, the dominant virus in any particular host at any one time will probably be different, but the swarm diversity is preserved – in part because the dominant type in any case may facilitate the replication of other, less fit virus types in any one host. Consequently, it is possible to see very different dominant viruses in maize and wheat grown in consecutive summer and winter growing seasons in the same field, as a result of independent selection by the host genotype of the most fit virus genotype.

Future Threat

MSV is rightly regarded as a significant potential threat to maize production outside of Africa: while the vectors do not occur outside of the current range, there is no obvious reason that they would not survive in other, climatically similar areas such as the southern USA and South America and Eurasia, and it is a distinct possibility that they could inadvertently spread or be deliberately taken there. If one or more vector species did become established, and were viruliferous, spread of MSV and its relatives into native grasses and cultivated maize and other cereals would be inevitable. It is worth noting here that as none of the maize varieties grown outside of Africa has any but the weakest resistance to MSV, the probability of severe economic consequences would be very high.

See also: Plant Resistance to Viruses: Geminiviruses.

Further Reading

Bosque-Pérez NA (2000) Eight decades of maize streak virus research. *Virus Research* 71: 107–121.

Boulton MI (2002) Functions and interactions of mastrevirus gene products. *Physiological and Molecular Plant Pathology* 60: 243–255.

Damsteegt V (1983) Maize streak virus. Part I: Host range and vulnerability of maize germ plasm. *Plant Disease* 67: 734–737.

Efron Y, Kim SK, Fajemisin JM, *et al.* (1989) Breeding for resistance to maize streak virus – A multidisciplinary team-approach. *Plant Breeding* 103: 1–36.

Fuller C (1901) *First Report of the Government Entomologist, Natal, 1899–1900.* http://www.mcb.uct.ac.za//msv/fuller.htm (accessed February 2008).

Harrison BD, Barker I, Bock K, *et al.* (1977) Plant viruses with circular single-stranded DNA. *Nature* 270: 760.

Martin DP, Willment J, Billharz R, *et al.* (2001) Sequence diversity and virulence in *Zea mays* of maize streak virus isolates. *Virology* 288: 247.

McLean AP (1947) Some forms of streak virus occurring in maize, sugarcane and wild grasses. *Science Bulletin of Department of Agriculture for Union of South Africa* 265: 1–39.

Palmer KE and Rybicki EP (1998) The molecular biology of mastreviruses. *Advances in Virus Research* 50: 183–234.

Storey HH (1925) The transmission of streak disease of maize by the leafhopper. *Balclutha mbila naudé. Annals of Applied Biology* 12: 422–443.

Zhang W, Olson NH, Baker TS, *et al.* (2001) Structure of the maize streak virus germinate particle. *Virology* 279: 471–477.

Relevant Website

http://www.mcb.uct.ac.za – The Maize Streak Virus Home Page at Online Resources of the University of Cape Town Department of Molecular and Cell Biology.

Mungbean Yellow Mosaic Viruses

V G Malathi and P John, Indian Agricultural Research Institute, New Delhi, India

Glossary

Agroinoculation Delivery of the viral genome through *Agrobacterium* inoculation.

Binary vector In these systems, the T-DNA region containing a gene of interest is contained in one vector and the *vir* region is located in a separate disarmed Ti plasmid. The plasmids co-reside in *Agrobacterium* and remain independent.

Nuclear localization signal Arrangement of basic residues like arginine or lysine in a protein that facilitates entry through nuclear pores.

Replicon Autonomously replicating DNA segments having an independent replication origin.
Rolling circle replication A mechanism for copying single-stranded circular genome by means of double-stranded intermediates.

History

The mungbean yellow mosaic virus (MYMV) and mungbean yellow mosaic India virus (MYMIV) infect a variety of leguminous crop plants and cause devastating yellow mosaic and golden mosaic diseases. A virus disease of mungbean (*Vigna radiata* (L.) Wilczek) exhibiting bright yellow mosaic symptoms was first observed in 1950s in Delhi, India by Nariani. The causal virus was easily transmissible to selected legume hosts through the whitefly vector. Nariani recorded the virus as mungbean yellow mosaic virus in 1960. Yellow mosaic viruses were reported subsequently from several hosts either as strains of MYMV or as separate entities. The annual loss due to yellow mosaic disease caused by these two viruses in blackgram, mungbean, and soybean together is estimated to be $300 million.

Taxonomy and Classification

The two viruses belong to the species *Mungbean yellow mosaic virus* and *Mungbean yellow mosaic India virus* (genus *Begomovirus,* family *Geminiviridae*). The type species *Mungbean yellow mosaic virus* is represented by a mungbean isolate from Thailand. The type species *Mungbean yellow mosaic India virus* is represented by a blackgram isolate

from Delhi, India. The members of the family *Geminiviridae* have circular, single-stranded DNA genome of 2.5–2.7 kbp and are distinguished from other viruses by their characteristic geminate particle (about 30×20 nm in size) morphology. In the family *Geminiviridae,* the genus *Begomovirus* is differentiated, among other criteria, from the genera *Mastrevirus, Curtovirus,* and *Topocuvirus* by the fact that its members are transmitted by whiteflies (*B. tabaci* Genn) in a persistent manner.

Geographic Distribution

MYMV occurs in Thailand, Pakistan, and in western and southern states of India. MYMIV occurs in northern, eastern, and central states of India, Pakistan, Nepal, and Bangladesh. Identity of yellow mosaic viruses infecting grain legumes in Bhutan and Sri Lanka is not yet elucidated.

Symptoms in Plants

As the name implies, the characteristic symptoms caused by the infection are yellow and golden mosaic in the leaves (**Figure 1**). Symptoms start as scattered small specks or yellow spots in the leaf lamina which enlarge to irregular yellow and green patches alternating with each other on matured leaves. The yellow areas increase, coalesce resulting in complete yellowing of leaves. Plants produce fewer flowers and pods. Pod size and seed size are reduced. In some of the varieties of blackgram, necrotic mottling is seen. Sometimes the green areas are raised and the leaves show puckering and reduction in size. In French bean, affected plants show downward leaf curling and corrugated leaf lamina instead of yellow mosaic.

Figure 1   Yellow and golden mosaic symptoms in different legumes in the field: (a) mungbean, (b) blackgram, (c) soybean, and (d) cowpea.

Virion or Particle Structure

Typical geminate particles measuring $(15-18) \times 30$ nm have been observed in leaf dip preparation of plant extracts of naturally infected plants.

In double antibody sandwich enzyme-linked immunosorbent assay (ELISA) the protein preparations from MYMV- and MYMIV-infected plants reacted to polyclonal and monoclonal antibodies to Indian and African cassava mosaic viruses, although A_{405} values obtained for field-infected plants were very low. The molecular weight of the coat protein (CP) of MYMV and MYMIV is ~ 28 kDa and consists of ~ 230 amino acids. There is one α-helix (WRKPRFY) at the N-terminus and the rest of the protein contains potential β-sheets. The recombinant MYMIV CP protein lacking the first N-terminus 22 amino acids was expressed as maltose-binding and glutathione S-transferase (GST) fusion protein. The expressed CP bound preferentially to single-stranded (ss) DNA rather than double-stranded (ds) DNA. The blackgram isolate of MYMV-Vig, CP was shown to interact with nuclear import factor, α importin of plants, suggesting nuclear import of CP via α-importin-dependent pathway.

Genome Organization

The MYMV and MYMIV have a genome organization (**Figure 2**) similar to Old World bipartite begomoviruses. The genome consists of two circular ssDNA components. They are designated as DNA-A and DNA-B components which are encapsidated separately. The ~ 2745 bp DNA-A component codes for the CP and for viral DNA replication and transcription proteins. The ~ 2616 bp DNA-B encodes for proteins for movement and nuclear localization. DNA-A replicates autonomously and is dependent on DNA-B for movement functions. Replication of DNA-B

component is dependent on DNA-A-encoded proteins and both components are essential for viral pathogenicity. DNA-A has two open reading frames (ORFs) in the viral sense, (ORF AV1-CP; AV2-pre-CP) and five in the complementary sense (ORF AC1-replication initiation protein, Rep; ORF AC2-transcription activator protein, TrAP; ORF AC3-replication enhancer protein, REn; ORFs AC4 and AC5). The role of ORFs AV2, AC4, and AC5 is yet to be deciphered. The ORF AV2 is present only in Old World begomoviruses. In DNA-B, there is one ORF in viral sense strand, ORF BV1-nuclear shuttle protein (NSP), and one in complementary sense, ORF BC1-movement protein (MP). The nucleotide sequence identity between the DNA-A component of the two viruses is only 80%, justifying their recognition as separate species.

One characteristic feature of MYMV occurring in a southern state (Tamil Nadu) of India is the occurrence of multiple DNA-B components along with only one DNA-A component. Between rightward and leftward ORFs, in both the DNA components, there is a noncoding intergenic region. Intergenic region contains a characteristic stem–loop structure which is conserved in all geminiviruses. It lies 29 nt downstream of TATA box. The structure consists of GC-rich stem and an invariant nonanucleotide sequence, TAATATTAC. The Rep protein cleaves this nonanucleotide sequence to initiate replication. In the intergenic region there are 5–8 nt sequences that occur, as repeats and are called as iterons. Iterons represent Rep protein-binding sites, which cleaves the plus DNA strand to initiate replication. The Rep-binding repeat sequence was identified to be ATCGGTGT which occurs as invert and tandem repeat before the TATA box and one copy is also present after the nonanucleotide sequence in MYMIV. In MYMV the iterons are identified as GGTGTAxxGGTGT (x any nucleotide). The arrangement of iterons in MYMV and MYMIV is unique and does not show lineage to other begomoviruses. The

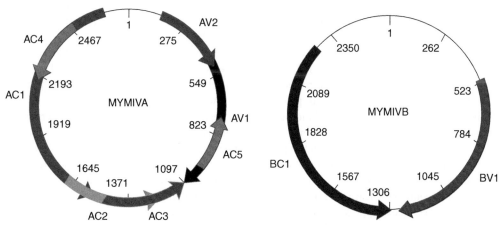

Figure 2 Genome organization of mungbean yellow mosaic India virus (MYMIV). Arrows represent open reading frames of virion and complementry sense. AV1, coat protein; AV2, pre-coat protein; AC1, replication intimation protein; AC2, transcription activator protein; AC3, replication enhancer protein; BC1, movement protein; BV1, nuclear shuttle protein.

term origin of replication (*ori*) refers to sequences from the tandem repeat of iteron, upstream of TATA box to the invariant nonanucleotide sequence. Intergenic region also contains promoters for rightward and leftward ORFs.

Within the intergenic region, there is a stretch of 180–210 nt sequence that is near identical between DNA-A and DNA-B components of bipartite begomoviruses; and this region called as common region (CR) is specific for a virus. In the case of MYMV and MYMIV, analysis of the CR of DNA-A and DNA-B components of more than 17 isolates revealed that there is considerable divergence in the whole CR in DNA-B compared to DNA-A component. The divergence in the CR is essentially in the origin of replication and ranges from 14% to 23% in the whole CR, 22% to 29% in the origin of replication among different isolates of MYMV and MYMIV.

Replication

The delicate mouth parts of the whitefly vector inject the virion particles in sieve tube cells while sucking the plant juice. From the geminate particles, ssDNA genome is released, whether the intact geminate particle or the genomic DNA enters the nucleus through the nuclear pore is debated. Once the ssDNA enters the nucleus, it is dependent on host DNA polymerase to synthesize a replicative dsDNA. It is the dsDNA which is the template both for transcription of various genes and for replication.

Investigations on geminiviruses have suggested that the viral DNA replicates in a rolling-circle mode (RCR). During RCR, Rep binds to specific iterons present in the CR and hydrolyzes the phosphodiester bond between the seventh and eighth residues of the invariant nonamer sequence 5′TAATATT↓ AC 3′ (arrow indicates site of cleavage). Rep remains bound covalently to the 5′ phosphate end and 3′ hydroxyl end thus generated becomes available for rolling-circle replication. After a full cycle of replication, the new origin is generated which is again hydrolyzed by Rep. Subsequently, Rep closes the nascent 3′ end of the DNA with the previously generated 5′ end. In this way one unit genome-length circular, ssDNA molecule, that is, the mature viral genome, is processed.

The Rep protein of MYMIV was overexpressed in *Escherichia coli*. The recombinant and refolded protein bound to CR, in a sequence-specific manner; binding of DNA-A was more efficient than DNA-B. The recombinant protein showed site-specific nicking/closing and type-1 topoisomerase activities. The cleavage function was especially upregulated by ATP, suggestive of ATP-mediated conformational changes required to cleave the nonanucleotides. A large oligomeric complex (approximately 24 mer) of Rep protein was shown to function as helicase. The recombinant Rep protein of MYMIV showed binding with recombinant pea, proliferating cell nuclear antigen (PCNA)

protein. The site-specific cleaving and closing activity and ATPase function of Rep were also impaired when bound with PCNA. There was a strong interaction between Rep and CP of MYMIV, the domain of interaction with the CP has been mapped to the central region of Rep. The activities of Rep were downregulated by the CP indicating how geminiviral DNA replication could be regulated by the CP.

Gene Expression and Transcription Regulation

The circular ssDNA genome of geminiviruses replicate through a double-stranded intermediate which is also the template for transcription for polymerase II. Transcription in these viruses is bidirectional and both viral and complementary strand encode different proteins.

Total RNA isolated from young leaves of blackgram seedlings agroinoculated with MYMV-Vig were subjected to transcript analysis by circularized reverse transcriptase-polymerase chain reaction (RT-PCR) method. The study showed that in DNA-A there are two major transcription units, one on viral sense and another one in complementary sense. Both are dicistronic, where rightward transcription unit is used for translation of ORF AV2/AV1 and leftward transcription unit for AC1/AC4. Likewise, there are two major transcripts in DNA-B, one rightward and one in leftward side, ORF BV1 and ORF BC1, are translated from these two transcripts respectively. The two major transcripts in both the DNA are driven by the promoter in the intergenic region and so the promoter is considered to be bidirectional. The promoter present between ORF AC1/AV2 is contained within 252 bp intergenic region in DNA-A. The promoter in DNA-B resides in the larger fragment of the intergenic region (957 bp) between ORFs BC1/BV1. Both the bidirectional promoters have direction-specific core elements located at an optimal distance from the transcription start site. Both rightward and leftward transcription are activated by the transcription factor AC2 encoded by the virus.

A new feature that was seen in MYMV-Vig for the first time is that the Rep protein showed synergistic activity with the AC2 protein and contributed to the activation of rightward promoters.

Besides these four transcripts, a fifth transcript of 1.4 kbp that hybridizes with ORF AC2 probe is predicted to be the transcript from which ORF AC2/AC3 will be translated. This transcript was driven by a strong monodirectional promoter upstream of ORF AC2.

Another unique feature was the transcription unit of the BC1 protein. There is a conserved leader-based intron which is not seen in any other begomoviruses. The 123 nt intron within BC1 transcript appears to have all the features of plant introns. Such an intron with consensus

5′ and 3′ splice sites (AG/GU,CAG/G) and one or more short ORFs (sORFs) appear at the same or nearby location in all the isolates of MYMV and MYMIV.

One characteristic feature observed in all transcripts except that of AC1 and BC1 is multiple transcription initiation sites, closely spaced (3–4 nt) from each other and multiple polyadenylation sites. Translation of ORFs AV1, AC4, and AC3 is predicted to be of leaky scanning type wherein the proximal 5′ AUG is in suboptimal context and is bypassed allowing the second product (AV1, AC4, and AC3) to be translated. Short ORFs present between AV2 and AV1 may also help in reinitiation of transcription of second product.

Infectivity of Cloned Components

Infectivity of blackgram isolate of MYMV, MYMV-[Vig] and blackgram, cowpea, mungbean and soybean isolates of MYMIV, MYMIV-[Bg], MYMIV-[Cp], MYMIV-[Mg], and MYMIV-[Sb] on leguminous hosts have been established by 'agroinoculation'. In this strategy dimers of DNA-A and DNA-B components are cloned in a binary vector, which is mobilized in *Agrobacterium tumefaciens*. The partial or complete tandem repeat constructs are amplified in *A. tumefaciens* in nutrient broth; the bacterial culture is used to deliver the viral inoculum. In all these cases, considerable divergence was seen in the origin of replication in DNA-B component, which did not impair their infectivity. An interesting feature was seen with the cowpea isolate; though it is not whitefly transmissible to hosts other than cowpea and French bean, through agro-inoculation the cowpea isolate was systemically infected and produced yellow mosaic symptoms in blackgram and mungbean. This adaptation to new hosts was maintained by the viral progeny isolated from agroinoculated plants, which could then be easily transmitted by whitefly to other hosts. However, the blackgram isolate of MYMIV did not produce yellow mosaic symptoms in cowpea even by agroinoculation. An improved agroinfection using one strain of *A. tumefaciens* was shown for MYMV-Vig. Partial tandem repeat construct of DNA-A and DNA-B was made in vectors having compatible replicons and introduced into the same *Agrobacterium* cells. When *Agrobacterium* cells having both constructs were inoculated, infectivity rate was 100% in blackgram.

Inoculation with different DNA-B components, KA22 and KA27 with DNA-A of MYMV-Vig, showed differences in symptoms severity. KA22 DNA-B (which is more closely related to MYMIV DNA-B) caused more intense mosaic symptoms with high viral DNA titer in blackgram. In contrast, KA27 DNA-B (closely related to DNA-B of Thailand isolate of MYMV) caused more intense symptoms and high viral DNA titer in mungbean. DNA-B is therefore considered as an important pathogenicity determinant of host range between blackgram and mungbean.

Host Range

The MYMV and MYMIV isolates have very narrow host range infecting only leguminous hosts. Whatever information is available on host range, they have been generated before the demarcation of isolates into two viruses. Whether differentiation into two species based on DNA-A nucleotide sequence will reflect on the biological properties need to be investigated freshly. The isolates from blackgram, mungbean, horsegram, and mothbean are easily transmissible to following leguminous species: mungbean (*V. radiata*), blackgram (*V. mungo*), mothbean (*V. aconitifolia* (Jacq) Marechal), soybean (*Glycine max*), phaemey bean (*Phaseolus lathyroides*), horsegram (*Macrotyloma uniflorum*) and black tapery bean (*P. acutifolius*), pigeonpea (*Cajanus cajan*), and French bean (*P. vulgaris*). These isolates are not transmissible to *Dolichos*. The transmission of these isolates to cowpea is inconsistent and contradictory results are reported. The cowpea isolate of MYMIV is transmissible only to cowpea (*V. unguiculata*), yard long bean (*V. unguiculata* (L.) Walp. f.sp. *sesquipedalis*), and French bean and not to any other host. The pigeonpea and soybean isolates could be transmitted to cowpea, though back transmission from cowpea to soybean or pigeonpea is inconsistent. Thus, it has been difficult to exactly assess the host range of the viruses as they are not sap transmissible and transmission by whitefly is dependent on the biotype of the vector used, its feeding preferences, and susceptibility of the genotypes tested. Weeds like *Brachiaria ramosa*, *Eclipta alba*, and *Xanthium strumarium*, and garden plant *Cosmos bipinnatus* were reported as hosts for MYMIV. Total nucleic acid extracted from above hosts showing symptoms of begomovirus infection did not hybridize with radiolabeled MYMIV DNA-B probe, indicating that the begomoviruses infecting these hosts are different. At present, MYMIV isolates are not transmissible to any other nonleguminous host.

Transmission and Virus–Vector Relationship

All isolates of MYMV and MYMIV are neither sap transmissible (except the Thailand isolate of MYMV) nor are they transmitted through seeds. MYMV and MYMIV are transmitted by whitefly, *B. tabaci* Genn, in a persistent circulatory manner. Young expanding leaves are better sources of inoculum than old leaves. A single whitefly can transmit the virus given an acquisition (AAP) and inoculation access period (IAP) of 24 h. Transmission percentage definitely increases with increased number of whiteflies. Female whiteflies are found to transmit more effectively and retain the virus for a longer period than male whiteflies. Starvation for a duration of 15 min, before acquisition and inoculation access period increases the transmission rate to 50%. Though the vector can acquire and transmit the virus immediately after an AAP and IAP of 10–15 min, the minimum period of 4–6 h is required. There is a latent period of more than 3 h inside the vector for transmission to occur. The whiteflies can retain the virus for 10 days and transmission rate decreases gradually. The period of retention may differ from one batch of insects to another batch. There is no transovarial transmission. No evidence exists till date for association of any specific biotype of the vector with the epidemic outbreak of the disease.

Ecology and Epidemiology

Bemisia tabaci is a polyphagous vector with wide host range; however, it shows strong host preferences and its feeding behavior is a major factor in deciding the active spread of the virus from one crop species to another. For example, the cowpea isolate of MYMIV does not spread from golden mosaic disease-affected cowpea plants to adjacent blackgram or mungbean fields in northern India.

Bemisia tabaci thrives best under hot and humid conditions that prevail in the tropical and subtropical area in the Indian subcontinent. The population of the vector is influenced by temperature, relative humidity, and rainfall.

Depending on vector population build-up, spread of the virus is gradual, cumulative, and is in the direction of prevalent wind. In northern India with the onset of monsoon rain (June–July), as the population of vector increases, rate of spread of virus increases. Whereas before monsoon rain *B. tabaci* populations are nonviruliferous, whiteflies may pick up the inoculum from self-perpetuating weeds harboring viruses. However, the role of any weed or crop plant specifically serving as reservoir for MYMV and MYMIV has not been ascertained. A serious outbreak of yellow mosaic disease caused by MYMV in mungbean occurred in northern Thailand in 1977, resulting in reduction in mungbean cultivation and shift in cropping pattern.

Management

The main components of management of the disease are cultivation of resistant genotypes, cultural practices to ensure weed-free or inoculum-free field, and judicious use of chemicals to reduce the vector population.

The genetics of resistance to MYMIV is understood and it appears to be governed by a single recessive gene or two complementary recessive genes in mungbean. Sources of resistance to both the viruses have been identified by rigorous screening in the field; some of the resistance sources identified are being used for breeding purposes.

Cultural practices like adjustment of sowing dates to avoid the disease, rouging the weeds and plants that get infected early in the season, inter or mixed cropping of mungbean or blackgram with nonhost plants like sorghum, pearl millet, and maize have been adopted and shown effective in bringing down the disease incidence. The vector whitefly is controlled by applying phorate, disulfoton granules @ 2 kg ha^{-1} during sowing. Spray of metasystox malathion @ 0.1% is known to effectively reduce the virus spread and disease incidence.

Biolistic delivery of RNAi constructs targeting CR sequences of MYMV-Vig isolate resulted in complete recovery from infection (68–77%) in blackgram seedlings, offering an alternative approach for the management of the disease. Possibility of using *Paecilomyces farinosus*, an endophyte of whitefly, as biocontrol agent is being explored.

Phylogenetic Relationship

Complete nucleotide sequence of DNA-A component of 28 MYMV and MYMIV isolates from Indian subcontinent and Thailand is available in the GenBank database. Nucleotide sequence of a limited number of yellow mosaic virus isolates infecting *Dolichos* and horsegram has also been completed. There are currently four different species comprising 'yellow mosaic legume viruses'. They are *Mungbean yellow mosaic virus, Mungbean yellow mosaic India virus, Horsegram yellow mosaic virus,* and *Dolichos yellow mosaic virus*. Percentage nucleotide identity for DNA-A and DNA-B component between the viruses of the different species is given in **Tables 1** and **2**. Comparison of nucleotide sequence of DNA-A component revealed 81% identity between MYMIV and MYMV isolates; both the viruses share only 60% identity with cowpea golden mosaic virus from Nigeria.

A phylogenetic tree based on complete nucleotide sequence (**Table 3**) of DNA-A component of selected

Table 1   Matrix of pairwise identity percentages of complete nucleotide sequences of DNA-A of selected begomoviruses

YMV isolates	MYMIV	MYMIV-Mg[Isl]	MYMV-Vig	MYMV	DoYMV	HgYMV
MYMIV	*	95.8	80.3	80.0	60.5	80.9
MYMIV-Mg[Isl]		*	81.3	81.0	61.7	82.3
MYMV-Vig			*	96.7	62.4	84.1
MYMV				*	62.4	84.0
DoYMV					*	62.0
HgYMV						*

Acronyms are as given in text; '*' indicates homologous comparisons. Matrix was generated using the CLUSTAL algorithms.

Table 2 Matrix of pairwise identity percentages of complete nucleotide sequences of DNA-B of selected begomoviruses

YMV isolates	MYMIV	MYMV-Vig	MYMV-Vig [Mad KA 21]	MYMV-Vig [Mad KA 27]	MYMV-Vig [Mad KA 28]	MYMV-Vig [Mad KA 34]	MYMV	HgYMV
MYMIV	*	91.8	92.3	69.6	92.4	92.0	69.5	66.9
MYMV-Vig		*	96.8	69.7	96.6	98.9	86.3	66.6
MYMV			69.5	95.0	69.8	69.4	*	65.9
MYMV-Vig [Mad KA 21			*	69.7	98.2	96.9		67.1
MYMV-Vig [Mad KA 27]				*	69.7	69.7		66.6
MYMV-Vig [Mad KA 28]					*	96.7		67.5
MYMV-Vig [Mad KA 34]						*		66.7
HgYMV								*

Acronyms are as given in text; '*' indicates homologous comparisons. Matrix was generated using the CLUSTAL algorithms.

Table 3 List of viruses with their accession numbers from GenBank database used for sequence analysis and phylogenetic comparison

Virus	Acronym	Accession Number
Bean calico mosaic virus	BCaMV	AF110189
Bean dwarf mosaic virus	BDMV	M88179
Bean golden yellow mosaic virus-[Mexico]	BGYMV-[MX]	AF173555
Bhendi yellow vein mosaic virus-[301]	BYVMV-[301]	AJ002453
Bhendi yellow vein mosaic virus-[Madurai]	BYVMV-[Mad]	AF241479
Cabbage leaf curl virus	CaLCuV	U65529
Cotton leaf curl Rajasthan virus	CLCuRV	AF363011
Cowpea golden mosaic virus-[Nigeria]	CPGMV-[NG]	AF029217
Indian cassava mosaic virus-[Maharashtra]	ICMV-[Mah]	AJ314739
Maize streak virus-A[Kom] (RANDOM)	MSV-A[Kom]	AF003952
Mungbean yellow mosaic India virus	MYMIV	AF126406, AF142440
Mungbean yellow mosaic India virus-[Bangladesh]	MYMIV-[BG]	AF314145
Mungbean yellow mosaic India virus-Cowpea	MYMIV-[Cp]	AF481865, AF503580
Mungbean yellow mosaic India virus-Mungbean	MYMIV-[Mg]	AF416742, AF416741
Mungbean yellow mosaic India virus-Mungbean [Akola]	MYMIV-Mg[Akol]	AY271893, AY271894
Mungbean yellow mosaic India virus-[Soybean]	MYMIV-[Sb]	AY049772, AY049771
Mungbean yellow mosaic India virus-Mungbean [Islamabad]	MYMIV-Mg [Isl]	AY269992
Mungbean yellow mosaic India virus-Mungbean [Nepal]	MYMIV-Mg [Nep]	AY271895
Mungbean yellow mosaic India virus-[Soybean TN]	MYMIV-[SbTN]	AJ416349, AJ420331
Mungbean yellow mosaic India virus-Mungbean [Islamabad]	MYMIV-[PK;Cp]	AJ512496
Mungbean yellow mosaic India virus-[MBK-A25]	MYMIV-[MBK-A25]	AY937175, AY937196
Mungbean yellow mosaic virus-Mungbean[Haryana]	MYMV-Mg [Har]	AY271896
Mungbean yellow mosaic virus-Soybean [Islamabad]	MYMV-Sb [Isl]	AY269991
Mungbean yellow mosaic virus-Soybean [Madurai]	MYMV-Sb [Mad] MYMV-[Sb]	AJ421642, AJ867554, AJ582267
Mungbean yellow mosaic virus–Vigna	MYMV-Vig[KA22]	AJ132575, AJ132574
Mungbean yellow mosaic virus-Vigna [KA27]	MYMV-Vig [KA:27]	AF262064
Mungbean yellow mosaic virus-Vigna [Maharashtra]	MYMV-Vig [Mah]	AF314530
Mungbean yellow mosaic virus-Thailand	MYMV-Thailand	D14703, D14704
Hosegram yellow mosaic virus	HgMV-[IN:Coi]	AJ627904, AJ627905
Dolichos yellow mosaic virus	DoYMV-[IN]	AY306241
Mungbean yellow mosaic virus–Vigna [Madurai]	MYMV-[KA:34]	AJ439057
Mungbean yellow mosaic virus–Vigna[Madurai]	MYMV-[KA:28]	AJ439058
Mungbean yellow mosaic virus–Vigna [Madurai]	MYMV-[KA:21]	AJ439059
Potato yellow mosaic virus-[Venezuela]	PYMV-[VE]	D00940
Soybean crinkle leaf virus-[Japan]	SbCLV-[JR]	AB050781
Tomato golden mosaic virus-Yellow Vein	TGMV-YV	K02029
Tomato leaf curl Bangalore virus-[Ban5]	ToLCBV-[Ban5]	AF295401
Tomato leaf curl New Delhi virus-Severe	ToLCNDV-Svr	U15015
Tomato yellow leaf curl Thailand virus-[2]	TYLCTHV-[2]	AF141922

begomoviruses revealed their distinct lineages. MYMV and MYMIV showed a clear dichotomy. Three viruses, MYMV, MYMIV, and DoYMV, occupy three branches separately (**Figure 3**).

The topology of the phylogenetic tree based on nucleotide sequence of DNA-B (**Figure 4**) component is different from DNA-A. There are three major branches: one major branch comprising DNA-B of MYMIV isolates and (four) DNA-B found associated with MYMV-Vig DNA-A. The second branch consisting of one isolate each of soybean (MYMV-Sb [Mad]B1) and blackgram (MYMV-Vig[Mad]KA27) from southern India and one mungbean isolate from Thailand (MYMV); the third branch is highly divergent from others consisting of one soybean isolate of MYMV-Sb[Mad]B2 and HgYMV. Recombination events have been identified in a mungbean

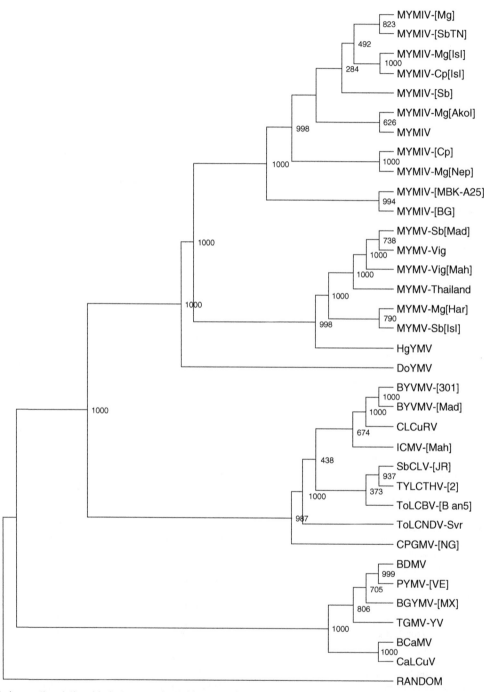

Figure 3 Phylogenetic relationship between selected begomovirus isolates based on complete nucleotide sequence of DNA-A. Dendogram was constructed using the neighbour-joining method with bootstraping (1000 replicates) in CLUSTAL X. Vertical distances are arbitary, horizontal distances are propotional to genetic distances.

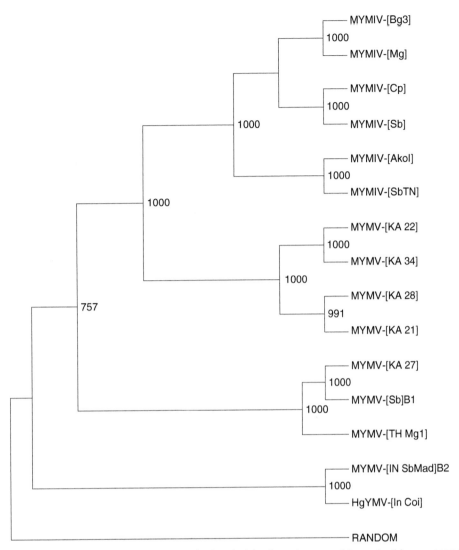

Figure 4 Phylogenetic relationship between yellow mosaic virus isolates based on complete nucleotide sequence of DNA-A dendogram was constructed using the neighbor-joining method with bootstraping (1000 replicates) in CLUSTAL X. Vertical distances are arbitary, horizontal distances are propotional to genetic distances.

isolate of MYMIV. Occurrence of multiple DNA-B components with MYMV DNA-A indicates how MYMIV DNA-B could have been captured by MYMV DNA-A during mixed infection of both MYMV and MYMIV.

A striking feature of DNA-A tree is the unique position occupied by MYMV and MYMIV. They do not cluster with typical begomoviruses of the Indian subcontinent origin, which is unusual. Begomoviruses of a specific geographical origin show high degree of conservation in CP gene, due to selection pressure exerted by the whitefly biotype. On the contrary, MYMV and MYMIV which are transmitted effectively by the same whitefly biotypes in Indian subcontinent like other begomoviruses, yet do not cluster with them. Due to their unique identity, MYMV and MYMIV are separately classified as 'legumoviruses'. They neither show host-related affinity as they are divergent from legume begomoviruses of both Old World (the

virus species *Cowpea golden mosaic virus* from Nigeria, *Soybean crinkle leaf virus* from Thailand) and New World (the virus species *Bean golden yellow mosaic virus*, *Bean dwarf mosaic virus*, and *Bean calico mosaic virus*). Probably yellow mosaic begomoviruses of legumes diverged from a common ancestor virus much earlier in evolution and might have gained their unique features while adapting to legume hosts during evolution in the Indian subcontinent.

See also: Tomato Leaf Curl Viruses from India.

Further Reading

Chaudhary NR, Malik PS, Singh DK, Islam MN, Kaliappan K, and Mukherjee SK (2006) The oligomeric Rep protein of Mungbean yellow mosaic India virus (MYMIV) is a likely replicative helicase. *Nucleic Acid Research* 34: 6362–6377.

Girish KG and Usha R (2005) Molecular characterization of two soybean infecting begomoviruses from India and evidence for recombination among legume infecting begomoviruses from Southeast Asia. *Virus Research* 108: 167–176.

Green SK and Kim D (eds.) (1992) *Mungbean Yellow Mosaic Disease. Proceedings of an International Workshop*, Bangkok, Thailand, 79pp. Taipei: AVRDC.

Malathi VG, Surendernath B, Naghma A, and Roy A (2005) Adaptation to new host shown by the cloned components of Mungbean yellow mosaic India virus causing golden mosaic in northern India. *Canadian Journal of Plant Pathology* 27: 439–447.

Malik PS, Kumar V, Bagewadi B, and Mukherjee SK (2005) Interaction between coat protein and replication initiation protein of Mungbean yellow mosaic India virus might lead to control of viral DNA replication. *Virology* 337: 273–281.

Mandal B, Varma A, and Malathi VG (1997) Systemic infection of *Vigna mungo* using the cloned DNAs of the blackgram isolate mungbean yellow mosaic geminivirus through agroinoculation and transmission of the progeny virus by whiteflies. *Journal of Phytopathology* 145: 503–510.

Nariani TK (1960) Yellow mosaic of mung (*Phaseolus aureus* L.). *Indian Phytopathology* 13: 24–29.

Nene YL (1968) A survey of the viral diseases of pulse crops in Uttar Pradesh. In: *First Annual Report, FG-In-358*, Pantnagar, India: Uttar Pradesh Agricultural University,

Shivaprasad PV, Akbergenov R, Trinks D, *et al.* (2005) Promoters, transcripts and regulatory proteins of Mungbean yellow mosaic geminivirus. *Journal of Virology* 79: 8149–8163.

Stanley J, Bisaro DM, Briddon RW, *et al.* (2005) Geminiviridae. In: Fauquet CM, Mayo MA, Maniloff J, Desselberger U,, and Ball LA (eds.) *Virus Taxonomy: Eighth Report of the International Committee on Taxonomy of Viruses*, San Diego, CA: Elsevier Academic Press.

Trinks D, Rajeswaran R, Shivaprasad PV, *et al.* (2005) Suppression of RNA silencing by a geminivirus nuclear protein AC2 correlates with transactivation of host gene. *Journal of Virology* 79: 2517–2527.

Usharani KS, Surendranath B, Haq QMR, and Malathi VG (2004) Yellow mosaic virus infecting soybean in northern India is distinct from the species infecting soybean in southern and western India. *Current Science* 86: 845–850.

Varma A, Dhar AK, and Mandal B (1992) MYMV-transmission and its control in India. In: Green SK and Kim D (eds.) *Mungbean Yellow Mosaic Disease. Proceedings of an International Workshop*, pp. 1–25. Bangkok, Thailand, Taipei: AVRDC.

Nanoviruses

H J Vetten, Federal Research Centre for Agriculture and Forestry (BBA), Brunswick, Germany

Glossary

Rep (replication initiator protein) It initiates replication only of its own DNA molecule and possesses origin-specific DNA cleavage, nucleotidyl transferase activity, and ATPase activity.

Satellite Satellites are subviral agents composed of nucleic acids; they depend for their multiplication on co-infection of a host cell with a helper virus. When a satellite encodes the coat protein in which its nucleic acid is encapsidated it is referred to as a satellite virus.

Introduction

Until the late 1980s, yellowing and dwarfing diseases of legumes and banana whose causal agents were persistently transmitted by aphids, not transmitted by sap and difficult to isolate, were generally thought to be caused by single-stranded RNA (ssRNA) viruses of the family *Luteoviridae*. In attempts to elucidate the etiology of these diseases, unusually small, icosahedral particles measuring only 18–20 nm in diameter were consistently isolated from infected plants. These particles did not contain one type of linear ssRNA but several circular ssDNA molecules, all of which were about 1 kb in size. Because of the disease symptoms (dwarfing) and the small size of the virions and genome components these viruses were referred to as nanoviruses. They differ from geminiviruses, the only other known group of ssDNA viruses of plants, in particle morphology, genome size, number and size of DNA components, genomic organization, mode of transcription, and vector species.

Meanwhile, four of these viruses, namely banana bunchy top virus (BBTV), faba bean necrotic yellows virus (FBNYV), milk vetch dwarf virus (MDV) and subterranean clover stunt virus (SCSV), have been formally described. Because of striking differences to other ssDNA viruses of plants, bacteria, and vertebrates, these viruses have recently been assigned to the family *Nanoviridae*. On the basis of differences in biology (host range, aphid vectors) and both genome size and organization, members of this family ('nanovirids') are subdivided into the genera *Nanovirus* and *Babuvirus*. Moreover, increasing evidence suggests that these viruses are not only quite variable in their biological and molecular properties but also that crops in tropical and subtropical countries of the Old World harbor further nano- and babuviruses, such as abaca bunchy top virus (ABTV) in the Philippines, cardamom bushy dwarf virus (CBDV) in India, and faba bean necrotic stunt virus (FBNSV) in Ethiopia and Morocco. BBTV occurs widely on Pacific Islands (including Hawaii), in Australia and Indochina but has an erratic geographic distribution in South Asia and Africa. SCSV and MDV have been reported only from Australia and Japan, respectively, whereas FBNYV appears to have a much wider geographic distribution (West Asia, Middle East, North and East Africa, and Spain). Nanovirids are not known to occur in the New World.

Particle Properties

Virions of the nanovirids are not enveloped, 17–20 nm in diameter, and presumably of an icosahedral T = 1 symmetry structure containing 60 subunits. Capsomeres may be evident, producing an angular or hexagonal outline (**Figure 1**). Virions are stable in Cs_2SO_4 but may not be stable in CsCl. The buoyant density of virions is about 1.24–1.30 g cm^{-3} in Cs_2SO_4, and 1.34 g cm^{-3} in CsCl. They sediment as a single component in sucrose rate-zonal and Cs_2SO_4 isopycnic density gradients.

Virions have a single capsid protein (CP) of about 19 kDa. No other proteins have been found associated with virions. Up to 12 distinct DNA components each of about 1 kb have been isolated from virion preparations of different species and their isolates. Each ssDNA component appears to be encapsidated in a separate particle.

Virions are strong immunogens. Most nanovirid species are serologically distinct from one another. However, antisera and some monoclonal antibodies (MAbs) to BBTV cross-react fairly strongly with ABTV and CBDV. On the contrary, antisera to FBNYV and SCSV cross-react weakly with SCSV and FBNYV, respectively, in Western blots and immunoelectron microscopy but not at all in double antibody sandwich enzyme-linked

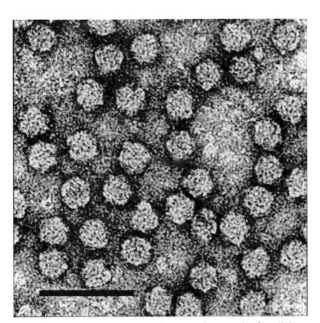

Figure 1 Negative constrast electron micrograph of particles of an isolate of *Faba bean necrotic yellows virus*. Scale = 50 nm. Courtesy of D. E. Lesemann and L. Katul.

immunosorbent assay (DAS-ELISA). However, MDV antigen reacts strongly not only with an antiserum to FBNYV but also with the majority of MAbs to FBNYV. Therefore, species-specific MAbs are required for the differentiation and specific detection of not only FBNYV and MDV (CP amino acid sequence identity of about 83%), but presumably also BBTV, ABTV, and CBDV.

Biological Properties

Economic Importance

Based on symptomatology and transmission characteristics, all diseases now known to have a nanovirid etiology were earlier suspected to be caused by luteoviruses. Unlike luteoviruses, however, nanovirids generally cause more severe symptoms. In many virus–host combinations, early infections lead to very severe effects and even premature plant death. The disease caused by BBTV is considered the most serious viral disease of banana worldwide. SCSV and FBNYV are also thought to be of great economic importance as they have caused serious diseases of subterranean clover in Australia and of faba bean in Egypt, respectively, leading to repeated crop failures.

Host Range

Individual species have narrow host ranges. FBNYV, MDV, and SCSV naturally infect a range of leguminous species, whereas BBTV has been reported only from *Musa* species and closely related species within the Musaceae, such as abaca (*M. textilis* Née) and *Ensete ventricosum* Cheesem. There are no confirmed non-*Musa* hosts of BBTV. Symptoms of BBTV include plant stunting, foliar yellowing, and most characteristic dark green streaks on the pseudostem, petioles, and leaves. All economically important natural hosts of FBNYV, MDV, and SCSV are legumes, in which these viruses generally cause plant stunting and a range of foliar symptoms, such as leaf deformations and chlorosis or reddening. Although FBNYV infects >50 legume species and only few nonlegume species (*Stellaria media*, *Amaranthus*, and *Malva* spp.) under experimental and natural conditions, major legume crops naturally infected by FBNYV are faba bean, lentil, chickpea, pea, French bean, and cowpea. Likewise, SCSV experimentally infects numerous legume species, but its economically important natural hosts include only subterranean clover, phaseolus bean, faba bean, pea, and medics. MDV is known to cause yellowing and dwarfing in Chinese milk vetch (*Astragalus sinicus* L.), a common green manure crop in Japan, as well as in faba bean, pea, and soybean.

Tissue Tropism and Means of Transmission

SCSV and FBNYV have been shown to replicate in inoculated protoplasts. All members of assigned species are restricted to the phloem tissue of their host plants and are not transmitted mechanically and through seeds. Apart from graft transmission, vector transmission had been the only means of experimentally infecting plants with nanovirids, until infectivity of purified FBNYV virions by biolistic bombardment was demonstrated.

Transmission by Aphids

Under natural conditions, all viruses are transmitted by certain aphid species, in which they can persist for many days or weeks without replicating in their vectors. Whereas only one aphid species (*Pentalonia nigronervosa*) has been reported as vector of BBTV, several aphid species transmit FBNYV, MDV, and SCSV. *Aphis craccivora* appears to be the major natural vector of these viruses as it is the most abundant aphid species on legume crops in the afflicted areas and was among the most efficient vectors under experimental conditions. Other aphid vectors of FBNYV are *Aphis fabae* and *Acyrthosiphon pisum* but in no case were *Myzus persicae* and *Aphis gossypii* able to transmit this virus. SCSV has been reported to be vectored also by *Ap. gossypii*, *M. persicae*, and *Macrosiphum euphorbiae*, but some of these accounts now appear questionable.

Transmission studies showed that aphids are able to transmit FBNYV and SCSV following short acquisition and inoculation access feeding periods of about 30 min each. Although viruliferous aphids often retain transmission ability for life, nanovirids do not multiply in their insect vectors. Together with the observation that longer acquisition and inoculation access feeding periods resulted in higher transmission rates, the strikingly long persistence of nanovirids in their insect vectors indicates that they are transmitted in a circulative persistent manner similar to that of luteoviruses.

For FBNYV it has been demonstrated that purified virions alone are not transmissible by its aphid vector, regardless of whether they are acquired from artificial diets or directly microinjected into the aphid's hemocoel. However, faba bean seedlings biolistically inoculated with intact virions or viral DNA developed symptoms typical of FBNYV infections and were efficient sources for FBNYV transmission by aphids. These observations together with results from complementation experiments suggest that FBNYV (and other nanovirids) require a virus-encoded helper factor for its vector transmission that is either dysfunctional or absent in purified virion preparations.

Genome Organization and Protein Functions

DNA Structure

Up to 12 distinct DNA components each of about 1 kb in size have been isolated from virion preparations of different nanovirid species and their isolates. The majority of

them seem to be structurally similar in being positive sense, transcribed in one direction, and containing a conserved stem–loop structure (and other conserved domains) in the noncoding region. Each coding region is preceded by a promoter sequence with a TATA box and followed by a polyadenylation signal (**Figure 2**). By analogy to the BBTV DNA-U3, -S, -C, -M, and -N, each of which has been shown to yield only one mRNA transcript, the majority of the nanovirid DNAs contain only one major gene. Only for BBTV DNA-R two mRNA transcripts were detected, a large one mapping to the major replication initiator protein (Rep)-encoding open reading frame (ORF) and a small one completely internal to the major *rep* ORF.

Viral DNA Types

The understanding of the genomic organization of nanovirids was initially complicated by the fact that several DNAs encoding different replication initiator (Rep) proteins had been found associated with BBTV, FBNYV, MDV, and SCSV isolates. However, it was soon demonstrated that only one of the Rep-encoding DNAs of each nanovirid is an integral part of its genome and required for the replication of the other DNAs that encode other types of viral proteins. In addition to being consistently associated with a nanovirid infection and capable of initiating replication of the other viral DNAs, master Rep-encoding DNAs share with the other DNAs of a nanovirid species, a highly conserved sequence encompassing the stem loop.

The genomic information of each nanovirid is distributed over at least six or eight molecules of circular ssDNA (**Table 1**). The fact that a typical set of six and eight distinct DNAs has been consistently identified from a range of geographical isolates of BBTV and FBNYV (and MDV), respectively, suggests that the babuvirus genome consists of six DNAs and the nanovirus genome of eight DNA components. DNA-R, -S, -C, -M, and -N have been identified from all four assigned species of the family *Nanoviridae* (**Table 1**). DNA-U1 is shared by all three nanovirus species but is absent from the BBTV genome. The apparent absence of DNA-U2 and -U4 in the SCSV genome appears to be due to the fact that these genome components have not been identified yet from this nanovirus. DNA-U1, -U2, and -U4 seem to be absent from the BBTV genome and specific components of the nanovirus genome. In contrast, DNA-U3 appears to be specific of the babuvirus genome and absent from the nanovirus genome. However, the question as to whether DNA-U3 and -U4, which potentially encode similar-sized proteins (∼10 kDa), are functionally distinct, remains to be determined (**Table 1**).

Integral Genome Segments

Despite the aforementioned circumstantial evidence, the number and types of ssDNA components constituting the

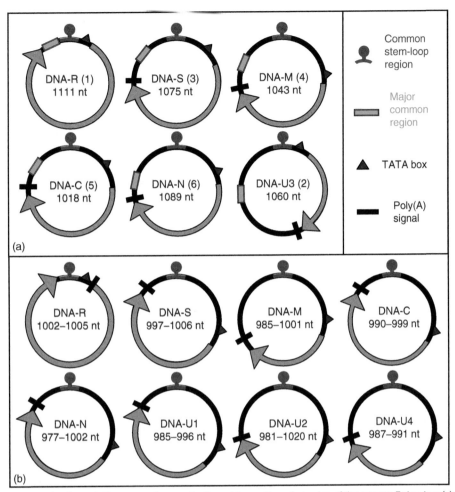

Figure 2 Diagram illustrating the putative genomic organization of the assigned species of the genera *Babuvirus* (a) and *Nanovirus* (b) and depicting the structure of the six and eight identified DNA components identified from the genomes of BBTV (a) and the three nanoviruses (b) FBNYV, MDV, and SCSV (see also **Table 1**). Each DNA circle contains its designated name and its size (range). Arrows refer to the location and approximate size of the ORFs and the direction of transcription. Note that DNA-U2 and -U4 have not been identified for SCSV.

integral parts of the nanovirid genome are still enigmatic. There has been only one (reported) attempt to use cloned DNAs to reproduce a nanovirid infection. Characteristic symptoms of FBNYV infection were obtained in faba bean, the principal natural host of FBNYV, following biolistic DNA delivery or agroinoculation with full-length clones of all eight DNAs that have been consistently detected in field samples infected with various geographical isolates of FBNYV. However, experimental infection with different combinations of fewer than these eight DNAs also led to typical FBNYV symptoms. Only five genome components, DNA-R, -S, -M, -U1, and -U2, were sufficient for inducing disease symptoms in faba bean upon agroinoculation. Symptomatic plants agroinoculated or bombarded with eight DNAs contained typical FBNYV virions; however, the virus produced after agroinoculation of cloned viral DNAs has not yet been transmitted by *Ap. craccivora* or *Ac. pisum*, two efficient aphid vectors of FBNYV.

Satellite-Like Rep DNAs

In addition to the putative genomic DNAs, a large number of additional DNAs encoding Rep proteins have been described from nanovirid infections. These DNAs are very diverse and phylogenetically distinct from the DNA-R of the nanovirids (**Figure 3**). They are structurally similar and phylogenetically closely related to nanovirid-like *rep* DNAs that have recently been found associated with some begomoviruses (e.g., ageratum yellow vein virus DNA 1 (AJ238493) and DNA 2 (AJ416153); cotton leaf curl Multan virus DNA 1 (AJ132344-5). However, due to the inclusion of an A-rich sequence within the intergenic region, the begomovirus-associated DNAs are larger (~1300 nt) than the nanovirid-associated DNAs (~930 to ~1100 nt). In contrast to the genomic DNA-R which encodes the only known Rep protein essential for the replication of the multipartite genome of the nanovirids, these additional *rep* DNAs are only capable of initiat-

Table 1   Key properties of assigned and tentative members of the genera *Babuvirus* and *Nanovirus* of the family *Nanoviridae*

	Genus *Babuvirus*			Genus *Nanovirus*			
	Assigned species	Tentative species		Assigned species			Tentative species
	BBTV	ABTV	CBDV	FBNYV	MDV	SCSV	FBNSV
Geographic distribution	Pacific Islands, Australia, southern Asia, Africa	Borneo, Philippines	India	Near East, North Africa, Ethiopia	Japan	Australia, Tasmania	Ethiopia, Morocco
Biological properties							
Major host plants	*Musa* spp.	*Musa* spp. (abaca, banana)	Large cardamom (*Amomum subulatum*)	Legumes	Legumes	Legumes	Legumes
Aphid vectors	*Pentalonia nigronervosa*	*Pentalonia nigronervosa*	*Micromyzus kalimpongensis*	*Aphis craccivora, A. fabae, Acyrthosiphon pisum*	*Aphis craccivora, Acyrthosiphon pisum*	*Aphis craccivora, A. gossypii* and other aphid spp.	*Aphis craccivora*
Virion properties							
Morphology	Icosahedral	No data	Icosahedral	Icosahedral	Icosahedral	Icosahedral	Icosahedral
Particle diameter (nm)	18–20	No data	17–20	18	18	17–19	18
Sedimentation coefficient	46 S	No data	No data	No data	No data	No data	No data
Density (g cm^{-3})	1.28	No data	No data	1.245	No data	1.24	No data
Capsid protein (kDa)	20	No data	No data	20	No data	19	20
Genome properties							
Number of components[a]	6 (9)	6	1	8 (12)	8 (12)	6 (8)	8
Component sizes (nts)	1018–1111	1013–1099	No data	985–1014	977–1022	988–1022	923–1003
Integral genome components[b]							
DNA-R (M-Rep, 33.1–33.6)[c]	+ (1)[d]	+ (1)	+ (1)	+ (2)	+ (11)	+ (8)	+ (2)
(U5, 5.0)	+ (1)	− (1)	− (1)	− (2)	− (11)	− (8)	− (2)
DNA-S (CP, 18.7–19.3)	+ (3)	+ (3)	No data	+ (5)	+ (9)	+ (5)	+ (5)
DNA-C (Clink, 19.0–19.8)	+ (5)	+ (5)	No data	+ (10)	+ (4)	+ (3)	+ (10)
DNA-M (MP, 12.7–13.7)	+ (4)	+ (4)	No data	+ (4)	+ (8)	+ (1)	+ (4)
DNA-N (NSP, 17.3–17.7)	+ (6)	+ (6)	No data	+ (8)	+ (6)	+ (4)	+ (8)
DNA-U1 (U1, 16.9–18.0)	−	−	No data	+ (3)	+ (5)	+ (7)	+ (3)
DNA-U2 (U2, 14.2–15.4)	−	−	No data	+ (6)	+ (7)	−	+ (6)
DNA-U3 (U3, 10.3)	+ (2)[e]	+ (2)[e]	No data	−	−	−	−
DNA-U4 (U4, 10 or 12.5)	−	−	No data	+ (12)	+ (12)	−	+ (12)
Additional (satellite-like)							
Rep-encoding DNAs	DNA-S1, -S2, -S3, -Y1, -W2	No data	No data	DNA-C1, -C7, -C9, and -C11	DNA-C1, -C2, -C3, and -C10	DNA-C2 and -C6	No data

[a] Numbers of identified genome components possibly forming the viral genome. Numbers in parentheses give the total number of distinct ssDNA components described from one or various isolates of each virus. For details see text.

[b] DNA(s) encoding proteins that are either functionally equivalent and/or share significant levels of sequence similarities were placed on the same line.

[c] Assigned and tentative functions of the protein encoded by the genome components is given in parentheses: master replication initiator protein (M-Rep), capsid protein (CP), cell-cycle link protein (Clink), movement protein (MP), putative nuclear shuttle protein (NSP), and proteins of unknown functions (U1–U4). The deduced molecular mass (in kDa) of the protein(s) encoded by each genome component is given in parenthesis (behind the protein designation).

[d] A plus (+) and dash (−) indicates as to whether a component (or ORF) encoding a similar protein has been identified from a virus species or not, respectively. Numbers in parentheses give the DNA-component numbering used originally by the research group studying this virus.

[e] A U3-encoding ORF has been identified only from several South Pacific isolates of BBTV, but not from South Asian isolates of BBTV and the two known ABTV strains.

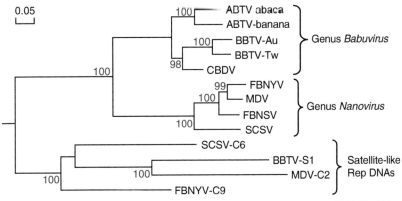

Figure 3 Neighbor-joining dendrogram illustrating the nucleotide sequence relationships in the DNA-R within and between the genera *Babuvirus* and *Nanovirus* of the family *Nanoviridae*. For comparison, four diverse representatives of the numerous satellite-like Rep-encoding DNAs frequently found associated with nanovirid infections were included to demonstrate their phylogenetic distinctness from the DNA-R of the nanovirids. DNA-R sequences used are those of members of the four assigned nanovirid species, *Banana bunchy top virus* (BBTV), *Faba bean necrotic yellows virus* (FBNYV), *Milk vetch dwarf virus* (MDV), and *Subterranean clover stunt virus* (SCSV), and the tentative species, abaca bunchy top virus (ABTV), cardamom bushy dwarf virus (CBDV), and faba bean necrotic stunt virus (FBNSV). Since some species appear to be particularly diverse, the DNA-R sequences of ABTV isolates from abaca and banana as well as BBTV isolates from Australia (Au) and Taiwan (Tw) representing the South Pacific and Asian groups, respectively, were also included in the comparison. Vertical branch lengths are arbitrary and horizontal distances are proportional to percent sequence differences. Sequence alignments and dendrograms were produced using DNAMAN (version 6, Lynnon Corporation, Quebec, Canada) which uses a CLUSTAL-type algorithm. The dendrograms were bootstrapped 1000 times (scores are shown at nodes).

ing replication of their cognate DNA but not of any heterologous genomic DNA. Since they are, moreover, only erratically associated with nanovirid infections, they are regarded as satellite-like DNAs that depend on their helper viruses for various functions, such as encapsidation, transmission, and movement.

There are no data as to whether these additional *rep* DNAs are of any biological significance to the helper virus. However, recent agroinoculation experiments with eight FBNYV DNAs or with the same eight FBNYV DNAs in combination with a satellite-like *rep* DNA (FBNYV-C11) suggest that FBNYV-C11 can reduce the number of symptomatic (42/77 vs. 21/74) and severely infected faba bean plants (16/42 vs. 2/21) and, thus, interfere with establishment of disease. This may be due to competition between the additional *rep* DNAs and the genomic nanovirid DNAs for factors required for replication, systemic movement, or encapsidation. It is also noteworthy that the protein encoded by another additional *rep* DNA (C1) of FBNYV was about ten times more active in an *in vitro* origin cleavage and nucleotidyl-transfer reaction than the master Rep protein of FBNYV. Moreover, from many nanovirid infections the additional *rep* DNAs were identified often prior to the master Rep-encoding DNA-R, suggesting that they attain higher concentrations than DNA-R in nanovirid-infected plants.

Proteins

In addition to the CP (about 19 kDa) coded for by the DNA-S transcript, at least 5–7 nonstructural proteins are encoded by the mRNA(s) transcribed from the genomic ssDNAs (**Table 1**, **Figure 2**). The large transcript from DNA-R encodes the master Rep protein (33.1–33.6 kDa). Although a second smaller transcript from the BBTV DNA-R contains a virion-sense ORF completely nested within the master Rep-encoding ORF, there is no experimental evidence that this small ORF potentially encoding a 5 kDa protein (U5) of unknown function is expressed. In addition, a small ORF similar in size and location has not been identified from the DNA-R of both nanoviruses and the tentative babuviruses ABTV and CBDV. DNA-C encodes a 19.0–19.7 kDa protein ('Clink') which contains a conserved LxCxE motif and has been shown to interact with plant proteins involved in cell-cycle regulation. A 12.7–13.7 kDa protein described from all four nanovirid species contains a stretch of 25–30 hydrophobic residues at its N terminus. This together with other experimental evidence obtained for the BBTV DNA-M-encoded protein (MP) indicates that it is involved in cell-to-cell movement of nanovirids. DNA-N encodes a 17.3–17.7 kDa protein which has been identified from all four nanovirids and proposed to act as a nuclear shuttle protein (NSP). Based on significant levels of amino acid sequence identities among some of the other nanovirid proteins, they appear to have similar but unknown functions and are provisionally referred to as U1–U4 proteins (**Table 1**).

The most conserved nanovirid proteins are the master Rep protein (54–97% identity) and NSP (41–91%), followed by the CP (20–84%), Clink (18–72%), and MP (14–76%). Consequently, the babuvirus BBTV shares significant levels of amino acid sequence identity with the nanoviruses only in the M-Rep (54–56%) and NSP (41–45%), whereas the amino acid sequence similarities between the two genera are negligible in the CP (20–27%), the MP (20–23%), and the Clink protein (18–23%). One of the least conserved proteins of the babuviruses and nanoviruses appears to be the protein U3 and U4, respectively.

Nanovirid Replication

Since the nanovirids DNAs and some of the biochemical events determined for nanovirid replication resemble those of the geminiviruses, their replication is also thought to be completely dependent on the host cell's DNA replication enzymes and to occur in the nucleus through transcriptionally and replicationally active double-stranded DNA (dsDNA) intermediates by a rolling-circle type of replication mechanism. Upon decapsidation of viral ssDNA, one of the first events is the synthesis of viral dsDNA with the aid of host DNA polymerase. As the virus DNAs have the ability to self-prime during dsDNA synthesis, it is likely that preexisting primers are used for dsDNA replicative form (RF) synthesis, as has been shown for BBTV. From these dsDNA forms, host RNA polymerase then transcribes mRNAs encoding the M-Rep and other viral proteins required for virus replication. Viral DNA replication is initiated by the M-Rep protein that interacts with common sequence signals on all the genomic DNAs. Nicking and joining within the conserved nonanucleotide sequence TAT/GTATT-AC by the Rep proteins of BBTV and FBNYV has been demonstrated *in vitro*. The nonanucleotide sequence is flanked by inverted repeat sequences with the potential to form a stem–loop structure, a common feature of every nanovirid DNA. Replication of the viral DNAs by the cellular replication machinery is enhanced by the action of Clink, a nanovirid-encoded cell-cycle modulator protein.

Relationships to Other Families of ssDNA Viruses

All Rep proteins of the nanovirid species have most of the amino acid sequence domains characteristic of Rep proteins of ssDNA viruses of other taxa, such as the families *Geminiviridae* and *Circoviridae*. However, the nanovirid Rep proteins differ from those of members of the family *Geminiviridae* in being smaller (about 33 kDa), having a slightly distinct dNTP-binding motif (GPQ/NGGEGKT), and in sharing amino acid sequence identities of only 17–22% with them. A particularly noteworthy feature of the nanovirid M-Rep protein is the lack of the Rb-binding (LxCxE) motif and, thus, the apparent absence of cell-cycle modulation functions from this protein. In contrast to some

geminiviruses, whose monopartite genome encodes a Rep protein with a conserved Rb-binding motif, the nanovirids have a separate DNA segment encoding an LxCxE-containing protein involved in cell-cycle regulation. Moreover, nanovirids are clearly distinct from geminiviruses in particle morphology (isometric vs. geminate) and dimensions (18–20 nm vs. 18 × 30 nm), genome size (6.5 or 8.0 vs. 2.6–3.0 or 5.0–5.6) and segments (6 or 8 vs. 1 or 2), mode of transcription (uni- vs. bidirectional) as well as in vector species (aphids vs. whiteflies, leafhoppers, or a treehopper). Although nanovirids and circoviruses share similar particle morphologies, the most notable differences between these two virus taxa are that the circoviruses infect vertebrates (pigs and birds) and have a monopartite genome which is only 1.8–2.0 kb in size and from which the *rep* and *cap* genes are bidirectionally transcribed. All of these viruses have a conserved nonanucleotide motif at the apex of the stem–loop sequence which is consistent with the operation of a rolling-circle model for DNA replication.

See also: Banana Bunchy Top Virus; Luteoviruses; Maize Streak Virus; Plant Resistance to Viruses: Geminiviruses.

Further Reading

Aronson MN, Meyer AD, Györgyey J, *et al.* (2000) Clink, a nanovirus encoded protein binds both pRB and SKP1. *Journal of Virology* 74: 2967–2972.

Boevink P, Chu PWG, and Keese P (1995) Sequence of subterranean clover stunt virus DNA: Affinities with the geminiviruses. *Virology* 207: 354–361.

Burns TM, Harding RM, and Dale JL (1995) The genome organization of banana bunchy top virus: Analysis of six ssDNA components. *Journal of General Virology* 76: 1471–1482.

Chu PWG and Helms K (1988) Novel virus-like particles containing circular single-stranded DNA associated with subterranean clover stunt disease. *Virology* 167: 38–49.

Chu PWG and Vetten HJ (2003) *Subterranean clover stunt virus*. AAB Descriptions of Plant Viruses, No. 396. http://www.dpvweb.net/dpv/showadpv.php?dpvno=369 (accessed June 2007).

Franz AW, van der Wilk F, Verbeek M, Dullemans AM, and van den Heuvel JF (1999) Faba bean necrotic yellows virus (genus *Nanovirus*) requires a helper factor for its aphid transmission. *Virology* 262: 210–219.

Karan M, Harding RM, and Dale JL (1994) Evidence for two groups of banana bunchy top virus isolates. *Journal of General Virology* 75: 3541–3546.

Katul L, Timchenko T, Gronenborn B, and Vetten HJ (1998) Ten distinct circular ssDNA components, four of which encode putative replication-associated proteins, are associated with the faba bean necrotic yellows virus genome. *Journal of General Virology* 79: 3101–3109.

Mandal B, Mandal S, Pun KB, and Varma A (2004) First report of the association of a nanovirus with foorkey disease of large cardamom in India. *Plant Disease* 88: 428.

Sano Y, Wada M, Hashimoto T, and Kojima M (1998) Sequences of ten circular ssDNA components associated with the milk vetch dwarf virus genome. *Journal of General Virology* 79: 3111–3118.

Timchenko T, de Kouchkovsky F, Katul L, David C, Vetten HJ, and Gronenborn B (1999) A single rep protein initiates replication of multiple genome components of faba bean necrotic yellows virus, a single-stranded DNA virus of plants. *Journal of Virology* 73: 10173–10182.

Timchenko T, Katul L, Aronson M, *et al.* (2000) Infcotivity of nanovirus DNAs: Induction of disease by cloned genome components of *Faba bean necrotic yellows virus. Journal of General Virology* 87: 1735–1743.

Timchenko T, Katul L, Sano Y, de Kouchkovsky F, Vetten HJ, and Gronenborn B (2000) The master rep concept in nanovirus replication: Identification of missing genome components and potential for natural genetic reassortment. *Virology* 274: 189–195.

Vetten HJ, Chu PWG, Dale JL, *et al.* (2005) Nanoviridae. In: Fauquet CM, Mayo MA, Maniloff J, Docselberger U, and Ball LA (eds.) *Virus Taxonomy: Eighth Report of the International Committee on Taxonomy of Viruses,* pp. 343–352. San Diego, CA: Elsevier Academic Press.

Wanitchakorn R, Hafner GJ, Harding RM, and Dale JL (2000) Functional analysis of proteins encoded by banana bunchy top virus DNA-4 to -6. *Journal of General Virology* 81: 299–306.

Necrovirus

L Rubino, Istituto di Virologia Vegetale del CNR, Bari, Italy
G P Martelli, Università degli Studi, Bari, Italy

Taxonomy and Classification

As the representative of a monotypic group, *Tobacco necrosis virus A* (TNV-A) was among the 16 groups of plant viruses described in 1971, and became the type species of the genus *Necrovirus* when it was established in 1995. Currently, *Necrovirus* is a genus in the family *Tombusviridae* and comprises seven definitive member species (i.e., *Beet black scorch virus* (BBSV), *Chenopodium necrosis virus* (ChNV), *Leek white stripe virus* (LWSV), *Olive latent virus 1* (OLV-1), *Olive mild mosaic virus* (OMMV), *Tobacco necrosis virus A* (TNV-A), and *Tobacco necrosis virus D* (TNV-D)) and two tentative species (i.e., Carnation yellow stripe virus (CYSN) and Lisianthus necrosis virus (LNV)).

Virion Properties

Necroviruses have very stable particles which resist temperatures in excess of 90 °C. TNV-A virions sediment a single component with a coefficient ($S_{20,w}$) of 118S and have buoyant density of 1.399 g ml^{-1} at equilibrium in CsCl.

Virion Structure and Composition

Virions are approximately 28 nm in diameter, have angular profile and a capsid made up of 60 copies of a trimer consisting of three chemically identical but independent protein subunits (A, B, and C) stabilized by Ca^{2+} ions, arranged in a $T = 3$ lattice. Subunit size ranges from 24–27 kDa (BBSV and LWSV, respectively) to 29–30 kDa (other viral species). The capsid has a smooth appearance as protein subunits lack the protruding domain proper of members of the majority of the other genera in the family *Tombusviridae.*

Virions encapsidate a molecule of single-stranded, positive-sense RNA, approximately 3.7 kb in size, constituting *c.* 19% of the particle weight. The genome of five definitive viral species has been fully sequenced: TNV-A, 3684 nt (accession NC001777); TNV-D, 3762 nt (NC003487, U62546); LWSV, 3662 nt (X94660), OLV-1, 3699 nt (X85989); and OMMV, 3683 nt (NC006939).

Genome Organization and Expression

The monopartite genome contains four open reading frames (ORFs) which, in the order from the 5′ to the 3′ terminus, code for replication-associated proteins (ORF1 and ORF1-RT), movement proteins (ORF2 and ORF3), and the coat protein (CP) (ORF4). Some species possess a fifth ORF either located in the 3′ terminal region (TNV-A) or in the middle of the genome (TNV-D^H and BBSV), partially overlapping the C-terminus of ORF1-RT and the N-terminal region of ORF2. The genome is very compact, having noncoding regions of limited size (**Figure 1**).

Genomic RNA acts as messenger for the translation of a protein of 22–24 kDa from ORF1. By translational readthrough of the UAG termination codon of ORF1, a protein of 82–83 kDa is synthesized (**Figure 1**), which contains, in the readthrough portion, the GDD motif of RNA-dependent RNA polymerases (RdRp). This protein, together with the expression product of ORF1, is indispensable for virus replication.

Genes downstream ORF1-RT are expressed via the synthesis of two subgenomic RNAs, 1.4–1.6 nt and 1.1–1.3 nt in size (**Figure 1**).

ORFs 2 and 3 are two small centrally located ORFs which, depending on the viral species, encode proteins of 7–8 and 6–7 kDa, respectively. Both of these proteins are

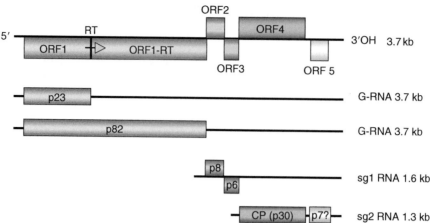

Figure 1   Genome organization and replication strategy of tobacco necrosis virus A.

involved in cell-to-cell transport but are dispensable for viral replication. LWSV ORF2 differs in size from that of other sequenced necroviruses, for it codes for a protein of 11 kDa. The additional central ORFs of TNV-D^H and BBSV code for a 7 kDa and a 5 kDa protein, respectively which, like the products of ORFs 2 and 3, are necessary for cell-to-cell movement.

ORF4 is the CP gene encoding a 24–30 kDa protein, the building block of the capsid, which is required also for efficient systemic virus spreading in the host. No experimental evidence is available of the expression of TNV-A ORF5.

Interspecific Relationships

Necroviruses are serologically distinguishable from one another. Some are unrelated (e.g., LWSV with TNV-A and TNV-D), whereas others share antigenic determinants that result in weak relationships. This is the case of TNV-A and TNV-D that are distantly related with OLV-1 (serological differentiation index from 6 to 9), and of OMMV or BBSV with certain TNV isolates.

The level of molecular similarity varies very much with the viral species. Thus, the RdRp of OMMV, a putative recombinant between TNV-D and OLV-1, shows 91.2% identity at the amino acid level with the comparable protein of OLV-1, whereas the identity level of CP is highest (86.2%) with that of TNV-D. Among the other sequenced species in the genus, the amino acid identity of RdRps and CPs is much lower, ranging from 32% to 34% and 35% to 50%, respectively.

Satellites

TNV-A and TNV-D activate the replication of a satellite virus (TNSV) present in nature as four biologically and

serologically distinct strains, supported by different viral isolates. TNSV has isometric particles *c.* 17 nm in diameter, sedimenting as a single component with a coefficient of 50S. Its capsid is constructed with 60 identical subunits *c.* 22 kDa in size, arranged in a $T = 1$ lattice, and contains a single-stranded, positive-sense RNA molecule 1239 nt in size. TNSV RNA accounts for *c.* 20% of the particle weight, and comprises a single ORF encoding the CP.

Whereas satellite viruses are not associated with any other member of the genus, BBSV supports the replication of a small, linear, single-stranded, noncoding satellite RNA 615 nt in size, which is encapsidated in the virions in monomeric or, more rarely, dimeric form. Monomers are thought to be produced from multimeric templates.

TNSV interferes to some extent with helper virus infections for its presence in TNV inocula reduces slightly the virus concentration and the size but not the number of local lesions. By contrast, BBSV satRNA enhances the aggressiveness of the helper virus for more lesions are produced on infected hosts when mixed inocula are used.

Transmission and Host Range

All necroviruses are readily transmitted by mechanical inoculation to experimental herbaceous hosts, which usually react with necrotic local lesions, not followed by systemic infection. Under natural conditions, infection is often restricted to the roots. Some members are transmitted through the soil either by the chitrid fungus *Olpidium brassicae* (TNV-A, TNV-D, BBSV), or without the apparent intervention of a vector (OLV-1). Particles of species transmitted by *O. brassicae* are acquired by the vector from the soil where they are released from roots of infected plants through sloughing off epidermal cell layers and/or following decay of plant debris. Virions are bound tightly to the plasmalemma and the axoneme (flagellum) of the fungal zoospores and are transported inside the zoospore cytoplasm when the flagellum is retracted prior to encystment that precedes penetration into the host. The same mechanism is exploited by TNSV to gain entrance into both vector and host. Since necroviruses and their satellites (TNSV and BBSV satRNA) multiply in the plant cells but not in the fungal plasmodium, zoospores released from infected roots are virus free. Transmission can therefore occur only if they come again in contact with and adsorb virus particles.

Seed transmission of necroviruses has not been reported, except for OLV-1, which was detected in the integuments and internal tissues of 82% olive seeds, and transmitted to 35% of the seedlings.

The natural host range and geographical distribution of necroviruses varies with the species. Thus, TNV-A and TNV-D are ubiquitous and infect a wide range of cultivated and wild plants. OLV-1 was recorded from olive in several Mediterranean countries, citrus in Turkey, and tulip in Japan, OMMV from olive in Portugal, BBSV from beet in China, and LWSV from leek in France.

Virus–Host Relationships

Necroviruses replicate very actively in their hosts, which translates into the production of a large number of virus particles in infected cells of all tissue types, including vessels. Virions are either scattered in the cytoplasm, gathered in bleb-like evaginations of the tonoplast into the vacuole, or arranged in crystalline arrays of various sizes. TNSV particles can also give rise to crystals that can be found in the same cells along with those of the helper virus. Cells infected by TNV and OLV-1 also contain two types of inclusions, that is, clumps of electron-dense amorphous material resembling accumulations of excess

CP and fibrous bundles made up of thin filaments with a helical structure. In OLV-1 infected cells, these bundles were identified as accumulations of the 8 kDa movement protein expressed by ORF2. The same protein and the 6 kDa movement protein coded for by ORF3 were detected by immunogold labeling near plasmodesmata. Cytoplasmic clusters of membranous vesicles with fibrillar material, derived from the endoplasmic reticulum or lining the tonoplast, were observed in cells infected by OLV-1 and LWSV.

See also: Carmovirus; Plant Virus Diseases: Ornamental Plants; Tombusviruses.

Further Reading

Castellano MA, Loconsole G, Grieco F, *et al.* (2005) Subcelluar localization and immunodetection of movement proteins of olive latent virus 1. *Archives of Virology* 150: 1369.

Gua L-H, Cao H-E, Li D-W, *et al.* (2005) Analysis of nucleotide sequences and multimeric forms of a novel satellite RNA associated with beet black scorch virus. *Journal of Virology* 79: 3664.

Lommel SA, Martelli GP, Rubino L, *et al.* (2005) Necrovirus. In: Fauquet CM, Mayo MA, Maniloff J, Desselberger U,, and Ball LA (eds.) *Virus Taxonomy: Eighth Report of the International Committee on Taxonomy of Viruses*, 926pp. San Diego, CA: Elsevier Academic Press.

Molnar H, Havelda Z, Dalmay T, *et al.* (1997) Complete nucleotide sequence of tobacco necrosis virus D^H and genes required for RNA replication and virus movement. *Journal of General Virology* 78: 1235.

Oda Y, Saeki K, Takahashu Y, *et al.* (2000) Crystal structure of tobacco necrosis virus at 2.25 Å resolution. *Journal of Molecular Biology* 300: 153.

Russo M, Burgyan J, and Martolli GP (1994) Molecular biology of Tombusviridae. *Advances in Virus Research* 44: 381.

Uyemoto JK (1981) Tobacco necrosis and satellite viruses. In: Kurstak E (ed.) *Handbook of Plant Virus Infections and Comparative Diagnosis*, 123pp. Amsterdam, The Netherlands: Elsevier/North Holland Biochemical Press.

Nepovirus

H Sanfaçon, Pacific Agri-Food Research Centre, Summerland, BC, Canada

Glossary

Odonstyle Anterior section of the stylet of a longidorid nematode. The odonstyle consists of a needle-like mouth spear used to penetrate plant root cells.

Odontophore Posterior section of the stylet of a longidorid nematode. The odontophore is located at the base of the nematode mouth. Stylet protactor muscles are attached to the ondotophore.

Triradiate lumen Posterior region of the lumen (or food channel) of the esophagus of a longidorid nematode. The triradiate lumen is located in a muscular bulb at the base of the esophagus. The radial muscles attached to the triradiate lumen are used for food ingestion.

Introduction and Historical Perspective

The genus *Nepovirus* was among the first 16 groups of viruses recognized by the International Committee on Taxonomy of Viruses (ICTV). The name stands for nematode-transmitted viruses with polyhedral particles. Although nematode transmission was one of the original defining characteristics of the genus, the primary criteria for inclusion of viruses in the genus are now the structure of the RNA genome and of the particles. As a result, not all viruses belonging to the genus *Nepovirus* are nematode transmitted (**Table 1**), and some nematode-transmitted polyhedral viruses have been reclassified in distinct genera (e.g., strawberry latent ringspot virus is now considered a tentative member of genus *Sadwavirus*).

Taxonomy and Relation to Other Viruses

The genus *Nepovirus*, along with the genera *Comovirus* and *Fabavirus*, belongs to the family *Comoviridae*. Nepoviruses are also related to the unassigned genera *Sadwavirus* and *Cheravirus* which include viruses previously considered as tentative nepoviruses. Common characteristics among members of these five genera include: small polyhedral particles, a bipartite positive-strand RNA genome, and a conserved arrangement of protein domains within the polyproteins encoded by RNA1 and -2. Nepoviruses are distinguished from viruses of the other four genera by their single large coat protein (CP). This classification is supported by phylogenetic comparisons of the RNA genomes. Nepoviruses are similar to members of the family *Picornaviridae* in that they share the same modular arrangement of replication proteins on the polyproteins, conserved motifs in these replication proteins, and a similar capsid structure. Nepoviruses have been divided into three subgroups based on the length and packaging of RNA2, sequence similarities, and serological properties (subgroups A, B, and C; see **Table 1**).

Virus Particle Structure

Nepoviruses have isometric particles of 26–30 nm in diameter with sharp hexagonal outlines (**Figure 1(a)**). Equilibrium centrifugation in CsCl of purified virus particles typically reveals the presence of three types of particles. T-particles (top component) sediment at 50S and do not contain an RNA molecule. In electron microscopy, these empty particles are penetrated by a negative stain. B-particles (bottom component) sediment at 115–134S and contain a single molecule of RNA1. In the case of subgroup A nepoviruses, B-particles can also contain two molecules of RNA2. M-particles (middle component) sediment at 86–128S and contain a single molecule of RNA2. M- and B-particles of subgroup C nepoviruses are often difficult to separate due to the larger size of RNA2.

Nepovirus particles contain 60 molecules of a single CP with an M_r of $(53–60) \times 10^3$. Tobacco ringspot virus (TRSV) (see **Table 1** for abbreviations) is the only nepovirus for which the atomic structure of the virus particle has been solved. The pseudo $T = 3$ structure was found to be very similar to that of comoviruses and picornaviruses. The single CP contains three functional domains with one β-barrel each. Each functional domain corresponds to one of the three smaller CPs found in picornaviruses. Comoviruses also share a similar structure with one large CP containing two β-barrels and one small CP with a single β-barrel. It was suggested that the CP(s) of nepoviruses, comoviruses, and picornaviruses have evolved from a common ancestor.

Genome Structure

The two RNA molecules of nepoviruses are polyadenylated at the 3′ end and are covalently linked to a small viral protein (VPg) at the 5′ end. Each RNA codes for one large polyprotein which is cleaved by a viral proteinase (Pro). RNA1 encodes replication proteins and can replicate independently of RNA2. RNA1 and -2 are both required for cell-to-cell movement of the virus. The identification of conserved sequence motifs and the characterization of cleavage sites recognized by the viral Pro have led to the definition of protein domains within the polyproteins. The genomic organization of a representative virus from each subgroup is shown in **Figure 2**. The RNA2-encoded polyprotein includes the domains for the CP and movement protein (MP) at its C-terminus as well as one (for subgroup A and B nepoviruses) or two (for ToRSV, a subgroup C nepovirus) additional protein domains at its N-terminus. The C-terminal portion of the RNA1-encoded polyprotein contains the domains for the putative helicase (also termed NTB protein because it contains a conserved nucleoside triphosphate-binding sequence motif), VPg, Pro, and RNA-dependent RNA polymerase (Pol). In the case of ToRSV, characterization of cleavage sites recognized *in vitro* by the viral Pro has resulted in the identification of two protein domains upstream of the NTB domain: X1 of unknown function and X2. X2 is a highly hydrophobic protein that shares conserved sequence motifs with the RNA1-encoded 32 kDa protein of comoviruses. An equivalent domain is present in the N-terminal region of the RNA1-encoded polyprotein of subgroup A and B nepoviruses (shown in orange in **Figure 2**). Further characterization of cleavage sites will be necessary to determine whether this region constitutes an independent protein domain in these viruses or whether it is included as part of a larger NTB protein (in the case of GFLV) or 1a protein (in the case of BRSV).

Table 1 Some properties of nepoviruses

Virus name	Abbreviation	Vector[a]	Database accession numbers			
			RNA1	RNA2	Satellite RNAs	
					Type B[b]	Type D
Subgroup A (RNA2 M_r 1.3–1.5 × 10[6])						
Arabis mosaic virus	ArMV	*Xiphinema diversicaudatum*	NC006057	NC006056	NC003523	NC001546
Arracacha virus A	AVA					
Artichoke Aegean ringspot virus	AARSV					
Cassava American latent virus	CsALV					
Grapevine fanleaf virus	GFLV	*X. index*; *X. italiae*[c]	NC003615	NC003623	NC003203	
Potato black ringspot virus	PBRSV		AJ616715			
Raspberry ringspot virus	RpRSV	*Longidorus elongatus*; *L. macrosoma*; *Paralongidorus maximus*	NC005266	NC005267		
Tobacco ringspot virus	TRSV	*X. americanum*	NC005097	NC005096		NC003889
Subgroup B (RNA2 M_r 1.4–1.6 × 10[6])						
Artichoke Italian latent virus	AILV	*L. apulus*; *L. fasciatus*		X87254		
Beet ringspot virus[d]	BRSV	*L. elongatus*	NC003693	NC003694		
Cocoa necrosis virus	CNV					
Crimson clover latent virus	CCLV					
Cycas necrotic stunt virus	CNSV		NC003791	NC003792		
Grapevine chrome mosaic virus	GCMV		NC003622	NC003621		
Mulberry ringspot virus	MRSV	*L. martini*			+	
Myrobalan latent ringspot virus	MLRSV					
Olive latent ringspot virus	OLRSV			AJ277435		
Tomato black ring virus	TBRV	*L. attenuatus*	NC004439	NC004440	NC003890	
Subgroup C (RNA2 M_r 1.9–2.2 × 10[6])						
Apricot latent ringspot virus	ALRSV			AJ278875		
Artichoke yellow ringspot virus	AYRSV					
Blackcurrant reversion virus	BRV	*Cecidophyopsis ribis* (mite)	NC003509	NC003502	NC003872	
Blueberry leaf mottle virus	BLMoV		U20622	U20621		
Cassava green mottle virus	CsGMV					
Cherry leaf roll virus	CLRV		Z34265	U24694		
Chicory yellow mottle virus	ChYMV				NC006452	NC006453
Grapevine Bulgarian latent virus	GBLV				+	
Grapevine Tunisian ringspot virus	GTRSV					
Hibiscus latent ringspot virus	HLRSV					
Lucerne Australian latent virus	LALV					
Peach rosette mosaic virus	PRMV	*X. americanum*; *L. diadecturus*[e]	AF016626			
Potato virus U	PVU					
Tomato ringspot virus	ToRSV	*X. americanum*; *X. bricolensis*; *X. californicum*; *X. rivesi*	NC003840	NC003839		

[a]Only cases for which a specific association with a nematode vector has been verified experimentally are reported here. Nonvalidated associations with nematode vectors have been reported for the following viruses: CLRV – *Xiphinema* sp., GCMV – *X. index*, PBRSV – *X. americanum*, PVU – *Longidorus* sp.

[b]The sequence accession number of satellite RNAs is given when known. In the case of MLRSV and GBLV, the + symbol indicates that a type B satellite is known to be associated with the virus but has not been sequenced.

[c]Transmission of GFLV by *X. italiae* has been reported in only one instance. Many other populations of *X. italiae* did not transmit the virus. Thus, *X. index* is considered the main vector of GFLV.

[d]BRSV was previously known as TBRV-S or TBRV-Scottish.

[e]Transmission of PRMV by *L. diadecturus* was reported for only one location in spite of the widespread distribution of the vector in North America. Thus, *X. americanum* is considered the main vector of PRMV.

In addition to the coding region, each RNA includes untranslated regions (UTRs) at its 5′ and 3′ ends. The 5′ UTR is 70–300 nt long while the 3′ UTR varies in length from 200–400 nt for subgroup A and B nepoviruses to 1300–1600 nt for subgroup C nepoviruses. A short conserved sequence is present at the immediate 5′ end of the RNAs of many but not all nepoviruses. In addition, conserved structural features have been identified in the 5′ UTRs of subgroup A and B nepovirus RNAs (i.e., a series of stem and loop structures). The presence of conserved sequences and/or structural motifs within the UTRs points to a possible role for these elements in the viral replication cycle. However, this needs to be confirmed experimentally. In addition to these short conserved motifs, RNA1 and -2 often share regions of complete or partial sequence identity in the UTRs (**Figure 2**). In subgroup B and C nepoviruses, the 3′ UTRs are identical between RNA1 and -2 while the 5′ UTRs share 68–100% sequence identity. For example, the 5′ UTRs of ToRSV RNA1 and -2 share a region of 100% sequence identity which extends into the coding region. On the other hand, the 5′ UTRs of the BRV RNAs share only 78% sequence identity and do not extend beyond the UTR. In subgroup A nepoviruses, the 5′ and 3′

UTRs of RNA1 and -2 share homology but are not identical (70–97% sequence identity). It was suggested that the extensive regions of sequence identity detected in the 5′ and 3′ ends of ToRSV RNAs are the result of recombination events occurring during replication of the viral RNAs. Experimental support for this suggestion was provided using pseudorecombinants consisting of GCMV RNA1 and BRSV RNA2. Sequencing of the viral progeny revealed that the 3′ UTR of GCMV RNA1 was transferred to BRSV RNA2 after three passages. In contrast, recombination was not readily observed between the 3′ UTRs of BLMV RNA1 and -2. It was suggested that selection rather than recombination played a role in the conservation of sequence identity in BLMV RNAs.

So far, the production of infectious cDNA clones has only been reported for subgroup A nepoviruses (GFLV, ArMV, RpRSV). Thus, reverse genetics is only possible for these viruses. Removal of the covalently linked VPg from purified viral RNAs either decreased (ArMV, RpRSV) or abolished (TRSV, ToRSV, BRSV) their infectivity, although it did not affect the ability of these RNAs to be translated *in vitro*. Possibly, the requirement for an intact VPg is an obstacle for the production of infectious clones for some nepoviruses.

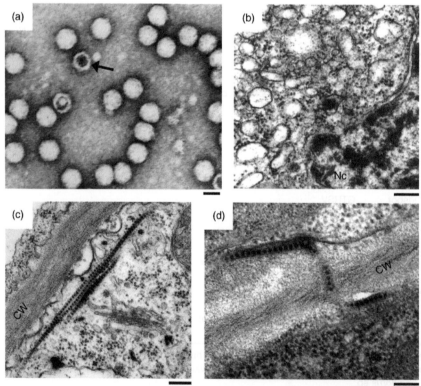

Figure 1 Electron micrograph depicting purified nepovirus particles and cytopathological structures typical of nepovirus-infected cells. (a) Purified ToRSV particles in negative staining. Note the empty particle (T-particle) which is penetrated by the negative stain (arrow). (b) Proliferation of membrane vesicles observed in the vicinity of the nucleus (Nc) in ToRSV-infected cells. (c) Tubular structures containing virus-like particles accumulating near the cell wall (CW) in PRMV-infected cells. (d) Tubular structure traversing the cell wall in ArMV-infected cells. Scale = 25 nm (a), 200 nm (b–d).

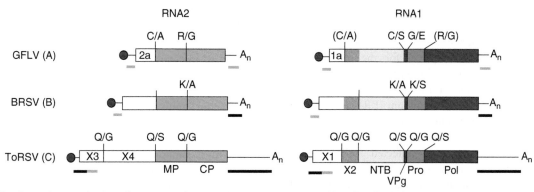

Figure 2 Genomic organization of representative nepoviruses of subgroup A (GFLV), B (BRSV), and C (ToRSV). Each RNA is represented with the covalently attached VPg (pink circle) and the polyA tail (A_n). The coding regions are represented by the boxes. Cleavage sites confirmed by *in vitro* processing experiments or by the detection of viral proteins in infected plants are indicated by the continuous vertical lines. Identified or putative (in parentheses) cleavage sites are indicated above each line when known. The function of each protein domain is indicated at the bottom of the figure. Thick bars below each RNA indicate regions with high degree of sequence identity between RNA1 and -2. The black portions indicate regions with 100% sequence identity and the gray portions represent regions with 75–83% sequence identity.

Regulated Polyprotein Processing

Nepoviruses encode a single proteinase which is responsible for processing the two polyproteins. Nepovirus proteinases are related to the 3C-Pro of picornaviruses. The catalytic triad consists of a histidine, aspartic acid, and cysteine. The proteinase also contains a substrate-binding pocket which determines its cleavage site specificity. A conserved histidine is found in the substrate-binding pocket of subgroup C nepovirus, comovirus, and picornavirus proteinases. The cleavage sites recognized by subgroup C nepovirus proteinases contain a glutamine, asparagine, or aspartate at the −1 position (**Table 2**). This is similar to the conserved glutamine or glutamate found at the −1 position of picornavirus cleavage sites. In contrast, the proteinases of subgroup A and B nepoviruses contain a leucine instead of a histidine in their substrate-binding pocket and recognize very different cleavage sites that have a lysine, cysteine, arginine, or glycine at the −1 position.

The RNA1-encoded polyprotein is cleaved predominantly intramolecularly (*cis*-cleavage) although intermolecular processing (*trans*-cleavage) of the N-terminal cleavage site has been reported for GFLV. The RNA2-encoded polyprotein is cleaved *in trans* by the proteinase. The processing cascade results in the release of mature proteins as well as stable processing intermediates containing two or more protein domains. These intermediate polyproteins accumulate in infected plant cells and may have different activities from the mature proteins. In ToRSV-infected cells, several polyprotein intermediates containing the NTB domain are detected in addition to the mature NTB protein. In BRSV-infected cells, an intermediate containing the Pro and Pol domains accumulates rather than the mature Pro and Pol proteins. This

Table 2 Identified cleavage sites in the polyprotein of nepoviruses

Virus	Cleavage site
Subgroup A	
ArMV	R/G
GFLV	R/G, C/A, C/S, G/E
RpRSV	C/A
TRSV	C/A
Subgroup B	
BRSV	K/A, K/S
GCMV	R/A
OLRSV	K/A
CNSV	K/S
Subgroup C	
BRV	D/S
BLMoV	N/S
CLRV	Q/S
ToRSV	Q/G, Q/S

suggests preferential recognition of some cleavage sites by the viral Pro. Slow release of mature proteins by processing of stable intermediates at suboptimal cleavage sites may provide a regulatory mechanism to control the accumulation of specific protein species during the replication cycle. In fact, the activity of the proteinase itself is regulated by its release from larger polyprotein precursors. Indeed, the mature proteinase of GFLV and ToRSV cleaves cleavage sites on the RNA2-encoded polyprotein more efficiently than the VPg-Pro precursor.

Viral RNA Replication

Infection of plant cells by nepoviruses results in membrane proliferation and the formation of cytoplasmic inclusion bodies which contain membrane vesicles (**Figure** 1(**b**)). The use of cerulenin, an inhibitor of *de novo* phospholipid synthesis, has confirmed that membrane proliferation is required for the replication of GFLV RNAs. Brefeldin A also inhibits GFLV replication suggesting a requirement for intact vesicle trafficking between the endoplasmic reticulum (ER) and the Golgi apparatus. The replication complex of two nepoviruses (GFLV and ToRSV) has been shown to co-localize with ER-derived membranes in infected cells. Double-stranded RNA replication intermediates, viral replication proteins, and replication activity are associated with the membrane-bound complexes. GFLV VPg antibodies were used to isolate membrane vesicles that had a rosette-like structure similar to that associated with picornavirus replication complexes.

In the case of ToRSV, the mature NTB protein or a larger intermediate polyprotein containing the NTB domain has been proposed to play a role in anchoring the replication complex to the membranes. The mature NTB and the intermediate NTB-VPg polyprotein are integral membrane proteins that co-fractionate with the replication complex. The NTB protein is targeted to ER membranes when expressed individually and contain two membrane-binding domains: a C-terminal transmembrane domain and a putative N-terminal amphipathic helix. The ToRSV X2 protein is also an ER-targeted integral membrane protein and may play a role in viral replication although its association with the replication complex in infected cells remains to be confirmed. Other replication proteins (Pro and Pol) are soluble when expressed individually but are found in association with the membrane-bound replication complex in infected cells. Thus, they are probably brought to the replication complex either as part of a larger polyprotein that includes the NTB domain or through protein–protein interaction with the viral membrane anchors. By analogy with picornaviruses, the VPg protein may act as a primer for viral replication, although this has not been demonstrated experimentally for nepoviruses. In the context of the NTB-VPg polyprotein, the VPg domain is translocated in the lumen of the membranes. Because replication presumably takes place on the cytoplasmic side of the membranes, it is unlikely that the luminally oriented VPg present in the NTB-VPg protein plays an active role in replication. Therefore, other intermediate precursors containing the VPg domain (e.g., VPg-Pro or VPg-Pro-Pol) probably act as donors for a replication-active VPg protein. In addition to RNA1-encoded replication proteins, the GFLV RNA2-encoded 2a protein is also associated with the replication complex. The 2a protein is required for the replication of RNA2 but not RNA1 and probably interacts with RNA1-encoded replication proteins.

Cell-to-Cell and Long-Distance Movement in the Plant

Nepovirus-infected cells are characterized by the presence of tubular structures containing virus-like particles in or near the cell wall (**Figures** 1(**c**) and 1(**d**)). These tubules are similar to the ones found in comovirus-infected cells and have been suggested to direct the cell-to-cell movement of intact virus particles. The viral MP is a structural component of the tubular structure. Expression of the GFLV MP alone is sufficient to induce the formation of empty tubular structures in intact plant cells or protoplasts. In the latter case, tubular extensions are found projecting from the surface of the protoplasts, a phenomenon also induced by comovirus MP. The GFLV MP is an integral membrane protein. The secretory pathway and the cytoskeleton are involved in the intracellular targeting of the GFLV MP from its site of synthesis (probably in association with the ER-bound replication complex) to specific foci in the cell wall where it assembles into tubules. The presence of specific sites of tubule formation within the cell wall suggests an interaction between MP and a cellular receptor but this has not been confirmed experimentally. By analogy with comoviruses, it is likely that an interaction between MP and CP is necessary to enable nepovirus cell-to-cell movement. Long-distance movement of TRSV occurs through the phloem resulting in the invasion of most tissues of the plant including meristematic tissues. The virus probably reaches the phloem through cell-to-cell movement from inoculated cells to phloem sieve tubes.

Host Range, Symptomatology, and Interaction of Nepoviruses with the Plant Post-Transcriptional Gene Silencing Pathway

Most nepoviruses have a wide host range that includes woody and herbaceous hosts. In nature, many hosts remain symptomless while others display symptoms such as necrotic or chlorotic rings which can appear as concentric rings (ringspots) or lines. Other symptoms can include leaf flecking and mottling, vein necrosis, plant stunting, and in some cases death. Common experimental hosts used for virus propagation are *Chenopodium quinoa* (in which most nepoviruses induce obvious symptoms), *C. amaranticolor, C. murale, Cucumis sativus, Nicotiana clevelandii, N. benthamiana, Petunia hybrida,* and *Phaseolus vulgaris*. The intensity of symptoms produced by nepovirus infection depends on the specific virus–host combination and to a large extent on environmental conditions. In many herbaceous hosts and in particular in *Nicotiana* species, symptoms develop on the inoculated leaves and on the first upper systemic leaves. Later in infection, new leaves remain free of symptoms although the virus is present. This phenomenon is termed recovery. Recovered leaves often contain somewhat reduced titer of the virus compared to symptomatic leaves and are resistant to secondary viral infection in a sequence-specific manner. This suggests that induction of the plant RNA silencing machinery plays a role in virus clearance and symptom recovery. It was recently shown that although recovery of *N. benthamiana* from necrotic symptoms induced by ToRSV is accompanied with induction of RNA silencing, the virus titer is not significantly reduced in recovered leaves. Thus, the relationship between symptom recovery and RNA silencing may be more complex than first envisaged. While many plant viruses encode potent suppressors of RNA silencing, analysis of two nepoviruses (TBRV and ToRSV) did not reveal significant silencing suppression activity.

Satellites

Two classes of satellite RNAs (satRNAs) have been found in association with some but not all nepoviruses (**Table 1**). SatRNAs depend on the helper virus for their replication and are encapsidated in virus particles. Type B satRNAs are 1100–1500 nt in length. They are linked to a VPg molecule at their 5′ end, polyadenylated at their 3′ end, and encode a nonstructural protein which is essential for their replication. The exact function of the encoded protein is not known but it has been suggested that it interacts with the viral replication complex. Sequences at the 5′ and 3′ ends of type B satRNAs are also important for their replication. The 5′ ends of type B satRNAs often have short sequence motifs that are identical or nearly identical to the 5′ ends of the viral RNAs and are likely recognized by the viral replication complex. One or several copies of type B satRNAs are packaged in the viral particles, either alone or together with one molecule of RNA2. As a general rule, type B satRNAs are replicated specifically by the virus with which they are associated although there are exceptions (e.g., the replication of GFLV satRNA is supported by ArMV). Type B satRNAs are usually found in low concentration in the field or in experimental systems and do not seem to affect significantly the replication of the helper virus or the symptomatology of the disease. For example, type B satRNAs have been found to be associated with only 15–17% of ArMV or GFLV isolates.

Type D satRNAs are less than 500 nt long. They are not linked to a VPg molecule or polyadenylated and do not encode a protein. Type D satRNAs are encapsidated as monomeric or multimeric linear molecules. A circular form of the molecule is present in infected cells and serves as the polymerase template for replication through a rolling-circle mechanism. The multimeric linear forms produced during replication are cleaved in an autocatalytic reaction which allows the release of linear monomers. These monomers are circularized to form new templates of positive or negative polarity. Type D satRNAs have been shown to either attenuate (TRSV) or intensify (ArMV) symptoms associated with the helper virus.

Diseases and Economic Considerations

Nepoviruses cause a wide range of diseases on a variety of crops including: grapevine (ArMV, GBLV, GCMV, GFLV, GTRSV, RpRSV, TBRV, ToRSV, TRSV), soft fruits such as strawberry, raspberry, blueberry, black currant, and red currant (ArMV, BLMV, BRV, CLRV, RpRSV, TBRV, ToRSV, TRSV), fruit trees such as peach, apricot, almond, cherry, plum, walnut, and apple (CLRV, MLRSV, PRMV, RpRSV, ToRSV), and horticultural crops including but not limited to hop, soybean, potato, beet, and tobacco (AILV, ArMV, AVA, BRSV, CGMV, PBRSV, and TRSV). Many nepoviruses also infect and induce diseases in ornamental species. Most nepoviruses are restricted geographically by the natural distribution of their nematode vector. An exception to this is GFLV which has been disseminated worldwide along with its vector. GFLV is the most significant nepovirus at the economic level and can reduce yield in grapevine by as much as 80%. Other nepoviruses can cause significant diseases where they occur.

Transmission

Many nepoviruses are transmitted by soil-inhabiting nematodes belonging to three closely related genera *Xiphinema*, *Longidorus*, or *Paralongidorus* in the order Dorylaimida, family Londigoridae (**Table 1**). The nematodes feed ectoparasitically on the roots using long mouth stylets. There is no evidence that nepoviruses replicate in the nematodes. The acquired viruses remain transmissible for varying periods of time (9 weeks for species *Longidorus* and up to 4 years for species *Xiphinema*), suggesting different modes of retention and release of the virus. Because of the restricted movement of the nematodes through the soil, the spread of nematode-transmitted nepoviruses through an infected field is slow and often occurs in patches. The interaction between nepoviruses and nematodes is usually specific with only one or two species of nematodes transmitting a given nepovirus. A notable exception is RpRSV which is transmitted by nematodes from the genera *Longidorus* and *Paralongidorus*. The viral determinant for the specificity of nematode transmission has been mapped to the CP in the case of GFLV. It is not known whether specific nematode receptors recognize the viral CP. Earlier studies suggested that carbohydrates may be involved in the retention of ArMV particles in its vector. However, further experiments will be required to determine if the viral CP has lectin properties. It has been suggested that pH changes may be involved in the release of viral particles from their site of retention within the nematode. Nepoviruses transmitted by *Longidorus* sp. are usually associated with the odontostyle, while viruses transmitted by *Xiphinema* sp. are found associated with the cuticle lining the lumen of the odontophore and the esophagus. Interestingly, while TRSV and ToRSV are both transmitted by *X. americanum*, immunofluorescence labeling of these viruses in the nematode vector revealed different sites of retention. TRSV is retained predominantly in the lining of the lumen of the stylet extension and the anterior esophagus, while ToRSV is localized only in the triradiate lumen.

Although the predominant vector for transmission of TRSV is a nematode, possible aerial vectors have been suggested including *Thrips tabaci* and *Epitrix hirtipennis* (flea beetle). The importance of these vectors in natural epidemics of TRSV-induced diseases needs to be confirmed. Other nematode-transmitted nepoviruses do not have known aerial vectors. BRV is transmitted by the erio-phyid gall mite of black currant (*Cecidophyopsis ribis*) and possibly other *Cecidophyopsis* species but not by nematodes. Plant-to-plant transmission can occur rapidly (in only 4 h). Virus particles have not been found inside mites, suggesting that the transmission may be nonpersistent or semipersistent. Two surface-exposed amino acid triplets are conserved between the CP of BRV and other mite-transmitted viruses from unrelated genera, suggesting that they may play a role in the interaction between the virus and its vector. However, this remains to be determined experimentally.

Seed transmission has been reported for most but not all nepoviruses. For example, BRV is apparently not seed transmitted. Infection can occur through the ovule or the pollen, although nepovirus-infected pollen may not compete effectively with healthy pollen. An exception to this is BLMoV and CLRV which are efficiently transmitted by pollen, even to the mother plant. Many nepoviruses are readily transmitted through grafting and mechanical inoculation. The propagation of infected seed and plant stocks plays an important role in the long-distance movement of nepoviruses. Because many nepoviruses have a wide host range including many common weeds, dormant weed seeds may constitute an important reservoir for the virus in the field.

Population Structures

Examination of nepovirus isolates recovered from different hosts or geographic locations has revealed a degree of sequence diversity of 2–20% at the nucleotide level.

Isolates may also differ in their serological properties. Several nepovirus isolates have arisen through recombination implying that mixed infections occur in nature. A detailed analysis of GFLV population structure in an infected vineyard demonstrated the presence of mixed infections and a high degree of recombination between the various isolates. Evidence for mixed infections was also provided from the analysis of an ArMV isolate which revealed the presence of two species of RNA2, each encoding a distinct polyprotein. Analysis of a number of CLRV isolates revealed that the degree of diversity was defined primarily by the host rather than the geographic location. This is an unusual situation among plant viruses and may reflect the fact that this virus is pollen transmitted.

Control

Nepoviruses are mainly controlled through the removal of infected plants and replanting with resistant cultivars (when available) or with virus-free, certified plant material. In the case of nematode-transmitted nepoviruses, soil can be fumigated with broad-range nematicides. However, this method is not always effective because nematode populations can occur at considerable depths in the soil (1 m or more). Further, nematicides are costly and toxic to the environment. Bait plants can be used to test for the presence of viruliferous nematodes in the soil. Careful weed control is recommended to eliminate potential reservoirs for further nepovirus infection. Since few resistant cultivars have been reported for nepoviruses, the usefulness of transgenic approaches has been investigated. Resistance to various nepoviruses has been engineered in herbaceous hosts using the coding region for the viral CP in the sense or anti-sense orientation. Other regions of the viral genome have also been used to engineer resistance to nepoviruses through the induction of RNA silencing. Resistance to GFLV has been reported in transgenic grapevines transformed with the CP coding region. The resistance was effective in a field situation heavily infected with viruliferous nematodes. It is of interest to note that the presence of susceptible and resistant GFLV-CP transgenic grapevine in the field did not increase the occurrence of recombination events in the virus population. It is likely that the next generation of transgenic lines will be aimed at increasing the efficiency of induction of sequence-specific RNA silencing using transgenes that contain only very small portions of the viral genome.

Nepovirus Research in the Future

The next phase of nepovirus research will undoubtedly address several fundamental questions. First, the function and mode of action of nepovirus proteins, in particular protein domains in the N-terminal region of the two polyproteins, require further investigation. Although a putative function has been assigned in some cases (e.g., the GFLV 2a protein was shown to play a role in RNA2 replication), in other cases the role of the protein in the virus replication cycle is unknown (e.g., ToRSV X1, X3, and X4). Second, the role played by host factors in the translation, replication, and cell-to-cell (or systemic) movement of nepoviruses needs to be characterized. Large-scale studies of the interaction of the plant and virus proteomes will be necessary to address this question. In addition, it will be useful to analyze the ability of nepoviruses to infect collections of plants mutated or silenced for the expression of specific genes. Third, further work is necessary to understand the specificity of nepovirus–vector interactions. These questions are not only important to satisfy scientific curiosity, but also for the rational design of alternative approaches for the control of nepoviruses.

See also: Plant Virus Diseases: Fruit Trees and Grapevine; Plant Virus Diseases: Ornamental Plants.

Further Reading

Andret-Link P, Schmitt-Keichinger C, Demangeat G, Komar V, and Fuchs M (2004) The specific transmission of grapevine fanleaf virus by its nematode vector *Xiphinema index* is solely determined by the viral coat protein. *Virology* 320: 12–22.

Brown JF, Trudgill DL, and Robertson WM (1996) Nepoviruses: Transmission by nematodes. In: Harrison BD and Murant AF (eds.) *The Plant Viruses, Vol. 5: Polyhedral Virions and Bipartite RNA Genomes*, pp. 187–209. New York: Plenum.

Chandrasekar V and Johnson JE (1998) The structure of tobacco ringspot virus: A link in the evolution of icosahedral capsids in the picornavirus superfamily. *Current Biology* 6: 157–171.

Gaire F, Schmitt C, Stussi-Garaud C, Pinck L, and Ritzenthaler C (1999) Protein 2 A of grapevine fanleaf nepovirus is implicated in RNA2

replication and colocalizes to the replication site. *Virology* 264: 25–36.

Harrison BD and Murant AF (1996) Nepoviruses: Ecology and control. In: Harrison BD and Murant AF (eds.) *The Plant Viruses, Vol. 5: Polyhedral Virions and Bipartite RNA Genomes*, pp. 211–228. New York: Plenum.

Jovel J, Walker M, and Sanfacon H (2007) Recovery of *Nicotiana benthamiana* plants from a necrotic response induced by a nepovirus is associated with RNA silencing but not with reduced virus titer. *Journal of Virology* 81: 12285–12297.

Laporte C, Bettler G, Loudes AM, et al. (2003) Involvement of the secretory pathway and the cytoskeleton in intracellular targeting and tubule assembly of grapevine fanleaf virus movement protein in tobacco BY-2 cells. *Plant Cell* 15: 2058–2075.

Le Gall O, Iwanami T, Karasev AV, et al. (2005) Family *Comoviridae*. In: Fauquet CM, Mayo MA, Maniloff J, Desselberger U, and Ball LA (eds.) *Virus Taxonomy: Eighth Report of the International Committee on Taxonomy of Viruses*, pp. 807–818. San Diego, CA: Elsevier Academic Press.

Mayo MA and Robinson DJ (1996) Nepoviruses: Molecular biology and replication. In: Harrison BD and Murant AF (eds.) *The Plant Viruses Vol. 5: Polyhedral Virions and Bipartite RNA Genomes*, pp. 139–185. New York: Plenum.

Murant AF, Jones AT, Martelli GP, and Stace-Smith R (1996) Nepoviruses: General properties, diseases and virus identification. In: Harrison BD and Murant AF (eds.) *The Plant Viruses, Vol. 5: Polyhedral Virions and Bipartite RNA Genomes*, pp. 99–137. New York: Plenum.

Rebenstorf K, Candresse T, Dulucq MJ, Buttner C, and Obermeier C (2006) Host species-dependent population structure of a pollen-borne plant virus, cherry leaf roll virus. *Journal of Virology* 80: 2453–2463.

Ritzenthaler C, Laporte C, Gaire G, et al. (2002) Grapevine fanleaf virus replication occurs on endoplasmic reticulum-derived membranes. *Journal of Virology* 76: 8808–8819.

Susi P (2004) Black currant reversion virus, a mite-transmitted nepovirus. *Molecular Plant Pathology* 5: 167–173.

Vigne E, Bergdoll M, Guyader S, and Fuchs M (2004) Population structure and genetic variability within isolates of grapevine fanleaf virus from a naturally infected vineyard in France: Evidence for mixed infection and recombination. *Journal of General Virology* 85: 2435–2445.

Zhang SC, Zhang G, Yang L, Chisholm J, and Sanfacon H (2005) Evidence that insertion of tomato ringspot nepovirus NTB-VPg protein in endoplasmic reticulum membranes is directed by two domains: A C-terminal transmembrane helix and an N-terminal amphipathic helix. *Journal of Virology* 79: 11752–11765.

Ophiovirus

A M Vaira and R G Milne, Istituto di Virologia Vegetale, CNR, Turin, Italy

Glossary

Bipartite nuclear targeting sequence Signal involved in nuclear translocation of proteins characterized by (1) two adjacent basic amino acids (Arg or Lys), (2) a spacer region of ten residues, (3) at least three basic residues (Arg or Lys) in the five positions after the spacer region.

Mononegavirales Order of viruses comprising species that have a non-segmented, negative-sense RNA genome. The order includes four families: *Bornaviridae*, *Rhabdoviridae*, *Filoviridae*, and *Paramyxoviridae*.

Introduction

The genus *Ophiovirus* does not fit into any existing virus family, and a new family (*Ophioviridae*) has been proposed but is not yet official. The genus comprises a number of viral species, all plant-infecting viruses, relatively recently discovered. In some cases ophioviruses are now known to be the cause of well-known and 'classical' major plant diseases, but in other cases the presence of the virus has not been linked to any specific symptom, because of invariably mixed infections with other viruses. Ophioviruses occur in monocots and dicots, in vegetables, ornamentals, and trees, in the New and Old World, suggesting a well-adapted group of viruses containing more species than the ones already discovered. Where identified, the vectors of ophioviruses have proved to be *Olpidium brassicae*, a soil-inhabiting fungus. Ophioviruses have been slow to emerge mainly because the virions are not easy to see in the electron microscope; indeed, the first ophiovirus particle was observed only in 1988 and its morphology understood in 1994. The generic name derives from the Greek word *ophis*, meaning snake, in reference to the serpentine appearance of the virus particles.

Taxonomy and Classification

There are currently five species in the genus *Ophiovirus*, and at least one tentative species (**Table 1**). The accepted species in the genus are: *Citrus psorosis virus* (CPsV), the type species, *Lettuce ring necrosis virus* (LRNV), *Mirafiori lettuce virus* (MiLV), *Ranunculus white mottle virus* (RWMV), and *Tulip mild mottle mosaic virus* (TMMMV). The family *Ophioviridae*, containing a single genus listing all the species has recently been proposed but is not official. To date, the genomes of three species, CPsV, LRNV, and MiLV, are fully sequenced.

Phylogenetic analysis of the complete RNA-dependent RNA polymerase (RdRp) domain of RWMV and of the RdRp core modules of CPsV, MiLV, and RWMV, provide evidence of a relationship with *Rhabdoviridae* and *Bornaviridae* species, both belonging to the *Mononegavirales*. However, ophioviruses appear to form a monophyletic group, separate from the other negative-stranded RNA viruses; and of course the *Ophiovirus* genome is multipartite, excluding membership in the *Mononegavirales*.

For species demarcation within the genus, different criteria have been considered: the different coat protein (CP) sizes, absence of, or distant serological relationship between the CPs, differences in natural host range and different number, organization, and/or size of genome segments. From available data, there is 94–100% identity among complete CP amino acid sequences of isolates belonging to the same species, in particular for MiLV and CPsV, and alignments between incomplete but considerable parts of CP sequences of two RWMV isolates also show almost 100% identity. The percentage of identity falls to 30–52% for interspecies alignments, with the exception of the identity between MiLV and TMMMV CP sequences, which is about 80%, confirming the closeness of these two viruses. They are currently considered as two different species at least because of different host ranges; more information on TMMMV sequences is needed prior to any revision of their taxonomic position.

The percentage of identity/similarity between CP amino acid sequences may also be proposed as a further tool for species demarcation when complete sequences are available. A phylogenetic tree based on alignment of representative CP amino acid sequences is shown in **Figure 1**.

There has been some further discussion on the position of CPsV, as it is presently the only ophiovirus with a woody natural host, and there is no evidence of soil transmission, contrary to the apparent rule for ophioviruses; Western blotting has shown no serological cross-reaction between CPsV coat protein and other ophiovirus CPs and there are several molecular differences, the main one being the presence of three, not four genome segments. These differences, if confirmed and reinforced, could lead to considering placement of CPsV in a different genus within the proposed family *Ophioviridae*.

Reverse transcriptase-polymerase chain reaction (RT-PCR) amplification, using degenerate primers, of a 136 bp fragment from the RdRp gene, located outside the sequence coding the polymerase domain, is currently the best tool for detecting and identifying species within the genus. Sequencing and phylogenetic analysis of the 45 amino acid string deduced from the amplified fragment of several isolates belonging to different species, fully supports the present species classification and gives an indication of the taxonomic positioning of newly diagnosed isolates. An ophiovirus in freesia provisionally

Table 1 Virus members in the genus *Ophiovirus*

Virus (alternative name)	Abbreviations	Accession numbers[a]
Citrus psorosis virus (Citrus ringspot virus)	CPsV (CRsV)	NC006314, NC006315, NC006316
Lettuce ring necrosis virus	LRNV	NC006051, NC006052, NC006053, NC006054
Mirafiori lettuce virus (Mirafiori lettuce big-vein virus)	MiLV (MLBVV)	NC004779, NC004781, NC004782, NC004780
Ranunculus white mottle virus	(RWMV)	AF335429, AF335430, AY542957
Tulip mild mottle mosaic virus	(TMMMV)	AY204673, AY542958
Freesia ophiovirus (freesia sneak virus)	(FOV-FreSV)	AY204676, DQ885455

[a]Complete genome sequences are reported when available.

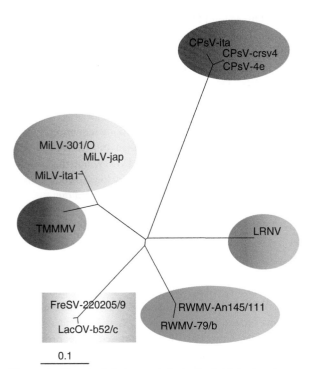

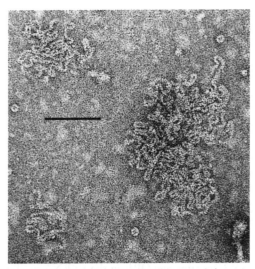

Figure 2 Negative contrast electron micrograph (uranyl acetate) of partially purified virion preparation from field lettuce showing big-vein symptoms. Note large and small ophiovirus particles. The bar represents 100 nm.

Figure 1 Unrooted phylogenetic tree of established and putative members of genus *Ophiovirus* based on their coat protein amino-acid sequences. The tree was generated by the neighbor-joining method and bootstrap values were estimated using 1000 replicates. All bootstrapping values were above 500. Branch lengths are related to the evolutionary distances. Viral isolates included in the analysis (accession numbers are reported when available): CPsV-4e (AAC41022), CPsV-CRSV4 (AAF00018), CPsV-ita (CAJ43825), MiLV-301/O (AAN60449), MiLV-ita1 (AAU12873), MiLV-jap (AAO49152), TMMMV (AAT08133), LRNV (AAT09112), RWMV-79/b (AAT08132) partial sequence, RWMV-an145/111 partial sequence, FreSV-220205/9 (DQ885455), LacOV-b52/c from *Lachenalia*, partial sequence. Ovals grouping isolates represent established viral species, the rectangle the proposed new species.

named freesia sneak virus (FreSV), considered a tentative species, was diagnosed and proposed as a new ophiovirus species owing to this procedure, when no other viral genome sequences were known; the complete CP sequence later obtained confirmed the hypothesis. A positive reaction with the genus-specific RT-PCR has recently been obtained for several samples from *Lachenalia*, a monocot in the family *Hyacinthaceae*, in this case the RT-PCR and subsequent analysis of the 45-amino-acid string suggest placing the isolate within the freesia sneak virus tentative species. Also in this case, subsequent CP gene sequencing showed nearly 100% identity with the homologous portion of FreSV at amino acid level, confirming this position.

Properties of Particles

The virions are naked filamentous nucleocapsids about 3 nm in diameter forming circularized structures of different lengths. The virions appear to form internally coiled circles that in some cases can collapse into pseudo-linear duplex structures (**Figure 2**). There is no evidence of an envelope. The 5′ and the 3′ ends of the RNAs have been structurally analyzed in the complete genome sequences available for CPsV and MiLV, to assess the probability of panhandle-like or similar structures. In CPsV, secondary structure predictions do not support significant complementarity between the termini of the viral RNAs to afford panhandle formation. On the contrary, in MiLV, the terminal sequences do carry partial inverted repeats, potentially allowing panhandles to form; the terminal sequences also contain palindromes potentially able to fold into stem-loops, so far not found for CPsV sequences. Thus, the apparent circularization of the particles of all ophioviruses remains incompletely explained.

Genome Organization and Replication

The ophiovirus genome is ssRNA, 11.3–12.5 kbp in size, divided into three or four segments (RNAs 1–4) and mainly negative sense. Positive-sense RNA is also encapsidated, to some extent with CPsV or in nearly equimolar amounts, with MiLV. The 3′ termini of the viral RNAs show sequences of 9–12 nt conserved within the RNAs of each species: A_7GUAUC for CPsV, $A_{4-6}UAAUC$ for MiLV, and $A_7GUAUCA$ and $A_3UA_3GUAUCA$ for LRNV; these may be involved in the recognition of the RNAs by the RdRp.

The genome organization of ophioviruses is as follows (**Figure 3** and **Table 2**). RNA1 is negative sense, shows short 5′ and 3′ untranslated regions, and contains two open reading frames (ORFs) separated by an AU-rich intergenic region. The putative product of ORF1 did not show any significant similarity with others available in databases and its function remains unknown. The product of the large ORF2 contains the core polymerase module with the five conserved motifs proposed to be part of the RdRp active site and shows low but significant amino acid sequence similarities with the L protein of several rhabdoviruses.

RNA2 contains one ORF in the negative strand in all three sequenced viruses, encoding a putative nonstructural protein with no significant sequence similarity to other known proteins and therefore of unknown function. In the amino acid sequence of the CPsV protein, two motifs bear similarity to a putative 'bipartite nuclear targeting sequence' considered to be a nuclear localization signal. An additional minor ORF, in the virus-sense strand, is present in MiLV RNA2, but to date there is no evidence for expression of the predicted 10 kDa protein.

RNA 3 contains one ORF in the vc RNA in all three viruses which has been identified as coding for the nucleocapsid protein.

An RNA4 has been found only in MiLV and LRNV. MiLV RNA4 contains two overlapping ORFs in the negative strand in different reading frames: the first putatively encodes a 37 kDa protein, but the second, overlapping the first by 38 nt, appears to lack an initiation

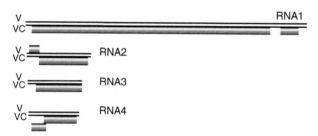

Figure 3 Genomic organization of ophioviruses. The four genomic RNAs are represented. The RNA4 is not described in all species, see text and **Table 2** for details; v-RNA and vc-RNA indicate the viral and the viral complementary RNAs, respectively. Boxes indicate ORFs.

codon. This second ORF has a theoretical coding capacity of 10.6 kDa and is suggested to be expressed by a +1 frameshift. A slippery sequence, GGGAAAU, can be recognized immediately in front of the UGA stop codon of the 37 kDa ORF. This unique feature of ORF positioning has been shown for two different isolates of MiLV, and therefore seems not to be caused by cloning artifacts.

RNAs 1, 2, and 3 of the other partially sequenced species are similar in size to those described. No RNA4 has been found in CPsV, RWMV, and TMMMV, but the possible existence of an RNA4, perhaps in very low concentration or very similar in size to RNA3, cannot be ruled out.

Analysis for the presence of subgenomic RNAs has not extensively been pursued for all species, even though several studies have used minus-strand and plus-strand probes in Northern blot hybridization using RNA extracted from infected tissues as well as virus particles, to obtain information regarding the polarity of encapsidated viral RNA. With CPsV in such studies no subgenomic RNAs were ever found for any of RNA1, 2, or 3.

A putative nuclear localization sequence present in the 280 kDa protein coded by RNA1 and in the 54 kDa protein coded by RNA2 in CPsV, and also observed in available analogous sequences of MiLV and RWMV, together with the polymerase similarities found with isolates from the genera *Nucleorhabdovirus* and *Bornavirus*, may suggest nuclear replication also for ophioviruses; in any case, gold immunolabeling and electron microscopy of RWMV-infected tissues showed accumulation of CP in the cytoplasm.

Viral Proteins

The most studied ophiovirus protein is the CP. There is only one CP, varying somewhat in size according to species. For CPsV, MiLV, and LRNV, this is 48.6, 48.5, and 48 kDa, respectively; Western blotting has given apparent

Table 2   Details of genomic organization of fully sequenced viruses inside the genus *Ophiovirus*

Virus	RNA1	RNA2	RNA3	RNA4
CPsV	8186 nt	1645 nt	1447 nt	
	2 ORFs in vc RNA	1 ORF in vc RNA	1 ORF in vc RNA	
	ORF1: 24 kDa	54 kDa protein	49 kDa	
	ORF2: 280 kDa proteins			
MiLV	7794 nt	1788 nt	1515 nt	1402 nt
	2 ORFs in vc RNA	1 ORF in vRNA:	1 ORF in vc RNA	2 ORFs in vc RNA:
	ORF1: 25 kDa	10 kDa protein	48.5 kDa	ORF1: 37 kDa
	ORF2: 263 kDa proteins	1 ORF in vc RNA:		ORF2: 10.6 kDa proteins
		55 kDa protein		
LRNV	7651 nt	1830 nt	1527 nt	1417 nt
	2 ORFs in vc RNA	1 ORF in vc RNA	1 ORF in vc RNA	1 ORF in vc RNA
	ORF1: 22 kDa	50 kDa protein	48 kDa protein	37 kDa protein
	ORF2: 261 kDa proteins			

sizes of 43 and 47 kDa for RWMV and TMMMV CPs, respectively. FreSV CP is 48.4 kDa. Ophiovirus CPs are relatively poor antigens. The antisera or, in one case, monoclonal antibodies, have been produced using purified virus preparations or recombinant protein. Western blots show that the CPs of RWMV, TMMMV, MiLV, and LRNV are slightly to moderately related; CPsV CP appears to be unrelated to the others. No clear information is available for FreSV. Genetic variability in the CP genes has been extensively studied in CPsV and MiLV.

Variability of the CP gene of CPsV was assessed serologically and by sequence analysis of two regions located in the 3′ and 5′ halves of the gene. Variability assessed by a panel of monoclonal antibodies to the protein resulted in 14 reaction patterns but no correlation was found between serogroups and specific amino acid sequences, field location, or citrus cultivar. Results from sequence analysis showed limited nucleotide diversity in the CP gene within the population. Diversity was slightly higher in the 5′ region. The ratio between nonsynonymous and synonymous substitutions (d_N/d_S) for the two regions indicated a negative selective pressure for amino acid changes, more intense at the 3′ end. When the entire CP sequences were considered, two clusters were identified, one comprising 19 isolates from Italy and an isolate from Spain and a second one containing only the Florida CPsV isolate.

Phylogenetic analysis of MiLV CP genes has revealed two distinct subgroups; however, this grouping was not correlated with symptom severity on lettuce or the geographic origin of the isolates; whether these two subgroups show different characteristics with regard to virulence in indicator plants or serological relationships remains to be determined; furthermore also in this case, a low value of d_N/d_S ratio was estimated for all MiLV isolates and also between the two subgroups, supporting a negative selection pressure for amino acid changes. In general, under natural conditions, genetic stability seems to be the rule rather than the exception.

The other viral protein, considered mainly for taxonomic purposes, is the putative RdRp, encoded by RNA1 in all species, described earlier.

No information is available regarding the synthesis or the function of other putative ORF products.

Pathogenicity and Geographic Distribution

Formerly a so-called citrus psorosis group of diseases included some of the most widespread graft-transmissible disorders of citrus, in some cases of undemonstrated etiology. Characterization of CPsV has shown that it is the cause of the great majority of psorosis symptoms but that a small but still undefined proportion of trees with

'psorosis-like' bark-scaling shows no evidence of carrying the virus, and the idea of a non-CPsV psorosis-like disease of unknown etiology is emerging. In the past, different kinds of symptoms were described: 'psorosis A', characterized by causing bark-scaling in trunk and limbs of infected field trees; 'psorosis B', causing rampant scaling of thin branches in field trees and chlorotic blotches in old leaves with gummy pustules in the underside; 'ringspot', characterized by presence of chlorotic blotches and rings in the old leaves of inoculated seedlings but apparently no specific symptoms on infected field trees or in other cases, chlorotic flecks and ringspots on leaves, and trunk and fruit symptoms (**Figure 4**). All these appear to be caused by CPsV but in the former 'psorosis group' there are other non-CPsV diseases with symptoms such as chlorotic leaf-flecking and oak-leaf patterns; while graft-transmissible, these diseases have not yet yielded

Figure 4 Bark scaling in citrus, typical of severe psorosis.

to further analysis. Psorosis is an ancient disease; bark-scaling of citrus was first observed in Florida and California in the 1890s. The disease has been brought under control in most advanced citrus-growing countries due to rigorous indexing and quarantine. In Argentina it remains a severe problem, in Mediterranean areas it is also reported, and in citrus-growing parts of Asia the disease may well be widespread, although rigorous testing for the presence or absence of CPsV has generally been lacking.

RWMV has been reported in two species, ranunculus (*Ranunculus asiaticus* hyb.) and anemone (*Anemone coronaria*) in northwest Italy since the 1990s. The pathogenic impact is uncertain as it was almost always found in mixed infection with other viruses commonly infecting the two species. The symptom description 'white mottle' present in its name derives from the bright white mottling symptoms consistently observed on *Nicotiana benthamiana* leaves mechanically infected by RWMV. In some cases the bright mottle has also been observed in naturally infected ranunculus, always in mixed infection. Indexing for the presence of RWMV in ranunculus crops done in 1996 showed an incidence of 2.5% among symptomatic plants; furthermore, the very few plants apparently infected only with RWMV did not show any distinctive symptoms. Infection of ranunculus seedlings by mechanical inoculation results in limited necrosis and deformation of stems and leaves. The pathogenic potential of RWMV is thus still unclear.

Tulip mild mottle mosaic disease, caused by TMMMV, is one of the most serious diseases in some bulb-producing areas of Japan (which produces bulbs on a large scale for the Southeast Asian market) and has been reported since 1979. Symptoms on tulips include color-attenuating mottle on flower buds and color-increasing streak on petals (**Figure 5**); mild chlorotic mottle and mosaic slightly appear along the leaf veins. TMMMV infection is up to now restricted to *Tulipa* species and cultivars, and has not been reported in other geographical areas.

Lettuce big-vein disease was first reported in the United States in 1934 and occurs in all major lettuce-producing areas in the word. The disease becomes serious during cooler periods of the year. The main symptoms are vein-banding in the leaves, due to zones parallel to the vein cleared of chlorophyll (**Figure 6**); there is associated leaf distortion, delayed head formation, and decreased head size.

The causal agent of the disease, long known to be soil-transmitted, was first identified as 'lettuce big-vein virus' (LBVV) (genus *Varicosavirus*); recently, re-evaluation of the etiology has been necessary as a second less easily detected virus, MiLV, was found in lettuces with big-vein symptoms. Following experimental inoculation of the two viruses together and separately, MiLV was shown to be the etiological agent of big-vein, while LBVV, now renamed as lettuce big-vein associated virus (LBVaV), apparently plays no part in the disease, although it is almost always present, and is, like MiLV, transmitted by *Olpidium*. Studies on symptom development in the field have confirmed that both viruses very commonly occur together in lettuce crops. MiLV likely occurs worldwide;

Figure 5   TMMMV symptoms on tulip (healthy tulip on the left). Courtesy of T. Morikawa, Toyama Agric. Res. Center, Tonami, Toyama, Japan.

Figure 6 Leaf of butterhead lettuce showing big-vein symptoms. Close-up showing vein-banding.

it is reported in California (USA), Chile, France, Germany, Italy, Spain, the Netherlands, England, Denmark, Japan, Australia, and New Zealand. To date, natural infection has been reported only in cultivars of *Lactuca sativa*.

LRNV is closely associated with lettuce ring necrosis disease, first described in the Netherlands and in Belgium as 'kring necrosis' and also in France as *maladie des taches orangées*, in the 1980s. It is an increasingly important disease of butterhead lettuce crops in Europe. Definitive proof that LRNV is the cause of ring necrosis is still awaited. In southern France the disease is observed primarily in winter lettuces (September–January) under plastic or glass, when the crop is maturing, the day length is short, and both light intensity and temperature are low. Symptoms depend on the lettuce type and environmental conditions, and mainly consist of necrotic rings and ring-like patterns on leaves, which may render the product unmarketable. LRNV is often found together with MiLV (and, of course, LBVaV) with which it shares host and vector. The virus has also been reported in California.

Although the ophiovirus isolated from freesia has not yet been recognized as an official new ophiovirus species, we would like to say a few words about the freesia disease. Necrotic disorders of freesia (*Freesia refracta* hyb., family Iridaceae), known as 'leaf necrosis' and 'severe leaf necrosis', were first described in freesia crops in the Netherlands before 1970, and a similar disease, named 'freesia streak', was reported in England and Germany in the same years. The 'severe leaf necrosis' appeared to be caused by mixed infection with the potyvirus freesia mosaic virus and a virus with varicosavirus morphology and mode of transmission, that has been tentatively named Freesia leaf necrosis virus. In recent years, in northwest Italy, a necrotic disease of freesia has spread and caused considerable economic losses, even though it was present since 1989. Both in Italy and in the

Netherlands, the disease seems to be linked to the presence of an ophiovirus. Typical symptoms are chlorotic spots and streaks at the leaf tips that expand downwards and turn necrotic, and may vary according to cultivar and climate. Freesia sneak virus has been proposed as the name of this new ophiovirus species.

Experimental Symptoms and Host Ranges

All ophiovirus are, though sometimes not easily, mechanically transmissible to a limited range of test plants, including Chenopodiaceae and Solanaceae, inducing local lesions and in some cases systemic mottle.

Cytopathology

Thin sections of *Nicotiana clevelandii* leaves mechanically infected by RWMV have been examined by electron microscopy but no distinctive inclusions were observed and no virus particles were seen. In classical thin sections, viral nucleic acids stain up well but protein coats are faintly contrasted, so it is not surprising that very thin nucleoprotein threads in random orientations should escape detection. After gold immunolabeling with RWMV-specific polyclonal antibodies, sections of the cytoplasm of parenchyma cells were seen in the EM to be clearly labeled, but nuclei, chloroplasts, mitochondria, and microbodies were unlabeled. No other studies on ophiovirus cytopathology are till now available.

Transmission, Prevention, and Control

Most ophiovirus species are soil-transmitted through the obligately parasitic soil-inhabiting fungus *Olpidium brassicae*; this has been proved for TMMMV, MiLV, and LRNV; FreSV is known to be soil-transmitted, and the freesia leaf necrosis complex (presumably containing FreSV and the varicosavirus freesia leaf necrosis virus) has been transmitted by *O. brassicae*. The virus–vector relationship was recognized as the *in vivo* type, at least for MiLV-*O. brassicae*; virus-free zoospores acquire the virus during the vegetative part of their life cycle in the roots of virus-infected host plants. Zoospores released from infected sporangia apparently carry the virus internally in their protoplasts and transmit it to healthy roots. The virus enters vector resting spores, and can survive for years in the soil in the absence of a growing host plant. Furthermore, disease control is difficult because of the lack of safe and effective treatments against the fungal vector. As ophioviruses occur in low concentration and are physically labile *in vitro*, it would be interesting to establish whether they multiply in the vector.

No information on natural vectors is available for RWMV. CPsV is commonly transmitted by vegetative propagation and no natural vectors have been identified; in some cases natural spread of psorosis in limited citrus areas has been reported, but the spatial patterns would suggest a hypothetic aerial vector instead of a soil-borne one. The data are however based on symptom observation, not on analysis for the spread of CPsV.

For disease control the use of resistant or tolerant crops may be the best choice. In Japan, the use of resistant tulip cultivars is the most important component of managing TMMMV disease, as it can be highly effective and has no deleterious effect on the environment; resistance assays have allowed researchers to identify highly resistant tulip lines and use them for breeding new resistant cultivars. For lettuce in soil-less cultivation, using ultraviolet (UV) sterilization of nutrients has shown good results in prevention of MiLV and LRNV infection, although for field lettuces the prospect is less good as no classical sources of resistance or tolerance have yet been identified. In the case of CPsV, control of the sanitary status of mother plants for producing propagating material is essential. Shoot-tip grafting *in vitro* associated with thermotherapy or somatic embryogenesis from stigma and style cultures have been successfully used to eliminate CPsV from plant propagating material. Several transgenic citrus lines exist carrying parts of the CPsV genome, and promising resistance may emerge from these.

See also: Plant Rhabdoviruses.

Further Reading

Derrick KS, Brlansky RH, da Graça JV, Lee RF, Timmer LW, and Nguyen TK (1988) Partial characterization of a virus associated with citrus ringspot. *Phytopathology* 78: 1298–1301.

García ML, Dal Bo E, Grau O, and Milne RG (1994) The closely related citrus ringspot and citrus psorosis viruses have particles of novel filamentous morphology. *Journal of General Virology* 75: 3585–3590.

Martín S, López C, García ML, et al. (2005) The complete nucleotide sequence of a Spanish isolate of citrus psorosis virus: Comparative analysis with other ophioviruses. *Archives of Virology* 150: 167–176.

Morikawa T, Nomura Y, Yamamoto T, and Natsuaki T (1995) Partial characterization of virus-like particles associated with tulip mild mottle mosaic. *Annals of the Phytopathological Society of Japan* 61: 578–581.

Roggero P, Ciuffo M, Vaira AM, Accotto GP, Masenga V, and Milne RG (2000) An ophiovirus isolated from lettuce with big-vein symptoms. *Archives of Virology* 145: 2629–2642.

Roggero P, Lot H, Souche S, Lenzi R, and Milne RG (2003) Occurrence of mirafiori lettuce virus and lettuce big-vein virus in relation to development of big-vein symptoms in lettuce crops. *European Journal of Plant Pathology* 109: 261–267.

Vaira AM, Accotto GP, Costantini A, and Milne RG (2003) The partial sequence of RNA 1 of the ophiovirus ranunculus white mottle virus indicates its relationship to rhabdoviruses and provides candidate primers for an ophio-specific RT-PCR test. *Archives of Virology* 148: 1037–1050.

Vaira AM, Accotto GP, Gago-Zachert S, et al. (2005) Ophiovirus. In: Fauquet CM, Mayo MA, Maniloff J, Desselberger U,, and Ball LA (eds.) *Virus Taxonomy: Eighth Report of the International Committee on Taxonomy of Viruses*, pp. 673–679. San Diego, CA: Elsevier Academic Press.

Van der Wilk F, Dullemans AM, Verbeek M, and Van den Heuvel JFJM (2002) Nucleotide sequence and genomic organization of an ophiovirus associated with lettuce big-vein disease. *Journal of General Virology* 83: 2869–2877.

Papaya Ringspot Virus

D Gonsalves, J Y Suzuki, and S Tripathi, USDA, Pacific Basin Agricultural Research Center, Hilo, HI, USA
S A Ferreira, University of Hawaii at Manoa, Honolulu, HI, USA

Published by Elsevier Ltd.

Introduction

The term papaya ringspot virus (PRSV) was first used in the 1940s to describe a viral disease of papaya. The name was used primarily to describe the ringspots that appeared on fruits from infected plants. Early investigations showed that the virus was transmitted by several species of aphids in a nonpersistent manner. That is, the aphid vector could acquire the virus in a short period of time while feeding on infected plants and likewise transmit the virus in a span of few seconds to less than a minute during subsequent feeding. In the same decade researchers from India and other places like Puerto Rico reported the occurrence of an aphid-transmitted disease of papaya; based on the symptoms on the leaves, it was identified as papaya mosaic virus. Work in the 1980s showed that the aphid-transmitted papaya mosaic virus and PRSV were really the same, and the name of PRSV was adopted. PRSV is a member of the family *Potyviridae*, a large and arguably the most economically important group of plant viruses. Today, the term papaya mosaic virus is reserved for a virus that is not aphid transmitted, belongs to the family *Potexviridae*, and causes the papaya mosaic disease which is seldom observed and not important commercially.

The systemic host range of PRSV is confined to plants in the families Caricaceae and Cucurbitaceae, with the primary economically important host being papaya and a range of cucurbits such as squash, watermelon, and melons.

It does cause local lesions on plants of the family Chenopodiaceae such as *Chenopodium quinoa* and *C. amaranticolor.* The disease on cucurbits was, early on, referred to as being caused by watermelon mosaic virus-1 (WMV-1). Later serological and molecular characterization showed that PRSV and WMV-1 are virtually identical. Based on their close relationship, a single name was adopted to unify both viruses into one group. The name PRSV was chosen due to its being named before WMV-1. To clarify host range, 'P' (PRSV-P) or 'P type' is used to designate virus infecting papaya and cucurbits, while 'W' (PRSV-W) or 'W type' refers to virus infecting cucurbits only. The virus symptoms on cucurbits are identical to those on Caricaceae. Leaves of infected plants show severe mosaic, and chlorosis, are deformed, and often exhibit shoestring-type

Figure 1 Symptoms of PRSV on papaya.

symptoms. The fruits are also often deformed and bumpy. In papaya, PRSV infection is characterized by mosaic and chlorosis symptoms on leaves, water-soaked streaks on the petiole, and deformation of leaves that can result in shoestring-like symptoms that resemble mite damage (**Figure 1**). The virus can cause deformation and ringspot symptoms on the fruit, hence the name PRSV. Commercial PRSV-resistant transgenic papaya expressing the coat protein (*CP*) gene of the virus has been used to control PRSV P in Hawaii, as will be discussed later.

General Properties of PRSV

The virus particles are flexuous rods about 760–800 nm × 12 nm with single RNA of about 10 326 b in length. Virus particles consist of 94.5% protein and 5.5% nucleic acid by weight. It has a single coat protein (CP) of about 36 kDa. Analysis of purified virus preparations that are stored show that the CP degrades to smaller proteins of *c.* 31–34 and 26–27 kDa proteins, possibly due to proteolytic degradation. The density of the virion in purified preparations is $1.32\,\mathrm{g\,cm^{-3}}$ in CsCl.

PRSV should not be confused with another potyvirus, papaya leaf distortion mosaic virus (PLDMV), which occurs in Okinawa and other parts of Asia, such as Taiwan. This virus causes very similar symptoms as PRSV on papaya and cucurbits but is serologically unrelated and its CP shares only 55–59% similarity to that of PRSV.

PRSV Genome

A genetic map of PRSV genome with polyprotein processing sites and products is presented in **Figure 2** and their possible functions in **Table 1**. Much of the knowledge on the genome of PRSV has been obtained from extensive work done by the laboratory of Dr. Shyi-Dong Yeh of National Chung-Hsing University in Taiwan. The genomic RNA of PRSV is 10 326 nt in length excluding the poly(A) tract and contains one large open reading frame that encodes a polyprotein of 3344 amino acids starting at nucleotide position 86 and ending at position 10 120. A VPg protein is linked to the 5′ end of the RNA while a poly(A) tract is at the 3′ end. The polyprotein is cleaved into proteins designated (name (size in M_r)): P1 (63K), helper component (HC-Pro, 52K), P3 (46K), cylindrical

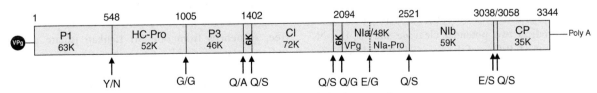

Figure 2 Genome map of PRSV. Vertical arrows indicate the proteolytic cleavage sites.

Table 1   PRSV proteins and their possible functions

Viral protein	Size (M_r)	Functions
P1	63K	Proteinase Cell-to-cell movement
HC-Pro	52K	Vector transmission Proteinase Pathogenicity Suppressor of RNA silencing Cell-to-cell movement
P3	46K	Unknown, but possible role in replication
6K1	6K	Unknown, but possible role in replication
CI	72K	Genome replication (RNA helicase) Membrane attachment Nucleic acid-stimulated ATPase activity Cell-to-cell movement
6K2	6K	Unknown, but possible roles in: • Replication • Regulation; inhibition of NIa nuclear translocation
NIaVPg	21K	Genome replication (primer for initiation of RNA synthesis)
NIaPro	27K	Major proteinase
NIb	59K	Genome replication (RNA-dependent RNA polymerase, RdRp)
CP	35K	RNA encapsidation Vector transmission Pathogenicity Cell-to-cell movement

inclusion protein (CI, 72K), nuclear inclusion protein a (NIa, 48K), nuclear inclusion protein b (NIb, 59K), coat protein (CP, 35K), as well as two other proteins 6K1 (6K) and 6K2 (6K). The cleaved proteins are arranged on the genome starting from the 5′ in order as: P1–(HC-Pro)–P3–6K1–CI–6K2–NIa–NIb–CP (**Figure 2**).

The cleavage proteins mentioned above have been identified by immunoprecipitation and dynamic precursor studies with PRSV-p as well as by extensive studies on proteolytic processing of polyproteins from other potyviruses. Three virus-encoded proteinases are responsible for at least seven cleavages: the P1 protein from N-terminus of the polyprotein autocatalytically liberates its own C-terminus, the HC-Pro also cleaves its own C-terminus, and NIa is responsible for *cis*- and *trans*-proteolytic processing to generate the CI, 6K, NIa, NIb, and CP proteins. NIa has also been shown to contain an internal cleavage site for delimitation of the genome-linked protein (VPg) and the proteinase (Pro) domains. Thus, the genomic organization and processing of the polyprotein of PRSV is similar to those of other potyviruses.

A rather interesting feature of the PRSV is that sequence analysis predicts two potential cleavage sites at the N-terminus of the CP. One of the sites (VFHQ/SKNF) predicts a CP of 33K and a NIb of 537 amino acids about 20 amino acids larger than those of other potyviruses. The second predicted cleavage site (VYHE/SRGTD) generates

a CP of 35K and an NIa of 517 amino acids. There is no firm evidence to suggest that only one cleavage site is used. If both sites are used in polyprotein processing, one would expect heterogeneous products. This may explain why the analysis of purified CP preparations that are stored frequently shows the major ∼36K form in addition to smaller CPs that are 2–5K smaller.

Sequence Diversity and Evolution

Knowledge of the sequence diversity among isolates of a virus has great implications in developing an effective virus disease management program and in understanding the origin and biology of the virus. Recently, numerous PRSV P and W sequences from virus isolated from different parts of the world have been reported in the sequence database. Amino acid and nucleotide sequence divergence among PRSV isolates differ by as much as 14%. These differences, interestingly, are considerably less than that found among isolates of other potyviruses, such as yam mosaic virus (YMV), that differ by as much as 28%. Although initial data from the USA and Australia suggested that there was little variation among PRSV isolates within these countries, more recent data from India and Mexico have suggested that the sequence variation between PRSV isolates in other countries may be greater than previously recognized. Heterogeneity in *CP* length ranging from 840 to 870 nt has been noted. The observed size differences in *CP* sequence occurred in multiples of three, preserving the reading frame between genes of different genomes and resulting in CPs of between 280 amino acids (Indian P isolate KA2) and 290 amino acids (VNW-38 from central Vietnam). Interestingly, the CP-coding region of all isolates from Thailand were 286 amino acids in length, while those from India and Vietnam demonstrated considerable heterogeneity in CP length, at 280–286 and 285–290 amino acids, respectively. The first 50 or so amino acids of the N-terminal region of the PRSV *CP* gene were found to be highly variable and all differences in CP length were confined to this region. The differences in this region that did exist consisted of conservative amino acid substitutions. The majority of the size differences occurred in one of two hypervariable regions and most were due to differences in the number of EK repeats.

A phylogenetic study based on *CP* sequences from 93 isolates of type P and W PRSV from different geographic locations was done by generating a phylogenetic tree using the neighbor-joining method. In the phylogenetic tree, sequences from one Sri Lankan isolate and two Indian isolates formed a sister cluster to the rest of the sequences. The other isolates formed two major lineages: I included all isolates from the Americas, Puerto Rico, Australia, and a few from South Asia; and II included isolates from Southeast Asia and the Western Pacific. Lineage I was the major of the two, containing three clusters of Brazilian isolates, two Indian isolates, and Australian, Mexican and US isolates. Within lineage I, the Brazilian and Mexican isolates were more diverse than the US and Australian isolates. Lineage II included all of the isolates from the Southeast Asia and Western Pacific, including China, Indonesia, Thailand, Vietnam, Taiwan, Japan, and the Philippines. However, the subclustering of isolates did not correlate well with their geographic origins; rather, they appeared to be a single mixed population with some well-defined subpopulations. These observations suggest that considerable movement of PRSV isolates has occurred among the Southeast Asian countries. Thai isolates of the P type diverged together, whereas PRSV W type diverged with other Southeast Asian isolates. Both P and W isolates of PRSV from Vietnam were intermingled with other Asian isolates. Sequence analysis showed that all Vietnamese isolates (except the P type from the southern part of the country) diverged from a common branch with P isolates from Japan and Taiwan while PRSV isolates from South Vietnam were diverged compared to those from the Philippines and seemed closely related to several W types from Thailand.

An interesting feature of PRSV is the origin of types P and W. As noted above, a number of viral diseases on cucurbits were historically associated with WMV-1 and not with any diseases of papaya. Did type P originate from W, or was it the reverse, or did they evolve independently? Evidences from various sources indicate that PRSV is primarily a pathogen of cucurbits, and that PRSV P originated from PRSV W. Work in Australia suggests that the recent outbreak of PRSV P came from the population of PRSV W already present in Australia. This suggestion is also supported by the diversity in cucurbit-infecting potyviruses that are phylogenetically related to PRSV.

Infectious Transcripts of Recombinant PRSV Help to Reveal Potential Determinants of Several Biological Characteristics

Determinants for Host Range Specificity

Studies utilizing the technique of producing infectious transcripts from recombinant viruses followed by bioassays of the transcripts have demonstrated that sequences of the PRSV genome responsible for determining papaya and cucurbit host specificity are not in the region of the *CP* gene. However, nucleotides 6509–7700 encoding the NIa gene and parts of the NIb gene were critical for papaya infection. Amino acids of the NIb gene of this region (nucleotides 7644–7700) between PRSV-P and PRSV-W type are identical, whereas sequence comparison of nucleotides 6509–7643 of four type P and two type

W viruses showed that two amino acids at positions 2309 (K → D) and 2487 (I → V) of PRSV are significantly different between papaya-infecting type P and non-papaya-infecting type W. Further point mutational studies in these sites indicated that these two amino acids located in the NIa proteinase are responsible for conferring the ability to infect papaya.

Determinants for Local Lesion Formation on *Chenopodium*

As noted earlier, PRSV causes local lesions on *C. amaranticolor* and *C. quinoa*. The severe strain PRSV HA strain from Hawaii causes local lesions on *C. quinoa* but a mild nitrous acid mutant of it, PRSV HA 5-1, does not. Recombinant infectious viruses were generated by exchanging genome parts between PRSV HA and PRSV HA 5-1. The study revealed that the pathogenicity-related region is present between nucleotide positions 950 and 3261 of the PRSV HA genome and mutations in the *P1* and *HC-Pro* genes resulted in the attenuation of PRSV HA symptoms and the loss of ability to produce local lesions on *C. quinoa*. The *HC-Pro* gene of PRSV is the major determinant factor for local lesion formation.

Determinants on Severity of Symptoms, Suppression of Gene Silencing, Infection of Transgenic Papaya

Virus–host interaction studies based on recombinant analyses between severe and mild strains of PRSV indicated that the *HC-Pro* gene plays an important role in viral pathogenicity and virulence and acts as a suppressor of the gene-silencing defense mechanism in the papaya host plant. In addition, the comparative reaction of recombinant PRSV with chimeric *CP* gene sequences showed that heterologous sequences and their position in the *CP* gene influences their pathogenicity on PRSV-resistant transgenic papaya.

An interesting phenomenon of strain-specific cross-protection was observed in papaya and horn melon provided by the mild strain HA 5-1 of PRSV. The PRSV mild mutant HA 5-1 provided 90–100% protection against the severe parental strain PRSV HA in greenhouse and field conditions. However, the degree of protection provided in horn melon by HA 5-1 against a PRSV type-W strain from Taiwan was only 20–30%. Studies on strain-specific cross-protection phenomenon indicated that the recombinant HA 5-1 carrying both the heterologous *CP* and the 3′ untranslated region (UTR) of the PRSV W from Taiwan significantly enhanced the protection against the Taiwan strain in cucurbits. However, chimeric HA 5-1 virus carrying either heterologous *CP* or heterologous 3′ UTR showed reduced effectiveness of protection against PRSV HA in papaya when compared to protection by the native mild HA 5-1.

Similar to other potyviruses, PRSV is also transmitted by insect vector aphids (*Myzus persicae* and *Aphis gossypii*) in a nonpersistent manner. Detailed studies with other potyviruses show that HC-Pro/virions interaction is essential for aphid transmission of potyviruses. Although not empirically tested, it would seem likely that aphid transmission of PRSV would similarly be governed by HC-Pro/virion interactions.

Pathogen-Derived Resistance for Controlling PRSV: The Hawaii Case

The control or management of PRSV has been approached through practices such as quarantine, eradication, avoidance by planting papaya in areas isolated from the virus, continual rogueing of infected plants, use of tolerant lines to reduce damage caused by PRSV, cross-protection through the use of mild virus strains, and resistance using the approach of 'pathogen-derived resistance'. The efforts to control PRSV in Hawaii are described because it involves all of the above practices. Ultimately, the most successful has been the 'pathogen-derived resistance' approach.

The state of Hawaii consists of eight main islands that are in rather close proximity to each other with the shortest and farthest distance between islands being 7 mile between Maui and Kahoolawe and 70 mile between Kauai and Oahu. Travels between the islands are prevalent with the exception of Kahoolawe, which is not inhabited, and Niihau, which is privately owned. PRSV was first detected in the 1940s on Oahu where Hawaii's papaya industry was located at that time. Efforts to control the virus on Oahu largely consisted of state officials and farmers continually monitoring for infected plants and rogueing them, especially in areas where the virus was not prevalent. However, by the late 1950s, PRSV was causing extensive damage, which caused the papaya industry to relocate to Puna on the island of Hawaii.

The relocation of the industry to Puna was timely and effective because Puna had an abundance of land that was suitable to grow Kapoho, a cultivar of excellent quality that adapted to the volcanic soil base there, allowing excellent drainage, had high rainfall and yet lots of sunshine, and the land there could be bought or leased at reasonable prices. By the 1970s, the Kapoho papaya grown in Puna accounted for 95% of the state's papaya production, making papaya the second most important fruit crop behind pineapple.

Despite strict quarantine on movement of papaya seedlings between islands, PRSV was discovered in the town of Hilo which was only about 18 miles away from the center of the papaya-growing area of Puna. However, PRSV was indeed discovered in Puna in May 1992 (**Figure 3(a)**) and the Hawaiian papaya industry would be forever changed. By 1995, a third of the papaya grow-ing area was completely infected and much of the rest of Puna had widespread infection (**Figure 3(b)**). By 1998, the production of papaya in Puna had dropped to 27 million pounds of papaya from 52 million pounds in 1992 when PRSV was discovered in Puna. In retrospect, the efforts of quarantine, monitoring and rogueing of infected plants in Hilo, and suppression efforts of PRSV in Puna all played key roles in helping Hawaii's papaya industry, because it gave researchers time to develop control measures for PRSV.

Research to develop tolerant varieties and cross-protection measures were started in the 1970s. Since resistance to PRSV has not been identified in *Carica papaya*, researchers have used tolerant germplasm in an attempt to develop papaya cultivars with acceptable PRSV tolerance and horticultural characteristics. However, tolerance to PRSV is apparently governed by a family of genes that is inherited quantitatively, which makes it technically difficult to develop cultivars of acceptable horticultural quality. Furthermore, the tolerant lines do become infected with PRSV, although fruit production continues still at a lower level. Indeed, in Thailand, the Philippines, and Taiwan, a number of tolerant lines have been developed and are used. However, Hawaii grows the small 'solo'-type papaya and efforts to introduce tolerance into acceptable 'solo' papaya cultivars have not been successful.

Efforts to use cross-protection were similarly started in the late 1970s to control PRSV in Hawaii. Cross-protection can be defined as the use of a mild strain of virus to infect plants that are subsequently protected against economic damage caused by a severe strain of the same virus. This practice has been used successfully for many years to minimize damage by citrus tristeza virus in Brazil, for example. In the early 1980s, a mild strain of PRSV (described above as PRSV HA 5-1) was developed through nitrous acid treatment of a severe strain, PRSV HA isolated from Oahu island. This mild strain was tested in Hawaii on Oahu island and showed good protection against damage by severe strains but produced symptoms that were very obvious on certain cultivars, such as Sunrise, especially in the winter months. This prominent symptom induction on certain cultivars and the logistics of mild strain buildup and inoculation of plants, among others, were factors that caused it not to be consistently used on the island of Oahu. There was no justification to use it on the island of Hawaii because PRSV was not yet found in Puna during the 1980s. Interestingly, the mild strain was used extensively for several years in Taiwan, but it did not afford sufficient protection against the severe strains from Taiwan and thus its use was abandoned after several years.

Transgenic Resistance

In the mid-1980s, an exciting development on tobacco mosaic virus (TMV) provided a rationale that resistance

Figure 3 (a) Healthy Puna papaya in 1992. (b) Severely infected papaya orchards in Puna in 1994. (c) Field trial of transgenic papaya. PRSV-infected nontransgenic papaya on left and PRSV-resistant transgenic papaya on right. (d) Commercial planting of transgenic papaya one year after releasing seeds of PRSV-resistant transgenic papaya. (e) Transgenic papaya commonly sold in supermarkets. (f) Risk of growing nontransgenic papaya still exists in 2005. Foreground is PRSV-infected nontransgenic papaya that are cut, and background shows healthy PRSV-resistant transgenic papaya.

to plant viruses could be developed by expressing the viral *CP* gene in a transgenic plant. This approach was called CP-mediated protection, and, at about the same time, a report introduced the concept of 'parasite-derived resistance'. The report on transgenic resistance to TMV set off a flurry of work in many laboratories to determine if this approach could be used for developing resistance to other plant viruses. Likewise, work was initiated in 1985 to use this approach for developing PRSV-resistant transgenic papaya for Hawaii.

Key requirements for successful development and commercialization of transgenic virus-resistant plants are the isolation and engineering of the gene of interest, vectors for mobilization into and expression of the gene in the host, transformation and subsequent regeneration of the host cells into plants, effective and timely screening of transformants, testing of transformants, and the ability to deregulate and commercialize the product.

The *CP* gene of the mild strain of PRSV was chosen as the 'resistance' gene because it had been recently cloned and it was of the PRSV P type. The gene was engineered into a wide host range vector that could replicate in *Escherichia coli* as well as in *Agrobacterium tumefaciens*, the bacterium used for one of the most widely used methods of plant transformation. The commercial cultivars Kapoho, Sunrise, and Sunset were chosen for transformation. Initially, transformation of papaya was attempted using the *Agrobacterium*–leaf piece approach, where leaf pieces would be infected with *Agrobacterium* harboring the *CP* gene and transformed cells would be regenerated via organogenesis into transgenic plants. The latter is the direct regeneration of cells from an organ such as the leaf. This approach did not work due to our failure to develop plants from leaf pieces. A shift to the transformation of somatic embryos via the biolistic (often referred to as the gene gun) approach resulted in obtaining of about a dozen transgenic papaya lines, four of which expressed the *CP* gene. In 1991, tests of the R_0 lines identified a transgenic Sunset that expressed the *CP* gene of PRSV HA 5-1, and showed resistance to PRSV from Hawaii. A field trial of R_0 plants was started in April 1992 on as the island of Oahu, and a month later PRSV was discovered at Puna in May 1992, as discussed above.

The Oahu field trial showed that R_0 plants of line 55-1 were resistant, and line 55-1 was further developed to obtain the cultivar 'SunUp' which is line 55-1 that has the *CP* gene in a homozygous state, and 'Rainbow' which is an F_1 hybrid of SunUp and the nontransgenic 'Kapoho'. SunUp is red-fleshed and Rainbow is yellow-fleshed. In 1995, SunUp and Rainbow were tested in a subsequent field trial in Puna and showed excellent resistance (**Figure 3(c)**). Due to its yellow flesh and good shipping qualities, Rainbow was especially preferred by the growers. Line 55-1 was deregulated by the US government and commercialized in May 1998. The deregulation also applied to plants that were derived from line 55-1. The timely commercialization of the transgenic papaya in 1998 was crucial since PRSV had decreased papaya production in Puna by 50% that year compared to 1992 production levels. The transgenic Rainbow papaya was quickly adopted by growers and recovery of papaya production in Hawaii was underway (**Figure 3(d)**). The transgenic papaya is sold throughout Hawaii (**Figure 3(e)**) and the mainland USA, and to Canada where it was deregulated in 2003. However, several challenges remain: coexistence, exportation of nontransgenic

papaya to Japan, deregulation of transgenic papaya in Japan, and the adoption of transgenic papaya in other countries that suffer from PRSV.

Hawaii still needs to grow nontransgenic papaya to satisfy the lucrative Japanese market as well as for production of organic papaya, for example. Interestingly, the islands of Kauai and Molokai do not have PRSV but grow only limited acreage of papaya. This situation illustrates the point that many factors influence decisions on the localities and crops that are grown. In Hawaii, Puna is the best place to grow papaya for the reasons mentioned above; there is a lot of land, farmers there are intuned to growing the crop, the region receives plenty of water, sunshine, and has a well-drained lava-based 'soil' structure, and there are high-quality cultivars adapted to the local growing conditions. The disadvantage is PRSV, but that disadvantage was overcome through the introduction of the PRSV-resistant Rainbow papaya that has good commercial attributes plus virus resistance. Puna accounts for 90% of Hawaii's papaya and, as of 2005, Rainbow represents 66% of the papaya grown in Puna. Growing nontransgenic papaya can be risky because PRSV is still around (**Figure 3(f)**), but judicious use of isolation from virus sources and constant rogueing can provide a means of raising nontransgenic papaya. However, a major market is Japan, which still has not deregulated the transgenic papaya. To maintain the lucrative Japanese market, Hawaii has to continue the exportation of nontransgenic papaya to Japan. The immediate solution is to concurrently grow nontransgenic and transgenic papaya, and subsequently to deregulate the transgenic papaya so it can be freely shipped into Japan. What approaches are being taken?

Currently, Japan accepts nontransgenic papaya but it needs to be free of 'contamination' by transgenic papaya. The Hawaii Department of Agriculture (HDOA) and the Japan Ministry of Agriculture, Forestry and Fisheries (MAFF) have agreed on an 'identity preservation protocol' (IPP), in which nontransgenic papaya can even be grown in close proximity (coexistence) to transgenic papaya in Puna, for example, and still be shipped to Japan. The protocol involves a series of monitoring and checkpoints in Hawaii that allow direct marketing of the papaya without delay while samples of the shipment are spot-checked by MAFF officials in Japan. The process has worked very well and allowed Hawaii to maintain its market share in Japan. This represents a practical case of 'coexistence' of transgenic and nontransgenic papaya and fruitful collaboration between governments (Hawaii and Japan) that provides mutual benefits to all parties.

The ideal situation would be, however, to freely export nontransgenic and transgenic papaya to Japan. To this end, efforts are underway to deregulate the transgenic papaya in Japan by obtaining approval from Japanese governmental agencies such as MAFF, the Ministry of Health, Labor, and Welfare (MHLW), and the Ministry of the Environment

(MOE). MAFF has provisionally approved Hawaii's transgenic Rainbow and SunUp papayas, and efforts to present the final documentation to all three agencies are nearing completion. Deregulation of the transgenic papaya in Japan not only would expand the Hawaiian transgenic market but would also be a good case study for evaluating the effectiveness of commercialization and marketing of fresh transgenic products, since the transgenic papaya in Japan would be labeled, and subsequently consumers there would be given an opportunity to make a personal choice between a transgenic and nontransgenic product. The implications of this opportunity are obvious given the current 'controversial' climate over genetically modified organisms (GMOs) in the world. It is indeed rare that the previously little known PRSV could perhaps provide an example that would help us to resolve such controversies.

Summary Remarks

PRSV has been thoroughly characterized and is a typical member of the family *Potyviridae*, arguably the largest and economically most important plant virus group. The complete genome sequence has been elucidated and infectious transcripts have provided a means to determine the genetic determinants of some important biological functions such as host range and virulence. Furthermore, pathogen-derived resistance has been used to control PRSV in Hawaii through the use of virus-resistant transgenic papaya. In the US, only three virus-resistant transgenic crops have been commercialized: squash, papaya, and potato. The transgenic virus-resistant papaya provides a potential means to test the global acceptance of GMOs while presenting a plausible approach to control a disease affecting papaya worldwide.

See also: Plum Pox Virus; Potato Virus Y; Plant Resistance to Viruses: Engineered Resistance; Watermelon Mosaic Virus and Zucchini Yellow Mosaic Virus.

Further Reading

Bateson M, Henderson J, Chaleeprom W, Gibbs A, and Dale J (1994) Papaya ringspot potyvirus: Isolate variability and origin of PRSV type P (Australia). *Journal of General Virology* 75: 3547–3553.

Fuchs M and Gonsalves D (2007) Safety of virus-resistant transgenic plants two decades after their introductions: Lessons from realistic field risk assessments studies. *Annual Review of Phytopathology* 45: 173–202.

Gonsalves D (1998) Control of papaya ringspot virus in papaya: A case study. *Annual Review of Phytopathology* 36: 415–437.

Gonsalves D, Gonsalves C, Ferreira S, et al. (2004) Transgenic virus resistant papaya: From hope to reality for controlling papaya ringspot virus in Hawaii. *APSnet Feature, American Phytopathological Society, Aug.–Sept.* http://www.apsnet.org/online/feature/ringspot/ (accessed January 2008).

Gonsalves D, Vegas A, Prasartsee V, Drew R, Suzuki JY, and Tripathi S (2006) Developing papaya to control papaya ringspot virus by transgenic resistance, intergeneric hybridization, and tolerance breeding. *Plant Breeding Reviews* 26: 35–78.

Yeh SD and Gonsalves D (1994) Practices and perspective of control of papaya ringspot virus by cross protection. In: Harris KF (ed.) *Advances in Disease Vector Research*, pp. 237–257. New York: Springer.

Yeh SD, Jan FJ, Chiang CH, et al. (1992) Complete nucleotide sequence and genetic organization of papaya ringspot virus RNA. *Journal of General Virology* 73: 2531–2541.

Pecluvirus

D V R Reddy, Hyderabad, India
C Bragard, Université Catholique de Louvain, Leuven, Belgium
P Sreenivasulu, Sri Venkateswara University, Tirupati, India
P Delfosse, Centre de Recherche Public-Gabriel Lippmann, Belvaux, Luxembourg

Glossary

Coiled-coil motif A protein structure in which two to six α-helices of polypeptides are coiled together like the strands of a rope.

Hetero-encapsidation Partial or full coating of the genome of one virus with the coat protein of a differing virus. Also termed transcapsidation or heterologous encapsidation.

Leaky scanning mechanism Mechanism by which the ribosomes fail to initiate translation at the first AUG start codon, and scan downstream for the next AUG codon.

Post-transcriptional gene silencing Mechanism for sequence-specific RNA degradation in plants.

t-RNA-like structure Structure mimicking a t-RNA.

Virus-like particles Consist of the structural proteins of a virus. These particles resemble virions meaning that they are not infectious.

History

Pecluviruses, responsible for the 'clump' disease in peanut (=groundnut, *Arachis hypogaea*), have been reported from

West Africa and the Indian subcontinent and contribute globally to annual losses estimated to exceed 38 million US dollars. Pecluviruses also cause economic losses also to other dicotyledonous crops such as pigeonpea, chilli, cowpea, and monocotyledonous crops such as wheat, barley, sorghum, pearl millet, foxtail millet, maize, and sugarcane.

Taxonomy and Classification

The genus *Pecluvirus* comprise two species: isolates from West Africa were grouped under the species name *Peanut clump virus* (African peanut clump virus), and those that occur in India were grouped under *Indian peanut clump virus*. Once the molecular features of the genomes of Indian peanut clump virus (IPCV) and peanut clump virus (PCV) were reported, the ICTV in 1997 assigned them to the newly established genus *Pecluvirus* (siglum from peanut clump). The two viruses differ in host range, antigenic properties, and genomic sequences (see below). Furthermore, the viruses can be distinguished from their geographical location.

Geographic and Seasonal Distribution

IPCV occurs in India in the states of Andhra Pradesh, Gujarat, Punjab, Rajasthan, and Tamil Nadu, and in Pakistan in the provinces of Sindh and Punjab. In Africa, PCV has been reported from Benin, Burkina Faso (Saria, Kamboinsé, Bobo Dioulasso, Niangoloko), Chad, Congo, Côte d'Ivoire, Niger (Sadoré, Maradi), Mali (Segou, Koutiala, Bamako), Gabon, Senegal (Bambey, Cap-Vert, Thies, Sine, Saloum, Pout, Mbour, Kirene), and Sudan. IPCV occurs mainly in the rainy season (July–November) in peanut crops. In post-rainy season (December–March), crops escape the disease. In West Africa, the majority of the peanut crops are grown during the rainy season.

Biological Properties

Host Range and Symptoms

PCV infects economically important crops such as peanut, sorghum, sugarcane, maize, pearl millet, finger millet, and cowpea, whereas IPCV infects peanut, cowpea, pigeon pea, chilli, wheat, barley, pearl millet, finger millet, sorghum, and maize. Both the viruses also infect several monocotyledonous weeds (*Cynodon dactylon, Cyperus rotundus*) that can play a vital role in the survival and dissemination.

Affected plants are conspicuous in the field as a result of their dark green appearance, stunting, and occurrence in patches. IPCV occurs at a high incidence in Rajasthan, India, where peanuts are grown on over 250 000 ha in rotation with irrigated winter wheat and barley and with rain-fed pearl millet during summer. Early-infected plants will not yield and even late-infected crops showed reduction in crop yields up to 60%.

The cereal hosts act as reservoirs of inoculum. The Hyderabad isolate of IPCV (IPCV-H) in young wheat plants up to 3 weeks old induced symptoms similar to rosette caused by soil-borne wheat mosaic virus (SBWMV). Diseased plants are stunted with poorly developed root system. Grain yield losses up to 58% were recorded. Wheat CV RR-21 infected with Durgapura isolate (IPCV-D) showed reduced growth without any overt symptoms on leaves. IPCV-infected barley plants were stunted and bushy, with chlorotic or necrotic leaves, and the majority of these plants died. Those plants that reached maturity produced poorly developed spikes. IPCV-H could also infect finger millet, foxtail, or Italian millet, pearl millet, and sorghum plants. In maize, IPCV-H is responsible for aerial biomass losses up to 33% and grain loss up to 36%.

PCV caused red mottle and chlorotic streaks or stripes on sugarcane with incidence up to 50%. The symptoms varied considerably depending on the cultivar. Yield reductions up to 6% were recorded and the sugar yield was reduced by 14%. In Niger, PCV caused yield losses to sorghum grain up to 62%, to sorghum straw up to 45%, and to pearl millet straw up to 15%, whereas no effect on millet grain yield was observed.

Various of PCV and IPCV were readily detected in the cells of roots, stems, and leaves of systemically infected hosts. PCV particles in wheat cells were found in the cytoplasm, near the nucleus or along the plasmalemma, and arranged in angled-layer aggregates. IPCV and PCV have wide experimental host ranges which include both dicotyledonous and monocotyledonous plants. *Nicotiana clevelandii* × *Nicotiana glutinosa* hybrid and *Phaseolus vulgaris* (Top Crop) are suitable for propagation and as assay/diagnostic hosts, respectively, for IPCV. The isolates of IPCV collected from clump-diseased peanut crops from different locations differed slightly in their host ranges. *Canavalia ensiformis* and *N. clevelandii* × *N. glutinosa* hybrid were found to be useful for distinguishing the IPCV isolates. *Chenopodium amaranticolor, Nicotiana benthamiana, N. glutinosa, P. vulgaris*, and *Triticum aestivum* are of diagnostic value for PCV. The symptoms induced in *C. amaranticolor* by various PCV isolates collected from Senegal, Burkina Faso, and Niger were shown to differ markedly. *Nicotiana benthamiana* and *P. vulgaris* are the propagation species of PCV.

Transmission

Pecluviruses can be transmitted by sap and through peanut seed (PCV up to 6% and IPCV up to 24%). IPCV is transmissible through the seed of pearl millet, finger millet, foxtail millet, wheat, and maize generally at rates of <2%. IPCV has been shown to be transmitted by the plasmodiophorid, *Polymyxa graminis. Sorghum arundinaceum* (bait plant) is a suitable host for testing vector transmission of PCV to roots. Convincing evidence for PCV transmission by *Polymyxa* is still lacking. The thick-walled resting spores of the vector probably carry the virus and contribute to its survival. *P. graminis* from tropical and subtropical regions differed in ribotype and temperature requirements. For North American and European *P. graminis* f. sp. *temperata*, optimum temperature for survival is between 15 and 20 °C as opposed to the narrow optimum temperature range (close to 30 °C) for the Indian *P. graminis* f. sp. *tropicalis. Polymyxa graminis* in the tropics has a wide host range, including monocotyledonous and dicotyledonous plants, as opposed to the narrow host range (largely restricted to monocotyledons) for *P. graminis* f. sp. *temperata*. The plasmodiophorid needs a cereal host for completing its life cycle.

Serological Diagnosis

PCV is highly immunogenic and its polyclonal antibodies did not react with IPCV, barley stripe mosaic (BSMV), tobacco mosaic (TMV), beet necrotic yellow vein, potato mop top (PMTV), and SBWMV viruses. Wide serological diversity exists among isolates of PCV and IPCV. Therefore serological tests have limitations to detect more than one isolate from a single antibody source. Antisera to different isolates of IPCV facilitated the grouping of isolates into three serotypes, IPCV-H, IPCV-D, and IPCV-Ludhiana (IPCV-L). All IPCV serotypes are serologically distinct from PCV isolates, and vice versa. Utilizing four different formats of enzyme-linked immunosorbent assay (ELISA) and a panel of monoclonal antibodies raised against a PCV isolate, a number of PCV isolates were grouped into five serotypes. Interestingly, one of the monoclonal antibodies reacted with IPCV-D in triple antibody sandwich ELISA.

Molecular Diagnosis

To detect pecluviruses in disease surveys, to eliminate virus-contaminated sources in quarantine, and to devise strategies for disease management, it is essential to utilize diagnostic tools that are highly sensitive and broadly specific. Nucleic acid probes (both radioactive and nonradioactive) for the conserved regions were found to be ideal for pecluvirus diagnosis. Initially probes derived from the 742 nt at the 3' end of IPCV-RNA1 were used. Subsequently, probes targeting the P14 gene and a probe (CGAGCCATAGAGCACGGTTGTGGG) derived from the conserved 3' terminal ends of both RNA1 and RNA2 of IPCV facilitated highly sensitive detection of a range

of IPCV and PCV isolates. Of the various methods tested, reverse transcription-polymerase chain reaction (RT-PCR) was found to be the most suitable one.

Molecular Properties

Virus Particles

ICPCV virions are nonenveloped rigid rods, 24 nm in diameter, with two predominant lengths of *c.* 250 and *c.* 180 nm. PCV particles contain two predominant rigid rods measuring *c.* 245 and *c.* 190 nm and 21 nm in diameter with a clear axial canal. They contain a single coat protein (CP) of 23 kDa.

Genome

The pecluvirus genome is bipartite. PCV and IPCV virions contain positive-sense, single-stranded RNAs (4% by weight) RNA1 is *c.* 5900 nt long and RNA2 is *c.* 4500 nt long. The RNAs of PCV showed little sequence identity with those of IPCV, with the exception for the 3' terminal 273 nt, which are conserved between the two RNAs (**Figure 1**).

Coding Sequences

RNA1 encodes two proteins involved in viral RNA replication (P131 and P191). P191 is a C-terminally extended form of P131 produced by translational readthrough of the UGA termination codon (P131) (**Table 1**). Additionally RNA1 codes for P15, a suppressor of posttranscriptional gene silencing (PTGS). The P15 open reading frame (ORF) is downstream of the P191 ORF and separated from it by a noncoding region of about 60 nt. The amino acid sequence of P191 contains methyltransferase, helicase, and RNA-dependent RNA polymerase domains. P15 on RNA1 is a cysteine-rich protein (CRP), translated from a relatively abundant sub-genomic RNA1 (**Figure 1**). P15 resembles CRPs of BSMV, poa semilatent virus, and SBWMV. These proteins have been suggested to act as regulatory factor during virus replication as well as for long-distance movement and contribute to virulence factor. RNA1 is able to replicate in the absence of RNA2 in protoplasts of tobacco BY-2 cells. However, both RNA1 and RNA2 are needed for infection. Experiments using enhanced 5' green fluorescent protein and 5'-bromouridine 5'-triphosphate labels have suggested that PCV replication complexes co-localize with endoplasmic reticulum green fluorescent bodies accumulating around the nucleus during infection. P15 does not act directly at sites of viral replication but intervenes indirectly to control viral accumulation levels. It acts as a suppressor of PTGS, though it shares no sequence similarities with previously described anti-PTGS molecules encoded by other

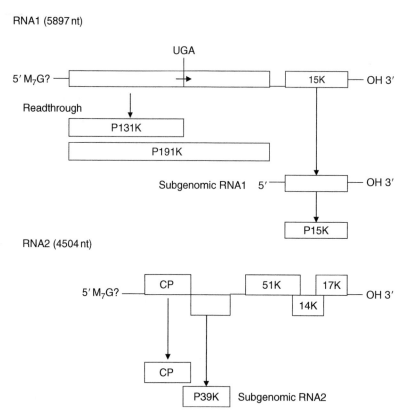

Figure 1 Illustrates the organization of the pecluvirus genome.

Table 1 Pecluvirus open reading frames, polypeptides, and their functions

Genomic RNA	ORF	M_r of polypeptide (kDa)	Function
RNA1	P131	131 ⎫	Methyltransferase, helicase, replicase
	P191	191 ⎭	
	P15	15	Suppressor of PTGS
RNA2	P23	23	Coat protein
	P39	39	Putative vector transmission factor
	P51	51 ⎫ Triple gene block	
	P14	14 ⎬	Virus movement
	P17	17 ⎭	

viruses. The P15 possesses four C-terminal proximal heptad repeats that can generate a coiled-coil interaction and is targeted to peroxisomes via a C-terminal SKL motif. Such a motif is conserved among pecluviruses from both Africa and India. It has been demonstrated that a coiled-coil motif is necessary for the anti-PTGS activity of P15, but the peroxisomal localization signal is not, although it is required for efficient intercellular movement of the virus.

RNA2 is relatively more complex and encodes five polypeptides (**Figure 1**). The 5′-proximal ORF (23K) encodes the CP. The following ORF (ORF2) encodes P39, a putative vector transmission factor, which is expressed by a leaky scanning mechanism *in vitro*. ORF2 starts 1 nt upstream of the first residue of UGA stop codon of the CP cistron. The remaining three ORFs encode P51,

P14, and P17 and form a triple gene block (TGB) (**Table 1**). TGB plays a role in virus movement. P51 is translated from a relatively abundant subgenomic RNA2, but subgenomic RNAs responsible for the synthesis of P14 and P17 have not yet been detected. However, analogy with other TGB-containing viruses suggests that both proteins are probably translated from a low-abundance subgenomic RNA with a 5′ terminus upstream of the P14 gene.

Noncoding Sequences

The 5′ and 3′ noncoding regions (NCRs) of RNA1 are about 130 and 300 nt in length, respectively (**Figure 1**). Those of the RNA2 are more diverse, between *c.* 390 and *c.* 500 nt in length. There is no distinct 5′ sequence feature common to all pecluviruses. RNA1 and RNA2 have similar 5′ NCRs except for six to seven nucleotides and these sequences are shared among pecluvirus species. The 3′ NCRs are *c.* 300 nt in length, and *c.* 100 terminal nucleotides are identical among pecluvirus RNAs sequenced so far. Such sequence similarity has enabled the development of a hybridization probe corresponding to the 3′-terminal 700 nt of IPCV-H. This probe detected all the currently known IPCV serotypes as well as an isolate of PCV.

The 3′ NCR of pecluviruses, as in the case of furoviruses, forms a t-RNA-like structure (TLS) that is capable of high-efficiency valylation and aids in the replication of both PCV RNAs. The internal NCRs in RNA1 are present between P191 and P15 (60 nt) and in RNA2 between P39 and P51 (145 nt).

Sequence Comparisons

The RNA1-encoded polypeptides of PCV and IPCV share identities ranging from 75% (P15) to 95% (readthrough part of P191) and show significant similarities with furoviruses (e.g., 56% identity with polymerase of SBWMV). The proteins encoded by RNA2 are 39% (P39) to 89% (P14) identical between species. The CPs are *c.* 60% identical and also have significant similarity (*c.* 30% identity) with the CP of BSMV (genus *Hordeivirus*). The TGB proteins resemble those of PMTV (genus *Pomovirus*).

The nucleotide sequence of IPCV-H RNA1, is similar to that of PCV and the polypeptides encoded by this are 60–95% identical. Comparison of the P15 gene shows a close relationship between IPCV isolates and a relatively high diversity among PCV isolates (**Figure 2**).

The five RNA2-encoded ORFs of IPCV-L are between 32% and 93% identical to those encoded by PCV RNA2. The partial nucleotide sequences of RNA2 of IPCV-H and -D showed that the polypeptides encoded by the two 5′-proximal ORFs (CP and P39) are similar to those of the IPCV-L serotype. A conserved motif 'F-E-X₆-W' is present near the CP C-terminus of all three IPCV serotypes and PCV, as in the CPs of other rod-shaped viruses (TMV and tobacco rattle virus).

The full-length sequence comparison between RNA2 of four isolates of PCV and two isolates of IPCV have revealed a high degree of variability in size (between 58% and 79%). Amino acid sequence alignments of each of the five ORFs of RNA2 showed that ORF4, encoding P14 of TGB, is highly conserved (90–98% identical), whereas the P39 encoded by ORF2 is less conserved (25–60% identical). The CP of eight isolates showed amino acid sequence identities between 37% and 89%. Phylogenetic comparisons, based on complete RNA2 sequences, showed that the eight isolates could be grouped into two distinct clusters with no geographical distinction between PCV and IPCV isolates. Phylogenetic tree topologies for

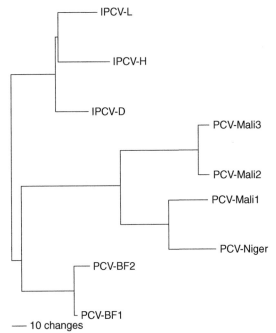

Figure 2 Maximum parsimony phylogenetic tree obtained by a heuristic search from 352 characters of the P15 ORF of IPCV and PCV. Multiple alignments of nucleotide sequences of the P15 ORF from different sources of IPCV (D, H, L serotypes) and PCV isolates from Mali, Burkina Faso (BF), and Niger were obtained using CLUSTAL W 2.08b with the suggested default settings. Phylogenetic analyses were performed using PAUP 4.0 beta 1 (Sinauer Associates, Inc., Sunderland, MA) (Dieryck and Bragard, unpublished data).

individual ORFs revealed an overall similarity with that obtained from complete RNA2 sequences, but the relative positions of individual isolates varied within each cluster. Further, such studies indicate that there is substantial divergence among the RNA2's of pecluviruses and suggest that different polypeptides have evolved differently, possibly due to different selection pressures.

Several PCV isolates propagated in *N. benthamiana* contain an RNA2 shorter than that of the type isolate. Partial characterization of two such isolates revealed that their RNA2's have undergone deletions in ORF2. The impact of deletions in ORF2, implicated in vector transmission, is to be established.

Assembly of Virus Particles

The origin of assembly sequence (OAS) positions has been identified in the ssRNA genomes of several rod-shaped viruses. By testing the ability of different RNA1 and RNA2 deletion mutants to be encapsidated *in vivo*, the RNA1 and RNA2 sequences required for assembly into PCV virions have been established. A putative OAS was mapped in the 5′-proximal part of the P15 gene of RNA2. Nevertheless, the nonencapsidation of subgenomic RNA that encodes P15 raises questions about the mechanism underlying the encapsidation process. Two sequence positions that could drive encapsidation of RNA2 have been identified. One is in the 5′-proximal CP gene and the other in the P14 gene near the RNA2 3′ terminus. No obvious sequence similarities between different assembly initiation sequences have been noted. The initiation of PCV assembly, like that of TMV, probably involves interaction of CP with a relatively short sequence which is presented by an RNA secondary structure in a special configuration.

Interestingly, the possible localization of an OAS in the CP gene of IPCV was realized as a result of formation of virus-like particles (VLPs) in *Escherichia coli* and *N. benthamiana*, confirming the results observed for PCV.

The assembly of VLPs in either bacteria or plants has been proposed as a way to protect and accumulate specific mRNA, as a means to study the molecular assembly of the capsid and for the production of oral vaccines. In genetically transformed *N. benthamiana* or *E. coli* cells, the cloned IPCV-H CP gene is expressed and assembled into VLPs. The monomer VLP size approximately corresponds to the one expected, according to the length of the encapsidated CP gene transcript RNA. Using immunocapture RT-PCR(IC-RT-PCR), such VLPs have been demonstrated to contain RNA encoding IPCV-H CP. When transgenic *N. benthamiana* expressing IPCV-H CP gene inoculated with the serologically distinct IPCV-L serotype, accumulated virus particles that contained both types of CP, the possibility of hetero-encapsidation in transgenic plants was suggested.

Epidemiology and Management

Introduction

The control of peanut clump depends upon the accurate and sensitive detection of IPCV and PCV in plants, seeds, and soil. Peanut clump disease is largely restricted to sandy soils and sandy loams. Detection is needed to identify infested fields and implement appropriate management strategies; second, to eliminate seed lots infected by the viruses and to assess the resistance of peanut, pearl millet, sorghum, and sugarcane breeding lines.

Life Cycle

Dicotyledonous hosts restrict the multiplication of the plasmodiophorid vector and hence are considered as fortuitous hosts that may not contribute to perpetuation of virus inoculum. Indeed, either virus-infected peanut roots or seed could transmit or establish the disease. Monocotyledonous plants such as maize, pearl millet, and sorghum are 'preferred' hosts of the vector and contribute to the build-up of vector inoculum potential in the soil. Seed of millets, maize, and wheat and rhizomatous grasses (e.g., *Cynodon dactylon*) are likely to contribute to the disease establishment in new areas by supporting the multiplication of both the virus and plasmodiophorid vector.

The role of rainfall and temperature in the dynamics of infection by IPCV-H and its vector were analyzed on various crops. Wheat followed by barley showed the highest virus incidence, although *P. graminis* was rarely observed in the roots of wheat and was not detected in those of barley. The roots of maize, pearl millet, and sorghum plants, colonized by *P. graminis*, showed the presence of the virus. Peanut is a systemic host for the virus but no *P. graminis* was found in its roots. High rainfall soon after summer months resulted in high incidences of the disease. Weekly rainfall of 14 mm is sufficient for the vector to initiate infection. Temperatures (27–30 °C) prevailing during the rainy season were found to be conducive to virus transmission. At temperatures below 23 °C, infection did not occur and the development of the plasmodiophorid was delayed. This appears to be the major reason for the absence of clump disease on crops raised during the post-rainy season in India.

Cultural Practices

Continuous cropping with fortuitous hosts such as peanut, cowpea, or pigeonpea is likely to reduce the plasmodiophorid population in soil. Seed-borne inoculum from dicotyledonous hosts does not aid in disease establishment. However, seed-borne inoculum from cereal hosts and rhizomatous grasses can contribute to disease establishment.

Initial experiments showed that application of soil biocides (e.g., dibromochloropropane, DD, fumigant nematicides that also have fungicidal action) and soil solarization were effective in reducing disease incidence. Nonetheless, they are not economical to adopt and additionally are hazardous to use.

The following cultural practices can reduce disease incidence of IPCV:

1. early planting of groundnut before the onset of the monsoon under judicious irrigation;
2. trap cropping with pearl millet, that is, sowing a pearl millet crop at a high density and then ploughing the entire crop, 2 weeks after germination, into the soil and then planting with peanut;
3. avoiding rotation of peanut with such highly susceptible crops as sorghum, wheat, and maize; and
4. maintain continuous cropping with dicotyledonous crops (peanut, pigeonpea, cowpea, and marigold) for at least three growing seasons.

The above-recommended measures are ecofriendly and economical and are practicable even under marginal farming conditions.

Host Plant Resistance

No resistance to IPCV was found in over 9000 cultivated *Arachis* germplasm lines. Resistance to IPCV was identified in a wild *Arachis* sp., but it is yet to be incorporated into cultivated peanut. Transgenic peanut lines carrying virus genes (CP and replicase) are currently being evaluated. However, they are unlikely to be available in the near future for cultivation.

Future Perspectives

The suspected role of P39 in vector transmission of the virus needs to be confirmed probably by mutational analysis to exploit this ORF in transgenic research. The presence of subgenomic RNAs encoding P14 and P17 in infected tissues needs to be verified. The cultural control measures tested for IPCV are worth exploiting in West Africa to minimize the impact of PCV.

The most economical way to control pecluviruses is by developing virus-resistant cultivars. The organizations involved must seek approvals from appropriate licensing agencies so that the transgenic plants will become available for cultivation.

See also: Furovirus.

Further Reading

Bragard C, Doucet D, Dieryck B, and Delfosse P (2006) Detection of pecluviruses. In: Rao GP, Kumar PL,, and Holguin-Peña RJ (eds.) *Characterization, Diagnosis & Management of Plant Viruses: Vegetable and Pulse Crops,* 1st edn., vol. 3, pp. 125–140. Houston: Studium Press.

Bragard C, Duncan GH, Wesley SV, Naidu RA, and Mayo MA (2000) Virus-like particles assemble in plants and bacteria expressing the coat protein gene of *Indian peanut clump virus. Journal of General Virology* 84: 267–272.

Delfosse P, Reddy AS, Thirumala Devi K, *et al.* (2002) Dynamics of *Polymyxa graminis* and *Indian peanut clump virus* (IPCV) infection on various monocotyledonous crops and groundnut during the rainy season. *Plant Pathology* 51: 546–560.

Dunoyer P, Pfeffer S, Fritsch C, Hemmer O, Voinnet O, and Richards KE (2002) Identification, subcellular localization and some properties of a cystein-rich suppressor of gene silencing encoded by peanut clump virus. *Plant Journal* 19: 555–567.

Fritsch C and Dollet M (2000) Genus *Pecluvirus.* In: Van Regenmortel MHV, Fauquet CM, Bishop DHL, *et al.* (eds.) *Virus Taxonomy: Seventh Report of the International Committee on Taxonomy of Viruses,* pp. 913–917. San Diego: Academic Press.

Hemmer O, Dunoyer P, Richards K, and Fritsch C (2003) Mapping of viral RNA sequences required for assembly of peanut clump virus particles. *Journal of General Virology* 84: 2585–2594.

Herzog E, Guilley H, Manohar SK, *et al.* (1994) Complete sequence of peanut clump virus RNA-1 and relationships with other fungus transmitted rod shaped viruses. *Journal of General Virology* 75: 3147–3155.

Herzog E, Hemmer O, Hauser S, Meyer G, Bouzoubaa S, and Fritsch C (1998) Identification of genes involved in replication and movement of peanut clump virus. *Virology* 248: 312–322.

Legrève A, Delfosse P, and Maraite H (2002) Phylogenetic analysis of *Polymyxa* species based on nuclear 5.8S and internal transcribed spacers ribosomal DNA sequences. *Mycological Research* 106: 138–147.

Legrève A, Delfosse P, Vanpee B, Goffin A, and Maraite H (1998) Differences in temperature requirements between *Polymyxa* sp. of Indian origin and *Polymyxa graminis* and *Polymyxa betae* from temperate areas. *European Journal of Plant Pathology* 104: 195–205.

Manohar SK, Dollet M, Dubern J, and Gargani D (1995) Studies on variability of peanut clump virus: Symptomatology and serology. *Journal of Phytopathology* 143: 233–238.

Miller JS, Wesley SV, Naidu RA, Reddy DVR, and Mayo MA (1996) The nucleotide sequence of RNA-1 of Indian peanut clump furovirus. *Archives of Virology* 141: 2301–2312.

Naidu RA, Miller JS, Mayo MS, Wesley SV, and Reddy AS (2000) The nucleotide sequence of Indian peanut clump virus RNA 2: Sequence comparisons among pecluviruses. *Archives of Virology* 145: 1857–1866.

Reddy AS, Hobbs HA, Delfosse P, Murthy AK, and Reddy DVR (1998) Seed transmission of Indian peanut clump virus (IPCV) in peanut and millets. *Plant Disease* 82: 343–346.

Reddy DVR, Mayo MA, and Delfosse P (1999) Pecluviruses. In: Granoff A and Webster RG (eds.) *Encyclopedia of Virology,* 2nd edn., vol. 2, pp. 1196–1200. New York: Academic Press.

Plant Reoviruses

R J Geijskes and R M Harding, Queensland University of Technology, Brisbane, QLD, Australia

Glossary

Open reading frame A sequence of nucleotides in DNA that can potentially translate as a polypeptide chain.

Icosahedral Having twenty equal sides or faces.

Capsid The protein shell that surrounds a virus particle.

Monocotyledon A flowering plant that has only one cotyledon or seed leaf in the seed.

Dicotyledon A flowering plant that has two cotyledons or seed leaves in the seed.

Transovarial Transmission from one generation to another through eggs.

Introduction

The family *Reoviridae* comprises a diverse group of viruses which can infect vertebrates, invertebrates, and plants. Despite their large host range, all members of the family *Reoviridae* share common properties including an icosahedral shaped virion and segmented double-stranded RNA (dsRNA) genome. The family *Reoviridae* consists of nine genera of which three genera, *Fijivirus*, *Oryzavirus*, and *Phytoreovirus* are plant-infecting reoviruses. These reoviruses generally replicate in both plant hosts and insect vectors. Infection of the insect vector is non-cytopathic and persists often throughout the life of the insect. Infection of the host plant is tissue specific and can cause severe disease. Fiji leaf gall disease, caused by Fiji disease virus (FDV), has caused yield losses of up to 90% in susceptible varieties of sugarcane in Australia. Rice ragged stunt virus (RRSV) is reported to reduce yield of rice by up to 100% in severe infections (generally 10–20%). Rice dwarf disease, caused by rice dwarf virus (RDV), can also cause significant losses as infected plants often fail to bear seeds. The genera of plant-infecting reoviruses are differentiated according to the number of genomic dsRNA segments and their electrophoretic profile, hosts, serological relationships, and capsid morphology (**Table 1**).

Taxonomy and Classification

Currently there are three genera of the family *Reoviridae* which are classed as plant-infecting reoviruses, *Fijivirus*, *Oryzavirus*, and *Phytoreovirus*. These reoviruses replicate both in plant hosts (except for one fijivirus: *Nilaparvata lugens* reovirus) and in their insect vectors (**Table 1**). Infection of the host plant is species specific, although the host range can often be extended under experimental conditions, and can produce various symptoms, including severe disease. The complete genome sequence has been obtained for a number of viruses and at least partial sequence information is now available for all plant reoviruses. This has allowed detailed comparisons within these genera and across all of the *Reoviridae*, thus providing a basis for the classification of these viruses into species and genera.

Within the genus *Fijivirus*, individual species have considerable similarities to the type species, Fiji disease virus. Classification into separate species is based on unique characteristics such as capacity to exchange genome segments, relatively high amino acid sequence similarity, serological cross-reaction, cross-hybridization

Table 1   Characteristics of plant reoviruses

Genus	Virus	dsRNA genome segments	Host	Vector/s
Fijivirus	Fiji disease virus (FDV)	10	Monocot (Gramineae)	Planthoppers: *Perkinsiella saccharicida, P. vastatrix, P. vitiensis*
	Rice black-streaked dwarf virus (RBSDV)	10	Monocot (Gramineae)	Planthoppers: *Laodelphax striatellus, Ribautodelphax albafascia, Unkanodes sapporona*
	Maize rough dwarf virus (MRDV)	10	Monocot (Gramineae)	Planthopper: *Ribautodelphax notabilis*
	Pangola stunt virus (PaSV)	10	Monocot (Gramineae)	Planthoppers: *Sogatella furcifera, S. kolophon*
	Mal del Rio Cuarto virus (MRCV)	10	Monocot (Gramineae)	Planthopper: *Delphacodes kuscheli*
	Oat sterile dwarf virus (OSDV)	10	Monocot (Gramineae)	Planthoppers: *Javesella pellucidia, J. discolour, J.dubia, J.obscurella, Dicranotropis hamata*
	Garlic dwarf virus (GDV)	10	Monocot (Liliaceae)	Planthopper: Unknown
	Nilaparvata lugens reovirus (NLRV)	10	No plant host reported	Planthoppers: *Nilaparvata lugens, Laodelphax striatellus*
Oryzavirus	Echinochloa ragged stunt virus (ERSV)	10	Monocot (Gramineae)	Planthoppers: *Sogatella longifurcifera, S. vibix*
	Rice ragged stunt virus (RRSV)	10	Monocot (Gramineae)	Planthopper: *Nilaparvata lugens*
Phytoreovirus	Wound tumor virus (WTV)	12	Dicots	Leafhoppers: *Agallia constricta, A.quadripunctata, Agalliopsis novella*
	Rice dwarf virus (RDV)	12	Monocot (Gramineae)	Leafhoppers: *Nephotettix cincticeps, N. nigropictus, Recillia dorsalis*
	Rice gall dwarf virus (RGDV)	12	Monocot (Gramineae)	Leafhoppers: *Nephotettix cincticeps, N. nigropictus, N. virescens, N. malayanus, Recillia dorsalis*

of RNA or cDNA probes, host species, and insect vector species. In addition to these commonly used identifiers, analysis of the available genome sequences has assisted in identification of *Fijivirus* species. Gross genome characteristics for fijivirus members include a genome size of approximately 29 kbp and a characteristically low G+C content of 34–36%. Unique, and highly conserved, 5′ and 3′ terminal sequences are present in different plant reovirus species; in all RNA segments, the 3′ terminal trinucleotide is conserved across all species within the genus *Fijivirus* (**Table 2**). Inverted repeats are found adjacent to the terminal sequences and these differ from those of other plant reoviruses.

Members of the genera *Oryzavirus* and *Phytoreovirus* have significant similarity to type members rice ragged stunt virus and wound tumor virus, respectively. Demarcation of species within the oryzaviruses and phytoreoviruses are primarily based on the ability to exchange genome segments although other characteristics, as mentioned for fijiviruses above, are also used. When available, genomic sequences are examined to reveal distinguishing features to support the classification. Oryzaviruses have a total genome size of approximately 26 kbp and specific 5′ and 3′ terminal sequences in all RNA segments: $^{5'}$GAUAAA—GUGC$^{3'}$. Phytoreoviruses have a total genome size of

approximately 25 kbp, a G+C content between 38% and 48% and specific 5′ and 3′ terminal sequences, ($^{5'}$GG(U/C)A—(U/C)GAU$^{3'}$) in all RNA segments.

Virion Structure and Genome Organization

Fijivirus

The virions have a complex double icosahedral capsid construction and consist of a capsid, a core, and a nucleoprotein complex. Virions are fragile structures and readily break down *in vitro* to give cores. The outer capsid is 65–70 nm in diameter with 12 'A' type spikes located at the vertices of the icosahedron. The inner core is about 55 nm in diameter, with 12 'B' type spikes located at the vertices. The viral nucleic acid is located at the center of the virus particle, within the inner core capsid. Each virion contains a single full-length copy of the genome. Fijivirus genomes contain ten dsRNA genomic segments varying from approximately 1.8–4.5 kbp (**Figure 1**). The total genome is approximately 29 kbp with a low G + C content of 34–36%. Highly conserved unique 5′ and 3′ terminal sequences are found on all RNA segments (**Table 2**). Segment-specific inverted repeats are found

Table 2 Conserved 5′ and 3′ sequences identified in fijiviruses

Virus	*5′ conserved sequence*	*3′ conserved sequence*[a]
Fiji disease virus	$^{5'}$AAGUUUUU—$^{3'}$	$^{5'}$—CAGCNNNN*GUC*$^{3'}$
Rice black-streaked dwarf virus	$^{5'}$AAGUUUUU—$^{3'}$	$^{5'}$—AGCUNN(C/U)*GUC*$^{3'}$
Maize rough dwarf virus	$^{5'}$AAGUUUUUU—$^{3'}$	$^{5'}$—U*GUC*$^{3'}$
Mal del Rio Cuarto virus	$^{5'}$AAGUUUUU—$^{3'}$	$^{5'}$—CAGCUNNN*GUC*$^{3'}$
Oat sterile dwarf virus	$^{5'}$AACGAAAAAAA—$^{3'}$	$^{5'}$—UUUUUUUUA*GUC*$^{3'}$
Nilaparvata lugens reovirus	$^{5'}$AGU—$^{3'}$	$^{5'}$—G*UUGUC*$^{3'}$

[a]Italicized trinucleotide is conserved in all fijivirus sequences reported to date.

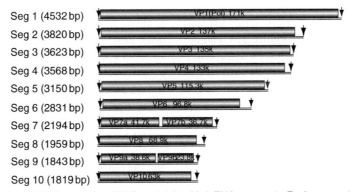

Seg 1 (4532 bp)
Seg 2 (3820 bp)
Seg 3 (3623 bp)
Seg 4 (3568 bp)
Seg 5 (3150 bp)
Seg 6 (2831 bp)
Seg 7 (2194 bp)
Seg 8 (1959 bp)
Seg 9 (1843 bp)
Seg 10 (1819 bp)

Figure 1 Genome organization of Fiji disease virus (FDV) containing 10 dsRNA segments. Each segment contains one ORF except for Seg 7 and Seg 9 which contain two ORFs. The arrows indicate the location of the 5′ and 3′ conserved sequences, respectively. Reproduced from Fauquet CM, Mayo MA, Maniloff J, Desselberger U, and Ball LA (2005) *Virus Taxonomy – Classification and Nomenclature of Viruses: Eighth Report of the International Committee on the Taxonomy of Viruses.* San Diego, CA: Elsevier Academic Press, with permission from Elsevier.

Table 3   Genome organization of FDV and predicted gene function

Segment	Size (bp)	Protein nomenclature	MW (kDa)	Predicted function (location)
S1	4532	VP1	170.6	RdRp (core)
S2	3820	VP2	137.0	Possible core protein (core)
S3	3623	VP3	135.5	Possible B spike (capsid)
S4	3568	VP4	133.2	Possible core protein (core)
S5	3150	VP5	115.3	Unknown
S6	2831	VP6	96.8	Unknown
S7	2194	VP7a	41.7	Unknown
		VP7b	36.7	Unknown
S8	1959	VP8	68.9	Possible NTP-binding (core)
S9	1843	VP9a	38.6	Structural protein (unknown)
		VP9b	23.8	Nonstructural
S10	1819	VP10	63.0	Outer capsid protein (capsid)

Reproduced from Fauquet CM, Mayo MA, Maniloff J, Desselberger U, and Ball LA (2005) *Virus Taxonomy – Classification and Nomenclature of Viruses: Eighth Report of the International Committee on the Taxonomy of Viruses.* San Diego, CA: Elsevier Academic Press, with permission from Elsevier.

adjacent to these terminal sequences. Segments 1–6, 8, and 10 are monocistronic, containing one open reading frame (ORF), while segments 7 and 9 each contain two ORFs. NLRV is the only fijivirus identified to date which differs from this structure, with one ORF on segment 7. The functions of proteins encoded by most ORFs are still unconfirmed; gene functions of segments 1–4 and 8–10 have been predicted based on protein expression studies or sequence similarities to related reoviruses (**Table 3**).

Oryzavirus

The virions have a double-shelled icosahedral capsid and consist of an outer capsid, an inner capsid, and a core. Virions are fragile and readily break down *in vitro* to give subviral core particles unless pre-treated. The outer capsid is 75–80 nm in diameter with 12 'A' type spikes located at the five fold axis of the icosahedron. The core capsid is about 57–65 nm in diameter, with 12 'B' type spikes. The viral nucleic acid is located at the center of the viral particle, within the core capsid. The virus genome consists of ten dsRNA segments ranging in size from 1162–3849 bp (RRSV) with a total length of 26 kbp (**Figure 2**). Genome segments 1–3, 5, 7–10 of RRSV each contain a single ORF, while segment 4 contains two ORFs. Segment 8 encodes a polyprotein which is cleaved into two proteins. **Table 4** summarizes the organization of the RRSV dsRNA segments and the predicted function of the encoded proteins.

Phytoreovirus

The virions have a double-shelled icosahedral capsid construction and consist of an outer capsid, a core capsid, and a smooth core. Virions are approximately 70 nm in diameter with 12 spikes located at the fivefold vertices of the icosahedron and generally remain intact when purified. WTV, the type member, possesses three protein shells: an outer

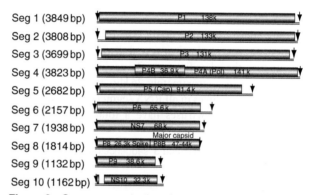

Figure 2   Genome organization of rice ragged stunt virus (RRSV) containing 10 dsRNA segments. Each segment contains one ORF except for Seg 4 which contains two ORFs. The arrows indicate the location of the 5′ and 3′ conserved sequences, respectively. Reproduced from Fauquet CM, Mayo MA, Maniloff J, Desselberger U, and Ball LA (2005) *Virus Taxonomy – Classification and Nomenclature of Viruses: Eighth Report of the International Committee on the Taxonomy of Viruses.* San Diego, CA: Elsevier Academic Press, with permission from Elsevier.

amorphous layer made up of two proteins, an inner capsid made up of two proteins, and a smooth core made up of three proteins that is about 50 nm in diameter. Each virion contains a single full-length copy of the genome. Phytoreoviruses have 12 segments of dsRNA which range in size from approximately 1–4.5 kbp with a total genome length of approximately 25 kbp (**Figure 3**) and a G+C content of 38–44%. Each segment of RDV contains a single ORF except for segment 12, which contains two ORFs. **Table 5** summarizes the organization of the RDV dsRNA segments and the putative function of the encoded proteins.

Replication and Gene Expression

The replication and gene expression of plant-infecting reoviruses is thought to be similar to that of other

Table 4 Genome organization of RRSV and predicted gene function

Segment	Size (bp)	Protein nomenclature	MW (kDa)	Predicted function (Location)
S1	3849	P1	137.7	Virion associated (B spike)
S2	3810	P2	133.1	Inner core capsid (core capsid)
S3	3699	P3	130.8	Major core capsid (core capsid)
S4	3823	P4a	141.4	RdRp
		P4b	36.9	(unknown)
S5	2682	P5	91.4	Capping enzyme
S6	2517	P6	65.6	Unknown
S7	1938	NS7	68	Nonstructural protein (unknown)
S8	1814	P8	67.3	Precursor polyprotein/protease
		P8a	25.6	Spike
		P8b	41.7	Major capsid protein
S9	1132	P9	38.6	Vector transmission (spike)
S10	1162	NS10	32.3	Nonstructural protein

Reproduced from Fauquet CM, Mayo MA, Maniloff J, Desselberger U, and Ball LA (2005) *Virus Taxonomy – Classification and Nomenclature of Viruses: Eighth Report of the International Committee on the Taxonomy of Viruses.* San Diego, CA: Elsevier Academic Press, with permission from Elsevier.

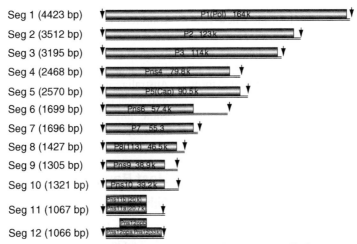

Figure 3 Genome organization of rice dwarf virus (RDV) containing 12 dsRNA segments. Each segment contains one ORF except for Seg 11 and Seg 12 which contain two ORFs. The arrows indicate the location of the 5′ and 3′ conserved sequences, respectively. Reproduced from Fauquet CM, Mayo MA, Maniloff J, Desselberger U, and Ball LA (2005) *Virus Taxonomy – Classification and Nomenclature of viruses: Eighth Report of the International Committee on the Taxonomy of Viruses.* San Diego, CA: Elsevier Academic Press, with permission from Elsevier.

reoviruses. The best described of these is bluetongue virus (BTV), type member of the genus *Orbivirus*. If the BTV model is accurate for the plant-infecting reoviruses, replication occurs after virions (or viral cores) are delivered into the host cell. Replication is initiated when the viral capsid layer is removed and the core enters the cytoplasm of the cell. The viral genome (10–12 segments) remains packaged in the central cavity of the viral core to ensure host cell defense responses to dsRNA are not activated. The core is biochemically active with RNA-dependent RNA polymerase (RdRp), capping enzyme, and helicase enzyme. The viral core contains a number of channels, the largest of which is at the fivefold axis of the icosahedral structure. Smaller channels allow the entry of nucleotides into the core which are required for transcription. The

large channel is located adjacent to the replicase/transcriptase complex which has helicase activity for the unwinding and rewinding of the dsRNA genome during transcription of negative RNA strand. The newly formed positive strand mRNA molecules are modified to form a Cap1 structure by the guanylyltransferase, nucleotide phosphohydrolase, and transmethylase activity of the capping enzyme prior to the extrusion of mRNA, from the major pore, into the cytoplasm. These mRNA molecules released into the cytoplasm can be translated to produce viral proteins. Nonstructural viral proteins aggregate to form inclusion bodies or viroplasms. The viroplasm is the site of most of the mRNA production and assembly of core proteins. The mRNA molecules, one of each segment, are assembled with these viral proteins to form new virus core

Table 5   Genome organization of RDV and putative gene function

Segment	Size (bp)	Protein nomenclature	MW (kDa)	Predicted function (location)
S1	4423	P1	170	RdRp (core)
S2	3512	P2	130	Capsid structural protein (outer capsid)
S3	3195	P3	110	Major core protein (core capsid)
S4	2468	Pns4	83	Nonstructural protein
S5	2570	P5	89	Guanylyltransferase (core)
S6	1699	Pns6	56	Nonstructural protein
S7	1696	P7	58	Nucleic acid binding protein (core)
S8	1427	P8	43	Major outer capsid protein
S9	1305	Pns9	49	Nonstructural protein
S10	1321	Pns10	53	Nonstructural protein
S11	1067	Pns11a	23	Nonstructural protein
		Pns11b	24	
S12	1066	Pns12	34	Nonstructural protein
		Pns12OPa	8	
		Pns12OPb	7	

particles. Once a copy of the mRNA is inside a new viral core, the negative strand is synthesized completing the replication of the dsRNA genome. The complete viral core containing dsRNA then moves to the periphery of the viroplasm where capsid proteins are assembled to form the complete new viral particle.

Control of gene expression of a multisegmented genome is complex and not fully understood. Each genome segment contained within the viral core is associated with a single replicase/transcription complex, located adjacent to the major pore in the vertices of the icosahedron, and is transcribed separately to make full-length positive sense RNA copies. The location of the replicase/transcription complex also restricts the number of genome segments to a maximum of 12. These 10–12 mRNAs are produced in different molar amounts based largely on segment size resulting in more copies of smaller mRNAs. This interaction between mRNA molecules of varying length and the replicase/transcription complex provides some control over the expression levels of individual virus genes. Translation of mRNA segments is largely independent of mRNA length although some segments are translated more efficiently resulting in a secondary method of control over expression. A third level of control results from the use of multiple or overlapping ORFs on one mRNA strand which are translated at different efficiencies. Lastly, some ORFs encode a polyprotein which must be processed to form functional proteins.

Distribution

Plant-infecting reoviruses are seasonally distributed as a result of plant host/crop cycles and presence of insect vector. Plant reoviruses have been isolated from every

continent but some genera are more widespread than others. Fijiviruses are the most widely distributed, which is not surprising given that they are the most numerous. Fijiviruses occur in Africa, Europe, South America, Asia, Australia, and South Pacific Islands. Oryzaviruses have only been isolated from the Indian subcontinent and Asia while phytoreoviruses have been isolated from North America, Asia, and Africa.

Host Range and Virus Transmission

Fijiviruses

The genus *Fijivirus* contains eight species whose members infect a range of monocotyledonous plants of the families Gramineae and Liliacae. Common plant hosts include the Gramineae: *Avena sativa, Oryza sativa, Saccharum officinarum, Zea mays,* and the Liliacae: *Allium sativum.* However, this natural host range can be extended significantly by experimental virus infection. Virus is transmitted by delphacid planthoppers (**Table 1**). Virus can be acquired in juvenile stages, replicates in the vector, and, following a two week latent period, is transmitted to plants in a persistent manner. No transovarial transmission of virus has been reported. In addition to transmission by insect vectors, mechanical transmission of the virus to susceptible hosts has been achieved for some members with difficulty.

Oryzaviruses

The genus *Oryzavirus* contains two species whose members infect monocotyledonous plants of the family Gramineae. Common plant hosts include *Oryza sativa* and *Echinochloa crus-galli.* However, this natural host range can be extended by experimental virus infection to include other economically important species such as *Hordeum vulgare, Triticum aestivum,* and *Zea mays.* Virus is transmitted by delphacid planthoppers (**Table 1**). An acquisition period of 3 h is required followed by a 9-day latent period prior to transmission at all life stages in an intermittent manner. No transovarial transmission or mechanical transmission of virus has been reported.

Phytoreoviruses

The genus *Phytoreovirus* contains two species whose members infect monocotyledonous plants of the family Gramineae and one species which infects dicotyledonous plants. Common plant hosts include the Gramineae: *Oryza sativa* and the dicot – *Melilotus officinalis.* However, the natural host range of the dicot-infecting WTV can be extended significantly by experimental virus infection. Virus is transmitted by cicadellid leafhoppers (**Table 1**). Virus can be acquired after a short feeding period, replicates in the vector, and, following a 10–20 day latent period, is transmitted to plants throughout the life of the vector. Transovarial transmission of virus has been reported. Attempts to transmit the virus to susceptible hosts by mechanical methods have been unsuccessful.

Pathogenicity

The pathogenicity of plant reoviruses is particularly interesting as most viruses replicate in both insects and plant hosts. Most of these viruses do not appear to cause any disease in the insect host and pathogenicity of these viruses is restricted to the plant host. The pathogenicity of fijiviruses varies considerably. Fiji leaf gall disease (caused by FDV) has been reported to cause losses of up to 90% in susceptible sugarcane varieties, while NLRV has no known plant host and, therefore, no pathogenicity. Oryzaviruses can also cause important yield losses. Rice ragged stunt disease (caused by RRSV) has been reported to cause losses of 10–20% but sometimes as high as 100% in severe infections of susceptible varieties. The pathogenicity of phytoreoviruses is much milder although rice dwarf disease (caused by RDV) can be severe as infected plants often fail to bear seeds. There is currently little information on the molecular basis for pathogenicity and it is not known if different isolates of the same virus cause diseases of varying severity.

Diagnosis and Control

Diagnosis of plant reovirus infections can be done on the basis of symptoms or by use of molecular tests. Symptoms vary in different virus/host complexes but symptoms such as plant stunting, increased numbers of side shoots, and tumors or gall formation in phloem tissue are commonly observed. Given the variability in time to symptom expression and symptom severity, alternative tests are often used. Molecular and serological tests have been developed to assist in the diagnosis of viral infection in nonsymptomatic plant material and vector insects. Serological tests are usually in enzyme-linked immunosorbent assay (ELISA) format and rely on polyclonal antisera raised against virions or expressed viral proteins. Recently, molecular tests such as reverse transcriptase-polymerase chain reaction (RT-PCR), which are faster and more specific than serology, have become the most common method of diagnosis. Species-specific primers are now commonly available as increasing numbers of plant reovirus genomes are being sequenced.

Control strategies for plant reoviruses can be focused on either the host plant or insect vector. Plant-based control through breeding to develop resistant plant species is most commonly utilized in combination with removal of susceptible varieties and infected plants which provide a source of inoculum. This approach has provided robust control of RDV in rice and FDV in sugarcane. Genetic engineering of plant hosts has also been explored as an alternative control strategy. Pathogen-derived resistance approaches using either coat protein or other viral genes to control RDV, RRSV, and FDV have not proved as successful as those used to control other RNA plant viruses. However, this may improve in the future as more information on virus infection and replication becomes available providing new resistance targets. Control of insect vector numbers with insecticide has provided some additional disease control. Unfortunately, chemical control appears to be of limited use in cases of high vector pressure. The current combination of diagnosis and control measures is already relatively effective and has resulted in reduced disease incidence and impact.

Future

Although our understanding of plant reoviruses is increasing, many of the molecular and biological properties of these viruses are still unknown. The complete sequence information is now available for a number of these viruses which will allow production of cDNA probes to further elucidate the infection and replication processes in both plant and insect cells. The potential to produce infectious clones also holds promise for detailed studies of both plant and insect host ranges and methods of resistance employed by nonhost species. This information combined with knowledge gained from comparison to animal reoviruses may assist in further development of control strategies for diseases caused by these viruses in plants.

Further Reading

Bamford DH (2000) Virus structures: Those magnificent molecular machines. *Current Biology* 10: R558–R561.
Fauquet CM, Mayo MA, Maniloff J, Desselberger U, and Ball LA (2005) *Virus Taxonomy – Classification and Nomenclature of Viruses: Eighth*

Report of the International Committee on the Taxonomy of Viruses. San Diego, CA: Elsevier Academic Press.
Hull R (2002) *Matthews Plant Virology.* London: Academic Press.
Mertens P (2004) The dsRNA viruses. *Virus Research* 101: 3–13.
Mertens PPC and Diprose J (2004) The Bluetongue virus core: A nano-scale transcription machine. *Virus Research* 101: 29–43.

Plant Rhabdoviruses

A O Jackson, University of California, Berkeley, CA, USA
R G Dietzgen, The University of Queensland, St. Lucia, QLD, Australia
R-X Fang, Chinese Academy of Sciences, Beijing, People's Republic of China
M M Goodin, University of Kentucky, Lexington, KY, USA
S A Hogenhout, The John Innes Centre, Norwich, UK
M Deng, University of California, Berkeley, CA, USA
J N Bragg, University of California, Berkeley, CA, USA

Introduction

The plant rhabdoviruses have distinctive enveloped bacilliform or bullet-shaped particles and can be distinguished based on whether they replicate and undergo morphogenesis in the cytoplasm or in the nucleus. Consequently, they have been separated into two genera, *Cytorhabdovirus* or *Nucleorhabdovirus*. More than 90 putative plant rhabdoviruses have been described although, in many cases, molecular characterizations necessary for unambiguous classification are incomplete or lacking. Recent analyses indicate that the eight sequenced plant rhabdoviruses have the same general genome organization as other members of the *Rhabdoviridae*, but that each encodes at least six open reading frames (ORFs), one of which probably facilitates cell-to-cell movement of the virus. Thus, plant rhabdoviruses have a number of similarities to members of other rhabdovirus genera, but they differ in several respects from rhabdoviruses infecting vertebrates.

Rhabdoviruses infect plants from a large number of different families, including numerous weed hosts and several major crops. Symptoms of infection vary substantially and range from stunting, vein clearing, mosaic and mottling of leaf tissue, to tissue necrosis. The most serious pathogens include maize mosaic virus (MMV), lettuce necrotic yellows virus (LNYV), rice yellow stunt virus (RYSV), also known as rice transitory yellowing virus (RTYV), eggplant mottled dwarf virus (EMDV), strawberry crinkle virus (SCV), potato yellow dwarf virus (PYDV), and barley yellow striate mosaic virus (BYSMV), which is synonymous with maize sterile stunt virus (MSSV), and wheat chlorotic streak virus (WCSV). A number of other rhabdoviruses also have disease potential that can be affected by agronomic practices, incorporation of genes for disease resistance, and control of insect vectors.

The spread of most plant rhabdoviruses is dependent on specific transmission by phytophagous insects that support replication of the virus, so their prevalence and distribution is influenced to a large extent by the ecology and host preferences of their vectors. Although some rhabdoviruses can be transmitted mechanically by abrasion of leaves, this mode of transmission does not contribute significantly to their natural spread due to the labile nature of the virion. Moreover, seed or pollen transmission of plant rhabdoviruses has not been described; thus, aside from vegetative propagation, direct plant-to-plant transmission is unlikely to be a major factor in the ecology or epidemiology of these pathogens.

This article focuses on recent findings concerning the taxonomy, structure, replication, and vector relationships of plant rhabdoviruses. More extensive aspects of plant rhabdovirus biology, specifically ecology, disease development and control, can be found in earlier reviews.

Taxonomy and Classification

The International Committee on Taxonomy of Viruses (ICTV) has used subcellular distribution patterns to assign plant rhabdoviruses to the genera *Cytorhabdovirus* and *Nucleorhabdovirus* (**Table 1**). Currently the ICTV has assigned eight virus species (BYSMV, *Broccoli necrotic yellows virus* (BNYV), *Festuca leaf streak virus* (FLSV), LNYV, *Northern cereal mosaic virus* (NCMV), *Sonchus virus* (SV), SCV, and *Wheat American striate mosaic virus* (WASMV)) to the genus *Cytorhabdovirus* and seven viruses (*Datura yellow vein virus* (DYVV), *Eggplant mottled dwarf virus* (EMDV),

Tablo 1   List of plant rhabdoviruses and their host and vector specificity

Virus	Host	Vector
Cytorhabdovirus		
Barley yellow striate mosaic virus (BYSMV)	M	P
[Maize sterile stunt virus]		
[Wheat chlorotic streak virus]		
Broccoli necrotic yellows virus (BNYV)	D*	A
Festuca leaf streak virus (FLSV)	M	
Lettuce necrotic yellows virus (LNYV)	D*	A
Northern cereal mosaic virus (NCMV)	M	P
Sonchus virus (SonV)	D*	
Strawberry crinkle virus (SCV)	D*	A
Wheat American striate virus (WASMV)	M	L
[Oat striate mosaic virus]		
Nucleorhabdovirus		
Cereal chlorotic mottle virus (CCMoV)	M	L
Datura yellow vein virus (DYVV)	D	
Eggplant mottled dwarf virus (EMDV)	D*	L
[Pittosporum vein yellowing virus]		
[Tomato vein yellowing virus]		
[Pelargonium vein clearing virus]		
Maize fine streak virus (MFSV)	M	L
Maize mosaic virus (MMV)	M	P
Potato yellow dwarf virus (PYDV)	D*	L
Rice yellow stunt virus (RYSV)	M	L
[Rice transitory yellowing virus]		
Sonchus yellow net virus (SYNV)	D*	A
Sowthistle yellow net virus (SYVV)	D	A
Taro vein chlorosis virus (TaVCV)	M	
Unassigned Plant Rhabdoviruses		
Asclepias virus	D	
Atropa belladonna virus	D	
Beet leaf curl virus	D	LB
Black current virus	D	
Broad bean yellow vein virus	D	
Butterbur virus	D*	
Callistephus chinensis chlorosis virus	D	
Caper vein yellowing virus	D	
Carnation bacilliform virus	D	
Carrot latent virus	D	A
Cassava symptomless virus	D	
Celery virus	D	
Chondrilla juncea stunting virus	D	
Chrysanthemum vein chlorosis virus	D	
Clover enation (mosaic) virus	D	
Colocasia bobone disease virus	D	P
Coriander feathery red vein virus	D*	A
Cow parsnip mosaic virus	D*	
Croton vein yellowing virus	D*	
Cucumber toad skin virus	D	
Cynara virus	D*	
Cynodon chlorotic streak virus	M	P
Daphne mezereum virus	D	
Digitaria striate virus	M	P
Euonymus fasciation virus	D	
Euonymus virus	D	
Finger millet mosaic virus	M	P
Gerbera symptomless virus	D	
Gloriosa fleck virus	D	
Gomphrena virus	D*	
Gynura virus	D	

Continued

Table 1   Continued

Virus	Host	Vector
Holcus lanatus yellowing virus	M	
Iris germanica leaf stripe virus	M	
Ivy vein clearing virus	D*	
Kenaf vein-clearing virus	D	
Laburnum yellow vein virus	D	
Launea arborescens stunt virus	D	
Lemon scented thyme leaf chlorosis virus	D	
Lolium ryegrass virus	M	
Lotus stem necrosis	D	
Lotus streak virus	D	A
Lucerne enation virus	D	A
Lupin yellow vein virus	D	
Maize Iranian mosaic virus	M	P
Maize streak dwarf virus	M	P
Malva sylvestris virus	D	
Meliotus (sweet clover) latent virus	D	
Melon variegation virus	D	
Mentha piperita virus	D	
Nasturtium vein banding virus	D	
Papaya apical necrosis virus	D	
Parsely virus	D*	
Passionfruit virus	D	
Patchouli mottle virus	D	
Peanut veinal chlorosis virus	D	
Pigeon pea proliferation virus	D	L
Pinapple chlorotic leaf streak virus	M	
Pisum virus	D*	
Plantain mottle virus	M	
Poplar vein yellowing virus	D	
Ranunculus repens symptomless virus	D	
Raphanus virus	D*	
Raspberry vein chlorosis virus	D	A
Red clover mosaic virus	D	
Sainpaulia leaf necrosis virus	D	
Sambucus vein clearing virus	D	
Sarracenia purpurea virus	D	
Sorghum stunt mosaic virus	M	L
Soursop yellow blotch virus	D	
Soybean virus	D	
Triticum aestivum chlorotic spot virus	M	
Vigna sinensis mosaic virus	D	
Viola chlorosis virus	D	
Wheat rosette stunt virus	M	P
Winter wheat Russian mosaic virus	M	P

Names in brackets are synonymous to those immediately above. Host: D, dicot; M, monocot. (*) indicates ability to be mechanically transmitted. Vectors: A, aphid, L, leafhopper; LB, lacebug; P, planthopper. Blank spaces indicate that no insect vector has been identified.

MMV, PYDV, RYSV, *Sonchus yellow net virus* (SYNV), and *Sowthistle yellow vein virus* (SYVV)) to the genus *Nucleorhabdovirus*. However, sufficient new information has been documented to justify provisional inclusion of the recently described maize fine streak virus (MFSV) and taro vein chlorosis virus (TaVCV) in the genus *Nucleorhabdovirus* (**Table 1**). Cereal chlorotic mottle virus (CCMoV) has also been provisionally included in the genus *Nucleorhabdovirus* based on its intracellular distribution and serology. The complete genomic sequences have been determined for three cytorhabdoviruses, LNYV, NCMV, and SCV and for five nucleorhabdoviruses, MFSV, MMV, RYSV, SYNV, and TaVCV. Phylogenetic analyses of these rhabdoviruses have confirmed their taxonomic classification. Most other plant rhabdoviruses have not been investigated in much detail beyond cursory infectivity studies, crude physicochemical analyses of virus particles, and electron microscopic observations of morphogenesis. Consequently, more than 75 putative rhabdoviruses await assignment to a genus (**Table 1**).

Particle Morphology and Composition

Plant rhabdoviruses are normally bacilliform after careful fixation (**Figure 1(a)**) and estimates of their sizes range from 45 to 100 nm in width and 130 to 350 nm in length. The outer layer consists of 5–10 nm surface projections that appear to be composed of G protein trimers that penetrate a host-derived membrane (**Figure 1(b)**). The

nucleocapsid core is composed of the genomic RNA, the nucleocapsid protein (N), the phosphoprotein (P), and the L polymerase protein (**Figures 1(a)** and **1(b)**). Rhabdovirus virions also contain a matrix protein (M) that interacts with the G protein to stabilize the particle. A sixth protein (sc4) is associated with the membrane fractions of SYNV particles but the presence of an sc4 derivative has not been found in virions of other plant rhabdoviruses.

The overall chemical composition (~70 % protein, 2 % RNA, 20–25% lipid, and a small amount of carbohydrate associated with the G protein) of the plant and animal rhabdoviruses is similar. The minus-sense RNA genomes of plant rhabdoviruses, which range in size from ~11 to 14 kb based on sedimentation, gel electrophoretic analyses, and genome sequencing, are slightly larger than those of most described animal rhabdoviruses. The lipids of plant and animal rhabdoviruses consist of fatty acids and sterols that are derived from sites of morphogenesis. The four sterols predominating in SYNV closely approximate sterols in the nuclear envelope, whereas those of NCMV, a cytorhabdovirus, are more typical of cytoplasmic membranes.

Genomic Structure and Organization

The consensus plant rhabdovirus genome deduced from the eight sequenced viruses is 3-ℓ-N-P-X-M-G-Y-L- t-5′ (**Figure 1(c)**). The N, P, M, G, and L genes appear in the same order as in other rhabdoviruses and their encoded proteins are thought to be functionally similar to the five proteins of vesicular stomatitis virus (VSV). A variable number of genes at the X site have been found in each of the sequenced viruses and some of these appear to be involved in movement. The Y sites between the G and L genes encode short ORFs of unknown functions that are present in the nucleorhabdovirus, RYSV, and the cytorhabdoviruses, NCMV and SCV.

Rhabdovirus ORFs are separated by intergenic or 'gene-junction sequences' that provide vital regulatory functions during transcription and replication (**Figure 2(a)**). The gene-junction sequences can be grouped into three elements consisting of (1) a poly (U) tract at the 3′ end of each gene, (2) a variable intergenic element that is not transcribed in the mRNAs, and (3) a short element complementary to the first 5 nt at the 5′ start site of each mRNA. In general, the gene-junction sequences of each virus are highly conserved, and those of the plant rhabdoviruses share substantial relatedness and differ mostly at element II. Slightly more limited divergence is noted when comparing the genomes of other families within the order *Mononegavirales*, suggesting that these regulatory sequences have been stringently conserved.

The coding regions are flanked by 3 leader (ℓ) and 5 trailer (**t**) noncoding sequences that represent recognition

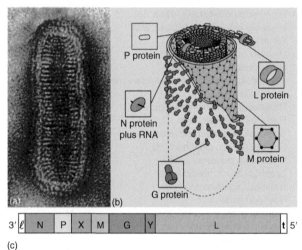

Figure 1 Electron micrograph, diagram, and genome organization of plant rhabdoviruses. (a) Transmission electron micrograph of a negative-stained virus showing the striated inner core, envelope, and glycoprotein spikes. (b) Architecture of the virus particle. The nucleocapsid core is composed of the minus-sense genomic RNA, the nucleocapsid protein (N), the phosphoprotein (P), and the polymerase protein (L). The matrix protein (M) is involved in coiling the nucleocapsid, attachment of the nucleocapsid to the envelope, and associations with the transmembrane glycoprotein (G). (c) Schematic representation of the negative-sense arrangement of genes encoded in the genomes of plant rhabdoviruses. The order of the genes is 3′-ℓ-N-P-X-M-G-Y-L-**t**-5′, where ℓ represents the leader RNA, **t** represents the trailer sequence, X denotes putative movement and undefined plant rhabdovirus genes, and Y shows the location of open reading frames of unknown function in the genomes of several plant and animal rhabdoviruses. Reprinted, with permission, from the Annual Review of Phytopathology, Volume 43, © 2005 by Annual Reviews.

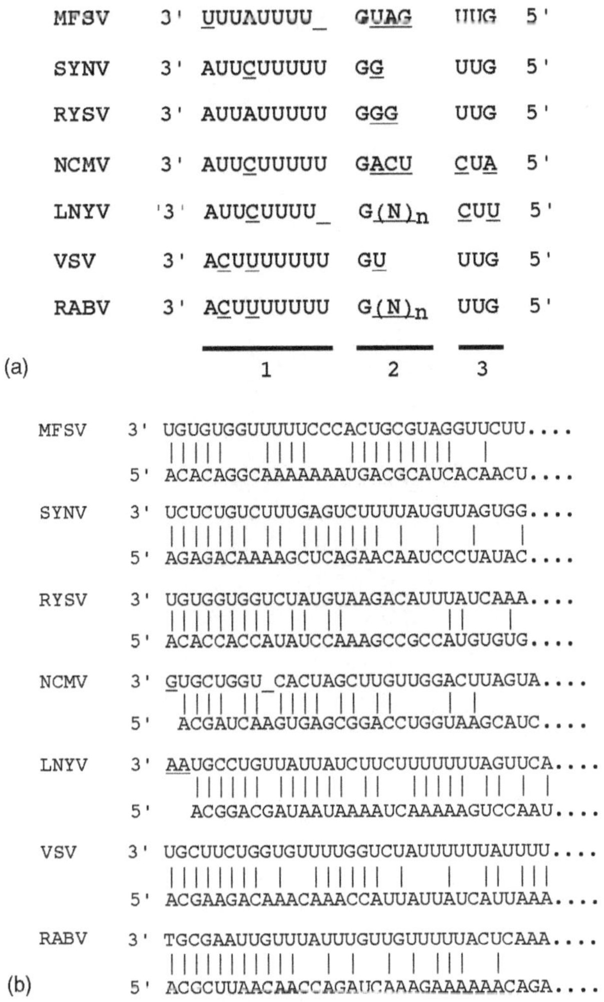

```
MFSV   3'  UUUAUUUU_   GUAG   UUG   5'

SYNV   3'  AUUCUUUUU   GG     UUG   5'

RYSV   3'  AUUAUUUUU   GGG    UUG   5'

NCMV   3'  AUUCUUUUU   GACU   CUA   5'

LNYV  '3'  AUUCUUUU_   G(N)n   CUU   5'

VSV    3'  ACUUUUUUU   GU     UUG   5'

RABV   3'  ACUUUUUUU   G(N)n   UUG   5'
```

(a)
```
            1        2     3
```

```
MFSV   3'  UGUGUGGUUUUUUCCCACUGCGUAGGUUCUU....
           |||||   ||||   ||||||||||  |
       5'  ACACAGGCAAAAAAAUGACGCAUCACAACU....

SYNV   3'  UCUCUGUCUUUGAGUCUUUUAUGUUAGUGG....
           |||||||| || |||||||| |  |   |
       5'  AGAGACAAAAGCUCAGAACAAUCCCUAUAC....

RYSV   3'  UGUGGUGGUCUAUGUAAGACAUUUAUCAAA....
           |||||||| || ||     || |   |
       5'  ACACCACCAUAUCCAAAGCCGCCAUGUGUG....

NCMV   3'  GUGCUGGU_CACUAGCUUGUUGGACUUAGUA....
           |||| ||_|||| || ||      | |
       5'  ACGAUCAAGUGAGCGGACCUGGUAAGCAUC....

LNYV   3'  AAUGCCUGUUAUUAUCUUCUUUUUUUAGUUCA....
           ||||||  ||||||| ||  ||||| || | |
       5'  ACGGACGAUAAUAAAAUCAAAAAGUCCAAU....

VSV    3'  UGCUUCUGGUGUUUUGGUCUAUUUUUUAUUUU....
           ||||||||| |  ||||||| |  |  || |||
       5'  ACGAAGACAAACAAACCAUUAUUAUCAUUAAA....

RABV   3'  TGCGAAUUGUUUAUUUGUUGUUUUUUACUCAAA....
           ||||||||||| | | | | ||| |
       5'  ACGCCUUAACAACCAGAUCAAAGAAAAAACAGA....
```
(b)

Figure 2 Comparisons of intergenic and terminal noncoding regions of rhabdovirus genomes. (a) Intergenic sequences separating the genes. (b) Complementary sequences at the 3′ and 5′ termini of the genomic RNAs. MFSV (maize fine streak virus), SYNV (sonchus yellow net virus), RYSV (rice yellow stunt virus), NCMV (northern cereal mosaic virus), LNYV (lettuce necrotic yellows virus), VSV (vesicular stomatitis virus), RABV (rabies virus). Modified from figures 2 and 3 of Tsai C-W, Redinbaugh MG, Willie KJ, Reed S, Goodin M, and Hogenhout SA (2005) Complete genome sequence and *in planta* subcellular localization of maize fine streak virus proteins. *Journal of Virology* 79: 5304–5314, with permission from American Society for Microbiology.

signals required for nucleocapsid assembly and regulation of genomic and antigenomic RNA replication. These sequences have short complementary termini and small amounts of common sequence relatedness (**Figure 2(b)**). However the plant rhabdovirus ℓ RNAs differ in sequence from the ℓ and t sequences of vertebrate rhabdoviruses and are considerably longer than those of VSV. The transcribed ℓ RNA of SYNV is polyadenylated and differs in this respect from the ℓ RNA of VSV and other known rhabdoviruses.

Properties of the Encoded Proteins

The most comprehensive biochemical analyses of the encoded proteins have been carried out with SYNV, LNYV, and RYSV. Overall, the plant rhabdovirus proteins have very little sequence relatedness to analogous proteins of animal rhabdoviruses, with the exception of the L protein, which has conserved polymerase motifs common to those of most rhabdoviruses. A description of these proteins and their probable functions is outlined below.

The nucleocapsid protein (N)

The N protein functions to encapsidate the viral genomic RNA and is a component of the viroplasms and of the polymerase complex (**Figure 1**). The N genes of the nucleorhabdoviruses, SYNV, MMV, MFSV, TaVCV, and RYSV, and the cytorhabdoviruses, LNYV, SCV, and NCMV have been sequenced. The SYNV, MFSV, MMV, and RYSV nucleorhabdovirus N proteins exhibit short stretches of sequence similarity, suggesting that these four viruses are closely related. These regions of the nucleorhabdovirus N proteins are not significantly related to those of the cytorhabdoviruses and have no extensive relatedness to vertebrate rhabdovirus N proteins.

Experiments conducted in plant and yeast cells have shown that SYNV N protein contains a bipartite nuclear localization signal (NLS) near the carboxy-terminus that is required for nuclear import, and biochemical studies have shown that the protein interacts *in vitro* with importin α homologs. Related nuclear localization sequences are also present in the MFSV and TaVCV N proteins, but this signal is lacking in the N protein of MMV and RYSV. During transient expression in plant cells, the SYNV N protein forms subnuclear foci that resemble the viroplasms found in infected plants, and coexpression of the N and P proteins results in colocalization of both proteins to subnuclear foci (**Figure 3**). These foci require homologous interactions of the N protein that are mediated by a helix-loop-helix motif near the amino-terminus. Interestingly, the SYNV subnuclear foci are distinct from those of nucleolar marker proteins, whereas foci formed during interactions of the MFSV N and P proteins appear to colocalize to the nucleolus.

The phosphoprotein (P)

Direct experiments showing phosphorylation of the P protein are available only for SYNV. The amino-terminal half of the SYNV P protein is negatively charged, as is the case for the other rhabdoviruses. In SYNV, the P protein is phosphorylated *in vivo* at threonine residues and hence differs from the VSV P protein, which is phosphorylated at serine residues. No discernable sequence relatedness is evident between the P protein of SYNV and those of other rhabdoviruses. However, the plant

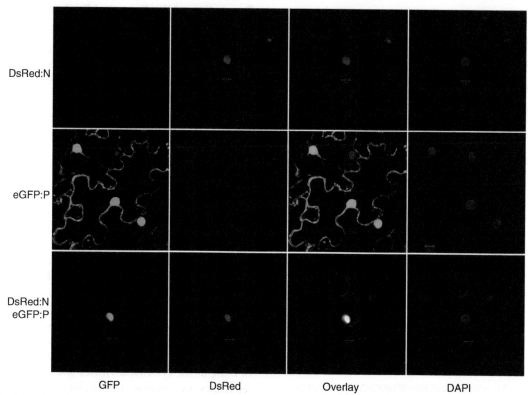

Figure 3 Subcellular localization of the N and P proteins of SYNV. The DsRed:N and eGFP:P fusion proteins were transiently expressed in *Nicotiana benthamiana* leaf tissue via infiltration with *Agrobacterium tumefacians* containing pGD vectors. The confocal micrographs show the subcellular localization of proteins at 3 days after infiltration. The top row depicts the individual expression of the DsRed:N fusion protein. The middle row shows fluoescence in cells expressing eGFP:P alone. The bottom row shows cells coexpressing the DsRed:N and eGFP:P proteins. Nuclei are identified by staining with DAPI (4′-6-diamino-2-phenylindole dihydrochloride). From Deng *et al.*, unpublished.

rhabdovirus P proteins have similar hydrophilic cores and the cytorhabdovirus P proteins overall are hydrophilic with similar isoelectric points. In addition, there is nearly 50% sequence identity between the P proteins of MMV and TaVCV. Although these results suggest that the provisional assignments of the P proteins of other plant rhabdoviruses are probably correct, additional data need to be accumulated to obtain a consensus of the functional activities and the biochemical interactions of the putative plant rhabdovirus P proteins.

The P protein is a component of the viral nucleocapsid core and the polymerase complex. The SYNV P protein forms complexes *in vivo* with the N and L proteins that are analogous to N:P and P:L complexes found in VSV-infected cells, and hence the P protein probably functions in SYNV polymerase recycling. Biochemical experiments have shown that the solubility of the SYNV N protein is increased during P protein interactions, so the P protein appears to have chaperone activity. Heterologous N:P protein complexes form by interactions of an internal region of the P protein with the amino-terminal helix-loop-helix region of N that overlaps the N:N protein binding site. The P protein also engages in homologous interactions that are mediated near the amino-terminus of the

P protein. Hence, the SYNV P protein has functions similar to those of the P proteins of other well-characterized vertebrate rhabdoviruses.

Reporter gene fusions show that the SYNV P protein, when expressed alone, accumulates in both the nucleus and the cytoplasm (**Figure 3**). The central third of the SYNV P protein is required for nuclear import, but other regions of the protein affect the import efficiency. Recent experiments indicate that the SYNV P protein binds directly to human importin β derivatives *in vitro* and, since the N protein has an NLS site and interacts with importin α, the N and P proteins have different mechanisms for nuclear import. Interestingly, sequence analyses show that the SYNV P protein does not have a bipartite NLS, whereas the P proteins of RYSV and MMV have a bipartite NLS, and both P proteins have a pronounced nuclear localization pattern. Hence, due to these differences between the SYNV and the RYSV and MMV proteins, it is likely that the viruses have diverged in their strategies of nuclear localization. In SYNV and MFSV, coexpression of the P protein with the N protein of the same virus results in colocalization of the complexes to subnuclear foci characteristic of viroplasms. In SYNV, formation of these foci requires interactions of P with the amino-terminus of the

N protein. However, the N and P interactions appear to be virus-specific because heterologous combinations of the SYNV and MFSV N and P proteins fail to form subnuclear foci, and the P protein continues to be expressed in both the nucleus and the cytoplasm.

The SYNV P protein also has a leucine-rich nuclear export signal located within the first third of the protein. Reporter proteins fused to the P protein are retained in the nucleus following treatment with Leptomycin B, an inhibitor of nuclear export. Interaction with the host nuclear export receptor XpoI provides further evidence for P protein nuclear export functions. These results and P protein mutagenesis experiments provide strong evidence suggesting that the SYNV P protein is involved in nuclear shuttling activities.

In addition to its role as a structural protein, the SYNV P protein shares many of the hallmarks of RNA-silencing suppressor proteins. These characteristics include suppression of reporter gene silencing in transgenic plants and the ability to bind small interfering (siRNAs) and single-stranded RNAs *in vitro*. Together, these activities clearly point to key roles of the P protein in nucleocapsid structure, replication, countering innate host defenses, and possibly intercellular movement.

Position X proteins

Like other plant viruses, plant rhabdoviruses must encode proteins to assist in cell-to-cell movement of virus derivatives through the plasmodesmata and their systemic transport through the vascular system. Considerable evidence for a movement function has been accumulated for proteins encoded at position X between the P and M genes (**Figure 1(c)**). The predicted secondary structures of several plant rhabdovirus proteins, including SYNV sc4, LNYV 4b, RYSV P3, MMV P3, and MFSV P4, have a distant relatedness to the TMV 30 K superfamily structural motifs. Additional evidence for a role of the position X-encoded proteins in cell-to-cell movement is their association with host and viral membranes during transient expression. Unpublished evidence indicates that the sc4 protein is phosphorylated, as is also the case with the TMV 30 K movement protein. The movement hypothesis for genes occupying the X position has been reinforced by experiments carried out with RYSV. With this nucleorhabdovirus, the P3 protein is able to *trans*-complement cell-to-cell movement of a movement-defective potato virus X in *Nicotiana benthamiana* leaves. The P3 protein also interacts with the N protein, the major component of nucleocapsids; hence, it could possibly facilitate movement of nucleocapsids through these interactions.

Despite the persuasive evidence for a movement function of genes encoded at position X, considerable diversity appears in the X ORF(s) of several plant rhabdoviruses. For example, although the TaVCV gene X codes for a protein of a size similar to the other X protein genes, the protein has no obvious sequence similarity to proteins in the 30 K superfamily. MFSV encodes two proteins between the putative P and M proteins, and both have different localization patterns from sc4 of SYNV. In addition, four small ORFs reside between the P and M genes of NCMV. Thus, additional studies need to be undertaken to clarify the functional activities of the 'unusual' X proteins.

The matrix protein (M)

The M proteins of plant rhabdoviruses are basic and are thought to function in nucleocapsid binding and coiling, and interactions with the G protein. Sequence alignments of the M proteins of several plant rhabdoviruses have not revealed extensively conserved motifs. Unpublished studies suggest that the SYNV M protein is phosphorylated *in vivo* at both threonine and serine residues. When expressed ectopically, the M proteins of SYNV and MFSV localize in the nucleus. A central hydrophobic region of the M protein is thought to mediate membrane–lipid interactions with the G protein during morphogenesis. In addition to their roles in viral morphogenesis, preliminary experiments indicate that rhabdovirus M proteins have important roles in host–virus interactions because they appear to be able to inhibit host gene expression.

Position Y ORFs

NCMV, RYSV, and SCV contain a short ORF at position Y that separates the genes encoding the G and L proteins (**Figure 1(c)**). Small, nonvirion ORFs preceding the L gene are also found in the genomes of some animal viruses, but the products of these ORFs are either nonstructural or have not been detected in infected cells. The three predicted plant rhabdovirus Y proteins are small (<100 amino acids) and do not share obvious sequence identity with each other or with the nonvirion genes of the animal rhabdoviruses. However, short stretches of the Y ORFs have limited relatedness to other negative-strand RNA virus proteins, suggesting that these regions of the genome may have originated by gene duplication or recombination.

The RYSV P6 protein at position Y is predicted to contain an aspartic protease motif (DTG) and has five potential phosphorylation sites (S/T-X-X-D/E). *In vitro* phosphorylation assays using a GST:P6 fusion protein have shown that P6 is phosphorylated at both serine and threonine residues. Although P6 could not be detected in total protein extracts from infected leaf tissue, immunoblots of purified virus and protein extracts from viruliferous leafhoppers suggest that P6 is associated with virions, so it may have a structural role in infection.

The glycoprotein (G)

The G protein forms the glycoprotein spikes of rhabdovirus virions (**Figure 1**). The plant rhabdovirus G proteins do not have extensive similarity, but they are more closely related to each other than to the G proteins of several vertebrate rhabdoviruses. The plant rhabdovirus G proteins share putative N-terminal signal sequences, a transmembrane anchor domain, and several possible glycosylation sites. In addition, the SYNV G protein contains a putative NLS near the carboxy-terminus that could be involved in transit to the inner nuclear membrane prior to morphogenesis. Several glycosylation inhibitors interfere with N-glycosylation of the SYNV G protein, and tunicamycin treatment blocks SYNV morphogenesis, leading to accumulation of striking arrays of condensed nucleocapsid cores that fail to bud through the inner nuclear membrane. Thus, the G protein has a prominent role in morphogenesis, and the available evidence suggests that glycosylation is required for interactions of the protein with coiled nucleocapsids.

The polymerase protein (L)

The L proteins of plant rhabdoviruses are present in low abundance within nucleocapsids and in infected cells. The L proteins are the most closely related of the rhabdovirus-encoded proteins and are positively charged with conserved polymerase domains and RNA-binding motifs. The L protein of SYNV is required for polymerase activity, because antibodies directed against the GDNQ (polymerase) motif inhibit transcription. Alignment of the L protein sequence with polymerases of several other nonsegmented negative-strand RNA viruses reveals conservation within 12 motifs. Phylogenetic trees derived from L protein alignments indicate that the nucleorhabdoviruses and cytorhabdoviruses cluster together in two clades separated from the vertebrate rhabdoviruses. This suggests that the plant rhabdoviruses have diverged less from each other than from the vertebrate rhabdoviruses.

Polymerase Activity

A viral RNA-dependent RNA polymerase is activated after treatment of LNYV and BNYV cytorhabdovirus virions with mild nonionic detergents, and this activity cosediments with loosely coiled nucleocapsid filaments that are released from virions. The transcribed products are complementary to the genome, as expected of mRNAs. Thus, the described polymerases of these plant cytorhabdoviruses appear to be similar to the extensively studied polymerases of the vesiculoviruses.

In contrast, no appreciable polymerase activity is evident in dissociated preparations of SYNV or other nucleorhabdovirus virions that have been analyzed. However, an active polymerase can be recovered from the nuclei of plants infected with SYNV. Polymerase activity is associated with a nucleoprotein derivative, consisting of the N, P, and L proteins, that cosediments with SYNV nucleocapsid cores. The polymerase complex can be

precipitated with P protein antibodies, but the activity of the complex is not inhibited by these antibodies. However, antibody inhibition experiments demonstrate that the L protein is required for polymerase activity. Kinetic analysis of transcription products also reveals that the complex is capable of sequentially transcribing a polyadenylated plus-sense leader RNA and polyadenylated mRNAs corresponding to each of the six SYNV-encoded proteins. Potential replication intermediates consisting of short incomplete minus-strand products homologous to the genomic RNA are also transcribed. These results thus provide a model whereby nucleorhabdovirus particles require polymerase activation by host components early in infection. In contrast, the polymerases of the cytorhabdoviruses appear to be present in an active form in virions and the released cores are capable of initiating primary transcription immediately upon uncoating *in vitro*.

Cytopathology and Replication

The plant rhabdoviruses vary profoundly in their sites of replication and morphogenesis, and those that replicate in the nucleus differ substantially from vertebrate rhabdoviruses that replicate and assemble in the cytoplasm. In plants, nucleorhabdoviruses replicate in the nucleus, bud in association with the inner nuclear membrane, and accumulate in enlarged perinuclear spaces formed between the inner and outer nuclear envelopes. Similar patterns normally occur in the majority of insect tissues, but MMV also buds in the outer membranes of the salivary glands and nerve cells of its leafhopper vector. Clearly, studies on insect cells need to be emphasized in future work to clarify aspects of insect transmission.

The limited evidence available indicates that the cytorhabdoviruses replicate in the cytoplasm, bud in association with the endoplasmic reticulum (ER), and accumulate in ER-derived vesicles (**Figure 4(a)**). Two slightly different variations in replication of LNYV and BYSMV have been proposed, based on extensive electron microscopic observations of infected cells. Indirect evidence suggests that a nuclear phase may be involved in LNYV replication because the outer nuclear membrane blisters and develops small vesicles that contain some virus particles. However, later in the life cycle, masses of thread-like viroplasms appear in the cytoplasm and these are located close to dense networks of the ER that appear during infection. These proliferated membranes form vesicles that may serve as sites for morphogenesis of the accumulating nucleocapsids. A similar scenario lacking a nuclear phase has been outlined for BYSMV. In this case, membrane-bound viroplasms appear in the cytoplasm and virus particles are found exclusively in association with cytoplasmic membranes that proliferate in close proximity to the viroplasms. Unfortunately, both of the cytorhabdovirus models

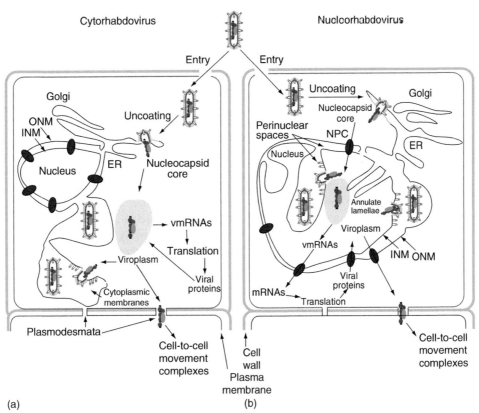

Figure 4 Models for cytorhabdovirus and nucleorhabdovirus replication in plant cells. Most rhabdoviruses are believed to enter plant cells during insect vector feeding and the nucleocapsid is thought to fuse with the endoplasmic reticulum (ER) and be liberated into the cytoplasm. Panels (a) and (b) provide contrasts between the cytorhabdovirus and the nucleorhabdovirus replication strategies, respectively. (a) Cytorhabdovirus replication model. The available information about cytorhabdovirus replication relies almost entirely on ultrastructural observations, and molecular or modern cytological information has not been obtained to extend these observations. After nucleocapsid release, primary and secondary rounds of transcription are followed by nucleocapsid accumulation in viroplasms to form dense masses that are associated with proliferated membrane vesicles. Morphogenesis occurs by budding of nucleocapsids into the ER. During the later stages of replication, large aggregates of bacilliform virions accumulate in pronounced vesicles that are thought to have originated from the ER. (b) Nucleorhabdovirus replication model. The nucleocapsid is thought to be transported into the nucleus through the nuclear pore complex and host components are thought to activate the nucleocapsids to initiate primary rounds of transcription to produce polyadenylated leader RNA and mRNAs for each of the viral proteins. The mRNAs are transported to the cytoplasm and translated, and the N, P and L proteins are transported through the nuclear pore complex into the nucleus. As the N, P, and L proteins increase in abundance, a switch occurs from primary transcription of mRNAs to a mode consisting of intermittent rounds of replication to produce antigenomic and genomic nucleocapsids, followed by secondary rounds of transcription to increase the pool of mRNAs. This phase of replication is regulated by a feedback mechanism that relies on the abundance of the core proteins for encapsidation of nascent leader RNAs. As replication progresses, the nuclei become greatly enlarged, and subnuclear viroplasms appear that consist of large masses of granular material that contain viral RNA and the N, P, and L proteins. Early during replication, some of the newly synthesized nucleocapsids are transported to the cytoplasm where they associate with movement proteins and are transported to other cells through the plasmodesmata. Late in replication, the M protein reaches sufficient concentration to coil the genomic nucleocapsids and mediate interactions with G protein patches at the inner nuclear membrane. During this process, virions undergo morphogenesis by budding through the inner nuclear membrane and accumulate in the perinuclear spaces. Reprinted, with permission, from the Annual Review of Phytopathology, Volume 43, © 2005 by Annual Reviews.

have been derived solely from ultrastructural observations, and none of these studies has utilized specific antibodies to identify individual virus proteins, viral-specific probes for *in situ* hybridization, or modern techniques of cell biology to probe replication.

No direct information is yet available about the early entry and uncoating events, but a model for nucleorhabdovirus infection predicts that, after entry into the cell during vector feeding or mechanical transmission,

rhabdovirus particles associate with the ER to release the nucleocapsid cores into the cytoplasm (**Figure 4(b)**). Released cores may then utilize the host nucleocytoplasmic transport machinery to recognize karyophylic signals present on the N and P proteins to facilitate nucleocapsid entry into the nucleus. During the early stages of infection, the virion-associated polymerase is probably activated by host components to produce an active transcriptase that copies the genomic RNA into capped and polyadenylated mRNAs that are transported to the cytoplasm and translated. The translated N and P proteins are imported into the nucleus by separate mechanisms using host importin α and β proteins, respectively.

After entry into the nucleus, the N, P, and L proteins probably participate in multiple rounds of mRNA transcription, and antigenomic and genomic RNA replication. As replication proceeds, the viroplasms form discrete foci that appear near the periphery of dramatically enlarged nuclei. During the early stages of infection, small amounts of the nucleocapsids are postulated to be exported to the cytoplasm by interactions of the nuclear export signals on the P protein with host nuclear export receptor proteins. These exported nucleocapsids then interact with movement protein homologs to mediate transport through the plasmodesmata to adjacent cells. As infection progresses, the M protein accumulates in the nucleus and reaches concentrations sufficient to downregulate transcription and participate in coiling of minus-sense RNA nucleocapsid cores. The coiled cores then associate with G protein at sites on the inner nuclear envelope that are located in close proximity to the viroplasms. During budding, numerous enveloped virions accumulate in perinuclear spaces between the inner and outer nuclear envelope where they may be ingested during vector feeding.

A recent discovery that may shed new light on the processes of nucleorhabdovirus replication and maturation has been noted during infection of transgenic *N. benthamiana* plants that express green fluorescent protein (GFP) targeted to the ER. During infections with either SYNV or PYDV, a substantial proportion of the GFP appears to be redistributed to form spherules within the nuclei. In the case of SYNV, the spherules colocalized with foci formed by the N protein. A model to explain this phenomenon is that the spherules contacting the viroplasms in SYNV-infected plants are derived from the ER and become redistributed to the inner nuclear membrane to serve as sites for replication and virion maturation.

Vector Relationships, Distribution, and Evolution

Plant pathogenic rhabdoviruses are highly dependent on arthropod vectors for their distribution between plants. Although some plant rhabdoviruses have no known vector, most well-characterized members are transmitted by insects in which they also multiply, so it is possible that the plant rhabdoviruses radiated from a primitive arthropod. Plant rhabdoviruses are most commonly transmitted by aphids (Aphididae), leafhoppers (Cicadellidae), or planthoppers (Delphaccidae) (**Table 1**). An incompletely characterized putative rhabdovirus, beet leaf curl virus (BLCV), reportedly has a heteropteran beet leaf bug (*Piesma quadratum*) vector, but more extensive molecular

and cytological analyses of this virus and other poorly characterized plant rhabdoviruses need to be carried out so their properties and vector relations can be clarified.

Vector–host relationships have profoundly affected plant rhabdovirus distribution and host range. For example, leafhoppers, planthoppers, and aphids are prevalent on both monocots and dicots, but the rhabdoviruses causing diseases of the Gramineae are all transmitted by leafhoppers or planthoppers. Except for PYDV and EMDV, which have leafhopper and planthopper vectors, respectively, dicot-infecting rhabdoviruses whose transmission has been investigated are transmitted by aphids. In all cases of insect transmission that have been carefully examined, rhabdoviruses are persistently transmitted in a propagative fashion, and in many cases, can be transmitted to vector progeny. Long latent periods are required before transmission occurs; insects often remain viruliferous throughout their lives, and transovarial passage has been observed through eggs and nymphs. In addition, strain-specific infection of tissue culture lines and explants combined with serological detection in vector cells provides unequivocal proof that rhabdoviruses replicate with high specificity in leafhopper and aphid vectors.

Several classical studies with PYDV in leafhoppers and SYVV in aphids, as well as recent studies with MMV in its planthopper vector *Peregrinus maidis*, have provided models for tissue-specific events in insect infection (**Figure 5**). After virus acquisition from plants, MMV initially accumulates in epithelial cells of the anterior part

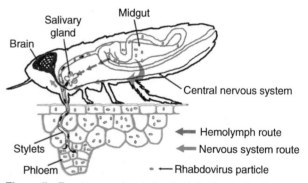

Figure 5 Events occurring during the infection cycle of rhabdoviruses in leafhopper vectors. Viruses are acquired during feeding on plant cells and move from the stylet to the midgut lumen of the digestive tract where they are hypothesized to invade epithelial cells by receptor-mediated endocytosis. From the epithelial cell layers, the virus moves into the nervous system, trachea, and hemolymph and spreads throughout the insect into the salivary glands and reproductive tissues. The salivary glands accumulate high levels of virus particles that are released by exocytosis, transported through the salivary canal, and transmitted to new host plants during subsequent feeding. Reproduced from figure 2a, Hogenhout SA, Redinbaugh MG, and Ammar E (2002) Plant and animal rhabdovirus host range: A bug's eye view. *Trends in Microbiology* 11: 264–271, with permission from Elsevier.

of the insect midgut and subsequently in nerve cells. Then, the virus appears in tracheal cells, hemocytes, muscles, and the salivary glands, and finally, in the fat cells, mycetocytes, and epidermal tissues. MMV infection is most extensive in the anterior portion of the gut, nerve cells, tracheal cells, and salivary glands of the planthopper. Based on this order of events, it is postulated that epithelial cells of the midgut are the first virus-entry sites, and that the virus quickly moves to the nervous system, trachea, and the hemolymph. From these tissues, MMV can be transmitted systemically to other tissues, including the salivary glands, which support high levels of virus accumulation.

MMV buds through the inner nuclear membrane in plant cells and has a similar pattern of morphogenesis in most cells of *P. maidis* tissues. However, MMV has also been observed to bud frequently from outer membranes in cells of nervous tissue and salivary glands of the planthopper, and similar observations have been made with other virus–vector combinations. Thus, the cellular budding site of MMV, and probably other rhabdoviruses, in insect hosts may be dependent on the cell type. Budding from outer cell membranes in salivary glands may be important because the process could allow release of virions into the saliva and permit introduction into plant cells during feeding.

Genetic experiments with PYDV have shown that highly efficient and inefficient leafhopper vectors can be selected. Continuous passage of PYDV by serial injection of insects can result in isolates that are unable to infect plants. Additional studies have shown that strains that have lost their capacity to be insect-transmitted can be recovered after protracted passage in plants. This phenomenon could provide a mechanism for evolution of vectorless rhabdoviruses, particularly in cases where infections were established in vegetatively propagated hosts.

Rhabdoviruses normally have the capacity to infect a greater range of plant hosts than the narrow range of species colonized by their insect vectors, because experimental host ranges usually can be extended considerably through mechanical transmission. In addition, plant rhabdoviruses have a wider insect host range than their natural insect vector hosts would indicate. For example, the majority of maize-feeding leafhoppers and planthoppers can acquire MFSV, but only one leafhopper, *Graminella nigrifrons*, can transmit this virus to maize. In addition, studies have shown that surrogate nonvectors of SCV injected with infected plant extracts can support replication and transmit virus to dicot hosts that do not support feeding by the native vector. Finally, cowpea protoplast infectivity experiments with the grass rhabdovirus FLSV and with SYNV show that both viruses are able to infect the legume protoplasts, but neither virus is able to infect cowpea plants. Together, these observations indicate that some plant rhabdoviruses have the ability to

infect cells of several distantly related hosts, but that the natural host specificity is determined by (1) the insect vector feeding range; (2) the ability of the virus to move through the insect vector into the salivary glands and into the plant; and (3) systemic movement in the infected plant.

During evolution, plant rhabdoviruses faced two major challenges of a fundamentally different nature brought about by the necessity to alternately infect plants and insects. In each host, the virus must utilize different entry methods and accommodate distinct cellular and defense mechanisms. Rhabdovirus acquisition by the vector probably necessitates attachment to specific receptors at the surface of cells in the digestive system, followed by active invasion of the reproductive organs, fat bodies, and salivary glands. Very different barriers must be circumvented to establish systemic infections of plants. In order to establish a primary infection focus, the cell wall must first be breached by mouthparts of the insect, the virus must be introduced into the plant cell, where it uncoats and initiates the replication cycle. To establish systemic infections in plants, the virus must move from cell to cell through very small plasmodesmatal connections, enter the vascular system and spread throughout the plant. Therefore, plant rhabdoviruses are anticipated to have evolved a number of sophisticated mechanisms to circumvent the barriers to infection of their insect and plant hosts.

Further Reading

Black LM (1979) Vector cell monolayers and plant viruses. *Advances in Virus Research* 25: 192–271.

Hogenhout SA, Redinbaugh MG, and Ammar E (2002) Plant and animal rhabdovirus host range: A bug's eye view. *Trends in Microbiology* 11: 264–271.

Jackson AO and Wagner JDO (1998) Procedures for plant rhabdovirus purification, polyribosome isolation, and replicase extraction. In: Foster G and Taylor S (eds.) *Plant Virology Protocols: From Virus Isolation to Transgenic Resistance*, vol. 81, ch. 7, pp. 77–97. Totowa, NJ: Humana Press.

Jackson AO, Dietzgen RG, Goodin MM, Bragg JN, and Deng M (2005) Plant Rhabdoviruses. *Annual Review of Phytopathology* 43: 623–660.

Krichevsky A, Kozlovsky SV, Gafney Y, and Citovsky V (2006) Nuclear import and export of plant virus proteins and genomes. *Molecular Plant Pathology* 7: 131–146.

Melcher U (2000) The 30K super family of viral movement proteins. *Journal of General Virology* 81: 257–266.

Sylvester ES and Richardson J (1989) Aphid-borne rhabdoviruses relationships with their vectors. In: Harris KF (ed.) *Advances in Disease Vector Research*, vol. 9, pp. 313–341. New York: Springer.

Tsai C-W, Redinbaugh MG, Willie KJ, Reed S, Goodin M, and Hogenhout SA (2005) Complete genome sequence and *in planta* subcellular localization of maize fine streak virus proteins. *Journal of Virology* 79: 5304–5314.

Tordo N, Benmansour A, Calisher C, *et al.* (2005) Family *Rhabdoviridae*. In: Fauquet CM, Mayo MA, Maniloff J, Desselberger U, and Ball LA (eds.) *Virus Taxonomy: Eighth Report of the International Committee on Taxonomy of Viruses*, pp. 623–644. San Diego, CA: Elsevier Academic Press.

Plum Pox Virus

M Glasa, Slovak Academy of Sciences, Bratislava, Slovakia
T Candresse, UMR GDPP, Centre INRA de Bordeaux, Villenave d'Ornon, France

Introduction

Plum pox virus (PPV), the agent responsible for the Sharka disease, belongs to the genus *Potyvirus*. The natural host range of this virus is restricted to *Prunus* spp. (stone fruits and ornamental trees). The infection of susceptible genotypes results in characteristic foliar and fruit symptoms and premature fruit drop. The wide geographical distribution of PPV includes most of Europe and the Mediterranean region as well as some countries in Asia and North and South America, although with widely different incidence levels in different countries. PPV is transmitted nonpersistently by more than 20 aphid species, by grafting and vegetative multiplication of infected plants, but is not seed-borne. To date, six strains/groups of PPV have been identified based on biological, serological, and molecular properties (M, D, Rec, C, EA, and W). Although many diagnostic tools are available for the sensitive and/or specific detection of PPV, its uneven distribution in infected woody hosts and its low titer outside of the active growth period significantly complicate its detection. In regions free of PPV, strict quarantine measures are usually enforced. In the quasi-absence of resistant fruit tree varieties, a mix of prophylactic approaches including the use of virus-free propagation material, eradication of diseased trees, and vector control is generally used in an effort to control the virus in regions where it has not reached an endemic status.

Economical Importance

PPV is considered as the most detrimental viral pathogen of stone fruit crops (peach, apricot, plum, Japanese plum). During approximately a century of recognized existence, PPV had a devastating effect on the European stone-fruit industry, mainly in the central and south European countries. Fruit trees infected with PPV cannot be cured and are often eliminated as a consequence of disease eradication or containment efforts. Although the infected trees are usually not stunted and do not die, fruit yields can be severely affected. Besides foliar symptoms, the virus often severely damages fruits, so that they have decreased weight and sugar content, overall lower gustative quality and often become blemished and are frequently unsuitable for consumption or processing. Premature fruit drop can also be observed and may reach 80–100% in susceptible cultivars. Consequently, traditional susceptible cultivars have to be replaced by less susceptible or tolerant cultivars, which are often of lower agricultural or gustatory quality. Infected propagation material and nursery stock (rootstock, budwood, scions) is not marketable and has to be destroyed. Because of its high potential impact on stone fruit crops, plum pox virus is listed as a quarantine pathogen in many parts of the world. In Europe for example, PPV is listed in the EC Plant Health Directive (Annex II of the European Union council directive 2000/29/EEC).

History and Geographical Distribution

The sharka disease was observed for the first time around 1917 on plum cv. Kjustendil in the village of Zemen in Bulgaria, near the Yugoslavian border. The disease is named according to the characteristic symptoms on fruits (sharka means pox in Bulgarian). Initially described on plum, the virus was observed in 1933 on apricot, in the 1960s on peach, and in the 1980s on sour and sweet cherry.

After World War I, the virus progressively spread to a large part of the European continent and Mediterranean basin, probably mostly as a consequence of exchange of infected propagation material. In recent years, the virus has been reported from China, South America (Chile, Argentina), and North America (USA, Canada). However, the prevalence of the disease differs from region to region, from the endemic occurrence observed in the central and eastern European countries, where the virus is well established, to local and more limited incidence observed in other countries where the virus was only introduced recently or where strict phytosanitary control measures have been enforced and have successfully retained the virus under some level of control.

Host Range and Symptomatology

Natural host range is restricted to species of the genus *Prunus*, including cultivated stone fruits – plum (*P. domestica*), Japanese plum (*P. salicina*), apricot (*P. armeniaca*), peach and nectarine (*P. persica*), almond (*P. amygdalus*). Although not infected by the majority of PPV isolates, sweet cherry (*P. avium*) and sour cherry (*P. cerasus*) are also natural hosts for some specific PPV isolates.

Symptoms vary depending on virus isolate, host/cultivar susceptibility, physiological status and age of the host, and environmental conditions, such as temperature.

Depending on the host, the symptoms may affect leaves, flowers, fruits, and stones. In recently infected trees, symptoms tend to be restricted to only some parts of tree but tend to generalize with time. Under field conditions, the symptoms on *Prunus* plants infected with PPV are often masked late in the season or during the warm period of the growing season.

Mixed infection with other viruses such as prunus necrotic ringspot virus (PNRSV) or prune dwarf virus (PDV) may further increase the severity of symptoms.

Foliar symptoms on plum consist generally of pale green chlorotic rings, spots, or patterns. Susceptible cultivars develop shallow ring or arabesque depressions on fruits, sometimes with brown or reddish necrotic flesh and gumming. Tolerant plum cultivars show no symptoms on fruits.

Infected apricots develop chlorotic or pale-green rings and lines on leaves, light-colored depressed rings on fruits, which may be severely deformed. Stones are marked with typical discolored rings.

Symptoms on susceptible peach genotypes are pronounced vein clearing, small chlorotic blotches, and distortions of the leaves. Color-breaking symptoms on the petals are observed in some varieties. Pale rings or diffuse band are visible on the skin of the fruits. In general, the symptoms on peach tend to be less visible in comparison with those on plums and apricots. Infection of almond is often symptomless or with limited foliar symptoms.

Characteristic symptoms on cherries consist of pale-green patterns and rings on leaves. Fruits may be slightly deformed, with chlorotic and necrotic rings and notched marks.

Premature dropping of fruits is frequently observed in the most susceptible varieties of the various hosts, in particular plum and apricot. Depending on host (cultivar) susceptibility, the losses may reach up to 100%.

PPV also infects wild and ornamental trees, such as myrobalan (*P. cerasifera*), Japanese apricot (*P. mume*), nanking cherry (*P. tomentosa*), Canada plum (*P. nigra*), American plum (*P. americana*), dwarf flowering almond (*P. glandulosa*), and blackthorn (*P. spinosa*), which may act as local reservoirs of the virus.

The experimental host range of PPV is large, with over 60 reported host species (such as *Chenopodium foetidum*, *Nicotiana benthamiana* and *N. clevelandii*, *Pisum sativum*, *Ranunculus arvensis*, *Senecio vulgaris*, *Stellaria media*, etc.) in eight families. However, no epidemiologically significant contribution of weeds and annual herbaceous plants to the spread of plum pox virus has been reported.

In woody hosts, the virus often shows an uneven distribution within the tree and full systemic invasion of a tree may require several years. Long-distance viral movement is restricted in some resistant (apricot, peach) and hypersensitive (plum) genotypes. The irregular distribution and translocation of the virus in the tree and the low titer outside the active growth period may complicate the detection of the virus when methods of insufficient sensitivity are used.

Virion Structure and Genome Properties

The flexuous filamentous viral particles are approximately 750×15 nm. Viral particles are composed of the single-stranded genomic RNA encapsidated by a single type of capsid protein subunit. The genome is of positive polarity and is 9741–9795 nucleotides in length. It has a polyadenylated 3′ end and a virus-encoded protein (VPg) covalently bound at its 5′ end. The genomic organization is typical of potyviruses, with a single open reading frame encoding a large polyprotein precursor (3125–3143 amino acids, *c.* 355 kDa) that is proteolytically processed by three virus-encoded proteinases (P1, HC-Pro, and NIaPro) to yield as many as ten mature functional proteins.

Strains/Groups

Initially, different strains or groups of isolates have been described on the basis of symptomatology in various experimental hosts. The most developed analysis of this kind relied on symptoms on *C. foetidum*, an experimental local-lesion host. PPV isolates were classified as yellow, intermediate, and necrotic strains depending on the type of local lesions they induced. However, this phenotypic characterization proved to be at least to some extent under environmental control and no relationships could be later established between this biological property and serological/molecular properties.

In the late 1970s, the existence of two serogroups of PPV was demonstrated using agar double diffusion assays employing polyclonal antibodies and a purified, formaldehyde-treated, suspension of undegraded viral particles. These two serogroups were named M (based on the type isolate Marcus from peach in Greece) and D (based on the type isolate Dideron from apricot in southeastern France). Later it was shown that both strains differ also in the coat protein (CP) mobility in denaturing polyacrylamide gel electrophoresis (PAGE) and in the presence or absence of an *Rsa* I restriction site in the region encoding the CP C-terminus.

On the basis of recent molecular and serological analysis, six strains/groups of PPV isolates sharing common biological, serological, and molecular properties are now recognized. This classification has also been validated by comparisons of complete genomic sequences of isolates representative of each of these strains (**Figure 1**).

PPV-M (from the type isolate Marcus). The isolates of this group are present in many European countries, but absent from the Americas and China. They are often associated with rapidly spreading epidemics in peach but are less frequently found on plums. Usually, PPV-M isolates are transmitted efficiently by aphids.

PPV-D (from the type isolate Dideron). PPV-D isolates are widespread in all areas where PPV has been reported, including the recent outbreaks in Asia and South and North America. Isolates infect all the susceptible *Prunus* species excluding cherries, but PPV-D isolates are less frequently associated with spreading epidemics in peach.

PPV-Rec (from 'recombinant'). PPV-Rec isolates are derived from a single homologous recombination event between PPV-M and PPV-D isolates, with a cross-over located in the 3' terminal part of the NIb coding region. This group of isolates was recognized only recently through the use of improved strain-typing methods. It was in fact found that a number of isolates originally described as PPV-M belong in fact to the PPV-Rec group. Isolates belonging to the PPV-Rec strain are widespread in several central and eastern European countries, frequently associated with plums and efficiently transmitted by aphids.

PPV-EA (from the type isolate El Amar) was originally isolated from apricots in the El Amar region of the Nile delta in Egypt in the late 1980s. It has also been observed in peach and in other regions of Egypt but has not been reported from outside this country thus far.

PPV-C (from 'Cherry') was first reported on sour cherry in Moldova and on sweet cherry in Italy. This group of cherry-adapted isolates seems sporadically present in some central and eastern European countries. PPV-C isolates are aphid transmissible and seem to be the only PPV isolates able to systemically infect cherry and *P. mahaleb*. PPV-C isolates are however also able to infect other *Prunus* species under experimental conditions.

PPV-W (from 'Winona') has been recently reported from a few infected plum trees in Canada. The PPV-W genome shows unique features, justifying its classification as a new strain but the epidemiology and other biological properties of this recently characterized strain remain to be determined.

Full-length genomic sequences have been determined for 18 PPV isolates representing each of the six recognized strains (**Table 1**). Moreover, partial sequences, focusing mainly on the 3' terminal part of genome, are available for a large number of isolates, making PPV one of the genetically best studied potyviruses. The comparison of complete genomic sequences revealed up to 27.7% nucleotide divergence between representative isolates of the PPV-M, D, Rec, EA, C, and W strains (**Table 2**).

Recombination seems to have played a significant role in the evolutionary history of PPV. The PPV-Rec strain derives from a single recombination event involving isolates belonging to the PPV-D and PPV-M strains, with a recombination breakpoint in the C-terminus of the NIb gene. In addition, analysis of complete genomic sequences has recently demonstrated that the PPV-D and PPV-M strains themselves share an ancestrally

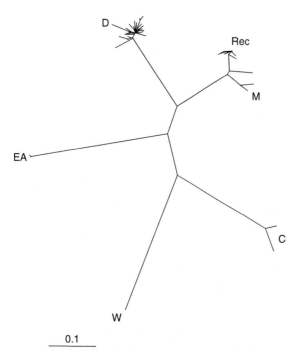

Figure 1 Unrooted phylogenetic tree computed using full-length coat protein gene sequences showing the relationships between isolates representative of all six known PPV strains.

Table 1   Origins and references of plum pox virus isolates for which complete genomic sequences are available

Isolate	Strain	Country of origin	Original host	Reference
PS	M	Yugoslavia	Peach	AJ243957
SK-68	M	Hungary	Plum	M92280
Dideron	D	France	Apricot	X16415
NAT	D	Germany	?	NC_001445
SC	D	?	?	X81083
PENN-1	D	USA	Peach	AF401295
PENN-2	D	USA	Plum	AF401296
PENN-3	D	USA	Peach	DQ465242
PENN-4	D	USA	Peach	DQ465243
Fantasia	D	Canada	Nectarine	AY912056
Vulcan	D	Canada	Peach	AY912057
48–922	D	Canada	Peach	AY912058
BOR-3	Rec	Slovakia	Apricot	AY028309
El Amar	EA	Egypt	Apricot	AM157175
El Amar	EA	Egypt	Apricot	DQ431465
SoC	C	Moldova	Sour cherry	Y09851
SwC	C	Italy	Sweet cherry	AY184478
W3174	W	Canada	Plum	AY912055

Table 2 Pairwise genetic distances calculated on the complete genomes of representative isolates of all six known PPV strains

	PPV-M	PPV-D	PPV-Rec	PPV-EA	PPV-C
PPV-D	0.127				
PPV-Rec	0.106	0.046			
PPV-EA	0.239	0.237	0.233		
PPV-C	0.257	0.266	0.260	0.277	
PPV-W	0.238	0.235	0.233	0.265	0.234

recombined 5′ part of their genome (5′ noncoding region, P1, HC-Pro, and N-terminus of P3). Other recently identified recombination events involve PPV-W, which shows a clear recombination signal with PPV-M in the region extending from the P1 coding region to the HC-Pro region and a divergent PPV-M isolate from Turkey (AY677114). It thus appears that at least four of the six recognized strains of PPV (among which the three most prevalent, PPV-D, -M, and -Rec) have recombination events in their evolutionary history. One remarkable consequence is that these four strains all share a similar genomic region corresponding to the C-terminus of their P1 and the N-terminus of their HC-Pro genes (genomic positions ~800–1450). On the other hand, PPV-C and PPV-EA appear to represent independent evolutionary lineages.

Assays that allow the discrimination between the various PPV strains include serological analysis with strain-specific monoclonal antibodies, restriction fragment length polymorphism (RFLP) analysis of polymerase chain reaction (PCR) fragments derived from the various genomic regions (CP, P3–6K1, and CI), PCR with strain specific primers and partial or complete sequence analysis. In addition, proper identification of recombinant isolates may require the use of several techniques targeting different parts of the viral genome or specific primers with binding sites located on both sides of the targeted recombination breakpoint. Strain-specific monoclonal antibodies have so far been obtained for isolates belonging to the PPV-M, PPV-D, PPV-EA, and PPV-C groups, but some isolates may show abnormal typing properties using either monoclonal antibodies or RFLP analysis techniques so that only PCR-based strain-specific assays or sequencing can provide unambiguous PPV strain identification.

Virus Spread

Natural spread of PPV occurs through aphids from neighboring infected reservoirs (cultivated, wild, or ornamental trees), while long-range movement of the virus is linked to international exchange of contaminated propagation material. The recent demonstration that aphids can acquire and transmit the virus from contaminated fruits indicates that fruit shipments could also represent a pathway for international virus movement. The virus is thus usually introduced in a new location by infected propagation material. Once established in the orchard or nursery, the virus is rapidly spread by aphid vectors. PPV is naturally transmitted by over 20 different aphid species in a nonpersistent manner (*Aphis craccivora*, *A. gossypii*, *A. hederae*, *A. spiraecola*, *Brachycaudus cardui*, *B. helichrysi*, *B. persicae*, *Myzus persicae*, *M. varians*, *Phorodon humuli*, *Rhopalosiphum padi*, etc.). Aphids generally transmit the virus over short distance, that is, next to infected sources, when migrating between plants and making test probes. The virus is transmitted in a nonpersistent manner so that aphids may acquire the virus very rapidly when probing infected plants with their stylets (piercing–sucking mouthpart). However, aphids can only transmit the virus for a short time after virus acquisition as the virus is retained in a viable state in the aphid's mouthparts for a period of only a few minutes to a few hours. The virus appears to be lost by the aphid the first time the aphid probes on the next plant and aphids do not remain infectious after moulting nor do they pass PPV onto their progeny.

The efficiency of transmission depends on the particular aphid species, virus isolate, and host species from which the virus is acquired and to which the virus is transmitted. As for other potyviruses, two viral proteins are known to be involved in the transmission mechanism, HC-Pro (helper component) and CP. The DAG amino acid motif is highly conserved in the N-terminus of the CP protein of aphid-transmissible PPV isolates and loss of transmissibility is correlated with its deletion.

Conflicting results have been reported in the past concerning seed transmission in some *Prunus* species, but recently more thorough studies have failed to demonstrate seed transmission of PPV in any of its woody hosts.

Ecology and Control

In the countries and regions where the virus is not yet present, strict quarantine measures need to be established to prevent the introduction of PPV through legal or illegal importation of infected fruit tree propagation material or fruits. In regions where the disease is present but still under control, a mix of prophylactic approaches is usually implemented, including the use of virus-free propagation material (rootstocks, budwoods), tight aphid vector control, regular inspection of orchards, and prompt eradication of infected plants. If the rate of infected trees in the prospected orchard exceeds a critical level (10–20% in most countries), whole orchards have to be removed.

In regions where eradication is no longer an option and where the spread of the disease is no longer under control (e.g., central and southeastern Europe), the cultivation of less susceptible or tolerant varieties is so far the only option that may allow to continue the production of susceptible fruit species. This practice, however, further contributes to the viral spread.

For PPV control in the long term, most European countries are placing strong emphasis on breeding programs aiming at the development of PPV-resistant varieties. Unfortunately, only a very limited number of natural sources of resistance to PPV have been identified within the *P. armeniaca* and *P. domestica* germplasm, and these are often characterized by low-quality fruits. In peach, despite extensive screening efforts, no resistant varieties are known but resistance has been identified in the wild relative *P. davidiana*. Breeding efforts are ongoing using these various resistance sources to introduce PPV resistance in commercially interesting varieties. Progress is however slowed by the long generation times, the length of the resistance tests, and the often polygenic nature of the resistances involved.

Development of genotypes with a hypersensitive response as an active defense mechanism against PPV is another promising way to produce resistant fruit trees, as was demonstrated for some plum varieties, but no hypersentivity sources are so far known in apricot and peach. In addition, it should be cautioned that in herbaceous crops, hypersensitive resistance to viruses has often proved to be nondurable as a consequence of genetic evolution of the virus.

Early attempts to use cross-protection with attenuated virus isolates, a technique successfully used to control some other potyviruses (zucchini yellow mosaic virus, papaya ringspot virus), have not met with success in the case of PPV.

Transgene-based resistance offers a complementary approach for the development of PPV-resistant stone fruit cultivars. Experimental herbaceous plants transformed with different regions of the PPV genome (i.e., P1, HC-Pro, CI, NIa, Nib, or CP) have been developed and have shown partial or complete resistance to PPV. A transgenic clone of *P. domestica* (C5), containing the PPV capsid protein gene, has been described as highly resistant to PPV in greenhouse and field tests, displaying characteristics typical of post-transcriptional gene silencing (PTGS) based resistance. Although this resistance can be partially overcome by graft inoculation, it provided effective protection against the natural spread of PPV in several field tests, even under conditions of high natural inoculation pressure.

See also: Papaya Ringspot Virus; Potato Virus Y; Watermelon Mosaic Virus and Zucchini Yellow Mosaic Virus.

Further Reading

EPPO (2004) EPPO Standards PM 7/32 Diagnostic Protocol for Plum pox potyvirus. *EPPO Bulletin* 34: 247–256.
EPPO (2006) A review of plum pox virus. *EPPO Bulletin* 36: 201–349.
Glasa M and Candresse T (2005) Plum pox virus. AAB Description of Plant Viruses. No. 410. http://www.dpvweb.net/dpv/showdpv.php?dpvno=410 (accessed July 2007).
Kegler H, Fuchs E, Gruntzig M, and Schwarz S (1998) Some results of 50 years of research on the resistance to plum pox virus. *Acta Virologica* 42: 200–215.
López-Moya JJ, Fernández-Fernández MR, Cambra M, and García JA (2000) Biotechnological aspects of plum pox virus. *Journal of Biotechnology* 76: 121–136.
Martinez-Gomez P, Dicenta F, and Audergon JM (2000) Behaviour of apricot (*Prunus armeniaca* L.) cultivars in the presence of sharka (*Plum pox potyvirus*): A review. *Agronomie* 20: 407–422.

Pomovirus

L Torrance, Scottish Crop Research Institute, Invergowrie, UK

Glossary

Spore balls Resting spores (cystosori) of *Spongospora subterranea* found in lesions or pustules (powdery scabs) on potato tubers; a single cystosorus comprises 500–1000 resting spores aggregated to form in a ball that is partially hollow, traversed by irregular channels.
Spraing Virus disease symptoms of internal brown lines and arcs in potato tuber flesh.

Introduction

Pomoviruses have tubular rod-shaped particles and tripartite genomes; they are transmitted by soil-borne zoosporic organisms belonging to two genera (*Polymyxa* and *Spongospora*) in the family Plasmodiophoraceae. Pomoviruses have limited host ranges, infecting species in a few families of dicotyledenous plants. Agriculturally important hosts include potato and sugar beet. There are four member species: *Potato mop-top virus, Beet soil-borne virus, Beet virus Q*, and *Broad bean necrosis virus*. Beet virus Q (BVQ) and beet soil-borne virus (BSBV) can occur in mixed infections with the benyvirus beet necrotic yellow vein virus (BNYVV).

Taxonomy and Classification

The classification of tubular rod-shaped viruses transmitted by soil-borne plasmodiophorid vectors was revised in 1998 to establish four genera: *Furovirus*, *Pomovirus*, *Benyvirus*, and *Pecluvirus*. The revision was prompted by new virus sequence information that revealed major differences in genome properties (number of RNA species, sequence, and genome organization). The genera are not assigned currently to any family. Pomovirus is a siglum from *Potato mop-top virus*, the type species.

Physical Properties of Particles

Pomovirus particles are hollow, helical rods, 18–20 nm in diameter, comprising multiple copies of a single major coat protein (CP; *c.* 19–20 kDa). The CP gene is terminated by a UAG (or UAA in broad bean necrosis virus (BBNV)) stop codon that is thought to be suppressible, readthrough (RT) of which would produce a fusion protein of variable mass (54–104 kDa). One or a few copies of the RT fusion protein are present at the extremity of potato mop-top virus (PMTV) particles thought to contain the 5′ end of the virus RNA. Pomovirus particles are fragile and particle size distribution measurements are variable; PMTV particles have predominant lengths of 125, 137, and 283 nm. PMTV particles sediment as three components with sedimentation coefficients ($S_{20,w}$) of 126, 171, and 236 S.

Genome Properties

Complete genome sequences are available for three member species and an almost complete sequence (without the 5′ and 3′ untranslated regions (UTRs)) is available for BBNV. Pomovirus genomes comprise three species of positive-sense single-stranded RNA of *c.* 5.8–6, 2.8–3.4, and 2.3–3.1 kbp (**Figure 1**). RNA-1 encodes the replicase proteins. It contains a large open reading frame (ORF) that is interrupted by a UGA stop codon (ORF1) (or UAA in BVQ and BSBV); the sequence continues in-phase to encode an RT protein (204–207 kDa). The ORF1 protein (145–149 kDa) contains methyltransferase and helicase motifs while the RT domain contains the GDD RNA-dependent RNA polymerase (RdRp) motif. Phylogenetic analysis reveals that the RdRps of the pomoviruses and soil-borne wheat mosaic furovirus share between 50% and 60% sequence identity.

The 5′ UTR of pomovirus RNAs contains the starting sequence $GU(A)_{1-4}(U)_n$ (except BVQ RNA-1 which begins with AUA). The RNAs are probably capped at the 5′ end since the RNA-1 ORF contains methyltransferase motifs associated with capping activity. The terminal 80 nucleotides of the 3′ UTR can be folded into a tRNA-like structure that contains an anticodon for valine. Both BSBV and PMTV RNAs were shown to be valylated experimentally.

The virus movement proteins are encoded on RNA-2 of PMTV (RNA-3 in BSBV, BVQ, and BBNV). Three 5′ overlapping ORFs encode a conserved module of movement proteins known as the triple gene block (TGB). TGB movement proteins are found in the genomes of other rod-shaped viruses (hordei-, beny-, pecluviruses) and in monopartite filamentous viruses in the genera *Potexvirus* and *Carlavirus*. The TGB proteins (TGB1, TGB2, and TGB3) are named according to their position on the RNA and have molecular masses of 48–53, 13, and 20–22 kDa, respectively. The TGB1 contains a deoxyribonucleotide triphosphate (dNTP) binding site and helicase motifs in the C-terminal half typical of all TGB1 sequences and an extended N-terminal domain found in hordei-like TGB1s that do not share obvious sequence identities. TGB1 binds RNA and is

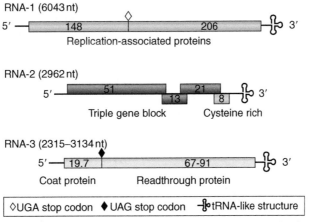

Figure 1 Diagram of the PMTV genome organization; boxes indicate open reading frames with the molecular masses of predicted protein products (kDa) indicated within.

thought to interact with genomic RNAs to facilitate movement. The sequence of the second TGB protein is the most conserved with BSBV, BVQ, and PMTV sharing 63–75% sequence identity and 49% identity with that of BBNV; there is little sequence identity among the TGB3 sequences. Analysis shows that TGB2 and TGB3 proteins contain two hydrophobic regions (predicted transmembrane domains) separated by a hydrophilic domain and these proteins are associated with intracellular membranes in infected plants.

In PMTV and BBNV, a fourth small ORF is predicted that encodes an 8 kDa cysteine-rich or 6 kDa glycine-rich protein, respectively, of unknown function whereas no such ORF is present in BSBV or BVQ. The 8 kDa cysteine-rich protein of PMTV is not needed for virus movement or infection of *Nicotiana benthamiana* and is not thought to function as silencing suppressor. Although a subgenomic RNA (sgRNA) that could encode the 8 kDa protein was detected in infected *N. benthamiana*, the protein is not readily detectable in extracts of infected leaves.

The CP and RT proteins are encoded on RNA-3 of PMTV (RNA-2 in BSBV, BVQ, and BBNV). The PMTV RNA-3 is of variable size, from 2315 to 3134 nt. Analysis of naturally occurring and glasshouse-propagated isolates revealed that deletions of *c.* 500–1000 nucleotides occur in both field and laboratory isolates and that they occur predominantly in the RT domain, particularly in the region toward the C-terminus. Deletions in this region are correlated with loss of transmission by the natural vector *Spongospora subterranea*. The first PMTV sequence to be published contained a shorter form of the CP–RT-encoding RNA and was designated as RNA-3, whereas in BSBV, BVQ, and BBNV the corresponding RNA is RNA-2. The RT domain of BVQ is shorter than the others and the RT coding sequence is followed by two additional ORFs for proteins of predicted mass of 9 and 18 kDa. Amino acid sequence comparisons reveal conserved motifs between these two proteins and domains in the C-terminal portions of BSBV and PMTV RT proteins which may indicate that they have arisen by degeneration of a larger ORF.

The occurrence of encapsidated deleted forms of RNA-2 and RNA-3 in natural and laboratory isolates of PMTV has been described and sequence analysis suggests that these PMTV RNAs contain sites that are susceptible to recombination possibly through a template switching mechanism. Variable base composition is found in natural isolates of BSBV in the sequence at the 3′ end of RNA-3 between the stop codon of the third TGB and the terminal tRNA-like structure.

Virus–Host Interactions and Movement

Cytoplasmic inclusions of enlarged endoplasmic reticulum (ER) and the accumulation of distorted membranes and small virion bundles can be seen by electron microscope examination of thin sections of BSBV- and BVQ-infected leaves. In PMTV-infected potato leaves, abnormal chloroplasts with cytoplasmic invaginations were seen in thin sections as well as tubular structures in the cytoplasm associated with the ER and tonoplast and in the vacuole.

PMTV does not require CP for movement and it is thought that TGB1 interacts with viral RNA forming a movement competent ribonucleoprotein (RNP) complex. Studies of transiently expressed PMTV TGB proteins fused to marker proteins such as green fluorescent protein or monomeric red fluorescent protein in epidermal cells of *N. benthamiana* have helped to elucidate events in intracellular trafficking and indicate that PMTV interacts with the cellular membrane recycling system. PMTV TGB2 and TGB3 were shown to co-localize on the ER (**Figure 2**) and in small motile granules that utilize the actin–ER network to reach the cell periphery and plasmodesmata (PD) and TGB3 contains a putative tyrosine sorting signal (Y-Q-D-L-N), mutation of which inhibits PD localization. TGB2 and TGB3 act together to transport GFP-TGB1 to the PD for movement into neighboring cells and TGB2 and TGB3 have the capacity to gate the PD pore. TGB2 co-localizes in endocytic vesicles with the Rab 5 ortholog Ara 7 (AtRabF2b) that marks the early endosomal compartment. Also, protein-interaction analysis revealed that recombinant TGB2 interacted with a tobacco protein belonging to the highly conserved RME-8 (receptor mediated endocytosis-8) family of J-domain chaperones, essential for endocytic trafficking.

Host Range, Geographical Distribution, and Transmission by Vector

Pomoviruses have a limited host range and are transmitted in soil by zoosporic plasmodiophorid vectors that have been classified as protists (**Figure 3**). Viruses that are transmitted by plasmodiophorid vectors include pomo-, peclu-, furo-, beny-, and bymoviruses and they are thought to be carried within the zoospores. The vector life cycle includes production of environmentally resistant thick-walled resting spores and viruliferous resting spores can survive in soil for many years.

BBNV has been reported only from Japan; it causes necrosis and stunting in broad beans and peas. It is mechanically transmitted by inoculation of sap to a few species including *Vicia faba*, *Pisum sativum*, and *Chenopodium quinoa*.

BVQ and BSBV are often found associated with BNYVV; BVQ is reported only from Europe whereas BSBV is found in sugar-beet-growing areas worldwide. No symptoms have been attributed to BSBV or BVQ alone in sugar beet. The viruses are thought to be transmitted in soil by *Polymyxa betae*. BVQ is mechanically

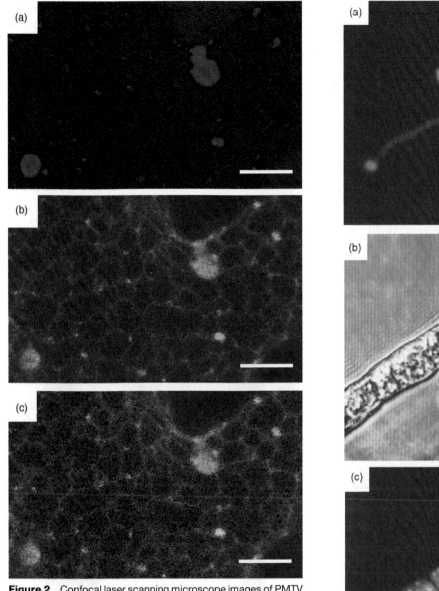

Figure 2 Confocal laser scanning microscope images of PMTV TGB2 and TGB3 fluorescent fusion proteins in epidermal cells of *Nicotiana benthamiana*. Monomeric red fluorescent protein (mRFP) tagged PMTV-TGB2 was transiently expressed with green fluorescent protein (GFP) tagged PMTV-TGB3, the proteins co-localized on membranes of the ER and in motile granules seen moving on the ER network. (a) Expression of mRFP-TGB2 (red channel); (b) expression of GFP-TGB3 (green channel); (c) merged image. Scale = 10 nm.

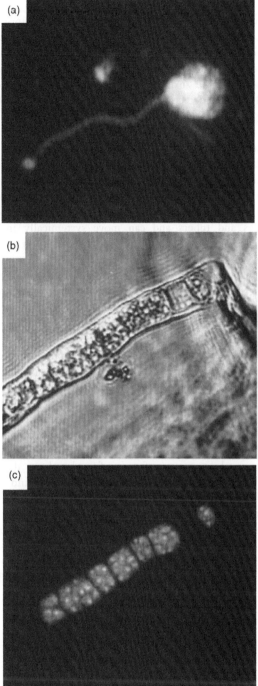

Figure 3 (a) Biflagellate zoopore of *Spongospora subterranea*. (b) Bright field and (c) fluorescence microscope images of zoosporangia in tomato root hair.

transmissible only to *C. quinoa* and BSBV to members of the Chenopodiaceae.

PMTV is found in potato-growing regions of Europe, North and South America, and Asia; virus incidence is favored by cool, wet growing conditions. It is transmitted in soil by *S. subterranea*, also a potato pathogen which causes the tuber blemish disease powdery scab. PMTV can be transmitted mechanically to members of the Solanaceae and Chenopodiaceae.

Serological Relationships, Diagnosis, and Control

The viruses are serologically distinct. Distant relationships have been reported between BSBV and BVQ; PMTV and soil-borne wheat mosaic furovirus; and PMTV, BBNV, and tobamoviruses. PMTV, BSBV, and

BVQ CPs contain a conserved sequence (SALNVAHQL) that reacts in Western blots with a monoclonal antibody (SCR70) produced against PMTV, but that is not exposed on intact particles. PMTV particles contain an immuno-dominant epitope at the N-terminus of the CP that is exposed at the surface along the sides of the particles and can be detected by monoclonal antibody SCR69 (**Figure 4**).

The viruses can be detected by serological tests (immunosorbent electron microscopy, enzyme-linked immunosorbent assay) and by assays based on the reverse transcriptase-polymerase chain reaction (RT-PCR) in leaves, roots, or tubers from naturally infected plants. PMTV is known to be erratically distributed in potato leaves and tubers and can move systemically in the absence of CP in potato leaves which raises a risk of false negative diagnosis. In addition, test plants grown in soil that has been previously air-dried can be used as indicators either by observing visual symptoms such as the PMTV-induced necrotic 'thistle-leaf'-shaped line patterns on leaves of *Nicotiana debneyi* or by conducting serological or RT-PCR analysis on the test plant roots or leaves.

PMTV causes an economically important disease affecting the quality of tubers grown for the fresh and processing markets. Tubers of sensitive potato cultivars that are infected from soil during the growing season develop spraing symptoms that include unsightly brown lines, arcs, or marks in the flesh sometimes accompanied by slightly raised external lines and rings (**Figure 5**). Potato plants grown from infected tubers display yellow markings or chevrons on the leaves and may have shortened internodes (mop-top) producing cracked or malformed tubers; both tuber quality and yield can be affected. However, the virus does not infect all plants grown from infected tubers. Haulm and tuber symptoms vary markedly with cultivar, and some cultivars are symptomlessly infected. Environmental conditions also affect disease incidence and severity, and PMTV incidence was shown to increase with annual rainfall. Soil temperatures of 12–17 °C and high soil moisture at tuber initiation favor powdery scab incidence.

PMTV can be established at new sites by planting infected tubers and once established, PMTV is a persistent problem, as the resting spore balls of the vector *S. subterranea* are long-lived and resistant to drought and agrochemicals. Viruliferous spore balls are spread readily to new sites by farm vehicles, contaminated seed tubers, wind-blown surface soil; motile zoospores can be spread through contaminated irrigation or drainage water. In

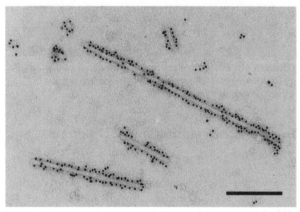

Figure 4 Electron micrograph of PMTV particles labeled with monoclonal antibody SCR69/gold conjugate.

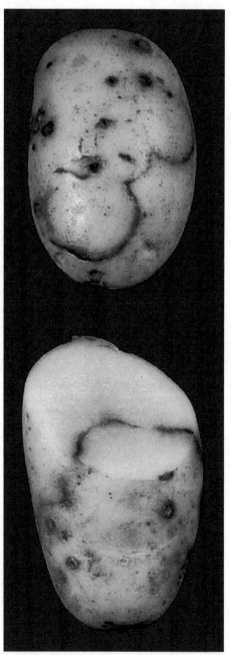

Figure 5 Symptoms of PMTV in potato tubers cv. Nicola.

certain areas, potatoes have become infected by PMTV 18 years after potatoes were last grown (the longest period recorded). There is no effective practicable means to control *S. subterranea* but decreased severity of powdery scab can be achieved by application of fluazinam to soil. In addition, disinfection of tubers with chemicals such as formaldehyde decreases virus incidence, although the efficacy of this treatment depends on the level of soil infestation where the tubers are planted and a combination of tuber and soil treatments may be more effective.

The best prospect for virus disease control is development of resistant cultivars but there are no known sources of PMTV resistance in commercial potato cultivars and most of the commercially grown cultivars are also susceptible to *S. subterranea*. However, plants transformed with virus transgenes (CP and a mutated form of TGB2) have exhibited resistance to PMTV with decreased virus accumulation and incidence in tubers of plants grown in infested soil.

See also: Benyvirus; Furovirus; Pecluvirus; Tobamovirus.

Further Reading

Arif M, Torrance L, and Reavy B (1995) Acquisition and transmission of potato mop-top furovirus by a culture of *Spongospora subterranea* f. sp. *subterranea* derived from a single cystosorus. *Annals of Applied Biology* 126: 493–503.

Haupt S, Cowan GH, Ziegler A, Roberts AG, Oparka KJ, and Torrance L (2005) Two plant-viral movement proteins traffic in the endocytic recycling pathway. *Plant Cell* 17: 164–181.

Koenig R and Loss S (1997) Beet soil-borne virus RNA1: Genetic analysis enabled by a starting sequence generated with primers to highly conserved helicase-encoding domains. *Journal of General Virology* 78: 3161–3165.

Koenig R, Pleij CWA, Beier C, and Commandeur U (1998) Genome properties of beet virus Q, a new furo-like virus from sugarbeet determined from unpurified virus. *Journal of General Virology* 79: 2027–2036.

Lu X, Yamamoto S, Tanaka M, Hibi T, and Namba S (1998) The genome organization of the broad bean necrosis virus (BBNV). *Archives of Virology* 143: 1335–1348.

Pereira LG, Torrance L, Roberts IM, and Harrison BD (1994) Antigenic structure of the coat protein of potato mop-top furovirus. *Virology* 203: 277–285.

Reavy B, Arif M, Cowan GH, and Torrance L (1998) Association of sequences in the coat protein/read-through domain of potato mop-top virus with transmission by *Spongospora subterranea*. *Journal of General Virology* 79: 2343–2347.

Rochon D'A, Kakani K, Robbins M, and Reade R (2004) Molecular aspects of plant virus transmission by olpidium and plasmodiophorid vectors. *Annual Review of Phytopathology* 42: 211–241.

Sandgren M, Savenkov EI, and Valkonen JPT (2001) The readthrough region of potato mop-top virus (PMTV) coat protein encoding RNA, the second largest RNA of PMTV genome, undergoes structural changes in naturally infected and experimentally inoculated plants. *Archives of Virology* 146: 467–477.

Savenkov EI, Sangren M, and Valkonen JPT (1999) Complete sequence of RNA1 and the presence of tRNA-like structures in all RNAs of potato mop-top virus, genus *Pomovirus*. *Journal of General Virology* 80: 2779–2784.

Zamyatnin AA, Solovyev AG, Savenkov EI, *et al.* (2004) Transient coexpression of individual genes encoded by the triple gene block of potato mop-top virus reveals requirements for TGBp1 trafficking. *Molecular Plant Microbe Interactions* 17: 921–930.

Potato Virus Y

C Kerlan, INRA, UMR1099 BiO3P, Le Rheu, France
B Moury, INRA – Station de Pathologie Végétale, Montfavet, France

Brief Description and Significance

Potato virus Y (PVY) was first recognized in 1931 as an aphid-transmitted member within a group of viruses associated with potato degeneration, a disorder known since the eighteenth century. PVY is the type species of the genus *Potyvirus*, one of the six genera in the family *Potyviridae*. PVY is naturally spread by vegetatively propagated material and by aphids in numerous species in a nonpersistent manner. Transmission by contact has also been reported. PVY has a wide host range and is highly variable with some host specificity. Genome sequences and reliable tools for detection and strain differentiation are available. Bioassays and serology have been largely developed.

PVY is one of the most damaging plant pathogens causing significant losses in four main crops around the world: potato, pepper, tomato, and tobacco. In surveys of viruses with worldwide economic importance, PVY was listed in the top-five viruses affecting field-grown vegetables. PVY was also found responsible for damages in petunias in Europe and in eggplant crops in India.

Efficient control strategies depending on the crop have been developed. However none of them seems capable to take into account PVY evolution and to suppress risks of new epidemics.

Viral Particle and Genome

PVY virions are nonenveloped, filamentous, flexuous rods, 730–740 nm long, 11–12 nm in diameter, with an axial canal 2–3 nm in diameter and helical symmetry. Assembly and disruption of PVY particles were studied in detail. Virions contain about 6% nucleic acid and a viral genome linked protein (VPg), but no lipid or other components.

The genome consists of one single-stranded, linear, positive-sense RNA molecule ($3.1–3.2 \times 10^{6}$ Da) with a polyadenosine sequence at the 3′ terminus. VPg is attached to the 5′ end via a phosphate ester linkage to Tyr[60]. A three-dimensional model structure of VPg was proposed. Complete genome sequences are available in databases for isolates from potato, tobacco, pepper, tomato, and *Solanum nigrum*. Hundreds of partial sequences of PVY isolates (more than 200 coat protein (CP) sequences) are also available.

PVY RNA is approximately 9700 nt in length excluding the poly(A) tail. As in all members of the genus *Potyvirus*, its single open reading frame is expressed as a large polyprotein (3061–3063 amino acids) autoproteolytically cleaved to yield ten functional proteins: P1 (284 aa), HC-Pro (456 aa), P3 (365 aa), 6K1 (52 aa), CI (634 aa), 6K2 (52 aa), VPg (188 aa), NIaPro (244 aa), NIb (521 aa), and CP (267 aa). There are two distal noncoding regions, 5′NTr (184 nt) and 3′NTr (from 326 to 333 nt) (**Figure 1**).

RNA synthesis is believed to occur in the cytoplasm. The replication complex comprises the proteins NIb, CI and VPg and possibly involves the proteins 6K1 and 6K2. The NIb protein is believed to be the RNA-dependent RNA polymerase.

The CP (29.95 kDa) consists of 267 amino acid residues. It comprises a core (218 aa), highly conserved in members of the genus *Potyvirus*, and two surface-exposed N-terminal (300 aa) and C-terminal (19 aa) regions which

are not required for virus assembly and maintenance of infectivity.

The helper component protein (HC-Pro) is supposed to have a biologically active dimeric form (with a subunit molecular mass of 58 kDa). It is indeed capable in the yeast two-hybrid system of self-interaction that can be drastically reduced by mutations in its cysteine-rich region. HC-Pro is involved in PVY accumulation in tobacco and is a suppressor of post-transcriptional gene silencing (PTGS).

Transmission of PVY

PVY is naturally spread by vegetatively propagated plant organs (tubers, cuttings) and by aphids. PVY has also been reported to be spread by plant-to-plant contact, for instance in tobacco and tomato crops in Southern America. It can also be transmitted by contact between sprouts of potato tubers during storage. Transmission by seed has been reported from *S. nigrum* and *Nicandra physaloides*. Transmission by pollen has never been proved for any host plant.

Unlike other potyviruses, PVY displays an unusually large range of aphid vectors. Aphids in 70 species, all in the family Aphidinae, were demonstrated to be able to transmit PVY, most of them with very low efficiencies compared to that of *Myzus persicae*. Apterae and alatae are vectors (**Figure 2**).

PVY is aphid-transmitted in a nonpersistent manner. Acquisition and inoculation periods are brief (a few seconds or minutes). Aphid stylets penetrate into the epidermal cell layer of the plants and puncture plant cell membranes. There is no discernible latent period.

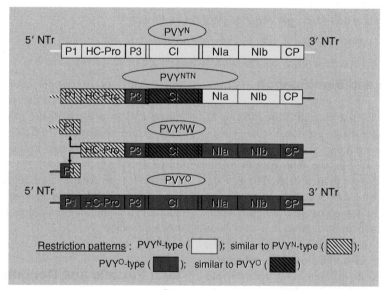

Figure 1 Schematic map of the PVY genome: strains PVY[O] and PVY[N], variants PVY[NTN] and PVY[N]W. Reproduced from Glais L, Tribodet M, and Kerlan C (2002) Genetic variability in Potato Polyvirus Y (PVY): Evidence that PVY[NW] and PVY[NTN] variants are single to multiple recombinants between PVY[O] and PVY[N] isolates. *Archives of Virology* 147: 363–378, with permission from Springer-verlag.

Figure 2 Aphid vector of PVY: winged form of *Aphis nasturtii*. Photograph: B. Chaubet, INRA, France.

PVY does not pass through insect moults and retention of the virus in the aphid usually lasts not more than 1 or 2 h. However longer retention periods (up to 17 h in *Aphis nasturtii*) were reported. Prior starvation of the aphids increases the efficiency of transmission though it does not affect the number of electrically recorded membrane punctures during acquisition periods.

PVY transmissibility is determined by both HC-Pro and CP proteins. All aphid-transmissible PVY isolates contain the 'DAG triplet' (Asp–Ala–Gly) in the CP N-terminal domain. PVY isolates having the sequence DAGE are also aphid-transmissible, unlike those of tobacco vein mottling virus. PVY was the first virus for which it was proved that a virus-induced component of the plant sap is needed for aphid transmission. Effectively, HC-Pro of an efficiently transmitted PVY isolate was shown to mediate the transmission of nonaphid-transmitted PVY isolates and of other potyviruses. Its activity can be blocked by antisera specific to PVY HC-Pro but not by antisera to HC-Pros of other viruses. Monoclonal antibodies (MAbs) to PVY HC-Pro were also produced. Loss of HC-Pro activity correlates with nonretention of virions on aphid stylets. Aphid-transmissible and -nontransmissible isolates differ by one or two amino acid substitutions: Gly^{35} to Asp, Lys^{50} to Glu or Lys^{50} to Asn, Ile^{225} to Val, Ser^{355} to Gly. Lys^{50} is part of a conserved 'KITC motif' (Lys–Ile–Thr–Cys). Changes in or around this motif can result in losses of aphid transmissibility. Conversely the reverse mutation Glu^{50} to Lys restores the helper function of a defective HC-Pro.

Host Plants

The natural host range comprises plants in nine families. It includes potato (*S. tuberosum* ssp. *tuberosum*), several species of native potatoes in the Andes namely *S. andigena*, numerous *Solanum* wild species, pepper (*Capsicum* spp.), tobacco (*Nicotiana* spp.), tomato (*Lycopersicon* spp.), eggplant (*S. melongena*), ornementals (*Petunia* spp., *Dahlia* spp.), perennial plants (*Physalis virginiana*, *P. heterophyla*), and a number of annual or perennial self-propagating plants. Many region-specific hosts were recorded such as *Cotula australis* in New Zealand or *Sorbaria tomentosa* in Himalaya.

Weeds in the family Solanaceae are often potential inoculum sources for PVY infections in tomato and pepper crops: *S. nigrum* and *S. dulcamara* in many countries; *S. chacoense* and *P. viscosa* in Southern America; *S. gracile*, *S. aculeatissimum*, *P. angulata*, *P. ciliosa*, and *P. floridana* in Florida. *P. virginiana* and *P. heterophyla* (perennial ground cherries) are overwintering hosts in Northern America. *S. dulcamara* in Western Europe or *Datura* spp. in Mediterranean countries are potential inoculum sources in potato crops as well as volunteer potatoes. Infected pepper and tomato plants, and seed and volunteer potatoes were also identified as potential sources of inoculum in tobacco crops in the USA, Canada, and Italy.

Host specificity clearly stated in the past is currently under reinvestigation. Most PVY isolates appear able to infect common tobacco and tomato cultivars. Conversely most isolates naturally infecting potato are unable to infect pepper cultivars systemically while pepper isolates are usually not detected in potato fields.

PVY is easily sap-transmitted and can be transmitted by stem- and tuber-grafting. Its experimental host range comprises plants in 495 species in 72 genera of 31 families. It includes 287 species in the family Solanaceae (among which 141 *Solanum* species and 70 *Nicotiana* species), 28 species of Amaranthaceae, 25 species of Fabaceae, 20 species of Chenopodiaceae, and 11 species of Asteraceae. A large part of these plants comprises only local lesion hosts. *Datura stramonium*, formerly reported as a host plant, was demonstrated to be totally immune to all strains tested in 1980–90. Some PVY isolates can infect the model plant *Arabidopsis thaliana*.

N. tabacum, *N. benthamiana*, *N. occidentalis* can be used as diagnostic species susceptible to all PVY strains (**Figure 3**). *N. tabacum* plantlets are often used as source and test plants for aphid transmission experiments. *N. tabacum* is a suitable host for virus purification. Yields of purification including a final step of cesium chloride gradient centrifugation usually vary from 10 to 25 mg kg^{-1} of tobacco leaves (**Figure 4**). Leaves of *N. tabacum* are the best virus-infected material to store. Antigenic properties can be retained for 1 year in freeze-dried crude extracts. Infectivity was reported to be preserved in freeze-dried material stored over calcium chloride at 4 °C for 15 years. However from our own experience numerous long-term stored isolates of PVY available in international collections are no longer infectious.

S. demissum Y, *S. demissum* PI 230579, and the hybrid *S. demissum* A6 are local lesion hosts. The 'A6 test' on detached leaves was commonly used in the past as a diagnostic tool for differentiating PVY from potato virus A (PVA), another potyvirus.

Cultivars of *N. tabacum*, potato and pepper, *Chenopodium amaranticolor* (**Figure 5**), *P. floridana*, some accessions of *S. brachycarpum* and *S. sparsipilum* are useful for distinguishing among PVY strains and pathotypes. Potato cultivars such as Bintje or Saco can be used to separate PVY from plants co-infected with PVA or potato virus X (PVX).

Serology

PVY is considered to be strongly immunogenic. Antisera with precipitin titers up to 1/4096 and MAbs have been

Figure 3 Typical mottle induced by a PVY isolate from potato on a leaf of *Nicotiana tabacum* cv *Xanthi* 15 days after mechanical inoculation. Photograph: C. Lacroix, INRA, France.

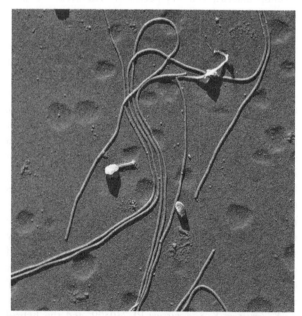

Figure 4 Purified suspension of PVY with a bacteriophage T$_4$ as internal calibration standard. Magnification 22 000×. Photograph: D. Thomas, CNRS-Rennes 1 University, France.

produced in rabbits or mice immunized with purified virus preparations or synthetic peptides, or by using DNA-based immunization, or phage display antibody technology. Numerous sources of antibodies and serological detection kits are available.

ELISA (standard DAS-ELISA and related protocols), dot-blot immunobinding assay, immunosorbent electron microscopy, latex and virobacterial agglutination tests, immunodiffusion in agar gels were intensively studied for use in strain differentiation and virus detection from plant material and from aphids.

Polyclonal antisera do not discriminate among PVY strains. MAbs allow to separate two main serogoups broadly corresponding to potato strains PVY^O and PVY^C from one part and PVY^N from the other part (see below). Specific MAbs to strain PVY^C or variants PVY^{NTN} and PVY^{N}W have also been produced, but their reliability still needs confirmation.

Relationships to Other Potyviruses

PVY is distantly serologically related to PVA and potato virus V (PVV), and to 17 other viruses in the genus *Potyvirus* including pepper mottle virus (PepMoV).

PVY, PepMoV, PVV, pepper yellow mosaic virus, pepper severe mosaic virus, wild potato mosaic virus and Peru tomato virus constitute a phylogenetic group distinguishable from other potyviruses including PVA.

Cytopathology

Virions are usually closely associated with inclusions in the cytoplasm of infected cells. They have also been observed within plasmodesmata or aligned on Golgi apparatus,

Figure 5 Host plants of PVY: local lesions induced by a PVY isolate from pepper on *Chenopodium amaranticolor* after mechanical inoculation. Photograph. L. Glais, INRA, France.

endoplasmic reticulum, and around mitochondria. Most PVY strains induce type-IV inclusions (i.e., pinwheels, scrolls, and short curved laminated aggregates). They consist of a single, nonglycosylated protein (67 kDa), serologically unrelated to the CP protein, or to host proteins. PVY also induces cytoplasmic rod-like amorphous inclusions, but except two isolates from Brazil does not induce large crystalline cytoplasmic and nucleolar inclusions.

Cytopathology in PVY infections has recently been intensively studied in potato and tobacco leaves, notably changes in fresh matter content, photosynthesis, and other metabolic activities. P1 protein was detected in association with cytoplasm inclusions in tobacco cells. Amorphous inclusions appear to be the primary site of HC-Pro accumulation. NIa protein was proved to accumulate in both cytoplasm and nucleus. PVY CP, HC-Pro, and RNA were found within chloroplasts of tobacco leaves suggesting they may alter the chloroplastic function as also proved in transgenic tobacco plants expressing PVY CP.

PVY in Potato

PVY has become in the last decade the most important virus in most growing areas for seed, ware, and processed potatoes. Serious outbreaks were, reported in the 2000s. Tuber quality can be severely affected due to necrosis or defects for processing potatoes. Reduction in size and number of harvested tubers can result in losses up to 80%.

Symptoms consist of mild to severe mottle, often associated with crinkling of the leaves. Yellowing and necrosis (vein necrosis and necrotic spots) frequently occur in the lower leaves. Symptoms also include collapse and dropping of intermediate leaves, which remain clinging to the stem (leaf drop). Secondarily infected plants (when mother tubers are infected) are stunted with crinkled and smaller leaflets (**Figure 6**). Necrosis on and around leaf veins, on petioles, stems, and tubers may occur in numerous cultivars. Some necroses on stems and petioles are called stipple-streak disease. The potato tuber necrotic ringspot disease (PTNRD) is characterized by particular patterns on leaves (sometimes oak-leaf necrosis) (**Figures 7(a)** and **7(b)** and stems, and superficial annular and arched necroses on tubers, first slightly protruding from tuber skin, then becoming dark brown with sometimes crackings of tuber skin (**Figures 7(c)** and **7(d)**).

Potato isolates have historically been divided into three main strain: PVY^O, PVY^N, and PVY^C. PVY^O and PVY^C are separated on the basis of hypersensitive reactions in potato cultivars bearing the resistance genes Ny_{tbr} and N_c (**Figure 8**). PVY^N differs from PVY^O and PVY^C in causing a veinal necrosis reaction in *N. tabacum* cv *Samsun* or cv *Xanthi* (**Figures 9(a)** and **9(b)**), but does not elicit any hypersensitive response in most potato cultivars. Two amino acids at positions 400 and 419 in the HC-Pro protein are involved

Figure 6 PVY on potato. Natural secondary infection by PVY in a potato field: crinkling, yellowing, and growth reduction of the leaflets. Photograph: V. Le Hingrat, FNPPPT, France.

in the induction of this necrotic reaction in tobacco. *P. floridana* is a third host for strain differentiation. PVY^C induces collapse and premature death of this plant. PVY^C was first characterized as nonaphid-transmissible, but many isolates of this strain were proved to be readily aphid-transmissible. Two former PVY^C isolates were later recognized as isolates of PVV.

Many unclassified variant isolates were reported in the past, notably isolates classified in a group called PVY^{An}. More recently PVY^Z and PVY^E were characterized as overcoming the resistance genes Ny_{tbr} and N_c. Two main variants have emerged for two decades: PVY^{NTN}, characterized by its ability to induce tuber necrosis, and PVY^{N}W which differs by its pathogenicity and its serotype O–C instead of serotype N. PVY$^{N:O}$ isolates described in the 2000s also share properties with both PVY^N and PVY^O and some of them induce tuber necrosis. PVY^O and PVY^N strains are distributed worldwide though PVY^N is yet a quarantine pathogen in Canada and the USA. The PVY^C strain is less frequent. The PVY^{NTN} variant has been identified in most potato-growing countries including the USA and Peru. The PVY^N-W variant has become prevalent in Poland and has been reported in several countries. PVY$^{N:O}$ was reported in Canada and the USA.

PVY potato isolates have been divided into three genetically distinct strains designated PVY^O, PVY^N, and PVY^{C1}. PVY^{C1} is a part of the genetically defined strain PVY^C or PVY^{NP} (nonpotato) which also includes isolates from pepper, tomato, and tobacco. PVY^{N}W, PVY$^{N:O}$, and the majority of PVY^{NTN} isolates are recombinants between PVY^N and PVY^O strains, with one to three recombination breakpoints (**Figure 1**). Several tuber necrosis-inducing isolates possess a PVY^N-type genome without any recombination breakpoint. Lastly, North American and European PVY^{NTN} isolates are separated on the basis of polymorphism in the 5′ NTr-P1 region.

Figure 7 PVY on potato. Symptoms associated with the potato tuber necrotic ringspot disease (PTNRD). (a) Yellowing and necroses on a basal leaf. (b) Necrotic oak-leaf pattern on an intermediate leaf. (c) Typical PTNRD symptoms on tubers of the cv *Monalisa*. (d) Atypical PTNRD induced by a PVY$^{N:O}$ isolate on a tuber of the cv *Yukon Gold*. Photographs: K. Charlet-Ramage, GNIS-INRA, France (a–c) and R. P. Singh Potato Research Centre, NB, Canada (d).

Figure 8 Hypersensitive reactions on potato cultivars 10 days after mechanical inoculation on cv *Desiree* containing the gene Ny_{tbr} inoculated by a PVY^{O} isolate (left) and on cv *Eersteling* containing the gene *Nc* inoculated by a PVY^{C} isolate (right). Photographs: C. Kerlan and J. P. Cohan, INRA, France.

Differences in aphid transmission between strains were reported. PVY^{N} isolates seem better transmitted than other PVY isolates with longer retention periods in aphids. Whatever the strain, *M. persicae* is clearly the most important vector in most potato-growing areas. However despite their low transmission efficiency, other species colonizing potatoes such as *A. nasturtii* (**Figure 2**), or visiting potatoes can also contribute to PVY spread. Some of these species,

notably cereal aphids or the pea aphid, were thought to be involved in PVY epidemics in potato crops (in Europe and the USA) due to their high abundance.

Control methods are first based on control of seed potatoes, breeding for resistance and quarantine regulations, especially regarding PVY^{N}. Sophisticated schemes of seed production include monitoring of aphid vectors, treatments by mineral oils against aphid transmission, eradication of weeds, and post-harvest detection tests using large-scale ELISA. Numerous molecular assays were also developed for PVY detection in leaves, potato tubers, and aphids, including cDNA hybridization, various RT-PCR protocols, and microarray technology. Breeding for potato resistance to PVY takes into account resistance to infection (which includes resistance to virus and to vectors, and is largely used), hypersensitivity resistance (HR) and extreme resistance (ER). HR and ER are based

on single dominant genes Ny and Ry, respectively. Ry genes (Ry_{sto}, Ry_{adg}), from *Solanum* wild species, map on chromosomes XI and XII, and confer broad-spectrum resistance, whereas HR genes Nc, Ny_{tbr} (which map on chromosome IV) (**Figure 10**) and Nz found in old potato cultivars, protect against PVY^C, PVY^O, and PVY^Z, respectively. The NIaPro protein of PVY is involved in the elicitation of Ry. The gene $Y-1$ involved in the HR reaction has been cloned. More than 20 cultivars containing Ry have been

Figure 9 PVY indexing on *Nicotiana tabacum* cv *Xanthi* (15 days after mechanical inoculation). (a) Typical vein necrosis and leaf distortion induced by a PVY^N isolate. (b) Typical vein clearing without any leaf distortion induced by a PVYO isolate. Photographs: C. Lacroix, INRA, France.

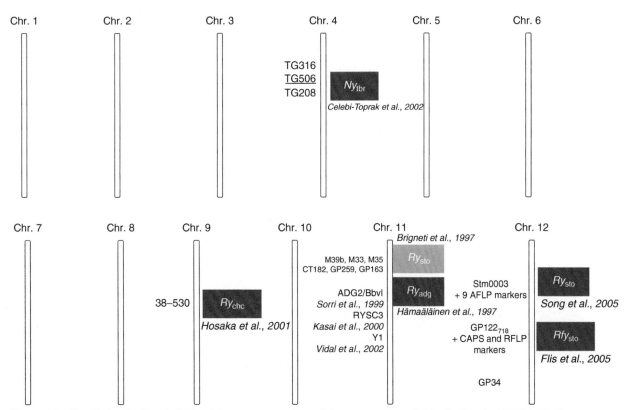

Figure 10 Genetic localization of PVY resistance genes on the potato genome. From S. Marchadour, FNPPPT-INRA, France (unpublished).

produced and this resistance has proved to be quite durable. Marker-assisted selection for resistance to PVY has been developed for genes Ry_{adg}, Ry_{chc}, and Ny_{tbr}. Genetically engineered resistance has been intensively studied and used to generate resistant transgenic plants from commercial potato cultivars. CP-mediated resistance was described as ER with often a broad spectrum.

PVY in Pepper

PVY is the causal agent of major diseases and production losses in pepper crops. In some situations it can affect 100% of the plants and can be the most important disease. PVY affects pepper production worldwide, while other pepper-infecting potyviruses are mostly restricted to particular continents and are not present in Europe. Symptoms are mostly visible on the vegetative parts of the plants but rarely on fruits. Depending on the pepper genotype and on the virus isolate, they consist of mosaic and vein banding on leaves or on necrotic symptoms on leaves, petioles, and stems (**Figure 11**).

Pepper isolates of PVY belong to the same group as the C phylogenetic group of PVY in potato, although they could constitute a separate C subgroup. One recently described exception consists of a O:C recombinant strain inducing veinal necrosis in pepper in Italy. No PVY isolates from the N or O phylogenetic groups have been described as epidemic in pepper. Pepper PVY isolates have been classified according to their pathotype relative to the numerous resistance genes and alleles used to control PVY in pepper crops (see further).

The most widespread and efficient way to control PVY in pepper is through the growing of cultivars carrying resistance genes. The first resistance genes used to control PVY were the $pvr2^1$ and $pvr2^2$ alleles, which control high-level resistance since no virus can be detected in inoculated organs of the plants. These genes have been exploited for decades in a large number of pepper cultivars and proved to be durable. However, PVY isolates virulent to the $pvr2^1$ allele were occasionally observed, especially in the Mediterranean basin and in tropical regions, while isolates virulent to the $pvr2^2$ allele are exceptional.

The three alleles $pvr2^+$ (susceptibility reference allele), $pvr2^1$ and $pvr2^2$ were used to classify the pepper isolates in three pathotypes: pathotype (0) isolates which can infect only plants with the $pvr2^+$ allele, pathotype (0,1) isolates which only infect plants carrying the $pvr2^+$ or $pvr2^1$ alleles and pathotype (0,1,2) isolates which are virulent toward all three $pvr2$ alleles.

It was shown that these alleles correspond to various copies of an isoform of the eukaryotic translation initiation factor 4E (eIF4E), and that amino acid substitutions in the central part of the viral protein genome-linked (VPg) of PVY determined virulence to both $pvr2^1$ and $pvr2^2$.

More recently, the dominant $Pvr4$ gene, originating from a *C. annuum* Mexican population, was introgressed into bell-pepper cultivars. This gene present in about 40% of European cultivars confers a high-level resistance to all tested isolates of PVY. This resistance shares common properties with hypersensitivity-based resistances. No isolate of PVY virulent to $Pvr4$ has been described so far, indicating a high durability of the resistance.

Other pepper resistances to PVY have been described (notably polygenic resistances from *C. annuum* and $Pvr7$ resistance from *C. chinense*).

Other control measures such as the use of oil sprays or physical barriers (polyethylene sheets and coarse nets) were also used for controlling spread of PVY in pepper crops in Israel and Florida.

PVY in Tomato

PVY can induce severe diseases in tomato crops but has long been considered a pathogen of secondary importance, inducing mild mosaic on the foliage only. Since the 1980s, new strains of PVY have arisen in Mediterranean countries, causing serious yield and quality loss of tomato fruits. These strains induce necrotic lesions on leaves and necrotic streaks on stems in all tomato varieties and frequently affected 100% of the plants, in greenhouses

Figure 11   PVY infection in a pepper field: (a) severe crinkling and distortion of the leaves. (b) Necrotic patterns on a leaf. Photographs: P. Gognalons, INRA, France (a) and M. Pepelnjak Slovenia (b).

as well as in open fields. Importantly, combination of PVY infections with other viruses such as cucumber mosaic virus can induce very severe diseases in tomato crops.

Data about the genetic diversity of PVY isolates from tomato are scarce. Isolates belonging to the C/pepper group of strains and to the recombinant NTN group have been observed in tomato crops. In laboratory tests, most of the isolates belonging to all phylogenetic groups of PVY induce systemic infections in tomato.

There are few reports about PVY resistance in tomato and related species. Accession PI247087 of *L. hirsutum* was described highly resistant to PVY. The resistance is efficient toward all tested tomato or pepper isolates of PVY, inhibits the multiplication of the virus in the inoculated organs and is controlled by a single recessive gene (named *pot-1*) which maps to a region on chromosome 3 syntenic to the *pvr2* locus in pepper. *Pot-1* was shown to belong to the same family of genes as *pvr2*. Strains of PVY virulent to the *pot-1* resistance can be selected during laboratory tests, a property that is controlled by a single amino acid substitution in the VPg of the virus. The *pot-1* gene was introduced into genotypes of *L. esculentum*, but varieties carrying this gene are not widespread.

PVY in Tobacco

PVY was reported to be the most damaging virus in tobacco crops. It causes height reductions and modifies the chemical composition of cured leaves, especially the nicotine content. Yield losses of 14–59% were reported with incidence of up to 100%. Symptoms on leaves are usually mild mottling but particular chlorotic patterns and necroses, notably the veinal necrosis disease, may also occur (**Figures 12(a)** and **12(b)**).

Three strain groups, M^SM^R, M^SN^R, N^SN^R, have been identified according to their reaction in tobacco cultivars resistant or susceptible to the root-knot nematode (*Meloidogyne incognita*). The NIb protein was found to elicit the hypersensitive response in these resistant tobacco cultivars. VAM-B refers to a resistance-breaking group of isolates overcoming the gene *va* originally found in the genotype Virginia A mutant and characterized by deletions at the Va locus. The VPg protein was suggested to be involved in overcoming resistance conferred by *va*.

Breeding for resistance is the main control method against PVY in tobacco production. Many European and American commercial cultivars display noteworthy levels of recessive resistance. The gene *va* has been extensively utilized, though its introgression is often associated with decrease of yield and cured leaf quality. In countries where early infection frequently occurs, additional measures consist of protecting seedbeds by fleece, eradication of weeds, and isolation from potato, tomato and pepper fields. Pathogen-derived resistance has been extensively studied in tobacco, although more for investigating the mechanisms of protection than in order to use this strategy in commercial tobacco cultivars.

Prospects

Strains

Current distinctions within the species *Potato virus Y* is not totally clear. The exact relationship between the necrotic strains from potato and from tobacco (PVY^N, N^SN^R, M^SN^R) still must be defined precisely, as the distinction between PVY^{NTN} and other isolates of the PVY^N strain. Is there any PVY^N isolate unable to

Figure 12   PVY infection on tobacco cv *Burley* in field trials in Southwestern France. (a) Chlorotic oak-leaf and halotic patterns. (b) Severe systemic necrosis. Photograph: D. Blancard, INRA, France.

induce potato tuber necrosis whatever the conditions? Are PVY$^{N:O}$ isolates different to those previously referred as PVY^{N}W isolates? Molecular classification in most cases correlates only partially with biological classifications since they are based on neutral markers.

Evolution

The large variability and genomic diversity of PVY makes its evolutionary story fascinating. Links between mutation or recombination events and evolution of PVY are thoroughly studied. Current knowledge indicates that PVY was present in pre-Columbian America and might have followed different evolutionary pathways as a result of co-evolution with various solanaceous hosts.

See also: Vector Transmission of Plant VirusesVector Transmission of Plant Viruses; Plant Resistance to Viruses: Engineered Resistance; Plant Virus Diseases: Economic Aspects; Potato Viruses; Tobacco Viruses.

Further Reading

Blancard D (1998) *Maladies du Tabac.* Paris: INRA.
Blancard D (1998) *Maladies de la Tomate.* Paris: INRA.
de Bokx JA and van der Want JPH (eds.) (1987) *Viruses of Potato and Seed-Potato Production,* Wageningen, The Netherlands: Pudoc.
Edwardson JR and Christie RG (1997) Potyviruses. In: *Florida Agricultural Experiment Station Monograph Series 18-II – Viruses Infecting Pepper and Other Solanaceous Crops,* pp. 424–524. Gainesville, FL: University of Florida.
Flis B, Hennig J, Strelczyk-Zyta D, Gebhardt C, and Marczewski W (2005) The *Ry-f$_{sto}$* gene from *Solanum stoloniferum* for extreme resistance to potato virus Y maps to chromosome XII and is diagnosed by PCR marker GP122$_{718}$ in PVY resistant potato cultivars. *Molecular Breeding* 15: 95–100.
Glais L, Kerlan C, and Robaglia C (2002) Variability and evolution of *Potato virus Y* (PVY), the type-member of the *Potyvirus* genus. In: Khan JA and Dijkstra J (eds.) *Plant Viruses as Molecular Pathogens,* pp. 225–253. Binghamton, NY: The Haworth Press.
Kasai K, Morikawa Y, Sorri VA, Valkonen JPT, Gebhardt C, and Watanabe KN (2000) Development of SCAR markers to the PVY resistance gene *Ry$_{adg}$* based on a common feature of plant disease resistance genes. *Genome* 43: 1–8.
Loebenstein G, Berger P, Brunt AA,, and Lawson RG (eds.) (2001) *Virus and Virus-Like Diseases of Potatoes and Production of Seed-Potatoes.* Dordrecht, The Netherlands: Kluwer Academic Publishers.
Milne RG (1988) The economic impact of filamentous viruses. In: Milne RG (ed.) *The Plant Viruses, Series 4: The Filamentous Plant Viruses,* pp. 331–407. New York: Plenum.
Moury B, Morel C, Johansen E, and Jacquemond M (2002) Evidence for diversifying selection in potato virus Y and in the coat protein of other potyviruses. *Journal of General Virology* 83: 2563–2573.
Moury B, Morel C, Johansen E, et al. (2004) Mutations in *Potato virus Y* genome-linked protein determine virulence toward recessive resistances in *Capsicum annuum* and *Lycopersicon hirsutum. Molecular Plant–Microbe Interactions* 17: 322–329.
Ruffel S, Dussault M-H, Palloix A, et al. (2002) A natural recessive resistance gene against potato virus Y in pepper corresponds to the eukaryotic initiation factor 4 E (eIF4E). *Plant Journal* 32: 1067–1075.
Ruffel S, Gallois J-L, Lesage M-L, and Caranta C (2005) The recessive potyvirus resistance gene *pot-1* is the tomato orthologue of the pepper *pvr2*-eIF4E gene. *Molecular Genetics and Genomics* 274(4): 346–353.
Shukla DD, Ward CW, and Brunt AA (1994) *The Potyviridae.* Wallingford, UK: CAB International.
Tribodet M, Glais L, Kerlan C, and Jacquot E (2005) Characterization of *Potato virus Y* (PVY) molecular determinants involved in the vein necrosis symptom induced by PVY^N isolates in infected *Nicotiana tabacum* cv. Xanthi. *Journal of General Virology* 86: 2101–2105.

Potexvirus

K H Ryu and J S Hong, Seoul Women's University, Seoul, South Korea

Glossary

Gene silencing A general term describing epigenetic processes of gene regulation, that is, a gene which would be expressed (turned on) under normal circumstances is switched off by machinery in the cell.
RNA interference (RNAi) A mechanism for RNA-guided regulation of gene expression in which double-stranded RNA inhibits the expression of genes with complementary nucleotide sequences.
Triple gene block (TGB) A specialized evolutionarily conserved gene module of three partially overlapping open reading frames involved in the cell-to-cell and long-distance movement of plant viruses.

Introduction

The genus *Potexvirus* (derived from *Potato virus X*) is one of nine genera in the family *Flexiviridae.* Most potexviruses are found wherever their hosts are grown. The genus *Potexvirus* contains a large number of members or tentative members. Host plants naturally infected with potexviruses may show mosaic, necrosis, ringspot, or dwarf symptoms or may be symptomless. The natural host range of individual viruses is usually restricted to a few plant species, although a few of the viruses can infect a wide range of plant species. Many members of the genus have been identified although only relatively recently. Most potexviruses are not transmitted by vertebrate, invertebrate, or fungal vectors; potato aucuba mosaic virus (PAMV), white clover mosaic virus (WClMV), and a few others are

Table 1   Virus species in the genus *Potexvirus*

Species	Virus abbreviation	Accession number[a]
Definitive species		
Alternanthera mosaic virus	AltMV	NC_007731
Asparagus virus 3	AV-3	
Bamboo mosaic virus	BaMV	NC_001642
Cactus virus X	CVX	NC_002815
Caladium virus X	CalVX	AY727533
Cassava common mosaic virus	CsCMV	NC_001658
Cassava virus X	CsVX	
Clover yellow mosaic virus	ClYMV	NC_001753
Commelina virus X	ComVX	
Cymbidium mosaic virus	CYMMV	NC_001812
Daphne virus X	DVX	
Foxtail mosaic virus	FoMV	NC_001483
Hosta virus X	HVX	AY181252
Hydrangea ringspot virus	HdRSV	NC_006943
Lily virus X	LVX	NC_007192
Mint virus X	MVX	NC_006948
Narcissus mosaic virus	NMV	NC_001441
Nerine virus X	NVX	NC_007679
Opuntia virus X	OVX	NC_006060
Papaya mosaic virus	PapMV	NC_001748
Plantago asiatica mosaic virus	PlAMV	NC_003849
Plantago severe mottle virus	PlSMoV	
Plantain virus X	PlVX	
Potato aucuba mosaic virus	PAMV	NC_003632
Potato virus X	PVX	NC_001455
Scallion virus X	ScaVX	NC_003400
Schlumbergera virus X	SVX	AY366207
Strawberry mild yellow edge virus	SMYEV	NC_003794
Tamus red mosaic virus	TRMV	
Tulip virus X	TVX	NC_004322
White clover mosaic virus	WClMV	NC_003820
Zygocactus virus X	ZVX	NC_006059
Tentative species		
Allium virus X	AVX	AY826413
Alstroemeria virus X	AlsVX	NC_007408
Artichoke curly dwarf virus	ACDV	
Barley virus B1	BarV-B1	
Boletus virus X	BolVX	
Centrosema mosaic virus	CenMV	
Chenopodium mosaic virus X	ChMVX	NC_008251
Discorea latent virus	DLV	
Lychnis symptomless virus	LycSLV	
Malva veinal necrosis virus	MVNV	
Nandina mosaic virus	NaMV	
Negro coffee mosaic virus	NeCMV	
Paris polyphylla virus X	PPVX	DQ530433
Parsley virus 5	PaV-5	
Parsnip virus 3	ParV-3	
Parsnip virus 5	ParV-5	
Patchouli virus X	PatVX	
Pepino mosaic virus	PepMV	NC_004067
Rhododendron necrotic ringspot virus	RoNRSV	
Rhubarb virus 1	RV-1	
Smithiantha latent virus	SmiLV	
Viola mottle virus	VMoV	
Zygocactus symptomless virus	ZSLV	

[a]Accession number in GenBank.

transmitted by aphids. The type species of the genus *Potexvirus* is *Potato virus X*. During the last few decades, potato virus X (PVX) has greatly contributed to our understanding of host resistance and gene silencing mechanisms, in particular because of the use of viral vectors in studying gene expression and RNA silencing in plants.

Taxonomy and Classification

The genus *Potexvirus* is one of nine genera in the family *Flexiviridae*. The type species of the genus is *Potato virus X*, and the name potexvirus was derived from the type species. The genus contains 55 members, of which 32 are definitive species and 23 tentative (**Table 1**). Some members listed as species and many members listed as tentative species have not been sequenced, and their taxonomic status is yet to be verified.

Different degrees of relationship exist among potexviruses, and phylogenetic analysis has revealed the presence of two major branches based on the composition of polymerase, coat protein (CP) and triple gene block (TGB) proteins.

Species demarcation criteria in the genus *Potexvirus* include the following: (1) members of distinct species have less than ~72% nucleotide or ~80% amino acid identity between their CP or replicase genes; (2) members of distinct species are readily differentiated by serology, and some virus strains can be differentiated with monoclonal antibodies; and (3) members of distinct species fail to cross-protect in common host plant species, and they usually have distinguishable experimental host ranges.

Virus Structure and Composition

Virions are flexuous filaments that are not enveloped, and measure about 470–580 nm in length and 13 nm in diameter, have helical symmetry and a pitch of 3.3–3.7 nm. Nucleocapsids are longitudinally striated. Virion M_r is about 3.5×10^6 with 5–7% nucleic acid content. Virions contain a monopartite, linear, single-stranded, positive-sense RNA that has a size range of 5.8–7.5 kbp. The 3′ terminus has a poly(A) tract and the 5′ terminus occasionally has a methylated nucleotide cap. CP subunits are of one type with a size in the range of 22–27 kDa. The genome is encapsidated in 1000–1500 CP subunits. The CPs of some strains of PVX and other viruses are glycosylated.

In some potexviruses such as bamboo mosaic virus (BaMV), a satellite RNA has been reported. The satellite RNA of 836 nt depends on BaMV for its replication and encapsidation. The BaMV satellite RNA (satBaMV) contains a single open reading frame (ORF) encoding a 20 kDa nonstructural protein.

Physicochemical Properties

Purified virion preparations sediment as one component with sedimentation coefficient of 115–130 S. The isoelectric point of the virion is about pH 4.4. The ultraviolet (UV) absorbance spectra of potexviruses have maxima at 258–260 nm and minima at 244–249 nm, with A_{max}/A_{min} ratios of 1.1–1.3, and the A_{260}/A_{280} ratio of purified preparation as 1.09–1.37. Physical properties of viruses in this genus are: thermal inactivation point 60–80 °C, longevity *in vitro* (several weeks to months), and dilution endpoint 10^{-5}–10^{-6}. Infectivity of sap does not change on treatment with diethyl ether. Infectivity is retained by the virions despite being deproteinized by proteases, phenols, or detergents.

The CPs of potexviruses are partly degraded during the purification and storage of virus preparations. The virions may dissociate by denaturing agents such as sodium dodecyl sulfate (SDS), urea, guanidine hydrochloride, acetic acid, and alkali.

Genome Structure and Gene Expression

The genome is a single-stranded RNA (ssRNA), 5.8–7.5 kbp in size and contains five ORFs, encoding the replicase, three putative protein components of movement proteins (MPs) called TGB, and the CP, from the 5′- to 3′- end in that order (**Figure 1**). Its genome organization resembles that of the genera *Allexivirus*, *Foveavirus*, *Mandarivirus*, and *Carlavirus*, but is distinguished from them by the presence of five ORFs and a small replication protein. The genome RNA is capped at the 5′-end and polyadenylated at the 3′-end. The 146–191 kDa viral replicase ORF is translated from the full-length genomic RNA. For expression of its 3′ proximal viral genes, the virus utilizes at least two subgenomic RNAs. The genome structure of PVX (6435 nt) is represented in **Figure 1**. PVX RNA has 83 nt 5′ untranslated region (UTR) at the 5′-terminus and 76 nt 3′ UTR followed by a poly(A) tail at the 3′-terminus. The 5′ UTR regulates genomic and subgenomic RNA synthesis and encapsidation, and the 3′ UTR controls both plus- and minus-strand RNA synthesis. Potexvirus-infected plants contain double-stranded RNAs (dsRNAs) whose molecular mass corresponds to that of genomic RNA and two subgenomic RNAs. Although potexviruses typically have five ORFs, some potexviruses (cassava common mosaic virus (CsCMV), narcissus mosaic virus (NMV), strawberry mild yellow edge virus (SMYEV), and WClMV) have a sixth smaller ORF located within the ORF5, with no known protein product.

ORF1 encodes a 166 kDa polypeptide that is presumed to be the viral replicase. It contains motifs of methyltransferase, helicase, and RNA-dependent RNA polymerase. ORFs 2, 3, and 4 form the TGB encoding

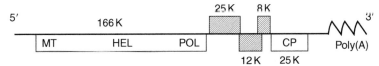

Figure 1 Genome organization of PVX, showing the typical genome structure present in the genus *Potexvirus*, family *Flexiviridae*. The 5′ proximal large ORF encodes an RNA-dependent RNA polymerase (viral replicase), three overlapping ORFs encode the putative MPs (TGBs), and an ORF encodes CP. Motifs in the replicase are methyltransferase (MT), helicase (HEL), and RNA-dependent RNA polymerase (POL).

polypeptides of 25, 12, and 8 kDa, respectively, which facilitate the virus cell-to-cell movement. ORF5 encodes the 25 kDa CP. The CP is required for virion assembly and cell-to-cell movement. The 25 kDa TGBp1 protein contains an NTPase-helicase domain but is not involved in RNA replication.

Complete genomic sequences have recently been obtained for many potexviruses (see **Table 1**).

Infectious cDNA clones have been reported for some potexviruses, such as PVX, BaMV, cymbidium mosaic virus (CymMV), zygocactus virus X (ZVX), and WClMV. In particular, PVX has been widely used as a vector for the study of gene silencing and small RNAs as well as for expression of foreign genes in plants.

Viral Transmission

Potexviruses are spread by mechanical contact and horticultural and agricultural equipment. None of the viruses have known invertebrate or fungal vectors. Two potexviruses (PAMV and WClMV) are transmitted by aphids. A few members are not transmitted by mechanical inoculation. Seed transmission may occur with some potexviruses, but is not common.

Cytopathology

The cytoplasm of potexvirus-infected cells contains fibrous, beaded, banded, or irregular aggregates of virus particles. Virus particles, frequently in large aggregates, occur in the cytoplasm and occasionally in the nuclei. The distribution of potexviruses in the infected plant is not tissue specific. Inclusions such as cytoplasmic laminated inclusion components are present in infected cells.

Host Range

The natural host range of individual members is mostly limited, although some species can infect a wide range of experimental hosts. Some of the viruses cause serious damage to their hosts while others cause little damage to infected hosts.

Symptomatology

Infection by most members of the genus *Potexvirus* is latent or causes only mild mosaic symptoms in the natural host. Symptoms vary cyclically or seasonally and may disappear soon after infection. Some viruses cause necrosis, ringspot, or dwarf symptoms in a wide range of plant species. If symptoms are evident, more severe symptoms such as mosaics appear in the early stages of infection. Some potexviruses such as PVX cause diseases that are of economic importance on their own; however, most of them are associated with more serious diseases when plants are co-infected with other viruses.

Serology

Virions are good immunogens. Most potexviruses are serologically related to several others, with the relationships varying from close to distant.

Geographical Distribution

Potexviruses are found wherever their hosts are grown, and so the geographical distribution of many species is restricted to only certain parts of the world.

Viral Epidemiology and Control

Most potexvirus-associated diseases are usually very mild or symptomless. The need for their control is often perceived as not important. However, crops such as potatoes, certain cactis, and some ornamental crops that may be infected by more damaging potexviruses require suitable control measures. Transgenic potato, tobacco, and orchid plants, which are resistant to infection by PVX and CymMV, have been developed.

See also: Allexivirus; Capillovirus, Foveavirus, Trichovirus, Vitivirus; Carlavirus; Flexiviruses; Plant Virus Vectors (Gene Expression Systems); Vector Transmission of Plant Viruses.

Further Reading

Adams MJ, Antoniw JF, Bar-Joseph M, *et al.* (2004) The new plant virus family *Flexiviridae* and assessment of molecular criteria for species demarcation. *Archives of Virology* 149: 1045–1060.

Fauquet CM, Mayo MA, Maniloff J, Desselberger U, and Ball LA (2005) In: *Virus Taxonomy, Classification and Nomenclature of Viruses,* *Eighth Report of the International Committee on the Taxonomy of Viruses*, 1101pp. New York: Academic Press.

Martelli GP, Adams MJ, Kreuze JF, and Dolja VV (2007) Family *Flexiviridae*: A case study in virion and genome plasticity. *Annual Review of Phytopathology* 45: 73–100 (online published).

Verchot-Lubicz J, Ye CM, and Bamunusinghe D (2007) Molecular biology of potexviruses: Recent advances. *Journal of General Virology* 88: 1643–1655.

Rice Yellow Mottle Virus

E Hébrard and D Fargette, IRD, Montpellier, France

G Konaté, INERA, Ouagadougou, Burkina Faso

Introduction

Rice yellow mottle virus (RYMV) is a member of the genus *Sobemovirus*. RYMV is present only on the African continent. It was first reported in Kenya in 1966 and in Côte d'Ivoire in 1974 in the irrigated rice fields. Since the early 1990s, RYMV is present everywhere in sub-Saharan Africa and in Madagascar where rice is grown. It affected all types of rice cultivations, including lowland, upland, rain-fed, floating, and mangrove rice. RYMV regularly induces severe yield losses to rice production ranging from 25% to 100%. In some regions, when epidemics are recurrently very severe, farmers abandoned their fields and eradicated the tropical forests in order to plant new rice fields. Highly susceptible cultivars have been eliminated by the disease. RYMV is now ranked as the main biotic threat to rice production in Africa.

Virion Properties

RYMV has icosahedral particles of 25 nm in diameter. The virions contain a single coat protein (CP) of 29 kDa, a genomic RNA (gRNA), and one subgenomic RNA (sgRNA) molecule. The capsid contains 180 copies of the CP subunit arranged with $T = 3$ symmetry (**Figure 1**). The structure of RYMV was determined by X-ray crystallography at 2.8 Å resolution and compared to the structure of southern cowpea mosaic virus (SCPMV) and sesbania mosaic virus (SeMV). Sobemovirus CP subunits are chemically identical but structurally not equivalent. Three types of CP subunits termed A, B, C are related by quasi-threefold axes of symmetry and are involved in different inter-subunit contacts. In the C-type subunits, a longer part of the N-terminus is ordered (residues 27–49), forming an additional β-strand named βA arm. The analysis of molecular determinants involved in the $T = 3$ assembly of SeMV demonstrated the major role of βA arm. Sobemovirus particles are stabilized by divalent cations, pH-dependent protein–protein interactions, and salt bridges between protein and RNA. Upon alkali treatment in presence of chelators, the capsid shell swells and becomes sensitive to enzymes and denaturants. In these conditions, RYMV is more stable than SCPMV. This property is likely due to the 3D swapping of the RYMV βA arm around the distal quasi-sixfold axes as found with SeMV whereas the SCPMV βA arm makes a U-turn and is localized around the nearby quasi-sixfold axes. Another reason for the better stability of RYMV is likely to be the strong RNA–protein interactions resulting from the presence of ordered RNA within the capsid shell. RYMV particles exist in three forms with different stability (1) an unstable swollen form dependent on basic pH but lacking Ca^{2+}, (2) a transitional compact form dependent on acidic pH, but also lacking Ca^{2+}, and (3) a stable compact form that is pH independent and contains Ca^{2+}. The compact form is highly infectious and probably required for virus movement and transport by vector.

Recently, molecular diversity of the CP was analyzed in relation to the capsid three-dimensional structure in order to identify which amino acids are involved in the differential recognition by certain monoclonal antibodies. The residues in position 178 and 180, despite their internal localization in the capsid, can modify the antigenic reactivity, with Mabs G and E allowing serotypes Sr3 and Sr5 to be differentiated.

Organization of the Genome

The genomic RNA is one single-stranded messenger-sense molecule, 4450 nucleotides in size. The 5′ terminus of the RNAs has a genome-linked protein (VPg), and lacks a poly(A) tail. RYMV often encapsidates, in addition to its genomic RNA, a viroid-like satellite RNA (satRNA)

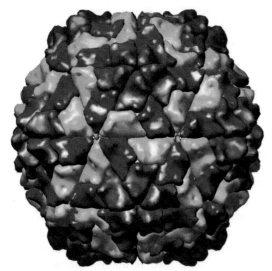

Figure 1 RYMV capsid structure. The capsid comprised 180 copies of one single type of polypeptide arranged in *T* = 3 quasi-equivalent symmetry. The icosahedral asymmetric unit contains three subunits: A (in blue), B (in red), and C (in green). Each subunit is involved in different inter-subunit contacts. This image was automatically generated from ViPER virus capsid PDB file 1F2N using the MultiScale extension to the Chimera interactive molecular graphics package. Reproduced from Rice yellow mottle virus In: *Characterization, Diagnosis and Management of Plant Viruses*, Vol. 4, Ch. 2, pp. 31–50, 2008. Houston, TX: Studium Press, with permission from Studium Press.

that is dependent on a helper virus for replication. RYMV satRNA has 220 nt and is the smallest naturally occurring viroid-like RNA known today. The RYMV satellite RNA is not involved in pathogenicity.

Like other sobemoviruses, the RYMV genome is organized in 4 overlapping open reading frames (ORFs). Two noncoding regions at the RNA extremities contain 80 and 289 nt, respectively. The ORF1 and the ORF4 are localized at the 5′ and 3′ extremities, respectively. An intergenic region of 54 nt separates the first two ORFs. The two coding regions ORF2a and ORF2b overlapped, ORF2b being situated in the −1 reading frame within ORF2a (**Figure 2**).

The ORF1 encodes a protein P1 which is dispensable for replication but is required for infectivity in plants. RYMV P1 exists in two forms of 18 and 19 kDa which probably result from degradation or post-translational modifications. P1 is required for cell-to-cell movement of the virus. Moreover, RYMV and cockfoot mottle virus (CoMV) P1 are suppressors of post-translational gene silencing in the nonhost plant *Nicotiana benthamiana*. The two P1 are able to suppress the initiation and the maintenance of silencing but the suppression of systemic silencing is stronger with RYMV P1 than with CoMV P1.

The polyproteins encoded by the RYMV ORF2 are presumably translated from leaky translation. The start codon of ORF2 appeared to be in a more favorable context than ORF1. The ORF2b is translated as a polyprotein fused with ORF2a after a-1 programmed ribosomal frameshifting mechanism. This phenomenon happened with *in vitro* efficiency from 26 to 29% for the CoMV. The presence of a heptanucleotide slippery sequence UUUAAAC and a predicted stem–loop structure are necessary. The ORF2a contains sequence motifs for the proteinase and the VPg and the ORF2b encodes the RNA-dependent RNA polymerase (RdRp). RYMV proteinase contain the consensus sequence $H(X_{32-35})[D/E](X_{61-62})TXXGXSG$ characteristic of serine proteinases, they cleave between E–T and E–S amino acid residues.

RYMV VPg is not absolutely required for infectivity but *in vitro* transcripts required capping to be infectious. The putative and known sobemovirus VPgs are characterized by the presence of a conserved W[A/G]D sequence followed by a D- and E-rich region. The position of sobemovirus VPg, like polerovirus VPg, differs from the genome arrangement VPg proteinase–polymerase characteristic of many RNA virus species including members of the families *Picornaviridae* and *Comoviridae*. The localization of the polymerase was predicted from the presence of the GDD motif and surrounding conserved motifs characteristics of RdRp. Little is known about the replication signals needed for initiation of plus and minus-strand synthesis in sobemoviruses.

The ORF4 is translated from the subgenomic RNA and encodes the CP (which appeared as a doublet of 28–29 kDa). RYMV CP is dispensable for replication but required for movement to long distance and perhaps from cell to cell. Encapsidation is likely to be essential for long distance movement. This movement was suggested to occur during the cell differentiation to vessels. The N-terminal sequence contains a putative bipartite nuclear localization signal (NLS). This highly basic region is thought to interact with viral RNA and to stabilize the virion.

Relationships of the Species with Other Taxa

The species *Rice yellow mottle virus* belongs to the genus *Sobemovirus* which is not assigned to any family. The genus contains 13 definitive species including *Southern bean mosaic virus*, the type species. The sobemovirus CPs are related to those of necroviruses (*Tombusviridae*), whereas the proteinase, VPg, and polymerase are related to those of poleroviruses and enamoviruses (*Luteoviridae*).

Biological Properties

The natural host range of RYMV is restricted to a few members of the Poaceae family, principally *Eragrostidae* and

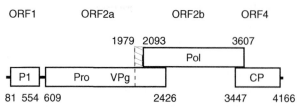

Figure 2 RYMV genomic organization. Positions of the ORFs are indicated in nucleotides. P1, proteinase (Pro), VPg, RNA-dependent RNA polymerase (Pol), and CP are labeled. The dotted line at nucleotide 1979 represents the frameshifting signal. The fusion point of the polyprotein P2a+b is unknown, and an AUG codon present at the beginning of the ORF2b (nucleotide 2093) is indicated by the vertical line. Reproduced from Rice yellow mottle virus In: *Characterization, Diagnosis and Management of Plant Viruses*, Vol. 4, Ch. 2, pp. 31–50, 2008. Houston, TX: Studium Press, with permission from Studium Press.

Oryzae sp. The most commonly cited species are *Oryza sativa*, *O. longistaminata*, and *Echinocloa colona*. A few additional Poaceae plants have been infected experimentally.

RYMV particles are present in all plant parts including roots and seeds. Particles are present in large amount in mesophyll cells, and in vascular tissues mainly in xylem and associated parenchyma cells. Particles were found individually, as aggregates, or as crystalline forms in the cytoplasm and the vacuoles of infected cells. In addition, the cytoplasm of infected cells contains fibrillar material located either in vesicles or distributed in diffused patches. Virus particles are also found in mature xylem, as well as inside the primary wall.

Cytological changes were found in the chloroplasts of infected mesophyll cells where the starch grains decreased in size and number. The chloroplasts form invaginations containing mitochondria, peroxysomes, and virus particles. Electron-dense materials were observed in the nuclei and associated with fibrillar elements in cytoplasmic vesicles. The most dramatic changes induced by RYMV occurred in the cell walls of parenchyma and mature xylem cells causing disorganization of the middle lamellae wall. From infected cells, a viral ribonucleoprotein complex moves from cell to cell to reach the vascular bundle sheath. Viral replication, encapsidation, and storage in vacuoles occur mainly in vascular parenchyma cells. The calcium linkage from pit membranes to virus particles during xylem differentiation could contribute to disruption of these membranes and facilitate systemic virus transport. In the upper leaves, virions are suspected to cross the pit membrane to infect new vascular cells and virions spread out by cell-to-cell movement through plasmodesmata. At this stage of infection, most replication occurs in mesophyll and vascular cells, and most of the virions accumulate in large crystalline patches.

Vector transmission is mainly due to chrysomelidae vectors, but other biting insects including the grasshopper *Conocephalus* have been reported to occasionally transmit the virus. Transmission is not persistent. The virus is present within the seed, although RYMV is not seed-transmissible. Abiotic transmissions through plant residues, irrigation water, gutation water, direct contact between plants, and contamination by infected agricultural tools have been observed. Experimentally, mechanical transmission is easily achieved.

Diagnostic and Identification

RYMV diagnosis was first based on symptoms. However, many factors such as mineral deficiencies can induce symptoms of yellow mottle on rice. Therefore, several tools for specific detection of RYMV were developed. First, polyclonal antibodies (PAbs) directed against different isolates were used in double-diffusion tests and in direct antibody sandwich enzyme-linked immunosorbent assay (DAS-ELISA). No cross-reactions with other sobemoviruses was observed, demonstrating the high specificity of these antibodies. Western-blot analysis can be used to detect RYMV with PAbs. These antibodies were also used in immunolocalization techniques. Immunoprinting tests derived from the direct tissue blotting technique were developed to analyze the distribution of the virus in entire leaves. Moreover, cytological detection of RYMV CP in rice tissues can be performed by immunofluorescence microscopy after glutaraldehyde/paraformaldehyde fixation and labeling with secondary antibodies linked to fluorescein. Polyclonal antibodies directed against P1 produced in *E. coli* are also available, and were used in Western-blot analyses. Monoclonal antibodies were also developed to study RYMV diversity. Indirect triple antibody ELISA (TAS-ELISA) with strain-specific MAbs allows the major serotypes to be differenciated.

In parallel to the serological methods, molecular tools were developed. The first full-length sequence was obtained after RNA extraction from purified virus. Several specific primers were defined to perform RT-PCR amplification and to produce infectious transcripts. Northern and Southern blot methods are available using a DNA probe obtained by PCR amplification of the CP gene (ORF4) from the infectious clone CIa. *In situ* hybridization tests of the RNA in rice tissues were also performed. In this case, the probe was DNA, obtained from the full-length genome of the clone CIa labeled randomly with digoxigenin-UTP. RT-PCR amplification of the CP gene followed by direct DNA sequencing progressively has replaced other molecular methods of analyses.

Epidemiological Aspects

Typical symptoms of RYMV are a mottle and a yellowing of the leaves. However, streaks, necrosis, and whitening

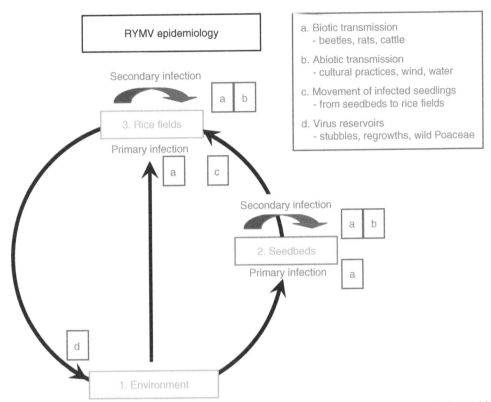

Figure 3 Descriptive model of RYMV epidemiology. 1. Environment. RYMV is present in the environment in rice stubble and regrowths and in wild Poaceae. 2. Rice seedbeds. Primary infection occurred via biotic transmission (beetles, rats, cows) whereas cultural practices contribute to secondary infection in seedbeds 3. Rice fields. Primary infection occurred via biotic transmission from the environment, and also when transplanting infected seedlings from seedbeds. Both biotic and abiotic transmissions contribute to secondary spread in rice fields. After harvesting, RYMV is perpetuated in contra-season in the environment through rice stubbles and regrowths and transmitted to wild Poaceae.

are sometimes observed with some cultivars and under specific growing conditions. Infection resulted in stunting of the plant, reduced tillering, poor panicle exertion, and sterility. Death of the plants of susceptible cultivars occurs after early infection. Severe yield losses ranging from 20% to 100% after RYMV infection have been reported in several countries. Most cultivars currently grown are susceptible to RYMV. Heavy infection is associated with reduced fertility.

RYMV is an emergent disease. It is thought that the virus was originally harbored in wild Poaceae and transferred only recently to cultivated rice. The rapid and intense spread of the virus was associated to the change of agricultural practices in response to increasing food demand. Very productive but susceptible *O. sativa* cultivars have been introduced from Asia which allows two crop cycles a year resulting in the maintenance of the inoculum all over the year. RYMV propagation probably occurs as follows: (1) a virus source made of volunteers, of wild Poaceae or plant residues act as primary inoculum; (2) a direct contamination of the new rice fields occur after mechanical transmission by man or biological infection by insects; (3) rice beds can also be infected by insect vector; (4) the new rice field originating from these infected seed-beds are contaminated by man during replanting; (5) further propagation of the virus is done by man, animals, wind, and irrigation water (**Figure 3**).

The epidemiology has been little studied and there are even uncertainties about the respective importance of biotic and abiotic transmission. Under these conditions, precise forecasting of RYMV infection is not possible, although growing intensification of rice cultivation in Africa will no doubt favor RYMV spread, unless durable resistant cultivars are introduced. Phytosanitary measures are sometimes advised although they are often economically not practicable and their impact to reduce virus spread is unknown. They include protection of seedbeds by nets, disinfection of tools used at replanting, destruction of volunteers and rice residues. Chemical control of the vectors is not economically feasible, is ecologically dangerous, and is most unlikely to be successful considering the large number of species involved in transmission.

Several breeding programs have been conducted in order to select and breed resistant cultivars. The genotype and the phenotype of two kinds of natural resistance have been characterized. Partial resistance is encountered in cultivars of *O. japonica* species with a delayed virus multiplication and symptom expression. Partial resistance is polygenic and a major quantitative trait locus has been identified on chromosome 12. High resistance has been identified in a limited number of accessions of *O. glaberrima* and *O. sativa indica* species. No symptoms are apparent and virus content is most often undetectable. This resistance is monogenic and recessive. The high resistance is controlled by the recessive gene *Rymv1* that maps on chromosome 4 and encodes the translation initiation factor eIF(iso)4G. Different alleles were identified in the sativa-resistant varieties and the *O. glaberrima* accessions. Compared to susceptible varieties, they are characterized by a point mutation or a small deletion in the conserved domain of the gene. The high resistance was efficient against representative isolates of the main virus strains. However, high resistance was overcome experimentally. The VPg was identified as the virulence factor. A single point mutation was sufficient to break the resistance. Transgenic plants with portions of the ORF2 expressing partial resistance have also been produced. This transgenic resistance is thought to involve a gene silencing mechanism. However, the transgenic lines showed a less effective, partial, and temporary resistance compared to natural resistances.

Diversity and Evolution

The diversity of RYMV was assessed by studying isolates from several countries where the disease has been observed. Isolates were serologically typed with monoclonal antibodies. Their CP was sequenced. Several isolates representative of the geographic and molecular diversity were fully sequenced. RYMV is a highly variable virus and analyses of the geographic distribution of the genetic diversity elucidated the process of evolution and of dispersal of the virus. RYMV showed a high level of population structure marked at the continental scale with three subdivisions: East Africa, Central Africa, and West Africa. The highest diversity was observed in East Africa, with a pronounced peak in Eastern Tanzania, and a decrease from the east to the west of the continent. This pattern suggests a westward expansion with a succession of founder effects and subsequent diversification phases. Accordingly, genetic diversity would be adversely affected by recurrent bottlenecks occurring along the route of colonization, resulting in the lowest diversity in the extreme west. This, together with accumulation of *de novo* mutations postdating population separation, provides an explanation for the genetic differences among strains across Africa.

See also: Cereal Viruses: Rice; Emerging and Reemerging Virus Diseases of Plants; Sobemovirus.

Further Reading

Abubakar Z, Ali F, Pinel A, *et al.* (2003) Phylogeography of rice yellow mottle virus in Africa. *Journal of General Virology* 84: 733–743.

Albar L, Bangratz-Reyser M, Hébrard E, *et al.* (2006) Mutations in the eIF(iso)4G translation initiation factor confer high resistance of rice to rice yellow mottle virus. *Plant Journal* 47: 417–426.

Bonneau C, Brugidou C, Chen L, *et al.* (1998) Expression of the rice yellow mottle virus P1 protein *in vitro* and *in vivo* and its involvement in virus spread. *Virology* 244: 79–86.

Brugidou C, Holt C, Yassi A, *et al.* (1995) Synthesis of an infectious full-length cDNA clone of rice yellow mottle virus and mutagenesis of the coat protein. *Virology* 206: 108–115.

Fargette D, Konate G, Fauquet C, *et al.* (2006) Molecular ecology and emergence of tropical plant viruses. *Annual Review of Phytopathology* 44: 235–260.

Fargette D, Pinel A, Abubakar Z, *et al.* (2004) Inferring the evolutionary history of rice yellow mottle virus from genomic, phylogenetic and phylogeographic studies. *Journal of Virology* 78: 3252–3261.

Fargette D, Pinel A, Halimi H, *et al.* (2002) Comparison of molecular and immunological typing of isolates of rice yellow mottle virus. *Archives of Virology* 147: 583–596.

Hébrard E, Pinel-Galzi A, Bersoult A, *et al.* (2006) Emergence of a resistance-breaking isolate of rice yellow mottle virus during serial inoculations is due to a single substitution in the genome-linked viral protein VPg. *Journal of General Virology* 87: 1369–1373.

Hébrard E, Pinel-Galzi A, Catherinot V, *et al.* (2005) Internal point mutations of the capsid modify the serotype of rice yellow mottle sobemovirus. *Journal of Virology* 79: 4407–4414.

Konaté G, Traoré O, and Coulibaly M (1997) Characterization of rice yellow mottle virus isolates in Sudano-Sahelian areas. *Archives of Virology* 142: 1117–1124.

Pinel A, Traoré O, Abubakar Z, Konaté G, and Fargette D (2003) Molecular epidemiology of the RNA satellite of rice yellow mottle virus in Africa. *Archives of Virology* 148: 1721–1733.

Qu C, Liljas L, Opalka N, *et al.* (2000) 3 D domain swapping modulates the stability of members of an icosahedral virus group. *Structure* 8: 1095–1103.

Sorho F, Pinel A, Traoré O, *et al.* (2005) Durability of natural and transgenic resistances in rice to rice yellow mottle virus. *European Journal of Plant Pathology* 112: 349–359.

Traoré O, Pinel A, Hébrard E, *et al.* (2006) Occurrence of resistance-breaking isolates of rice yellow mottle virus in West and Central Africa. *Plant Disease* 90: 256–263.

Traoré O, Traoré M, Fargette D, and Konaté G (2006) Rice seedbed as a source of primary infection by rice yellow mottle virus. *European Journal of Plant Pathology* 115: 181–186.

Sequiviruses

I-R Choi, International Rice Research Institute, Los Baños, The Philippines

Glossary

Helper component A viral-encoded protein that mediates the transmission of viruses by vectors.
Semipersistent transmission A mode of vector-mediated transmission of plant viruses in which viruses are usually acquired within few minutes, and retained up to several days by vectors. Viruses do not multiply in vectors and often require a helper component during semipersistent transmission.

Introduction

Sequiviridae is a relatively newly recognized family of viruses, although the viruses classified into this family had been reported already in the early 1970s. Viruses belonging to the family *Sequiviridae* were often referred to as 'plant picorna-like viruses' due to their similarities to picornaviruses in virion morphology and genome structure. They have a monopartite single-stranded RNA (ssRNA) genome encapsidated in isometric particles. They infect plants and are usually transmitted by insect vectors in nature. Special attention has been given to two of the members, maize chlorotic dwarf virus (MCDV) and rice tungro spherical virus (RTSV), since they are involved in major diseases of important cereal crops. Chlorotic dwarf disease is considered to be the second most important viral disease of maize in the USA, and tungro disease is the most serious threat to rice production in South and Southeast Asia. One of the unique biological features found in these viruses is that most of them act as or require a helper virus for transmission by insect vectors. The name of the family comes from the Latin word *sequi* which means to follow or accompany, in reference to the dependent insect transmission of parsnip yellow fleck virus (PYFV), one of the type members in this family.

Taxonomy

The family *Sequiviridae* is divided into two genera, *Sequivirus* and *Waikavirus*. At present the genus *Sequivirus* has two species *Parsnip yellow fleck virus* (type species) and *Dandelion yellow mosaic virus*. Three species, *Rice tungro spherical virus* (type species), *Maize chlorotic dwarf virus*, and *Anthriscus yellows virus* are classified as members of the genus *Waikavirus* (**Table 1**). In addition, lettuce mottle

virus is considered as a tentative member of the genus *Sequivirus*.

Viruses belonging to the genus *Sequivirus* infect mesophyll and epidermal cells and are able to be transmitted by mechanical inoculation and insect vectors, while those belonging to the genus *Waikavirus* are usually limited in phloem tissue and transmitted only by insect vectors. Sequiviruses are dependent on a helper virus for their transmission by insect vectors. Waikaviruses are independently transmitted by insect vectors and presumably encode a helper component in their genomes. The viruses in the family *Sequiviridae* are primarily classified on the basis of their biological and physical characteristics, but conspicuous differences are also found in their genome features. The genomes of MCDV and RTSV are approximately 12 kb in length and polyadenylated at the 3' end, whereas that of PYFV is about 10 kb and devoid of the 3' poly(A) region.

Properties of Virions

Viruses belonging to *Sequiviridae* have nonenveloped isometric particles of approximately 30 nm in diameter (**Figure 1**). The sedimentation coefficients of the virions are 153–159S for sequiviruses and 175–183S for waikaviruses. The buoyant density of PYFV virions in CsCl is 1.49 g ml^{-1}, and that of waikaviruses is 1.51–1.55 g ml^{-1}. Genome sequences and immunodetection of the viruses belonging to the family *Sequiviridae* indicate that the virions consist of three capsid proteins (CPs). The sizes of CPs range from 22 to 35 kDa, depending on virus species and isolates. The virion of RTSV (Philippine-type strain A) consists of three proteins of 22.5 (CP1), 22 (CP2), and 33 (CP3) kDa. The predicted molecular masses for the CP of MCDV (Tennessee (TN) isolate) are 22, 23, and 31 kDa, and those of PYFV (P-121 isolate) are 22.5, 26, and 31 kDa.

Genome Structure

The genomes of viruses belonging to the family *Sequiviridae* are positive-sense, monopartite ssRNA. The length of genomes varies significantly, ranging from approximately 10 kb for sequiviruses to 12 kb for waikaviruses. The genomes contain a large open reading frame (ORF) encoding a polyprotein presumably proteolytically processed by viral-encoded protease(s) during translation (**Figure 2**). The large ORF in the genome of PYFV (P-121 isolate) putatively encodes a polyprotein consisting of 3027 amino

Table 1   Virus members in the family *Sequiviridae*

Genus/virus	Genome size (kb)	Geographical distribution	Major natural host	Transmission vector
Sequivirus				
Parsnip yellow fleck virus	9.9	Europe	Parsnip, hogweed, cow parsley	Aphids (*Cavariella aegopodii, C. pastinacae*)
Dandelion yellow mosaic virus	10.0[a]	Europe	Lettuce, dandelion	Aphids (*Acyrthosiphon solani, Myzus ornatus, M. ascalonicus, M. persicae*)
Waikavirus				
Rice tungro spherical virus	12.2	Asia	Rice, *Oryza* species	Green leafhoppers (*Nephotettix virescens* and four other species)
Maize chlorotic dwarf virus	11.8	USA	Maize, Johnson grass	Deltocephaline leafhopper (*Graminella nigrifrons*)
Anthriscus yellow virus	10.6[a]	Eurasia, UK	Cow parsley	Aphid (*Cavariella aegopodii*)

[a]Estimated size (nucleotide sequence not determined).

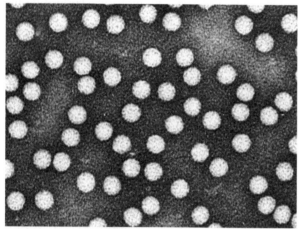

Figure 1   Virions of RTSV. The estimated size of particles is approximately 30 nm.

acid residues with the predicted molecular mass of about 336 kDa. The large ORF in the genomes of waikaviruses encodes a polyprotein of approximately 400 kDa, which has about 3440–3470 amino acid residues. In addition to the ORF for the polyprotein, short ORFs were also identified near the 3′ and the 5′ ends of MCDV and RTSV genomes. Two RNA species which seemingly correspond to subgenomic transcripts from the short ORFs locating near the 5′ end of the RTSV genome (strain A) were detected from infected plants. However, the lengths and locations of the short ORF in the genomes of waikaviruses vary considerably among viruses and isolates, and the presence of products translated from these ORFs was not confirmed in plants. The 5′-untranslated region (UTR) in the genome of PYFV (P-121 isolate) is about 0.28 kb in length, while those of waikaviruses are longer, about 0.43 kb in MCDV and 0.52 kb in RTSV. The 5′-UTR in the genomes of MCDV and RTSV has several AUG sequences upstream of the putative polyprotein start site, and appeared to form extensive secondary structures. As observed in the genome of

picornaviruses, such secondary structures may serve as the internal ribosomal entry site to avoid the interference on translation by upstream AUG sequences. No stable secondary structures are recognized in the 5′-UTR of the PYFV genome; however, several short stretches of pyrimidines (UCUCUY) are present in the region. The 3′-UTR in the genomes of waikaviruses are polyadenylated and unusually long. The length of 3′-UTR in the genomes of MCDV and RTSV are approximately 1.0 and 1.2 kb, respectively, provided that the short ORFs near the 3′ end of genomes are not translated. The 3′-UTR in the genome of PYFV is approximately 0.5 kb in length. Unlike the genomes of waikaviruses, that of PYFV is not polyadenylated, but a part of 3′-UTR is likely to form a stem–loop structure which is similar to that found in the genomes of flaviviruses.

Properties of Viral Proteins

Based on the protein sequences predicted from the large ORF, and the actual sequences of the N-termini of CPs, it was predicted that the polyproteins of viruses belonging to the family *Sequiviridae* are cleaved into at least seven proteins through proteolytic maturation. It appears that the N-terminal half of the polyprotein contains regions for a leader protein and three CPs. Protein regions showing similarities to helicase, protease, and RNA-dependent RNA polymerase (RdRp) of picornaviruses and comoviruses are recognized in the central to carboxyl (C)-terminal regions of the polyproteins. The arrangement of functional domains in the polyproteins of the viruses belonging to *Sequiviridae* shows significant similarity to that of picornaviruses, indicating that the strategies of genome replication and expression for the viruses of *Sequiviridae* might be analogous to those of picornaviruses.

Comparative sequence analysis indicated that the polyproteins of the viruses belonging to the family *Sequiviridae* contain a protease region similar to the 3C cysteine prote-

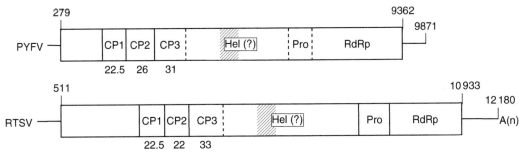

Figure 2 Genome structure of PYFV (P-121 isolate) and RTSV (strain A). The ORFs for polyprotein are indicated with rectangles. Horizontal lines at both ends of the ORFs represent the 5'- and 3'-UTR of the genome. Solid vertical lines dividing the ORF correspond to the positions of proteolytic cleavage sites in the polyproteins when translated, while the positions corresponding to predicted cleavage sites are indicated with dotted lines. Positions of translation start and stop codons for the polyproteins and the length of the entire genomes are indicated with the numbers above the genomes. Numbers below the regions encoding CP1–CP3 are the predicted molecular weight (kDa) for the respective CP. Shadowed areas represent the approximate region encoding NTP-binding domain. The regions in the ORF are shown with Hel (?) for putative NTPase/helicase, Pro for 3C-like cysteine proteinase, and RdRp for RNA-dependent RNA polymerase. A(n) indicates the polyadenylation at the 3' end of the RTSV genome.

ase of picornaviruses and other viral proteases resembling the 3C protease such as the 24 kDa protease of cowpea mosaic virus (CPMV) and the NIa protease of tobacco etch virus. For example, considerable similarity was found between the region delimited by the amino acid residues 2643 and 2853 in the polyprotein of RTSV (strain A) and a region in the 24 kDa protease of CPMV. Immunodetection using an antiserum raised against a protein containing the putative protease region of RTSV indicated that the size of mature protease in infected plants is approximately 35 kDa. The sequence context of cleavage site for the 3C cysteine protease and the results from *in vitro* translation from partial genomic RNA templates defined the region of the protease in the RTSV polyprotein to be from amino acid residues 2527–2852. The RTSV protease presumably acts *in cis* to cleave itself from the adjacent protein regions, but *trans*-cleavage by the RTSV protease was also observed to occur between the protease and the putative helicase regions *in vitro*. Based on the sequence alignment among the 3C-type proteases, amino acid residues such as His^{2680}, Glu^{2717}, Cys^{2811}, and His^{2830} in the RTSV polyprotein were predicted to constitute the catalytic triad or substrate-binding pocket of the cysteine protease. Substitutions of these amino acid residues abolished or drastically reduced the proteolytic activity, substantiating their critical roles in the cysteine protease. Conserved amino acid residues which may constitute the catalytic triad were also identified in the protease regions of MCDV and PYFV, although the involvement of the individual residues in the protease activity has not been experimentally demonstrated.

The C-terminal region in the polyproteins of viruses belonging to the family *Sequiviridae* shows extensive similarity in amino acid sequence to RdRp of picornaviruses and comoviruses. The RdRp region of RTSV (strain A) was defined in the region between amino acid residues 2853 and 3473 with the conserved YGDD motif at amino acid residues 3270–3273. The RdRp region of MCDV (isolate TN)

has the conserved YGDD motif at amino acid residues 3238–3241. In addition, motifs such as DYSXFDG (amino acid residues 3129–3135 in the MCDV polyprotein) and PSGX$_3$TX$_3$NS (amino acid residues 3189–3200) were identified to be conserved among the RdRp regions of MCDV, CPMV, and tomato black ring virus (TBRV). Multiple alignment of the RdRp region of PYFV (isolate P-121) with those of CPMV, TBRV, and poliovirus showed that they share several conserved motifs including sequences YGDD (amino acid residues 2629–2632 in the PYFV polyprotein), PSGX$_3$TX$_3$NS (amino acid residues 2580–2591), and FLKR (amino acid residues 2684–2687).

The central region flanked by the CP and the protease regions in the polyproteins of viruses belonging to the family *Sequiviridae* contains sequence motifs GX$_4$GKS and DD, which are conserved among proteins with NTP-binding domain. The NTP-binding domain of the central polyprotein regions shows extensive similarity to the corresponding domains in the 58 kDa protein of CPMV and the 2C proteins of picornaviruses. The 58 kDa protein and the 2C protein presumably function as a nucleoside triphosphatase (NTPase)/helicase required for the initiation of negative-strand RNA synthesis. Antibodies specific to protein segments from the central region of the MCDV polyprotein reacted with three protein species in extracts from infected plants. The sizes of the proteins detected were smaller than predicted from the intact central region, indicating that the protein in the central region might be processed into smaller proteins.

It appeared that the polyproteins of viruses belonging to *Sequiviridae* have a region for putative leader protein(s) at the N-terminal region. Although the predicted sizes of the putative leader proteins are about 40 kDa in PYFV, and 70–78 kDa in waikaviruses, the sizes of proteins detected with the antisera specific to the putative leader proteins are significantly smaller than predicted. An antiserum specific to the putative RTSV leader protein detected a protein of about 32 kDa in extracts from infected plants. Meanwhile, proteins with apparent sizes of 50, 35, and 25 kDa in the extracts from plants infected with MCDV (severe (S) strain) reacted with the antibody specific to the putative MCDV leader protein. These results suggest that the putative leader proteins of waikaviruses may be processed post-translationally in plants. The function associated with the leader protein region is still unclear. However, such leader protein found in the polyprotein of aphthoviruses in the family *Picornaviridae* has a protease activity which cleaves itself autocatalytically from the polyprotein. It was proposed that the 35 kDa protein of MCDV is generated from the leader protein region through autoproteolysis at the putative cleavage site ALVRLFHGSAE (amino acid residues 150–160), while the 25 kDa protein results from the cleavage at Q^{445}/S^{446} and Q^{686}/S^{687} by a cysteine protease. The 25 kDa protein may function as the helper component in the insect transmission of MCDV since the protein was observed to accumulate in vector insects after feeding on plants infected with MCDV. This observation is consistent with the result from a serological blocking experiment showing that the helper component is not the virion or CP.

Three consecutive regions of CPs are present in the N-terminal half of the polyproteins of the viruses belonging to the family *Sequiviridae*. The proteolytic cleavage sites for the RTSV CPs determined by N-terminal amino acid sequencing were mapped to Q^{644}/A^{645}, Q^{852}/S^{853}, and Q^{1055}/D^{1056} for the junctions between leader protein/CP1, CP1/CP2, and CP2/CP3, respectively. The context of these cleavage sites suggests that the CPs are processed *in trans* by a cysteine protease. However, the cleavage at these sites with the RTSV protease was not detected *in vitro*. The difference in the reactivity among the CPs of RTSV with the antibody raised against the virus particles indicated that the 33 kDa CP (CP3) is the major antigenic determinant on the surface of particles, although the structural details of virus particles have not been elucidated yet. The size of CP3 of RTSV (Philippine isolate) in the crude extract from infected plants detected by the antibody specific to CP3 appeared to be 40–42 kDa, markedly larger than that detected from purified virus preparation or that expected from the genome sequence. It is likely that the size of CP3 in infected cells is larger due to post-translational modification, but the modified moiety appeared to be cleaved off probably by the treatment with cellulolytic enzymes during virion purification.

Phylogenetic Relationships

Comparison of amino acid sequences revealed that the sequence similarities of the MCDV (isolate TN) polyprotein to those of RTSV (strain A) and PYFV (isolate P-121) were 51% and 35%, respectively. Overall similar-

ity in genome organization and the presence of several conserved motifs in the polyproteins indicate that the viruses of the family *Sequiviridae* are closely related to picornaviruses and comoviruses. The NTP-binding domain appeared to be the most conserved region in the polyproteins among the viruses belonging to the family *Sequiviridae*, picornaviruses and comoviruses. The NTP-binding domain in the polyprotein of MCDV (TN isolate) showed significant sequence similarity to those of viruses such as RTSV (strain A, 79%), PYFV (isolate P-121, 55%), CPMV (46%), hepatitis A virus (HAV, 47%), and poliovirus (45%). The regions of RdRp also show significant similarities to one another among members of the family *Sequiviridae*, picornaviruses and comoviruses. For instance, the sequence similarities in the RdRp region of MCDV (TN isolate) to those of related viruses are 75% for RTSV, 50% for PYFV, 48% for CPMV, 44% for TBRV, and 40% for HAV. Phylogenetic analysis based on the sequences of NTP-binding domains indicates that PYFV and RTSV are not noticeably similar to each other compared to their relatedness to picornaviruses and comoviruses. However, phylogenetic analysis based on the regions of RdRp suggests that PYFV and RTSV are more closely related to each other than to picornaviruses and comoviruses. Unlike the region of RdRp and NTP-binding domain, the CPs of viruses belonging to the family *Sequiviridae* show no sequence similarities to those of comoviruses, and only limited similarities to those of some picornaviruses. The 26 kDa CP of PYFV and the 22.5 kDa CP of RTSV contain amino acid sequences resembling those in VP3 of encephalomyocarditis virus and human rhinovirus 14. The closer relatedness of nonstructural protein regions in members of the family *Sequiviridae* to those of comoviruses, and the similarity in the CP sequences among PYFV, RTSV, and some picornaviruses imply that the family *Sequiviridae* may fit taxonomically between *Comoviridae* and *Picornaviridae*.

Variation of Isolates and Strains

Isolates of PYFV are largely divided into two groups. One is the parsnip serotype which includes isolates from parsnip, celery, and hogweed, while those from carrot and cow parsley belong to the other group, the *Anthriscus* serotype. The two groups of isolates are distinguishable by reciprocal immunodiffusion tests with antisera raised against isolates belonging to either group. In addition to the difference in natural hosts, the artificial inoculation of test plants with the respective isolates showed evident difference in host ranges between the two groups of isolates, although minor differences in host range and symptoms on certain plants were also observed among the isolates within each serotype.

Examination of nucleotide sequences of the RTSV CP revealed broad genotypic variation among and within geographic isolates, although the relationship of genotypic variation with the pathogenicity is not understood yet. Isolates of RTSV from the Philippines and Malaysia show about 95% sequence similarity, while those from Bangladesh and India differ from the Philippine isolate by about 15%. The CP3 of the Indian isolate was distinguishable from that of the Philippine isolate in electrophoretic mobility and the response to cellulolytic enzyme. Genotypic survey on the CP sequences of RTSV field isolates collected from various sites in the Philippines and Indonesia indicated that a high degree of genetic diversity exists among the field isolates, and that infections with mixed genotypes in single sites are not uncommon. Phylogenetic analysis based on the CP sequences of the RTSV field isolate suggested that the clustering of genotypes found in the Philippines sites was significantly different from that found in the Indonesia sites, indicating geographic isolation of RTSV populations. Strain Vt6 of RTSV was found to possess enhanced virulence, showing infectivity to some rice cultivars which the type strain A may not be able to infect. Amino acid sequence of strain Vt6 was approximately 95% identical to that of strain A, with greater dissimilarity in the leader protein region and the putative small ORF found near the 3′ end.

Few isolates of MCDV with distinctive biological and genotypic characteristics have been reported. The S isolate of MCDV was observed to produce more pronounced symptoms than the type (T) isolate. The mild (M1) isolate usually exhibits mild symptoms by itself, but it develops severe symptoms by synergistically interacting with other MCDV isolates. The deduced amino acid sequences of isolates S and T show 99.5% identity, while that of isolate M1 has only 61% identity to that of isolate T. In fact, antisera raised against isolate T react strongly with isolate S, but not with isolate M1. Isolate TN of MCDV is also significantly divergent from isolate T, showing only 60% of amino acid sequence identity. The low levels of amino acid sequence identity among the isolates of MCDV raised the possibility that they may represent distinct virus species.

Interactions between Viruses

Viruses belonging to the family *Sequiviridae* are often detected in plants infected with other viruses, for instance, cow parsley infected with PYFV and anthriscus yellows virus (AYV), and rice plants infected with rice tungro bacilliform tungrovirus (RTBV) and RTSV. Such mixed infections appear to be the outcome of the dependent insect transmission of one virus on the other. Incidences of mixed infection in lettuce with dandelion yellow mosaic virus and lettuce mosaic virus were also reported, but their relationships in aphid-mediated transmission are still unclear.

PYFV is transmitted by mechanical inoculation, but AYV is not. Both viruses are also transmitted by aphids in a semipersistent manner. However, the aphid-mediated transmission of PYFV is dependent on AYV. PYFV is transmitted by aphids from plants infected with both viruses or by aphids previously fed on plants infected with AYV. Aphids seem to retain PYFV and AYV for up to 4 days. Even though AYV cannot infect parsnip, PYFV can be transmitted to parsnip by aphids fed on other plants infected with AYV and PYFV. Therefore, AYV apparently acts only as the helper virus for aphids to acquire PYFV, and is not necessary for the infection process of PYFV.

Rice tungro disease is caused by the interaction between RTSV and RTBV. Both viruses are transmitted by green leafhoppers (GLHs) in a semipersistent manner. GLHs transmit RTSV and RTBV simultaneously or singly from source plants infected with both viruses. RTSV is independently transmitted by GLHs, but RTBV can be transmitted by GLHs which feed on plants infected with RTSV. Although GLHs retain RTSV for only 3–4 days, their ability to acquire and transmit RTBV may persist for 7 days. Neutralization of RTSV-viruliferous GLHs with anti-RTSV immunoglobulin markedly reduced the ability to transmit RTSV but still retained the ability to acquire and transmit RTBV. These observations indicate that RTSV is the helper virus for the GLH-mediated transmission of RTBV, but the helper function is associated with factors other than RTSV virions or CP. Rice plants infected with RTBV alone exhibit symptoms such as stunted growth, yellow to yellow-orange discoloration of leaves, and reduced tillering. The symptoms become more severe when plants are simultaneously infected with RTBV and RTSV, despite the fact that RTSV alone does not cause conspicuous symptoms except occasional slight stunted growth (**Figure 3**). Such synergistic effects of RTSV on symptom development are also observed when RTSV co-

Figure 3 Synergistic effects between RTSV and RTBV on the symptom development in rice. Rice plants shown (left to right) were infected with RTBV alone (RTBV), both RTBV and RTSV (RTBV + RTSV), RTSV alone (RTSV), or not infected (healthy). Plants infected with both RTBV and RTSV were stunted and showed severe discoloration of leaves.

infects plants with viruses such as rice grassy stunt virus and rice ragged stunt virus.

See also: Cereal Viruses: Maize/Corn; Cereal Viruses: Rice; ; ; Vector Transmission of Plant Viruses.

Further Reading

Azzam O, Yambao MLM, Muhsin M, McNally KL, and Umadhay KML (2000) Genetic diversity of rice tungro spherical virus in tungro-endemic provinces of the Philippines and Indonesia. *Archives of Virology* 145: 1183–1197.

Chaouch-Hamada R, Redinbaugh MG, Gingery RE, Willie K, and Hogenhout SA (2004) Accumulation of maize chlorotic dwarf virus proteins in its plant host and leafhopper vector. *Virology* 325: 379–388.

Elnagar S and Murant AF (1976) Relations of the semi-persistent viruses, parsnip yellow fleck and anthriscus yellows, with their vector, *Cavariella aegopodii*. *Annals of Applied Biology* 84: 153–167.

Hemida SK and Murant AF (1989) Host ranges and serological properties of eight isolates of parsnip yellow fleck virus belonging to the two major serotypes. *Annals of Applied Biology* 114: 101–109.

Isogai M, Cabauatan PQ, Masuta C, Uyeda I, and Azzam O (2000) Complete nucleotide sequence of the rice tungro spherical virus genome of the highly virulent strain Vt6. *Virus Genes* 20: 79–85.

Reddick BB, Habera LF, and Law MD (1997) Nucleotide sequence and taxonomy of maize chlorotic dwarf virus within the family *Sequiviridae*. *Journal of General Virology* 78: 1165–1174.

Shen P, Kaniewska M, Smith C, and Beachy RN (1993) Nucleotide sequence and genomic organization of rice tungro spherical virus. *Virology* 193: 621–630.

Turnbull-Ross AD, Mayo MA, Reavy B, and Murant AF (1993) Sequence analysis of the parsnip yellow fleck virus polyprotein: Evidence of affinities with picornaviruses. *Journal of General Virology* 74: 555–561.

Sobemovirus

M Meier, A Olspert, C Sarmiento, and E Truve, Tallinn University of Technology, Tallinn, Estonia

Glossary

−1 Ribosomal frameshifting Event occurring during translation elongation when the ribosome shifts its frame for reading the mRNA exactly one position in the upstream direction.

Icosahedral particle Spherical viral particle that is a polyhedron having 20 faces.

Leaky scanning mechanism Mechanism during translation initiation for escaping the first start codon, which occurs when the first AUG resides in a very poor context and therefore only some ribosomes initiate translation at that point.

Polycistronic RNA Contains the genetic information to translate more than one protein.

Satellite RNA (satRNA) Subviral agent consisting of RNA that becomes packaged in protein shells made from coat protein of the helper virus and whose replication is dependent on that virus.

***T* = 3 particle** Icosahedral virus particle that contains three chemically identical coat protein monomers in the icosahedral asymmetric unit; $T = 3$ particle contains 180 coat protein molecules.

VPg Protein that is attached to the 5′ end of viral genomic RNA.

Introduction

It was proposed in 1969 that single-component-RNA beetle-transmitted viruses be placed into a southern bean mosaic virus group. Since 1995, this group has been recognized by the International Committee on Taxonomy of Viruses (ICTV) as the genus *Sobemovirus* (sigla from *so*uthern *bean mo*saic *virus*) unassigned to any family. The establishment of this virus group was based on similarities in particle morphology, capsid stabilization, sedimentation coefficients, sizes of protein subunits and genomic RNA, features in mode of vector transmission, and distribution of the particles within the cell. In the *Eighth Report of the International Committee on Taxonomy of Viruses*, 13 virus species were accepted as definite species of the genus *Sobemovirus* and four tentative species were proposed (**Table 1**).

In addition, two viruses presently not recognized by the ICTV are closely related to the sobemoviruses. Nucleotide sequence comparison of the polymerase, VPg, and coat protein (CP) genes of papaya lethal yellowing virus (PLYV) shows high homology to sobemoviruses (about 41–51% with lucerne transient streak virus (LTSV), southern bean mosaic virus (SBMV), southern cowpea mosaic virus (SCPMV), and cocksfoot mottle virus (CfMV)). Snake melon asteroid mosaic virus (SMAMV) RNA-dependent RNA polymerase (RdRp) fragment possesses 71% amino acid sequence similarity to rice yellow mottle virus (RYMV) RdRp.

The host range of sobemoviruses is usually narrow; individual viruses can naturally infect plants from one family only. The exception is sowbane mosaic virus (SoMV) that infects plants from the families Chenopodiaceae, Vitaceae, and Rosaceae. Some sobemoviruses (SBMV, SCPMV, SoMV) are distributed throughout the world; others are limited to one continent or even to one country (**Table 1**).

Sobemoviruses are readily transmitted mechanically. RYMV, for example, is efficiently transferred from plant

Table 1   Viruses of the genus *Sobomovirus* and their biological properties

Virus	Abbr.	Distribution	Natural host	Insect vector	Transmission Mechanical	Seed
Definitive species						
Blueberry shoestring virus	BSSV	USA (Maine, Michigan, New Jersey, Virginia), Canada (New Brunswick, Ontario, Quebec)	*Vaccinium corymbosum, V. angustifolium*	*Masonaphis pepperi* (aphid)	Yes	No
Cocksfoot mottle virus	CfMV	Europe (France, Germany, Norway, Russia, UK), New Zealand, Japan	*Dactylis glomerata, Triticum aestivum*	*Lema melanopus, L. lichensis* (beetles)	Yes	No
Lucerne transient streak virus	LTSV	Australia (Victoria, Tasmania), New Zealand, Canada	*Medicago sativa*	ND	Yes	No
Rice yellow mottle virus	RYMV	Africa (Benin, Burkina Faso, Cameroon, Chad, Côte d'Ivoire, Gambia, Ghana, Guinea, Guinea Bissau, Kenya, Liberia, Madagascar, Mali, Malawi, Mauritania, Mozambique, Niger, Nigeria, Rwanda, Senegal, Sierra Leone, Tanzania, Togo, Uganda)	*Oryza sativa, O. longistaminata*	*Chaetocnema pulla, Sesselia pusilla, Trichispa sericea* (beetles)	Yes	No
Ryegrass mottle virus	RGMoV	Japan, Germany	*Lolium multiflorum, Dactylis glomerata*	ND	Yes	ND
Sesbania mosaic virus	SeMV	India (Andra Pradesh)	*Sesbania grandiflora*	ND	Yes	ND
Solanum nodiflorum mottle virus	SNMoV	Australia (Queensland, New South Wales)	*Solanum nodiflorum, S. nitidibaccatum, S. nigrum*	*Epilachna sparsa, E. doryca australica,*		
		E. guttatopustulata, Psylliodes sp. (beetles), *Cyrtopeltis nicotianae* (mirid)	Yes	No		
Southern bean mosaic virus	SBMV	USA (Arkansas, California, Lousiana), North and South America (Brazil, Colombia, Mexico), Africa (Côte d'Ivoire, Morocco), Europe (France, Spain), Iran	*Phaseolus vulgaris*	*Ceratoma trifurcata, Epilachna variestis* (beetles)	Yes	Yes
Southern cowpea mosaic virus	SCPMV	USA (Wisconsin), Africa (Botswana, Ghana, Kenya, Nigeria, Senegal), Asia (India, Pakistan)	*Vigna unguiculata*	*Ootheca mutabilis* (beetle)	Yes	Yes
Sowbane mosaic virus	SoMV	USA, Canada, Central and South America, Europe (Bulgaria, Czech Republic/ Slovakia, Croatia, France, Hungary, Italy, Moldova), Japan, Australia (Queensland, New South Wales, Victoria, Tasmania)	*Chenopodium* spp., *Atriplex subrecta, Spinacia oleracea, Vitis* sp., *Prunus domestica, Alisma plantago-aquatica, Danae racemosa*	*Myzus persicae* (aphid), *Liriomyza langei* (leafminer), *Circulifer tenellus* (leafhopper), *Halticus citri* (fleahopper)	Yes	Yes
Subterranean clover mottle virus	SCMoV	Australia (New South Wales, South Australia, Tasmania, Victoria, Western Australia)	*Trifolium subterraneum*	ND	Yes	Yes
Turnip rosette virus	TRoV	UK	*Brassica campestris, B. nigra*	*Phyllotreta nemorium* (beetle)	Yes	ND

Continued

Table 1 Continued

| Virus | Abbr. | Distribution | Natural host | Insect vector | Transmission | |
					Mechanical	Seed
Velvet tobacco mottle virus	VTMoV	Australia (Northern Territory, Queensland, South Australia)	*Nicotiana velutina*	*Cyrtopeltis nicotianae* (mirid), *Epilachna* spp. (beetle)	Yes	No
Tentative species						
Cocksfoot mild mosaic virus	CMMV	Europe (Czech Republic/ Slovakia, Denmark, France, Germany, Norway, UK), Canada (Ontario)	*Phleum pratense, Dactylis glomerata, Agrostis stolonifera, Bromus mollis, Festuca pratensis, Poa trivialis, Triticum aestivum*	*Myzus persicae* (aphid), *Lema melanopus* (beetle)	Yes	No
Cynosurus mottle virus	CnMoV	Europe (Germany, UK, Ireland), New Zealand	*Cynosurus cristatus, Agrostis tenuis, A. stolonifera, Lolium perenne X L. multiflorum*	*Lema melanopus* (beetle), *Rhopalosiphum padi* (aphid)	Yes	ND
Ginger chlorotic fleck virus	GCFV	India, Malaysia, Mauritius, Thailand	*Zingiber officinale*	ND	Yes	ND
Rottboellia yellow mottle virus	RoMoV	Nigeria	*Rottboellia cochinchinensis*	ND	Yes	ND

to plant by farming operations, donkeys, cows, grass rats, wind-mediated leaf contacts, soil, etc. In addition, sobemoviruses are transmitted by vectors. The most common vectors are different species of beetles that transmit sobemoviruses in a semipersistent manner. However, blueberry shoestring virus (BSSV) and SoMV are transmitted by aphids, SoMV also by leafminers and leafhoppers, and velvet tobacco mottle virus (VTMoV) by mirids. Several viruses in the genus are seed-transmissible (**Table 1**).

Genome Organization and Replication

The full-length genomic nucleotide sequences have been determined for members of nine sobemovirus species. Their genome sizes vary from 4.0 to 4.5 kb. The genomic (as well as subgenomic) RNA has a viral genome-linked protein (VPg) covalently bound to its 5' end. The 3' terminus of the genomic RNA is nonpolyadenylated.

All the sequenced sobemoviruses have a polycistronic positive-sense single-stranded RNA (ssRNA) genome that consists of four open reading frames (ORFs) (**Figure 1**). The genome is compact, as for most viruses all ORFs overlap. ORFs 1, 2a, and 2b are all translated from the genomic RNA. The initiation of translation from the genomic RNA is facilitated at least in case of CfMV by the translational enhancer in the 5' untranslated region (UTR) of the genome. Translation of ORF1 and ORF2a occurs via a leaky scanning mechanism. ORF2b is expressed as a fusion protein with

ORF2a through a −1 ribosomal frameshift mechanism. Previously, it was reported that some sobemoviruses express the RdRp from a single in-frame polyprotein, not via a −1 translational frameshift. Recently, it has been demonstrated, however, that the difference between the two kinds of genomic organization resulted from a single erroneous extra nucleotide in the genomic sequences of these sobemoviruses that RdRp was thought to be expressed without the −1 ribosomal frameshifting mechanism. Thus, all sequenced sobemovirus species possess similar genomic organization.

A genome 3'-proximal ORF is translated from the subgenomic RNA (sgRNA). The sgRNA has been detected in sobemovirus-infected tissues as well as in virus particles. In addition to the genomic and sgRNA, some sobemoviruses (LTSV, RYMV, subterranean clover mottle virus (SCMoV), Solanum nodiflorum mottle virus (SNMoV), VTMoV) encapsidate a circular viroid-like satellite RNA (satRNA). The sizes of these circular satRNAs range from 220 to 390 nt, the 220 nt RYMV satRNA being the smallest known naturally occurring circular RNAs.

The replication of sobemoviruses uses is poorly understood. The genomic RNA of incoming virus particles is probably uncoated by the co-translational disassembly mechanism that is followed by RNA replication. Little is known about the signals needed for the replication. The 5' terminal nucleotides of the genomic RNA are ACAAAA for SCPMV, ACAA for RYMV, ACAAA for LTSV and ryegrass mottle virus (RGMoV), ACAAAA for SCMoV, CACAAAA for Sesbania mosaic virus (SeMV) and SBMV,

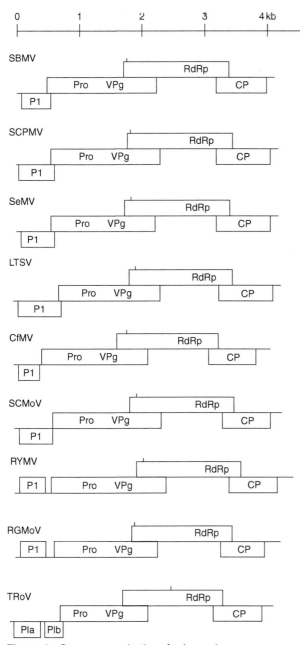

Figure 1 Genome organization of sobemoviruses.

and CAAAA for turnip rosette virus (TRoV). For all these viruses, the 5′ ACAAA or ACAAAA sequence is also present upstream from the CP translation initiation codon, indicating the possible 5′ terminus of sgRNA. This sequence is also characteristic of 5′ termini of polero-, diantho-, and barnaviruses. Due to the conservation of this sequence, it or its complementary sequence in (−)-strand has been predicted to function in viral RNA replication by promoting or enhancing the binding of viral RdRp. Different from other sobemoviruses, such motif is present neither at the 5′ end nor upstream from the CP translation initiation codon in CfMV genome. In addition,

all sobemoviruses contain a polypurine tract 5′-aAGgAAA near the beginning of their genomic RNA. Nearly nothing is known about signals at the 3′ end of the genomic RNA essential for the initiation of the synthesis of the genomic minus strand. A potential tRNA-like structure has been attributed to the 3′ end of some sobemoviruses, but no experimental data are available on that.

Mutation and recombination rates associated with the replication by sobemoviral RdRps are largely uncharacterized. No intra- or interspecies recombinant sobemoviruses have been described so far. Based on the phylogenetic analysis of RYMV sequences, it has been concluded that RYMV has evolved in the absence of recombination events. No recombinants were also detected between CfMV and RGMoV in doubly infected plants. However, several defective interfering (DI) RNA molecules have been cloned from CfMV-infected plants containing the 5′ end of the genomic RNA linked to 850–950 nt of the 3′ terminus. SatRNAs have several interesting interactions with the replication of the helper sobemoviruses. For example, LTSV supports the replication of satRNA of SNMoV but SNMoV does not replicate LTSV satRNA. At the same time, LTSV satRNA replication is supported by CfMV, SBMV, SoMV, and TRoV, whereas this support is host dependent.

Gene Products and Their Functions

P1

P1 is encoded by the 5′ terminal ORF of the viral genomic RNA and its translation occurs with poor efficiency, as the translation initiation context is suboptimal. The molecular masses of different P1s range between 11.7 and 24.3 kDa and, surprisingly, there is no similarity between the P1 amino acid sequences, making this region the most variable one in the genome of sobemoviruses.

The most-studied P1 proteins are the ones of RYMV, SCPMV, and CfMV. All these proteins are required for systemic infection and are dispensable for viral replication. The P1 of SCPMV has also been shown to be nonessential for viral assembly. In addition, RYMV P1 is described as an important pathogenicity determinant. Both CfMV P1 and RYMV P1 act as suppressors of RNA silencing in *Nicotiana benthamiana*, a nonhost species. P1 of CfMV binds ssRNA in a sequence-independent manner and it does not bind double-stranded small interfering RNAs (siRNAs).

The transient gene expression of ORF1 from CfMV and from SCMoV demonstrated that P1 – as a green fluorescent protein (GFP) fusion protein – is involved in movement. However, the cell-to-cell movement of P1, independent of other viral components, was very limited.

Little is known about the subcellular localization of P1. In the case of CfMV, the P1 is coupled with cellular membranes and/or heavily aggregated when overexpressed in insect cells.

It is worth mentioning that the 5′ terminal half of the genomes of sobemoviruses and poleroviruses is similar in organization. Moreover, poleroviral P0, encoded by the 5′ terminal ORF, shares many features with sobemoviral P1: it acts as suppressor of RNA silencing, it is the most divergent protein of the viral genome, and it has a poor translation initiation context.

Polyprotein

Translation of the P2a or P2a2b polyprotein takes place from sobemoviral genomic RNA via a leaky scanning mechanism. The comparison of the sequences surrounding the initiation codons for ORF1 and ORF2a/ORF2a2b of sobemoviruses with the consensus sequences established for plant mRNAs shows that the sequence surrounding the ORF2a initiation codon is always in a more favorable context for translation by plant ribosomes.

In vitro translation of sobemoviral RNAs as identified a protein with a molecular mass of about 100–105 kDa. This protein is encoded by partially overlapping ORF2a and ORF2b due to the signal for −1 ribosomal frameshifting. The −1 ribosomal frameshift is needed to regulate the production of sobemoviral RdRp which is encoded by ORF2b. Although no sobemoviral RdRp has been molecularly characterized yet, the presence of the highly conserved GDD motif with its surrounding (characterized as SGSYCTSSTNX$_{19–35}$GDD) feature for RdRps of positive-strand ssRNA viruses has been identified by computer-based sequence analysis of sobemovirus genomes. In the case of CfMV, the consensus signal for −1 ribosomal frameshift event has been shown to consist of a slippery sequence UUUAAAC and a stem–loop structure located 7 nt downstream of it. The slippery sequence UUUAAAC, followed by a simple stem–loop structure, is absolutely conserved for all sequenced sobemoviruses. In a wheat germ extract, CfMV-derived −1 ribosomal frameshifting takes place with an efficiency of *c.* 10%. It has been proposed that the C-terminal processing products of ORF2a-encoded protein regulate the efficiency of the frameshifting.

ORF2a encodes the N-terminal part of the polyprotein that contains at least a viral serine protease and a VPg. The position of the VPg in between the viral protease and replicase (Pro-VPg-RdRp) is unique to sobemo-, polero-, enamo-, and barnaviruses. It is proposed that sobemoviral VPgs exist in a 'natively unfolded state' lacking both secondary and tertiary structures. The only conserved amino acid sequence element observed among sobemoviral VPgs is a WAD or WGD motif followed by a D- or E-rich region.

The proposed consensus amino acid sequence for the catalytic triad of the serine proteases of sobemo-, polero-, enamo-, and barnaviruses is H(X$_{32–35}$)[D/E](X$_{61–62}$) TXXGXSG. The glycine and histidine residues downstream from the catalytic residues (H181, D216, S284 for SeMV) are suggested to be the site for substrate binding. The crystal structure of the protease domain has been determined for SeMV at 2.4 Å resolution. The structure exhibits the characteristic features of trypsin fold.

Mutations of active site residues of H181A, D216A, or S284A render the SeMV protease inactive. The SeMV serine protease domain lacking the N-terminal 70 amino acids is inactive *in trans*. According to *in silico* analysis, the N-termini of the sobemoviral polyprotein sequences (except LTSV that lacks the domain respective to N-termini of other sobemoviral polyproteins) show the presence of high-propensity membrane helices. Interestingly, the presence of the VPg domain at the C-terminus of the SeMV protease domain lacking the N-terminal 70 amino acids restores the protease activity both in *cis* and in *trans*. Furthermore, the substitution of conserved W43 in SeMV VPg to phenylalanine or the deletion of the entire VPg domain abolishes the proteolytic processing of the viral polyprotein.

The N-terminal sequencing of SBMV, CfMV, and SeMV VPgs attached to viral genomes as well as other approaches have indicated that the polyprotein is processed at E/T or E/N sites between the protease, VPg, and RdRp domains. Additionally, A/V cleavage site was identified at the N-terminus of SeMV polyprotein by mass spectrometric analysis. As the alanine and valine residues are small enough to be accommodated in the active site without steric hindrance, it is suggested that the specificity of sobemoviral proteases depends not only on the sequence but also on the conformation of the polypeptide.

Virion Structure and Coat Protein

The virions of members of the genus have an icosahedral capsid roughly the size of 30 nm, which is assembled according to $T = 3$ symmetry. The capsid contains 180 molecules of a single ∼30 kDa CP, which is translated from an sgRNA. The single-stranded genomic RNA and sgRNA together with VPg are packaged inside the virion. The three-dimensional structures of SCPMV, SeMV, RYMV, RGMoV, and CfMV virions have been determined utilizing X-ray crystallography. Each icosahedral unit of the virion comprises of three quasi-equivalent subunits A, B, and C, which have minor differences in conformation (**Figure 2(a)**). The A subunits group at fivefold axes while pairs of B and C subunits meet at threefold axes. In addition to protein–protein and protein–RNA interactions each icosahedral unit is stabilized by three calcium-binding sites located between subunits AB, BC, and CA.

The CP is divided into two domains: C-terminal S (shell) domain, which has an eight-strand jellyroll β-sandwich topology, common to nonenveloped icosahedral viruses, and N-terminal R (random) domain, which is disordered in subunits A and B, but is partially structured in subunit C. The S domain is the building block of the virion, whereas the R domain is involved in the regulation

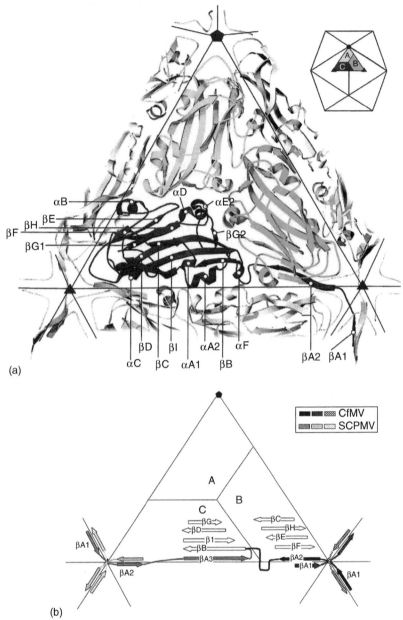

Figure 2 (a) Arrangement of CP molecules in CfMV capsid. (b) A schematic representation of the N-terminal arm in sobemoviruses. Reproduced from Tars K, Zeltins A, and Liljas L (2003) The three-dimensional structure of cocksfoot mottle virus at 2.7 Å resolution. *Virology* 310: 288–289, with permission from Elsevier.

of the capsid structure. The primary sequences of CPs among the members of the genus are quite different. However, the three-dimensional structures are nearly identical, for example root mean square difference between RYMV and SCPMV is 1.4 Å. Regardless of that, two different structures of the R domain in C subunit have been found (**Figure 2(b)**). In SCPMV and SeMV the N-terminus of subunit C, makes a U-turn and extends toward the three-fold axis nearest to C, where it makes a β-structure together with R domains from analogous C subunits. In RYMV and CfMV there is no U-turn, instead the N-terminal arm of subunit C extends toward subunit B and makes a similar

structure at the distal threefold axes closest to subunit B. When the R domain of SCPMV or SeMV CP is removed, only $T = 1$ particles are formed, indicating the importance of the region in $T = 3$ particle formation.

Due to its complex nature, very little is known about the mechanism of capsid formation. Yet, there is evidence suggesting that the virion assembly could be nucleated by AB dimers at icosahedral fivefold axes since pseudo $T = 2$ SeMV particles comprise of groups of A and B subunits. Studies with SCPMV, SeMV, and RYMV particles demonstrate that the stability of the virions depends greatly on pH and the availability of calcium ions. Upon alkaline pH or removal of the cations the virus particles swell and become less stable. Mutation analysis of SeMV CP calcium-binding sites demonstrates that cation-mediated interactions are mainly needed for particle stability. The R domain of all sobemovirus CPs is rich in basic amino acid residues and contains an arginine-rich motif (ARM). Studies with SCPMV and SeMV CP demonstrate that ARM is essential for RNA encapsidation but not for particle formation. However, the presence of RNA enhances the overall stability of capsids. The N-terminus of SCPMV CP witholds a potential to form an α-helix and can interact with membranes *in vitro*. It is proposed that the R domain of all sobemoviral CPs contains a nuclear localization signal.

The functions of sobemoviral CPs in viral life cycle besides capsid formation are not fully understood. Studies with full-length clones of SCPMV and RYMV display that the CP is dispensable for virus replication but systemic virus movement is completely abolished in the absence of CP. Sobemoviral CPs have also been reported to complement the long-distance movement of taxonomically distinct plant viruses.

Subcellular Localization, Short- and Long-Distance Movement

Sobemoviral particles are found mainly in mesophyll and vascular tissues, but also in epidermal, bundle sheath, and guard cells. The quantity of particles present is in correlation with the severity of symptoms. In vascular tissues there are reports of virus particles in both xylem and phloem. CfMV, SCPMV, and SBMV virus particles have been found in phloem companion cells, whereas RYMV particles have been detected predominantly in xylem. RYMV particles accumulated in xylem parenchyma cells and vessels; additionally association with intervascular pit membranes was observed. For RYMV, the common belief is that the virus is transported between xylem cells through pit membranes.

Subcellularly virus particles are found at least in cytoplasm, vacuoles, and nuclei. Virus particles in cytoplasm or vacuoles are known to form crystalline structures, sometimes particles are found in vesicles. No particles have been found in mitochondria and chloroplasts, but the latter are noted to form finger-like extrusions in infected cells. Studies with RYMV suggest that vacuoles of xylem parenchyma cells become the storage compartments for virions in late phase of infection. It is proposed that swollen and less compact virions coexist in the cytoplasm, whereas vacuoles with their lower pH and higher Ca^{2+} concentration contain compact virions.

Studies with SCPMV, SBMV, and RYMV emphasize that cell-to-cell and vascular movement of sobemoviruses are two distinct processes, whereas the long-distance movement is dependent on the correct capsid formation.

Pathology, Economic Importance, Resistance

Sobemoviral infections can cause a variety of disease symptoms: mild or severe chlorosis and mottling, stunting, necrotic lesions, vein clearing, sterility. Subcellularly, sobemoviruses form crystalline arrays and tubules in the cytoplasm, some of which are enveloped in endoplasmic reticulum-derived vesicles. It has been observed that nucleolus of the RYMV-infected cell enlarges. Probably the most dramatic change induced by RYMV occurs in the cell walls of parenchyma and mature xylem cells, where middle lamellae of the wall are disorganized. RGMoV has been reported to induce apoptotic cell death in oat leaves. The outcomes of these histopathological changes range from symptomless infections of sobemoviruses to severe diseases and death of plants. RYMV infection also causes important changes in the abundance of many host proteins. For instance, the expression levels of several defense- and stress-related proteins like superoxide dismutase and different heat shock proteins increase several times.

Several sobemoviruses are economically important pathogens. RYMV causes one of the most damaging and rapidly spreading diseases of rice in Africa. Yield losses fluctuate between 10% and 100%, depending on plant age prior to infection, susceptibility of the rice variety, and environmental factors. PLYV, causing serious chlorosis, is responsible for an important disease of papaya in northeast Brazil. SCMoV decreases seed and herbage production in Australia. Over time, SCMoV-infected pastures become weedy and unproductive. SBMV infections in common bean leads to the mosaic and distortion of pods and reduced size and number of seeds.

Natural resistance to sobemoviruses has been detected at least for CfMV in cocksfoot, for CnMoV in *Cynosurus cristatus*, for RYMV in rice and *Oryza glaberrima*, for SBMV in beans, for SCMoV in subterranean clover, and for SCPMV in cowpea. The molecular mechanisms conferring resistance have only been described for RYMV in *Oryza* species. Namely, the recessive resistance gene *Rymv-1* encodes the eukaryotic translation initiation factor eIF(iso)4G whose interaction with viral VPg is responsible for the high-resistance trait. In parallel, a quantitative trait locus (QTL) is described conferring partial resistance against several RYMV isolates, but the molecular mechanisms responsible for that trait are unknown. Pathogen-derived transgenic resistance against RYMV has been achieved by transforming plants with constructs expressing either RdRp or CP sequences of the virus.

Phylogenetic Relationships

Within the genus *Sobemovirus*, phylogenetic analysis of different proteins indicates that three species – SeMV,

SBMV, and SCPMV – are very closely related to each other. Also LTSV and SCMoV as well as CfMV and RYMV cluster into two corresponding subgroups within the genus (**Figure 3**).

When the sobemoviral sequences are compared to all other sequences available, it is evident that the 5' terminus of the sobemoviral genomes together with ORF1 are unrelated to any other known viral genera (**Figure 3**). The middle part of the genomes (encoding for the successive domains of Pro-VPg-RdRp), however, is similar to those of genera *Polerovirus* and *Enamovirus* from the family *Luteoviridae*. In contrast, the 3' part of

the sobemoviral genomes encoding for the CP is more closely related to CP genes of the genus *Necrovirus* from the family *Tombusviridae* (**Figure 3**). These similarities indicate that possibly early recombination events have played an important role during the evolution of sobemoviruses, luteoviruses, and tombusviruses. This possibility is further supported by the existence of poinsettia latent mottle virus (PnLMoV) whose sequence (AJ867490) shows a close relationship to poleroviruses within the first three quarters of its genome (encoding P1 and polyprotein), but rather to sobemoviruses in the last quarter (encoding CP gene) (**Figure 3**).

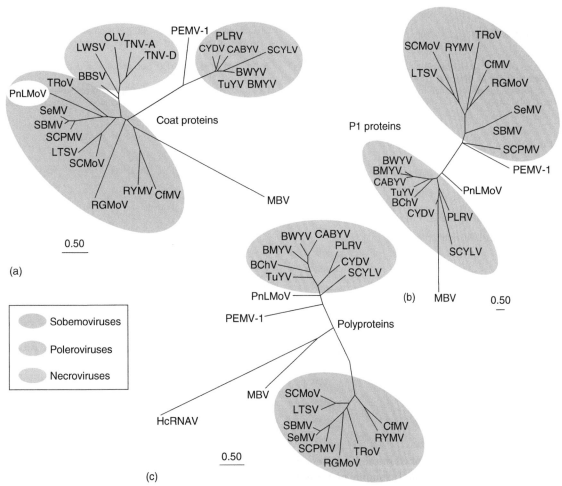

Figure 3 Unrooted phylogram of sobemoviral proteins and the respective proteins of related viruses using maximum-likelihood method. (a) Coat proteins, (b) P1 proteins, and (c) polyproteins. The bar shows the scale of branch length. The protein sequences were aligned with ClustalW and the phylogenetic trees were constructed using PHYLIP PROML 3.6.1 and visualized using Treetool. Viruses (with GenBank accession number) involved in the analyses presented in this figure were the following. Sobemoviruses: CfMV, cocksfoot mottle virus (DQ680848); LTSV, lucerne transient streak virus (U31286); RYMV, rice yellow mottle virus (AJ608206); RGMoV, ryegrass mottle virus (AB040446); SeMV, Sesbania mosaic virus (AY004291); SBMV, southern bean mosaic virus (AF0558871); SCPMV, southern cowpea mosaic virus (M23021); SCMoV, subterranean clover mottle virus (AF208001); TRoV, turnip rosette virus (AY177608). Poleroviruses: BChV, beet chlorosis virus (AF352024); BMYV, beet mild yellowing virus (X83110); BWYV, beet western yellows virus (AF473561); CYDV, cereal yellow dwarf virus (Y07496); CABYV, cucurbit aphid-borne yellows virus (X76931); PLRV, potato leafroll virus (D00530); ScYLV, sugarcane yellow leaf virus (AF157029); TuYV, turnip yellows virus (X13063). Enamovirus: PEMV-1, pea enation mosaic virus 1 (L04573). Barnavirus: MBV, mushroom bacilliform virus (U07551). Necroviruses: BBSV, beet black scorch virus (AF452884); LWSV, leek white stripe virus (X94560); OLV-1, olive latent virus 1 (X85989); TNV-A, tobacco necrosis virus A (M33002); TNV-D, tobacco necrosis virus D (D00942). Unclassified viruses: PnLMoV, poinsettia latent mottle virus (AJ867490); HcRNAV, Heterocapsa circularisquama RNA virus (AB218609).

A virus belonging to the single species *Mushroom bacilliform virus* of the family *Barnaviridae* (genus *Barnavirus*) has a genomic organization similar to sobemoviruses (except that it lacks ORF1 of sobemoviruses) and its Pro-VPg-RdRp and CP genes are related to the same sequences of different sobemoviruses. In addition, putative protease and RdRp of a positive-sense ssRNA virus infecting a marine dinoflagellate *Heterocapsa circularisquama*, HcRNAV (Heterocapsa circularisquama RNA virus, AB218609), are similar to those of the sobemo- and poleroviruses as well as of PnLMoV and MBV.

See also: Barnaviruses; Necrovirus; Rice Yellow Mottle Virus.

Further Reading

Albar L, Bangratz-Reyser M, Hébrard E, Ndjiondjop M-N, Jones M, and Ghesquière A (2006) Mutations in the eIF(iso)4G translation initiation factor confer high resistance of rice to rice yellow mottle virus. *Plant Journal* 47: 417–426.

Aus dem Siepen M, Pohl JO, Koo BJ, Wege C, and Jeske H (2005) Poinsettia latent mottle virus is not a cryptic virus, but a natural polerovirus–sobemovirus hybrid. *Virology* 336: 240–250.

Gayathri P, Sateshkumar PS, Prasad K, Nair S, Savithri HS, and Murthy MRN (2006) Crystal structure of the serine protease domain of Sesbania mosaic virus polyprotein and mutational analysis of residues forming the S1-binding pocket. *Virology* 346: 440–451.

Hull R and Fargette D (2005) Genus *Sobemovirus*. In: Fauquet CM, Mayo MA, Maniloff J, Desselberger U, and Ball LA (eds.) *Virus Taxonomy: Eighth Report of the International Committee on Taxonomy of Viruses*, pp. 885–890. San Diego, CA: Elsevier Academic Press.

Kouassi NK, N'Guessan P, Albar L, Fauquet CM, and Brugidou C (2005) Distribution and characterization of rice yellow mottle virus: A threat to African farmers. *Plant Disease* 89: 124–133.

Mäkelainen K and Mäkinen K (2005) Factors affecting translation at the programmed −1 ribosomal frame-shifting site of cocksfoot mottle virus RNA *in vivo*. *Nucleic Acids Research* 33: 2239–2247.

Meier M and Truve E (2007) Sobemoviruses possess a common CfMV-like genomic organization. *Archives of Virology* 152: 635–640.

Qu C, Liljas L, and Opalka N (2000) 3D domain swapping modulates the stability of members of an icosahedral virus group. *Structure* 8: 1095–1103.

Tamm T and Truve E (2000) Sobemoviruses. *Journal of Virology* 74: 6231–6241.

Tars K, Zeltins A, and Liljas L (2003) The three-dimensional structure of cocksfoot mottle virus at 2.7 Å resolution. *Virology* 310: 288–289.

Tenuivirus

B C Ramirez, CNRS, Paris, France

Glossary

RNA silencing Evolutionarily conserved mechanism in many eukaryotes to target and degrade aberrant RNA molecules. It constitutes an antiviral defense in plants and insects.

Introduction

The tenuiviruses were first described in the Fifth Report of the International Committee on Taxonomy of Viruses in 1982 as nonenveloped plant viruses possibly possessing a negative single-stranded (ss) RNA genome. Their name comes from the Latin 'tenuis', (thin, fine, weak) which refers to the structure of the viral particle as seen by electron microscopy (**Figure 1**). Epidemics of rice stripe virus (RSV) and rice hoja blanca virus (RHBV) cause important yield losses in rice-growing areas of Asia and the former USSR and of tropical America, respectively. Tenuiviruses exhibit unique properties that make them different from other plant viruses. Some properties of tenuiviruses are the following:

1. The peculiar flexuous viral particles have a thread-like morphology and can adopt circular forms (**Figure 1(a)**).
2. The viruses are persistently transmitted by a particular species of planthopper in a circulative, propagative manner. For some of the members of the genus, it has been demonstrated that the virus multiplies both in the host plant and in the insect vector. Multiplication of the virus in the vector may have deleterious effects on the insect. The viruses can be transmitted transovarially by viruliferous female planthoppers to their offspring, and through sperm from viruliferous males.
3. The genome of tenuiviruses is multisegmented and composed of ssRNAs that have either negative or ambisense polarity (**Figure 2**).
4. An RNA-dependent RNA polymerase (RdRp) is associated with the viral particle.
5. A nonstructural protein of 16–22 kDa accumulates in large amounts in infected plants, forming large inclusions (**Figure 1(b)**).
6. It has been observed for some tenuiviruses that the mRNAs are synthesized via cap-snatching (**Figure 3**).
7. Tenuiviruses infect plants of the family Poaceae.

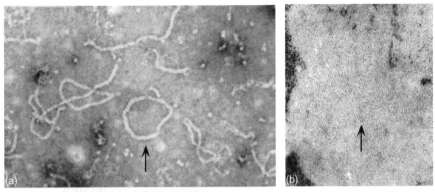

Figure 1 (a) Micrograph of purified ribonucleoproteins (RNPs) of rice hoja blanca virus (RHBV): An RNP is indicated by the arrow. (b) Ultrathin section of a rice leaf cell infected with RHBV. The arrow indicates viral inclusion bodies inside the nucleus with cytopathic effects.

Type Species and Other Species in the Genus

Rice stripe virus (RSV) is the type species of the genus *Tenuivirus*. Other species in the genus are *Echinochloa hoja blanca virus* (EHBV), *Maize stripe virus* (MSpV), *Rice grassy stunt virus* (RGSV), *Rice hoja blanca virus* (RHBV), and *Urochloa hoja blanca virus* (UHBV). Tentative species are Brazilian wheat spike virus (BWSpV), European wheat striate mosaic virus (EWSMV), Iranian wheat stripe virus (IWSV), rice wilted stunt virus, and winter wheat mosaic virus.

Virion Properties

Morphology

The particles also referred as ribonucleoproteins (RNPs) are thin filaments that may appear circular or spiral shaped (**Figure 1(a)**). The RNPs are 3–10 nm in diameter, with lengths proportional to the sizes of the RNAs they contain. No envelope has been observed.

Physical and Physicochemical Properties

RNP preparations can be separated into four or five components by sucrose density gradient centrifugation. The buoyant density of the RNP in CsCl when centrifuged to equilibrium is 1.282–1.288 g cm^{-3}. RNA constitutes 5–12% of the particle weight.

Components

The tenuivirus genome is composed of 4–6 noncapped ssRNA segments; the approximate size of the genome of the type member RSV is 17 kbp (**Figure 2**). The largest segment (RNA1, ~9 kbp) of RSV, MSpV, and RHBV is

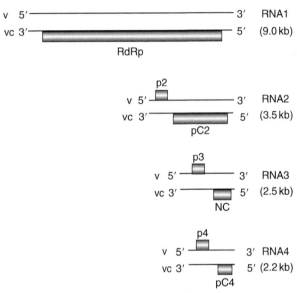

Figure 2 Schematic representation of the genome organization of RSV. RdRp, RNA-dependent RNA polymerase; NC, nucleocapsid protein; p4, nonstructural protein, which accumulates in viral inclusion bodies; v, viral sense RNA; vc, viral complementary RNA.

of negative polarity and encodes the RdRp. Segments 2 (RNA2, 3.3–3.6 kbp), 3 (RNA3, 2.2–2.5 kbp), and 4 (RNA4, 1.9–2.2 kb) of RSV, MSpV, and RHBV are ambisense. Segment 5 (1.3 kbp) detected in virions of MSpV and of EHBV is of negative polarity. A fifth RNA segment has also been reported for some isolates of RSV. RGSV RNA1, 2, 5, and 6 are homologous to RNA1, 2, 3, and 4, respectively, of other tenuiviruses, whereas RNA3 (3.1 kbp) and 4 (2.9 kbp) are ambisense and unique to RGSV. Subgenomic RNAs (sgRNAs) of different sizes and of either polarity are detected; they serve as mRNA for the synthesis of the viral proteins (**Figure 3**).

The RNPs contain a nucleocapsid (NC) protein of 34–35 kDa, and small amounts of a 230 kDa protein which is co-purified with RNPs of RSV, RHBV and RGSV. This

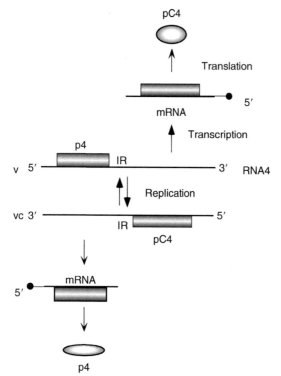

Figure 3 Schematic representation of the replication, transcription, and translation strategies of the ambisense RNA segments of tenuiviruses. Cap and nonviral nucleotides of host origin used as primers by the RdRp as a result of cap-snatching are represented by ''. IR, intergenic region.

protein is the RdRp, associated with filamentous RNPs. The RNA polymerase activity of RHBV is capable of replicating and transcribing the RNA segments *in vitro*.

Genome Organization, Encoded Proteins, Replication and Transcription

The 5′ and 3′ terminal sequences (for about 20 nt) are complementary to each other: they can base-pair and give rise to circular RNPs. The terminal 8 nt (5′ ACACAAAG) and their complement are conserved between tenuiviruses and viruses of the genus *Phlebovirus* of the family *Bunyaviridae*. Several RNA segments encode two proteins in an ambisense arrangement (**Figure 2**). The RdRp and NC protein (32 Kda) are encoded by the viral complementary (vc)RNA1 and vcRNA3, respectively. The p3 (24 Kda) encoded by the viral (v)RNA3 is a suppressor of RNA silencing as demonstrated for RHBV in both plant and insect hosts. A major nonstructural protein (p4; 22 Kda) that accumulates in infected plants is encoded by vRNA4. The sizes of other RSV proteins shown in **Figure 2** are 23 kDa (p2), 94 kDa (pC2), 24 kDa (p3), and 32 kDa (pC4).

For MSpV, RHBV, and RSV, the mRNAs are synthesized via cap-snatching (**Figure 3**). The 5′ end of the mRNAs contains 10–17 nonviral nucleotides and are capped, these extra sequences are derived from host cell mRNAs that are taken or snatched by the RdRp and used as primers to initiate mRNA synthesis. Intergenic noncoding regions located between the open reading frames (ORFs) can in certain cases adopt hairpin structures. The cap-snatching mechanism has been observed for mRNA synthesis of influenza viruses and viruses of the families *Bunyaviridae* and *Arenaviridae*.

Antigenic Properties

The NC proteins of RSV and MSpV are serologically related, as are the NC and p4 proteins of RSV and RGSV. Likewise, the NC proteins of RHBV, EHBV, and UHBV are serologically related. The NC protein of RSV reacts weakly with antibodies made to virion preparations of RGSV or RHBV.

Biological Properties

Host Range

Plant hosts of tenuiviruses all belong to the family Poaceae.

Cytopathology

In infected plants large inclusion bodies (**Figure 1(b)**) are observed which contain the major nonstructural protein p4.

Symptoms

Infected plants exhibit chlorotic stripes and yellow stippling on the leaf blade. Rice plants infected with RHBV at an early stage of development are stunted and may develop necrosis.

Insect Vector and Kind of Transmission

Each virus species is transmitted by a particular species of planthopper in a circulative, propagative manner and can be transmitted transovarially by viruliferous female planthoppers to their offspring, and through sperm from viruliferous males. The principal vectors of the species are *Laodelphax striatellus* for RSV, *Tagosodes cubanus* for EHBV, *Peregrinus maidis* for MSpV, *Nilaparvata lugens* for RGSV, *Tagosodes oryzicolus* for RHBV, and *Caenodelphax teapae* for UHBV. Known vectors of the tentative species are *Sogatella kolophon* for BWSpV, *Javesella pellucida* for EWSMV, and *Ukanoes tanasijevici* for IWSV.

Criteria for Species Demarcation and Phylogenetic Relationships between Species in the Genus

The criteria for species demarcation are given as follows:

1. Vector specificity: transmission by different species of vector.
2. Different sizes and/or numbers of RNA segments.
3. Host range: abilities to infect different key plant species.
4. Amino acid (aa) sequence identity of less than 85% between corresponding gene products.
5. Nucleotide sequence identity of less than 60% between corresponding noncoding intergenic regions.

An example of species discrimination is that between RSV and MSpV. RSV is transmitted by *L. striatellus* and infects 37 graminaceous species including rice and wheat. MSpV is transmitted by *P. maidis* and infects maize, occasionally sorghum, and a few other graminaceous plants but not rice or wheat. RSV isolates have genomes of four RNA segments and the MSpV genome consists of five segments.

An example of difficult species demarcation is the group of the hoja blanca viruses (RHBV, EHBV, and UHBV). They have different vectors, different sizes, and numbers of RNA segments, different hosts, and the nucleotide sequence identity of their intergenic regions is less than 60%. However, the aa sequences of the four proteins on RNAs 3 and 4 are about 90% identical between RHBV, EHBV, and UHBV. Since four of five criteria are met, these viruses could be considered as distinct species that possibly separated recently and are now diverging with little field contact between them.

Phylogenetic analysis of the sequence data from RNA3 and RNA4 of the tenuiviruses shows that RHBV, EHBV, and UHBV are related and form a group distinct from RSV and MSpV (**Figure 4**).

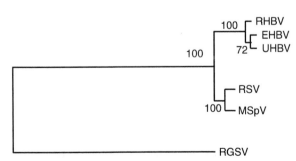

Figure 4 Phylogenetic tree showing the relationships between tenuiviruses on the basis of the nucleotide sequences of the ORFs on RNAs 3 and 4 (RNAs 5 and 6 for RGSV). Shown is the percentage bootstrap support at each node. RHBV, *Rice hoja blanca virus;* EHBV, *Echinochloa hoja blanca virus;* UHBV, *Urochloa hoja blanca virus;* RSV, *Rice stripe virus;* MSpV, *Maize stripe virus;* RGSV, *Rice grassy stunt virus.*

Relation to Other Taxa

Viruses of the genus *Tenuivirus* share several properties with those of the genus *Phlebovirus* of the family *Bunyaviridae.* The multisegmented genomes of tenui- and phleboviruses contain negative-sense and ambisense components. The 5′ and 3′ terminal complementary sequences of viruses of either genus can base-pair and could give rise to circular RNPs. The terminal 8 nt (5′ ACACAAAG) and their complement are conserved between tenui- and phleboviruses. Synthesis of mRNA by the viral RdRp follows a cap-snatching mechanism for viruses of the two genera. Tenuiviruses and most genera in the family *Bunyaviridae* infect their insect vectors as well as their plant hosts. The different number of genome components and the apparent lack of an enveloped viral particle distinguish tenuiviruses from viruses in the family *Bunyaviridae.*

Acknowledgments

The author is grateful to F. Morales for the electron microscope pictures and to Anne-Lise Haenni for useful discussions and critical reading of the manuscript.

Further Reading

de Miranda JR, Muñoz M, Wu R, and Espinoza AM (2001) Phylogenetic placement of a novel tenuivirus from the grass *Urochloa plantaginea.* *Virus Genes* 22: 329–333.

de Miranda JR, Muñoz M, Wu R, Hull R, and Espinoza AM (1996) Sequence of rice hoja blanca tenuivirus RNA-2. *Virus Genes* 12: 231–237.

de Miranda JR, Ramirez BC, Muñoz M, *et al.* (1997) Comparison of Colombian and Costa Rican strains of rice hoja blanca tenuivirus. *Virus Genes* 15: 191–193.

Estabrook EM, Suyenaga K, Tsai JH, and Falk BW (1996) Maize stripe tenuivirus RNA 2 transcripts in plant and insect hosts and analysis of pvc2, a protein similar to the Phlebovirus virion membrane glycoproteins. *Virus Genes* 12: 239–247.

Falk BW and Tsai JH (1998) Biology and molecular biology of viruses in the genus Tenuivirus. *Annual Review of Phytopathology* 36: 139–163.

Haenni A-L, de Miranda JR, Falk BW, *et al.* (2005) *Tenuivirus.* In: Fauquet CM, Mayo MA, Maniloff J, Desselberg U,, and Ball LA (eds.) *Virus Taxonomy: Eighth Report of the International Committee on Taxonomy of Virus,* pp. 717–723. San Diego, CA: Elsevier Academic Press.

Huiet L, Feldstein PA, Tsai JH, and Falk BW (1993) The maize stripe virus major non-capsid protein messenger RNA transcripts contain heterogeneous leader sequences at their 5′ termini. *Virology* 197: 808–812.

Nguyen M, Ramirez BC, Golbach R, and Haenni A-L (1997) Characterization of the *in vitro* activity of the RNA-dependent RNA polymerase associated with the ribonucleoproteins of rice hoja blanca tenuivirus. *Journal of Virology* 71: 2621–2627.

Ramirez BC, Garcin D, Calvert LA, Kolakofsky D, and Haenni A-L (1995) Capped non-viral sequences at the 5′ end of the mRNA of rice hoja blanca virus RNA4. *Journal of Virology* 69: 1951–1954.

Ramirez BC and Haenni A-L (1994) Molecular biology of tenuiviruses, a remarkable group of plant viruses. *Journal of General Virology* 75: 467–475.

Ramirez BC, Lozano I, Constantino L-M, Haenni A-L, and Calvert LA (1993) Complete nucleotide sequence and coding strategy of rice hoja blanca virus RNA4. *Journal of General Virology* 74: 2463–2468.

Ramirez BC, Macaya G, Calvert LA, and Haenni A-L (1992) Rice hoja blanca virus genome characterization and expression *in vitro*. *Journal of General Virology* 73: 1457–1464.

Shimizu T, Toriyama S, Takahashi M, Akutsu K, and Yoneyama K (1996) Non-viral sequences at the 5′-termini of mRNAs derived from virus-sense and virus-complementary sequences of the ambisense RNA segments of rice stripe tenuivirus. *Journal of General Virology* 77: 541–546.

Toriyama S, Kimishima T, and Takahashi M (1997) The proteins encoded by rice grassy stunt virus RNA5 and RNA6 are only distantly related to the corresponding proteins of other members of the genus *Tenuivirus*. *Journal of General Virology* 78: 2355–2363.

Toriyama S, Kimishima T, Takahashi M, Shimizu T, Minaka N, and Akutsu K (1998) The complete nucleotide sequence of the rice grassy stunt virus genome and genomic comparisons of the genus *Tenuivirus*. *Journal of General Virology* 79: 2051–2058.

Tobacco Mosaic Virus

M H V van Regenmortel, CNRS, Illkirch, France

The Beginnings of Virology

Tobacco mosaic virus occupies a unique place in the history of virology and was in the forefront of virus research since the end of the nineteenth century. It was the German Adolf Mayer, working in the Netherlands, who in 1882 first described an important disease of tobacco which he called tobacco mosaic disease. He showed that the disease was infectious and could be transmitted to healthy tobacco plants by inoculation with capillary glass tubes containing sap from diseased plants. Although Mayer could not isolate a germ as the cause of the disease, he did not question the then prevailing view that all infectious diseases were caused by microbes and he remained convinced that he was dealing with a bacterial disease.

About the same time, in St. Petersburg, Dmitri Ivanovsky was studying the same disease and he reported in 1892 that when sap from a diseased tobacco plant was passed through a bacteria-retaining Chamberland filter, the filtrate remained infectious and could be used to infect healthy tobacco plants. Ivanovsky was the first person to show that the agent causing the tobacco mosaic disease passed through a sterilizing filter and this gave rise to the subsequent characterization of viruses as filterable agents. A virology conference was held in 1992 in St. Petersburg to celebrate the centenary of this discovery. Although Ivanovsky is often considered one of the fathers of virology, the significance of his work for the development of virology remains somewhat controversial because all his publications show that he did not really grasp the significance of his filtration experiments. He remained convinced that he was dealing with either a small bacterium or with bacterial spores and never appreciated that he had discovered a new type of infectious agent.

Following in the footsteps of Mayer, Beijerinck in Delft, Holland, again showed in 1898 that sap from tobacco plants infected with the mosaic disease was still infectious after filtration through porcelain filters. He also demonstrated that the causative agent was able to diffuse through several millimeters of an agar gel and he concluded that the infection was not caused by a microbe.

Beijerinck called the agent causing the tobacco mosaic disease a *contagium vivum fluidum* (a contagious living liquid), in opposition to a *contagium vivum fixum*. In those days the term *contagium* was used to refer to any contagious, disease-causing agent, while the term *fixum* meant that the agent was a solid particle or a cellular microbe. On the basis of his filtration and agar diffusion experiments, Beyerinck was convinced that the agent causing tobacco mosaic was neither a microbe nor a small particle or corpuscle (meaning a small body or particle from the Latin *corpus* for body). He proposed instead that the disease-causing agent, which he called a virus, was a living liquid containing a dissolved, nonparticular and noncorpuscular entity.

Lute Bos has claimed that Beijerinck's introduction in 1898 of the unorthodox and rather odd concept of a *contagium vivum fluidum* marked the historic moment when virology was conceived conceptually. However, it is clear that his definition of a virus as a soluble, living agent not consisting of particles certainly does not correspond to our modern view of what a virus is. It is Beijerinck's insistence that a virus is not a microbe and his willingness to challenge the then widely held view that all infectious diseases are caused by germs that make many people regard his contribution to the development of virology as more important than that of Ivanovsky. In 1998, a meeting held in the Netherlands commemorated the centenary of the work of Beijerinck on TMV. About the same time a meeting was held in Germany to honor the contribution of Loeffler and Frosch who, also in 1898, had shown that the agent causing the foot-and-mouth disease of cattle was able to pass through a Chamberland-type filter, in the same way as TMV. In addition the German workers also established that their disease-causing agent was not able to go through a finer grain Kitasato filter, from which they concluded that the agent, which was multiplying within the host, was corpuscular and not soluble as claimed by Beijerinck. Loeffler, however, continued to believe that his pathogen was a very small germ or spore, invisible in the light microscope, that was unable to cross the small pores of a Kitasato filter.

There is no doubt that none of these 'fathers' of virology realized the nature of the filterable pathogenic agents they were investigating and that the term virus which they used did not have the meaning it has today. It took another 30 years before chemical analysis eventually revealed what viruses are actually made of.

Physical and Chemical Properties of TMV

A major change in our perception of what viruses are occurred in 1935 when Stanley, working at the Rockefeller Institute in Princeton, obtained needle-like crystals of TMV that were infectious and consisted of protein. He used methods that were being developed at the time for purifying enzymes and was greatly helped by a bioassay developed by Holmes in 1929 that made it possible to quantify the amount of TMV present in plant sap. In this assay, extracts containing the virus are rubbed on the leaves of *Nicotiana glutinosa* tobacco plants which leads to a number of necrotic lesions proportional to the amount of virus present.

Stanley's demonstration that TMV was a crystallizable chemical substance rather than a microorganism, a discovery that earned him the Nobel Prize in 1946, had a profound impact on the thinking of biologists because it suggested that viruses were actually living molecules able to reproduce themselves. For a while, it seemed that viruses closed the gap between chemistry and biology and might even hold the key to the origin of life.

Stanley had originally described TMV as a pure protein, but in 1936 Bawden and Pirie showed that the virus contained phosphorus and carbohydrate and actually consisted of 95% protein and 5% RNA. TMV thus became the first virus to be purified and shown to be a complex of protein and nucleic acid. Following the development of ultracentrifuges in the 1940s, ultracentrifugation became the standard method for purifying TMV and many other viruses.

In 1939, TMV became the first virus to be visualized in the electron microscope and in 1941, TMV particles were shown to be rods about 280 nm long and 15 nm wide. Subsequently, X-ray analysis established that the rods were hollow tubes consisting of a helical array of 2130 identical protein subunits with 16(1/3) subunits per turn and containing one molecule of RNA deeply embedded in the protein subunits at a radius of 4 nm (**Figure 1**). The length of the TMV particle is controlled by the length of the RNA molecule which becomes fully coated with protein and is thereby protected from nuclease attack.

The viral protein subunits can also aggregate on their own, without incorporating RNA, to form rods very similar to TMV particles but in this case the rods are of variable length (**Figure 2**). It is possible to degrade the virus particles with acetic acid or weak alkali and to obtain in this way dissociated protein subunits and viral RNA, the latter being rapidly degraded by nucleases. In 1956 Gierer and Schramm in Tübingen, Germany, and Fraenkel-Conrat in Berkeley, California, showed that if the viral RNA was obtained by degrading the particles with phenol or detergent, intact RNA molecules were obtained that were infectious and could produce the same disease as intact virus. This was the first demonstration that the nucleic acid component of a virus was the carrier of viral infectivity and that it possessed the genetic information that coded for the viral coat protein.

In 1960 the TMV coat protein was the first viral protein to have its primary structure elucidated when its sequence was determined simultaneously in Tübingen and in Berkeley and found to consist of 158 amino acids. Subsequently, the coat protein sequence of numerous TMV mutants obtained by treating the virus with the mutagenic agent nitrous acid was also determined and these sequence data helped to establish the validity of the genetic code that was being elucidated in the early 1960s in Nirenberg's and Ochoa's laboratories. One of the changes induced by the action of nitrous acid on viral RNA is the nucleotide conversion of cytosine (C) to uracil (U). This changes the codon CCC for the amino acid proline to the codons UCC or CUC for serine and leucine, respectively. In addition, these two codons can be further converted to UUU which codes for phenylalanine. When the coat protein sequences of various TMV mutants were determined, it was found that most of the exchanged amino acids could be attributed to C→U nucleotide conversions, which confirmed the validity and universality of the proposed genetic code.

A particularly interesting mutation present in mutant Ni 1927 was the exchange proline to leucine at position 156 in the sequence which was found to greatly decrease the chemical stability of the virus. It had been known for many years that the virus was resistant to degradation by the enzyme carboxypeptidase and that this enzyme released only about 2000 threonine residues from each virus particle. The C-terminal residue in the coat protein is in fact threonine and the enzyme degradation data were interpreted at the time to mean that the virus particle contained 2000 subunits. However, when the proline at position 156 was replaced by leucine in the mutated protein, the enzyme was able to degrade the protein far beyond the C-terminal residue. The location of a proline near the C-terminus together with the presence of an acetyl group on the N-terminus are actually responsible for the remarkable resistance of the virus to exopeptidase degradation. The same exchange at position 156 in mutant Ni 1927 also led to an interesting discovery in immunology (see below).

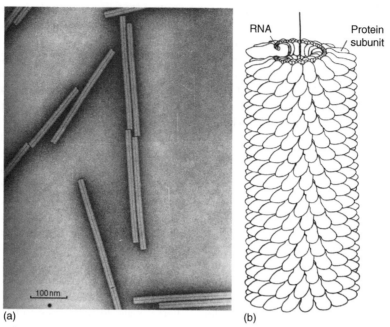

Figure 1   (a) Electron micrograph of negatively stained TMV particles. (b) Diagram of a TMV particle showing about one-sixth of the length of a complete particle. The protein subunits form a helical array with 16(1/3) units per turn and the RNA is packed at a radius of about 4 nm from the helix axis. From Klug A (1999) The tobacco mosaic virus particle: Structure and assembly. *Philosophical Transactions of the Royal Society: Biological Sciences* 354: 531–535.

Self-Assembly of TMV Particles

Dissociated coat protein molecules of TMV are able to assemble into different types of aggregates depending on the pH and ionic strength (**Figure 2**). The disk aggregate is a two-layer cylindrical structure with a sedimentation coefficient of 20 S, where each layer consists of a ring of 17 subunits compared with the 16(1/3) molecules present in each turn of the assembled helix. The disks are able to form stacks which can be either polar or nonpolar, depending on the relative orientation of adjacent disks (**Figure 2**). Short stacks of disks were shown to be nonpolar by immunoelectron microscopy, using monoclonal antibodies that reacted with only one end of the viral helix but were able to bind to both ends of the stacked disks. Whereas short polar stacks of disks can be transformed via lock-washer intermediates into helices (**Figure 2**), this is not possible for nonpolar stacks which cannot be incorporated as such during the elongation process that leads to the formation of RNA-containing virus particles.

It is generally accepted that initiation of TMV assembly from its RNA and protein components requires a 20 S protein aggregate, which could be either a disk or a short, lock-washer type proto-helix. The surface of this aggregate constitutes a template which is recognized by a specific viral RNA sequence. The nucleation of the assembly reaction was found to occur by a rather complex process involving a hairpin structure on the viral RNA, known as the origin of assembly and located about 1000 nucleotides

from the 3′ end, which is inserted through the central hole of the 20 S aggregate. The nucleotide sequence of the stem–loop hairpin is responsible for the preferential incorporation of viral RNA rather than foreign RNA. Elongation occurs by addition of more protein subunits which pull up more RNA through the central hole. This process requires that the 5′ tail of the RNA loops back down the central hole and is slowly 'swallowed up' while the 3′ tail continues to protrude freely from the other end. This unexpected mechanism was visualized by electron micrographs taken in Hirth's laboratory in Strasbourg which revealed how particles were growing with both 3′ and 5′ RNA tails protruding initially from the same end of the particle. The elongation is thus bidirectional and occurs faster along the longer 5′ tail by incorporation of 20 S aggregates and more slowly toward the 3′ end through incorporation of 4 S protein. The elucidation of the assembly process was greatly facilitated by the sequencing of the viral RNA which was completed in 1982. TMV RNA consists of 6395 nucleotides and was the first genome of a plant virus to be sequenced completely.

Virus Disassembly

Since TMV particles are extremely stable *in vitro*, it was not at all obvious how the RNA managed to be released from the particles in order to start the virus replication cycle. Using a cell-free translation system,

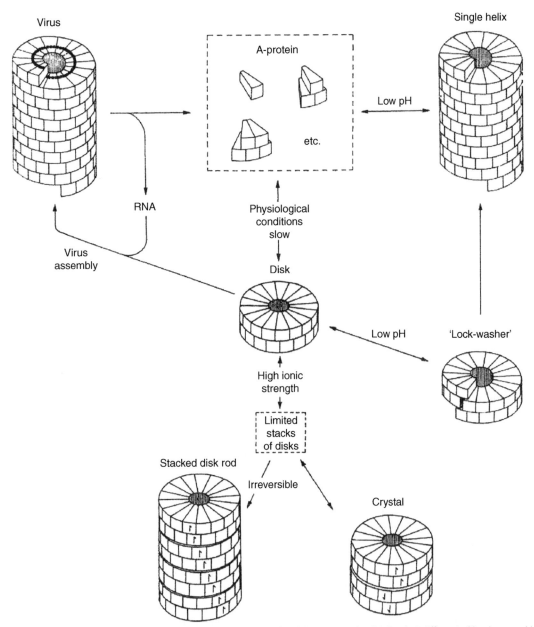

Figure 2 Diagram showing the different polymorphic aggregates of TMV coat protein obtained at different pH values and ionic strengths. Note that both polar and nonpolar stacked disks can be obtained. The 'lock-washer' is not well defined and represents a metastable transitory state. Adapted from Durham ACH and Klug A (1972) Structures and roles of the polymorphic forms of tobacco mosaic virus protein. *Journal of Molecular Biology* 67: 315–332, with permission from Elsevier.

Wilson discovered that the disassembly of TMV particles is initiated when the end of the particles containing the 5' terminus of the RNA becomes associated with ribosomes. This leads to viral subunits becoming dislodged from the particles while the 5' terminal open reading frame in the RNA is being translated by the ribosomes, a process known as co-translational disassembly. This mechanism allows the coat protein subunits to protect the RNA from enzymatic degradation until the particle has reached a site in the infected cell where translation can be initiated.

Antigenicity of TMV

The antigenic properties of TMV have been studied extensively for more than 60 years and these studies have given us much information on how antibodies recognize proteins and viruses. TMV is an excellent immunogen and antibodies to the virus are readily obtained by immunization of experimental animals. When the sequence of TMV coat protein became available, it was possible for the first time to locate the antigenic sites or epitopes of a viral protein at the molecular level. Initial studies focused on two antigenic regions of the coat protein, the C-terminal region (residues 153–158) situated at the surface of virus particles and the disordered loop region (residues 103–112) located in the central hole of the particles and accessible to antibodies only in dissociated protein subunits (**Figure 3**). The C-terminal hexapeptide coupled to bovine serum albumin was used by Anderer in Tübingen to raise antibodies and the resulting antiserum was found to precipitate the virus and neutralize its infectivity. Since both natural peptide fragments and synthetic peptides were used in this work, Anderer and his colleagues should be credited with the discovery that synthetic peptides can elicit antibodies that neutralize the infectivity of a virus. Only when similar results were obtained with animal viruses 15 years later, did the potential of peptides for developing synthetic vaccines become clear.

It has been known since the 1950s that intact TMV particles and dissociated coat protein subunits harbor different types of epitopes recognized by specific antibodies. Certain epitopes present only on virions are constituted by residues from neighboring subunits that are recognized as a single entity by certain antibodies; other epitopes arise from conformational changes in the protein that result from intersubunit bonds. Both these types of epitopes which depend on the protein quaternary structure and are absent in dissociated protein subunits have been called neotopes. Another type of epitope known as cryptotope occurs on the portion of the protein surface that is buried in the polymerized rod and becomes accessible to antibodies only in the dissociated subunits.

The mapping of neotopes and cryptotopes on the surface of TMV coat protein was greatly simplified once monoclonal antibodies (Mabs) became available. Many continuous epitopes were identified at the surface of dissociated coat protein subunits by measuring the ability of peptides to react with Mabs or with antisera (**Figure 3**). These epitopes were found to correspond to regions of the protein shown by X-ray crystallography to possess a high segmental mobility. This correlation between antigenicity and mobility along the peptide chain was also found to exist in other proteins and was used to develop algorithms for predicting the location of epitopes in proteins from their primary structure.

The surface of the protein subunits accessible at either end of the virus particle is different and the one located near the 5′ terminus end of the RNA harbors the two helices corresponding to residues 73–89 and 115–135 (**Figure 3**). Many Mabs specific for this surface have been obtained and in addition to reacting with both ends of nonpolar stacked disks, they were found to block TMV disassembly by sterically preventing the interaction between RNA and ribosomes. It has been suggested that if such antibodies could be expressed in plants, they might be able to control viral infection.

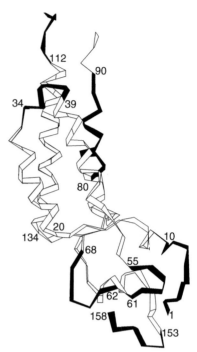

Figure 3 Backbone of the TMV coat protein subunit based on crystallographic data. Residues 94–106 have been omitted because this region, located on the inside of the particle, is disordered. The N and C termini of the protein are located on the outer surface of TMV particles. The position of seven continuous epitopes (residues 1–10, 34–39, 55–61, 62–68, 80–90, 108–112, and 153–158) is indicated by solid lines. Reproduced from Al Moudallal Z, Briand JP, and Van Regenmortel MHV (1985) A major part of the polypeptide chain of tobacco mosaic virus is antigenic. *EMBO Journal* 4: 1231–1235, with permission from Nature Publishing Group.

Another interesting immunological phenomenon was discovered when TMV antibodies were analyzed for their ability to react with certain TMV mutants. It was found that all rabbits immunized with TMV induced the formation of heterospecific antibodies, that is, antibodies that were unable to react with the TMV immunogen but recognized the mutant Ni 1927 quite well. When all the antibodies in a TMV antiserum capable of reacting with TMV were removed by cross-absorption with the virus, it was found that the depleted antiserum still reacted strongly with this mutant which had a single proline→leucine exchange. Apparently the removal of the proline at position 156 exposes binding sites for both peptidases and antibodies that are normally out of reach in the wild-type protein structure. The induction of heterospecific antibodies by immunization with TMV is another illustration of the difference between the antigenicity and immunogenicity of proteins. TMV particles possess the immunogenic capacity of eliciting heterospecific antibodies that react with the Ni 1927 mutant but do not have the antigenic capacity of reacting with these antibodies. The reverse situation where an antigenic peptide or protein is able to react with a particular antibody but is unable to induce the same type of antibody when used as immunogen is a more commonly observed phenomenon which greatly hampers the development of synthetic peptide vaccines.

Many of the viruses that are currently classified in the genus *Tobamovirus* were initially considered to be strains of TMV on the basis of similar particle morphology and ability to cross-react with TMV antibodies. Antigenic relationships between different tobamoviruses were quantified using a parameter known as the serological differentiation index (SDI) which is the average number of twofold dilution steps separating homologous from heterologous antiserum titers. A close correlation exists between the antigenic distance of two viruses expressed as SDI values and the degree of sequence difference in their coat proteins. In general, when two viruses differ antigenically by an SDI value larger than 4, they are considered to belong to separate species. This is valid only if relatedness is measured with polyclonal antisera containing antibodies to a range of different epitopes since comparisons made with Mabs specific for a single epitope will emphasize antigenic similarities or differences depending on which particular Mab is used.

Relationships between individual tobamoviruses are nowadays usually assessed by comparisons between viral genomes. Phylogenetic studies have shown that tobamoviruses are very ancient and co-evolved with their angiosperm hosts, which means that they are at least 120 million years old.

Replication and Cell-to-Cell Movement of TMV

TMV-infected tissues contain four viral proteins: the 126 kDa and 183 kDa proteins of the replicase complex, the 30 kDa movement protein, and the 17.6 kDa coat protein. TMV replication is initiated by the translation of the viral RNA to produce the replicase proteins and this leads to the synthesis of minus-sense and plus-sense copies of the RNA. Translation of the viral coat protein gene then occurs leading to the assembly of progeny virus particles from full-length genomic RNA and coat protein. The translation of the coat protein and movement protein is controlled by the production of two separate subgenomic mRNAs, a mechanism later found to occur in many other viruses but first demonstrated with TMV.

The mechanism that allows plant viruses to move from cell to cell in their hosts has remained a mystery for many years. The rigid cellulose-rich walls of plant cells impede intercellular communication which occurs only through tubular connections known as plasmodesmata. However, plasmodesmata are too narrow to allow the passage of virus particles. Studies with the 30 kDa TMV movement protein showed that this protein changed the size exclusion limit of the plasmodesmata, allowing the virus to move through them in the form of a thin, less than 2 nm wide, ribonucleoprotein complex composed of the genomic RNA and the movement protein. Movement proteins have subsequently been found in many other plant viruses and shown to possess RNA-binding properties.

Studies with TMV also revealed how plant viruses encode a suppressor to combat the post-transcriptional gene silencing reaction that plants use to fight virus infection.

Biotechnology Applications of TMV

TMV-resistant plants have been obtained by transforming them with a DNA copy of the TMV coat protein gene. This coat protein-mediated resistance is due to the ability of transgenically expressed coat protein to interfere in transgenic cells with the disassembly of TMV particles. Coat protein-mediated resistance has subsequently been obtained with many positive-sense RNA plant viruses.

A portion of the TMV RNA leader sequence, called omega, was shown to enhance the translation of foreign genes introduced into transgenic plants. This translational enhancer has been incorporated successfully in many gene vectors for a variety of applications.

In another type of application, TMV particles have been used as surface carriers of foreign peptide epitopes for constructing recombinant vaccines and producing them in plants. It was found that the N- and C-termini of the TMV coat protein as well as the surface loop area corresponding to residues 59–65 were able to accept foreign peptide fusions without impairing the ability of the resulting chimeric virus to infect plants systemically. Several experimental vaccines against viral and parasitic infections that are based on genetically engineered TMV particles produced in tobacco are currently under development.

In conclusion, it is clear that a series of historical accidents together with the fact that large amounts of remarkably stable TMV particles could be obtained from infected plants, allowed this virus to play a major role in the development of virology. In addition, studies of TMV also contributed significantly to the advancement of molecular biology and to our understanding of the physicochemical and antigenic properties of macromolecules.

See also: Plant Virus Vectors (Gene Expression Systems); Tobamovirus; Vaccine Production in Plants.

Further Reading

Al Moudallal Z, Briand JP, and Van Regenmortel MHV (1985) A major part of the polypeptide chain of tobacco mosaic virus is antigenic. *EMBO Journal* 4: 1231–1235.

Bos L (1999) Beijerinck's work on tobacco mosaic virus: Historical context and legacy. *Philosophical Transactions of the Royal Society, London, Series B* 534: 675–685.

Calisher CH and Horzinek MC (eds.) *100 Years of Virology. The Birth and Growth of a Discipline*, pp. 1–220. Vienna: Springer.

Creager ANH (ed.) (2002) *The Life of a Virus. Tobacco Mosaic Virus as an Experimental Model*, pp. 1–398. Chicago, IL: University of Chicago Press.

Durham ACH and Klug A (1972) Structures and roles of the polymorphic forms of tobacco mosaic virus protein. *Journal of Molecular Biology* 67: 315–332.

Harrison BD and Wilson TMA (1999) *Tobacco Mosaic Virus: Pioneering Research for a Century. Philosophical Transactions of the Royal Society, London, Series B* 354: 517–685.

Hirth L and Richards KE (1981) Tobacco mosaic virus: Model for structure and function of a simple virus. *Advances in Virus Research* 26: 145–199.

Mahy BWJ and Lvov DK (eds.) (1993) *Concepts in Virology. From Ivanovsky to the Present*, pp. 1–438. Langhorne, PA: Harwood Academic Publishers.

Scholthof K-BG (2004) Tobacco mosaic virus: A model system for plant biology. *Annual Review of Phytopathology* 42: 13–34.

Scholthof K-BG, Shaw JG, and Zaitlin M (1999) *Tobacco Mosaic Virus: One Hundred Years of Contributions to Virology*, pp. 1–256. St. Paul, MN: American Phytopathologial Society Press.

Van Helvoort T (1991) What is a virus? The case of tobacco mosaic disease. *Studies in History and Philosophy of Science* 22: 557–588.

Van Regenmortel MHV (1999) The antigenicity of TMV. *Philosophical Transactions of the Royal Society, London, Series B* 354: 559–568.

Van Regenmortel MHV and Fraenkel-Conrat H (eds.) (1986) *The Plant Viruses. Vol. 2: The Rod-Shaped Plant Viruses*, pp. 1–180. New York, NY: Plenum.

Tobamovirus

D J Lewandowski, The Ohio State University, Columbus, OH, USA

Glossary

Origin of assembly Stem–loop structure that is the site of initiation of virion assembly.

Pseudoknot An RNA structure with base pairing between a loop and other regions of the RNA.

Introduction

Early research in the late 1800s on the causal agent of the mosaic disease of tobacco led to the discovery of viruses as new infectious agents. Thus tobacco mosaic virus (TMV), the type species of the genus *Tobamovirus*, became the first virus to be discovered, and subsequently has had a significant role in many fundamental discoveries in virology. The first quantitative biological assay for plant viruses was the use of *Nicotiana glutinosa* plants, which produce necrotic local lesions when inoculated with TMV and many tobamoviruses. The resistance gene *N* that confers this hypersensitive response-type resistance to TMV was the first resistance gene against a plant virus to be cloned and characterized. TMV was the first virus to be purified and crystallized, which led to the discovery of the nucleoprotein nature of viruses and determination of the atomic structure of the coat protein and the virion. TMV was the first virus to be visualized in the electron microscope, confirming the predicted rigid rod-shaped virions. The genetic material of TMV was shown to be RNA, a property previously thought to be restricted to DNA. The first viral protein for which an amino acid sequence was determined was the coat protein of TMV. TMV was the first virus to be mutagenized and the subsequent determination of coat protein sequences from a number of strains and mutants helped to establish the universality of the genetic code. Methods of infecting plant protoplasts with viruses were developed with the tobacco–TMV system, creating a synchronous system to study events in the infection cycle. The TMV 30 kDa protein was the first viral protein shown to be required for virus movement.

Taxonomy and Classification

The genus *Tobamovirus* has not been assigned to a family. Currently, there are 23 recognized species within the genus *Tobamovirus* (**Table 1**). Several recently sequenced viruses are tentative members of new species (**Table 1**). Although tobamoviruses comprise one of the more intensively studied plant virus genera, taxonomy has often been confusing. Historically, plant viruses with rigid virions of approximately $18 \times 300 \, \text{nm}^2$ and causing various diseases were all designated strains of TMV. Thus, many viruses originally referred to as TMV strains are now recognized as belonging to separate species. For example, the tobamovirus that was referred to as the tomato strain of TMV, and is approximately 80% identical to TMV at the nucleotide sequence level, is actually tomato mosaic virus. One criterion for distinguishing the members of separate tobamovirus species is a nucleotide sequence difference of at least 10%.

Table 1 Definitive and tentative members of the genus *Tobamovirus*

Cucumber fruit mottle mosaic virus	CFMMV
Cucumber green mottle mosaic virus	CGMMV
Frangipani mosaic virus	FrMV
Hibiscus latent Fort Pierce virus	HLFPV
Hibiscus latent Singapore virus	HLSV
Kyuri green mottle mosaic virus	KGMMV
Maracuja mosaic virus	MaMV
Obuda pepper virus	ObPV
Odontoglossum ringspot virus	ORSV
Paprika mild mottle virus	PaMMV
Pepper mild mottle virus	PMMoV
Ribgrass mosaic virus	RMV
Sunn-hemp mosaic virus	SHMV
Sammons' Opuntia virus	SOV
Tobacco latent virus	TLV
Tobacco mild green mosaic virus	TMGMV
Tobacco mosaic virus	TMV
Tomato mosaic virus	ToMV
Turnip vein-clearing virus	TVCV
Ullucus mild mottle virus	UMMV
Wasabi mottle virus	WMoV
Youcai mosaic virus	YoMV
Zucchini green mottle virus	ZGMMV
Tentative members	
Brugmansia mild mottle virus	
Cucumber mottle virus	
Cactus mild mottle virus	
Streptocarpus flower break virus	
Tropical soda apple mosaic virus	

Virus Structure and Composition

Tobamovirus virions are straight tubes of approximately $18 \times 300 \, nm^2$ with a central hollow core 4 nm in diameter. Virion composition is approximately 95% protein and 5% RNA. For TMV, approximately 2100 subunits of a single coat protein are arranged in a right-handed helix around a single genomic RNA molecule, with each subunit associated with three adjacent nucleotides. Protein–protein associations are the essential first event of virion assembly. Coat protein subunits assemble into several types of aggregates. Coat protein monomers and small heterogeneous aggregates of a few subunits are collectively referred to as 'A-protein'. The equilibrium between A-protein and larger aggregates is primarily dependent upon pH and ionic strength. Purified coat protein and viral RNA can assemble into infectious particles *in vitro*. Larger aggregates are disks composed of two individual stacked rings of coat protein subunits, and protohelices. Protohelices contain approximately 40 coat protein subunits arranged in a spiral around a central hollow core, similar to the arrangement within the virion. A sequence-specific stem–loop structure in the RNA, the origin of assembly (OAS), initiates encapsidation and prevents defective packaging that could result from multiple independent initiation events on a single RNA molecule. Virion assembly initiates as the primary loop of the OAS is

threaded through a coat protein disk or protohelix with both ends of the RNA trailing from one side. The conformation of the coat protein protohelix changes as the RNA becomes embedded within the groove between the two layers of subunits. Elongation is bidirectional, proceeding rapidly toward the 5' end of the RNA as the RNA loop is extruded through the elongating virion and additional coat protein disks are added. There is disagreement about the mechanism of elongation toward the 3' terminus of the RNA, but it appears that this slower process involves the addition of smaller protein aggregates.

Subgenomic mRNAs containing the OAS are encapsidated into shorter virions that are not required for infectivity. The OAS is located within the open reading frame (ORF) for the movement protein of most tobamoviruses. The level of accumulation of a particular subgenomic mRNA containing the OAS determines the relative proportion of that particular virion species. Thus, all tobamovirus virion populations contain a small percentage of movement protein subgenomic mRNAs. In some tobamoviruses, including cucumber green mottle mosaic virus, hibiscus latent Singapore virus (HLSV), kyuri green mottle mosaic virus, maracuja mosaic virus, sunn-hemp mosaic virus, zucchini green mottle mosaic virus, and cactus mild mottle virus, the OAS is located within the coat protein ORF. Thus, these so-called subgroup 2 tobamoviruses produce a significant proportion of small virions that contain coat protein subgenomic mRNA. Hybrid nonviral RNAs containing an OAS will also assemble with coat protein into virus-like particles of length proportional to that of the RNA.

Genome Organization

The genome of tobamoviruses consists of one single-stranded positive-sense RNA of approximately 6300–6800 nt (**Figure 1(a)**). There is a methylguanosine cap at the 5' terminus, followed by an AU-rich leader 55–75 nt in length. The 3' nontranslated end of the RNA consists of sequences that can be folded into a series of pseudoknot structures, followed by a tRNA-like terminus. The hibiscus-infecting tobamoviruses HLSV and hibiscus latent Fort Pierce virus contain a polyA stretch between the 3' end of the coat protein ORF and the tRNA-like structure. The tRNA-like terminus can be aminoacylated *in vitro*, and in most cases specifically accepts histidine. The exception is the 3' terminus of SHMV, which accepts valine and appears to have arisen by a recombination event between a tobamovirus and a tymovirus.

Four ORFs that are contained within all tobamovirus genomes (**Figure 1(a)**) correspond to the proteins found in infected tissue. Two overlapping ORFs begin at the 5' proximal start codon. Termination at the first in-frame stop codon produces a 125–130 kDa protein. A 180–190 kDa protein is produced by readthrough of this leaky termination codon approximately 5–10% of the time. The remaining proteins are expressed from individual 3′ co-terminal subgenomic mRNAs, from which only the 5′ proximal ORF is expressed (**Figures 1(c) and 2(a)**). The next ORF encodes the 28–34 kDa movement protein, which has RNA-binding activity and is required for cell-to-cell movement. The 3′ proximal ORF encodes a 17–18 kDa coat protein. A subgenomic mRNA containing an ORF for a 54 kDa protein that encompasses the readthrough domain of the 180–190 kDa ORF has been isolated from infected tissue, although no protein has been detected.

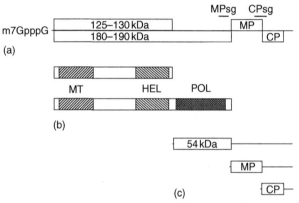

Figure 1 Tobamovirus genome organization and gene expression strategy. (a) Tobamovirus genome organization. ORFs designated as open boxes. Nontranslated sequences designated as lines; positions of subgenomic promoters are marked. (b) Nonstructural proteins involved in tobamovirus replication. Functional domains shared with other viruses within the 'alphavirus supergroup' are designated as hatched boxes. (c) Subgenomic mRNAs with 5′ proximal ORF labeled. MP, movement protein; CP, coat protein; MT, methyltransferase; HEL, helicase; POL, polymerase; MPsg, MP subgenomic mRNA; CPsg, CP subgenomic mRNA.

Within the protein-coding regions of the genome, there are nucleotide sequences that also function as *cis*-acting elements for subgenomic mRNA synthesis, virion assembly, and replication. Gene expression from subgenomic mRNAs is regulated both temporally and quantitatively. The movement protein is produced early and accumulates to low levels, whereas the coat protein is produced late and accumulates to high levels. The regulatory elements for subgenomic mRNA synthesis are located on the genome-length complementary RNA overlapping the upstream ORF (**Figure 1(a)**). There is limited (40%) sequence identity between the TMV movement protein and coat protein subgenomic promoters. The TMV movement protein subgenomic promoter is located upstream of the movement protein ORF, flanking the transcription initiation site. Unlike the movement protein subgenomic promoter, full activity of the coat protein subgenomic promoter requires sequences within the coat protein ORF.

Viral Proteins

The tobamovirus 125–130/180–190 kDa proteins are involved in viral replication, gene expression, and movement. Both are contained in crude replicase preparations, and temperature-sensitive replication-deficient mutants map to these ORFs. The 125–130/180–190 kDa proteins contain two functional domains common to replicase proteins of many positive-stranded RNA plant and animal viruses (**Figure 1(b)**). The N-terminal domain has methyltransferase and guanylyltransferase activities associated with capping of viral RNA. The second common domain is a proposed helicase, based upon conserved sequence motifs. The readthrough domain of the 180–190 kDa protein has sequence motifs characteristic of RNA-dependent RNA polymerases. Both proteins are necessary for

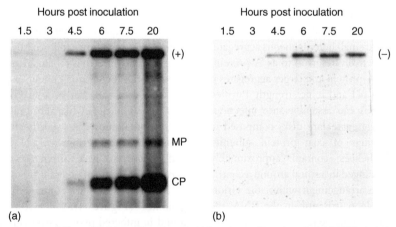

Figure 2 Time course of accumulation of TMV positive-stranded (a) and negative-stranded (b) RNAs in tobacco protoplasts. Total RNA was extracted from tobacco suspension cell protoplasts transfected with TMV *in vitro* transcripts at the time points indicated and analyzed by Northern blot hybridization. (+), TMV genomic RNA; MP, movement protein subgenomic RNA; CP, coat protein subgenomic RNA; (–), negative-stranded complement to the genomic RNA.

efficient replication, although the TMV 126 kDa protein is dispensable for replication and gene expression in protoplasts. The 125–130 kDa protein (or sequences within this region) of the 180–190 kDa protein are required for cell-to-cell movement. Additionally, these multifunctional proteins are symptom determinants, as mutations in mild strains map to these ORFs.

The 28–34 kDa movement protein has a plasmodesmatal binding function associated with its C-terminus and a single-stranded nucleic acid-binding domain associated with the N-terminus. The movement protein–host interaction determines whether the virus can systemically infect some plant species. Although principally a structural protein, the coat protein is also involved in other host interactions. Coat protein is required for efficient long-distance movement of the virus. Coat protein is also a symptom determinant in some susceptible plant species and an elicitor of plant defense mechanisms in other plant species.

Interactions between Viral and Host Proteins

Available evidence suggests that the interactions of viral proteins with host factors are important determinants of viral movement and host ranges. Amino acid substitutions in the movement protein and 125–130/180–190 kDa proteins can alter the movement function in different hosts. Some viruses, including tobamoviruses, can assist movement of other viruses that are incapable of movement in a particular plant species. These interactions suggest that there are more precise associations of viral proteins with host factor(s) than with viral RNA. Additionally, precise coat protein–plant interactions are required for movement to distal positions within the plant. The helicase domain of the 130–190 kDa proteins elicits the *N* gene-mediated resistance in *N. glutinosa*.

Virus Replication

Virions or free viral RNA will infect plants or protoplasts. Because tobamoviruses have a genome consisting of messenger-sense RNA that is infectious, one of the first events is translation of the 5′-proximal ORFs to produce the proteins required for replication of the genomic RNA and transcription of subgenomic mRNAs. When virions are the infecting agent, the first event is thought to be co-translational disassembly, in which the coat protein subunits at the end of the virion surrounding the 5′ end of the RNA loosen, making the RNA available for translation. Ribosomes then associate with the RNA, and translation of the 126/183 kDa ORFs is thought to displace coat protein subunits from the viral RNA. After the formation of an active replicase complex, a complementary negative-strand

RNA is synthesized from the genomic positive-strand RNA template. Negative-strand RNA serves as template for both genomic and subgenomic mRNAs. Negative-strand RNA synthesis ceases early in infection, while positive-strand RNA synthesis continues. This results in an asymmetric positive- to negative-strand RNA ratio. Early in infection, genomic RNA functions as template for negative-strand RNA synthesis and as mRNA for production of the 126/183 kDa proteins. Later in the infection cycle, most of the newly synthesized genomic RNA is encapsidated into virions. Subgenomic mRNAs transcribed during infection function as mRNA for the 3′ ORFs. Within cells of an infected leaf, replication proceeds rapidly between approximately 16 and 96 h post infection within a cell, then ceases. Even though the infected cells become packed with virions, these cells remain metabolically active for long periods. During the early stages of infection of an individual cell, the infection spreads through plasmodesmatal connections to adjacent cells. This event requires the viral movement protein that modifies plasmodesmata to accommodate larger molecules and the 126/183 kDa proteins. Movement through plasmodesmata does not require the coat protein. A second function of the movement protein appears to be binding to the viral RNA to assist its movement through the small plasmodesmatal openings. The movement protein also appears to associate with the cytoskeleton. As the virus spreads from cell to cell throughout a leaf, it enters the phloem for rapid long-distance movement to other leaves and organs of the plant, a complex process that requires the coat protein.

cis-Acting Sequences

The 5′ nontranslated region contains sequences that are required for replication. This region is an efficient translational leader. The 3′ nontranslated region contains *cis*-acting sequences that are involved in replication. Certain deletions within the pseudoknots are not lethal, but result in reduced levels of replication. Exchange of 3′ nontranslated elements between cloned tobamovirus species has resulted in some lethal and nonlethal hybrids, suggesting a requirement for sequence specificity and/or secondary structure. The 3′ nontranslated region appears to be a translational enhancer, both in the viral genome and when fused to heterologous reporter mRNAs. Sequences encoding the internal ORFs for the movement and coat proteins are dispensable for replication. Duplication of the subgenomic promoters results in transcription of an additional new subgenomic mRNA. Heterologous tobamovirus subgenomic promoters inserted into the viral genome are recognized by the replicase complex and transcribed. Foreign sequences inserted behind tobamovirus subgenomic mRNA promoters have been expressed to high levels in plants and protoplasts.

Satellite Tobacco Mosaic Virus

Satellite tobacco mosaic virus (STMV), a tobamovirus-dependent satellite virus, was isolated from *Nicotiana glauca* plants infected with tobacco mild green mosaic virus (TMGMV). The STMV genome consists of one single-stranded positive-sense RNA of 1059 nt. The 240 3′ nucleotides share approximately 65% sequence identity with TMGMV and TMV, contain pseudoknot structures, and have a tRNA-like terminus. No sequence similarity with any tobamovirus exists over the remainder of the genome. Two overlapping ORFs that are expressed in *in vitro* translation reactions are present in the genomic RNA of most STMV isolates. The 5′-proximal ORF encodes a 6.8 kDa protein that has not been detected *in vivo* and is not present in all STMV isolates. The second ORF encodes a 17.5 kDa coat protein that is not serologically related to any tobamovirus coat protein. The 17 nm icosahedral virions are composed of a single STMV genomic RNA encapsidated within 60 STMV coat protein subunits. Replication of natural populations of STMV is supported by other tobamoviruses, but at lower levels than with the natural helper virus, TMGMV. The host range of STMV parallels that of the helper virus.

See also: Plant Virus Diseases: Economic Aspects; Tobacco Mosaic Virus.

Further Reading

Gibbs AJ (1977) Tobamovirus group. CMI/AAB Descriptions of Plant Viruses, No. 184. http://www.dpvweb.net/dpv/showadpv.php?dpvno=184 (accessed July 2007).

Hull R (2002) *Matthews' Plant Virology,* 4th edn., New York: Academic Press.

Lewandowski DJ (2005) *Tobamovirus* genus. In: Fauquet CM, Mayo MA, Maniloff J, Desselburger U,, and Ball LA (eds.) *Virus Taxonomy, Classification and Nomenclature of Viruses: Eighth Report of the International Committee on Taxonomy of Viruses*, pp. 1009–1014. San Diego, CA: Elsevier Academic Press.

Pogue GP, Lindbo JA, Garger SJ, and Fitzmaurice WP (2002) Making an ally from an enemy: Plant virology and the new agriculture. *Annual Review of Phytopathology* 40: 45–74.

Scholthof KB (2005) Tobacco mosaic virus, a model system for plant virology. *Annual Review of Phytopathology* 42: 13–34.

Scholthof KBG, Shaw JG,, and Zaitlin M (eds.) (1999) *Tobacco Mosaic Virus: One Hundred Years of Contributions to Virology*. St. Paul, MN: APS Press.

Van Regenmortel MHV and Fraenkel-Conrat H (eds.) (1986) *The Plant Viruses, Vol. 2: The Rod-Shaped Plant Viruses*. New York: Plenum.

Tobravirus

S A MacFarlane, Scottish Crop Research Institute, Dundee, UK

Taxonomy and Characteristics

The genus *Tobravirus* is comprised of three species, the type species *Tobacco rattle virus* (TRV) together with *Pea early-browning virus* (PEBV) and *Pepper ringspot virus* (PepRSV), which was previously referred to as the CAM strain of TRV. The genus has not been assigned to a virus family. Tobraviruses have a genome of two, positive-sense, single-strand RNAs that are packaged separately into rod-shaped particles. In some situations, the larger genomic RNA (RNA1) can cause a systemic infection in the absence of the second, smaller RNA (RNA2) and without the formation of virus particles. Tobraviruses are transmitted between plants by root-feeding nematodes of the genera *Trichodorus* and *Paratrichodorus*, and in some plant species are also seed transmitted.

Virus Particle Production and Structure

Tobravirus RNA1 is encapsidated into the L (long) particle with a length of 180–215 nm, depending on virus species, and RNA2 is encapsidated into S (short) particles which range in length from 46 to 115 nm, depending on virus isolate (**Figure 1**). Both L and S particles have an apparent diameter of 20–23 nm, depending on the technique used to examine them. *In vitro* translation experiments using RNA extracted from purified virus preparations, as well as studies with the TRV SYM isolate, which has an unusual genome structure, showed that some, if not all, tobravirus subgenomic (sg)RNAs are also encapsidated, in particles of various lengths. The tobraviruses encode a single coat protein (CP), molecules of which assemble in a helical arrangement around a central cavity with a diameter of 4–5 nm, and with a distance of 2.5 nm between successive turns of the helix. *In vitro* reconstitution experiments suggested that virus particle formation initiates at the 5′ end of the viral RNA, although the encapsidated sgRNAs do not carry the 5′ terminal part of the virus genomic RNAs. Peptide mapping showed that the major antigenic regions of the CP are, in descending order of strength, the C-terminal 20 amino acids (aa), 5 aa in the central region of the protein and 5 aa at the N-terminus. This and other spectroscopic analyses suggest that the N- and C-termini are exposed on the outer surface of the particle, while the central region is exposed in the central canal (where interactions of the CP with the viral RNA take place).

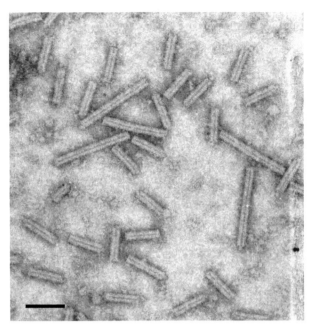

Figure 1 Electron micrograph of long and short particles of TRV isolate RH. Scale = 100 nm.

The C-terminal domain appears to be unstructured and plays a role in interactions with other virus proteins that are involved in nematode transmission of the virus. Particularly with TRV, there is significant amino acid sequence difference between the CP of different virus isolates, resulting in many different serotypes of this virus.

M-Type and NM-Type Infections

In early studies with tobraviruses (usually with TRV), the infectivity of fractionated and purified L and S particles was examined, showing that L particles were infectious (producing local and systemic symptoms on particular hosts) but that S particles were not. However, plants infected with L particles did not contain virus particles, whereas plants infected with both L and S particles did contain (both) virus particles. This was explained by sequencing, which revealed that RNA1 encodes proteins for RNA replication and movement, whereas RNA2 encodes the CP. Because infections with RNA1 do not produce virus particles, they were very difficult to maintain by repeat inoculation of sap extracts, and were referred to as nonmultiplying (NM-type) infections. Infections derived from both RNA1 and RNA2 and producing virus particles were easily passaged and were referred to as multiplying (M-type) infections. Subsequently, it was found that extraction with phenol of RNA from NM-infected plants in fact produced highly infectious preparations.

Early observations also suggested that NM-type infections caused more severe symptoms than M-type infections and moved only slowly up the plant, giving rise to the theory that unencapsidated RNA1 could spread systemically only from cell to cell via plasmodesmata. However, experiments with TRV and PEBV carrying defined mutations in the CP gene demonstrated unequivocally that unencapsidated virus RNAs moved very rapidly via the vascular system in both *Nicotiana benthamiana* and *Nicotiana clevelandii* plants. In addition, with wild-type (encapsidated) virus, systemic infection always included both RNA1 and RNA2 but with the (unencapsidated) CP mutants RNA2 occasionally became separated from RNA1 and was not detected in some systemic leaves. One PEBV mutant carried only a small (28 aa) deletion at the C-terminus of the CP but did not produce virus particles or indeed any detectable CP in infected plants. Nevertheless, this mutant moved systemically as quickly as did wild-type virus and in this case RNA2 did not become separate from RNA1. This suggests, perhaps, that some form of CP but not incorporated into particles, which can be present at very low levels, ensures the coordinated transport of RNA1 and RNA2 through the phloem.

In another study, plants were infected with TRV RNA1 and two types of RNA2, one wild type and the second encoding a CP in which the 15 aa at the C-terminus were replaced with three nonviral residues. Both RNA2 species were encapsidated by the CP that they encoded but neither appeared to be encapsidated by the CP from the other RNA2. Apparently, although RNA1 is encapsidated *in trans*, RNA2 is only encapsidated *in cis*. The mechanism for this is not known but may be linked to the role of CP in the coordinated systemic movement of RNA1 and RNA2.

Genome Structure and Expression

The viral RNAs have a 5′ methylated guanosine cap. The 3′ end of the RNAs is not polyadenylated but folds into a tRNA-like pseudoknot structure that, in contrast to some other virus RNAs, cannot be aminoacylated. RNA1 is from 6.8 to 7 kb in size, and RNA2 varies between isolates from 1.8 to ∼4 kb (**Figure 2**). Complete sequences have been obtained for RNA1 from four isolates of TRV, two isolates of PEBV, and one isolate of PepRSV. In addition, complete sequences have been obtained for RNA2 from 15 isolates of TRV, three isolates of PEBV, and one isolate of PepRSV. (Information on RNA sequences can be obtained from the eighth ICTV report.) Other than the sequencing, almost no further molecular studies have been carried out on PepRSV.

RNA Sequences

The larger RNA (RNA1) is highly conserved in nucleotide sequence between different isolates of the same virus (e.g., 99% identity between RNA1 of TRV isolates SYM and ORY), but there is much less identity between isolates of the different tobravirus species (e.g., 62% identity

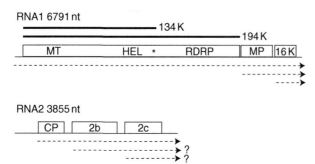

RNA1 6791 nt

RNA2 3855 nt

Figure 2 Genome diagram and expression strategy of TRV isolate PpK20. Open boxes denote virus genes. The solid lines above the RNA1 genes shows the location of the two, overlapping replicase genes. The 134K protein contains methyltransferase (MT) and helicase (HEL) motifs. The C-terminal part of the 194K protein contains an RNA-dependent RNA polymerase (RDRP) motif. The asterisk denotes the 134K gene termination codon where readthrough translation to produce the 194K protein occurs. The movement protein (MP) gene is also known as the 29K or 1a gene. The cysteine-rich 16K gene is also known as the 1b gene. For RNA2, CP denotes the coat protein gene. The dashed lines beneath each RNA denote the 3′ co-terminal subgenomic RNAs that are known or suspected to exist.

between RNA1 of TRV isolate SYM and PEBV isolate SP5). The smaller RNA (RNA2) varies considerably both in terms of overall nucleotide sequence identity as well as protein-coding capacity between isolates of the same virus (TRV or PEBV), although phylogenetic analysis shows that the CPs of the different tobraviruses, which are encoded by RNA2, are more closely related to each other than to CPs from viruses in other genera. The tobraviruses are most closely related to tobamoviruses, in terms of particle structure, overall gene organization, and viral protein sequence homologies.

RNA1 has a 5′ noncoding region (NCR) of between 126 and 202 nt, and a 3′ NCR of 459–255 nt. RNA2 has a much larger 5′ NCR of 470–710 nt, and a 3′ NCR of 780–392 nt. There is almost no sequence homology between the 5′ NCRs of the tobraviruses; however, the 25 nt at the 3′ terminus of the TRV and PEBV RNAs are identical and there is 70% identity over the next 140 nt. Consequently, when recombinants were made between the RNA2 of TRV PpK20 and PEBV SP5, the replicase complex of both viruses could replicate RNA2 molecules carrying the 3′ NCR from either virus. However, RNA2 carrying the TRV 5′ NCR could only be replicated by a TRV replicase complex, and RNA2 carrying the PEBV 5′ NCR could only be replicated by a PEBV replicase complex.

Expression of Virus Genes

The 5′ proximal gene of RNA1, encoding viral replicase proteins, is expressed by direct translation of the genomic RNA. The two other RNA1-encoded genes (1a and 1b) are expressed from sgRNAs that start with the sequence AUA (within a conserved motif GCAUA) and that are co-terminal with the 3′ end of the genomic RNA.

The 5′ proximal gene of RNA2 of almost all tobravirus isolates encodes the CP. However, the CP gene is located at least 470 nt downstream of the 5′ terminus of RNA2, and this region contains numerous (at least six) AUG codons upstream of the translation initiation codon of the CP gene, that are inhibitory to CP expression. Experiments to delete parts of the 5′ NCR of PEBV RNA2 showed that only a very small amount of CP is translated *in vitro* from full-length (genomic) RNA2 but that translation is increased 25-fold when the 5′ NCR is removed. Thus, even though the CP gene is the 5′ proximal gene on RNA2, it (and the other genes on RNA2) is expressed from an sgRNA. Uniquely, the CP gene of TRV SYM is located near to the 3′ end of RNA2, downstream of another gene. The sgRNA for this CP is much shorter than genomic RNA2 and is encapsidated into a clearly identifiable VS (very short) virus particle.

A stem–loop sequence is found upstream of tobravirus sgRNA start sites, and is particularly conserved for the CP sgRNA. The introduction of mutations into this structure (8 bp stem, 4 nt loop) upstream of the PEBV CP gene showed that the stem–loop forms an essential part of the sgRNA promoter, and that the structure of the stem rather than its actual sequence is important for sgRNA synthesis *in planta*. The promoter regions for the TRV and PEBV CP genes could be used as a cassette and moved to other locations in RNA2 to express nonviral genes.

Recombination in Tobravirus RNAs

The large variation in the overall length of RNA2 from different tobravirus isolates reflects the fact that most are recombinant molecules, where the 3′ part of RNA2 has been replaced by sequences from the 3′ part of RNA1. The recombination junction can occur at any position in RNA2 that is downstream of the CP gene, and the recombinant may retain none, some, or even both of the other genes (2b and 2c) that might be considered RNA2 specific. The region of RNA1 that is transferred to RNA2 always includes the 3′ NCR, often includes the 1b gene immediately upstream of the 3′ NCR, and may occasionally include part of the 1a gene that is located upstream of the 1b gene. This means that many of these recombinant isolates carry two, probably both functional, copies of the 1b gene, one on RNA1 and the second on RNA2. The mechanism for recombination is not known but is speculated to be caused by template switching by viral replicase from RNA1 to RNA2 during minus-strand RNA synthesis. Sequences that resemble the 5′ terminus of tobravirus genomic RNAs (rich in A and U residues) are often found at or near the recombination junction, and conceivably could facilitate the recombination event. It is also not completely clear when tobravirus recombination occurs, though the feeling is that repeated passage of virus by mechanical inoculation in glasshouse-grown plants, bypassing the nematode transmission process, either encourages recombination or reduces selection against the survival of recombinants. Nevertheless, at least one known recombinant isolate, TRV PaY4, which was cloned after only a very limited period of multiplication in the glasshouse, did retain its ability to be nematode transmitted.

A different recombination process also occurs in which the CP and/or 2b gene in RNA2 of some TRV isolates appears to have been derived from PEBV. For this reason, serological analysis is not always successful in discriminating between the different tobraviruses. Attempts to reproduce this recombination by co-inoculating TRV and PEBV to plants in the glasshouse were not successful, suggesting that this may be only a rare occurrence, or may be stimulated by particular environmental conditions. Nevertheless, one report showed that 30% of the TRV isolates recovered from fields in the coastal bulb-growing region in the Netherlands were TRV/PEBV recombinants.

Viral Proteins

RNA1

RNA1 encodes proteins for virus RNA replication and intraplant movement (local and systemic). The 5′ half of RNA1 encodes a large (134–141 kDa) protein that contains methytransferase and helicase domains and is expected to be part of the viral replicase complex. Readthrough translation of the stop codon of this protein produces a larger protein (194–201 kDa) that contains motifs associated with RNA-dependent RNA polymerase proteins. *In vitro* translation experiments showed that the opal (UGA) translation termination codon of the TRV 134 kDa protein is suppressed by tRNAs that incorporate tryptophan or cysteine at this position. Downstream of the replicase genes is the 1a gene encoding a 29–30 kDa cell-to-cell movement protein. Disruption of this gene prevents accumulation of TRV in inoculated leaves but can be overcome by co-inoculation with tobacco mosaic virus (TMV) or to transgenic plants expressing the TMV 30K movement protein. The C-terminal 1b gene encodes a 12–16 kDa cysteine-rich protein that is involved in seed transmission of PEBV in pea, pathogenicity of TRV and PEBV, and in cultured *Drosophila* cells suppresses RNA silencing. Deletion of the 1b gene prevents systemic movement of TRV in *N. benthamiana*, which can be recovered by co-expression, from TRV RNA2, of other cysteine-rich proteins from PEBV, soil-borne wheat mosaic virus, and barley stripe mosaic virus, as well as by the 2b silencing suppressor protein of cucumber mosaic virus.

RNA2

RNA2 encodes proteins for virus particle formation and transmission by nematodes. Several studies have identified the CP as being involved in the nematode transmission process, and deletion of part of the C-terminal domain of the CP prevented the transmission of PEBV without affecting virus particle formation. Infectious, nematode-transmissible clones of three TRV isolates and one PEBV isolate have been sequenced. In addition to the CP, these RNAs all carry two other genes encoding the 2b and 2c proteins, both of which (for PEBV) have been detected by Western blotting in extracts of virus-infected plants. Mutation studies showed that the 2b gene is necessary for transmission of TRV and PEBV, whereas the 2c protein was only required for transmission of PEBV. This difference may reflect the different species of nematode that transmitted the particular virus isolates used in these studies, rather than a clear mechanistic difference between TRV and PEBV transmission. A further, small open reading frame, encoding a putative 9 kDa protein immediately downstream of and in-frame with the CP gene, is present in PEBV-TpA56 and TRV TpO1. For PEBV, mutation of this gene greatly reduced nematode transmission frequency, although the protein was not detected by Western blotting.

The 2b proteins of the different tobravirus isolates share some amino acid sequence homology with each other, ranging from only 11% identity (TRV Umt1 vs. TRV PaY4) to 99% identity (TRV Umt1 vs. TRV OR2). They also range in size from 238 aa (TRV PaY4) to 354 aa (TRV PpK20). The TRV PaY4 2b protein has been show to act *in trans*, with nematode transmission of a 2b-mutant virus being complemented by co-infection with wild-type virus. However, this appears to be isolate specific, as the TRV PaY4 2b mutant was complemented only by wild-type TRV PaY4 and not by wild-type TRV PpK20.

The 2b protein may influence nematode transmission by more than one mechanism. In a microscopy study of the roots of *N. benthamiana* plants infected by PEBV, large numbers of particles of the wild-type virus were found in all regions of the root tip, which is where the vector nematodes preferentially feed, whereas virus particles of a mutant deleted for the 2b and 2c genes were present in roots only in much lower numbers. A similar enhancement was found in the efficiency of invasion of roots in *N. benthamiana*, as well as in leaves of *Arabidopsis thaliana*, when TRV carrying the 2b gene was compared with that of TRV lacking the 2b gene.

The 2b protein also physically interacts with the TRV CP (as examined by yeast two-hybrid (Y2H) analysis) and the virus particle (as examined by immunoelectron microscopy). One suggestion is that the 2b protein acts as a bridge to trap virus particles to specific sites of retention on the nematode esophageal cuticle. Aggregates of virus particles have been observed using the electron microscope to collect in this part of the nematode but the co-location of the 2b protein has not yet been demonstrated. It appears that interaction between the CP and 2b protein stabilizes the 2b protein, as in plants infected with wild-type TRV PaY4, carrying both PaY4 CP and PaY42b genes, the 2b protein was detected by Western blotting. However, with a recombinant TRV carrying the TRV PpK20 CP gene and the TRV PaY4 2b gene, the 2b protein could not be detected, possibly because these proteins from two different isolates cannot interact with one another. This is reinforced by results from the Y2H study, where the TRV PpK20 CP and PpK20 2b proteins were found to interact. Similarly, the PaY4 CP and PaY4 2b proteins interacted; however, the PpK20 CP did not interact with the PaY4 2b protein and the PaY4 CP did not interact with the PpK20 2b protein.

Although the 2c protein is involved in transmission of PEBV, little else is known about this protein. Amino acid sequence homologies between the 2c proteins of different TRV isolates range from almost none to over 95% identity. In a Y2H assay, the TRV PpK20 2c protein interacted with the PpK20CP, and removal of the CP C-terminal flexible domain did not affect the interaction.

Tobraviruses as Gene Expression/ Silencing Vectors

As RNA1 encodes all the proteins necessary for tobravirus replication and movement, the RNA2 can be modified to carry nonviral sequences without greatly affecting virus infection. Expression vectors have been constructed from all three tobraviruses, in which a duplicate CP promoter sequence is inserted downstream of the native CP gene followed by restriction sites to allow the cloning of other sequences. Together, the tobraviruses can infect a wide range of plant species, often without causing particularly severe symptoms, features which increase their utility as expression vectors.

Plants infected with tobraviruses often undergo a rapid recovery in which infection symptoms and virus levels fall dramatically, although the plants do not become free of virus, most particularly in the meristem regions. However, these plants have developed a strong resistance to further infection by the same virus, most likely by an RNA silencing-based mechanism. Although not well understood, it seems that tobraviruses are potent triggers of RNA silencing but do not encode a strong silencing suppressor protein to counteract this host defense activity. A consequence of this is that when a host plant sequence is inserted into the virus genome, very strong silencing is initiated that targets expression of the host gene itself, a process known as virus-induced gene silencing (VIGS). TRV has become one of the most widely used VIGS vectors for studies of plant gene function, and recent work has shown that PEBV is also a very effective VIGS vector for studies in pea.

Diseases Caused by Tobraviruses

TRV is found in many regions (in Europe, North America, Japan, and Brazil) and has a particularly wide host range, infecting more than 100 species in nature and more than 400 species when tested in the glasshouse, although not all of these infections are systemic. As tobraviruses are transmitted by soil-inhabiting nematodes, infection may be limited in the field to the roots. Weeds may play an important role in the maintenance and spread of tobraviruses, with *Capsella bursa-pastoris*, *Senecio vulgaris*, *Stellaria media*, and *Viola arvensis* being the most commonly found weed hosts of TRV. Virus transmission in seed of these plants was also reported. Many crop plants are infected by TRV, the major diseases being those of potato and ornamental bulbs (narcissus, gladiolus, tulip, lily, and crocus). The symptoms of TRV infection in potato are the formation of arcs and flecks of brown corky tissue in the tuber which is referred to as spraing, and which can make the tuber unfit for sale. Both M-type and NM-type infections can produce spraing symptoms, the biochemical basis for which is not known. It was thought that potato cultivars that did not show spraing symptoms were resistant to TRV; however, recent work has shown that some infections can be symptomless. Nevertheless, even these symptomless infections lead to significant reductions in tuber yield and tuber quality. Inclusion of tubers carrying symptomless infection could have major consequences for the production and distribution of seed potatoes.

PEBV has been reported in several European countries (UK, Netherlands, Italy, Belgium, Sweden) as well as Algeria and Morocco. In the field, it infects mainly legumes, including pea, faba bean, French bean, lupin, and alfalfa. Several weeds and other crop plants may be infected, although often only in the roots.

PepRSV has only been reported from Brazil, where it infects pepper, tomato, and artichoke, as well as local weed species.

See also: Nepovirus; Tobamovirus.

Further Reading

Harrison BD and Robinson DJ (1978) The Tobraviruses. *Advances in Virus Research* 23: 25–77.
MacFarlane SA (1999) Molecular biology of the tobraviruses. *Journal of General Virology* 80: 2799–2807.
MacFarlane SA and Robinson DJR (2004) Transmission of plant viruses by nematodes. In: Gillespie SH, Smith GL,, and Osbourne A (eds.) *SGM Symposium 63: Microbe–Vector Interactions in Vector-Borne Diseases*, pp. 263–285. Cambridge: Cambridge University Press.
Robinson DJ (2005) Tobravirus. In: Fauquet CM, Mayo MA, Maniloff J, Desselberger U, and Ball LA (eds.) *Virus Taxonomy: Eighth Report of the International Committee on Taxonomy of Viruses*, pp. 1015–1019. San Diego, CA: Elsevier Academic Press.
Visser PB, Mathis A, and Linthorst HJM (1999) Tobraviruses. In: Granoff A and Webster R (eds.) *Encyclopedia of Virology*, 2nd edn., pp. 1784–1789. London: Elsevier.

Tomato Leaf Curl Viruses from India

S Chakraborty, Jawaharlal Nehru University, New Delhi, India

Glossary

Pseudorecombinants New strains of a virus that result from the reassortment of genome nucleic acids during the replication of viruses with divided genomes in mixed infections.

Recombinant A new strain/species of a virus that occurs as a result of the breakage and renewal of covalent links in a nucleic acid chain for rearrangement of nucleic acids in the chain.

Synergism The association of two or more viruses acting at the same time, which enhances symptom severity.

Introduction

Whitefly-transmitted geminiviruses cause epidemics in vegetable, staple, and fiber crops. The diseases are generally associated with local or regional whitefly (*Bemisia tabaci* Gennadius) infestations. They cause enormous economic losses in several crops in the Tropics, which provide ideal conditions for the perpetuation of viruses and the insect vector. Intensive agricultural practices necessitated by the ever-increasing demands of a rapidly growing population and the introduction of new genotypes, cropping pattern, and crops have further aggravated the situation. For example, the continuous cultivation of crops (such as cotton, tomato, pepper, beans, soybean, and melon which are susceptible to the viruses and are attractive hosts for the whiteflies) certainly account for some of the increase in the severity and the vast spread of diseases caused by geminiviruses.

Tomato leaf curl disease (ToLCD) is the most devastating disease of tomato, affecting a large area under cultivation; it can be on the scale of an epidemic. Incidence of this disease was first reported from northern India in 1948. During the 1950s, incidence was reported from central India, followed by occurrence of the disease in the main tomato-growing regions of southern India. Since then, this disease has emerged as a major threat to tomato cultivation, and incidence has increased after introduction of high-yielding tomato varieties during the late 1960s. The disease is ubiquitous with the crop and has been observed in all the tomato-producing areas of the country. Leaf curl disease of tomato is so serious that the ability of small farmers to cultivate tomato in

several major production areas, especially during the peak of whitefly infestation season, has been eliminated.

Yield loss varied with the age of plants at the time of infection and varieties being tested. The virus affects all stages of the growth causing 17.6–99.7% loss in yield, depending on the stage of the crop at the time of infection. When plants get infected early, within 20 days after transplanting, they remain stunted and produce few or no fruits and the yield loss may reach up to 100%. However, if the plants are infected at 35 and 50 days after transplanting, the yield loss is reduced to 74.1% and 28.9%, respectively. This indicates that earlier the infection, higher is the yield loss.

Symptoms

Leaf curl disease of tomato is characterized by severe stunting of the plants with upward and downward rolling and crinkling of the leaves. Infected plants exhibit intervenal yellowing, vein clearing, and crinkling and puckering of the leaves, sometimes accompanied by inward rolling of the leaf margins. The disease induces severe stunting, bushy growth, and partial or complete sterility, depending on the stage at which infection has taken place (**Figure 1**).

Transmission

Under natural conditions, whiteflies transmit the tomato leaf curl viruses (ToLCVs) from infected to healthy plants. Even a single whitefly can transmit the virus. Minimum acquisition access period and inoculation access period of 30 min each is required for successful transmission to occur. Pre-acquisition and pre-inoculation starving of the vector results in higher levels of transmission. ToLCV can persist up to 10 days after acquisition in a single adult whitefly. Females are more efficient transmitters than the males. The virus is also transmitted by dodder (*Cuscuta reflexa* Roxb.). Under artificial conditions, grafting can transmit the disease. It is also known that some of the isolates of ToLCVs are also sap transmissible under laboratory conditions.

Epidemiology

Like other vector-borne diseases affecting crop species, the factors contributing to ToLCD buildup are: (1) availability of virus inoculum and the vector around the fields;

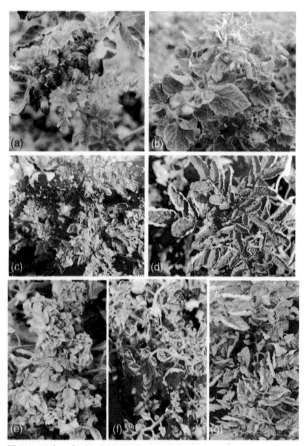

Figure 1 (a–g) Variation of symptoms induced by begomoviruses infecting tomatoes in India.

fall, temperature). It was observed that the disease progresses from February to June, when the dry and hot season with low humidity prevail. The incidence may even reach up to 100% during these months. These types of weather conditions favor whitefly multiplication and disease spread. The whitefly population and ToLCV incidence remain comparatively less during winter and rainy seasons. The tropical climate in southern India allows year-round tomato cropping, which, together with the presence of perennial host plants for both ToLCVs and *B. tabaci*, enables an easy carry-over of ToLCD between growing seasons. Whitefly biotype B can effectively transmit the virus. Overall, factors like long persistence of the virus in the vector, efficient transmission by the biotype, cultivation of tomato throughout the year, and abundance of weed hosts are the contributing factors for the high incidence of the disease under natural conditions.

Host Range

ToLCV strains/species are easily transmitted by grafting and through vector to a wide range of weeds and cultivated crops. Molecular detection using ToLCV-specific primers has led to identification of a range of plant species that can harbor ToLCVs in India. ToLCVs can infect crops such as *Lycopersicon esculentum*, *L. peruvianum*, *L. hirsutum*, *L glandulosum*, *L. pimpinellifolium*, *Capsicum annuum*, *Nicotiana tabacum*, *Vigna unguiculata*, and *Luffa cylindrica*. The viruses perpetuate on many weed hosts, viz. *Acanthospermum hispidum*, *Ageratum conyzoides*, *Blainvella rhomboids*, *Euphorbia hirta*, *Fraveria hirta*, *Parthenium hysterophorus*, *Malvastrum coromandalinum*, and *Croton bonplandianum*.

Genome Organization

Full-length genome of isolates of ToLCVs have been cloned and sequenced. Three types of genomic DNA (DNA-A, DNA-B, and DNA-β) have been found to be associated with begomoviruses causing ToLCD in India.

Apparently, the isolates from southern India have a monopartite genome (DNA-A) associated with DNA-β while both monopartite (DNA-A with DNA-β) and bipartite (DNA-A and DNA-B) begomoviruses have been found to infect tomatoes in northern India. The genome organization of DNA-A (ranges between 2739 and 2759 bp) resembles other begomoviruses having two open reading frames (ORFs) on the viral strand (AV1 and AV2) and four on the complementary strand (AC1, AC2, AC3, and AC4). Genome organization of DNA-B (ranges between 2656 and 2686 bp) and DNA-β (genome size between 1344 and 1376 bp) also resembles other begomoviruses having two ORFs (one on the viral strand,

(2) the movement of viruliferous vectors into the freshly sown field; (3) the susceptibility of the variety that builds up vector population and allows establishment of the virus; (4) weather parameters favoring vector population buildup; and (5) vector biotype which can effectively transmit the disease. Interaction of these factors leads to epidemic outbreak. The most important factors responsible for the epidemic are source of inoculum and vector. In India, wherever tomato is grown continuously, there is increase in leaf curl disease incidence. Virus inoculum in the weeds and continuous cropping contribute to rapid disease buildup. Weather parameters, both at macro- and micro-level, affecting the developmental stages of the plant and the life cycle of the vector are important. Tomato is grown in different agroclimatic zones in India, which makes prediction of the outbreak of epidemics more difficult. However, there is definitely a correlation between vector population and disease incidence.

Maximum temperature and rainfall play an important role for spread of the disease in southern India, while minimum temperature and minimum relative humidity influence the whitefly population in the north. ToLCD incidence depends on weather conditions (humidity, rain-

BV1, and the other on the complementary strand, BC1) and one on the complementary strand (βC1), respectively.

Diversity of Tomato-Infecting Begomoviruses

ToLCD appears to be caused by a complex of several viruses based on symptom variations on different indicator hosts. During the 1980s, based on symptoms produced in a particular tomato cultivar, ToLCVs were divided into five groups: (1) severe leaf curl with thickening of veins, (2) severe symptom with enation, (3) screw pattern of leaf arrangement, (4) vein purpling and leaf curl, and (5) exclusively downward curling of leaves. Variability was

subsequently also found in the epitope profiles of ToLCVs collected from Karnataka, with groupings suggesting that the tomato crop and some neighboring weed species were hosts to the same ToLCV strains/species. Species status for begomoviruses, however, cannot be conferred based on symptom type or epitope profile.

In accordance with the ICTV *Geminiviridae* study group guidelines, nucleotide sequence identity of DNA-A, and genome organization (see **Table 1**), the following five Indian ToLCV species are to be demarcated: *Tomato leaf curl Bangalore virus*, *Tomato leaf curl Gujarat virus*, *Tomato leaf curl Karnataka virus*, *Tomato leaf curl New Delhi virus*, and *Tomato leaf curl Pune virus*. Except for *Tomato leaf curl Pune virus*, four species have been characterized in detail based on their biological and molecular properties (**Figure 2**).

Table 1   GenBank accession numbers of selected begomoviruses DNA-A, DNA-B and DNA-β sequences used for analysis

| Species | Virus name | Accession numbers | | Abbreviation |
		DNA-A	DNA-B	
DNA-A and DNA-B				
Tomato leaf curl Bangalore virus	Tomato leaf curl Bangalore virus-[Ban1]	Z48182		ToLCBV-[Ban1]
	Tomato leaf curl Bangalore virus-[Kerala]	DQ887537		ToLCBV-[Ker]
	Tomato leaf curl Bangalore virus-[Kolar]	AF428255		ToLCBV-[Kol]
	Tomato leaf curl Bangalore virus-[Ban5]	AF295401		ToLCBV-[Ban5]
	Tomato leaf curl Bangalore virus-[Ban4]	AF165098		ToLCBV-[Ban4]
	Tomato leaf curl Bangalore virus-[Ban AVT1]	AY428770		ToLCBV-[Ban AVT1]
Tomato leaf curl Bangladesh virus	Tomato leaf curl Bangladesh virus	AF188481		ToLCBDV
Tomato leaf curl Gujarat virus	Tomato leaf curl Gujarat virus-[Mirzapur]	AF449999		ToLCGV-[Mir]
	Tomato leaf curl Gujarat virus-[Vadodara]	AF413671		ToLCGV-[Vad]
	Tomato leaf curl Gujarat virus-[Varanasi]	AY190290	AY190291	ToLCGV-[Var]
	Tomato leaf curl Gujarat virus-[Nepal]	AY234383		ToLCGV-[Nepal]
Tomato leaf curl Joydebpur virus	Tomato leaf curl Joydebpur virus	DQ673859		ToLCJoV
Tomato leaf curl Karnataka virus	Tomato leaf curl Karnataka virus-[Bangalore]	U38239		ToLCKV-[Ban]
	Tomato leaf curl Karnataka virus-[Janti]	AY754812		ToLCKV-[Janti]
Tomato leaf curl New Delhi virus	Tomato leaf curl New Delhi virus-[Lucknow]	Y16421	X89653	ToLCNDV-[Luc]
	Tomato leaf curl New Delhi virus-[Mild]	U15016		ToLCNDV-[Mild]
	Tomato leaf curl New Delhi virus-[Severe]	U15015	U15017	ToLCNDV-[Svr]
	Tomato leaf curl New Delhi virus-[Pakistan]	AF448058	AY150305	ToLCNDV-[PK]
	Tomato leaf curl New Delhi virus-[Pakistan-Islamabad]	AF448059	AY150304	ToLCNDV-[PK-IS]
Tomato leaf curl Malaysia virus	Tomato leaf curl Malaysia virus-[India]	DQ629102		ToLCMYV-[IN]
Tomato leaf curl Pune virus	Tomato leaf curl Pune virus	AY754814		ToLCPV
Tomato leaf curl Sri Lanka virus	Tomato leaf curl Sri Lanka virus	AF274349		ToLCSLV
DNA-β				
Tomato leaf curl beta-[Aurangabad]		EF095958		ToLCB-[Aur]
Tomato leaf curl beta-[Bangalore]		AY428768		ToLCB-[Ban]
Tomato leaf curl beta-[Chinthapalli]		AY43855		ToLCB-[Chi]
Tomato leaf curl beta-[Coimbatore]		AY438560		ToLCB-[Coi]
Tomato leaf curl beta-[Jabalpur]		AY230138		ToLCB-[Jab]
Tomato leaf curl beta-[New Delhi]		AJ542490		ToLCNDB
Tomato leaf curl beta-[Pune]		AY838894		ToLCB-[Pune]
Tomato leaf curl beta-[Rajasthan]		AY438558		ToLCB-[Raj]
Tomato leaf curl beta-[Varanasi]		AY438559		ToLCB-[Var]

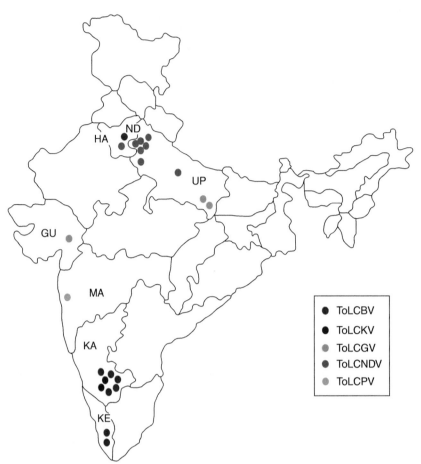

Figure 2 Distribution of five species of tomato-infecting begomoviruses in India. Presence of ToLCVs was located based on availability of full-length DNA-A sequences in GenBank.

Tomato Leaf Curl New Delhi Virus

During the mid-1990s, two isolates of this virus were reported from New Delhi and Lucknow. In addition to the severe isolate, a mild isolate was also described from New Delhi. It has a bipartite genome with DNA-A and DNA-B. However, occurrence of DNA-β has also been observed. Infectivity of the cloned DNAs has been demonstrated.

Tomato Leaf Curl Gujarat Virus

Three isolates from Varanasi, Vadodara, and Mirzapur belong to this species. Among them, Varanasi strain has been characterized in detail. DNA-A alone is infectious but DNA-B increases symptom severity. Association of DNA-β has also been observed with this mono-bipartite species under natural conditions. Tomato leaf curl Gujarat virus (ToLCGV) is also sap transmissible to tomato, pepper, *N. benthamiana*, and *N. tabacum*. Unexpectedly, ToLCGV-Var DNA-A (AY190290) and DNA-B (AY190291) share a common region (CR) of 155 bp that is only 60% identical which were cognate pair of components that cause severe disease of tomato under field conditions.

Tomato Leaf Curl Bangalore Virus

This virus was reported for the first time from Bangalore, southern India. Several isolates referred to as tomato leaf curl virus Ban1, Ban3, Ban4, Ban5, and Kolar belong to this species. It contains a monopartite DNA-A genome, and a satellite DNA-β molecule was observed to be present with tomato leaf curl Bangalore virus (ToLCBV) infection. Pathogenicity of cloned DNAs has not been demonstrated.

Tomato Leaf Curl Karnataka Virus

This sap-transmissible virus has been isolated from Bangalore. DNA-A alone is infectious and produces typical leaf curl symptoms on tomato. Association of DNA-β has also been observed.

Tomato Leaf Curl Pune Virus

This is the most recently reported new virus from India, whose only DNA sequences are available in GenBank. Tomato leaf curl Pune virus (ToLCPV) contains a

DNA-β molecule, which is quite distinct from all other known ToLCVs.

In addition, viruses such as tomato leaf curl Bangladesh virus, tomato leaf curl Malaysia virus, and tomato leaf curl Joydebpur virus are also known to infect tomatoes in India.

Molecular Relationships among ToLCVs

Relationships among the type member of the five species (viz. ToLCBV, ToLCGV, tomato leaf curl Karnataka virus (ToLCKV), tomato leaf curl New Delhi virus (ToLCNDV), and ToLCPV) revealed that there is a great degree of diversity among the Indian tomato-infecting begomoviruses. To examine the diversity of the sequences, phylogenetic trees were generated of the 18 full-length DNA A sequences together with representative sequences present in GenBank (**Figure 3**). The trees constructed using either neighbor joining or most parsimonious methods for full-length sequences were all similar. Members of four different species are closely related and form a

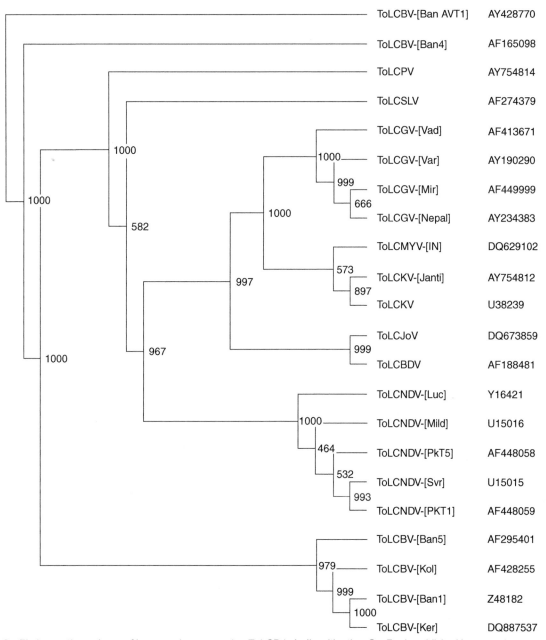

Figure 3 Phylogenetic analyses of begomoviruses causing ToLCD in India with other GenBank-published begomovirus sequences. Numbers at nodes indicate the bootstrap value out of 1000 replicates. The tree was generated with the PHYLIP programs using full-length DNA-A sequence data of Indian ToLCVs and other selected ToLCVs of the SE Asia. The full names of the viruses can be found in **Table 1**.

well-knitted cluster with the exception of ToLCMYV-[IN] which was grouped with ToLCKV-Janti and ToLCKV. Full-length DNA-A sequences of Indian ToLCVs formed five different clusters, which generally had 72–82% nucleotide identity between clusters. However, one cluster formed by ToLCKV, belonging to a species known to have arisen through recombination, shared up to 88.3% with ToLCBVs.

Four out of the five sequences typed as ToLCBV shared 94.2–99.5% nucleotide identity with each other. ToLCPV also fell into the ToLCBV group, but had only 88% nucleotide identity to previously published ToLCBVs. ToLCBDVs originated from Bangladesh are distinct and form a different cluster while tomato leaf curl Sri Lanka virus (ToLCSLV) is close to ToLCPV and ToLCBV.

Comparison of DNA-A Sequence

Representatives of five species share varied degree of identity among each other (66–88%). Minimum nucleotide identity (66%) was observed with ToLCGV and ToLCNDV while maximum identity of 88% was observed between ToLCBV and ToLCPV. Among the five species, ToLCKVs share more than 75% homology with all the species (**Table 2(a)**).

Comparison of DNA-B Sequence

GenBank database revealed the presence of four full-length DNA-B sequences associated with ToLCD in India. DNA-B of ToLCVs is highly conserved among each other in comparison to DNA-A. All the DNA-Bs are closely related (80–97%) (**Table 2(b)**). ToLCNDV, ToLCNDV-[Luc], and ToLCGV-[Var] have the same pairwise comparison profile all along their genomes.

Comparison of DNA-β Sequence

DNA-β molecules associated with ToLCVs share a less degree of homology as compared to DNA-A and DNA-B (**Table 2(c)**), indicating their uniqueness. So far nine full-length sequences have been reported with ToLCD in India and a wide range of 63–96% identity was observed among them. ToLCB-Varanasi share 96% with ToLCB-Aurangabad, followed by ToLCB-Jabalpur having 94% identity with ToLCB-Chinthapalli and by ToLCB-Coimbatore having 93% with ToLCBB. Two DNA-β sequences available from ToLCD-infected samples from Pune share 96% identity with each other.

Table 2   Percent identity (nucleotide) among begomoviruses causing TLCD in India

(a) DNA-A

	ToLCBV	ToLCGV-Var	ToLCKV	ToLCNDV-Svr	ToLCPV
ToLCBV	–	72	82	72	88
ToLCGV-Var		–	84	66	79
ToLCKV			–	76	83
ToLCNDV-Svr				–	72
ToLCPV					–

(b) DNA-B

	ToLCGV-Var	ToLCNDV-Luc	ToLCNDV 1	ToLCNDV 2
ToLCGV-Var	–	84	85	85
ToLCNDV-Luc		–	80	80
ToLCNDV 1			–	97
ToLCNDV 2				–

(c) DNA-β

	ToLCB-Var	ToLCB-Aur	ToLCB-Chi	ToLCB-Jab	ToLCB-Ban	ToLCB-Coi	ToLCB-Pun	ToLCB-Raj	ToLCB-Nd
ToLCB-Var	–	96	64	65	64	63	63	65	64
ToLCB-Aur		–	64	64	64	63	63	64	65
ToLCB-Chi			–	94	77	77	79	71	70
ToLCB-Jab				–	77	77	78	69	69
ToLCB-Ban					–	93	78	68	68
ToLCB-Coi						–	78	67	68
ToLCB-Pun							–	69	69
ToLCB-Raj								–	80
ToLCB-Nd									–

Comparison of Common Regions

The replication of the B-component by the Rep protein of the A-component is possible because of the existence of the so-called CR, a short stretch of ±200 nt that usually is highly conserved between the two molecules. The CRs of the DNA-A (CRAs) of A-components of several isolates of a particular species are very closely related, with 89–99% identity. The CRAs of different species share 52–77% identity with each other. Sequence identities between the CRs of ToLCGV-[Var] DNA-B and genomic components of ToLCNDV are very high: it is 83% identical to ToLCNDV-[Svr] DNA-A CR and 86% identical to ToLCNDV DNA-B CR, indicating that these sequences could pertain to isolates of the same species. However, the CRs of DNA-A and DNA-B of ToLCGV-[Var] are only 61% identical. The lowest sequence identity (52%) was observed between the CR of ToLCGV-[Var] DNA-A and the CR of ToLCNDV DNA-B. A multiple alignment of the CRs and inter-cistronic regions of the five species causing ToLCD in India revealed that all the isolates of ToLCNDV, ToLCGV, and ToLCKV have similar or very related iteron sequences (GGTGT-XX/X-GGAGT) while ToLCBV and ToLCPV have similar iteron sequences (GGTGG-XX/X-GGTGG) (**Table 3**). The exception is the ToLCNDV-Mld and ToLCNDV-Luffa isolates whose iteron sequence is GGCGT-CT-GGCGT. The major difference between the CRs of the DNA-A among these viruses is found in the spacer sequence between the two iterons that varies between 2 and 3 nt. It was also observed that a third identical iteron is present at the 5' end of the CRs. The CRs of DNA-A of ToLCGV-[Var], compared with DNA-B of ToLCGV-[Var], DNA-B of ToLCNDV, and DNA-A of ToLCNDV-Svr, share more than 85% identity in the first 80 nt of their CRs, whereas the region between the TATA box and the hairpin loop is below 50%.

Recombination

Tomato-infecting begomoviruses from India appear to have a great capacity to recombine. The amount of material that viruses may exchange ranges from small fragments of a few nucleotides to very large fragments of 2000 nt or more. When DNA-A was compared, no recombination was observed among ToLCGV-[Var] with other two isolates of the same species, ToLCGV-[Vad] or ToLCGV-[Mir]. Remarkably, a short possible recombination (150 nt) was observed between ToLCGV-[Var] DNA-A and ToLCNDV at the 3' end of AC1. With ToLCKV, ToLCGV share high nucleotide sequence identity for their IR (approximately the first 100 nt), the 5' end of AV1 (200 nt), and a long stretch of 1350 nt from the 5' end of AC3 to the 5' end of the IR (**Figure 4**). This indicates that, at these sites, recombination events possibly took place between these two tomato-infecting viruses or with a third unknown virus. Putative recombination sites among isolates of ToLCBV have been identified as AV1, AV2, AC1, and IR of the viral that may account of variability in strains/species.

Pseudo-Recombination and Synergism

Under natural conditions, mixed virus infections in a single plant possess biological and epidemiological implications. For the first time, synergism between two distinct species of begomoviruses infecting tomatoes in India was observed that results in an increase in viral DNA and symptom severity. Also, the occurrence of a more virulent pseudo-recombination between two distinct species may explain the sudden breakdown of resistant tomato cultivars and the development of epidemics in tomato-growing areas in India. Recently, the association of both ToLCGV-[Var] and ToLCNDV-Svr components in a single severely infected tomato plant under natural conditions has been detected. Also, based on coat protein (CP) gene sequences, associations of ToLCGV and ToLCBV, ToLCBV and ToLCKV, and ToLCBV and ToLCNDV have been observed in a single plant. The synergistic role of ToLCGV DNA-A and ToLCNDV-Svr DNA-A, resulting in a much higher level of ToLCNDV-Svr DNA-A in turn, helped to a more efficient replication of the B-components, and particularly the ToLCGV DNA-B component, resulting onto greater DNA accumulation in the systemically infected leaves, and consequently more severe symptoms. Since ToLCNDV-Svr and ToLCGV-[Var] belong to the same genus, infect the same hosts, and are transmitted by the same whitefly vector, they are therefore more likely to co-infect the same plants. The synergism between the two viruses will increase the amount of both the viruses in the systemically infected leaves and increase chances of transmission, if need be. As a consequence, doubly infected plants have a considerable potential as sources of inoculum for both viruses, and whiteflies feeding on such plants would, therefore, more easily acquire and transmit both viruses to virus-free plants. This information provides another source of geminivirus biodiversity.

Table 3 Rep binding sequences of ToLCVs from India

Virus species	DNA-A	DNA-B
ToLCBV	GGTGG-AAT-GGTGG	
ToLCGV	GGTGT-ATT-GAGT	GGTGT-CT-GGTGT
ToLCKV	GGTGT-ACT-GGAGT	
ToLCNDV-Svr	GGTGT-CT-GGAGT	GGTGT-CT-GGTGT
ToLCNDV-Mild	GGCGT-CT-GGCGT	
ToLCPV	GGTGG-AAC-GGTGG	

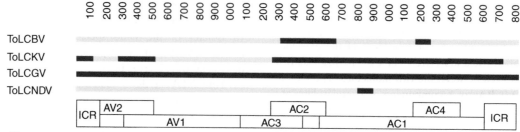

Figure 4 Diagrammatic representation of recombinant fragments between tomato leaf curl Gujarat virus-[Varanasi] and another three biologically characterized ToLCVs originating from India. Each line represents a linearized begomovirus genome (in the sense orientation starting from the origin of replication, in the CR) and each presence of red segments indicates homologous stretches of sequences to ToLCGV-[Var] at the strain level (>89%) within the genome of other Asian geminiviruses. Positions of the regions of the genome are represented at the top of the figure and a representation of a linearized geminivirus genome is represented at the bottom.

Exchange of genomic components of the members of two distinct species of begomoviruses causing ToLCD in India can form infectious pseudo-recombinants. Transcomplementation between ToLCNDV-Svr DNA-A and ToLCGV DNA-B resulted in a more severe symptom phenotype as compared to wild type. ToLCGV DNA-B can also transcomplement ToLCNDV-Mld (mild isolate) and ToLCKV. The viable nature of these pseudo-recombinants was attributed to the highly conserved nature of AC1 and CR and also to identical iteron sequences. A highly specific interaction between the *Rep* protein of ToLCNDV (mild and severe strains) with their cognate iteron sequences also demonstrated the intimate relationship between these elements and the consequence in terms of DNA accumulation and symptoms. This led to the concept that matching N-Rep and iterons is required for efficient replication and consequently a severe symptomatology, and it also was supportive of the species concept where most of the time, with the exceptions of the recombinant in these regions, these two elements are different between species. Although the perfect match between the N-Rep sequence and the iteron sequences is probably vital in most of the cases, it may not be fatal in that particular one. This suggests that there must be other cooperative factors that are shared between the Rep of ToLCNDV-Mild and ToLCGV-[Var] B component CR, to the point that they compensate a nonmatching iteron interaction.

Replication and Pathogenesis

A highly specific interaction between the *Rep* protein of ToLCNDV (mild and severe strains) and their cognate iteron sequences also demonstrated the intimate relationship between these elements and the consequence in terms of DNA accumulation and symptoms. Two strains of ToLCNDV, viz. severe and mild, share 94% sequence identity on the basis of symptoms on tomato and tobacco. The studies demonstrated that the amino acid at position 10 in Rep protein coupled with a change in the binding site

sequence may determine the replication of viral DNA. Change of Asp10 to Asn in Rep protein of the mild strain accompanied by exchange of the 13-mer binding site (making it identical to the severe strain) altered its replication, leading to increased accumulation of viral DNA. In addition, the modified mild strain could replicate heterologous strain DNA-B, indicating that the interaction of Rep protein with its binding site may be essential for replication of viral DNA.

Mutational analysis of ToLCNDV-Svr virion-sense ORFs has been carried out to assign the function. Plants inoculated with infectious DNA which contained deletions in AV2 developed very mild symptoms and accumulated only low levels of both single-stranded (ss-) and double-stranded (ds-) viral DNA, whereas inoculated protoplasts accumulated both ss- and dsDNA to wild-type levels, showing that AV2 is required for efficient viral movement. Mutations in the CP caused a marked decrease in ssDNA accumulation in plants and protoplasts while increasing dsDNA accumulation in protoplasts. The results demonstrated that multiple functions provided by AV2, BV1, and BC1 are essential for viral movement, and that changes in A-component virion-sense mRNA structure or translation affect viral replication.

The role of the movement protein (MP) and nuclear shuttle protein (NSP) in the pathogenicity of ToLCNDV, a bipartite begomovirus, has been elucidated using either potato virus X (PVX) expression vector or by stable transformation of gene constructs under the control of the 35S promoter in *N. tabacum*. No phenotypic changes were observed in any of the three species when the MP was expressed from the PVX vector or constitutively expressed in transgenic plants. Expression of the ToLCNDV NSP from the PVX vector in *N. benthamiana* resulted in leaf curling that is typical of the disease symptoms caused by ToLCNDV in this species. However, expression of NSP from PVX in *N. tabacum* and *L. esculentum* resulted in a hypersensitive response (HR), suggesting that the ToLCVDV NSP is a target of host defense responses in these hosts. The NSP, when expressed as a transgene under the control of the 35S promoter, resulted in necrotic lesions in expanded leaves that initiated from a point and then spread across the leaf. The necrotic response was systemic in all the transgenic plants. N-terminus of NSP is required for the HR. These findings demonstrate that the ToLCNDV NSP is a pathogenicity determinant as well as a target of host defense responses. The necrosis in transgenic tobacco plants is systemic, as it starts from a point on the fully emerged leaf and spreads over the lamina and to other leaves. Thus, ToLCNDV NSP is an avirulence determinant that interacts with the product of a resistance gene encoded by a host defense system, possibly an R gene product, triggering a host defense response involving an HR in *N. tabacum* and tomato.

Management

Until the late 1990s, the main control method employed against ToLCD was intensive use of insecticides targeted at viruliferous immigrant adult *B. tabaci* that spread ToLCVs into and within tomato crops. Identification of sources of resistance is difficult as the leaf curl disease syndrome is caused by different tomato begomoviruses. In the recent past, however, three high-yielding ToLCD-resistant tomato varieties have been developed that can be grown successfully with minimal insecticide use. Artificial screening through whiteflies has been carried out to identify resistant/tolerant varieties from different *Lycopersicon* spp., viz., *L. peruvianum*, *L. peruvianum* f. *glandulosum*, *L. peruvianum* f. *regulare*, *L. esculentum*, *L. chilense*, *L. hirsutum* f. *glabratum*, and *L. pimpinellifolium*. Gene governing resistance has also been mapped in tomato cultivar, H-24 (a derivative from *L. hirsutum* f. *glabratum*) on chromosome 11 against TYLCV and ToLCBV. In order to manage the deployment of this valuable resource, improve the efficacy with which further ToLCV-resistant material is screened, and investigate resistant-genotype/virus interactions, an improved understanding of the diversity and distribution of ToLCVs present will help in developing strategies for ecofriendly management of ToLCD. Agrobacterium-mediated inoculation of cloned DNAs will certainly provide better tool for identification of R genes in tomatoes in the future. Pathogen-derived resistance needs more attention in order to develop broad-spectrum resistance against Indian ToLCVs.

See also: Potato Viruses.

Further Reading

Chakraborty S, Pandey PK, Banerjee MK, Kalloo G, and Fauquet CM (2003) *Tomato leaf curl Gujarat virus*, a new begomovirus species causing a severe leaf curl disease of tomato in Varanasi, India. *Phytopathology* 93: 1485–1495.

Chatchawankanphanich O and Maxwell DP (2002) Tomato leaf curl Karnataka virus from Bangalore, India, appears to be recombinant begomovirus. *Phytopathology* 92: 637–645.

Chatterji A, Chatterji U, Beachy RN, and Fauquet CM (2000) Sequence parameters that determine specificity of binding of the replication-associated protein to its cognate in two strains of *Tomato leaf curl virus-New Delhi*. *Virology* 273: 341–350.

Chowda Reddy RV, Colvin J, Muniyappa V, and Seal S (2003) Diversity and distribution of begomoviruses infecting tomatoes in India. *Archives of Virology* 150: 845–867.

Green SK and Kalloo G (1994) Leaf curl and yellowing viruses of pepper and tomato: An overview. *Technical Bulletin No. 21*. Tainan, Republic of China: Asian Vegetable Research and Development Center.

Muniyappa V, Venkatesh HM, Ramappa HK, *et al.* (2000) Tomato leaf curl virus from Bangalore (ToLCV-Ban4): Sequence comparison with Indian ToLCV isolates, detection in plants and insects, and vector relationships. *Archives of Virology* 145: 1583–1598.

Padidam M, Beachy RN, and Fauquet CM (1995) Tomato leaf curl from India has a bipartite genome and coat protein is not essential for infectivity. *Journal of General Virology* 76: 25–35.

Vasudeva RS and Sam Raj J (1948) A leaf curl disease of tomato. *Phytopathology* 38: 364–369.

Tombusviruses

S A Lommel and T L Sit, North Carolina State University, Raleigh, NC, USA

Glossary

Modular evolution Evolution of viral genomes involving the combination of common genes/gene families.

Movement protein A plant viral protein that potentiates viral cell-to-cell movement through plasmodesmata.

Origin of assembly Unique RNA sequence and structure that specifically binds CP to initiate capsid assembly.

Polycistronic The characteristic of a given RNA containing more than one gene (open reading frame).

Quasiequivalence Capsid protein subunits arranged so that they are in somewhat equivalent environments with respect to their adjacent subunits.

(Silencing) suppressor Virally encoded product that inhibits a stage in the host RNA silencing pathway.

Subgenomic RNAs Less-than-full-length RNAs that are produced during replication, usually to express internal open reading frames.

Introduction

The *Tombusviridae* is a relatively large and diverse family of single-stranded, positive-sense, RNA plant viruses with common morphological, structural, molecular, and genetic features. Due to their small size and extremely high virus titer in experimental hosts, viruses in the family are particularly well characterized in terms of virion structure, replication, gene expression strategies, local and systemic movement, suppression of host gene silencing and associated satellite viruses, and defective interfering RNAs.

The family is constituted based on a unifying phylogenetic and biological feature. The RNA-dependent RNA polymerase of tombusviruses is highly conserved in terms of sequence identity, genomic structure, and gene expression and function. Biologically, tombusviruses share the property of being primarily soil transmitted, often without a biological vector, and accumulate to high levels in the roots of infected plants. Beyond these constants the family is remarkably diverse in biology, pathology, host range, and genome organization.

Many of the viruses now comprising the family *Tombusviridae* have been studied for a number of decades. Viruses like tomato bushy stunt virus (TBSV), carnation ringspot virus (CRSV), carnation mottle virus (CarMV), and cymbidium ringspot virus (CymRSV) were first described and virions purified and characterized in the 1940s. The taxonomic and phylogenetic relatedness of many of the viruses now comprising this family was not resolved until the mid-1980s, when genomes of these viruses were first cloned and sequenced.

Taxonomy, Phylogeny, and Evolution

The family *Tombusviridae* of plant viruses is composed of the genera *Aureusvirus*, *Avenavirus*, *Carmovirus*, *Dianthovirus*, *Machlomovirus*, *Necrovirus*, *Panicovirus*, and *Tombusvirus*, with more than 43 species and 15 tentative species recognized. Several genera are represented by many species whereas three genera are monotypic. The type species of the genus *Aureusvirus* is *Pothos latent virus*. This genus is quite similar to the genus *Tombusvirus* but is distinguished by having significantly different sized movement and silencing suppressor proteins. The type species of the monotypic genus *Avenavirus* is *Oat chlorotic stunt virus*. This species constitutes a separate genus because the genome organization is intermediate between those of the genera *Carmovirus* and *Tombusvirus*. Furthermore its capsid protein (CP) is significantly larger than those found in other genera whose CPs have a protruding (P) domain. *Carnation mottle virus* is the type species of the genus *Carmovirus*. This genus is distinguished by having two small proteins associated with virus movement and a CP with a P domain. The genus *Dianthovirus*, of which *Carnation ringspot virus* is the type species, has the most

dramatic taxonomic distinction: its genome is split into two segments. *Maize chlorotic mottle virus* is the type and monotypic species of the genus *Machlomovirus*. This genus is structurally quite similar to the genus *Panicovirus* but contains an additional open reading frame (ORF) at the 5′ end of its genome which nearly completely overlaps the polymerase. The genus *Necrovirus* is represented by *Tobacco necrosis virus A*. The genome organization and expression strategy are quite similar to those in the carmoviruses, but the necroviruses have the phylogenetically distinct CPs lacking a P domain. The genus *Panicovirus* is also represented by a single species, *Panicum mosaic virus*. This virus is distinguished by having carmovirus-like movement proteins (MPs), a CP without a P domain, as well as several accessory genes (**Figure 1**).

This family serves as an excellent example supporting the concept of modular evolution of viruses. The polymerase is the sole shared module binding the family. Within the limitations of the viral polymerase the family has taken great liberties with the arrangement and expression of the various genes on its polycistronic RNA. While it is true that all members of the family have a $T = 3$ icosahedral virion, different genera achieve this structure using two phylogenetically distinct CP modules. At this time, it appears that this family has acquired at least three phylogenetically different cell-to-cell movement modules. The sources of the virus suppressor of host gene silencing appear to be equally diverse. In addition, various species within a particular genus have acquired additional unique accessory modules, possibly for transmission by fungal vectors.

The minimal viral polymerases of the *Tombusviridae* lack any identifiable helicase motif, and do not have a nucleotide triphosphate binding motif. They do, however, have the canonical glycine–aspartate–aspartate (GDD) motif found in most RNA polymerases. All *Tombusviridae* polymerases are similarly expressed from an interrupted ORF by one of two translational regulatory mechanisms: terminator readthrough or -1 ribosomal frameshifting (**Figure 1**). The *Tombusviridae* polymerase is also phylogenetically related to that in the genus *Luteovirus* and, to a lesser extent, *Enamovirus* but not with the genus *Polerovirus* in the family *Luteoviridae*. Polymerases of tombusviruses belong to the Sindbis-like superfamily of RNA polymerases, but they are often categorized in a separate subgroup because of their reduced size and missing motifs.

The virions of tombusviruses are morphologically similar. They all form icosahedral $T = 3$ symmetry particles approximately 30–38 nm in diameter. Virions are formed from 180 copies of a single CP subunit ranging from 25 to 48 kDa in size. Virions parse into two morphological subclasses the first of which have rough or granular surfaces and the second smooth surfaces. The smooth particles, such as those found in the genera *Machlomovirus*, *Necrovirus*, and *Panicovirus*, have smaller CPs and lack the P domain found in the larger CPs (**Figure 2**).

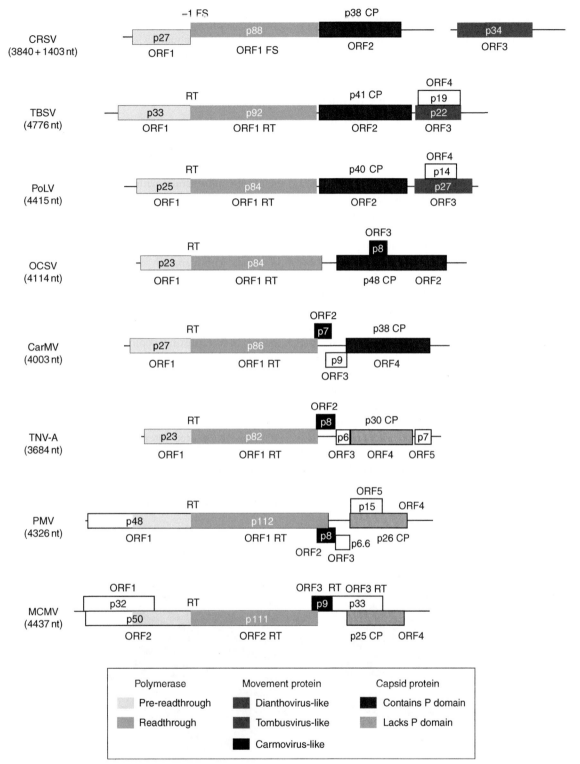

Figure 1 Genomic organization of the type species for each genus in the family *Tombusviridae*. The genomes are aligned vertically at the site of the polymerase readthrough event. Boxes represent known and predicted ORFs with the sizes of the respective proteins (or readthrough products) indicated, for example, p27 for 27 kDa protein. Similarly, colored boxes represent proteins with extensive sequence conservation and function. RT, translational readthrough of termination codon; −1 FS, -1 ribosomal frameshifting event; CP, capsid protein; CRSV, carnation ringspot virus (genus *Dianthovirus*); TBSV, tomato bushy stunt virus (genus *Tombusvirus*); PoLV, pothos latent virus (genus *Aureusvirus*); OCSV, oat chlorotic stunt virus (genus *Avenavirus*); CarMV, carnation mottle virus (genus *Carmovirus*); TNV-A, tobacco necrosis virus A (genus *Necrovirus*); PMV, panicum mosaic virus (genus *Panicavirus*); MCMV, maize chlorotic mottle virus (genus *Machlomovirus*).

There are also common elements in the viral genome structures of tombusviruses. They are all small, single-stranded, positive-sense RNA viruses, encoding four to six ORFs on a single RNA molecule, with the exception of the dianthoviruses, which have a split or bipartite genome. In addition, members of the *Tombusviridae* all rely on the generation of one or more subgenomic RNAs (sgRNAs) to express genes from the polycistronic viral genome.

Virion Structure

Although the underlying size and structure of the virions are similar, the members can be broadly subdivided into two groups based on the presence or absence of a P domain at the c-terminus of the CP. Genera containing P domains (*Aureusvirus, Avenavirus, Carmovirus, Dianthovirus,* and *Tombusvirus*) produce virions that are 32–38 nm in diameter and display a granular surface. These CP subunits range in size from 37 to 48 kDa. The X-ray crystal structures of TBSV and turnip crinkle virus (TCV; *Carmovirus*) revealed the discrete organization of each CP subunit. Generally, the CP subunit can be divided into four domains: the RNA-binding (R) domain is located at the N-terminus followed by the arm (a), shell (S), and P domains (66, 35, 67, and 110 amino acids, respectively, in TBSV). The R domain contains many basic residues and is found in the interior of the virion. Cryoelectron microscopy reconstructions have further revealed the presence of internal ordered cages of RNA intertwined with CP residues from the R domains beneath the virion surface. This internal scaffold may play a role in directing specific packaging of viral RNA and

formation of the icosahedral virions. The virion structure is primarily formed by the globular S domains (stabilized by a pair of Ca^{2+} per subunit) which are composed of two sets of four-stranded antiparallel β-sheets. The S domain is also the most highly conserved region of the CP subunit. There is a flexible hinge of five residues between the S and P domains that allows the CP subunit to adopt different configurations by varying the angle between the domains. This feature of the CP subunit overcomes the structural constraints imposed by the icosahedral morphology. P domains (containing antiparallel β-sheet structures in a jellyroll conformation, with one six-stranded β-sheet and one four-stranded β-sheet) of adjacent CP subunits dimerize to produce 90 projections leading to the granular surface texture. Genera lacking the P domain (*Machlomovirus, Necrovirus,* and *Panicovirus*) produce virions that are 30–32 nm in diameter with a smooth surface similar to viruses in the genus *Sobemovirus.* These shorter CP subunits range from 25 to 29 kDa in size.

Due to the quasi-equivalent nature of the CP subunits when arranged in an icosahedron, each subunit can take on one of three distinct conformations, termed A, B, and C (respectively, blue, red, and green in **Figure 2**). The A and B conformations differ in the angle between the S and P domains and their arrangement on the virion surface (A: fivefold axes, B: threefold axes). The C conformation differs from A and B in that their R/a domains intertwine to form an ordered internal structure around threefold axes of symmetry termed the β-annulus. The three different CP subunit conformations pack together as either AB or CC dimers in the virion particle. For TCV, virion assembly initiates with three CP dimers and viral RNA

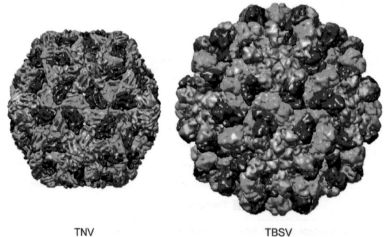

TNV   TBSV

Figure 2 Virion images of tombusviruses based on X-ray crystal structures. Left image is tobacco necrosis virus (TNV) at 2.25 Å. The capsid protein of this virus does not contain a protruding domain and is representative of species in the genera *Machlomovirus,* *Panicovirus,* and *Necrovirus.* Right image is TBSV at 2.9 Å. The capsid protein of this virus contains a protruding domain and is representative of species in the genera *Avenavirus, Aureusvirus, Carmovirus, Dianthovirus,* and *Tombusvirus.* For both images, the individual subunits are colored according to their various conformations: A (blue), B (red), and C (green). From Shepherd CM, Borelli IA, Lander G, *et al.* (2006) VIPERdb: A relational database for structural virology. *Nucleic Acids Research* 34 (Database Issue): D386–D389, VIPERdb.

followed by formation of the virion shell according to structural constraints imposed by CP/RNA interactions. In TCV, the origin of assembly RNA sequence (which specifically initiates the interaction with CP subunits) is contained within a 186 nt region at the 3′ end of the CP ORF. Assembly studies with TCV also showed that only RNAs equal to or smaller than 4.35 kbp in size could be packaged, suggesting very strict size limitations for icosahedral virions.

Virions of tombusviruses have an M_r of $\sim 8.2 – 8.9 \times 10^6$ and produce a single, well-defined band upon centrifugation with sedimentation coefficients ranging from 118 to $140 S_{20,w}$. Virion densities range from 1.34 to $1.36 \, g \, cm^{-3}$ in CsCl gradients. Virions are stable at acidic pH, but expand above pH 7 and in the presence of ethylenediaminetetraacetic acid (EDTA). Virions are resistant to elevated temperatures although thermal inactivation usually occurs above 80 °C. Due to the lack of a lipid membrane, virions are insensitive to organic solvents and nonionic detergents.

Most genera produce virions containing a single molecule of positive-sense, linear single-stranded RNA (ssRNA) ranging in size from 3.7 to 4.8 kb depending on the genus. The genomic RNA constitutes about 17% of the particle weight. The exception is the genus *Dianthovirus* where virions contain two genomic RNAs: a large RNA-1 of $\sim 3.9 \, kb$ and a smaller RNA-2 of 1.5 kb. The genomic RNAs lack a 5′ cap structure and the 3′ ends are not polyadenylated nor do they form tRNA-like structures. Satellite RNAs (satRNAs) of nonviral origin are associated with several genera. Satellite viruses (which code for their own CP) are also associated with several genera but these are readily distinguishable due to their significantly smaller dimensions. Both small RNAs modify the symptoms of their respective helper viruses.

Genome Organization and Replication Strategy

The dimensions of the icosahedral virions constrain the genome sizes of the various members of the family *Tombusviridae*. This forces the members to adopt several strategies for expression of their genetically compact, polycistronic genomes. Their unifying genomic feature is the presence of an interrupted ORF for the virally encoded RNA-dependent RNA polymerase at the 5′ end of the genome. The interruption is manifested as either an amber termination codon which may be read through (most members) or a −1 ribosomal frameshifting signal (genus *Dianthovirus* only). In either case, this leads to the direct translation of two ORFs with identical amino termini from the genomic RNA (**Figure 1**). These polymerase subunits share a high degree of sequence similarity within the family. Notably, they do not contain a helicase-type motif and the genomes do not encode a helicase-like ORF.

ORFs downstream of the polymerase are expressed via 3′ co-terminal sgRNAs with the resident ORFs produced with an assortment of translational strategies such as ribosomal scanning and frameshifting. Aside from the single CP ORF mentioned previously, the genomes of *Tombusviridae* also encode one or more MPs that are involved in cell-to-cell spread of the virus. These range in size from the 34–35 kDa version found in the genus *Dianthovirus* to the 22 kDa product found in the genus *Tombusvirus* and down to the 6–9 kDa pair found in the genus *Carmovirus*. In most cases, the MPs can potentiate the local movement of unencapsidated viral RNA but systemic infection generally requires encapsidation. The exception is for members of the genus *Tombusvirus* which can infect systemically without CP albeit at a reduced rate. The other prominent protein product with a clearly defined function is the genus *Tombusvirus* p19 protein which is a potent suppressor of post-transcriptional gene silencing (PTGS). p19 binds double-stranded short interfering RNAs (siRNAs) that are generated by a Dicer-like RNase as part of the host RNA-silencing response to viral infection. This binding prevents incorporation of the siRNAs into the RNA-induced silencing complex (RISC) to prevent further cleavage of viral RNAs. Although other members of the family may not encode a unique suppressor protein, it is known that other proteins can have suppressor activity aside from their primary functions. This is the case for TCV where the CP is the viral suppressor of PTGS. Various additional ORFs are also encoded with undefined functions. One other method utilized by members of the family *Tombusviridae* for gene expression is genome segmentation which is only employed by the genus *Dianthovirus* where RNA-2 is monocistronic and encodes the MP.

Since the genomic RNAs are positive sense, they are directly infectious as unencapsidated RNAs. However, the lack of a 5′ cap structure and a 3′ poly(A) tail precludes the normal circularization procedure employed by mRNAs for translation. Instead, a cap-independent mechanism involving a long-distance RNA–RNA interaction between sequence elements in the 5′ and 3′ noncoding regions has been demonstrated in TBSV and tobacco necrosis virus-A (TNV-A) and proposed in the case of red clover necrotic mosaic virus (RCNMV; genus *Dianthovirus*) RNA-1. The direct translation readthrough strategy ensures that the longer ORF (containing the catalytic site of the polymerase) is present in only 5–10% of the translation products since readthrough is inefficient. Translation of the virally encoded polymerase subunits is followed by localization to and proliferation of cellular membranes. The location of viral replication varies dependent on the particular virus and is specified by the pre-readthrough portion of the polymerase. In all cases studied to date, the polymerase anchors on membranes, causing proliferations. The advent of yeast-based replication systems for TBSV, cucumber necrosis virus (CNV; genus *Tombusvirus*), and CymRSV has provided valuable insights into the host components utilized by the viral replication complex.

Once the polymerase has been assembled, it binds to the 3' terminus to initiate minus-strand synthesis. If full-length copies are generated, these serve as templates for synthesis of progeny genomic RNAs. This synthesis is highly asymmetric with positive-stranded RNA accumulating at much higher levels than negative-stranded RNA. Occasionally, positive-strand synthesis will be terminated prematurely which leads to the formation of templates for plus-strand sgRNA synthesis. This premature termination is controlled by long-distance RNA–RNA interactions which occur in *cis* (for instance, in TBSV) or in *trans* (for instance, RCNMV). Generally, sgRNAs are produced later in the infection cycle and behave as monocistronic mRNAs. However, the sgRNAs also demonstrate various translational strategies such as ribosomal scanning in the case of p19 expression which is nested within the p22 ORF in TBSV. In this case, the p22 start codon is suboptimal and occasionally read through by ribosomes before translation at the optimal p19 start codon.

During replication, defective interfering (DI) RNAs (which are not packaged into virions) arise in viruses belonging to some genera. These DI RNAs are derived from the viral genome and are replicated to very high titers which lead to their interfering with viral genomic replication levels. DI RNAs have been used extensively to delimit critical sequence elements required for RNA replication.

Transmission, Host Range, and Epidemiology

The natural host range of a given species is relatively narrow; however, the experimental host range tends to be broad. Members can infect either monocotyledonous or dicotyledonous plants, but no species can infect both. Natural infections can be limited to or are often concentrated in the root system. Many species induce a necrosis symptom in the foliar parts of the plant. All species are readily transmitted by mechanical inoculation and by host vegetative propagation and some may be transmitted by contact and through seeds. A number of species are readily detected in the soil, surface waters, rivers, lakes, and even the ocean. Transmission by the chytrid fungi in the genus *Olpidium* and beetles have also been reported for members of several genera. Most, if not all, members can be transmitted through the soil either dependent on, or independent of, a biological vector. This rather unusual soil and water mode of transmission is unique to this plant virus family and is based on the unusually robust constitution of the virion.

Geographical distribution of particular species varies from wide to restricted. Most species occur in temperate regions although legume-infecting carmoviruses and a tentative member of the genus *Dianthovirus* have been recorded from tropical areas.

In nature, viruses in the family *Tombusviridae* can be spread in a variety of ways, including seed and pollen transmission, mechanical transmission, vegetative propagation and grafting, growth in infected soil, and by vectors such as fungi, thrips, and beetles. As mentioned above, one hallmark of members of the family *Tombusviridae* is the amount of virus present in the roots of infected plant tissue, and the ease of transmission through the soil, whether that transmission is dependent on a biological vector or not. For viruses that can be transmitted by fungi, there is a considerable amount of specificity required between the virus and the fungal vector. For example, CNV (*Tombusvirus*), melon necrotic spot virus (*Carmovirus*), and cucumber leaf spot virus (*Aureusvirus*) are transmitted by the root-inhabiting chytrid fungus *Olpidium bornovanus*, whereas TNV (*Necrovirus*) is transmitted exclusively by *Olpidium brassicae*.

Pathogenesis

For the species in which it has been studied, the virus titers and pathogenic effects are at least as high, if not higher, in the root system of the host as in the other parts of the plant. This is consistent with the robust nature of the virions and that viruses in the family are transmitted through the soil.

Many tombusviruses generate DI RNAs that can affect pathogenesis. DI RNAs are essentially deletion mutants of the viral genome which are generated by errors in replication such as rearrangement or recombination. Several species such as TBSV and TCV produce DI RNAs during viral infection. For TBSV the presence of DI RNAs results in attenuation of symptoms, whereas for TCV the presence of DI RNAs intensifies symptoms. Many species, including both TBSV and TCV, are also known to harbor satRNAs. These small RNAs are generally not derived from viral sequences, but they depend on the helper virus for replication and packaging. As with DI RNAs, they can attenuate TBSV or intensify symptoms (TCV).

The origin of nonhomologous satRNAs is unclear at present, but like DI RNAs, satRNAs can be generated from recombination events with other satRNAs, the viral genome, DI RNAs, or host sequences. Both DI RNAs and satRNAs have been useful molecular tools for studying viral replication and recombination. Since both satRNAs and DI RNAs are dependent on the host or parent virus for replication, they must have retained any sequence or structural signals required for recognition by the viral polymerase, and therefore can help determine exactly what those signals are.

Recombination can facilitate viral evolution as well as repair of viral genomes, and so it is an important factor in the virus life cycle and it can have profound effects on pathogenesis. In the family *Tombusviridae*, recombination in the genera *Carmovirus* and *Tombusvirus* has been shown to repair damaged or deleted 3′ ends of virus-associated RNAs, as well as generate new satRNAs or DI RNAs. Recombination can occur between both homologous and nonhomologous sequences, probably by a copy-choice mechanism in which the viral replicase jumps from one template to another during the replication process.

Plants infected with tombusviruses display a distinctive cytopathology, observed by electron microscopy as dense-staining features. These features are known as multivesicular bodies, and are believed to be sites of viral replication as well as accumulation of exceedingly high concentrations of progeny virions. To date, cell biology studies have shown that species can replicate on, remodel, and proliferate membranes of peroxisomes, mitochondria, and the cortical ER. There appears to be no correlation between a specific organelle membrane used for replication and taxa.

See also: Carmovirus; Luteoviruses; Machlomovirus; Necrovirus; Vector Transmission of Plant Viruses.

Further Reading

Fauquet CM, Mayo MA, Maniloff J, Desselberger U,, and Ball LA (eds.) (2005) *Virus Taxonomy: Eighth Report of the International Committee on Taxonomy of Viruses*. San Diego, CA: Elsevier Academic Press.

Knellor EL, Rakotondrafara AM, and Miller WA (2006) Cap-independent translation of plant viral RNAs. *Virus Research* 110: 63–75.

Martelli GP, Gallitelli D, and Russo M (1988) Tombusviruses. In: Koenig R (ed.) *The Plant Viruses, Polyhedral Virions with Monopartite RNA Genomes*, vol. 3, pp. 13–72. New York: Plenum.

Miller WA and White KA (2006) Long-distance RNA–RNA interactions in plant virus gene expression and replication. *Annual Review of Phytopathology* 44: 447–467.

Morris TJ and Carrington JC (1988) Carnation mottle virus and viruses with similar properties. In: Koenig R (ed.) *The Plant Viruses, Polyhedral Virions with Monopartite RNA Genomes*, vol. 3, pp. 73–112. New York: Plenum.

Nagy PD and Pogany J (2006) Yeast as a model host to dissect functions of viral and host factors in tombusvirus replication. *Virology* 344: 211–220.

Nagy PD and Simon AE (1997) New insights into the mechanisms of RNA recombination. *Virology* 235: 1–9.

Robbins MA, Reade RD, and Rochon DM (1997) A cucumber necrosis virus variant deficient in fungal transmissibility contains an altered shell protein domain. *Virology* 234: 138–146.

Rochon D, Kakani K, Robbins M, and Reade R (2004) Molecular aspects of plant virus transmission by olpidium and plasmodiophorid vectors. *Annual Review of Phytopathology* 42: 211–241.

Rossman MG and Johnson JE (1989) Icosahedral RNA virus structure. *Annual Review of Biochemistry* 58: 533–573.

Russo M, Burgyan J, and Martelli GP (1994) Molecular biology of *Tombusviridae*. *Advances in Virus Research* 44: 381–428.

Scholthof HB (2006) The tombusvirus-encoded P19: From irrelevance to elegance. *Nature Reviews Microbiology* 4: 405–411.

Shepherd CM, Borelli IA, Lander G, *et al.* (2006) VIPERdb: A relational database for structural virology. *Nucleic Acids Research* 34 (Database Issue): D386–D389.

White KA and Nagy PD (2004) Advances in the molecular biology of tombusviruses: Gene expression, genome replication, and recombination. *Progress in Nucleic Acid Research and Molecular Biology* 78: 187–226.

Tospovirus

M Tsompana and J W Moyer, North Carolina State University, Raleigh, NC, USA

Glossary

Ambisense genome Viral RNA genome with open reading frames in both the viral- and viral complementary (vc) sense on the same genome segment.

Envelope Membrane-like structure that packages genome segments.

IGR The intergenic region is the untranslated, A-U rich region found between the two open reading frames on the S and M RNA segments.

Negative sense genome Viral RNA genome that codes for proteins in the vc sense. Transcription of vc mRNA is required for translation of viral proteins.

Nucleocapsid Viral RNA encapsidated in the nucleoprotein.

Protoplast Plant cell lacking its cell wall.

RNP Ribonucleoprotein complex consisting of the viral RNA genome segment, nucleoprotein, and a small number of polymerase molecules.

Virion Quasispherical structure containing the viral genome and bounded by a membrane-like envelope.

History

Diseases now known to be caused by tomato spotted wilt virus (TSWV) were first reported in 1915 and were shown to be of viral etiology by 1930. This taxon of plant viruses was categorized as a monotypic virus group consisting of a single virus (TSWV) until the report of impatiens necrotic spot virus (INSV) in 1991.

Thus, most of the characteristics which define the genus *Tospovirus* were obtained through investigation of TSWV even after the discovery of additional viruses in the genus (**Table 1**). Biological investigations beginning in the 1940s revealed a virus that had an unusually large host range and occurred in nature as a complex mixture of phenotypic isolates. However, it was one of the least stable viruses and most difficult plant viruses to mechanically transmit. Although the enveloped virions were observed in the 1960s, molecular characterization and elucidation of the genome organization were not completed until the early 1990s. The virus was shown to be vectored by thrips in the 1930s and later transmitted in a persistent manner. Thrips were demonstrated to be a host for replication of the virus and that replication was required for transmission in the early 1990s. Later it was recognized that limited, localized replication may occur in thrips that does not result in the thrips becoming viruliferous. Advances in gene function and cellular biology have been limited due to the absence of a robust *in vitro* plant or thrips cell culture system, and lack of an efficient reverse genetics system. However, limited progress has been made utilizing gene expression systems and classical viral genetics.

Taxonomy and Classification

Tospoviruses constitute the only genus of plant-infecting viruses in the family *Bunyaviridae*, however, these viruses share many molecular characteristics typical of other

Table 1   List of *Tospovirus* species[a, b, c]

Tospovirus species	Abbreviation
Groundnut bud necrosis virus (Peanut bud necrosis virus)	GBNV
Groundnut ringspot virus	GRSV
Groundnut yellow spot virus (Peanut yellow spot virus)	GYSV
Impatiens necrotic spot virus	INSV
Tomato chlorotic spot virus	TCSV
Tomato spotted wilt virus	TSWV
Watermelon silver mottle virus	WSMoV
Zucchini lethal chlorosis virus	ZLCV
Tentative *Tospovirus* species	
Capsicum chlorosis virus (Gloxinia tospovirus) (Thailand tomato tospovirus)	CACV
Chrysanthemum stem necrosis virus	CSNV
Iris yellow spot virus	IYSV
Groundnut chlorotic fan-spot virus	GCFSV
Physalis severe mottle virus	PhySMV
Watermelon bud necrosis virus	WBNV

[a]http://www.ncbi.nlm.nih.gov/ICTVdb.
[b]Whitfield AE, Ullman DE, and German TL (2005) Tospovirus–thrips interactions. *Annual Review of Phytopathology* 43: 459–489.
[c]Synonyms are indicated inside parentheses.

members of this virus family. They have an enveloped virion containing the viral genome which is distributed among three RNA segments that replicate in a manner consistent with that of other negative strand viruses. All three segments have highly conserved, complementary termini resulting in a pan-handle structure and genes with functions similar to those of viruses in other genera are located in similar locations on the genome. However, the genome organization is distinct from the other genera. The small (S) and middle (M) segments each encode two genes in opposite or ambisense polarity.

Classification of a *Tospovirus* population as a distinct species (virus) is based upon the similarity of sequence between the nucleocapsid genes of the respective viruses. This is in contrast to the system used to differentiate viruses in other genera which traditionally relied on serological neutralization of infectivity or other biological properties (hemagluttination) mediated by the glycoproteins. Tospovirus isolates with greater than 90% nucleotide similarity in the nucleocapsid gene are classified as isolates of the same species (virus). Serologically related isolates with 80–90% sequence identity are subjectively classified as strains or as distinct species depending on other criteria. Isolates with less than 80% identity are classified as distinct species.

Geographic Distribution

TSWV, the type member of the tospoviruses is found worldwide in temperate regions in association with its thrips vector. The wide host-range of TSWV and its thrips vector is consistent with the geographic distribution. Other tospoviruses have more well-defined distribution. For example, GBNV, WBNV, and WSMoV, that are transmitted by *Thrips palmi*, a thrips species found only in the subtropics are only known to occur in Southeast Asia. Another anomaly is INSV. While INSV is reported to occur around the world, it is almost entirely limited to greenhouse-grown floral crops.

Host-Range and Virus Propagation

TSWV has one of the most diverse host-ranges of any plant-infecting virus. The virus infects over 925 plant species belonging to 70 botanical families, both monocots and dicots. In addition, TSWV infects approximately ten thrips species. Important economic plants susceptible to TSWV include tomato, potato, tobacco, peanut, pepper, lettuce, papaya, and chrysanthemum. Other tospoviruses (e.g., IYSV) have much narrower host-ranges and thus the broad host-range of TSWV is not characteristic of the genus. These viruses can be transmitted mechanically or by their thrips vector, but are not transmitted transovarially, by plant seeds or pollen. Purified RNA preparations are not infectious. There are no robust plant or insect culture systems for tospoviruses. However, plant and insect protoplasts have been successfully inoculated.

Virion Properties

Tospovirus virions are quasispherical, enveloped particles 80–120 nm in diameter (**Figure 1**). Two viral coded glycoproteins, G_N and G_C, are embedded in the viral envelope and form surface projections 5–10 nm long. Ribonucleoprotein (RNP) particles consisting of the viral RNA encapsidated in the nucleoprotein (nucleocapsid), and a small number of polymerase molecules are contained within the envelope. Nucleocapsids are pseudocircular due to noncovalent bonding of the complementary RNA termini. Intact virions as well as carefully prepared RNPs retrieved from sucrose or $CsSO_4$ gradients are infectious. There are several reports that TSWV and INSV isolates, while infectious, are defective for virion formation.

Genome Properties

Tospoviruses have a single-stranded, tripartite RNA genome with segments designated as L, M, and S in order of decreasing size (**Figure 2**). The termini of each

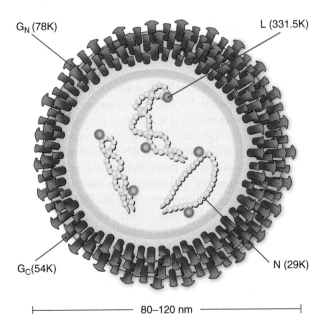

G_N (78K)   L (331.5K)

G_C(54K)   N (29K)

|—————— 80–120 nm ——————|

Figure 1 Tospovirus quasispherical virion particles. The S, M, and L RNA genomic segments are encapsidated by the nucleoprotein, are in association with L protein molecules, and form pan-handle structures due to the complementarity of their 5′ and 3′ ends. The glycoproteins G_N and G_C are embedded within the viral envelope.

of the RNA segments consist of an eight nucleotide sequence (5′ AGAGCAAU 3′) that is strictly conserved among all tospoviruses. The remaining untranslated region at the termini also has a high degree of complementarity. Base pairing at the termini between the inverted complementary sequences supports a pan-handle structure that most likely serves as a promoter for replication. The L RNA is 8.9 kbp and codes for the L or RdRp protein in the viral complementary (vc) sense (**Figure 2**). The M and S RNAs are in ambisense orientation. The M RNA is 4.8 kbp and codes in the viral sense for the nonstructural protein NSm and for the G_N/G_C precursor glycoprotein in the vc sense. The S RNA is 2.9 kbp and codes in the viral sense for the nonstructural protein NSs and the nucleocapsid protein in the vc sense (**Figure 2**).

TSWV M and S RNA IGRs have variable lengths, are A–U rich, and are the most hypervariable regions of the genome. The 5′ and 3′ ends of the IGRs are conserved, separated by variable sequences, deletions, and insertions. In addition, highly conserved sequences are embedded within the S RNA IGR. A 33 nucleotide (nt) duplication occurring in the S RNA IGR of some isolates has been correlated with loss of competitiveness in mixed infections of isolates with and without the duplication. A 31nt conserved sequence, with significantly higher GC-content compared to the remaining S RNA IGR, has also been found in some TSWV isolates. The IGRs of the M and S segments have high inclination for base pairing thought to be involved in initiation and termination of transcription. There is speculation that the termination of transcription is dictated by a conserved nucleotide sequence (CAACUUUGG) in the center of the S and M RNA IGR or that it is due to a secondary structure highly stabilized in the 31nt region referred to above.

Full length molecules of the M and S RNAs are found in infected tissue and purified virions in both the viral and vc sense (approximate ratio of 10:1), consistent with ambisense segments from other viruses. Defective interfering RNAs (DIs) associated with attenuated symptom expression and increased replication rate are also frequently observed. DIs from the L ORF in TSWV infected tissue are the result of and associated with attenuated infectivity. Deletions, frameshift, and nonsense mutations in the G_N/G_C ORF have been shown to interfere with thrips transmissibility and virion assembly. Recently, frameshift and nonsense mutations with unknown effect have also been identified in the N ORF. The formation of DIs is favored by repeated mechanical passage in certain plant hosts, high inoculum concentration, and low temperatures. Available evidence supports the hypothesis that secondary structure rather than sequence is the primary determinant of the site of deletion. There is also a high frequency of DIs that maintain the original reading frame resulting in translation of truncated proteins whose existence was confirmed in nucleocapsid preparations.

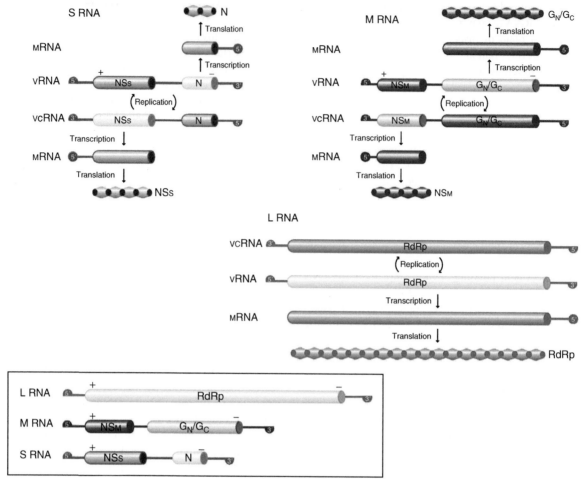

Figure 2 Tospovirus ambisense genome organization (inset in figure) and expression strategy. Positive (+) and negative (−) sense ORFs are dark and light shaded tubes respectively. Proteins from the S and M segments are translated from subgenomic mRNAs which are capped with 10–20 nt of non-viral orgain at the 5′ end.

Protein Properties

The 331.5 kDa L protein encoded by the L RNA has been identified as the putative RdRp, through sequence homology with other members of the *Bunyaviridae* and identification of sequence motifs characteristic of polymerases. RdRp activity has been associated with detergent-disrupted TSWV virion preparations. The 33.6 kDa NSm protein encoded by the M RNA has been shown to induce tubule structures in plant protoplasts and insect (*Spodoptera* and *Trichoplusia*) cells. Induction of tubules in plants, ability to change the size exclusion limit of plasmodesmata, an early expression profile and complementation of cell-to-cell and systemic movement in a movement-defective tobacco mosaic virus vector is evidence that NSm is the TSWV movement protein and that it supports long-distance movement of viral RNAs. In thrips, NSm does not aggregate into tubules, indicating that this protein might not have any function in the vector's life cycle. It is also known that NSm specifically interacts with the N protein, the

At-4/1 intra- and intercellular trafficking plant protein and binds single-stranded RNA in a sequence-nonspecific manner. An NSm homolog is absent in the animal infecting *Bunyaviridae*. The 127.4 kDa G_N/G_C precursor glycoprotein also coded by the M RNA contains a signal sequence that allows its translation on the endoplasmic reticulum. Proteolytic cleavage of the polyprotein does not require other viral proteins. The M_rs of G_N and G_C is 78 kDa and 54 kDa respectively. Evidence for the involvement of the glycoproteins in thrips transmission is provided by: (1) their interaction with proteins of the thrips vector, (2) their association with the insect midgut during acquisition, (3) the loss of thrips transmissibility of envelope-deficient mutants, (4) the presence of a glycoprotein sequence motif that is characteristic for cellular attachment domains and (5) the observation that only reassortants with the M RNA of a thrips-transmissible isolate rescue thrips transmissibility. Specifically, G_N is involved in virus binding and/or entry in thrips midgut cells, whereas G_C is a possible fusion protein playing a significant role in pH-dependent virus entry. It is also believed that these proteins are implicated in virion assembly. The NSs protein encoded on the S RNA is 52.4 kDa and accumulates to high levels as loose aggregates or paracrystalline arrays of filaments. NSs has RNA silencing suppressor activity, affects symptom expression in TSWV-infected plants, and is not present in the mature virus particle. The N protein, also encoded by the S RNA, ranges in size from 29 to 3 kDa depending on the virus. This protein encapsidates the viral RNA segments, is highly abundant, and is the predominant protein detected in serological assays. A 'head-to-tail' interaction of the nucleoprotein N terminus (aa 1–39) with the C terminus (aa 233–248) results in multimerization.

Replication

Replication of viral RNA and assembly of virions occurs in the cytoplasm of both plant and insect cells. Tospovirus replication, however, has mainly been described based on plant infection. Upon entry into the plant cell, the virus loses its membrane and releases infectious nucleocapsids into the cytoplasm. In thrips cells, infection by tospoviruses is accommodated by binding of the viral surface glycoproteins to a host cell receptor(s) (possibly a 50 kDa and/or a 94 kDa protein). This is followed by release of infectious nucleocapsids into the cytoplasm, through fusion between the viral and thrips membranes possibly initiated by low pH. Depending on the concentration of N protein, the viral RNA is either transcribed or replicated. At low N concentrations, the polymerase transcribes mRNAs that are translated into the virus proteins. Translation of proteins from the S and M ambisense RNAs occurs from subgenomic mRNAs (**Figure 2**). The S and M subgenomic mRNAs are capped at the 5' terminus with 10–20 nucleotides of nonviral origin indicating that tospoviruses utilize a cap-snatching mechanism to regulate transcription. Leader sequences of alfalfa mosaic virus have also been detected as caps of TSWV mRNA in mixed infections of the two viruses. The TSWV transcriptase has a reported preference for caps with multiple base complementarity with the viral template. Upon increase of N protein concentration, the polymerase switches its mode to replication with the viral RNA serving as the template. Replicated viral RNAs form RNPs that can presumably associate with the NSm protein for movement through plasmodesmata to adjacent plant cells through tubular structures. Alternatively, RNPs form new virions by associating with the glycoproteins and budding through the Golgi membranes. Virions are initially double-membraned, but soon coalesce and form groups of virions with a single membrane surrounded by another membrane.

Pathogenicity and Cytopathology

Tospoviruses are noted for the severity of the diseases they cause in plants. Symptoms are highly variable, depending on the virus, the virus isolate, the host plant, time of the year, and environment, and are thus of little diagnostic value. Chlorosis, necrosis, ring or line patterns, mottling, silvering, and stunting often appear on inoculated and systemically infected leaves. Systemic invasion of plants is frequently nonuniform. Stems and petioles may exhibit necrotic lesions. Observed symptoms often mimic disease and injury caused by other biotic and abiotic stresses. Infection of younger plants results in severe stunting and high mortality rates. TSWV has been shown to affect more severely *Datura*, *Nicotiana*, and *Physalis* plants under a specific temperature regime (daytime, 29 + 2 °C, nighttime, 24 + 3 °C). The effect of tospovirus infection on thrips has been controversial, due to the confounded effects of plant host, virus, and environment on the insect vector and the genetic variability of thrips and virus populations. TSWV infection of *F. occidentalis* provided evidence that thrips exhibit an immune response to the virus. Recent work with TSWV-infected *F. fusca* reared on infected foliage indicated a direct effect of the virus on thrips resulting in reduced fitness. The same study showed that the plant infection status and the TSWV isolate have also an effect on the insect, explaining the variable results obtained from independent studies of virus pathogenicity on the insect vector.

Tospoviruses induce characteristic cytopathic structures that are host and virus-isolate dependent. In addition to virions, inclusions of viroplasms consisting of the NSs or N protein may be abundant in the cytoplasm. NSs may aggregate in loose bundles (e.g., TSWV) or in highly ordered paracrystalline arrays (e.g., INSV). Excess N protein occurs in granular electron dense masses. NSs and nucleocapsid protein inclusions have been observed in infected plant and insect cells. NSm protein induces tubule structures in plant protoplasts and insect cells.

Transmission and Epidemiology

Tospoviruses are transmitted from plant to plant by at least ten thrips species in the genera *Frankliniella*, *Scirtothrips*, and *Thrips*. Among the more common vectors are *Frankliniella occidentalis*, *F. fusca*, *F. schultzei*, *F. intonsa*, *F. bispinosa*, *Thrips palmi*, *T. setosus*, and *T. tabaci*. Thrips feed on the cytoplasm of plant cells. The contents of infected cells are ingested and the virus is transported along the lumen of the digestive tract to the midgut, the primary binding and entry site into the insect cells. The brush border of the midgut lumen is the first membrane barrier that the virus encounters. The virus replicates in the midgut and crosses the basement membrane into the visceral muscle cells. The virus subsequently enters the primary salivary glands. It has been hypothesized that the virus moves from the midgut to the salivary glands through infection of ligament-like structures, or when there is direct contact between membranes of the visceral muscles and the primary salivary glands during the larval stages of development. A less plausible hypothesis is that the virus infects the salivary glands after entry and circulation in the hemocoel. Viral inoculum is introduced into plants in the insect saliva coincident with feeding on the plant by adult thrips.

The process of successful acquisition occurs only by larvae and acquisition rates decrease as larvae develop, affecting adult vector competency. Vector competency is also determined by the thrips' feeding preference on a particular host, the uniformity of distribution of virus in plant cells, the rate of virus replication in the midgut, and the extent of virus migration from the midgut to the visceral muscle cells and the salivary glands. In some instances the virus can be acquired by adult thrips and infects midgut cells, but is unable to spread further possibly due to the formation of an age-dependent midgut barrier (e.g., basal lamina). Research has shown the existence of thrips transmitters with detectable levels of virus, nontransmitters with detectable virus, and nontransmitters with no detectable virus, supporting multiple sites for vector specificity between tospoviruses and thrips. Evidence for replication of the virus in the insect vector is based on the accumulation of NSs and the visualization of viral inclusions in midgut epithelial cells, muscle cells, and the salivary glands. Although the virus is maintained transtadially throughout the life of the insect, there is no evidence for transovarial transmission. Thus, each generation of thrips must acquire the virus during the larval stages.

The primary dispersal of tospoviruses is by adult thrips and dissemination of infected somatic tissue in vegetatively propagated crops. These viruses are thought to move long distances in thrips carried by wind currents. They may also survive in commercial agricultural systems in weeds that serve as a bridge between crops. Infected summer weeds (e.g., in NC *I. purpurea*, *I. hederacea*, *M. verticillata*, *A. palmeri*, *C. obtusifolia*, *R. scabra*, *Ambrosia artemisiifolia* L., *Polygonum pensylvanicum* L., *and Chenopodium album* L.) are the principal source for spread of TSWV to winter annual weeds, from which the virus is spread to susceptible crops in spring. Secondary spread within a crop can only occur in crops that concomitantly support virus infection and reproduction of the vector as only the larval stage can acquire the virus for transmission. Transmission through plant seed and pollen has not been conclusively demonstrated. The emergence of these viruses as serious pathogens in crops has been attributed to the increased prevalence of *F. occidentalis* as an agricultural pest on a worldwide basis.

Genetics and Evolution

The knowledge base for genetics and evolution of tospoviruses has been derived almost exclusively from TSWV. TSWV has a characteristic ability to adapt to new or resistant hosts and to lose phenotypic characters following repeated passage in experimental hosts, especially *Nicotiana benthamiana*. The virus occurs in plants as a heterogeneous mutant population with one or two predominant haplotypes and 9–21 rare haplotypes. Recent research shows that natural TSWV variants evolve in nature through recombination, random genetic drift, and mutation. Intergenomic recombination is important for the genesis and evolution of ancestral TSWV lineages. Genetic drift during thrips transmission and mutation concurrent with virus population growth, shape the genetic architecture of the most recently evolved lineages. The existence of single viral strains as mutant populations and recombination in ancestral viral lineages arm TSWV with a unique genetic reservoir for causing disease and spreading in epidemic proportions in nature. Additional research at the species level supports a distinct TSWV geographical structure and the occurrence of species-wide population expansions. TSWV is also known to use reassortment of genome segments to adapt to resistant hosts under specific laboratory conditions. The determinants of adaptation to resistance in tomato and pepper have been mapped to the M and S RNAs respectively. Little is known about the thrips–tospovirus coevolution and the genetic diversity of the thrips vector itself. The altered status of *Thrips tabaci* as a TSWV vector is one of the very few likely examples of coevolution between tospoviruses and their insect vector.

Detection and Diagnosis

Tospoviruses have certain unique biological properties that are useful for diagnosis. These viruses can be mechanically transmitted by gently rubbing inoculum on plants dusted with carborundum. *Nicotiana glutinosa* L., *Chenopodium quinoa* Wild., and garden petunia give characteristic lesions that progress as spots or concentric zones, and sometimes as lethal necrosis. Tospoviruses can also be identified by electron microscopy of leaf-dip preparations on thin sections of infected plants. Additional techniques for identification are based on the enzyme-linked immunosorbent assay (ELISA) using polyclonal and monoclonal antibodies, and detection of viral-specific nucleic acids using ribo- and cDNA-probes. The reverse transcription-polymerase chain reaction (RT-PCR) is the most powerful and commonly used technique for detecting small amounts of tospovirus RNA. Real time RT-PCR has been successfully used to detect and quantify TSWV in leaf soak and total RNA extracts from infected plants and thrips.

RT-PCR with degenerate primers can detect five distinct tospovirus species. Tissue selection and sampling strategy are critical factors in TSWV diagnosis and detection regardless of the technique. Because, TSWV titer varies throughout the plant and does not spread uniformly throughout plants that are 'systemic' hosts, sampling strategies should be validated in each situation.

Prevention and Control

Tospoviruses cause significant economic losses annually, due to suppressed growth, yield, and reduced quality. These viruses can be partially managed in well-defined cropping systems such as glasshouses by obtaining uninfected plant propagules, implementing a preventative thrips control program in high risk areas, together with constant monitoring of production areas for thrips and infected plants. However, these strategies are costly and require intensive management. Control in field crops is problematic due to the array of external sources of inoculum. Vector control is generally ineffective against the introduction of virus from external sources, due to thrips' high fecundity, ability to develop insecticide resistance, and to infest many TSWV-susceptible crops. Some measure of control can be achieved using thrips-proof mesh tunnels in the field and reflective mulches. Cultural practices such as utilization of virus-tested planting stock, careful selection of planting dates, removal of cull piles and weeds, rotation with nonsusceptible crops, prevention of planting TSWV-susceptible crops adjacent to each other, reduced in-field cultivation to avoid movement of thrips from infected sources, can reduce the spread of tospoviruses. In peanuts, higher plant density, planting from early until late May and application of selected insecticides have reduced the incidence of TSWV. In flue-cured tobacco, early-season treatment with activators of plant defenses and insecticides have also significantly reduced TSWV incidence.

Deployment of resistant cultivars has provided benefits in only three of the crops infected by TSWV. Although little is known about the benefits of host resistance against most of the tospoviruses, TSWV defeated nearly every resistance gene deployed against it in many crops. Single-gene resistance is available for TSWV in a limited number of tomato (*Sw-5*) and pepper (*tsw*) cultivars. Naturally occurring, resistance-breaking isolates of TSWV have been recovered from pepper and tomato cultivars containing their respective resistance genes. A co-dominant cleaved amplified polymorphic sequence (CAPS) marker has been developed for TSWV marker-assisted selection in pepper. 'Field' resistance has been reported for some peanut varieties. Progress has been made in understanding the genetic basis of the ability of TSWV to overcome single gene resistance by mapping determinants to specific segments of the TSWV genome and characterizing the selection process. Pathogen-derived resistance utilizing the N and NSm genes has been effective in some greenhouse and field tests; however, isolates have been obtained that overcome nucleocapsid mediated resistance. Best suppression of TSWV epidemics has been achieved with the integrated use of moderately resistant cultivars, chemical, and cultural practices. The impact of these viruses on agricultural production is in large part due to the absence of durable forms of resistance in the affected crops or other highly effective control measures.

Future Perspectives

The last decade has been characterized by exciting developments in our understanding of tospovirus molecular biology, evolution, and virus–host relationships. Further progress in understanding replication and gene function requires the development of efficient reverse genetics, plant and insect culture systems for tospoviruses. In addition, effective management of these viruses will depend on a deeper understanding of thrips' genetic diversity, virus–thrips coevolution and the changes in viral and thrips population dynamics upon exertion of specific selective forces. Such understanding can be acquired only through integrated research at the interface of virology, entomology, and ecology.

Further Reading

Adkins S (2000) Tomato spotted wilt virus-positive steps towards negative success. *Molecular Plant Pathology* 1: 151–157.

Best RJ (1968) Tomato spotted wilt virus. In: Smith KM and Lauffer MA (eds.) *Advances in Virus Research*, pp. 65–145. New York: Academic Press.

de Avila AC, de Haan P, Kormelink R, *et al.* (1993) Classification of tospoviruses based on phylogeny of nucleoprotein gene sequences. *Journal of General Virology* 74: 153–159.

German TL, Ullman DE, and Moyer JW (1992) Tospoviruses: Diagnosis, molecular biology, phylogeny, and vector relationships. *Annual Review of Phytopathology* 30: 315–348.

Goldbach R and Peters D (1994) Possible causes of the emergence of tospovirus diseases. *Seminars in Virology* 5: 113–120.

Prins M and Goldbach R (1998) The emerging problem of tospovirus infection and nonconventional methods of control. *Trends in Microbiology* 6: 31–35.

Sin SH, McNulty BC, Kennedy GG, and Moyer JW (2005) Viral genetic determinants for thrips transmission of tomato spotted wilt virus. *Proceedings of the National Academy of Sciences, USA* 102: 5168–5173.

Tsompana M, Abad J, Purugganan M, and Moyer JW (2005) The molecular population genetics of the tomato spotted wilt virus (TSWV) genome. *Molecular Ecology* 14: 53–66.

Ullman DE, Meideros R, Campbell LR, Whitfield AE, and Sherwood JL (2002) Thrips as vectors of tospoviruses. In: Plumb R (ed.) *Advances in Botanical Research*, pp. 113–140. London: Elsevier.

Ullman DE, Sherwood JL, and German TL (1997) Thrips as vectors of plant pathogens. In: Lewis TL (ed.) *Thrips as Crop Pests*, pp. 539–565. London: CAB International.

Whitfield AE, Ullman DE, and German TL (2005) Tospovirus–thrips interactions. *Annual Review of Phytopathology* 43: 459–489.

Tymoviruses

A-L Haenni, Institut Jacques Monod, Paris, France
T W Dreher, Oregon State University, Corvallis, OR, USA

Glossary

Icosahedron A solid having 20 faces and 12 vertices.
Open reading frame Region of a genome that can be translated into a protein.
Phylogeny The complete evolutionary history of group of animals, plants, bacteria, viruses, etc.
Pseudoknot RNA structure formed by base-pairing between nucleotides within the loop subtending a stem and nucleotides outside of this loop.
sgRNA Subgenomic RNA, derived from a viral RNA genome during replication, serves as mRNA for the expression of genes that are translationally silent in the genomic RNA.
Vein-clearing Yellowing along the leaf veins.

The Family and Its Distinguishing Features

The members of the family *Tymoviridae* are presented in **Table 1**. Like the family itself, the genus *Tymovirus* derives its name from the type species, *Turnip yellow mosaic virus*. Turnip yellow mosaic virus (TYMV) was first isolated in 1946 and is by far the most intensively studied member of the family. Indeed, TYMV is one of the best-characterized plant viruses. The family was recently created in recognition of the close relationships between the genera *Tymovirus*, *Marafivirus*, and the founding member of newly created genus *Maculavirus*, grapevine fleck virus (GFkV). Poinsettia mosaic virus (PnMV) is currently a family member unassigned to a genus. Complete genome sequences are available for the type species of each of the three genera of the *Tymoviridae* and for PnMV.

The members of the family *Tymoviridae* are characterized by their icosahedral, nonenveloped, ~29 nm virions, that can readily be visualized by negative-staining electron microscopy (EM). Infections produce a characteristic mixture of filled, infectious virions and empty or near-empty capsids. Regular surface features, representing prominent peaks formed by pentamers and hexamers of coat protein (CP) molecules, are evident by EM (**Figure 1**).

The genomes of members of the *Tymoviridae* are composed of a single positive-stranded RNA generally 6.0–6.5 kb long, although the genome of GFkV is 7.5 kb

long. Subgenomic (sg) RNAs of 1 kb or less are associated with CP expression. All *Tymoviridae* genomes have a distinctive skewed nucleotide composition that is rich in C residues (32–50%). The unifying characteristic of the genome design of all *Tymoviridae* members is the presence of a long open reading frame (ORF) that covers most of the genome and encodes the replication polyprotein with identifiable domains: methyltransferase, papain-like proteinase, helicase, and RNA-dependent RNA polymerase (RdRp) in order N- to C-terminal (**Figure 2**). The CP ORF is situated downstream of the polyprotein ORF, to which it is fused in the marafivirus and PnMV genomes. Close familial relationships are easily discerned from alignments of the sequences of the RdRp and CP genes.

Most viruses in the family have narrow host ranges. Many of the tymoviruses have been isolated from noncrop hosts and have thus far not presented major disease threats to crops. Marafiviruses are associated with significant crop losses, perhaps resulting from their more effective transmission by flying insects.

Tymoviruses

Properties and Distinguishing Characteristics

Although virtually all information about tymoviruses has been derived from studies on TYMV, it is considered to be generally applicable to the other members of the genus. Virions of tymoviruses contain genomic RNAs 6.0–6.7 kb long and produce a single 3′ collinear sgRNA less than 1 kb in length encoding the ~20 kDa CP. Their genomes include a distinctive 16-nt-long 'tymobox' sequence, and the genomes of most species possess a 3′ tRNA-like structure (TLS) that can be efficiently esterified with valine. The genomes encode three ORFs (**Figure 2**), two of which almost completely overlap: the overlapping protein (OP) and replication protein (RP) ORFs, which begin at AUG codons separated by only 4 nt. Tymoviruses replicate in all major tissues of their host plants, and accumulate to high levels (more than $0.1 \, \mathrm{mg \, g^{-1}}$ leaf tissue). Both filled and empty or near-empty particles accumulate. They are readily transmitted mechanically under laboratory conditions, and are spread over limited distances by beetle vectors in nature. Infection produces mosaic symptoms and a distinctive cytopathy that is evident upon EM observation by the appearance of small vesicles on the surface of chloroplasts, together with vacuolation and clumping of chloroplasts.

Table 1   Members of the family *Tymoviridae*

Species name	Virus abbreviation	Complete sequence accession no.
Genus *Tymovirus*		
Andean potato latent virus	APLV	
Belladonna mottle virus	BeMV	
Cacao yellow mosaic virus	CYMV	
Calopogonium yellow vein virus	CalYVV	
Chayote mosaic virus	ChMV	AF195000
Clitoria yellow vein virus	CYVV	
Desmodium yellow mottle virus	DYMoV	
Dulcamara mottle virus	DuMV	AY789137
Eggplant mosaic virus	EMV	J04374
Erysimum latent virus	ErLV	AF098523
Kennedya yellow mosaic virus	KYMV	D00637
Melon rugose mosaic virus	MRMV	
Okra mosaic virus	OkMV	EF554577
Ononis yellow mosaic virus	OYMV	J04375
Passion fruit yellow mosaic virus	PFYMV	
Peanut yellow mosaic virus	PeYMV	
Petunia vein banding virus	PetVBV	
Physalis mottle virus	PhyMV	Y16104
Plantago mottle virus	PlMoV	AY751779
Scrophularia mottle virus	SrMV	AY751777
Turnip yellow mosaic virus	TYMV	J04373, X16378, X07441
Voandzeia necrotic mosaic virus	VNMV	
Wild cucumber mosaic virus	WCMV	
Tentative members		
Anagyris vein yellowing virus	AVYV	AY751780
Nemesia ring necrosis virus	NeRNV	AY751778
Genus *Marafivirus*		
Bermuda grass etched-line virus	BELV	
Maize rayado fino virus	MRFV	AF265566
Oat blue dwarf virus	OBDV	U87832
Citrus sudden death-associated virus	CSDaV	AY884005, DQ185573
Tentative members		
Grapevine asteroid mosaic-associated virus	GAMaV	
Grapevine rupestris vein feathering virus	GRVFV	AY706994
Genus *Maculavirus*		
Grapevine fleck virus	GFkV	AJ309022
Tentative member		
Grapevine red globe virus	GRGV	
Unassigned virus		
Poinsettia mosaic virus	PnMV	AJ271595

Capsid Structure

Virions of tymoviruses are highly stable ~29 nm, $T = 3$ icosahedra formed by 180 copies of the single CP, arranged as 12 pentamers and 20 hexamers. These groupings form the vertices with fivefold and sixfold symmetry that constitute the surface peaks visible upon high-resolution EM observation (**Figure 1**). Intersubunit stabilization is provided primarily by hydrophobic protein–protein contacts, allowing the formation of stable shells that appear to be devoid of RNA. These empty or near-empty capsids can account for about one-third of the particles present in infected tissues and are readily identifiable by internal staining in negative-contrast EM (**Figure 1**). They sediment as the 'top component' at 45–55S in CsCl density gradients, and readily separate from the 'bottom com-ponent', the infectious 110–120S virions that contain the genomic RNA. Minor components of intermediate density contain a range of subgenomic-size RNAs that can be translated to yield CP. Roughly equimolar amounts of genomic and sgRNAs are encapsidated, but the precise disposition of sgRNA in the various particles has not been determined. Like many other viral CPs, tymoviral CPs fold into eight-stranded, β-barrel, jelly-roll structures. However, unlike some other CPs, there is no positively charged domain at the N terminus for interaction with RNA and charge neutralization during packaging. Charge neutralization is thought to be provided by polyamine molecules (mostly spermidine) associated with the RNA. The crystal structures of empty and infectious particles have been determined for TYMV and physalis mottle virus. Consis-

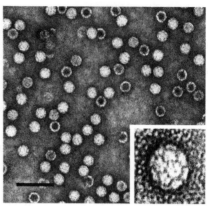

Figure 1 Tymovirus particle structure. A cartoon of the arrangement of subunits into pentamers and hexamers to build the $T = 3$ icosahedron is shown at left. A negative-contrast electron micrograph of virions and 'empty' particles of belladonna mosaic tymovirus is shown at right, with the high-magnification inset showing the prominent surface structure. Scale = 100 nm. Image courtesy Dr. D.-E. Lesemann. Reproduced from Mayo MA, Dreher TW, and Haenni A-L (2000) Genus *Tymovirus* In: van Regenmortal MHV, Fauquet CM, Bishop DHL, *et al.* (eds.) *Seventh Report of the International Committee on Taxonomy of Viruses.* New York: Academic Press, with permission from Elsevier.

tent with the dominance of protein–protein interactions in particle stabilization, empty and infectious particles have very similar structures.

Packaging signals in tymoviral RNAs have yet to be identified, though RNA recruitment may involve conserved hairpins in the 5′ untranslated region (UTR) and interaction of C-rich segments of the RNA with CP at low pH. Localized areas of low pH (5–6) have been postulated to arise at the surfaces of photosynthesizing chloroplasts. Tymoviral replication occurs in characteristic vesicles that form at the chloroplast surfaces (see below), and capsid formation is most active near these vesicles, suggesting a possible coupling between replication and encapsidation during infection. Despite much effort, there has been no success in developing a cell-free packaging system. These efforts have focused on the possible role of 'artificial top component' (ATC) capsids as decapsidation and encapsidation intermediates. ATCs are protein shells devoid of RNA that can be made from infectious virions by treatments such as freeze–thawing and exposure to high pH and pressure. They are similar in structure to infectious virions, but lack a capsomere of six CP molecules that is ejected during the treatment, allowing RNA escape.

Genome Organization

Tymoviral genomic RNAs possess a 5′-m7GpppG cap and terminate at the 3′ end in -CC(A) except for dulcamara mottle virus (DuMV) that is in most cases part of a TLS. The typical tymoviral TLS is just over 80 nt long, has a distinctive pseudoknot close to the 3′ terminus, and is a close structural mimic of cellular tRNA^{Val}. A valine-specific anticodon is present, and the 3′ terminus can be specifically aminoacylated with valine. The valylated RNAs of some, though not all tymoviruses, can form a

tight complex *in vitro* with the GTP-bound form of translation elongation factor eEF1A. Most molecules of encapsidated RNA lack the terminal A residue in the -CC(A) end, which is thought to be added by the host tRNA-specific CCA-nucleotidyltransferase at the beginning of infection. The main roles of the TYMV TLS are thought to be (1) as a 3′ enhancer of translation initiation, (2) as a regulator of the onset of RNA replication by modulating access by the polymerase to the 3′ end, and (3) in maintaining an intact-CCA 3′ end. A minority of tymoviral genomic RNAs have 3′ UTRs that lack the typical valine-specific TLS (DuMV, erysimum latent virus, and nemesia ring necrosis virus, NeRNV).

Tymoviral RNAs act as cap-dependent messenger RNAs. Of the three ORFs, two (OP and RP) are expressed directly from the genomic RNA, whereas the third (CP) is expressed from the 5′-capped sgRNA (**Figure 2**). Although the sequence contexts around the OP and RP initiation codons vary, 4 nt always separates these closely spaced AUGs. The RP ORF covers most of the viral genome, and encodes the ∼200 kDa RP, the only viral protein essential for supporting viral RNA replication in protoplasts. This ORF encodes discrete domains (**Figure 2**) that are discernable by virtue of sequence relationships to similar domains encoded by a wide variety of positive-stranded RNA viral genomes. Particularly the RdRp domain is well conserved and appropriate for phylogenetic comparisons. Closely related, yet distinct, tymoviral genomes have RdRp domains with about 70% nucleotide sequence identity. The RP is translated as a precursor protein that subsequently undergoes maturation cleavage catalyzed by the papain-like proteinase domain. The single known cleavage separates the RdRp domain from the remainder of the RP; thus, for TYMV, the precursor p206 is processed to yield p141 and p66.

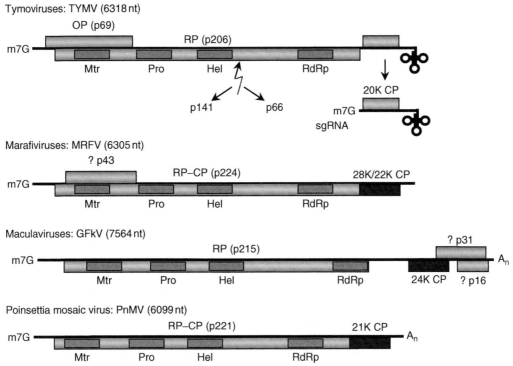

Figure 2 Genomes of the type members of the genera of the family *Tymoviridae* and of PnMV. All genomes have a 5′-m7GpppG cap, but the 3′ terminal structures vary. TYMV has a valine-specific tRNA-like structure (cloverleaf), GFkV and PnMV have a poly(A) tail, while MRFV has neither. The known or predicted (?) expressed ORFs are indicated with the molecular weight (K, kDa) of the predicted protein. The RPs all possess methyltransferase (Mtr), papain-like proteinase (Pro), helicase (Hel), and RdRp domains. The TYMV RP is cleaved as indicated, and similar cleavages are expected for the other viruses. The TYMV OP is the viral movement/RNAi suppressor protein. The TYMV CP is expressed from a sgRNA, and all other CPs are probably also expressed from sgRNAs (not shown). The 28 kDa MRFV CP (whose true size is likely to be closer to 25 kDa) may be produced by proteolytic cleavage from the RP–CP fusion protein, and an analogous event may occur with PnMV. Expression of the MRFV p43 and GFkV p31 and p16 has not yet been validated.

The OP ORF almost entirely overlaps the RP ORF. Its length and the sequence of the encoded protein are highly variable. OPs range between 49 and 82 kDa, and have only 25–40% amino acid sequence identity between the most closely related pairs of viruses. OP expression is needed for establishment of infection and for spread of the virus in plants. The OP is a suppressor of the host RNA silencing antiviral response and is also believed to be the viral movement protein. It is a substrate for ubiquitin-dependent degradation by the proteasome. Long-distance movement of the virus in plants requires expression of the CP.

The CP ORF initiates close to the end of the RP ORF, sometimes even a little upstream. CP sequences are variably conserved among tymoviruses, with 36–86% amino acid sequence identity between the most closely related pairs of viruses.

Tymoviral genomes possess a highly conserved sequence, the 16 nt tymobox (–GAGUCUGAAUUGCUUC–) with small variations in some viruses, especially wild cucumber mosaic virus, just upstream of the CP ORF. The tymobox and its associated 'initiation box' sequence CAA(U/G) positioned 8 or 9 nt downstream of the tymo-box are believed to serve as the core elements of the sgRNA promoter. The tymobox overlaps with the 3′ end of the RP ORF, resulting in the presence of the tripeptide – ELL – near the C terminus of all RPs.

Replication Cycle

The ultrastructural changes that reflect viral replication activity and that result in the distinctive pathology of the chloroplasts have been particularly well described for TYMV. As is typical of positive-stranded RNA viruses, the replication cycle is completed entirely in the cytoplasm, and RNA replication occurs in association with membranes, specifically the chloroplast outer membrane. Vesicles 50–80 nm in diameter form as invaginations of the two outer membranes of chloroplasts. They form before the appearance of virions, which are initially found close to vesicle clusters and later throughout the cytoplasm; in some cases, empty capsids accumulate in nuclei. The RdRp has been localized to zones on chloroplasts that are rich in vesicles, and this localization depends on protein–protein interaction between the proteinase and RdRp domains of the mature RPs p141 and p66, respectively.

RNA replication occurs via the production of full-length minus strands, whose synthesis in TYMV is directed by the 3′ terminal −CCA serving as promoter and initiation site. No other minus-strand promoter elements have been identified in the 3′ UTR. As mentioned above, the tymobox appears to function with the initiation box as the core of the promoter directing sgRNA synthesis by internal initiation on the minus strand. The sgRNA of TYMV has a m7GpppA 5′ terminus.

Studies on the functions of viral proteins and of *cis*-acting sequences in the genomic RNA have been greatly facilitated by the use of 'infectious clones', that is, molecularly cloned cDNA versions of the genome from which infectious RNA can be derived.

Infection and Transmission

Tymoviruses cause chlorotic mosaic, vein-clearing, and mottling symptoms, generally without strong stunting. Host ranges are mostly narrow, and to date, no tymoviruses infecting monocot plants have been isolated. Tymoviruses are transmitted over limited distances by chrysomelid beetles, and some are weakly seed-transmissible. They can also be transmitted mechanically.

Phylogenetic Relationships and Species Demarcation

In the past, serology using antisera raised against intact virus was the main criterion in classifying tymoviruses and in distinguishing between species. This is no longer the most convenient approach, however, and it has in fact been shown that distinct viruses whose genomic and CP sequences have identities less than 80% and 90%, respectively, can appear to be serologically identical. This seems to be due to similar or identical dominant epitopes within otherwise distinct CPs.

The study of relationships among tymoviruses and to other viruses will undoubtedly in the future rely on the interpretation of sequences derived from the genomic RNA. The phylogenetic trees based on the sequences of the CP and the RdRp (**Figure 3**) show similar relationships, both among the tymoviruses and to the other members of the family *Tymoviridae*. This suggests that recombination between these coding regions has not been a strong evolutionary force among the *Tymoviridae*. However, the genome of NeRNV, which has a tobamoviral-type TLS capable of aminoacylation with histidine, indicates that recombination can occur and shape tymoviral genome evolution. The complete genome sequences are available for 13 tymoviruses or tentative tymoviruses (**Table 1**).

Alignments based on the RP protein sequences indicate that the capilloviruses, carlaviruses, trichoviruses, and potexviruses, all members of the *Flexiviridae*, are the next most closely related virus groups. More distant

sequence relationships indicate that members of the family *Tymoviridae* belong to the alpha-like virus group.

Marafiviruses

Properties and Distinguishing Characteristics

Three current members of the genus *Marafivirus* are viruses principally infecting monocots (grasses), while another infects citrus and the two tentative marafiviruses were isolated from grapevine. Plant host ranges are narrow. The genomes are 6.3–6.8 kb long, with a single ORF encoding an RP/CP fusion protein covering most of the genome. Two forms of the same CP (∼25 and ∼21 kDa) are found in virions. A single 3′-collinear sgRNA (<1 kb) encoding the smaller form of the CP is produced during infection. A slightly modified form of the tymobox sequence, termed 'marafibox', is present in the genome near the junction between the RdRp- and capsid-coding sequences. Marafiviruses are phloem-limited and not mechanically transmissible. The grass-infecting marafiviruses are transmitted by cicadellid leafhoppers in a persistent-propagative manner, involving virus replication in the insect.

A distinctive feature of the marafiviruses is the presence of two CPs in the virions, a major CP ∼21 kDa form and a minor N-terminally extended 22–28 kDa form. The latter is thought to arise by proteolytic release from the polyprotein, whereas the former is expressed from the sgRNA. The production of this sgRNA is probably under the control of the marafibox, a 16 nt sequence (−CA(A/G)GGUGA AUUGCUUC−) very closely related to the tymobox, and the −CA(A/U)-initiation box located 8 nt downstream. The roles of the two forms of CP are unclear.

Virion Structure, Genome Organization, Replication Cycle, and Phylogenetic Relationships

Marafiviruses produce virions and empty capsids that are similar to those produced by tymoviruses when observed by negative-staining EM. However, due to the phloem-limited replication of marafiviruses, the yield of virus is low, limiting the usefulness of EM for viral identification. Complete genome sequences are available for four confirmed or tentative marafiviruses (**Table 1**).

Marafivirus genomic RNAs appear to be 5′-capped but they lack the 3′-TLS that is distinctive of tymoviruses. Apart from maize rayado fino virus (MRFV), marafivirus genomes possess a 3′ poly(A) tail. Almost the entire genome of marafiviruses is devoted to encoding a large 224–240 kDa polyprotein, which includes replication-associated and CP domains (**Figure 2**). The relationships between the CP and RdRp domains of the marafiviruses and other members of the family *Tymoviridae* are indicated in **Figure 3**. The polyprotein is believed to be proteolytically processed

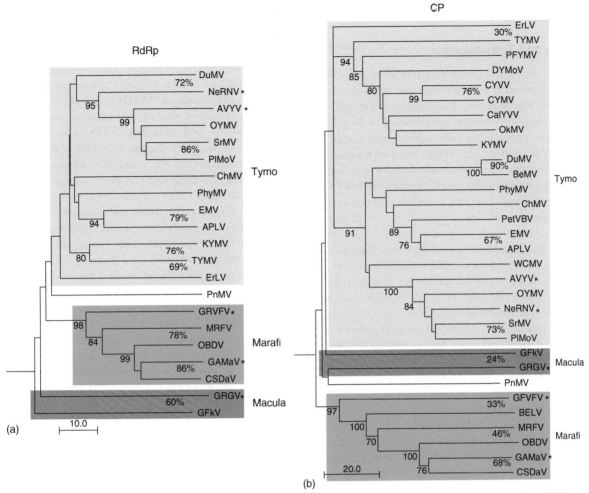

Figure 3 Phylogenetic trees showing the relationships between members of the family *Tymoviridae* based on neighbor-joining amino acid alignments of (a) the core RdRp domain and (b) the CP sequences. Bootstrap values (%) are indicated at well-supported nodes, and percent identities taken from pairwise alignments are indicated between selected sequences. Branch lengths reflect the number of amino acid differences. Viruses from each of the three genera are separately shaded as marked. Viruses with tentative taxonomic status are marked with an asterisk.

by the viral proteinase at two sites, between the helicase and RdRp domains as in tymoviruses, and between the RdRp and CP domains. Although the only experimentally mapped polyprotein cleavage site in a member of the family *Tymoviridae* is that occurring in TYMV, amino acid sequence alignments have discerned probable cleavage sites occurring immediately downstream of Gly-Gly or Gly-Ala dipeptides at the helicase–RdRp and RdRp–CP junctions in marafivirus polyproteins.

At the present state of relative paucity in marafivirus genome sequences, the taxonomic implications of evident variation in genome organization are uncertain. In addition to lacking a poly(A) tail, MRFV is the only marafivirus possessing a significant overlapping ORF, encoding a putative 43 kDa protein, whose expression *in vivo* has not been verified. Related but strongly interrupted coding sequences have been discerned overlapping the methyltransferase domains of other marafiviral genomes and, based on

sequence similarities to tymoviral OP, it has been postulated that these sequences represent variously degenerate versions of the tymoviral OP. The loss of OP during evolution may explain the phloem limitation of marafiviruses. Citrus sudden death-associated virus (CSDaV) is the only marafivirus with an additional ORF near the 3′ end of the genome. This ORF almost completely overlaps the CP ORF and could encode a 16 kDa protein.

The three grass-infecting marafiviruses, MRFV, OBDV, and Bermuda grass etched-line virus, are serologically related, but are not serologically cross-reactive with tymoviruses or more distantly related viruses. Despite the indications of genome variability mentioned above, the marafiviruses do form a phylogenetic group with interspecies relationships similarly close to those between the various tymoviruses (**Figure 3**). Sequence relationships among the RP and CP ORFs clearly link the marafiviruses to the other members of the *Tymoviridae*. Like the tymoviral

CPs, marafivirus CPs lack a cluster of basic amino acids at the N terminus.

Infection and Transmission

The best-studied marafiviruses are MRFV and OBDV. They cause stunting and chlorotic leaf spots or streaks, general leaf discoloration, or enations on leaf and stem veins. Cytological symptoms are restricted to the phloem and adjacent parenchyma cells, with clearly discernible hyperplasia and hypertrophy. These viruses are not mechanically transmissible or transmissible through seed, but are vectored in nature by leafhoppers. Significantly, MRFV and OBDV are maintained for long periods and through moults in leafhopper vectors, and transmission can occur from viruliferous vectors over a period of several weeks. OBDV replicates in the aster leafhopper (*Macrosteles fascifrons*), and the transmission behavior of MRFV is also consistent with replication in the cicadellid leafhopper (*Dalbulus maidis*). MRFV is commonly associated and co-transmitted with maize stunt spiroplasma or maize bushy stunt mycoplasma in causing economically significant outbreaks of corn stunt syndrome. OBDV can also be co-transmitted with a mycoplasma. The presence of virus in the leafhopper vectors is not associated with detectable symptoms or loss of reproductive fitness.

The most economically significant marafivirus is probably CSDaV. This virus is believed to be the causative agent of the citrus sudden death syndrome, which has had devastating effects on orange trees in Brazil. Preliminary evidence suggests transmission by aphids. The grapevine-infecting tentative marafiviruses, grapevine asteroid mosaic-associated virus and grapevine rupestris vein-feathering virus, have no known insect vector.

Maculaviruses

Properties and Distinguishing Characteristics

Recent sequencing of the complete genome of GFkV supported the creation of the new genus *Maculavirus*. The distinguishing characteristics were the length of the GFkV genome, the absence of a tymobox/marafibox, the lack of a fused RP/CP ORF as present in marafivirus genomes, and the lack of phylogenetic clustering with either tymoviruses or marafiviruses. Grapevine red globe virus (GRGV) is a tentative member of the genus. Both viruses infect only *Vitis* species (grapevine and relatives). In common with marafiviruses, these maculaviruses are restricted to the phloem and are not sap transmissible. No insect vectors have been identified. As for all other members of the family *Tymoviridae*, maculaviruses produce ~29 nm particles with prominent surface features observable by EM. Top component and infectious bottom component particles are produced. GFkV infections are cytologically distinct from marafivirus infections through

the presence of severely modified mitochondria ('multivesiculate bodies').

The genome of GFkV (7564 nt) is the longest among the members of the family *Tymoviridae*, and has the highest cytosine content (49.8%). The arrangement of RP and CP ORFs is similar to that of tymoviruses (**Figure 2**). The RP ORF encodes a 215 kDa protein with replication-associated domains that present sequence relatedness to the tymoviral and marafiviral RPs. The amino acid sequence of the RdRp domain shows the closest relationship to sister genera (59% and 67% identities with the RdRps of a tymovirus and marafivirus, respectively).

The 24.5 kDa GFkV CP has 23–31% sequence identity with CPs of tymoviruses and marafiviruses, and like other CPs of members of the family *Tymoviridae* it lacks a cluster of basic amino acids at the N terminus. The GFkV genome encodes two additional proteins from ORFs that partially overlap the CP ORF in the 3′ region of the genome. No functions for these putative proteins are predicted from sequence comparisons. Close relationship between GFkV and GRGV is rather weakly supported by sequence comparisons (**Figure 3**).

The GFkV genome is probably capped at the 5′ end and has a poly(A) tail at the 3′ end, in common with most marafiviruses and PnMV. It is the only known member of the *Tymoviridae* to lack a tymobox- or marafibox-like sequence. Nevertheless, two or more sgRNAs, ~1.3 and ~1.0 kb in length, are present in infected tissues. These RNAs appear to be packaged in both top- and bottom-component particles, in the latter case apparently together with the genomic RNA.

Poinsettia Mosaic Virus

Properties and Distinguishing Characteristics

PnMV possesses properties characteristic of both the tymoviruses and marafiviruses. The ~28 nm virus particles have the EM structure typical of the *Tymoviridae* and separate into typical top components containing the sgRNA and infectious bottom components. Like tymoviruses, PnMV is produced in high yield during infection, is not phloem-limited, and can be mechanically transmitted in the laboratory. Replication is associated with chloroplast cytopathy, though not with the appearance of the replication-associated vesicles typical of tymovirus infection. The properties and coding arrangement of the 6.1 kb RNA are more closely aligned with those of the marafiviruses (**Figure 2**). The genomic RNA is polyadenylated, contains a marafibox-related putative sgRNA promoter, and possesses a single long ORF that encodes an RP/CP fusion protein (p221). No OP ORF is present. The 21–24 kDa PnMV CP is presumably expressed from the 0.65 kbp sgRNA; the involvement of potential cleavage of the RP/CP fusion protein to produce a second CP variant is not known. Although PnMV sequences are more closely related to the marafiviruses than tymoviruses, the relationships appear to be insufficiently close to warrant inclusion in the genus *Marafivirus* (**Figure 3**).

In commercial poinsettia cultivation, PnMV is transmitted by vegetative propagation; the virus is not transmissible via seed or pollen. The natural host range is restricted to *Euphorbia* sp., especially poinsettia, *E. pulcherrima*. Symptoms appear seasonally, varying from inapparent to light mottling. PnMV is often associated with poinsettia cryptic virus (family *Partitiviridae*) and with a phytoplasma that is responsible for the desirable freebranching phenotype.

See also: Flexiviruses.

Further Reading

Bradel BG, Preil W, and Jeske H (2000) Sequence analysis and genome organisation of poinsettia mosaic virus (PnMV) reveal closer relationship to marafiviruses than to tymoviruses. *Virology* 271: 289–297.

Canady MA, Larson SB, Day J, and McPherson A (1996) Crystal structure of turnip yellow mosaic virus. *Nature Structure Biology* 3: 771–781.

Ding SW, Howe J, Keese P, *et al.* (1990) The tymobox, a sequence shared by most tymoviruses: Its use in molecular studies of tymoviruses. *Nucleic Acids Research* 18: 1181–1187.

Dreher TW (2004) *Turnip yellow mosaic virus*: Transfer RNA mimicry, chloroplasts and a C-rich genome. *Molecular Plant Pathology* 5: 367–375.

Edwards MC, Zhang Z, and Weiland JJ (1997) Oat blue dwarf marafivirus resembles the tymoviruses in sequence, genome organization, and expression strategy. *Virology* 232: 217–229.

Francki RIR, Milne RG,, and Hatta T (eds.) (1985) *Atlas of Plant Viruses, Vol. I.* Boca Raton, FL: CRC Press.

Hirth L and Givord L (1988) Tymoviruses. In: Koenig R (ed.) *The Plant Viruses*, pp. 163–212. New York: Plenum.

Jakubiec A, Notaise J, Tournier V, *et al.* (2004) Assembly of *Turnip yellow mosaic virus* replication complexes: Interaction between the proteinase and polymerase domains of the replication proteins. *Journal of Virology* 78: 7945–7957.

Kadaré G, Rozanov M, and Haenni A-L (1995) Expression of the turnip yellow mosaic virus proteinase in *Escherichia coli* and determination of the cleavage site within the 206 kDa protein. *Journal of General Virology* 76: 2853–2857.

Koenig R, Pleij CW, Lesemann DE, Loss S, and Vetten HJ (2005) Molecular characterization of isolates of anagyris vein yellowing virus, plantago mottle virus and scrophularia mottle virus: Comparison of various approaches for tymovirus classification. *Archives of Virology* 150: 2325–2338.

Maccheroni W, Alegria MC, Greggio CC, *et al.* (2005) Identification and genomic characterization of a new virus (*Tymoviridae* family) associated with citrus sudden death disease. *Journal of Virology* 79: 3028–3037.

Martelli GP, Sabanadzovic S, Abou-Ghanem Sabanadzovic N, Edwards MC, and Dreher T (2002) The family *Tymoviridae*. *Archives of Virology* 147: 1837–1846.

Mayo MA, Dreher TW, and Haenni A-L (2000) Genus *Tymovirus* In: van Regenmortal MHV, Fauquet CM, Bishop DHL, *et al.* (eds.) *Seventh Report of the International Committee on Taxonomy of Viruses.* New York: Academic Press

Morch MD, Boyer JC, and Haenni A-L (1988) Overlapping open reading frames revealed by complete nucleotide sequencing of turnip yellow mosaic virus genomic RNA. *Nucleic Acids Research* 16: 6157–6173.

Sabanadzovic S, Ghanem-Sabanadzovic NA, Saldarelli P, and Martelli GP (2001) Complete nucleotide sequence and genome organization of grapevine fleck virus. *Journal of General Virology* 82: 2009–2015.

Weiland JJ and Dreher TW (1989) Infectious TYMV RNA from cloned cDNA: Effects *in vitro* and *in vivo* of point substitutions in the initiation codons of two extensively overlapping ORFs. *Nucleic Acids Research* 17: 4675–4687.

Umbravirus

M Taliansky, Scottish Crop Research Institute, Dundee, UK
E Ryabov, University of Warwick, Warwick, UK

Glossary

Fibrillarin An abundant nucleolar protein that participates in the processing and modification of rRNAs and a methyltransferase.

Nucleolus A prominent subnuclear domain and the site of transcription of ribosomal RNA (rRNA), processing of the pre-rRNAs, and biogenesis of pre-ribosomal particles, and also participates in other aspects of RNA processing and cell function.

Phloem Plant vascular system used for rapid long-distance transport of assimilates and macromolecules.

Plasmodesmata Plant-unique intercellular membranous channels that span plant cell walls linking the cytoplasm of adjacent cells.

Introduction

Umbraviruses differ from other plant viruses in that they do not encode a coat protein (CP) and, thus no conventional virus particles are formed in infected plants. The name of the genus *Umbravirus* is derived from the Latin *umbra*, which means a shadow or an uninvited guest who comes with an invited one. This name reflects the way in which umbraviruses depend for survival in nature on an assistor virus, which is always a member of the family *Luteoviridae*. For transmission between plants, the CP of the luteovirus forms aphid-transmissible hybrid virus particles encapsidating umbraviral RNA. In nature, each umbravirus is associated with one particular luteovirus, although in experiments transcapsidation is not needed for umbravirus accumulation within infected plants because functions such as protection and movement of the virus RNA do not require the presence of the luteovirus or its CP. Moreover, under experimental conditions, mechanical transmission of umbraviruses can take place without the aid of an assistor virus. This implies that umbraviruses encode some product(s) that functionally compensate for the lack of a CP (see below).

Currently the genus *Umbravirus* includes seven distinct virus species: *Carrot mottle virus* (CMoV), *Carrot mottle mimic virus* (CMoMV), *Groundnut rosette virus* (GRV), *Lettuce speckles mottle virus* (LSMV), *Pea enation mosaic virus-2* (PEMV-2), *Tobacco mottle virus* (TMoV), and *Tobacco bushy top virus* (TBTV). These viruses together with the tentative members of the genus *Umbravirus* are listed in **Table 1**. Some of these viruses have been known since the early days of plant virology. The first to be described was TMoV, reported from Zimbabwe and Malawi in 1945. The most economically important umbravirus is GRV, which is endemic throughout sub-Saharan Africa. Sporadic, unpredictable outbreaks of groundnut rosette disease (actually caused by a satellite RNA of GRV) cause severe crop losses. Pea enation mosaic disease outbreaks are also sporadic and localized, but losses of nearly 90% in peas and up to 50% in field beans have been reported. One or both of the carrot-infecting umbraviruses probably occurs worldwide

wherever carrots are grown, but they are uncommon in commercial crops because the insecticides used to control carrot fly also control the aphid vectors of CMoV and CMoMV. LSMV is reported only from California, USA.

Virus Properties

Although no virions are formed in plants infected with umbraviruses unaccompanied by the assistor virus, the infectivity of CMoV and GRV in buffer extracts of infected leaves is surprisingly stable: infectivity remains for several hours at room temperature and is resistant to treatment with ribonuclease. Infectivity is, however, sensitive to a treatment with organic solvents suggesting that lipid-containing structures are involved in the protection of umbravirus RNA. Indeed, in plants infected with a number of umbraviruses, including CMoV, enveloped structures ~50 nm in diameter were observed. However, an infective fraction from GRV-infected tissue contained complexes with a buoyant density of $1.34-1.45 \text{ g cm}^{-3}$ consisting of filamentous ribonucleoprotein (RNP) particles, composed of the umbraviral ORF3 protein (see below) and virus RNA, embedded in a matrix (**Figure 1**). The relationship between the enveloped and filamentous structures is unclear.

Genome Organization and Expression

The genome of umbraviruses consists of a single linear segment of positive-sense, single-stranded RNA (ssRNA) (**Figure 2**). The complete genomic RNA sequences of CMoMV, GRV, PEMV-2, and TBTV are comprised of

Table 1   Virus members of the genus *Umbravirus*

Virus	Abbreviation	Sequence	Accession no.	Assistor virus
Carrot mottle mimic virus	(CMoMV)	Complete genomic RNA	U57305	Carrot red leaf virus (CaRLV)
Carrot mottle virus	(CMoV)			Carrot red leaf virus (CaRLV)
Groundnut rosette virus	(GRV)	Complete genomic RNA	Z69910	Groundnut rosette assistor virus (GRAV)
		Satellite RNAs	Z29702–Z29711	
Lettuce speckles mottle virus	(LSMV)			Beet western yellows virus (BWYV)
Pea enation mosaic virus-2	(PEMV-2)	Complete genomic RNA	U03563	Pea enation mosaic virus-1 (PEMV-1)
		Satellite RNA	U03564	
Tobacco bushy top virus	(TBTV)	Complete genomic RNA	AF402620	Unknown
Tobacco mottle virus	(TMoV)	Partial genomic RNA	AY007231	Tobacco vein distorting virus (TVDV)
Tentative members in the genus				
Sunflower crinkle virus	(SuCV)			Unknown
Sunflower yellow blotch virus	(SuYBV)			Unknown
Tobacco yellow vein virus	(TYVV)			Tobacco yellow vein assistor virus (TYVAV)

4201, 1019, 1253, and 4152 nt, respectively. There is no polyadenylation at their 3' termini and there is no information about modifications of their 5' termini. At the 5' end, a very short noncoding sequence precedes ORF1, which encodes a putative 31–37 kDa protein. ORF2, which overlaps the 3' end of ORF1, potentially encodes a 63–65 kDa protein, but lacks an AUG initiation codon. It is likely that ORF1 and ORF2 are translated as a single 94–98 kDa protein by a −1 translational frameshifting mechanism; the sequence associated with frameshifting events in several animal and plant viruses is found in the region at the 3' end of the ORF1. The predicted amino acid sequence of the ORF2-encoded product has similarities with the sequences of RNA-dependent RNA polymerases (RdRps) of viruses in the families *Tombusviridae* and *Luteoviridae* and contains all eight conserved motifs of RdRp of positive-strand RNA viruses. A short untranslated stretch of nucleotides separates ORF2 from ORF3 and ORF4, both of which almost completely overlap in different reading frames and encode 26–29 kDa products. The ORF4 contains motifs characteristic of the cell-to-cell movement proteins (MPs) of plant viruses, in particular those of cucumoviruses. The ORF3 products from different umbraviruses possess up to 50% homology to each other but show no significant similarity to any other viral or nonviral product. The ORF3 and OFR4 products are likely to be translated from subgenomic RNA(s) – 3' terminal RNAs of the appropriate size were detected in GRV-infected plant tissue. It should be pointed out that the sequenced genomes of umbraviruses do not encode a potential CP(s). The essential role of the ORF3–ORF4 block in movement of umbraviruses within plants has been experimentally demonstrated and is discussed below.

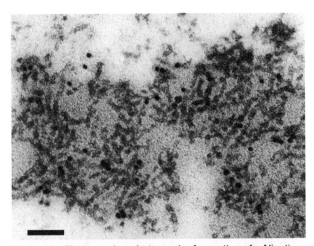

Figure 1 Electron microphotograph of a section of a *Nicotiana benthamiana* cell expressing GRV ORF3 protein. The section shows a complex of filamentous RNP particles embedded in an electron-dense matrix. The section was labeled by *in situ* hybridization with an RNA probe specific for viral RNA. Scale = 100 μm.

Replication

Replication of umbravirus RNA presumably involves the ORF1/ORF2-encoded RdRp. Leaves of plants infected with umbraviruses contain abundant double-stranded RNA (dsRNA) including a major species of about 4.4–4.8 kbp corresponding in size to that expected for a double-stranded form of the viral genomic RNA which may be an umbravirus RNA replication intermediate. No other details of the replication mechanism have been elucidated.

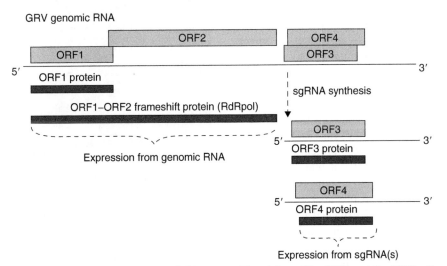

Figure 2 Organization and expression strategy of the GRV genome. Translation of the ORF1 and the ORF2 with a frameshift event results in production of the ORF1–ORF2 frameshift protein; subgenomic RNA(s) (sgRNA(s)) synthesis is required for expression of the ORF3 and ORF4. Solid lines represent RNA molecules; gray boxes represent open reading frames (ORFs); black boxes represent translation products.

Satellite RNA

Satellite RNAs are associated with some umbraviruses. In the case of GRV, satellite RNA is found in all naturally occurring isolates, and is primarily responsible for the symptoms of groundnut rosette disease. GRV satellite RNA is an ssRNA of about 900 nt which relies on GRV for its replication and, more unusually, is also required for the groundnut rosette assistor virus (GRAV)-dependent aphid transmission of GRV. Thus, unlike most virus satellite RNAs, it is essential for the biological survival (though not the replication) of its helper virus. The role of the satellite RNA in the transmission process is to mediate transcapsidation of GRV RNA by GRAV protein to form stable aphid-transmissible hybrid virus particles. Although different GRV satellite RNA variants contain up to five potential ORFs, none of the ORFs is essential for any of the functions and biological properties that have been ascribed to GRV satellites. In contrast, the satellite RNA that is associated with some isolates of PEMV-2 is not required for transcapsidation of PEMV-2 RNA by the CP of its assistor virus PEMV-1 or for aphid transmission of the hybrid particles, and other umbraviruses, such as CMoV, do not have satellite RNAs, yet are transcapsidated by their assistors and thus transmitted by aphids. The reasons for these differences have not been explained.

Cell-to-Cell Movement Function

The highly conserved ORF4-encoded proteins of umbraviruses exhibit significant sequence similarity with cell-to-cell MPs of other plant viruses, in particular the 3a proteins of cucumoviruses. Therefore, it has been suggested that this protein is involved in cell-to-cell trafficking of umbravirus RNA through plasmodesmata. This suggestion has been confirmed by a number of genetic, cytological, and biochemical approaches. By using the gene replacement strategy, it was demonstrated that that the GRV ORF4 protein could functionally replace the MPs of unrelated viruses, Potato virus X (PVX) (all the products encoded by the triple gene block and the CP) and cucumber mosaic virus (CMV) (the 3a MP and the CP). Localization of the GRV ORF4 protein has much in common with localization of MPs of other plant virus groups. The green fluorescent protein (GFP)-tagged GRV ORF4 protein targeted to plasmodesmata (**Figure 3(a)**). Also, this protein formed extended tubular structures on the surface of protoplasts infected either with GRV or with the heterologous virus expressing the GFP-tagged GRV ORF4. The GRV ORF4 protein binds to RNA *in vitro* in a noncooperative manner which may make the viral RNA of the RNP complex accessible for translation and replication.

Phloem-Dependent Long-Distance Movement Function

One of the striking features of umbraviruses is their ability to move long distance within the plant without having a conventional CP. Involvement of CPs in long-distance movement of viral infection has been shown for a number of plant viruses, including tobacco mosaic virus (TMV) which utilizes the CP exclusively for its long-distance but not cell-to-cell movement function. By using the gene replacement strategy it was demonstrated that the ORF3 proteins of GRV, PEMV-2, and ToMV were able to functionally substitute for the TMV CP in the long-distance movement process. The hybrid TMV mutants expressing the umbraviral ORF3 proteins were able to move rapidly through the phloem. It was also shown that specific mutations in the ORF3 protein of PEMV-2 abolish the ability

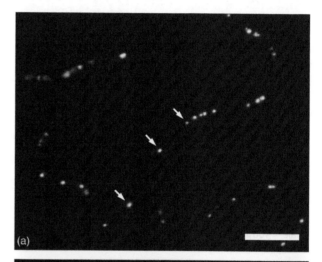

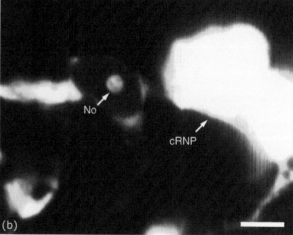

Figure 3 Confocal laser scanning images showing the localization of (a) GRV ORF4 protein fused to GFP to plasmodesmata (shown by arrows) and (b) GRV ORF3 protein fused to GFP to the nucleolus (No) and cytoplasmic inclusions containing ORF3 protein RNP particles (cRNP). Scale = 25 μm (a), 10 μm (b).

of this virus to move long distance without affecting its cell-to-cell spread within inoculated leaves.

It should be noted that the mechanisms of the long-distance movement facilitated by umbraviral ORF3 proteins is different from those mediated by suppressors of RNA silencing such as the 2b protein of CMV and the HC-Pro protein of potyviruses. Rather than supporting virus infection by suppressing the host RNA-mediated response, the ORF3 protein seems to protect RNA by binding to viral RNA. It was shown that the ORF3 proteins of GRV, PEMV-2, and ToMV increase the stability of viral RNA and protect it from RNases. Immunogold-labeling and *in situ* hybridization experiments showed that the GRV ORF3 protein accumulated in cytoplasmic granules consisting of filamentous RNP particles composed of the ORF3 protein and viral RNA (**Figure 1**). It has been suggested that these particles may be a form in which viral RNA moves long-distance through the phloem. Also the ORF3-containing RNP complex may protect viral RNA from nucleases and the RNA silencing machinery.

Involvement of the Nucleolus in Umbravirus Systemic Infection

In addition to cytoplasmic granules, the GRV ORF3 protein was also found in nuclei, predominantly targeting nucleoli (**Figure 3(b)**). Sequence analysis of the ORF3 proteins of umbraviruses revealed the presence of both a nuclear localization signal (NLS) and a nuclear export signal (NES), the functional roles of which in nuclear import and export of the GRV ORF3 has been confirmed experimentally by genetic analysis. Functional analysis of ORF3 protein mutants revealed a correlation between the ORF3 protein nucleolar localization and its ability to form RNP particles and transport viral RNA long distances. It was also shown that the ORF3 protein interacts with a nucleolar protein, fibrillarin, redistributing it from the nucleolus to the cytoplasm, and that such an interaction is absolutely essential for umbravirus long-distance movement through the phloem.

Interaction with Assistor Virus

Although umbraviruses accumulate and spread very efficiently within infected plants, they depend on the assistance of luteoviruses for their survival in nature, as they require encapsidation by luteoviral CP for horizontal transmission by aphids. In turn, umbraviruses facilitate the movement of phloem-limited luteoviruses to and from the phloem, as well as the cell-to-cell movement of luteoviruses between mesophyll and epidermal cells. It was shown that the ability to promote cell-to-cell movement of luteoviruses is a unique feature of the umbraviral ORF4 MP. Moreover, the ORF4 MP of GRV can facilitate cell-to-cell movement of *Potato leafroll virus* even when it is expressed from heterologous PVX or CMV genomes. In most instances, the luteovirus partner does not depend on the umbravirus infection for its survival in nature. A complex consisting of PEMV-1 (the genus *Enamovirus*, family *Luteoviridae*) and PEMV-2 is a notable exception. Unlike other members of the family *Luteoviridae*, PEMV-1 on its own lacks the ability to move, even through the phloem; both long-distance and cell-to-cell movement functions of PEMV-1 are provided by the umbraviral component of the complex, PEMV-2. Such a strong mutual dependence and adaptation between umbraviruses and luteoviruses suggest a long co-evolution, which has resulted in establishing a range of interactions from facultative coexistence (GRV/GRAV) to complete dependence (PEMV-1/PEMV-2).

Similarity with Other Taxa

Amino acid sequence comparisons showed that the putative RdRp encoded by CMoMV, GRV, PEMV-2, and TBTV belong to the so-called supergroup 2 of RNA polymerases, as do those of viruses in the genera *Carmovirus*, *Necrovirus*, *Machlomovirus*, and *Tombusvirus*. Since these enzymes are the only universally conserved proteins of positive-strand RNA viruses, the genus *Umbravirus* might be considered to be in or close to the family *Tombusviridae*.

Host Range

Individual umbraviruses are confined in nature to one or two host plant species. For example, groundnut is the only known natural host of GRV and the entire rosette disease complex (GRV, its satellite RNA and GRAV). Experimental host ranges of umbraviruses are broader but still restricted. They usually induce a leaf mottle and/or mosaic in infected plants.

Transmission

In nature, umbraviruses are transmitted by aphids, but only from plants that are infected also with an assistor luteovirus. The mechanism of the dependent transmission is the encapsidation of the umbraviral RNA by the CP provided by the assistor virus. Hence, the transmission of the dependent umbravirus occurs in the same persistent (circulative, nonpropagative) manner as that of the assistor-luteovirus.

Prevention and Control

For the avoidance of pea enation mosaic disease, it is recommended that pea or faba bean crops should be sited away from alfalfa and clover fields. Early sowing and close spacing can reduce groundnut rosette disease incidence, probably by inhibiting the landing response of the vector. This approach is also effective against tobacco rosette disease.

However, the best control approach is to use resistant cultivars if they are available. Resistance to GRV controlled by two independent recessive genes has been found and groundnut lines possessing this resistance have been developed.

A possibility for the future is the deployment of transgenic forms of resistance. Strategies for engineering transgenic resistance against umbraviruses include the transformation of plants with translatable or nontranslatable sequences from the umbraviruses themselves or their satellite RNAs.

See also: Luteoviruses; Plant Virus Vectors (Gene Expression Systems).

Further Reading

Demler SA, Borkhsenious ON, Rucker DG, and de Zoeten GA (1994) Assessment of the autonomy of replicative and structural functions encoded by the luteo-phase of pea enation mosaic virus. *Journal of General Virology* 75: 997–1007.

Kim SH, Ryabov EV, Brown JW, and Taliansky M (2004) Involvement of the nucleolus in plant virus systemic infection. *Biochemical Society Transactions* 32: 557–560.

Nurkiyanova KN, Ryabov EV, Kalinina NO, et al. (2001) Umbravirus-encoded movement protein induces tubule formation on the surface of protoplasts and binds RNA incompletely and noncooperative. *Journal of General Virology* 82: 2579–2588.

Ryabov EV, Robinson DJ, and Taliansky ME (1999) A plant virus-encoded protein facilitates long distance movement of heterologous viral RNA. *Proceedings of the National Academy of Sciences, USA* 96: 1212–1217.

Ryabov EV, Robinson DJ, and Taliansky M (2001) Umbravirus-encoded proteins that both stabilize heterologous viral RNA *in vivo* and mediate its systemic movement in some plant species. *Virology* 288: 391–400.

Taliansky M, Roberts IM, Kalinina N, et al. (2003) An umbraviral protein, involved in long-distance RNA movement, binds viral RNA and forms unique, protective ribonucleoprotein complexes. *Journal of Virology* 77: 3031–3040.

Taliansky ME and Robinson DJ (2003) Molecular biology of umbraviruses: Phantom warriors. *Journal of General Virology* 84: 1951–1960.

Watermelon Mosaic Virus and Zucchini Yellow Mosaic Virus

H Lecoq and C Desbiez, Institut National de la Recherche Agronomique (INRA), Station de Pathologie Végétale, Montfavet, France

Glossary

Mild-strain cross-protection A plant systemically infected by a mild virus strain will not develop additional symptoms when inoculated by a severe strain of the same virus. Most often, the severe strain does not multiply in the cross-protected plant.

Filiformy A leaf deformation symptom in which the leaf blade is drastically reduced but not the veins, giving a shoestrings aspect to the leaf.

Transmission propensity A measure of vector importance quantifying the natural ability of a species to inoculate a plant with a virus under conditions that allow vectors to move and feed freely.

History and Taxonomy

Watermelon Mosaic Virus

Mosaic diseases of cucurbit crops were first reported in the 1920s and the early literature contains a diversity of names for viruses or virus diseases that were only partially characterized. In 1940, Milbrath reported a severe watermelon (*Citrullus lanatus*) mosaic disease in California, but the nomenclature of the causal agent remained controversial until 1979. In 1965, Webb and Scott compared ten isolates from southern USA of what was then called the watermelon mosaic virus (WMV) complex. Based on cross-protection experiments, serological relationships and host range reactions they divided them into two groups: WMV1 and WMV2. WMV1 and WMV2 were considered as different viruses, also different from a watermelon mosaic isolate reported from South Africa in 1960 by van Regenmortel. Further work by Milne and Grogan in 1969 increased the confusion by concluding that WMV1 and WMV2 were strains of the same virus. In 1979, Purcifull and Hiebert definitively clarified the situation by demonstrating that WMV1 and WMV2 were indeed serologically distinct entities, and that WMV1 was closely related to papaya ringspot virus (PRSV). Today, WMV1 is considered as the W strain of PRSV, while WMV2 is referred to as watermelon mosaic virus (WMV). In addition, the same authors showed that a watermelon mosaic virus isolate from Morocco was a third serological entity. This isolate is considered as a distinct virus, Moroccan watermelon mosaic virus (MWMV), to which probably also belongs the South African isolate.

So, from the initial watermelon mosaic virus complex emerged three different virus species: *Watermelon mosaic virus*, *Papaya ringspot virus*, and *Moroccan watermelon mosaic virus*.

Zucchini Yellow Mosaic Virus

An apparently new cucurbit virus was isolated in 1973 from a zucchini squash (*Cucurbita pepo*) plant in Northern Italy, and Lisa *et al.* described this virus as belonging to a new potyvirus species, *Zucchini yellow mosaic virus*, in 1981. In 1979, many melon (*Cucumis melo*) crops were devastated in southwestern France by an apparently new virus disease, whose causal agent was tentatively named muskmelon yellow stunt virus; very rapidly, it appeared that it was a strain of zucchini yellow mosaic virus (ZYMV). Within a few years, ZYMV was reported in many countries on the five continents, and in this regard, ZYMV appears as a typical example of an emerging plant virus.

Classification

Based on particle morphology, aphid transmissibility, serological relationships, ability to induce pinwheel cytoplasmic inclusions in host cells, genome organization, and nucleotide sequences, WMV and ZYMV were identified as members of the genus *Potyvirus*, family *Potyviridae*. Several other potyviruses have been shown to infect cucurbit crops including PRSV-W, MWMV, zucchini yellow fleck virus (ZYFV), melon vein banding mosaic virus (MVBMV), telfairia mosaic virus (TeMV), turnip mosaic virus (TuMV), clover yellow vein virus (ClYVV), and bean yellow mosaic virus (BYMV). However, these viruses have either only limited geographic distribution or minor economical incidence.

Molecular analyses based on the coat protein (CP) coding sequence revealed that cucurbit-infecting potyviruses belong to several 'clusters' of closely related species: ZYMV and WMV belong to a cluster that also contains mostly legume-infecting potyviruses, whereas PRSV, MWMV, and ZYFV are grouped in a 'PRSV-like' cluster containing mostly cucurbit-restricted viruses (**Figure 1**).

Symptoms

Watermelon Mosaic Virus

WMV induces a diversity of symptoms according to the isolate and the host cultivar. On leaves, symptoms are mosaics, vein banding, more or less severe leaf deforma-

tions, and filiformy. Some isolates induce discoloration and deformation on fruits of zucchini squash susceptible cultivars while other isolates do not affect fruit and yield quality. Mosaic and discoloration are also observed on leaves and fruits of some melon cultivars (**Figure 2**).

Zucchini Yellow Mosaic Virus

Since its first detection, ZYMV was recognized as a virus causing extremely severe symptoms leading to complete yield losses in the case of early contamination. This severity of symptoms was probably an important factor for the rapid identification of ZYMV soon after its first outbreaks, in many countries. In melon, leaf symptoms include vein clearing, yellow mosaic, leaf deformation, occasionally with blisters and enations. There is often a severe plant stunting. Some ZYMV isolates can induce a rapid and complete wilt in cultivars possessing the *Fn* gene. On fruits, a diversity of symptoms are observed: external mosaic or necrotic cracks, internal marbling, and hardening of the flesh. Seeds are occasionally severely deformed and have poor germination rates. In zucchini squash, symptoms are very severe on leaves with mosaic, yellowing, leaf distortion and sometimes very severe filiformy. Fruits are generally severely misshaped with prominent knobs and are of course unmarketable (**Figure 3**). In cucumber (*Cucumis sativus*) and watermelon, mosaic and deformations are generally observed on leaves and fruits.

Synergism

Synergism has been observed between WMV or ZYMV and other viruses infecting cucurbits which can be expressed either by increase in virus multiplication rates or by more severe symptoms. Significant increase in cucumber mosaic virus (CMV) multiplication rate was observed in cucumber plants co-infected by ZYMV. CMV could also partially overcome a resistance in cucumber when in mixed infection with ZYMV. In grafted cucumbers, double infection by CMV and ZYMV induces a severe and rapid wilting reaction which is not observed with infections by CMV or ZYMV alone. When WMV or ZYMV are in mixed infection with the polerovirus, cucurbit aphid-borne yellows virus (CABYV), CABYV multiplication rate and symptom intensity are increased. As for other potyviruses, these synergetic effects could be related to the strong potyvirus gene silencing suppressor.

Geographic Distribution

Both WMV and ZYMV are now widely distributed in the major cucurbit production areas worldwide. Their geographic distributions are broadly overlapping and frequent mixed infections are observed in the fields.

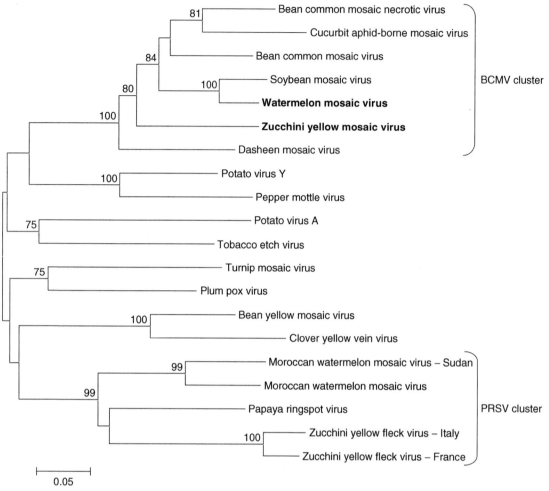

Figure 1 Relationships based on CP amino acid sequences, of cucurbit-infecting potyviruses. Branch lengths indicate the molecular divergence of viruses (the scale bar represents 0.05 mutations per residue). Figures at some nodes represent bootstrap values (in %), indicating the robustness of each node. Only values above 75% are indicated.

For unknown reasons, WMV appears to be rare or absent in cucurbits in subtropical or tropical areas. For instance, in exhaustive surveys conducted in Nepal, Sudan, and French West Indies, no WMV was detected, although ZYMV was relatively abundant. In Florida, WMV is frequent in northern and central counties but is not detected in southern counties, whereas ZYMV can be found throughout the state. This cannot be related to a lack of potential WMV vectors or reservoirs because both are abundant in tropical or subtropical areas.

ZYMV is present worldwide in almost all countries where cucurbits are grown, under temperate, Mediterranean, subtropical, and tropical climatic conditions. It affects highly mechanized cropping systems (such a glasshouse crop production in Northern Europe) as well as more traditional agroecosystems (such as flood-irrigated crops on the Nile banks). ZYMV has been reported in very remote areas including semi-desertic regions or islands.

Host Range

WMV has a relatively wide experimental host range for a potyvirus. It infects over 170 species in 26 mono- or dicotyledonous families. Besides cucurbits, WMV causes mosaic diseases in legumes (pea, broad bean) and orchids (vanilla, *Habenaria radiata*) and infects many weeds that can serve as alternative hosts. Generally, naturally infected weeds do not present evident symptoms of viral infection.

ZYMV has a relatively narrow host range. In natural conditions, it infects mostly cultivated or wild cucurbits but also a few flower species (*Delphinium, Althea*) or weeds.

Diagnostic Method

The confusion that was prevalent in the early descriptions of watermelon mosaic diseases was mainly due to

Figure 2 Mosaic symptoms on a leaf and fruit of a melon plant infected by a moderately severe WMV isolate.

Figure 3 Severe mosaic and deformations on leaves and fruits of a zucchini squash plant infected by ZYMV. Reproduced from Astier S, Albouy J, Maury Y, and Lecoq H (2001) *Principes de virologie végétale: Génome, pouvoir pathogène, écologie*. Paris: INRA Editions, with permission from QUAE.

the convergence of symptoms caused by WMV, PRSV, and MWMV in cucurbits and to the lack of proper diagnostic tools. Symptoms and virus particle morphology were clearly insufficient to differentiate these viruses. The production of specific polyclonal antisera, and the development of simple serological tests, such as the gel double-diffusion test in agar containing sodium dodecyl sulfate and sodium azide (SDS-ID), brought a major contribution to the proper diagnosis of cucurbit potyviruses. In particular, this method contributed to the rapid and unequivocal identification of ZYMV in several countries soon after its first observation. Now, double antibody sandwich enzyme-linked immunosorbent assay (DAS-ELISA) is generalized and commercial kits are available for WMV and ZYMV. Recently, dipstick serological tests based on the lateral flow technique have been developed that allow an easy and rapid diagnosis of ZYMV in the fields.

Monoclonal antibodies (MAbs) have been produced for ZYMV and WMV. They proved to be very useful to study the serological variability and to differentiate ZYMV and WMV subgroups. They have also been used successfully to analyze virus interactions, and in particular cross-protection efficiency and specificity.

Many partial nucleotide sequences are now available for ZYMV and WMV, particularly in the CP coding region, which allowed the development of specific primers for each virus. However, in routine testing, DAS-ELISA seems to be more reliable than RT-PCR.

Vector Relationships

WMV is transmitted by at least 35 aphid species in 19 genera. Fewer aphid species were tested for their ability to transmit ZYMV, and 11 were identified as ZYMV vectors. *Aphis craccivora, Aphis gossypii, Macrosiphum euphorbiae*, and

Myzus persicae are efficient WMV and ZYMV vectors. Some aphid species were shown to be poor or nonvectors of WMV and ZYMV, which suggests some level of specificity in the virus–vector interaction.

WMV and ZYMV are transmitted on the nonpersistent mode: they are acquired and transmitted during very short probes (a few seconds to minutes), and their retention period in the vector is relatively short (a few hours).

WMV and ZYMV as typical potyviruses require the presence of a virus-encoded helper component (HC-Pro) protein for transmission. HC-Pro from WMV and ZYMV are interchangeable and both mediate efficiently the transmission of purified virions of both species.

Several ZYMV isolates that have lost aphid transmissibility have been characterized, and a unique feature for this virus is that single amino acid mutants have been identified in the three domains important for transmission. ZYMV-NAT has an A to T substitution in the DAG motif in the CP, ZYMV-PAT a T to A substitution in the PTK motif of HC-Pro, and ZYMV-R1A a K to E substitution in the KLSC motif of HC-Pro. These mutants led to the identification of an interaction between the HC-Pro and CP through their PTK and DAG domains.

The nontransmissible isolate ZYMV-NAT (having the DTG motif in the CP) could be transmitted by aphids from plants infected concomitantly by a transmissible isolate of PRSV. This occurred through heteroencapsidation, a phenomenon by which ZYMV RNA is completely or partially encapsidated by PRSV CP, which is functional in aphid transmission. An aphid nontransmissible isolate deficient for HC-Pro can also be transmitted by aphids when in mixed infection with an isolate that has a functional HC-Pro. The transmissible isolate provides its functional HC-Pro to mediate the transmission of the deficient isolate. These two mechanisms can contribute to the maintenance, in natural conditions, of variants which have lost their vector transmissibility.

An interesting interaction has been observed between ZYMV and *Aphis gossypii*, an aphid vector colonizing cucurbit crops. *Aphis gossypii* lives longer and produces more offspring on ZYMV-infected than on noninfected plants. In addition, more alatae are produced on infected plants, which may stimulate the spread of ZYMV. These phenomena might be related to the observed changes in phloem exudates' composition (free amino acids, sugars) in virus-infected plants.

Epidemiology

Virus Sources

WMV has not been described as seed borne in cucurbits or other crops, but may be transmitted through vegetative propagation in vanilla. ZYMV seed transmission remains controversial. It has been reported in squash but other studies repeatedly failed to observe seed transmission. This is an important issue, since ZYMV seed transmission would be the simplest explanation for the rapid ZYMV dissemination throughout the world in the 1980s. If ZYMV seed transmission does occur, it is at a low rate and may be only for some strains or hosts. Another possible way for long-distance dissemination of ZYMV is through the globalization of vegetable production and trade. It has been shown that ZYMV-infected fruits imported from Central America into Europe could be very efficient virus sources for aphids.

Virus sources from which epidemics will initiate could be overwintering weeds or crops infected during the previous crop season. In tropical and subtropical regions, cucurbit crops or weeds grow all year round, and viruses could easily move from an old infected crop or cucurbit weed to a young planting. In more temperate regions, non-cucurbit weeds were found to be efficient reservoirs for WMV but not for ZYMV. Winter protected-crops could contribute to ZYMV overwintering in Mediterranean regions and residential gardens were found to be important sources of ZYMV in California.

Efficient Vectors

Many potential aphid vector species have been identified for WMV and ZYMV. A study was conducted to compare the vector capacity of two aphid species, one colonizing cucurbits (*A. gossypii*), the other not (*A. craccivora*). Two parameters were used. Transmission efficiency was measured in the laboratory with single aphids exposed in sequence to an infected plant and then to four healthy plants. Transmission propensity was measured by arena tests (more representative of natural conditions) in which aphids could move between plants and feed without interference.

It was shown that *A. craccivora* had both a higher efficiency and propensity to disseminate ZYMV than *A. gossypii*. This highlights the importance of noncolonizing transient vector species in the epidemiology of ZYMV.

Pattern of Spread

The same general pattern is observed for WMV and ZYMV. The first contaminations occur generally shortly after planting, depending upon the availability of virus sources in the environment. The secondary virus spread then occurs when aphid flights, particularly of noncolonizing species, spread the virus from the primary infection foci to the rest of the crop. The diseased plants are often distributed in large patches that rapidly extend and join each other, leading to the complete contamination of the crop within a few weeks. Simultaneously, the infected crop will serve as a source of virus to contaminate weeds or nearby young plantings. Epidemic development curves have an overall S-shape, generally fitting well with the logistic model.

Control Methods

Prophylactic Measures

Prophylactic measures are intended to prevent or limit the contact of viruliferous aphids with cultivated plants. They are not specific for a particular virus and are generally efficient for all aphid-borne viruses. These include careful weeding near plantings and avoiding overlapping crops in the same area to reduce virus and aphid sources near new plantings. Plastic mulches have a repelling action on aphids and significantly delay WMV and ZYMV spread. However, they confer only a temporary protection that is limited to the early stages of the crop, because their efficiency decreases when plant growth covers their surface. Row covers of different types (unwoven, perforated plastics, etc.) can also be used; they physically prevent winged aphids from reaching the plants, but they must be removed to allow insect pollination necessary for cucurbits. Both methods have a major drawback: they require a lot of plastic material that farmers must dispose in an environmentally sound way after the crop cycle. Insecticides applications have generally been found inefficient in limiting WMV and ZYMV spread. This is to relate to the large number of winged aphids that land on the plants and to the rapidity of the transmission process. Oil applications can delay virus spread when inoculum pressure is moderate. When applicable, a 1 month crop-free period has been shown to be efficient in limiting ZYMV virus spread.

Cross-Protection

Although the mechanism of cross-protection has not been fully elucidated (it probably relies on the gene silencing machinery), this method has been developed at a commercial level to protect cucurbit crops against ZYMV. The principle is simple (**Figure 4**): when a mild virus isolate (i.e., that has no significant impact on commercial yield) is inoculated to young seedlings, it protects the plant from subsequent contaminations by severe isolates of the same virus. The mild strain ZYMV-WK is a natural variant of a severe aphid nontransmissible isolate. Although efficient against most ZYMV isolates, ZYMV-WK does not protect against very divergent isolates such as those from Réunion Island, indicating some specificity in the protection. A single amino acid change (R to I) in the FRNK conserved domain of the HC-Pro is responsible for symptom attenuation of ZYMV-WK. A complete technological package (mild strain production, quality control protocols, inoculation machines) has been developed to implement commercially ZYMV cross-protection. Mild WMV isolates have been reported that could also have a potential for cross-protection.

Resistant Cultivars

The use of virus-resistant cultivars is probably the easiest and cheapest way to control plant viral diseases at the farmer's level. Breeding for resistance still relies mainly upon searching for resistance characters in germplasm collections and introgression of the resistance gene(s) into commercially acceptable cultivars. Considerable efforts have been made to look for resistance to WMV and ZYMV in genetic resources, and some WMV- or ZYMV-resistant commercial cultivars are now available. Some resistance genes confer complete and durable resistance (such as the *zym* gene in cucumber), while others confer only partial resistances or may be overcome by virus evolution (such as the *Zym* gene in melon). An interesting situation is observed for ZYMV resistance in squash. Although the resistance level was high in the original accession of *Cucurbita moschata* in which the resistance was identified, when transferred through interspecific crosses to zucchini squash (*C. pepo*), the resistance phenotype was different: ZYMV multiplies but the plants present only very mild symptoms (a phenomenon called tolerance). However, tolerance appears not to be stable since aggressive variants of the virus (i.e., causing severe symptoms in tolerant plants) may emerge in these plants. A single amino acid change in the P3 gene is sufficient to confer this aggressive phenotype. However, the aggressive variants are counter-selected when in competition with common ZYMV isolates in susceptible cucurbits. This genetic load associated with aggressiveness could be a factor that will make the tolerance durable. In melon, a resistance to WMV and ZYMV transmission by *A. gossypii*

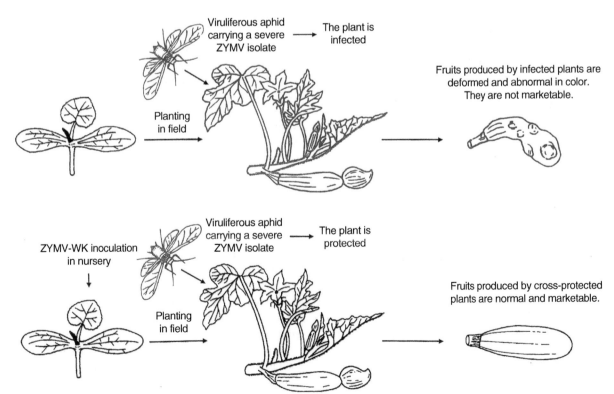

Figure 4 Representation of mild strain cross-protection as applied with ZYMV-WK in zucchini squash. Reproduced from Astier S, Albouy J, Maury Y, and Lecoq H (2001) *Principes de virologie végétale: Génome, pouvoir pathogène, écologie.* Paris: INRA Editions, with permission from QUAE.

was found to be governed by the single dominant gene *Vat*. This gene is present in many commercial cultivars, but confers only limited protection in the fields, probably because WMV and ZYMV are also transmitted by many other aphid species in natural conditions.

In the last two decades, attempts were made to obtain WMV and ZYMV transgenic resistant plants using the pathogen-derived resistance approach. Different constructs were tested to obtain resistant WMV or ZYMV plants and the best results were obtained with the full-length CP gene and with ribozymes. Freedom II, a transgenic squash hybrid containing the WMV and ZYMV CP genes, was released in the USA in 1995, as the first virus-resistant transgenic crop to be commercially cultivated in the world. It proved to have a very efficient resistance to WMV and ZYMV in field conditions. Similar biotech crops are presently grown mainly in southeastern USA, particularly during late summer and fall when WMV and ZYMV inoculum pressure is high.

Variability

Only limited biological variability has been reported for WMV. This concerns mainly differences in symptom intensity, host range, or aphid transmissibility. In contrast, from its first description, ZYMV appeared to have a very important biological diversity. When collections of field isolates were compared, an important variability was revealed in host range and symptoms on susceptible hosts, with isolates producing mild or atypical mosaic symptoms, necrosis, or wilting reactions. When these isolates were inoculated to melon or squash varieties possessing resistance genes, different pathotypes could be differentiated. Important variability has also been observed in aphid transmissibility and several aphid nontransmissible or poorly transmissible isolates have been described.

Limited serological variability was observed in WMV and ZYMV when polyclonal antibodies were used. However, the development of monoclonal antibodies against WMV and ZYMV allowed the characterization of serotypes closely correlated to the molecular variability.

The molecular variability of ZYMV and WMV was assessed mostly on the CP region, particularly the N-terminal part of the CP that is known for potyviruses to be highly variable and is frequently used for molecular studies. More than 200 partial CP sequences are now available for ZYMV. Strains from some regions, for example, Réunion Island and Singapore, are highly divergent molecularly, whereas isolates from other regions are more closely related, and fall into three main clusters without geographic structure; more recent sequence data available for Asian strains (China, Korea, Taiwan) indicate that some of those strains tend to locate between previously described groups or clusters. Molecular analyses have

allowed, in a few cases, tracking the putative origins of ZYMV strains emerging in a new area.

Genetics and Evolution

WMV and ZYMV genome organization is very similar to that of other potyviruses sequenced so far. The single-stranded, positive-sense RNA genome (9.6 kbp for ZYMV, 10 kbp for WMV excluding the polyadenylated 3′ extremity) is translated as a single polyprotein that is self-cleaved in 10 functional proteins.

The full-length sequence revealed that WMV is related to the legume-infecting soybean mosaic virus (SMV); however, the P1 protein is 135 amino acids longer than that of SMV, and the N-terminal half of P1 shows no relation to SMV but is 85% identical to another legume-infecting potyvirus, bean common mosaic virus (BCMV). This suggests that WMV has emerged through an ancestral recombination event between an SMV-like and a BCMV-like potyvirus. SMV and BCMV have narrow host ranges, mostly restricted to legumes, while WMV is one of the potyviruses with the broadest host range, including monocots and dicots. The impact of its recombinant nature on WMV biological properties remains unknown and speculative.

Partial sequence data also indicate the presence of intraspecific recombinants in WMV: the molecular variability was structured in three groups based on the CP sequence; when other parts of the genome were considered, several isolates switched groups, indicating that recombination may have taken place. By sequencing the full-length genome of putative recombinants, recombination points were characterized in different parts of the genome.

In the case of ZYMV, sequence analysis softwares suggest that recombination might have taken place, but the situation is not as clear-cut as for WMV.

Besides recombination, viral genomes also evolve by mutations that take place during replication. In the case of ZYMV, the biological consequences of several point mutations that emerged in natural conditions were assessed by molecular studies: sequences of closely related strains with different biological properties were compared, and the mutations observed – particularly non-silent mutations located in conserved domains – were introduced by site-directed mutagenesis in an infectious cDNA of ZYMV in order to check if the mutation alone is sufficient to induce the difference in biological properties.

Molecular studies have thus shown that single point mutations in HC-Pro or P3 of ZYMV were important for symptoms. Similarly, single mutations in HC-Pro or CP greatly affect aphid transmissibility. For unknown reasons, some ZYMV isolates such as ZYMV-E15 seem to be particularly prone to mutations and many variants (mild, aphid nontransmissible, aggressive, or virulent) have been derived from this isolate.

Biotechnological Application

The development of infectious cDNA clones of ZYMV brought the possibility of using ZYMV as a biotechnological tool to produce proteins of pharmaceutical or crop protection interest. The gene coding for the protein of interest can be inserted in the ZYMV genome either between the P1 and HC-Pro domains or between the NIb and CP domains. For optimal protein production, there is the possibility to use mild clones that have the mutation in the FRNK motif of HC-Pro so that they produce mild symptoms and do not affect plant growth. The protein may be produced in the edible cucurbit fruits, and therefore used directly for oral administration. Several molecules of pharmaceutical interest have been efficiently produced through this technology: the human interferon-alpha 2, antiviral and antitumor proteins MAP30 and GAP31, and a mite allergen. This technology also provided a way to produce large amounts of nucleocapsid proteins of five tospoviruses that could be used for immunization and production of specific antibodies. Finally, the *bar* gene coding for a phosphinothricin acetyltransferase that confers resistance to herbicides based on glufosinate ammonium has been inserted into the mild ZYMV expression vector. In weed-infested plots, this construct efficiently protected inoculated zucchini squash plants from damage caused by a herbicide treatment that completely destroyed the weeds. In addition, these plants were protected against severe ZYMV isolates. It is interesting to see that ZYMV, which emerged as the most damaging cucurbit virus in the last decades, can be manipulated and used for the benefit of human health and agriculture.

See also: Papaya Ringspot Virus; Plant Antiviral Defense: Gene Silencing Pathway; Plant Resistance to Viruses: Engineered Resistance; ; Plant Virus Diseases: Economic Aspects; Plant Virus Vectors (Gene Expression Systems); Vector Transmission of Plant Viruses; Vegetable Viruses.

Further Reading

Arazi T, Slutsky SG, Shiboleth YM, *et al.* (2001) Engineering zucchini yellow mosaic potyvirus as a non-pathogenic vector for expression of heterologous proteins in cucurbits. *Journal of Biotechnology* 87: 67–82.

Astier S, Albouy J, Maury Y, and Lecoq H (2001) *Principes de virologie végétale: Génome, pouvoir pathogène,* écologie. Paris: INRA Editions.

Desbiez C and Lecoq H (1997) Zucchini yellow mosaic virus. *Plant Pathology* 46: 809–829.

Desbiez C and Lecoq H (2004) The nucleotide sequence of watermelon mosaic virus (WMV, *Potyvirus*) reveals interspecific recombination between two related potyviruses in the 5′ part of the genome. *Archives of Virology* 149: 1619–1632.

Fuchs M, Ferreira S, and Gonsalves D (1997) Management of virus diseases by classical and engineered protection. *Molecular Plant Pathology On-Line.* http://www.bspp.org.uk/mppol/1997/0116fuchs (accessed October 2007).

Lecoq H (1998) Control of plant virus diseases by cross protection. In: Hadidi A, Kheterpal RK, and Koganezawa H (eds.) *Plant Virus Disease Control,* pp. 33–40. St. Paul, MN: APS Press.

Lecoq H (2003) Cucurbits. In: Loebenstein G and Thottapilly G (eds.) *Viruses and Virus-Like Diseases of Major Crops in Developing Countries,* pp. 665–687. Dordrecht: Kluwer Academic Publishers.

Lisa V and Lecoq H (1984) *Zucchini Yellow Mosaic Virus. CMI/AAB Descriptions of Plant Viruses No. 282.* Kew, UK: Commonwealth Mycological Institute.

Purcifull DE and Hiebert E (1979) Serological distinction of watermelon mosaic virus isolates. *Phytopathology* 69: 112–116.

Purcifull D, Hiebert E, and Edwardson J (1984) *Watermelon Mosaic Virus 2. CMI/AAB Descriptions of Plant Viruses No. 293.* Kew, UK: Commonwealth Mycological Institute.

PLANT VIRUS DISEASES

Cereal Viruses: Maize/Corn

P A Signoret, Montpellier SupAgro, Montpellier, France

Introduction

Maize is the main cereal crop in the world regarding total yield; it is grown on about 120 Mha. Maize has been grown for millennia in Central America. From a plant mainly used as human food, maize is now a main component for feeding animals but is still a major staple food crop in sub-Saharan Africa and America.

Viruses can cause important diseases of maize worldwide. While some viruses are widespread, others are localized. Some maize viruses can be sporadic and devastating causing severe yield losses, others may occur each year but losses are relatively minor. The average yield of maize in several African countries is only about a third of the world's average, and some viral diseases of maize are one of the major factors responsible for this low productivity. When available, resistant or tolerant hybrids provide the most effective means to control maize viruses. A pathogen-derived resistance strategy was developed for some viruses such as maize dwarf mosaic virus (MDMV) and maize streak virus (MSV). The major viruses affecting maize are transmitted by leaf- or planthoppers (16) but four viruses have aphids, two have mites, and one has beetles as vector. Populations of leafhopper and planthopper vectors can be controlled by systemic insecticides applied as seed dressing or at sowing time. Herbicide eradication of Johnson grass, an overwintering host of some viruses, had a significant impact on disease control. In some cases, crop rotation and selection of planting date result in significant decreases in virus infection.

Maize Virus Diseases

Barley Yellow Dwarf (MAV-PAS-PAV), Cereal Yellow Dwarf (GPV-RPS-RPV), RMV-SGV

All the virus species of the family *Luteoviridae* in this subsection cause similar symptoms (see **Table 1** for a list of relevant virus names). Usually infected cultivars express discoloration of the leaves, reddening and yellowing, but some present no symptoms. More or less red or yellow bands are observed mainly on the borders of the older leaves. Compared to the healthy plants, possible grain yields are lower by 15–20%. Seeds coated with a systemic insecticide have to be used in areas where the disease is frequently encountered. A few maize inbred lines tolerant to barley yellow dwarf virus are available. Maize hybrids are in general tolerant.

Chloris Striate Mosaic Virus

This virus found in 1963 in Australia causes striate mosaic in *Chloris gayana* Kunth and other grass species and in cereals including barley, maize, and oats. It is transmitted by the leafhopper *Nesoclutha pallida* (Evans). Natural hosts are several grasses, and barley, oats, maize, and wheat. Infected plants develop chlorotic grayish white striation on leaves; notching and curling of leaves may also occur. Maize is occasionally infected but most hybrids are resistant.

Cynodon Chlorotic Streak

Cynodon chlorotic streak virus (CCSV) can cause a serious disease in the Mediterranean region. Symptoms on maize are characterized by the development on the youngest leaves of yellow streaks which become larger and can coalesce to form yellow stripes (**Figure 1(e)**). Finally the whole leaf turns yellow with some parts becoming reddish. If contamination occurs at a young stage, the plants are dwarfed and die early. The virus is transmitted by planthoppers and occurs naturally in Bermuda grass (*Cynodon dactylon* (L.) Pers.) and maize. The level of CCSV infection in maize is directly related to the level of CCSV infection and planthopper vector populations in Bermuda grass. As many as 70–84% virus infected plants have been observed for some years in southern France. Seed dressing with imidachloprid allowed a good control of the vector population.

High Plains Virus

High Plains virus (HPV), a previously unknown pathogen, was found in 1993 to infect maize in US High Plains states. It has also been found in other countries. The virions are flexible, thread-like particles. Analysis of the nucleic acid shows four bands of double-stranded RNA (dsRNA) after electrophoresis. A great variety of symptoms may be expressed in maize, including spots or flecks along vascular bundles, purpling or reddening of the leaf margins or sectors of the leaf, and stunting. The virus is transmitted obligately by the wheat curl mite *Aceria tosichella* (Keifer) that also transmits wheat streak mosaic virus (WSMV). HPV can be seed transmitted. The host range includes barley, maize, and wheat in addition to several weed species. Three dominant genes for resistance have been identified and mapped in maize, and they provide a high degree of resistance.

Table 1   Taxonomic position of maize viruses

Virus names	Genus	Family
Barley yellow dwarf-MAV virus	*Luteovirus*	*Luteoviridae*
Barley yellow dwarf-PAS virus	*Luteovirus*	*Luteoviridae*
Barley yellow dwarf-PAV virus	*Luteovirus*	*Luteoviridae*
Barley yellow striate mosaic virus	*Cytorhabdovirus*	*Rhabdoviridae*
Brome mosaic virus[b]	*Bromovirus*	*Bromoviridae*
Cereal yellow dwarf-GPV virus	*Polerovirus*	*Luteoviridae*
Cereal yellow dwarf-RPS virus	*Polerovirus*	*Luteoviridae*
Cereal yellow dwarf-RPV virus	*Polerovirus*	*Luteoviridae*
Chloris striate mosaic virus	*Mastrevirus*	*Geminiviridae*
Corn lethal necrosis complex		
Corn stunt complex		
Cynodon chlorotic streak[a]		
High Plains virus[a]		
Indian peanut clump virus	*Pecluvirus*	
Johnson grass mosaic virus	*Potyvirus*	*Potyviridae*
Maize chlorotic dwarf virus	*Waikavirus*	*Sequiviridae*
Maize chlorotic mottle virus	*Machlomovirus*	*Tombusviridae*
Maize dwarf mosaic virus	*Potyvirus*	*Potyviridae*
Maize fine streak virus	*Nucleorhabdovirus*	*Rhabdoviridae*
Maize Indian Fiji-like virus[a]		
Maize Iranian mosaic virus[a]		
Maize mosaic virus	*Nucleorhabdovirus*	*Rhabdoviridae*
Maize mottle/chlorotic stunt virus[a]		
Maize necrotic streak virus[a]		
Maize rayado fino virus	*Marafivirus*	*Tymoviridae*
Maize rough dwarf virus	*Fijivirus*	*Reoviridae*
Maize sterile stunt virus[a]		
Maize streak virus	*Mastrevirus*	*Geminiviridae*
Maize stripe virus	*Tenuivirus*	*Tenuiviridae*
Mal de Rio Cuarto virus	*Fijivirus*	*Reoviridae*
Pennisetum mosaic virus[a]		
Rice black-streaked dwarf virus	*Fijivirus*	*Reoviridae*
RMV		*Luteoviridae*
SGV		*Luteoviridae*
Sugarcane mosaic virus	*Potyvirus*	*Potyviridae*
Wheat American striate mosaic virus	*Cytorhabdovirus*	*Rhabdoviridae*
Wheat streak mosaic virus	*Tritimovirus*	*Potyviridae*

[a]Unassigned virus.
[b]Not described, virus with low or no agronomical importance.

Johnson Grass Mosaic Virus

Johnson grass mosaic virus (JGMV) was first reported in Australia in maize and sorghum and later in the US and South America. Johnson grass (*Sorghum halepense* (L.) Pers.), maize, and some other plants of the Poaceae are natural hosts. The virus causes mosaic, mottling, and ringspot symptoms on maize and grass hosts, which is particularly evident on young leaves. Since the symptoms of JGMV and MDMV are very similar, serology or polymerase chain reaction (PCR) may be necessary to distinguish between the two viruses. The type strain from Australia can infect oats, which is not susceptible to the other graminaceous potyviruses. The perennial host Johnson grass allows virus survival between seasons in rhizomes. Resistant or tolerant hybrids provide the best control option.

Maize Chlorotic Dwarf Virus

Confirmed reports of maize chlorotic dwarf virus (MCDV) infection are limited to the US, where the virus was discovered in 1969. The extent of distribution of MCDV in the US is second to that of MDMV. The virus is associated with the occurrence of indigenous Johnson grass (*S. halepense* (L.) Pers.), its overwintering host, and the black-faced leafhopper, *Graminella nigrifrons* (DeLong and Mohr), its principal vector. Symptoms on susceptible maize are shortening of the upper internodes or proportionate stunting of the internodes, chlorotic clearing or banding of the tertiary leaf veins, and red or yellow discoloration of upper leaves. Infected leaves are also more difficult to tear off a plant. On resistant maize, MCDV induces mainly chlorotic clearing or banding of tertiary veins without marked stunting or discoloration.

Figure 1 (a) Long continuous stripes along the veins of sweet corn caused by maize mosaic virus. (b) Severe dwarfing, stripping of upper leaves and 'leek aspect' of maize rough dwarf virus-infected maize plant. (c) Symptoms of maize streak virus on maize. (d) Maize plants infected with Mal de Rio Cuarto virus showing severe dwarfing and leaves distorted and ragged. (e) Yellow stripe and dwarfing on maize infected by cynodon chlorotic streak virus.

The diagnostic symptom of MCDV infection is vein banding. MCDV's host range is limited to species of the Poaceae. MCDV may complex with MDMV in infections of maize and Johnson grass in the same field or plant and with sugarcane mosaic virus strain MDB in maize. Biological tests for MCDV include transmission by the leafhopper in a semipersistent manner and vascular-puncture inoculation (VPI) to susceptible plants with expression of the diagnostic symptoms of vein clearing or banding of the tertiary leaf veins. Confirmation by serological methods is often necessary. No molecular tests for MCDV have been developed.

Agronomical control of MCDV is possible by early planting of maize to escape peak vector populations and by herbicide eradication of overwintering Johnson grass. However, the latter treatment may result in increased MCDV incidence in susceptible hybrids due to mass migration of viruliferous leafhoppers from dying Johnson grass. Maize inbreds with high levels of tolerance to MCDV have been developed. Virus resistance appears inherited mostly as a dominant to partially dominant trait. Resistant Caribbean germplasm has been identified and the inbred line Oh1VI highly resistant to MCDV has been developed. Commercial hybrids with various degrees of resistance or tolerance and with agronomically satisfactory characteristics are available. The systemic insecticide carbofuran which remains toxic to the leafhopper for up to 55 days after application at planting time is very effective and can control the disease. As a result the economic impact of the virus has decreased and it no longer poses a significant threat to US maize production.

Maize Chlorotic Mottle Virus

Maize chlorotic mottle virus (MCMV) was first reported from Peru in 1973, and was later found in central USA, Hawaii, Mexico, and Argentina. Symptoms vary from a mild chlorotic mottling to severe yellowing, necrosis, and plant death depending on the maize variety and plant age when infected. Plants may be stunted with short, malformed, partially filled ears, and male inflorescences may be shortened with sparse spikes. Under natural infection, 10–15% yield reductions can be observed. Maize is the only natural host of MCMV, and the experimental host range is limited to Poaceae. Laboratory transmission of MCMV can be obtained with six species of chrysomelid beetles, but only *Diabrotica* spp., which have a New World distribution and are mostly tropical, are natural vectors. In areas of South America where maize is grown continuously, MCMV spreads from older plants to younger plants via adult and larval beetles. Control would require a halt in continuous cultivation or management of the beetles. Most maize cultivars are susceptible to MCMV but a new variety (N211) is resistant. Transgenic maize shows milder symptoms from MCMV infection. MCMV is sometimes found as part of the corn lethal necrosis complex.

Maize Dwarf Mosaic Virus

MDMV was first reported on corn in 1965 from Ohio, USA. It is considered to be distributed worldwide. Symptoms of MDMV in corn vary with the stage of host development at time of infection, genotype, and virus strain. Initial leaf symptoms are chlorotic spots and short streaks followed by typical mosaic or mottle. When the plants develop, flecks, streaks, and rings appear on new leaves. Ear formation and development are arrested, resulting in losses in grain yield. MDMV is widely adapted to species of the Poaceae infecting over 66% of 293 species tested. Maize and sorghum are important agricultural hosts. More than 15 species of aphids transmit this virus in a nonpersistent manner. MDMV can be mechanically transmitted; it survives by overwintering in the rhizomes of infected Johnson grass. Crop loss due to MDMV ranges from 9% to 72% for susceptible dent corn hybrids. Losses are generally important in seed production or sweet corn hybrid fields.

Many commercial corn MDMV-resistant hybrids are available. A major gene for resistance to MDMV is a dominant gene (*mdm1*) located on chromosome 6. The control of Johnson grass helps to control disease.

Maize Fine Streak Virus

Maize fine streak virus (MFSV) was isolated in 1999, from sweet corn grown in Georgia, USA. Symptoms in the field include fine chlorotic streaks along the major leaf veins, similar to symptoms associated with maize rayado virus (**Figure 1(c)**). Maize is so far the only identified natural host of MFSV. The virus is transmitted by the leafhopper *Graminella nigrifons* (DeLong & Mohr) in a persistent manner. MFSV is readily transmitted mechanically by kernel vascular-puncture inoculation. Maize lines resistant to maize mosaic virus (MMV: Hi31 and Hi34) are also resistant to MFSV.

Maize Indian Fiji-Like Virus

Symptoms include severe plant dwarfing, dark green leaves, enations on the lower leaf surface, and small malformed ears. All of these symptoms are similar to those produced by Mal de Rio Cuarto virus (MRCV) and maize rough dwarf virus (MRDV) on maize. This unreported reovirus was found in southern India, where its incidence ranged from 4% to 61%.

Maize Iranian Mosaic Virus

First observed in the Fars Province of Iran, it is now distributed in other parts of the country. Maize Iranian mosaic virus (MIMV) symptoms consist of long chlorotic lines and stripes along the veins and sheaths, stunting of the plants, and abortion of ears when infected early. MIMV is serologically distinct from others rhabdoviruses infecting Poaceae plants. In polyacrylamide gel electrophoresis, six proteins corresponding to proteins L, G, N, NS, M, and M2 were detected. MIMV is a new virus in the genus *Nucleorhabdovirus* that may be distantly related to MMV. Delphacids are vectors of MIMV. The natural host range includes maize, oat, rice, and wheat. The disease incidence in maize fields can be as high as 80% depending on sowing date and variety. Late planting effectively controls the virus.

Maize Mosaic Virus

MMV causes a serious disease of maize in tropical and subtropical areas of Africa, the Americas, and Australia. It was first reported in 1914 in Hawaii. MMV can occur in mixed infections with maize stripe virus (MSpV). Initial leaf symptoms on sweet corn inoculated with MMV are long stripes, colored light green to yellow, along the midrib (**Figure 1(a)**). These stripes elongate to form distinct, chlorotic stripes between and along the veins extending on the whole leaf. Stripes can appear on sheath and husk. Stunting of plants is common. Mosaic patterns do not occur despite the disease name. Only species of the Poaceae are known to be hosts of MMV. It occurs naturally in itchgrass, maize, and *Setaria vulpiseta* (Lam.) Roem. & Schult. MMV is transmitted in a persistent-propagative manner by the corn planthopper *Peregrinus maidis* (Ashmead), the sole vector. Transmission is by both nymphs and adult males and females. MMV has been shown to be inoculated by vascular puncture of maize kernels (VPI). Several contact and systemic insecticides can control *P. maidis*. Host-plant resistance is likely to be the most effective means of control. Tropically adapted parental inbred lines of maize resistant to MMV have been developed and sweet corn cultivars have been bred and released, both by the University of Hawaii. From Caribbean and Mascarene germ plasm, high resistance was found in line Hi40 and resistance to some levels was found in lines 37–2, A211, and Mp705.

Maize Mottle Chlorotic Stunt Virus

This virus was first reported, in 1937, in East Africa and called maize mottle virus (MMotV). Maize mottle chlorotic stunt virus (MMCSV) is restricted to tropical African countries and the adjacent islands. The virions are spherical, *c.* 40 nm in diameter. MMCSV can be found in mixed infections with maize mosaic and maize streak viruses. Symptoms on maize appear as chlorotic mottling with mild to severe chlorosis of tertiary veins and stunting. Young mottled leaves fail to support themselves in a normal upright position. Tassel abortion can occur in severe cases. Maize is the only known natural host of the virus which is transmitted by several *Cicadulina* spp. in a persistent manner. Disease control can be achieved by vector control or by growing resistant cultivars, which are now available. Most maize streak virus-resistant cultivars have also moderate to high levels of resistance to MMCSV.

Maize Necrotic Streak Virus

Maize necrotic streak virus (MNeSV) was discovered in 2000 in maize samples from the US state of Arizona. The virus is of little agronomic importance. Virions are isometric, *c.* 32 nm in diameter. MNeSV appears to be a member of the family *Tombusviridae*. Distinctive symptoms of MNeSV infection are initially chlorotic spots and streaks that become spindle shaped. They later coalesce into long chlorotic bands that become translucent and necrotic around the edges. Plant stalks show a chlorosis that becomes necrotic, a symptom especially distinctive for

MNeSV. There is no known aerial vector – the virus is transmitted through the soil. Maize lines Oh 1VI OSU23i and Mo17 are highly resistant to MNeSV.

Maize Rayado Fino Virus

Maize rayado fino virus (MRFV) was first described in 1969 in Costa Rica and El Salvador. MRFV is widespread and becoming increasingly important in tropical areas of the Americas. It is the type member of the genus *Marafivirus*, and is the only known indigenous virus of maize in Mesoamerica. Initially, infection causes fine chlorotic dots or stipples, which coalesce into chlorotic stripes at the base and along the veins of young leaves. As the plant grows older, the symptoms become less conspicuous and may disappear when the plant reaches maturity. In locally adapted Central American varieties, the virus incidence varies from 0% to 20%, but may reach 100% in more susceptible foreign or newly developed cultivars. Maize, teosinte, and the perennial *Zea* spp. are the only known natural hosts. In nature, MRFV is transmitted exclusively by the corn leafhopper vector *Dalbulus maidis* (DeLong and Wolcott) in a persistent manner. The natural host range of MRFV may be limited by the ability of the vector to feed on different species. It has been proposed that MRFV, its insect vector, and the maize host co-evolved in a triad in which the parasitic members (insect and virus) displayed highly specialized interactions. It has been suggested that MRFV originated in Mexico or Guatemala, from where it spread in the region.

MRFV is one of the three major components of the corn stunt disease complex. MRFV is endemic in many areas of Central and South America, where the virus and *D. maidis* reach epidemic levels. The use of systemic insecticides does not reduce virus incidence. Crop rotation, mixed plantings, and selection of planting dates result in significant decreases in virus infection. The Guatemalteco variety introduced in Ecuador in 1962 is used as a source of virus tolerance. Production of transgenic plants resistant to MRFV and their evaluation for their susceptibility are in progress in Costa Rica.

Maize Rough Dwarf Virus

This virus was discovered in Europe when corn hybrids were introduced from the US after World War II. In 1949, in Italy, a severe outbreak threatened the maize cultivation, lowering the yield by ~40%. Later, MRDV was reported in several European countries, where it had the potential to be economically damaging. In China, a similar disease was found to be caused by rice black-streaked dwarf virus (RBSDV), not by MRDV. In young field-grown corn, symptoms caused by MRDV are dark green color of leaves, stunting and irregular swellings of veins (enations) along the lower surfaces of leaves and sometimes also of leaf sheaths, ligules, and husks. The enations are rough to the touch, hence the disease name. Plants are stunted with increased girth giving the plant a 'leek' aspect (**Figure 1 (b)**). Short chlorotic streaks develop on mature leaves and coalesce into yellowish green stripes parallel to the veins. Later the leaves can turn reddish. Tassels are sterile. The root system is reduced, roots develop swellings, resulting in their frequent splitting. MRDV has a fairly wide experimental host range including only a few species of Poaceae, among which maize is the only one of economic importance. Oats, wheat, and several grasses can be infected naturally. The planthopper *Laodelphax striatellus* (Fallén) is the natural vector and the only known winter host of MRDV. However, in northern Italy, oats were reported as the first overwintering plant. Planthoppers can be controlled either by spraying grasses around the fields with an insecticide in the spring, by seed treatment with imidachlopride, or by row application of an insecticide at sowing time (e.g., carbofuran). Some maize hybrids are less susceptible to MRDV, but the resistance can be overcome if the vector population is very high.

Maize Sterile Stunt Virus

Maize sterile stunt virus (MSSV) was first reported from Australia in 1977. It is a strain of barley yellow striate mosaic virus (BYSMV). The virus causes severe stunting, sterility, purple coloration, and top necrosis in a small number of maize genotypes. MSSV is transmitted in a persistent manner by the delphacid planthopper *Sogatella longifurcifera* Esaki and Ishihara and inefficiently by *Sogatella kolophon* (Kirk) and *P. maidis* (Ashm.). Natural hosts of MSSV are barley, maize, triticale, and wheat.

Maize Streak Virus

Maize streak was first recorded in 1901 in South Africa and is the most widespread and important disease of maize in sub-Saharan Africa and adjacent islands. Maize streak symptoms are characterized by the development of chlorotic spots and streaks in longitudinal lines on leaves. There is a progressive increase in the number and length of streaks that occur on new leaf tissue. The streaks often fuse laterally, resulting in narrow broken chlorotic stripes, which can in some cases cover the entire leaf. In very susceptible maize cultivars, severe chlorosis occurs leading to stunted plant and premature death. MSV is transmitted by several species of leafhoppers in the genus *Cicadulina*, with *C. mbila* (Naudé) being the most common vector. It transmits MSV in a circulative nonpropagative manner. Streak symptoms have been described in numerous species of wild grasses. Late plantings are generally more severely infected than early ones. The major source of MSV infection on maize is a previously infected maize

field. Therefore, planting of maize close to a previous maize field must be avoided. Carbamate insecticides (such as carbofuran) can effectively prevent MSV transmission but prices of chemicals and equipment for spraying are often prohibitive for smallholder farmers. The development of insect resistant plants is under study.

Maize Stripe Virus

MSpV has often been confused with MMV because of their similarity in vector specificity, mode of transmission, symptomatology, and host range. It often occurs in mixed infection with MMV. The virus is commonly found in most parts of the subtropics and the tropics where the vector occurs. MSpV is found to infect maize, sorghum, and itchgrass (*Rottboellia exaltata* L.). It can be transmitted experimentally to barley, oats, rye, and triticale. Initial symptoms on sweet corn are fine chlorotic stipplings and narrow stripes similar to that of MMV. Later, continuous chlorotic stripes develop with varying width and intensity which show a 'brushed out' appearance, and apical bending occurs. Infected plants are often stunted and die prematurely. The virus is transmitted by *P. maidis* (Ashm.), also a vector of MMV, in a persistent-propagative manner. Both nymphs and adults transmit the virus. Control of the maize stripe disease is largely dependent on chemical control of *P. maidis*.

Mal de Rio Cuarto Virus

The Mal de Rio Cuarto was first reported in maize fields of the Rio Cuarto County in Argentina, at the end of the 1960s. MRCV is a major constraint to maize production in Argentina. It spread in the endemic area of the province of Cordoba, but also appeared in the central region. MRCV initially reported as a strain of MRDV but is now considered to be a separate species within the genus *Fijivirus*. In field-infected maize, the symptoms show up within 4–5 weeks. Plants are dwarfed, showing fine chlorotic flecks on secondary and tertiary veins. Stems are usually flattened with shortened internodes and leaves from the upper third may be stiff, distorted, and ragged (**Figure 1(d)**). The root system is fragile, and greatly reduced. The ears are reduced and malformed. Natural dispersion of the virus occurs by the planthopper *Delphacodes kuscheli* Fennah in a propagative manner. Apart from maize, MRCV can infect several monocots, winter small grains (barley, oats, rye, and wheat), spring–summer grains (millet and sorghum), and several spring–summer annual and perennial weeds. The initial inoculum source is its vector that feeds on winter cereals like oat and wheat, where it acquires the virus and migrates to maize when these cereals become senescent. Early sowing of maize helps to prevent high virus infection. Systemic insecticides (carbofuran, imidachlopride) applied as seed coating protect maize seedlings from the vector. Some commercial hybrids with high tolerance levels are now available.

Pennisetum Mosaic Virus

A potyvirus was isolated from whitegrass (*Pennisetum flaccidum* Griseb) in North China, in the early 1980s and was later named pennisetum mosaic virus (PenMV). This virus infects maize and sorghum naturally. Symptoms on maize are not specific when compared to other members of the sugarcane mosaic virus (SCMV) subgroup in the genus *Potyvirus*. The virus is less important, in terms of yield loss, than the other prevalent viruses of the SCMV subgroup.

Rice Black-Streaked Dwarf Virus

The causal agent of characteristic symptoms on maize (stunting and rough white line veins of leaves) was initially identified as MRDV in China in the late 1970s. Subsequently, molecular studies demonstrated that rice black-streaked virus was the cause of the disease. These two viruses are quite similar in host range, serology, morphology, and the genome sequence. Reverse transcription-polymerase chain reaction can discriminate between RBSDV and MRDV. Host range is limited to species in the Poaceae: oats, hordeum, oryza, triticum, and zea. This virus is transmitted by the leafhopper *L. striatellus* in a persistent manner. RBSDV is managed mainly by control of leafhoppers.

Sugarcane Mosaic Virus

SCMV was first detected in sugarcane in 1919 and in maize in 1963, both in the US. SCMV occurs throughout the world. Infected maize plants develop distinct mosaic symptoms which are especially clear on the lower part of the younger leaves. Very susceptible cultivars react with a strong chlorosis, sometimes together with a red striped pattern. It is often impossible to distinguish SCMV and MDMV infections on the basis of symptoms. In Southern Europe, MDMV is the most prevalent virus. Maize inbred lines resistant to the virus have been described and transgenic maize expressing the coat protein gene of SCMV is also available.

Wheat Streak Mosaic Virus

The symptoms consist of linear arrays of small, yellow-ringed eyespots. On some germplasm, the spots and streaks later form mosaics, or mottle patterns. WSMV is mite transmitted. Three independent genes controlling resistance to WSMV have been identified in maize. Resistance in certain maize lines is not effective against all strains of WSMV. Late planting of corn adjacent to maturing wheat should be avoided.

Maize Diseases Caused by Complex of Pathogens

Corn Lethal Necrosis Complex

This disease is caused by a synergistic interaction between MCMV and certain members of the family *Potyviridae* (MDMV-A, SCMV-MDB, JGMV, and WSMV). Maize plants infected at early stages develop leaf chlorosis followed by necrosis. Plants are stunted, produce small deformed ears, and die prematurely. Only terminal leaves show symptoms with a late infection. Crop losses of up to 90% have been reported.

Corn Stunt Complex

Corn stunt disease complex, also known by the names achaparramiento, maize stunt, and red stunt, is caused by two or more of the following agents: MRFV, corn stunt spiroplasma (CSS), and maize bushy stunt phytoplasma (MBSP). The corn stunt complex pathogens appear to be restricted to the Americas. Early symptoms of CSS consist of yellowish streaks in the youngest leaves. Later, much of the leaf turns purple. Infected plants are stunted and will often have more ears, but the ears are smaller. Maize plants infected with MBSP show stunted growth and reduced grain production, ear proliferation, and greater tillering. The leafhopper vector *D. maidis* (DeLong & Wolcott) can simultaneously transmit CSS, MBS, and MRFV. Although *D. maidis* is the primary vector, other leafhoppers transmit one or more of the corn stunt complex pathogens under experimental conditions.

See also: Cereal Viruses: Rice; Cereal Viruses: Wheat and Barley; Maize Streak Virus.

Further Reading

Lapierre H and Signoret PA (2004) *Viruses and Virus Diseases of Poaceae Gramineae,* 857pp. Versailles, France: INRA.

Lübberstedt T, Ingvardsen C, Melchinger AE, *et al.* (2006) Two chromosome segments confer multiple potyvirus resistance in maize. *Plant Breeding* 125: 352–356.

Redinbaugh MG, Jones MW, and Gingery RE (2004) The genetics of virus resistance in maize (*Zea mays* L.). *Maydica* 49: 183–190.

Cereal Viruses: Rice

F J Morales, International Center for Tropical Agriculture, Cali, Colombia

Glossary

Ambisense A viral nucleic acid that can be 'read' in both directions to be translated into viral encoded proteins.

Capsid Protein envelope of the viral genome (nucleic acid).

Circulative A virus that passes through the intestine of an insect vector into the hemolymph, and eventually reaches the salivary glands of the insect to be injected back into a plant.

Etiology Studies the cause of disease.

Inclusion Virus particles or viral components found in infected plant cells.

Incubation Period of time that a virus needs inside its vector before it can be transmitted to a susceptible host.

ORF Any sequence of DNA or RNA that can be translated into a protein.

Persistent A virus that is not lost by a vector after feeding in a series of healthy plants, without having access to a virus-infected plant (virus source).

Pinwheel A cytoplasmic protein inclusion induced by potyviruses.

Propagative A virus that multiplies in its vector.

Semi-persistent Virus is lost after a vector feeds on a series of virus-free plants.

Single-stranded Viruses consisting of only one strand of viral RNA or DNA.

Transovarial transmission Of a virus from a female insect vector to her progeny.

Transtadial A virus that is retained by an insect vector after each developmental stage until it reaches adulthood.

Virion A mature virus particle (capsid protein, nucleic acid, and any other constituents).

Viroplasm Amorphous aggregation of viral components.

Introduction

Rice (*Oryza sativa* L.) is the second most extensive food crop grown in the world (*c.* 154 million ha) after wheat, and is the main food crop grown in the tropics. Half of the world's population (over 3 billion people) depends on rice, particularly in Asia, where annual per capita rice consumption may be as high as 170 kg in countries such as Vietnam, as compared to 9 kg in the United States. Asia produces almost 90% of all the rice cultivated in the world (136 million ha), mainly China and India, where rice cultivation and domestication seemed to have started over 5000 years ago. Africa cultivates approximately 9 million ha, and Latin America 6.8 million ha of rice. Nigeria and Brazil are the main producers of rice in Africa and Latin America, respectively, with average cultivated areas of 3.8 million ha, each.

Given the history and extensive cultivation of rice around the world, including Europe (565 000 ha) and the United States (1 355 000 ha), it is not surprising to note that this crop has several pest and disease problems of significant socioeconomic importance. Among the various biotic constraints of rice production, viruses constitute an important group of plant pathogens, particularly in Asia. Viral diseases are also considered major constraints to rice production in Africa and Latin America, even though only two viruses are considered pathogens of economic importance outside Asia.

Rice is also affected in the tropics by numerous insect pests, particularly by several species of leafhoppers and planthoppers (Homoptera:Auchenorrhyncha) that cause direct feeding damage and, more important, transmit economically important plant viruses. This article describes the main viral diseases of rice in the world; their causal viruses and vectors; and the different disease management strategies implemented to date.

Rice Dwarf

Rice dwarf was first described in Japan, in 1883. A Japanese rice grower was the first person to demonstrate in 1894, the relation between this disease and the presence of leafhoppers in affected rice fields. Rice dwarf is also known to occur in Korea, China, Nepal, and the Philippines. In 1885, K. Takata reported that the leafhopper *Recilia dorsalis* was 'responsible' for the disease. In 1900, the leafhopper *Nephotettix cincticeps* was thought to be the cause of rice dwarf. In 1899, *N. cincticeps* individuals captured in a disease-free area could not induce disease in healthy rice plants, unless these individuals had been previously allowed to feed on rice dwarf-affected rice plants. Despite this evidence suggesting the existence of an unknown causal agent, it was not until 1960 that the first electron micrographs of an isometric virus *c.* 70 nm in diameter were obtained. The suspected virus was shown in 1966–70, to contain double-stranded (ds) RNA, divided in 12 different segments (4.354–0.828 kb) with a total genome size of 25.13 kb. The causal virus was classified in 1972 as *Rice dwarf virus*, a species of the genus *Phytoreovirus*, in the family *Reoviridae*. The virus replicates in the cytoplasm of infected cells, and the dsRNA genome is encapsidated in super-coiled form.

Rice dwarf virus (RDV) is transmitted by the leafhoppers *N. cincticeps* (main vector in Japan, Korea and China), *Nephotettix nigropictus* (formerly *N. apicalis*; the main vector in Nepal and the Philippines), *Nephotettix virescens*, and *R. dorsalis* (formerly *Inazuma dorsalis*). Few individuals of these species are active or potential vectors, and they acquire the virus by feeding for up to a day on infected plants. The incubation period of the virus in a potential vector ranges from 12 to 25 days ('propagative' virus), and the leafhopper vector remains viruliferous for life ('persistent' mode of transmission). Viruliferous individuals require from a few minutes to half an hour to transmit the virus to a healthy plant, and there is transovarial transmission of RDV to the progeny of viruliferous female vectors. Some nymphs infected through the egg die prematurely. The average lifespan of viruliferous *N. cincticeps* females is 12.1 days, and of virus-free females is 16.6 days. RDV is not mechanically transmissible and it is not transmitted by sexual seed or pollen.

RDV induces stunting and chlorotic specks in infected rice plants. RDV's host range is restricted to the Gramineae, including the genera *Alopecurus*, *Avena*, *Echinochloa*, *Oryza*, *Panicum*, *Paspalum*, *Poa*, and some cultivated species, namely, barley, millet, oats, rye, and wheat. The disease usually appears in the field after transplanting, suggesting an early infection in the seedbed stage. Early infection generally results in severe stunting, shortening of internodes, and small, rosette-like tillers. The root system of diseased plants does not develop well either, and infected plants usually do not produce panicles or seed. Diseased plants remain green in contrast to virus-free plants, which naturally turn yellow when mature. Virions are found in phloem, cytoplasm, and in cell vacuoles. Viroplasms can be observed in infected cells. Genetic resistance to RDV has been observed in some rice varieties, but current efforts are directed toward the development of genetically modified (transgenic) rice plants possessing resistance to the virus. Complementary cultural practices and other integrated disease management strategies are also recommended.

Rice Black Streaked Dwarf

Rice black streaked dwarf was first reported to affect rice in Japan in the early 1950s, and it also spreads in China and Korea. This disease was shown to be different from rice dwarf, and to be associated with the planthopper *Laodelphax striatellus*. The causal agent was isolated in 1969 and further characterized in 1974 as a spherical virion *c.* 60 nm in diameter, belonging to the reovirus group. Rice black streaked dwarf virus (RBSDV) is currently classified as a member of a species of the genus *Fijivirus*, family *Reoviridae*. It has 10 dsRNA segments coding for the RNA polymerase, core protein, nonstructural components, spikes, major outer shell, and other proteins of unknown function. Three major proteins (130, 120, and 56 kDa) and three minor proteins (148, 65, and 51 kDa) have been detected in RBSDV virions. The 56 kDa protein is the main component of the outer capsid shell. Most of the genome segments are monocistronic, and replication occurs in the cytoplasm of phloem cells. Inclusion bodies of unknown nature are observed in malformed tissue. Some degree of genomic variability among RBSDV isolates has been observed.

RBSDV is transmitted by the delphacid vector *L. striatellus*, and occassionally by *Unkanodes sapporona* and *Unkanodes albifascia*, in a persistent, circulative, and propagative manner. The proportion of active transmitters of *L. striatellus* is about 30%. Shortest acquisition feeding period was 30 min; and incubation periods ranging from 4 to 35 days in the vector *L. striatellus* have been reported. The virus is not transmitted congenitally. *Laodelphax striatellus* breeds on rice, wheat, and barley, but not on maize; whereas *U. sapporona* favors maize, wheat, and barley, but does not breed on rice. Besides rice, RBSDV infects oats, barley, wheat, maize, rye, and species of *Agrostis*, *Alopecurus*, *Digitaria*, *Echinochloa*, *Eragrostis*, *Glyceria*, *Lolium*, *Panicum*, and *Poa* in the Gramineae. Perpetuation of the virus between seasons occurs through overwintering planthoppers. Main symptoms associated with this disease are: stunting, darkening of leaves, leaf malformation, and waxy swellings along veins of the abaxial leaf surface, which later form dark tumors. Infected plants usually produce excessive tillers.

Resistance to RBSDV has been observed under field conditions. There is at least one dominant gene involved in the resistance to RBSDV, but some modifying genes may also be involved in the outcome of the interaction between the plant genotype and the amount of virus inoculated. Some wild rice species have exhibited moderate resistance to the virus. Cultural practices aimed at reducing vector populations, such as sowing dates that do not coincide with peak populations of the vector, are recommended.

Rice Gall Dwarf

Rice gall dwarf was first observed in Thailand, in the early 1980s. This disease is also present in China, Malaysia, and Thailand. The causal virus belongs to a species in the genus *Phytoreovirus*, family *Reoviridae*. Virions are icosahedral (60–70 nm) and possess 12 dsRNA genome segments and up to seven structural proteins (45–160 kDa). Rice gall dwarf virus (RGDV) is confined to phloem tissues of the plant host. Largest genome segment has 4.4 kb, and the smallest, 0.6 kb. Virions are found in the phloem, cytoplasm, and cell vacuoles in low concentrations, together with viroplasms containing immature virions.

RGDV is transmitted by the leafhoppers *N. nigropictus*, *R. dorsalis*, *N. cincticeps*, *N. malayanus*, and *N. virescens*, in a persistent, circulative, and propagative manner. RGDV is transmitted transovarially, is acquired rapidly from infected rice plants, and has an incubation period of 10–20 days in the insect vector. Main symptoms in rice are stunting, translucent gall formation on the underside of the leaves and on leaf sheaths (enations), and permanent dark green leaves. Infected plants produce few tillers, and form poorly developed panicles. Other grasses susceptible to this virus include oats, barley, rye, wheat, Italian grass (*Lolium* sp.), and *Oryza rufipogon*. Japanese grass (*Alopecurus aequalis*) is an important reservoir of RGDV; and maize also seems to be a host of RGDV in China.

Cultural practices aimed at breaking the cycle of the leafhopper vectors that move from spring to autumn plantings; and the elimination of ratoon plantings from over-wintered rice are recommended.

Rice Ragged Stunt

Rice ragged stunt was first reported from the Philippines and Indonesia in 1976. It is currently distributed in Bangladesh, China, India, Japan, Malaysia, Sri Lanka, Taiwan, and Thailand. The causal virus was isolated in the early 1980s, and shown to belong to a species of the genus *Oryzavirus*, family *Reoviridae*. Rice ragged stunt virus (RRSV) has icosahedral, double-shelled particles (75–80 nm) with conspicuous spikes. Oryzaviruses have a genome of 10 linear dsRNA segments with a total size of 26.6 kb. RRSV particles are composed of five major structural proteins (33–120 kDa) and three nonstructural proteins (31, 63, and 88 kDa). The dsRNA segments contain a single ORF each, except S4, which contains two ORFs. RRSV is serologically related to Echinochloa ragged stunt virus.

RRSV is transmitted by the delphacid planthoppers *Nilaparvata lugens* (brown planthopper) and *N. bakeri*, in a persistent, circulative, and propagative manner. RRSV is not transovarially transmitted. RRSV infects rice and other *Oryza* species, causing stunting, enations, ragged (serrated) leaves, and suppression of reproductive structures. Infected rice plants remain green when healthy plants have senesced. RRSV also infects maize and wheat. Virions are found in phloem enations; and infected cells show the presence of viroplasms and cell wall proliferation.

RRSV persists in rice and in the brown planthopper, and, consequently, continuous rice planting perpetuates the virus and its vectors. Planthoppers migrate from the tropics to temperate areas in the summer time. Resistance to RRSV has been identified in some breeding lines; but molecular breeding techniques have been preferred to genetically modify rice genotypes for resistance to the virus.

Rice Tungro

Tungro means 'degenerate growth' in Ilocano, a Philippine language. Tungro-like symptoms were first observed in the Philippines, Indonesia, Malaysia, and Thailand in the early 1960s, under different names, such as 'mentek', 'penyakit habang', 'penyakit mera', and 'yellow orange leaf'. Severe outbreaks of tungro disease occurred in the late 1960s, following the introduction of early maturing, fertilizer-responsive, and high-yielding rice varieties. The disease has been reported from other Asian countries, namely, India, Bangladesh, and China. Tungro was first associated with the presence of leafhoppers in affected rice fields, but one of the causal agents could only be visualized in 1967, as isometric particles, 30–35 nm in diameter. Between 1967 and 1970, at least four different strains of the tungro 'virus' were identified, and this number increased to eight strains by 1976. In 1978, Indonesian scientists reported the observation of both isometric (30 nm) and bacilliform (35 × 350 nm) particles in rice plants affected by the 'penyakit habang' disease (tungro) in Indonesia. Severe symptoms were associated with both particles, whereas rice plants infected only by the bacilliform virus showed moderate symptoms. Plants infected only by the isometric virus remained symptomless. The isometric particles were transmitted by the leafhopper *N. virescens*, but the bacilliform virus could not be transmitted in the absence of the isometric virus. This constituted the first demonstration that tungro is a synergistic disease, and that the isometric virus acts as an 'assistor' or 'helper' virus for the bacilliform virus that causes main symptoms associated with tungro disease. In 1979, *N. nigropictus* was also shown to be an occasional vector of the isometric virus, but not of the bacilliform virus. The shortest acquisition feeding for *Nephotettix impicticeps* is 5–30 min, and there is no incubation period. Therefore, the virus does not persist in the vector; and there is no trans-stadial or transovarial passage of the virus in the insect vector.

Rice tungro disease is thus caused by two different viruses: rice tungro spherical virus (RTSV), a positive-sense RNA virus, and rice tungro bacilliform virus (RTBV), a circular dsDNA virus. The nucleotide sequence of rice tungro bacilliform badnavirus (RTBV) DNA was determined in 1991. The circular genome has 8.3 kb and one strand contains four open reading frames (ORFs). One ORF encodes a protein of 24 kDa (P24). The other three ORFs potentially encode proteins of 12, 194, and 46 kDa. Comparative analyses with retroviruses suggested that the 194 kDa polyprotein is proteolytically cleaved to yield the virion coat protein, a protease, and replicase (reverse transcriptase and RNase H), characteristic of retro-elements. The DNA sequence of RTBV suggests that RTBV is a pararetrovirus. RTBV is currently classified as a member of a species of the genus *Badnavirus*, family *Caulimoviridae*.

The 12 433 nucleotide (nt) sequence of RTSV contains a large ORF, capable of encoding a viral polyprotein of 390.3 kDa. Two viral subgenomic RNAs of *c.* 1.2 and 1.4 kb, respectively, were detected in RTSV-infected leaf tissues. There are at least three capsid protein subunit cistrons near the N-terminus of the large ORF. The C-terminal half of the large ORF revealed conserved protein sequence motifs for a viral RNA polymerase, proteinase, and a putative nucleoside triphosphate (NTP)-binding protein. The sequence motifs are arranged in a manner that resembles those of picorna-like viruses. RTSV is the type species of the genus *Waikavirus*, family *Sequiviridae*.

Rice tungro bacilliform virus isolates from Bangladesh, India, Indonesia, Malaysia, and Thailand were compared with the type isolate from the Philippines. Restriction endonuclease maps revealed differences between the isolates, and cross-hybridization showed that they formed two groups, one from the Indian subcontinent and a second one from Southeast Asia.

Diseased leaves become yellow or orange-yellow, and malformed, starting with the tip of the lower leaves. Young infected leaves may show mottling and chlorotic stripes parallel to the veins. Root development is poor and the reproductive stage may be delayed. Affected panicles are usually small and sterile. Symptom expression varies considerably depending on the variety affected and environmental conditions. Yield losses depend upon plant age at the time of infection, and the plant genotype. *Eleusine indica, Echinochloa colonum, E. crusgalli, Oryza* spp., *Paspalum* sp., *Setaria* sp., sorghum, and wheat have also shown to be susceptible to tungro disease.

Several sources of resistance to RTSV have been identified at the International Rice Research Institute (IRRI). Promising lines have been identified and some of the superior lines resistant to tungro have been released as varieties. IR20, IR26, and IR30 were the first tungro-resistant cultivars released by IRRI followed by IR28, IR29, IR34, IR36, IR38, and IR40 during the 1970s. Some cultivars resistant to *N. virescens*, such as IR54, IR56, IR60, and IR62, were released in the 1980s, followed by PSBRc4, PSBRc10, PSBRc18, and PSBRc28 during the 1990s. Recently, five tungro-resistant lines showing resistance to RTSV were released as cultivars in Indonesia and the Philippines. Utri Merah, Balimau Putih, Habiganj DW8, and *O. rufipogon* served as donors of resistance genes.

Rice Stripe

Rice stripe is the second oldest viral disease of rice, having been first observed in the 1890s, in Japan. In 1931, the disease was associated with the presence of the leafhopper *L. striatellus*, but its causal agent could not be isolated. The disease continued to spread in Japan in the early 1950s,

due to the introduction of new cultural practices (seeding of rice in wheat fields and early plantings) that expanded the temporal availability of young rice plants to the leafhopper vector. Rice stripe eventually emerged in Korea, Taiwan, China, and southwestern regions (Vladivostok) of the former USSR. The causal agent eluded detection until 1975, when Japanese scientists isolated thin (3–8 nm), coiled, filamentous particles of different lengths (2110–510 nm) from rice stripe-affected plants. In 1989, rice stunt virus (RSV) was shown to consist of four ssRNA components, similar to the genome of maize stripe virus (MSpV) isolated in 1981 in the United States. These viruses, together with rice hoja blanca virus (RHBV), isolated in Colombia in 1982, formed the 'rice stripe virus group' in 1988. Further molecular work led to the creation of the genus *Tenuivirus* in 1995. Tenuiviruses consist of up to six segments of linear, negative-sense, and ambisense ssRNA. RSV is believed to have four to five segments, depending on the isolate. Sizes are *c.* 9 kb for RNA 1 (negative polarity); 3.3–3.6 kb for RNA 2 (ambisense), 2.2–2.5 kb for RNA 3 (ambisense); 1.9–2.2 kb for RNA 4 (ambisense); and 1.2 kb for RNA 5 (negative polarity). The viral genome encodes structural (32–35 kDa) and nonstructural proteins (230 kDa), both of which are useful for diagnostic purposes. The nucleocapsid protein is encoded by the 5′ proximal region of the virion-complementary sense strand of RNA 3. Some proteins are translated from subgenomic RNAs.

RSV is transmitted by *L. striatellus*, the 'smaller brown planthopper', *Terthron albovittatum, Unkanodes sapporonus*, and *U. albifascia* (formerly *Ribautodelphax albifascia*). The virus multiplies in the insect vector (propagative) and there is transovarial passage of the virus to the progeny. The proportion of active vectors of *L. striatellus* can be as high as 54%. The shortest virus acquisition feeding period may be less than 30 min, and the incubation period in the vector is 5–10 days, although longer incubation periods have been reported for *L. striatellus* (21 days) and *U. albifascia* (26 days). The virus persists in its planthopper vectors for life ('persistent' transmission). RSV is not transmitted manually or through seed. The main hosts of *L. striatellus* are wheat, barley, and Italian rye in spring, and upland rice, *Echinochloa crusgalli*, and *Digitaria* spp. in summer.

Initial symptoms induced by RSV are failure of emerging leaves to unfold normally. Affected leaves may show chlorosis, and chlorotic stripes running parallel to the leaf veins. Affected leaves may eventually develop necrosis and die. In infected, older leaves, a chlorotic mottle may appear. The tillering capacity is often reduced, and panicles may fail to produce seed. RSV also infects at least 35 other species of plants in the Gramineae, including oats, barley, rye, sorghum, wheat, and maize.

All the Japanese paddy rice cultivars tested as seedlings proved susceptible, whereas most Japanese upland rice cultivars are highly resistant or immune. Outside Japan, most *indica* rice cultivars were resistant, but *japonica* rice cultivars were susceptible. Resistance in Japanese upland varieties was found to be controlled by two complementary dominant genes (St_1 and St_2). Multiple alleles of the latter gene may be responsible for the resistance associated with *indica* cultivars. Partial dominant resistance has been identified, linked to another dominant gene (St_3). Resistance to the virus is not necessarily linked to resistance to the vector. Transgenic rice cultivars (coat protein mediated resistance) possessing resistance to RSV have already been developed.

Rice Grassy Stunt

Rice grassy stunt was first observed in 1963 in the Philippines, where the characteristic stunting was initially referred to as 'yellow dwarf' or 'rice rosette'. This disease is now present in China, India, Indonesia, Japan, Malaysia, Sri Lanka, Taiwan, and Thailand. The association of this disease with the brown planthopper, *N. lugens*, was made in 1964. The causal agent of rice grassy stunt was isolated in the early 1980s, as a member of a distinct species of the genus *Tenuivirus*. Hence, rice grassy stunt virus (RGSV) shares most of its physicochemical characteristics with RSV. However, RGSV has an additional (sixth) ssRNA segment (ambisense), and its RNA 3 (3.1 kb) and RNA 4 (2.9 kb) are different from the corresponding segments of other tenuiviruses. RGSV RNA 1 (9.8 kb), 2 (4.1 kb), 5 (2.7 kb), and 6 (2.6 kb) are homologous to RNAs 1, 2, 3, and 4, respectively, of other tenuiviruses. Both the capsid and non-structural proteins of RSV and RGSV are serologically related. A tentative tenuivirus, rice wilted stunt virus, originally reported in 1981 from Taiwan, is often considered a synonym of RGSV.

RGSV is transmitted by the planthoppers *Nilaparvata bakeri*, *N. lugens*, and *N. muiri*, in a persistent manner. RGSV multiplies in the vector, but it is not transmitted congenitally. Main symptoms consist of severe stunting, excessive tillering, erect growth, and short, narrow chlorotic leaves. Young leaves may show mottling and/or stripes. Inclusions are present in infected cells. Other hosts of RGSV include: *E. colonum*, *Cynodon dactylon*, *Cyperus rotundus*, *Leersia hexandra*, and *Monochoria* sp.

Genetic resistance to RGSV has been identified in some rice varieties. A single dominant gene conferring resistance to RGSV was identified in *Oryza nivara*. Some sources of RGSV resistance are susceptible to the planthopper vectors.

Rice Hoja Blanca

Rice hoja blanca ('white leaf') was first observed in the Cauca Valley of Colombia, South America, in 1935. By 1940, the disease was present in most rice-producing regions of Colombia. The disease was detected in Panama in 1952, in Cuba in 1954, in Venezuela in 1956, in Costa Rica in 1958, in El Salvador in 1959, in Guatemala in 1960, and in Nicaragua, Honduras, and the Dominican Republic in 1966. In South America, rice hoja blanca also occurs in Ecuador and Peru. The virus was detected in southern United States in the late 1950s, but it is not a problem there. Until the end of the twentieth century, hoja blanca was the only known viral disease of rice in the Americas.

The etiology of rice hoja blanca remained elusive until 1983, when virus-like particles isolated from infected rice plants in Colombia, were shown to be similar to those described in Japan for RSV. The causal virus was finally characterized as a member of the rice stripe virus group created in 1988, which also include MSpV. In 1995, these viruses became species of the genus *Tenuivirus*. The main physicochemical characteristics of these viruses have been described above, but RHBV virions consist of four species of ssRNA, with a total genome of 17.6 kb. RNA 1 is 9.8 kb and encodes the viral polymerase. RNA 2 (3.5 kb) has the capacity to code for two proteins in an ambisense manner, and RNA 3 (2.3 kb) encodes two proteins in an ambisense manner, including the capsid protein (35 kDa). RNA 4 (1.9 kb) has the same basic arrangement as RNAs 2 and 3, and codes for a nonstructural protein (23 kDa). Echinochloa hoja blanca virus (EHBV) was initially considered a strain of *Rice hoja blanca virus*, because it coexists with this virus in most affected rice fields, but it is now considered a distinct, serologically related tenuivirus.

RHBV is transmitted in a persistent, circulative, and propagative manner by the planthopper *Tagosodes orizicolus* (previously *Sogatodes oryzicola*). The virus also infects the insect vector and passes transovarially to the progeny of viruliferous females (80–95% efficiency). However, only 5–15% of the wild *T. orizicolus* population has the ability to transmit the virus in nature. Incubation periods of 6–36 days have been reported for RHBV in its planthopper vector. *Tagosodes cubanus*, the planthopper vector of EHBV, has been reported to transmit RHBV from rice to *E. colonum* but not from rice to rice, or from *Echinochloa* to rice, unless the viruliferous insects are forced to feed on the test plants. Colonies containing over 90% active vectors can be developed by crossing known male and female vectors. If left unattended, these colonies revert to their natural 5–15% of active vectors, suggesting that viruliferous individuals are at a biological disadvantage with respect to nontransmitters. The acquisition feeding period observed for at least 50% of the potential vectors to transmit RHBV is 15 min, with an optimum of 1 h. The incubation period is 6–9 days for individuals who acquire the virus transovarially, but may be up to 36 days for individuals acquiring the virus as nymphs or adults (males usually have a shorter life span and must acquire the virus transovarially). Nonviruliferous *T. orizicolus* females lay more eggs than viruliferous females, which suggested a deleterious effect of RHBV in its insect vector.

RHBV induces small chlorotic spots at the base of infected leaves, which then show longitudinal chlorotic stripes. The following leaves may be completely chlorotic (hence the name 'white leaf'). Plants affected at an early stage are usually stunted, and may eventually die. Malformation of grains or sterility of the panicle can be observed in systemically infected rice plants that reach the reproductive stage. RHBV has been shown to infect *Digitaria* spp., *Leptochloa* spp., oats, barley, rye, and wheat. Massive screenings of rice cultivars took place in Cuba, Venezuela, and Colombia in the late 1950s and early 1960s. Most commercial long-grain rice varieties are susceptible. Resistance to RHBV was identified in short-grain *japonica* varieties and hybrids with *indica* types. These hybrids have been used to develop resistant rice cultivars, but their level of resistance ultimately depends on virus pressure. Some varieties possessing high levels of RHBV resistance have been developed in recent years in Colombia, which seem to be the result of combining different mechanisms of resistance, not necessarily against the virus. Transgenic plants that show moderate levels of resistance to the causal virus have been developed. RHBV epidemics are cyclic, often separated by several years of low disease incidence. The deleterious effect of the virus in viruliferous females may account for this cyclical phenomenon.

Rice Yellow Stunt

Rice yellow stunt (previously known as 'rice transitory yellowing') was first reported from southern Taiwan in 1960. The disease was initially believed to be caused by the lack of aeration in paddy fields. The characteristic foliar yellowing, stunting, and root rot were once blamed for the destruction of over 2000 ha of rice in the 1960s. In 1965, a virus was first suspected as the causal agent, and the leafhopper *N. nigropictus* (previously *N. apicalis*) was implicated as the vector. The name 'transitory yellowing' was given to this disease because affected rice plants often recovered from the yellowing symptoms at later stages of growth. The causal virus was isolated in 1986 and shown to be a bullet-shaped rhabdovirus approximately 190 × 95 nm. *Rice yellow stunt virus* is currently characterized as a species of the genus *Nucleorhabdovirus*, family *Rhabdoviridae*. Nucleorhabdoviruses multiply in the nucleus of infected cells, forming large granular inclusions. Virus maturation occurs at the inner nuclear envelope, where rice yellow stunt viruses (RYSVs) acquire their lipid envelope. The virus contains a single molecule of negative-sense ssRNA about 13.5 kb in size. A modular organization of the genome shows three highly conserved blocks, encoding structural and nonstructural proteins, membrane proteins, movement protein, and the polymerase.

RYSV is transmitted by the leafhoppers *N. nigropictus*, *N. cincticeps*, and *N. virescens* in a persistent and propagative manner. The virus does not pass congenitally to the progeny. Warm winters and continuous rice plantings increase virus incidence. *Japonica* varieties are more susceptible to RYSV than *indica* types. Highly resistant rice varieties have been selected since the early 1960s to manage this disease.

Rice Yellow Mottle

Rice yellow mottle was first observed in 1966 in Kenya and is currently the most important viral disease of rice in sub-Saharan Africa. The disease also affects rice in Madagascar. Yield losses ranging from 25% to 100% are not uncommon in West Africa. The causal agent was described in 1974 as an isometric virus about 30 nm in diameter, possessing positive-sense ssRNA (1.4×10^6 M). The RNA has 4450 nt organized into four ORFs, three of which overlap (except ORF 1). ORF4 codes for the 26 kDa capsid protein. This protein is required for cell-to-cell and long-distance movement. Rice yellow mottle virus (RYMV) is currently classified as a member of the species *Rice yellow mottle virus* in the unassigned genus *Sobemovirus*.

RYMV is transmitted by chrysomelid beetles (Coleoptera): *Sesselia pusilla*, *Chaetocnema pulla*, *C. dicladispa*, *Trichispa sericea*, *Dicladispa (Hispa) viridicyanea*, and *D. gestroi*, in a semi-persistent manner. The virus is lost when beetles molt; it does not multiply in the vector and is not congenitally transmitted. RYMV is transmitted by mechanical means but is not seed-borne. Rice yellow mottle disease is characterized by stunting, crinkling, mottling, yellow streaks, malformation of panicles, reduced tilllering, and sterility. RYMV also infects other *Oryza* species, and the grasses *Acroceras zizanioides*, *Dinebra retroflexa*, *Eragrostis aethipica*, *E. ciliaris*, *E. namaquensis*, *E. tenella*, *Echinochloa colona*, *Ischaemum rugosum*, *Panicum repens*, *P. subalbidum*, *Phleum arenarium*, *Sacciolepis africana*, and *Setaria longiseta*.

Most African rice cultivars are highly susceptible to RYMV. Two types of resistance have been identified in *Oryza*: polygenic resistance in *O. sativa*, *japonica*, which slows down virus dissemination; and monogenic resistance in *Oryza glaberrima*, which prevents infection. Transgenic rice plants showing 'enhanced' resistance to RYMV have also been developed. The epidemiology of RYMV is poorly understood, specially regarding the identification of primary inoculum sources, and propagation of the virus in the field. Resistance-breaking isolates of RYMV have been recently detected in West and Central Africa.

Rice Stripe Necrosis

Rice stripe necrosis first emerged in Ivory Coast, West Africa, in 1977. The disease induces chlorotic striping, malformation, and necrosis of leaves, besides stunting and reduced tillering. The causal virus was shown to consist of rod-shaped virions of two predominant lengths (110–160 and 270–380 nm), and 20 nm wide. The virus can be mechanically transmitted to *Chenopodium amaranticolor* (local lesions), but is not seed-borne. The virus was transmitted through contaminated soil, which led to the identification of the fungus *Polymyxa graminis* as the vector. The causal virus, rice stripe necrosis virus (RSNV), was tentatively classified as a member of the genus *Furovirus*. However, its relatively low incidence and economic importance, when compared to rice yellow mottle virus, relegated RSNV to the condition of a minor pathogen. In the meantime, RSNV slowly disseminated into the main rice-growing areas of West Africa, where farmers and scientists alike mistook the disease for a soil or physiological problem. It was not until 1991, when a new disease of rice emerged in the Eastern Plains of Colombia, South America, that research conducted at CIAT, Palmira, Colombia, demonstrated that the new disease was in fact 'rice stripe necrosis', probably introduced from West Africa in contaminated rice seed. The Colombian isolate of RSNV had rod-shaped particles with a bi-modal length of 260 and 360 nm, and a particle width of 20 nm. These particles were observed in the cytoplasm of infected rice cells, affecting mainly mitochondria and the endoplasmic reticulum.

The virus was isolated as four components of ssRNA (6.3, 4.6, 2.7, and 1.8 kb), and a capsid protein subunit of 22.5 kDa. At least two distinct replication-associated domains of RSNV RNA1 show sequence identities of >78% with the corresponding domains of Beet necrotic yellow vein virus (BNYVV-RNA 1). As in the case of BNYVV, the RNAs of RSNV terminate in a 3'-poly(A) tail. RSNV's particle length, number of RNA segments, polyadenylation of the RNAs 3'-end, and partial homology to BNYVV suggest that RSNV is a species of the genus *Benyvirus*. Partial sequence obtained from the Colombian isolate's RNA 1 segment also showed a 51% similarity to the corresponding region of BNYVV, member of the type species of the genus *Benyvirus*. The vector of the Colombian RSNV isolate was molecularly characterized as a type II isolate of the fungus *P. graminis*.

Since the 1970s, when RSNV was first reported from West Africa, it was evident that rice varieties reacted differentially to RSNV. Some IRAT lines (8, 9, and 13) showed low virus incidences or escaped disease. Screening of the main rice varieties grown in Colombia showed that most of these cultivars were susceptible. Some breeding lines and some Asian varieties exhibited moderate resistance levels. *Oryza glaberrima*, a related species planted for human consumption in West Africa, was observed to possess a high level of resistance to RSNV in that region and in Colombia. The disease seems to spread more rapidly in mechanized and irrigated rice plantings, but the main virus-vector dissemination mechanism is undoubtedly the use of seed produced in affected fields, where the fungus and the virus persist for many years. Rice stripe necrosis is currently found throughout West Africa, and in four neighboring countries of Latin America: Colombia, Ecuador, Panama, and Costa Rica.

Rice Necrosis Mosaic

Rice necrosis mosaic was first observed in Japan in 1959. The causal virus was observed in 1968 to have filamentous particles of two different lengths (275 and 550 nm) and 13 nm in width. Rice necrosis mosaic virus (RNMV) is currently classified as a member of the species *Rice necrosis mosaic virus* of the genus *Bymovirus*, family *Potyviridae*. Virions contain two molecules of positive-sense ssRNA. RNA 1 is 7.5–8.0 kb, and RNA 2 is 3.5–4.0 kb, and both RNAs are polyadenylated at their 3'-ends. The capsid protein is located in the 3'-proximal region of RNA 1. Virions have a single capsid protein subunit of *c.* 33 kDa. As other species in the family *Potyviridae*, RNMV induces cylindrical (pinwheels) inclusions in the cytoplasm of infected cells.

RNMV is transmitted by mechanical means and through soil by the vector fungus *P. graminis*. The main symptoms are initial mottling of the lower leaves, which gives rise to streaks and irregular foliar patches. Affected leaves turn yellow. Although susceptible rice plants are not markedly stunted, the number of tillers is usually reduced. Necrotic lesions can appear on leaf sheaths and the base of culms. Grain production is significantly reduced in diseased plants. The virus seems to be limited to *O. sativa*.

Viruses transmitted by *P. graminis* are generally disseminated through contamination of agricultural tools and machinery; irrigation water; and seed contaminated with soil particles containing viruliferous fungal propagules (cystosori). These propagules can remain viable in contaminated soils for over 10 years. Soil used in seedbeds should be kept free of the vector. Resistance to this disease has been identified in some rice varieties.

See also: Cereal Viruses: Maize/Corn; Cereal Viruses: Wheat and Barley; Plant Reoviruses; Plant Rhabdoviruses; Rice Yellow Mottle Virus; Tenuivirus; Vector Transmission of Plant Viruses.

Further Reading

APS (1992) *Compendium of Rice Diseases,* 62pp. St. Paul, MN: APS Press.

Brunt AA, Crabtree K, Dallwitz MJ, *et al.* (1996) *Viruses of Plants,* 1484pp. Wallingford: CAB International.

Fauquet CM, Mayo MA, Maniloff J, Desselberger U,, and Ball LA (eds.) (2005) . *Virus Taxonomy: Eighth Report of the International Committee on Taxonomy of Viruses,* 1259pp. San Diego, CA: Elsevier Academic Press.

IRRI (1969) *The Virus Diseases of the Rice Plant,* 354pp. Los Baños, The Philippines: The International Rice Research Institute.

Ling KC (1972) *Rice Virus Diseases,* 142pp. Los Baños, The Philippines: The International Rice Research Institute.

Morales FJ and Niessen AI (1983) Association of spiral filamentous virus-like particles with rice hoja blanca. *Phytopathology* 73: 971–974.

Morales FJ, Ward E, Castaño M, Arroyave J, Lozano I, and Adams M (1999) Emergence and partial characterization of rice stripe necrosis virus and its fungus vector in South America. *European Journal of Plant Pathology* 105: 643–650.

Wilson MR and Claridge MF (1991) *Handbook for the Identification of Leafhoppers and Planthoppers of Rice,* 142pp. Bristol, UK: Natural Resources Institute and C.A.B. International.

Cereal Viruses: Wheat and Barley

H D Lapierre and D Hariri, INRA – Département Santé des Plantes et Environnement, Versailles, France

Glossary

Guttation Phenomenon linked to the natural movement of the fluid from the apoplastic areas to the leaf surface through stomata under conditions that allow gutate to form.

Pathotype A pathotype is a virus strains that exhibits a differential interaction with the host, usually increased virulence.

Pyramiding genes A combination of different resistance genes introduced in one line or cultivar.

Strain A virus isolate called strain exhibits specific interactions in front of several plant species. A strain may also be charaterized by particular properties of the viral particles.

Introduction

Most of the viruses that infect barley also infect wheat. Of the 57 viruses infecting wheat and or barley (**Table 1**), ten have been discovered less than 30 years ago and new strains or pathotypes are regularly reported. A few apparently nonpathogenic viruses belonging to the genera *Metavirus, Pseudovirus,* and *Endornavirus* have been characterized but are not described here. Only the most damaging viruses and some emerging viruses are described.

Agropyron Mosaic Virus

This virus of low agronomical importance on wheat is believed to be transmitted by the cereal rust mite *Abacarus hystrix* (Nalepa). In USA, agropyron mosaic virus (AgMV) may occur in wheat fields infected with wheat streak mosaic virus (WSMV). In the presence of the two viruses, a yield loss of 85% has been estimated.

Arabis Mosaic Virus

In Switzerland, arabis mosaic virus (ArMV) has been detected in fields of winter barley. Partial stunting, without yellowing, can occasionally be observed. Barley infection is achieved by the nematode *Xiphinema diversicaudatum* (Micoletzky).

Aubian Wheat Mosaic Virus

This emergent virus was observed in some winter wheat fields from northern and southern France. A similar virus is present in the UK. The rod-shaped particles of this virus are detected by enzyme-linked immunosorbent assay (ELISA). *Polymyxa graminis* cystosori are detected in the roots of infected plants but aubian wheat mosaic virus (AWMV) is not serologically related to the *Polymyxa*-transmitted viruses already known. The only known natural hosts of this virus belong to the genus *Triticum*. AWMV is mechanically transmitted to wheat and to some eudicots but mosaic symptoms were observed only on wheat. All the cultivars of bread and durum wheat inoculated were infected. In wheat fields, yellow patches of diseased plants are observed from the beginning of the year till heading stage. Most of the cultivars are slightly stunted showing variable mosaic symptoms. Nucleic acid sequences of the genome and natural transmission of the virus are under investigation.

Barley Stripe Mosaic Virus

This virus is no longer a global agronomical problem to barley. In the past, up to 60% yield loss has been reported. Pollen transmission of barley stripe mosaic virus (BSMV) is the main way of contamination. Immunological control of the seeds allows growing virus-free cultivars. BSMV is less frequently found on wheat. Its recent detection in Turkey means this virus should be monitored closely. With its tripartite genome and the high seed transmission, BSMV remains a model for studies of virus/plant interactions.

Table 1 Viruses infecting naturally wheat and barley in different parts of the world

Viruses	Genera	Families
Agropyron mosaic virus [a]	Rymovirus	Potyviridae
Arabis mosaic virus [b]	Nepovirus	Comoviridae
Aubian wheat mosaic virus [a, l]		
Barley mild mosaic virus [b, c]	Bymovirus	Potyviridae
Barley stripe mosaic virus [d]	Hordeivirus	
Barley yellow dwarf virus-MAV [d, e]	Luteovirus	Luteoviridae
Barley yellow dwarf virus-PAS [d, e]	Luteovirus	Luteoviridae
Barley yellow dwarf virus-PAV [d, e]	Luteovirus	Luteoviridae
Barley yellow mosaic virus [b, c]	Bymovirus	Potyviridae
Barley yellow streak mosaic virus [d, l]		
Barley yellow striate mosaic virus [d]	Cytorhabdovirus	Rhabdoviridae
Brazilian wheat spike virus [a, l]		
Brome mosaic virus [d]	Bromovirus	Bromoviridae
Brome streak mosaic virus [f]	Ttritimovirus	Potyviridae
Cereal mosaic virus [g, l]		
Cereal pseudorosette virus [g, l]		
Cereal yellow dwarf virus-GPV [d, e]	Polerovirus	Luteoviridae
Cereal yellow dwarf virus-RPS [d, e]	Polerovirus	Luteoviridae
Cereal yellow dwarf virus-RPV [d, e]	Polerovirus	Luteoviridae
Chinese wheat mosaic virus [a, h]	Furovirus	
Chloris striate mosaic virus [d]	Mastrevirus	Geminiviridae
High Plains virus [d, l]		
Indian peanut clump virus [d]	Pecluvirus	
Iranian wheat stripe virus [a, l]		
Maize Iranian mosaic virus [a, l]		
Maize rough dwarf virus [a, l]	Fijivirus	Reoviridae
Maize sterile stunt virus [d, l]		
Maize streak virus [d]	Mastrevirus	Geminiviridae
Maize white line mosaic virus [a, l]		
Mal de Rio Cuarto virus [d]	Fijivirus	Reoviridae
Northern cereal mosaic virus [d]	Cytorhabdovirus	Rhabdoviridae
Peanut clump virus [i]	Pecluvirus	
Rice black streaked dwarf virus [d]	Fijivirus	Reoviridae
RMV [d, e]		Luteoviridae
SGV [d, e]		Luteoviridae
Soil-borne cereal mosaic virus [a, h]	Furovirus	
Soil-borne wheat mosaic virus [a, h]	Furovirus	
Soil-borne wheat mosaic virus Marne strain [b, h]	Furovirus	
Triticum mosaic virus [a, l]		
Wheat American striate mosaic virus [a]	Cytorhabdovirus	Rhabdoviridae
Wheat dwarf virus B and W strains [a, b, l]	Mastrevirus	Geminiviridae
Wheat Eqlid mosaic virus [a, l]		
Wheat rosette stunt virus [g, l]		
Wheat spindle streak mosaic virus [a, j]	Bymovirus	Potyviridae
Wheat spot mosaic [d, k, l]		
Wheat streak mosaic virus [a]	Tritimovirus	Potyviridae
Wheat yellow head virus [a, l]		
Wheat yellow leaf virus [d]	Closterovirus	Closteroviridae
Wheat yellow mosaic virus [a, j]	Bymovirus	Potyviridae
Winter wheat Russian mosaic virus [g, l]		

[a]Virus infecting wheat.
[b]Virus infecting barley.
[c]See section on bymoviruses of barley.
[d]Virus infecting wheat and barley.
[e]See section on yellow dwarf viruses.
[f]Virus detected in barley volunteers.
[g]See section on Northern cereal mosaic virus.
[h]See section on furoviruses.
[i]See section on Indian peanut clump virus.
[j]See section on bymoviruses of wheat.
[k]See section on High Plains virus.
[l]Unassigned virus.

Barley Yellow Streak Mosaic Virus

This virus has been detected in several US states and Canada. The agronomical importance of barley yellow streak mosaic virus (BaYSMV) has not been evaluated. Nearly all plants in a field may be infected resulting in yield loss up to 100%. Infected plants are stunted and show chlorotic streaks and stripes parallel to the leaf veins. Virus particles are filamentous *c.* 64 nm in diameter and of varying lengths between 127 and 4000 nm. Particles have an outer lipid-like envelope. From purified particles, a dominant protein of ~32 Da and several single-stranded RNA (ssRNA) species of ~13–11 kbp have been isolated. This unassigned virus which resembles some insect and animal viruses probably belongs to a new genus.

BaYSMV is transmitted by the brown wheat mite *Petrobia latens* and is retained in eggs of mites which are the overwintering reservoir. No resistant barley cultivars are known.

Barley Yellow Striate Mosaic Virus

The distribution of this virus was first described in the Mediterranean areas. Recently, barley yellow striate mosaic virus (BYSMV) has been detected in East European countries. More than 10% yield loss in susceptible cultivars of durum wheat is suspected. Symptoms consist of leaf chlorotic striations, dwarfing, and excess tillering, and heads may be confined in the boot. Awns are short or broken and show mosaic. Wheat and barley plants die in spring in the case of an early infection in the autumn. BYSMV is transmitted in a persistent manner mainly by the planthopper *Laodelphax striatellus* (Fallén). The virus can be transmitted transovarially. Limited molecular tools have been used to compare BYSMV with other serologically related viruses (digitaria striate virus, maize sterile stunt virus, northern cereal mosaic virus (NCMV), wheat rosette stunt virus (WRSV)). Small grain species and several grasses are hosts of BYSMV. Durum wheat is generally more susceptible to BYSMV than bread wheat. Genetic basis of susceptibility/tolerance in wheat is under study.

Brazilian Wheat Spike Virus

This virus found in Brazil seems to be of low incidence on wheat crops. Young leaves of infected plants are entirely chlorotic, older leaves show bright stripes. Spikes of affected plants are pale yellow to whitish, containing empty grains. Particle morphology and cytology indicate that Brazilian wheat spike virus (BWSpV) could be a tenuivirus.

Brome Mosaic Virus

Brome mosaic virus (BMV) disease has been reported on wheat and triticale. Infected plants show yellow streak and stunting. The virus is present in the guttation fluid of the leaves. Beetles are considered to be the main vectors. Mechanical inoculation by animal and machinery traffic probably also occur.

Bymoviruses of Barley

Barley Yellow Mosaic Virus, Barley Mild Mosaic Virus

Bymoviruses have bipartite ssRNA genomes and each segment (RNA1 and RNA2) carries a single open reading frame (ORF) which encodes a polyprotein. The coat protein (CP) gene is located at the C-terminus of the RNA1.

Barley yellow mosaic virus (BaYMV) and barley mild mosaic virus (BaMMV) are major pathogens of winter barley in Europe and East Asia. Yellow discolored patches of different sizes are often the first sign of field infection in early spring. First symptoms are chlorotic or pale green spots and streaks along the leaf veins and can be observed in December or after snow break in February or March. Generally, the third leaves show as rolled young leaves. The bymoviruses of barley significantly reduce height and number of fertile tillers and losses caused by these viruses vary from 10% to 90%. When temperature rises, the new leaves remain green but show distinct pale green streaks. BaYMV and BaMMV can be found, either separately or together. In general, high yield reductions can be anticipated when the winter is very severe (**Figures 1** and **2**).

Different biological or/and serological variants of BaYMV, BaMMV, and wheat spindle streak mosaic virus (WSSMV) have been reported in Europe. In Europe, the resistant genes *rym4* and *rym5* have been overcome by BaYMV 2 and BaMMV Sillery strain (BaMMV-Sil). In France, the presence of several pathotypes of BaYMV and BaMMV able to overcome at least seven of 15 known genes has been reported. These pathotypes, in slow progression, do not pose an agronomic problem for the moment. In Japan, seven strains of BaYMV and two strains of BaMMV have been described on the basis of pathogenicity toward barley cultivars. A Korean strain of BaMMV (BaMMV-Kor) differing biologically and serologically from the Japanese and European isolates and several biological variants of BaYMV in China have also been recognized.

The vector of BaYMV/BaMMV is *P. graminis* Ledingham, a soil organism belonging to the family Plasmodiophoraceae.

The only effective means of controlling these viruses is through the use of resistant cultivars. Resistant genes have been described and localized in the barley genome. Some

Figure 1 Symptoms of barley yellow mosaic virus pathotype 2 (BaYMV 2) in winter barley.

Figure 2 Symptoms of barley mild mosaic virus Sillery pathotype (BaMMV-Sil) in barley.

Figure 3 Symptoms of wheat spindle streak mosaic virus in winter bread wheat.

of them confer complete immunity against BaYMV and/ or BaMMV while others only delay the appearance of symptoms. Pyramiding genes have been introduced in barley cultivars and other genes are studied in various *Hordeum* species.

Bymoviruses of Wheat

Wheat Spindle Streak Mosaic Virus, Wheat Yellow Mosaic Virus

WSSMV and wheat yellow mosaic virus (WYMV) were reported in America/Europe and Asia, respectively. These two viruses are major pathogens of bread and durum wheat, rye, and triticale (**Figure 3**). In the case of WSSMV, it was demonstrated that its genome associated to viral CP is internalized by *P. graminis*. In France, a new serotype (sII) of the WSSMV has been reported. Depending on the wheat

cultivars, WSSMV resistance is controlled by a single dominant gene or by two pairs of alleles, which show complementary effects.

Furoviruses

Chinese Wheat Mosaic Virus, Soil-Borne Cereal Mosaic Virus, Soil-Borne Wheat Mosaic Virus

Several wheat furoviruses infecting cereal species have been described in the worldwide: soil-borne wheat mosaic virus (SBWMV) in wheat, rye, and barley; soil-borne cereal mosaic virus (SBCMV) in wheat and rye (**Figure 4**); and Chinese wheat mosaic virus (CWMV) in wheat. These viruses are transmitted by *P. graminis*. SBWMV, the type member of the genus *Furovirus*, occurs mostly on winter wheat in America, Europe, Asia, and Africa. Based on genomic organization and biological characteristics, it was proposed that the American SBWMV, CWMV, European SBCMV, and Japanese SBWMV are four

Figure 4 Symptoms of soil-borne cereal mosaic virus in winter wheat.

strains of the same virus species. SBWMV is internalized by *P. graminis* in the form of viral RNAs associated to movement proteins. Genomes of these viruses are divided into two ssRNA species that are individually encapsidated. The furoviruses of wheat often causes a 'rosetting' disease, meaning that the plants are severely stunted. The number of tillers and kernel weight are also reduced. The disease usually results in a 10–30% yield loss, but may cause up to 80% yield loss in seriously infested fields. Resistance to SBCMV is controlled by a single locus *Sbm* on 5DL chromosome. Recently a bulk segregant analysis demonstrated that the SBWMV resistance gene was also on chromosome 5DL. The relationship between the two loci is being investigated. This resistance limits the movement of the viruses from the roots to leaves.

Soil-borne wheat mosaic virus Marne strain (SBWMV-Mar) was isolated from barley cv. Esterel in France (Marne department). This furovirus, in contrast to common French isolates of SBCMV, is mechanically transmitted to barley and to oats. Nucleotide and amino acid sequence analyses revealed that the French virus infecting barley is closely related to a Japanese isolate of SBWMV (SBWMV-JT), which was originally isolated from barley.

High Plains Virus

This virus, discovered in USA in 1993, has been identified in two other continents. High Plains virus (HPV) is similar or identical to wheat spot mosaic virus described since 1952. Severe symptoms are shown by wheat plants when infected by HPV. The host range of HPV is large including barley, maize, and many grasses. The virus is transmitted obligatorily by the mite *Aceria tritici* (Shevtchenko) that also transmits WSMV. Two sources of resistance to WSMV *RonL* and *Wsm1* are not effective on HPV.

Indian Peanut Clump Virus

Found in five states of India and in Pakistan, Indian peanut clump virus (IPCV) is one of the few viruses infecting both Poaceae and Eudicot species. Several Poaceae species are important reservoirs of this virus which induces a severe disease in peanut. Like barley and maize, wheat is severely affected by IPCV. Early infected plants are dark green, stunted, and generally die. Spikes of late infected plants are malformed and produce shriveled seeds. IPCV is transmitted by the soil-borne vector *P. graminis* Ledingham and at a low level by wheat seeds. Peanut clump virus from West Africa is closely related to IPCV. Its agronomical importance on wheat and barley is unknown.

Iranian Wheat Stripe Virus

The agronomical importance of this virus was not estimated but is probably not high in the Fars Province of Iran where it was discovered. Chlorosis, whitening, and striping are observed in infected plants which are dwarfed. Iranian wheat stripe virus (IWSV) is transmitted in a persistent-propagative manner by the planthopper *Unkanodes tanasijevici* (Dlabola). From the comparison of genomic nucleic acid sequences, it has been suggested that a common ancestor between IWSV and RHBV/EHBV tenuiviruses may exist in the New World.

Maize Streak Virus – B Strain

Three strains of MSV have been distinguished on the basis of nucleic acid sequences. Strain A is associated to maize. Strain B (MSV-B) has been characterized in wheat, barley, and grasses in South Africa and in Tasmania. MSV-B is more aggressive on wheat and barley than on maize. Recombinants between strain A (from maize) and strain B have been characterized. Agronomical importance and epidemiology of strain B have to be investigated.

Northern Cereal Mosaic Virus

Several viruses such as WRSV from China, cereal mosaic virus, and winter wheat Russian mosaic virus (WWRMV) from east Russia are probably isolates or strains of the NCMV described in Japan. On winter wheat associated with conditions of high viral inocula in the autumn, yield losses may reach 80%. The infected wheat plants show stunting and prolific tillers. The leaves develop longitudinal yellow and green stripes. Plants infected at the seedling stage die during winter. When infection occurs later, the plant develops thickened stems and foliage. Grains are frequently shriveled. The virus is transmitted in a persistent-propagative manner mainly by *L. striatellus*. Small grains and several grasses are natural hosts of NCMV. Maize has been reported to be resistant to NCMV. Late sowing in autumn and eradication of volunteer plants are recommended.

Particle size, antigenic properties, and vector specificity of these cytorhabdoviruses are very similar or identical. Cereal pseudorosette virus (CPV) is transmitted also by *L. striatellus*. Since its particles are shorter than those of NCMV, CPV is considered a distinct species.

Triticum Mosaic Virus

This virus, discovered in 2006 in Kansas, has been found in areas where WSMV and HPV are present. The gene RonL giving resistance to WSMV is not efficient on triticum mosaic virus (TriMV). The Wsm1 source of resistance is active on TriMV at 19 °C but not at 25 °C. Symptoms induced by TriMV are similar to those of WSMV. In the future, serological diagnosis may become useful to distinguish TriMV from other viruses commoly found in wheat.

Wheat American Striate Mosaic Virus

Reported from Canada and the north-central regions of US, this virus has a limited importance because it is usually present only on the borders of wheat fields. Wheat American striate mosaic virus (WASMV) has a relatively narrow natural host range including barley and maize. Fine chlorotic parallel streaks, brown necrotic streaking of culms and glumes are relatively good indicators of WASMV infection. Susceptible cultivars are stunted and most ears are sterile. *Endria inimica* (Say), the main leafhopper vector, transmits WASMV in a persistent-propagative manner. Early planting favors autumn infection. Several cultivars such as TAM107 are tolerant.

Wheat Dwarf Virus

In continental areas of Europe and its borders, this virus may be very severe on winter wheat, durum wheat, triticale, rye, and barley. Two strains of wheat dwarf virus (WDV) have been characterized. Overall nucleic acid sequences of the wheat-adapted strain (WDV-W) which does not infect barley showed 16% divergence with the barley-adapted strain (WDV-B) which does not infect wheat. The monoclonal antibody MAB 3C10 developed in Germany detects only WDV-B. The two strains have several common hosts. In winter wheat, the leaves show large yellowing bands, the plants are stunted, and an excess of tillering is frequent in contrast to infection by

yellow dwarf viruses (YDVs). In winter, barley plants are dwarfed, leaves are shortened, and start yellowing. As in wheat, early infected plants die. *Psammotettix* is the only known leafhopper vector of this virus. Both nymphs and adults of *Psammotettix alienus* (Dahlbom) transmit WDV in a persistent-circulative manner. Only local populations of *P. alienus* carry the viral inoculum to new crops. Depending on the climatic conditions, one, two, or three vector generations may be observed during the warm active periods. In autumn, flying insects are not seen when the maximum temperature is below 10 °C. *Psammotettix alienus* survives the winter in the egg stage on several Poaceae species. Fallows, regrowths, and volunteers play a major role in the constitution of the autumn viral inoculum. Herbicide sprays and tillage of fallows and of preceding susceptible crops reduce the size of *P. alienus* populations. Sowing as late as possible is also recommended. Insecticide-dressed seed offers good protection. Depending on the evolution of *P. alienus* populations, protection of crops can be achieved using one or two pyrethroid sprays.

Wheat Eqlid Mosaic Virus

This emergent virus found in the Fars Province of Iran is transmitted by the root aphid *Forda marginata* (Koch). It has been suggested that wheat Eqlid mosaic virus (WEqMV) should be included in a new genus of the family *Potyviridae*.

Wheat Streak Mosaic Virus

Agronomical importance of this virus is high mainly in European continental areas where analysis of the genome sequence suggested that this virus probably originated in the Fertile Crescent. WSMV-infected plants show mosaic symptoms and develop a rosette appearance. Symptoms on winter wheat usually appear only in spring at temperatures above 10 °C. Infected plants are stunted with sterile heads. Yield losses can be as high as 100%. WSMV is transmitted by the wheat curl mite (WCM) *Aceria tosichella* (Keifer). The virus is acquired during nymphal stages and is retained up to several weeks even following molting. Prevailing winds play an important role in long-distance dispersion of mites. Seed transmission of WSMV in wheat was estimated from 0.5% to 1.5%. The host range of WSMV is very wide including maize. Prevention of the infection of WSMV in winter wheat in the autumn is achieved by elimination of volunteers 2 weeks before planting.

Wsm1 is a gene of *Thinopyrum intermedium* translocated in wheat. Other partial amphiploids (Zhong1, Zhong 2) are resistant to WSMV and WCM. Recently, two wheat origins from Iran (Adl Cross and 4004) have been reported to carry two distinct genes providing complete resistance

Figure 5 Symptoms of wheat streak mosaic virus in winter wheat.

to WSMV symptoms. Five genes of resistance to WCM detected in *Aegilops* have been transferred to wheat. Introduction of WSMV replicase (NIa) or CP gene into transgenic wheat confers virus resistance (**Figure 5**).

Wheat Yellow Head Virus

This virus detected in USA is of limited agronomical importance. The infected plants show yellow heads and mosaic symptoms of flag leaves. Its vector is not known but wheat yellow head virus (WYHV) can be transmitted by vascular puncture inoculation to maize and to *Nicotiana clevelandii* plants. In infected plants, a major protein of 32–34 kDa is detected. Analysis of the amino acid sequences of this protein demonstrated that WYHV is closely related to rice hoja blanca virus, a tenuivirus.

Wheat Yellow Leaf Virus

Agronomical importance of this virus, detected in Japan, China, and Italy, is limited. The virus causes diffuse chlorotic flecks or interveinal chlorosis and severe yellowing. All small grain crops are susceptible to wheat yellow leaf virus (WYLV). In Japan, the weed *Agropyron tsukushiense* (Ohwi) is probably an important reservoir of the virus. *Rhopalosiphum maidis* (Fitch) and *Rhopalosiphum padi* (L.) transmit WYLV in a semi-persistent manner. Resistant Chinese cultivars have been reported.

Yellow Dwarf Viruses

Barley Yellow Dwarf-MAV, -PAS, -PAV; Cereal Yellow Dwarf-GPV, -RPS, -RPV; RMV; SGV

Different viruses of the family *Luteoviridae* infecting Poaceae (barley yellow dwarf-MAV, -PAS, -PAV; cereal yellow dwarf-GPV, -RPS, -RPV) and two other viruses not assigned to a genus are collectively called YDs. The distribution of these viruses is dependent on environmental conditions. BYDV-PAV is probably the most widespread. In fields mainly infected by BYDV-PAV, yield reduction from 5% to 20% and nearly 40% have been reported in susceptible wheat and barley cultivars, respectively. Symptoms in infected fields occur in patches. Yellowing (mostly in barley) or reddening (mostly in wheat) are characteristic of YDV infection. These symptoms appear at the two- to three-leaf stage in barley and often after heading in wheat. YDVs are transmitted in a circulative-persistent manner by one or several aphid species. The capsid of YDVs includes a CP and a read-through protein, both needed for aphid transmission.

In temperate areas, YDV vectors have two main flight periods. At the beginning of autumn, winged aphids leave summer hosts (ripening maize, perennial grasses, regrowths, and volunteers). These aphids are attracted to low plant density areas in young winter cereal fields. Depending on the temperature, one to several cycles of virus infection occur before aphids are killed by low temperatures. In oceanic and in other warm areas, low concentration of apterous aphids is maintained during winter. In spring, aphids leave their hosts and infect spring crops and then maize crops. Perennial grasses maintain a low but permanent reservoir of YDVs and aphids.

Risk assessment systems have been developed. In Australia, forecasts of vector incidence are based on the temperature and rainfall in late summer/early autumn. In France, a decision-support system based on temperature-driven simulations of aphid populations was proposed. In UK, a computer-based decision-support system associating the number of aphids found in suction traps and the numbers of foci of YDV infection per unit area of crop was developed. Using these systems, a risk index allows rationalization of foliar pesticide usage (usually synthetic pyrethroids). Treating seeds with an aphicide is necessary in some areas when early sowing of susceptible cultivars is decided or when the pressure of YDV inocula is known to be regularly high. Imidacloprid (nicotinic agonist) or fipronil (chloride channel agonist) gives a good protection during several weeks.

Only minor sources of resistance are known in wheat. Several sources of tolerance have been detected in other Triticinae genera. Several lines derived from crosses between wheat and *Thinopyrum* sp. are very tolerant to YDVs. The basis for *Thinopyrum*-derived resistance to CYDV-RPV is associated to resistance via inhibition of viral systemic infection.

In barley, a major gene of resistance mapped to chromosome 3H, Ryd2, has been introduced into present-day cultivars. The level of protection conferred by this gene varies according to strains of BYDV-PAV and has a low efficiency for CYDV-RPV. Recently, a new major gene

Ryd3 mapped to chromosome 6 has been characterized. In resistant plants, these genes tend to increase seed yield compared to healthy controls.

Recent data on the mechanism and genetics of transmission by aphids and evidence of the glycosylation of virus CP open new avenues to control these viruses in the future.

See also: Barley Yellow Dwarf Viruses; Cereal Viruses: Maize/Corn; Cereal Viruses: Rice; Furovirus; Maize Streak Virus; Nepovirus; Pecluvirus; Plant Reoviruses; Plant Resistance to Viruses: Engineered Resistance; Plant Rhabdoviruses; Tenuivirus.

Further Reading

Chain F, Riault G, Trottet M, and Jacquot E (2007) Evaluation of the durability of the barley yellow dwarf virus-resistant Zhong ZH and TC14 wheat lines. *European Journal of Plant Pathology* 117: 35–43.

Hariri D, Meyer M, and Prud'homme H (2003) Characterization of a new barley mild mosaic virus pathotype in France. *European Journal of Plant Pathology* 109: 921–928.

Lapierre H and Signoret P (eds.) (2004) *Viruses and Virus Diseases of Poaceae (Gramineae)*, 857pp. Versailles, France: INRA.

Li H, Conner R, Chen Q, *et al.* (2005) Promising genetic resources for resistance to wheat streak mosaic virus and the wheat curl mite in wheat – *Thinopyrum* partial amphiploids and their derivatives. *Genetic Resources and Crop Evolution* 51: 827–835.

Malik R, Smith CM, Brown-Guedira GL, Harvey TL, and Gill BS (2003) Assessment of *Aegilops tauchii* for resistance to biotypes of wheat curl mite (Acari: Eriophyidae). *Journal of Economic Entomology* 96: 1329–1333.

Niks RE, Habeku ß A, Bekele B, and Ordon F (2004) A novel major gene on chromosome 6H for resistance of barley against barley yellow dwarf virus. *Theoretical and Applied Genetics* 109: 1536–1543.

Seddas P and Boissinot S (2006) Glycosylation of beet western yellows proteins is implicated in the aphid transmission of the virus. *Archives of Virology* 151: 967–984.

Stein N, Perovic D, Kumlehn J, *et al.* (2005) The eukaryotic translation initiation factor 4E confers multiallelic recessive bymovirus resistance in *Hordeum vulgare* (L.). *Plant Journal* 42: 912–922.

Wiangjun H and Anderson JM (2004) The basis for *Thinopyrum*-derivative resistance to cereal yellow dwarf virus. *Phytopathology* 94: 1102–1106.

Emerging Geminiviruses

C M Fauquet and M S Nawaz-ul-Rehman, Danforth Plant Science Center, St. Louis, MO, USA

Glossary

Geminivirus A class of plant viruses with a genome composed of single-stranded DNA encapsidated in geminate particles.
Pandemic An epidemic at the regional or worldwide level.
Satellite An infectious molecule that depends on a helper virus for crucial functions.

Introduction

Geminiviruses, which are circular single-stranded DNA (ssDNA) plant viruses, are divided into four different genera (*Curtovirus*, *Mastrevirus*, *Topocuvirus*, and *Begomovirus*). All geminiviruses are transmitted by insects; the most devastating begomoviruses are transmitted by whiteflies (*Bemisia tabaci*). In the last 20 years, we have witnessed in many parts of the world the emergence of newly discovered or previously recorded geminivirus diseases. These emergences are attributable to several factors, including an explosion of whitefly populations in many parts of the world, an increase in human activities, and a large development of global trade introducing new plant hosts, vectors, and viruses in different ecosystems. As there are too many geminivirus emerging diseases to be discussed here, we will restrict our description to the three most important diseases caused by begomoviruses that will illustrate the impact of the various factors influencing geminivirus emergences in the world.

The Tomato Yellow Leaf Curl Disease

Geographical Distribution

The tomato yellow leaf curl virus (TYLCV) was originally reported/isolated from Israel and was described in the 1960s, mostly infecting tomatoes and causing very severe symptoms and huge losses. In Spain and Italy, tomato production increased considerably in the 1980s, and closely related viruses were identified in these countries which were first mistakenly identified as new strains of TYLCVs, although the Israeli virus was subsequently identified in these countries. Furthermore, recombinants between local viruses and TYLCV were found and are now prevalent in several regions of both countries. TYLCV was later exported to several countries around the Mediterranean region and worldwide. In 1994, TYLCV was identified in the Caribbean islands and in 1997 was identified for the first time in the USA. Since then, the virus has been exported worldwide (**Figure 1**) and is present in the following countries: USA (California, Georgia, New Mexico, Arizona), Mexico, Puerto Rico, Dominican Republic, Cuba, Tunisia, Morocco, Egypt, Spain, Italy, France, Greece, Turkey, Sudan, India, China, Japan, and Australia.

Symptoms and Yield Losses

The symptoms induced by TYLCV are typically a leaf curling of the leaves with different levels of yellowing (**Figure 2**). However in a single field of infected tomato plants a variety of symptoms can be observed from green to purple leaf curling, thickening of the veins and the stems, and various levels of stunting. In some instances, the leaf surface is reduced a minimum, with large veins and thick lamina. When plants are infected at an early stage, the plant fail to produce fruits and will stay stunted. Losses are in general very high, and depending on the earliness of the infection and on the level of resistance of the variety can reach 100%. TYLCV is considered a major threat for farmers growing tomatoes, both for commercial or subsistence purposes.

Causal Agent with Classification

TYLCV belongs to the genus *Begomovirus*, in the family *Geminiviridae*. Begomoviruses are either monopartite and bipartite, meaning that their genome is composed of a single or two molecules of circular ssDNA and TYLCV is a monopartite virus. This feature does not explain the prevalence of TYLCV in the world, as there are many monopartite geminiviruses in the Old World that do not show the characteristics of a new virus emergence. Regarding the genome organization, TYLCV is a typical monopartite begomovirus coding for six open reading frames (ORFs). A total of five strains of TYLCV, called TYLCV-Iran, -Gezira, -mild, -Oman, and -Israel, have been identified, but it is mostly the Israeli strain that has, so far, spread all over the world.

Epidemiology

Begomoviruses are not seed transmitted, nor mechanically transmitted (with a few exception in lab conditions); they are only transmitted by the whitefly *Bemisia tabaci*. In the 1960s, TYLCV was found in local weeds in Israel, serving as virus reservoir, when tomatoes were planted in the field. In the 1980s in Spain, the year-round cultivation of tomatoes under plastic screen houses promoted the spread of the disease, but beans were also found as a host of the virus. Over the years, it was found that the privileged mode of spreading of the disease in the world was via international trade of tomato plantlets, but it was demonstrated that the green tissue attached to the fruits (sepals, stems) was a very efficient virus reservoir for local whiteflies. In several instances (Dominican Republic and Florida), it was observed that TYLCV became the prevalent virus, displacing the local geminiviruses previously infecting tomatoes. This could be due to the fact that TYLCV is very aggressive and could become prevalent because of a better fitness to the host. In addition to the trade impact, the explosion of the B-type population of whiteflies can explain the worldwide spread of TYLCV. This particular B-type population of whiteflies is very ubiquitous and infects more than 1000 different plant hosts, providing the insect with enormous resources for winter survival and local adaptation.

Control

Geminiviruses are in general very difficult to control. The most efficient way is the use of virus-resistant varieties when they are available, acceptable, and affordable. The cost of such varieties is limiting their use to industrial tomato plantations and therefore most of the tomato fields in poor farmers fields are extremely susceptible. Another limiting factor is the taste and shape of virus-resistant tomatoes that do not match local requirements and encourage the use of susceptible varieties. Controlling the insect, by chemical means, is practically impossible and very expensive. Agricultural practices, although efficient in certain conditions, have been of little help to control TYLCV in a practical manner.

The African Cassava Pandemic

Geographical Distribution

The cassava geminivirus pandemic in Africa started in the early 1990s and is still going on today. It started in Uganda in 1994, invaded several countries in East Africa around Lake Victoria, before crossing the mountains in Rwanda-Burundi, to infect the entire Congo Basin and invade up to Gabon and the south of Cameroon (**Figure 3**). Cassava plants were already infected with African cassava mosaic virus (ACMV) and other viruses, but became superinfected with an Ugandan strain of East African cassava mosaic virus (EACMV-UG), which displaced other viruses.

Symptoms and Yield Losses

Symptoms due to the EACMV-UG are extreme reduction of the surface leaf, yellowing of the lamina, and downcurling of the leaves. The entire infected plants look yellow compared to healthy plants, with reduced

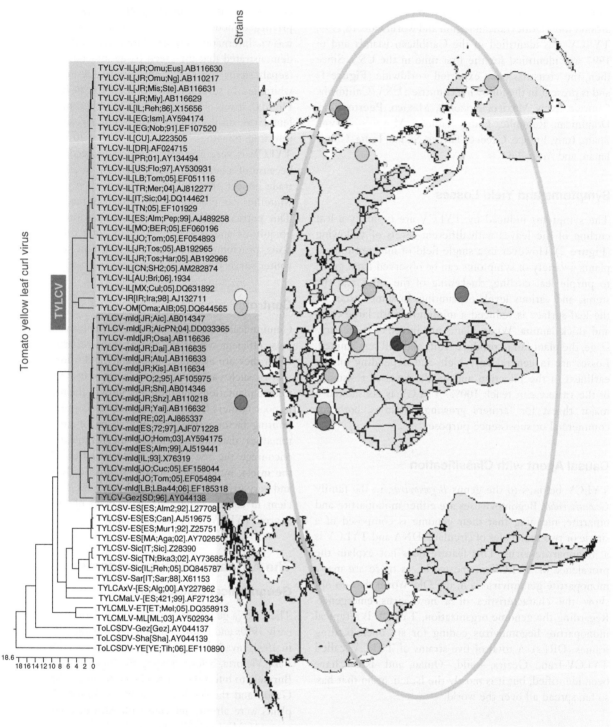

Figure 1 World map on which each of the 44 members representing five begomovirus strains of the TYLCV species have been indicated. The abbreviated name and accession numbers of the isolates are listed in phylogenetic tree in the upper part of the diagram. The phylogenetic tree has been built using their complete A component sequence. The Clustal V algorithm of the program MegAlign from DNAStar has been used and distances in percentage difference are indicated on the left. The tree shows a partition in six major clusters, one for each of the six designated viruses, TYLCV, tomato yellow leaf curl Sardinia virus (TYLCSV), tomato yellow leaf curl Axarquia virus (TYLCAxV), tomato yellow leaf curl Malaga virus (TYLCMalV), tomato yellow leaf curl Mali virus (TYLCMLV), and tomato leaf curl Sudan virus (ToLCSDV). These six species constitute the so-called TYLCV cluster of the Old World begomoviruses. The individual viruses composing the TYLCV cluster are positioned on the world map, as filled circles of various colors representing their pertaining to one of the six specific TYLCV strains, as indicated in the colored boxes at the bottom of the tree. On the world map, the individuals pertaining to the TYLCV species are shown with filled circles.

size and extremely limited foliage. Furthermore, in many instances, plants are infected with both EACMV-UG and ACMV, and these two viruses synergize to cause a very dramatic syndrome, called 'candlestick syndrome'. In this

instance, the stem is thick and turns purple, the leaves are even more reduced in size, with very thick veins and almost no lamina (**Figure 4(d)**). Cassava is normally propagated using the stem, but if the plants have been co-infected with ACMV and EACMV-UG, the stems cannot be used anymore and cassava cannot be propagated to the next season. This caused dramatic cassava shortages in Uganda between 1994 and 2000, estimated to losses amounting to 6 million tons per year, and famine and death for thousands of people.

Causal Agent with Classification

ACMV and EACMV-UG are both members of the genus *Begomovirus*, in the family *Geminiviridae*, similarly to TYLCV above. These viruses however are bipartite viruses, with a second DNA molecule coding for two movement proteins. Furthermore, EACMV-UG is a recombinant between ACMV and EACMV, with a fragment of 550 nt from the core part of the coat protein of ACMV integrated into the EACMV genome. Although there is no direct proof that this recombinant has any specific advantage over the nonrecombinant EACMV, it is a fact that the EACMV-UG strain is present in all the plants that show severe symptoms

Figure 2 Typical symptoms of TYLCV on tomato in a field in Jordan. Courtesy of Mohammed Abhary.

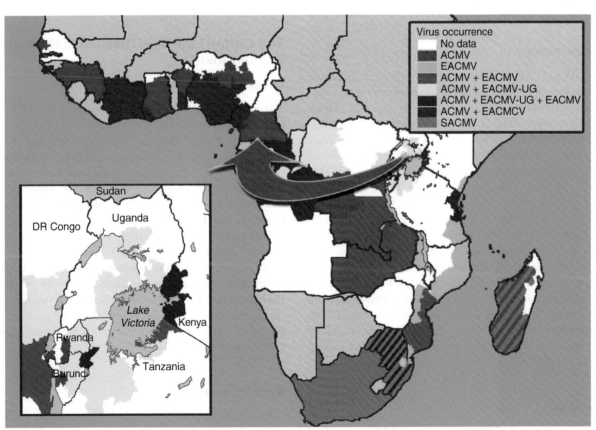

Figure 3 Distribution of cassava mosaic geminiviruses in Africa obtained from CMD surveys in Africa, 1998–2004. ACMV, African cassava mosaic virus; EACMV, East African cassava mosaic virus; EACMV-UG, East African cassava mosaic virus – Uganda; EACMCV, East African cassava mosaic Cameroon virus; SACMV, South African cassava mosaic virus.

Figure 4 Typical symptoms of cassava infected with ACMV (b) and EACMCV from Cameroon (c), and synergism between the two viruses expressing the 'candlestick' symptoms (d), compared to the control healthy cassava leaves (a).

and is strictly correlated with the development of the pandemic across the African continent. In addition, these two viruses do synergize to cause a dramatic 'candlestick' symptom, preventing the cuttings from these infected plants to grow for the next generation. The explanation for this phenomenon resides in the fact that each virus codes for a very strong and differential gene silencing suppressor, and when both suppressors are present, the infected plant collapses from the resulting dual virus infection. This is the first time that such synergism between two geminiviruses has been found.

Epidemiology

ACMV and EACMV-UG are spread by the whitefly *Bemisia tabaci* and through the cassava cuttings, as there is no virus-clean material propagation system in place in Africa. The explosion of the cassava mosaic pandemic in Africa is currently explained by a conjunction of several mechanisms. (1) The synergism between the two viruses certainly played a key role, as it translates into a huge increase in viral DNA in the tip of the plants by 30–100 times, thereby enhancing the whitefly transmission capacity. (2) The recombinant fragment could provide an advantage to EACMV-UG, as it has been shown that the whitefly transmission epitopes are coded by this fragment and that there has been co-adaptation as far as the rate of transmission is concerned, between the local whitefly and viruses. (3) Finally, there has been a tremendous increase in whitefly populations adapted to cassava in Uganda, where the epidemic started. The conjunction of these three elements is believed to be at the triggering and maintenance of the cassava mosaic disease pandemic in Africa.

Control

Similar to the TYLCV, the only available method of control of the cassava mosaic disease is the use of virus-resistant cassava cultivars. Fortunately, such plants were

made available to farmers in Uganda by the end of the 1990s, and the use of these plants allowed to restore the cassava production in Uganda. This however did not stop the pandemic that is still going on in West Africa. Furthermore, many of the farmers in Uganda, and elsewhere, are now going back to their preferred cassava plant material that is not virus resistant, but highly appreciated for their organoleptic and processing qualities. Consequently, more virus-resistant cassava cultivars, acceptable to farmers, are needed.

The Cotton Leaf Curl Disease in the Indian Subcontinent

Geographical Distribution

The leaf curl disease of cotton (*Gossypium hirsutum* L.) was first observed in Multan, Pakistan, in 1967. The disease reappeared in 1987 in its epidemic form in most of the cotton-growing areas of Pakistan. In 1992, the disease was observed throughout the central and southern parts of Punjab in Pakistan. During 1997–98, the disease was established in the border areas of India (Rajasthan, Haryana, and Punjab), joining the southern Punjab of Pakistan. The most devastating virus known as cotton leaf curl Multan virus (CLCuMV) is attributed to huge losses throughout the Indian subcontinent. The estimated direction of movement for the disease is from the center of Pakistan to northern and southern districts of India. Nowadays, the disease can be observed in all primary or secondary cotton-growing areas of the Indian subcontinent (**Figure 5**).

Symptoms and Losses

Cotton leaf curl disease (CLCuD) is not a seed-borne disease, but is only transmitted through whiteflies. After 2–3 weeks of virus inoculation by the insects, vein thickening on young leaves can be observed. Later, upward or

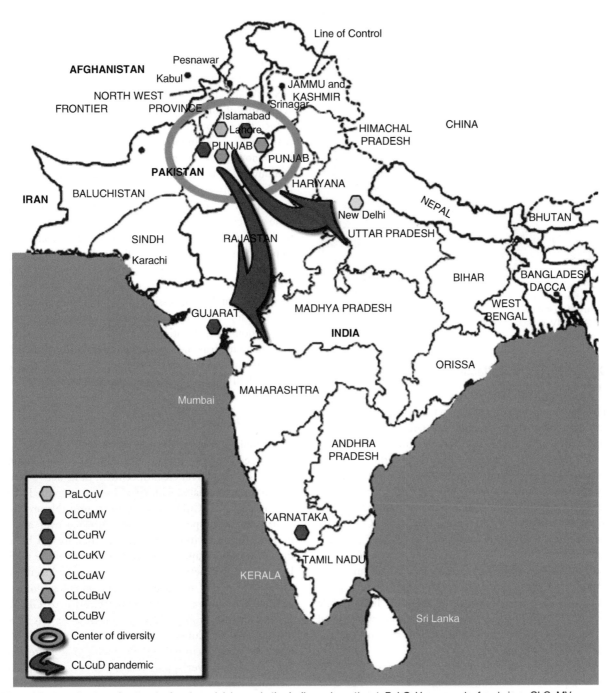

Figure 5 Localization of cotton leaf curl geminiviruses in the Indian subcontinent. PaLCuV, papaya leaf curl virus; CLCuMV, cotton leaf curl Multan virus; CLCuRV, cotton leaf curl Rajasthan virus; CLCuKV, cotton leaf curl Kokhran virus; CLCuAV, cotton leaf curl Allahabad virus; CLCuBuV, cotton leaf curl Burawalia virus; CLCuBV, cotton leaf curl Bangalore virus.

downward curling of leaves starts and a cup-shaped leaf enation appears at the underside of the leave (**Figure 6**). The infected plants remain stunted throughout their life cycle and may lose 100% of the total yield. Disease spread can be observed with the wind direction. During the last two decades, the disease caused heavy cotton yield losses which reached up to US$ 5 billion in Pakistan in 1992–97. Recently, the introduction of relatively resistant varieties has selected a more aggressive virus in 2002 in central Pakistan, called cotton leaf curl Burawalia virus (CLCu-BuV), and this strain is now prevailing in all cotton-growing areas of Pakistan.

Figure 6 Downward leaf curl symptoms of cotton induced by the CLCuMV in association with the cotton leaf curl Multan beta satellite.

Causal Agent and Classification of Cotton Leaf Curl Disease

The begomoviruses associated with CLCuD are monopartite begomoviruses with an associated ssDNA satellite molecule known as DNA-β (beta satellite).

According to sequence analysis of the available sequences in gene bank, there are viruses belonging to seven different species involved in the etiology of CLCuD named as, CLCuMV, cotton leaf curl Kokhran virus (CLCuKV), cotton leaf curl Rajasthan virus (CLCuRV), cotton leaf curl Allahabad virus (CLCuAV), cotton leaf curl Burawalia virus (CLCuBuV), cotton leaf curl Bangalore virus (CLCuBV), and papaya leaf curl virus (PaLCuV). Cotton leaf curl Bangalore virus is only present in the Bangalore district of southern India, while viruses belonging to the other six species are widespread throughout the cotton-growing areas of the north of the Indian subcontinent.

The area of maximum diversity for CLCuD is in central parts of Pakistan, hosting viruses belonging to five species of cotton leaf curl viruses (CLCuVs) and one of PaLCuV.

CLCuD is in reality caused by a complex of viruses associated with virus satellites. A monopartite geminivirus, belonging to different species of viruses, is typically associated with a beta satellite (DNA-β) or a DNA-1 satellite. If DNA-β's have only been found associated with geminiviruses, DNA-1 satellites have originally been found associated to nanoviruses. Both satellite genomes are encapsidated by the geminivirus capsid protein, and thereby both are transmitted by whiteflies. Full-length clones of several CLCuVs DNA-A were unable to reproduce the symptoms in cotton and other different hosts such as tobacco. Infectious clones of DNA-1 were also not capable of inducing symptom development of CLCuVs in the inoculated plants.

With the discovery of DNA-β, it was proved that both DNA-A and DNA-β are necessary to induce typical symptoms of CLCuD, while DNA-1 is not necessary to induce the symptoms and is seldom found in association with the disease complex. At present, the origin of DNA-β is unknown but it can be transreplicated by diverse geminiviruses associated with the CLCuD complex. The DNA-β is only ±1300 nt long and codes only for one ORF called βC1. The gene βC1 codes for a strong suppressor of gene silencing, thereby providing a tremendous advantage to the cognate helper virus. DNA-1 satellite codes for a Rep protein that allows this satellite to be replicated on its own, but the satellite depends on the helper virus for all the other functions, such as gene-silencing suppression, movement, and transmission. DNA-β is dependent on DNA-A for replication, movement, and transmission. The genome organization of DNA-A is typical of any other begomovirus and has no similarity with its associated satellite molecules. Both DNA-β and DNA-1 are approximately half the size of DNA-A, that is, 1350 nt.

Epidemiology

Cultivated cotton or upland cotton (*G. hirsutum*) was introduced into the Indian subcontinent from southern Mexico during the nineteenth century. In its native place, cotton is not associated with virus diseases as in the Indian subcontinent. CLCuD complex most probably evolved from some wild species of cotton growing over a long period of time or from other weed hosts present before the cultivation of upland cotton in this part of the world. In 1991, a new variety of cotton (S12) was introduced from Texas, because of its excellent fiber qualities; unfortunately, this new variety was hypersensitive to CLCuD and the disease propagated very quickly after this introduction. The CLCuD threat was greatly increased due to intensive cotton cultivation, monocropping pattern of agriculture, and overlapping seasons of other crop plants. CLCuVs can infect a number of diverse hosts in the families Malvaceae and Solanaceae. The virus infection was not highly noticed at the early beginning of the epidemic, when perhaps it could have been possible to eradicate the disease. The disease is transmitted through whitefly in a persistent manner. The disease spread is directly correlated with the spread of the whitefly population, along the prevailing wind direction.

Control

Controlling the whitefly vector population is an important aspect to control CLCuD. Insecticides used to control whitefly play a vital role in the control of disease, but still cannot completely eliminate the disease. Over the years, breeders have selected several varieties of cotton with various levels of resistance, but either this level of resistance was unsatisfactory or has been broken by the appearance of new strains of the virus, such as the CLCuBuV from Burawalia. Recently RNA interference (RNAi) has been used and shown to be efficient, in model plants, to control the disease. However, there is still a need to do more work to control the disease in crop plants. In order to create a broad-spectrum resistance to control the disease complex, there is a need of integrated efforts of conventional and genetic engineering approaches.

Conclusion

The three examples of geminivirus emergence presented here provide examples of the importance of different factors that favor the emergence of geminiviruses under very different conditions and environments. In the case of TYLCV, it is a relatively simple case with a single monopartite geminivirus that is transported, through human-based international trade, to various places in the world, and where this very effective virus overcomes existing local viruses and prevails. Although we do not know the molecular and biological basis for the prevalence of TYLCV, it is clear that this virus has a better fitness in tomato and other hosts, compared to local geminiviruses, whether they be bipartite New World geminiviruses or monopartite Old World geminiviruses.

In the case of cassava mosaic geminiviruses, the situation is more complex as it involves at least two different geminiviruses, a new whitefly population, better adapted to cassava, as well as human participation to move the viruses, and probably the whiteflies, in new environments over chains of mountains, where the disease can again explode. It is therefore the apparent conjunction of various biological and human-based activities that promoted the pandemic in Africa. On the biological side, at least three elements can be related to the emergence: (1) the encountering of two geminiviruses with differential and combinatorial gene silencing suppressors that promoted synergism; (2) the occurrence of an apparent successful recombination between these two viruses; and (3) the adaptation of a new population of whiteflies to cassava. The pandemic has been able to travel eastward more than 3500 km in 12 years, and this would not have happened without human intervention. It remains to be seen if the pandemic will continue in West Africa or if it will fade away in this new ecological zone.

The CLCuD case is even a more complex situation where human intervention by introducing new cotton varieties, highly susceptible to local unknown viruses and satellites, as well as the absence of human intervention to stop the disease, triggered the CLCuD pandemic in the Indian subcontinent, costing billions of dollars to poor farmers and impacting Pakistan economy to about 30% for years. On the virological side, this pandemic revealed a new concept, that is, the crucial role of satellites like the DNA-β for inducing symptoms and finally for allowing the whole pandemic. It is the association of a single DNA-β to a variety of helper geminiviruses that is causing the disease and not a single monopartite geminivirus. It is now obvious that recombination of both DNA-β and its geminivirus helpers is an important evolutionary mechanism to create new diversity to overcome virus resistance, revealing the power of geminiviruses and their satellites to adapt to new ecological situations.

The common factor among these three examples is that human intervention, or lack of intervention, is providing viruses and their satellites opportunities to promote new disease emergences in combination with ecological changes, which are also linked to human impact. These dramatic pandemics, costing billions of dollars in losses and thousands of lives, offer however opportunities to the scientists to better understand the biological causes of these emerging diseases. Taking advantage of these global scale 'experiments', it may be possible to identify the relevant biological factors that could in future allow such diseases to be controlled.

See also: Nanoviruses; Plant Resistance to Viruses: Geminiviruses; Tomato Leaf Curl Viruses from India.

Further Reading

Briddon RW (2003) Cotton leaf curl disease, a multicomponent begomovirus complex. *Molecular Plant Pathology* 4: 427–434.

Czosnek H (ed.) (2007) *Tomato Yellow Leaf Curl Virus Disease.* New York: Springer.

Legg JP and Fauquet CM (2004) Cassava mosaic geminiviruses in Africa. *Plant Molecular Biology* 56: 585–599.

Mansoor S, Briddon RW, Zafar Y, and Stanley J (2003) Geminivirus disease complexes: An emerging threat. *Trends in Plant Science* 8: 128–134.

Polston JE and Anderson PK (1997) Emergence of whitefly transmitted geminiviruses in tomato in the Western Hemisphere. *Plant Disease* 81: 1358–1369.

Rojas MR, Hagen C, Lucas WJ, and Gilbertson RL (2005) Exploiting chinks in the plant's armor: Evolution and emergence of geminiviruses. *Annual Review of Phytopathology* 43: 361–394.

Emerging and Reemerging Virus Diseases of Plants

G P Martelli and D Gallitelli, Università degli Studi and Istituto di Virologia Vegetale del CNR, Bari, Italy

Introduction

A number of RNA plant viruses constitute new threats to economically relevant vegetable crops, grapevine, and citrus. These are listed among new disease-causing agents, and are therefore denoted 'emerging or re-emerging' being often new viruses or virus strains responsible for serious diseases. The emerging plant viruses and their known vectors are quickly expanding in new areas as a consequence of the increasing international trade and the large numbers of people traveling in many countries. Once introduced, the fitness of these viruses to new agroclimatic conditions increases, leading to the development of severe epiphytotics.

Vegetable Viruses

Pepino Mosaic Virus

Geographical distribution

Pepino mosaic virus (PepMV) was originally described in Peru on pepino (*Solanum muricatum*) and found to infect tomato and related wild species symptomlessly, in experimental trials. Since 1999, PepMV outbreaks have been reported almost simultaneously in many European countries (Austria, Bulgaria, Finland, France, Germany, Hungary, Italy, Netherlands, Norway, Poland, Slovakia, Spain, Sweden, Switzerland, Ukraine, and UK) where it is considered an emerging pathogen of tomato glasshouses. PepMV was also found in Canada, USA (Arizona, California, Colorado, Florida, Oklahoma, Texas), Ecuador, and Chile. After discovering the PepMV occurrence in tomato grown in Europe and North America, a survey carried out in central and southern Peru and Ecuador demonstrated that the virus was present in Peruvian tomato crops as well as in the following wild *Lycopersicon* species: *L. chilense, L. chmilewskii, L. parviflorum, L. peruvianum*, and in *L. pimpinellifolium* in Ecuador.

Disease symptoms and yield losses

In pepino, the virus causes yellow mosaic in young leaves, whereas affected tomato plants show a wide array of symptoms. These include stunting of the whole plant, bubbling of the leaf surface, interveinal chlorosis, mosaic and green striations on stem and sepals. Foliar symptoms resemble hormonal herbicide damage, while lower leaves show necrotic lesions that resemble damage caused by water that dripped onto the plant. Fruits from early infected tomato plants may show blotchy ripening and gold marbling. Ripe fruits develop yellow speckles and spots that make them unmarketable. Observations by Dutch scientists indicate that symptoms are more readily seen during autumn and winter months, and are masked during warmer months. Yield losses caused by PepMV are probably below 5%, although surveys in glasshouse-grown tomatoes reported crop reductions up to 40%. The disease spreads very rapidly and crop losses may be significant if early action is not taken to eliminate infected sources. Although Peruvian virus strains seem to be little virulent, new strains of higher virulence could arise through recombination. Because of the adverse effects on quality and yield, PepMV is becoming one of the most important tomato pathogens in Europe.

Causal agent and classification

PepMV (genus *Potexvirus*, family *Flexiviridae*) has helically constructed semirigid filamentous particles with a modal length of 508 nm and a diameter of 11 nm. Sequencing of a number of strains has shown that virions contain a single molecule of linear, positive-sense, single-stranded RNA (ssRNA) 6410–6425 nt in size. The genome organization is typical of the genus *Potexvirus*, with five open reading frames (ORFs): ORF1, encoding a putative replicase of 164 kDa; ORFs 2–4, coding for the triple gene block proteins 1–3 (TGBp) of 26 kDa, 14 kDa, and 9 kDa, respectively; and ORF5, encoding the 25 kDa coat protein (CP). Phylogenetic analyses carried out on replicase, TGBp1, and CP amino acid sequences revealed that PepMV is closely related to narcissus mosaic virus (NMV), scallion virus X (SVX), cymbidium mosaic virus (CymMV), and potato aucuba mosaic virus (PAMV) (**Figure 1**). Tomato isolates of PepMV can be distinguished from the original isolate from pepino on the basis of symptomatology, host range, and sequence data. On the other hand, tomato isolates from Europe appear very similar to each other sharing nucleotide sequence identities higher than 99% and in the range of 95–96% with PepMV from *L. peruvianum* from Peru. Nucleotide sequence comparisons between US and the European tomato isolates show only 79–82% identity. Gene-for-gene comparison between sequenced isolates suggests that TGBp1 and TGBp3 are more suitable than either the replicase or CP gene products for discriminating virus isolates.

Epidemiology

PepMV is transmitted by contact and readily spread mechanically by contaminated hands, tools, shoes, clothing,

Figure 1   (a) Strong symptoms of PepMV on tomato; (b) symptom of CYSDV on melon; (c) severe vein yellowing on cucumber infected by CVYV; (d) swelling at the graft union and death of the scion of a vine grafted on Kober 5BB, infected by *Grapevine leafroll-associated virus 2;* (e) a citrus grove affected by sudden death disease; and (f) a citrus plant killed by sudden death disease. (a) Courtesy of Dr. E. Moriones. (b) Courtesy of Dr. E. Moriones. (c) Courtesy of Dr. I. M. Cuadrado. (e, f) Courtesy of Dr. J.Bovè.

and plant-to-plant contact. The virus is thought to remain viable in dry plant material for as long as 3 months. Seed transmission or surface seed contamination in tomato is suspected but not demonstrated. In the UK, the virus is frequently found in imported fruits. Infection of other solanaceous crops such as eggplant, tobacco, and potato has only been observed in experimental trials, while infection in pepper has not been demonstrated. Cucumber can be artificially infected but the virus does not appear to spread systemically in the plant.

Alternative hosts that may serve as virus reservoirs were studied in Spain, where native plants with virus-like symptoms growing in or around tomato fields were collected and analyzed for the presence of PepMV. As many as 18 weed species were found to be infected.

Because of the high similarity of tomato virus isolates, a common origin seems likely although where this origin is located is still unclear. The extent of PepMV distribution in Peru, together with the fact that many of the wild Peruvian *Lycopersicon* populations sampled were isolated and had not been manipulated by man, led to the conclusion that this virus has been present in the region for a long time and that other factors might be involved in its spread. *Myzus persicae* does not transmit PepMV. In southern Spain a faster spread of PepMV was observed in bumble-bee pollinated greenhouses than in those with no pollinator insects.

Control

Recommended control strategies for PepMV focus on sanitation. Use of certified seed lots, complete removal (including roots) of infected plants, limited access to affected rows, and sanitation of clothing and tools are all critical. The increasing concern caused by this new disease has led the European Commission to authorize member countries to take measures to prevent the spread of PepMV within the European Union.

Criniviruses

Criniviruses are whitefly-transmitted RNA viruses that cause plant diseases with increasingly important yield impact, consequent to the sudden explosion of whitefly populations in temperate regions in the last 10–20 years. All these viruses induce yellowing symptoms in their hosts, are generally phloem-limited, nonmechanically transmissible, and have large ssRNA genomes. Cucurbit yellow stunting disorder virus (CYSDV), tomato infectious chlorosis virus (TICV), and tomato chlorosis virus (ToCV) are important emerging criniviruses.

Cucurbit Yellow Stunting Disorder Virus

Geographical distribution

CYSDV was first observed in the southeastern coast of Spain, on melon and cucumber grown under plastic.

The virus was also found in the Canary Islands, Egypt, France, Israel, Jordan, Lebanon, Mexico, Morocco, Portugal, Saudi Arabia, Syria, Texas, Turkey, and the United Arab Emirates.

Disease symptoms and yield losses
Protected crops of cucumber and melon show severe yellowing that starts as interveinal mottle of older leaves to develop into complete yellowing of the leaf lamina, except for the veins, followed by rolling, brittleness, and stunting. On the whole, symptoms are virtually indistinguishable from those caused by beet pseudo yellows virus (BPYV). Zucchini can also be infected but symptoms have not been described. Disease incidence in protected crops can easily reach 100%. In Spain and many other countries, CYSDV is considered the prevailing virus of protected cucurbit crops.

Causal agent and classification
CYSDV (genus *Crinivirus*, family *Closteroviridae*) has a narrow host range limited to the family *Cucurbitaceae* and is confined to phloem tissues. Virions are flexuous filaments with lengths between 750 and 800 nm. Genome consists of two molecules of ssRNA of plus-sense polarity designated RNA-1 and -2. RNA-1 is 9123–9126 nt long and contains five ORFs with papain-like protease, methyltransferase, RNA helicase, and RNA-dependent RNA polymerase domains in the first two overlapping ORFs, a small 5 kDa hydrophobic protein, and two further downstream ORFs potentially encoding proteins 25 and 22 kDa in size, respectively. RNA-2 is 7976 nt long and contains the hallmark ORFs of the family *Closteroviridae*, encoding, in the order, a heat shock protein 70 (HSP70) homolog, a 59 kDa protein, the major (CP) and the minor (CPm) CP. In the 3'-terminal region, RNA-1 contains an ORF potentially encoding a protein of 25 kDa which has no homologs in any databases, and RNA-2 has an unusually long 59 noncoding region. Subgenomic RNAs were detected in CYSDV-infected plants, suggesting that they serve for the expression of internal ORFs. CYSDV can be divided into two divergent groups of isolates. One group is composed of isolates from Spain, Lebanon, Jordan, Turkey, and North America, the other of isolates from Saudi Arabia. Nucleotide identity between isolates of the same group is greater than 99%, whereas identity between groups is about 90%.

Epidemiology
Natural hosts of CYSDV are restricted to cucurbits: watermelon, melon, cucumber, and zucchini. *Cucurbita maxima* and *Lactuca sativa* are experimental host plants. The life cycle of CYSDV is dependent on its vector, the whitefly *Bemisia tabaci*, as viral outbreaks are associated with heavy infestations of whitefly biotypes A, B, and Q. Transmission of CYSDV by biotype B is more effcient than by biotype A, whereas biotype Q transmits as efficiently as biotype B.

Trialeurodes vaporariorum was displaced *by B. tabaci* as the dominant whitefly along the southeastern coast of Spain when CYSDV took over BPYV as the agent of yellowing diseases of cucurbits. Acquisition periods of 18 h or more and inoculation periods of 24 h or more seem necessary for high transmission rates of CYSDV, which can persist for at least 9 days in the vector. The virus is not known to be seed-borne.

Control
Control of CYSDV consists in controlling *B. tabaci*, and on eliminating infection sources. Chemical control of *B. tabaci* has not been effective in preventing the spread and in reducing the incidence of the disease because the vector has a wide host range and quickly develops resistance to most of existing insecticides. Roguing infected plants and weeds that can act as hosts for the vector and removal of overwintering crops prior to the emergence of adult whiteflies may prove useful. This helps if applied over large areas and where there is no continuous cropping in glasshouses, which are the sites of whitefly survival and the source of virus spread throughout the year. Growing plants under physical barriers such as low-mesh tunnels may also have a positive effect. No resistant cultivars are currently available commercially but experimental evidence for delayed viral infection and decreased symptom severity in accessions of *Cucumis sativus* has recently been obtained.

Tomato Infectious Chlorosis Virus

Geographical distribution
TICV infections have been reported from California, France, Greece, Indonesia, Italy, Japan, North Carolina, Spain, and Taiwan. This virus may also be present in the Czech Republic but the record has not been confirmed.

Disease symptoms and yield losses
Symptomatic tomato plants in open field and greenhouses exhibit interveinal yellowing in older leaves, followed by generalized yellowing. Symptoms can be confused with nutritional disorders (i.e., magnesium deficiency), pesticide toxicity, or natural senescence as older leaves may also turn red. Necrosis and occasional upward rolling of the leaves have also been reported. On the whole, infected plants are less vigorous and with fruits that may show delayed ripening. There is no information on the effects of TICV on artichoke and lettuce crops or ornamental species known to be natural hosts. Disease incidence may vary from one or few plants to severe outbreaks, depending on the abundance of whitefly populations. In California and Greece, disease incidence between 80% and 100% was reported.

Causal agent and classification
The natural hosts of TICV (genus *Crinivirus*, family *Closterovidae*) include members of the families *Chenopodiaceae*,

Compositae, Ranunculaceae, and *Solanaceae* where the virus is confined to phloem tissue. Virions are flexuos filaments with lengths between 750 and 800 nm. Genome consists of two molecules of ssRNA of positive polarity, denoted RNA-1 and RNA-2, whose nucleotide sequence has partially been determined. Both RNAs contain the hallmark ORFs of the family *Closteroviridae.*

Epidemiology

TICV is spread by the whitefly *T. vaporariorum* but not by *B. tabaci* and it is not mechanically transmitted. Besides tomato, natural infections have been detected in *Callistephus chinensis, Chenopodium album C. murale, Cynara cardunculus,* lettuce, *Physalis ixocarpa* (tomatillo), globe artichoke, *Nicotiana glauca, Petunia hybrida, Picris echioides, Ranunculus* sp., and *Zinnia elegans.*

Control

TICV has the potential to cause significant losses to tomato and other naturally susceptible crops if it becomes established. Control largely depends on the efficacy of treatments and strategies to limit populations of *T. vaporariorum.* Tomato seedlings for planting should come from disease-free stocks. The general management strategies outlined for CYSDV can also be adopted for TICV.

Tomato Chlorosis Virus

Geographical distribution

ToCV infections have been reported from the US (Colorado, Connecticut, Florida, Louisiana), Europe (France, Greece, Italy, Portugal, Spain), Morocco, Israel, Puerto Rico, South Africa, and Taiwan.

Disease symptoms and yield losses

Symptoms are very similar to those induced by TICV, that is, progressive yellowing of the whole plant. Apparently, there are no estimates of yield losses, although since its discovery, the virus represents a serious problem for tomato production in many parts of the world.

Causal agent and classification

ToCV (genus *Crinivirus,* family *Closteroviridae*) virions are flexuous filaments encapsidating two molecules of positive-sense ssRNA denoted RNA-1 and RNA-2, whose complete nucleotide sequence has been determined. RNA-1 consists of 8595 nt organized into four ORFs and encodes replication-associated proteins. RNA-2 is 8244 nt long and putatively encodes nine ORFs comprising in the order, the HSP70 homolog, a 59 kDa protein, CP, and CPm, which may be involved in determining the unique, broad vector transmissibility of the virus. Phylogenetically, ToCV is closely related to sweet potato chlorotic stunt virus (SPCSV) and CYSDV.

Epidemiology

ToCV is transmitted by *T. vaporariorum, B. tabaci* biotypes A and B, and *T. abutilonea.* Besides tomato, the natural host range includes pepper, *Datura stramonium, Physalis wrightii, Solanum nigrum,* and *Z. elegans.*

Control

ToCV has the potential to cause significant losses to tomato if it becomes established. Control depends on the efficacy of treatments and strategies to limit populations of its broad range of vectors. The management strategies outlined for CYSDV apply also for ToCV.

Cucumber Vein Yellowing Virus

Geographical distribution

Cucumber vein yellowing virus (CVYV) has been recorded form Greece, Israel, Jordan, Portugal, Spain, Sudan, and Turkey.

Disease symptoms and yield losses

In cucumber, CVYV causes pronounced vein clearing deformation of the leaves followed by a generalized chlorosis and necrosis of affected plants. Fruits show light to dark green mottling. Nonparthenocarpic cucumbers are symptomless carriers of CVYV while parthenocarpic cucumbers develop severe symptoms. Stunting was also observed in cucumber and melon and sudden death in protected melons in Portugal. In watermelon, symptoms are often light or not expressed, but splitting of the fruits has occasionally been observed. In zucchini, symptoms vary from chlorotic mottling to vein yellowing of the leaves, but symptomless infections have also been recorded. Symptoms severity may be increased by synergistic reactions with other viruses. Yield losses have not been quantified. CVYV could present a threat to cucurbits grown outdoors or under glasshouses, although its actual impact in the complex pathosystem affecting cucurbit crops in the Mediterranean and subtropical regions has not been determined.

Causal agent and classification

CVYV (genus *Ipomovirus,* family *Potyviridae*) has filamentous particles and is transmitted by *B. tabaci* but, unlike other whitefly-transmitted viruses, it can also be transmitted by mechanical inoculation. CVYV genome consists of a single molecule of plus-sense ssRNA of 9751 nt (ALM 23 isolate) containing most hallmarks of the genome of members of the family *Potyviridae.* The absence of a coding region for the helper component-proteinase seems to be a distinctive trait of CVYV. Two CVYV strains have been recognized from Israel and Jordan, that is, CVYV-Is and CVYV-Jor. The two strains induce similar vein-clearing symptoms in cucumber and melon, but CVYV-Jor infections in cucumber are more severe. CVYV is more closely related to *Sweet potato mild mottle virus* (SPMMV) than any other species in the family *Potyviridae*.

Epidemiology

CVYV naturally infects cucumber, melon, watermelon squash, and zucchini and several weed species are also natural hosts of the virus (e.g., *Ecballium elaterium*, *Convolvulus arvensis*, *Malva parviflora*, *Sonchus oleraceus*, *Sonchus asper*, and *Sonchus tenerrimus*). All experimental hosts belong to the family *Cucurbitaceae* and include *Cucurbita moschata*, *Cucurbita foetidissima*, and *Citrullus colocynthis*. CVYV is semipersistent in its white-fly vector (*B. tabaci*) which retains the virus for less than 6 h. Therefore, individuals moving to nonhost plants may not remain viruliferous long enough to transmit the virus. *Aphis gossypii* and *M. persicae* are not vectors. Whether CVYV is seedborne has not been determined.

Control

CVYV management is mainly based on the use of virus-free stock plants as well as of *B. tabaci* populations as already outlined for CYSDV. Several detection methods based on molecular hybridization and polymerase chain reaction (PCR) have been developed and used for screening *Cucurbitaceae* germplasm that can show some degree of resistance to CVYV.

Graft Incompatibility in Grapevines

Geographical Distribution

Cases of graft union disorders have been documented from Europe ('Kober 5BB incompatibility', 'Syrah decline'), California, New Zealand, Italy, Australia, and Chile ('Young vine decline'), and again California ('Roostock stem lesions', 'Necrotic union', and 'Stem necrosis'), but are likely to occur also in other viticultural countries.

Disease Symptoms and Yield Losses

The increased use of grapevine clonal material is revealing unprecedented and widespread conditions of generalized decline that develop dramatically in certain scion–rootstock combinations. Newly planted vines grow weakly, shoots are short, leaves are small sized, with margins more or less extensively rolled downwards, and the vegetation is stunted. The canopy shows autumn colors off season so that leaves turn reddish in red-berried varieties or yellow in white-berried varieties much earlier than normal. A prominent swelling forms at the scion/rootstock junction ('Kober 5BB imcompatibility', 'Young vine decline') sometimes accompanied by necrosis at the graft union ('Necrotic union'), and variously extended necrotic lesions develop on the rootstock stem ('Roostock stem lesion', 'Stem necrosis'). Severely affected vines decline and may die within 1 or 2 years. Syrah decline is a severe disease characterized by early reddening of the leaves and swellings with grooves and deep cracks at the graft union. Appearance of graft union disorders depends more on the rootstock than on the scion. For instance, European grape varieties grafted on tolerant rootstocks (e.g., Freedom, Harmony, Salt creek, 03916, 101-14) exhibit a green canopy and perform well, whereas varieties grafted on susceptible roostocks (e.g., Kober 5BB, 5C, 1103P, 3309) develop a discolored canopy, decline, and may die.

Causal Agents and Classification

An ordinary strain of grapevine leafroll-associated virus 2 (GLRaV-2) is consistently associated with Kober 5BB incompatibility in Europe, appears to be involved in California's young vine decline, and was detected in diseased Chilean and Argentinian grapes. GLRaV-2, a definitive member of the genus *Closterovirus* (family *Closteroviridae*), has flexuous filamentous particles *c.* 1600 nm long an RNA genome 15 528 nt in size made up of nine ORFs. A virus originally detected in cv. Redglobe in California called grapevine rootstock stem lesion-associated virus (GRSLaV) proved to be a molecular and biological variant of GLRaV-2 (GLRaV-2 RG). Other molecular variants of GLRaV-2 were reported from New Zealand, Chile, and Australia in association with young vine decline conditions. Based on the differential responses of a panel of 18 rootstocks, up to five different graft-transmissible agents inducing incompatibility were detected in California. Of these, only GLRaV-2 RG, the putative agent of roostock stem lesion, was identified. The agents of necrotic union and stem necrosis are unknown. Equally unkown is the agent of Syrah decline, although there is circumstantial evidence that grapevine rupestris stem pitting-associated virus (GRSPaV) may have a bearing on its aetiology. GRSPaV, a definitive member of the genus *Foveavirus* (family *Flexiviridae*), has filamentous particles *c.* 730 nm in length, a genome 8726 nt in size comprising five or six ORFs, and occurs in nature as a family of molecular variants.

Transmission

GLRaV-2 and GRSPaV have no known vectors, but GRSPaV is pollen- and seed-borne. Infected propagative material is the major means for dissemination of both viruses.

Control

Use of certified virus-free scionwood and rootstocks is recommended. Currently known graft incompatibility agents can be eliminated with reasonable efficiency by heat therapy, meristem tip culture, or a combination of the two. If scionwood is infected, the use of sensitive rootstocks is to be avoided.

Sudden Death of Citrus

Geographical Distribution

Citrus sudden death (CSD) has only been reported from the State of Saõ Paulo in Brazil, where it has already killed about 1 million trees. It has the potentiality for spreading to other Brazilian States and neighboring countries.

Disease Symptoms and Yield Losses

CSD is a destructive disease first observed in 1999 in Brazil on sweet orange and mandarin trees grafted on Rangpur lime (*Citrus limonia*) or Volkamer lemon (*Citrus volkameriana*). Outward symptoms of infected trees and modifications of the bark anatomy at the bud union resemble very much those elicited by citrus tristeza, the major difference being that CSD develops on trees grafted on tristeza-resistant rootstocks. Symptoms of CSD are characterized by a generalized discoloration of the leaves which are pale green initially, then turn yellowish and abscise. With time, defoliation becomes more intense, the trees do not push new vegetation, and the root system decays. A rapid decline and death of the plant ensues, due also to phloem degeneration in the graft union region. The phloem of the susceptible rootstocks Rangpur lime and Volkamer lemon shows a characteristic yellow strain. Until their sudden collapse, infected trees bear a normal crop.

Causal Agent and Classification

All trees affected by CSD host a strain of CTV and, with 99.7% association, a spherical virus denoted citrus sudden death-associated virus (CSDaV). This virus is a tentative member of the genus *Marafivirus* (family *Tymoviridae*), has isometric particles about 30 mn in diameter, a ssRNA genome 6805 nt in size with a high cytosine content (37.4%), encompassing two ORFs. ORF1 codes for a large polyprotein with a predicted M_r of 240 kDa which is processed to yield the replication-associated proteins (methyltransferase, helicase, and RNA-dependent RNA polymerase), a papain-like protease, and two CP subunits 21 and 22 kDa in size, respectively. One of the CP subunits is produced from the cleavage of the C-terminus of the polyprotein, the other is translated from a subgenomic RNA. ORF2 codes for a putative protein 16 kDa in size, with some relationship with a movement protein of viruses belonging to the sister genus *Maculavirus* (family *Tymoviridae*). It has not been ultimately established whether CTV or CSDaV are the causal agents of CSD, but their consistent association with the disease suggests that both may have a bearing on its etiology.

Epidemiology

CSD is transmitted by grafting and can be disseminated with propagation material. Natural spreading is by aphids (*Toxoptera citricida*) with a temporal and spatial pattern similar to that of citrus tristeza. CSDaV was detected in *T. citricida* that had fed on infected trees and was transmitted experimentally to healthy plants. Since marafiviruses are not aphid-borne, it was hypothesized that CSDaV may use CTV as a helper virus.

Control

Production and distribution of healthy scion material to be grafted onto tolerant rootstocks (Cleopatra mandarin, Swingle citrumelo) can help restraining the spread of CSD in newly established groves. Grafting the affected trees above the graft union with seedlings of tolerant roostocks allows their recovery.

See also: Plant Virus Diseases: Economic Aspects; Plant Virus Diseases: Fruit Trees and Grapevine.

Further Reading

Boubals D (2000) Le dépérissement de la Syrah. Compte-rendu de la réunion du Groupe de Travail National. *Progrés Agricole et Viticole* 117: 137–141.

Célix A, López-Sesé A, Almarza N, Gómez-Guillamón ML, and Rodríguez-Cerezo E (1996) Characterization of cucurbit yellow stunting disorder virus, a *Bemisia tabaci*-transmitted closterovirus. *Phytopathology* 86: 1370–1376.

Duffus JE, Liu HY, and Wisler GC (1996) Tomato infectious chlorosis virus – A new clostero-like virus transmitted by *Trialeurodes vaporariorum*. *European Journal of Plant Pathology* 102: 219–226.

Greif C, Garau R, Boscia D, et al. (1995) The relationship of grapevine leafroll-associated virus 2 with a graft incompatibility condition of grapevines. *Phytopathologia Mediterranea* 34: 167–173.

Lecoq H, Desbiez C, Delécolle B, Cohen S, and Mansour A (2000) Cytological and molecular evidence that the whitefly-transmitted cucumber vein yellowing virus is a tentative member of the family *Potyviridae*. *Journal of General Virology* 81: 2289–2293.

Legin R and Walter B (1986) Etude de phénomènes d'incompatibilité au greffage chez la vigne. *Progrés Agricole et Viticole* 103: 279–283.

Maccheroni W, Allegria MC, Greggio CC, et al. (2005) Identification and genomic characterization of a new virus (*Tymoviridae* family) associated with citrus sudden death disease. *Journal of Virology* 79: 3028–3037.

Meng B and Gonsalves D (2003) Rupestris stem pitting associated virus of grapevines: Genome structure, genetic diversity, detection, and phylogenetic relationship to other plant viruses. *Current Topics in Virology* 3: 125–135.

Roman MP, Cambra M, Juares J, et al. (2004) Sudden death of citrus in Brazil: A graft-transmissible bud union disease. *Plant Disease* 88: 453–467.

Rubio L, Soong J, Kao J, and Falk BW (1999) Geographic distribution and molecular variation of isolates of three whitefly-borne closteroviruses of cucurbits: Lettuce infectious virus, cucurbit yellow stunting disorder virus, and beet pseudo-yellows virus. *Phytopathology* 89: 707–711.

Soler S, Prohens J, Díez MJ, and Nuez F (2002) Natural occurrence of *Pepino mosaic virus* in *Lycopersicon* species in Central and Southern Peru. *Journal of Phytopathology* 150: 49–53.

Uyemoto JK, Rowhani A, Luvisi D, and Krag R (2001) New closterovirus in 'Redglobe' grape causes decline of grafted plants. *California Agriculture* 55(4): 28–31.

Van der Vlugt RAA, Cuperus C, Vink J, Stijger ICMM, Lesemann DE, Verhoeven JTJ, and Roenhorst JW (2002) Identification and characterization of Pepino mosaic potexvirus in tomato. *Bulletin OEPP/EPPO Bulletin* 32: 503–508.

Verhoeven JTJ, van der Vlugt RAA, and Roenhorst JW (2003) High similarity between tomato isolates of *Pepino mosaic virus* suggests a common origin. *European Journal of Plant Pathology* 109: 419–425.

Wisler GC, Duffus JE, Liu H-Y, and Li RH (1998) Ecology and epidemiology of whitefly-transmitted closteroviruses. *Plant Disease* 82: 270–279.

Legume Viruses

L Bos, Wageningen University and Research Centre (WUR), Wageningen, The Netherlands

Glossary

Ecology Highly complex interaction of organisms with each other and with their physical environment and its study.

Epidemiology (quantitative ecology) Rapid quantitative spread of parasites and pathogens in populations of their host(s) and its study.

Vector Organism or other agent that can spread a virus or any parasite and introduce it to or into other organisms including plants so that attack or infection results.

Virus identification The acts of (1) describing the particular identity or independence of a new virus (i.e., virus characterization) and (2) recognizing a virus as an entity earlier described, classified, and named, with the aim of etiological disease diagnosis.

Introduction

Legume viruses and those of any other type of crop must be dealt with in relation to the crop concerned. Initially, plant viruses drew attention in crops because they were injurious to plants and reduced crop productivity. Legumes were among the first crops studied for virus infections. Also, plants were for long the sole medium to study the viruses and distinguish between them from the view point of the species they infected the symptoms they produce. Most plant viruses still derive their names from crops on which they were first detected and from the symptoms they caused, for example, bean yellow mosaic virus (BYMV). Plants were also used as differential hosts for separating viruses from mixtures by selective passage. Natural hosts affect the evolution of the viruses that infect them. Selected test or indicator plants soon helped for detecting viruses in plants and in vector organisms, and later in fractions during purification. Laboratory studies of viruses for their intrinsic properties made virological interest gradually shift from their effects on plants to molecular biology. But plants remain indispensable for propagating most plant viruses, and basic information on plant viruses must merge in relation to crops such as legumes for elucidating the biological and societal impact of viruses, and showing how to combat them in agricultural practice.

Legumes

Legumes are botanical species in the pod-bearing family Leguminosae or Fabaceae (order Fabales) of which the subfamily Papilionoideae is the largest and economically most important. Root-nodule bacteria (*Rhizobium* spp.) that bind nitrogen from the air make legumes do well under poor growing conditions and improve soil fertility. Legumes are protein-rich sources of food and fodder of high nutritive value to man and animals. The seeds of grain legumes serve as a meat substitute in developing countries. Those of soybean and groundnut contain valuable oil for industrial processing, and they are increasingly produced on large holdings as cash crops for export, as of soybean in Brazil. Green seeds or whole pods and sometimes green leaves or sprouted seeds are also eaten as a vegetable. Some tropical legume species are grown for their edible swollen roots. Clovers are forage or fodder crops, or serve as green manure and cover crops used in crop rotation for maintaining soil fertility. Leguminous tree species are often grown in the tropics for shading and sometimes for timber.

Plants, including leguminous species, keep playing important roles in studying plant viruses for their effects on plants and crops, and also for their separation from natural mixtures, their propagation, and even their detection. Common bean (French bean or bush bean; *Phaseolus vulgaris*) and cowpea (*Vigna unguiculata*) are still used as test plants for detecting several viruses or measuring their infectivity by the local lesions they produce on uniform opposite primary leaves either on plants or detached.

History of Legume Virology

The history of research on legume viruses illustrates how, through trial and error, plant virology has evolved from its middle ages concentrating on their effects on plants and crops into the present high-technology era dominated by molecular biology of the viruses themselves. Legumes were among the first crops studied for virus diseases. Some early examples were bean common mosaic (1917, 1921) (**Figure 1**) and soybean mosaic (1921). Viruses were originally distinguished by their hosts or host ranges and by symptoms, and later also by their ways of spread, as in seed (bean common mosaic potyvirus, BCMV, 1919; **Figure 2**) and by specific insects such as aphids or beetles. Later, further distinction was by the so-called 'physicochemical properties' studied in expressed sap, especially since 1930 when sap transmission was facilitated with abrasives. Viruses of annual legumes, such as peas and beans, were also found in perennial legumes such as clovers, and viruses described from legumes and those from nonlegumes increasingly appeared to cross-infect their respective hosts. For example, beet curly-top curtovirus (BCTV) was found to cause severe disease in beans, and BYMV often infected gladiolus. With information accruing in the 1930s, efforts to devise keys for the recognition of viruses of legumes were first made in 1939 and 1945, and order slowly began to emerge.

After Wold War II, viruses were increasingly recognized as worldwide production constraints, and legume virus workers were among the first to standardize the techniques for identifying the viruses dealt with (1960).

An International Working Group on Legume Viruses (IWGLV) was established in 1961 for the exchange of seeds of test plants and antisera and of information. New techniques such as serology and electron microscopy were advocated for further laboratory characterization of the viruses themselves, and collaboration between researchers across national barriers was stimulated. A resulting tentative list of viruses reported from naturally infected leguminous plants (1964) animated further worldwide assemblage of information in computerized form as by the Australian Virus Identification Data Exchange project (VIDE). Its microfiche publication on *VIDE Viruses of Legumes* was soon followed by a printed version (1983). Similar books on viruses of plants in Australia (1988) and of tropical plants (1990) were succeeded by *Viruses of Plants* (1996), which was also distributed on the Internet. It later contributed to the database of the International Committee on Taxonomy of Viruses (ICTV), but it all began with IWGLV.

The IWGLV has in 2001 merged with the International Working Group on Vegetable Viruses, set up in 1970 with similar means and objectives and partially dealing with the same crops. The major aim remains surveying of the crops for viruses and their economic importance, understanding their ecology, and learning how they can be controlled for safeguarding crop productivity. But first of all comes virus identification.

Legume Viruses

In early plant virology viruses were thought to be rather specific in host range, for example, limited to Leguminosae or even to single species. Examples of narrow host specialization were BCMV and soybean mosaic virus (SMV) on beans (**Figure 1**) and soybean, respectively. They are highly seed transmitted and are introduced into crops every year by the seed that is sown, whereupon

Figure 1 Dramatic effect of common mosaic, caused by bean common mosaic potyvirus (BCMV) in some cultivars of common bean (*Phaseolus vulgaris*). The disease was described during the decade starting 1910 as one of the first recognized, now globally distributed legume virus diseases. (Left, resistant cultivar.)

Figure 2 BCMV in common bean with primary infection from the seed in plant at left and secondary infection by aphids in adjacent plant at right.

they are further spread from infected seedlings by aphids (**Figure 2**). Alternative hosts then are not needed for survival from one growing season to another. The aphid-transmitted wisteria vein mosaic potyvirus (WVMV) has so far been found worldwide in a few ornamental *Wistera* spp. only, possibly from a single vegetative origin.

Several viruses, of legumes, soon appeared to be much more polyphagous than supposed before. BYMV, for example, infects several annual legumes, but is prevalent in ornamental gladiolus as well, and occurs in other Iridaceae and some 14 other plant families. The related cowpea aphid-borne potyvirus (CABMV) also causes woodiness disease in passionfruit, and BCMV has meanwhile been found naturally in at least 26 leguminous species.

Some 'nonlegume viruses', other than the beet virus BCTV, also infecting legumes are the potyviruses beet mosaic virus (BtMV), lettuce mosaic virus (LMV), and turnip mosaic virus (TuMV). Extremely polyphagous viruses like the aphid-borne cucumber mosaic cucumovirus (CMV) and the soil-borne and fungus-transmitted tobacco necrosis necrovirus (TNV) include legumes among their many natural hosts (**Figure 3**). Presumed host barriers are increasingly transgressed, and novel types of disease emerge in cases of often haphazard new encounters between plants and viruses – as through the global movement of plant species and cultivars and of viruses and their strains distributed worldwide in plant propagation materials. Legume viruses are therefore increasingly difficult to define as a special category, but over 14% of the plant viruses still derive their name from legumes.

Most taxonomic groups of plant viruses are represented in legumes, and a 1991 monograph by Edwardson and Christie lists 279 different legume species reported naturally infected by 171 viruses then thought to be distinct. Perennial legumes, such as clovers, host many viruses, often in mixtures, and in plants the viruses may interact, either directly or via their host. BYMV, for example, helps multiply CMV and increase disease severity as in cowpea stunt. Groundnut rosette in Africa is caused by a complex of groundnut rosette umbravirus (GRV) together with an associated satellite RNA and groundnut rosette assistor luteovirus (GRAV), that helps aphid transmission. In nature, leguminous hosts and the viruses infecting them vary greatly and the result of infection may range from symptomless infection to severe disease up to premature plant death. Symptoms may often be mistaken for mere physiological disturbances as by mineral deficiency (**Figure 4**). Concise literature on virus diseases of specific legume crops is given by the compendia of crop diseases published by the American Phytopathological Society, St. Paul, MN, USA. These and other reviews up till 1996 have been listed by Bos in 1996.

Variation of Legume Viruses

BCMV was one of the first viruses of which 'strains' were detected on new cultivars of common bean bred for resistance. The same was later found for SMV on soybean where resistance breeding was also prominent. Genetic change of crops is known to exert selection pressure on the viruses allowing these to co-evolve. Of BCMV, at least seven pathogenicity genes have been postulated

Figure 3   Severe pod necrosis in common bean caused by the soil-borne and sap- and fungus-transmitted tobacco necrosis necrovirus (TNV).

Figure 4   Beginning chlorosis in top of faba-bean plants with bean leafroll luteovirus (BLRV), often mistaken for a mere effect of mineral deficiency.

and four of them interact with the bean's susceptibility or resistance genes in a gene-for-gene relationship.

Genetic adaptation to plant species and host cultivars where the viruses happen to land may also account for the wide variation that exists within the virus genera, explaining the evolution of virus species. Legume potyviruses (genus *Potyvirus*, now part of the family *Potyviridae*) have many variants and intermediates as between and around BCMV, SMV, and BYMV. Their expanding number made biological properties, such as host ranges, symptoms, cross-protection, and serology, increasingly inadequate for distinction and specific detection, and a semblance of continuity started to blur the originally postulated boundaries. It suggested progressive evolution through mutation, recombination, and selective adaptation in niches often created by man. Molecular genetics, recording percentages of nucleotide and amino acid sequence identities, has helped artificially to draw boundaries between different 'viruses' and devise phylogenetic dendrograms (**Figure 5**). This allows more final identification but entails increasing reliance on molecular techniques. Similar evolutionary developments have been observed among, for example, legume luteoviruses and begomoviruses.

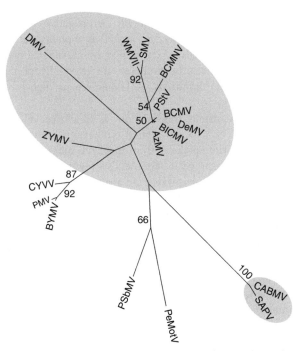

Figure 5 Phylogenetic tree of the legume-infecting viruses in the family *Potyviridae* based on analysis of the nucleotide sequences of the 3'-end noncoding region (NCR) of their ss (+) RNA genome. Branch lengths are proportional to genetic distance. Reproduced from Berger PH, Wyatt SD, Shiel PJ, Silbernagel MJ, Druffel K, and Mink GI (1997) Phylogenetic analysis of the *Potyviridae* with emphasis on legume-infecting potyviruses. *Archives of Virology* 142: 1979–1999, with permission from Springer-Verlag.

Ecology

What happens to the legume viruses in legume crops is a matter of an intriguingly complex ecology. Factors to consider are: (1) the viruses already mentioned, (2) the sources of infection, (3) the vectors that spread the viruses, (4) the crops that are subject to infection and differ in genetic vulnerability, and finally (5) the growing conditions, including soil, climate, and the cropping system that influence host plants and vectors. These factors may capriciously interact, and the outcome is often hard to predict. Several of the factors listed largely depend on the grower's decisions, as for his/her choice of place and time of cultivation and the choice of crop and cultivar and of cultural practices. Legume viruses illustrate the widest known array of ecological relationships between viruses and plants. The quantitative involvement of the factors mentioned is the subject of epidemiology. Comprehension of the ecology of the viruses and of their epidemiology explaining to what extent and with what speed the viruses spread is essential for devising ways and means to control the viruses or limit their effects.

Sources of Infection

Many legume viruses are seed-borne, and commercial seed lots, even when usually only partially infested, provide infected seedlings throughout a newly sown crop. For BCMV (**Figure 2**) and SMV in common bean and soybean, respectively, these are the almost only, but efficient and early, within-crop sources of infection. BYMV that is seed-transmitted in yellow lupin and at low rates in pea and faba bean, has other sources of infection as well. Pea seed-borne mosaic potyvirus (PSbMV), which caused considerable concern in the pea-growing and -processing industry in the USA during the late 1960s and 1970s, is now known to often occur in seed of several other legumes worldwide. Notorious and vast other sources of infection for annual legumes are nearby perennial clovers and medics in pastures or grown as pure crops. BYMV readily moves from clovers (**Figure 6**) and also from nonleguminous ornamental gladiolus to bean when grown nearby. Infection of wild plants, as by CMV, pea early-browning tobravirus (PEBV) and TNV, is often symptomless and the wild sources of infection may then be hard to find.

Means and Ways of Spread

By contact

For legumes, spread from one plant to another when plants touch or are touched, seems to be limited to white clover mosaic potexvirus (WClMV), TNV, and the beetle-transmitted red clover mottle comovirus (RCMV). They occur in plants in high concentration and have very stable virions. RCMV may even be spread with a lawn mower.

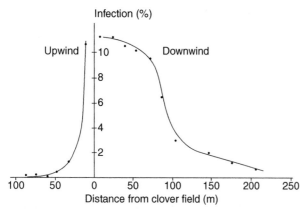

Figure 6 Spread of bean yellow mosaic potyvirus (BYMV) that is nonpersistently transmitted by aphids from red clover into field of common bean, influenced by prevailing wind. Reproduced from Hampton RO (1967) Natural spread of viruses infectious to beans. *Phytopathology* 57: 476–481, with permission from American Physopathological Society.

Some stable plant viruses, such as the fungus-transmitted red clover necrotic mosaic dianthovirus (RCNMV) may also be released into soil and water from decaying roots and may then get mechanically into plants when these are wounded. Tomato bushy stunt tombusvirus (TBSV), reported from *Robinia pseudoacacia* and *Wisteria floribunda*, may be transmitted directly from living roots.

Via seed

Many legume viruses (69 of 171 listed) have been reported to be distributed via seed, although often in such low percentages (below 1% or less) that infected seedlings easily escape notice. Such seed may still spread viruses over long distances to places where they did not occur before. Many legume viruses have already been spread worldwide in seed, and they continue to threaten legume crop improvement in developing countries. However, rates of seed-lot infestation with infected seeds may be high as of PSbMV where up to 90% of the seeds of infected mother plants may contain the virus, so that crops grown from such seed will directly suffer economically. Most seed-transmitted viruses of legumes are carried in the embryo, which remains infective for the seed's life. The legume viruses that in their plant hosts are limited to the phloem, as a rule, do not pass via the seed. Most mechanically transmissible viruses are detectable for a while in the seed coats from where they usually cannot reach the seedling.

By living organisms

Most spread of legume viruses is by living organisms, called vectors. It usually is rather, if not highly, specific, and most viruses have complicated relationships with their vectors so that viruses transmitted by one type of vector usually are not transmitted by any other type, and the vectors have complex ecologies of their own as well.

Seed-borne viruses, with the exception of cryptoviruses which have no known vector, have either an aerial or subsoil vector for further dissemination.

Aboveground, most legume viruses are readily insect transmitted.

1. 'Nonpersistent and noncirculative transmission', as of many carla-, caulimo-, cucumo-, and fabaviruses, and the large group of potyviruses that are all readily transmissible in expressed sap, is by mere mouthpart contamination and usually is not specific. It is rapid but over short distances only (**Figure 6**). These viruses usually move in zones from a nearby other crop or do so spotwise from infected seedlings or weeds.

2. 'Persistent transmission' particularly holds for phloem-limited viruses that usually are not sap-transmissible. Examples are the luteoviruses of bean leafroll (BLRV) (**Figure 4**) and soybean dwarf (SbDV) by aphids, the mastrevirus of chickpea chlorotic dwarf (CpCDV) and the curtovirus BCTV in common beans by leafhoppers, the bean golden mosaic begomovirus (BGMV) by whiteflies, and the tospoviruses of tomato spotted wilt (TSWV) and groundnut bud necrosis (PBNV) in groundnut by thrips. Their relationship with the vectors is highly specific, that is, circulative. Some viruses even multiply in their vector (propagative transmission) and transovarially pass to offspring. The vectors then remain viruliferous for extended periods or for life. Propagative viruses are viruses of animals (insects) as well as of plants. Persistently transmitted plant viruses may be transferred over very long distances. Leafhoppers are strong flyers and they may cover hundreds of kilometres in high-level wind currents.

3. 'Semipersistent transmission' as of clover yellows closterovirus (CYV) and related criniviruses often is by whiteflies. Virus uptake is from the sieve tubes, and the virus is only adsorbed in the insect's foregut.

4. A variety of viruses that are artificially transmissible in plant sap is naturally transmitted by beetles (*Coleoptera*) including blister beetles (bean pod mottle comovirus; BPMV), lady beetles (southern bean mosaic sobemovirus; SBMV), leaf beetles (cowpea mosaic comovirus; CPMV; and BPMV) and some weevils (broad bean mottle bromovirus; BBMV). These have biting mouthparts, and virus acquisition and introduction by these vectors is immediate and after a single bite. Virus retention in the beetles is in the hemolymph for days or weeks, and specificity of transmission is due to specific inactivation of most of the viruses in the regurgitant fluid that is produced by the beetles during feeding.

5. Exceptional among legume viruses is the possibly persistent transmission of pigeon pea sterility mosaic virus (PPSMV) in India by the eryophyid mite *Aceria cajani* that is invisible to the naked eye and mostly moves passively as by wind.

Within the soil, a number of legume viruses are transmitted by either nematodes or fungi.

1. The trichodorid nematodes that spread PEBV are of low mobility. Patches of infected plants in fields of pea and common bean hardly enlarge. The transmission resembles semipersistence, and retention in the vector may be for months especially at low temperatures. Disease recurrence the next season, as well as long-distance dissemination of the viruses, is especially explained by high rates of seed transmission and by occurrence in weeds. Trichodorid nematodes and the viruses they transmit do best on sandy soils or in sandy patches.

2. (a) Chytrid fungi (*Olpidium brassicae* and *O. bornovanus*) transmit the necrovirus TNV (**Figure 3**) and the dianthovirus RCNMV, respectively. These viruses are carried externally on the fungal zoospores, and transmission is nonpersistent. The virus-carrying spores move for short distances in soil water or over longer distances in irrigation water. (b) The plasmodiophorid fungus *Polymyxa graminis* spreads peanut clump pecluvirus (PCV) and broad bean necrosis pomovirus (BBNV). The viruses are carried within the hard-walled resting spores of the fungus and may persist there for many years. These spores may be transferred over long distances in drainage water and much farther in dry soil on seeds, vegetative propagation material, tools and transport vehicles, and in wind-blown dry soil and seeds. Fungus-transmitted viruses are favored by rainfall and irrigation.

The Role of Man

Viruses are nearly always and often invisibly present in or around crops or in wild species. Whether disease results greatly depends on susceptibility and sensitivity (vulnerability) of the crop and thus on the crop species or cultivar that the growers choose. A change of farming system may also create new niches for viruses and their vectors. Large-scale commercial cropping of groundnut in Northern Nigeria, for example, led to a dramatic epidemic of groundnut rosette in 1975. Dense populations of the vectoring aphids resulted from their survival and multiplication on volunteer groundnut plants and on numerous other plant species available throughout the year due to increased irrigation. In Brazil, the acreage of soybean for the world market tremendously increased and its growing period extended, creating vast areas of an excellent food and breeding plant for the whitefly *Bemisia tabaci*. This led to the enormous upsurge of bean golden mosaic virus (BGMV) in common bean. Where soybean cultivation had also expanded in traditional bean-producing areas, the seed-borne SMV, not usually infecting bean, also showed up in that crop because of high infection pressure from soybean and the cultivation of susceptible bean varieties. In the Latin Americas and other parts of the world, whitefly-

transmitted viruses have also increased enormously in soybean and a range of crops including many nonlegumes such as cotton, tobacco and tomato. New niches may also result from the introduction of alien crops or new cultivars, and alien viruses inadvertently introduced in germplasm for breeding programs and in commercial seed may well be pathogenic to local crops or crop cultivars.

Most of the many seed-borne viruses of legumes have already shown up everywhere. Legume improvement programs in developing countries often entailed new diseases by viruses introduced in germplasm. In developed countries the introduction of virus-resistant cultivars, as of bean and soybean, has evoked the emergence of resistance-breaking virus strains. Continuing agricultural modernization and further internationalization of trade and traffic are therefore bound to lead to newly emerging diseases. In fact, most if not all virus diseases of crops are 'man made'.

Epidemiology

Accurate quantitative information about epidemic development of virus infection in crops usually is scarce. Epidemic disease development in crops when beginning with a limited number of infected plants is known to be a polycyclic process and to proceed according to an S-curve. This also holds for legume viruses (**Figure 7**; curve *V*). The speed of development greatly depends on distance between source and subject of infection, on the number of infection sources, on vector population density

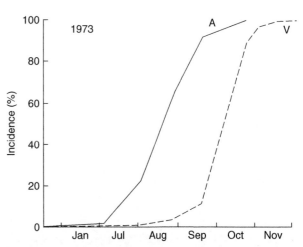

Figure 7 The increase in incidence with time of subterranean clover red leaf virus (a strain of soybean dwarf virus, SbDV, an unassigned virus of the family *Luteoviridae*) (**V**), and of its aphid vector *Aulacorthum solani* (**A**) in crop of faba bean in Tasmania. Reproduced from Thresh JM (1983) Progress curves of plant virus disease. *Advances in Applied Biology* 8: 1–85, with permission from Elsevier.

(**Figure** 7; curve *A*) and efficiency, and on crop and cultivar susceptibility (or resistance). When seeds are major if not sole primary sources of infection as of BCMV and SMV in bean and soybean, respectively, initial disease incidence determined by seed infestation below 0.1% mostly does not lead to incidences at the time of harvest that cause appreciable loss. In resistant cultivars, virus multiplication in individual plants usually is lower than in susceptible ones so that it takes longer for plants to become efficient secondary sources of infection. Epidemic development in crops of such cultivars is then delayed and crops are resistant as a whole.

Persistently transmitted viruses such as the luteoviruses usually move in from far away and many plants may become infected around the site where the viruliferous aphids land and move about. Then, infection may soon evenly invade an entire crop. Nematode-transmitted viruses like PEBV in pea and bean occur spotwise depending on irregularity of soil type and nematode distribution. Plants usually are invaded with such viruses early when the nematodes become active, and the spots may slowly enlarge over the years if the soil permits the nematodes to extend their territory.

Economic Importance

Crop loss depends on incidence of diseased plants, on time of infection during crop development, and on symptom severity. Many viruses of legumes are highly contagious and often rapidly reach all plants of a crop. Losses then are high if the symptoms are severe as of faba bean with faba bean necrotic yellows nanovirus (FBNYV) where plant death often leads to total crop failure (e.g., **Figure 8**). Early infections usually are most severe, particularly when symptoms accumulate as of the phloem-limited

luteoviruses viruses in chickpea, faba bean (**Figure 4**), pea, and several other legumes. These viruses primarily cause degeneration of the sieve tubes and impede transport in vascular bundles. This then is followed by a cascade of secondary and tertiary symptoms such as poor root growth and leaf yellowing, plant stunting, and even premature death. Late infections usually are less harmful than early ones because plant susceptibility and sensitivity often decrease toward plant maturity. Necrosis directly affects yield and quality particularly when the pods are involved. Examples are certain genetic host/virus combinations of soybean with SMV and of bean with BCMV and BCMNV, of bean with TNV (**Figure 3**), and of the same and pea with PEBV.

The effects of viruses on yield are often overlooked or neglected. Plant yellowing (**Figure 4**) may be mistaken for noninfectious mineral deficiency, but final losses may still be dramatic. Chlorophyl defects, as in various mosaics and yellows diseases (**Figures 1** and **4**), always reduce assimilation capacity and adversely affect yield. Even when visible symptoms are absent, root nodulation, plant vitality and productivity may considerably be diminished, and plant susceptibility and sensitivity to other pathogens increased. In perennial clovers, longevity of stands may be reduced by increased infection by root rot fungi and by raised sensitivity to drought and winter injury. One virus may also support another virus. In cowpea stunt, caused by a mixture of CMV and blackeye cowpea mosaic virus (BCMV or a variant), experiments have shown that together they caused a reduction in seed yield of 86%, whereas single infections only led to seed losses of 14 and 2.5%, respectively.

In practice, severe crop losses have been reported in legumes. The dramatic 1975 epidemic of groundnut rosette in Northern Nigeria occurred on over 1 million ha of groundnut and destroyed an estimated 0.7 million ha worth over $250 million. The same year, the whitefly-transmitted BGMV had become the main production constraint in the traditional cultivation of bean in Brazil. In Latin America, where beans is one of the main staple foods, more than 2.5 million ha was under attack by the virus during the 1990s. The related bean dwarf mosaic begomovirus (BDMV) showed up in Argentina in 1981 and at least one million ha previously planted with bean in South America was abandoned because of the risk of total yield loss. In 1992 in the Nile Valley, Egypt, the then newly recognized FBNYV wiped out entire crops of faba bean on *c.* 16 000 ha in the Beni Suef Governorate (**Figure 8**).

Financial losses that mostly are not taken into account but may be substantial are the costs of hygienic measures to be taken by growers, and of the more expensive price of seed of resistant cultivars and of seed certified for low rates of infection. Nationwide, costs of research, education, extension, and quarantine must also be considered.

Figure 8 Total crop loss in faba bean (*Vicia faba*) over vast area in Egypt in 1992 caused by faba bean necrotic yellows nanovirus (FBNYV).

Control

Virus-diseased plants cannot be cured once they are infected, and control nearly always is preventive. Either one or, mostly, a combination of mainly hygienic measures are to be taken based on profound knowledge of the ecology and epidemiology of the viruses in their relation to crop ecology. But this, first of all, requires proper disease diagnosis, that is, reliable identification of the causal virus for allowing a goal-directed strategy of control.

Often highly effective, even when applied singly, is the avoidance or removal of the sources of infection. For example, soil infested with soil-borne and nematode- or fungus-transmitted viruses, such as PEBV, TNV, or PCV, should not be used for growing peas, bean, and groundnut. Commercial seed used for growing crops must be relatively free of seed infected with seed-borne viruses, and possibly be certified for low rates of infection. Certification schemes have been developed for the production of commercial seed of legume crops that often harbor aphid-transmitted seed-borne viruses such as BCMV and SMV. The commercial seed must be produced under relatively vector-free conditions away from sources of infection, while infected plants must be regularly removed. The harvested product must be tested and certified for low rates of virus infection so that infection of crops to be grown from the seed is likely to remain below damage thresholds. For most legumes, commercial seed should not contain more than 0.1–0.5% of infected seeds. Germplasm used for breeding purposes and as basic material for the production of commercial seed, must be multiplied under vector-proof conditions from seed of plants that have been tested individually for virus freedom, and suspected plants must be removed. Annual legume crops must not overlap with nearby older crops or with other perennial crops such as clovers that may harbor several legume viruses (**Figure 6**).

Epidemic buildup must be prevented by avoiding high densities of vector insects by choosing growing conditions or a season with less vector insects. Systemic insecticides are helpful to reduce the spread of persistently insect-transmitted viruses such as the legume luteoviruses. Virus dissemination over long distances must be avoided or prevented also by the use of clean seed, which is especially important for legumes because of the many viruses that are seed-borne in them.

The use of resistant cultivars, and breeding for resistance, has been the most widely used means of managing virus disease incidence and severity, and has a long tradition for legumes. It should be an ongoing effort because of the selection pressure that new cultivars exert on viruses for developing new resistance-breaking strains. Breeding for resistance to seed transmission is also possible as found in beans and lupin for BCMV and CMV, respectively. For developing countries, International Agricultural Research Institutes (IARCs) as CIAT in Cali, Colombia, ICARDA in Aleppo, Syria, ICRISAT in Hyderabad, India, and IITA in Ibadan, Nigeria, that are mandated for legume crops, are instrumental in improving resistance of important leguminous food and fodder crops by breeding, often in concert with national institutes in their respective regions. They also provide information and survey the respective crops in their regions of outreach, study the identity and ecology of the viruses that are important there, and develop techniques and means for their proper identification and for testing of plant genetic materials for resistances to the viruses concerned. The institutes have immense collections of germplasm in their gene banks that are also sources of genes for resistance to viruses. Genetic engineering is increasingly replacing original methods of gene transfer by pollination.

The ever-modernizing agriculture, including the continuing change of crop genetic makeup as well as of farming systems, however, keeps entailing new risks. Even resistance of cultivars practically never means immunity, and resistance hardly ever is durable. For the many seed-borne viruses of legumes, in particular, the continuing globalization of trade and traffic create a special hazard. Certification of commercial seed, which must be grown in the open with a view to the immense quantities needed, and quarantine of germplasm, which cannot handle individual seeds of the large quantities that are distributed worldwide, is unlikely to ever guarantee absolute freedom from the many known and still unknown and often latent seed-borne viruses.

Most virus control in crops, therefore, involves interference through crop management with the ecology of the viruses. Most measures are imperfect and are likely to entail new problems. Hence, man will never be ready to combat viruses in legumes and whatever other crops. Since the mostly preventive measures of control must be taken outside the field-grown crops, for example, at the source of infection, and what individual growers are doing or neglecting has a bearing on the health of crops of other growers, public interests are at stake. That is why public institutions, governments, and international organizations must remain involved as for regulation, quarantine, certification, advice, teaching, and continuing research. Such research, as on legume viruses, is of great relevance to human society. The study of legume viruses provides an outstanding example of the dynamics and complexity of viruses in their relation to crops.

See also: Luteoviruses; Plant Resistance to Viruses: Engineered Resistance; Plant Resistance to Viruses: Geminiviruses.

Further Reading

Berger PH, Wyatt SD, Shiel PJ, *et al.* (1999) Phylogenetic analysis of the *Potyviridae* with emphasis on legume-infecting potyviruses. *Archives of Virology* 142: 1979.

Berger PH, Wyatt SD, Shiel PJ, Silbernagel MJ, Druffel K, and
    Mink GI (1997) Phylogenetic analysis of the *Potyviridae* with emphasis
    on legume-infecting potyviruses. *Archives of Virology* 142:
    1979–1999.
Bos L (1970) The identification of three new viruses isolated from
    Wisteria and *Pisum* in the Netherlands, and the problem of variation
    within the potato virus Y group. *Netherlands Journal of Plant
    Pathology* 76: 8.
Bos L (1983) Plant virus ecology: The role of man, and the involvement
    of governments and international organizations. In: Plumb RT and
    Thresh JM (eds.) *Plant Virus Epidemiology; the Spread and Control of
    Aphid-Borne Viruses*, 7pp. Oxford: Blackwell.
Bos L (1992) New plant virus problems in developing countries: A corollary
    of agricultural modernization. *Advances in Virus Research* 41: 349.
Bos L (1996) *Research on Viruses of Legume Crops and the
    International Working Group on Legume Viruses; Historical Facts and
    Personal Reminiscences*, 151pp. Aleppo, Syria: International
    Working Group Legume Viruses.
Bos L, Hagedorn DJ, and Quantz L (1960) Suggested procedures for
    international identification of legume viruses. *Tijdschr. Plantenziekten*
    66: 328.
Boswell KF and Gibbs AJ (1983) *Viruses of Legumes 1983. Descriptions
    and Keys from VIDE*, 139pp. Canberra: Research School of
    Biological Sciences, The Australian National University.
Edwardson RE and Christie RG (1986) *Viruses Infecting Forage
    Legumes,* monograph No.14, 742pp + appendices. Gainesville:
    University of Florida Agricultural Experiment Station.
Edwardson RE and Christie RG (1991) *CRC Handbook of Viruses
    Infecting Legumes*, 504pp. Boca Raton, FL: CRC Press.
Hampton RO (1967) Natural spread of viruses infectious to beans.
    Phytopathology 57: 476.
Jones RAC (2000) Determining 'threshold' levels for seed-borne virus
    infection in seed stocks. *Virus Research* 71: 171.
Summerfield RJ and Roberts EH (eds.) (1985) *Grain Legume Crops.*
    859pp. London: Collins.
Thresh JM (1983) Progress curves of plant virus disease. *Advances in
    Applied Biology* 8: 1–85.
Van der Maesen LJG and Sadikin S (eds.) (1989) *Pulses. Plant
    Resources of South-East Asia No. 1.* 105pp. Wageningen: Pudoc.
Weiss F (1945) Viruses described primarily on leguminous
    vegetable and forage crops. *Plant Disease Reporter, Supplement*
    155: 32.

Plant Virus Diseases: Economic Aspects

G Loebenstein, Agricultural Research Organization, Bet Dagan, Israel

Introduction

Estimates on crop losses due to pests and diseases are lacking, with respect to both up-to-date data and accuracy. Data from different countries obtained by a variety of methods should be taken only as a rough guide. Nevertheless, the available information sheds light on the tremendous loss to the food and fiber supply for the ever-growing world population.

A team of German crop scientists, with support of the European Crop Protection Association, published a comprehensive study of pest-induced crop losses on major food and cash crops. In the four principal crops (rice, wheat, barley, and maize) losses due to diseases ranged between 10% and 16% of potential production which translates into approx. US$64 × 10 billion in the years 1988–90. The losses increased to US$84 × 10 billion if four additional crops were included (potatoes, soybeans, cotton, and coffee) that together occupy half the world's cropland, with harvests worth US$300 billion in 1988–90. The study did not cover some important food crops of developing countries, such as cassava, millet, and sorghum, and horticultural crops, which are often heavily affected by virus diseases. The study found that pests and diseases accounted for pre-harvest losses of 42% of the potential value of output during 1988–90, with 15% attributable to insects and 13% each to weeds and pathogens. An additional 10% of the potential yield was lost post-harvest.

Yield losses due to plant virus infections are estimated to range between 10% and 15%, while losses due to fungal pathogens are estimated at 40–60%, the rest being caused by bacterial and phytoplasma pathogens and nematodes. Again, data should be viewed with caution, as many survey data are estimates based on educated guesses. On their face value these data do not agree with the previous data. Apparently, the selection of crops was different and fewer viral diseases were considered.

Yield losses caused by 11 different virus diseases in various crops, including bananas, cassava, cacao, maize, groundnuts varied from 0% up to 100%. In a survey conducted in eight African countries, cassava mosaic disease yield losses were estimated to reach 30–40%.

Economic losses are not only due to reduced vigor and growth resulting in yield losses, but may also affect product quality, as for example deformations in zucchini yellow mosaic virus (ZYMV)-infected squash or plum pox virus-infected peaches. In addition, major economic aspects to be considered are the costs to produce virus-tested propagation material, eradication programs, vector control, and breeding for resistance.

It seems that spread of viruses and subsequent economic losses are more prevalent in tropical and subtropical regions with continuous vegetation throughout the year. The continuous vegetation, annual and perennial, enables persistence of viruses and their vectors. In cooler areas winter temperatures are not favorable for annual vegetation and prevalence of vectors. Similarly, in regions with a dry hot summer, continuous herbaceous vegetation is interrupted which also reduces vectors and subsequent spread of plant viruses.

Economic losses due to viruses are often higher in vegetatively propagated crops such as citrus, cassava, and potato, than in seed-propagated ones. In many plants, viruses are not transmitted through seeds, a mechanism developed through evolution whereby the plant ensures that the next generation starts healthy. In vegetatively propagated crops, if cuttings, seed tubers, bud wood, bulbs, runners, etc. are taken from infected plants the virus will inevitably infect the next generation. This is in addition to spread of the virus by vectors, which may occur in both cases.

During the last decade whitefly transmitted viruses are on the increase in many crops, including vegetables, often becoming a major constraint.

This article is restricted to virus diseases in horticultural and plantation crops where virus diseases cause severe economic losses, many of them in developing countries.

Citrus Tristeza Virus

Citrus tristeza virus (CTV) is probably the most destructive citrus virus in the world. About 30 million trees on sour orange rootstocks were lost in Brazil and Argentina in the years 1940–60, 6.6 million trees in Venezuela in the 1980s, and an estimated 10 million trees in Florida and other Caribbean Basin countries. In these regions CTV was and is still being spread by the brown citrus aphid *Toxoptera citricida*. In Spain CTV killed about 10 million trees and several million more were lost in California, Israel, and other areas. In these areas the melon aphid, *Aphis gossypii*, is spreading CTV. When both vectors are present, as in Florida, spread of CTV is greatly enhanced. Yield losses have been documented in several places, for example, in Jamaica. At two locations totaling 1159 acres, losses over 5 years were more than US$4 million. These costs did not include the costs for removing dead trees and replanting. However, even though millions of trees in Brazil and Spain were lost it did not affect production of citrus in both countries in the long run. Thus, in Brazil production increased from 1.7 million t in 1960 to more than 20 million t in the 1990s of the previous century. Apparently, replanting new orchards on tolerant rootstocks, together with budwood control, and improved horticultural technologies made it possible to overcome the severe damages by CTV in the old plantations.

Various strains or isolates of CTV are known. Mild strains cause little if any symptoms on most commercial citrus and stem pitting on Mexican lime while severe strains cause decline and death on citrus grafted on sour orange rootstocks; seedling yellow strains cause stunting and yellowing of leaves while orange and grapefruit stem pitting strains will stem-pit sweet orange varieties and grapefruit, respectively. Stem pitting isolates are the most severe strains of CTV and cannot be controlled by the use of tolerant rootstocks.

CTV is a member of the monopartite genus *Closterovirus*, with a positive single-stranded RNA genome encapsidated in a flexuous particle about 2000 nm in length. The virus can be detected by ELISA or by biological indexing on Mexican lime.

The best way to control CTV and other graft-transmissible diseases of citrus is by a mandatory certification program. This program includes a 'clean stock program' for production and maintenance of pathogen-free propagation stock, and 'certification programs' for maintenance and distribution of virus-free material for commercial use. These publicly operated programs are generally self-supportive. Thus, in Florida a fee of US$2 is charged for registration of validated or parent trees.

Cross-protection with mild CTV strains is widely practiced in Brazil, Australia, and South Africa. The control of CTV strains causing stem pitting on sweet orange and grapefruit in Brazil is, by far, the largest and most successful use of cross-protection. By 1980, over 8 million trees of Pera sweet orange were cross-protected. Cross-protection to control CTV in Pera sweet orange in Brazil is still practiced.

An eradication program for tristeza is in operation in California. During 2000–01 the Central California Tristeza Eradication Agency surveyed approximately 31 400 acres and conducted up to 375 000 individual ELISA assays. This program delayed undoubtedly the spread of the disease but is expensive since it also includes the costs of removing infected trees. In Israel a tristeza suppression program was in operation between 1970 and 1977, whereby 300 000 tests were made using indicator plants and electron microscopy and 1700 CTV-infected trees were detected and removed. Between 1979 and 1980, 1.25 million trees were indexed by ELISA and only 0.13% of the tested trees were found to be infected. Apparently the eradication program delayed the outbreak of a tristeza pandemic in Israel by at least 15–20 years.

Cassava Mosaic Disease

Cassava is a low-cost carbohydrate food predominantly grown by resource-poor farmers in developing countries. Virus diseases seriously impede cassava production, especially in Africa, where it is grown in diverse climates, from coastal regions to semiarid zones of the Sahel. Virus diseases are regarded as the major constraint on cassava production.

Several similar but distinct whitefly transmitted geminiviruses cause cassava mosaic disease (CMD) in Africa. They can occur singly or in combination. The most important of them are African cassava mosaic virus (ACMV) and East African cassava mosaic virus (EACMV), ascribed to the genus *Begomovirus*. From 1988 to the present, a major pandemic of an unusually severe form of CMD is spreading throughout East and Central Africa, causing massive losses and affecting food supplies. A distinct strain of EACMV – the Ugandan strain – is the most significant threat to cassava production in Africa.

Yield losses with individual cassava cultivars in different African countries range from 20% to 95%, and continent-wide losses have been estimated at 12–23 million t of fresh tuberous roots per year, worth about US$1200–2300 million. Yield losses in Uganda at the peak of the recent EACMV-pandemic were immense and farmers had to abandon cultivation of cassava in the worst affected areas. It was assumed that each year an area of *c.* 60 000 ha of cassava yielding 600 000 t worth US$60 000 was lost due to CMD.

Symptoms of CMD are mainly green or yellow mosaic. Severe chlorosis is often associated with premature leaf abscission.

CMD is transmitted by the whitefly *Bemisia tabaci*. Virus dissemination between fields and over long distances is mainly through the use of infected stem cuttings as planting material. The disease is caused by several geminiviruses.

CMD can be detected by serological gel-diffusion tests, different ELISA and immunosorbent electron microscopy. Monoclonal antibodies enable cassava mosaic virus isolates to be differentiated though none exists so far for detecting specifically the Ugandan strain of EACMV. Polymerase chain reaction (PCR) with specific nucleotide sequence primers however detects all different cassava mosaic viruses, including EACMV.

For control of CMD, the main emphasis has been on the development of resistant varieties. Phytosanitation, involving the selection of cuttings from healthy plants, rouging of diseased plants, and production of virus-free tissue-cultured planting material are also practiced. Planting date can be adjusted to avoid times when populations of whiteflies are high. Use of insecticides to restrict spread of CMD by controlling the whitefly vector was not effective.

Virus Diseases of Potato

Virus diseases are a major constraint in potato production and often reduce yields by more than 50%. Due to their economic importance much research was devoted to them in the first half of the twentieth century, and new techniques such as serology, meristem cultures, and electron microscopy were developed to control them.

At least 37 viruses naturally infect cultivated potatoes, with potato virus Y (PVY), potato leafroll virus (PLRV), potato virus S (PVS), potato virus M (PVM), potato virus X (PVX), potato aucuba mosaic virus (PAMV), and potato mop-top virus (PMTV) being the most widespread and damaging ones. In addition, potato spindle tuber viroid

(PSTV) and phytoplasma diseases are also causing major crop losses in potato.

PLRV is of great economic importance. It is widespread worldwide and in plants grown from infected tubers (secondary infection) yields may be reduced by 33–50%. Tubers from infected plants are small to medium sized. Even greater losses are observed when PLRV occurs together with PVX or PVY, reaching 40–70%. Yield losses due to PVY may reach 10–80%, especially if the virus occurs with PVX. A severe group of PVY strains designated PVY^{NTN} present in Europe, Japan, and the US cause a damaging disease in which tubers develop superficial rings that later are sunken and necrotic. They often become more conspicuous during storage, and can affect 90% of tubers of susceptible cultivars. At present there is no serological or other assay, which can distinguish between PVY^{NTN} and other PVY strains.

Yields of a crop in which all plants are secondarily infected with PVX and PVS will be 5–15% lower than normal. PVS and PVM are common in Eastern European countries. PVM can result in tuber yield losses of 40–75%. PVM occurs in complex with other viruses (especially PVS). PVS isolates may reduce tuber yield by 3–20%. Plants from infected stocks that have been propagated for several generations may suffer more severe losses. Thus, for example, potato yields in Kazakhstan are extremely low, averaging 9 t ha^{-1} during the years 1993–95. In a pilot scheme conducted by the Agricultural Institute of Astana, Kazakhstan, within a USAID project, microtubers obtained from meristems of virus-tested potatoes were used and resulted in a low infection rate in the elite seed tubers. These elite seed gave an increase in yield of about 90% when compared with yields of commercial fields in the vicinity.

Yields in former Soviet Union countries are low, averaging 11.26 t ha^{-1} during 1993–95 as there is no reliable system for providing certified tuber seeds, and farmers use seeds from their own fields. Low yields are to a great extent due to virus diseases carried over in the planting material. Thus, Loshitsky and Belorusky 3 N cultivars developed by the Belarusian Research Institute for Potato Growing (BRIP) yielded 30.0–50.0 t ha^{-1} in experimental fields, while Belarus' average potato yields range between 12.2 and 16.3 t ha^{-1}. Supposing that only 20% of this difference is due to virus infection in the planting material, and extrapolating this over the total area of potato in the former Soviet Union countries of about 6 million ha, producing on an average 70 million t of potato an annual loss of about 18 million t seems to be a conservative estimate.

Costs of seed potato certification schemes should also be considered. The aim of these schemes is to produce true-to-type disease-free (mainly from viruses) potato seed tubers. These schemes nowadays often include *in vitro*-derived plantlets for the initial increase of pathogen-tested clonal selections, including production of minitubers.

Prices for certified tuber seeds range between US$350 and 500 per ton while prices for ware potatoes range between US$170 and 180 per ton. If about 3 t seed potato is needed to plant 1 ha, this amounts to an additional expense of about US$450 when certified seeds are used compared to the farmer using his own crop for next year's planting. In fact the difference is presumably higher since the farmer might use the small, not marketable tubers for planting.

Virus Diseases of Sweet Potato

Sweet potato, *Ipomoea batatas*, is the seventh most important food crop in the world in terms of production. They are grown on about 9 million ha, yielding ~140 million t, with an average yield of about 15 t ha^{-1}. They are mainly grown in developing countries, which account for over 95% of world output. Sweet potato is a 'poor man's crop', with most of the production done on a small or subsistence level. Sweet potato produces more biomass and nutrients per hectare than any other food crop in the world. It is well suited to survive in fertile tropical soils and to produce tubers without fertilizers and irrigation and is one of the crops with an important role in famine relief. Thus, for example, across East Africa's semiarid, densely populated plains, thousands of villages depend on sweet potato for food supply. The Japanese used it when typhoons demolished their rice fields. Yields differ greatly in different areas or even fields in the same region. Thus, the average yield in African countries is about 4.7 t ha^{-1}, with yields of 8.9, 4.3, 2.6, and 6.5 t ha^{-1} in Kenya, Uganda, Sierra Leone, and Nigeria, respectively. The yields in Asia are significantly higher, averaging 18.5 t ha^{-1}. China, Japan, Korea, Thailand, and Israel have the highest yields with about 20, 24.7, 20.9, 12, and 33.3 t ha^{-1}, respectively. In South America the average yield is 12.2 t ha^{-1}, with Argentina, Peru, and Uruguay in the lead with 18, 11, and 10 t ha^{-1}, respectively. For comparison, the average yield in the US is 16.3 t ha^{-1}.

These differences in yields are mainly due to variation in quality of the propagation material. Sweet potatoes are vegetatively propagated from vines, root slips (sprouts) or tubers, and farmers in African and other countries often take vines for propagation from their own fields year after year. Thus, if virus diseases are present in the field they will inevitably be transmitted with the propagation material, resulting in a decreased yield. Often these fields are infected with several viruses, thereby compounding the effect on yields. In China, on average, losses of over 20% due to sweet potato virus diseases are observed mainly due to sweet potato feathery mottle virus (SPFMV) and sweet potato latent virus (SwPLV). The infection rate in the Shandong province reaches 5–41%. In countries where care is taken to provide virus-tested planting material, for instance in the USA and Israel, markedly higher yields of 16.3 and 30 t ha^{-1}, respectively, are obtained.

The sweet potato virus diseases (SPVDs) caused by the interaction of SPFMV and sweet potato chlorotic stunt virus (SPCSV) or sweet potato sunken vein virus (SPSVV) (possibly a variant of SPCSV) is the most important virus (complex) disease in East Africa. It can cause losses of 50–80%, especially in Uganda and Kenya, though in another study from Uganda losses were much smaller probably due to relatively high levels of virus resistance in their varieties. In a 3-year field study in Cameroon, SPVD reduced root yields by 56–90% in susceptible varieties. Yield reductions of 78% due to SPVD have been reported from field trials in Nigeria. In Israel in a 2-year field experiment, yield reductions of ~50% were observed in plots planted with SPVD-infected cuttings while infection by SPFMV or SPSVV alone were minor. Cucumber mosaic virus (CMV) can cause a complete failure of the crop when infected sweet potatoes carry SPSVV. However, CMV is unable to infect healthy sweet potato.

At present the best way to control virus diseases in sweet potato is to supply the grower with virus-indexed propagation material. Such programs are operating in Israel and in the Shandong province of China. In Israel, as a result of planting virus-tested material, yields increased at least by 100%, while in China increases ranged between 22% and 92%. The payoff to the farmer has been high and in Israel use of certified material, prepared by special nurseries, is common practice, costing ~3.0 US cents per cutting. Farmers buys about 30% of material needed to plant their fields and fill in the rest from cutting of vines grown from these plants. Yields are high and stable when virus-tested propagation material is used. In African countries such programs are operating only on a limited scale, because sweet potatoes are grown mainly as a food security crop, and not as a commercial one.

Breeding programs might be a future answer and such programs are in operation in Uganda, providing SPVD resistance. It will have to be seen if these improved cultivars will retain their resistance. Thus, several clones obtained from the International Potato Center (CIP) that were claimed to be resistant to SPFMV were found to be susceptible when Israeli and Ugandan virus isolates were tested.

Cacao Swollen Shoot Disease

The disease occurs in West Africa and was first reported in Ghana in 1936. Cacao swollen shoot disease is one of the most economically important viral diseases and has contributed to the drastic decline of cocoa production in Ghana. Since 1936, nearly 200 million cacao trees have been cut out from about 130 000 ha. From 1945 to 1950 yield of beans fell by 86% almost proportional to the killing of trees. In 1953 the 50 million infected trees represented a capital depreciation of £25 million since the discovery of the disease.

Cacao swollen shoot disease virus (CSSV) belongs to the genus *Badnavirus* and has bacilliform virions of 113 nm length and 28 nm width. A range of mealybug species, with *Planococcoides njalensis* being the most important one, transmit CSSV.

Attempts were made to control the disease by use of insecticides and biological control of the mealybugs, breeding for resistant cocoa cultivars, and protection with mild strains, but the main practice is by roguing of infected and their neighboring trees.

The eradication policy of infected trees enforced by the British colonial authorities in the Gold Coast (now Ghana) in the late 1940s met with serious opposition and was a major factor contributing to the rise of the political party led by Kwame Nkruma, who accused the authorities of attempting to destroy farmers' income.

Whitefly-Transmitted Geminiviruses in Tomato

These viruses such as tomato yellow leaf curl virus (TYLCV) have become a limiting factor in tomato production. In the USA whitefly-transmitted geminiviruses (WTGs) appeared in the early 1990s and losses are estimated to reach 20%, while in Cuba, Mexico, Guatemala, Costa Rica, and Brazil yield losses ranged between 30% and 100%. Losses in the Dominican Republic during 1989–95 were estimated at US$50 million.

Chemical control of the whitefly vector is a possibility, but the large whitefly populations make this option inefficient. Physical barriers such as greenhouses protected with fine net (50-mesh) screens are effective while 'floating barriers' of perforated polyethylene sheets stretched over tomato fields prevent the landing of whiteflies. These measures, however, increase the costs of production markedly. The best approach to control WTG is through the use of resistant or tolerant cultivars. Breeding programs are in progress, but so far no resistant or tolerant cultivars have been obtained.

Plum Pox Virus Sharka Disease of Stone Fruits

Plum pox symptoms were first observed in plums in Bulgaria between 1915 and 1918. Between 1932 and 1960 the disease moved north and east from Bulgaria into Yugoslavia, Hungary, Romania, Albania, Czechoslovakia, Germany, and Russia. The disease was observed mainly in plums and apricots and since the 1980s also in peaches.

Virus infection can cause considerable losses. About 100 million stone fruit trees in Europe are currently infected, and susceptible cultivars can result in 80–100% yield losses. In Eastern and Central Europe, sensitive plum varieties can exhibit premature fruit drop and bark splitting. Some sweet cherry fruits develop chlorotic and necrotic rings and premature fruit drop.

Plum pox does not kill trees but it makes the fruit unmarketable and reduces yield by 20–30%. The fruits drop before maturity and are unfit for use as they are bitter and unsweet. In Bulgaria, losses in 1968 were estimated to be 30 000 t. In the last three decades, the average fruit yield in the Czech Republic dropped by 80%, and the total number of plum trees was reduced from 18 million to 4 million. Emilia-Romagna region of Italy, between 1998 and 2002, 69 000 stone fruit trees were removed due to Sharka infection; and US$450 000 is being spent annually for their removal and replanting. In Spain since 1988, 1.5 million trees have been removed at a cost of US $17 million. In France, plum pox virus (PPV) is mainly present in the southeast part of the country. All trees showing symptoms are eliminated and if infection levels are above 10%, the whole orchard is destroyed. These methods have resulted in the destruction of about 27 000 trees in 1992 (of which 550 were found to be infected) and in 1993 100 ha (mainly peach) were eliminated. From 1973 to 1990, it is estimated that 91 000 trees were destroyed in France.

PPV is a potyvirus transmitted by aphids in a nonpersistent manner. However much of the spread in Europe can be attributed to movement of infected nursery materialfrom the Balkans.

See also: Emerging and Reemerging Virus Diseases of Plants; Plant Resistance to Viruses: Engineered Resistance; Plant Virus Diseases: Fruit Trees and Grapevine; Plant Virus Vectors (Gene Expression Systems); Vector Transmission of Plant Viruses.

Further Reading

Loebenstein G and Thottappilly G (2003) *Virus and Virus-Like Diseases of Major Crops in Developing Countries.* The Netherlands: Kluwer Academic Publishers.

Orke EC, Dehne HW, Schonbeck F, and Weber A (1994) *Crop Production and Crop Protection: Estimated Losses in Major Food and Cash Crops.* Amsterdam: Elsevier.

Van der Zaag DE (1987) Yield reduction in relation to virus infection. In: de Bolas JA and van der Want JPH (eds.) *Viruses of Potatoes and Seed-potato Production,* pp. 146–150. Wageningen, The Netherlands: Pudoc.

Waterworth HE and Hadidi A (1998) Economic losses due to plant diseases. In: Hadidi A, Khetarpal RH,, and Koganezawa H (eds.) *Plant Virus Disease Control,* pp. 1–13. St. Paul: American Pythopathological Society Press.

Relevant Website

http://faostat.fao.org – FAOSTAT, Food and Agriculture Organization of the United Nations (FAO).

Plant Virus Diseases: Fruit Trees and Grapevine

G P Martelli, Università degli Studi and Istituto di Virologia vegetale CNR, Bari, Italy
J K Uyemoto, University of California, Davis, CA, USA

Introduction

Grapevine (*Vitis* spp.), rosaceous fruit tree (stone and pome fruits), and nut crop (walnut, hazelnut) varieties are propagated by grafting scions onto rootstocks which, in turn, are either clonally propagated or derive from seedlings. Both members (scion and rootstock) of the grafted plant are therefore liable to carry viruses if they come from infected sources, or can be infected in the field. Remarkably, the causal agents of a number of diseases of these different plant species share the same epidemiological behavior and/or taxonomic position as, for example, soil-borne nepoviruses and pollen-borne ilarviruses and cherry leaf-roll virus (CLRV).

Virus Diseases of Grapevines

Disease Symptoms and Yield Losses

There are four major virus diseases of grapevines, that is, infectious degeneration-decline, leafroll, rugose wood, and fleck, which differ symptomatologically and in the type of the causal agents.

Infectious degeneration affects European grapes (*Vitis vinifera*) and American rootstocks and is characterized by two distinct syndromes, 'infectious malformations' and 'yellow mosaic' caused by distorting and chromogenic virus strains, respectively. Leaves and shoots of vines infected by distorting virus strains are more or less severely malformed (**Figure 1(a)**), bunches are smaller and fewer than normal, and berries ripen irregularly, are small-sized, and set poorly. Yellow mosaic-affected vines show bright chrome yellow discolorations that may affect leaves (**Figure 1(g)**) shoots, tendrils, and inflorescences. Leaves and shoots show little malformation, but bunches are small and few. Symptoms of 'decline' resemble those of 'infectious malformations' but affected European grape varieties decline and may die. Crop losses due to these diseases can exceed 60–70%.

Leafroll elicits an early discoloration of the interveinal tissues of leaves of infected European vines which turn purple red in red-berried cultivars (**Figure 1(i)**) and yellowish in white-berried cultivars; both interveinal tissue colorations develop against a background of green primary veins. Discolorations are usually accompanied by downward rolling of the leaf margins and thickening of the blade. In white-berried cultivars of *V. vinifera*, the symptoms are similar, but the leaves become chlorotic to yellowish, instead

of reddish. Bunches mature late and irregularly. Yield is decreased by 15–20% in average and rooting ability, graft take, and plant vigor are adversely affected as well as the quality of grapes and musts (sugar and protein content, aromatic profile, soluble solids, titratable acidity). Infection of American *Vitis* species and rootstock hybrids is symptomless, except for a variable decrease in vigor.

Rugose wood is a complex disease in which four different syndromes are recognized, that is, Rupestris stem pitting (RSP), Kober stem grooving (KSG), Corky bark (CB), and LN33 stem grooving (LNSG). Infected vines are less vigorous than normal, may show delayed bud opening in spring and a swelling above the bud union, which reflects a marked difference between the relative diameter of scion and rootstock. The bark above the graft union may be exceedingly thick and corky. Some vines decline and die within a few years from planting. The woody cylinder is marked by pits and/or grooves that can show on the scion, the rootstock (**Figure 1(j)**), or both, their severity varying with the scion/stock combination. Under cool and wet climates wood symptoms are milder or absent. No specific symptoms are seen on the foliage, although certain cultivars show rolling, yellowing, or reddening of the leaves similar to those induced by leafroll. Bunches may be fewer and smaller than normal and the crop is reduced on average by 20–30%.

Fleck is a complex consisting of several diseases ('fleck', 'asteroid mosaic', 'rupestris necrosis', and 'rupestris vein feathering') and viruses (grapevine redglobe virus, GRGV) that cause latent or semilatent infections in *V. vinifera* and most American *Vitis* species and rootstock hybrids. Only *Vitis rupestris* reacts to the different diseases with differential symptoms. Although the elusive nature of the complex hinders the assessment of its economic impact, adverse influence on vigor, rooting ability of rootstocks, and on graft take have been reported.

Geographical Distribution

All diseases are ubiquitous. There is not a single viticultural country where surveys were carried out in which one or more of the viruses involved in their etiology have not been found.

Causal Agents and Classification

Several different viruses are involved in the etiology of, or are associated with each single disease.

Figure 1   Symptoms induced by: (a) Distorting strain of grapevine fanleaf virus (grapevine). (b) Double infections by apple mosaic virus (yellow banding) and prunus necrotic ringspot virus (shredding) (cherry). (c) American plum line pattern virus (plum). (d) Cherry rasp leaf virus (cherry). (e) Apple stem pitting virus (pear). (f) Apple stem grooving virus (Virginia crab). (g) Chromogenic strain of grapevine fanleaf virus (grapevine). (h) Plum pox virus (apricot). (i) Grapevine leafroll disease (grapevine). (j) Grapevine rugose wood disease (grapevine). (e, f) Courtesy of L. Giunchedi.

Infectious degeneration and decline are two diseases caused by nematode-borne viruses with isometric particles *c.* 30 nm in diameter and bipartite RNA genome 2.2–2.8 kb (RNA-1), and 1.5–2.4 kb (RNA-2) in size. Of the 16 viruses recovered from infected grapevines, 15 belong to different species of the genus *Nepovirus* (family *Comoviridae*): (1) subgroup A: *Arabis mosaic virus* (virus: ArMV), *Grapevine deformation virus* (GDefV), *Grapevine fanleaf virus* (GFLV), *Raspberry ringspot virus* (RpRSV), *Tobacco ringspot virus* (TRSV); (2) subgroup B: *Artichoke Italian latent virus* (AILV), *Grapevine Anatolian ringspot virus* (GARSV), *Grapevine chrome mosaic virus* (GCMV), *Tomato black ring virus* (TBRV); (3) subgroup C: *Blueberry leaf mottle virus* (BLMoV), CLRV, *Grapevine Bulgarian latent virus* (GBLV), *Grapevine Tunisian ringspot virus* (GTRV), *Peach rosette mosaic virus* (PRMV), *tomato ringspot virus* (ToRSV). The species *Strawberry latent ringspot virus* (SLRSV) is a member of the genus *Sadwavirus*. Infectious degeneration is caused by European nepoviruses (ArMV, GFLV, GCMV, RpRV, TBRV) and SLRSV, whereas the agents of decline disease are American nepoviruses (BLMoV, PRMV, TRSV, ToRSV).

The agents of leafroll are filamentous viruses with a monopartite RNA genome and a length of 1400–2200 nm, classified in the family *Closteroviridae*. Grapevine leafroll-associated virus 2 (GLRaV-2), the only representative of the genus *Closterovirus*, has a genome 15 528 nt in size. GLRaV-1 (17 647 nt), GLRaV-3 (17 917 nt), and GLRaV-5

are members of the genus *Ampelovirus*. GLRaV-4, GLRaV-6, GLRaV-8, and GLRaV-9 are tentative species in the same genus, whereas GLRaV-7 is unassigned to the family.

Viruses of the rugose wood complex have filamentous particles 730–800 nm long, encapsidating a monopartite RNA genome, and are classified in two genera of the family *Flexiviridae*. The genus *Vitivirus* comprises grapevine virus A (GVA) (7349 nt), the putative agent of 'Kober stem grooving', grapevine virus B (GVB) (7599 nt) and grapevine virus D (GVD); these two viruses may be involved in the etiology of 'Corky bark' rupestris stem pitting-associated virus (RSPaV) (8726 nt), a member of the genus *Foveavirus*, is the putative agent of the homonymous disease.

The family *Tymoviridae* comprises all agents associated with the fleck complex, that is, viruses with isometric particles *c.* 30 nm in diameter showing a prominent surface structure and a monopartite RNA genome. Grapevine fleck virus (GFkV, 7564 nt), the causal agent of fleck, and GRGV are both members of the genus *Maculavirus*, whereas grapevine asteroid mosaic-associated virus (GAMaV) and grapevine vein feathering virus (GVFV) are tentative members of the genus *Marafivirus*.

Transmission

All diseases are graft transmissible and persist in the propagating material which is responsible for their long-distance dissemination. Spread at a site varies with the disease.

Infectious degeneration and decline are two soil-borne diseases. Dorylaimoid nematodes have been identified as vectors of some of their agents. In particular, GFLV is transmitted by *Xiphinema index*; ArMV and SLRSV by *X. diversicaudatum*; ToRSV by *X. americanum sensu stricto*, *X. rivesi*, and *X. californicum*; PRMV by *X. americanum sensu lato* and *Longidorus diadecturus*; TRSV by *X. americanum sensu stricto*; TBRV by *L. attenuatus*; and RpRSV by *Paralongidorus maximus*. Seed transmission has been reported for PRMV, ToRSV, BLMoV, GCMV, and GFLV; however, grapevine is not seed-propagated in commercial viticultural practices.

Some of the viruses of the leafroll and rugose wood complexes are transmitted in a nonspecific semipersistent manner by pseudococcid mealybugs (i.e., *Planococcus ficus* (GVA, GVB, GLRaV-3), *Pl. citri* (GVA, GLRaV-3), *Pseudococcus longispinus* (GVA, GVB, GLRaV-3, GLRaV-5, GLRaV-9), *Ps. affinis* (GVA, GVB, GLRaV-3), *Ps. comstocki* (GVA, GLRaV-3), *Ps. calceolariae* (GLRaV-3), *Ps. viburni* (GLRaV-3), *Ps. maritimus* (GLRaV-3), *Heliococcus bohemicus* (GVA, GLRaV-1, GLRaV-3), *Phenacoccus aceris* (GLRaV-1)) and by soft scale insects (i.e., *Neopulvinaria innumerabilis* (GVA, GLRaV-1, GLRaV-3), *Pulvinaria vitis* (GLRaV-1, GLRaV-3), and *Parthenolecanium corni* (GLRaV-1)). None of these viruses are seed-borne, contrary to RSPaV, which is pollen-borne and transmitted through seeds. None of the viruses of the fleck complex have known vectors.

Control

Use of clonally selected and sanitized propagative material is the best preventive method currently available for controlling all diseases. Heat therapy, meristem tip culture, and somatic embryogenesis are effective, though to different extents, for the elimination of parenchyma (nepoviruses) and of phloem-restricted (closteroviruses, vitiviruses, foveaviruses, maculaviruses) viruses. Restraining field re-infection of virus-free stocks is, however, difficult because no ultimate control of soil- or airborne vectors is possible. Attempts for introduction of transgenic resistance to some nepoviruses (GFLV, ArMV, GCMV), vitiviruses (GVA and GVB), and closteroviruses (GLRaV-2 and GLRaV-3) are in progress.

Virus Diseases of Stone Fruits

Disease Symptoms and Yield Losses

Stone fruits are affected by several diseases caused by recognized viruses and by unidentified graft-transmissible pathogens which, for brevity sake, are not addressed in this article.

Symptoms of virus infections depend on the pathogen, the host plant, and the environmental conditions. Variants of prunus necrotic ringspot virus (PNRSV) cause different diseases such as calico or bud failure in almond, necrotic ringspot or rugose mosaic in cherry, tatter leaf in cherry (**Figure 1(b)**) and peach, and mule's ear in peach. Variants of prune dwarf virus (PDV) cause yellows in sour cherry, blind wood and narrow leaf in sweet cherry, and gummosis in apricot. Co-infections by PNRSV and PDV cause peach stunt. Peach yields are drastically reduced (30–60%). Other viruses inducing yellow line pattern or vein netting symptoms in plum are Danish plum line pattern, a strain of PNRSV, European plum line pattern, a strain of apple mosaic virus (ApMV), and American plum line pattern virus (APLPV) (**Figure 1(c)**). In Mediterranean countries, mosaic symptoms in almond trees involve PNRSV, PDV, and/or ApMV.

Sharka, induced by plum pox virus (PPV), is characterized by chlorotic/necrotic ring pattern or mottling of the leaves, color breaks of flower petals, distortion of fruit shape, and rings or blotches of their surface (**Figure 1(h)**). Infected trees drop fruit prematurely. Because of this and fruit alterations, the crop can be completely lost. Peach mosaic virus (PcMV) and cherry mottle leaf virus (CMLV) cause delayed bud-break, stunted shoot growth, leaves with chlorotic spots and vein feathering and deformed rough skinned fruits (PcMV) or chlorotic spots over an uneven leaf surface with shredded leaf margins (CMLV). Cherry is resistant to PcMV, while CMLV causes latent infections in peach. Little cherry disease caused by little cherry virus 2 (LChV-2) affects shape and size of the fruits which lack deep color in dark varieties, and are flavorless. Canopy is light green and tree growth reduced. Black Beaut plums (*Prunus salicina*) infected by plum bark necrosis-stem pitting associated virus (PBNSPaV) are short-lived, exhibit gummosis and necrotic bark on branches and trunks, and severe stem pits.

Stem pitting is induced by different ToRSV strains in *Prunus*. In early spring, canopies of affected trees appear light green, progressing to general chlorosis, leaves are drooping and drop prematurely. The basal portion of trunks develops thick, spongy textured bark and pits and grooves on the woody cylinder, and trees decline. ToRSV-infected prune trees with a brown line at the graft union exhibit poor growth, sparse canopy, and in chronic infections develop limb dieback. PRMV causes rosettes of shoots with shortened internodes. Affected trees are dark green, stunted, and fruitless.

Other nematode-borne stone fruit diseases are: apricot bare twig and unfruitfulness caused by SLRSV, decline and death by cherry rasp leaf virus (CRLV), which also elicits the formation of enations on the underside of leaves (**Figure 1(d)**). In Europe, sweet cherry trees with rasp leaf disease are infected by ArMV, CLRV, raspberry ringspot virus (RpRSV), or SLRSV either alone or in combination with PDV. CLRV can also cause rapid decline of infected cherry trees on Colt rootstock. In Germany, TBRV-infected peach trees develop chlorotic spots and distortion of the leaves, dieback of scaffold branches, and tree death. Infections in peach by SLRSV are associated with willow leaf rosette (Italy), court-noué (France), and shoot dwarfing (Germany). In plum, RpRSV infections are associated with mild symptoms, but combined with PPV symptoms are severe and infected trees decline (Poland). Myrobalan latent ringspot virus (MyLRSV) induces stunting, sparse canopy, and decline in Myrobalan B plum rootstocks (France). Recently, a new putative cheravirus named stocky prune virus (StPV) was detected in stunted, low-yielding trees in France.

Apple chlorotic leaf spot virus (ACLSV) causes dark green, sunken mottle in leaves of the peach cv. Okinawa, is associated with a fruit disorder known as pseudopox in apricot and plum, and, in cherry, it induces fruit necrosis or fruit distortion when in combination with PNRSV. Bark split in plum and graft-incompatibility in apricot and peach have also been associated with infections by different ACLSV strains.

Apricot latent virus (ApLV), latent in apricot, elicits symptoms of asteroid spot disease in peach. Cherry green ring mottle virus (CGRMV) severely impacts sour cherry and Japanese flowering cherry trees, whereas cherry virus A (CVA) is always found in co-infections with other viruses; thus, specific symptoms have yet to be defined.

Geographical Distribution

Due also to symptomless infections in different *Prunus* species, ApMV, PDV, PNRSV, ACLSV, CGRMV, and, presumably, CVA and CLRV have now a worldwide distribution. PPV occurs through most of Europe, the Mediterranean region, and, more recently, became established in Chile, USA (Pennsylvania), Canada (Ontario), and India. PcMV is endemic in southwestern United States and in Northern Mexico. LChV-2 is present in North America, Europe, and Japan. PBNSPaV was identified in USA (California), Italy, Turkey, Morocco, Jordan, and Serbia. ToRSV and CRLV are endemic in North America and PRMV is restricted to Michigan (USA) and neighboring Ontario (Canada).

Causal Agents and Classification

Causal agents of stone fruit virus diseases belong to a wide array of genera and families, the properties of some of which, that is, the genera *Nepovirus* (*Comoviridae*), *Ampelovirus* (*Closteroviridae*), and *Foveavirus* (*Flexiviridae*), have been outlined above. This applies to the nepoviruses ToRSV, PRMV, ArMV, CLRV, RpRSV and TBRV, the ampeloviruses LChV-2 and PBNSPaV, and the foveaviruses ApLV and CGRMV.

ApMV, PNRSV, PDV, and APLPV (genus *Ilarvirus*, family *Bromoviridae*) have quasi-isometric particles 25–35 nm in diameter and a tripartite RNA genome 1.0–1.3 kb (RNA-1), 1.0–1.2 kb (RNA-2), and 0.7–0.9 kb (RNA-3) in size. CRLV, the type species of the recently established genus *Cheravirus*, has isometric particles *c.* 30 nm in diameter and a bipartite RNA genome 7030 nt (RNA-1) and 3315 nt (RNA-2) in size. ACLSV, PcMV, and CMLV (genus *Trichovirus*) have filamentous particles 720–760 nm in length and a 2.2–2.4 kb monopartite RNA genome. The capillovirus CVA has filamentous particles of undetermined length and a monopartite RNA genome 7383 nt in size while PPV, the only potyvirus known to infect stone fruits, has also filamentous particles *c.* 750 nm long, encapsidating a monopartite RNA genome 9741 nt in size.

Transmission

Ilarviruses and CLRV are pollen- and seed-borne, giving rise to infected seedlings. Horizontal spread of PNRSV and PDV requires also the sequential activities of honeybees (*Apis mellifera*) and thrips (*Thrips tabaci* or *Frankliniella occidentalis*). Spread via root grafts between neighboring trees have been reported in sweet cherry orchards in Washington (USA) for CLRV and in nurseries for ApMV. Several aphid species transmit PPV in a nonpersistent manner to account for localized spread, and movement in infected stocks for long-distance spread. The apple mealybug *P. aceris* vectors LChV-2. Similarly, the ampelovirus PBNSPaV is putatively vectored by mealybug species. PcMV is vectored by the bud mite, *Eriophyes insidiosus* and CMLV by *E. inaequalis*. Natural spread by vectors has not been confirmed for ACLSV, CGRMV, and CVA; however, root grafts are known to occur or suspected. ToRSV and CRLV are vectored by the dagger nematode, *Xiphinema americanum sensu stricto* and PRMV by *X. americanum sensu lato* and *Longidorus diadecturus*.

Control

Because infected orchard trees cannot be cured, effective control strategies are production and use of clean stocks and programs for pathogen assays to identify pretested propagation materials. To this aim, rules and regulations have been promulgated in various European and North American countries to enact some form of nursery tree fruit improvement schemes. Although planting clean stocks is an essential first step in prevention and control of virus diseases, in some instances additional steps may be required. With nepoviruses, diseased sites will require soil fumigation, use of resistant rootstocks, and/or broad leaf weed control.

Virus Diseases of Pome Fruits
Disease Symptoms and Yield Losses

Some of the viruses infecting stone fruit trees (ACLSV, ApMV, CRLV, and ToRSV) are also pathogenic to pome fruits. ApMV and tulare apple mosaic virus (TAMV) induce yellow mottling of the leaves which is outstanding on the spring vegetation but tends to fade away when the temperature rises. A yield reduction of 20–40% has been associated with infection by severe ApMV strains. ACLSV is the agent of different syndromes, such as *Malus platicarpa* dwarf, leaf deformation of several ornamental apple species, russet ring of apple fruits, and ring pattern mosaic of the leaves of pear and quince. Sensitive apple and quince cultivars can be heavily damaged; for example, yield of apples can be reduced in excess of 20%.

Apple stem pitting virus (ASPV) infections are latent in the great majority of commercial apple cultivars, whereas, in sensitive rootstocks, this virus elicits stem grooving, epinasty, that is, a marked curling of the leaves caused by necrosis of the main vein, and chlorotic/necrotic mottling of the leaves, followed by decline. Some of its strains are also responsible for diseases that affect the leaves (vein yellowing, necrotic spots) (**Figure 1(e)**) and fruits (stony pit) of pear, and the leaves (sooty ring spot), branches (necrotic grooves and bark necrosis), and fruits (deformation and stony pits) of quince.

Like ASPV, apple stem grooving virus (ASGV) infects latently ungrafted cultivars and rootstocks of apple, pear, and quince. However, a severe disease that leads to decline and death of the plants, known as 'apple junction necrotic pitting' or 'top working', develops when infected scions are used for grafting, especially on Virginia Crab, which reacts. Grafted plants react with stem grooving (**Figure 1(f)**), and necrosis of the tissues at the graft union (brown line) which leads to a rapid decline and death.

A somewhat similar disorder called 'union necrosis and decline' is induced by ToRSV in apples grafted on the rootstock MM.106. 'Flat apple', a disease induced by CRLV, is characterized by deformation of the fruits, reduced growth of lateral branches and upward rolling of the leaves.

Geographical Distribution

Except for TAMV, CRLV, and ToRSV, which are largely confined to the USA, all other pome fruit-infecting viruses (ApMV, ACLSV, ASGV, and ASPV) have a worldwide distribution.

Causal Agents and Classification

Affiliation and properties of ACLSV, ApMV, TAMV, CRLV, and ToRSV are as reported above. ASPV and ASGV are the type species of the genus *Foveavirus* and *Capillovirus*, respectively. Both viruses have filamentous particles *c.* 640 nm (ASGV) or *c.* 800 nm (ASPV) long, and a monopartite RNA genome 6495 nt (ASGV) and 9306 nt (ASPV) in size.

Transmission

ASPV, ASGV, and ACLSV have no known vector. Long-distance dissemination occurs via infected propagative material and nursery productions, whereas spread in apple orchards may take place through root grafts. The epidemiology of ilarviruses (ApMV, TAMV), the nepovirus ToRSV, and the cheravirus CRLV is as reported above.

Control

To control pome fruit viruses the same strategies used for stone fruit viruses can be implemented.

Virus Diseases of Walnut

Disease Symptoms and Yield Losses

Walnut blackline is a trunk union malady in English walnut scions propagated on rootstocks of several *Juglans* species. In California, blackline disease developed in English walnut trees propagated on seedlings of northern California black walnut, the hybrids Paradox and Royal, and Chinese wingnut (*Pierocarya stenoptera*).

In California, blackline diseased trees exhibit poor tree vigor and shoot growth, limb dieback, and yellow, drooped leaves. Rootstocks often produce sucker shoots. The main diagnostic symptoms are comprised of necrotic tissues embedded in bark tissue and into the woody cylinder at the scion-rootstock junction displayed as union blackline in trees of English walnut on black walnut rootstock or tissue necrosis extending downward from the scion-rootstock junction into Paradox rootstock. In Europe, grafted CLRV-infected trees develop a similar decline pattern. In addition, leaves develop chlorotic spots, rings, and yellow line patterns. Such leaf symptoms have not been observed in California.

Geographic Distribution

Walnut blackline disease has been reported in the USA (California and Oregon) and in Europe (Bulgaria, England, France, Hungary, Italy, Romania, and Spain).

Causal Agent and Classification

The incitant of walnut blackline disease is CLRV, genus *Nepovirus*.

Transmission

CLRV spreads by top-working trees with infected scions during cultivar conversion or in nature by pollen infecting healthy walnut trees during the flowering period. The virus is seed-borne and gives rise to infected seedlings.

Control

Use clean sources of scion wood and rootstocks. In orchards requiring supplemental pollination, pollen should be collected from CLRV-free trees.

Virus Disease of Hazelnut

Disease Symptoms and Yield Losses

Hazelnut mosaic diseased trees may develop leaves with chlorotic rings, flecking, and a variety of line patterns. Virus-infected trees may be symptomless also. Nut yields may be halved compared to production on healthy trees.

Geographical Distribution

The disease is reported in commercial orchards in Europe (Bulgaria, Italy, Turkey, Spain, Georgia) and likely occurs in other European countries. In the USA, hazelnut mosaic was detected in breeding lines and clonal germplasm importations, but not in commercial orchards.

Causal Agents and Classification

A complex of ilarviruses (ApMV, PNRSV, or TAMV) has been associated with diseased trees. In the USA, only ApMV has been identified in hazelnut trees.

Transmission

Hazelnut seeds harvested from ApMV-infected trees give rise to infected seedlings. ApMV is not known to be pollen-transmitted or vectored by insects. In contrast, PNRSV is known to be pollen- and seed-borne in several *Prunus* species.

Control

Propagate plants for planting from clean stocks.

See also: Nepovirus; Citrus Tristeza Virus; Flexiviruses; Tymoviruses; Ilarvirus; Plum Pox Virus.

Further Reading

Andret-Link P, Laporte C, Valat L, *et al.* (2004) Grapevine fanleaf virus: Still a major threat to the grapevine industry. *Journal of Plant Pathology* 86: 183–195.

Bovey R, Gärtel W, Hewitt WB, Martelli GP, and Vuittenez A (1980) *Virus and Virus-Like Diseases of Grapevines.* Lausanne: Editions Payot.

Desvignes JC, Boyé R, Cornaggia D, and Grasseau N (1999) *Maladies à Virus des Arbres Fruitiers.* Paris: Editions Ctifl.

Jones AL and Aldwinckle HS (eds.) (1990) *Compendium of Apple and Pear Diseases.* St. Paul, MN: APS Press.

Krake LR, Scott NS, Rezaian MA, and Taylor RH (1999) *Graft-Transmissible Diseases of Grapevines.* Collingwood, VIC: CSIRO.

Martelli GP (ed.) (1993) *Detection and Diagnosis of Graft-Transmissible Diseases of Grapevines.* Rome: FAO Publication Division.

Martelli GP and Boudon-Padieu E (2006) Directory of infectious diseases of grapevines. *Options Méditérranéens, Series B* 55: 11–201.

Mink GI (1992) Ilarvirus vectors. *Advances in Disease Vector Research* 9: 261–281.

Nemeth M (1986) *Virus, Mycoplasma, and Rickettsia Diseases of Fruit Trees.* Dordrecht, The Netherlands: Martinus Nijhoff.

Ogawa JM, Zehr EI, Bird GW, Ritchie DF, Uriu K,, and Uyemoto JK (eds.) (1995) *Compendium of Stone Fruit and Diseases.* St. Paul, MN: APS Press.

Teviotdale BL, Michailides TJ, and Pscheidt JW (2002) *Compendium of Nut Crop Diseases in Temperate Zones.* St. Paul, MN: APS Press.

Walter B (ed.) (1997) *Sanitary selection of the grapevine. Protocols for Detection of Viruses and Virus-Like Diseases Les Colloques, No. 86.* Paris: INRA Editions.

Walter B, Boudon-Padieu E, and Ridé M (2000) *Maladies à virus, bacteries et phytoplasmes de la vigne.* Bordeaux: Editions Féret.

Wilcox WF, Gubler WD,, and Uyemoto JK (eds.) (2007) *Compendium of Grape Diseases.* St. Paul, MN: APS Press.

Plant Virus Diseases: Ornamental Plants

J Engelmann and J Hamacher, INRES, University of Bonn, Bonn, Germany

Glossary

Chlorotic symptoms Degradation or depletion of chlorophyll leads to light green coloration of plant parts. Characteristic patterns such as ring spots, line patterns, and mosaics occur during virus infections.

Deformations Virus-induced growth abnormalities leading to reduced growth of cells, twisted or curled leaves or a stunted appearance of the whole plant.

Flower breaking Discoloration of the flowers often leading to an uneven distribution or total loss (bleaching) of pigments.

Generative propagation Propagation via pollen and seeds, nonuniform daughter generations are obtained.

Micropropagation Art and science of plant multiplication *in vitro*. The process includes many steps – stock plant care, explant selection, and sterilization, and media manipulation – to obtain proliferation, rooting, acclimation, and growing on of liners.

Necrotic symptoms These symptoms occur when cells die, giving rise to dull grayish, brown, or black coloration of the respective plant tissues. Necrotic lesions are often surrounded by darkly stained cells.

Phylogenetic origin The evolutionary relatedness among groups of organisms (e.g., species and populations). Phylogenetics treats a species as a group of lineage-connected individuals over time.

Serological methods Diagnostic methods based on the specific reaction of antibodies raised in rabbits, mice, or chicken by immunization with plant viruses. Methods are, e.g., ELISA, direct immunoblots, Western blots, or lateral flow tests.

Subgenomic RNA Short copy of a genomic RNA often existing as a single additional translatable RNA for capsid protein within a viral capsid. One of the translation strategies of RNA-viruses to express their whole genetic information.

Vector transmission Most plant viruses are vectored by arthropods, often hemipteran insects. For persistently transmitted viruses, the virus is ingested, passes through the gut wall into the haemolymph, and then moves to the salivary glands where it can potentially be transmitted to other plants with the saliva. Long acquisition times of more than 1 h, a latency period of several days between uptake and release of the virus, and a long inoculation access of 1 h to several hours is characteristic. Persistently transmitted viruses, have two subclasses termed circulative if there is no multiplication in the insect vector and propagative if there is. For nonpersistently transmitted viruses, the virus is restricted to the tips of the insect stylet, where salivary duct and food canal coalesce. Nonpersistent viruses are efficiently transmitted after relatively brief (<5 min) acquisition and inoculation access periods. Semipersistent transmission is intermediate between nonpersistent and persistent. In this kind of transmission the virus moves to the foregut of the insect. Acquisition and inoculation access periods are longer than with nonpersistent transmission.

Vegetative propagation All kinds of plant propagation where parts of the plant are used to generate new plantlets, aiming uniform daughter generations.

Introduction

Virus diseases are of great economic importance in ornamentals because most of them are propagated vegetatively. This kind of plant propagation, either by tissue culture (micropropagation) or by scions, bulbs, rhizomes, or other tissues, is the most economic method to propagate plants maintaining their uniformity. Unfortunately, by this method viruses are propagated and spread from the mother stock to the next generation as well. After, for example, vegetatively propagated trailing petunias flooded the market in the 1990s, problems with tobamoviruses, mainly tobacco mosaic virus (TMV), arose quickly, compared to seed-propagated petunias.

The risk of virus infections in ornamental cultures is also raised by additional factors:

- introduction and establishment of novel virus vectors (e.g., *Thrips palmi*, *Frankliniella occidentalis*, *Bemisia tabaci*) and/or novel virus species into production areas;
- cross-breeding of novel, potentially virus-infected wild types of plants;
- introduction of novel exotic genera or species to broaden the range of ornamentals;
- cultural practices in closed irrigation and fertilization (nutrient solution) systems;
- worldwide trade with ornamentals; and
- production of plant material in countries representing different standards of production and/or other virus pressure.

As the number of ornamental species and varieties is very high and their phylogenetic as well as geographic origins vary considerably, the number of infecting agents, especially viruses, varies accordingly, giving rise to an immense range of virus–plant pathosystems. Viruses of ornamental plants have very different host ranges, many of them being ubiquitous, such as tomato spotted wilt virus (TSWV), impatiens necrotic spot virus (INSV), or cucumber mosaic virus (CMV). Some of them, however, are specialized and infect only certain species or genera of plants, such as, for example, pelargonium flower break virus (PFBV) or angelonia flower mottle virus (AFMoV). The following section gives an overview of the most common as well as some recently detected viral diseases of economically important ornamentals.

Virus Diseases in Diverse Ornamentals

Virus Diseases of Pelargonium

All pelargonias are often named geraniums. Pelargonias originated from the South African region. The main ornamental forms are obtained by cross-breeding of species, thus representing hybrids. Three forms of pelargonium are of main importance in floriculture: *Pelargonium grandiflorum* hybrids (pot plants), *P. peltatum* hybrids (ivy-leaved hanging pelargonias), and *P. zonale* hybrids (upright growing), the latter two forms being of highest economical importance as bedding or balcony plants. As pelargonias are usually propagated by cuttings or tissue culture, viral diseases endanger production and quality of that crop. In pelargonia, many virus symptoms evolving in late winter and early spring will disappear during summer and may reappear in the next season.

Specific pelargonium viruses

The most prevalent virus in Europe is by far PFBV, a virus assigned to the genus *Carmovirus*, family *Tombusviridae*. This icosahedral virus spreads via draining water, via contaminated pollen carried by thrips vectors, possibly by allowing virus from the surface of infected pollen to enter through feeding wounds, and most efficiently by vegetative propagation of infected plant material. Symptoms vary according to time of the year and cultural practice: pink flowering varieties show flower break (**Figure 1**), leaf symptoms vary from symptomless to light mottling and to ring spotting (**Figure 2**) as well as line patterns. The virus occurs more often in *P. zonale* than in *P. peltatum* hybrids (**Figure 3**).

There has been considerable confusion about some pelargonium viruses causing very similar ring spot and line pattern symptoms. Investigations of these viruses on the molecular level solved the initial disarrangement: four different virus species have been characterized which are proposed to form a new genus of plant viruses named *Pelarspovirus*.

Pelargonium line pattern virus (PLPV) is a spherical virus, 30–32 nm in diameter, belonging to the family *Tombusviridae*, which has many traits of a carmovirus, but is different in one type of protein not expressed by carmoviruses and transcribes only one subgenomic RNA instead of two. It was reported initially in *P. zonale* from Great Britain. Most infections seem to be symptomless, but chlorotic rings and line patterns may develop (**Figure 4**). Investigations from Spain report that PLPV represents the most common symptomless virus disease of pelargonias. It can be transmitted mechanically and by vegetative propagation.

Pelargonium ring spot virus (PelRSV) produces symptoms of pronounced ring and line patterns resembling

Figure 3   Leaf mottling on *P. peltatum* (PFBV). By Joachim Hamacher.

Figure 1   Flower breaking on *P. zonale* induced by PFBV. By Joachim Hamacher.

Figure 2   Ring spots on *P. zonale* induced by PFBV. By Joachim Hamacher.

Figure 4   Chlorotic ring spots and line patterns induced by PLPV. By Joachim Hamacher.

those induced by PLPV. PelRSV could, however, be clearly distinguished from PLPV by serology and nucleotide sequence as well as by the size of double-stranded RNAs (dsRNAs) produced in *Chenopodium quinoa*.

A further virus belonging to the family *Tombusviridae*, pelargonium chlorotic ring pattern virus (PCRPV) was isolated in Italy as geranium virus isolate 57. It was characterized and proposed to belong to the new virus genus mentioned above. This virus also leads to distinct chlorotic ring pattern symptoms (**Figure 5**).

Symptoms of pelargonium leaf curl virus, assigned to the genus *Tombusvirus*, were first reported in *P. zonale* from Germany as early as 1927. The symptoms are characterized by the appearance of small white-yellowish spots on leaves, which enlarge, becoming round flecks or stellate spots with necrotic centers (**Figure 6**). In later infection stages leaf crinkling, leaf splitting, and plant degeneration or dwarfing occurs (**Figure 7**). Leaf symptoms will disappear in summer. The virus can be transmitted by water or nutrient solutions in soil-free substrates, mechanically, though with some difficulty, with plant sap, and most efficiently by cuttings from infected symptom-free sources.

Pelargonium zonate spot virus (PZSV), a virus now assigned to the new genus *Anulavirus*, family *Bromoviridae* has been reported from Italy (Apulia). PZSV causes concentric chrome-yellow bands in the leaves of *P. zonale*. Besides mechanical transmission, it can be transmitted by seeds and by pollen in *Nicotiana glutinosa* and *Diplotaxis erucoides*.

Pelargonium vein clearing virus (PelVCV) syn. eggplant mottled dwarf virus (EMDV), assigned to the genus *Cytorhabdovirus*, is reported from southern Italy. It causes mild vein clearing in *P. peltatum* and *P. zonale*. Transmission occurs via mechanical transmission and by vegetative propagation.

An often observed anomaly of pelargonium is named yellow net vein disease. It is graft transmissible and leads to obvious veinal chlorosis (**Figure 8**), persisting throughout the year. A causal virus or other agent has not been isolated or identified up to now.

Figure 5 Chlorotic ring patterns on *P. peltazonale* induced by PCRPV. By Joachim Hamacher.

Figure 7 Leaf crinkling and curling on *P. zonale* caused by PLCV. By Joachim Hamacher.

Figure 6 Whitish spots on PLCV-infected *P. zonale*. By Joachim Hamacher.

Figure 8 Yellow net vein symptoms on *P. zonale*, pathogen unknown. By Joachim Hamacher.

Pelargonium nonspecific viruses

Some nepoviruses also infect pelargonias, mainly in North American varieties: tomato ringspot virus (ToRSV), tobacco ring spot virus (TRSV), and tomato black ring virus (TBRV) lead to ring spotting and chlorotic flecking and stippling (**Figure 9**). These viruses cause quarantine diseases because of their potential to infect fruit trees and their transmissibility by nematodes, pollen, and seeds.

A considerable threat for floriculturists are two tospoviruses: TSWV and INSV – the first infecting a great number of vegetables and ornamentals, the latter mainly ornamentals. TSWV infects *P. peltatum*, producing ring spots and line patterns (**Figure 10**). INSV infects pelargonias very rarely. Both viruses are transmitted by thrips, mainly the western flower thrips (*Frankliniella occidentalis*, Pergande).

CMV is a very ubiquitous virus with the widest known host range of all plant viruses, exceeding 1000 host plant species. It occurs with many different strains, may infect pelargonias and leads to pronounced flecking (**Figure 11**),

asymmetry of leaves as well as breaking of the characteristic brown horseshoe zone in *P. zonale*. It can be transmitted by many aphid species in a nonpersistent manner.

Virus Diseases of Solanaceae

Petunia, calibrachoa

When vegetatively propagated trailing petunias (surfinias) came on the market, problems with viruses, unknown in seed-propagated petunias, arose. More than 150 different virus species may infect petunias, but only some of them infect petunias naturally. Petunias are vigorously growing plants requiring a balanced fertilization, water supply, and culture maintenance. If these requirements are not achieved, symptoms suspicious of viral origin may arise very soon. This has to be considered when monitoring the crops.

Specific petunia viruses

A new virus disease of petunias has been identified recently in Brazil. The pathogen has been named petunia vein banding virus (PetVBV) and is assigned to the genus *Tymovirus*. The virus occurred in mixed infections with another spherical virus (possibly PVCV) and produced local chlorotic and necrotic spots and systemic vein banding on petunias when inoculated as single pathogen.

Petunia vein clearing virus (PVCV), genus *Petuvirus*, eludes detection by immunological methods most time of the year, since its dsDNA genome is integrated in the host genome of some Petunia varieties. The genome expression is activated only in late winter/early spring or by stress. During symptom development, few icosahedral virus particles measuring about 45 nm in diameter can be observed by electron microscopy (**Figure 12**). Symptoms are veinal chlorosis with shrunken leaf veins (**Figure 13**) sometimes becoming necrotic.

Figure 9 Chlorotic stippling of *P. zonale* induced by TBRV. By Joachim Hamacher.

Figure 10 Chlorotic line patterns and rings induced by TSWV on *P. peltatum*. By Joachim Hamacher.

Figure 11 Chlorotic flecking on *P. peltatum* induced by CMV (holes were punched in the left-hand leaf for analytical purposes). By Dietrich E. Lesemann.

Viruses nonspecific of petunias

The most frequently detected virus in trailing petunias is TMV. Symptoms vary according to variety, fertilization, and developmental stage of the plant. Veinal chlorosis (**Figure 14**) and necrosis (**Figure 15**), mottling (**Figure 16**),

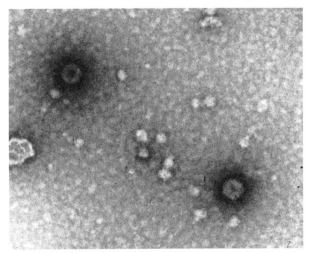

Figure 12 Spherical particles of PVCV from infected petunia showing darkly staining centers. Bar represents 50 nm. By Joachim Hamacher.

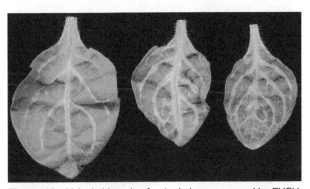

Figure 13 Veinal chlorosis of petunia leaves caused by PVCV. By Dietrich E. Lesemann.

Figure 14 Leaf vein chlorosis of young petunia plants induced by TMV. By Joachim Hamacher.

blistering with dark green areas as well as leaf curl and stunting of whole plants may appear. Flowers may also express color deviations as observed with bicolored petunia varieties (**Figure 17**). Tomato mosaic virus (ToMV)-infected plants show similar symptoms, although symptom expression is less severe. Mixed infections of both viruses occur as well. Two other tobamoviruses, tobacco mild green mosaic virus (TMGMV) and turnip vein clearing virus (TVCV) have been detected in petunias, both leading to severe necrotic symptoms in petunias (**Figure 18**).

Transmission of tobamoviruses with contaminated plant sap is very easy and results in high infection rates, as petunias are vigorously growing plants with soft leaves and many hairs, which break when plants are handled.

Potato virus Y (PVY) is the type virus of the genus *Potyvirus* and often found in petunias. The main symptoms

Figure 15 Necrotic leaf spots and veins of petunia caused by TMV. By Joachim Hamacher.

Figure 16 Mottling of TMV-infected petunia leaves. By Joachim Hamacher.

are mottling of leaves and color deviations (mottling) in purple flowering plants, also occurring in mixed infections with CMV (**Figure 19**). Another potyvirus infecting

Figure 17   Flower breaking in bicolored petunia varieties with TMV infection. By Joachim Hamacher.

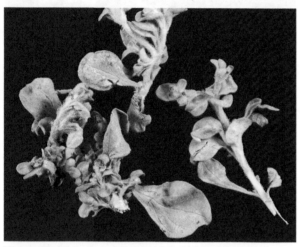

Figure 18   Crippling of TMGMV-infected petunia scions (necrosis and deformations). By Joachim Hamacher.

Figure 19   Flower mottle of purple *Surfinia* caused by a mixed infection with PVY and CMV. By Dietrich E. Lesemann.

petunias, which induces flower breaking is turnip mosaic virus (TuMV). Potyviruses are aphid transmissible in a nonpersistent manner.

Tospoviruses may infect petunias locally but represent no real danger, as vegetative propagation will not transmit these viruses effectively. Nonflowering petunias, interestingly, may be used to monitor the occurrence of tospoviruliferous thrips, because they develop brown to black local lesions at the edges of thrips-feeding scars when infection has occurred.

Two viruses of the genus *Cucumovirus*, CMV and tomato aspermy virus (TAV) may infect petunias, leading to mosaic, flecking (**Figure 20**) or mottled leaves and TAV to flower deformation. Both viruses are aphid transmissible in a nonpersistent manner.

Alfalfa mosaic virus (AMV) produces marbling of leaves (**Figure 21**) and flower breaking in petunia.

Some nepoviruses (ArMV, TBRV, ToRSV, and TRSV) are reported to infect petunias, producing ring spots, line patterns, or tip necrosis. The symptoms will disappear in the course of infection.

Figure 20   Bright yellow leaf spots of CMV-infected Surfinia. By Joachim Hamacher.

Tobacco necrosis virus (TNV), tobacco rattle virus (TRV), and tobacco streak virus (TSV), were detected to infect petunias, but seem to appear only sporadically.

Some viruses naturally infecting petunias are of minor importance as they do not evoke obvious symptoms. *Calibrachoa mottle virus* (CbMV), assigned as a new virus species mainly infecting *Calibrachoa* sp., does not produce symptoms in petunias. Broad bean wilt virus 1 (BBWV1) is another latent virus of petunia.

Calibrachoa or 'Million Bells'

The name of calibrachoa goes back to the nineteenth century Mexican botanist and pharmacologist Antonio de Cal y Bracho. Million Bells are solanaceous bedding and balcony plants with growing popularity.

The most frequently detected virus of calibrachoas has been characterized recently, named calibrachoa mottle virus (CbMV) and has been tentatively assigned to the genus *Carmovirus*. Symptoms evolving after infection vary from light mottling to chlorotic blotching (**Figure 22**), but some varieties may remain symptomless.

Tobamoviruses play an important role in calibrachoas. ToMV is of greater importance than TMV, as it leads to pronounced chlorotic ring spots and line patterns

(**Figure 23**), whereas TMV remains symptomless or induces only light mottling.

Some other viruses have been diagnosed in calibrachoas such ArMV, TRSV, BBWV1, INSV, PMMoV, RMV, CVB, PVY, and other potyviruses, as well as PNRV and TSV.

Virus Diseases of Balsaminaceae

Impatiens New Guinea hybrids and *I. walleriana* play a prominent role as bedding and balcony plants and may be hosts for quite a lot of virus diseases. The prevalent virus of *Impatiens* spp. is INSV. The symptoms observed after infection with INSV are necrotic spots (**Figure 24**), ring spots or mottling of leaves, and necrotic flecking of stems. On *I. New Guinea* hybrids the discolorations are often accompanied by foliar deformations (**Figure 25**). Besides vegetative propagation, virus transmission of

Figure 22 Chlorotic leaf blotching and mottling of *Calibrachoa* infected by CbMV. By Joachim Hamacher.

Figure 21 Marbling of petunia leaves induced by AMV. By Joachim Hamacher.

Figure 23 Chlorotic line patterns in ToMV-infected *Calibrachoa* leaves. By Joachim Hamacher.

tospoviruses within crops is mainly due to feeding of *Frankliniella occidentalis* (Pergande) the western flower thrips. Other thrips species may transmit as well but are of minor importance (**Table 1**). TSWV occurs by far less frequently on impatiens than INSV, but induces very similar necrotic symptoms.

Tobamoviruses are as well quite common in impatiens. TMGMV and TVCV have been diagnosed in *I. New Guinea* hybrids. The symptoms are very similar and comprise stunting, leaf deformation, or die back of young plants (**Figure 26**), leaf necrosis and deformation as well as foliar reddening on older plants. Pink or orange flowering varieties may exhibit flecking of flowers (**Figure 27**). Black line patterns, resembling symptoms induced by INSV or TSWV, have also been observed on some varieties. The viruses may be symptomless on mature plants. Both viruses may be transmitted by mechanical injury

Figure 24 Necrotic spots on leaves of INSV-infected *Impatiens New Guinea* hybrid (NGI). By Joachim Hamacher.

Figure 25 Leaf deformation and necrotic spots due to INSV infection of NGI. By Joachim Hamacher.

when handling the plants, by contaminated substrate and pots as well as by irrigation.

Leaf narrowing and rugged leaf borders as well as split petals and color breaking are typical symptoms of infections with CMV (**Figures 28** and **29**). Several aphid species may transmit the virus in a nonpersistent manner within seconds.

Further virus species infecting impatiens are TuMV, AMV (aphid transmissible in a nonpersistent manner as well), and TSV, a seed-borne ilarvirus, which probably may be transmitted with contaminated pollen via thrips feeding.

TuMV leads to dark areas at the base or tip of the leaves of some varieties of New Guinea hybrids as well as to asymmetry of the leaves. AMV-infected plants show slight leaf necrosis or cloudy yellowish discoloration of the leaves of green-leaved varieties (**Figure 30**). TSV is reported to occur quite frequently. It induces only slight mosaic on the lower side of the leaves or remains symptomless.

An hitherto not characterized virus on both species of cultivated impatiens occurs rather frequently in Europe and produces chlorotic to yellow concentric ring spots and line patterns (**Figures 31** and **32**). It is graft transmissible between *I. New Guinea* hybrids and *I. walleriana*. In plants exhibiting such symptoms, few spherical virus-like particles, with about 30 nm diameter could be observed by electron microscopy in newly infected tissue (**Figure 33**). A provisional name for the pathogen could be impatiens chlorotic line pattern agent.

Virus Diseases of Scrophulariaceae

Bacopa, Diascia, Nemesia, Angelonia

Bacopa (*Sutera cordata*), *Nemesia* (*N. fruticans* hybrids), *Diascia* (*D. vigilis*), and *Angelonia* (*A. angustifolia*) are genera of the family Scrophulariaceae, continuously gaining popularity.

Sutera cordata (diffusa), also known as *Bacopa*, is often infected by potyviruses. Lettuce mosaic virus (LMV) was detected by electron microscopy as well as by enzyme-linked immunosorbent assay (ELISA). Little and dented leaves, leaf curl, and marginal foliar necroses (**Figure 34**) were observed, but most frequently no symptoms were induced at all. Other filamentous viruses of bacopas are PVY and CVB. BBWV 1 was detected sporadically in bacopa, leading to mottling and flecking of leaves. Tobamoviruses, such as ToMV and TMV, tospoviruses (INSV and TSWV), as well as viruses belonging to the family *Bromoviridae* (AMV and CMV), have also been detected in bacopa.

Nemesia ring necrosis virus (NeRNV) has only recently been described as a new virus and assigned to the genus *Tymovirus*. Before identification as a new species, infections with the respective symptoms have been thought to be induced by scrophularia mottle virus (ScrMV), because of its strong immunological cross-reaction with that virus. NeRNV is widespread,

Table 1 Virus acronyms, species, and genera

Acronym	Virus/viroid species	Virus/viroid genus	Virus/viroid family
AFMoV	Angelonia flower mottle virus*	Carmovirus	Tombusviridae
AMV	Alfalfa mosaic virus	Alfamovirus	Bromoviridae
AnFBV	Angelonia flower break virus	Carmovirus	Tombusviridae
ApMV	Apple mosaic virus	Ilarvirus	Bromoviridae
ArMV	Arabis mosaic virus	Nepovirus	Comoviridae
BBWV 1	Broad bean wilt virus 1	Fabavirus	Comoviridae
BCMV	Bean common mosaic virus	Potyvirus	Potyviridae
BWYV	Beet western yellows virus	Polerovirus	Luteoviridae
BYMV	Bean yellow mosaic virus	Potyvirus	Potyviridae
CRSV	Carnation ringspot virus	Dianthovirus	Tombusviridae
CbMV	Calibrachoa mottle virus*	Carmovirus	Tombusviridae
ClYMV	Clover yellow mosaic virus	Potexvirus	Flexiviridae
CMV	Cucumber mosaic virus	Cucumovirus	Bromoviridae
CSNV	Chrysanthemum stem necrosis virus	tent. Tospovirus	Bunyaviridae
CSVd	Chrysanthemum stunt viroid	Pospiviroid	Pospiviroidae
CVB	Chrysanthemum virus B	Carlavirus	Flexiviridae
CymMV	Cymbidium mosaic virus	Potexvirus	Flexiviridae
CymRSV	Cymbidium ringspot virus	Tombusvirus	Tombusviridae
CypCSV	Cypripedium chlorotic streak virus	tent. Potyvirus	Potyviridae
CypVY	Cypripedium virus Y	Potyvirus	Potyviridae
EMDV	Eggplant mottled dwarf virus syn. Pelargonium vein clearing virus	Nucleohabdovirus	Rhabdoviridae
DMV	Dahlia mosaic virus	Caulimovirus	Caulimoviridae
DVNV	Dendrobium vein necrosis virus	tent. Closterovirus	Closteroviridae
INSV	Impatiens necrotic spot virus	Tospovirus	Bunyaviridae
LMV	Lettuce mosaic virus	Potyvirus	Potyviridae
LSV	Lily symptomless virus	Carlavirus	Flexiviridae
LVX	Lily virus X	Potexvirus	Flexiviridae
MNSV	Melon necrotic spot virus	Carmovirus	Tombusviridae
NeRNV	Nemesia ring necrosis virus*	Tymovirus	
OFV	Orchid fleck virus	Rhabdovirus	Rhabdoviridae
ORSV	Odontoglossum ringspot virus	Tobamovirus	
PCRPV	Pelargonium chlorotic ring pattern virus*	Pelarspovirus (proposed)	Tombusviridae
PelRSV	Pelargonium ring spot virus*	Pelarspovirus (proposed)	Tombusviridae
PelVCV	Pelargonium vein clearing virus syn. Eggplant mottled dwarf virus	Cytorhabdovirus	Rhabdoviridae
PetVBV	Petunia vein banding virus	Tymovirus	
PFBV	Pelargonium flower break	Carmovirus	Tombusviridae
PMMoV	Pepper mild mottle virus	Tobamovirus	
PLCV	Pelargonium leaf curl virus	Tombusvirus	Tombusviridae
PLPV	Pelargonium line pattern virus*	Pelarspovirus (proposed)	Tombusviridae
PNRSV	Prunus necrotic ringspot virus	Ilarvirus	Bromoviridae
PVCV	Petunia vein clearing virus	Petuvirus	Caulimoviridae
PVY	Potato virus Y	Potyvirus	Potyviridae
PZSV	Pelargonium zonate spot virus	Anulavirus (proposed)	Bromoviridae
ReTBV	Rembrandt tulip breaking virus	Potyvirus	Potyviridae
RMV	Ribgrass mosaic virus	Tobamovirus	
SLRSV	Strawberry latent ringspot virus	Sadwavirus	Comoviridae
SMV	Soybean mosaic virus	Potyvirus	Potyviridae
SrMV	Scrophularia mottle virus	Tymovirus	
TAV	Tomato aspermy virus	Cucumovirus	Bromoviridae
TBV	Tulip breaking virus	Potyvirus	Potyviridae
TBRV	Tomato black ring virus	Nepovirus	Comoviridae
TBSV	Tomato bushy stunt virus	Tombusvirus	Tombusviridae
TCBV	Tulip chlorotic blotch virus syn. Turnip mosaic virus	Potyvirus	Potyviridae
TEV	Tobacco etch virus	Potyvirus	Potyviridae
TMGMV	Tobacco mild green mosaic	Tobamovirus	
TMV	Tobacco mosaic virus	Tobamovirus	
TNV	Tobacco necrosis virus	Necrovirus	Tombusviridae
ToMV	Tomato mosaic virus	Tobamovirus	
ToRSV	Tomato ringspot virus	Nepovirus	Comoviridae
TRSV	Tobacco ringspot virus	Nepovirus	Comoviridae
TRV	Tobacco rattle virus	Tobravirus	
TSV	Tobacco streak virus	Ilarvirus	Bromoviridae

Continued

Table 1   Continued

Acronym	Virus/viroid species	Virus/viroid genus	Virus/viroid family
TSWV	*Tomato spotted wilt virus*	*Tospovirus*	*Bunyaviridae*
TulMV	*Tulip mosaic virus*	*Potyvirus*	*Potyviridae*
TuMV	*Turnip mosaic virus*	*Potyvirus*	*Potyviridae*
TVCV	*Turnip vein clearing virus*	*Tobamovirus*	
VanMV	*Vanilla mosaic virus*	*Potyvirus*	*Potyviridae*
VeLV	*Verbena latent virus*	*Carlavirus*	*Flexiviridae*

List of acronyms, virus names, and viral genera.
Names with an asterisk do not appear in the ICTV virus name list.
Adapted from Fauquet CM, Mayo MA, Maniloff J, Desselberger U, and Ball LA (eds.) (2005) *Virus Taxonomy: Eighth Report of the International. Committee on Taxonomy of Viruses*. San Diego, CA: Elsevier Academic Press.

Figure 26   Leaf necrosis and distortion of NGI caused by TMGMV. By Joachim Hamacher.

Figure 28   Leaf narrowing and curling of red-leaved NGI infected with CMV. By Joachim Hamacher.

Figure 27   Flower mottle of pink-flowering NGI caused by infection with TMGMV. By Joachim Hamacher.

Figure 29   Flower breaking and deformation of bicolored NGI induced by CMV. By Joachim Hamacher.

leading to foliar concentric necrotic ring spots and line patterns as well as to sporadical flecking of flowers (**Figures 35** and **36**) in Nemesia, and in Diascia to black discoloration and line patterns, as well as dwarfing of the whole plant or parts of the plant (**Figures 37** and **38**).

Infections of nemesia with tospoviruses lead to necrotic areas and leaf narrowing (**Figure 39**), and in mixed infections with NeRNV and TSWV to bleaching of flowers and necrotic flecks and ring spots (**Figures 40** and **41**).

Figure 30 Cloudy yellow mottling of green-leaved NGI infected with AMV. By Joachim Hamacher.

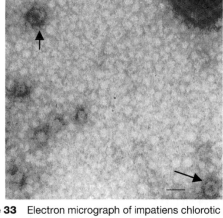

Figure 33 Electron micrograph of impatiens chlorotic line pattern virus-like particles (arrowheads) in plant sap of *I. walleriana* exhibiting symptoms of chlorotic line pattern. Bar represents 35 nm. By Joachim Hamacher.

Figure 31 Chlorotic line pattern on leaves of NGI infected by an unknown virus (impatiens chlorotic line pattern agent). By Joachim Hamacher.

Figure 34 LMV-infected bacopa with dented leaves and marginal necrosis. By Joachim Hamacher.

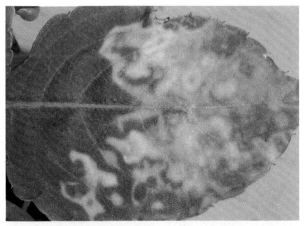

Figure 32 Marbling of *Impatiens walleriana* leaves infected with an unknown virus (impatiens chlorotic line pattern agent) after graft transmission from NGI. By Joachim Hamacher.

Figure 35 Blotches and ring spots on flowers of nemesia infected with NeRNV. By Joachim Hamacher.

A new virus disease of *Angelonia* sp. has recently been proved to be induced by a new putative member of the genus *Carmovirus*, for which the names AFMoV or angelonia flower break virus (AnFBV) respectively have been proposed. The virus (~30 nm in diameter, **Figure 42**) leads to dark blotching of flowers (**Figure 43**). The virus occurs very frequently in angelonias.

Virus Diseases of Verbenaceae

Verbenas are traded as upright-, trailing-, and ground-covering forms (e.g., Tapien and Temari). Virus infections of verbenas occur frequently and are induced by a lot of virus species. AFMoV has only recently been detected to occur in varieties of verbenas. It may be latent or leads to distinct flecking of the plants, depending on the respective variety, and to degeneration when co-infected with INSV (**Figures 44** and **45**).

NeRNV has been reported from the UK to induce necrotic and chlorotic flecking, whereas clover yellow mosaic virus (ClYMV) leads to darkly staining and necrotic spots on the base of leaves as well as on stems in trailing verbenas.

TSWV, INSV, CMV, BBWV1, AMV, and TMV have as well been diagnosed in verbenas. INSV and TSWV cause necrotic flecking, ring spotting, and line patterns (**Figure 46**). CMV and BBWV1 lead to mottling, necrosis, and little leaves (**Figures 47** and **48**). TMV causes yellow mottling, necrotic leaf borders, or die-back of leaves (**Figure 49**).

More viruses have been reported to infect verbenas: ApMV, TSV, ArMV, TRSV, ToRSV, CVB, VeLV, CRSV, MNSV, TEV, and a strain of BYMV, PNRV, RMV, and ToMV (for virus names see **Table 1**).

Figure 37   Black rings and lines on NeRNV-infected diascia leaves. By Joachim Hamacher.

Figure 36   Necrotic concentric ring spots on leaf of nemesia infected with NeRNV. By Joachim Hamacher.

Figure 38   Dwarfed shoots of diascia infected with NeRNV. By Joachim Hamacher.

Figure 39 Necrotic symptoms and leaf narrowing on INSV-infected nemesia. By Joachim Hamacher.

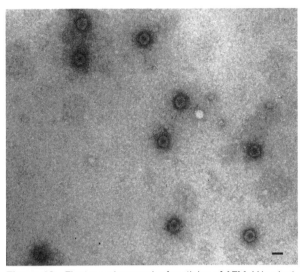

Figure 42 Electron micrograph of particles of AFMoV in plant sap from infected angelonia. Bar represents 30 nm. By Joachim Hamacher.

Figure 40 Flower bleaching of nemesia co-infected with NeRNV and TSWV. By Joachim Hamacher.

Figure 43 Darkly colored flecks on flowers (flower mottle) of angelonia infected with AFMoV. By Joachim Hamacher.

Figure 41 Necrotic ring spots and flecks on nemesia leaves co-infected with NeRNV and TSWV. By Joachim Hamacher.

Figure 44 Chlorotic spots on leaves of verbena co-infected with AFMoV and INSV. By Joachim Hamacher.

Figure 45 Necrotization of lower leaves of verbena co-infected with AFMoV and INSV. By Joachim Hamacher.

Figure 48 Light mottling and little leaves of BBWV1-infected verbena. By Joachim Hamacher.

Figure 46 Necrotic lines and spots on TSWV-infected verbena. By Joachim Hamacher.

Figure 49 TMV-infected verbena, yellow mottling of leaves, and necrotization of leaf margins. By Joachim Hamacher.

Virus Diseases of Compositae

Osteospermum and dimorphotheca, dahlia, and chrysanthemum

Compositae or Asteraceae include plant species from all over the world

The South African species *Osteospermum* and *Dimorphotheca*, vernacular name, African daisy or cape daisy, are closely

Figure 47 Necrotic spots and mottling on leaves of BBWV1-infected verbena. By Joachim Hamacher.

related genera and consist of about 70 (*Osteospermum*) and 2 (*Dimorphotheca*) species of evergreen shrubs, half shrubs, or annual plants. Osteospermums are relatively new to most gardeners. They have gained considerable popularity as summer bedding, balcony, and potted plants in the last decade. The frequently occurring virus disease in both species is caused by LMV which leads to foliar mottling, while some isolates also induce flower breaking. In some cases plants remain symptomless.

The tospoviruses TSWV and INSV may disturb plant growth, leading to a dwarfed appearance, deformed leaves, and chlorosis (**Figures 50** and **51**).

Infections with CMV also lead to deformation of plant parts: leaf narrowing and dwarfing in combination with chlorosis are typical symptoms (**Figure 52**).

Dahlias are native of Mexico and Latin America and belong to a genus with only 30 species. The very important bedding and cut flower forms go back to three species

D. coccinea Cav., *D. pinnata* Cav., and *D. juarezii* hort. The latter most probably being already a garden form. The modern cultivars result from breeding in West European countries.

Dahlias are frequently infected by dahlia mosaic virus (DMV), a spherical dsDNA virus which mostly induces chlorotic oak leaf patterns (**Figure 53**) and other chlorotic symptoms and may induce stunting of susceptible varieties. The virus is spread via aphids in a nonpersistent or semipersistent way.

TSWV and INSV are among the commonly occurring viruses of dahlias. Symptoms may vary according to virus and variety but most often these viruses induce necrotic ring spots becoming concentric (**Figures 54** and **55**). Oak leaf patterns or discrete necrotic lines along the midrib may also appear in young leaves of certain cultivars.

Symptoms induced by CMV vary according to the developmental stage of the plant. They comprise light mosaic and typical leaf narrowing, also called 'fern leaf'. In some cultivars oak leaf patterns may prevail (compare **Figure 53**).

Figure 50   Dwarfing of TSWV-infected osteospermum (left), noninfected control (right). By Joachim Hamacher.

Figure 52   CMV infection of osteospermum: leaf narrowing of plant parts. By Joachim Hamacher.

Figure 51   Leaf narrowing, chlorosis, and stunting of TSWV-infected osteospermum plant. By Joachim Hamacher.

Figure 53   Chlorotic oak leaf pattern on DMV-infected dahlia. By Joachim Hamacher.

TSV appears rather frequently in dahlias. It does not induce typical or severe symptoms, but represents a potential danger for other crops.

The former genus *Chrysanthemum* encompasses about 200 species, originating predominantly from the Mediterranean region and Western Asia but some come from South Africa. *Chrysanthemum indicum* hybrids (*Dendranthema indicum* and *D. x grandiflorum*) originate from China to Japan. The genus *Chrysanthemum* is now differentiated into 14 genera, the economically most important of which are *Argyranthemum* (species *A. frutescens*), traded as potted flowering shrubs) and *Dendranthema* (species *Dendranthema x grandiflorum*), traded as cut plants or potted plants. *Chrysanthemum* is one of the leading ornamentals in the international market.

Dendranthema hybrids are most frequently infected by chrysanthemum stunt viroid (CSVd). The disease originated in the USA but reached Europe in the 1950s with

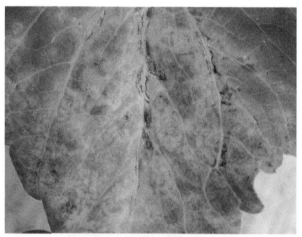

Figure 54   Chlorotic concentric rings and necrotic lines on TSWV-infected dahlia. By Joachim Hamacher.

infected cultivars exported to England. About 70% of the infected plants are systemically stunted with dwarfed leaves (**Figure 56**) and scrubby flowers, exhibiting floral bleaching of red colored cultivars. Color deviations of leaves are not always obvious, but may include pale, upright young leaves. Sometimes, leaf spots or flecks, often associated with leaf distortions ('crinkling'), are also observed. Symptoms are variable and highly dependent on environmental conditions, especially temperature and light. As CSVd is very stable *in vitro*, sap transmission during handling and cutting may occur. Detection of the viroid may be done by visual inspection, grafting onto indicator varieties, bidirectional electrophoresis, molecular hybridization, as well as reverse transcriptase polymerase chain reaction (RT-PCR) and nested RT-PCR, but not by immunological methods or electron microscopy.

A common virus of chrysanthemums is chrysanthemum virus B (CVB), a carlavirus. It leads to very mild leaf mottling or vein clearing in some cultivars; some infected varieties show slight loss of flower quality, and a few varieties sometimes develop brown necrotic streaks on the florets. Many cultivars become entirely infected, often without visible symptoms. CVB occurs in many other ornamental plants with rather long incubation periods and produces rather mild symptoms. Besides vegetative propagation, the virus can be transmitted in a nonpersistent manner by several aphids.

Tospoviruses (TSWV and INSV) infect chrysanthemums quite often, inducing irregular chlorotic or necrotic spots (**Figure 57**), mild leaf deformations, primarily on young leaves, as well as browning or wilting of shoots. As tospoviruses are a major threat to growers, elimination of infected plants has to be executed early and rigidly.

TAV is fairly important in chrysanthemums, as it may cause heavy symptoms like breaking, dwarfing, as well as distortion of the flowers, and foliar chlorotic spots or ring

Figure 55   Necrotic ring spots on dahlia infected with TSWV. By Joachim Hamacher.

Figure 56   Dwarfed and stunted chrysanthemum infected with CSVd (left) noninfected control (right). By Joachim Hamacher.

Figure 58 Tulips exhibiting flower breaking induced by TBV. By Franz J. Nienhaus.

Figure 57 Necrotic flecks of TSWV-infected chrysanthemum leaves. By Rainer Wilke.

spots. Most infected cultivars, however, do not show leaf symptoms.

Other virus diseases induced by AMV, BBWV1, BWYV, LMV, SMV, TMV, ToMV, ToRSV, and TSV have been reported for *Chrysanthemum* spp. but seem to be of minor importance.

Virus Diseases of Liliaceae

Tulipa, Lilium

Taxonomically, Liliaceae are assigned to the order Liliiflorae of the monocotyledonous plants. We concentrate on virus diseases of the genera *Tulipa* and *Lilium*, as they represent large genera with economically very important ornamentals.

Tulips originated from Central Asia and include about 100 different species. Tulips are propagated via bulbs, which develop at the base of the bulb of the previous year.

More than 20 different viruses are reported to infect tulips, the most important of which are TBV, TNV, and TRV.

Tulip breaking virus (TBV) is the most frequently encountered virus in tulips. It is assigned to the genus *Potyvirus* and can be transmitted by aphids (e.g., *Myzus persicae* and *Aphis fabae*) in a nonpersistent manner. TBV affects color breaking of flowers particularly in late-flowering pink, purple, and red cultivars, while white- and yellow-flowered cultivars are not affected. Breaking symptoms have been described as bars, stripes, streaks, featherings, or flames of different colors on petals (**Figure 58**). The color variation is caused by local fading, intensification, or accumulation of pigments in the upper epidermal layer after development of the normal flower color. Mottling or striping of the leaves

also occurs. The infection causes loss of vigor and poor flower production. TBV played an important role in the Dutch 'tulipomania' in the seventeenth century, in that it increased the value of tulips with decorative flower breaking and led to wild speculations with astronomical prices for one variegated tulip bulb. At that time, the undesirable viral cause of the spectacular flower breaking was not yet known.

Tobacco necrosis virus D (TNV-D) causes the so-called Augusta disease, named after the variety in which it was first detected. The virus is transferred to the roots specifically adhering to the zoospores of the chytridiomycete *Olpidium brassicae*. This particularly happens when soil temperatures rise above 9 °C. Infected plants do not necessarily show symptoms, since the virus may be confined to the roots. The disease may suddenly appear when the plants are planted out during frost or when planted early at high temperatures after storage on a standing ground. Stunting and distortion of the shoot and leaves are the most severe symptoms but streaking or angular or elliptical spots are more typical symptoms. Fine necrotic lines in the tepals are of considerable value for differential diagnosis.

TRV (genus *Tobravirus*) is another common virus of tulips with an exceptionally wide host range of cultivated and wild plants. A vector transmission by soil-inhabiting nematodes of the genera *Trichodorus* and *Paratrichodorus* may spread the virus in the field. Symptoms of the disease include chlorotic flecks, oval lesions and streaks, appearing early in the season. Streaks of darker color may occur on tepals of red flowering varieties, whereas those streaks appear translucent in yellow or white flowers. Plants may become stunted and flowers sometimes deformed.

Tulip virus X (TVX) has been reported in tulips from Great Britain and from Japan. It causes chlorotic or necrotic streaks in leaves and streaks of intensified color mainly at the margins of tepals. The tepals may also become bleached or necrotic, resembling symptoms caused by TNV. TVX is a member of the genus *Potexvirus*.

Mechanisms of spread within field or glasshouse plots are unclear but may rely on sap transmission.

Other viruses of tulips are nepoviruses: ArMV, TBRV, TRSV; a cucumovirus: CMV; several potyviruses: LMoV, ReTBV, TCBV, TulMV; a carlavirus: LSV; a tombusvirus: TBSV; and a tobamovirus: TMV (see **Table 1**).

Three viruses are common in lilies, these are lily mottle virus (LMoV), lily virus X (LVX), and lily symptomless virus (LVS).

LMoV is a potyvirus and was formerly thought to be a strain of TBV, but serological, host range, and molecular investigations have set it apart from TBV as an own species. It can be found wherever lilies are grown. Symptoms develop as chlorotic mottle to stripe-mosaic (**Figure 59**) and leaves may be twisted or show narrowing. When young plants are infected, severe yellowing of leaves or browning of veins in stems may occur. As LMoV is a potyvirus its transmission is affected by aphids in a nonpersistent manner. The virus can also infect tulips.

LVS does not induce specific symptoms in many cultivars, but reduced growth, small flowers, and lower bulb yields are recorded. Foliar vein clearing or intercostal light green stripes may develop as well. Other natural hosts are tulip and *Alstroemeria*. The virus can be transmitted nonpersistently by aphids.

LVX is a potexvirus, which normally does not lead to pronounced symptoms in most cultivars, but enhances symptoms of LSV. When symptoms appear, the plants show faint chlorotic spots and sometimes necrotic lesions. The virus can be sap-transmitted by mechanical inoculation onto herbaceous test plants.

Other viruses, reported to naturally infect lilies are ArMV, TRSV, ToRSV, CMV, TRV, TSWV, TMV, and SLRSV.

Virus Diseases of Orchidaceae

The family Orchidaceae contains about 800 genera with more than 25 000 species. Centers of origin are probably the Malaysian region with *c*. 12 000 species followed by the tropics of America with about 10 000 species. Genera with high economic importance are *Cattleya*, *Cymbidium*, *Dendrobium*, *Odontoglossum* hybrids, and *Phalaenopsis*.

Lawson summarizes 27 different viruses of orchids in a detailed review. The most important and often detected viruses are cymbidium mosaic virus (CyMV) and odontoglossum ringspot virus (ORSV). Many rhabdoviruses or rhabdovirus-type viruses (nine different viruses) have been reported for orchids as well as potyviruses, such as BYMV, dendrobium mosaic virus syn, bean common mosaic virus (BCMV), cypripedium virus Y (CypVY), cypripedium chlorotic streak virus (CypCSV), TuMV and vanilla mosaic virus (VanMV), tospoviruses TSWV and INSV, a tentative closterovirus: dendrobium vein necrosis virus (DVNV), ToRSV, cymbidium ringspot virus (CyRSV), and CMV.

CyMV is a potexvirus and as such easily sap-transmissible. It infects a wide range of orchids and leads to necrotic flecks or streaks (**Figures 60** and **61**), black necrotic spots or line patterns (**Figure 62**) on *Cymbidium*, *Phalaenopsis*, and in mixed infections together with DVNV

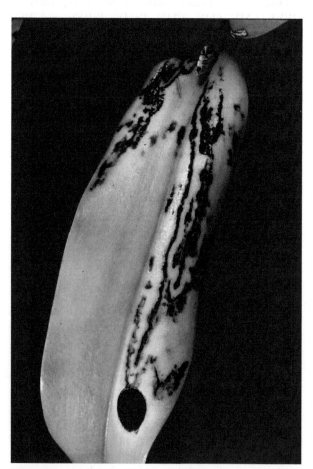

Figure 60 Black line patterns on CymMV infected Cattleya leaves. By Dietrich E. Lesemann.

Figure 59 Intensive streaking of lily flower caused by co-infection of LMoV and LSV. By Martin Bucher.

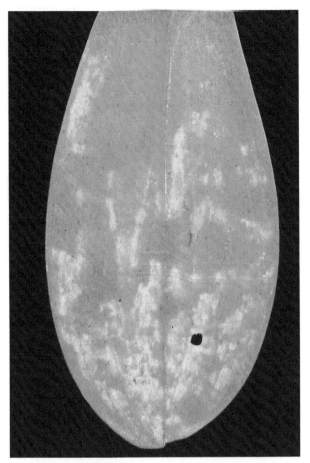

Figure 62 Cattleya leaf with black streaking caused by CymMV. By Dietrich E. Lesemann.

Figure 61 Necrotic flecks on CymMV infected Phalaenopsis leaves. Dietrich E. Lesemann.

in *Dendrobium* (**Figures 63** and **64**). Flower necrosis induced in *Cattleya, Laelia, Cymbidium, Phalaenopsis, Epidendrum,* and *Vanda* has also been shown to be caused by CyMV (**Figure 65**).

ORSV (syn. TMV-O) is a tobamovirus and leads to a range of different symptoms in many orchid species (**Figures 66–68**). Flower breaking in violet *Cattleya* varieties with streaks of intensified pigmentation on tight buds have been observed as well as fine necrotic stripes (**Figure 69**). Symptoms on leaves comprise chlorotic and dark concentric ringspots in *Odontoglossum grande.* The virus is easily sap-transmissible and spread occurs via contaminated cutting tools.

Virus Control in Ornamentals

The ornamental industry is a very important branch of agriculture around the world with high amounts of investment and rising competition among growers. Outbreaks of diseases have shattered the ornamental industry more than once. Especially epidemics of viral or bacterial etiology may be disastrous for the reputation and standing of the enterprise struck by such diseases.

As direct chemical control strategies have proven to be ineffective in plants, only prophylactic measures to prevent or hinder infection and spread of viruses in the respective cultures are effective. Traditional resistance cross-breeding as a reliable measure to combat virus diseases, however, is of minor importance, for on one hand cross-breeding of resistance genes is time consuming, and on the other hand the market demands a rapid change or enlargement of the range of varieties. Hygienic measures and, in many cases, meristem culture and thermotherapy are the essential prophylactic strategies to ensure virus-free cultures. The success of transgenic virus-resistant plants as a practicable alternative is strongly dependent on the legislative and the acceptance in the respective countries. To date the most promising strategy for the producer is the establishment of virus-free mother plants and elite stocks as well as continuous monitoring of production stocks. This demands staff skilled in horticultural production, as virus symptoms appear very varied and may be confounded with symptoms provoked by environmental stress or faulty cultural practices. Specialized diagnosis labs have to know the peculiarities of sampling and time of the year, in which virus testing is practicable, because virus concentrations vary considerably during the course of the year in many ornamental

Figure 63   Cattleya flower with black streaking induced by CymMV. By Dietrich E. Lesemann.

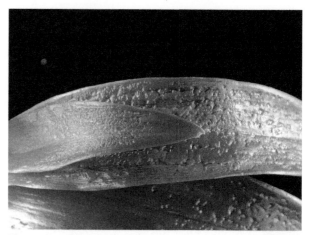

Figure 65   Dendrobium leaves with sunken necrotic spots and necrotic stripes caused by a mixed infection of CymMV and DVNV. By Joachim Hamacher.

Figure 66   Chlorotic and necrotic ring spots and line patterns on ORSV-infected cymbidium leaves. By Dietrich E. Lesemann.

Figure 64   Dendrobium leaves with black spots and chlorotic flecks caused by a co-infection of CymMV and DVNV. By Joachim Hamacher.

species. They tend to be highest in late winter/early spring in countries with moderate climate, and nearly disappear in summer and autumn, thus disabling secure diagnosis. To ensure virus-free production, the European and American plant protection organizations European and Mediterranean Plant Protection Organization (EPPO/ OEPP) and US Department of Agriculture-Animal and Plant Health Inspection Service (USDA-APHIS), for example, have established continuously updated guidelines for the production of virus-free ornamentals. This is achieved by quarantine measures and well-regulated production requirements with highest phytosanitary

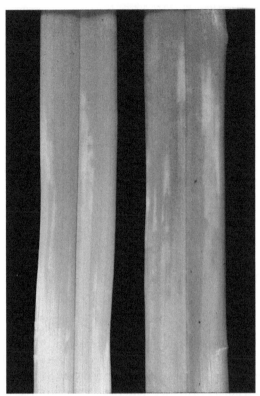

Figure 67   Chlorotic striping on cymbidium leaves infected with ORSV. By Dietrich E. Lesemann.

Figure 69   Chlorotic flecking on cattleya flowers infected with ORSV. By Dietrich E. Lesemann.

Figure 68   Black streaking caused by ORSV on cymbidium leaves. By Dietrich E. Lesemann.

standards. These include regular testing of candidate and elite stocks in specialized diagnostic labs that are either part of the horticultural enterprises or autonomous. Plant clinics and plant protection services contribute to the program as well. Routine testing is done by ELISA or comparable immunological tests. Molecular techniques with higher sensitivity than that of the conventional testing methods are gaining wider fields of application. Upon occurrence of symptoms (of unknown cause), electron microscopy may be of great help to clarify the viral etiology of observed symptoms. Regular visual inspections of production stocks (monitoring) ensure high quality and freedom from diseases of the products.

See also: Papaya Ringspot Virus; Plant Antiviral Defense: Gene Silencing Pathway.

Further Reading

Brunt A, Crabtree K, Dallwitz M, Gibbs A, and Watson L (1996) *Viruses of Plants*. In: *Descriptions and Lists from the VIDE Database*, 1484pp. Wallingford, UK: CAB International.

Daughtrey ML and Benson DM (2005) Principles of plant health management for ornamental plants. *Annual Review of Phytopathology* 43: 141–169.

Daughtrey ML, Robert L, Wick RL,, and Peterson JL (eds.) (1995) *Compendium of Flowering Potted Plant Diseases*. St. Paul, MN: APS-Press.

Fauquet CM, Mayo MA, Maniloff J, Desselberger U,, and Ball LA (eds.) (2005) *Virus Taxonomy: Eighth Report of the International Committee on Taxonomy of Viruses.* San Diego, CA: Elsevier Academic Press.

Hammond J (ed.) (2002) ISHS Acta Horticulturae 568: *X International Symposium on Virus Diseases of Ornamental Plants* (CD-ROM format only).

Hull R (ed.) (2001) *Matthews Plant Virology.* San Diego, CA: Academic Press.

Loebenstein G, Hammond J, Gera A, Derks T,, and van Zaayen A (eds.) (1996) ISHS Acta Horticulturae 432: IX *International Symposium on Virus Diseases of Ornamental Plants.* (CD-ROM format only).

Loebenstein G, Lawson RH,, and Brunt AA (eds.) (1995) *Virus and Virus-Like Diseases of Bulb and Flower Crops,* 543pp. New York: Wiley.

Mokrá V, Brunt AA, Derks T,, and van Zaayen A (eds.) (1994) ISHS Acta Horticulturae 377: VIII *International Symposium on Virus Diseases of Ornamental Plants* (CD-ROM format only).

Pirone TP and Perry KL (2002) Aphids: Non-persistent transmission. *Advances in Botanical Research* 36: 1–19.

Ullman DE, Meideros RB, Campbell LR, Whitfield AE, Sherwood JL, and German TL (2002) Thrips as vectors of Tospoviruses. *Advances in Botanical Research* 36: 113–140.

Relevant Websites

http://www.dpvweb.net – Descriptions of plant viruses (Association of Applied Biologists).

http://www.agdia.com – List of ornamental plant samples of Agdia.

http://www.ebi.uidaho.edu – University of Idaho, VIDE Database.

Potato Viruses

C Kerlan, Institut National de la Recherche Agronomique (INRA), Le Rheu, France

Glossary

Primary infection (primarily infected plants) Potato plants become infected during the growing season.
Secondary infection Infection results from an infected mother tuber.

Introduction

At least 38 potato viruses have been described, several of them being classified by ICTV as tentative species or possible strains of one species. They can be divided into three groups according to their importance and their distribution in potato-growing areas in the world. They include well-known viruses, and recently discovered, often poorly characterized viruses. Eight viruses are major pathogens causing severe damages in potato crops (**Table 1**). Many potato viruses occur only in Latin America. None of these viruses is economically important; some of them have been found only once in one cultivar (**Table 2**). The third group comprises viruses that occur in other parts of the world and are either of only local importance or without any significance (**Table 3**).

Potato Viruses: One Century

Viral diseases are thought to have become a threat to potato soon after its introduction in Europe. The phenomenon called potato degeneration was described in the eighteenth century. The leafroll disease was mentioned in the 1750s and the causal agent described in 1916 was one of the first viruses identified. The main species involved in the mosaic complex were identified in 1920s and 1930s, that is, potato virus M, potato virus X, potato virus Y, and potato virus A. Other important potato viruses were only discovered in the 1950s (tobacco rattle virus, potato virus S) or mid-1960s (potato mop-top virus). In the last four decades, an ever-increasing number of viruses were found infecting potato in Latin America, especially in Andean valleys where the potato originates from. Lastly, numerous viruses or strains more common in other host plants were found to infect potato occasionally. Some of them, for instance tomato spotted wilt virus, cause emerging diseases and could become important in the future. We may also point out that a contact-transmitted virus such as pepino mosaic virus, never detected in potato fields so far, was shown to be capable of infecting many important potato cultivars and hence could become a threat in the future.

Since the turn of the twentieth century, potato producers in many countries have recognized that certain potato-growing areas were better than others for the production of seed potatoes. Seed-potato production programs were developed, based during 1940–50 on clonal selection, then on basic seed production involving virus eradication by thermotherapy and meristem culture, and later on specific multiplication systems such as propagation *in vitro* or by stem cuttings or from true potato seeds (TPSs). Planting of certified seed potatoes with low levels of virus infections has now become the main element in the fight against viral diseases. Breeding for virus resistance has been intensively used at least since 1925 for PLRV, and sources of resistance in many wild *Solanum* spp. have been exploited. Three types of resistance were described in the 1950s and reliable knowledge on their genetic background and inheritance

Table 1   Major potato viruses

Species Acronym First mention	Genus	Economical importance[a]	Geographical distribution[a]	Spread[a]	Natural host range Other main hosts	Variability[a]
Potato virus Y PVY 1931	Potyvirus	Very high	Worldwide	Aphids NP[c], contact?	Wide Tobacco, tomato, pepper	Strains Y^O, Y^C, Y^N Variants Y^{NTN}, Y^NW, $Y^{N:O}$ Serotypes PVY^{O-C}, PVY^N
Potato leafroll virus PLRV 1916	Polerovirus	Very high	Worldwide	Aphids P[c]	Narrow Solanum spp., tomato, ulluco	Tomato yellow top and Solanum yellows strain
Potato virus X PVX 1931	Potexvirus	High	Worldwide	Contact	Narrow Tomato, pepper	Strain groups 1, 2, 3, 4 Strains HB, CP Serotypes PVX^O, PVX^A
Potato virus A PVA 1932	Potyvirus	Moderate	Worldwide[b]	Aphids NP[c]	Only potato	Four strain groups Three serotypes
Potato virus S PVS 1952	Carlavirus	Moderate	Worldwide	Aphids NP[c], contact	Narrow Pepino	Strains PVS^O, PVS^A
Potato virus M PVM 1923	Carlavirus	Moderate	Worldwide[b]	Aphids NP[c], contact	Narrow Mainly in Solanaceae	Strain PVM-ID
Tobacco rattle virus TRV 1946	Tobravirus	High	Worldwide	Nematodes	Weeds, flower bulbs, beet, tobacco, lettuce, spinach, etc.	Many strains
Potato mop-top virus PMTV 1966	Pomovirus	Rather high	Mainly in cooler climates	Fungus	Only potato	PMTV-S, PMTV-T

[a]In potato.
[b]Not found in Andean countries.
[c]Viruses are transmitted by aphids in a nonpersistent (NP), persistent (P) manner.
Species in bold italic letters: the virus (at least one strain) can induce tuber necrosis.

was available in the 1970s. Now many resistance genes have been mapped in the potato genome, which has permitted breeding based on molecular markers. Lastly a remarkable catalog of drastic measures has been established in order to avoid the introduction of quarantine viruses. Technical guidelines for safe movement and utilization of potato germplasm have recently been published under authority of relevant international specialized institutions.

Main Traits of the Viral Diseases in Potatoes

Symptomatology and Damages

Secondary infection symptoms are often more severe than those of primary infections, resulting in emergence delays and growth reductions of plants. Combinations of viruses also result in severe symptoms. Common symptoms are mosaic, mottle, or other (bright or pale green) discolorations, crinkling, rolling, or distortion of leaflets, leaf dropping, stunting, and deformation of the whole plant. Various necrotic patterns may occur on leaves, stems, and tubers (both external and internal).

Size, number, aspect, and content of tubers can be affected. Yield losses greatly vary depending on virus, strain, and time of infection. They can reach up to 90%. Some infections can result in defects in processed products such as blackening of tuber flesh. Lastly viruses inducing tuber necrosis (strains of at least 11 species) can make tubers totally unmarketable as ware or processed potatoes.

Spread of Potato Viruses

Potato viruses are efficiently transmitted by aphids, either by species colonizing potatoes or by itinerant species migrating from other crops and weeds. PLRV and PVY are model viruses for studying the mechanisms of persistent and nonpersistent transmission by aphids. PLRV was the first virus of the family Luteoviridae for which passage of the virus through the intestine and not the hindgut

Table 2 Potato viruses only found in Latin America

Species Acronym First report	Genus	Economical importance[a]	Geographical distribution[a]	Spread[a]	Natural host range Other main hosts	Variability[a]
Andean potato latent virus APLV 1966	Tymovirus	Very low	Bolivia, Colombia, Ecuador, Peru	Contact, flea beetle, TPS[c]	Only potato and ulluco	Serotypes Hu, CCC, Col-Caj
Andean potato mottle virus APMoV 1977	Comovirus	Unknown	Chili, Ecuador, Peru, Brazil	Contact, flea beetle?	Only potato	Strains B, C, H
Aracacha virus B-oca strain AVB-O 1981	Nepovirus?	Unknown	Bolivia, Peru	TPS	Only potato and oca	Oca and potato strains
Potato deforming mosaic virus PDMV 1985	Begomovirus?	High in one cultivar	Southern Brazil	Whitefly	potato, S. chacoense, S. sisymbrifolium	Not reported
Potato rough dwarf virus PRDV 1995	Carlavirus	Very low	Argentina, Uruguay	Unknown	Only potato	Strain of PVP
Potato virus P PVP 1993	Carlavirus	Locally in two cultivars	Brazil	Aphids	Only potato	PRDV strain
Potato virus T PVT 1977	Trichovirus	Unknown	Bolivia, Peru	Contact, TPS	Mashua, oca, potato, ulluco	Not reported
Potato virus U PVU 1983	Nepovirus	None	Locally in Peru (Comas Valley)	Nematode	Unknown	None

Virus	Genus		Distribution	Transmission	Host range	Related viruses
Potato yellow mosaic virus PYMV 1986	*Begomovirus*	Unknown	Venezuela	Whitefly	Narrow Tomato	Not reported
Potato yellow vein virus PYVV 1954	*Crinivirus*	Locally high	Colombia, Ecuador, Peru	Whitefly	Potato and several weeds	Not reported
Potato yellowing virus PYV 1991?	*Alfamovirus?*	Unknown	Peru, Chili, Bolivia	Aphid SP[b], TPS	Only potato	Not reported
Potato 14R virus P14R 1996	*Tobamovirus?*	None	Peru	Unknown	Only potato	None
Solanum apical leaf curl virus SALCV 1983	*Begomovirus?*	Very low	Peru	Unknown	Narrow *Physalis peruviana, Solanum nigrum*	Not reported
Sowbane mosaic virus SoMV 1997	*Sobemovirus*	Isolated once from a Mexican line	Unknown (Mexico?)	Unknown	Narrow *Chenopodium* spp.	None
Tobacco ringspot virus TRSV 1977	*Nepovirus*	Low	Peru	Nematode? TPS	Potato, arracacha, oca	TRSV-Ca & PBRSV
Wild potato mosaic virus WPMV 1979	*Potyvirus*	None	Peru	Aphid NP[b]	Narrow Pepino, tomato, wild potato	Not reported

[a] In potato.
[b] Viruses are transmitted by aphids in a nonpersistent (NP), persistent (P), and propagative (Pr) manner.
[c] TPS means true potato seeds.

Table 3 Viruses of limited significance to potato production

Species Acronym First report	Genus	Economical importance[a]	Geographical distribution[a]	Spread[a]	Natural host range Other main hosts	Variability[a]
Alfalfa mosaic virus AMV 1940	Alfamovirus	Only locally	Uncommon	Aphids NP[b], pollen, TPS[c]	Wide Clover, tobacco, tomato, pepper	Calico and tuber necrosing strains
Beet curly top virus BCTV 1954	Curtovirus	Only locally	Arid and semiarid regions	Leafhoppers	Bean, beet, spinach, tomato, pepper, many cucurbits	Not reported
Cucumber mosaic virus CMV 1958	Cucumovirus	Very low	UK, Egypt, India, Saudi Arabia	Aphids NP	Many cucurbits, pepper, tomato, etc.	Serotypes I, II
Eggplant mottled dwarf virus EMDV 1987	Nucleo-rhabdovirus	Very low	Iran	Aphids? contact?	Narrow Eggplant, tomato	Not reported
Potato aucuba mosaic virus PAMV 1961	Potexvirus	Low	Uncommon	Aphids NPH[b], contact	Narrow Clover, tomato, pepper, tobacco	Mild and severe strains
Potato latent virus PotLV 1996	Carlavirus	Unknown	North America	Aphid, whitefly?	Only potato	Not reported
Potato virus V PVV 1986	Potyvirus	Low	Bolivia, Peru, North Europe	Aphids NP	Narrow Tomato	Isolates Gl, AB, UF

Virus	Genus		Distribution	Transmission	Host range	Strains/Notes
Potato yellow dwarf virus PYDV 1 937	*Nucleo-rhabdovirus*	None	Canada, North USA, Saudi Arabia, South Russia?, Ukraine?	Leafhoppers P[b], Pr	Narrow Ox-eye daisy, periwinkle, clover	Two strains
Tobacco mosaic virus TMV 1977	*Tobamovirus*	None or low	China, India, Saudi Arabia, Andean countries	Contact	Wide Many crops and woody plants	Not reported
Tobacco necrosis virus TNV 1935	*Necrovirus*	Low	Europe, North America	Fungus	Bean, tulip, tobacco, cucumber, pea, lettuce	Strain D
Tobacco streak virus TSV 1981	*Ilarvirus*	Low	Brazil, Peru	Thrips?, TPS? pollen?	Wide	Not reported
Tomato black ring virus TBRV 1950	*Nepovirus*	Rather low	Europe	Nematodes, seeds, TPS	Very wide Grape, fruit trees, vegetables, weeds…	Potato bouquet and pseudo-aucuba strains, two serogroups,
Tomato mosaic virus TMV 1977	*Tobamovirus*	None	Hungary	Contact	Rather wide Pepper, tomato	TMV potato strain
Tomato spotted wilt virus TSWV 1938	*Tospovirus*	Locally high	South America, Australia, India, Europe, South Africa	Thrips	Very wide Tomato, pepper, pea, groundnut, soyabean, tobacco, etc.	Not clearly described

[a] In potato.
[b] Viruses are transmitted by aphids in a nonpersistent (NP), persistent (P), and propagative (Pr) manner; NPH means that aphid transmission requires a helper virus.
[c] TPS means true potato seeds.
Species in bold italic letters: the virus (at least one strain) can induce tuber necrosis.

was demonstrated. Other insects are vectors of less important viruses, notably leafhoppers, thrips, and whiteflies (**Tables 2** and **3**). PVX and some PVS strains are transmitted by contact between foliage or sprouts, between tubers when cutting, and also through farm implements. Several viruses are soil-borne, either transmitted by nematodes (TRV) or by fungi (PMTV, tobacco necrosis virus). Many Andean viruses can be transmitted by TPSs or by pollen. Inoculum sources are in most cases potato itself (volunteer or secondarily infected plants).

Control Methods

Control of seed potatoes

Production of certified seed potatoes nowadays is based on a catalog of measures more or less drastic and sophisticated depending on the country. It is undoubtedly a specialized and costly scheme involving the following main steps: (1) multiplication of virus-free prebasic and basic seeds under conditions where infection risks are minor, for example, first *in vitro* and in vector-proof greenhouses or plastic tunnels, then in isolated fields in areas known for low virus disease pressure; (2) field inspections involving roguing of diseased potatoes and eradication of all sources of inoculum; (3) treatments against vectors, notably insects; (4) specific growing techniques taking into account risks associated to the vector pressure such as early dates of plantation, haulm killing and use of barrier crops; (5) postharvest controls involving large-scale laboratory tests, commonly using ELISA and sometimes molecular techniques. The final aim is the quality grading of the commercial categories of seed potatoes, standards being dependent on each country but usually infection levels should not exceed 10% for all viruses together.

Resistance

Three types of resistance to viruses in potato are commonly distinguished: field resistance or resistance to infection, hypersensitivity (HR), and extreme resistance (ER). Resistance to infection is a polygenically inherited resistance appraised by the number of progeny tubers that are infected. It comprises resistance to virus itself and resistance to vectors. It is stated to be effective against all strains of the virus involved. Scores of resistance levels depend strongly on environmental conditions, time of infection, inoculum, and vector pressure. HR and ER are governed by single dominant genes coded N and R, respectively (**Figure 1**). ER is an enhanced form of HR, resulting in no obvious symptoms and no systemic infection. Conversely, HR is associated with local necrotic lesions on inoculated leaves but also in most virus–cultivar combinations with systemic necrotic reactions (spots, lines, arches) extending to lower and upper leaves, stems, and tubers. The virus though unevenly distributed can be detected in all these organs. However, except in special

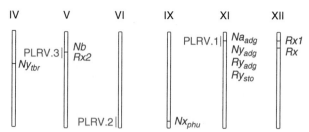

Figure 1 Genomic mapping of resistance genes (in blue) and QTL (in red) to four major viruses (PVY, PLRV, PVX, PVA) in potato. From S. Marchadour, FNPPPT-INRA, France (unpublished).

cases it is not transmitted to the daughter plants. N genes are strain-specific in contrast to R genes, each of which confers resistance to all strains of one or two viruses.

Breeding programs for resistance to infection to the main viruses have been well established in numerous countries producing potatoes. ER is also a frequently employed resistance, notably in countries where the use of certified seeds provides no satisfactory control. R genes providing protection against some major viruses (PVX, PVY, PVA, PVM) have been found in several wild *Solanum* species such as *S. stoloniferum*, *S. andigena*, *S. chacoense*, and *S. acaule*. They have been used in breeding programs since the mid-1940s and have appeared to provide sustainable resistance. N genes were present in the pool of old potato cultivars and hence have been easily incorporated into new cultivars. Potato is easily transformed by genetic engineering and pathogen-derived resistance against some major viruses (PVY, PLRV, PVX, PMTV) has been intensively studied, though without any significant applied use.

Quarantine measures

Quarantine measures taken against the risk of introducing foreign viruses (species or strains) in one state or continent have demonstrated their full efficiency. Indeed, so far no virus or strain listed as a quarantine pathogen is known to have spread outside its original region. Technical recommendations needed to test any exported or imported material (tubers, cuttings, true potato seeds, pollen, *in vitro* plantlets) have been established. These regulated viruses can be diagnosed by bioassays on selected indicator plants and, for the majority of them, by ELISA. Antibodies to potato virus U and potato yellow vein virus (PYVV) are not available. All potato viruses, except PYVV, potato-deforming mosaic virus and solanum apical leaf curl virus, are sap-transmissible. Molecular techniques or electron microscopy can also be used.

The Major Potato Viruses

Eight viruses are both widespread and damaging and, thus, of great concern in seed potato production (**Table 1**).

Figure 2 (a) Secondary infection by PVY: mottle and crinkle of the leaflets; stunting of the plant. (b) Primary infection by PVY: mosaic on an apical leaf of cv. Bintje. (c) Primary infection by PVY: typical yellowing with green halos on a basal leaf of cv. Nicola. (d) Oak-leaf patterns and necrotic spots induced by PVY on leaflets of cv. Nicola. (e) PTNRD typical symptoms on tubers of cv. Nicola harvested in a field in Southern France. (a) Reproduced by permission of Y. Le Hingrat, FNPPPT, France. (b) Photograph: Christelle Lacroix, INRA, France. (c–e) Photographs: K. Charlet-Ramage, GNIS-INRA, France.

Potato Virus Y

PVY has become in the last decade the most important potato virus. It affects most cultivars and causes major yield losses of up to 80% and even more in the case of the potato tuber necrotic ringspot disease (PTNRD) (**Figure 2(a)**).

PVY isolates infecting potato have been divided into three strains (PVY^O, PVY^C, PVY^N), and many pathotypes and variants. Currently, PVY^O and PVY^N are the main widespread strains. PVY^C is less frequent (less than 5% of PVY isolates in France). Variants PVY^{NTN}, PVY^NW, and $PVY^{N:O}$ are also widespread. Most isolates of these variants are recombinants between PVY^O and PVY^N. Strains PVY^O, PVY^C, and PVY^N are differentiated on the basis of their reaction on *Nicotiana tabacum* and on potato cultivars carrying the hypersensitive genes Ny_{tbr} and Nc. PVY^{N}W isolates differ from other PVY^N isolates in their virulence and in belonging to serotype O-C.

Typical symptoms associated with PVY infections are mosaic, crinkle, necrotic patterns, and stunting of the plant in the case of PVY^O secondary infections (**Figure 2(b)**). Infection by PVY^N is often symptomless or associated with a mild mosaic (**Figure 2(c)**), only sometimes with a severe mottle. Yellowing and green halos can be observed on basal leaves (**Figure 2(d)**). Necroses (on or around leaf veins, on petioles, stems, and tubers) are related to hypersensitivity reactions in numerous cultivars (**Figure 2(e)**). Necrosis of basal leaves which remain clinging to the stem (leaf drop) are typical of PVY^O primary infections. PVY^C induces stipple-streaks on stems and brown spots around eyes of tubers. PVY^{NTN}, and some PVY^{N}W and PVY$^{N:O}$ isolates, can induce PTNRD.

Aphid transmission is undoubtedly the most important means of PVY spread in the fields. However, transmission by plant-to-plant contact has been consistently reported by several authors. Under experimental conditions, 70 aphid species were able to transmit PVY, *Myzus persicae* being the most efficient vector. This species and other aphids colonizing potato fields have long been seen as the main natural vectors. However, due to their abundance in potato fields during many PVY outbreaks, itinerant species multiplying in other crops (cereals, peas, etc.) have also been said to play a significant role in spread of PVY. PVY has a wide host range but besides potato no other host has so far been shown to act as a significant inoculum source, at least in the main potato-growing areas.

Control of PVY in seed fields requires use of mineral oil treatments to prevent aphid transmission. Forecasting and simulation models of PVY spread have been developed in many countries. At least 20 cultivars possess durable immunity to PVY though none of them is economically important. However, Ry_{sto} and Ry_{adg} genes are more and more introgressed into breeding programs.

Potato Leafroll Virus

Although PLRV has been decreasing in incidence, it is yet a serious threat in most potato-growing areas. It can induce heavy yield losses, often more than 50%, sometimes up to 90%. Harvested tubers are small, whereas mother tubers often stay hard and unrotten. Spindling sprouts might occur in old cultivars. In some currently widespread cultivars, tuber quality is reduced due to the occurrence of necrotic spots distributed within the flesh (net necrosis) (**Figure 3(a)**). The virus is restricted to phloem. It induces formation of callose in sieve tubes blocking sugar transport and making the leaflets becoming hard and brittle. Cracking sounds resulting from squeezing such leaflets are a good indicator PLRV infections. It also induces phloem necrosis in petioles and stems. Photosynthesis and chlorophyll content are also reduced. Typical symptoms of secondary infection are an erect habit, plant stunting, and pale yellow foliage (**Figure 3(b)**). Leaflets roll upward, whereby lower leaves are dry, brittle, and leathery and may become necrotic. Primary infections are often latent, especially late infections. In most susceptible cultivars, apical leaves are erected and pale yellow (top yellowing) with sometimes purple or red margins.

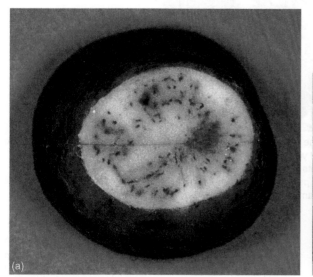

Figure 3   (a) Net-necrosis induced by PLRV in a tuber of cv. Russet Burbank. (b) Secondary infection by PLRV: erect habit, plant stunting and pale yellow foliage. (a) Photograph: J. Martin, LNPV, France. (b) Reproduced by permission of Y. Le Hingrat, FNPPPT, France.

PLRV is not transmitted by sap inoculation. At least ten aphid species were shown to be capable of transmitting it. PLRV does not multiply in the insect. Natural spread of PLRV in potato crops is mainly due to *M. persicae*, the most efficient vector though with large differences depending on aphid biotype and PLRV isolate. Other aphid species colonizing potatoes are poor vectors. However, *Aulacorthum solani* was shown to play a role in early infections in some PLRV outbreaks in Western Europe in the 1970s. *Capsella bursa-pastoris*, *Datura stramonium*, and *Physalis floridana* can act as virus reservoirs.

PLRV does not show any remarkable variability except that there have been reports on striking differences between isolates in aphid transmission rates. Differences between isolates can be established on the basis of symptom severity (stunting) in *P. floridana*. This host is also the common plant used to test the transmission efficiency of aphids.

In seed-potato producing countries, the use of efficient insecticide treatments has led to a considerable decrease of PLRV incidence. Many cultivars display high levels of field resistance. Hovever, there is no commercial cultivar carrying *R* or *N* genes though ER and HR to PLRV were shown to be present in several pools of *Solanum* wild species. Two genetically engineered cultivars resistant to PLRV were registered in the USA but do not seem to be widespread.

Potato Virus X

PVX was in the past the most common virus in potatoes. It often induces a mild mosaic and mottle, sometimes severe mosaic, rugosity, crinkling of the leaflets, and also tuber necrosis in some cultivars. Yield reduction is usually low, often 10–15% but can be over 50% in the case of infections by severe strains or of mixed infections with PVY and PVA.

PVX has been separated into four strain groups (X^1, X^2, X^3, X^4) that differ in virulence to HR genes *Nx* and *Nb*. PVX^3 is the commonest strain at least in Europe. PVX^4 overcomes both *Nx* and *Nb* but is fortunately rare. PVX_{CB} the common strain in Peru, belongs to group X^2. The strain HB discovered in Bolivia in 1978 overcomes these HR genes and also the *Rx* gene. Amino acid changes in its coat protein are responsible for its virulence. Both PVX_{HB} and PVX_{CP} belong to the serotype PVX^A which is distinct from the serotype PVX^O. Specific monoclonal antibodies and nucleic acid probes have been produced for diagnosis of isolates of each serotype. Reliable indicator cultivars for strain differentiation can be: Desiree or Pentland Crown (*nx*, *nb*) which are susceptible to all strains, King Edward (*Nx*, *nb*) resistant to X^1 and X^3, Bintje (*nx*, *Nb*) resistant to X^1 and X^2, Maris Piper or Pentland Dell (*Nx*, *Nb*) resistant to X^1, X^2, and X^3. *Gomphrena globosa*, *C. amaranticolor*, and *C. quinoa* are other PVX indicator hosts.

PVX is efficiently controlled by use of certified seeds. Moreover, numerous important European and American cultivars (Ranger Russet, Saturna, Cara) harboring Rx_{adg} or Rx_{acl} are PVX immune. Numerous cultivars also contain *Nx* and *Nb*. PVX_{HB} is not a real threat due to its lack of fitness.

Potato Virus A

PVA is widespread except in some countries where it is rare (Poland) or absent (Andes). It is frequent in susceptible cultivars such as Desiree and Russet Burbank. In the last decade it has become more prevalent in some countries. Yield losses may be negligible or amount to 40%. Infection is often symptomless even in the most susceptible cultivars. Typical symptoms are a very mild mosaic, often transient or only discernible in cloudy weather conditions. Leaflets may be shiny, rough, and slightly rippled. Crinkling occurs in mixed infections with PVX. Combination with PVY also results in a severe disease.

PVA has a limited host range. Other hosts such as tobacco and cherry tomato do not play any significant role in PVA epidemiology. *M. persicae* is the main vector. *M. euphorbiae*, some *Aphis* species, and *Brachycaudus helichrysi* were reported as potential vectors.

PVA isolates have been divided into three serogroups using monoclonal antibodies and into four strain groups based on HR or systemic response in several potato cultivars (notably King Edward). Some isolates were proved to have lost aphid transmissibility. Mutations in proteins 6K2 and VPg were shown to be responsible for particular avirulence in *Nicandra physaloïdes*.

Virus concentration in infected plants is low and purification of virions is difficult. *N. tabacum* can be used as a propagation host. Virions are weakly immunogenic and specific antisera without any cross-reactivity with PVY are difficult to prepare. Commercial ELISA kits are available, though not always specific. RT-PCR protocols were also reported for the detection in potato leaves and dormant tubers. In the past *Solanum demissum* A6 and *S. demissum* A were frequently used for differentiating PVA from PVY.

A large part of old and current cultivars is either immune or hypersensitively resistant to PVA (notably the widespread cultivars Bintje, Eersteling, Spunta, Kennebec, Charlotte, Monalisa, Katahdin, King Edward).

Potato Virus S and Potato Virus M

PVS and PVM are distantly related to each other and to other potato viruses of the genus *Carlavirus* (PVP, PRDV; recently shown to be strains of the same species). They were widespread in the past infecting all plants of cultivars such as King Edward which were later virus-freed by meristem culture. They are currently quasi-eradicated in Western Europe but are yet widespread in Eastern Europe, notably in Poland, where PVS incidence may reach 100% in some cultivars. Yield losses do not exceed 15%, except in some Polish cultivars infected by PVM. Co-infections by both viruses are frequent. ELISA is routinely used for detection in leaves and tubers. Both viruses are controlled in seed-potato schemes.

PVS

Infections are usually symptomless though leaf mottling may be observed in particular genotypes (**Figure 4**). PVS^O and PVS^A are two strain groups which induce local and systemic reactions on *Chenopodium* spp., respectively. PVS^A can cause severe reactions in potato. Full-length genomes of at least three isolates were sequenced. A broad variability was shown throughout the whole genome. The 'Central European' variant (PVS-CP) which systemically infects *Chenopodium* spp. was genetically closely related to PVS^O but only distantly related to PVS^A.

Both PVS^O and PVS^A can be naturally transmitted by contact and by several aphid species, especially *M. persicae*, *Aphis frangulae*, *A. nasturtii*, *A. fabae*, and *Rhopalosiphum padi*. Natural host range is restricted to several species in the families Solanaceae and Chenopodiaceae. Tomato is immune to PVS^O, but symptomlessly infected by PVS^A. *Nicotiana debneyi* can be used for isolation from plants doubly infected with PVM. Specific anti-PVS^A monoclonal antibodies have been produced. Detection by radioactive and nonradioactive probes as well as by PCR was reported. Many cultivars harbor the Ns_{adg} gene and are efficiently protected. Conversely, resistance in cultivar Saco is polygenic and recessive.

PVM

PVM usually induces very slight mottles and mild abaxial rolling of leaflets (**Figure 5**). Main vectors are aphid species colonizing potato crops. Some isolates that are not transmitted by aphids have been described. PVM-ID is a serologically distinct strain. Potato is the main host naturally infected. Tomato (symptomless infection) and potato cultivar Saco are good hosts for propagation and separation from PVS. The main aphid vectors are *M. persicae*, *A. nasturtii*, and *A. frangulae*. Some strains may also be transmitted by contact. Monogenic dominant resistance was found in *S. gourlayi* (*Gm* gene), *S. megistacrolobum*, and *S. stoloniferum* (HR genes).

Tobacco Rattle Virus

TRV is a widespread virus infecting numerous weeds and crops. It causes a serious disease in potato, called spraing or stem mottle, mainly in areas with light and peaty soils. Spraing or corky ringspot denotes one type of necrotic arcing in the tuber flesh (**Figure 6(a)**) and on the tuber surface (**Figure 6(b)**). Stem mottle refers to the fact that only one single shoot may be stunted with leaves showing a mosaic and sometimes transient yellow chevrons. Spraing-affected tubers often give rise to virus-free progeny plants, sometimes to infected plants with symptomless daughter tubers. Lastly, some cultivars are symptomless carriers of TRV.

TRV particles are tubular straight rods varying in length depending on the isolate. Normal particle-producing isolates (called M-type) have two genomic RNAs and are readily transmitted by nematodes and by sap inoculation. NM-type isolates have only RNA-1, do not produce particles, are probably not transmitted by nematodes and are not easily sap-transmitted. Some of the best characterized strains were originally obtained from potato: PRN (potato ring necrosis) now used as the type strain although it seems to have lost its transmissibility by nematodes; Oregon strains, which include the variant Oregon Yellow; PSG, and PLB. Monoclonal antibodies to strain PLB have been

Figure 4 Infection by PVS: mottle observed on a plantlet from imported material tested in the Potato Quarantine Station in France. Photograph: J. Martin, LNPV, France.

Figure 5 Infection by PVM: soft rolling of leaflets and pale green color on an infected plant (right) versus an healthy plant (left). Reproduced by permission of Y. Le Hingrat, FNPPPT, France.

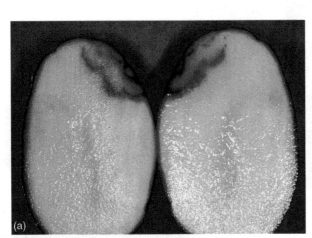
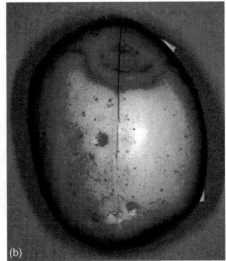

Figure 6 (a) Typical necrotic arches in the tuber flesh (spraing) induced by TRV. (b) Necrotic arches on tubers associated with spraing induced by TRV. (a, b) Photographs: J. Martin, LNPV, France.

produced. Complete genomic sequences of many potato strains are available in databases, namely for PLB and PSG.

At least 12 nematode species of the genera *Paratrichodorus* and *Trichodorus* are natural vectors, with a high specificity between virus strain and vector species. Adults and juveniles can transmit and retain virions for many months, particles being attached to the esophageal wall. Acquisition and inoculation periods can be short (1 h). TRV does not seem to multiply in its vector and to be transmitted through TPS.

Serological detection of TRV from potato is unreliable. Detection is based on the use of bait and indicator plants (*P. floridana*, *N. tabacum*, *C. quinoa*) but primarily on molecular techniques (RT-PCR and real-time PCR). However, none of these techniques is fully reliable.

Control methods are based on the use of certified seeds and of cultivars more or less resistant to spraing. Tolerant cultivars multiplying the virus without displaying any spraing cannot be recommended since they can contribute to spread of the virus. The use of nematicides for soil treatments is no longer permitted in most countries.

Potato Mop-Top Virus

PMTV can cause a serious disease in potato crops in Northern Europe, especially in cultivars such as Saturna. It is also present in the Andes, Japan, China, and Canada. PMTV induces necroses in potato tubers, both external and internal sinuous, often parallel, lines, also arcs (**Figure 7**). Blotching, skin cracking, and malformation can develop, mainly in cool conditions before harvest or during storage. Typical symptoms on leaflets after secondary infection are yellow or necrotic chevrons and blotches. Mop-top denotes bunched foliage on shortened stems. Virus does not invade all daughter tubers.

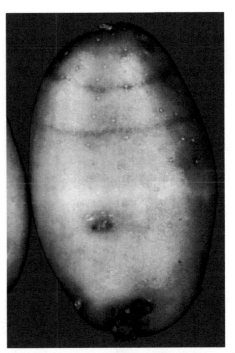

Figure 7 Necrotic sinuous lines induced by PMTV on a tuber of cv. Kerpondy. Photograph: C. Kerlan, INRA, France.

PMTV is vectored by the fungus *Spongospora subterranea* f. sp. *subterranea*, the causal agent of potato powdery also. Virus can persist in long-live resting spores of its vector. Strains differing in transmission and virulence have been reported. PMTV-T from Scotland is the type strain.

Potato is usually the only host plant. Many weed species such as *C. quinoa* may be natural hosts in Andean countries. *C. amaranticolor* and *N. debneyi* are local lesion hosts. *N. benthamiana* can be used as a bait plant and a suitable propagation host. ELISA, RT-PCR, and real-time

Figure 8 Calico symptoms induced by AMV on potato: bright yellow blotching of leaflets. Photograph: L. F. Salazar, CIP, Peru. Reproduced by permission of C. Jeffries, SASA, UK.

Figure 9 Infection by PAMV: bright yellow spots, flecks and blotches on cv. Ulster Premier. Reproduced by permission of C. Jeffries, SASA, UK.

Figure 10 Infection by TNV: necrotic lesions with crackings of the tuber skin. Reproduced by permission of C. Jeffries, SASA, UK.

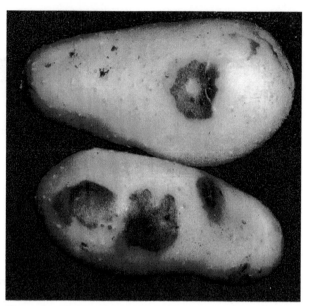

Figure 11 External necroses on tubers induced by TSWV. Photograph: J. Martin, LNPV, France.

PCR have been successfully used for detection, though results are inconsistent due to erratic distribution of the virus in infected plants. Virions are rod-shaped fragile particles similar to those of TRV. Polyclonal and monoclonal antibodies have been raised though the virus is difficult to purify. Antisera have also been produced using recombinant coat protein as antigen.

Control methods include breeding for tolerance to spraing, but no natural sources of resistance are available so far. Coat protein-derived resistance has been successfully tested in greenhouse conditions.

Viruses with Limited Significance

Many viruses of limited significance in potato may cause spectacular symptoms on potato foliage or tubers (**Figures 8–11**). Only TSWV was associated with significant losses in several countries with incidences sometimes reaching up to 90%. Its spread results from invasion of some species of thrips vectors.

See also: Carlavirus; Necrovirus; Nepovirus; Pomovirus; Potexvirus; Sobemovirus; Tobamovirus; Tobravirus; Tospovirus; Vector Transmission of Plant Viruses; Luteoviruses; Plant Virus Diseases: Economic Aspects; Potato Virus Y; Tymoviruses.

Further Reading

Brunt AA, Crabtree K, Dallwitz MJ, Gibbs AJ,, and Watson L (eds.) (1996) *Viruses of Plants. Descriptions and Lists from the VIDE Database.* Wallingford, UK: CAB International.

de Bokx JA and Van der Want JPH (eds.) (1987) *Viruses of Potato and Seed-Potato Production,* 2nd edn. Wageningen, The Netherlands: Pudoc.

Delleman J, Mulder A, Peeters JMG, Schiper E,, and Turkensteen LJ (eds.) (2005) *Potato Diseases.* Den Haag, The Netherlands: NIVAP Holland.

Gebhardt C and Valkonen J (2001) Organisation of genes controlling disease resistance in the potato genome. *Annual Review of Phytopathology* 39: 79–102.

Jeffries CJ (1998) *FAO/IPGRI Technical Guidelines for the Safe Movement of Germplasm, No. 19. Potato.* Rome: Food and Agriculture Organization of the United Nations/International Plant Genetic Resources Institute.

Loebenstein G, Berger P, Brunt AA,, and Lawson RG (eds.) (2001) *Virus and Virus-Like Diseases of Potatoes and Production of Seed-Potatoes.* Dordrecht, The Netherlands: Kluwer.

Lorenzen JH, Meacham T, Berger PH, *et al.* (2006) Whole genome characterization of *Potato virus Y* isolates collected in the western USA and their comparison to isolates from Europe and Canada. *Archives of Virology* 151: 1055–1074.

Radcliffe EB and Ragsdale EB (2002) Aphid-transmitted potato viruses: The importance of understanding vector biology. *American Journal of Potato Research* 79: 353–386.

Ratke W and Rieckman W (1991) *Maladies et ravageurs de la pomme de terre.* Gelsenkirchen-Buer, Germany: Verlag Th. Mann.

Salazar LF (1966) *Potato Viruses and Their Control.* Lima, Peru: International Potato Center.

Singh RP (1999) Development of the molecular methods for potato virus and viroid detection and prevention. *Genome* 42: 592–604.

Solomon-Blackburn RM and Barker H (2002) A review of host-major gene resistance to potato viruses X, Y, A and V in potato: Genes, genetic and mapped locations. *Heredity* 86: 8–16.

Solomon-Blackburn RM and Barker H (2002) Breeding virus resistant potatoes (*Solanum tuberosum*): A review of traditional and molecular approaches. *Heredity* 86: 17–35.

Valkonen J (1994) Natural genes and mechanisms for resistance to viruses in cultivated and wild potato species (*Solanum* spp.). *Plant Breeding* 112: 1–16.

Valkonen J (1997) Novel resistances to four potyviruses in tuber-bearing potato species, and temperature-sensitive expression of hypersensitive resistance to *Potato virus Y. Annals of Applied Biology* 130: 91–104.

Tobacco Viruses

S A Tolin, Virginia Polytechnic Institute and State University, Blacksburg, VA, USA

Glossary

Transient virus vector The use of a virus, such as tobacco mosaic virus, for the expression of a foreign gene following insertion of the sequence at a site in the viral genome.

Virus vector A biological organism, such as an aphid, leafhopper, thrips, whitefly, or fungus, that is capable of specific horizontal transmission of a virus from one plant to another.

Introduction

Tobacco has played a role in agricultural history and in the history of virology. Twenty plant virus names begin with tobacco, indicating that tobacco was a host of great agricultural and economic importance, was susceptible to a large number of viruses, and was the object of early virus research. Early in the last century, when viruses were first discovered and named, research programs emphasized controlling diseases of tobacco, many of which were caused by viruses. This article discusses the role of the tobacco in virology, viruses as causes of tobacco diseases, and some of the major milestones in virology achieved with tobacco viruses.

Tobacco is in the genus *Nicotiana* (family Solanaceae), created in 1565 and named after a French promoter of tobacco, Jean Nicot von Villemain. Although the genus is quite diverse and contains about 100 species, only two species have been extensively cultivated as commercial crops. *Nicotiana rustica* was originally grown by Native Americans in the eastern United States, and was the first tobacco species introduced to England and Portugal. *Nicotiana tabacum* was grown in Mexico and South America, and was introduced to the early Spanish explorers. With the cultivation of the species by the Virginia colony in Jamestown, *N. tabacum* became the preferred tobacco type, and is the predominant species today.

Tobacco is the most widely grown nonfood crop in the world, and is thought to have been used by people in the Americas for smoking and chewing since 1000 BC. It played a major role in the colonization of the Americas by Europeans, and was an economic driver for nearly four centuries. Although always a controversial and exotic commodity, tobacco became established in Europe before coffee, tea, sugar, and chocolate. In spite of the human health risks of tobacco products, tobacco continues to be grown throughout the world.

Three major types of the *N. tabacum* predominate, each with different characteristics depending on the final use of the tobacco product. The most common type is Virginia or flue-cured tobacco, which is used for cigarette, pipe, and chewing tobacco. It is grown worldwide, with the USA and China as major producers. Burley tobacco, the second-most popular type, is air-cured and used in chewing tobacco, and is flavored and blended for American-type cigarettes and pipe tobacco. Burley tobacco is also widely grown, but in not as many countries as is Virginia tobacco. A third type, Oriental or Turkish tobacco, is a small-leafed tobacco and is sun-cured and used in English blends. The name is derived from the eastern Mediterranean area where it is grown. Oriental cultivars Samson from Turkey and Xanthium from Greece have been widely used in virus research.

Viruses as Causal Agents of Tobacco Diseases

There are 20 viruses that were either first isolated from tobacco or contain tobacco as the host in their name (TEV, TWV) (**Table 1**). Chronologically these reports span over a century. Seven viruses naturally infect tobacco, but were first isolated from other crops and named from them (**Table 2**). The viruses are discussed individually, but some are synergistic in mixed infections and cause symptoms more severe than either virus alone. Satellites viruses and RNAs associated with tobacco viruses were the first to be recognized as subviral entities (**Table 3**). The greatest recent new activity with tobacco viral diseases has been in understanding the complexity of the viruses with similarities to tobacco leaf curl virus (family *Geminiviridae*) and their worldwide distribution.

Tobacco in Virology History

Tobacco was used as a host plant for virology research leading to many fundamental discoveries, only a few of which are highlighted here. *Nicotiana* species widely used for virus propagation and host range tests are *N. clevelandii*, *N. tabacum*, *N. glutinosa*, *N. megalosiphon*, *N. sylvestris*, and, more recently, *N. benthamiana*. These species are reported as being experimentally susceptible to 267, 231, 201, 93, 57, and 141 viruses, and insusceptible to 80, >200, 175, 35, 35, and 40 viruses, respectively. Responses of tobacco species and cultivars to mechanical inoculation are used for virus and strain identification. Cytopathic effects and inclusion bodies induced by plant viruses were first observed by light microscopy with X-bodies of TMV and nuclear and cytoplasmic inclusions of TEV and PVY. Details of plant virus replication processes were advanced by studies using tobacco leaf protoplasts. Tobacco was also the first plant to be genetically transformed by recombinant DNA techniques, and the first in which pathogen-derived resistance was demonstrated by the insertion of the coat protein gene from TMV. Tobacco transformed with noncoding viral sequences of TEV led to the discovery of RNAi and viral-induced gene silencing (VIGS).

Viruses Causing Economically Important Diseases of Tobacco

Tobamoviruses

The causal agent of a mosaic disease of tobacco, tobacco mosaic virus (TMV, genus *Tobamovirus*) became the first virus to be purified, crystallized, and characterized. Milestones in virology history that have been accomplished with TMV have been well documented, and will not be repeated here. The virus has a rigid, rod-shaped particle constructed of a single capsid protein arranged helically around a single strand of RNA encoding four proteins, functions of which are well characterized in viral replication and movement. Transient vectors can be constructed from TMV, permitting insertion and expression of foreign genes for new products, or for labeling virus for pathogenicity and cellular biology experiments.

The mosaic disease occurs worldwide wherever tobacco is grown. The stability of the virus in tobacco products, its highly infectious nature, and efficient inoculation by mechanical abrasion contributes to the difficulty tobacco farmers have had in controlling this virus. In certain cultivars of burley and ornamental tobaccos, the *N* gene for resistance has provided durable resistance for nearly 75 years, considered remarkable for an RNA virus, as resistance-breaking strains have not developed. Use of the *N* gene in Virginia or flue-cured tobacco, however, has not been successful because of poor-quality traits associated with linked genes. Efforts to resolve this using hybrids have only been moderately successful because systemic necrosis can develop in field-infected plants and cause more serious loss than that caused by mosaic.

Other tobamoviruses infecting tobacco are tomato mosaic virus (ToMV) and tobacco mild green mosaic virus (TMGMV). ToMV and TMV are seed-borne on tomato seed, but reportedly not on tobacco seed. However, production of tobacco seedlings in float trays in greenhouses suggests anecdotally that TMV-infested seed may provide inoculum source for seedling infections. Instances of spread of ToMV from tomato seedlings to tobacco seedlings have been observed. ToMV causes more severe symptoms than does TMV. TMGMV naturally infects *N. glauca*, and is often associated with an icosahedral satellite virus, satTMV (**Table 3**).

Bromoviruses

Certain viruses of tobacco are the type members of three genera of the family *Bromoviridae*, namely cucumber mosaic virus (CMV, genus *Cucumovirus*), alfalfa mosaic virus (AMV, genus *Alfamovirus*), and tobacco streak virus (TSV, genus *Ilarvirus*). These viruses continue to occur in tobacco worldwide. Disease outbreaks are sporadic, but are of economic significance in fields infected at a high inci-

Table 1 Chronology of discovery and naming of tobacco viruses

Virus name	Virus acronym	Genus/family	Year described	Crop of origin	Country	Investigator	Vector taxon
Tobacco mosaic virus	TMV	Tobamovirus	1892, 1896 1914	Tobacco Tobacco	USSR USA	Iwanowski Allard	None
Tobacco etch virus	TEV	Potyvirus/Potyviridae	1921	Datura spp.	USA	Blakeslee	Aphid
Tobacco ringspot virus	TRSV	Nepovirus / Comoviridae	1927	Tobacco	USA (VA)	Fromme	Nematode
Tobacco mild green mosaic virus	TMGMV	Tobamovirus	1929	N. glauca	Canary Islands	McKinney	None
Tobacco rattle virus	TRV	Tobravirus	1931	Tobacco	Germany	Boning	Nematode
Tobacco leaf curl virus	TLCV	Begomovirus / Geminiviridae	1931	Tobacco	Tanzania	Storey	Whitefly
Tobacco necrosis virus	TNV	Necrovirus Tombusviridae	1935	Tobacco	UK	Smith & Bald	Fungus
Tobacco streak virus	TSV	Ilarvirus /Bromoviridae	1936	Tobacco	USA (WI)	Johnson	Thrips
Tobacco yellow dwarf virus	TYDV	Mastrevirus/ Geminiviridae	1937	Tobacco	Australia	Hill	Leafhopper
Tobacco mottle virus + Tobacco vein distorting virus	TMoV + TVDV	Umbravirus + Luteovirus	1946	Tobacco	Zimbabwe	Smith	Aphid
Tobacco stunt virus	TStV	Varicosavirus	1950	Tobacco	Japan	Hidaka	Olpidium brassicae
Tobacco wilt virus	TWV	Potyvirus/Potyviridae	1959	Solanum jasminoides	India		Aphid
Tobacco bushy top virus	TBTV	Umbravirus	1962	Tobacco	Rhodesia		Aphid
Tobacco yellow vein virus + Tobacco yellow vein assistor virus	TYVV + TYVAV	Umbravirus + Luteoviridae	1972	Tobacco	Malawi	Adams & Hull	Aphid
Tobacco vein mottling virus	TVMV	Potyvirus/Potyviridae	1972	Tobacco	USA (NC)	Gooding	Aphid
Tobacco necrotic dwarf virus	TNDV	Luteoviridae	1977	Tobacco	Japan	Kubo	Aphid
Tobacco apical stunt virus	TASV	Begomovirus/ Geminiviridae	1999	Tobacco	Mexico	Brown	Whitefly
Tobacco curly shoot virus	TCSV	Begomovirus/ Geminiviridae	2002	Tobacco	China (Yunnan)	Zhou	Whitefly

Table 2　Additional viruses reported to naturally occur in tobacco

Virus name	Virus acronym	Genus/family	Year described	Crop of origin	Country	Investigator	Vector taxon
Cucumber mosaic virus	CMV	Cucumovirus/ Bromoviridae	1916	Cucumber	USA	Doolittle Jagger	Aphid
Tomato mosaic virus	ToMV	Tobamovirus	1919	Tomato	USA	Clinton	None
Tomato spotted wilt virus	TSWV	Tospovirus/ Bunyaviridae	1930	Tomato	India	Samuel	Thrips
Alfalfa mosaic virus	AMV	Alfamovirus/ Bromoviridae	1931	Alfalfa	USA	Weime	Aphid
Potato virus Y	PVY	Potyvirus/Potyviridae	1931	Potato	USA	Smith	Aphid
Eggplant mottled dwarf virus	EMDV	Nucleorhabdovirus/ Rhabdoviridae	1969	Eggplant	Italy	Martelli	Cicadellid leafhopper
Eggplant mosaic virus – tobacco strain	EMV-T	Tymovirus/ Tymoviridae	1996	Tobacco	Brazil	Ribiero	Chrosomelid beetle

Table 3　Satellite viruses and nucleic acids associated with tobacco viruses

Satellite name	Helper virus	Year described	Crop of origin	Investigator – country	Characteristics
Tobacco necrosis satellite virus	TNV	1960	Tobacco	Kassanis – UK	18 nm, $T = 1$
Tobacco mosaic satellite virus	TMV	1986	N. glauca	Dodds – US	17 nm, $T = 1$; 1059 nt
Tobacco bushy top virus satellite RNA	TBTV	2001	Tobacco	China	Linear, 650–800 nt
Tobacco ringspot virus satellite RNA	TRSV	1972	Bean	Schneider	Circular, 359 nt, viroid-like
Cucumber mosaic virus satellite RNA	CMV	1976	Tobacco	Kaper – Italy	Linear, 330–390 nt
Tobacco leaf curl virus DNA β	TLCV	2003	Tobacco	Zhou – China	Circular, 1300–1350 nt
Tobacco curly shoot virus DNA β	TCSV	2005	Tobacco	Li – China	Circular

dence. All three viruses have very wide host ranges, are known to infect a number of weeds and native plants naturally, and are seed transmitted in many of their host plants. CMV and AMV are transmitted in a stylet-borne, nonpersistent manner by many species of aphids. TSV, however, is spread by contact and through pollen carried by thrips.

Common properties include a tripartite, single-stranded RNA (ssRNA) genome, and encapsidation of a fourth subgenomic RNA encoding the capsid protein. Particles have icosahedral symmetry, but AMV forms multiple bacilliform particles and TSV forms quasi-icosahedrons. CMV capsids are of $T = 3$ symmetry.

Cucumber mosaic and alfalfa mosaic viruses

Symptoms caused by CMV are similar to those of TMV. CMV is an important tobacco pathogen in Asia and in certain European countries. In recent surveys in Greece, burley, Virginia, and oriental types all had a high incidence of CMV, even up to 100% infected plants in some fields. Symptoms of AMV on tobacco are a bright yellowing of leaf areas, broad rings, and mosaic on young leaves. AMV incidence was lower, but was highest in tobacco growing near alfalfa fields, a source of the virus.

Many fundamental discoveries of genome structure, organization, and encapsidation occurred with AMV and CMV in the 1960s and 1970s using tobacco as the main test species. Sucrose density-gradient centrifugation enabled separation of the multiple particles of AMV, leading to experiments to demonstrate the requirement of RNA1, RNA2, and RNA3 plus either RNA4 or free coat protein for infectivity. Genetic experiments were conducted by mixed inoculations with RNA species from biologically distinct strains. Similar experiments were conducted with CMV, which required separation on cesium chloride gradients based on density to demonstrate separate encapsidation of RNA1, RNA2, and RNA3 + RNA4 in a third capsid. Infectivity required RNAs 1, 2, and 3, but not RNA4 or coat protein. A fifth, small RNA species, initially termed RNA5 or CARNA5 (CMV-associated RNA), was later shown to be the first satellite RNA species. These satellite RNAs can either ameliorate or exacerbate symptoms in crops. Severe lethal necrosis symptoms are associated with CMV satellite RNA in tomato in Italy and tobacco in India.

Tobacco streak virus

TSV is rarely reported on tobacco, but is an important pathogen of many annual and vegetatively propagated crops and wild species worldwide. In tobacco, TSV causes necrotic lines along veins of leaves as virus moves systemically from the point of inoculation. Symptoms develop on a few leaves, and then plants recover. Incidence of infected tobacco plants is usually low. Thrips are recognized as a vector of TSV. TSV belongs to subgroup 1 of the genus *Ilarvirus*, which is divided into six subgroups based on serological relatedness.

Potyviruses

Three viruses of the genus *Potyvirus* (family *Potyviridae*), tobacco etch virus (TEV), potato virus Y (PVY), and tobacco vein mottling virus (TVMV), are among the most important viral pathogens of tobacco. All are aphid transmitted, with flexuous filamentous particles *c.* 13×750 nm containing an ssRNA genome of about 10 kb. Successful aphid transmission depends upon a DAG sequence in the N-terminus of the coat protein and a viral sequence encoding a helper component (HC). The genome is expressed as a polyprotein, which is cleaved by viral-encoded proteases (Pro) into a single structural and several nonstructural proteins. Both TEV and PVY occur nearly worldwide and have wide host ranges. Diseases caused by potyviruses are managed in part through resistant cultivars such as TN86, a burley tobacco resistant to TVMV, TEV, and PVY. Several burley varieties have partial or high resistance to one or more viruses, as well as resistance to fungi and nematodes.

Tobacco etch virus

TEV has a rather narrow host range, but is found in solanaceous weeds including *Datura* sp., from which it was first isolated in the 1920s. The virus is common only in North and South America. Isolates vary in symptom severity and ability to cause necrosis or severe versus mild etch in tobacco. TEV is also known to cause severe symptoms in pepper and tomato crops in the USA, Central America, and the Caribbean. The few isolates that have been sequenced show very little diversity, and are indistinguishable serologically.

TEV has also been a model virus in the elucidation of many host–virus interactions and viral genome expression. The $5'$ leader of TEV genomic RNA is used in many constructs for genetic engineering, as it contains an internal ribosome entry site and directs efficient translation of uncapped mRNA. TEV was the first virus demonstrated to encode proteases, including the autocatalytic small nuclear inclusion proteinase and a protease linked to helper component (HC-Pro).

Transgenic resistance to TEV was the first to be due to RNA interference and gene silencing, as nontranslatable sequences of the capsid protein-encoding region of the viral genome were more effective than the capsid protein translatable sequences. This observation led to the discovery RNA interference as well as the silencing suppression function by the HC-Pro region of the genome.

Potato virus Y

PVY, first isolated from potato in the 1930s as one of a complex of viruses associated with the century-old potato degeneration, is one of the most studied plant viruses and is the type member of the genus *Potyvirus*. PVY remains an important pathogen of potato and also of tobacco. Severe epidemics of PVY in tobacco have been reported recently in China. In Greece, PVY is one of the most prevalent viruses isolated from seedbeds and fields. Symptoms associated with PVY are vein clearing and/or top necrosis in oriental tobacco, and yellowing with veinal necrosis in Virginia and burley tobacco. PVY has a wide host range among the Solanaceae, including 70 species of *Nicotiana* and is also found in annual and perennial weeds in 15 or more plant families. A number of strains have been described both on potato and on tobacco, distinguishable by inoculation to differential hosts and by sequence analysis.

Other potyviruses

TVMV was not described until the 1970s, when it emerged as a new virus in the burley tobacco-growing regions of the USA, namely Kentucky, North Carolina, Tennessee, and Virginia. Severe losses were recorded until resistance to this virus, and partial resistance to PVY and TEV, was incorporated into burley tobacco cultivars. A virus isolated from tobacco in Nigeria was identified as being closely related to the potyvirus pepper veinal mottle virus, and both reacted with TVMV antibody in electron microscope serology, but not enzyme-linked immunosorbent assay (ELISA). However, this reaction alone is not sufficient to prove virus identity. Only one additional potyvirus, tobacco wilt virus (TWV), has been reported from tobacco, and it is only known in India.

Tospoviruses

Tomato spotted wilt virus (TSWV), a member of the genus *Tospovirus* (family *Bunyaviridae*), is a serious pathogen on tobacco in many regions of the world. This thrips-transmitted virus has a very wide host range infecting over 800 species in 90 or more plant families, both monocots and dicots, and causes significant yield losses in a number of economically important crops. The virus has spherical, membrane-bound particles, 80–120 nm, with a three-segment, ambisense genome encapsided in the same particle. Symptoms in tobacco infected early are severe leaf necrosis and stunting. Several management approaches have involved weed reservoir and vector management. In Georgia in the southeastern US, successful management has been accomplished with compounds to induce resistance combined with imidochloprid insecticide applications in seedling trays and at transplanting. Yield losses were reduced significantly.

Geminiviruses (Family *Geminiviridae*)

Mastreviruses

Tobacco yellow dwarf virus (TYDV) was described in Australia in 1937, and later determined to be one of only three dicot-infecting members of the genus *Mastrevirus* (family *Geminiviridae*). It is yet to be recorded outside of Australia where it causes severe dwarfing and downward leaf curl in tobacco, and a lethal, necrotic disease in bean (syn. bean summer death virus). The virus is transmitted persistently by the brown leafhopper *Orosius argentatus*, but not by other leafhoppers nor mechanically. TYDV also infects several summer annual and autumn–spring growing plants, providing a continuous succession of host plants for the virus, and a source of virus for tobacco. Like other mastreviruses, TYDV has a single-stranded DNA genome, single component, and has coding regions in both virion sense and complementary sense sequences.

Begomoviruses

The tobacco leaf curl disease was first described in Tanzania in the early 1930s, in which infected tobacco plants were severely crinkled, curled, and dwarfed. The disease was successfully transmitted by the whitefly *Bemisia tabaci*, but not mechanically, suggesting that a virus was the causal agent. Similar disease symptoms in tobacco were noted as early as 1902 in South Africa, in 1912 in the Netherlands East Indies, in the 1930s in India, and in Brazil and Venezuela prior to 1950. Positive association of the tobacco leaf curl disease with a begomovirus was not demonstrated until DNA analyses and polymerase chain reaction (PCR) became available. The name tobacco leaf curl virus (TLCV, genus *Begomovirus*, family *Geminiviridae*) is now accepted for this virus.

However, new viruses are being described worldwide in association with severely diseased tobacco. There are numerous recent reports of TLCV-like viruses from Asia, including Japan, China, and India, as well as from Dominican Republic and Cuba. Those viruses are now accepted as members of the genus *Begomovirus*: tobacco leaf curl Japan virus (TbLCJV), tobacco leaf curl Kochi virus (TbLCKoV), tobacco leaf curl Yunnan virus (TbLCYNV), tobacco leaf curl Zimbabwe virus (TbLCZV), and tobacco curly shoot virus (=tobacco leaf curl – China, TbCSV). Sequences have also been reported for tobacco leaf curl virus – Karnataka 1 and 2 from India. Tentative begomovirus members include tobacco apical stunt virus, tobacco leaf curl India virus, and tobacco leaf rugose virus – Cuba.

All TLCV-like viruses have a single-stranded, monopartite DNA genome, and have only the DNA-A component. TbLCYNV was the first virus to be shown to have a DNAβ component, a type of satellite DNA widely associated with the virus in the Yunnan Province of China and with TbCSV (**Table 3**). Like other satellites, DNA β molecules have no sequences in common with their helper virus, DNA-A, which replicates independently of DNA β. With TbCSV, DNA β affects pathogenicity, with increasing symptom severity in the presence of specific molecules.

Viruses Causing Diseases of Tobacco of Lesser or Minor Importance

Nepoviruses

Tobacco ringspot virus (TRSV) caused a severe disease of tobacco in Virginia in the 1920s. Symptoms initially were severe necrotic rings with green centers, and a pattern of necrotic lines described as an 'oak-leaf' pattern. Tobacco plants expressed symptoms on three to four leaves developing following inoculation, but not on later-developing leaves. This apparent recovery was the first phenomenon to be described as acquired immunity. In recent research, recovery from symptoms induced by TRSV was the first to be attributed to plant defenses and to involve RNAi. The ssRNA genome is bipartite and encapsidated in separate icosahedral particles, typical of members of the family *Comoviridae*. Additional small RNAs found in the 1970s to also be encapsidated in TRSV particles were the first satellite RNAs to be discovered (**Table 3**) and later shown to have hammerhead, ribozyme-like structures capable of self-cleavage. These structures are also found in certain viroids.

TRSV is the type species of the genus *Nepovirus*. It is classified in subgroup A of this genus based on RNA2 size and sequence, and on serological relationships. The genus is named from the property of being nematode transmitted (by *Xiphenema* spp.) and having polyhedral particles. The virus is spread naturally in North America by these nematodes as well as by possible arthropod vectors. It has a very wide host range and is found naturally in many weeds and woody species. Additionally, TRSV has been disseminated worldwide in vegetative propagules of perennial crops and ornamentals, but does not appear to spread in the absence of the nematode vector.

Artichoke yellow ringspot virus (AYRSV) is known to infect tobacco naturally, causing chlorotic blotches, rings, and lines, and to spread naturally in Greece, Italy, and

Eurasia. It is reported to spread via pollen to seeds or seed parent plants of *N. clevelandii*, *N. glutinosa*, and *N. tabacum*. Serological evidence indicates that AYRSV is a distinct species, and RNA2 properties place it in the subgroup C of nepoviruses, which includes tomato ringspot and cherry leaf roll viruses.

Tobraviruses

Tobacco rattle virus (TRV, genus *Tobravirus*), first described on tobacco in the 1930s, is now seldom reported on tobacco crops. Symptoms of rattle on tobacco are systemic necrotic flecks and line patterns, and death or stunting of shoots. It was known as *Mauche, ratel, Streifen und Kräuselkrankheit* in Germany, its geographic origin. Distribution today is limited to Europe, Japan, New Zealand, and North America. The host range of TRV is very wide, as it infects over 400 species of both dicots and monocots, many of which are weeds and other wild plants. TRV causes important diseases of potato, pepper, and various ornamentals.

Biological vectors of TRV are nematodes in the genera *Paratrichodorus* and *Trichodorus* (Trichodoridae). *Nicotiana clevelandii* is recommended as a propagative host for TRV, and as a trap plant for testing nematode transmission characteristics. Transmissibility is mediated by read-through proteins encoded on RNA-2 of the bipartite, ssRNA genome which extend from the surface of rigid rod-shaped particles of various sizes. This region of RNA-2 has been exploited for expression of introduced genes, and the virus is now widely used as a virus-induced gene silencing (VIGS) vector for genetic experiments.

Necroviruses

Tobacco necrosis virus (TNV) is no longer associated with diseases of tobacco, but rather with bean stipple streak and tulip necrosis diseases. TNV is classified in the genus *Necrovirus* (family *Tombusviridae*). It is one of the few viruses transmitted by the fungus *Olpidium brassicae* and often occurs in irrigated soils and in greenhouses. Virus particles adsorb to the surface of resting spores, on which they do not survive for long periods. Symptoms on tobacco are localized necrosis, with very little systemic movement within infected plants. Electron microscopy of purified TNV revealed 26–28 nm isometric particles of $T = 3$ symmetry. Some cultures also had 17 nm particles, which were a satellite virus (satTNV) encoding a capsid protein that assembled into a $T = 1$ particle. satTNV is found in Europe and North America in association with bean, but not with tobacco.

Varicosaviruses

Tobacco stunt virus (TStV) is reported only from Japan, where it was found in the 1950s to cause severe stunting and necrosis in tobacco. This virus was among the first to be shown to be transmitted by a fungus, *O. brassicae* (order Chytridiales), in which it is borne internally in resting spores and persists for 20 years or more. Once soil is infested with spores containing virus, tobacco planted into the soil will become infected and show symptoms within 2 weeks. The RNA genome of TStV is encapsided as two segments in straight, rod-shaped, rather labile, particles. The genome is now classified as being negative sense and single-stranded, with genome organization like that of rhabdoviruses. Taxonomically, TStV is in the genus *Varicosavirus*, whose name comes from the enlarged (varicose) veins induced in lettuce by the virus now known as lettuce big-vein associated virus. Recent molecular analyses suggest that TStV is a strain of this virus, rather than a distinct species.

Luteoviruses and Umbraviruses

Diseases caused by luteoviruses alone

Tobacco necrotic dwarf virus (TNDV) of the genus *Enamovirus* (family *Luteoviridae*) causes severe disease symptoms in some tobacco fields in Japan. It is transmitted by the aphid *Myzus persicae* in a persistent manner, but not mechanically. Early infection causes stunting and premature yellowing or death of young leaves of tobacco, but distribution is apparently limited.

Tobacco rosette and other diseases with dependent transmission

An aphid-transmitted disease known as tobacco rosette was widespread in Zimbabwe in sub-Saharan Africa in the 1930s. Smith in 1946 discovered that the sap-transmissible tobacco mottle virus (TMoV) could only be transmitted by aphids if plants were co-infected with tobacco vein distorting virus (TVDV), making tobacco rosette the first disease complex associated with dependent transmission. TVDV, a member of the genus *Enamovirus*, is the assistor virus for TMoV, a presumed member of the genus *Umbravirus* based on sap transmissibility and other biological properties. Sequence data of one of the two recognized strains of TMoV confirmed this assignment in 2001.

Two other diseases of tobacco have complex etiology presumed to involve an umbravirus–luteovirus complex. Tobacco yellow vein virus (TYVV), a tentative species of the genus *Umbravirus*, and tobacco yellow vein assistor virus, a tentative member of the family *Luteoviridae*, were described as causing a tobacco disease with chlorotic vein-banding and leaf malformation symptoms in Malawi. Tobacco bushy top disease occurring in Rhodesia since the early 1960s is known to be caused by TVDV and a different umbravirus, tobacco bushy top virus (TBTV), and is characterized by pale green, mottling, and extreme shoot proliferation. TBTV and TVDV have recently been detected in China in tobacco showing severe witches' broom symptoms, the etiology of which had been presumed to be phytoplasma. Sequence data from these viruses confirm their classification. The significance of this observation is that this is the first incidence of an umbravirus–luteovirus complex reported in China, and first outside of sub-Saharan Africa.

Tymoviruses

A virus was isolated from naturally infected tobacco in Brazil in 1996, and described as eggplant mosaic virus-tobacco strain (EMV-T), the first report of a virus of the genus *Tymovirus* (family *Tymoviridae*) from tobacco. The virus caused mosaic symptoms, but did not reduce plant productivity. It was experimentally transmitted by sap and by the chrysomelid beetle, *Diabrotica speciosa*. As this is the only report of its incidence in tobacco, EMV-T is of only minor importance in tobacco in a limited geographic region. EMV is a pathogen of eggplant and other solanaceous vegetables, and was limited in distribution to Trinidad and Tobago. It is closely related to Andean potato latent virus, but is considered a distinct species. Genomic relationships between EMV and EMV-T have not been reported.

Nucleorhabdoviruses

Eggplant mottled dwarf virus (EMDV, genus *Nucleorhabdovirus*, family *Rhabdoviridae*) was found in tobacco plants in Greece. Plants showed severe stunting, leaf deformation, and vein clearing, and were not infected by other common tobacco viruses. Incidence was at less than 0.01%, mainly in plants near the edges of the field suggesting a weed reservoir for the virus, which has a cicadellid leafhopper vector, *Agallia vorobjevi*. EMDV has been known in the Mediterranean area since 1969 and was reported from tobacco in Italy in 1996 as tobacco vein yellowing virus, now a synonym of EMDV. EMDV infects other solanaceous crops, and appears to be spreading to a wider geographical range from Spain to the Middle East.

See also: Alfalfa Mosaic Virus; Cucumber Mosaic Virus; Emerging and Reemerging Virus Diseases of Plants; Ilarvirus; Plant Resistance to Viruses: Engineered Resistance; Plant Virus Diseases: Economic Aspects; Potato Virus Y; Tobacco Mosaic Virus; Vector Transmission of Plant Viruses; Virus Induced Gene Silencing (VIGS).

Further Reading

Adams MJ and Antoniw JF (2005) DPVweb: An open access Internet resource on plant viruses and virus diseases. *Outlooks on Pest Management* 16: 268–270.

Adams MJ, Antoniw JF, and Fauquet CM (2005) Molecular criteria for genus and species discrimination within the family. *Potyviridae Archives for Virology* 150: 459–479.

Brunt AA, Crabtree K, Dallwitz MJ, Gibbs AJ, Watson L, and Zurcher EJ (eds.) (1996 onward) Plant viruses online: Descriptions and lists from the VIDE Database. Version: 20 Aug. 1996. http://image.fs.uidaho.edu/vide/ (accessed May 2007).

Chatzivassiliou IK, Efthimiou K, Drossos E, Papadopoulou A, Poimenidis G, and Katis NI (2004) A survey of tobacco viruses in tobacco crops and native flora in Greece. *European Journal of Plant Pathology* 110: 1011–1023.

Cui XF, Xie Y, and Zhou XP (2004) Molecular characterization of DNAβ molecules associated with tobacco leaf curl Yunnan virus. *Journal of Phytopathology* 152: 647–650.

Harrison BD and Wilson TMA (1999) Milestones in the research on tobacco mosaic virus. *Philosophical Transactions of the Royal Society of London, Section B* 345: 521–529.

Lindbo JA and Dougherty WG (2005) Plant pathology and RNAi: A brief history. *Annual Review of Phytopathology* 43: 191–204.

Mansoor S, Briddon RW, Zafir Y, and Stanley J (2003) Geminivirus disease complexes: An emerging threat. *Trends in Plant Science* 8: 128–134.

Paximadis M, Idris AM, Torres-Jerez I, Villareal A, Rey MEC, and Brown JK (1999) Characterization of tobacco geminiviruses in the Old and New World. *Archives of Virology* 144: 703–717.

Ratcliff F, Martin-Hernandez AM, and Baulcombe DC (2001) Tobacco rattle virus as a vector for analysis of gene function by silencing. *Plant Journal* 25: 237–245.

Sasaya T, Ishikawa K, Kuwata S, and Koganezawa H (2005) Molecular analysis of coat protein coding region of tobacco stunt virus shows that it is a strain of *Lettuce big-vein virus* in the genus Varicosavirus. *Archives of Virolology* 150: 1013–1021.

Shew HD and Lucas GB (eds.) (1991) *Compendium of Tobacco Diseases*, 96pp. St. Paul, MN: APS Press.

Simon AE, Roossinck MJ, and Havelda Z (2004) Plant virus satellite and defective interfering RNAs: New paradigms for a new century. *Annual Review of Phytopathology* 42: 415–437.

Syller J (2003) Molecular and biological features of umbravirus, the unusual plant viruses lacking genetic information for a capsid protein. *Physiological and Molecular Plant Pathology* 63: 35–46.

Xie Y, Jiang T, and Zhou X (2006) Agroinoculation shows tobacco leaf curl Yunnan virus is a monopartitie begomovirus. *European Journal of Plant Pathology* 115: 369–375.

Relevant Website

http://www.dpvweb.net – DPVWeb Home Page.

Vegetable Viruses

P Caciagli, Istituto di Virologia Vegetale – CNR, Turin, Italy

Glossary

Incubation The time period from the initial infection (inoculation) to the manifestation of the disease (symptoms). For virus diseases of vegetables, incubation can vary between 2 and 40 days, most commonly between 1 and 3 weeks.

Meristem A tissue in plants consisting of undifferentiated cells, found in zones of the plant where growth (cell division) can take place (meristematic dome). The term 'meristem tip' is used to denote the meristem dome together with 1–2 primordial leaves and measuring between 0.1 and 0.5 cm in height.

Roguing The practice of removing diseased (or abnormal) plants from a crop.

Vector An organism capable of transmitting a pathogen from one plant host to another by establishing specific relationships with the pathogen itself.

Vegetables

The plant species/subspecies cultivated as vegetables are close to 200 and belong to more than 30 botanical families, representing, with fruits, about 35% of the total food derived from plants (including cereals, vegetable oils, and sugar crops).

Viruses

According to different estimates, 3–5% of overall vegetable production is lost to virus infections, but losses can be occasionally very high, and are normally higher where pest control is lower, for example, in developing countries.

Keeping in mind that viruses of potato, sweet potato, cassava, and legumes are not considered in this article, as they are dealt with elsewhere in this encyclopedia, a total of about 200 viruses (excluding cryptic viruses, satellites, and viroids) have been reported as naturally infecting 67 of the 129 vegetables considered here. These viruses belong to 30 genera in 15 families and to nine genera unassigned to families; eight viruses are in three families, but have not been assigned to genera, and three more viruses have not been assigned to any taxa (**Table 1**).

As tomatoes, watermelons, cabbages, onions, cucumbers, and eggplants make up almost 50% of the total vegetable production (legumes, potatoes, sweet potatoes, and cassava excluded), it is no wonder that more than 90 viruses have been found in these crops, with about 50 viruses in tomatoes, the most cultivated vegetable of all (about 13% of the total) (**Table 2**).

Ecology and Epidemiology

Factors normally considered in epidemiological studies are hosts, pathogens, and environment. To these, we can add vectors and human action.

Host

Among possible reactions of a plant to virus attack, only susceptibility and tolerance (plant systemically infected with very mild or no visible effects) favor the spread of viruses. It is evident, from both the number of viruses isolated and the number of vegetables found naturally infected, that most vegetables escape virus infections, for a number of different reasons, including resistance, possibly as a consequence of centuries of human selection for good-looking products. In fact, only 38 vegetables are infected, under natural conditions, by more than one virus, and only 25 by more than two viruses.

Pathogen

The main virus features that affect epidemiology are variability, host range, and transmission pathways. The higher the variability of a virus, the higher is its chance to adapt to new hosts and environments. Most vegetable viruses (as most plant viruses) have a single-stranded RNA (ssRNA) genome, the genome most prone to errors during replication, as it lacks the proofreading mechanisms that are present in DNA replication. Although it has some drawbacks, this feature ensures a high level of variability to viruses with a single genome component. Viruses with a divided genome can also count on natural reassortment of multiple-component genomes to increase their variability. This can strongly influence the host range of the virus. Out of 76 viruses infecting more than one vegetable crop, only 12 naturally infect more than five crops, with eight of them having an ssRNA divided genome (viz., alfalfa mosaic virus, arabis mosaic virus, cucumber mosaic virus, lettuce infectious yellows virus,

Table 1   Taxonomic assignment of vegetable viruses, their numbers per genus, and their vectors

Genome type	Family	Genus	No. of virus species	Vectors
Single-stranded DNA	*Geminiviridae*	*Begomovirus*	35	Whiteflies
		Curtovirus	2	Leafhoppers
		Mastrevirus	1	Leafhoppers
	Nanoviridae	*Babuvirus*	1	Aphids
Double-stranded DNA RT	*Caulimoviridae*	*Badnavirus*	4	Mealybugs
		Caulimovirus	1	Aphids
Double-stranded RNA	*Reoviridae*	*Fijivirus*	1	Planthoppers
Single-stranded, negative-sense RNA	*Rhabdoviridae*	*Cytorhabdovirus*	2	Aphids
		Nucleorhabdovirus	2	Aphids
			1	Leafhoppers
		Unassigned in the family	2	Aphids
			1	Leafhoppers
			2	Not known
	Bunyaviridae	*Tospovirus*	3	Thrips
	Unassigned	*Ophiovirus*	2	Fungi
		Varicosavirus	1	Fungi
Single-stranded, positive-sense RNA	*Sequiviridae*	*Sequivirus*	2	Aphids
	Comoviridae	*Comovirus*	3	Beetles
		Fabavirus	1	Aphids
		Nepovirus	11	Nematodes
	Potyviridae	*Ipomovirus*	1	Whiteflies
		Potyvirus	40	Aphids
		Unassigned in the family	1	Aphids
	Luteoviridae	*Polerovirus*	3	Aphids
		Unassigned	2	Aphids
	Tombusviridae	*Aureusvirus*	1	Fungi
		Carmovirus	3	Fungi
		Necrovirus	1	Fungi
		Tombusvirus	5	Fungi
	Bromoviridae	*Alfamovirus*	1	Aphids
		Cucumovirus	2	Aphids
		Ilarvirus	5	(Thrips)
	Tymoviridae	*Tymovirus*	4	Beetles
	Closteroviridae	*Closterovirus*	3	Aphids
		Crinivirus	5	Whiteflies
	Flexiviridae	*Allexivirus*	6	Mites
		Carlavirus	8	Aphids
		Potexvirus	9	None known
	Unassigned	*Benyvirus*	1	Plasmodiophorids
		Ourmiavirus	2	None known
		Pomovirus	1	Plasmodiophorids
		Sobemovirus	1	None known
		Tobamovirus	7	None known
		Tobravirus	2	Nematodes
		Umbravirus	3	Aphids
	Unassigned family	Unassigned genus	1	Plasmodiophorids
			2	Not known

radish mosaic virus, squash mosaic virus, tobacco rattle virus, and tomato black ring virus); six of these have vegetable hosts in more than one botanic family. Viruses of vegetables with a divided genome are only 30% of the total. The analysis of the host-range done for this work does not include all cultivated plants and, furthermore, the host range of viruses is not usually restricted to cultivated plants, but nevertheless the number of crops naturally infected is a good indicator of the host-range width. The host range is also strictly interconnected to the

ways viruses are transmitted. Viruses, even those infecting centenary trees, need to be able to pass from one susceptible host to another. The means by which viruses move from host to host can be summarized as (1) vectors (flagellate protists: plasmodiophorids, fungi, nematodes, and arthropods); (2) seed and pollen; (3) vegetative propagation; and (4) mechanical transmission.

1. Vectors of the viruses infecting vegetables are shown in **Table 1** (column 5). Transmission by vectors involves

Table 2 The top vegetables in the world (excluding potato, sweet potato, cassava, and legumes), their production, and the number of viruses found in each crop in natural infection

Vegetable	Family	Production[a]	Viruses[b]
Tomato	Solanaceae	126.1	53
Watermelon	Cucurbitaceae	97.1	14
Cabbages	Brassicaceae	66.7	2
Onion	Alliaceae	62.1	4
Yam	Dioscoreaceae	46.7	3
Cucumber	Cucurbitaceae	41.6	20
Plantain banana	Musaceae	33.0	2
Eggplant	Solanaceae	30.4	8
Pepper	Solanaceae	27.2	28
Carrot (and turnip)	Apiaceae	23.8	13
Lettuce and chicory	Asteraceae	21.9	21
Squash, pumpkin	Cucurbitaceae	20.4	3
Cauliflower and broccoli	Brassicaceae	17.7	3
Garlic	Alliaceae	14.2	9
Spinach	Chenopodiaceae	12.8	10
Taro	Araceae	11.3	3
Asparagus	Asparagaceae	6.6	6
Okra	Malvaceae	5.3	2
Leek	Alliaceae	1.8	3
Artichoke	Asteraceae	1.4	14
Total		881	193[c]

[a]Data, expressed in millions of tons, refer to year 2004; from data made available by the Food and Agriculture Organization, United Nations.
[b]Number of viruses detected from natural infections; the same virus may infect more than one vegetable.
[c]Total of different virus species detected in vegetables.

all epidemiological factors: the virus, the vector, the host plant, the environment, in both the biotic and abiotic aspects, and man-action, in the form of movement of vectors, viruses, and host plants around the world.

2. Out of the 134 viruses that infect vegetables and have been studied for seed transmission, only 33 are seed transmitted in some plants. However, only 15 (listed in **Table 3**) are seed-transmitted in vegetables. Nevertheless, seed transmission is of paramount importance, as it can start virus infection in a newly planted crop. All viruses known to be seed-transmitted in vegetables have single-stranded, positive-sense RNA genome.

3. Viruses that are vegetatively propagated are listed in **Table 4** together with the number of viruses reported from natural infection in each crop. With the exception of artichoke, the number of virus detected is linked to the distribution of crops in terms of both the total production and cultivation areas outside the center of origin.

4. Mechanical transmission is particularly important for viruses (for instance, tobamo- and potexviruses) that reach high concentration in their vegetable hosts, particularly in the epidermis and hairs, and maintain infectivity

for a long time in the environment. The risk is particularly high for crops frequently manipulated during the growing season, as man and his tools can act as carriers. Mechanical transmission due to repeated cuts for harvesting are known for zucchini yellow mosaic virus in zucchini squash, and for radish mosaic virus in rocket (eruca).

The case of tobacco mosaic virus (TMV) is worth a special mention, as an example of a 'seed-mechanical' transmission: TMV can contaminate the tegument of a tomato seed and from here infect the seedling during its growth. A special case is also the pollen-mediated transmission of tobacco streak virus and pelargonium zonate spot virus. These viruses can be mechanically inoculated by thrips that rub contaminated pollen on the leaf surface while moving on a host plant.

Environment

The environment has an influence on virus spread through both biotic and abiotic factors. Among biotic factors, the most important is wild vegetation that can act as reservoir for viruses, and fauna that may include potential vectors. Among abiotic factors, soil has an influence on vectors of soil-borne viruses. Light, sandy soils favor the spread of nematode-borne viruses, while heavy, water-retaining soils favor the spread of plasmodiophorid- and fungal-borne viruses. Rain and humidity, wind, and, most of all, temperature has a strong influence on virus epidemiology, because of their effects on arthropod vectors. High humidity and moderate temperatures favor multiplication of aphids, while strong winds and rains impair their long-distance movements. Higher temperatures are more suited for whiteflies and leafhoppers. So, it is not by chance that whitefly-borne viruses are more common in areas with subtropical climate, and that the global warming is accompanied by an increase of their importance and diffusion areas.

Control Measures

Control of virus and virus diseases in vegetables, as in most crops, is indirect and based upon a number of strategies for preventing virus infection. Seeds should be as virus-free as possible, using certified seeds whenever possible, disinfecting seeds of crops susceptible to tobamoviruses by a number of simple, but effective chemical treatments, when certified seed is not available. Multiplication material of vegetatively propagated species (**Table 4**) need to be selected and virus-free vegetative stock prepared and maintained. For freeing plant material from viruses one can use (1) heating of dormant plant parts (thermotherapy) and growing small shoot tips removed immediately after heat treatment; (2) meristem tip culture, particularly efficient for eliminating viruses

Table 3 Viruses transmitted through vegetable seeds

Family	Virus (genus)	Vegetable	Species	Family
Comoviridae	Squash mosaic virus (Comovirus)	Watermelon	Citrullus lanatus (C. vulgaris)	Cucurbitaceae
		Squash, pumpkin	Cucurbita maxima	Cucurbitaceae
	Chicory yellow mottle virus (Nepovirus)	Parsley	Petroselinum crispum subsp. tuberosum	Apiaceae
		Chicory	Cichorium intybus	Asteraceae
	Beet ringspotvirus (Nepovirus)	Mangolds (beet roots, fodder beet)	Beta vulgaris	Chenopodiaceae
		Swiss chard	Beta vulgaris var. cicla	Chenopodiaceae
	Tobacco ringspot virus (Nepovirus)	Lettuce	Lactuca sativa	Asteraceae
	Tomato black ring virus (Nepovirus)	Celery	Apium graveolens	Apiaceae
		Lettuce	Lactuca sativa	Asteraceae
		Tomato	Solanum lycopersicon	Solanaceae
Potyviridae	Celery latent virus (Potyvirus?)	Celery	Apium graveolens	Apiaceae
	Lettuce mosaic virus (Potyvirus)	Lettuce	Lactuca sativa	Asteraceae
Tombusviridae	Tomato bushy stunt virus (Tombusvirus)	Pepper	Capsicum annuum	Solanaceae
		Tomato	Solanum lycopersicon	Solanaceae
	Cucumber leaf spot virus (Aureusvirus)	Cucumber	Cucumis sativus	Cucurbitaceae
Bromoviridae	Alfalfa mosaicvirus (Alfamovirus)	Pepper	Capsicum annuum	Solanaceae
	Spinach latent virus (Ilarvirus)	Spinach	Spinacia oleracea	Chenopodiaceae
Unassigned to family	Cucumber green mottle mosaic virus (Tobamovirus)	Squash, pumpkin	Cucurbita maxima	Cucurbitaceae
		Watermelon	Citrullus lanatus (C. vulgaris)	Cucurbitaceae
	Pepper mild mottle virus[a] (Tobamovirus)	Pepper	Capsicum annuum	Solanaceae
	Tobacco mosaic virus[a] (Tobamovirus)	Pepper	Capsicum annuum	Solanaceae
		Tomato	Solanum lycopersicon	Solanaceae
	Tomato mosaic virus[a] (Tobamovirus)	Tomato	Solanum lycopersicon	Solanaceae

[a]Transmission of tobamoviruses is not a true seed transmission as a result of embryo infection, but a 'seed-mechanical' transmission; TMV can contaminate the tegument of a seed and from here infect the seedling during its growth.

from plants of the family Araceae, for example, may be preceded by or associated with heat treatment (very effective for garlic, for example); (3) chemical treatment with antiviral substances, like ribavirin, either alone or with DHT; usually disappointing, it may work in association with the other two methods. Seedlings should be protected before and soon after transplanting as plants are usually very susceptible at this stage. After transplanting, it is important to prevent virus spread from outside and within the crop. Contamination from the soil can be avoided by removing debris from previous crop (particularly for tobamo-, tombus-, and carmoviruses) or by crop rotation, alternating crops susceptible to different viruses; crop rotation may help also in controlling vectors of limited mobility (nematodes, fungi, plasmodiophorids). Crops can be protected against airborne vectors by physical means, either excluding the vectors (nets and covers of various types, wind-breakers) or repelling them (mulching with repellent effect, e.g., UV-reflective silver plastic), or by chemical treatments. Biologic control of vectors is not very efficient, particularly against introduction of viruses into the crop, as efficient vectors can be occasional visitors of the crop rather than regular colonizers. Control of weeds and volunteer plants from the previous crop is also very important in reducing early infections. When a vegetable is grown on a wide territory

crop after crop, a 'crop-free period' can be very efficient in preventing virus infections (e.g., carrot virus Y in carrots in southern Australia). As for avoiding the introduction of viruses in areas (islands, countries, continents) where they are not present yet, most agriculturally advanced countries set quarantine rules aimed at excluding specific viruses (or their vectors) or limiting their further spread; setting and applying quarantine measures can be quite complex, but it may be worth particularly for viruses transmitted through seed or in dormant vegetative parts. In order to control the spread of viruses within the crops, one needs to reduce the movement of viruses. For vector-borne viruses, a number of chemical treatments are available for controlling the different vectors. Hygiene of operator and implements (by washing with a solution of 3% trisodium orthophosphate, for example) is particularly important for viruses readily mechanically transmissible (tobamo-, potex-, carmoviruses, but also comoviruses) in crops that need frequent tending. Roguing can be effective only if virus incidence is still low (e.g., less than 10%), the incubation is short, and the movement of the virus from plant to plant, by whatever means, is slow. When roguing plants hosting vectors, care should be taken not to disperse them while removing plants. A further possibility for protecting a vegetable crop from virus damage is artificial infection of the plants with a strain of the virus causing only mild disease

Table 4 Vegetatively propagated vegetables from which viruses have been reported, and the number of viruses reported in natural infections

Species	Common name	Family	No. of viruses reported
Cynara scolimus	Artichoke	Asteraceae	14
Allium sativum	Garlic	Alliaceae	9
Asparagus officinalis	Asparagus	Asparagaceae	6
Allium ascalonicum	Shallot	Alliaceae	5
Colocasia esculenta or C. antiquorum	Taro	Araceae	4
Ullucus tuberosus	Olluco	Basellaceae	4
Canna edulis	Achira	Cannaceae	3
Dioscorea spp.	Yam	Dioscoreaceae	3
Arracacia xanthorrhiza	Arracacha	Apiaceae	2
Musa spp.	Plantain banana	Musaceae	2
Tropaeolum tuberosum	Mashua	Tropaeolaceae	2
Rorippa (Nasturtium) officinale or R. officinale × R. microphyllum	Water cress	Brassicaceae	1
Bambuseae (tribe)	Bamboo shoot	Poaceae	1

symptoms (cross-protection) or associating a satellite to a virus to the same aim (e.g., cross-protection is used for protecting cucurbits from damages caused by zucchini yellow mosaic virus and tomatoes from tomato mosaic virus; co-infection with a satellite can be used to reduce damages by cucumber mosaic virus in pepper and tomato). Both methods have a number of objections, based on ecological and practical considerations, so they are not recommended as general practice, but still can reduce virus damages at least for some time in some crops. Another method of choice to avoid virus infection of vegetables is resistance, to vectors in few instances (aphids, whiteflies, thrips) and to viruses in many cases. For analysis of resistance to viruses, either natural or engineered, readers are referred to specific articles of this encyclopedia. **Table 5** lists some of the crops, and their viruses, for which natural resistance is commercially available. Transgenic lines of vegetables have been developed that resist to very damaging viruses, but they are not listed here because, for a number of reasons, transgenic seeds are not available to the majority of growers yet.

A Few Special Cases

A few viruses infecting vegetables and the symptoms they cause in their hosts are hereafter described, chosen as examples among the many vegetable virus diseases.

Table 5 Vegetables and viruses for which natural resistance of different degrees has been incorporated in commercially available cultivars

Species	Common names	No. of viruses reported	Genetic (natural) resistance available to
Lactuca sativa	Lettuce	20	Lettuce mosaic virus (Potyvirus)
Brassica oleracea var. capitata	Cabbage	2	Turnip mosaic virus (Potyvirus)
Spinacia oleracea	Spinach	10	Cucumber mosaic virus (Cucumovirus)
Cucumis sativus	Cucumber	20	Cucumber mosaic virus (Cucumovirus), papaya ringspot virus (Potyvirus), watermelon mosaic virus (Potyvirus), zucchini yellow mosaic virus (Potyvirus)
Cucurbita maxima	Squash, pumpkin	3	Cucumber mosaic virus (Cucumovirus), papaya ringspot virus (Potyvirus), watermelon mosaic virus (Potyvirus), zucchini yellow mosaic virus (Potyvirus)
Cucurbita pepo	Zucchini, marrows, pumpkin, squash	14	Cucumber mosaic virus (Cucumovirus), papaya ringspot virus (Potyvirus), watermelon mosaic virus (Potyvirus), zucchini yellow mosaic virus (Potyvirus)
Capsicum annuum	Pepper	27	Cucumber mosaic virus (Cucumovirus), pepper mottle virus (Potyvirus), potato virus Y (Potyvirus), tobacco etch virus (Potyvirus), tobacco mild green mosaic virus (Tobamovirus), tobacco mosaic virus (Tobamovirus), tomato spotted wilt virus (Tospovirus)
Solanum lycopersicon	Tomato	53	Tobacco mosaic virus (Tobamovirus), tomato mosaic virus (Tobamovirus), tomato spotted wilt virus (Tospovirus)
Solanum melongena	Eggplant (aubergine)	8	Cucumber mosaic virus (Cucumovirus), tomato mosaic virus (Tobamovirus)

For details on the structure and properties of the viruses mentioned, the reader is referred to the specific articles of this encyclopedia.

Cucumber Mosaic Virus

Cucumber mosaic virus (CMV) is a positive-sense ssRNA virus with a tripartite genome encapsidated in isometric particles measuring about 30 nm in diameter. CMV has a very wide host range (more than 800 species in 70 botanic families). Natural infections have been reported from at least 10 vegetable species and in a large number of weeds, the most important virus reservoir. In many of these weeds CMV is not associated to any symptoms and is efficiently seed-transmitted. The virus is transmitted by several aphid species in a nonpersistent manner: infectivity is retained by vectors for a few hours and is lost with molt. Symptoms of CMV in vegetables like tomato plants infected at an early stage are chlorotic local lesions between the secondary leaf vein. Necrosis, first in the form of brown leaf areas, and later as brown lines along petioles and stems, proceeds toward the bottom of the plant, which can die in a few weeks after infection. Fruits, if present, are misshaped, sun-burnt, and necrotic. Plants infected at a mature stage normally show growth reduction, bushy appearance, and leaf deformation (shoestring-like leaves). Fruits have reduced size and do not ripen. Depending on a number of factors, including the presence of satellite RNAs, plants may have normal-looking leaves but show necrosis of fruits (losses are quite reduced in this case) or show typical, but stronger symptoms (in this case, losses can be very high). Symptoms of CMV infection in 'pepper' plants depend on the virus strain and on the plant age at the moment of infection. In young pepper plants CMV causes leaf yellowing and narrowing, and necrotic symptoms, on both the foliage and the fruit. If older plants are infected, CMV symptoms can be mild and appear on lower, yellowing leaves as green ring spots or oak-leaf patterns, but also as a mild mosaic, with a general dull appearance. Cucurbits infected by CMV, particularly zucchini, squash, and melons, show strong mosaic symptoms and leaf narrowing, and distortion (**Figure 1**). Fruits are pitted and misshapen, unmarketable. To worsen things, mixed infections with other viruses are common. Celery and parsley infected by CMV show leaf yellowing and necrosis. 'Lettuce' infected by CMV shows intense mosaic, vein chlorosis and, frequently, vein browning and necrosis when temperature drops below 13 °C. CMV infection of spinach causes a variety of symptoms including stunting, yellowing, and mottling of the older leaves and malformation of the younger leaves.

Garlic Viruses

Because of exclusive vegetative propagation, by cloves or, in some cultivars, by inflorescence bulbils, garlic is

Figure 1 Symptoms of cucumber mosaic virus infection on zucchini leaves. Courtesy of V. Lisa.

particularly prone to accumulate complex virus infections. The most known virus-associated syndrome in garlic is mosaic. Symptoms include various degrees of leaf discolorations, most pronounced in the younger leaves, chlorotic mottling, striping, and streaking. Plants are also stunted compared to healthy looking plants under the same conditions. Most probably, two potyviruses are responsible for mosaic in garlic: one is a garlic strain of onion yellow dwarf virus (OYDV-G), the other a specific strain of leek yellow stripe virus (LYSV-G). Garlic clones, first freed from OYDV and LYSV and then mechanically inoculated with either virus, have been shown to suffer yield reduction from 17% to 60% depending on the virus–cultivar combinations. Yield losses are worse in case of mixed infections or chronic infections (vegetatively propagated material originally infected). Two aphid-borne carlaviruses, garlic common latent virus (GarCLV) and shallot latent virus (SLV), also known as garlic latent virus (GarLV), have also been identified as parts of the garlic virus complex. A number of allexiviruses (genus *Allexivirus*; family *Flexiviridae*) have been identified in recent years in garlic. Although none of them has been associated with any particular symptom, some of them cause yield effects, particularly in mixed infections with OYDV and LYSV. Garlic dwarf virus (GDV) (family *Reoviridae*; genus *Fijivirus*), originally isolated in France, causes well-defined symptoms of dwarfing and leaf thickening.

Tobamoviruses

Tobamoviruses are possibly the most-studied viruses of plants. Tobamoviruses have rod-shaped particles 300×18 nm, each containing a single molecule of positive-sense, ssRNA. Particles are very stable. No biological vector is known, but because of high stability and high concentration in plant tissues, tobamoviruses are readily transmitted mechanically during crop tending. They are also seed-transmitted but, as mentioned before, in a manner very different from true seed transmission. Tobacco mosaic virus (TMV) (from which the name of the genus is derived) and tomato mosaic virus (ToMV) are the most known, but other viruses, pepper mild mottle virus (PMMoV) and tobacco mild green mosaic virus (TMGMV), infect vegetables in the family Solanaceae and can be differentiated by biological and serological tests. TMV, ToMV, and PMMoV commonly infect peppers, causing chlorotic mosaic, leaf distortion, sometimes systemic necrosis, and defoliation, depending on the usual factors: plant cultivar and age, virus strain, light intensity, and temperature. Fruits are disfigured, with discolored or necrotic areas. PMMoV usually causes milder symptoms on leaves but is more severe on fruits. TMGMV has been found occasionally in pepper. Tomato plants are mostly infected by ToMV, but occasionally also by TMV. The most characteristic symptoms are mottled areas of light and dark green on the leaves. Plants infected at an early stage of growth are yellowish and stunted. Leaves may also be malformed, narrowed, although not as much as with CMV, or showing enations (outgrowths) on the lower leaf lamina. High temperature can mask leaf symptoms. Fruits can be from almost normal to misshapen and be reduced in size and number, showing uneven ripening, corky or necrotic rings, internal browning. TMV and ToMV can infect eggplant occasionally, causing mild symptoms on both leaves and fruits.

Tomato Spotted Wilt Virus

Tomato spotted wilt virus (TSWV) is the type member of the genus *Tospovirus* in the family *Bunyaviridae*. Virus particles are roughly isometric, 80–90 nm in diameter, enveloped in a lipoprotein membrane, contain three single-stranded linear RNAs, one of negative and two of ambisense polarity, associated with a nucleoprotein to form the nucleocapsid. TSWV is transmitted by thrips in a circulative, propagative manner. It can infect hundreds of species in 70 botanical families. The first symptoms of TSWV infection in tomato are chlorotic spots, 3–4 mm in diameter, on apical leaves, rapidly turning to bronze and dark-brown (necrosis). If infected when young, the plant will eventually die. If infected later, dark brown streaks also appear on stems and leaf petioles; growing tips are greatly stunted, and usually severely affected with systemic necrosis. Fruits, mostly reduced in size, will display characteristic symptoms – immature fruit have mottled, light green rings with raised centers that develop to orange and red discoloration patterns on mature fruits (**Figure 2**). TSWV-infected pepper shows symptoms similar to those described for tomatoes. Plants infected at a very early stage are usually severely stunted and yield no fruit. Plants infected later show chlorotic or necrotic rings on leaves and stems. Some cultivars react

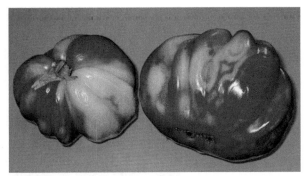

Figure 2 Symptoms of tomato spotted wilt virus infection on tomato fruits. Courtesy of the late P. Roggero.

with flower and leaf drop. Fruits develop necrotic or discolored spots or rings. Lettuce and endive infected when young turn yellow, collapse, and die. Older plants develop marginal wilting, yellowing, and necrosis of the leaves. Typical is the bending of midribs as a consequence of symptoms appearing only on one half of the leaf. The plant may also look twisted. Endive appears to be more susceptible than lettuce. Chicory infected by TSWV has a stunted, bushy, yellowish look, with chlorotic necrotic spots along the midribs and secondary veins. Eggplants and artichokes can also be naturally infected by TSWV, but only occasionally.

Turnip Mosaic Virus

Turnip mosaic virus (TuMV) is a typical member of the genus *Potyvirus* in the family *Potyviridae*. It has nonenveloped, filamentous virions, usually flexuous, with a modal length of 720 nm, containing a monopartite, positive-sense ssRNA genome. It is transmitted by several species of aphids in a nonpersistent manner. It has a relatively wide host range; naturally, infections occur mainly in the family Brassicaceae and in few weeds outside this family. All *Brassica* species are susceptible to TuMV. The virus also infects lettuce, watercress, radish, and rocket, and it is present all over the world. Young leaves of infected cauliflower plants show chlorotic ringspots followed, as the leaf ages, by yellow or brownish spots surrounded by circular or irregular necrotic rings. Generally, these occur in the vicinity of the leaf veins. Leaf blight appears in infected sections when many lesions coalesce. Outer leaves of infected cabbages may show ringspot pattern or appear more uniformly necrotic, but, as the leaves age, a chlorotic color almost replaces the normal green tissue. In white cabbage, the occurrence of internal necrotic lesions, sunken and coalescent, has been correlated with the presence of field symptoms of turnip mosaic virus. This postharvest disorder may result in the loss of large quantities of stored cabbage. Symptom expression is temperature sensitive, with most pronounced symptoms at temperatures ranging from 22 to

Figure 3 Leaves of a tomato plant infected by tomato yellow leaf curl virus (right) compared to leaves of a healthy plant (left).

30 °C, and is very much masked below 17 °C. Other *Brassica* species, like turnip, broccoli, and Chinese cabbage, react to TuMV infection with different degrees of mosaic, necrosis, leaf distortion, and reduced growth.

Tomato Yellow Leaf Curl Virus and Related Begomoviruses

Tomato yellow leaf curl virus (TYLCV) is a member of the genus *Begomovirus* in the family *Geminiviridae*. It has geminate particles measuring 20 × 30 nm and encapsidating a monopartite genome of circular, ssDNA 2.8 kbp in size. The virus is transmitted by the whitefly *Bemisia tabaci* in a circulative manner, but it is not mechanically transmissible. Seed transmission has not been reported. TYLCV has a relatively narrow natural host range. Infected tomato plants are stunted, with branches and petioles tending to assume an erect position. Leaflets of the infected plants are smaller than those of healthy ones and upward curled with margins more or less yellow (**Figure 3**). Flowers have a normal appearance, but fruit production is strongly reduced. TYLCV was first described in Israel and it is now known to be present in the Mediterranean basin, in Egypt and Sudan, in Iran, and in the Caribbean basin. A number of monopartite begomoviruses, transmitted by *B. tabaci*, inducing very similar symptoms on tomato plants, with some differences in the host range and significant differences in their sequences, have been described in recent years and classified as distinct species of the genus *Begomovirus* (e.g., *Tomato yellow leaf curl Sardinia virus*). Similar symptoms on tomato plants are induced by other begomoviruses, named in the same way, but are actually bipartite viruses (e.g., tomato yellow leaf curl China, -Iran, -Thailand virus); some of these viruses may be mechanically transmissible with difficulties. A few other begomoviruses, named tomato leaf curl someplace virus, where someplace is the name of the area (state, country) of origin, may induce some yellowing of tomato leaves, despite the name. Molecular diagnostics are therefore the only tools for identification of begomoviruses inducing yellow leaf curl symptoms in tomato plants.

Figure 4 Symptoms of zucchini yellow mosaic virus infection on zucchini fruits. Courtesy of the late P. Roggero.

Zucchini Yellow Mosaic Virus

Zucchini yellow mosaic virus (ZYMV) is a typical member of the genus *Potyvirus* (family *Potyviridae*) (see TuMV). It is transmitted by several species of aphids in a nonpersistent manner and its natural host range is essentially limited to members of the family Cucurbitaceae. ZYMV is a 'recent' virus – it was first detected in 1973 in southern Europe, the first epidemics were reported in 1979–80, and in 1996 it was already present practically in all the areas where cucurbits are grown. We still lack an acceptable explanation for the rapid worldwide diffusion of the virus. Early symptoms in infected zucchini appear as vein clearing of fine leaf veins, followed by a general yellowing of the leaf with dark green areas. Later on, the mosaic becomes stronger, and vein banding appears. Leaves can be deformed to shoestring appearance, but the most affected are fruits, that are severely malformed and often develop longitudinal cracks (**Figure 4**). Similarly, squash and pumpkin plants develop knobby areas on the fruits resulting in prominent deformations. In watermelons, as in melons, ZYMV induces yellow mosaic, severe malformations of leaves and fruits, that often display radial or longitudinal cracks. On cucumber, symptoms of ZYMV infection are less severe than on the other cucurbit crops; mosaic of variable intensity, followed by dark green vein banding and discoloration of fruits.

See also: Cucumber Mosaic Virus; Diagnostic Techniques: Plant Viruses; Flexiviruses; Plant Resistance to Viruses: Engineered Resistance; Plant Resistance to Viruses: Geminiviruses; Plant Virus Diseases: Economic Aspects; Tobamovirus; Tospovirus; Vector Transmission of Plant Viruses.

Further Reading

Conti M, Gallitelli D, Lisa V, *et al.* (1996) *I principali Virus delle Piante Ortive.* Milan: Bayer.

Jones JB, Jones JP, Stall RE, and Zitter TA (eds.) (1991) *Compendium of Tomato Diseases.* St. Paul, MN: The American Phytopathological Society.

Pernezny K, Roberts PD, Murphy JF, and Goldberg NP (eds.) (2003) *Compendium of Pepper Diseases.* St. Paul, MN: The American Phytopathological Society.

Ryder EJ (1999) *Lettuce, Endive and Chicory.* Cambridge: CABI Publishing.

Thresh JM (ed.) (2006) *Advances in Virus Research 67: Plant Virus Epidemiology.* San Diego, CA: Academic Press.

Zitter AT, Hopkins DL, and Thomas CE (eds.) (1996) *Compendium of Cucurbit Diseases.* St. Paul, MN: The American Phytopathological Society.

Relevant Website

http://vegetablemdonline.ppath.cornell.edu – Cornell Plant Pathology Vegetable Disease Web Page.

FUNGAL VIRUSES

Ascoviruses

B A Federici, University of California, Riverside, CA, USA
Y Bigot, University of Tours, Tours, France

Glossary

Apoptosis Genetically programmed cell death.
Apoptotic bodies Cell vesicles resulting from apoptosis.
Caspase Protease that activates a major portion of programmed cell death.
Endoparasitic wasps Species of insect parasites belonging to the order Hymenoptera, which lay their eggs in insects where the wasp larvae develop.
Per os infection Infection by feeding.
Programmed cell death Genetically programmed cascade proteases and nucleases that cleave DNA and proteins within a cell leading to its death.
Reniform Shaped like a kidney.
Transovarial transmission Transmission of virus inside the egg.
Virion-containing vesicles Vesicles containing virions formed by ascoviruses by rescue of apoptotic bodies induced by ascovirus infection.

Introduction

The family *Ascoviridae* is one of the newest families of viruses, established in 2000 to accommodate several species of a newly recognized type of DNA virus that attacks larvae of insects of the order Lepidoptera. Viruses of this family produce large, enveloped virions, measuring 130 nm in diameter by 300–400 nm in length, and when viewed by electron microscopy have a reticulated appearance. They are typically bacilliform or reniform in shape, and contain a circular double-stranded DNA genome that, depending on the species, ranges from ~120 to 185 kbp. Whereas the virions of ascoviruses are structurally complex like those of other large DNA viruses that attack insects, such as those of iridoviruses (family *Iridoviridae*) and entomopoxviruses (family *Poxviridae*), they differ from these in two significant aspects. First, ascoviruses are transmitted from diseased to healthy lepidopteran larvae or pupae by female endoparastic wasps when these lay eggs in their hosts. Second, ascoviruses have a unique cell biology and cytopathology in which shortly after infecting a cell, they induce apoptosis and then rescue the developing apoptotic bodies and convert these into virion-containing vesicles. This aspect of viral reproduction apparently evolved to disseminate virions to the larval blood where they could contaminate the ovipositors of female wasps so that the virus could be transmitted to new hosts. Ascoviruses appear to occur worldwide, wherever there are endoparasitic wasps and larvae of species belonging to the lepidopteran family Noctuidae. However, as these viruses have been discovered relatively recently and their signs of disease are not commonly known in the scientific community, relatively few ascovirus species have been described.

History

The first ascoviruses were discovered during the late 1970s in southern California where they were found causing disease in larvae of moths belonging to the lepidopteran family Noctuidae. Diseased larvae were recognized by the presence of blood that was very white and opaque, in marked contrast to the blood of healthy larvae which is translucent and slightly green (**Figure 1**). The color and opacity of the blood in diseased larvae was shown to be due to the presence of high concentrations of vesicles that contained virions (**Figure 2**). The white blood and virion-containing vesicles are diagnostic for the disease, and the name for this group, ascoviruses (derived from the Greek *asco* meaning 'sac'), was chosen on the basis of the latter characteristic. Since the discovery of the first ascovirus, ascoviruses have been isolated as the cause of disease in many species of noctuid larvae. In addition, an ascovirus that attacks the pupal stage of a species belonging to the family Yponomeutidae was discovered in the 1990s in France.

Distribution and Taxonomy

With respect to distribution, ascoviruses have been reported from the United States, Europe, Australia, and Indonesia, and it is highly probable that they occur worldwide. This is because their most common hosts, larvae of lepidopteran species belonging to the family Noctuidae, the largest family of the order Lepidoptera, as well as their most common vectors, endoparasitic wasps of the families Braconidae and Ichnuemonidae, are distributed throughout the world. Although only a few ascovirus species have been described to date, there are probably many, including variants, that occur worldwide. Thus, given the common occurrence of their hosts and vectors, it is possible that

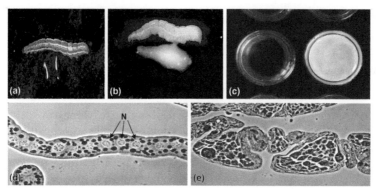

Figure 1   Major characteristics of the disease typically caused by ascoviruses in lepidopteran larvae. (a) and (b) Healthy and ascovirus-infected larvae, respectively, of the cabbage looper, *Trichoplusia ni*, infected with TnAV. Note the opaque white blood in the infected larva. (c) Spot plate containing blood from healthy (left) and infected larvae (right). (d, e) Sections through lobes of fat body from a healthy and infected larva, respectively, of the fall armyworm, *Spodoptera frugiperda*, infected with SfAV ascovirus. Note the greatly hypertrophied cells in the fat body of the infected larva. The cells in most of this tissue have already cleaved into viral vesicles. N, nuclei.

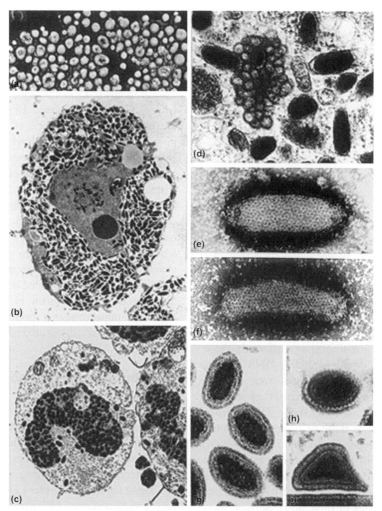

Figure 2   Structural and morphological characteristics of ascovirus virions and virion-containing vesicles. (a) Wet mount preparation viewed with phase microscopy of blood from a *Spodoptera frugiperda* larva infected with SfAV. The spherical refractile bodies are virion-containing vesicles. (b, c) Electron micrographs of ultrathin sections through viral vesicles produced by the trichoplusia (TnAV) and spodoptera (SfAV) ascoviruses, respectively. (d) Matrix of the occlusion body produced by the spodoptera (SfAV) ascovirus. The occlusion consists of virions, protein, and small spherical vesicles. (e, f) Negatively stained virions of the spodoptera (SfAV) and trichoplusia (TnAV) ascoviruses, respectively. Note the reticulate appearance of the virions. (g, h) Electron micrographs of ultrathin cross sections through inner particles of the spodoptera (SfAV) ascovirus after formation (g) and during envelopment (h). (i) Ultrathin cross section through a fully developed virion of the trichoplusia (TnAV) ascovirus.

Table 1 Members in the genus *Ascovirus* belonging to the family *Ascoviridae*

Species name	Virus abbreviation	Accession number
Recognized species		
Spodoptera frugiperda ascovirus 1a	SfAV-1a	[AM3988432]
Trichoplusia ni ascovirus 2a	TnAV-2a	[AJ312707]
Heliothis virescens ascovirus 3a	HvAV-3a	
Diadromus pulchellus ascovirus 4a	DpAV-4a	[AJ279812]
Tentative species		
Spodoptera exigua ascovirus 5a	SeAV-5a	
Spodoptera exigua ascovirus 6a	SeAV-6a	
Helicoverpa armigera ascovirus 7a	HaAV-7a	
Helicoverpa punctigera ascovirus 8a	HpAV-8a	

ascoviruses are very common insect viruses. That they have not been discovered in large numbers, for example, like baculoviruses (family *Baculoviridae*), is probably because they cause a chronic disease with few easily detectable signs, making it difficult for individuals not familiar with the disease to recognize diseased larvae in field populations.

At present, five species of ascoviruses are officially recognized based on a combination of properties including the relatedness of key genes coding for the DNA polymerase and major capsid protein, the degree to which their genomic DNA cross-hybridizes under conditions of low stringency, their lepidopteran host range, and their tissue tropism (**Table 1**). The type species is the *Spodoptera frugiperda ascovirus* (SfAV-1a), with the other species being *Tricoplusia ni ascovirus* (TnAV-2a), *Heliothis virescens ascovirus* (HvAV-3a), and *Diadromus pulchellus ascovirus* (DpAV-4a). The Arabic numeral reflects the order in which each species was formally recognized, whereas the lower case letter indicates the type species of the variants. Variants from the type species are recognized by different consecutive lower case numbers; for example, TnAV-2b and 2c would represent two different variants of TnAV-2a recognized subsequently. Herein, these viruses are referred to by their acronyms without the numerical and lower case suffix.

Ascoviruses have been isolated from many more insect species than those listed above, but these isolates have turned out to be variants of known ascoviruses, and therefore they have not been named after the host from which they have been isolated. For example, ascoviruses related to TnAV and HvAV have been isolated from noctuid species such as *Autographa precationis*, *Helicoverpa zea*, *Helicoverpa armigera*, and *Helicoverpa punctigera*; however, they do not bear the name of their host of isolation. What this implies is that ascoviruses belonging to the TnAV and HvAV species have a broad and overlapping host range among different noctuid species, although this has only been tested experimentally to a limited extent.

Virion Structure and Composition

Depending on the species, the virions of ascoviruses are either bacilliform or reniform in shape, with complex symmetry, and very large, measuring about 130 nm in diameter by 300–400 nm in length. The virion consists of an inner particle surrounded by an outer envelope (**Figure 2**). The inner particle is complex containing a DNA/protein core as well as an apparent internal lipid bilayer surrounded by a distinctive layer of protein subunits. Thus, the virion appears to contain two lipid membranes: one associated with the inner particle and the other forming the lipid component of the envelope. In negatively stained preparations, virions have a distinctive reticulate appearance, which is thought to be due to superimposition of subunits on the surface of the internal particle with those in the envelope.

As indicated by the size and complexity of the virions, the genome of ascoviruses is large, and consists of a single molecule of double-stranded circular DNA. The genomes of three species have been sequenced, the type species SfAV, TnAV, and HvAV. The SfAV genome is 157 kbp and codes for at least 120 proteins (**Figure 3**), whereas the TnAV 2a genome is slightly larger, 174 kbp, and codes for at least 134 proteins. The genome of HvAV is 186 kbp and codes for approximately 180 potential proteins. Based on gel analyses, ascovirus virions contain at least 12 structural polypeptides ranging in size from 12 to 200 kDa. In addition to proteins and the DNA genome, the presence of an envelope as detected by electron microscopy, as well as experiments with detergents and organic solvents, indicate that virions contain a substantial lipid component. And, as in other enveloped viruses of eukaryotes, it is likely that the virion also contains carbohydrate in the form of glycoproteins, though none have been identified.

Transmission and Ecology

One of the most interesting features of ascoviruses is that their transmission from host to host appears to be dependent on their being vectored by female endoparasitic wasps belonging to the families Braconidae and Ichneumonidae (order Hymenoptera). Ascoviruses are extremely difficult to transmit *per os*, with typical infection rates averaging less than 15% even when larvae are fed as many as 10^5 virion-containing vesicles in a single dose. In

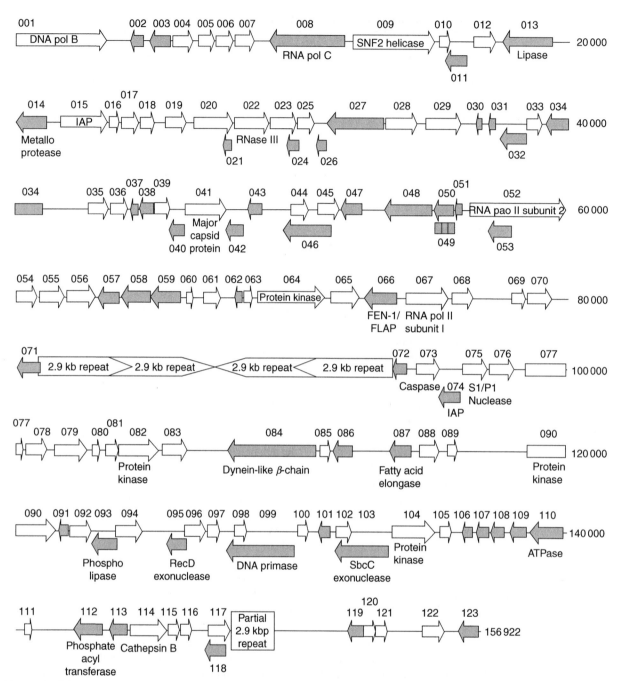

Figure 3 Schematic illustration in linear form of the circular DNA 156 922 base pair genome of the SfAV 1a ascovirus, the ascovirus type species, isolated from a larva of the fall armyworm, *Spodoptera frugiperda*. The illustration identifies the relative positions of key genes including those coding for the DNA polymerase, major capsid protein, executioner caspase, as well as several genes coding for proteins involved in lipid metabolism. White and dark arrows represent, respectively, ORFs in forward and reverse orientations along the genome. From Bideshi DB, Demattei MV, Rouleux-Bonnin F, *et al*. (2006) Genomic sequence of spodoptera frugiperda ascovirus 1a, an enveloped, double-stranded DNA insect virus that manipulates apoptosis for viral reproduction. *Journal of Virology* 80: 11791.

contrast to this, infection rates for caterpillars injected with as few as 10 virion-containing vesicles are typically greater than 90%. Moreover, experiments with parasitic wasps show that they can effectively transmit ascoviruses to their noctuid hosts. For example, when females are allowed to lay eggs in ascovirus-infected noctuid cater-

pillars, thereby contaminating their ovipositor, and then allowed to lay eggs in healthy larvae, the majority of the latter contract ascovirus disease. Interestingly, though the parasite eggs hatch in their infected noctuid hosts, the parasite larvae die as the ascovirus disease develops in the caterpillar. Under field conditions, the prevalence of ascovirus disease in caterpillars is correlated with high rates of parasitization by endoparasitic wasps. When wasps from these populations are collected in the field and allowed to oviposit in healthy caterpillars reared in the laboratory, the latter often exhibit ascovirus disease within a few days. Thus, laboratory and field studies provide sound evidence that the primary mechanism for the transmission of ascoviruses attacking noctuid larvae is through being vectored mechanically by parasitic wasps. No evidence has been found in the lepidopteran hosts for transovum or transovarial transmission.

In the case of DpAV, the association of the virus with its wasp and caterpillar hosts is much more intimate. DpAV DNA is carried in wasp nuclei as a circular molecule, and small numbers of virions are produced in the oviducts of females. However, the virus does not cause noticeable pathology in the wasp host. The females lay eggs in the pupal stage of the lepidopteran host, *Acrolepiopsis assectella*, introducing small numbers of ascovirus virions along with the wasp eggs. These virions invade lepidopteran host cells, replicate, and initiate destruction of major host tissues. The wasp larva then emerges from the egg and feeds on the host tissues and ascovirus virions. The DpAV genome is carried by both male and female wasps, where it is apparently transmitted from generation to generation transovarially. These observations make ascoviruses the only known group of viruses pathogenic to insects primarily dependent on vectors for their transmission.

Now that the characteristics of the disease are known, field studies in the southeastern United States and California are beginning to show that ascoviruses are probably the most common type of virus to occur during most of the year in populations of several important noctuid pests, including the cabbage looper, *T. ni.*, fall armyworm, *S. frugiperda*, and the corn earworm, *H. zea*. Prevalence rates range from 10% to 25%, depending on the species and time of the year, with the highest rates of infection, as noted above, being correlated with high levels of parasitization. In South Carolina, ascovirus infection rates as high as 60% have been reported in populations of noctuid larvae at the end of summer.

Host Range

The experimental host range of ascoviruses varies with the viral species. TnAV, HvAV, and SeAV have a broad host range and are capable of replication in a variety of noctuid species, as well as in selected species belonging to other families of the order Lepidoptera. Alternatively, the experimental host range of SfAV is limited to other species of the genus *Spodoptera*. DpAV can replicate in hymenopteran and lepidopteran hosts closely related to its natural host species, *A. assectella*. To propagate virus in the laboratory, all ascoviruses can be grown in their larval or pupal hosts.

To infect caterpillars, they are injected with virus in the fourth or early fifth instar, and virion-containing vesicles are harvested from the blood 5–7 days later.

Pathology and Pathogenesis

Signs of Disease

The signs of ascovirus disease are very subtle, and this probably accounts for why ascoviruses were discovered only recently. The most obvious sign of disease within 24 h of infection is a decrease in the normal rate of feeding. The feeding rate continues to slow as the disease progresses, and as a result larvae fail to gain weight or advance in development. Healthy larvae, particularly in the early stages of development, will easily quadruple their weight and size in a period of 3 or 4 days, whereas ascovirus-infected larvae cease to grow and may actually lose weight. This feature of ascovirus disease is almost impossible to detect in infected larvae in the field. However, it is easily noticed under laboratory conditions when infected and healthy larvae are reared side by side over a period of a few days. A second feature easily noted in the laboratory is that ascovirus diseases are chronic, though usually fatal. When infected during early stages of development, ascovirus-diseased larvae often survive for 2 or 3 weeks beyond the time at which most healthy larvae have completed their development and pupated. Signs of disease other than these are minor, but include the inability to completely cast the molted cuticle, a bloated thoracic region, and a white or creamy discoloration and hypertrophied appearance of the larval body at advanced stages of disease development.

Cytopathology and Cell Biology

In comparison to all other known viruses, the most unique property of ascoviruses is the unusual cytopathology that leads to the formation of the virion-containing vesicles. This process resembles apoptosis, and recent studies of the SfAV genome have shown that it encodes an executioner caspase, synthesized 9 h after infection, which by itself is capable of inducing apoptosis (**Figure 4**).

At the cellular level *in vivo*, the disease begins with extraordinary hypertrophy of the nucleus accompanied by invagination of sections of the nuclear envelope, followed by a corresponding enlargement of the cell. Cells typically grow from 5 to 10 times the diameter of uninfected cells. As the nucleus enlarges, the nuclear envelope ruptures and disintegrates into fragments. At about this stage, the cell plasmalemma begins to invaginate along 'planes' toward the now anucleate cell center. Concomitantly, sheets of membrane form closely adjacent to mitochondria that accumulate along the planes. As this process continues, the membrane sheets coalesce and join the

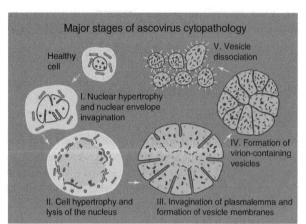

Figure 4 Major stages of cellular pathogenesis caused by a typical ascovirus, a process that resembles apoptosis. After infection, the nucleus enlarges and the nuclear membrane invaginates, and then lyses. Subsequently, the plasma lemma of the cell invaginates and coalesces with cytoplasmic membranes, apparently formed *de novo*, thereby dividing the cell into a cluster of virion-containing vesicles. These vesicles dissociate and are liberated into the blood as the basement membrane of infected tissues degenerates. Virion assembly becomes apparent as the nuclear membrane lyses, and continues throughout all subsequent stages of vesicle formation.

invaginating plasmalemma, thereby cleaving the cell into a cluster of 20 to more than 30 vesicles, ranging in size from 5 to 10 µm in diameter. This aspect of ascovirus cellular pathology resembles the formation of apoptotic bodies during apoptosis. However, rather dissipate as the cell dies, the developing apoptotic bodies are rescued by the virus and progress to form vesicles in which virions continue to assemble. These virion-containing vesicles, also referred to as viral vesicles, typically remain in the tissue until the basement membrane ruptures, though on occasion cell hypertrophy can be so great that the enlarging cell erupts out through the basement membrane of the infected tissue, releasing large fragments of the infected cell directly into the blood. Analysis of both the SfAV and TnAV genomes shows that, unlike many other large DNA viruses, ascoviruses encode several lipid-metabolizing enzymes that are likely involved in the process of converting developing apoptotic bodies into virion-containing vesicles.

Although the process by which viral vesicles are cleaved from cells varies among different ascoviruses, the histopathology is similar among virtually all viruses. Vesicles accumulate in the tissues where they are formed, but as these tissues degenerate during disease progression, the basement membrane of infected tissues deteriorates and ruptures, allowing the vesicles to spill out into the blood. There they accumulate reaching concentrations as high as $10^7 - 10^8$ vesicles ml$^-$ within 3–4 days of infection. There is some evidence that viral replication proceeds within the vesicles as they circulate in the blood, and thus

this tissue must also be considered one of the tissues attacked by ascoviruses. If fact, because such high concentrations of viral vesicles are found in the blood, this tissue could be considered a major site of infection, particularly if it is eventually shown that these viruses continue to replicate in the vesicles as they circulate in the blood.

Despite the chronic nature of the disease caused by ascoviruses, virion-containing vesicles are present in the blood within 2 or 3 days of infection. When the virus replicates in cells *in vitro*, the vesicles are formed within 12–16 h of infection. The rapid development and circulation of the viral vesicles in the blood probably evolved to enhance transmission of the virus by parasitic wasps.

Tissue Tropism

The cytopathology of ascoviruses is consistent among different viral species; however, considerable variation occurs with respect to the tissues attacked, that is, in which replication occurs. TnAV, HvAV, and SeAV exhibit a relatively broad tissue tropism infecting the tracheal matrix, epidermis, fat body, and connective tissue. Differences exist between these species in that some HvAV variants infect the epidermis much more extensively than TnAV variants, whereas some of the latter can also replicate more extensively in fat body cells, but appear only to do this when larvae are infected early in their development. Alternatively, the type species, SfAV, and its variants have a very narrow tissue tropism, with the fat body being the primary site of infection. DpAV occurs in the nuclei of all tissues of its wasp host, but appears to only produce progeny in ovarial tissues. In its lepidopteran pupal host, it attacks and replicates in a wide variety of tissues.

Replication and Virion Assembly

Although there have been few biochemical studies of viral DNA replication or protein synthesis, studies carried out with ascoviruses *in vivo* and *in vitro* show that progeny virions first appear about 12 h after infection. Virion assembly is initiated after the nucleus ruptures, and occurs prior to and during the cleavage of the cell into viral vesicles. The first recognizable structural component of the virion to form is the multilaminar layer of the inner particle. Based on its ultrastructure, this layer consists of a unit membrane and an exterior layer of protein subunits. As the multilaminar layer assembles, a dense nucleoprotein core aggregates on the interior surface. This process continues until the inner particle is complete. After formation, the inner particle is enveloped by membranes within the cell or vesicle. These membranes are apparently synthesized *de novo*. Thus, the assembly of the virions is reminiscent of that in other viruses with complex virions, such as the ridoviruses, herpesviruses, and poxviruses, where the virions differentiate after association of the precursors of virion structural components.

After formation, the virions of the TnAV ascovirus accumulate toward the periphery of the vesicle where they often form inclusion bodies, that is, aggregations of virions (**Figure 2**). In SfAV, occlusion bodies are formed in which the virions are actually occluded in a 'foamy' vesicular matrix that consists of a mixture of protein and minute spherical vesicles. When viewed with phase microscopy, these viral inclusion and occlusion bodies are phase bright, and are largely responsible for the highly refractile appearance of the vesicles. Ascoviruses do not typically form the types of occlusion bodies characteristic of other types of DNA insect viruses, such as baculoviruses and entomopoxviruses.

Origin and Evolution

The subject of viral evolution over millions of years has received relatively little study due to the lack of a fossil record. Moreover, viruses are considered polyphyletic, and thus most of the more than 70 families of viruses are thought to have originated independently. In this regard, ascoviruses may provide a unique opportunity to obtain insights into virus evolution over long periods. Phylogenetic comparisons of ascovirus genes sequenced to date including those coding DNA polymerase and major capsid protein as well as several enzymes indicate that these viruses evolved from a lepidopteran iridovirus (family *Iridoviridae*). Iridoviruses, in turn, appear to have originated from phycodnavirsues (family *Phycodnaviridae*), which attack certain ciliates and algae. On the other end of the evolutionary scale, ascovirus virions are structurally and morphologically similar to the particles formed by ichnoviruses of the family *Polydnaviridae*. Ichnovirus particles are produced in the reproductive tracts of endoparastic wasps of the family Ichneumonidae, and the wasp vector and host of DpAV is a member of this family. Thus, there is a reasonable possibility that ascoviruses and ichnoviruses are related phylogenetically, and share a common ancestor. This possibility is currently under investigation, and should be resolved over the next several years through a comparative analysis of the molecular evolution of genes of ascoviruses and ichnoviruses, after more structural genes from the latter viruses have been cloned and sequenced. A major question to be addressed is whether the DpAV represents an early ascovirus branch that evolved from an iridovirus or is representative of an ascovirus branch that eventually led to the origin of ichnovirus particles. With respect to the ichnoviruses and bracoviruses, recent data on the DNA contained by the particles of these putative viruses suggest that these are not viruses after all, but rather are an unusual highly evolved type of organelle that evolved from DNA viruses, which are used by endoparasitic wasps to suppress the internal defense responses of their insect hosts.

Future Perspectives

As present, too little is known about ascoviruses to assess whether they are or will turn out to be of economic importance. Their poor infectivity *per os* makes it highly unlikely they will ever be developed as viral insecticides, especially given the successful advent of insect-resistant transgenic crops. However, as more entomologists become familiar with the disease caused by ascoviruses, it may be shown that in habitats rarely treated with chemical insecticides, such as transgenic crops, these viruses are responsible for significant levels of natural pest suppression, particularly where parasitic wasps are abundant. Such findings would encourage even greater emphasis on the development of biological control and other more environmentally sound methods of pest control. With respect to the cell biology of viral vesicle formation, ascoviruses provide an interesting model for how apoptosis can be manipulated at the molecular level. Additionally, study of the unusual process by which ascoviruses rescue the developing apoptotic bodies to form viral vesicles could lead to insights into how cells manipulate the cytoskeleton and mitochondria. Finally, it is possible that viral vesicles will provide a unique anucleate cellular system for studying the replication of a complex type of enveloped DNA virus *in vitro*.

Further Reading

Asgari S (2006) Replication of Heliothis virescens ascovirus in insect cell lines. *Archives of Virology* 151: 1689.

Asgari S, Davis J, Wood D, Wilson P, and McGrath A (2007) Sequence and Organization of the Heliothis virescens ascovirus genome. *Journal of General Virology* 88: 1120–1132.

Bideshi DB, Demattei MV, Rouleux-Bonnin F, et al. (2006) Genomic sequence of spodoptera frugiperda ascovirus 1a, an enveloped, double-stranded DNA insect virus that manipulates apoptosis for viral reproduction. *Journal of Virology* 80: 11791.

Bideshi DB, Tan Y, Bigot Y, and Federici BA (2005) A viral caspase contributes to modified apoptosis for virus transmission. *Genes and Development* 19: 1416.

Bigot Y, Rabouille A, Doury G, et al. (1997) Biological and molecular features of the relationships between Diadromus pulchellus ascovirus, a parasitoid hymenoptera wasp (*Diadromus pulchellus*) and its lepidopteran host, *Acrolepiopsis assectella*. *Journal of General Virology* 78: 1149.

Cheng XW, Wang L, Carner GR, and Arif BM (2005) Characterization of three ascovirus isolates form cotton insects. *Journal of Invertebrate Pathology* 89: 193.

Federici BA (1983) Enveloped double-stranded DNA insect virus with novel structure and cytopathology. *Proceedings of the National Academy of Sciences, USA* 80: 7664.

Federici BA and Bigot Y (2003) Origin and evolution of polydnaviruses by symbiogenesis of insect DNA viruses in endoparasitic wasps. *Journal of Insect Physiology* 49: 419.

Federici BA, Bigot Y, Granados RR, *et al.* (2005) Family *Ascoviridae*. In:
    Fauquest CM, Mayo MA, Maniloff J, Desselberger U, and Ball LA
    (eds.) *Virus Taxonomy. Eighth Report of the International Committee
    on Taxonomy of Viruses*, pp. 269–274. San Diego, CA: Elsevier
    Academic Press.
Federici BA and Govindarajan R (1990) Comparative histopathology of
    three ascovirus isolates in larval noctuids. *Journal of Invertebrate
    Pathology* 56: 300.
Federici BA, Vlak JM, and Hamm JJ (1990) Comparison of virion
    structure, protein composition, and genomic DNA of three ascovirus
    isolates. *Journal of General Virology* 71: 1661.

Govindarajan R and Federici BA (1990) Ascovirus infectivity and the
    effects of infection on the growth and development of Noctuid larvae.
    Journal of Invertebrate Pathology 56: 291.
Pellock BJ, Lu A, Meagher RB, Weise MJ, and Miller LK (1996)
    Sequence, function, and phylogenetic analysis of an ascovirus DNA
    polymerase gene. *Virology* 216: 146.
Stasiak K, Renault S, Demattei MV, Bigot Y, and Federici BA (2003)
    Evidence for the evolution of ascoviruses from iridoviruses. *Journal of
    General Virology* 84: 2999.
Wang L, Xue J, Seaborn CP, Arif BM, and Cheng XW (2006) Sequence
    and organization of the Trichoplusia ni ascovirus 2c (*Ascoviridae*)
    genome. *Virology* 354: 167.

Barnaviruses

P A Revill, Victorian Infectious Diseases Reference Laboratory, Melbourne, VIC, Australia

Glossary

Casing A layer of peat moss placed on top of the growing beds to encourage sporophore formation.
Sporophore Mushroom fruiting body.

Introduction

In the USA in 1948, a disease of the cultivated mushroom *Agaricus bisporus* was discovered on a property in Pennsylvania that had a major impact on the mushroom industry and fungal pathology in general. It was characterized by poor colonization of compost by mycelium and misshapen sporophores with long thin stipes and small globular caps producing a drumstick-like appearance, or they were thickened with a barrel-like appearance. The poor colonization of the compost and casing by infected mycelium often produced characteristic bare patches on the growing beds and reduced yields. The disease was named La France disease and a virus was implicated as a possible cause in 1960, after it was shown that the disease could be transmitted to healthy cultures by hyphal anastamosis. In 1962, three different virus-like particles were identified, two of which were spherical (25 and 29 nm), and the third was a 19 nm × 50 nm elongated or bacilliform particle with rounded ends. Subsequently a 34–36 nm particle with a double-stranded (ds)RNA genome (La France infectious virus (LFIV)) has been identified as the causal agent of La France disease.

The bacilliform virus particle was of particular interest as almost all mycoviruses identified to that point had a spherical or isometric morphology. Bacilliform 19 nm × 48 nm virus-like particles were subsequently identified in the ascomycete *Microsphaera mougeotti*, and 17 nm × 35 nm bacilliform particles were also observed in the deuteromycete *Verticilium fungicola*, itself a pathogen of *A. bisporus*. However no relationship with the mushroom bacilliform virus (see below) and these bacilliform particles has been established.

Originally named mushroom virus 3 (MV3), the virus was subsequently renamed mushroom bacilliform virus (MBV). The viral genome was identified as positive-sense single-stranded (ss)RNA, and the virus was classified as the type member of the genus *Barnavirus* in the family *Barnaviridae*. The family name derives its roots from Bacilliform RNA virus. MBV remains the only barnavirus identified to date.

Virion Properties

The MBV virion M_r is 7.1×10^6, with a buoyant density in Cs_2SO_4 of $1.32 \, \text{g cm}^{-3}$. Virions are stable between pH 6 and 8 and ionic strength of 0.01 to 0.1 M phosphate.

Virion Structure and Composition

MBV has a bacilliform or bullet-shaped morphology, with particles generally 19 nm × 50 nm in size (**Figure 1**). Virions contain a single major capsid protein (CP) of 21.9 kDa and there are approximately 240 molecules in each capsid.

Virions encapsidate a single linear molecule of a positive-sense ssRNA, 4.0 kb in size. The complete 4009 nt sequence is available (GenBank accession No. U07551). The RNA has a 5′-linked VPg and lacks a poly(A) tail. RNA constitutes about 20% of virion weight.

Genome Organization and Expression

The RNA genome (4009 nt) contains four major and three minor open reading frames (ORFs) and has 5′- and

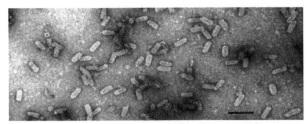

Figure 1 Electron micrograph *Mushroom bacilliform virus* particles. Scale = 100 nm. Reprinted from *Virus Taxonomy: Seventh Report of the International Committee on Taxonomy of Viruses*, Wright PJ and Revill PA, The *Barnaviridae*, copyright 2004, with permission from Elsevier.

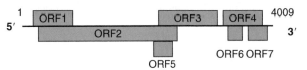

Figure 2 The MBV genome arrangement.

3′-untranslated regions (UTRs) of 60 and 250 nt, respectively (**Figure 2**). ORFs 1–4 encode polypeptides of 20, 73, 47, and 22 kDa, respectively. The deduced amino acid sequence of ORF2 contains three conserved chymotrypsin-related serine protease sequence motifs. Blast searches of the deduced ORF2 amino acid sequence show similarity to serine proteases encoded by plant sobemoviruses. ORF2 also encodes the VPg. ORF3 contains the $GX_3TX_3NX_nGDD$ amino acid sequence shared by the putative RNA-dependent RNA polymerases (RdRps) of positive-sense ssRNA viruses and has similarity to the RdRps of sobemoviruses, enamoviruses, and poleroviruses. ORF4 encodes the CP. ORFs 5–7 encode 8, 6.5, and 6 kDa polypeptides, respectively. The polypeptides potentially encoded by ORFs 1, 5, 6, and 7 show no significant similarity to known polypeptides. The negative strand of MBV contains seven small ORFs of unknown significance. These potentially encode polypeptides ranging from M_r 6.5K to M_r 10.5K.

The genome arrangement and transcription/translation strategies of MBV are strikingly similar to those of a number of plant viruses, particularly poleroviruses and sobemoviruses. MBV probably also uses similar strategies to express its gene products, including leaky ribosomal scanning for expression of ORF2, ribosomal frameshifting for expression of the RdRp, and subgenomic RNA for expression of the CP. Of these, only subgenomic RNA has been confirmed *in vivo*. In a cell-free system, genomic-length RNA directs the synthesis of major 21 and 77 kDa polypeptides and several minor polypeptides of 18–60 kDa. The full-length genomic RNA and a sgRNA (0.9 kb) encoding ORF4 (CP) are found in infected cells. Virions accumulate singly or as aggregates in the cytoplasm. However the MBV life cycle has yet to be determined.

Evolutionary Relationships

The MBV genome sequence has no similarity with any other mycovirus genome characterized to date. However the deduced ORF2 and ORF3 amino acid sequences share striking similarity with those of plant viruses, particularly

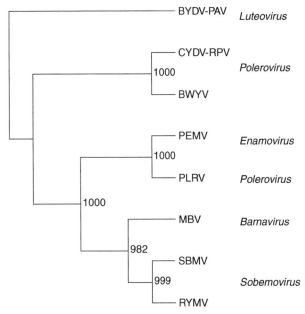

Figure 3 Neighbor-joining tree of the MBV RdRp compared to RdRps of a number of plant viruses. Sequences were aligned using Clustal X (1000 bootstrap replicates) and the tree was constructed with Treeview. The tree was rooted using the outgroup (BYDV-PAV, barley yellow dwarf virus). CYDV-RPV, cereal yellow dwarf virus-RPV; SBMV, southern bean mosaic virus; RYMV, rice yellow mottle virus; PEMV, pea enation mosaic virus; BWYV, beet western yellows virus; PLRV, potato leafroll virus. The GenBank accession numbers of the sequences used in the analysis were PLRV (D00530), CYDV-RPV (NC004751), BWYV (NC004756, PEMV (NC003629), RYMV (NC001575), SBMV (DQ875594), MBV (U07557), BYDV-PAV (EF043235).

sobemoviruses, poleroviruses, and enamoviruses (**Figure 3**). This, together with the similarity of the MBV and sobemovirus/polerovirus genome arrangements, suggests that MBV may have shared a common ancestor with these plant virus groups.

Transmission and Host Range

MBV is transmitted horizontally through infected mycelium and it is yet to be determined if the virus can be transmitted in spores. There is no known insect vector. Although morphologically similar viruses to MBV have been identified in the field agaric, *A. campestris*, it is unknown whether these particles are related to MBV. Consequently MBV remains the only barnavirus identified to date.

See also: Fungal Viruses; Luteoviruses; Sobemovirus.

Further Reading

Goodin MM, Schlagnhaufer B, and Romaine CP (1992) Encapsidation of the La France disease specific double-stranded RNAs in 36 nm isometric virus-like particles. *Phytopathology* 82: 285–290.

Moyer JW and Smith SH (1976) Partial purification and antiserum production to the 19 × 50 nm mushroom virus particle. *Phytopathology* 66: 1260–1261.

Moyer JW and Smith SH (1977) Purification and serological detection of mushroom virus-like particles. *Phytopathology* 67: 1207–1210.

Revill PA, Davidson AD, and Wright PJ (1994) The nucleotide sequence and genome organization of mushroom bacilliform virus: A single-stranded RNA virus of *Agaricus bisporus* (Lange) Imbach. *Virology* 202: 904–911.

Revill PA, Davidson AD, and Wright PJ (1998) Mushroom bacilliform virus RNA: The initiation of translation at the 5′-end ofthe genome and identification of the VPg. *Virology* 249: 231–237.

Revill PA, Davidson AD, and Wright PJ (1999) Identification of a subgenomic mRNA encoding the capsid protein of mushroom bacilliform virus, a single-stranded RNA mycovirus. *Virology* 260: 273–276.

Romaine CP and Schlagnhaufer B (1991) Hybridization analysis of the single-stranded RNA bacilliform virus associated with La France disease of *Agaricus bisporus*. *Phytopathology* 81: 1336–1340.

Tavantzis SM, Romaine CP, and Smith SH (1980) Purification and partial characterization of a bacilliform virus from *Agaricus bisporus*: A single-stranded RNA mycovirus. *Virology* 105: 94–102.

Tavantzis SM, Romaine CP, and Smith SH (1983) Mechanism of genome expression in a single-stranded RNA virus from the cultivated mushroom *Agaricus bisporus*. *Phytopathology* 106: 45–50.

Wright PJ and Revill PA (2004) The *Barnaviridae*. In: Fauquet CM (ed.) *Virus Taxonomy: Seventh Report of the International Committee on Taxonomy of Viruses*, London: Elsevier Academic Press.

Chrysoviruses

S A Ghabrial, University of Kentucky, Lexington, KY, USA

Glossary

Hyphal anastomosis The union of a hypha with another resulting in cytoplasmic exchange.

Mycoviruses Viruses that infect and multiply in fungi.

Viruses with multipartite genomes The essential genome is divided among several genomic segments (segmented genome) that are either separately encapsidated in identical capsids (multicomponent viruses, i.e., chrysoviruses) or jointly enclosed in a single particle (e.g., reoviruses).

Introduction

The discovery in the late 1960s and early 1970s of polyhedral virus particles in many of the industrial strains of *Penicillium chrysogenum* used for penicillin production generated considerable interest in the study of *Penicillium* viruses. It was surmised then that virus infection might be responsible for the instability of some of these strains. *Penicillium chrysogenum virus* (PcV), the type species of the genus *Chrysovirus*, was one of the first mycoviruses to be extensively studied at the biochemical, biophysical, and ultrastructural levels. Although the results of these earlier studies on PcV properties were mostly in agreement, they differed in their explanation of the nature of genome complexity. Although PcV and the related *Penicillium*

viruses penicillium brevicompactum virus (PbV) and penicillium cyaneo-fulvum virus (Pc-fV) have similar isometric particles, 35–40 nm in diameter, and are serologically related, there was confusion as to whether they contain three or four double-stranded RNA (dsRNA) segments. Because none of these viruses was characterized at the molecular level until recently, these three *Penicillium* viruses were originally grouped under the genus *Chrysovirus* and provisionally placed in the family *Partitiviridae* with the assumption that their genomes are bipartite, with dsRNA1 encoding the RNA-dependent RNA polymerase (RdRp) and dsRNA2 encoding the major capsid protein (CP). The additional dsRNAs (dsRNAs 3 and/or 4), like those of some partitiviruses, were presumed to be defective or satellite dsRNAs. The classification of the genus *Chrysovirus* was recently reconsidered because the complete nucleotide sequence and genome organization of each of the four monocistronic dsRNA segments associated with PcV virions and with the chrysovirus Helminthosporium victoriae 145S virus were recently reported. Based on the consistent simultaneous presence of their four dsRNA segments, the existence of extended regions of highly conserved terminal sequences at both ends of all four segments, sequence comparisons and phylogenetic analysis, it became clear that PcV and related viruses should not be classified with the family *Partitiviridae*. This led to the creation of the new family *Chrysoviridae* to accommodate the isometric dsRNA mycoviruses with multipartite genomes. The name *Chryso* is derived from the specific epithet of *Penicillium chrysogenum*, the fungal host of the type species.

Virion Properties

The buoyant densities of virions of members in the family *Chrysoviridae* are in the range of 1.34–1.39 g cm^{-3} and their sedimentation coefficients $S_{20,w}$ (in Svedberg units) are in the range of 145S to 150S. Generally, each virion contains only one of the four genomic dsRNA segments. However, purified preparations of PcV and Pc-fV can contain minor distinctly sedimenting components that include empty particles and replication intermediates.

Chrysovirus particles possess virion-associated RNA-dependent RNA polymerase (RdRp) activity, which catalyzes the synthesis of single-stranded RNA (ssRNA) copies of the (+) strand of each of the genomic dsRNA molecules. The *in vitro* transcription reaction occurs by a conservative mechanism, whereby the released ssRNA represents the newly synthesized plus strand.

Genome Organization

The virions of members of the family *Chrysoviridae* contain four unrelated linear, separately encapsidated, monocistronic dsRNA segments (2.4–3.6 kbp in size; **Table 1**). the largest segment, dsRNA1, codes for the RdRp and dsRNA2 codes for the major CP. The dsRNA segments 3 and 4 code for proteins of unknown function. The genomic structure of PcV, the type species of the genus *Chrysovirus*, comprising four dsRNA segments, is schematically represented in **Figure 1**. The earlier conflicting

reports on whether PcV contains three or four segments were recently explained when studies on cDNA cloning and sequencing of the viral dsRNAs were completed. Although the dsRNA extracted from purified virions is resolved into three bands by agarose gel electrophoresis (**Figure 2**, lane EB), northern hybridization analysis using cloned cDNA probes representing the four dsRNA segments shows clearly that each of the four segment has unique sequences (**Figure 2**). Because dsRNAs 3 and 4 differ in size by only 74 bp (**Table 1**), they co-migrate when separated by agarose gel electrophoresis. Previous studies on sequencing analysis and *in vitro* coupled transcription–translation assays showed that each of the four dsRNAs is monocistronic, as each dsRNA contains a single major open reading frame (ORF) and each is translated into a single major product of the size predicted from its deduced amino acid sequence. Thus, the fact that PcV virions contain four distinct dsRNA segments has clearly been established.

Unlike PcV, dsRNAs 3 and 4 from other chrysoviruses, helminthosporium victoriae 145SV (Hv145SV), amasya cherry disease associated chrysovirus (ACDACV) and cherry chlorotic rusty spot associated chrysovirus (CCRSACV) are clearly resolved from each other when purified dsRNA preparations are subjected to agarose gel electrophoresis. As shown in **Table 1**, dsRNAs 3 and 4 from these viruses are significantly different in size. Assignment of numbers 1–4 to PcV dsRNAs was made according to their decreasing size. Following the same criterion used for PcV, the dsRNAs associated with Hv145SV, CCSRACV, and

Table 1   List of members and tentative members in the family *Chrysoviridae*

Virus species	Abbreviation	DsRNA segment no. (length in bp, encoded protein; size in kDa)	GenBank accession no.
Helminthosporium victoriae 145S virus	Hv145SV	1 (3612; RdRp,125)	NC_005978
		2 (3134; CP, 100)	NC_005979
		3 (2972; chryso-P4, 93)	NC_005980
		4 (2763; chryso-P3, 81)	NC_005981
Penicillium brevicompactum virus	PbV	Four dsRNA segments; no molecular data	
Penicillium chrysogenum virus	PcV	1 (3562; RdRp,129)	NC_007539
		2 (3200; CP, 109)	NC_007540
		3 (2976; chryso-P3 101)	NC_007541
		4 (2902; chryso-P4, 95)	NC_007542
Penicillium cyaneo-fulvum virus	Pc-fV	Four dsRNA segments; no molecular data	
Tentative members			
Agaricus bisporus virus 1 or LaFrance isometric virus	AbV-L1	L1 (3396; RdRp, 122)	X94361
	LFIV	L5 (2455; unknown, 82)	X94362
Amasya cherry disease associated chrysovirus	ACDACV	1 (3399; RdRp, 124)	AJ781166
		2 (3128; CP, 112)	AJ781165
		3 (2833; chryso-P4, 98)	AJ781164
		4 (2498; chryso-P3, 77)	AJ781163
Cherry chlorotic rusty spot associated chrysovirus	CCRSACV	1 (3399; RdRp, 124)	AJ781397
		2 (3125; CP, 112)	AJ781398
		3 (2833; chryso-P4, 98)	AJ781399
		4 (2499; chryso-P3, 77)	AJ781400

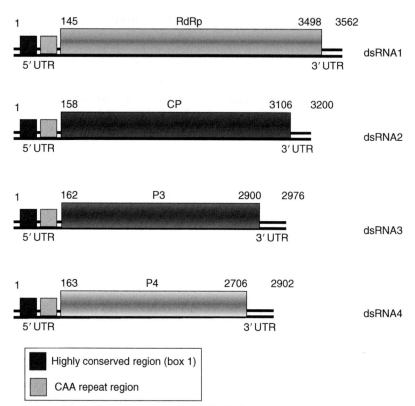

Figure 1 Genome organization of penicillium chrysogenum virus (PcV), the type species of the genus *Chrysovirus*. The genome consists of four dsRNA segments; each is monocistronic. The RdRp ORF (nt positions 145–3498 on dsRNA1), the CP ORF (nt positions 158–3106 on dsRNA2), the p3 ORF (nt positions 162–2900 on dsRNA3), and the p4 ORF (nt positions 163–2706 on dsRNA4) are represented by rectangular boxes. Adapted from Ghabrial SA, Jiang D, and Castón RJ (2005) *Chrysoviridae*. In: Fauquet CM, Mayo MA, Maniloff J, Desselberger U, and Ball LA (eds.) *Virus Taxonomy: Eighth Report of the International Committee on Taxonomy of Viruses,* pp. 591–595. London: Academic Press, with permission from Elsevier.

ACDACV were accordingly assigned the numbers 1–4. Sequence comparisons, however, indicated that dsRNAs 3 of Hv145SV, CCSRACV, and ACDACV are in fact the counterparts of PcV dsRNA4 rather than dsRNA3. Likewise, dsRNA4 of these three chrysoviruses are the counterparts of PcV dsRNA3. Since PcV was the first chrysovirus to be characterized at the molecular level and to avoid confusion, the protein designations P3 and P4 as used for PcV will be adopted and referred to as chryso-P3 and chryso-P4. Thus, whereas the chryso-P3 protein represents the gene product of PcV dsRNA3, it comprises the corresponding gene product of Hv145SV dsRNA4, and so on.

Except for CCRSACV and ACDACV dsRNAs, the 5′ UTRs of chrysovirus dsRNAs are relatively long, between 140 and 400 nucleotides in length. In addition to the strictly conserved 5′- and 3′-termini, a 40–75 nt region with high sequence identity is present in the 5′ UTR of all four dsRNAs (box 1; **Figure 3**). A second region of strong sequence similarity is present immediately downstream from 'box 1' (**Figure 3**). This consists of a stretch of 30–50 nt containing a reiteration of the sequence 'CAA'. The $(CAA)_n$ repeats are similar to the enhancer elements

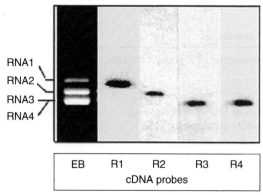

Figure 2 Northern hybridization analysis of PcV dsRNA segments. Virion dsRNAs were separated on a 1.5% agarose gel, transferred onto Hybond-N$^+$ membrane and hybridized with ^{32}P-labeled probes prepared by random-primer labeling of cloned cDNA to PcV dsRNA1 (R1), dsRNA2 (R2), dsRNA3 (R3), and dsRNA4 (R4). EB, ethidium bromide-stained virion dsRNAs separated on a 1.5% agarose gel. Reproduced from Jiang D and Ghabrial SA (2004) Molecular characterization of penicillium chrysogenum virus: Reconsideration of the taxonomy of the genus chrysovirus. *Journal of General Virology* 85: 2111–2121, with permission from Society for General Microbiology.

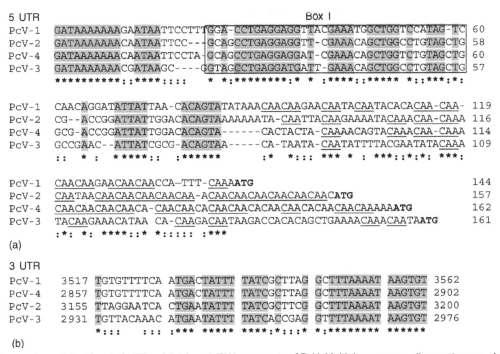

Figure 3 Comparison of the 5′ and 3′ UTRs of the four dsRNA segments of PcV. Multiple sequence alignments were obtained using CLUSTAL X (and some manual adjustments) with the nucleotide sequences of the 5′ UTR (a) and the 3′ UTR (b). The (CAA) repeats are underlined in (a). Asterisks signify identical bases at the indicated position (shaded) and colons specify that three out of four bases are identical at the indicated positions. Reproduced from Jiang D and Ghabrial SA (2004) Molecular characterization of penicillium chrysogenum virus: Reconsideration of the taxonomy of the genus chrysovirus. *Journal of General Virology* 85: 2111–2121, with permission from Society for General Microbiology.

present at the 5′ UTRs of tobamoviruses. Although the 5′ UTR of CCRSACV and ACDACV contain the 'CAA' repeat region upstream of the translation initiation codon, the 'box 1' region is significantly shortened. Furthermore, these viruses do not share with PcV and Hv145SV the strictly conserved terminal 8 and 7 nucleotides, respectively at the 5′ and 3′ ends (**Table 2**). The discrepancies in the length and features of the 5′ and 3′ UTRs of CCRSACV and ACDACV dsRNAs compared to those of PcV and Hv145SV could be due to the cloning procedure used, which may not have allowed for the exact termini to be cloned. This, however, seems unlikely because of the strong similarities of the termini and internal sequences of the 5′ and 3′ UTRs of all four CCRSACV and ACDACV dsRNA segments. Alternatively, the divergence in the features of the 5′ and 3′ UTRs of ACDACV and CCRSACV may be because they represent plant rather than fungal chrysoviruses (see section on 'Biological properties').

Genome Expression and Replication

Chrysovirus dsRNA1s Code for RdRp

The largest dsRNA segment (dsRNA1) of the chrysoviruses so far sequenced contains a single large ORF

coding for RdRp. The calculated molecular mass of chrysovirus RdRps ranges from 122 to 129 kDa (**Table 1**). These values are consistent with those estimated by SDS-PAGE of the *in vitro* translation products of full-length transcripts derived from cDNAs to dsRNA1s of PcV and Hv145SV. Examination of the deduced amino acid sequence of the RdRp ORF reveals the presence of the eight conserved motifs characteristic of RdRps of dsRNA viruses of simple eukaryotes (**Figure 4**). A comparison of the conserved motifs of chrysovirus RdRps with those of totiviruses and partitiviruses reveals that the RdRps of chrysoviruses are more closely related to those of totiviruses than to those of partitiviruses (**Figure 4**). This conclusion is also supported by published results of phylogenetic analysis of RdRp conserved motifs and flanking sequences of chrysoviruses and viruses in the families *Totiviridae* and *Partitiviridae* (see section on 'Evolutionary relationships').

Chrysovirus dsRNA2s Code for CP

The second largest dsRNA segment (dsRNA2) of chrysoviruses so far sequenced contains a single large ORF coding for CP. The calculated molecular mass of chrysovirus CPs ranges from 100 to 112 kDa (**Table 1**). The predicted size of PcV CP (109 kDa) is similar to that

Table 2   Comparison of the nucleotide sequences at the 5′ and 3′ termini of chrysovirus dsRNAs[a]

	5′-terminus	3′-terminus
Hv145SV		
dsRNA1	5′-GAUAAAAAGAAAAA-U..	.. UUAGGACUUUAAGUGU-3′
dsRNA2	5′-GAUAAAAACAAAAAU..	.. UUCGGACUUUAAGUGU-3′
dsRNA3	5′-GAUAAAAACAGAAAU..	.. UUCGGACUUUAAGUGU-3′
dsRNA4	5′-GAUAAAAACAGAAAU..	.. UGCGGACUUUAAGUGU-3′
PcV		
dsRNA1	5′-GAUAAAAAAAGAAUAA..	.. GCUUUAAAAUAAGUGU-3′
dsRNA2	5′-GAUAAAAAACAAUAA..	.. GCUUUAAAAUAAGUGU-3′
dsRNA3	5′-GAUAAAAAACGAUAA..	.. GCUUUAAAAUAAGUGU-3′
dsRNA4	5′-GAUAAAAAACGAUAA..	.. GUUUUAAAAUAAGUGU-3′
CCRSACV		
dsRNA1	5′-GAAAUUAUGGUUUUUG..	.. AUUGUCAAUAAUAUGC-3′
dsRNA2	5′-GAAAUUAUGGUUUUUG..	.. GUGUUGAUAUAUAUGC-3′
dsRNA3	5′-GAAAUUAUGGUUUUUG..	.. GGUUAUAACUAUAUGC-3′
dsRNA4	5′-GAAAUUAUGGAUUUUG..	.. AUGUGUAACUAUAUGC-3′

[a]Identical nucleotides in the same position are shaded.

	1	2	3		4		5	
SsRV1	LLGRA(61)	WCVNGSQND(47)	KL-EHG-KTRAIFACDTRSY	(47)	LDFDDFNSHHS(45)		TLPSGHRGTTIVNSVLNAAYI(14)	
Hv190SV	LQGRY(61)	WCVNGSQNA(42)	KL-ENG-KDRAIFACDTRSY	(47)	LDYDNFNSQHS(45)		TLMSGHRATTFTNSVLNAAYI 14)	
SsRV2	LQGRA(64)	WAVNGSQSG(46)	KL-EHG-KTRAIFACDTLNY	(47)	LDYDDFNSHHS(46)		TLMSGRRGTTYISSVLNEVYL(14)	
UmVH1	LYGRG(66)	WLVSGSSAG(55)	KLNETGGKARAIYGVTLWHY	(47)	YDYPDFNSMHT(64)		GLYSGDRDTTLINTLLNIAYA(20)	
	* **	* *:**:	** * *** :::: *		:*::*** *:		********** :: *:*** **	
Hv145SV	LLGRR(73)	WMTKGSLVS(56)	KLNENGHKDRVLLPGGLLHY	(44)	YDWANFNVQHS	(49)	GLYSGWRGTTWDNTVLNGCYM(20)	
PCV	LVGRG(74)	WLTKGSLVY(60)	K-YEVGKKDRTLLPGTLVHF	(44)	YDWADFNEQHS	(49)	GLYSGWRGTTWINTVLNFCYV(19)	
CCRSACV	LVGRR(73)	WLTKGSTVY(65)	KLNECGYKDRTLLPGSLFHY	(44)	FDWANFNAFHS	(49)	GLYSGWRGTSFLNSVLNSCYT(19)	
	*:**:	*:****:*:	*::* * ***:**** * *:		:***:** :**		********** *:*** **	
FpV1			SDRDGILKQRPVYAVDDLFL	(47)	IDWSGFDQRLP	(72)	GVPSGMLNTQFLDSFGNLFLL	(19)
RhsV-717			SKRDGTLKVRPVVAVDELFL	(47)	IDWSGYDQRLP	(71)	GVPSGMLLTQFLDSFGNLYLI	(19)
AhV1			SKRD-NLKVRPVYNAPMIYI	(47)	IDWSRFDHLAP	(92)	GVPSGILMTQFIDSFVNLTIL	(19)
HaV			SKIT-KLKVRPVYNAPMLFL	(47)	FDYSRFDQLAP	(101)	GVPSGIFMTQILDSFVNLFIF	(19)
BCV3			ADLREKTKVRGVWGRAFHYI	(48)	LDWSSFDSSVT	(50)	GIPSGSYYTSIVGSVVNRLRI	(15)
FsV1			SPRD-DPKTRLAWIYPSEML	(47)	LDFSSFDTKVP	(61)	GVPSGSWWTQLVDSVVNWILV	(14)
			: :: :*:*::: :		:*:* :*		*:*** : *: :*::*:	

	6		7		8
SsRV1	LHTGDDVYIRA	(18)	RINPAKQSVGFGTGEFLRM	(8)	GYLARSVASFVSGNW
Hv190SV	LHAGDDVYLRL	(18)	RMNPTKQSIGYTGAEFLRL	(8)	GYLCRAIASLVSGSW
SsRV2	IHVGDDVYLGV	(18)	RMNPMKQSVGHTSTEFLRL	(8)	GYLARAVASTISGNW
UmV-H1	LCHGDDIITVH	(18)	KGQESKLMIDHKHHEYLRI	(9)	GCLARCVATYVNGNW
	:: ***:		: :* *::: *:**		*:*:* *: ::*:*
Hv145SV	DQGGDDVDQEF	(18)	EATKSKQMIG-RNSEFFRV	(8)	ASPVRGLATFVAGNW
PCV	DHGGDDIDLGL	(18)	KANKWKQMFGTR-SEFFRN	(8)	ASPTRALASFVAGDW
CCRSACV	DHGGDDIDGGI	(18)	EAQKIKQMIGID-SEFFRI	(8)	GSATRALARFVSGNW
	*:*****:* :		:* * ***:* : *****		:*::*:** **:*:*
FpV1	FIMGDDNSAFT	(26)	SKTKSIITTLRHKIETLSY	(8)	RPIGKLVAQLCFPER
RhsV-717	FIMGDDNSIFT	(26)	SKTKSVITTLRSKIETLSY	(8)	RDVEKLIAQLVYPEH
AhV1	FIMGDDNVIFT	(26)	NISKSAVTSIRRKIEVLGY	(8)	RSISKLVGQLAYPER
HaV	FIQGDDNLVFY	(26)	SPDKSWITRLRTKIEVLGY	(8)	RDVSKLIATLAYPER
BCV3	YTQGDDSLIGE	(20)	NPDKTEYSTDPGYVTFLGR	(8)	RSLDKCLRLLMFPEY
FsV1	RVLGDDS-AFM	(21)	SDEKSISVEDATELKLLGV	(8)	RETEEWFKLALYPEG
	:: ***:		: *: : : ::: *::		* ::: : :**

Figure 4   Comparison of the conserved motifs of RdRps of selected isometric dsRNA viruses. Numbers 1–8 refer to the eight conserved motifs characteristic of RdRps of dsRNA mycoviruses. The amino acid positions corresponding to conserved motifs 1 and 2 for the RdRps of viruses in the family *Partitiviridae* are not well defined and, therefore, they are not presented. Multiple sequence alignments were obtained using the CLUSTAL X program with RdRp amino acid sequences of the following viruses. Upper set: viruses in the family *Totiviridae*: sphaeropsis sapinea RNA virus 1 (SsRV-1), SsRV-2, helminthosporium victoriae 190S virus (Hv190SV), and ustilago maydis virus H1 (UmV-H1). Middle set: viruses in the family *Chrysoviridae*: Hv145SV PcV and CCRSACV. Lower set: viruses in the family *Partitiviridae*: fusarium poae virus 1 (FpV1), rhizoctonia solani virus-717 (RhsV-717), atkinsonella hypoxylon virus 1 (AhV1), heterobasidion annosum virus (HaV), beet cryptic virus 3 (BCV3), and fusarium solani virus 1 (FsV1). Asterisks signify identical residues (shaded) at the indicated positions; colons signify highly conserved amino acid residue within a column; numbers in parentheses correspond to the number of amino acid residues separating the motifs.

estimated by SDS-PAGE of purified PcV virions as well as that determined for the *in vitro* translation product of full-length transcript of dsRNA2 cDNA. Direct evidence that PcV dsRNA2 encodes CP was provided by amino acid sequencing of a tryptic peptide derived from a gradient-purified PcV capsid.

Chryso-P3 Shares a 'Phytoreo S7 Domain' with Core Proteins of Phytoreoviruses

PcV dsRNA3 codes for its chryso-P3 protein, whereas Hv145SV, ACDACV, and CCRSACV dsRNA4s encode the corresponding chryso-P3s. Although the function of chryso-P3 is not known, sequence analysis and database searches offer some clues. ProDom database searches reveal that chryso-P3 sequences share a 'phytoreo S7 domain' with a family consisting of several phytoreovirus P7 proteins known to be viral core proteins with nucleic acid binding activities. The consensus for the three chrysoviruses is [X(V/I)V(M/L)P(A/M)G(C/H)GK(T/S)T-(L/I)]. Phytoreovirus P7 proteins bind to their corresponding P1 (transcriptase/replicase) proteins, which bind to the genomic dsRNAs. It is of interest, in this regard, that the N-terminal regions of all chryso-P3s (encompassing the amino acids within positions 1–500) share significant sequence similarity with comparable N-terminal regions of the putative RdRps encoded by chrysovirus dsRNA1s. A multiple alignment of a portion of the N-terminal region sequences of chrysovirus P3s and RdRps is shown in **Table 3** to demonstrate the level of similarity among the N-terminal sequences of these proteins. The regions in the dsRNA1-encoded proteins with high similarity to chryso-P3 occur upstream of the eight highly conserved motifs characteristic of RdRps of dsRNA viruses of simple eukaryotes. The significance of this sequence similarity to the function of chryso-P3 is not known for certain, but one may speculate that the N-terminal region of these proteins may play a role in viral RNA binding and packaging.

Chryso-P4 is Virion Associated as a Minor Protein

Present evidence, based on amino acid sequencing of a tryptic peptide derived from gradient-purified PcV virions, strongly supports the conclusion that PcV chryso- P4 is virion-associated as a minor protein. The chryso-P4 encoded by chrysoviruses contains the motif PGDGXCXXHX. This motif (I), along with motifs II (with a conserved K), III, and IV (with a conserved H), form the conserved core of the ovarian tumor gene-like superfamily of predicted cysteine proteases. Multiple alignments showed that motifs I–IV are also present in other viruses including AbV-1, a tentative member of the family *Chrysoviridae*. Whether the RNAs of these viruses indeed code for the predicted proteases remains to be investigated.

Replication of Chrysoviruses

There is very limited information on how chrysoviruses replicate their dsRNAs. The virion-associated RdRp catalyzes *in vitro* end-to-end transcription of each dsRNA to produce mRNA by a conservative mechanism. Purified virions containing both ssRNA and dsRNA have been isolated from *Penicillium* spp. infected with PcV or Pc-fV, which may represent replication intermediates.

Virion Structure

Virions are isometric, nonenveloped, 35–40 nm in diameter (**Figure 5**). The capsid structure of PcV was recently determined at relatively moderate resolutions (~2.5 nm) using cryotransmission electron microscopy combined

Table 3   Comparison of the amino acid sequences of the N-terminal regions of chryso-P3 and corresponding regions of RdRp proteins

Viral protein	Amino acid sequence[a]
PcV-chryso-P3	(102) LYGVVMPMGHGKTTLAQEEGWIDCDSLI (129)
Hv145S-chryso-P3	(82) LYTVVMPAGCGKTTIANEFNCIDVDDLA (109)
ACDACV-chryso-P3	(63) LFAIVLPAGCGKSTLCRKYGYLDIDECA (90)
CCRSACV-chryso-P3	(63) LFAIVLPAGCGKSTLCRKYGYLDIDECA (90)
PcV-RdRp	(84) LFAVIMPSGCGKTTLARTYGMVDVDELV (111)
Hv145S-RdRp	(60) LFAIILPAGTGKTYLAKKYGFIDVDKCV (87)
ACDACV-RdRp	(62) LFAIVMPGGTGKTRWAREYGLVDVDELV (89)
CCRSACV-RdRp	(62) LFAIVMPGGTGKTRWAREYGLVDVDELV (89)
	** : ** : * . * . *** . : ** . ** **** . . .

[a]Asterisks signify identical or similar residues (shaded) at the indicated position; colons signify at least six identical residues within a column; single dots signify 50% identical residues at the indicated position.

with three-dimensional image reconstruction. The outer surfaces of full particles of PcV, viewed along a five-, three-, and twofold axis of symmetry are shown in **Figure 6**. The capsid comprises 60 protein subunit monomers arranged on a $T=1$ icosahedral lattice. The outer diameter of the capsid is 406 Å and the average thickness of the capsid shell is 44 Å. At this low to moderate resolution, some features are distinguished on the relatively smooth topography of full and empty particles. The $T=1$ full capsid is formed by 12 slightly outward protruding pentamers making an underlying cavity. The pentamers are rather complex; on the outer surface, they are formed by five connected, elongated, ellipsoid-like structures surrounded by another five smaller similar structures that are neither connected to each other, nor intercalated with inner ellipsoid-like structures. No holes are evident

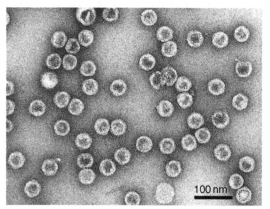

Figure 5 Negative contrast electron micrograph of particles of penicillium chrysogenum virus, the type species of the genus *Chrysovirus* in the family *Chrysoviridae*.

in the shell. The outer surface of the empty capsid is essentially identical to that of the full capsid except for the presence of five pores around the inner, elongated, ellipsoid-like structures at the fivefold axis, and three small pores around the threefold axis (not shown). There are significant differences between empty and full capsids on the inner surface around the five- and threefold positions, suggesting conformational changes. Although conformational changes have been characterized in related cores, the structural changes that are observed between empty and full PcV particles are considered unique.

The PcV capsid with its genuine $T=1$ lattice is the exception among dsRNA viruses whose capsids have '$T=2$' layers, which is the typical architecture for dsRNA viruses. PcV has the largest coat protein making a $T=1$ shell. Examination of the PcV coat protein sequence reveals a possible relationship with the canonical $T=2$ layers. Sequence analysis indeed suggests that the PcV capsid subunit falls into two similar domains that are likely to build a similar fold. Even though they are covalently linked, this unusual building unit could resemble the regular 120-subunit capsid. In this situation, PcV capsid might be considered a pseudo-$(T=2)$ structure. The coat protein of the chrysoviruses Hv145SV and CCRSACV share some of the repeated segments only in their amino-terminal half. The structures of the Hv145SV and CCRSACV capsids have yet to be elucidated.

Biological Properties

There are no known natural vectors for the recognized chrysoviruses PcV, PbV, Pc-fV, and Hv145SV. They are

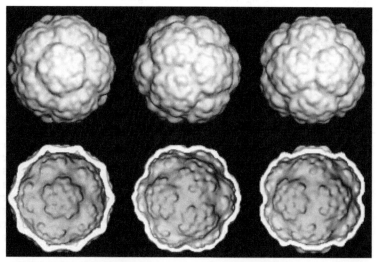

Figure 6 Three-dimensional structures of full PcV capsids. Surface-shaded representations of the outer surfaces of full PcV capsids viewed along a five-, three-, and twofold (left to right, upper row) axis of icosahedral symmetry. Models with the front half of the protein shell removed viewed along the five-, three-, and twofold axis are shown (left to right, lower row). Courtesy of R. J. Castón.

transmitted intracellularly during cell division and sporogenesis (vertical transmission), and following hyphal anastomosis (cell fusion) between compatible fungal strains (horizontal transmission).

Unlike the *Penicillium* chrysoviruses, which are associated with latent infections of their hosts, all other known chrysoviruses occur in mixed infections with other mycoviruses (or possibly plant viruses) and are associated with disease phenotypes of their hosts. The chrysovirus Hv145SV, which together with the totivirus Hv190SV co-infect the plant pathogenic fungus *Helminthosporium* (*Cochliobolus*) *victoriae*, is associated with a debilitating disease of the fungal host. The role of Hv145SV in disease development, however, is not yet clear.

ACDACV and CCRSACV are associated with two diseases of cherry, the Amasya cherry disease (ACD) and cherry chlorotic rusty spot (CCRS) disease. Both ACD and CCRS diseases are associated with a complex pattern of virus-like dsRNAs. Symptomatologically, both diseases are indistinguishable, a conclusion that was further supported by similar PAGE profiles for their associated dsRNAs. Furthermore, the sequences of at least six of these dsRNAs are essentially identical. Four of the ACD- and CCRS-associated dsRNAs comprise the genomic dsRNAs of the chrysoviruses ACDACV and CCRSACV, respectively. The etiology of these two cherry diseases is unknown and it has yet to be determined whether the dsRNAs associated with the diseases represent the genomes of plant or fungal viruses. In addition to the chrysovirus dsRNAs, the mixture of dsRNAs associated with the disease also contain a partitivirus and and larger dsRNAs with similar sizes to totiviruses. Whether the chrysoviruses alone or in combination with other dsRNA viruses play a role in disease development has yet to be elucidated.

The tentative chrysovirus AbV-1, also designated La France isometric virus (LFIV) induces a serious disease of cultivated mushroom (named La France disease). The AbV1 virions, isolated from diseased fruit bodies and mycelia, are isometric 34–36 nm in diameter and co-purify with nine dsRNA segments (referred to as disease-associated dsRNAs). The sizes of the dsRNA segments vary from 3.6 to 0.78 kbp, at least three of which are believed to be satellites. AbV1 represents a multiparticle system in which the various particle classes appear to have similar densities. Interestingly, phylogenetic analysis of RdRp conserved motifs of AbV1, encoded by dsRNA segment L1, and other dsRNA mycoviruses showed that AbV1 is closely related to the multipartite chrysoviruses. Although there is convincing evidence that infection with AbV-1 alone is essential and sufficient for induction of the La France disease, another mushroom virus, the mushroom bacilliform virus (MBV), the type species of the family *Barnaviridae*, is commonly found in mixed infections with AbV1 in diseased mushrooms. MBV is apparently not essential for disease development since no obvious pheno-

typic changes were observed in an *Agaricus* culture singly infected by MBV. However, synergistic interactions between AbV1 and MBV in doubly infected mushrooms cannot be ruled out.

Evolutionary Relationships among Chrysoviruses

The *Penicillium* chrysoviruses PcV, Pc-fV, and PbV are serologically related and have similar biochemical and biophysical properties. Although molecular data is only available for PcV, the three viruses could be considered as strains of the same virus for all practical purposes. The fact that these closely related viruses occur in different fungal species suggested that transmission by means other than hyphal anastomosis may occur in nature, since hyphal fusion between different fungal species is doubtful. Horizontal transmission of fungal viruses in nature, however, has yet to be demonstrated, and in the case of viruses of *Penicillium* species may not need to occur since the viruses replicate in parallel with their hosts and are carried intracellularly during the vegetative growth of the host (vertical transmission). Furthermore, the viruses are efficiently disseminated via the asexual spores (conidia) of *Penicillium* species. It seems feasible, however, that virus infection arose early in the phylogeny of *P. brevi-compactum*, *P. chrysogenurn*, and *P. cyaneo-fulvum* before they diverged and that the resident virus remained associated with them during their subsequent evolution.

BLAST searches of PcV RdRp amino acid sequence showed that it has significantly high sequence similarity (40% identity and 57% aa sequence similarity) to the RdRps encoded by the chrysoviruses Hv145SV, ACDACV, and CCRSACV. High similarities (BLAST hits of e^{-16} or lower) were also found to the RdRps of the tentative chrysovirus agaricus bisporus virus 1 (AbV-1) as well as to the totivirus ustilago maydis virus H1 (UmV-H1) and the giardiavirus trichomonas vaginalis virus (TVV). Still high similarity hits can be obtained with the RdRps of several members of the family *Totiviridae*. Interestingly, no significant hits were evident with any of the viruses in the family *Partitiviridae*, another validation for the removal of chrysoviruses from the family *Partitiviridae* and their placement in the newly created family *Chrysoviridae*. The conclusion that chrysovirus RdRps are more closely related to those of totiviruses than to those of partitiviruses is consistent with the results shown in **Figure 4**, where the RdRp conserved motifs of members of the three families are compared. Phylogenetic analysis (**Figure 7**), based on the complete nucleotide sequences of the RdRps of members of the three families, provides further confirmation to this conclusion.

BLAST searches of the deduced amino acid sequence of dsRNA2 ORF showed significant high similarity

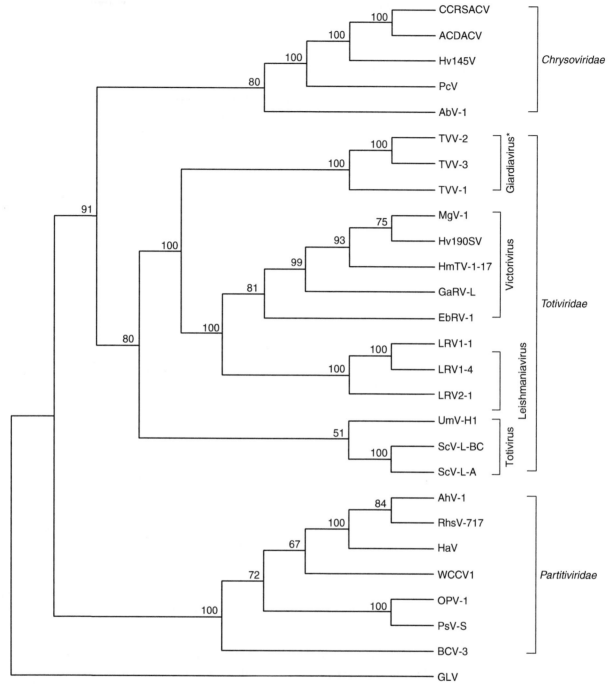

Figure 7 Neighbor-joining phylogenetic tree constructed based on the complete amino acid sequences of RdRps of selected isometric dsRNA viruses. The RdRp sequences were derived from aligned deduced amino acid sequences of members of the families *Chrysoviridae*, *Partitiviridae*, and *Totiviridae* using the program CLUSTAL X. See **Table 1** for names and abbreviations of chrysoviruses. The following viruses in the family *Totiviridae* were included in the phylogenetic analysis (abbreviations in parenthesis): trichomonas vaginalis virus-1 (TVV-1), TVV-2, and TVV-3, magnaporthe grisea virus 1 (MgV-1), helminthosporium victoriae 190S virus (Hv190SV), Helicobasidium mompa *totivirus 1-17* (HmTV-1-17) gremmeniella abietina RNA virus L (GaRV-L), eimeria brunetti RNA virus 1 (EbRV-1), leishmania RNA virus 1-1 (LRV1-1), LRV1-4, LRV2-1, ustilago maydis virus H1 (UmV-H1), saccharomyces cerevisiae virus L-A (ScV-L-A), and ScV-L-BC. The following viruses in the family *Partitiviridae* were included: atkinsonella hypoxylon virus 1 (AhV-1), rhizoctonia solani virus-717 (RhsV-717), heterobasidion annosum virus (HaV), white clover cryptic virus 1 (WCCV-1), ophiostoma partitivirus 1 (OPV-1), penicillium stoloniferum virus S (PsV-S), and beet cryptic virus 3 (BCV-3). The phylogenetic tree was generated using the program PAUP*. Bootstrap numbers out of 1000 replicates are indicated at the nodes. The tree was rooted with the RdRp of Giardia lamblia virus, the type species of the genus *Giardiavirus* in the family *Totiviridae*, which was included as an outgroup. * Note that TVV-1, TVV-2, and TVV-3 are tentative members of the genus *Giardiavirus*.

hits (3e⁻62) to Hv145SV CP (29% identity and 50% amino acid sequence similarity) and ACDACV and CCRSACV CPs (24% identity and 46% similarity). It is interesting that the region of high sequence similarity between the CPs of these chrysoviruses is limited to the N-terminal half of the proteins (aa 19–560 of PcV CP; data not shown). This finding may have implications when considering the structural organization of chrysovirus capsids.

Future Perspectives

Functions of Chryso-P3 and Chryso-P4

Present evidence suggests that both chryso-P3 and chryso-P4 are virion associated and that they may play a role in RNA transcription, RNA binding, and packaging. Because infectivity assays are not amenable for chrysoviruses, no direct evidence is currently available in support of the conclusion that chrysovirus segments 3 and 4 and their gene products are essential for virus infection and replication. However, based on the consistent co-presence of the four dsRNA segments, the existence of extended regions of highly conserved terminal sequences at both ends of all four segments, sequence comparisons and phylogenetic analysis, it is abundantly clear that the structural features of chrysoviruses are typical of RNA viruses with multipartite and multicomponent genomes. The dsRNA pattern of PcV virions isolated from different strains of *Penicillium chrysogenum* has remained unchanged throughout the years since PcV was first isolated. This is also true for other chrysoviruses isolated from different *Penicillium* species and from various strains of *Helminthosporium victoriae*. The co-presence of all four segments in different fungal species and strains harboring chrysoviruses and the stability of the dsRNA patterns support the contention that all four segments are essential for infection and that none is defective or satellite in nature.

An alternative approach to infectivity assays that should be applicable to chrysoviruses is to transform virus-free fungal isolates with full-length cDNA clones of viral dsRNAs. In a recent preliminary study involving transformation of virus-free *H. victoriae* protoplasts with individual as well as different combinations of full-length cDNAs of the four dsRNAs of Hv145SV, productive yield of ssRNA transcripts corresponding to all four dsRNA segments was only generated in transformants containing DNA copies of all four dsRNAs. Although dsRNA synthesis was not launched in these transformants, the results suggest that dsRNA segments 3 and 4 and their gene products are necessary for accumulation and stability of the viral transcripts.

Etiology of the CCRS Disease and Potential Occurrence of Plant Chrysoviruses

The etiology of the CCRS disease remains a mystery. Although there is circumstantial evidence for a fungal etiology for the disease, no fungal pathogen was ever isolated from diseased trees. Although it is true that no plant chrysoviruses have been identified to date, there is no reason to exclude this possibility. As a matter of fact, it is not yet determined whether the partitivirus cherry chlorotic rusty spot associated partitivirus (CCRSAPV), which is co-isolated along with CCRSACV from diseased cherry trees, is a plant or a fungal partitivirus. Phylogenetic analysis based on RdRp conserved motifs placed CCRSAPV in a cluster with a mixture of plant and fungal partitiviruses. This finding raised the interesting possibility of horizontal transfer of members of the family *Partitiviridae* between fungi and plants. This possibility is reasonable because some of the viruses in this phylogenetic cluster have fungal hosts that are pathogenic to plants. As more chrysoviruses from a wider range of fungal (and possibly plant) hosts are isolated and characterized, the reality of plant chrysoviruses may become apparent as well as the need to reconsider the taxonomy of the family *Chrysoviridae*.

See also: Fungal Viruses.

Further Reading

Castón JR, Ghabrial SA, Jiang D, *et al.* (2003) Three-dimensional structure of Penicillium chrysogenum virus: A double-stranded RNA virus with a genuine *T* = 1 capsid. *Journal of Molecular Biology* 331: 417–431.

Covelli L, Coutts RHA, Di Serio F, *et al.* (2004) Cherry chlorotic rusty spot and Amasya cherry diseases are associated with a complex pattern of mycoviral-like double-stranded RNAs. Part I: Characterization of a new species in the genus *Chrysovirus*. *Journal of General Virology* 85: 3389–3397.

Ghabrial SA, Jiang D, and Castón RJ (2005) *Chrysoviridae*. In: Fauquet CM, Mayo MA, Maniloff J, Desselberger U,, and Ball LA (eds.) *Virus Taxonomy: Eighth Report of the International Committee on Taxonomy of Viruses*, pp. 581–590. London: Academic Press.

Ghabrial SA, Soldevila AI, and Havens WM (2002) Molecular genetics of the viruses infecting the plant pathogenic fungus *Helminthosporium victoriae*. In: Tavantzis S (ed.) *Molecular Biology of Double-Stranded RNA: Concepts and Applications in Agriculture, Forestry and Medicine*, pp. 213–236. Boca Raton, FL: CRC Press.

Jiang D and Ghabrial SA (2004) Molecular characterization of Penicillium chrysogenum virus: Reconsideration of the taxonomy of the genus *Chrysovirus*. *Journal of General Virology* 85: 2111–2121.

Dicistroviruses

P D Christian, National Institute of Biological Standards and Control, South Mimms, UK
P D Scotti, Waiatarua, New Zealand

Glossary

Cohort A group of similar individuals.
Dipteran A member of the insect order Diptera: true flies.
Hemipteran A member of the insect order Hemiptera: true bugs (including aphids).
Hymenopteran A member of the insect order Hymenoptera: wasps and bees.
Intergenic region Region between the two open reading frames in the dicistrovirus genome.
Lepidopteran A member of the insect order Lepidoptera: moths and butterflies.
Orthopteran A member of the insect order Orthoptera: crickets and grasshoppers.
Penaeid A shrimp from the family Penaeidae.
Picorna-like Viruses that are ostensibly like members of the *Picornaviridae*; but the term is generally used to refer to any small (*c.* 30 nm in diameter) icosahedral viruses with single-stranded RNA genomes.
Polyprotein A protein that is cleaved after synthesis to produce a number of smaller functional proteins.
Vertical transmission Transmission of virus directly from an infected mother to her offspring.
VPg A virally encoded protein covalently linked to the 5′ end of the viral genome.

Introduction

After the early 1960s, it became apparent that invertebrates, as well as vertebrates and plants, played host to a number of small (<40 nm in diameter) icosahedral viruses with RNA genomes. The initial descriptions of many of these viruses involved little more than their physical and biochemical characteristics such as diameter, density, and S-value. By the 1970s, many new invertebrate small RNA viruses were being isolated and described and even minimal characterization made it clear that new families of viruses were emerging from this assemblage of small RNA-containing viruses of invertebrates, along with apparent members of existing virus families. Two major families of viruses recognized in the late 1970s and early 1980s were the *Tetraviridae* and *Nodaviridae*. Any of the other viruses were simply considered to be 'invertebrate picornaviruses' or 'small RNA viruses of insects'.

The properties of many of the yet-unclassified viruses were found to be very similar to those of the mammalian picornaviruses. In particular, the size of the virions (*c.* 30 nm), the composition of the capsids (three major proteins of around 30 kDa), and single-stranded, positive-sense RNA genomes all suggested these were invertebrate picornaviruses and this was very much the prevalent feeling – until 1998. At this point, the first full genome sequence of one such invertebrate virus, drosophila C virus (DCV), was published and surprisingly revealed a genome organization strikingly different from the picornaviruses, and indeed quite different from any other viruses known at that time. During the next several years, the genomes of a number of insect small RNA-containing viruses were sequenced and it became clear that two organizational paradigms existed. The first group became the *Dicistroviridae* while the second has become the (currently) unassigned genus *Iflavirus*.

Taxonomy and Classification

The family *Dicistroviridae* currently comprises 12 species, most of them in the only genus recognized so far, *Cripavirus* (**Table 1**). There are a number of other potential candidates for the family but these have yet to be accepted as species by the International Committee on Taxonomy of Viruses. For the purposes of this article, we will limit our discussion to only those species shown in **Table 1**.

Biophysical Properties

Dicistroviruses appear roughly spherical under the electron microscope in negative stained preparations with particle diameters of approximately 30 nm and no envelope (**Figure 1**). The mature virions contain three major structural proteins, VP1, VP2, and VP3 of between 28 and 37 kDa, although taura syndrome virus (TSV) does appear to have one larger structural protein of 56 kDa. In many mature virion preparations, one or more minor structural components can be present which are larger than the major capsid proteins and are presumed to be the precursor(s) of structural proteins. In some viruses, a fourth smaller structural protein (VP4) – of between 4.5 and 9 kDa – is also present in the virion. A summary of the biophysical properties of dicistrovirus virions and the size and composition of the genome is given in **Table 2**.

Table 1   Members of the virus family *Dicistroviridae*. Isolate and vernacular names are shown in brackets. Accession number for whole genome sequences are also given. A recently suggested taxonomy that places ABPV, KBV, SINV-1, and TSV as unassigned species in the family is followed

Genus	Species (isolate name)	Accession number	Abbreviation
Cripavirus	*Cricket paralysis virus* [type species] (cricket paralysis virus)	[AUF218039]	CrPV
	Aphid lethal paralysis virus (aphid lethal paralysis virus)	[AUF536531]	ALPV
	Black queen-cell virus (black queen-cell virus)	[AUF183905]	BQCV
	Drosophila C virus (drosophila C virus)	[AUF014388]	DCV
	Himetobi P virus (himetobi P virus)	[AUB017037]	HiPV
	Plautia stali intestine virus (plautia stali intestine virus)	[AUB006531]	PSIV
	Rhopalosiphum padi virus (rhopalosiphum padi virus)	[AUF022937]	RhPV
	Triatoma virus (triatoma virus)	[AUF178440]	TrV
Unassigned species in the family	*Acute bee paralysis virus* (acute bee paralysis virus)	[AUF150629]	ABPV
	Kashmir bee virus (Kashmir bee virus)	[AUY452696]	KBV
	Solenopsis invicta virus-1 (solenopsis invicta virus-1)	[AUY634314]	SINV-1
	Taura syndrome virus (taura syndrome virus)	[AUF277675]	TSV

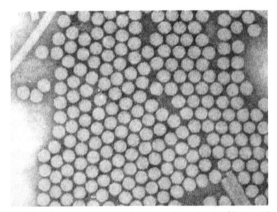

Figure 1   Negative stained electron micrograph of isometric particles CrPV showing rod-shaped particles of tobacco mosaic virus (diameter 18 nm) in the upper left- and lower right-hand corners. Electron micrograph supplied courtesy of Carl Reinganum.

The virions exhibit icosahedral, pseudo $T = 3$ symmetry and are composed of 60 protomers. Each of the protomers is composed of a single molecule of each of the structural proteins VP2, VP3, and VP1. The protomers are arranged so that the molecules of VP1 are set around the fivefold axes (**Figure 2**), with VP4 lying inside the virion below the molecules of VP1. Where the molecules of VP1 come together the surface of the virion shows a slightly raised crown – similar to that on the surface of poliovirus and other picornaviruses (**Figure 3**). In contrast to poliovirus, the surface of cricket paralysis virus (CrPV) does not show the characteristic deep canyon around the fivefold axes – which in the case of the former is where the receptor-binding site is known to be. The mature virions have a buoyant density in neutral CsCl of between 1.34 and 1.39 g cm^{-3} and sedimentation coefficients that range between 153S and 167S. For those viruses where physico-chemical stability has been assessed, for example, CrPV, the virions are stable at pH 3.0 and are resistant to treatment with detergents and organic solvents such as ether and chloroform.

The virions contain a single molecule of linear, positive-sense, single-stranded RNA (ssRNA) of approximately 9000–10 000 nt in size. Structural studies have not revealed any ordered structure to the RNA within the virion. Terminal modifications to the RNA include a covalently linked protein at the 5′ end of the genome (referred to as the VPg) and a polyA tract at the 3′ end.

Organization of the Dicistrovirus Genome

The single-stranded genomes possess a 5′ untranslated region (5′ UTR) of 500–800 nt followed by two open reading frames (ORFs) of *c.* 5500 and 2600 nt. The ORFs are separated by an untranslated region of ∼190 nt, commonly referred to as the intergenic region (IGR) (**Figure 4**).

Table 2   Summary of some biophysical properties of dicistroviruses

Virus	Molecular weight of major capsid proteins (kDa)[a]	Buoyant density in CsCl (g ml^{-1})	Particle diameter (nm)
Acute bee paralysis virus	35, 33, 24, 9	1.34	30
Aphid lethal paralysis virus	34, 32, 31 (41)	1.34	27
Black queen-cell virus	34, 32, 29, 6	1.34	30
Cricket paralysis virus	35, 34, 30 (43)	1.37	27
Drosophila C virus	31, 30, 28, 9 (37)	1.34	27
Himetobi P virus	37, 33, 28	1.35	29
Kashmir bee virus	41, 37, 25, 6	1.37	30
Plautia stali intestine virus	33, 30, 26, 5	n.d.	30
Rhopalosiphum padi virus	31, 30, 28 (41)	1.37	27
Solenopsis invicta virus-1	n.d.	n.d.	31
Taura syndrome virus	55, 40, 24 (58)	1.34	31
Triatoma virus	39, 37, 33	1.39	30

[a]Minor virion components are shown in brackets. These are presumed to be precursors of VP4–VP3.

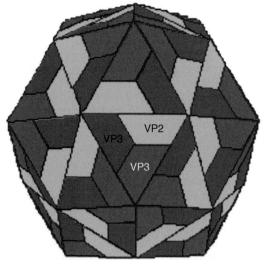

Figure 2   Diagram showing the surface packing of the coat proteins (VP1, VP2, VP3) of CrPV. Reproduced with permission from Fauquet CM, Mayo MA, Maniloff J, Desselberger U, and Ball LA (eds.) (2005) *Virus Taxonomy: Eighth Report of the International Committee on Taxonomy of Viruses.* San Diego, CA: Elsevier Academic Press.

The virion proteins (VPs) have been shown, by direct sequence analysis, to be encoded by an ORF proximal to the 3′ end, while the more 5′ ORF encodes protein(s) that have sequence motifs lying in the order (5′–3′) Hel-Pro-Rep, a feature common among a large number of other positive-sense RNA viruses such as picornaviruses, comoviruses, sobemoviruses, caliciviruses, sequiviruses, and potyviruses.

Where present, the sequence coding for the small virion protein VP4 is in the ORF encoding the structural proteins, between the region coding for the capsid proteins VP2 and VP3 (as it is also in the iflaviruses). VP4 is cleaved from VP3 during the maturation of the virion. The VPg is encoded in the nonstructural protein-encoding region, and in most dicistroviruses there are multiple copies of the VPg coding sequence which show some degree of heterogeneity.

Virus Replication and Genome Expression

The mechanism of virus entry into susceptible cells is unknown. Initiation of protein synthesis coincides with the shutdown or downregulation of host cell protein synthesis. During infection, a large number of precursor proteins are produced which are then cleaved to produce an array of smaller polypeptides. In the case of CrPV (one of the few dicistroviruses for which a permissive cell culture system exists), the structural proteins are produced in supramolar excess relative to the nonstructural proteins. Like many positive-sense RNA viruses, no subgenomic RNAs (sgRNAs) are produced during the infection cycle.

The absence of sgRNAs indicates that the translation of ORF2 could be initiated by an internal ribosome entry site (IRES) similar to the mechanism known from picornaviruses. Experimental studies did indeed demonstrate this and that both the 5′ UTR and IGR of several dicistroviruses including CrPV, plautia stali intestine virus (PSIV), and rhopalosiphum padi virus (RhPV) can act as IRES elements to direct initiation of translation in either *in vitro* translation systems or in cultured invertebrate cells.

However, perhaps the most unique feature of the replicative strategy of dicistroviruses is the fact that translation of the structural proteins from IGR–IRES element does not require the presence of a methionine codon. Computer modeling has shown that the IGRs of all dicistroviruses have predictable stem–loop structures. While there are two basic types among the dicistroviruses (**Figure 5**), these are fundamentally the same with six stem–loops – the most 3′ of which forms a pseudoknot with the codons involved in translation initiation.

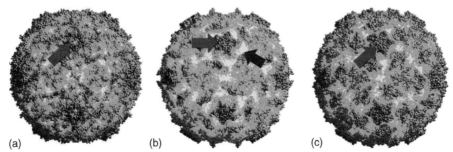

Figure 3 The surface structure of (a) CrPV, (b) poliovirus, and (c) rhinovirus 14. The red arrows show the raised crown around the fivefold axes and in poliovirus the black arrow indicates the deep canyon around this raised crown. Courtesy of John Tate.

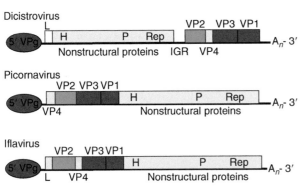

Figure 4 Diagrammatic representation of the genomic arrangement of the dicistroviruses, picornaviruses, and iflaviruses. The helicase (H), protease (P), and replicase (Rep) domains of the nonstructural proteins are indicated.

The initiation of translation is thought to be mediated by the codon which forms the part of the pseudoknot with stem loop VI, immediately upstream of the triplet which encodes the first amino acid of the mature VP2 (**Table 3**). In most cases, the initiation codon is CCU (Pro) while the first codon of VP2 encodes an alanine residue (**Table 3**). It has been shown experimentally that the glutamine at the 5′ end of PSIV VP2 can be replaced with any other amino acid to produce a mature protein in an *in vitro* translation system.

Where cell culture systems are available, pulse-chase studies have shown that translation of dicistrovirus genomes results in the production of polyproteins that are then cleaved to produce the structural and nonstructural proteins. For the virion proteins of some dicistroviruses, the cleavage sites between proteins in the structural polypeptide have been experimentally determined and are shown in **Table 3**. These sites show some degree of conservation and point toward the involvement of cysteine proteases. While the viruses themselves encode cysteine protease-like peptides in the nonstructural region there is also some evidence, for CrPV at least, that host-cell-encoded proteases may also be involved in the processing of viral polypeptides.

Host Range

To date, all members of the *Dicistroviridae* have been isolated from invertebrates, generally from a single species, or at most three closely related species. The hymenopteran viruses, acute bee paralysis virus (ABPV) and black queen-cell virus (BQCV), are known only from honeybees (*Apis mellifera*) but Kashmir bee virus (KBV) has also been isolated from the Asiatic hive bee *Apis dorsata*. It should also be noted that ABPV and KBV have been identified in the parasitic mite *Varroa destructor*, although there is no clear evidence that the virus is capable of replicating in this host. Solenopsis invicta virus-1 (SINV-1) is only known to infect the ant species, *Selonopsis invicta*.

Dicistroviruses found in homopterans and hemipterans (which includes aphids and true bugs, respectively) have slightly broader host ranges. RhPV has been isolated from laboratory and field populations of the aphids *Rhopalosiphum padi*, *R. maidis*, *R. rufiabdominalis*, *Schizaphis graminum*, *Diuraphis noxia*, and *Metapolophium dirrhodum*. Aphid lethal paralysis virus (ALPV) has only been isolated from aphids, but in this case from only three species: *R. padi*, *M. dirrhodum*, and *Sitobian avenae*. Himetobi P virus (HiPV) and PSIV are found in true bugs rather than aphids with HiPV having been isolated from the leafhoppers *Laodelphax striatellus*, *Sogatella furcifera*, and *Nilaparvata lugens*, and PSIV from the brown-winged green bug, *Plautia stali*. Triatoma virus is also a virus of hemipterans and has been found in the hematophagous triatomine bug, *Triatoma infestan*, which is also a vector of the protozoan agent that causes Chagas' disease in South America.

DCV is the only dicistrovirus with a host range restricted to dipterans and has been isolated from *Drosophila melanogaster* and the sibling species *D. simulans*. TSV is a virus of penaeid shrimps and has been isolated from a number of species including *Litopenaeus vannamei*, *L. stylirostris*, *Metapenaeus ensis*, and *Penaeus monodon*. The majority of the dicistroviruses have relatively restricted host ranges and, at most, have only been isolated from insects of a single order. CrPV is the striking exception.

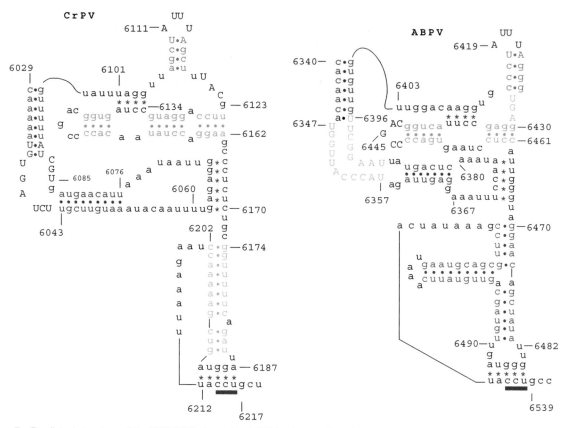

Figure 5 Predicted structure of the IGR–IRES elements of dicistroviruses. The structure represented by that of CrPV is shared by most dicistroviruses with the exception of ABPV, KBV, SINV-1, and TSV that have a structure similar to that represented by ABPV. The stem loop VI pseudoknot stuctures are shown in pink for CrPV and in purple for ABVP. The first translated codon is underlined in blue. Adapted with permission from Nobuhiko Nakashima.

Table 3   Some properties of the translation and processing of the structural polyprotein of the dicistroviruses. Only cleavage sites that have been empirically deduced from sequencing of the virion proteins are shown

Virus	First residue[a]	Initiation codon	Cleavage VP2/VP4	Cleavage VP4/VP3	Cleavage VP3/VP1
Cricket paralysis virus	A	CCU	IYAQ/AASE	LFGF/SKPT	n.d.
Acute bee paralysis virus	n.d.	CCU	VTMQ/INSK	IFGW/SKPR	ASMQ/INLA
Aphid lethal paralysis virus	A	CCU	n.d.	n.d.	n.d.
Black queen-cell virus	A	CCU	MLAQ/AGLK	LFGF/SKPL	MVAG/SNSG
Drosophila C virus	A	CCU	n.d.	MLGF/SKPT	IVAQ/VMGE
Himetobi P virus	A	CUA	AREQ/VNLN	APGF/KKPD	STAQ/EQAN
Kashmir bee virus	n.d.	CCU	n.d.	n.d.	n.d.
Plautia stali intestine virus	Q	CUU	LILQ/SGET	AFGF/SKPQ	LTLQ/SGDT
Rhopalosiphum padi virus	A	CCU	n.d.	THGW/SKPL	SIAQ/VGTD
Solenopsis invicta virus-1	n.d.	CCU	n.d.	n.d.	n.d.
Taura syndrome virus	A	CCU	n.d.	MFGF/SKDR	PSTH/AGLD
Triatoma virus	n.d.	CUC	n.d.	ALGF/SKPL	PIAQ/VGFA

[a]The first residue of the mature VP2.

Originally isolated from the field crickets *Teleogryllus oceanicus* and *T. commodus*, CrPV has subsequently been isolated from a further 20 species belonging to five taxonomic families: *Orthoptera, Hymenoptera, Lepidoptera, Hemiptera,* and *Diptera*. Interestingly, while a number of other RNA-containing viruses are known from lepidopterans (moths and butterflies), that is, the iflaviruses and tetraviruses, as well as a number of uncharacterized viruses, CrPV is the only dicistrovirus isolated from lepidopterans – in fact from ten lepidopteran species.

All of the above records refer to the natural host range of the viruses. To a certain extent, studies on the experimental host range of dicistroviruses are limited and those that have been carried out have not substantially extended the known host ranges. Again the exceptions are CrPV and DCV. CrPV replicates in a number of established insect cell lines including those from *Drosophila*, the hemipteran *Agallia constricta*, and the lepidopterans *Pieris rapae*, *Plutella xylostella*, *Spodoptera ornithogalli*, and *Trichoplusia ni*. In addition to insect cell lines, CrPV has also been found to replicate readily in larvae of the greater waxmoth, *Galleria mellonella*. This is an easy insect to rear and maintain and virus yields can be very high. Apart from CrPV, the only other dicistrovirus shown to replicate in cultured cells is DCV, which multiplies in several *Drosophila* cell lines (some DCV isolates also replicate in the greater waxmoth). Reports that TSV can replicate in some mammalian cell lines have never been substantiated and may simply be attributable to the production of a cytopathic effect in the absence of virus replication.

Pathology and Transmission

The names of several of the dicistroviruses imply that virus infection can produce a noticeable disease symptom, that is, CrPV, ABPV, ALPV, and BQCV (potential queens die as propupae or pupae in hive cells in which the walls turn black). However, the majority of virus–host interactions produce no noticeable disease. Although BQCV kills some queens in their cells, most infected larvae, pupae, and workers appear to be completely unaffected.

Nevertheless, while disease symptoms may not be evident, in most of the cases studied, dicistrovirus infections reduce the life span of the infected individual. Or, more precisely, the presence of a dicistrovirus in a cohort (group of similar individuals) of insects coincides with a reduced life span relative to an uninfected cohort. This point is explicitly made since no studies have yet been undertaken where an invertebrate host has been nonlethally sampled for virus through its life span. In many other virus–host systems, for example, plants, mammals, or even fish and amphibians, it is possible to sample for virus without killing the host so the progress of infection can be measured at the level of the individual. With invertebrates, this has not been done and would generally be very difficult or impossible. The experimental approach has therefore been to introduce the virus into an uninfected cohort and compare the effects to a control cohort.

Despite these practical limitations, a number of studies have demonstrated the routes of virus transmission. BQCV has been detected in the eggs, larvae, and offspring of queens that were found to be infected – indicating that the virus is vertically transmitted. Similar findings have been made with ALPV which can be vertically transmitted in the aphid host, *R. padi*; this infection subsequently results in reduced longevity and fecundity. It has also been shown that ALPV RNA can be detected in the developing embryos inside infected females.

DCV is also vertically transmitted although virus is not present in the cytoplasm of the egg but is associated with the chorion on the egg surface. Infection presumably occurs when emerging larvae ingest the virus since they do not become infected if the chorion is removed. A similar phenomenon is found with CrPV and *T. oceanicus* where surface sterilization/dechorionation of eggs with dilute hypochlorite blocks the transmission of the virus to emerging nymphs.

Evidence of horizontal transmission is in many respects even more difficult to obtain than data on vertical transmission. However, experimental feeding studies indicate that dicistroviruses can be transmissed horizontally. In the case of triatoma virus (TrV) and its host *T. infestans*, infected insects excrete virus in their feces and since triatomine bugs are coprophagous and TrV is infectious *per os*, the virus can be readily spread. With DCV, it has been found that when uninfected males are placed with uninfected females, these males become infected (and vice versa). In fact, even if uninfected female flies are placed on media on which infected males have been allowed to feed for several hours, the females become infected. Such manipulations with other virus and invertebrate host systems are not as easy as those involving *Drosophila*, as it is difficult to obtain colonies free of viruses and the insects are not as easy to manipulate in the laboratory.

Evidence of vectors playing a role in dicistrovirus transmission is limited to honeybee viruses and their parasitic mite, *Varroa destructor*. It has been known for some time that the prevalence of a number of honeybee viruses increases in the presence of varroa mites. Initially, it was thought that the stress caused by varroa infestation induced the viruses to replicate. However, recently it has been shown that ABPV and KBV can be detected in mites implicating them more strongly in transmission of the viruses. In contrast, BQCV, although present at high frequency in adult honeybees, has not been detected in mites, suggesting that they play little or no role in transmission.

Geographic and Strain Variation

Some dicistroviruses infect specific hosts with a wide geographic distribution or have been found to infect a number of species spread over a large geographic range. Several studies have looked at strain variation between geographic isolates of dicistroviruses using a variety of techniques. CrPV and DCV possess a range of biological and serological characteristics that show differentiation between geographical isolates. With CrPV, two major serogroups have been identified that separate Australian and New Zealand isolates from North American isolates. With DCV, isolates from different localities varied both in their pathogenicity and virus yield after injection into virus-free flies. Additional genetic studies on CrPV and DCV have utilized ribonuclease T1 fingerprinting and subsequently polymerase chain reaction restriction endonuclease analysis. Estimates of the maximum nucleotide divergence between isolates of CrPV and DCV were about 10%. In the case of CrPV, it was found that that the North American isolates were quite distinct from the antipodean isolates, a finding which reflects the serological data.

More recently, a number of molecular studies with the honeybee dicistroviruses and TSVs have been undertaken to determine the levels of genetic variation between isolates of these viruses. In most instances, slightly different regions of the virion protein coding regions (usually regions of VP3) have been used, which makes direct comparison between studies quite difficult. Nevertheless, ABPV isolates show levels of nucleotide identity between 90% and 100%, but isolates from the same geographic region are more closely related than isolates from more distant locations. One such study has revealed that viruses isolated from different central Europe regions were more similar to each other than to isolates from North America or from the UK. Similar patterns are also evident for BQCV and KBV with nucleotide identity within species at 90–100%. To put these values in perspective, the region used for the KBV studies is *c.* 75% identical to the same region from ABPV.

For TSV, the situation is slightly different, and it has been found that levels of nucleotide identity between isolates from North and Central America and Asia range from 95% to 100%. These lower levels of diversity may indicate that TSV has rapidly spread into many of the regions and hosts where it is now found – a hypothesis that is supported to some extent by the rapid emergence of the disease over the last two decades. There is also some evidence that there are serological differences between isolates.

Relationships within the Family

The taxonomy of the dicistroviruses currently recognizes only one genus, *Cripavirus*, into which most of the species are placed. However, comparisons of the coding sequences of the dicistroviruses reveal that they are only distantly related. For instance, the amino acid identity between the structural polyproteins of different species ranges from 19% to 66% while the amino acid similarity ranges from 37% to 80%. Using these data, the phenogram presented as **Figure 6** shows the pattern of relationships and reveals only two closely related pairs of species: CrPV and DCV

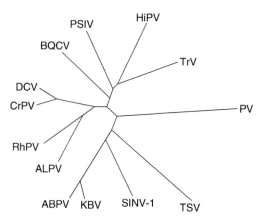

Figure 6 Phenogram showing the relationships among members of the family *Dicistroviridae,* constructed from the amino acid identity of the structural proteins encoded in ORF2. The phenogram was constructed using the neighbor-joining algorithm of the MEGA software. The sequence from Poliovirus type 3 [L23844] was used as an outgroup for the analysis. Branch lengths are drawn approximately to scale.

(which also share a serological relationship) and ABPV and KBV. All other species are only distantly related. While the genus *Cripavirus* is currently proposed to include all but ABPV, KBV, SINV-1, and TSV, this genus will probably be divided and new genera established to accept the currently unassigned species listed above.

Similarity with Other Taxa

The dicistroviruses share various properties with a number of other positive-sense ssRNA virus genomes. Historically, these viruses have been collectively grouped as the picornavirus superfamily which has been considered to include the virus families *Comoviridae, Picornaviridae, Potyviridae, Sequiviridae, Caliciviridae,* and more recently the unassigned genera *Iflavirus, Sadwavirus,* and *Cheravirus* and the recently created family *Marnaviridae.* In all of the groups mentioned above, the gene order for the nonstructural proteins is the same, viz. Hel-Pro-Rep (as in **Figure 4**). When compared to other positive-sense viruses, these genes appear to be more closely related to each other than to genes from other viruses, for example, coronaviruses, toroviruses, and tetraviruses. However, there is a subset of the above taxa which shares a larger set of properties: for example, isometric virus particles with pseudo $T = 3$ symmetry (formed of three units of eight-stranded β-barrel); the absence of sgRNAs during replication; and the presence of a 3–4 kDa VPg. On the basis of these properties, the grouping would exclude the *Caliciviridae* (sgRNAs, large VPg, and true $T = 3$ symmetry) and the *Potyviridae* (large VPg, rod-shaped particles with helical symmetry).

There are also a large number of other isolates of invertebrate picorna-like viruses that share some properties with dicistroviruses, for example, single-stranded postive-sense RNA genomes, isometric virions of around 30 nm in diameter, and up to three virion proteins of around 30 kDa. Further studies will undoubtedly reveal additional structural and organizational paradigms among the yet uncharacterized picorna-like viruses of invertebrates besides increasing the number of known dicistroviruses.

Further Reading

Bailey L and Ball BV (1991) *Honey Bee Pathology,* 2nd edn. Sidcup: Harcourt Brace Jovanovich.

Fauquet CM, Mayo MA, Maniloff J, Desselberger U,, and Ball LA (eds.) (2005) *Virus Taxonomy: Eighth Report of the International Committee on Taxonomy of Viruses.* San Diego, CA: Elsevier Academic Press.

Miller LK and Ball LA (eds.) (1998) *The Insect Viruses.* New York: Plenum Press.

Fungal Viruses

S A Ghabrial, University of Kentucky, Lexington, KY, USA
N Suzuki, Okayama University, Okayama, Japan

Glossary

Hyphal anastomosis The union of a hypha with another resulting in cytoplasmic exchange.

Hypovirulence Attenuated fungal virulence mediated by virus infection, mitochondrial defects, or mutations in the fungal genome.

Mycoviruses Viruses that infect and multiply in fungi.

Vegetative incompatibility A genetically controlled self/nonself recognition system in fungi that determines the ability to undergo hyphal anastomosis.

Introduction

Relative to plant virology, animal virology, or bacterial virology, fungal virology is new. In 1948, an economically important disease of the cultivated mushroom, *Agaricus bisporus,* characterized by malformed fruiting bodies and serious yield loss, was first reported in a mushroom house owned by the La France Brothers of Pennsylvania. The disease was called 'La France disease', and similar diseases were reported shortly thereafter from Europe, Japan, and Australia. Different designations, such as 'X-disease', 'watery stripe', 'brown disease', and 'dieback' were given to basically the same disease as the La France. The significance of the 1948 report lies in the fact that it led to the discovery of fungal viruses. In 1962 Hollings noted the presence of at least three types of virus particles in diseased mushroom sporophores. This was the first report of virus particles in association with a fungus and is regarded as the dawn of mycovirology. The subsequent discovery that dsRNA of mycoviral origin was responsible for interferon-inducing activities of cultural filtrates of several species of *Penicillium* spp. greatly stimulated the search for fungal viruses.

Fungi, like other living organisms, can be infected by a number of viruses, and mycoviruses are found in all the major groups of fungi. Although mycoviruses are widely prevalent, only those infecting a limited number of fungal host species have been studied, for example, the yeast *Saccharomyces cerevisiae,* edible mushroom, and phytopathogenic fungi. Given the predicted vast number of fungal species (approximately 10 000 known species and many more unknown species), it is expected that a greater number of unrecognized mycoviruses occur in nature. Support for this idea comes from recent extensive searches of field fungal isolates that showed relatively high frequencies of virus infection, for example, approximately 65%, 20%, and 2–28% of *Helicobasidium mompa, Rosellinia necatrix,* and *Cryphonectria parasitica* isolates were found to be infected, respectively.

Biological Properties

Host Range

The natural host range of mycoviruses is likely to be restricted to the same or closely related vegetative compatibility groups that allow lateral transmission. Until recently, there were no known experimental host ranges for fungal viruses because of lack of suitable infectivity assays. Experimental host ranges for some mycoviruses, however, were recently demonstrated and shown to extend to different vegetative compatibility groups and even to different genera. For example, the prototype mycoreovirus (mycoreovirus 1) can replicate and induce phenotypic alterations in different vegetative compatibility groups of the chestnut blight fungus,

Cryphonectria parasitica, similar to those exhibited by the original virus-containing strain. Furthermore, CHV1-EP713, the type member of the family *Hypoviridae*, can replicate in and confer hypovirulence onto members of several other fungal genera, for example, *Endothia gyrosa* and *Valsa ceratosperma*, in addition to its natural host, *C. parasitica*.

Symptom Expression

Although the majority of the viruses that infect phytopathogenic fungi have been reported to be avirulent, phenotypic consequences of infections with mycoviruses can vary from symptomless to severely debilitating, and from hypovirulence to hypervirulence. Mycoviruses that attenuate the virulence of plant pathogenic fungi provide excellent model systems for basic studies on development of novel biological control measures and for dissecting the mechanisms underlying fungal pathogenesis. In general, infections due to mycoviruses are both symptomless and persistent. Latency benefits the host for survival, and persistence helps the virus in the absence of extracellular modes of transmission. To ensure their retention, some mycoviruses have evolved to bestow selective advantage to their host (e.g., the killer phenotypes in yeasts and smuts). Because of their ability to secrete killer toxins, yeast killer strains have been utilized by the brewing industry to provide protection against contamination with adventitious sensitive strains. The genes encoding the smut killer toxins have been used for the development of novel transgenic approaches to control the corn smut.

Macroscopic symptoms caused by fungal viruses are a consequence of alterations in complex physiological processes that involve interactions between host and virus factors. Several virally encoded proteins are identified as symptom determinants including the papain-like protease, p29, of some hypoviruses. This protein acts to repress host pigmentation and conidiation regardless of whether it is expressed from host chromosomes or the homologous virus genome. Another interesting example is the proteins encoded by the totivirus *Helminthosporium victoriae* 190S virus. Co-expression of the capsid and RNA-dependent RNA polymerase (RdRp) proteins results in empty capsid production and phenotypic changes similar to those induced in virus-infected isolates, suggesting that viral replication is not required for symptom development. Host factors involved in symptom expression are not well studied except for host genes involved in the killer phenomenon in yeast infected with the totivirus L-A and associated satellite dsRNAs. Transcriptome analysis was performed for only limited virus/natural host combinations. Of the 2200 *C. parasitica* genes, 13.4% are either upregulated or downregulated upon infection with the prototype severe strain of the hypovirus CHV1, while only 7.5% are altered in their transcription levels by infection with a mild strain of CHV1. One-half of the genes responsive to the infection with the latter virus are commonly altered in transcription by the severe strain, which generally causes greater magnitude of transcriptional changes.

Transmission

Mycoviruses lack an extracellular phase to their life cycles. They are transmitted intracellularly during cell division, sporogenesis, and cell fusion. Lateral (horizontal) transmission usually occurs only between individuals within the same species, which belong to the same or closely related vegetative compatibility groups. Vegetative compatibility is governed genetically. Mycoviruses may be eliminated during sexual spore formation. Although the totiviruses and narnaviruses that infect the yeasts are effectively transmitted via ascospores, the mycoviruses infecting the ascomycetous filamentous fungi are eliminated during ascospore formation. Whereas the ssRNA and dsRNA mushroom viruses are transmitted efficiently via basidiospores, the virus-containing strains of the basidiomycete *H. mompa* are cured during the sexual sporulation processes. Therefore, whether a mycovirus is transmitted through sexual spores depends on the host/virus combination involved. Whereas mycovirus transmission through asexual spores occurs frequently, its rate varies greatly depending on the combination of viral and host strains. It was recently shown that the papain-like protease p29, encoded by CHV1, might play a role in virus transmission since it enhances the transmission of the homologous virus (CHV1) as well as heterologous viruses (mycoreoviruses) in *C. parasitica* conidia.

While rare in nature, interspecies transmission has been reported between members within the same genus including *Cryphonectria*, *Sclerotinia*, and *Ophiostoma* that share the same habitats. It remains unknown whether interspecies barrier is overcome by physical contacts or by vectors. Experimental evidence for vector transmission of fungal viruses, however, is lacking.

Mixed Infections

Mixed infections with two or more unrelated viruses and accumulation of defective dsRNA and/or satellite dsRNA are common features of mycovirus infections. Examples include the sphaeropsis sapinea totiviruses SsRV1 and SsRV2, saccharomyces cerevisiae totiviruses ScV-L-A and ScV-LB/C, and penicillium stoloniferum partitiviruses PsV-S and PsV-F.

There are a number of known examples of mixed infections with plant or animal viruses where one virus either interferes with or enhances the replication of the other one. As a consequence, reduction or increase in symptom severity may arise. Recent extensive searches for fungal viruses confirmed relatively high mixed infection rates in some phytopathogenic fungi like *C. parasitica*, *R. necatrix*, and *H. mompa*. In contrast to mixed virus infections in plants or animals, possible interactions between co-infecting mycoviruses in single hosts is little studied because of the limitation in experimental manipulation of viruses and hosts. Recent studies suggest synergistic interactions between a hypovirus (phylogenetically related to potyviruses and viruses belonging to the picorna-like superfamily) and a mycoreovirus through a hypovirally encoded protein. In this case, one-way synergism is observed in that only hypovirus infection transactivates the replication of a mycoreovirus, which has a replication strategy distinct from that of hypoviruses.

Fungal Virus Taxonomy

A list of the virus families and genera into which mycoviruses are classified is included in **Table 1**. Members of some fungal virus families, for example, the families *Narnaviridae*, *Chrysoviridae*, and *Hypoviridae*, infect only fungi, while members in other families, for example, the families *Metaviridae*, *Pseudoviridae*, *Reoviridae*, *Totiviridae*, and *Partitiviridae*, infect fungi, protozoa, plants, or animals. Except for the rhizidiomyces virus with dsDNA genome (the only member in the genus *Rhizidiovirus*), almost all other mycoviruses have RNA genomes and many have dsRNA genomes. Viruses with (−)-strand RNA or ssDNA genomes have yet to be found in fungi.

dsRNA Mycoviruses

Mycoviruses with dsRNA genomes represent the majority of the fungal viruses so far reported. With the exception of the mycoreviruses (genus *Mycoreovirus*, family *Reoviridae*), which have spherical double-shelled particles 80 nm in diameter, the dsRNA mycoviruses are typically isometric particles 25–50 nm in diameter. They are classified, based on the number of genome segments, into three families, *Totiviridae*, *Partitiviridae*, and *Chrysoviridae*. Viruses in the family *Totiviridae* have nonsegmented dsRNA genomes coding for a CP and an RdRp. At present, three genera have been placed in this family: *Totivirus*, *Giardiavirus*, and *Leishmaniavirus*. Viruses in the genus *Totivirus* infect fungi, whereas those belonging to the latter two genera infect parasitic protozoa. At least two distinct RdRp expression strategies have been reported for totiviruses: (1) those that express their RdRp as a fusion protein (CP-RdRp or Gag-Pol) by ribosomal frameshifting, such as the yeast L-A and the viruses that infect parasitic protozoa and (2) those that synthesize RdRp as a separate nonfused protein by an internal initiation mechanism (e.g., a coupled termination–reinitiation mechanism), as proposed for Hv190SV and others that

infect filamentous fungi. Phylogenetic analysis of CP or RdRp sequences of totiviruses reflects these differences, and separate phylogenetic clusters can be generated. Hv190SV and other totiviruses that infect filamentous fungi are closer to each other than to viruses infecting yeast, smut fungi, and protozoa. The fact that independent alignments of CP and RdRp sequences give similar phylogenetic relationships supports the conclusion that totiviruses infecting filamentous fungi should reside in a genus of their own. The genus *Victorivirus* has been proposed to include the totiviruses that infect filamentous fungi with Hv190SV, as the type species (**Table 1**). The genomes of partitiviruses and chrysoviruses consist of two and four segments, respectively.

The unclassified dsRNA mycovirus agaricus bisporus virus 1 (AbV1), also designated La France isometric virus, causes a serious disease of cultivated mushroom (named La France disease). AbV1 is of special interest because of its historical and economic importance. The AbV1 virions, isolated from diseased fruit bodies and mycelia, are isometric 36 nm in diameter and co-purify with nine dsRNA segments (referred to as disease-associated dsRNAs). The size of dsRNA segments varies from 3.6 to 0.78 kbp, three of which are believed to be satellites. It is not clear at present whether the nine dsRNA segments are encapsidated individually, in various combinations, or all nine segments are packaged in single particles. Based on the size of the particles, cesium sulfate gradient profile, and results of dsRNA and protein analyses of the gradient fractions, it is highly unlikely that all dsRNAs are packaged together in single particles. More realistically, AbV1 represents a multiparticle system in which the various particle classes have similar densities. Interestingly, phylogenetic analysis of the conserved motifs of AbV1 RdRp, encoded by dsRNA segment 1, and other dsRNA mycoviruses showed that AbV1 is closely related to the multipartite chrysoviruses.

ssRNA Viruses

There are a number of mycoviruses with apparent ssRNA genomes that do not code for capsid proteins and exist more or less predominantly as dsRNA 'replicative' forms in their hosts. Because of lack of true virions, these viruses were easier to isolate and study as their dsRNA forms, and some were grouped with dsRNA viruses (e.g., family *Hypoviridae*). However, there is ample evidence at present that many of these viruses replicate and express their genomes like (+)-strand RNA viruses and that the lineage of their RdRp and helicase genes are within the lineages of (+)-strand RNA viruses. The simplest types of these viruses include members of the genera *Narnavirus* and *Mitovirus* (family *Narnaviridae*), whose RNA genomes code only for RdRp and the viruses exist as RNA/RdRp nucleoprotein complexes. The corresponding dsRNAs can be isolated from infected

Table 1 List of viral families and genera into which fungal viruses are classified

Family/genus/virus species[a]	Virus abbreviation	No. of segments	Accession number
Double-stranded DNA genome			
Unassigned			
Rhizidiovirus			
Rhizidiomyces virus	RhiV	1	
RNA reverse transcribing genome			
Pseudoviridae			
Pseudovirus			
Saccharomyces cerevisiae Ty1 virus	SceTy1V	1	M18706
Hemivirus			
*Saccharomyces paradoxus Ty5 virus**	SceTy5V	1	U19263
Metaviridae			
Metavirus			
*Saccharomyces cerevisiae Ty3 virus**	SceTy3V	1	M34549
Double-stranded RNA genome			
Reoviridae			
Mycoreovirus			
Cryphonectria parasitica mycoreovirus 1 (9B21)	MyRv1–9B21	11	AY277888; AY277889; AY277890; AB179636; AB179637; AB179638; AB179639; AB179640; AB179641; AB179642; AB179643
Totiviridae			
Totivirus			
Saccharomyces cerevisiae virus L-A (L1)	Sc V-L-A	1	J04692; X13426
Victorivirus[b]			
Helminthosporium victoriae 190SV	Hv190SV	1	U41345
Partitiviridae			
Partitivirus			
Atkinsonella hypoxylon virus	AhV	2	L39125; L39126; L39127
Chrysoviridae			
Chrysovirus			
Penicillium chrysogenum virus	PcV	4	AF296439; AF296440; AF296441; AF296442
Single-stranded (+) RNA genome			
Narnaviridae			
Narnavirus			
Saccharomyces 20S narnavirus	ScNV-20S	1	M63893
Mitovirus			
Cryphonectria mitovirus 1	CMV1	1	L31849
Barnaviridae			
Barnavirus			
Mushroom bacilliform virus	MBV	1	
Hypoviridae[c]			
Hypovirus			
Cryphonectria hypovirus 1-EP713	CHV-1/EP713	1	M57938
Endornavirus[c]			
*Phytophthora endornavirus 1**	PEV1	1	AJ877914
Unassigned			
Botrytis virus F	BVF	1	AF238884
Sclerotinia sclerotiorum debilitation-associated RNA virus	SsDRV	1	AY147260
Diaporthe RNA virus	DRV	1	AF142094
Sclerophthora macrospora virus A	SmV A	2	AB083060; AB083061
Sclerophthola macrospora virus B	SmV B	1	AB012756

[a]The type species of the specified genus are listed. Otherwise, a fungal virus member of the genus is listed and marked with an asterisk.
[b]Proposed new genus in the family *Totiviridae*.
[c]Grouped with dsRNA viruses in the Eighth Report of ICTV.

tissues, usually in lesser molar amounts than the genomic ssRNA. Phylogenetic analysis of RdRps of members of the family *Narnaviridae* along with those of other fungal viruses and related taxa indicate a distant relationship between members of the family *Narnaviridae* and bacteriophages belonging to the family *Leviviridae*.

Lack of true virions is characteristic of two other groups of classified viruses, those belonging to the family

Hypoviridae and the genus *Endornavirus*. Although these viruses were grouped with dsRNA viruses in the latest ICTV Report, their genome organization and expression strategies are indicative of (+)-strand RNA viruses. Viruses in the family *Hypoviridae* are phylogenetically related to the (+)-strand RNA viruses in the family *Potyviridae* (picorna-like virus supergroup). Comparisons of hypovirus-conserved motifs of RdRp, helicase, and protease with those of members of the family *Potyviridae* suggest that viruses in the genus *Bymovirus* are the closest relatives to hypoviruses. Phytophthora endornavirus (PEV1) is the only nonplant virus in the genus *Endornavirus* of plant viruses. Although *Phytophthora* species and other members of the class Oomycetes have many biological properties in common with fungi, they are currently classified, based on sequence similarities, in a protist group known as the Stramenopiles. Endornaviruses are believed to have evolved from an alpha-like virus that has lost its capsid gene. This is consistent with the recent finding that RdRps of PEV1 and other endornaviruses cluster with those of families and genera in the alpha-like virus superfamily of (+)-strand RNA viruses.

The mushroom bacilliform virus (MBV; genus *Barnavirus*, family *Barnaviridae*) is the only mycovirus known to have bacilliform virions. MBV has a (+)-strand RNA genome that contains seven open reading frames (ORFs), three of which encode a putative chymotrypsin-like serine protease, a putative RdRp, and a CP. The polypeptides encoded by the remaining four ORFs have no homology to known proteins. Amino-acid sequence comparisons of the putative protease and RdRp suggest that MBV is evolutionarily related to sobemoviruses and poleroviruses. Although double infections of cultivated mushrooms with MBV and ABV1 are of common occurrence, the role of MBV in the ensuing dieback disease of cultivated mushroom remains unknown.

Unassigned ssRNA Viruses

It is worth noting that a relatively large number of mycoviruses remain unassigned including some well-characterized ones like botrytis virus F (BVF), sclerotinia sclerotiorum debilitation-associated RNA virus (SsDRV), diaporthe RNA virus (DRV), and sclerophthora macrospora viruses A and B (SmV A and SmV B). BVF has flexuous rod-shaped particles comparable in size and morphology to ssRNA plant 'potex-like' viruses. Amino-acid sequence identities of the conserved helicase and RdRp regions and the coat protein genes are greatest to those of potex-like viruses. The main difference between BVF and these plant viruses is the lack of a movement protein. Although particle morphology along with amino-acid sequence similarities of both the replicase and coat protein genes support the classification of BVF in the family *Flexiviridae*, it is obvious that the mycovirus BVF is distinct enough to belong to a new genus in this family. It is proposed that a

new genus tentatively designated '*Mycoflexivirus*' is created to include BVF (**Table 1**). The genome of SsDRV contains a single ORF encoding a protein with significant sequence similarity to the replicases of the 'alphavirus-like' super-group of (+)-strand RNA viruses. The SsDRV-encoded putative replicase protein contains the conserved methyl transferase, helicase, and RdRp domains characteristic of the replicases of potex-like plant viruses (flexiviruses) and BVF. Although phylogenetic analysis of the conserved RdRp motifs verified that SsDRV is closely related to BVF and to the allexiviruses in the family *Flexiviridae*, SsDRV is distinct enough from these viruses, mainly based on the lack of coat protein and movement protein, to justify the creation of yet another genus in the family *Flexiviridae*.

DRV is another naked RNA mycovirus that is associated with hypovirulence of its fungal host. It has two large ORFs present in the same reading frame, which are most likely translated by readthrough of a UAG stop codon in the central part of the genome. The longest possible translation product has a predicted molecular mass of about 125 kDa, which shows significant homology to the nonstructural proteins of carmoviruses of the (+)-strand RNA virus family *Tombusviridae*. Interestingly, transcripts derived from full-length cDNA clones were infectious when inoculated to spheroplasts and the transfected isolates exhibited phenotypic traits similar to the naturally infected isolate.

Sclerophthora macrospora virus A (SmV A) found in *S. macrospora*, the pathogenic fungus responsible for downy mildew of gramineous plants, is a small icosahedral virus containing three segments of (+) ssRNAs (RNAs 1, 2, and 3). Whereas RNA 1 contains the RdRp motifs, RNA 2 codes for a capsid protein. RNA 3 is a satellite RNA. Whereas the deduced amino acid sequence of RdRp shows some similarity to RdRps of members of the family *Nodaviridae*, the amino acid sequence of the viral CP shows similarity to those of members in the family *Tombusviridae*. The capsid of SmV A is composed of two capsid proteins, CP 1 and CP 2, both encoded in ORF2. CP 2 is apparently derived from CP 1 via proteolytic cleavage at the N-terminus. The genome organization of SmV A is distinct from those of other known fungal RNA viruses, and suggests that SmV A should be classified into a new genus of mycoviruses.

Sclerophthora macrospora virus B (SmV B), which is also found in *S. macrospora*, has small icosahedral, monopartite virions containing a (+) ssRNA genome. The viral genome has two large ORFs: ORF1 encodes a putative polyprotein containing the motifs of chymotrypsin-related serine protease, and ORF2 encodes a capsid protein. The genome arrangement of SmV B is similar to those belonging to the genera *Sobemovirus*, *Barnavirus*, and *Polerovirus*. The putative domains for the serine protease, VPg, RdRp, and the CP are located in this order from the 5' terminus to the 3' terminus. SmV B, however, is distinctive since its genome has only two

ORFs. The genome organization of the barnavirus MBV, on the other hand, resembles that of poleroviruses. These results suggest that SmV B, like SmV A, should also be classified into a new genus of mycoviruses.

The mycoviruses belonging to the families *Pseudoviridae, Metaviridae, Reoviridae* (genus *Mycoreovirus*), *Totiviridae, Partitiviridae, Chrysoviridae, Narnaviridae,* and *Barnaviridae* are discussed in more detail elsewhere in this encyclopaedia.

Replication and Gene Expression Strategy

Replication cycles of fungal viruses are not well studied except for a few cases including the *Saccharomyces cerevisiae* L-A virus. For dsRNA fungal viruses including members of the *Totiviridae, Partitiviridae, Reoviridae,* and *Chrysoviridae*, virus particles or subviral particles, containing RdRp, are believed to play pivotal roles in RNA transcription and replication. Replication of naked RNA mycoviruses represented by members of the family *Hypoviridae* may occur in infection-specific, lipid-membranous vesicles presumed to contain viral RdRp and RNA helicase. These vesicles are able to synthesize *in vitro* both plus and minus RNA at a ratio of 1:8. The narnavirus RNA, encoding only a single protein (RdRp), is associated with RdRp rather than being encapsidated. These RNA/RdRp complexes seem to play a key role in RNA replication.

Fungal viruses, like many RNA viruses of plants and animals, employ noncanonical translational strategies for expressing their genomes. These include −1 (*Totiviridae*) and +1 frameshifting (*Totiviridae, Pseudoviridae, Metaviridae*), termination-coupled initiation (*Totiviridae, Hypoviridae*), and IRES-mediated initiation (*Totiviridae*). Furthermore, a readthrough of a termination codon strategy is proposed for translation of the 3′ proximal ORF of DRV. A noncanonical mechanism may also be required for efficient translation of mRNA of narnaviruses, which lack poly(A) tails. The (CAA)n repeats found at the 5′ UTR of chrysoviruses are implicated in translation augmentation, as observed for the 5′ UTR sequence of tobacco mosaic virus. Translation of the viral genes that are regulated by these mechanisms is considered critical for virus viability.

Recent Technical Advances in Fungal Virology

Fungal virology has been thwarted by many constraints on manipulation of fungal viruses. Many, if not all, plant and animal viruses can be inoculated into individuals/tissue cultures of plant and animal hosts. Some assays with those hosts allow quantitative detection of biologically active viruses. As for mycoviruses, it is rather rare to be able to inoculate into host fungi because of experimental limitations, which often makes the etiology of mycoviruses difficult to establish. Fungal cells have rigid cell walls and are usually difficult to digest for preparation of cell-wall-free protoplasts ready for transformation or transfection. Even if protoplasts are made, their maintenance like animal cell culturing is not possible. Furthermore, fungal hosts usually have self/nonself recognition systems operating at inter- and intraspecies levels. Intraspecies barriers are based on vegetative incompatibility/compatibility that is governed genetically. This is often regarded as one of the host defense barriers that inhibits virus transfer between individuals. To overcome these barriers, a few methods are available. The prototypic hypovirus cryphonectria parasitica hypovirus 1 (CHV1) is the first for which a reverse genetics is established. Infection with different CHV1 strains can be launched either from cDNA integrated into host chromosomes or *in vitro*-synthesized RNA viral cDNA. It is noteworthy that via bombardment of mycelia, not protoplasts, the infectious CHV1 cDNA clone can be integrated into chromosomes of fungi other than the natural host *C. parasitica*. cDNA-based transfection systems are now available for three other species of RNA viruses: *Diaporthe RNA virus, Saccharomyces cerevisiae 20S narnavirus,* and *Saccharomyces cerevisiae 23S narnavirus*.

Mycoviruses in general lack infectivity as purified virions. However, all members of the new genus *Mycoreovirus* including mycoreovirus 1 (MyRV1), MyRV2, and MyRV3 were found to be infectious as purified particles when applied to fungal protoplasts. It is of interest in this regard that treatment of purified virions with trypsin or chymotrypsin was not required for infectivity. Protoplast fusion provides an alternative approach to introduce mycoviruses into vegetatively incompatible fungal strains that are incapable of hyphal anastomosis. Intra- and interspecies virus transfer via protoplast fusion has been reported in *Aspergillus* spp. This method is particularly useful for viruses for which infectious particles or cDNA-derived RNA are unavailable. Another recent revelation is that monokaryotic strains are able to serve as an intermediate virus transmitter between different mycelial incompatibility groups within the same species of *R. necatrix*.

To complete Koch's postulates, virus curing is as important as virus inoculation, because the virus must be back inoculated into an isogenic, virus-free strain with the same genetic background as the original virus-infected strain. Virus-free isolates may be obtained from germlings of asexual spores if virus transmission through spores is less than 100%. Alternatively, virus-free strains may be isolated by hyphal tip culturing, as in the case for *H. mompa* and *R. necatrix*. This technique is applicable to fungi infected with a virus that is transmitted to 100% of the asexual spores or for fungi that produce little or no spores.

Future Perspectives

Yeast as a (Model) Host to Study Viral Replication

S. cerevisiae has provided an excellent system to investigate virus assembly and replication of the dsRNA totivirus L-A and the ssRNA narnaviruses that infect yeast. With the robust yeast genetics, a number of host factors involved in totivirus replication were identified, many of which are related to translation events. Furthermore, the yeast provides an 'artificial' viral host model system to explore host genes affecting viral replication on a genomewide basis. Genetic screens of a collection of 4500–4800 single-gene deletion yeast strains (Yeast Knockout strain collection) have been successfully conducted for identifying host factors involved in replication and recombination with two different plant ssRNA viruses: brome mosaic virus and tomato bushy stunt virus (TBSV). Each screen led to the identification of approximately 100 host genes (approximately 1.8% of the entire yeast genes) that affect virus replication. Interestingly, the replication of the two viruses in yeast is affected by a different set of genes. A similar approach was employed to identify genes affecting the recombination of TBSV. This type of use of the yeast system can be expanded to a (+)-strand RNA animal virus and a (−)-strand RNA vertebrate virus. It may be surprising that the yeast has yet to be used for viruses infecting filamentous fungi. The yeast should be able to serve as a model host for mycoviruses other than those that naturally infect yeast.

Host Defense against Fungal Viruses

Host defense responses against viruses have not been explored intensively. RNA silencing is regarded as one of the host defense strategies of eukaryotes to molecular parasites including viruses, and operates in a number of fungi including important model and phytopathogenic fungi. However, no direct evidence is shown for RNA silencing that target mycoviral RNA. Recently, the hypovirus p29 was shown to be a suppressor of RNA silencing targeting a transgene that functions in both plant and fungal cells. This suggests that RNA silencing may function as an antiviral mechanism in fungal cells. The observation that p29 enhances the replication of a heterologous virus supports this idea. There are genetic elements involved in RNA silencing, for example, that are conserved widely from fungi to vertebrates. Functional roles of these factors in RNA silencing as antiviral reactions will need to be elucidated. Unraveling the mechanism by which the hypovirus p29, or other mycovirus-encoded RNA silencing suppressors, may block the RNA silencing pathway will be an interesting challenge.

Role of Mycoviruses in Plant–Fungal Mutualistic Associations

The question of whether mycoviruses are involved in the mutualistic interactions between endophytic fungi and their host plants is of considerable interest because of the attractive beneficial features of these associations and because of the common occurrences of fungal viruses in all major groups of fungi. This question was recently addressed in an intriguing report that presented evidence for a dsRNA mycovirus being involved in the mutualstic interaction between a fungal endophyte (*Curvularia protuberata*) and a tropical panic grass. This association allows both organisms to grow at high soil temperatures. The virus in question, which was designated curvularia thermal tolerance virus (CThTV), has unusual genome organization with an unknown genome expression strategy. CThTV has apparently a bipartite genome (RNA 1 and RNA 2), but no evidence that these RNAs are packaged in the 27-nm isometric particles isolated from the fungal host. Although many questions pertinent to the fungal endophyte, CThTV, and the veracity of the evidence for viral etiology remain unanswered, this report will undoubtedly stimulate the search for mycoviruses in other mutulastic fungal endophytes. In this regard, it is noteworthy that the well-characterized mutualistic endophyte, *Epichloë festucae*, was found to harbor a totivirus, but no phenotypes were associated with virus infection.

Mycovirus as Biocontrol Agents and as Tools for Fundamental Studies

The hypovirulence phenotype in the chestnut blight fungus (*Cryphonectria parasitica*) is an excellent and well-documented example for a mycoviral-induced phenotype that is currently being exploited for biological control. The debilitating disease of *Helminthosporium victoriae*, the causal agent of Victoria blight of oats, and the disease phenotype of the Dutch elm disease fungus *Ophiostoma novo-ulmi* are examples of pathogenic effects of dsRNA fungal viruses. An understanding of the molecular basis of disease in these fungal-virus systems would provide excellent opportunities for development of novel biocontol strategies of plant pathogenic fungi. Mycoviruses also continue to serve as versatile tools to study the virulence of host fungi, as recognized in studies with the hypovirus/ *Cryphonectria parasitica* system, in which substantial advances in our understanding of the molecular basis of hypovirulence have been made.

See also: Barnaviruses; Hypovirulence; Hypoviruses; Mycoreoviruses; Narnaviruses; Partitiviruses of Fungi; Ustilago Maydis Viruses; Yeast L-A Virus.

Further Reading

Buck KW (1986) Fungal virology-an overview. In: Buck KW (ed.) *Fungal Virology*, pp. 1–84. Boca Raton, FL: CRC Press.

Buck KW (1998) Molecular variability of viruses of fungi. In: Bridge PD, Couteaudier Y, and Clarkson JM (eds.) *Molecular Variability of Fungal Pathogens*, pp. 53–72. Wallingford: CAB International.

Ghabrial SA (1994) New developments in fungal virology. *Advances in Virus Research* 43: 303–388.

Ghabrial SA (1998) Origin, adaptation and evolutionary pathways of fungal viruses. *Virus Genes* 16: 119–131.

Ghabrial SA (2001) Fungal viruses. In: Maloy O and Murray T (eds.) *Encyclopedia of Plant Pathology*, vol. 1, pp. 478–483. New York: Wiley.

Goodin MM, Schlagnhaufer B, and Romaine CP (1992) Encapsidation of the La France disease-specific double stranded RNAs in 36 nm isometric viruslike particles. *Phytopathology* 82: 285–290.

Hillman BI and Suzuki N (2004) Viruses in the chestnut blight fungus. *Advances in Virus Research* 63: 423–472.

Howitt RLJ, Beever RE, Pearson MN, and Forster RLS (2001) Genome characterization of Botrytis virus F, a flexuous rod-shaped mycovirus resembling plant 'potex-like' viruses. *Journal of General Virology* 82: 67–78.

Márquez LM, Redman RS, Rodriguez RJ, and Roossinck MJ (2007) A virus in a fungus in a plant: Three-way symbiosis required for thermal tolerance. *Science* 315: 513–515.

McCabe PM, Pfeiffer P, and Van Alfen NK (1999) The influence of dsRNA viruses on the biology of plant pathogenic fungi. *Trends in Microbiology* 7: 377–381.

Nuss DE (2005) Hypovirulence: Mycoviruses at the fungal–plant interface. *Nature Reviews Microbiology* 3: 632–642.

Sun L-Y, Nuss DL, and Suzuki N (2006) Synergism between a mycoreovirus and a hypovirus mediated by the papain-like protease p29 of the prototypic hypovirus CHV1-EP713. *Journal of General Virology* 87: 3703–3714.

Tavantzis S (ed.) (2001) *Fungal dsRNA Elements: Concepts and Application in Agriculture, Forestry and Medicine*. Boca Raton, FL: CRC Press.

Van der Lende TR, Duitman EH, Gunnewijk MGW, Yu L, and Wessels JGH (1996) Functional analysis of dsRNAs (L1, L3, L5, and M2) associated with isometric 34-nm virions of *Agaricus bisporus* (White Button Mushroom). *Virology* 217: 88–96.

Wickner RB (1996) Double-stranded RNA viruses of *Saccharomyces cerevisiae*. *Microbiological Reviews* 60: 250–265.

Hypovirulence

N K van Alfen and P Kazmierczak, University of California, Davis, CA, USA

Glossary

Ascomycete A division of fungi whose members produce spores in a saclike structure called the ascus.

Canker A dead section of bark on the branches or main trunks of trees.

Hyphae Vegetative growth structures of the fungus.

Hypovirulence A reduction of virulence.

Fungal Viruses

Hypovirulence is a phenomenon that can occur in virus-infected fungi. Although fungal viruses are now known to be common and widespread within all groups of fungi, their discovery was fairly recent. It was not until the early 1960s that viruses were found in fungi, the first discovery being of viral particles that were isolated from the edible mushroom *Agaricus bisporus*. Fungal viruses became more widely known after it was discovered that strains of *Penicillium* spp., which were strong inducers of interferon, were found to be infected with a double-stranded RNA (dsRNA) virus.

Most descriptions of fungal viruses suggest that they are not normally associated with clearly defined symptoms. The lack of known infectivity cycles of fungal viruses is a major barrier to the study of these viruses since it limits the ability to demonstrate Koch's postulates. Transmission of fungal viruses is known to naturally occur only during reproduction by the formation of spores and other such structures and by cytoplasmic fusion between hyphae of different strains of the same fungus. In this respect, transmission of fungal viruses is more similar to the transmission of plasmids than it is to the transmission of other viruses, that is, the infectious agent never leaves the cytoplasmic environment. With a few exceptions, the genomic material of known fungal viruses is dsRNA, and in viruses where sequences are known, they contain a virus-associated RNA-dependent RNA polymerase (RdRp). Because fungal virus genomes are clearly viral in origin and affinity, the apparent lack of a viral-like infectivity cycle has led researchers to postulate an ancient origin for fungal viruses.

The most extensively studied fungal viruses cause symptoms that can generally be classified into three general groups of symptoms: (1) killer phenotype, (2) hypovirulence, and (3) debilitation. The killer systems consist of a helper totivirus and associated satellite dsRNAs, which encode a protein toxin that is secreted and is lethal to strains of the same fungus lacking the virus. The best studied examples of killer systems are those found in the common baker's yeast, *Saccharomyces cerevisiae*. Viruses that reduce the virulence of plant pathogenic fungi either directly, or by debilitation, are the subject of this article.

Hypovirulence

A reduction of the virulence of a plant pathogenic fungus is known as hypovirulence. Generally, anything that reduces the ability of a pathogenic fungus to grow normally will concomitantly reduce its virulence. So, by this general definition, viruses that debilitate a pathogen cause hypovirulence. It is useful, however, to distinguish this type of debilitating hypovirulence from a symptom of virus infection that is more directly associated with reducing virulence expression and/or other developmental processes of a fungus without significant effects on fungal growth. Virus-caused hypovirulence of both types has been described in diverse groups of plant pathogenic fungi. These mycoviruses that have hypovirulence as an associated symptom may have either ssRNA or dsRNA genomes and are found in the virus families *Chrysoviridae*, *Hypoviridae*, *Narnaviridae*, *Partitiviridae*, *Reoviridae*, and *Totiviridae*.

Perturbation of Virulence/Development

Other than the viruses associated with the killer phenotype of yeast, the most extensively studied fungal virus is cryphonectria hypovirus 1 (CHV1) that infects the ascomycete *Cryphonectria parasitica*. This fungus is the causal agent of chestnut blight and is the prime example of a virus that is able to cause the hypovirulence symptom without associated debilitation. Chestnut blight was first reported in North America during the summer of 1904 in the New York City Zoological Park. It was probably imported on nursery stock of Chinese or Japanese chestnuts, *Castanea mollissima* and *Castanea crenata*, respectively. The disease rapidly spread to the American chestnut, *Castanea dentata*, growing in surrounding woodlands and forests. Within 50 years, the disease spread throughout the range of the highly susceptible native tree. Few American chestnuts now survive in their natural range because of this disease. Prior to the blight, the American chestnut was the most common tree of the Eastern deciduous forest and was found on over 200 million acres across eastern North America. This is one of the most devastating diseases of recorded history.

Chestnut blight was also accidentally introduced into Europe, but during its spread in Italy it was noticed that chestnut trees in some orchards were recovering from the disease, that is, the cankers caused by the fungus stopped spreading. When the bark was peeled back, the fungal growth was found to be confined to the uppermost layers

of the bark with no penetration occurring into the living tissues of the tree. The isolates of the fungus recovered from these cankers were of low virulence in assays, so were called hypovirulent by the French plant pathologist, J. Grente. Using genetically marked virulent and hypovirulent strains of the fungus, Van Alfen and colleagues demonstrated that a cytoplasmic element found in the hypovirulent strain was responsible for the symptoms and could be transferred cytoplasmically to the virulent strain, converting it into a hypovirulent strain. This cytoplasmic element was later found to be a fungal virus that is a member of the type species of the family *Hypoviridae*. **Figure 1** shows the phenotypic differences between virulent and hypovirulent isolates of *C. parasitica* on both the chestnut tree and in culture.

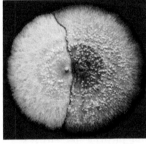

Figure 1 Phenotypic changes in *C. parasitica* caused by the hypovirus CHV1. The photograph on the right is a tree that is infected with a virulent strain of the fungus. The photograph below shows this strain in culture. In addition to producing orange/brown pigments, virulent strains produce numerous conidia in asexual fruiting bodies, both in culture and on trees. The tree on the left is infected with a fungus containing CHV1. Strains of the fungus infected with this virus typically produce superficial healing cankers that do not kill the tree. Virus-infected strains of the fungus, as shown growing in culture, are not pigmented; they remain white and produce very few asexual conidia in culture. Canker photograph courtesy of Linda Haugen, University of Georgia, http://www.forestryimages.org. Culture photograph reproduced from McCabe PM, Pfeiffer PL, and Van Alfen NK (1999) The influence of dsRNA viruses on the biology of plant pathogenic fungi. *Trends in Microbiology* 7: 377–381, with permission from Elsevier.

C. parasitica is also host to viruses classified in several other families of viruses: these include the *Reoviridae*, *Narnaviridae*, and *Chrysoviridae*. A hypovirulent phenotype of *C. parasitica* has also been described in strains infected by the virus NB631, a mitochondrial virus classified in the family *Narnaviridae*. This virus reduces virulence and growth of the fungus somewhat, but sporulation is normal. Two reoviruses of *C. parasitica*, C18 and 9B21 have been isolated. The best characterized of the two, 9B21, reduces virulence of the fungus significantly, but sporulation and pigmentation are not significantly affected. A very unusual feature of 9B21 is that purified particles of the virus are able to infect fungal protoplasts.

Hypovirulence symptoms caused by viruses have been reported in many plant pathogenic fungi. Generally these viruses remain uncharacterized and in most cases the virulence of the fungus is reduced in association with a general debilitation of the infected fungus.

Structure and Classification of CHV1

Once it was demonstrated that a cytoplasmic element caused transmissible hypovirulence of the chestnut blight fungus there was a search for a fungal virus as the potential cause. This search culminated in the report of dsRNA associated with hypovirulent strains of the fungus. Further studies of the dsRNA showed that it was associated with fungal vesicles and that no detectable capsid was associated with this dsRNA, although RdRp activity was associated with the vesicles. The dsRNA of this 12.7 kb virus was cloned, sequenced, and an infectious clone of the virus was transformed into *C. parasitica*, confirming that this virus is responsible for transmissible hypovirulence. Based on the symptoms caused by the virus it has formally been named CHV1. This virus is the type member of the family *Hypoviridae* and was the first virus family described whose members have no capsid. Comparing the genome of CHV1 with other known viruses suggested that it is most similar to plant potyviruses and other picorna-like viruses.

Further study of the nature of the RdRp associated with fungal vesicles supported the hypothesis that the genome of this virus is derived from the replicative form

of an ssRNA virus that lost its capsid during evolution. The coding strand of CHV1 contains two open reading frames: ORF A encodes a polyprotein that is processed into two polypeptides, p29 and p40 and ORF B which encodes a polyprotein from which only a single 48 kDa polypeptide p48 has thus far been shown to be autocatalytically processed. The sequences contained within ORF B indicate that it also contains RNA polymerase and helicase domains. **Figure 2** shows a schematic diagram of the genomic organization and expression strategy of CHV1.

Transmission of CHV1

There is no evidence that a typical infection cycle exists in the relationship of CHV1 with its fungal host. In order for infection of new hosts to occur the virus is transmitted through cytoplasmic exchange processes of the fungus. This cytoplasmic exchange occurs during fusion of hyphae (vegetative growth structures of the fungus) that happens frequently in fungal colonies and can occur between related strains of the fungus. In fungi, this hyphal fusion is controlled by vegetative incompatibility (*vic*) genes of the fusing strains; fusions most easily occur when alleles of the *vic* genes are the same in the two fungi. In *C. parasitica*, there are six known *vic* genes that control hyphal anastomosis. The more differences that occur in the *vic* loci, the less likely that transfer of CHV1 will occur. This vegetative incompatibility reaction is a defense in *C. parasitica* and other fungi against the spread of viruses. CHV1 spreads within a population with greater success if that population has few differences in alleles at the *vic* loci. Since strains of the fungus that contain CHV1 are incapable of sexual mating, infection by the virus suppresses allelic recombination at the *vic* loci and, as a result, may increase the effectiveness of virus spread within a population over time.

Very little is known about movement of the virus within the host fungus. However, the rate of movement of dsRNA within the fungus has been measured, and is rapid. The average rate of movement of the dsRNA within a fungal colony was found to be approximately 16 mm d^{-1}. This is 3–4 times faster than the colony growth rate

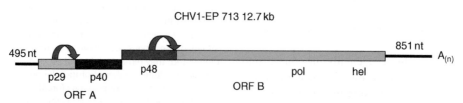

CHV1-EP 713 12.7 kb

Figure 2 The genomic organization and expression strategy of CHV1–EP713, the prototypic strain of CHV1. Redrawn from Hillman BI and Suzuki N (2004) Viruses of the chestnut blight fungus, *Cryphonectria parasitica*. In: Maramorosch K and Shatkin AJ (eds.) *Advances in Virus Research*, vol. 63, pp. 423–472. San Diego, CA: Academic Press.

during the same time period. Transmission electron microscope studies on the ultrastructure of hyphal anastomosis in *C. parasitica* reported that the dsRNA containing vesicles move from hypovirulent to virulent strains within 4–6 h after anastomosis. The incompatibility reaction controlled by the *vic* genes is an imperfect defense, since virus transfer can occur despite allelic differences in *vic* genes and virus transfer has been reported between different species of *C. parasitica*. There is clearly much left to be learned about virus movement within and between fungi.

The only other means by which the virus is known to spread is through asexual spores of the fungus. Although the virus suppresses developmental processes of the fungus, it does not completely eliminate asexual sporulation. Many fewer asexual spores are produced by CHV1-infected strains of the fungus, but those asexual spores that are produced may contain the virus.

Hypovirulence as Developmental Perturbation

Hypovirulence caused by CHV1 is of interest not only because it provides a means of biological control for an important plant disease, but also because it offers the potential to understand critical control point(s) for development in a filamentous fungal pathogen. The perturbation of development, without fungal growth being affected, suggests that the virus interferes with development in specific ways that leave normal vegetative growth functions undisturbed. Understanding how the virus acts to perturb development in this fungus may provide insights of value on how to reduce the economic impact of filamentous fungi.

The symptoms of CHV1 infection of *C. parasitica* are a reduction of asexual sporulation, reduced pigmentation, female sterility, and hypovirulence (**Figure 1**). Careful growth measurements show little or no adverse effects of viral infection on fungal vegetative growth in culture. While there have been some claims that CHV1 reduces vegetative growth, these growth effects are likely an artifact of prolonged growth of the fungus in culture. We have found similar reduced growth rates in culture if some strains of the fungus, particularly EP713, are kept in culture too long. Such slow growth mutations are common in cultured fungi, and in our experience, are independent of virus infection. To prevent the accumulation of such growth mutations in prolonged culture, we routinely single-spore EP155 to select for a normal growth strain, and then reintroduce CHV1 by anastomosis to recreate EP713 with normal vegetative growth characteristics. Routine single-spore isolation is necessary to prevent the slower growth mutant strains from developing in culture.

Ultrastructural examination of virus-infected strains of the fungus show no significant cytopathology in CHV1-containing cells. The only differences observed between infected and uninfected cells are unique vesicles that are described as being located in association with unique Golgi in areas devoid of ribosomes and other cellular components. Recent cytological fractionation studies have isolated these small vesicles that accumulate in CHV1-infected strains and have shown that they are associated with the *trans*-Golgi network. These vesicles contain the virus and its replication-associated enzymes. **Figure 3** shows the accumulation of these vesicles in the CHV1-infected strain, EP802.

Two general experimental approaches have been used to understand the molecular basis of how CHV1 causes symptoms in *C. parasitica*: (1) seek evidence that the virus perturbs normal fungal signal transduction pathways, and (2) identify commonalities in genes and proteins differentially expressed between normal and virus-infected strains of the fungus. A significant body of literature has been generated regarding how internal signal transduction pathways are affected in virus-infected strains, but, to date, there is no evidence of direct virus effects on signal transduction. As expected, signal transduction differs temporally and quantitatively between virus-infected and noninfected strains of the fungus.

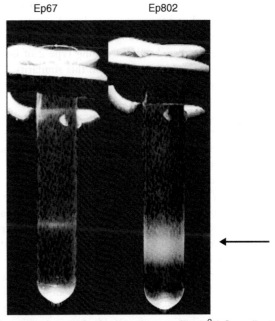

Ep67   Ep802

Figure 3 Subcellular fractionation on a Ficoll/^{2}H$_2$O gradient showing vesicle accumulation in CHV1-infected strain EP802 on the right and the isogenic virus free strain EP67 on the left. Gradient fractions were analyzed by Western and kex2 activity to show peak kex2 activity, AP-1μ, dsRNA, and viral helicase all cofractionate within this band. The endoprotease kex2 and the subunit for the AP-1 adaptor protein are markers for the *trans*-Golgi network.

The most reasonable explanation of the published observations is that perturbation of signal transduction is the consequence of virus downregulation of development, rather than the cause. There is in fact no evidence that the symptoms are caused by direct virus perturbation of signal transduction and no testable hypothesis has emerged from these studies as to how the virus may be affecting signal transduction to cause the observed symptoms.

The approach that sought commonalities between differentially expressed proteins and genes associated with virus infection is leading to an understanding of how CHV1 may be causing symptoms. Given that the symptom of virus infection is a lack of development in infected strains, significant differences in expression between the two strains were expected and observed. Early two-dimensional gel and differential hybridization studies identified a number of protein and gene expression differences between the infected and uninfected strains. A relatively small number of these genes/proteins that were highly expressed in uninfected but not in infected strains were characterized to seek commonalities between them.

Three of the genes highly expressed in uninfected strains, and not expressed in CHV1-infected strains, encoded sex pheromones. Two of the genes contain an 83bp ORF that encodes an identical 23-amino-acid peptide. This small peptide is similar to fungal lipopeptide sex pheromones: a C-terminal CAAX box with an asparagine residue 8–11 amino acids upstream of the box. CAAX is a prenylation signal, and farnesyl groups have been detected on the pheromones in other fungi. It was found that these peptides were expressed only in mating type 2 of uninfected strains of the fungus. Based on this similarity and loss of sexual mating upon deletion of the genes, the genes were named *mf2–1* and *mf2–2*. Deletion of *mf2–2* also resulted in a significant reduction in production of asexual fruiting bodies in culture and thus reduced numbers of asexual spores. Transcription run-on studies confirmed that these genes were transcriptionally downregulated in CHV1-infected cells.

Because *mf2–1* and *mf2–2* were so similar in structure to yeast pheromones, investigations were carried out to isolate the pheromone precursor gene of the opposite mating type. The gene isolated was shown to be similar in structure to the *S. cerevisiae* α-factor pheromone. There is a single copy of this gene expressed only by mating type 1 strains; it was named *mf1–1*. When the gene was deleted from *C. parasitica*, conidia or mycelial fragments normally used as spermatia in mating were sterile. Expression of this gene is significantly downregulated in CHV1-infected strains.

Differential protein expression studies identified two secreted proteins that are highly expressed by uninfected strains, but are downregulated in virus-infected cells. One of these is an extracellular laccase and the other an extracellular structural protein that was named cryparin. This later protein is a hydrophobin, a class of cell-surface proteins widely found in fungi and lichens. All hydrophobins are located on hyphal surfaces and confer hydrophobic properties to the surfaces, but they appear to have evolved different functions in various fungi. In *C. parasitica*, cryparin accumulates in fruiting bodies and is necessary for the eruption of the fungal fruiting bodies through the bark of the host tree. **Figure 4** shows the results of the deletion of cryparin.

The role of the laccase enzyme downregulated in CHV1-infected strains is not known. Knockout mutants did not exhibit any detectable phenotype, but did reveal the existence of multiple laccase enzymes expressed by the fungus. Laccases have been implicated in a number of roles related to fungal development and virulence including degradation of lignin, formation of fruiting bodies, and pigment production.

The isolation and characterization of a number of the gene products downregulated in virus-infected cells was done to determine if there were any commonalities among them that could point to how the virus acts to cause symptoms. All of the gene products characterized are secreted into the extracellular environment and three of the proteins have sequences that suggest they are post-translationally processed by the same secretion pathway. Each protein is processed by an endoprotease after the signal peptide is cleaved, and the recognition signals suggest that they are processed by the same enzyme.

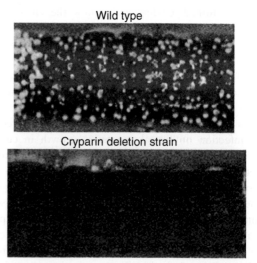

Wild type

Cryparin deletion strain

Figure 4 Stromal pustule (asexual fruiting bodies) eruption on chestnut wood. Sterile stem pieces of chestnut wood were inoculated with EP67 (upper panel) and the cryparin deletion strain (lower panel). Adapted from Kazmierczak P, Kim DH, Turina M, and Van Alfen NK (2005) A hydrophobin of the chestnut blight fungus, Cryphonectria parasitica, is required for stromal pustule eruption. *Eukaryotic Cell* 4: 931–936 with permission from American Society for Microbiology.

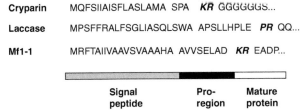

Cryparin	MQFSIIAISFLASLAMA SPA	*KR*	GGGGGGS...
Laccase	MPSFFRALFSGLIASQLSWA APSLLHPLE	*PR*	QQ...
Mf1-1	MRFTAIIVAAVSVAAAHA AVVSELAD	*KR*	EADP...

Signal peptide	Pro-region	Mature protein

Figure 5 Amino acid N-terminal sequences derived from cloned mRNAs of three developmentally regulated host genes downregulated by CHV1. The signal peptide directs the protein to the ER for secretion and into the Golgi network for processing of the proprotein by the kex2 endoprotease at the residues lysine/arginine or proline/arginine which results in the mature form of the secreted protein.

Figure 5 shows the preproproteins of Mf1–1, the fungal mating pheromone of mating type 1 strains, cryparin, and laccase. These three proteins have a signal peptide that presumably is cleaved early in the secretion process, followed by a propeptide region that terminates with a recognition sequence for processing by a kex2-like serine protease. These processing signals suggest that these three proteins are secreted via a kex2-like protein processing and secretion pathway. Kex2 processing is a post-Golgi function and the kex2 enzyme is a standard marker for *trans*-Golgi vesicles.

The processing of some secreted proteins by the kex2 pathway is highly conserved in eukaryotes. A similar pathway has been found in plants and animals and is utilized for the processing of specific proteins. It is generally not utilized for the secretion of most proteins. The specific role of this processing pathway in secretion appears to vary between organisms. It was first discovered in yeast because it is involved in the processing and secretion of the viral encoded killer protein. This pathway was later found to be necessary for the secretion of the yeast alpha sex pheromone. The fact that three randomly isolated developmentally regulated host proteins, which are downregulated by the virus, are all processed by the same enzyme during secretion is an important clue in understanding how the virus perturbs development. Research has shown that not all secreted proteins of *C. parasitica* are affected by the virus; the secreted rennin-like endoprotease, endothiapepsin, for instance does not have a kex2 processing signal sequence nor it is downregulated by the virus.

The possibility that the virus directly affects secretion of some proteins, and thereby causes at least some of the symptoms associated with infection, is supported by evidence that the virus replicates using *trans*-Golgi vesicles. The vesicle fraction that contains the dsRNA genome of the virus and the viral encoded RNA polymerase and helicase co-purifies with kex2 enzymatic activity. This vesicle fraction has also been shown to contain cryparin, one of the proteins downregulated by the virus.

The vesicle fraction in CHV1-infected cells that contains the virus dsRNA, viral encoded proteins, and kex2 activity are present in much greater amounts in virus-containing cells than in uninfected cells. This proliferation of vesicles in *C. parasitica* is typical of the effect of RNA viruses on their hosts, that is, the use of host membrane systems for replication by many RNA viruses cause a proliferation of these membranes within the host cells. We have estimated that in CHV1-infected cells there is at least a fivefold increase in this specific vesicle fraction when compared with noninfected cells. As previously mentioned, evidence from microscopic and cell fractionation observations suggest that CHV1 effects on membranes is specific to these *trans*-Golgi vesicles.

Recent research from our laboratory has shown that the CHV1 encoded protease p29 specifically integrates into the membranes of this *trans*-Golgi vesicle fraction. This protein is a close relative of the plant potyviral protease HC-Pro. HC-Pro is a papain-like protease that is autoprotelytic and which facilitates aphid transmission and promotes potyviral genome amplification. P29 is encoded on ORF A of CHV1 and is translated as part of the polyprotein p69 from which it is autocatalytically cleaved. Transformation of p29 into *C. parasitica* causes a loss of pigmentation, reduction in asexual sporulation, and suppression of host laccase expression, and there is evidence that it may be involved in suppression of RNA silencing by the host. Enhancement of both dsRNA accumulation and transmission of the virus through asexual spores are also functions attributed to CHV1 p29.

Recent studies show that p29 integrates into the *trans*-Golgi vesicle membranes of the host and that no other viral elements are required for this integration. While within the membrane p29 is fully susceptible to proteolytic digestion suggesting that it is primarily on the cytoplasmic side of the membrane. Deletion analysis of p29 showed that the C-terminal sequences of p29 mediate the membrane association. These transformation studies also confirmed the previous studies that showed p29 to be responsible for causing the reduced asexual sporulation and loss of pigment symptoms of virus infection. Further studies of the effect of p29 membrane integration on virus-caused host membrane proliferation and symptom production are in progress.

Based on the evidence available to date, CHV1 utilizes *trans*-Golgi vesicles of its host for replication and perhaps movement. *trans*-Golgi vesicles are critical for the final processing and secretion of proteins through the cytoplasmic membrane of the fungus, and for transport of proteins to certain cellular compartments. Studies of the viral protein p29 clearly link this protein to these vesicles and to causing some of the virus infection symptoms. The simplest hypothesis for how the virus is able to cause these and perhaps additional symptoms is that use of *trans*-Golgi vesicles for virus replication disrupts normal function of the vesicles and interferes with protein secretion or compartmentalization. The proprotein signal sequences directing kex2 endoprotease processing in a number of viral downregulated host proteins, such as the sex pheromone, cryparin, and laccase, also point to the *trans*-Golgi network as being a key to understanding the cause of virus symptom induction. The role of kex2 processing, a *trans*-Golgi vesicle function, in fungal development and its perturbation by the virus are currently under investigation.

There are precedents for RNA virus disruption of vesicle trafficking. Protein 3A of poliovirus specifically inhibits endoplasmic reticulum (ER)-to-Golgi traffic in mammalian cells, causing proteins otherwise destined for export to accumulate in ER-derived membranes. Many other viruses have been shown to have dramatic effects on the host membrane systems and a number of nonstructural virus proteins have been shown to interact with cellular factors responsible for the targeting and docking of membranes within the various organelles and membrane systems of virus hosts. Although no vesicle membrane proteins have been isolated yet from *C. parasitica* that are involved in vesicle function, the viral protein p29 which has an extensive cytoplasmic domain upon integration into vesicle membranes is a likely candidate for interaction with any host factors that direct cellular membrane traffic. An understanding of how the virus specifically targets these *trans*-Golgi membrane vesicles, and how this targeting affects normal function of these vesicles, will likely lead to a better understanding of virus symptom production, and provide new insights into critical host developmental controls.

See also: Fungal Viruses; Hypoviruses.

Further Reading

Boland GJ (2004) Fungal viruses, hypovirulence, and biological control of *Sclerotinia* species. *Canadian Journal of Plant Pathology* 26: 6–18.

Choe SS, Dodd DA, and Kirkegaard K (2005) Inhibition of cellular protein secretion by picornaviral 3A proteins. *Virology* 337: 18–29.

Ghabrial SA (1994) New developments in fungal virology. *Advances in Virus Research* 43: 303–388.

Ghabrial SA (1998) Origin, adaptation and evolutionary pathways of fungal viruses. *Virus Genes* 16: 119–131.

Hillman BI and Suzuki N (2004) Viruses of the chestnut blight fungus, *Cryphonectria parasitica*. In: Maramorosch K and Shatkin AJ (eds.) *Advances in Virus Research, vol.* 63, pp. 423–472. San Diego, CA: Academic Press.

Jacob-Wilk D, Turina M, and Van Alfen NK (2006) Mycovirus cryphonectria hypovirus 1 elements cofractionate with *trans*-Golgi network membranes of the fungal host *Cryphonectria parasitica*. *Journal of Virology* 80: 6588–6596.

Kazmierczak P, Kim DH, Turina M, and Van Alfen NK (2005) A hydrophobin of the chestnut blight fungus, *Cryphonectria parasitica*, is required for stromal pustule eruption. *Eukaryotic Cell* 4: 931–936.

Marsh M (2005) *Current Topics in Microbiology and Immunology, Vol. 285: Membrane Trafficking in Viral Replication*. Heidelberg: Springer.

McCabe PM, Pfeiffer PL, and Van Alfen NK (1999) The influence of dsRNA viruses on the biology of plant pathogenic fungi. *Trends in Microbiology* 7: 377–381.

McCabe PM and Van Alfen NK (2002) Molecular basis of symptom expression by the *Cryphonectria* hypovirus. In: Tavantzis SM (ed.) *dsRNA Genetic Elements*, pp. 125–144. Washington, DC: CRC Press.

Milgroom MG and Cortesi P (2004) Biological control of chestnut blight with hypovirulence: A critical analysis. *Annual Review of Phytopathology* 42: 311–338.

Nuss DL (2005) Hypovirulence: Mycoviruses at the fungal–plant interface. *Nature Reviews Microbiology* 3: 632–642.

Turina M, Zhang L, and Van Alfen NK (2006) Effect of *Cryphonectria hypovirus 1* (CHV1) infection on Cpkk1, a mitogen-activated protein kinase kinase of the filamentous *Cryphonectria parasitica*. *Fungal Genetics and Biology* 43: 764–774.

Villanueva RA, Rouille Y, and Dubuisson J (2005) Interactions between virus proteins and host cell membranes during the viral life cycle. *International Review of Cytology – A Survey of Cell Biology* 245: 171–244.

Hypoviruses

D L Nuss, University of Maryland Biotechnology Institute, Rockville, MD, USA

Glossary

Anastomosis The fusion of fungal hyphae resulting in exchange of cytoplasmic material and hypovirus transmission.

Hypovirulence Virus-mediated attenuation of fungal virulence.

Vegetative incompatibility A system controlled by at lease six genetic loci that determines the ability of two fungal strains to undergo anastomosis.

Introduction

The discovery of hypoviruses, a group of RNA viruses that reduce the virulence (hypovirulence) of the chestnut blight fungus *Cryphonecria parasitica*, has stimulated intensive research into the potential of using fungal viruses for biological control of fungal diseases. Documented examples of virus-mediated hypovirulence have been reported for fungal diseases of plants that range from trees to turfgrass and involve mycoviruses, which include representatives from the *Totiviridae*, *Chrysoviridae*, *Reoviridae*, *Narnaviridae*, and *Hypoviridae* (the hypoviruses). However, the hypoviruses remain the most thoroughly studied of the hypovirulence-associated viruses, primarily due to several significant advances in hypovirus molecular biology.

A classic example of the havoc that can result from the introduction of an exotic organism, the North American chestnut blight epidemic, first reported in 1905, caused the destruction of millions of mature chestnut trees by 1950. *Cryphonectria parasitica* was subsequently introduced into Italy during the 1930s, threatening European chestnut forests and orchards. However, for reasons still to be completely understood, the European chestnut blight epidemic was much less severe than that witnessed in North America. One clear contribution to this reduced severity was the prevalence of *C. parasitica* strains exhibiting a reduced virulence phenotype (hypovirulence), first described by an Italian forest pathologist in the 1950s. French investigators subsequently showed that the hypovirulence phenotype was transmissible following anastomosis (fusion of hyphae) between vegetatively compatible *C. parasitica* strains, implicating a cytoplasmic genetic element as the causative agent of the phenotype. The observation that a hypovirulent strain could produce a curative effect when inoculated onto existing cankers on diseased trees stimulated a successful government-sponsored biological control program using hypovirulent *C. parasitica* strains for management of chestnut blight in French chestnut plantations. Recent studies in Switzerland also support the view that natural hypovirulent strains retard disease progression in European forest ecosystems.

In 1977, Peter Day and co-workers at the Connecticut Agricultural Experiment Station reported that hypovirulent *C. parasitica* strains harbor double-stranded (ds) RNAs, providing the first indication of the nature of cytoplasmic elements responsible for the phenotype. Subsequent surveys of dsRNAs associated with different North American and European hypovirulent strains revealed considerable variations in concentration, number, and size of dsRNA components. By the late 1980s, it was clear that a detailed molecular analysis of the dsRNAs associated with a single hypovirulent strain was required to bring some measure of order to the mounting confusion generated by such surveys. This resulted in the cloning and complete sequence determination of the prototypic hypovirus, now designated CHV1-EP713, in 1991. This milestone was followed in 1992 by the construction of an infectious full-length (12 712 bp) cDNA clone of CHV1-EP713 RNA by Choi and Nuss. This development furnished direct evidence that hypoviruses are indeed the causative agents responsible for transmissible hypovirulence and provided the means for facile manipulation of the hypovirus genome. Current interest in hypoviruses extends past biological control potential to their utility as unique experimental tools for probing fundamental

processes underlying fungal pathogenesis and mycovirus–fungal host interactions.

Taxonomy and Genetic Organization

Hypoviruses are classified within the family *Hypoviridae*, consisting of the genus *Hypovirus* and four species designated *C. parasitica* hypovirus 1–4 (CHV1–CHV4). Hypovirus taxonomy is not based on virus structure, since this group of viruses does not encode a coat protein, but on genome organization, sequence similarity and symptom expression. Hypovirus genetic information is found predominantly as dsRNA associated with membrane vesicles ranging in diameter from 50 to 80 nm. As observed for fungal viruses generally, hypoviruses exhibit no extracellular phase in their life cycle. Infections cannot be initiated by inoculation with an infected cell extract or enriched fractions. Instead, these viruses are transmitted by cytoplasmic mixing as a result of fusion (anastomosis) between vegetatively compatible strains or to a variable degree in asexual spores (conidia).

The hypovirus species designations CHV1–CHV4 were assigned in the order in which their genome sequence was completed. The primary nucleotide sequence for the coding strand of the prototypic member of the genus *Hypovirus*, CHV1-EP713, specifies two large open reading frames (ORFs) designated ORF A and ORF B (**Figure 1**). A second closely related member, CHV1-Euro7, has the same organization and shares approximately 90% identity at the nucleotide level with CHV1-EP713. The type member of the CHV2 species, CHV2-NB58, shares only about 60% nucleotide sequence identity with CHV1-EP713 and lacks a portion of ORF A that, for the CHV1 species, encodes a functional *cis*-acting cysteine protease. Sequenced members of the CHV3 and CHV4 species, CHV4-GH2 and CHV4-SR2, respectively, are several kbp shorter than the CHV1 and CHV2 species, ~9.2–9.8 kbp versus 12.5–12.7 kbp, contain a single large ORF rather than two ORFs and are both more distantly related phylogenically to species CHV1 than CHV1 is to CHV2.

Hypovirus Gene Expression Strategy

Although hypovirus genetic information is readily recovered from infected cultures as linear dsRNA, the absence of a discrete virus particle and an extracellular infection phase presents some difficulties in precisely defining the hypovirus genome. Synthetic copies of the coding strand are infectious by electroporation into fungal spheroplasts and phylogenetic analyses suggest a common ancestry with the positive strand RNA plant potyviruses. Thus, one could consider hypoviruses as having a positive strand RNA genome and the dsRNA as representing accumulated replicative form RNA. Irrespective of this complication, direct analysis of hypovirus dsRNAs, cDNA cloning studies, and *in vitro* translational analyses has provided the following view of genetic organization and expression strategies for the prototypic hypovirus CHV1-EP713 shown in **Figure 2**. One strand contains a 3′-poly A tail while the complementary strand contains a 5′-poly U tract. All of the CHV1-EP713 coding information appears to reside within two contiguous ORFs on the 12 712 nt long polyadenylated strand. ORF A encodes two polypeptides, p29 and p40, that are released from a polyprotein precursor, p69, by an autoproteolytic event between Gly-248 and Gly-249, mediated by a

cysteine-like protease catalytic domain located within p29. ORF B has the capacity to encode a polyprotein of 3165 amino acids and contains unmistakable RNA-dependent RNA polymerase and helicase motifs. Proteolytic processing of only a portion of the ORF B polyprotein has been elucidated in the form of the autoproteolytic release of a 48 kDa protein, p48, from the N-terminus. This cleavage event occurs between Gly-418 and Ala-419 and is catalyzed by essential residues Cys-341 and His-388 within p48. The junction between ORF A and ORF B is defined by the sequence 5′-UAAUG-3′, in which the UAA portion clearly serves as the termination codon for ORF A and the AUG portion is

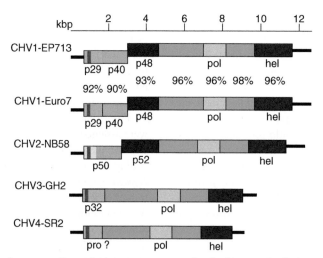

Figure 1 Genetic organization of sequenced hypovirus genomes representing the four species that comprise the genus *Hypovirus*, family *Hypoviridae*. Amino acid identity levels for coding regions of the two sequenced members of the CHV1 species, CHV1-EP713 and CHV1-Euro7, are indicated between representations of the two viral genomes. Protein coding regions homologous to CHV1-EP713 encoded p29, p40, p48, polymerase and helicase are color coded. The magenta regions represent a short conserved cysteine rich domain. Note that the genomes of CHV3-GH2 and CHV4-SR2 contain a single ORF. Modified from Dawe AL and Nuss DL (2001) Hypoviruses and chestnut blight: Exploiting viruses to understand and modulate fungal pathogenesis. *Annual Review of Genetics* 35: 1–29, with permission from Annual Reviews.

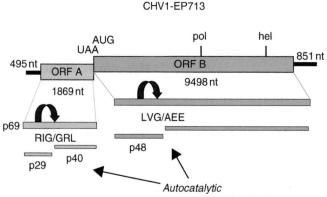

Figure 2 Expression strategy for prototypic hypovirus CHV1-EP713. The CHV1-EP713 coding strand consists of 12 712 nucleotides excluding a poly(A) tail, and contains two major coding domains designated ORF A and ORF B. Details are discussed in the text. Modified from Dawe AL and Nuss DL (2001) Hypoviruses and chestnut blight: Exploiting viruses to understand and modulate fungal pathogenesis. *Annual Review of Genetics* 35: 1–29 with permission from Annual Reviews.

thought to serve as the initiation codon for ORF B. While the mechanism involved in ribosome transition through the junction is not known, this unusual pentanucleotide sequence is found at the ORF A/ORF B junction for all confirmed CHV1 species. There is clearly a need for additional fine detailed mapping of the processing cascades for hypovirus-encoded polyproteins.

The genome of CHV2-NB58, type strain of species CHV2, also consists of a two ORF configuration with a UAAUG junction (**Figure 2**). However, ORF A lacks the p29 papain-like catalytic or cleavage sites and directs the translation of a 50 kDa protein product. ORF B of CHV2-NB58 does contain a p48 homolog, p52. The N-terminal portion of the single ORF of the CHV3-type strain CHV3-GH2 contains a protease, p32, with similarity to p29. A putative protease domain has been identified at the N-terminal portion of the CHV4-type strain CHV4-SR2, but protease cleavage has not been demonstrated.

In most hypovirus-infected *C. parasitica* isolates, the full-length viral dsRNA is accompanied by a constellation of shorter dsRNA species. These ancillary dsRNAs appear to be generated by internal deletion events, are replicated only in the presence of the full-length viral RNA, and are not associated with any function or phenotypic effect.

Hypovirus–Host Interactions

The phenotypic changes that are associated with hypovirus infection are not limited to hypoviurlence. Additional hypovirus-mediated phenotypic changes can include altered colony morphology, female infertility, reduced asexual sproulation (conidiation), and reduced pigmentation. The pleiotropic nature of these changes suggests that hypoviruses might perturb one or several cellular signaling pathways. Consistent with this view, alterations in G-protein, cyclic AMP-mediated-, mitogen-activated protein kinase-, and calcium/calmodulin/inositol trisphosphate-dependent-signaling pathways have been reported for hypovirus-infected fungal strains.

The construction of an ordered expressed sequence tag (EST) library representing approximately 2200 *C. parasitica* genes and development of a corresponding *C. parasitica* cDNA microarray platform has provided a view of host transcriptional responses to hypovirus infection. The modulation of transcript accumulation for approximately 13.4% of the 2200 cDNAs was shown following CHV1-EP713 infection. These transcriptional profiling studies also resulted in the identification of a subset of hypovirus responsive genes regulated through the G-protein signaling pathway and revealed a linkage between viral and mitochondrial hypovirulence.

Hypovirus-encoded symptom determinants and important replication elements have been mapped (**Figure 3**) by a combination of approaches that include (1) the construction of recombinant chimeras from hypoviruses that differ in their influence on host phenotype, (2) mutagenesis of a hypovirus infectious cDNA clone, and (3) cellular expression of viral coding domains independent of virus infection.

In analogy with plant viruses, CHV1-EP713 and CHV1-Euro7 can be viewed as severe and mild hypovirus isolates, respectively. Although these two CHV1 isolates cause quite different phenotypic changes in their fungal host, they share a high level of sequence similarity that has allowed the construction of viable chimeric viruses to begin mapping the determinants responsible for the differences in phenotypic changes.

Differences in colony morphologies were found to map to a region extending from a position just downstream of the p48 coding domain (map position 3575) to map position 9879, with clear indications of multiple discrete determinants. More specifically, the region extending from position 3575 to 5310 was able to confer a CHV1-EP713-like colony morphology when inserted into a CHV1-Euro7 genetic background. The CHV1-EP713

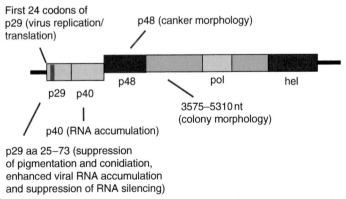

Figure 3 Emerging map of CHV1-EP713 symptom determinants and essential/dispensable replication elements as described in the text. Adapted from Dawe AL and Nuss DL (2001) Hypoviruses and chestnut blight: Exploiting viruses to understand and modulate fungal pathogenesis. *Annual Review of Genetics* 35: 1–29, with permission from Annual Reviews.

p48 coding region was found to be a dominant determinant contributing to suppression of asexual spore formation on the canker face.

The chimeric hypoviruses also proved to be very useful reagents when coupled with a pathway specific promoter/reporter system to map viral determinants responsible for altering G-protein/cAMP-mediated signaling. A common undesired side effect of hypovirus-mediated virulence attenuation is a significantly reduced ability of the fungal host to colonize and produce spores on the corresponding plant host. This reduces the ability of hypovirulent fungal strains to persist and spread through the ecosystem. Thus, from a practical perspective, a better understanding of the nature of viral symptom determinants and their relative effects on specific regulatory pathways and expression of gene clusters provides the means for a more rational approach for engineering hypoviruses that exhibit a desired balance between virulence attenuation and ecological fitness.

Additional insights into the functional role of viral coding regions were indirectly provided during efforts to develop hypoviruses as gene expression vectors. The nucleotide sequence corresponding to the first 24 codons of p29 was found to be required for viral replication, while the remaining 598 codons of ORF A, including all of the p40 coding region, was found to be dispensable. Substantial alterations were also tolerated in the pentanucleotide UAAUG that contains the ORF A termination codon and the overlapping putative ORF B initiation codon. For example, replication competence was maintained following either a frameshift mutation that caused a two-codon extension of ORF A or a modification that produced a single-ORF genomic organization. Further characterization of p40 revealed a role as an accessory function in viral RNA amplication with a functional domain extending from Thr(288) to Asn(313).

Expression of the CHV1-EP713 encoded papain-like protease, p29, in the absence of virus infection was shown to cause a subset of phenotypic changes exhibited by CHV1-EP713-infected strains, for example, a white phenotype (reduction in orange pigmentation), reduced asexual sporulation and a slight reduction in the production of fungal laccase activity. By deleting all but the first 24 N-terminal codons of p29 in the context of the CHV1-EP713 infectious cDNA clone (mutant virus Δp29), it was also possible to show that the p29 protein is dispensable for viral replication and to demonstrate a near restoration of orange pigment production and a moderate increase in conidiation levels relative to wild-type CHV1-EP713-infected fungal colonies. Deletion of p29 had no effect on virus-mediated virulence attenuation. A gain-of-function analysis involving progressive repair of the Δp29 mutant was also devised to map the p29 symptom determinant domain to a region extending from Phe-25 to Gln-73. When expressed from a chromosomally integrated cDNA copy, p29 elevated RNA accumulation and vertical transmission (through conidia) of the Δp29

mutant virus to levels observed for wild-type CHV1-EP713. Additional mutational studies indicated a linkage between p29-mediated changes in host phenotype in the absence of virus infection and p29-mediated in *trans*-enhancement of viral RNA accumulation and transmission.

The multifunctional nature of p29 was recently extended to include suppressor of RNA silencing both in the natural fungal host and in a heterologous plant system. This activity was predicted based on similarities between p29 and the well-characterized potyvirus-encoded suppressor of RNA silencing HC-Pro and has provided the first circumstantial evidence that RNA silencing in fungi may serve as an antiviral defense mechanism, a well-established function in plants.

Prospects for Biological Control

Chestnut blight cankers on American chestnut trees can be controlled by the direct application of hypovirus-infected hypovirulent *C. parasitica* strains to the canker margin, provided that the hypovirulent and resident virulent strains are vegetatively compatible. Anastomosis between the two strains results in hypovirus transmission and conversion of the canker to the hypovirulent phenotype, thereby preventing additional expansion of the canker and promoting formation of new callus tissue. However, since treated trees do not become resistant to new infections and North American *C. parasitica* populations generally exhibit a very diverse vegetative compatibility structure, this form of treatment is labor intensive and has not proved to be practical for effective control of chestnut blight in a North American forest ecosystem.

High levels of VC diversity and the severe phenotypic characteristics of the hypoviruses that have been introduced for biocontrol are thought to be two of the main contributing factors to the inability of introduced hypoviruses to spread through North America *C. parasitica* populations. In this regard, transgenic hypovirulent strains that contain a nuclear copy of the hypovirus cDNA exhibit novel hypovirus transmission properties that have been predicted to reduce these limitations. Hypovirus RNA is not transmitted to ascospores during mating. However, the hypovirus cDNA present in transgenic hypovirulent strains is inherited by a portion of the ascospore progeny, followed by the production of cDNA-derived cytoplasmically replicating viral RNA. Additionally, since the progeny represent a spectrum of vegetative compatibility groups, launching of the cytoplasmic viral RNA into these new vegetative compatibility groups is predicted to expand vegetative dissemination of hypovirus genetic information.

Three field trials with CHV1-EP713 transgenic strains have confirmed hypovirus transmission to ascospore progeny in a forest setting and also exposed the requirement for significant improvements in formulation and delivery methods. These trials also confirmed predictions that the CHV1-EP713 hypovirus performs poorly as a biological control agent because it severely reduces the ability of the fungal host to colonize, expand, and produce asexual spores on chestnut tissue. In this regard, the availability of an infectious cDNA clone for the mild hypovirus CHV1-Euro7 has provided the means to construct a transgenic hypovirulent strain that combines the properties of enhanced colonization and spore production with a novel mode of hypovirus transmission to ascospore progeny.

The infectious hypovirus cDNA clones have also been used to expand hypovirus host range to include several other pathogenic fungi that include the *Eucalyptus* canker pathogen *Cryphonectria cubensis* and the fruit tree pathogens *Phomopsis G-type* and *Valsa ceratosperma*. Hypovirus-based management of these economically important fungal diseases in controlled agricultural settings may not face the same problems of formulation and application that has been encountered for control of chestnut blight in a complex forest ecosystem.

See also: Fungal Viruses; Hypovirulence.

Further Reading

Dawe AL and Nuss DL (2001) Hypoviruses and chestnut blight: Exploiting viruses to understand and modulate fungal pathogenesis. *Annual Review of Genetics* 35: 1–29.

Hillman BI and Suzuki N (2004) Viruses of the chestnut blight fungus, *Cryphonectria parasitica*. *Advances in Virus Research* 63: 423.

Nuss DL (2005) Hypovirulence: Mycoviruses at the fungal–plant interface. *Nature Reviews Microbiology* 3: 632.

Segers GC, van Wezel R, Zhang X, Hong Y, and Nuss DL (2006) Hypovirus papain-like protease p29 suppresses RNA silencing in the natural fungal host and in a heterologous plant system. *Eukaryotic Cell* 5: 896.

Mycoreoviruses

B I Hillman, Rutgers University, New Brunswick, NJ, USA

The 8th Report of the International Committee for the Taxonomy of Viruses (ICTV) which lists 12 genera of reoviruses that infect mammals, invertebrates, and plants, was the first to also include the fungus-infecting genus *Mycoreovirus*. Members of many of the reovirus genera in the 8th Report replicate in organisms representing more than one kingdom or phylum: for example, all plant reoviruses replicate in and persistently infect their insect vectors, and similarly many of the mammalian reoviruses replicate in their invertebrate vectors. Three fungal virus species have been confirmed to be members of the genus *Mycoreovirus* of the family *Reoviridae*. Two of these viruses were isolated from the filamentous ascomycete fungus *Cryphonectria parasitica*, and the other is from the soil-borne fungus *Rosellinia necatrix*, also an ascomycete but representing a different fungal family.

All of the mycoreoviruses cause disease in their infected hosts, resulting in greatly reduced virulence, reduced growth in culture, reduced laccase accumulation, and reduced sporulation relative to wild-type, virus-free cultures of the same genetic background (**Figure 1**). The mycoreoviruses are most closely related to the genus *Coltivirus* of tick-borne reoviruses (**Figure 2**). Both Colorado tick fever virus (CTFV) and the closely related Eyach virus (EyaV) cause mammalian diseases, and both viruses replicate in their arthropod vectors as well as in their mammalian hosts. The coltiviruses are not well understood at the molecular level and so have shed only a little light on the study of mycoreoviruses. The genomic sequences of both CTFV and EyaV have been published, and they have been found to be more closely related to the genus *Orthoreovirus*, which includes the common human pathogen mammalian reovirus (MRV), than to members of the other two genera of the family *Reoviridae* (*Orbivirus* and *Rotavirus*) that have been well studied at the structural and molecular levels.

Filamentous fungi are valuable subjects for investigation of eukaryotic viruses. This has been especially true for ascomycetes, which usually are haploid throughout their vegetative phases and thus readily amenable to tools of classical genetics and to relatively simple gene knockout and knockdown experiments. *Cryphonectria parasitica* has been exceptional as a host for examination of fungal viruses. The fungus caused the greatest pandemic of a tree species in recorded history, killing 3–4 billion American chestnut trees. The fungus is very stable in culture, showing a consistent morphology upon continued maintenance and subculture, is easily transformed and transfected, and can be examined through classical genetics. Because of its historical importance as a pathogen and the potential for its control using natural or genetically engineered viruses, more viruses that result in stable morphological changes and/or changes in virulence have been identified and characterized in this fungus than in any other.

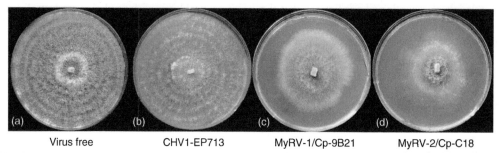

Figure 1 Morphologies of isogenic *Cryphonectria parasitica* colonies infected with three mycoviruses. (a) Uninfected colony, strain EP155; (b) strain EP155 infected with hypovirus CHV-1/EP713; (c) strain EP155 infected with mycoreovirus MyRV-1/Cp9B21; (d) Strain EP155 infected with mycoreovirus MyRV-2/CpC18.

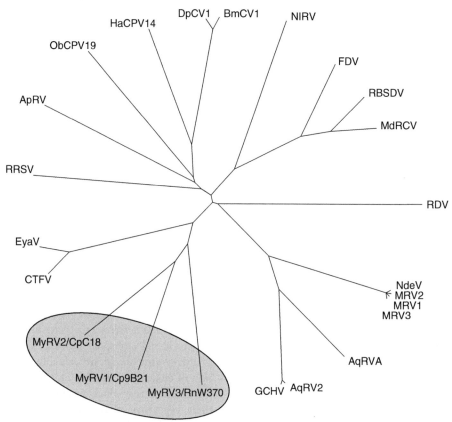

Figure 2 Unrooted neighbor-joining tree based on ClustalW alignments of complete deduced amino acid sequences of RNA-dependent RNA polymerase genes of the reoviruses most closely related to the mycoreoviruses (shaded).

Structure–Function Relationships

Relatively little is known of the details of mycoreovirus structure, but all indications from electron micrographs of negatively stained virus particles and genome sequence analysis are that it belongs to the orthoreovirus subgroup. The orthoreoviruses are distinct from the orbiviruses and rotaviruses in that they have identifiable pentameric turrets on top of each fivefold axis of the core particle, and these have been demonstrated to be involved in capping of nascent mRNA. The protein that was experimentally demonstrated to be the mycoreovirus guanylyltransferase is homologous to turret proteins of orthoreoviruses that have the same function, supporting the grouping of mycoreoviruses within the orthoreovirus subgroup of the *Reoviridae*.

Relatively little is known about the movement of fungal viruses in general within the mycelium and during horizontal or vertical virus transmission. Viruses that have been investigated are generally found in hyphal tip cells and move with the growing mycelium; however, location of mycoreoviruses within the mycelium has not been investigated. The steps in the infection process of the mammalian orthoreovirus have been investigated in considerable detail. Virus entry into cells via vesicles occurs by receptor-mediated endocytosis, whereupon the μ1 protein, which is myristoylated at its N-terminus and inserted into the inner membrane of the vesicle cleaves autocatalytically, resulting in pore formation in the vesicle membrane and subsequent virus release into the cytoplasm. Both of the *C. parasitica* mycoreoviruses have homologs of the orthoreovirus μ1 protein, with strong myristoylation signals at the amino termini of the corresponding proteins and putative sites for autocatalytic cleavage. This suggests that a similar mechanism of egress from vesicles is a component of the infection cycle of these reoviruses. Interestingly, although the Rosellinia virus, MyRV-3/RnW370, contains a homolog of the μ1 protein, it does not contain a glycine residue at the penultimate position of the N-terminus of this deduced protein, and based on its sequence it has a very low predicted probability of myristoylation. It does, however, contain the N/P putative cleavage motif at a similar position and in a similar environment as those in the other viruses.

It is relatively easy to cure *C. parasitica* and *R. necatrix* from their associated mycoreoviruses. In *C. parasitica*, single asexual spores are usually virus free, and in *R. necatrix*, cultures initiated from excised hyphal tips (the terminal 2–8 cells) are also virus free. These are in contrast to cultures infected with the well-studied hypovirus of *C. parasitica*, cryphonectria hypovirus-1 (CHV-1), in which most or sometimes all conidia contain virus and the virus-infected cultures cannot be cured by hyphal tip isolation. It is likely that details of mycoreovirus movement within the mycelium as well as horizontal and vertical transmission are quite different from those properties of hypoviruses, which have no capsid protein, are more closely related to positive-sense single-stranded RNA viruses such as plant potyviruses and animal picornaviruses, and whose replication is associated with the fungal *trans*-Golgi network.

Taxonomy and Nomenclature

Naming of fungal viruses is often problematic because infectivity studies are very difficult with fungal viruses and have not been done with most. In nature, fungal viruses do not exit completely from an infected isolate and enter an uninfected one exogenously. Instead, they move from one isolate to another only after the hyphae of the two isolates have fused (anastomosed), in which case the contents of one or usually more cells from the infected and uninfected isolate are mixed. Mixed infections are common and symptomless infections are the norm for fungal viruses. For these reasons, fungal virus names usually contain reference not only to the host genus and species, but to the host strain of

origin as well. The genus name *Mycoreovirus* was a natural one to use because it is descriptive . For the sake of simplicity and consistency with other viral nomenclature, species are numbered progressively as they are described. The fungal species and isolate in which a particular virus was identified is provided after the species number. In this system, the first mycoreovirus species described was designated *Mycoreovirus-1*. The only isolate of this virus species identified to date was from *C. parasitica* strain 9B21, so the virus is designated mycoreovirus-1/Cp9B21 or MyRV-1/Cp9B21. The reason that MyRV-2 represents a separate virus species even though the two were isolated from the same fungal species and only a few miles away is because it shares much less sequence similarity at both the nucleotide and amino acid levels (<50%) than expected for two viruses in a single species. Surprisingly, both of these species are monotypic: these two virus isolates, each representing a different species, are the only two mycoreoviruses isolated from *C. parasitica*, even though thousands of isolates infected with members of the family *Hypoviridae* have been identified worldwide. In contrast, dsRNA or virus from dozens of isolates of *Rosellinia necatrix* from different parts of Japan have been isolated, but all are closely related to each other, indicating that they represent strains of a single virus species, MyRV-3.

Genome Structures, Organizations, and Relationships

The three fungal reoviruses examined to date appear to have 11 segments of dsRNA that are required for infection. Isolates of one of the viruses, MyRV-3, have been found to have either 12 or 11 segments (see below). There is significant sequence similarity among the larger segments of the three mycoreovirus species and between homologous segments of the mycoreoviruses and their closest relatives, the coltiviruses, but this similarity becomes less apparent in the middle segments and is not apparent at all in the small segments. This feature is common with other members of the family *Reoviridae*, in which the more distant relationships are generally revealed in only the large segments. Each of the 11 required segments of mycoreoviruses appears to contain a single open reading frame. **Figure 3** Segments 1–5 of the mycoreoviruses are homologous to segments 1–5 of the coltiviruses (segments 4 and 5 of MyRV2-CpC18 are transposed relative to the others). Based largely on similarity with other reoviruses, their predicted functions are: S1: RNA-dependent RNA polymerase (core protein); S2: dsRNA-binding (core protein); S3: guanylyltransferase (turret protein); S4: Myristoylated membrane penetration protein (outer capsid); S5: cytoskeleton-interacting (core). Segment 6 of the mycoreoviruses is predicted to encode a nucleic acid binding core protein and is homologous to segment

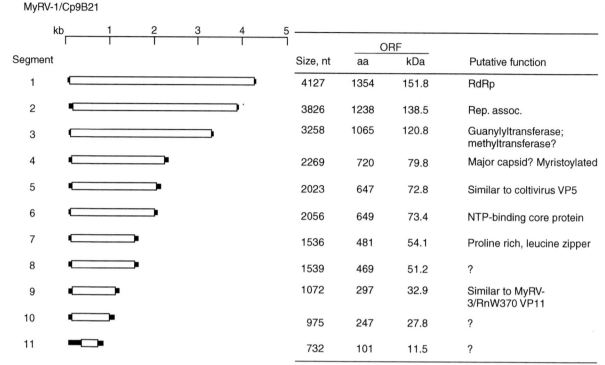

Figure 3 Diagram of the genome segments of mycoreovirus-1/Cp9B21 shown with segment size, deduced open reading frame (ORF) in number of amino acid residues and predicted protein product size, and putative protein function. Only the guanylyltransferase function of segment 3 has been demonstrated experimentally.

10 of the two coltiviruses. Segment 7 of MyRV-1/Cp9B21 encodes a proline-rich protein with similarity to several viral and nonviral proline-rich domains. Homologies among the smaller mycoreovirus proteins (from S8–11 or 12) and other reovirus deduced proteins, including those of coltiviruses, are unknown.

One of the indications of close relationship among reoviruses is conserved terminal sequences. This also is a good indication of whether or not pseudorecombinants can be generated by co-infection with two related viruses. In the case of the mycoreoviruses, the three species have different conserved terminal sequences, consistent with their taxonomic separation. Furthermore, although co-infection of a single colony with the two reoviruses has been achieved, pseudorecombinants have not yet been recovered from these doubly infected isolates.

Mycoreovirus-1/Cp9B21

Although MyRV-1/Cp9B21 was not the first of the two *Cryphonectria* reoviruses identified, fungal cultures infected with this virus proved to be more stable and easily studied than those infected with MyRV-2/CpC18 (see below), so early molecular investigations have focused on this virus. When grown on solid media in petri dishes (e.g., a defined complete medium or potato dextrose agar (PDA)), MyRV-1/Cp9B21-infected colonies are deep orange in color and have little aerial hyphae compared to their uninfected counterparts. As with other *C. parasitica* viruses, the phenotype of the infected culture has very little to do with the host isolate but is determined almost entirely by the virus.

Fungal isolates infected with MyRV-1/Cp9B21 are much less virulent than uninfected isolates, and are among the most debilitated of any virus-infected *C. parasitica* cultures studied to date. Although the virus accumulates to reasonably high concentrations in infected colonies, it is transmitted very poorly through conidia, at rates of only 2–5%. This may account in part for the rarity of the virus in nature.

Recently, double infections of the well-characterized hypovirus CHV-1/EP713 and MyRV-1/Cp9B21 were examined. In these analyses, it was found that presence of the hypovirus increased both the concentration and vertical transmission rate through conidia of the reovirus, but that the reovirus did not affect the concentration or transmission rate of the hypovirus. Furthermore, transgenic expression of only the hypovirus protein p29 also resulted in increased accumulation and transmission of MyRV-1/Cp9B21. This is consistent with the prediction that p29 serves as a suppressor of RNA silencing (interference) during hypovirus infection, and that this effect acts *in trans* to support enhanced mycoreovirus replication.

Expression of MyRV-1/Cp9B21 Gene Products

Functional analysis of the MyRV-1/Cp9B21 genome was initiated by cloning the 11 individual segments into a baculovirus expression vector and expressing them in insect cells. All 11 segments were expressed, resulting in 11 identifiable protein products on polyacrylamide gels. Only one of the proteins, the segment 3 product, has been studied functionally. This protein was found to be active in autoguanylylation assays, confirming it as the viral guanylyltransferase. Deletion and site-directed mutational analysis determined that the amino acid sequence EPAGYHPRPSIVVPHYFVFR constituted the catalytically active site of the MyRV-1/Cp9B21 guanylyltransferase. The Hx_8H motif was identified as absolutely conserved in all three members of the genus *Mycoreovirus*, as well as in the structurally related genera *Coltivirus*, *Orthoreovirus*, *Aquareovirus*, *Cypovirus*, *Dinovernavirus*, *Oryzavirus*, and *Fijivirus*. In all of the above genera in which the guanylyltransferase has been identified functionally, the Hx_8H motif has been found within the sequence. The core consensus sequence for the guanylyltransferase within this group of genera was a/vxxHxxxxxxxxHhyf/lvf, with only the H residues being absolutely conserved.

Mycoreovirus-2/CpC18

The first virus that was tentatively identified as a mycoreovirus was MyRV-2/CpC18. Like many of the virus-infected *C. parasitica* cultures, the one that was found to contain this virus was isolated from a canker on an American chestnut tree. The circumstances of the discovery were somewhat unusual in that only one of 36 fungal isolates from that particular canker was virus infected; the rest were virus free. Although it is not unusual to find mixed fungal infections within a canker or to isolate virus-containing and virus-free cultures from one canker, the ratio of 1/36 is extraordinary. The phenotype of the infected culture, designated C-18 (the 18th isolate from canker C on that particular tree) was distinct from the phenotype of MyRV-1/Cp9B21-infected cultures described above in that it was light brown in color and had more aerial mycelium (**Figure 1**). The two viruses have very similar, dramatic negative effects on fungal virulence.

A major difference between the two *C. parasitica* mycoreoviruses is their stability and transmissibility. MyRV-2/CpC18 is easily lost upon subculture of the fungus, while this has never been observed with MyRV-1/Cp9B21. Furthermore, MyRV-2/CpC18 is extremely difficult to transmit from an infected isolate to an isogenic uninfected isolate by hyphal anastomosis, a property that is not seen with other *C. parasitica* viruses. It is presumed that these properties are related, but their reasons have not been elucidated. Both MyRV-1/Cp9B21 and MyRV-2/CpC18 have been transmitted to uninfected fungal isolates by inoculating *C. parasitica* protoplasts with purified virus particle preparations and allowing the protoplast to regenerate cell walls, form a hyphal network, and grow into a single colony. The ability to infect protoplasts efficiently using purified virus particle preparations is an interesting and useful feature of *C. parasitica* reoviruses. This has allowed for infection of different genotypes of *C. parasitica* regardless of vegetative incompatibility group and potential for transmission by hyphal anastomosis.

Mycoreovirus-3/RnW370

White rot is a root disease of fruit trees that can be limiting to production. Control of the fungus that causes the disease, *Rosellinia necatrix*, by chemical means is difficult and not economically feasible. Pathologists in Japan, where the disease is particularly severe, have sought to use virus-infected strains to control the disease, leading to the identification of several viruses. This plant/fungus interaction represents an interesting contrast to the chestnut/*C. parasitica* interaction. As a root disease, there are challenges and opportunities for biological control of a fungal pathogen with viruses that do not apply to aerial diseases. One of the viruses under investigation for biocontrol of *R. necatrix* is the mycoreovirus MyRV-3, which causes reduced virulence of the fungus. Unlike the monophyletic *C. parasitica* reoviruses, different strains of MyRV-3 have been isolated from a variety of *R. necatrix* strains from around Japan.

Of the *R. necatrix* viruses, the virus isolate that has been most thoroughly characterized is MyRV-3/RnW370. In surprising contrast to the *C. parasitica* viruses, MyRV-3/RnW370 was found to contain 12 rather than 11 segments. However, the presence of 12 segments is not a consistent feature of all MyRV-3 isolates. Examination of different virus-infected isolates of *R. necatrix* shows that they may have either 12 or 11 segments. Experiments to investigate virus composition and transmission have been performed on hyphal tip cultures from infected *R. necatrix* isolates, resulting in demonstration that these viruses behave like the *Cryphonectria* mycoreoviruses.

When only 11 segments are present in MyRV-3 isolates, segment 8 is the one that is absent from the full complement. With most reoviruses, sequence conservation among species and genera is evident in the larger segments, but much less so in the smaller segments, and this is true in the mycoreoviruses and related genera. Consistent with this general trend, sequence comparison between the 12 segments of MyRV-3 and the 12 segments of the two coltiviruses has suggested nothing about possible function of the apparently dispensable segment 8.

It is intriguing to think that perhaps segment 8 is vestigial for reoviruses in fungi and is required only in another host, past or present. This would be similar to the leafhopper-transmitted phytoreovirus, wound tumor virus (WTV), in which deletion mutations in any of three segments may be found upon successive serial, insect-free virus passage or long-term virus maintenance in plants, whereupon resulting mutant viruses become defective in their transmission properties and incapable of replicating in their leafhopper vectors. An apparent major difference between MyRV-3 and WTV is that in MyRV-3 there is no evidence for remnants of segment 8; it appears to be either present or entirely absent. In the well-characterized mutants of WTV, the deleted segments are not completely gone, but shorter segments containing the two termini and varying amounts of adjacent sequence remain, ensuring that a total of 12 segments remain. Whether this represents a difference between the dsRNA segment sorting and packaging mechanisms of phytoreoviruses and mycoreoviruses is not known. In this line of inquiry, the close association of fungi with mites in natural settings may be significant to the evolutionary biology of mycoreoviruses: it may be no coincidence that their closest relatives are the tick-borne coltiviruses, with ticks and mites both in the arachnid subclass Acari. Unfortunately, mites are very difficult experimental subjects and cell cultures are not currently available. Furthermore, much less is known about coltivirus gene function than is known about many of the other members of the family *Reoviridae* that are pathogenic to humans, making it more difficult to pursue this line of research from a strictly bioinformatic standpoint.

Effects of Mycoreoviruses on Fungal Gene Expression

Considerable information has been amassed on the impact of CHV-1, a positive-sense RNA virus, on its fungal host. In contrast, studying the mechanisms of mycoreovirus infection of fungi is in its infancy. The first study addressing these questions was done by microarray analysis using mRNA isolated from isogenic fungal isolates of *C. parasitica* infected with either MyRV-1 or MyRV-2, and comparing results to the same strain that was uninfected or infected with either of two different CHV-1 strains, and with fungal mutants defective in virulence characteristics. To date, these experiments have been performed only on EST-based arrays representing ~20% of the total *C. parasitica* gene complement. Overall, there was consistency in the effects of the two *C. parasitica* mycoreoviruses on host gene expression: MyRV-1 infection resulted in differential expression of 6.5% of the genes on the array, whereas MyRV-2 infection affected expression of 5.8% of those genes. As might be expected based on their phenotypes, similar but distinct suites of genes were up- or downregulated in isogenic fungal isolates infected with the two reoviruses. Approximately 60% of the genes whose expression was affected were the same whether infection was by MyRV-1 or MyRV-2, and all but one of those genes were altered in the same direction. Some of these groups of genes are in common with those that are differentially regulated in cultures infected with the unrelated hypoviruses, but there are predictable differences. For example, hypovirus infection of *C. parasitica* results in female infertility, whereas mycoreovirus infection does not, and this is reflected in the expression of two genes predicted to be involved in the *C. parasitica* mating response. Both *mf2-1*, which encodes the fungal pheromone precursor, and *Csp12*, which encodes a homolog of the yeast *Ste12*-like transcription factor, were substantially downregulated in hypovirus-infected fungal isolates, which are female sterile, whereas there was much less effect on expression of these genes in either of the mycoreovirus-infected *C. parasitica* isolates. Virus is transmitted to ascospores at a rate of ~50% or less when the female parent is infected, but there is no virus transmission to ascospore progeny if the male parent in a mating is infected. Sequencing the complete genome of *Cryphonectria parasitica* is now underway. This will allow for the complete set of genes to be represented in an oligo array for more thorough investigation of differential gene expression.

See also: Fungal Viruses; Hypoviruses.

Further Reading

Enebak SA, Hillman BI, and MacDonald WL (1994) A hypovirulent *Cryphonectria parasitica* isolate with multiple, genetically unique dsRNA segments. *Molecular Plant-Microbe Interactions* 7: 590–595.

Hillman BI, Supyani S, Kondo H, and Suzuki N (2004) A reovirus of the fungus *Cryphonectria parasitica* that is infectious as particles and related to the *Coltivirus* genus of animal pathogen. *Journal of Virology* 78: 892–898.

Hillman BI and Suzuki N (2004) Viruses of *Cryphonectria parasitica*. *Advances in Virus Research* 63: 423–472.

Kanematsu S, Arakawa M, Oikawa Y, et al. (2004) A reovirus causes hypovirulence of *Rosellinia necatrix*. *Phytopathology* 94: 561–568.

Mertens P and Hillman BI (2005) Genus mycoreovirus. In: Fauquet CM, Mayo MA, Maniloff J, Desselberger U,, and Ball LA (eds.) *Virus Taxonomy: Eighth Report of the International Committee on Taxonomy of Viruses*, pp. 556–560. San Diego, CA: Elsevier Academic Press.

Nibert ML and Schiff LA (2001) Reoviruses and their replication. In: Knipe DM and Howley PM (eds.) *Fields Virology*, 2nd ed., vol. 2, pp. 1679–1729. Philadelphia, PA: Lippincott Williams and Wilkins.

Nuss DL (2005) Hypovirulence: Mycoviruses at the fungal-plant interface. *Nature Reviews Microbiology* 3: 632–642.

Osaki H, Wei CZ, Arakawa M, et al. (2002) Nucleotide sequences of double-stranded segments from hypovirulent strain of the white root rot fungus *Rosellinia necatrix*: Possibility of the first member of the *Reoviridae* from fungus. *Virus Genes* 25: 101–107.

Supyani S, Hillman BI, and Suzuki N (2006) Baculovirus expression of the 11 *Mycoreovirus*-1 genome segments and identification of the guanylyltransferase-encoding segment. *Journal of General Virology* 88: 342–350.

Suzuki N, Supyani S, Maruyama K, and Hillman BI (2004) Complete genome sequence of *Mycoreovirus 1*/Cp9B21, a member of a new genus in the family *Reoviridae* isolated from the chestnut blight fungus, *Cryphonectria parasitica*. *Journal of General Virology* 85: 3437–3448.

Wei CZ, Osaki H, Iwanami T, Matsumoto N, and Ohtsu Y (2003) Molecular characterization of dsRNA segments 2 and 5 and electron microscopy of a novel reovirus from hypovirulent isolate, W370, of the plant pathogen *Rosellinia necatrix*. *Journal of General Virology* 84: 2431–2437.

Wei CZ, Osaki H, Iwanami T, Matsumoto N, and Ohtsu Y (2004) Complete nucleotide sequences of genome segments 1 and 3 of Rosellinia anti-rot virus in the family *Reoviridae*. *Archives of Virology* 149: 773–777.

Narnaviruses

R Esteban and T Fujimura, Instituto de Microbiología Bioquímica CSIC/University de Salamanca, Salamanca, Spain

Glossary

Ribozyme RNA with a catalytic activity.

Introduction

The narnaviruses 20S RNA and 23S RNA (ScV20S and ScV23S, respectively) are positive-strand RNA viruses found in the yeast *Saccharomyces cerevisiae*. Currently only these two viruses are ascribed to the genus *Narnavirus* of the family *Narnaviridae*. Like most fungal viruses, they have no extracellular transmission pathway. They are transmitted horizontally by mating, or vertically from mother to daughter cells. It is believed that the high frequency of mating or hyphal fusion that occurs in the host life cycle makes an extracellular route of transmission dispensable for the viruses. The thick cell wall of fungi may also form a formidable barrier. The lack of extracellular transmission may, in turn, explain two prominent features found in narnaviruses. First, they are persistent viruses and do not kill the host cells. If their infection caused damages or disadvantages to the host, then the viruses might have perished during the course of evolution because of the lack of an escape route. Second, because there is no extracellular phase, the viruses do not need to form virions to protect their RNA genomes in the extracellular environment. In addition, they do not need machinery to ensure exit or reentry to a new host. The lack of a virion structure may sound peculiar to those who are familiar with infectious viruses. Considering that viruses are selfish parasites, however, it will be natural for them to shed genes or functions unnecessary for their existence. Consequently, the genomes of narnaviruses are simple and small: they only encode a single protein, the RNA-dependent RNA polymerase (RdRp). This may contribute to their persistence by reducing a number of viral proteins that might interfere with metabolism vital for the host. The simplicity of their RNA genomes encoding a single protein, together with the recent development of 20S and 23S RNA virus launching systems from yeast expression vectors, makes narnaviruses a good model system to investigate replication and the molecular basis for intracellular persistence of RNA viruses.

Historical Background

20S RNA was first described in 1971 as a single-stranded RNA (ssRNA) species accumulated in yeast cells transferred to 1% potassium acetate, a standard procedure to induce sporulation in yeast under nitrogen-starvation conditions. Because of its mobility relative to 25S and 18S rRNAs, the species was named 20S RNA. Later it was found, however, that the accumulation of 20S RNA was not related with the sporulation process because haploid cells that do not sporulate also accumulate 20S RNA under nitrogen-starvation conditions. It was also found that 20S RNA is a cytoplasmic genetic element. The realization of 20S RNA as a viral entity, however, had to wait several years, until the characterization of 20S RNA by cloning and sequencing in 1991. 23S RNA was reported first time in 1992. Both viruses were placed in the genus *Narnavirus* of the new family, *Narnaviridae* (naked RNA virus), a taxonomic group that appeared for the first time in the seventh edition of the International Committee on Taxonomy of Viruses (ICTV). The other genus in the family is *Mitovirus*, whose members are found in mitochondria of fungi, many of them pathogenic to plants. The members of the family have small RNA genomes (2–3 kb) that encode single proteins, their RdRps, and reside either in the cytoplasm (members of the genus *Narnavirus*, narnaviruses) or in the mitochondria (mitoviruses) of the host.

Viral Genomes

Many laboratory strains of *S. cerevisiae* harbor 20S RNA virus and fewer strains contain 23S RNA virus. Both viruses are compatible in the same host. The presence of 20S and 23S RNA viruses does not render phenotypic changes to the host. Under nitrogen-starvation conditions, the amounts of the viral genomes become almost equivalent to those of rRNAs (>100 000 copies/cell; **Figure 1**). In contrast, vegetative growing cells contain much lower amounts of the viral RNAs (5–20 copies/cell). **Figure 2(a)** shows the genome organization of narnaviruses. Both 20S

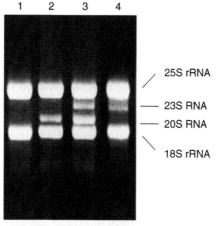

Figure 1   Agarose gel electrophoresis of RNA extracted from virus-free and virus-infected nitrogen-starved yeast cells. RNA from virus-free yeast cells (lane 1) or cells infected with 20S RNA virus alone (lane 2), 23S RNA virus alone (lane 4), or both viruses together (lane 3), was separated in an agarose gel and visualized by ethidium bromide staining. The positions of the 20S and 23S RNAs together with the rRNAs are indicated to the right.

and 23S RNAs are small (2514 and 2891 nt, respectively) and each genome encodes a single protein: a 91 kDa protein (p91) by 20S RNA and a 104 kDa protein (p104) by 23S RNA. The 5′ untranslated regions in both RNAs are extremely short: 12 nt in the case of 20S RNA and only 6 nt in 23S RNA. These RNAs lack poly(A) tails at the 3′ ends and have perhaps no 5′ cap structures. The same RNA can serve as template for translation and also for negative-strand synthesis. The antigenomic (or negative-strand) RNAs have no coding capacity for protein and are present at much lower copy numbers compared to the genomic (or positive-strand) RNAs under the induction conditions. The double-stranded forms of 20S and 23S RNAs are known and called W and T, respectively. These double-stranded RNAs (dsRNAs) accumulate when the cells are grown at 37 °C, a rather high temperature for yeast (the optimal temperature for growth is about 28 °C). These dsRNAs are not intermediates of replication but by-products. Replication proceeds from a positive strand to a negative strand and then to a positive strand.

The proteins encoded in the viral genomes are not processed to produce smaller fragments with distinct functional domains. Both proteins contain amino acid motifs well conserved among RdRps from positive-strand and dsRNA viruses (**Figure 2(a)**) In addition, p91 and p104 share stretches of amino acid sequences (denoted by 1–3 in **Figure 2(a)**) in the same order throughout the molecules, indicating a close evolutionary relationship between these two viruses. Remarkably, their RdRp consensus motifs are most closely related to those of RNA bacteriophages such as Qβ.

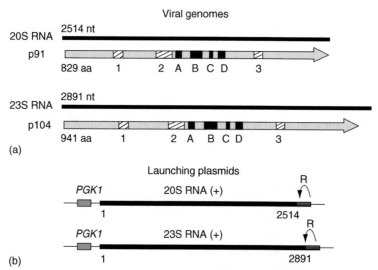

Figure 2   Genomic organization of 20S and 23S RNA viruses (a) and their launching plasmids (b). (a) Diagrams of 20S and 23S RNAs and the proteins encoded by them, p91 and p104, respectively. A–D represents motifs conserved among RdRps from positive strand and dsRNA viruses and 1–3 indicates amino acid stretches conserved between p91 and p104. (b) The complete cDNA of 20S or 23S RNA genome is inserted downstream of the constitutive *PGK1* promoter in a yeast expression vector in such a way that positive strands are transcribed from the promoter. The HDV ribozyme (R) is fused directly to the 3′ end of the viral genome.

Ribonucleoprotein Complexes as Viral Entities

Yeast is also a natural host for dsRNA totiviruses, called L-A and L-BC. Like narnaviruses, totiviruses have no extracellular transmission pathway. However, these viruses have *gag* and *pol* genes, and their dsRNA genomes are encapsidated into intracellular viral particles. In contrast, 20S and 23S RNA viruses have no capsid genes to form virion structures. Then, how do these viruses exist inside the cell and establish a persistent infection without a protective coat? Earlier studies demonstrated that 20S RNA migrated as 'naked RNA' in sucrose gradients. Furthermore, deproteination with phenol had no apparent effect on its mobility. Because protein provides a large part of the molecular mass in virions, these data clearly indicate that narnaviruses lack a virion structure. When specific antibodies against their RdRps became available, however, it was realized that each RdRp is associated with its RNA genome and this interaction is specific; that is, p91 is associated only with 20S RNA and p104 only with 23S RNA. These ribonucleoprotein complexes reside in the cytoplasm and are not associated with the nucleus, mitochondria, or intracellular membranous structures. Further studies indicated that most of the positive strands of 20S and 23S RNA viruses under induction conditions are associated with their own RdRps in a 1:1 stoichiometry. It is not known whether host proteins are present in the complexes. These complexes are called 'resting complexes' to distinguish them from the 'replication complexes' described in the following section. Negative strands are present at much lower amounts compared to positive strands, and available data indicate that they also form complexes with their own RdRps. These findings suggest that the formation of ribonucleoprotein complexes between the viral RNA and its RdRp is important for the life cycles of 20S and 23S RNA viruses.

Replication Intermediates

Lysates prepared from virus-induced cells have an RdRp activity. The activity is insensitive to actinomycin D or α-amanitin, thus independent of a DNA template. The majority of *in vitro* products are positive strands of 20S RNA. Synthesis of negative strands accounts for a small fraction of the RNA products compared to that of positive strands, thus reflecting the high positive/negative-strand ratio in the lysates. There is no, or very little, *de novo* synthesis *in vitro*. Therefore, radioactive nucleotides are unevenly distributed into 20S RNA positive-strand products with more incorporation into the 3′ end region. Replication complexes that synthesize 20S RNA positive strands have a ssRNA backbone and migrate in native agarose gels as a broad band corresponding to an ssRNA in the size ranging from 2.5 to 5 kbp long. These complexes consist of a full-length negative-strand template (2.5 kbp) and a nascent positive strand of less than unit-length, probably held together by the polymerase machinery. Deproteination with phenol converts them to dsRNA. Therefore, W dsRNA is not a replication intermediate but a byproduct. It is likely that the high temperature (37 °C) for growth may destabilize replication complexes, thus resulting in the accumulation of dsRNA. Upon completion of RNA synthesis *in vitro*, the positive-strand products as well as the negative-strand templates are released from replication complexes. It is likely that, in the cell, the released negative strands are immediately recruited to another round of positive-strand synthesis, because the majority of negative strands in lysates are present in replication complexes engaging in the synthesis of positive strands. Interestingly, both positive and negative strands released from replication complexes are associated with protein. Because replication complexes contain at least one p91 molecule per complex, p91 is a good candidate for the protein.

Generation of Narnaviruses *In Vivo*

As mentioned earlier, the presence of narnaviruses does not render phenotypic changes to the host. This has hindered studies on replication or virus/host interactions using yeast genetics. This obstacle has been overcome by recent developments in generating 20S and 23S RNA viruses *in vivo* from a yeast expression vector (**Figure 2(b)**). In either case, the complete viral cDNA was inserted in the vector downstream of a constitutive promoter in such a way that positive strands can be transcribed from the promoter. The 3′ end of the viral sequence was directly fused to the hepatitis delta virus (HDV) antigenomic ribozyme. Therefore, intramolecular cleavage by the ribozyme will create transcripts *in vivo* having the 3′ termini identical to the viral 3′ end. The efficiency of virus launching is high. The 20–70% of the cells transformed with the vector generated the virus. The primary transcripts expressed from the vectors have nonviral sequences (about 40 nt) at the 5′ ends. The generated viruses, however, possessed the authentic viral 5′ ends without the extra sequences. It is likely that the 5′ nonviral extension was eliminated by a 5′ exonuclease. Using these launching systems, it has been demonstrated that each RdRp is essential and specific for replication of its own viral RNA. p91 is essential for 20S RNA replication and does not substitute p104 for replication of 23S RNA virus. Similarly, p104 is essential for 23S RNA replication and does not support 20S RNA replication. Because negative strands cannot be decoded to the RdRps, vectors in which the viral cDNAs were reversed failed to generate the virus. These negative-strand-expressing vectors, however, successfully generated narnaviruses, if active polymerases were provided in *trans* from a second vector. Therefore, both 20S and 23S RNA viruses can be generated from either positive or negative strands expressed from a vector.

cis-Acting Signals for Replication

20S and 23S RNA genomes share the same 5 nt inverted repeats at the 5′ and 3′ termini (5′-GGGGC...GCCCC-OH). Extensive analysis was done modifying each nucleotide at the 3′ ends. It was found that the third and fourth C's from the 3′ termini are essential for replication in both viruses. While the 3′ terminal and penultimate C's can be eliminated or changed to other nucleotides without affecting virus generation, the generated viruses recovered the wild-type C's at the termini. Therefore, the consecutive four C's at the 3′ terminus are essential for these viruses (**Figure 3**). In contrast, the G at position 5 from the 3′ end

is dispensable for replication in both viruses. 23S RNA virus requires an additional 3′ *cis*-signal for replication. The stem–loop structure proximal to the 3′ end contains a mismatched pair of purines in the stem (**Figure 3**). This mismatched pair is essential for replication but the virus tolerates any combination of purines at this position. On the other hand, changing the purines to pyrimidines or eliminating one of the purines at the mismatched pair blocked virus generation. The distance between the mismatched pair and the 3′ terminal four C's and/or their spatial configuration appears to be critical, because shortening or increasing the length of the stem between the two sites by more than 1 bp abolished virus launching. It is not known whether 20S RNA virus has a similar *cis*-signal in the stem–loop structure proximal to the 3′ end. The G at position 5 from the 3′ end is located at the bottom of the stem structure. This G, as mentioned earlier, can be

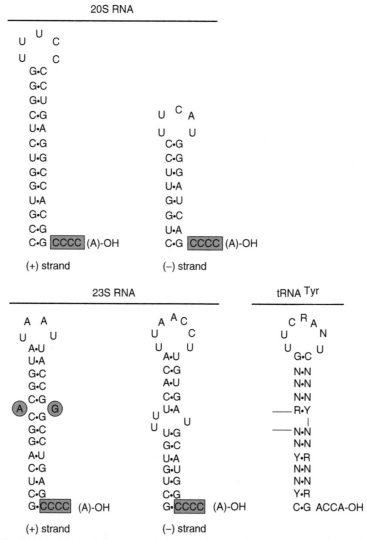

Figure 3 Comparison of the 3′ terminal secondary structures in the positive (+) and negative (−) strands of 20S and 23S RNA viruses, with the top half domain of tRNA^Tyr. The nontemplated A residues at the viral 3′ termini are indicated by parenthesis. The consecutive four C's essential for replication are boxed (green). A second *cis*-signal (the mismatched pair of purines) present in the positive strand of 23S RNA virus is circled (green). Y, R, and N stand for pyrimidine, purine, and any base, respectively.

changed to another nucleotide without impairing replication, as long as the modified nucleotide is hydrogen-bonded at the bottom of the stem.

The negative strands of 20S and 23S RNA viruses also possess four consecutive C's at the 3' ends. Using the two-vector system mentioned above, it has been found that the third and fourth C's from the 3' end are essential for replication. Similar to the positive strands, the 3' terminal and penultimate C's can be eliminated or changed to other nucleotides without affecting virus generation and the generated viruses recovered the wild-type four C's at the 3' ends. Therefore, the consecutive four C's at the 3' end of the negative strand are again a *cis*-signal for replication. The 5' ends of viral positive strands have not been analyzed extensively. Elimination of the 5' terminal G or changing it to other nucleotide had no effect on virus generation and the generated viruses recovered this G at the 5' ends.

cis-Signals for Formation of Ribonucleoprotein Complexes

Narnaviruses, as mentioned earlier, exist as ribonucleoprotein complexes in the host cytoplasm. In the absence of the HDV ribozyme, RNA transcribed from the launching vectors failed to generate viruses because of the presence of nonviral extensions at the 3' end. The transcripts, however, can be decoded to viral polymerases and the polymerases can form complexes *in vivo* with the transcripts, thus providing an assay system to analyze *cis*-signals for formation of ribonucleoprotein complexes. By immunoprecipitation with antiserum specific to p104, it has been found that the bipartite 3' *cis*-signal for replication (more specifically, the mismatched pair of purines and the third and fourth C's from the 3' end) is essential for 23S RNA positive strand to form complexes with the polymerase. This highlights the importance of formation of ribonucleoprotein complexes for the life of narnaviruses. The cytoplasm is filled with host RNAs, including a great variety of mRNAs. Formation of complexes between the viral polymerase and its template RNA will therefore facilitate replication and increase its fidelity by discriminating against nonviral RNAs as templates. The importance of the third and fourth C's from the 3' end for formation of complexes with p104 suggests that these nucleotides are in close contact with p104. Because the 23S RNA virus has no coat protein to protect the viral RNA, it is likely that such interaction protects the 3' ends of 23S RNA from exonuclease cleavage. In the case of 20S RNA virus, a similar *in vivo* assay indicates that the 3' *cis*-signal for replication (in particular the third and fourth C's from the 3' end) is also important for formation of ribonucleoprotein complexes with p91. When isolated resting complexes of 20S RNA virus were analyzed *in vitro*, however, it was found that p91 interacts with 20S RNA not only at the 3' end but also at the 5' terminal region of

the molecule. The 5' binding site is located at the second stem–loop structure from the 5' end. Computer-predicted analysis indicates that the 5' and 3' termini of 20S RNA (and 23S RNA) are brought together into close proximity by a long-distance RNA/RNA interaction (**Figure 4**). This may allow a single molecule of p91 to interact simultaneously with both ends of 20S RNA genome in a resting complex. This protein/RNA interaction may provide a clue to understand the molecular basis of narnaviruses persistence (see below).

Narnavirus Persistence in the Host

mRNA degradation in yeast, like in other eukaryotes, is initiated by shortening the 3' poly(A) tail followed by decapping at the 5' end. Then the decapped mRNA is degraded by the potent Xrn1p/Ski1p 5' exonuclease as well as by a 3' exonuclease complex called exosome. The RNA genomes of narnaviruses, as mentioned earlier, have no 3' poly(A) tails and perhaps no cap structures at the 5' ends, thus resembling intermediates of mRNA degradation. This suggests that these RNA genomes are vulnerable to the exonucleases involved in mRNA degradation. In fact, the copy numbers of 20S and 23S RNAs increase several-fold in strains having mutations in *SKI* genes such as *SKI2*, *SKI6*, and *SKI8*. These mutations were originally identified by their failure in lowering the copy numbers of L-A dsRNA totivirus and its satellite RNA M. It is known that the *SKI2*, *SKI6*, and *SKI8* gene products are components or modulators of the exosome. These observations suggest that the 3' end of the viral genome is constantly nibbled by 3' exonucleases. Therefore, one of the reasons for narnaviruses to form ribonucleoprotein complexes may be to protect their 3' ends from exonuclease cleavage. The fact that the third and fourth C's from the 3' end are important to form complexes in both viruses fits this hypothesis because binding of the RdRp to these nucleotides would block progression of the exonuclease and protect the internal region. As described earlier, mutations introduced at the terminal and penultimate positions at the 3' end had no deleterious effects on virus launching and the generated viruses recovered the wild-type sequences. This suggests that the terminal and penultimate positions at the 3' ends are not only vulnerable to cleavages but also accessible to the repair machinery. The 3' ends of these viruses may undergo constant turnover at these positions.

As regards the 5' end, the first four nucleotides in both 20S and 23S RNAs are consecutive G's (**Figure 4**). It is known that oligo G tracts inhibit progression of the Xrn1/Ski1 5' exonuclease. Furthermore, these consecutive G's are buried at the bottom of a long stem structure in both viruses. These features thus suggest that 20S and 23S RNAs by themselves are quite resistant to the 5' exonuclease. The initiation codon of p91 is located in the middle of the long stem structure proximal to the 5' end. If p91 binds to this

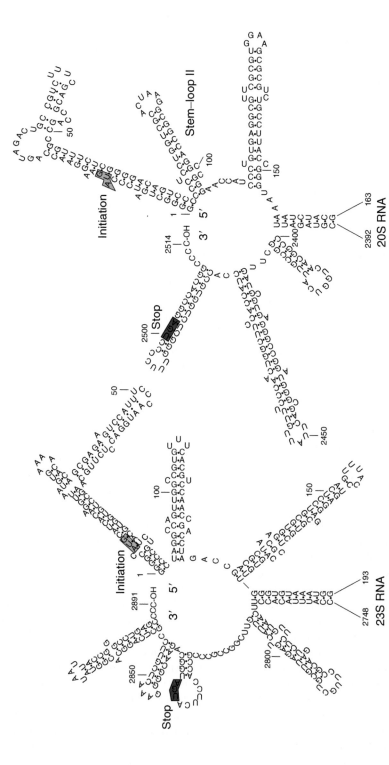

Figure 4   Secondary structures at the 5′ and 3′ end regions of 23S and 20S RNA positive strands, as predicted by the MFOLD program. The AUG initiation codons (green) and the stop codons (red) for p104 and p91 are boxed. Stem-loop II where the 5′ binding site for p91 is located is shown. Note that about 150 nt from the ends in each viral genome there are inverted repeats of 8–12 nt long that bring both 5′ and 3′ ends to a close proximity.

stem in the ribonucleoprotein complex, then such a stable binding may interfere with translation of new p91 molecules from the RNA. In this context, it may make sense that the 5′ binding site of p91 in the complexes is located at the second stem–loop structure from the 5′ end. By binding simultaneously to the 3′ end and also to the region close to the 5′ end of the same RNA molecule, p91 may stabilize the long-distance RNA–RNA interactions that bring the 5′ and 3′ ends of the RNA into proximity, and thus helps the RNA to form an organized structure in resting complexes.

It is not known whether the 3′ end repair is carried out by the replicase machinery during the replication process, or by host enzymes. However, the following evidence favors the latter case. The 3′ terminal structures of 20S and 23S RNAs resemble a half of tRNA, the so-called 'top-half' domain, consisting of the acceptor stem and T stem (**Figure 3**). The domain provides the determinants necessary for specific interactions with tRNA-related enzymes such as the tRNA nucleotidyltransferase (CCA-adding enzyme). This raises the possibility that 20S and 23S RNAs nibbled at the 3′ ends by 3′ exonucleases are repaired to the wild-type sequences by the CCA-adding enzyme. The fact that 15–30% of both positive and negative strands of 20S and 23S RNAs possess an unpaired A at the 3′ ends supports this possibility. Furthermore, that the 3′ repair is confined to the terminal and penultimate positions is consistent with the catalytic activity expected for the CCA-adding enzyme. Given that narnaviruses are persistent viruses, the underlying mechanism(s) to maintain the integrity of the viral 3′ ends will, therefore, have considerable significance for a long-term infection.

See also: Fungal Viruses; Yeast L-A Virus.

Further Reading

Buck KW, Esteban R, and Hillman BI (2005) *Narnaviridae*. In: Fauquet CM, Mayo MA, Maniloff J, Desselberger U,, and Ball LA (eds.) *Virus Taxonomy: Eighth Report of the International Committee on Taxonomy of Viruses*, pp. 751–756. San Diego, OA: Elsevier Academic Press.

Esteban LM, Rodríguez-Cousiño N, and Esteban R (1992) T double-stranded (dsRNA) sequence reveals that T and W dsRNAs form a new RNA family in *Saccharomyces cerevisiae*: Identification of 23S RNA as the single-stranded form of T dsRNA. *Journal of Biological Chemistry* 267: 10874–10881.

Esteban LM, Fujimura T, García-Cuéllar MP, and Esteban R (1994) Association of yeast viral 23 S RNA with its putative RNA-dependent, RNA polymerase. *Journal of Biological Chemistry* 269: 29771–29777.

Esteban R and Fujimura T (2003) Launching the yeast 23S RNA narnavirus shows 5′ and 3′ *cis*-acting signals for replication. *Proceedings of the National Academy of Sciences, USA* 100: 2568–2573.

Esteban R and Fujimura T (2006) Yeast narnavirus replication. In: Hefferon KL (ed.) *Recent Advances on RNA Virus Replication*, pp. 171–194. Trivandrum, India: Research Ringspot.

Esteban R, Vega L, and Fujimura T (2005) Launching of the yeast 20S RNA narnavirus by expressing the genomic or anti-genomic viral RNA *in vivo*. *Journal of Biological Chemistry* 280: 33725–33734.

Fujimura T and Esteban R (2004) Bipartite 3′ *cis*-acting signal for replication in yeast 23 S RNA virus and its repair. *Journal of Biological Chemistry* 279: 13215–13223.

Fujimura T, Solórzano A, and Esteban R (2005) Native replication intermediates of the yeast 20S RNA virus have a single-stranded RNA backbone. *Journal of Biological Chemistry* 280: 7398–7406.

Kadowaki K and Halvorson HO (1971) Appearance of a new species of ribonucleic acid synthesized in sporulation cells of *Saccharomyces cerevisiae*. *Journal of Bacteriology* 105: 826–830.

Matsumoto Y and Wickner RB (1991) Yeast circular RNA replicon: Replication intermediates and encoded putative RNA polymerase. *Journal of Biological Chemistry* 266: 12779–12783.

Rodríguez-Cousiño N, Esteban LM, and Esteban R (1991) Molecular cloning and characterization of W double-stranded RNA, a linear molecule present in *Saccharomyces cerevisiae*: Identification of its single-stranded RNA form as 20S RNA. *Journal of Biological Chemistry* 266: 12772–12778.

Solórzano A, Rodríguez-Cousiño N, Esteban R, and Fujimura T (2000) Persistent yeast single-stranded RNA viruses exist *in vivo* as genomic RNA polymerase complexes in 1:1 stoichiometry. *Journal of Biological Chemistry* 275: 26428–26435.

Wejksnora PJ and Haber JE (1978) Ribonucleoprotein particle appearing during sporulation in yeast. *Journal of Bacteriology* 134: 246–260.

Wesolowski M and Wickner RB (1984) Two new double-stranded RNA molecules showing non-Mendelian inheritance and heat inducibility in *Saccharomyces cerevisiae*. *Molecular and Cellular Biology* 4: 181–187.

Widner WR, Matsumoto Y, and Wickner RB (1991) Is 20S RNA naked? *Molecular and Cellular Biology* 11: 2905–2908.

Partitiviruses of Fungi

S Tavantzis, University of Maine, Orono, ME, USA

Introduction

The vast majority of mycoviruses (viruses infecting fungi) are 'cryptoviruses', that is, they are not usually associated with an overt pathology of their fungal hosts. So, it is not surprising that the first mycovirus paper, by M. Hollings published in *Nature* (London) in 1962, was about virus-like particles (VLPs) thought to be the cause of a serious dieback disease of the cultivated mushroom *Agaricus bisporus*. Following Hollings' groundbreaking paper, progress in mycovirology was slow, as in addition to their cryptic nature, fungal viruses can be transmitted only through intracellular routes. Most of the characterized mycoviruses possess double-stranded RNA (dsRNA) genomes but this high proportion might be biased since most of the published data refer to fungal screenings for the presence of dsRNA followed by transmission electron microscopy or virus characterization studies.

Mycoviruses are classified into nine families (*Barnaviridae, Chrysoviridae, Hypoviridae, Metaviridae, Narnaviridae, Partitiviridae, Pseudoviridae, Reoviridae, Totiviridae*), and one genus (*Rhizidiovirus*) not classified into a specific family. Viruses belonging to the family *Partitiviridae* usually cause cryptic infections in fungi and plants, and are classified into three genera, *Partitivirus, Alphacryptovirus*, and *Betacryptovirus*. The name *Partitiviridae* originates from the Latin *partitus*, which means 'divided' and refers to the fact that the genome of these viruses is bipartite (separately encapsidated) and consists of two monocistronic dsRNAs that are similar in size.

Taxonomy and Classification

This section focuses on the genus *Partitivirus*, which consists of viruses that infect filamentous fungi. The genera *Alphacryptovirus* and *Betacryptovirus* include viruses that are found in plants. The recognized species of the genus *Partitivirus* are listed in **Table 1** whereas tentative members are shown in **Table 2**. The sequence accession numbers refer to the RNA-dependent RNA polymerase (RdRp) genes of the respective partitiviruses.

Virion Properties

The virion M_r estimates range from 6×10^6 to 9×10^6 and the $S_{20,w}$ values (in Svedberg units) range from 101 to 145, whereas the $S_{20,w}$ values of particles lacking nucleic acid range from 66 to 100S. Virion buoyant density in CsCl is $1.34-1.37 \mathrm{\,g\,cm}^{-3}$ for particles with nucleic acid and $1.29-1.30 \mathrm{\,g\,cm}^{-3}$ for particles devoid of nucleic acid. Purified virion preparations contain, in addition to mature virions and empty virions, sedimenting and density components that are thought to be replicative intermediates containing single-stranded RNA (ssRNA) and particles with both ssRNA and dsRNA.

Virion Structure and Composition

Viruses of the genus *Partitivirus* have isometric particles with icosahedral symmetry ranging from 30 to 35 nm in diameter (**Figure 1**). Negatively stained virions that are devoid of nucleic acid have capsids with dark centers as

Table 1   Virus members in the genus *Partitivirus*

Virus	Abbreviation	Accession number
Agaricus bisporus virus 4	AbV-4	
Aspergillus ochraceous virus	AoV	
Atkinsonella hypoxylon virus	AhV	L39125
Discula destructiva virus 1	DdV-1	AF316992
Discula destructiva virus 2	DdV-2	AY033436
Fusarium poae virus	FpV-1	AF047013
Fusarium solani virus 1	FsV-1	
Gaeumannomyces graminis virus 019/6-A	GgV-019/6-A	
Gaeumannomyces graminis virus T1-A	GgV-T1-A	
Gremmeniella abietina RNA virus MS1	GarV-MS1	AY08993
Helicobasidium mompa virus	HmV	
Heterobasidion annosum virus	HaV	
Penicillium stoloniferum virus S	PsVS	AY156521
Rhizoctonia solani virus 717	RhsV-717	AF133250

Table 2   Tentative members of the genus *Partitivirus*

Virus	Abbreviation	Accession number
Ceratocystis polonica partitivirus	CpPV	
Ceratocystis resinifera partitivirus	CrPV	
Diplocarpon rosae virus	DrV	
Gremmeniella abietina RNA virus MS2	GaRV-MS2	AY615211
Helicobasidium mompa partitivirus	HmPV	
Ophiostoma partitivirus 1	OPV-1	AM087202
Penicillium stoloniferum virus F	PsV-F	AY758336
Phialophora radicicola virus 2-2-A	PrV-2-2-A	
Pleurotus ostreatus virus	PoV	AY533038
Rosellinia necatrix virus 1-W8	RnV-1-W8	AB113347

they are penetrated by stain (Figure 1). The virion capsid is not enveloped and consists of 12 capsomers and 120 capsid protein (CP) subunits with a molecular mass ranging from 57 to 76 kDa. Virion-associated RNA polymerase activity is attributed to RdRp with sizes ranging from 77 to 86 kDa in fungal partitiviruses sequenced to date.

Antigenic Properties

Antisera to purified virus preparations appear to contain antibodies to the virion RdRp polypeptide. This argument is supported by two lines of evidence: (1) antiserum to the rhizoctonia solani virus 717 (RhsV 717) reacted to both virion proteins CP and RdRp and (2) antiserum to the aspergillus ochraceous virus (AoV) reacted to the AoV CP and RdRp as well as the RdRp of the 'slow' component (PsV-S) of penicillium stoloniferum virus complex. Unlike the respective CPs, the RdRp polypeptides of AoV and PsV-S share a high degree of similarity, and this is congruent with the phylogenetic analysis of CPs and RdRps of members of the genus *Partitivirus* which shows a higher rate of evolutionary changes occurring in the *CP* genes (**Figure 4 (b)**) as compared to the rate in the *RdRp* genes (**Figure 4(a)**).

Genome Organization

Virions contain two separately encapsidated linear dsRNAs, dsRNA1 and dsRNA2, that are 1.5–2.4 kbp in size. The dsRNA components of a particular virus are of similar size (differing by 20–200 bp). Additional dsRNAs (satellite or defective) may be present in purified fungal partitivirus particles. The larger segment, dsRNA1, codes for an RdRp, whereas the smaller segment, dsRNA2, codes for a CP (**Table 3**). The 5′ ends of the coding strands of dsRNA1 and dsRNA2 are often highly conserved, and usually contain inverted repeats that may form stable stem–loop structures (**Figure 2**).

It has been hypothesized that partitiviruses may have evolved from totiviruses by dividing their genomes between two dsRNA segments. Like the totiviruses, the partitivirus genome consists of two genes (CP and RdRp) but, unlike the totiviruses, these genes are located on separate dsRNAs (bipartite genomes). In contrast to partitiviruses, many totiviruses express their RdRp as a CP–RdRp fusion protein. Totiviruses, such as the helminthosporium victoriae 190S virus (Hv190SV), that express RdRp as a separate nonfused protein might be evolutionary intermediates between ancestral totiviruses and partitiviruses.

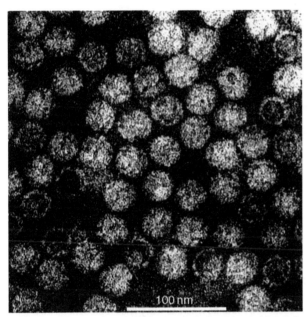

Figure 1 Electron micrograph of negatively stained virions of Rhizoctonia solani virus 717 (RhsV 717), a representative species of the genus *Partitivirus*. Adapted from Tavantzis SM and Bandy BP (1988) Properties of a mycovirus from *Rhizoctonia solani* and its virion-associated RNA polymerase. *Journal of General Virology* 69: 1465–1477, with permission from Society for General Microbiology.

Genome Expression and Virus Multiplication

As described above, the bisegmented genome of fungal partitiviruses consists of two dsRNA segments of similar size.

Table 3 Genome organization of the RhsV 717 partitivirus

Genome	ORF	Frame	Nucleotide coordinates	Amino acids	Mol. mass (kDa)	Putative function	Kozak sequence	Inverted repeats[a]
dsRNA1 (2363 bases)	1	2	86–2275	730	85.8	RdRp[b]	502–510	736–747
dsRNA2 (2206 bases)	1	1	79–2130	683	76.4	CP	None	None

[a]Inverted repeats longer than 12 bases.
[b]The Rhs 717 partitivirus RdRp (dsRNA 1, ORF 1) contains the dsRNA RdRp motifs according to Bruenn (1993). The residue coordinates of these motifs are as follows: Motif I, 242-LVTGT-246; Motif II, 315-FLKSFPTMM-323; Motif III, 363-ARKPECCIMYG-373; Motif IV, 397-IDWSGYDQRL-406; Motif V, 479-GVPSGMLLTQFLDSFGNLY-LII-500; Motif VI, 519-FIMGDDNSIF-528; Motif VII, 569-IETLSYRC-576; Motif VIII, 584-DVEK-587.
Reproduced from Strauss EE, Lakshman DK, and Tavantzis SM (2000) Molecular characterization of the genome of a partitivirus from the basidiomycete *Rhizoctonia solani*. *Journal of General Virology* 81: 549–555, with permission from Society for General Microbiology.

```
dsRNA1  - GUAGUCUUUUAGUAUCGAUCCCUCGACUCUCGACCGCACUAAAUCUCAUC  - 50
          : :::::::::::::::::::::::::::          :::::::::::::
dsRNA2  - G-AGUCUUUUAGUAUCGAUCCCUCGACU- - - - -CGCACUAAAUCUCAUU  - 43

dsRNA1  - GUUAUUACGAACGAACUCUCUUCAAUCAACACACAAUGCUCUACAACUUC  - 100
          ::: :::::::::: ::::::  :  ::   ::   ::::   :
dsRNA2  - GUUUUUAAGAACAAACUCUCAACUCUCGCAACCUGAUGCCUUCGCCAAAG  - 93
```

Figure 2 Homology at the 5′ ends of the coding strands of the two genomic segments of the RhsV 717. If a 6-base gap is allowed in dsRNA2, the two dsRNAs are highly conserved. The identity drops to 65% around the region that includes the translation initiation sites. AUG codons are shown in bold letters and the CAA motifs (possibly involved in translational enhancement) are double-underlined. Identical positions are shown with colons beneath the sequence. The underlined bases are complementary and are capable of forming stable stem–loop structures. Adapted from Strauss EE, Lakshman DK, and Tavantzis SM (2000) Molecular characterization of the genome of a partitivirus from the basidiomycete *Rhizoctonia solani*. *Journal of General Virology* 81: 549–555, with permission from Society for General Microbiology.

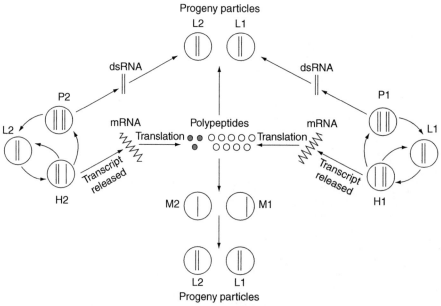

Figure 3 Model for replication of penicillium stoloniferum S virus (PsV-S). The open circles represent capsid protein subunits and the closed circles represent RNA polymerase subunits. Solid lines represent RNA strands whereas wavy lines represent mRNA. Adapted from Seventh Report of the International Committee on Taxonomy of Viruses, http://www.virustaxonomyonline.com/virtax/lpext.dll/vtax/agp-0013/dr06/dr06-fg/dr06-fig-0002, with permission from Elsevier.

The larger segment, dsRNA1, encodes an RdRp, whereas the smaller component, dsRNA2, encodes a CP. Northern blot analysis of purified virions as well as total RNA from partitivirus-infected mycelial tissue revealed no subgenomic RNA species in the form of discrete RNA bands with sizes smaller than that of full-length dsRNA1 or dsRNA2. This is consistent with analysis of the coding potential of these genomic RNAs, showing that the long open reading frames encoding the RdRp and CP polypetides cover the entire length of dsRNA1 and dsRNA2, respectively.

Mycoviruses, including members of the genus *Partitivirus*, have been difficult to transfect using an inoculum consisting of purified virions. Thus, methods of synchronous viral infection and multiplication have not been available for studying partitivirus gene expression or virion replication *in vivo*. The recent transfection of the rosellinia necatrix virus W8 (RnV-W8) into protoplasts of its fungal host is a positive development with regard to

unveiling the life cycle of fungal partitiviruses. Current knowledge of partitivirus gene expression, dsRNA replication, and virus multiplication is based on studies of RdRp activity in purified virus fractions, virus-related RNA content in infected mycelia, and *in vitro* translation studies of denatured dsRNA1 or dsRNA2 or full-length cDNA clones representing dsRNA1 or dsRNA2.

Populations of virions include particles of different $S_{20,w}$ values depending on their RNA content. Thus, purified virus preparations consist of mature virions (L1 or L2) containing a single molecule of dsRNA, virus particles with a single molecule of ssRNA corresponding to the plus-strand of the respective dsRNAs (M1, M2), a heterogeneous subpopulation of heavy (H) particles representing various replication stages and containing particles with a dsRNA genomic component and ssRNA tails or one molecule each of dsRNA and ssRNA; particles with two molecules of dsRNA (P) one of which is a

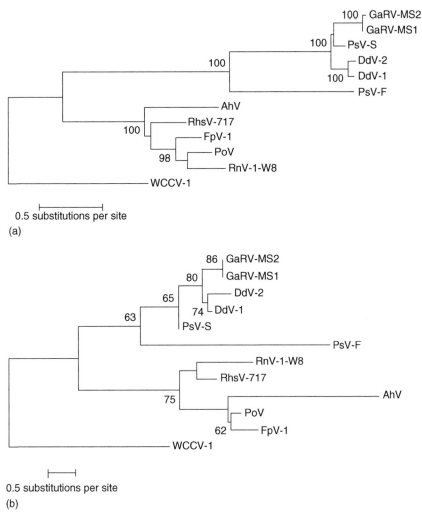

Figure 4 A neighbor-joining phylogram showing phylogenetic relationships among fungal partitivirus (a) RdRp and (b) CP amino acid sequences. The JTT model of amino acid evolution was applied in phylogenetic reconstruction using the software MEGA3.1. The tree was outgroup-rooted with the respective proteins of white clover cryptic virus 1 (WCCV-1), a plant partitivirus. Numbers above branches represent bootstrap support values and bar represents amino acid substitutions per site. See **Tables 1** and **2** for abbreviations of fungal partitivirus names.

product ('P') of replication (**Figure 3**). Data from *in vitro* RdRp activity studies in purified preparations of PsV-S and RhsV 717 are congruent with the replication scheme presented in **Figure 3**. However, experimental evidence for a number of the steps depicted in this replication scheme is currently lacking, and its verification awaits *in vivo* studies involving transfection of host mycelia with purified virions (or cDNAs representing dsRNAs 1 and 2) resulting in synchronous virus replication.

Phylogenetic Relationships among Fungal Partitiviruses

Phylogenetic analyses of the predicted amino acid sequences of fungal partitivirus RdRps and CPs (**Figures 4(a)** and **4(b)**, respectively) show that these viruses are closely related. Both analyses suggest that fungal partitiviruses form two

clusters. The RdRp clusters have 100% bootstrap support (**Figure 4(a)**), whereas the CP clusters are also well supported (**Figure 4(b)**). These results are in agreement with recently reported phylogenetic analyses showing that the evolutionary rate of CPs is higher than that of the RdRps, and are in congruence with data suggesting a cross-serological relationship between the RdRps of AoV and PsV-S.

Transmission of Partitiviruses

Like most mycoviruses, partitiviruses may be transmitted by anastomosis between hyphae of genetically related fungal genotypes. Moreover, the ineffectiveness of hyphal tip isolation to eliminate partitivirus particles or dsRNA in conjunction with the detection of partitiviruses in apical hyphal sections by electron microscopy suggest that these viruses are transmitted vertically by sexual spores (basidiospores or rarely ascospores) as well as asexual spores (conidiospores) derived from virus-infected fruit bodies. The proportion of spores containing partitivirus dsRNAs varies widely even within the same species and apparently depends on the combination of fungal host and virus genotypes. For example, in different isolates of the basidiomycete *Heterobasidion annosum*, 3–55% of germinated conidia, and 10–84% of germinated basidiospores, contained partitivirus dsRNAs.

Partitivirus Pathogenesis

Partitiviruses occur in all parts of an infected mycelial mat, but virion concentrations are higher in the older hyphal sectors than in hyphal tips. This adaptation to multiply mainly in areas of the fungal mycelium that are less involved in fungal growth might be one of the reasons that partitiviruses, although they replicate to relatively high levels, are usually associated with a symptomless infection of their respective fungal hosts. Recently, the latency of a partitivirus infection was demonstrated by direct experimental evidence in the case of ascomycete *Rosellinia necatrix* and the RnV-1-W8 partitivirus. Protoplasts of *R. necatrix* were transfected with purified particles of RnV-1-W8. The resulting mycelium contained transmissible RnV-1-V8 virions but showed no observable symptoms associated with the presence and replication of this virus. In con-

trast, in the *H. annosum* study (see previous section), germination frequency of basidiospores was significantly reduced ($P < 0.05$) by the presence of partitivirus dsRNA in the parental fruit bodies. Use of isogenic fungal lines may be needed to verify these results.

See also: Fungal Viruses; Chrysoviruses; Totiviruses.

Further Reading

Bruenn JA (1993) A closely related group of RNA-dependent RNA polymerases from double-stranded RNA viruses. *Nucleic Acids Research* 21: 5667–5669.

Ghabrial SA (1998) Origin, adaptation and evolutionary pathways of fungal viruses. *Virus Genes* 16: 119–131.

Ghabrial SA, Buck KW, Hillman BI, and Milne RG (2005) *Partitiviridae*. In: Fauquet CM, Mayo MA, Maniloff J, Desselberger U,, and Ball LA (eds.) *Virus Taxonomy: Eighth Report of the International Committee on Taxonomy of Viruses*, pp. 581–590. San Diego, CA: Elsevier Academic Press.

ICTVdB Management (2006) 00.049.0.01. Partitivirus. In: Büchen-Osmond C (ed.) *ICTVdB – The Universal Virus Database, version 4*, New York: Columbia University.

Kim JW, Choi EY, and Kim YT (2006) Intergeneric relationship between the aspergillus ochraceous virus F and the penicillium stoloniferum virus S. *Virus Research* 120: 212–215.

Strauss EE, Lakshman DK, and Tavantzis SM (2000) Molecular characterization of the genome of a partitivirus from the basidiomycete *Rhizoctonia solani*. *Journal of General Virology* 81: 549–555.

Tavantzis SM and Bandy BP (1988) Properties of a mycovirus from *Rhizoctonia solani* and its virion-associated RNA polymerase. *Journal of General Virology* 69: 1465–1477.

Prions of Yeast and Fungi

R B Wickner, H Edskes, T Nakayashiki, F Shewmaker, L McCann, A Engel, and D Kryndushkin, National Institutes of Health, Bethesda, MD, USA

Published by Elsevier Ltd.

Glossary

Nonchromosomal gene A gene that segregates 4+:0 in meiosis and can be transferred by cytoplasmic mixing, in contrast to chromosomal genes that segregate 2+:2- in meiosis, and are not transferred by cytoplasmic mixing.

Nonsense suppressor tRNA A mutant transfer RNA that recognizes a translational stop codon and inserts an amino acid thus allowing the peptide chain to continue.

Introduction and History

The word prion, meaning 'infectious protein' without need for a nucleic acid, was coined to explain the properties of the agent producing the mammalian transmissible spongiform encephalopathies (TSEs), although 25 years later there remains some debate if the TSEs are indeed caused by prions. The yeast and fungal prions were identified by their unique genetic properties which were unexpected for any nucleic acid replicon, but specifically predicted for an infectious protein.

[PSI] was described by Brian Cox in 1965 as a nonchromosomal genetic element that increased the efficiency of a weak nonsense suppressor transfer RNA (tRNA). [URE3] was described by Francois Lacroute as a nonchromosomal gene that relieved nitrogen catabolite repression, allowing expression of genes needed for utilizing poor nitrogen sources even when a good nitrogen source was available. The [Het-s] prion was described in 1952 by Rizet as a nonchromosomal gene needed for heterokaryon incompatibility in *Podospora anserina*. Each of these elements was later found to be a prion. The [PIN] prion was discovered in 1997 by Derkatch and Liebman in their studies of *de novo* generation of the [PSI] prion.

Genetic Signature of a Prion

Viruses of yeast and fungi generally do not exit one cell and enter another, but spread by cell–cell fusion, as in mating or heterokaryon formation. Infectious proteins (prions) should likewise be nonchromosomal genetic elements. To distinguish prions from nucleic acids, three genetic criteria were proposed (**Figure 1**): (1) If a prion can be cured, it can reappear in the cured strain at some low frequency. (2) Overproduction of the protein capable of being a prion should increase the frequency of the prion arising *de novo*. (3) If the prion produces a phenotype by the simple inactivation of the protein, then this phenotype should resemble the phenotype of mutation of the gene encoding the protein, which gene must be needed for prion propagation.

All three criteria were satisfied by [PSI] and [URE3], strongly indicating, perhaps proving, that they were prions. The [Het-s] prion of *P. anserina* and the [PIN] prion of *Saccharomyces cerevisiae* were likewise proved by application of the same genetic criteria, but because their prion form produces the phenotype, rather than the absence of the normal form, they do not satisfy criterion (3).

Self-Propagating Amyloid as the Basis for Most Yeast Prions

The finding that Sup35p, Ure2p, and HET-s were protease resistant and aggregated in prion-containing cells, that these proteins (and particularly their prion domains (**Figure 2**)) would form amyloid *in vitro* indicated that a self-propagating amyloid (**Figure 3**) was the basis of these prions. This was confirmed by the finding that the corresponding prions were transmitted by introduction of amyloid formed *in vitro* from the recombinant proteins (see below). However, in some cases, self-modifying enzymes have the potential to become prions (see below).

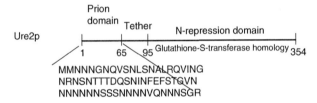

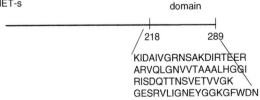

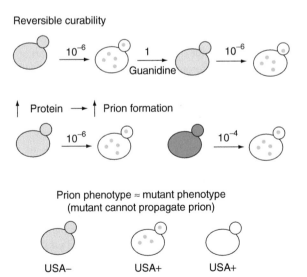

Figure 1 Genetic signature of a prion. Among nonchromosomal genetic elements, those with the three properties shown here must not be nucleic acid replicons and are almost certainly prions. However, only prions for which the prion form of the protein is inactive (such as [URE3] or [PSI]) will have property (3). The [Het-s] and [PIN⁺] prions are active forms of the HET-s protein and Rnq1 protein, respectively and have properties (1) and (2), but not (3).

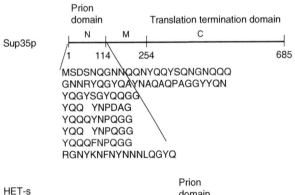

Figure 2 Prion domains. The prion domains of Ure2p, Sup35p, and HET-s are largely unstructured in the native form and in β-sheet in the prion form.

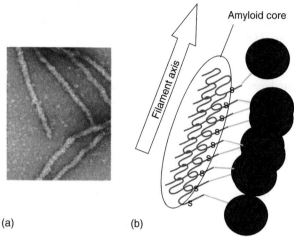

(a)   (b)

Figure 3 Amyloid. (a) Electron micrograph of amyloid of Ure2p. (b) Model of the structure of Ure2p amyloid. Red, prion domain; green, tether; blue, glutathione-*S*-transferase-like nitrogen regulation domain. Amyloid of recombinant HET-s, Sup35p, or Ure2p is infectious for yeast transmitting [Het-s], [PSI+], or [URE3], respectively. Amyloid of Ure2p is proposed to have the parallel in-register β-sheet structure shown based on the solid-state NMR studies of a fragment of the Ure2p prion domain, and the fact that the prion domain can be shuffled and still be a prion (see text).

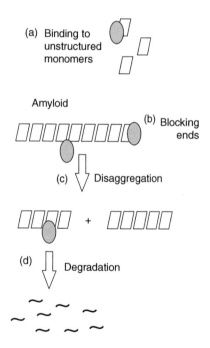

Figure 4 Chaperones and prions. Possible roles of chaperones in prion propagation are diagrammed.

Chaperones and Prions

Chaperones of the Hsp40, Hsp70, and Hsp104 groups, as well as Hsp90 co-chaperones, have been found to be clearly involved in prion propagation. Millimolar concentrations of guanidine are known to cure each of the amyloid-based prions, and the mechanism of action of guanidine curing has been shown to be specific inhibition of Hsp104. It is believed that at least one function of these chaperones is to break large amyloid filaments into smaller ones which can then be distributed at cell division to both daughter cells and insure the inheritance of the prion. Overexpression of some chaperones cure yeast prions, perhaps by solubilizing the filaments or perhaps by binding to the ends of filaments and preventing their elongation with new monomers. There is considerable specificity in which chaperone is needed for which prion and which chaperone can cure which prion by overexpression. The detailed mechanisms of chaperone action on prions (and amyloids in general) remain to be elucidated, but it is clear that they play an important role in these phenomena (**Figure 4**).

The Species Barrier and Prion Variants

Scrapie, a prion disease of sheep, only infects goats after a long incubation period, and subsequent goat-to-goat transmission has a much shorter incubation period. This is called the species barrier, and this barrier can in some cases be absolute, as appears to be the case between sheep and humans. The same phenomenon has now been documented in yeast, where [PSI] prions formed by the Sup35p of one species will not be transmitted to the Sup35p of another species, even though the other species' Sup35p can itself form a prion.

A single protein sequence can form several prions that are distinguishable, in yeast, by the intensity of their phenotype and the stability of their propagation. These are called 'prion variants' and are believed to reflect different amyloid structures. Paradoxically, a similar phenomenon, long documented in the mammalian TSEs, was used as an argument against the protein-only model. Elucidation of the structure of different prion variants, and the mechanism of their faithful propagation, in some cases across species barriers, remains an important problem.

The bovine spongiform encephalopathy epidemic in the UK has brought the species barrier and prion variant phenomena together. It is clear that the 'height' of the species barrier is a function of the prion variant. Collinge views the species barrier as a reflection of the degree of overlap of possible variants of the prion proteins of the two species. If they have few common amyloid conformers (prion variants) then the barrier will be high. If each sequence can adopt nearly all of the amyloid conformers of the other, there will be little species barrier.

Formation of Prions by Sup35p and Ure2p Homologs

The C-terminal domain of Sup35 is conserved in eukaryotes with a human homolog capable of complementing the

S. cerevisiae protein. All Sup35 proteins have N-terminal extensions, however, with limited or no sequence homology between species. N-terminal sequences from some species related to *S. cerevisiae* are capable of forming a [PSI+]-like prion. Ure2p is limited to the ascomycete yeasts. As with Sup35p, Ure2 proteins have a conserved C-terminal domain and a variable N-terminal domain that in general is rich in Asn and/or Glu residues. Ure2p homologs of some *Saccharomyces* yeasts can propagate [URE3] in *S. cerevisiae*.

Prion Generation, and [PIN]: A Prion That Gives Rise to Prions

One of the lines of evidence that showed [PSI] was a prion of Sup35p was that overproduction of Sup35p increased the frequency with which [PSI] arises *de novo*. However, it was found that in some strains, overproduction of Sup35p did not yield detectable emergence of [PSI]-carrying clones. Another nonchromosomal genetic element, named [PIN] for [PSI]-inducibility, was found necessary. [PIN] is a self-propagating amyloid form of the Rnq1 (rich in Asn (N) and Gln (Q)) protein, and it promotes *de novo* generation of [URE3] as well as [PSI].

Transfection with Amyloid of Recombinant Proteins

Amyloid filaments formed *in vitro* from recombinant HET-s protein, Sup35p, or Ure2p can efficiently transform cells to the corresponding [Het-s], [PSI], or [URE3] prion. In some cases it was shown that the soluble form or non-specific aggregates of the protein were ineffective. This argues that the respective amyloids are not by-products or a dead-end stage of these prions, but are themselves the infectious material. All infectious Ure2p amyloids are larger than about 40-mer size. Amyloid formed *in vitro* is capable, for at least [PSI] and [URE3], of transmitting any of several prion variants. This implies that the amyloids can have any of several structures, a fact demonstrated by solid-state nuclear magnetic resonance (NMR) for amyloid of the Alzheimer's disease peptide, Aβ.

Shuffling Prion Domains and Amyloid Structure

The prion domains (**Figure 2**) of Ure2p and Sup35p are quite rich in Asn and Gln residues, and nearly the entire sequence of Rnq1p, the basis of the [PIN] prion, is rich in these amino acids. However, many Q/N-rich proteins are not capable of being prions. Thus, it was assumed that specific sequences in the known prion domains were important for prion formation. The Sup35 prion domain

has octapeptide repeats much like those in PrP, and deletion or duplication of these showed substantial effects on prion generation. In addition, single amino acid changes in the prion domain of Sup35p blocked prion propagation.

To critically test whether the Ure2p prion domain had sequences essential for prion development, the entire Q/N-rich region (residues 1–89) was randomly shuffled (without changing the amino acid content) and each of five shuffled sequences were inserted into the chromosome in place of the normal prion domain. Surprisingly, each of these five shuffled sequences could support prion generation and propagation, although one was rather unstable. Each protein with the shuffled sequence could also form amyloid (**Figure 3**) *in vitro*. This showed that it was the amino acid content of the Ure2p prion domain that determines prion formation, and that sequence plays only a minor role.

Similarly, five shuffled versions of the Sup35p prion domain were each inserted in place of the normal sequence. Again, all five shuffled versions allowed formation and propagation of a [PSI]-like prion. It is likely that the effects of deletion or duplication of the octapeptide repeats observed on prion formation or propagation were due to changes in the length or composition of the prion domain. It appears that repeats *per se* are not important for prion generation or propagation.

Shuffleable Prion Domains Suggests Parallel In-Register β-Sheet Structure

Amyloids are β-sheet structures, but there are at least three kinds of β-sheets. Antiparallel β-sheets have adjacent peptide chains oriented in opposite directions: N→C next to C→N. This results in pairing of largely nonidentical residues. A β-helix also involves pairing of largely nonidentical residues, although they are within the same peptide chain. A parallel β-sheet pairs identical residues if it is in-register, but in principle, one could have an out-of-register parallel β-sheet, in which case nonidentical residues would be bonded to each other.

Prion propagation (and amyloid propagation in general) is a very sequence-specific process. For example, a single amino acid change (at residue 138) can block propagation of scrapie in tissue culture. Humans are polymorphic at PrP residue 129 with roughly equal numbers of alleles encoding M and V. Either M/M or V/V individuals can get Creutzfeldt–Jakob disease (a human prion disease), but M/V heterozygotes cannot. Similarly, a single amino acid change in the prion domain of Sup35p can block propagation of [PSI] from the normal sequence, but the mutant Sup35p can itself become a prion nonetheless. Thus, if a prion amyloid has an antiparallel, parallel out-of-register, or β-helix structure, there must be some form of complementarity between bonded residues. Shuffling such a sequence would be expected to destroy the complementarity. In contrast, shuffling the sequence of a parallel in-register β-sheet would still leave identical residues paired. This suggests that prion domains that can be shuffled without destroying their prion-forming ability are forming parallel in-register β-sheets. Indeed Ure2p$^{10-39}$, a fragment of the prion domain, forms amyloid with a parallel in-register β-sheet structure, as does the Sup35 prion domain.

Biological Roles of Prions: A Help or a Hindrance?

In an attempt to discern whether yeast prions are an advantage or disadvantage to their host organism, cell growth of isogenic [PSI^+] and [psi^-] strains have been carried out under a variety of conditions. To what extent the various growth conditions tested represent the normal yeast habitat seems unknowable, although [psi^-] was an advantage under far more conditions than was [PSI^+].

An alternative approach was to compare the frequency with which [PSI^+] or [URE3] was found in wild strains to those of several 'selfish' RNA and DNA viruses and replicons known in *S. cerevisiae*. In any organism, an infectious element (such as a virus) may be widely distributed in spite of it causing disease in its host because the infection process overcomes and outraces the loss of infected individuals from negative selection. Certainly an infectious element, which is an advantage to its host, will quickly become widespread as selection and infection operate in the same direction. In fact, while the mildly deleterious RNA and DNA viruses and plasmids of yeast are easily found in wild strains neither [URE3] nor [PSI^+] was found in any of the 70 wild strains examined. This indicates that [URE3] and [PSI^+] produce disease in their hosts, and a rather more severe disease than the mild nucleic acid replicons.

The [Het-s] prion of *Podospora* appears to carry out the normal fungal function of heterokaryon incompatibility, thought to be a protection against the sometimes debilitating fungal viruses. Indeed, as one would expect for a prion with a function for the cell, 80% of wild *Podospora* isolates carry [Het-s], confirming its beneficial effects.

Unlike [PSI^+] and [URE3], [PIN^+] is found in wild strains at a frequency similar to that of the parasitic RNA and DNA viruses and plasmids. This suggests that [PIN^+] is at least not as severe a pathogen as are [URE3] and [PSI^+].

Enzyme as Prion

While most of the known prions involve amyloids, the word prion (infectious protein) is more general, requiring only that transmission be by protein alone. If an enzyme is made as an inactive precursor that needs the active form of the same enzyme for its activation, then such a protein can be a prion. The vacuolar protease B (Prb1p) of *S. cerevisiae* can be such a prion in a mutant lacking protease A, which normally activates its precursor. Cells initially carrying only the inactive precursor remain so unless the active enzyme is introduced. Once a cell has some active enzyme, the autoactivation process can continue indefinitely. It is likely that other examples of this type of phenomenon will be found among the many protein kinases, methylases, acetylases, and other modifying enzymes that are known. Indeed, a protein kinase of *P. anserina* appears to be able to become a prion in this manner, producing a nonchromosomal genetic element called 'crippled growth'.

The advent of yeast prions has propelled the prion field forward, and is giving us insight into the broader field of amyloids and how they interact with cellular components. There is already evidence for a number of new prions, mostly in yeasts and fungi, in part because they are so well suited to genetic studies, and in part because their frequent natural mating or heterokaryon formation (in fungi) results in complete mixing and exchange of cellular proteins.

Further Reading

Aigle M and Lacroute F (1975) Genetical aspects of [URE3], a non-Mendelian, cytoplasmically inherited mutation in yeast. *Molecular and General Genetics* 136: 327–335.

Chan JCC, Oyler NA, Yau W-M, and Tycko R (2005) Parallel β-sheets and polar zippers in amyloid fibrils formed by residues 10–39 of the yeast prion protein Ure2p. *Biochemistry* 44: 10669–10680.

Chernoff YO, Lindquist SL, Ono B-I, Inge-Vechtomov SG, and Liebman SW (1995) Role of the chaperone protein Hsp104 in propagation of the yeast prion-like factor [psi$^+$]. *Science* 268: 880–884.

Coustou V, Deleu C, Saupe S, and Begueret J (1997) The protein product of the *het-s* heterokaryon incompatibility gene of the fungus *Podospora anserina* behaves as a prion analog. *Proceedings of the National Academy of Sciences, USA* 94: 9773–9778.

Cox BS (1965) PSI, a cytoplasmic suppressor of super-suppressor in yeast. *Heredity* 20: 505–521.

Derkatch IL, Bradley ME, Hong JY, and Liebman SW (2001) Prions affect the appearance of other prions: The story of [PIN]. *Cell* 106: 171–182.

Jung G, Jones G, and Masison DC (2002) Amino acid residue 184 of yeast Hsp104 chaperone is critical for prion-curing by guanidine, prion propagation, and thermotolerance. *Proceedings of the National Academy of Sciences, USA* 99: 9936–9941.

King CY and Diaz-Avalos R (2004) Protein-only transmission of three yeast prion strains. *Nature* 428: 319–323.

Maddelein ML, Dos Reis S, Duvezin-Caubet S, Coulary-Salin B, and Saupe SJ (2002) Amyloid aggregates of the HET-s prion protein are infectious. *Proceedings of the National Academy of Sciences, USA* 99: 7402–7407.

Masison DC and Wickner RB (1995) Prion-inducing domain of yeast Ure2p and protease resistance of Ure2p in prion-containing cells. *Science* 270: 93–95.

Paushkin SV, Kushnirov VV, Smirnov VN, and Ter-Avanesyan MD (1997) *In vitro* propagation of the prion-like state of yeast Sup35 protein. *Science* 277: 381–383.

Ross ED, Edskes HK, Terry MJ, and Wickner RB (2005) Primary
sequence independence for prion formation. *Proceedings of the
National Academy of Sciences, USA* 102: 12825–12830.
Tanaka M, Chien P, Naber N, Cooke R, and Weissman JS (2004)
Conformational variations in an infectious protein determine prion
strain differences. *Nature* 428: 323–328.

Wickner RB (1994) [URE3] as an altered *URE2* protein: Evidence for a
prion analog in *S. cerevisiae. Science* 264. 500–569.
Wickner RB, Edskes HK, Ross ED, *et al.* (2004) Prion genetics:
New rules for a new kind of gene. *Annual Review of Genetics*
38: 681–707.

Retrotransposons of Fungi

T J D Goodwin, M I Butler, and R T M Poulter, University of Otago, Dunedin, New Zealand

Glossary

Retrotransposition The process by which a
retrotransposon replicates.
Retrotransposon A eukaryotic mobile genetic
element that can replicate via the reverse
transcription of an RNA intermediate and the
insertion of the resulting DNA into the host genome.
Reverse transcriptase A DNA polymerase that can
use both RNA and DNA as a template.

Introduction

Retrotransposons are mobile genetic elements. They are
generally regarded as selfish or parasitic entities and
appear as inserts within the genomes of their hosts.
They can spread through the host genome by copying
their RNA transcripts into DNA and inserting these new
DNA copies back into the host chromosomes. As this
process, retrotransposition, is replicative it results in
an increase in the copy number of the retrotransposons.
Over time this can result in retrotransposons reaching
very high copy numbers and making up a large proportion of the genome. As an example, there are more
than 500 000 copies of the L1 retrotransposon in the
human genome and these make up about 17% of human
DNA.

Much of what is known about retrotransposons and
their interactions with their hosts has been derived
from studies with fungi. Fungi are a diverse group of
eukaryotes of great ecological, economic, medical, and
scientific importance. They range from tiny obligate
intracellular pathogens, such as *Encephalitozoon cuniculi*,
to large, multicellular organisms that can have a mass
in excess of 100 kg, such as *Bridgeoporus nobilissimus*.
They include such organisms as the bakers' and brewers'
yeast *Saccharomyces cerevisiae*, genetic model organisms
such as *Neurospora crassa*, plant pathogens such as the
causal agent of rice blast disease *Magnaporthe grisea*,

human pathogens such as *Candida albicans* and *Cryptococcus
neoformans*, sources of antibiotics such as *Penicillium chrysogenum*, and mycorrhiza-forming fungi such as *Glomus
intraradices*. The fungi are classified into six divisions:
Ascomycota, Basidiomycota, Chytridiomycota, Glomer-
omycota, Microsporidia, and Zygomycota. The great
majority of well-known fungi, such as *S. cerevisiae,
C. albicans*, and *N. crassa* are ascomycetes. The next largest
group is the basidiomycetes, which includes such organisms as the button mushroom *Agaricus bisporus*, the pathogenic yeast *C. neoformans*, as well as rusts and smuts. The
chytrids are unusual fungi that produce flagellated motile
spores. This group includes the frog pathogen *Batracho-
chytrium dendrobatidis*. The zygomycetes are characterized
by zygospores formed during sexual reproduction. They
include *Rhizopus oryzae*, which is commonly found on
decaying vegetable matter and can cause fatal infections
in humans. The glomeromycetes include the arbuscular
mycorrhizal fungi such as *G. intraradices* and *Gigaspora
margarita*. These species form symbiotic associations
with the roots of many land plants, the fungus supplying
the plant with nutrients obtained from the surrounding
soil, in exchange for carbohydrates. The last fungal division, the microsporidia, consists of obligate intracellular
parasites of animals. These are atypical fungi in that they
lack mitochondria and peroxisomes. They include the
human pathogen *Encephalitozoon cuniculi*.

Much of the contribution of fungi to our knowledge of
retrotransposons has come from the yeast *S. cerevisiae*. The
Ty elements found in this species were among the first
retrotransposons to be identified and characterized.
Indeed, the term retrotransposon was coined following
the demonstration that Ty elements replicate (transpose)
through an RNA intermediate, in a series of steps essentially the same as that for retroviruses. The widespread
use of *S. cerevisiae* as a model for many features of eukaryotic life, due to its ease of culture and experimental
manipulation, rapid growth, readily accessible genetics,
and small genome size, meant that many tools were
available to assist in the analysis of Ty elements. Numerous features that have made *S. cerevisiae* attractive for the

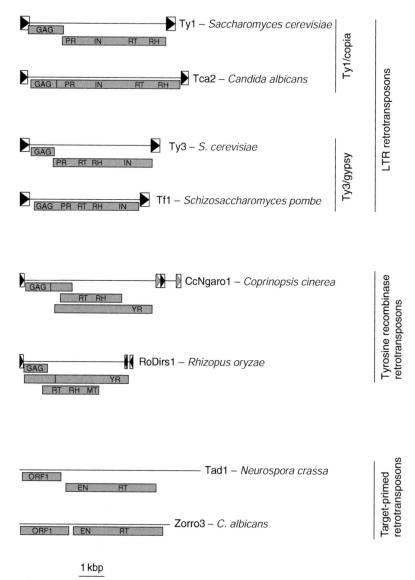

1 kbp

Figure 1 Structures of different types of retrotransposons. The lengths of the elements are indicated by the horizontal lines. Repeat sequences are shown as boxed triangles. Open reading frames are depicted as shaded boxes, with the approximate positions of the various protein-coding domains indicated: EN, endonuclease; IN, integrase; MT, putative methyltransferase; PR, protease; RH, ribonuclease H; RT, reverse transcriptase; YR, tyrosine recombinase. The groups to which the elements belong are shown on the right.

study of retrotransposons are also shared with other fungi. In particular, the general small size of fungal genomes has meant that many complete genome sequences, from an evolutionary diverse array of fungi, have been determined in recent years and many more will become available in the near future. Analyses of these genome sequences are ongoing but have already contributed much to our understanding of retrotransposons. Most importantly, the analysis of a diverse range of fungi has uncovered many types of retrotransposons, and interesting interactions between the retrotransposons and their hosts, that are not observed in *S. cerevisiae*, making it clear that study of a wide range of species will be essential for an in-depth understanding of fungal retrotransposons. Here we will outline many of the

features of retrotransposons, with special emphasis on the fungal elements. We also describe the impact of these elements on the fungal genome and the ways in which the interactions between the retrotransposons and their hosts have evolved to permit the long-term survival of the elements while minimizing the deleterious impact on their hosts.

Types of Retrotransposons

Retrotransposons come in a wide variety of forms with diverse sequences and diverse replication mechanisms (**Figures 1** and **2**). The one feature common to all retrotransposons is a sequence encoding reverse transcriptase (RT), the enzyme responsible for copying the RNA transcripts into DNA. The RTs of all known retrotransposons are homologous, indicating that they share a common origin. Phylogenetic analyses based upon alignments of RT sequences (**Figure 2**), together with comparisons of other structural and mechanistic features, have been used to classify retrotransposons into various groups. At the

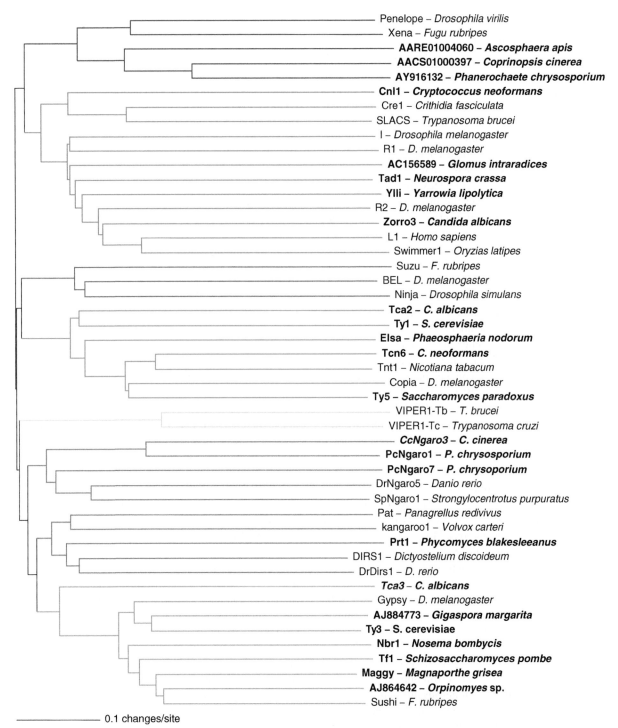

Figure 2 Relationships among retrotransposons. This tree is based on an alignment of RT sequences from a wide variety of retrotransposons. The various groups are indicated in different colors: red, Penelope-like retrotransposons; blue, target-primed retrotransposons; purple, BEL-like LTR retrotransposons; green, Ty1/copia LTR retrotransposons; yellow, VIPER elements; orange, tyrosine recombinase retrotransposons; khaki, Ty3/gypsy LTR retrotransposons. Elements from fungi are in boldface.

broadest level many retrotransposons can be classified into one of two groups. Members of the first group consist of a protein-coding internal region flanked by generally noncoding long terminal repeats (LTRs) and are known as LTR retrotransposons. The Ty elements of *S. cerevisiae* are all LTR retrotransposons. A typical LTR retrotransposon is 5–6 kbp long and contains two protein-coding open reading frames (ORFs), *gag* and *pol* (**Figure 1**). The *gag* ORF encodes a protein, Gag, which forms the major structural component of a virus-like particle (VLP) within which the reverse transcription reactions take place. The *pol* ORF encodes a polyprotein bearing various enzymatic domains: aspartic protease (PR), the RT, ribonuclease H (RH), and integrase (IN). PR is responsible for processing the initial *gag* and *pol* ORF products (see further ahead) into their functional domains. RT copies the RNA transcripts into DNA, assisted by RH. IN is involved in the insertion of the nascent DNA copy into the host genome. The LTRs flanking the protein-coding regions contain signals regulating transcription of the element, and, due to their redundant nature, they allow a full-length DNA copy of the element to be constructed from somewhat less-than-full-length RNA transcripts (see further ahead). LTR retrotransposons are themselves subdivided into further groups, named the Ty1/copia, Ty3/gypsy, and BEL groups, after founding elements from fungi and insects. The members of these groups can be distinguished from each other by sequence comparisons and the order in which the enzymatic domains appear within the *pol* gene. In Ty1/copia elements the *pol* domain order is PR-IN-RT-RH. In BEL and Ty3/gypsy elements (with a few exceptions) the order is PR-RT-RH-IN.

The second major group of retrotransposons are distinguished from LTR retrotransposons by the absence of LTRs, by differences in the set of encoded proteins, and a completely different replication mechanism (see further ahead). This group of retrotransposons has been given various names, including non-LTR retrotransposons, poly-A retrotransposons, long interspersed nuclear elements (LINEs), and retroposons. However, none of these names is ideal: the 'non-LTR' term refers to a feature all these elements lack, rather than one they all share; the 'poly-A' term refers to a run of A residues found at the 3′ end of many of the first elements of the group to be characterized, but subsequently found to be lacking in many other elements; the LINE terminology is not ideal, as many of the members of this group are not interspersed, but are found only in specific regions of the genome. Finally, the term 'retroposon' has been used in the literature to refer to numerous different groupings of retrotransposons and retrotransposon-like elements, so its usefulness has been lost. It has recently been suggested, however, that the elements of this group be referred to as the target-primed (TP) retrotransposons, after their insertion mechanism (target-primed retrotransposition – see

further ahead), which is distinct from that of LTR retrotransposons and is likely to be shared by all members of the group. This is the terminology used in this article.

A typical TP retrotransposon consists of one or two protein-coding regions flanked by 5′ and 3′ untranslated regions (**Figure 1**). All elements encode an RT and an endonuclease (EN). The TP retrotransposons can be divided into 15 or more clades, defined by sequence comparisons and structural distinctions. In members of what are believed to be the older clades, the EN-coding sequence lies downstream of RT and the EN is similar in sequence to certain restriction endonucleases. In the more recently evolved elements, the EN-coding sequence lies upstream of RT and is related to eukaryotic apurinic/apyrimidinic (AP) endonucleases. An RH-coding sequence also makes a sporadic appearance among these latter elements. In the elements with two ORFs, RT and EN are always encoded by the downstream ORF. The upstream ORF, ORF1, is generally poorly conserved. The function of the protein product of ORF1 is unknown, although in some elements ORF1p has been shown to have RNA-binding and nucleic acid-chaperone activities.

In addition to these two major categories of retrotransposons, several other types of retrotransposons have been described. These elements are generally rarer than the LTR and TP retrotransposons and so have not been characterized in as much detail. The first of these additional groups, the tyrosine recombinase (YR) retrotransposons, bear RT- and RH-coding sequences related to those of LTR retrotransposons, but they lack coding sequences for PR and IN. Instead they encode a tyrosine recombinase, related to the recombinase of bacteriophage lambda (**Figure 1**). Like LTR retrotransposons, they also bear flanking repeats, but these are distinct from typical LTRs in that they are either 'split direct repeats' or have a complex inverted-repeat structure. The DIRS1 element from the slime mold *Dictyostelium discoideum* was the first YR retrotransposon to be described. The second group, the Penelope-like elements (PLEs; named after an element from *Drosophila virilis*), are similar to TP retrotransposons in that they encode an RT and an EN. The EN of PLEs is, however, unrelated to that of TP retrotransposons, but is similar to the Uri endonucleases found in group I introns. PLEs are often found associated with variable terminal repeat sequences in both direct and inverted orientations. Their replication mechanism is not well understood, but may be an unusual form of target-primed retrotransposition. The final category of retrotransposons, the vestigial interposed retroelement (VIPER)-like elements, have been found in only a very small number of species (trypanosomes) to date. They bear RT and RH genes somewhat similar to those of LTR retrotransposons, but otherwise very little is known regarding their conserved features.

Distribution of Retrotransposons

Retrotransposons have only been found in eukaryotic genomes (with the exception of a few rare elements integrated into the genomes of eukaryotic viruses). While RT-encoding elements (group II introns and retrons) do appear in prokaryotes, these elements are not mobile genetic elements in the same sense as the eukaryotic retrotransposons. Among eukaryotes, retrotransposons have been found in almost all genomes that have been examined in detail. However, the diversity of retrotransposons found within particular species varies enormously. For instance, examples of nearly every type of known retrotransposon have been identified in the genome of the zebrafish *Danio rerio*, while only one active retrotransposon, the L1 TP retrotransposon, is known in humans and

no retrotransposons of any sort are found in the complete genome sequence of the malaria parasite *Plasmodium falciparum*. Similarly, the distribution of different types of retrotransposons also displays great variation. For instance, members of the Ty3/gypsy class of LTR retrotransposons appear in plants, animals, fungi, and various protists, whereas BEL-like LTR retrotransposons have only been found in animals to date.

Within the fungi, retrotransposons are widespread and abundant (**Table 1**). Most have been found in ascomycetes and basidiomycetes. Only a few have been identified in zygomycetes, glomeromycetes, chytrids, and microsporidia: this is, however, likely to be simply the result of there being less data available for species from these divisions, rather than low numbers of retrotransposons, as those species that have been examined often contain abundant

Table 1 Types of retrotransposons identified in various fungal phyla

Phylum	Type of retrotransposon	Examples (host)[a]	Accession no.
Ascomycota	LTR–Ty1/copia	Ty1 (*Saccharomyces cerevisiae*)	M18706
		Tca2 (*Candida albicans*)	AF050215
		Elsa (*Phaeosphaeria nodorum*)	AJ277966
	LTR–Ty3/gypsy	Ty3 (*S. cerevisiae*)	M34549
		Tf1 (*Schizosaccharomyces pombe*)	M38526
		Maggy (*Magnaporthe grisea*)	L35053
	Target–primed	Tad1 (*Neurospora crassa*)	L25662
		Zorro3 (*C. albicans*)	AF254443
	YR	—	
	Penelope	Unnamed (*Ascosphaera apis*)	AARE01004060
Basidiomycota	LTR–Ty1/copia	Tcn6 (*Cryptococcus neoformans*)	Retrobase[b]
	LTR–Ty3/gypsy	MarY1 (*Tricholoma matsutake*)	AB028236
	Target-primed	Cnl1 (*C. neoformans*)	Retrobase[b]
	YR	CcNgaro1 (*Coprinopsis cinerea*)	AACS01000194
	Penelope	Unnamed (*Phanerochaete chrysosporium*)	AY916132
Chytridiomycota	LTR–Ty1/copia	Unnamed (*Orpinomyces* sp.)	AJ864659
	LTR–Ty3/gypsy	Unnamed (*Orpinomyces* sp.)	AJ864642
	Target-primed	Unnamed (*Orpinomyces* sp.)	AJ864661
	YR	—	
	Penelope	—	
Glomeromycota	LTR–Ty1/copia	—	
	LTR–Ty3/gypsy	Unnamed (*Gigaspora margarita*)	AJ884773
	Target-primed	Unnamed (*Glomus intraradices*)	AC156589
	YR	—	
	Penelope	—	
Microsporidia	LTR–Ty1/copia	—	
	LTR–Ty3/gypsy	Nbr1 (*Nosema bombycis*)	DQ444465
	Target-primed	—	
	YR	—	
	Penelope	—	
Zygomycota	LTR–Ty1/copia	—	
	LTR–Ty3/gypsy	Unnamed (*Rhizopus oryzae*)	AACW02000049
	Target-primed	Unnamed (*R. oryzae*)	AACW02000090
	YR	Prt1 (*Phycomyces blakesleeanus*)	Z54337
	Penelope	Unnamed (*R. oryzae*)	AACW01000082

[a]A dash indicates that no element of a particular class has been identified in the phylum to date.
[b]Retrobase (http://biochem.otago.ac.nz).

elements. Of the LTR retrotransposons, members of the Ty1/copia group have been identified in a large number of ascomycetes, many of the basidiomycetes that have been examined, and also in one species of chytrid (*Orpinomyces* sp.). None has yet been identified in zygomycetes, glomeromycetes, or microsporidia. Ty3/gypsy elements have been identified in all fungal divisions. They are abundant in many ascomycetes and basidiomycetes and have also been found in microsporidia (*Nosema bombycis* and *Spraguea lophii*), glomeromycetes (*Glomus intraradices* and several species of Gigaspora), chytrids (*Orpinomyces* sp.) and zygomycetes (*R. oryzae*). In many fungi, Ty3/gypsy elements are more abundant than the Ty1/copia elements, suggesting that the former elements have been somewhat more successful in colonizing and persisting in fungal genomes. No BEL-like LTR retrotransposons have been identified, despite the large amounts of sequence data available, suggesting that this group is absent from fungi.

Target-primed retrotransposons are also widespread in fungi with numerous elements again found in ascomycetes and basidiomycetes and examples also identified in zygomycetes (*R. oryzae*), glomeromycetes (*G. intraradices* and some Gigaspora species), and chytrids (*Orpinomyces* sp.). None has yet been identified in microsporidia. Despite their widespread occurrence, TP elements are notably absent from several species for which complete genome sequences are known. In particular, they are absent from *S. cerevisiae* and another model yeast, *Schizosaccharomyces pombe*. This absence from model species has meant that the contribution of fungi to our understanding of the replication mechanisms and other features of TP retrotransposons has been much less than that for LTR retrotransposons.

YR retrotransposons have a more sporadic distribution in fungi than LTR and TP retrotransposons. Examples have been identified in basidiomycetes (*Phanerochaete chrysosporium* and *Coprinopsis cinerea*) and zygomycetes (*Phycomyces blakesleeanus* and *R. oryzae*). None has yet been identified in other divisions. For ascomycetes at least, with the vast amount of sequence data available, this suggests that they are either absent from this division or extremely rare. Sequences related to Penelope retrotransposons are also rare in fungi, but nevertheless, such sequences are evident in at least three fungal divisions: basidiomycetes (*P. chrysosporium* and *C. cinerea*), zygomycetes (*R. oryzae*), and ascomycetes (*Ascosphaera apis*). No elements similar to VIPER have been detected in fungi to date.

Overall, the distribution of the various types of retrotransposons in fungi parallels their distribution throughout eukaryotes in general; that is, TP retrotransposons and members of the Ty1/copia and Ty3/gypsy groups of LTR retrotransposons are common occurrences and found in most species. The types of retrotransposon that are rare in general, such as the YR and Penelope-like retrotransposons,

are also rare in fungi. Nevertheless, these latter elements are still widespread, in the sense that they are found in species from several different divisions. Their widespread distribution and a general high level of diversity in their sequences, suggests that these elements have a long history in fungi, probably dating back to the last common ancestor of all fungi. In this case, their current sporadic distribution is likely to be the result of the frequent loss of these elements from evolving lineages. It is not clear why these elements should be more frequently lost than TP and LTR retrotransposons. One possibility is that they might not replicate so efficiently and so never reach such high copy numbers and are thus more prone to stochastic loss through deletion or point mutation.

Retrotransposon Replication

LTR Retrotransposons

The replication of LTR retrotransposons is well understood due to its extensive similarities with retroviral replication and the thorough characterization of the Ty elements of *S. cerevisiae*. While specific features may vary from element to element, the general features of Ty1 element replication likely apply to all LTR retrotransposons. The first step in Ty1 replication is transcription of an integrated element. The Ty1 LTRs contain regulatory sequences which cause transcription to begin within the left LTR sequence, proceed all the way through the internal region, and terminate within the right LTR, resulting in a terminally redundant RNA. This RNA is then exported to the cytoplasm where it is translated. In Ty1 (and in all other LTR retrotransposons) the *pol* gene is located downstream of *gag* and is translated at a much lower level. This downregulation of *pol* is achieved by having it positioned in a different translational reading frame from the *gag* ORF, with the 5′ end of *pol* having a small overlap with the 3′ end of *gag*. All translation begins with the *gag* ORF and usually stops at the *gag* termination codon. The *pol* ORF is translated only when the ribosome undergoes a rare, programmed frameshift near the end of the *gag* ORF. When this occurs, a Gag–Pol fusion protein is produced. Next, the abundant Gag proteins and the rare Gag–Pol fusion proteins assemble into a hollow VLP. The N-terminus of the Gag protein lies on the exterior of the particle and the C-terminus lies on the inside. The fusion of the Pol protein to the C-terminus of Gag in the Gag–Pol fusion ensures that the Pol enzymes are packaged into the interior of the particle. Also packaged inside the particle are the retrotransposon RNA (usually two copies) and a tRNA (initiator Met tRNA) which will act as a primer during the reverse transcription step. The walls of the particle are porous to nucleotides and other small molecules.

The first step in the reverse transcription process is the annealing of the primer tRNA to a complementary sequence in the retrotransposon RNA (the minus-strand primer-binding site (PBS)) that lies just downstream of the left LTR. The 3′ OH group of the tRNA is then used as a primer by the RT to initiate minus-strand DNA synthesis. Minus-strand DNA synthesis then proceeds to the end of the molecule, the 5′ end of the mRNA, where it temporarily halts. The RNA in the RNA/DNA hybrid is then removed by RH. This allows the nascent minus-strand DNA to anneal to the complementary LTR sequence at the 3′ end of the RNA. Minus-strand DNA synthesis then resumes and proceeds to the 5′ end of the template. Next, the RNA of the RNA/DNA hybrid is again degraded by RH. A short purine-rich segment of RNA immediately upstream of the right LTR, known as the polypurine tract, is, however, resistant to removal by RH. This sequence remains bound to the minus-strand DNA and its 3′ OH group is used by RT to prime plus-strand DNA synthesis. This continues as far as the 5′ end of the minus-strand DNA template. The nascent plus-strand is then dissociated from the 5′ end of the minus-strand DNA, possibly by displacement by further plus-strand DNA synthesis initiated from a second poly-purine tract lying in the central region of the element. The displaced plus-strand DNA can then anneal to the 3′ end of the minus-strand DNA at their complementary sequences. A full-length double-stranded DNA copy of the retrotransposon can then be formed by extension of each strand to the end of its template and removal of any RNA nucleotides remaining from the primer sequences. Finally, in a series of steps, which are not so well understood, the new double-stranded DNA molecule associates with IN, exits the VLP, and enters the cell nucleus. Here, the DNA/protein complex associates with the chromosomal DNA and IN mediates the insertion of the retrotransposon into the host genome, completing the replication process.

While most of these steps are likely to be essentially the same in most LTR retrotransposons, many minor variations are known. For instance, in elements such as Tf1 and Tf2 from *S. pombe* (and closely related elements present in other fungi and also in vertebrates) the primer for minus-strand DNA synthesis is not a tRNA, but rather the 5′ end of the elements own mRNA which is complementary to the PBS. The two sequences anneal to each other and the resulting structure is recognized by the element's RH and cleaved to produce a free 3′ OH which is then used as a primer. As another example, in the element Tca2 from *C. albicans*, the *gag* and *pol* ORFs are in the same phase and separated by a stop codon, rather than being separated by a frameshift (**Figure 1**). Expression of the Gag–Pol fusion protein in this element presumably occurs via the occasional suppression of the *gag* stop codon. In other elements, for example, Tf1 and

Tf2 (**Figure 1**), the *gag* and *pol* ORFs are fused into a single ORF and translated together. The correct stoichiometry of the Gag and Pol proteins is subsequently obtained by the preferential degradation of Pol.

Target-Primed Retrotransposons

The replication of TP retrotransposons is not as well understood as that of LTR retrotransposons, at least partly due to the lack of a suitable model system in an experimentally tractable microorganism. Nevertheless, much has been learned from the study of TP elements in insect and mammalian systems. Replication begins with transcription initiated from an internal promoter in the 5′ untranslated region. Transcription proceeds to the 3′ end of the element to produce a full-length RNA. The RNA is exported to the cytoplasm where it is translated. In the case of elements with two ORFs, the first ORF is translated much more efficiently than the second ORF. Following translation, the proteins remain associated with the RNA to form a RNA/protein complex. The complex moves to the nucleus and associates with the chromosomal DNA. The EN produces a single-stranded nick in the host chromosome and the resulting 3′ OH group is used as a primer by RT to synthesize minus-strand DNA directly into the insertion site. The following steps are not well understood but presumably include the nicking of the second strand of the target site by EN, the use of the resulting 3′ OH group to prime plus-strand synthesis, the completion of both strands of DNA synthesis, and the sealing of the ends of the retrotransposon to the host chromosomes. Although fungal elements have, as yet, contributed little to our understanding of TP retrotransposition, several TP elements have recently been identified in yeasts that may be developed into useful experimental systems for analyzing the process in more detail. These elements include Zorro3 from *C. albicans* and Ylli from *Yarrowia lipolytica*.

Other Retrotransposons

Very little is known about the replication of the other types of retrotransposons and most of this is based upon analyis of sequence data rather than direct experimental evidence. Briefly, replication of the YR retrotransposons is thought to proceed via an RNA which is copied into a circular, double-stranded DNA by the actions of RT and RH. The circular DNA is then integrated into the host chromosome by recombination mediated by the tyrosine recombinase. It is likely that PLEs integrate via a mechanism related to target-primed retrotransposition. Nothing is known about the replication of the VIPER-type of retrotransposon. The recent identification of YR and Penelope-like retrotransposons in fungi, and the likelihood that more will be identified in the near future as more fungal genome sequence data are obtained, may soon permit the development of systems to characterize these elements in more detail.

Retrotransposons and Fungal Genomes

Retrotransposons can have many and varied effects on the genomes of their hosts. They can make up a significant proportion of the host genome. In humans, the L1 TP retrotransposon alone makes up about 17% of the genome. In some plants, up to as much as 90% of the genome is made up by retrotransposons. Obviously, such an abundance of retrotransposons will have a profound effect on the functioning of the genetic material. In the fungi that have been analyzed in depth, the proportion of the genome made up by retrotransposons is generally lower than the figures quoted above. Figures for fungi include 3.1% for *S. cerevisiae*, 1.3% for *C. albicans*, 1.1% for *S. pombe*, 5.4% for *M. grisea* (all ascomycetes), and 7.6% for *Microbotyrum violaceum* (a basidiomycete). It should be noted that many of the fungi that have been analyzed in depth have been chosen in part because of their small genome size (10–50 Mbp). However, fungi with much larger genomes exist. For instance, among the glomeromycetes the genome size of *Gigaspora margarita* has been estimated at ∼740 Mbp and that of *Scutellospora gregaria* at >1000 Mbp. It is possible that fungi with such large genomes might contain very large numbers of retrotransposons. On the other hand, some fungi have much lower numbers of retrotransposons. For instance, the microsporidian *E. cuniculi* has a very small genome (∼2.9 Mbp) which contains no identifiable retrotransposons at all. Likewise, the yeast *Pichia farinosa* appears to contain no retrotransposons, despite these elements being abundant in closely related species.

In addition to making up a significant fraction of many fungal genomes, retrotransposons can act as powerful mutagens. For instance, they can insert into or adjacent to genes thus altering gene structure and/or regulation. They can create new copies of genes via the occasional accidental copying of gene transcripts into DNA and the insertion of these back into the genome. The resulting sequences will often be nonfunctional (retropseudogenes) but occasionally a new copy will be inserted in an intact form, adjacent to a promoter sequence where it will be expressed and perform some useful function. Retrotransposons can move their flanking sequences around the genome, potentially creating new arrangements of exons in genes and/or altering promoter sequences. The retrotransposon sequences themselves may occasionally be adopted to perform functions useful for the host. In addition, by providing regions of sequence similarity dispersed throughout the genome, retrotransposons can promote recombination events between otherwise nonhomologous

regions, leading to chromosomal rearrangements. For example, the breakpoints of many chromosomal translocations in *S. cerevisiae* have been found to coincide with Ty elements. While the majority of mutational events associated with retrotransposons are likely to be neutral or deleterious to the hosts, occasional events will result in beneficial rearrangements. This suggests that the relationship between retrotransposons and their hosts should not be considered as strictly the same as that of parasite and host, but more as one with mutual benefits. In this regard, it is also of interest to note that, since the long-term survival of a retrotransposon is dependent on the long-term success of its host, many retrotransposons have developed strategies that act to minimize the damage they cause. For instance, some retrotransposons direct their integration to specific areas of the genome where they are least likely to cause deleterious mutations. As an example, the integration of Ty3 occurs very precisely 1–4 bp upstream of tRNA genes, areas where the insertion of the retrotransposon is unlikely to have a negative impact. Likewise, Ty5 integration is directed to areas of silent chromatin.

Not only can retrotransposons have profound effects on their hosts, but the hosts can also have profound effects on their retrotransposons. This is usually in the form of mechanisms to eliminate active retrotransposons, thus minimizing their potential mutagenic effects. Active retrotransposons can be eliminated in various ways. For instance, in addition to random point mutations, LTR retrotransposons can be inactivated by recombination between the two LTRs of a single element. This results in the internal region being excised and just a single LTR remaining at the original insertion site. In some species retrotransposons can be silenced by RNA interference or through methylation of repeat sequences. One of the most interesting mechanisms for eliminating active retrotransposons to have been identified in fungi is a process called RIP (repeat-induced point mutation). This was first found in *N. crassa* and similar processes have since been identified in a variety of other filamentous ascomycetes. RIP acts during the sexual cycle and efficiently identifies and mutates repeat sequences that are greater than about 400 bp in length and share more than about 80% sequence identity. In *N. crassa* RIP produces a large number of C-to-T and G-to-A mutations in both copies of the repeated sequence. RIP progressively mutates the repeated sequences over successive sexual cycles until they no longer are sufficiently similar to be recognized (i.e., <80% identity). This process very effectively inactivates retrotransposons. Its efficiency is demonstrated by the finding that in the strain of *N. crassa* whose genome was sequenced, numerous retrotransposon relics were identified but not a single active element remains. RIP in *N. crassa* is one of the more extreme examples of the extents to which fungi may go to eliminate active transposable elements. Its actions not only eliminate active retrotransposons but also appear to prevent gene duplication of any sort, thus eliminating this pathway as a means for evolving new gene functions. This is illustrated by the finding that in *N. crassa* the most closely related paralogous genes were duplicated >200 million years ago, likely predating the origins of RIP. Although RIP very effectively inactivates retrotransposons, an active element, the Tad1 TP retrotransposon, has nevertheless been identified in some strains of *N. crassa*. It is thought that active copies of this element have evaded RIP by persisting in strains that have not recently gone through a sexual cycle and by occasionally spreading via anastomoses between different asexual lineages. Analyses of these active Tad elements may have revealed a previously unsuspected importance of the asexual phase in the biology of *N. crassa*.

One of the most interesting features that becomes apparent when the sets of retrotransposons in different fungal genomes are compared is the great variation. Different species vary in the proportion of the genome made up by retrotransposons (as outlined above), in the diversity of retrotransposons that they harbor (e.g., in *S. pombe* the only known retrotransposons are two very closely related members of the Ty3/gypsy group of LTR retrotransposons, whereas species such as *C. albicans* or *C. neoformans* harbor a much wider range of elements), in the distribution of elements along the chromosomes (e.g., in *S. pombe* the elements seem to be fairly evenly distributed along the chromosomes, whereas in *C. neoformans* elements were found to be greatly concentrated in putative centromeric regions), and in whether or not there are many active elements (e.g., *S. cerevisiae* contains numerous intact elements, whereas nearly all elements in *N. crassa* are heavily mutated). In most cases the reasons for the variation have not been determined. No doubt some of the variation is simply due to chance, but the remainder likely reflects fundamental differences between the host species and the elements that they contain. Determining the reasons behind the variation will lead to important insights into the evolution of fungal genomes.

Uses of Fungal Retrotransposons

Fungal retrotransposons have been put to numerous uses. For instance, Ty elements have been employed as insertional mutagens and gene-tagging systems in *S. cerevisiae*. As another example, retrotransposons have been employed in population genetics studies of various species, such as the ascomycete *M. grisea* and the basidiomycete *Chondrostereum purpureum*. Probably the most important uses of fungal retrotransposons, however, are as models for elements from other species. Fungi are eukaryotes and contain many of the same cellular features as higher organisms, including man. However, the rapid growth, simple lifecycle, and ease of experimental manipulation of many fungi mean that it is possible to use them for studying aspects of eukaryotic cells that would be difficult or impossible in other organisms. The fact that fungi contain retrotransposons closely related to elements of profound importance, such as retroviruses and TP retrotransposons, makes them ideal for studying many aspects of retrotransposons and the retrotransposition process.

Further Reading

Daboussi M-J and Capy P (2003) Transposable elements in filamentous fungi. *Annual Review of Microbiology* 57: 275–299.
Galagan JE and Selker EU (2004) RIP: The evolutionary cost of genome defense. *Trends in Genetics* 20: 417–423.
Goodwin TJD and Poulter RTM (2000) Multiple LTR-retrotransposon families in the asexual yeast *Candida albicans*. *Genome Research* 10: 174–191.
Goodwin TJD and Poulter RTM (2001) The diversity of retrotransposons in the yeast *Cryptococcus neoformans*. *Yeast* 18: 865–880.
Kim JM, Vanguri S, Boeke JD, Gabriel A, and Voytas DF (1998) Transposable elements and genome organization: A comprehensive survey of retrotransposons revealed by the complete *Saccharomyces cerevisiae* genome sequence. *Genome Research* 8: 464–478.
Roth J-F (2000) The yeast Ty virus-like particles. *Yeast* 16: 785–795.

Totiviruses

S A Ghabrial, University of Kentucky, Lexington, KY, USA

Glossary

Mycoviruses Viruses that infect and multiply in fungi.
Pseudoknot A secondary structure in viral mRNA that slows movement of the ribosome and may cause a frameshift that allows entry to an alternative reading frame during translation.

Ribosomal frameshifting Ribosomes switching reading frame on an mRNA, in response to the presence of a slippery site and/or a pseudoknot, to synthesize a protein or a polyprotein from two overlapping reading frames.

Introduction

The discovery of the killer phenomenon in the 1960s in yeast (*Saccharomyces cerevisiae*) and in the smut fungus (*Ustilago maydis*) eventually led to the discovery of the isometric double-stranded (ds) RNA mycoviruses with nonsegmented genomes, currently classified in the genus *Totivirus* (family *Totiviridae*). Killer strains of yeast or smut secrete a protein toxin to which they are immune, but which is lethal to sensitive cells. The precursor to the killer toxin is encoded by a satellite dsRNA, which is dependent on a helper virus with nonsegmented dsRNA genome (totivirus) for encapsidation and replication. Unlike the helper totiviruses associated with the yeast and smut killer systems, other members of the family *Totiviridae*, including the viruses that infect filamentous fungi (members of the newly proposed genus *Victorivirus*) and those that infect parasitic protozoa (members of the genera *Giardiavirus* and *Leishmaniavirus*), are not known to be associated with killer phenotypes. Purified preparations of some of these viruses, however, can contain dsRNA species suspected of being satellite or defective dsRNAs.

The isometric dsRNA totiviruses that infect fungi and protozoa are unique among dsRNA viruses in that their genomes are undivided, whereas the genomes of all other dsRNA viruses are segmented. The yeast and smut totiviruses and associated killer systems, as well as the totiviruses infecting parasitic protozoa, are discussed elsewhere in this encyclopedia. In addition to examining the similarities and differences among members of the family *Totiviridae*, this article will focus on the totiviruses that infect filamentous fungi, a group of viruses that are phylogenetically more closely related to each other than to other totiviruses and utilize a different strategy to express their genomes.

Taxonomy and Classification

The family *Totiviridae* encompasses a broad range of viruses characterized by isometric particles, ~40 nm in diameter, that contain a monosegmented dsRNA genome coding for a capsid protein (CP) and an RNA-dependent RNA polymerase (RdRp). At present, three genera are recognized: *Totivirus*, *Giardiavirus*, and *Leishmaniavirus*. Viruses currently placed in the genus *Totivirus* infect yeast, smut fungi, or filamentous fungi, whereas those in the latter two genera infect parasitic protozoa. Two distinct RdRp expression strategies have been reported for species in the family *Totiviridae*: those that express RdRp as a fusion protein (CP–RdRp) by ribosomal frameshifting, such as saccharomyces cerevisiae virus L-A and the viruses that infect parasitic protozoa; and those that synthesize RdRp as a separate nonfused protein by an internal initiation mechanism, as shown for Hv190SV and proposed for all of the other totiviruses that infect filamentous fungi. Although *Helminthosporium victoriae 190S virus* (Hv190SV) is the only species recognized by the International Committee on Taxonomy of Viruses (ICTV) as a member of the genus *Totivirus*, the complete nucleotide sequences of several tentative totiviruses that infect filamentous fungi have recently been reported (**Table 1**). Sequence and phylogenetic analyses have demonstrated that these viruses have many properties in common with Hv190SV and that they are more closely related to each other than to the viruses infecting the yeast and smut fungi. A new genus (genus *Victorivirus*) was recently proposed to accommodate Hv190SV and the related viruses infecting filamentous fungi, with Hv190SV as the type species. The name 'Victorivirus' is derived from the specific epithet of *H. victoriae*, the host of the proposed type species.

In addition to fungal and protozoal hosts, totiviruses may also have crustacean hosts. A nonsegmented dsRNA virus with isometric particles, designated penaeid shrimp infectious myonecrosis virus (IMNV), was recently isolated from diseased penaeid shrimp and tentatively assigned to the family *Totiviridae*. IMNV is the causal agent of the shrimp myonecrosis disease characterized by necrosis of skeletal muscle, particularly in the distal abdominal segments and tail fan. Phylogenetic analysis based on the RdRp region of viruses in the totivirus family suggests that giardia lamblia virus (GLV; genus *Giardiavirus*) is the closest relative to IMNV. Interestingly, both IMNV and GLV, unlike all other members of the family *Totiviridae*, are infectious as purified virions. IMNV is presently unclassified and it has yet to be determined whether it is a member of a novel genus in the family *Totiviridae* or whether IMNV is a member of a novel family of dsRNA viruses that infect invertebrate hosts. The complete nucleotide sequences of several members and tentative members of the family *Totiviridae* have been published and the GenBank accession numbers are listed in **Table 1**.

Virion Properties

The buoyant densities in CsCl of virions of members of the totivirus family range from 1.36 to 1.43 g cm^{-3}, and the sedimentation coefficients of these virions range from 160S to 190S (S_{20w} in Svedberg units). Particles lacking nucleic acid sediment with apparent sedimentation coefficients of 90–105S. Isolates of ScV-L-A and UmV-H1 may have additional components, containing satellite or defective dsRNAs, with different sedimentation coefficients and buoyant densities. Virion-associated RdRp activity can be detected in all totiviruses examined to date. Protein kinase activity is associated with Hv190SV virions; capsids contain phosphorylated forms of CP.

Table 1 List of virus members and tentative members in the family *Totiviridae*, length of their genomes, size of encoded gene products and GenBank accession numbers

| Virus[a] | Abbreviation | Genome length (nt) | Size (kDa) | | GenBank accession no. |
			CP	RdRp/CP–RdRp[b]	
Genus: *Totivirus*					
Saccharomyces cerevisiae virus L-A	ScV-L-A	4579	76	171[b]	NC_003745
Saccharomyces cerevisiae virus L-BC	ScV-L-BC	4615	78	176[b]	NC_001641
Ustilago maydis virus H1	UmV-H1	6099	81	201[b]	NC_003823
Genus: *Victorivirus*					
Helminthosporium victoriae 190S virus	Hv190SV	5179	81	91	NC_003607
Chalara elegans RNA virus 1	CeRV1	5310	81	96	NC_005883
Coniothyrium minitans RNA virus	CmRV	4975	81	93	NC_007523
Epichloe festucae virus 1	EfV-1	5109	80	90	AM261427
Gremmeniella abietina RNA virus L	GaRV-L	5133	81	90	NC_003876
Helicobasidium mompa totivirus 1-17	HmTV-1-17	5207	83	93	NC_005074
Magnaporthe grisea virus 1	MgV-1	5359	77	91	NC_006367
Sphaeropsis sapinea RNA virus 1	SsRV-1	5163	89	92	NC_001963
Sphaeropsis sapinea RNA virus 2	SsRV-2	5202	83	91	NC_001964
Genus: *Giadiavirus*					
Giardia lamblia virus	GLV	6277	98	210[b]	NC_003555
Trichomonas vaginalis virus 1	TVV-1	4648	74	160[b]	U57898
Trichomonas vaginalis virus 2	TVV-2	4674	79	162[b]	NC_003873
Trichomonas vaginalis virus 3	TVV-3	4844	79	156[b]	NC_004034
Genus: *Leishmaniavirus*					
Leishmania RNA virus 1–1	LRV1-1	5284	82	98	NC_002063
Leishmania RNA virus 1–4	LRV1-4	5283	83	99	NC_003601
Leishmania RNA virus 2–1	LRV2-1	5241	78	88	NC_002064
Unclassified					
Penaeid shrimp myonecrosis virus	IMNV	7560	99	196[b]	NC_007915
Eimeria brunetti RNA virus 1	EbRV	5358	83	98	NC_002701

[a]The names of the ICTV-recognized virus species are written in italics.
[b]RdRp is expressed or proposed to be expressed as a CP–RdRp fusion protein.

Virion Structure and Composition

The totiviruses have isometric particles, approximately 40 nm in diameter, with icosahedral symmetry (**Figure 1**). The capsids are single-shelled and encompass a single major polypeptide. The capsids consist of 120 CP subunits of molecular mass in the range of 76–98 kDa. The capsid structures of three members of the family *Totiviridae* have been determined, at least one (the yeast L-A virus) at near atomic resolution using X-ray crystallography, and the other two (UmV-H1 and Hv190SV) at moderate resolutions (~1.4 nm) using cryo-transmission electron microscopy combined with three-dimensional image reconstruction. In all cases, the capsids of the fungal totiviruses are made up of 60 asymmetric CP dimers arranged in a 'T = 2' layer. Compared to the yeast L-A capsid, the Hv190SV capsid shows relatively smoother outer surfaces. The quaternary organization of the Hv190SV particle, however, is remarkably similar to the yeast L-A and the cores of the larger dsRNA viruses

of plants and animals: the A-subunits cluster around the fivefold axis and B-subunits around the threefold axis.

The ScV-L-A CP removes the 5' cap structure of host mRNA and covalently attaches it to the histidine residue at position 154. The decapping activity is required for efficient translation of viral RNA. The published yeast L-A capsid structure reveals a trench at the active site of decapping. The decapping activity has yet to be demonstrated for any other totivirus CPs.

Although a single gene encodes the capsid of Hv190SV, like other totiviruses, the Hv190SV capsid comprises two closely related major CPs, either p88 and p83 or p88 and p78. The capsids of all other totiviruses so far characterized appear to contain only a single major CP. Interestingly, HmTV-1-17, a totivirus infecting a filamentous fungus, has similar capsid heterogeneity to that of Hv190SV. It would be of interest to determine whether capsid heterogeneity and post-translational modification (phosphorylation and proteolytic processing) of the primary CP is a common feature of totiviruses infecting filamentous fungi (members of the tentative genus *Victorivirus*). Purified Hv190S virion preparations contain two types of particles, 190S-1 and 190S-2, which differ slightly in sedimentation rates (190S-1 is resolved as a shoulder on the slightly faster sedimenting component 190S-2) and capsid composition. The 190S-1 and 190S-2 virions are believed to represent different stages in the virus life cycle. The 190S-1 capsids contain p88 and p83, occurring in approximately equimolar amounts, and the 190S-2 capsids comprise similar amounts of p88 and p78. p88 and p83 are phosphoproteins, whereas p78 is nonphosphorylated (**Figure 2**). Totivirus virions encapsidate a single molecule of dsRNA, 4.6–6.3 kbp in size. Some totiviruses may additionally contain satellite dsRNAs or defective dsRNAs, which are encapsidated separately in capsids encoded by the totivirus genome.

Genome Organization and Expression

In general, the genome organization of the totiviruses infecting fungi and protozoa are similar: each virus genome contains two large open reading frames (ORFs); the 5′ proximal ORF encodes a CP and the 3′ ORF encodes an RdRp (see the genome organization of Hv190SV as an example; **Figure 3**). Except for LRV2-1 and UmV-H1, the RdRp ORF overlaps the CP ORF and is

in the −1 frame (ScV-L-A, ScV-L-BC, Hv190SV, TVV-2, and GLV) or in the +1 frame (LRV1-1 and LRV1-4) with respect to the CP ORF. The RdRp ORF of LRV2-1 does not overlap the CP ORF, and is separated from it by a stop codon. The Umv-H1 genome contains only a single ORF that encodes a polyprotein that is predicted to be auto-catalytically processed by a viral papain-like protease to generate the CP and RdRp proteins.

The totiviruses express their RdRps either as CP–RdRp (gag-pol-like) fusion proteins or as separate nonfused proteins. Expression of RdRp as a CP–RdRp fusion protein via −1 ribosomal frameshifting has been well documented for ScV-L-A. Virion-associated CP–RdRp has been detected as a minor protein in the capsids of ScV-L-A, ScV-L-BC, TVV, and GLV. Although CP–RdRp fusion proteins were neither detected *in vivo* nor associated with virions of LRV1-1, LRV1-4, or LRV2-1, expression of RdRp as a fusion protein by +1 ribosomal frameshifting or ribosomal hopping (LRV2-1) has been proposed.

The overlap region between ORF1 and ORF2 of ScV-L-A (130 nt), LRV1-1 and LRV1-4 (71 nt), and

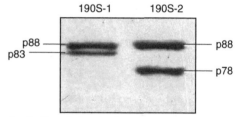

Figure 2 Sodium dodecyl sulfate-polyacrylamide gel electrophoresis (SDS-PAGE) analysis of Hv190SV sedimenting components 190S-1 and 190S-2. Purified preparations of Hv190SV contain two types of particles, 190S-1 and 190S-2, which differ slightly in sedimentation rates. The 190S-1 capsids contain p88 and p83, occurring in approximately equimolar amounts, and the 190S-2 capsids comprise similar amounts of p88 and p78. The capsid proteins p88 and p83 are phosphorylated, whereas p78 is nonphosphorylated.

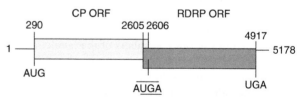

Figure 3 Genome organization of *Helminthosporium victoriae 190S virus*, the type species of the newly proposed genus *Victorivirus*. The dsRNA genome encompasses two large overlapping open reading frames (ORFs) with the 5′ ORF encoding a capsid protein (CP) and the 3′ ORF encoding an RNA-dependent RNA polymerase (RdRp). Note that the termination codon of the CP ORF overlaps the initiation codon of the RdRp ORF in the tetranucleotide sequence AUGA. Adapted from Ghabrial SA and Patterson JL (1999) *Encyclopedia of Virology*, 2nd edn., pp. 1808–1812. New York: Academic Press, with permission from Elsevier.

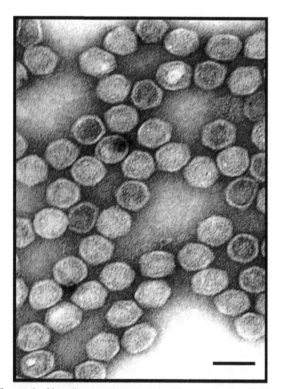

Figure 1 Negative contrast electron micrograph of particles of an isolate of *Helminthosporium victioriae 190SV*, the type species of the newly proposed genus *Victorivirus*. Scale = 50 nm.

GLV (122 nt) contains the structures necessary for ribosomal frameshifting including a slippery site and a pseudoknot structure to promote fusion of ORF1 and ORF2 *in vivo*. Although the overlap region in TVV is short (14 nt), it contains a potential ribosomal slippage heptamer.

The overlap regions in the dsRNA genomes of Hv190SV and Hv190S-like viruses (members of the newly proposed genus *Victorivirus*), on the other hand, are of the AUGA type where the initiation codon of the RdRp ORF overlaps the termination codon of the CP ORF (**Figure 3**), suggesting that expression of RdRp occurs by a mechanism different from translational frameshifting. For Hv190SV, the stop codon of the CP ORF (nucleotide position 2606-UGA-2608) overlaps with the start codon (nucleotide position 2605-AUG-2607) for the RdRp ORF in the sequence AUGA (**Figure 3**). The complete nucleotide sequences of Hv190SV and eight other putative members of the family *Totiviridae* infecting filamentous fungi (members of the genus *Victorivirus*) have been reported and the sequences deposited in the GenBank, with Hv190SV being biochemically and molecularly best characterized of these (**Table 1**).

The 5′ end of the plus strand of all totiviruses dsRNAs examined to date is uncapped and the 3′ end is not poly-adenylated. The 5′ end of Hv190SV dsRNA is uncapped and highly structured and contains a relatively long (289 nucleotides) 5′ leader with two minicistrons. These structural features of the 5′ untranslated region (UTR) of Hv190SV dsRNA suggest that the CP-encoding ORF1 (with its AUG present in suboptimal context according to the Kozak criteria) is translated via a cap-independent mechanism. The 5′-UTR of the leishmaniavirus LRV-1-1 functions as an internal ribosome entry site (IRES). Translation of the uncapped giardiavirus GLV mRNA in *Giardia lamblia* is initiated on a unique IRES element that contains sequences from a part of the 5′-UTR and a portion of the capsid coding region.

The UGA codon at position 2606–2608 of the Hv190SV genomic plus strand was verified by site-directed mutagenesis to be the authentic stop codon for ORF1 (**Figure 3**). The RdRp-encoding downstream ORF2 of Hv190SV is in a −1 frame with respect to ORF1 (**Figure 3**) and is expressed via an internal initiation mechanism (a coupled termination–initiation mechanism is proposed). The Hv190SV RdRp is detectable as a separate, virion-associated component, consistent with its independent translation from ORF2. The tetranucleotide AUGA overlap region, or a very similar structure, is characteristic of the overlap region of all putative members of the totivirus family that infect filamentous fungi (victoriviruses).

The initial report on the molecular characterization of IMNV, a monosegmented dsRNA virus infecting penaeid shrimp and tentatively assigned to the family *Totiviridae*, concluded that the viral genome encompasses two nonoverlapping ORFs with ORF1 encoding a polyprotein comprised of a putative RNA-binding protein and a CP. The coding region of the RNA-binding protein was located in the first half of ORF1 and contained a dsRNA-binding motif in the first 60 amino acids. The second half of ORF1 encoded a CP, as determined by amino acid sequencing, with a molecular mass of 99 kDa. ORF2 encoded a putative RdRp with the eight conserved motifs characteristic of totiviruses. Phylogenetic analysis based on the RdRp clustered IMNV with GLV, the type species of the genus *Giardiavirus* in the family *Totiviridae*. Important novel features of the genome organization of IMNV, however, were most recently uncovered that has significant bearing on how the viral proteins are expressed. These features include two encoded '2A-like' motifs, which are likely involved in ORF1 polyprotein 'cleavage', a 199 nt overlap between ORF1 and ORF2, and the presence a 'slippery heptamer' motif and predicted RNA pseudoknot in the region of ORF1–ORF2 overlap. The latter features probably allow ORF2 to be translated as a fusion with ORF1 by '−1' ribosomal frameshifting. Although the generation of CP as a polyprotein and the potential involvement of encoded '2A-like' peptides (GDVESNPGP and GDVEENPGP) in processing of the polyprotein to release the major CP represent novel features not shared by totiviruses, the potential expression of RdRp as a CP–RdRp fusion protein via ribosomal frameshifting is a common strategy utilized by totiviruses for expressing their RdRps. Experimental evidence for the presence of the putative CP–RdRp fusion protein in purified virions or in infected tissues, however, is lacking.

Virus Replication Cycle

Limited information is available on the replication cycle of totiviruses and has mainly been derived from *in vitro* studies of virion-associated RNA polymerase activity and the isolation of particles representing various stages in the replication cycle. In *in vitro* reactions, the RNA polymerase activity associated with virions of the fungal totiviruses ScV-L-A, UmV-H1, and Hv190SV, isolated from lag-phase cultures, catalyzes end-to-end transcription of dsRNA, by a conservative mechanism, to produce mRNA for CP, which is released from the particles. Purified ScV-L-A virions, isolated from log-phase cells, contain a less-dense class of particles, which package only plus-strand RNA. In *in vitro* reactions, these particles exhibit a replicase activity that catalyzes the synthesis of minus-strand RNA to form dsRNA. The resultant mature particles, which attain the same density as that of the dsRNA-containing virions isolated from the cells, are capable of synthesizing and releasing plus-strand RNA.

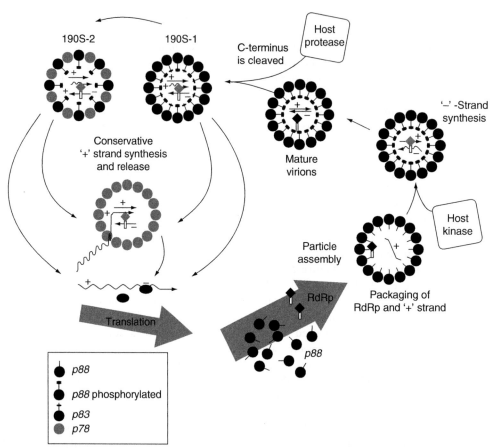

Figure 4 Life cycle of Hv190SV. Mature virions contain a single dsRNA molecule and their capsids are composed entirely or primarily of the capsid protein (CP) p88. Virions representing different stages of the virus life cycle can be purified from the infected fungal host *Helminthosporium victoriae* including the well-characterized 190S-1 and 190S-2 virions. These two types of virions differ in sedimentation rate, phosphorylation state, and CP composition; 190S-1 capsids contain p88 and p83, whereas the 190S-2 capsids contain p88 and p78 (p88 is the primary translation product of the CP gene; p83 and p78 represent post-translational proteolytic processing products of p88 at its C-terminus). p88 and p83 are phosphorylated, whereas p78 is nonphosphorylated. The virions with phosphorylated CPs (p88+p83) have significantly higher transcriptase activity *in vitro* than those containing the nonphosphorylated p78. Transcription occurs conservatively and the newly synthesized plus-strand RNA is released from the virions. Phosphorylation of CP is catalyzed by a host kinase, and is proposed to play a regulatory role in transcription/replication. A host-encoded protease catalyzes the proteolytic processing of phosphorylated p88; this occurs in two steps, leading first to p83 (the generation of the 190S-1 virions) and then to p78 (190S-2 virions). The conversion of p88-p83-p78 is proposed to play a role in the release of the plus-strand RNA transcripts from virions. The released plus-strand RNA is the RNA that is translated into CP and RNA-dependent RNA polymerase (RdRp) and packaged in capsids assembled from the primary translation product p88. It is not known whether p88 is phosphorylated before or after assembly. Synthesis of minus-strand RNA occurs on the plus-strand RNA template inside the virion; phosphorylation may be involved in turning on the replicase activity. Adapted from Ghabrial SA and Patterson JL (1999) *Encyclopedia of Virology*, 2nd edn., pp. 1808–1812. New York: Academic Press, with permission from Elsevier.

A proposed life cycle of Hv190SV is depicted in **Figure 4**. Host-encoded protein kinase and protease have been shown to be involved in post-translational modification of CP. Phosphorylation and proteolytic processing are proposed to play a role in the virus life cycle; phosphorylation of CP may be necessary for its interaction with viral nucleic acid and/or phosphorylation may regulate dsRNA transcription/replication. Proteolytic processing and cleavage of a C-terminal peptide, which leads to dephosphorylation and the conversion of p88 to p78, may play a role in the release of the plus-strand RNA transcripts from virions (**Figure 4**).

Biological Properties

There are no known natural vectors for the transmission of the fungal totiviruses. They are transmitted intracellularly during cell division, sporogenesis, and cell fusion. Although the yeast totiviruses are effectively transmitted via ascospores, the totiviruses infecting the ascomycetous filamentous fungi are essentially eliminated during ascospore formation. The leishmaniaviruses are not infectious as purified virions and are propagated during cell division. The giardiavirus GLV and the unclassified IMNV, on the other hand, are infectious as purified virions. Successful transfection of the protozoa *Giardia lamblia* has also been accomplished via electroporation with plus-strand RNA transcribed *in vitro* from GLV dsRNA. Whereas IMNV causes a myonecrosis disease in its crustacean host, GLV is associated with latent infection of its flagellated protozoan human parasite *G. lamblia*. GLV is released into the medium without lysing the host cells and the extruded virus can infect many virus-free isolates of the protozoan host.

There are no known experimental host ranges for the fungal totiviruses because of the lack of suitable infectivity assays. As a consequence of their intracellular modes of transmission, the natural host ranges of fungal totiviruses are limited to individuals within the same or closely related vegetative compatibility groups. Furthermore, mixed infections with two or more unrelated viruses are common, probably as a consequence of the ways by which fungal viruses are transmitted in nature. Dual infection of yeast with ScV-L-A and ScV-L-BC and the filamentous fungus *Sphaeropsis sapinea* with SsRV-1 and SsRV-2 are examples of mixed infections involving totiviruses. Dual infection of *H. victoriae* with Hv190SV and the chrysovirus Hv145SV is an example of a mixed infection involving two unrelated viruses belonging to two different families, a totivirus and a chrysovirus.

With the exception of IMNV, an unclassified virus tentatively assigned to the family *Totiviridae*, which causes a myonecrosis disease in shrimp, totiviruses are generally associated with symptomless infections of their hosts. A possible exception to this rule is Hv190SV, since mixed infections with Hv190SV and Hv145SV are associated with a debilitating disease of the fungal host. The disease phenotype and the two viruses are transmitted via hyphal anastomosis and diseased isolates are characterized by reduced growth, excessive sectoring, aerial mycelial collapse, and generalized lysis (**Figure 5**). The disease phenotype of *H. victoriae*, the causal agent of Victoria

blight of oats, is of special interest not only because diseased isolates are hypovirulent, but also because of the probable viral etiology. The disease phenotype was also transmitted by incubating fusing protoplasts from virus-free fungal isolates with purified virions containing both the Hv190S and 145S viruses. The frequency of infection and stability of the newly diseased colonies, however, were very low, and verification of transmission was based on virus detection by immune electron microscopy. A systematic molecular approach based on DNA-mediated transformation of a susceptible virus-free *H. victoriae* isolate with full-length cDNA clones of viral dsRNAs is currently being pursued to verify the viral etiology of the disease phenotype and elucidate the roles of the individual viruses in disease development. A transformation vector for *H. victoriae* based on the hygromycin B resistance marker was constructed and used to transform an *H. victoriae* virus-free isolate with a full-length cDNA clone of Hv190SV dsRNA. The Hv190SV cDNA was inserted downstream of a *Cochliobolus heterostrophus GPD1* promoter and upstream of an *Aspergillus nidulans trpC* terminator signal. The hygromycin-resistant transformants expressed the Hv190SV CP, as determined by Western-blotting analysis. Transformation of a normal virus-free fungal isolate with a full-length cDNA of Hv190SV dsRNA conferred a disease phenotype. Symptom severity varied among the transformants from symptomless to severely stunted and highly sectored (**Figure 6**). Symptom severity correlated well with the level of viral capsid accumulation. Integration of the viral genome in the host DNA was verified by Southern blot hybridization analysis and expression of the viral dicistronic mRNA was demonstrated by northern analysis. Like natural infection, the primary translation product of the *CP* gene (p88) was phosphorylated and proteolytically processed to generate p83 and p78. The CP assembled into virus-like particles indistinguishable in appearance from

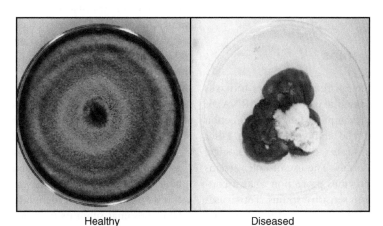

Healthy   Diseased

Figure 5 Colony morphology of virus-free (healthy), virus-infected (diseased) isolates of *Helminthosporium victoriae*. The colonies were grown for a week on potato dextrose agar medium at room temperature. The healthy colony shows uniform mycelial growth that extends to the entire plate, whereas the diseased colony is stunted and shows excessive sectoring.

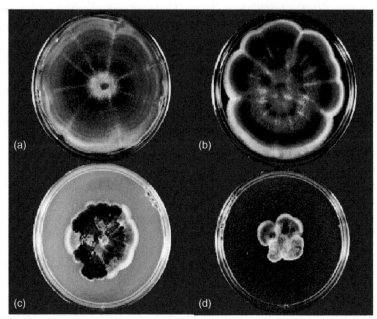

Figure 6 Colony morphology of a virus-free *Helminthosporium victoriae* isolate transformed with a full-length cDNA clone of the totivirus Hv190SV dsRNA. Colonies transformed with vector alone (a) or with recombinant vector containing full-length cDNA to Hv190SV dsRNA (b–d) were grown for a week at room temperature. Note that colonies (b)–(d) show increasingly more severe symptoms (reduced growth and sectoring).

empty capsids normally associated with virus infection. The empty capsids accumulated to significantly higher levels than in natural infections. Although the RdRp was expressed and packaged, no dsRNA was detected inside the virus-like particles or in the total RNA isolated from mycelium. Despite inability to launch dsRNA replication from the integrated viral cDNA, the demonstration that the ectopic 190SV cDNA copies were transcribed and translated and that the resultant transformants exhibited a disease phenotype provides convincing evidence for a viral etiology for the disease of *H. victoriae*. The role of the Hv145SV in disease development, however, has yet to be elucidated.

Virus–Host Relationships

The yeast killer system, comprised of a helper totivirus (ScV-L-A) and associated satellite dsRNA (M-dsRNA), is one of the very few known examples where virus infection is beneficial to the host. The ability to produce killer toxins by immune yeast strains confers an ecological advantage over sensitive strains. The use of killer strains in the brewing industry provides protection against contamination with adventitious sensitive strains. Totiviruses maintain only the genes that are essential for their survival (RdRp and CP), but make efficient use of host proteins. The host cells have evolved to support only a defined level of virus replication, beyond which virus

infection may become pathogenic. Because of amenability to genetic studies, the yeast–virus system has provided significant information on the host genes required to prevent viral cytopathology. A system of six chromosomal genes, designated superkiller (or SKI) SKI2, SKI3, SKI4, SKI6, SKI7, and SKI8, negatively control the copy number of the totivirus ScV-L-A and its satellite M-dsRNAs. The only crucial function of these genes is to block virus multiplication. Mutations in any of these SKI genes lead to the development of the superkiller phenotype as a consequence of the increased copy number of M-dsRNA. The SKI genes affect primarily the initiation of translation rather than the stability of mRNA, and are thus part of a cellular system that specifically blocks translation of nonpolyadenylated mRNAs (like the plus-strand transcripts of totiviruses). About 30 chromosomal genes, termed MAK genes (for maintenance of killer), are required for stable replication of the satellite M-dsRNA. Only three of these MAK genes are necessary for the helper virus (ScV-L-A) multiplication. Mutants defective in any of 20 MAK genes show a decreased level of free 60S ribosomal subunits. Since the *mak* mutations affecting 60S subunit levels are suppressed by *ski* mutations, and since the latter act by blocking translation of nonpolyadenylated mRNAs, the level of 60S ribosomal subunits is critical for translation of non-polyadenylated mRNAs.

The Hv190S totivirus that infects the plant pathogenic fungus *H. victoriae* utilizes host-encoded proteins (a protein kinase and a protease) for post-translational modification of its CP. Phosphorylation and proteolytic processing of CP may play a role in regulating transcription and the release of plus-strand transcripts from virions (**Figure 4**).

The *H. victoriae*-virus system is well characterized and provides a useful model system for studies on virus–host interactions in a plant pathogenic fungus. A major attribute of this system is the fact that the virus-infected *H. victoriae* isolates exhibit a disease phenotype (**Figure 5**), which is rare among fungal viruses. Modulation of fungal gene expression and alteration of phenotypic traits as a consequence of mycovirus infection are little understood, with the exception of the chestnut blight fungus–hypovirus system. It was previously demonstrated that the fungal gene *Hv-p68* is upregulated as a result of virus infection and proposed that upregulation of this gene might play a role in virus pathogenesis (**Figure 7**). Hv-p68, a novel alcohol oxidase/RNA-binding protein, belongs to the large family of FAD-dependent GMC oxidoreductases with 67–70% sequence identity to the alcohol oxidases of methylotrophic

yeasts. Hv-p68, however, shows only limited methanol-oxidizing activity and its expression is not induced in cultures supplemented with methanol as the sole carbon source (**Figure 7**). The natural substrate for Hv-p68 is not known, but the structurally similar alcohol oxidases are known to oxidize primary alcohols irreversibly to toxic aldehydes. Overexpression of Hv-p68 and putative accumulation of toxic intermediates was proposed as a possible mechanism underlying the disease phenotype of virus-infected *H. victoriae* isolates. Overexpression of Hv-p68 in virus-free fungal isolates, however, resulted in a significant increase in colony growth and did not induce a disease phenotype. Thus, overexpression of Hv-p68 *per se* is not sufficient to induce the disease phenotype in the absence of virus infection. If the function of *Hv-p68* is similar to that of the homologous alcohol oxidases of methylotrophic yeasts or other filamentous fungi, that is, irreversible oxidation of primary alcohols into aldehydes, then overexpression of Hv-p68 could lead to an accumulation of toxic aldehydes and development of disease phenotype. The finding that colonies overexpressing the Hv-p68 protein did not exhibit the disease phenotype and grew more rapidly than the nontransformed wild-type suggests that accumulation of toxic aldehydes did not occur, and that such aldehydes were probably assimilated into carbohydrates via the xylulose monophosphate pathway.

Hv-p68, which co-purifies with viral dsRNA (mainly that of Hv145SV), is a multifunctional protein with alcohol oxidase, protein kinase, and RNA-binding activities. Hv-p68 is detectable as a minor component of Hv190SV capsid. Recent evidence strongly suggests that Hvp68 is the cellular protein kinase responsible for phosphorylation of the capsid proteins. The RNA-binding activity of Hv-p68 has been demonstrated by gel mobility shift and northwestern blot analysis. Furthermore, the RNA-binding domain of Hv-p68 was mapped to the N-terminal region that contains the ADP-binding domain. Because dsRNA-binding proteins are known to sequester dsRNA and suppress antiviral host defense mechanisms, it is feasible that overexpression of the dsRNA-binding protein Hv-p68 may lead to the induction of the disease phenotype by suppressing host defense. This idea is consistent with the finding that overexpression of Hv-p68 led to enhancement in the accumulation of Hv145S dsRNA. The role of Hv145SV in the development of the disease phenotype, however, is not yet clear. Furthermore, it is curious that Hvp68 co-purifies with Hv145S dsRNA since the latter is predicted to be confined to the viral capsids where virus replication takes place. Recent results, however, suggest that a significant proportion of the Hv145V dsRNA does not appear to be encapsidated. Considering the multifunctional nature of the Hv-p68 protein, additional studies are needed to determine whether or not Hv-p68 upregulation has a role in viral pathogenesis.

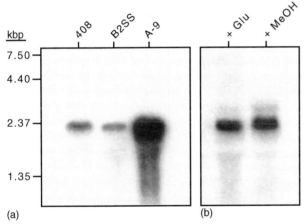

(a)   (b)

Figure 7 Northern analysis of Hv-p68 mRNA transcript levels. (a) Total RNA (25 μg) isolated from the virus-free isolates 408 and B-2ss and the virus-infected isolate A-9 were electrophoresed on a formaldehyde/ agarose gel, blotted, and hybridized under high stringency conditions with a radiolabeled probe for Hv-p68. Hv-p68 mRNA, 2.25 kbp in size, was detected in RNA samples from all three *H. victoriae* isolates. The level of Hv-p68 transcript in cultures of the virus-infected isolate A-9, however, was at least 15-fold higher than that for the virus-free isolates. (b) Total RNA (30 μg) isolated from cultures of the virus-free isolate B-2ss grown for 3 days in minimal medium supplemented with either glucose (Glu) or methanol (MeOH). Similar amounts of Hv-p68 mRNA were detected in total RNA isolated from fungal cultures supplemented with either glucose or methanol, as a carbon source. Reproduced from Soldevila AI and Ghabrial SA (2001) A novel alcohol oxidase/RNA-binding protein with affinity for mycovirus double-stranded RNA from the filamentous fungus *Helminthosporium* (*Cochliobolus*) *victoriae*. *Journal of Biological Chemistry* 276: 4652–4661, with permission from the American Society for Biochemistry and Molecular Biology, Inc.

Evolutionary Relationships among Totiviruses

Sequence comparison analysis of the predicted amino acid sequences of totivirus RdRps indicated that they share significant sequence similarity and characteristically contain eight conserved motifs. This sequence similarity was common to all totiviruses so far characterized including the totiviruses that infect the yeast, smut, and filamentous fungi, as well as those infecting parasitic protozoa. As indicated earlier, the viruses infecting filamentous fungi (members of the newly proposed genus *Victorivirus*) express their RdRp separate from the CP by an internal initiation mechanism, whereas the other members of the family *Totiviridae* express their RdRps as fusion proteins (CP–RdRp), mainly via a ribosomal frameshifting mechanism. Phylogenetic analysis based on multiple alignments of amino acid sequences of

totivirus RdRp conserved motifs (**Figure** 8) reflects these differences as the viruses infecting filamentous fungi (victoriviruses) are most closely related to each other and form a distinct large well-supported cluster (bootstrap value of 94%). Likewise, phylogenetic trees based on the CP sequences showed similar topology to those based on RdRp sequences (**Figure** 9). The fact that independent alignments of CP and RdRp sequences give similar phylogenetic relationships (**Figures** 8 and 9) further supports the creation of a new genus to accommodate the viruses infecting filamentous fungi.

The leishmaniaviruses LRV1, LRV2, and LRV4 (percent sequence identities of 46%) are most closely related to each other and form a discrete cluster with 100% bootstrap value. This is also true for the two yeast viruses ScV-L-A and ScV-LBC (RdRp identity of 32%), members of the genus *Totivirus*. UmV-H1, the third member in

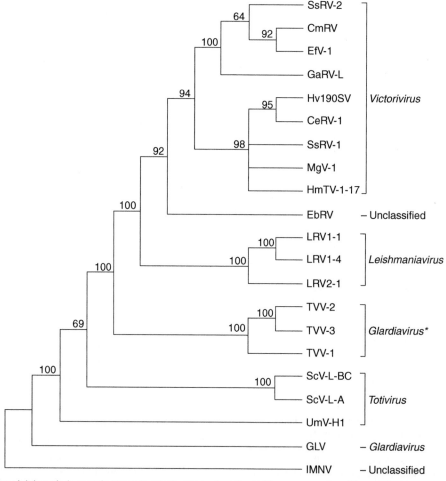

Figure 8   Neighbor-joining phylogenetic tree constructed based on the RdRp conserved motifs and flanking sequences. The RdRp sequences were derived from aligned deduced amino acid sequences of members of the family *Totiviridae* using the program CLUSTAL X. Motifs 1–8 and the sequences between the motifs, as previously designated by Ghabrial SA in 1998, were used. See **Table 1** for virus name abbreviations and GenBank accession numbers. The phylogenetic tree was generated using the program PAUP*. Bootstrap numbers out of 1000 replicates are indicated at the nodes. The tree was rooted with the RdRp of the penaeid shrimp infectious myonecrosis virus (IMNV), an unclassified virus tentatively assigned to the family *Totiviridae*, which was included as an outgroup.

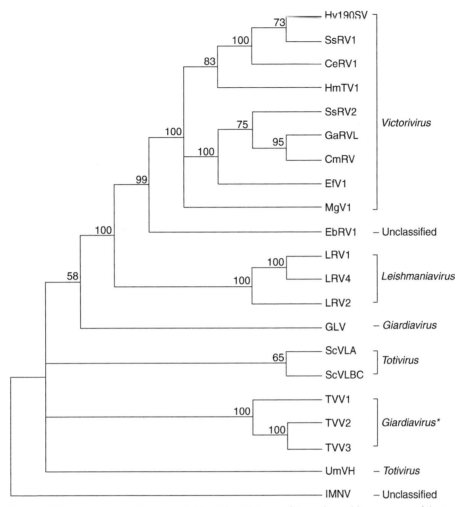

Figure 9 Neighbor-joining phylogenetic tree constructed based on the complete amino acid sequences of the capsid proteins of viruses in the family *Totiviridae*. The CP multiple sequence alignment was performed with the program CLUSTAL X and the phylogenetic tree was generated using the program PAUP*. See **Table 1** for virus name abbreviations and GenBank accessions numbers. Bootstrap numbers out of 1000 replicates are indicated at the nodes. The tree was rooted with the CP of the penaeid shrimp infectious myonecrosis virus (IMNV), an unclassified virus tentatively assigned to the family *Totiviridae*, which was included as an outgroup.

the genus *Totivirus* forms a phylogenetic clade distinct from the cluster of the two yeast viruses. This may reflect the difference in RdRp expression strategy between these viruses since UmV-H1 expresses its RdRp via the generation of a polyprotein followed by proteolytic processing to release the viral protein, whereas the yeast viruses express their RdRps as CP–RdRp fusion proteins via a −1 ribosomal frameshifting. It is of interest that Hv190SV and related viruses (victoriviruses) are phylogenetically more closely to the leishmaniaviruses than to the yeast viruses (**Figures 8** and **9**). It was previously hypothesized that leishmaniaviruses and the fungal totiviruses are of old origin having existed in a single-cell-type progenitor prior to the divergence of fungi and protozoa. The finding that the unclassified Eimeria brunetti RNA virus 1 (EbRV-1), which infects an apicomplexan parasitic protozoa, is more closely related to members of the new genus *Victorivirus* than to protozoal viruses is consistent with this idea.

EbRV-1 may represent an intermediate progenitor in the evolution of 'victoriviruses'.

It is worth noting that the classification of TVV1, TVV2, and TVV4 as tentative members of the genus *Giardiavirus* is not supported by phylogenetic analyses based on the totivirus CPs or RdRps since they do not cluster with GLV, the type species of the genus *Giardiavirus*. The trichomonas vaginalis viruses are most closely related to each other as they form a distinct phylogenetic cluster with 100% bootstrap support (**Figures 8** and **9**). It would probably be justifiable to create a new genus in the family totiviridae to accommodate these viruses.

Taxonomic considerations of totiviruses may benefit greatly from elucidating the capsid structure of representatives of the genera *Leishmaniavirus* and *Giardiavirus* (including the tentative members TVVs). Furthermore, resolving the capsid structure of the unclassified viruses EbRV-1 and IMNV may prove useful for their taxonomic placement. Information on the genome organization and RdRp expression strategy of EbRV-1 is lacking and such information would be crucial in determining whether a novel genus in the family *Totiviridae* should be created for EbRV-1. IMNV has some features that distinguishes it from totiviruses including that it expresses its CP as a polyprotein that undergoes autoproteolytic processing. It is likely that IMNV would be classified as member of a novel family of dsRNA viruses that infect invertebrate hosts.

See also: Fungal Viruses; Ustilago Maydis Viruses; Yeast L-A Virus.

Further Reading

Castón JR, Luque D, Trus BI, *et al.* (2006) Three-dimensional structure and stoichiometry of *Helminthosporium victoriae190S* totivirus. *Virology* 347: 323–332.

Cheng J, Jiang D, Fu Y, Li G, Peng Y, and Ghabrial SA (2003) Molecular characterization of a dsRNA totivirus infecting the sclerotial parasite *Coniothyrium minitans*. *Virus Research* 93: 41–50.

Ghabrial SA (1998) Origin, adaptation and evolutionary pathways of fungal viruses. *Virus Genes* 16: 119–131.

Ghabrial SA and Patterson JL (1999) *Encyclopedia of Virology*, 2nd edn., pp. 1808–1812. New York: Academic Press.

Ghabrial SA, Soldevila AI, and Havens WM (2002) Molecular genetics of the viruses infecting the plant pathogenic fungus *Helminthosporium victoriae*. In: Tavantzis S (ed.) *Molecular Biology of Double-Stranded RNA: Concepts and Applications in Agriculture, Forestry and Medicine*, pp. 213–236. Boca Raton, FL: CRC Press.

Huang S and Ghabrial SA (1996) Organization and expression of the double-stranded RNA genome of *Helminthosporium victoriae* 190S virus, a totivirus infecting a plant pathogenic filamentous fungus.

Proceedings of the National Academy of Sciences, USA 93: 12541–12546.

Icho T and Wickner RB (1989) The double-stranded RNA genome of yeast virus L-A encodes its own putative RNA polymerase by fusing two open reading frames. *Journal of Biological Chemistry* 264: 6716–6723.

Nibert ML (2007) '2A-like' and 'shifty heptamer' motifs in penaeid shrimp infectious myonecrosis virus, a monosegmented double-stranded RNA virus. *Journal of General Virology* 88: 1315–1318.

Nomura K, Osaki H, Iwanami T, Matsumoto N, and Ohtsu Y (2003) Cloning and characterization of a totivirus double-stranded RNA from the plant pathogenic fungus, *Helicobasidium mompa* Tanaka. *Virus Genes* 23: 219–226.

Park Y, James D, and Punja ZK (2005) Co-infection by two distinct totivirus-like double-stranded RNA elements in *Chalara elegans* (*Thielaviopsis basicola*). *Virus Research* 109: 71–85.

Preisig O, Wingfield BD, and Wingfield MJ (1998) Coinfection of a fungal pathogen by two distinct double-stranded RNA viruses. *Virology* 252: 399–406.

Romo M, Leuchtmann A, and Zabalgogeazcoa I (2007) A totivirus infecting the mutualistic fungal endophyte *Epichloé festucae*. *Virus Research* 124: 38–43.

Soldevila A and Ghabrial SA (2000) Expression of the totivirus Helminthosporium victoriae190S virus RNA-dependent RNA polymerase from its downstream open reading frame in dicistronic constructs. *Journal of Virolology* 74: 997–1003.

Soldevila AI and Ghabrial SA (2001) A novel alcohol oxidase/RNA-binding protein with affinity for mycovirus double-stranded RNA from the filamentous fungus *Helminthosporium* (*Cochliobolus*) *victoriae*. *Journal of Biological Chemistry* 276: 4652–4661.

Tuomivirta TT and Hantula J (2005) Three unrelated viruses occur in a single isolate of *Gremmeniella abietina* var. *abietina* type A. *Virus Research* 110: 31–39.

Wickner RB, Wang CC, and Patterson JL (2005) Totiviridae. In: Fauquet CM, Mayo MA, Maniloff J, Desselberger U,, and Ball LA (eds.) *Virus Taxonomy: Eighth Report of the International Committee on Taxonomy of Viruses*, pp. 581–590. San Diego, CA: Elsevier Academic Press.

Ustilago Maydis Viruses

J Bruenn, State University of New York, Buffalo, NY, USA

Glossary

Kex2p. Kexin, furin A membrane-bound proteinase in the Golgi that processes some secreted proteins in many eukaryotes.

Ustilago maydis Smut fungus. A plant pathogen infecting maize, wheat, oats, and barley that can adopt either yeast or mycelial form.

Introduction

The *Ustilago maydis* viruses belong to the genus *Totivirus* within the family *Totiviridae*. They are a related group of viruses that, like many fungal and protozoan viruses, exist as permanent passengers in their host cells, to which they are not deleterious. These are double-stranded RNA (dsRNA) viruses with similar structure and life cycle, derived from a very ancient ancestor but still possessing recognizable sequence similarities in their common protein, the viral RNA-dependent RNA polymerase (RdRp). They are not naturally infectious and are passed from cell to cell by meiosis or mitosis, although they will, inefficiently, infect protoplasts. In some cases, they provide a selective advantage to the host, encoding cellular toxins that kill cells of the same or related species that lack the virus. These killer toxins are apparently derived from cellular genes that have been co-opted by resident viruses. In some cases (as discussed below) cellular ancestors of these toxins still exist. When multiple species of genomic and satellite dsRNA exist in the fungal and protozoan dsRNA viruses, they are separately encapsidated. This is no disadvantage to viruses that do not depend on an infectious cycle for propagation.

Taxonomy

The fungal and protozoan dsRNA viruses generally fall into three groups: the totiviruses, the chrysoviruses, and the partitiviruses, all of which are closely related as judged by the sequence of their RdRps (**Figure 1**). In this scheme, only the giardiaviruses (which are presently classified as a separate genus within the family *Totiviridae*) are, rather inappropriately, marooned between the chrysoviruses and the partitiviruses. However, the *Ustilago maydis* viruses are placed closest to the *Saccharomyces* viruses, with which they have much in common. The three different *Ustilago maydis* viruses (UmVP1H1, UmVP6H1, and UmVP1H2) for which sequence information exists are more closely related to each other than to any other fungal viruses. This is now a common phenomenon among the fungal viruses, in which several similar dsRNA viruses may share the same host cell: ScV-L1 and ScV-La (*Saccharomyces cerevisiae*), DdV1 and DdV2 (*Discula destructiva*), and GaV-L1 and GaV-L2 (*Gremmeniella abietina*). However,

there are cases in which two unrelated dsRNA viruses occupy the same cell (HvV145S and HvV190S; SsV1 and SsV2). Aside from the similarities in RdRp sequence, the *Ustilago maydis* and *Saccharomyces* viruses share a number of other characteristics.

Genome Structure

The totiviruses are dsRNA viruses with a single defining property: a single segment of dsRNA encodes all necessary viral functions. All RNA viruses of eukaryotes have the same problem of expression: eukaryotic ribosomes generally initiate only in one place on an mRNA. Encoding more than one protein on a single RNA requires some method of obviating this limitation. The totiviruses have adopted a number of strategies for this purpose: programmed ribosomal frameshifting (ScV, GLV, LRV, TVV), stop and go translation, using overlapping termination and initiation codons (HvV190S, CmRV or CmV), internal initiation (SsV or SsRV), and proteolytic processing (UmV).

The UmV dsRNAs fall into three categories: H (heavy), M (medium), and L (light). The H segments that have been partially or completely sequenced code for viral capsid and RdRp polypeptides. The M segments encode secreted polypeptides, some of which are toxins that kill susceptible cells (killer toxins). The L segments have no significant open reading frames, and are exact copies of the 3′ portions of M plus strands.

The one UmV H segment entirely sequenced (P1H1) has a single open reading frame encoding a polypeptide of 1820 amino acids with an N-terminal capsid and a C-terminal RdRp sequence. There is a putative papain-like protease between the capsid and RdRp domains, so that generation of the capsid polypeptide appears to occur by endoproteolytic cleavage. The proportion of cap and cap-pol fusion protein in ScV is quite critical for capsid assembly; too much of either results in loss of the virus. If this is the case in UmV, the proportions must be controlled by the efficiency of endoproteolytic cleavage; with some low but finite probability, the protease must fail, generating a fusion protein containing the RdRp domain. Alternatively, the RdRp domain might be packaged as a single domain, as it must be in the partitiviruses. In either case, this is a rather sloppy way of doing things, since a large proportion of the C-terminal domain must be discarded as useless. However, it is no more wasteful than the processing of the poliovirus polyprotein.

There are three main isolates of *Ustilago maydis* whose viral dsRNAs have been characterized – P1, P4, and P6, named after the killer toxins that some viral segments encode (see further ahead). Some isolates have only three viral dsRNAs (H1, M1, L1), while others may have as many as seven segments. A summary of the known coding regions of these dsRNAs is shown in **Table 1**. In all cases, only the

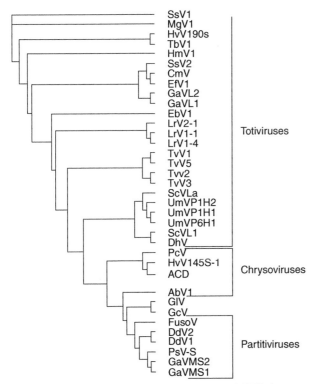

Figure 1 Cladogram of fungal and protozoan dsRNA virus RdRps. For clarity, statistics have been omitted, but each branch is supported by at least 500/1000 bootstraps. This is a Treeview rendering of a Clustal X alignment of complete RdRp sequences. Only a few of the known partitiviruses are shown due to space limitations.

Table 1   Protein products of the UmV dsRNAs

dsRNA	Length (bp)	Product	Length (aa)	Cleavage	Function
P1H1	6099	cap-pol	1820	None	RdRp
		cap	?	Endoprotease	capsid
P1M1	1504	prepropp	256		prepropp
		P1M1 alpha	103	sp, Kex2p	?
		P1M1 beta	83	Kex2p	?
P1M2	1034	KP1 pptoxin	291		pptoxin
		KP1 alpha	120	sp, Kex2p	?
		KP1 beta	117	Kex2p	KP1 toxin
P4M2	1006	KP4 pptoxin	128		pptoxin
		KP4 toxin	105	sp	KP4 toxin
P6M2	1234	KP6 pptoxin	219		pptoxin
		KP6 alpha	79	sp, Kex2p	KP6 alpha
		KP6 beta	81	Kex2p	KP6 beta

RdRp, RNA-dependent RNA polymerase; Prepropp, prepropolypeptide; sp, signal peptidase; Kex2p, Kex2p secretory proteinase; pptoxin, preprotoxin; ?, unknown.

viral plus strand has an open reading frame, and the predicted protein products have been demonstrated by *in vitro* protein synthesis, and/or by isolation *in vivo*. Presumably all totivirus mRNAs, like their genomic plus strands, are uncapped and lack polyA. This has only been directly demonstrated for ScV.

The *Ustilago maydis* Killer Toxins

The KP1, KP4, and KP6 toxins are encoded by P1M2, P4M2, and P6M1. The KP4 toxin is the best characterized of these. The KP4 toxin is synthesized as a 127-amino-acid preprotoxin from which the amino-terminal 22 amino acids are removed by signal peptidase during secretion. The resulting toxin is an unglycosylated peptide of 105 amino acids and 11.1 kDa. The KP4 toxin is the only killer toxin known to be not processed by Kex2p. Its three-dimensional structure is known (**Figure 2**). It is a monomer with five disulfides stabilizing a structure with seven beta-strands and three alpha-helices, consisting of a single beta-sheet with five antiparallel strands and two antiparallel alpha-helices at about a 45° angle to the beta-strands. Its structure is somewhat similar to that of the SMK toxin, even though the latter consists of a heterodimer. The SMK toxin is encoded by the nuclear genome of *Pichia farinosa*, rather than by a viral dsRNA, and it has essentially no primary structure similarity to the KP4 toxin. SMK is synthesized as a preprotoxin which is proteolytically processed to produce the heterodimer, so we can view the KP4 toxin as a preprotoxin that has lost its Kex2p sites (or SMK as a preprotoxin that has gained Kex2p sites).

The latter interpretation seems more likely, because there are other fungal proteins encoded by nuclear genomes that are highly similar to the KP4 toxin (**Figure 3**). The two proteins shown aligned with the KP4 toxin are from *Aspergillus oryzae* (BAE57860) and from *Gibberella zeae*

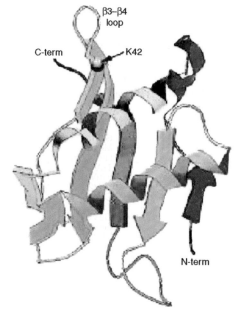

Figure 2   Structure of the KP4 toxin shown as a ribbon rendering. Figure courtesy of Tom Smith.

(XP_380238). These are both predicted proteins without known function. However, both are predicted (by the SignalP 3.0 Server) to be secreted proteins, and their mature peptides are predicted to begin with the same amino-terminal amino acid sequence as the KP4 toxin, LGINCR, or with an extra amino-terminal K (in the *A. oryzae* peptide). None of these proteins has predicted Kex2p cleavage sites (xyKR, xyRR, or xyPR, where x is hydrophobic and y is anything), and the KP4 toxin is known to be processed only by signal peptidase, alone among fungal killer toxins. Five of the ten cysteines in the KP4 toxin are conserved in both of these proteins, which share 25% and 36% sequence identity with the KP4 toxin, and a sixth cysteine may also be shared. The

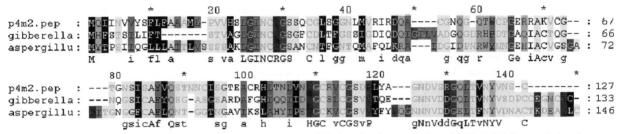

Figure 3 Comparison of KP4 (p4m2.pep) with cellular homologs. This is a Gendoc rendering of a Clustal X alignment showing physical–chemical similarities (color) and similarity (small letters in consensus line) or identity (capital letters in consensus line).

A. oryzae peptide has eight total cysteines and the *G. zeae* peptide 9. Hence each would have to be missing at least one of the five disulfides present in the KP4 toxin, and if they shared the KP4 structure would have to be stabilized by other interstrand interactions. K42 is known to be critical for KP4 toxin function, but this residue is not in a region conserved in the three proteins (**Figure 3**). The SMK toxin, with a structure somewhat similar to that of KP4, has only four disulfides. The striking similarity of these three proteins, all secreted, all lacking any N-linked sites for glycosylation, all processed only by signal peptidase, implies that the KP4 toxin was captured by the virus from a cellular transcript, since the *G. zeae* peptide, the *A. oryzae* peptide, and SMK are all encoded by nuclear genes. This evolutionary process is more obvious with the ScV k1 and k2 toxins, whose viral genes preserve the polyA they derived from the 3′ end of cellular messengers as internal portions of the viral plus strands.

The KP4 toxin is the only one of the UmV killer toxins about which we have some knowledge of mode of action. The KP4 toxin interferes with the function of calcium channels both in susceptible fungal cells and in mammalian cells. Calcium transport and signaling in fungi are now known to be necessary for many cellular processes and to involve the products of dozens of genes. The fungal calcium channel consists of a typical alpha-subunit (encoded by CCH1 in *S. cerevisiae*) and at least one more polypeptide (encoded by MID1 in *S. cerevisiae*). Presumably, the target for KP4 is located in the cell wall or cell membrane, since its effects are reversible, but it need not be the calcium channel itself. The SMK toxin is known to target a cell membrane ATPase in *S. cerevisiae*, which is susceptible to the SMK toxin but not to KP4. KP4 seems to act by interfering with calcium-regulated signal transduction pathways, thereby preventing cell growth and division. Resistance to KP4, which is encoded by a nuclear gene, might involve an alteration of any component of these signal transduction pathways.

The KP6 toxin is composed of two polypeptides, KP6-alpha and-beta, which are processed from a single precursor protein (a prepropolypeptide) of 219 amino acids encoded on P6M2. The preprotoxin is cleaved by signal peptidase after residue 19; by Kex2p cleavages after

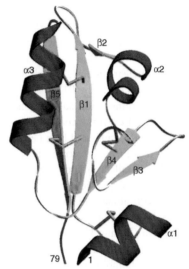

Figure 4 Structure of KP6-alpha shown as a ribbon rendering. Courtesy of the American Society for Biochemistry and Molecular Biology.

residues 27, 108, and 138; and by Kex1p cleavage after residue 137 to yield two polypeptides, alpha of 79 amino acids (8.6 kDa) and beta of 81 amino acids (9.1 kDa). Alpha and beta are not necessarily found as a multimer and can be added separately to susceptible cells to reconstitute killer activity; both are necessary for activity. Unlike the KP4 toxin, KP6 has irreversible effects on susceptible cells, actually killing cells rather than inhibiting growth. The structure of KP6-alpha is known (**Figure 4**). The monomer is a four-stranded antiparallel beta-sheet with two alpha-helices on one side, a small alpha-helix at the N-terminus, and a separate beta-strand, with four disulfide bonds, involving all eight cysteines in the polypeptide. In the crystal, the polypeptide forms trimers held together by salt bridges, and the trimers sit on each other forming an hour-glass shape, suggestive of an ion pore. However, expression of the KP6-beta peptide alone in susceptible cells is lethal, suggesting a mode of action more like that of colicins or diphtheria toxin, in which one of the toxin polypeptides is introduced into the cell by the action of one or more other toxin polypeptides and acts lethally within the cell. KP6-beta also has six cysteines, suggesting a structure with three disulfides. The two monomer nature of the KP6 toxin is reminiscent of the *S. cerevisiae* k1 toxin, which also has two monomers derived from a single prepropolypeptide by signal peptidase, Kex2p, and Kex1p processing. However, in the k1 toxin, the alpha and beta monomers are covalently bound by intermolecular disulfides, which are absent in KP6. The toxin is properly processed and secreted from *S. cerevisiae*, although there are some differences; none of the *Ustilago maydis* killer toxins (or other secreted proteins, apparently) are glycosylated, and processing of mutant precursors is different in yeast and *Ustilago maydis*. The KP6-alpha polypeptide has one possible site for N-linked glycosylation, which is used in yeast but not in *Ustilago maydis*. Unlike the KP4 toxin, the KP6 toxin polypeptides have no significant similarity to any proteins currently in the databases. Resistance to the KP6 toxin maps to the nuclear genome, as with KP4.

The KP1 toxin is encoded by P1M2 and translated as a preprotoxin of 291 amino acids, subsequently cleaved by signal peptidase and Kex2p to produce mature polypeptides alpha (120 amino acids) and beta (117 amino acids). Only beta is required for toxin activity. KP1-beta, like the other *Ustilago maydis* killer polypeptides, has an unusual number of cysteines (six) which, if all were involved in disulfides as in KP6-alpha and in KP4, would indicate three disulfides. P1M2 is different from the toxin encoding dsRNAs in UmVP4 and UmVP6 (P4M2 and P6M1). The 3' noncoding regions of the plus strands of P4M2 and P6M1 are entirely homologous to the L segments in UmVP4 and UmVP6, respectively, while the L segment in UmVP1 is derived from P4M1. P4M1 does encode a prepropolypeptide, whose processed polypeptides are secreted but have no detectable relation to the KP1 toxin (**Table 1**). Nothing is presently known of the mechanism of action of the KP1 toxin, and none of the UmVP1M1 or UmVP1M2 polypeptides have any homologs in the databases. Resistance to the KP1 toxin may be nuclear or cytoplasmic, hinting that some immunity may result from expression of one or more viral dsRNAs, as with the *S. cerevisiae* k1 toxin.

Replication and Transcription

Like all dsRNA viruses, UmV has a capsid-associated transcriptase that makes the viral plus strands. However, little is known of the nature of the transcripts or the mechanism of transcription. One peculiarity of this system is the presence of dsRNAs (L, 354–355 bp) entirely homologous to the 3' ends of the plus strands of their cognate dsRNAs (M). Presumably, like all viral dsRNAs, these are the result of replication of plus strands within nascent viral particles. It is not clear if the source of the L plus strands is internal transcription initiation or processing of M plus strands. Normal-sized L plus strands (with exactly the predicted 5' ends) are produced either by the viral transcriptase or by RNAP II transcription of cDNA clones in *Ustilago maydis* or heterologous systems, implying that a cleavage event must be responsible for production of the L segments.

This cleavage must be either an inherent property of the cognate M plus strand or cleavage by a ubiquitous nuclease activity recognizing a universal feature of the M plus strands. No ribozyme activity can be demonstrated in the cognate M plus strands *in vitro*. However, all the cognate M plus strands have a peculiar predicted secondary structure, in which the predicted L plus strand includes a very long predicted hairpin stem (23 perfect base pairs in P4M2) immediately following the predicted cleavage site (which is highly conserved). These should make ideal substrates for the recently described cellular machinery (dicer) responsible for siRNA production, which is ubiquitous in eukaryotes.

The function of the L segments is obscure: production of toxin proceeds perfectly well from cDNA constructs lacking the L encoding region, in either homologous or heterologous systems, and the P1M2 segment (encoding the KP1 toxin) has no cognate L segment. The L segments must include a packaging signal, but it is not clear what prevents its inclusion in an M plus strand sequence lacking the 3' L sequence, a situation that does exist with P1M2. In the closely related virus ScV, a sequence as small as 18 bases serves as the packaging signal.

Transcription in dsRNA viruses may be either semiconservative (resulting in displacement of the parental plus strand) or conservative (*de novo* synthesis of plus strands). Transcription in reovirus and ScV is conservative and in bacteriophage phi6 is semiconservative. An interesting difference in the structure of the viral RdRp may be responsible for this. The phi6 RdRp has an insertion of some 20 amino acids within the N-terminal F motif, an insertion that is absent in the reovirus RdRp (and absent in all the other known RdRp and reverse transcriptase structures). None of the totivirus, chrysovirus, or partitivirus RdRps has a phi6-like insertion; most may therefore be conservative transcriptases. However, UmVP1M2 has been reported to be semiconservatively transcribed.

Viral Structure

Like other fungal viruses, UmV has a small icosohedral capsid of about 41 nm in diameter, with a sedimentation coefficient of 172S and a buoyant density in CsCl of 1.42 g ml^{-1}. All dsRNA viruses so far characterized in detail (bluetongue virus, or BTV; reovirus, or Reo; and ScV, the *S. cerevisiae* virus) are icosohedral with an inner (or complete) capsid structure of 120 copies of a capsid protein. This is a highly unusual arrangement in viruses, which generally behave according to the Caspar and Klug rules for assembly of icosohedral viruses, which call for the simplest capsids to consist of 60 copies of a single polypeptide ($T = 1$) or 180 copies of a single polypeptide ($T = 3$). However, the dsRNA viruses do not really violate this stricture, since they have 60 copies of a capsid polypeptide in one configuration and 60 copies in another, so that, in effect, they have 60 copies of an asymmetric dimer, in which the two monomers share portions with identical shape and portions with different shape. The *Ustilago maydis* viruses obey this generalization, although, like all the fungal and protozoan viruses, they have a single capsid shell, not two, as in BTV and Reo. One variation of this arrangement exists in the *Penicillium chrysogenum* virus (PcV), in which the two monomers are actually part of a single polypeptide chain, the two halves of which probably share regions of identical structure.

This unique capsid structure must play an important role in properly arranging the viral polymerase within the particle to allow transcription and replication of the genomic dsRNA. In ScV and UmV, this is accomplished by making the polymerase as the carboxy-terminal portion of a fusion protein in which the N-terminal is the capsid polypeptide. In ScV, two copies of this fusion protein are incorporated into each viral particle. Each viral particle has one copy of a single species of viral dsRNA. In UmV and ScV (as well as the other totiviruses), this single species encodes both capsid polypeptide and RdRp. When additional dsRNAs are present (e.g., toxin-encoding species), these are also separately encapsidated, and are considered satellite dsRNAs: they are completely dependent on the capsid/RdRp-encoding dsRNA for their transcription, replication, and packaging. Satellite dsRNAs may be present in more than one copy per virion. This is a considerably simpler strategy than pursued by the reoviruses, orbiviruses, and cystoviruses, all of which package multiple species of dsRNA in each viral particle and as many as 12 copies of the viral RdRp in each.

Genetic Engineering

The production of killer toxins by a plant pathogen (*Ustilago maydis*) has excited some interest in using the toxin to genetically engineer resistance to the fungus in its host crop plants. Since it is possible to produce both KP6 and KP4 toxins in heterologous systems using cDNA expression

vectors, transgenic maize (KP6 and KP4) and wheat (KP4) plants have been constructed that express the toxins. These plants do secrete active killer toxins and exhibit some resistance to toxin-sensitive species of *Ustilago*. Since the toxins are harmless to animals when eaten, this may provide a new method of preventing the periodic smut infestations that occasionally decimate susceptible grain crops.

See also: Yeast L-A Virus.

Further Reading

Bruenn J (2002) The double-stranded RNA viruses of *Ustilago maydis* and their killer toxins. In: Tavantzis SM (ed.) *dsRNA Genetic Elements. Concepts and Applications in Agriculture, Forestry, and Medicine*, pp. 109–124. Boca Raton, FL: CRC Press.

Bruenn J (2003) A structural and primary sequence comparison of the viral RNA dependent RNA polymerases. *Nucleic Acids Research* 31: 1821–1829.

Bruenn J (2004) The *Ustilago maydis* killer toxins. In: Schmitt MJ and Schaffrath R (eds.) *Microbial Protein Toxins*, pp. 157–174. Heidelberg: Springer.

Cheng RH, Caston JR, Wang G-J, et al. (1994) Fungal virus capsids: Cytoplasmic compartments for the replication of double-stranded RNA formed as icosahedral shells of asymmetric Gag dimers. *Journal of Molecular Biology* 244: 255–258.

Gage MJ, Rane SG, Hockerman GH, and Smith TJ (2002) The virally encoded fungal toxin KP4 specifically blocks L-type voltage-gated calcium channels. *Molecular Pharmacology* 61: 936–944.

Gu F, Khimani A, Rane S, Flurkey WH, Bozarth RF, and Smith TJ (1995) Structure and function of a virally encoded fungal toxin from *Ustilago maydis*: A fungal and mammalian Ca^{2+} channel inhibitor. *Structure* 3: 805–814.

Jiang D and Ghabrial SA (2004) Molecular characterization of Penicillium chrysogenum virus: reconsideration of the taxonomy of the genus Chrysovirus. *Journal of General Virology* 85: 2111–2121.

Kang J, Wu J, Bruenn JA, and Park C (2001) The H1 double-stranded RNA genome of *Ustilago maydis* virus-H1 encodes a polyprotein that contains structural motifs for capsid polypeptide, papain-like protease, and RNA-dependent RNA polymerase. *Virus Research* 76: 183–189.

Li N, Erman M, Pangborn W, Duax WL, et al. (1999) Structure of *Ustilago maydis* killer toxin KP6 α-subunit: A multimeric assembly with a central pore. *Journal of Biological Chemistry* 274: 20425–20431.

Naitow H, Tang J, Canady M, Wickner RB, and Johnson JE (2002) L-A virus at 3.4 Å resolution reveals particle architecture and mRNA decapping mechanism. *Nature, Structural Biology* 9: 725–728.

Soldevila AI and Ghabrial SA (2000) Expression of the Totivirus Helminthosporium victoriae 190S virus RNA-dependent RNA polymerase from its downstream open reading frame in dicistronic constructs. *Journal of Virology* 74: 997–1003.

Wickner RB (1996) Double-stranded RNA viruses of *Saccharomyces cerevisiae*. *Microbiological Reviews* 60: 250–265.

Yeast L-A Virus

R B Wickner, National Institutes of Health, Bethesda, MD, USA
T Fujimura and R Esteban, Instituto de Microbiología Bioquímica CSIC/University of Salamanca, Salamanca, Spain

Glossary

Decapping Removal of the methylated GMP in 5′–5′ linkage at the 5′ end of most eukaryotic mRNAs.
Ribosomal frameshifting Ribosomes changing reading frame on an mRNA, in response to a special mRNA structure, to synthesize a protein from two overlapping reading frames.

Introduction

The L-A virus of bakers/brewers yeast *Saccharomyces cerevisiae* is one of several RNA viruses infecting this organism, each of which spreads via the cell–cell fusion of mating, rather than by the extracellular route. The totivirus L-A, like members of the very similar L-BC family of viruses, is a single-segment 4.6 kbp double-stranded RNA (dsRNA) virus encapsidated in icosahedral particles with a single major coat protein called Gag. The 20S and 23S RNA replicons are naked cytoplasmic single-stranded RNA (ssRNA) replicons except for their bound RNA-dependent RNA polymerases.

The L-A virus serves as the helper virus for any of several smaller satellite dsRNAs, called M dsRNAs, each encoding a secreted protein toxin and immunity to that toxin, producing the 'killer' phenomenon. Killer strains can eliminate some of the competition by this means, although only about 10% of wild strains harbor a killer dsRNA, suggesting there are costs to carrying this replicon. The killer phenotype was used to study the genetics of M dsRNAs and the helper L-A. Several functional variants of L-A were defined based on their interactions with different M dsRNAs and with the host, and host mutants affected in virus expression or propagation were also examined.

History

In 1963, Makower and Bevan reported that some yeast strains secrete a toxin that kills other yeasts This led to the identification by Bevan and by Fink of cellular dsRNAs and later viral particles correlated with the killer phenomenon. Studies of the chromosomal genes affecting the killer system revealed that the Kex2 protease, identified by its requirement for toxin secretion, was the long-sought proinsulin-processing enzyme. The Mak3 *N*-acetyltransferase, whose acetylation of Gag is needed for viral assembly, established consensus sequences for such enzymes. The loss of the L-A virus in *mak3* mutants revealed a second dsRNA species of the same size, called L-BC, and unrelated to the killer system.

Virion Structure

L-A virions have icosahedral symmetry, but contain 120 Gag monomers per particle, in apparent violation of the rules of quasi-equivalence. In fact, each virion is composed of 60 asymmetric dimers of Gag (**Figure 1**), a feature that is common to the cores of all dsRNA viruses that have been characterized. One type of Gag molecule makes contact with the icosahedral fivefold and twofold axes, but not with the threefold axes. The second type of Gag finds itself in contact with the threefold axis, but not with either the five- or twofold axes (**Figure 1**). The two environments of Gag lead to two distinct conformations, suggesting that Gag may be more flexible than some other coat proteins. The L-A virion has more volume per nucleotide than do ssRNA or dsDNA viruses. These facts may reflect the requirement that the dsRNA moves inside the particle and is transcribed by the RNA-dependent RNA polymerase that is fixed to the inner virion wall (see below). Pores at the fivefold axes are assumed to allow entry of nucleotides and exit of (+) strand transcripts, but retention of the dsRNA genome. A trench on the outside contains His154, the central active site residue of the mRNA-decapping activity to which the 7-methyl-GMP structure becomes covalently attached (**Figure 2**). A layered density observed for the viral dsRNA may reflect the dsRNA's rigid structure and the fact that it is forced to press against the inner capsid wall.

Genome Organization

The single segment of the L-A genome is a 4.6 kbp dsRNA with two long open reading frames (ORFs) (**Figure 3**). The 5′ ORF encodes the major coat protein, called Gag in analogy with retroviruses, while the longer 3′ ORF, called Pol, encodes the RNA-dependent RNA polymerase, and has homology with similar enzymes of other ssRNA and dsRNA viruses. Pol is expressed only as a fusion protein

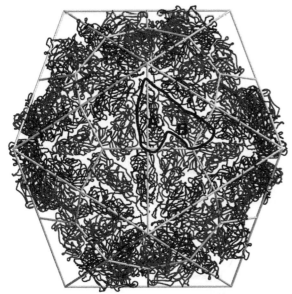

Figure 1 Wire diagram of the L-A virus capsid. The major coat protein, Gag, is found in two nonequivalent positions: 'A' molecules contact the fivefold and twofold axes, while 'B' molecules contact the threefold axes. Reproduced from Naitow I, Tang J, Canady M, Wickner RB, and Johnson JE (2002) L-A virus at 3.4 Å resolution reveals particle architecture and mRNA decapping mechanism. *Nature Structural Biology* 9: 725–728, with permission from Nature Publishing Group.

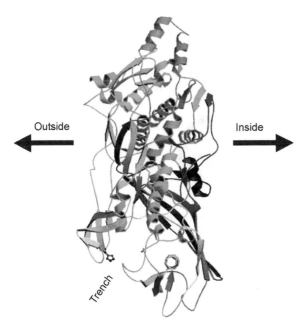

Figure 2 Ribbon diagram of a single Gag molecule. The trench on the outer surface includes His154, the Gag residue to which 7meGMP is covalently attached by the decapping activity.

whose amino end is a nearly complete Gag molecule. The fusion is carried out by a −1 ribosomal frameshift event in the region of overlap of the two ORFs. An RNA pseudoknot, in the region of overlap of the ORFs, slows

ribosome progression at a point on the mRNA (of the form X XXY YYZ (0 frame indicated)) where bonding of the A-site and P-site tRNAs to the mRNA is nearly as good in the −1 frame as in the 0 frame. About 1% of ribosomes slip into the −1 frame, and, continuing translation, make the Gag–Pol fusion protein.

Replication Cycle

Transcription

Viral particles have an RNA-dependent RNA polymerase that is due to the Pol part of the Gag–Pol fusion protein. Transcription is a conservative reaction, meaning that the parental strands remain together after the synthesis of a new (+) strand. The L-A (+) strand transcripts are extruded from the particles into the cytoplasm where they serve as both mRNA and as the species that is packaged by coat proteins to make new viral particles. However, (+) strand transcripts of the smaller M dsRNA or of deletion mutants of L-A are often retained within the particle where they may be replicated. This is called 'headful replication' because the volume of the particle seems to be the determinant of how many dsRNA molecules may accumulate in each particle.

RNA Packaging

The RNA packaging site (**Figure 3**) is a stem–loop structure about 400 nucleotides from the 3′ end of the L-A (+) strand. This stem–loop has an essential A residue protruding on the 5′ side of the stem. The loop sequence is also important, but the stem sequence is not critical, as long as the stem structure can form. The packaging site on the RNA is recognized by the proximal part of the Pol domain of the Gag–Pol fusion protein. The Gag part of Gag–Pol is part of the capsid structure, so the Gag–Pol fusion protein structure assures packaging of viral (+) strands in new particles.

RNA Replication

The (−) strand synthesis reaction is called replication (**Figure 4**). This reaction involves recognition of the internal binding site on the L-A (+) strands by Pol, followed by interaction with the now nearby 3′ end and initiation of new RNA (−) chains. Multiple rounds of RNA synthesis can proceed within the viral particle in the 'headful replication' process discussed above.

Viral Translation

The translation apparatus is the battleground of RNA viruses with their hosts. Poliovirus protease cleaves eIF4G so that host-capped mRNAs cannot be translated

L-A dsRNA virus of *S. cerevisiae*

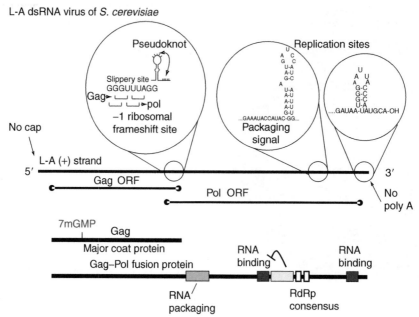

Figure 3   Genome organization. The L-A (+) strand encodes Gag and Pol in overlapping reading frames. The mRNA lacks 5′ cap and 3′ polyA structures. The location of the packaging site on the RNA and the region of the Pol protein segment (green box) that recognizes the packaging site are indicated. The RNA sites for replication, and the parts of Pol with homology to other RNA-dependent RNA polymerases (RdRp) are indicated. The cryptic *in vitro* RNA binding site (blue box) is inhibited by the area shown by the yellow box.

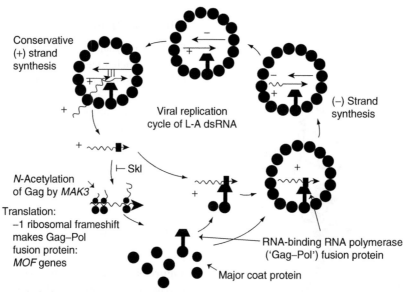

Figure 4   Replication cycle of the L-A virus. Both (+) and (–) RNA strand synthesis occur within the viral particles, but at different stages of the cycle. RNA (+) strands (transcripts) are extruded from the particles and serve as both mRNA and the species packaged to make new particles. Translation of viral mRNA is blocked by Ski proteins. N-terminal acetylation of Gag by Mak3p is necessary for viral assembly.

but its own IRES-containing mRNAs are used. Influenza virus steals caps from host mRNAs.

L-A (+) strand transcripts lack both the 5′ cap and 3′ polyA structures typical of eukaryotic mRNAs, and so are at a distinct disadvantage. Mutations resulting in a relative deficiency of 60S ribosomal subunits selectively lose the

ability to express cap-minus, polyA-minus mRNAs such as those of L-A. The *SKI* genes encode a series of proteins whose function is to specifically block the expression of non-polyA mRNAs. The Ski proteins have effects on both translatability of non-polyA mRNAs and on their stability. The *SKI1/XRN1* gene encodes a 5′→3′ exoribonuclease specific for uncapped mRNA. As a defense against its degradation of viral mRNA, L-A's Gag protein has a decapping activity that produces, from cellular mRNAs, 'decapitated decoys', uncapped RNAs that serve as alternative targets for the exonuclease action. In the absence of this decapping activity, viral mRNAs are not expressed, but deletion of the *SKI1/XRN1* gene restores viral mRNA expression.

L-A Genetics

Several natural variants of L-A have been described. The ability of an L-A variant to support M dsRNA replication was first called [HOK] (=helper of killer) and then H when it was found to be a property of L-As such as L-A-H, L-A-HN, or L-A-HNB. Making several chromosomal *MAK* genes dispensable for M propagation was named [B] (=bypass) and is found on L-A-HNB. Ability to exclude L-A-H was named [EXL] (=exclusion) and then just E on L-A-E. Insensitivity to the action of [EXL] was called [NEX] (=nonexcludable), and then shortened to N as in L-A-HN or L-A-HNB. The molecular basis of these interactions has resisted study because of the current inability to obtain L-A from a cDNA clone. M dsRNA can be supported from a cDNA clone of L-A, but the L-A virus has not been shown to be regenerated from these transcripts.

Other RNA Replicons in Yeast: L-BC, 20S RNA, 23S RNA

20S RNA was discovered as an RNA species which appeared when cells were placed under conditions that induce meiosis and spore formation, namely near-starvation for nitrogen and provision of acetate as a carbon source. It was shown that 20S RNA is an independent replicon, and can be made independent of the sporulation or meiosis processes. 20S RNA encodes its own RNA-dependent RNA polymerase which is bound to the otherwise naked cytoplasmic RNA. The mechanism of 20S RNA replication control by culture conditions remains to be elucidated.

23S RNA is a related, but independent yeast replicon which was found as a dsRNA form (called T). 23S RNA also encodes its RNA-dependent RNA polymerase. Both 20S RNA and 23S RNA can be launched from cDNA clones to form replicating virus.

L-BC is a totivirus, related to L-A but independent of it. Its dsRNA is essentially the same size as that of L-A although its copy number is usually about tenfold lower.

Thus it was not detected until chromosomal mutants that lose L-A were examined. L-BC does not interact with the killer system, and confers no obvious phenotype, so its genetics has not been extensively explored.

See also: Narnaviruses; Totiviruses.

Further Reading

Bevan EA, Herring AJ, and Mitchell DJ (1973) Preliminary characterization of two species of dsRNA in yeast and their relationship to the 'killer' character. *Nature* 245: 81–86.

Bostian KA, Elliott Q, Bussey H, Burn V, Smith A, and Tipper DJ (1984) Sequence of the preprotoxin dsRNA gene of type I killer yeast: Multiple processing events produce a two-component toxin. *Cell* 36: 741–751.

Dinman JD, Icho T, and Wickner RB (1991) A −1 ribosomal frameshift in a double-stranded RNA virus of yeast forms a gag-pol fusion protein. *Proceedings of the National Academy of Sciences, USA* 174–178.

Esteban R, Fujimura T, and Wickner RB (1989) Internal and terminal cis-acting sites are necessary for *in vitro* replication of the L-A double-stranded RNA virus of yeast. *EMBO Journal* 8: 947–954.

Fujimura T, Esteban R, Esteban LM, and Wickner RB (1990) Portable encapsidation signal of the L-A double-stranded RNA virus of *S. cerevisiae*. *Cell* 62: 819–828.

Fujimura T, Ribas JC, Makhov AM, and Wickner RB (1992) Pol of gag-pol fusion protein required for encapsidation of viral RNA of yeast L-A virus. *Nature* 359: 746–749.

Fuller RS, Brake A, and Thorner J (1989) Intracellular targeting and strructural conservation of a prohormone-processing endoprotease. *Science* 246: 482–486.

Icho T and Wickner RB (1989) The double-stranded RNA genome of yeast virus L-A encodes its own putative RNA polymerase by fusing two open reading frames. *Journal of Biological Chemistry* 264: 6716–6723.

Leibowitz MJ and Wickner RB (1976) A chromosomal gene required for killer plasmid expression, mating, and spore maturation in *Saccharomyces cerevisiae*. *Proceedings of the National Academy of Sciences, USA* 73: 2061–2065.

Masison DC, Blanc A, Ribas JC, Carroll K, Sonenberg N, and Wickner RB (1995) Decoying the cap-mRNA degradation system by a dsRNA virus and poly(A)-mRNA surveillance by a yeast antiviral system. *Molecular and Cellular Biology* 15: 2763–2771.

Naitow H, Canady MA, Wickner RB, and Johnson JE (2002) L-A dsRNA virus at 3.4 angstroms resolution reveals particle architecture and mRNA decapping mechanism. *Nature, Structural Biology* 9: 725–728.

Sommer SS and Wickner RB (1982) Yeast L dsRNA consists of at least three distinct RNAs; evidence that the non-Mendelian genes [HOK], [NEX] and [EXL] are on one of these dsRNAs. *Cell* 31: 429–441.

Tercero JC and Wickner RB (1992) *MAK3* encodes an N-acetyltransferase whose modification of the L-A *gag* N-terminus is necessary for virus particle assembly. *Journal of Biological Chemistry* 267: 20277–20281.

Vodkin MH, Katterman F, and Fink GR (1974) Yeast killer mutants with altered double-stranded ribonucleic acid. *Journal of Bacteriology* 117: 681–686.

Wickner RB (2006) Viruses and Prions of Yeast, Fungi and Parasitic Microorganisms. In: Knipe DM and Howley PM (eds.) *Fields Virology*, 5th edn. Philadelphia, PA: Lippincott Williams and Wilkins.

SUBJECT INDEX

Notes

Cross-reference terms in italics are general cross-references, or refer to subentry terms within the main entry (the main entry is not repeated to save space). Readers are also advised to refer to the end of each article for additional cross-references - not all of these cross-references have been included in the index cross-references.

The index is arranged in set-out style with a maximum of three levels of heading. Major discussion of a subject is indicated by bold page numbers. Page numbers suffixed by T and F refer to Tables and Figures respectively. *vs.* indicates a comparison.

This index is in letter-by-letter order, whereby hyphens and spaces within index headings are ignored in the alphabetization. Prefixes and terms in parentheses are excluded from the initial alphabetization.

To save space in the index the following abbreviations have been used

CJD - Creutzfeldt–Jakob disease

CMV - cytomegalovirus

EBV - Epstein–Barr virus

HCMV - human cytomegalovirus

HHV - human herpesvirus

HIV - human immunodeficiency virus

HPV - human papillomaviruses

HSV - herpes simplex virus

HTLV - human T-cell leukemia viruses

KSHV - Kaposi's sarcoma-associated herpesvirus

RdRp - RNA-dependent RNA polymerase

RNP - ribonucleoprotein

SARS - severe acute respiratory syndrome

TMEV - Theiler's murine encephalomyelitis virus

For consistency within the index, the term "bacteriophage" has been used rather than the term "bacterial virus."

Printed and bound by CPI Group (UK) Ltd, Croydon, CR0 4YY

03/10/2024

01040312-0019